MW00603468

Archaea

MOLECULAR AND CELLULAR BIOLOGY

Archaea

MOLECULAR AND CELLULAR BIOLOGY

EDITED BY

RICARDO CAVICCHIOLI

School of Biotechnology and Biomolecular Sciences
The University of New South Wales
Sydney, NSW, Australia

ASM
PRESS

Washington, DC

Copyright © 2007   ASM Press
American Society for Microbiology
1752 N St., N.W.
Washington, DC 20036-2904

Library of Congress Cataloging-in-Publication Data

Archaea : molecular and cellular biology/edited by Rick Cavicchioli.
    p. ; cm.
  Includes bibliographical references and index.
  ISBN-13: 978-1-55581-391-8 (hardcover : alk. paper)
  ISBN-10: 1-55581-391-7 (hardcover : alk. paper) 1. Archaebacteria—Molecular aspects. I. Cavicchioli, Ricardo.
  [DNLM: 1. Archaea—genetics. 2. Archaea—physiology. 3. Biotechnology. 4. Evolution, Molecular. 5. Genome,
Archaeal. QW 52 A669 2007]

QR82.A69A735 2006
579.3′21—dc22

2006034575

All rights reserved
Printed in the United States of America

10 9 8 7 6 5 4 3 2 1

Address editorial correspondence to ASM Press, 1752 N St., N.W., Washington, DC 20036-2904, U.S.A.

Send orders to: ASM Press, P.O. Box 605, Herndon, VA 20172, U.S.A.
Phone: 800-546-2416; 703-661-1593
Fax: 703-661-1501
E-mail: books@asmusa.org
Online: estore.asm.org

Cover figure: *Pyrodictium* cells after freeze-etching, exhibiting an S-layer with p6 symmetry (courtesy of Reinhard Rachel, Universität Regensburg) (see Chapter 14, Fig. 16A).

CONTENTS

CONTRIBUTORS

Michael W. W. Adams
Department of Biochemistry & Molecular Biology, Davison Life Sciences Complex, Green Street, University of Georgia, Athens, GA 30602-7229

Sonja V. Albers
Department of Microbiology, Groningen Biomolecular Sciences and Biotechnology Institute, University of Groningen, Kerklaan 30, 9751 NN Haren, The Netherlands

Alexandre Ambrogelly
Departments of Molecular Biophysics and Biochemistry, Yale University, New Haven, CT 06520-8114

Kimberly Anderson
Center of Marine Biotechnology, University of Maryland Biotechnology Institute, 701 E. Pratt Street, Baltimore, MD 21202

Stephen D. Bell
MRC Cancer Cell Unit, Hutchison MRC Research Centre, Hills Road, Cambridge CB2 2XZ, United Kingdom

Yan Boucher
Department of Chemistry and Biomolecular Sciences, Macquarie University, Sydney, NSW 2109, Australia

Alexander Machado Cardoso
Departments of Molecular Biophysics and Biochemistry, Yale University, New Haven, CT 06520-8114

Bonnie Chaban
Department of Microbiology and Immunology, Queen's University, Kingston, Ontario K7L 3N6, Canada

Chung-Jung Chou
Department of Chemical and Biomolecular Engineering, North Carolina State University, EB-1, Box 7905, 911 Partners Way, Raleigh, NC 27695-7905

Harald Claus
Institut für Mikrobiologie und Weinforschung, Johannes Gutenberg-Universität, Becherweg 15, D-55099 Mainz, Germany

Beatrice Cobucci-Ponzano
Institute of Protein Biochemistry, National Research Council, Via P. Castellino 111, 80131 Naples, Italy

Michael J. Danson
Centre for Extremophile Research, Department of Biology and Biochemistry, University of Bath, Bath BA2 7AY, United Kingdom

Patrick P. Dennis
Division of Molecular and Cellular Biosciences, National Science Foundation, 4201 Wilson Blvd., Arlington, VA 22230

Kieran C. Dilks
Biology Department, University of Pennsylvania, Philadelphia, PA 19104

Arnold J. M. Driessen
Department of Microbiology, Groningen Biomolecular Sciences and Biotechnology Institute, University of Groningen, Kerklaan 30, 9751 NN Haren, The Netherlands

Elena Evguenieva-Hackenburg
Institut für Mikrobiologie und Molekularbiologie, Universität Giessen, Heinrich-Buff-Ring 26-32, 35392 Giessen, Germany

James G. Ferry
Department of Biochemistry and Molecular Biology,
The Pennsylvania State University, University Park,
PA 16802

Patrick Forterre
Unité Biologie Moléculaire du Gène chez les
Extremophiles, Institut de Génétique et Microbiologie,
UMR CNRS 8621, Université Paris-Sud, 91405
Orsay, France

Shuya Fukai
Department of Biological Information, Graduate
School of Bioscience and Biotechnology, Tokyo
Institute of Technology, Yokohama-shi, Kanagawa
226-8501, Japan

Simonetta Gribaldo
Unité Biologie Moléculaire du Gène chez les
Extremophiles, Institut Pasteur, 25 rue du Dr. Roux,
75724 Paris Cedex 15, France

Dennis W. Grogan
Department of Biological Sciences, University of
Cincinnati, Cincinnati, OH 45221-0006

Stephanie Herring
Departments of Molecular Biophysics and
Biochemistry, Yale University, New Haven,
CT 06520-8114

Michael Hohn
Departments of Molecular Biophysics and
Biochemistry, Yale University, New Haven,
CT 06520-8114

David W. Hough
Centre for Extremophile Research, Department of
Biology and Biochemistry, University of Bath,
Bath BA2 7AY, United Kingdom

Ryo Izuka
Laboratory of Bio-Analytical Chemistry, Graduate
School of Pharmaceutical Sciences, The University of
Tokyo, 7-3-1 Hongo, Bunkyo-ku, Tokyo 113-0033,
Japan

Ken F. Jarrell
Department of Microbiology and Immunology,
Queen's University, Kingston, Ontario K7L 3N6,
Canada

Francis E. Jenney, Jr.
Department of Biochemistry & Molecular Biology,
Davison Life Sciences Complex, Green Street,
University of Georgia, Athens, GA 30602-7229

Kyle A. Kastead
Department of Biochemistry and Molecular Biology,
The Pennsylvania State University, University Park,
PA 16802

Robert M. Kelly
Department of Chemical and Biomolecular
Engineering, North Carolina State University,
EB-1, Box 7905, 911 Partners Way, Raleigh,
NC 27695-7905

Peter J. Kennelly
Department of Biochemistry, Mail Stop 0308,
Virginia Polytechnic Institute and State University,
Blacksburg, VA 24061

Arnulf Kletzin
Institute of Microbiology and Genetics, Darmstadt
University of Technology, Schnittspahnstrasse 10,
D-64287 Darmstadt, Germany

Gabriele Klug
Institut für Mikrobiologie und Molekularbiologie,
Universität Giessen, Heinrich-Buff-Ring 26-32,
35392 Giessen, Germany

Helmut König
Institut für Mikrobiologie und Weinforschung,
Johannes Gutenberg-Universität, Becherweg 15,
D-55099 Mainz, Germany

Wil N. Konings
Department of Microbiology, Groningen
Biomolecular Sciences and Biotechnology Institute,
University of Groningen, Kerklaan 30, 9751 NN
Haren, The Netherlands

Lakshmi Krishnan
Institute for Biological Sciences, 100 Sussex Drive,
National Research Council of Canada, Ottawa,
Ontario K1A 0R6, Canada

Henry J. Lamble
Centre for Extremophile Research, Department of
Biology and Biochemistry, University of Bath,
Bath BA2 7AY, United Kingdom

Si-houy Lao-Sirieix
MRC Cancer Cell Unit, Hutchison MRC Research
Centre, Hills Road, Cambridge CB2 2XZ,
United Kingdom

Paola Londei
Department of Medical Biochemistry, Biology and Physics (DIBIFIM), Università degli Studi di Bari (Policlinico), Bari, Italy

Anita Marchfelder
Molekulare Botanik, Universität Ulm, Albert-Einstein-Allee 11, 89069 Ulm, Germany

Victoria L. Marsh
MRC Cancer Cell Unit, Hutchison MRC Research Centre, Hills Road, Cambridge CB2 2XZ, United Kingdom

Marco Moracci
Institute of Protein Biochemistry, National Research Council, Via P. Castellino 111, 80131 Naples, Italy

Yuko Nakamura
Department of Biological Information, Graduate School of Bioscience and Biotechnology, Tokyo Institute of Technology, Yokohama-shi, Kanagawa 226-8501, Japan

Sandy Y. M. Ng
Department of Microbiology and Immunology, Queen's University, Kingston, Ontario K7L 3N6, Canada

Osamu Nureki
Department of Biological Information, Graduate School of Bioscience and Biotechnology, Tokyo Institute of Technology, Yokohama-shi, Kanagawa 226-8501, Japan

Arina D. Omer
Department of Biochemistry and Molecular Biology, University of British Columbia, Vancouver, BC V6T 1Z3, Canada

Hiroyuki Oshikane
Department of Biological Information, Graduate School of Bioscience and Biotechnology, Tokyo Institute of Technology, Yokohama-shi, Kanagawa 226-8501, Japan

Sotiria Palioura
Department of Molecular Biophysics and Biochemistry, Yale University, New Haven, CT 06520-8114

Giuseppe Perugino
Institute of Protein Biochemistry, National Research Council, Via P. Castellino 111, 80131 Naples, Italy

Mechthild Pohlschröder
Biology Department, University of Pennsylvania, Philadelphia, PA 19104

Carla Polycarpo
Departments of Molecular Biophysics and Biochemistry, Yale University, New Haven, CT 06520-8114

Reinhard Rachel
Lehrstuhl für Mikrobiologie, Universität Regensburg, Universitätsstraße 31, D-93053 Regensburg, Germany

Lennart Randau
Departments of Molecular Biophysics and Biochemistry, Yale University, New Haven, CT 06520-8114

John N. Reeve
Department of Microbiology, Ohio State University, Columbus, OH 43210

Frank T. Robb
Center of Marine Biotechnology, University of Maryland Biotechnology Institute, 701 E. Pratt Street, Baltimore, MD 21202

Mosè Rossi
Institute of Protein Biochemistry, National Research Council, Via P. Castellino 111, 80131 Naples, Italy

Juan Carlos Salazar
Departments of Molecular Biophysics and Biochemistry, Yale University, New Haven, CT 06520-8114

Rachel Samson
Department of Microbiology, Ohio State University, Columbus, OH 43210

Kelly Sheppard
Departments of Molecular Biophysics and Biochemistry, Yale University, New Haven, CT 06520-8114

Dieter Söll
Departments of Molecular Biophysics and Biochemistry, Yale University, New Haven, CT 06520-8114

Kevin R. Sowers
Center of Marine Biotechnology, University of Maryland Biotechnology Institute, 701 E. Pratt Street, Baltimore, MD 21202

G. Dennis Sprott
Institute for Biological Sciences, 100 Sussex Drive, National Research Council of Canada, Ottawa, Ontario K1A 0R6, Canada

Sabrina Tachdjian
Department of Chemical and Biomolecular Engineering, North Carolina State University, EB-1, Box 7905, 911 Partners Way, Raleigh, NC 27695-7905

Michael Thomm
Universität Regensburg, Lehrstuhl für Mikrobiologie, Universitätsstrasse 31, D-93053 Regensburg, Germany

Carl R. Woese
Department of Microbiology, University of Illinois, Urbana, IL 61801

Masafumi Yohda
Department of Biotechnology and Life Science, Tokyo University of Agriculture and Technology, 2-24-16 Naka-cho, Koganei, Tokyo 184-8588, Japan

Ying Yuan
Departments of Molecular Biophysics and Biochemistry, Yale University, New Haven, CT 06520-8114

Yvan Zivanovic
Institut de Génétique et Microbiologie, UMR CNRS 8621, Université Paris-Sud, 91405 Orsay, France

PREFACE

The *Archaea* are clearly recognizable as a unique and interesting group of organisms for many important reasons. They have distinct molecular characteristics that clearly distinguish them from the *Bacteria* and the *Eucarya*, and evolutionary studies highlight the quintessential role that the *Archaea* have played in shaping all life on Earth. Many archaea are extremophiles and have been responsible for rewriting the textbooks with regard to the degree that an organism can tolerate, or even thrive, in abiotic extremes. As a result, the combined implications of their evolutionary role and their capacity to thrive in environmental extremes have expanded the horizons for the astrobiology community in searching for extraterrestrial life. Properties that are characteristic of or unique to the *Archaea* span from fundamental biochemical signatures (e.g., the glycerol-1-phosphate lipid backbone) and metabolic pathways (e.g., methanogenesis) to fundamental properties of their "lifestyle" (e.g., no pathogenic archaea are known). While the *Archaea*, *Bacteria*, and *Eucarya* represent coherent lineages, they are also chimeric in part. Archaea tend to have information-processing systems (e.g., DNA replication, transcription, and translation) in common with eucarya, while they share other features (e.g., metabolism) with bacteria. These characteristics provide important fuel for debates about the course of evolution. They also offer practical benefits for advancing studies on eucaryal organisms, in particular, metazoans, which are often more complex and difficult to study, by providing analogous biological targets for a broad range of studies (e.g., protein X-ray structures for eucaryal homologs from archaeal hyperthermophiles).

Given their serious importance to biology, one might be surprised that suprisingly few books cover the biology of the *Archaea* to any real extent. The best introductory microbiology textbook that covers the *Archaea* is *Brock: Biology of Microorganisms*, in which

archaea are covered accurately, although necessarily briefly due to the scope of topics covered. *Archaea: Molecular and Cellular Biology* is pitched at practitioners in the field, and as a useful resource for those less familiar with the *Archaea* but who are charged with teaching the topic, and it is intended to be accessible to postgraduate and advanced level undergraduates who will, we might hope, be inspired to become the next generation of "archaeaologists." Each chapter covers essential background, focuses on the discoveries of the twenty-first century, and concludes with a view of what is expected to arise in the next five years. The book aims to be the authoritative reference source for the many disciplines interested in the *Archaea*.

In many fields, in particular, in the early years of growth, terminology is coined by different research groups and gradually evolves into a sea of descriptions. The development of the *Archaea* as a field of study is no exception. We have adopted the conventions proposed by Carl R. Woese, Otto Kandler, and Mark L. Wheelis in their article "Towards a natural system of organisms: proposal for the domains Archaea, Bacteria, and Eucarya," published in the *Proceedings of the National Academy of Sciences of the United States of America*, 1990 (**87**:4576–4579), and Norman Pace in his essay "Time for a change," published in *Nature*, 2006 (**441**:289). This avoids complications with the use of terms such as "archaebacteria" and "eubacteria," which in essence inappropriately describe two different forms of "bacteria," and the term "prokaryote."

While the *Archaea* have often been appreciated for the qualities listed above (e.g., their extremophilic nature), molecular ecological studies have highlighted the diversity and ubiquity of archaea on the planet. It may indeed be impossible to identify a type of habitat where archaea are completely absent. Traditionally, the taxonomy and physiology of the *Archaea* have been described by referring to halophiles, hyperthermophiles, and methanogens. While this serves a purpose, members of the *Archaea* know no boundaries and are unified by their fundamental cellular properties. This book embraces this knowledge and is dedicated to fully describing the molecular cell biology of the *Archaea*.

The core of the book, Chapters 3 to 18, describes the key cellular processes such as DNA replication, transcription, translation, lipids, and metabolism, while including a depth and completeness that includes unique features of aminoacyl-tRNA synthesis, signal transduction, and posttranslational modification (to mention but a few of the topics). Chapter 2 is a large chapter that covers general characteristics and important model organisms. By broadly overviewing the ecology and physiological diversity of the archaeal world, the chapter provides a reference point for contemplating the biology of the cell. Chapters 19 to 21 cover genomics, functional genomics, and molecular genetics. Although relevant aspects of these topics are integrated into preceding chapters, chapters 19 to 21 provide a depth and focus that highlight the evolutionary implications, the state-of-the-art, and the strengths and limitations of each of the fields. The book concludes with two chapters that cover biotechnological and biomedical applications of archaea and their cellular products.

Chapters 2 to 23 have been written and edited with the aim of capturing each author's individual expertise, while generating a reference text with a synthesized style and tone. It is indeed not merely a collection of conference proceedings. The exception to this general style is Chapter 1, a personalized account by Carl Woese of the "history of the *Archaea* movement." In his chapter, "The *Archaea*: an Invitation to Evolution," Carl offers a provocative and inspiring account of how the *Archaea* were discovered, delves into the implications of archaeal distinctiveness for the origin and evolution of life, and describes the scientific and social hurdles that needed to be overcome to bring the field forward to where it is today, a time in which the *Archaea*, whose existence could not have been guessed only a short time ago, have come of age.

It is my hope that the many hours of effort that preceded the publication of this volume will have not been in vain, that the volume in hand will be the first place active researchers go to for information on the *Archaea*, and that the younger generation of biologists will be inspired to try their hand at these organisms, whose secrets we have only begun to unlock and which offer so much in the way of fundamental insights and practical benefits.

It would be ungracious of me not to acknowledge the help I have received along the way from the many people who provided assistance. It is impossible to cite everyone, but, in particular, I thank Greg Ferry for constructive and encouraging words (and great coffee) who helped to initiate the book and drive it along, and Kevin Sowers for supporting my efforts (in particular, with Guinness) during sabbatical in his laboratory. Interactions with authors have been uplifting experiences, which I think illustrates that those in the *Archaea* field share a common excitement and joy of discovery, and who certainly know how to pull together when it counts. Sincere thanks to Carl Woese for letting me see inside his world, and for not only sharing rare insight but timely, inspiring, and warm encouragement. Thanks to Greg Payne who has provided the critical and smooth link to the rest of the team at ASM Press—here's to you all, and to the first edition of *Archaea: Molecular and Cellular Biology*.

Ricardo Cavicchioli

Archaea: Molecular and Cellular Biology
Edited by Ricardo Cavicchioli
© 2007 ASM Press, Washington, D.C.

Chapter 1

The Archaea: an Invitation to Evolution†

CARL R. WOESE

BEGINNINGS

The discovery of the archaebacteria was serendipitous, but not unexpected. In the late 1960s I had begun assembling the program for inferring (organismal) genealogical relationships through rRNA sequence comparisons (46). It was time to determine the structure of the universal phylogenetic tree (whatever it might be). Molecular evolution had been on the scene for the better part of a decade, and a universal framework within which to study evolution from the molecules on up was needed. Our initial emphasis was necessarily on the microbial world (bacteria, in particular), for almost nothing was known about microbial phylogenies.

My objective in establishing the phylogenetic program, however, was not to refine bacterial taxonomy per se, but to restore an evolutionary perspective/spirit to biology. This time the focus would be on the evolution of the cell itself, in particular, the evolution of its translation mechanism (and the information-processing systems in general) (38, 41). A reductionist discipline of molecular biology had deliberately ignored evolution—asserting that a fundamental understanding of life could be acquired without considering its evolution! This was unacceptable to me, anathema: a Biology that is not fundamentally evolutionary is *not* Biology.

The cell's translation mechanism (and the other cellular information-processing systems) are far too complex to have arisen in anything approaching their modern "fully evolved" state (38, 41). Therefore, each must have come from far more primitive beginnings

and so have undergone profound and telling evolutionary changes. To reconstruct even the final stages in this evolutionary progression was to come face to face with biological organization. (Biological organization *cannot* be understood apart from its evolution. All biological form is ultimately complex dynamic *process*—process involuted beyond our current capacity to understand it. In an important sense biological form is four dimensional.)

How exactly to go about determining a universal phylogenetic tree was, then, the question. Since the 1950s the idea that an enormous phylogenetic record lay buried in the sequences of macromolecules had been there—a record that could be read through comparative sequence analysis (6, 49). At that point, however, the approach had been used solely with proteins—in part, because nucleic acid sequencing had yet to be developed; in part, because most biologists had convinced themselves that proteins were the only way in which this could be done. The enterprise had even been labeled "protein taxonomy" initially (6). Having worked with ribosomes, I thought the approach might work with the ribosomal RNAs if only we had the proper method for characterizing them.

Fortunately, that method came along in 1965—Sanger's oligonucleotide-cataloging approach to RNA sequencing (25). The Sanger method and ribosomal RNA seemed the perfect combination: the method was powerful (yielded far more sequence information than anything that had preceded it), and rRNAs were ubiquitous and relatively easy to isolate; their sequences were quite highly conserved (10). The universal phylogenetic (genealogical) tree was only the first step in a journey, however: a world map, if you will, that helps to locate relics of the evolutionary past. I guess I didn't realize how long it was going to take just to *find* that map!

---

†In memory of Wolfram Zillig, a founder of the archaeal revolution.

Carl R. Woese • Department of Microbiology, University of Illinois, Urbana, IL 61801.

Several years were spent in tuning the Sanger method to suit our needs. By the early 1970s we were finally up and running, albeit very slowly. Initially we groped our way tentatively into the darkness of bacterial phylogeny, starting with garden variety microorganisms—*Escherichia coli* (here we had only to improve on what others had already done) (33); *Aerobacter aerogenes* (46); *Bacillus* (half a dozen or so species) (45); and various others. Almost immediately it became apparent that we were going to need help to cover the full range of the bacteria—at least in any reasonable time frame. Many organisms would just be too difficult for us to grow. Enlisting the aid of the experts, that is, those experts who were willing to grow their organisms in the proper radioactive medium, was essential. Such people were hard to find, but once the right constellation of experts started to form, we were on our way in earnest—characterizing rRNAs from mycoplasmas, other pathogens, a host of clostridia, many lactobacilli, bacteroides, a representative collection of photosynthetic bacteria, and, of course, key species of what would become the archaebacteria, and so on.

Of all the numerous suggestions we had gotten for organisms to study, the one I solicited from my colleague at the University of Illinois Department of Microbiology, Ralph Wolfe, turned out to be the most important. Ralph was in the process of working out the biochemistry of methanogenesis, which made it natural for him to suggest we characterize the methanogens. It was not the organisms per se, but the evolutionary conundrum they posed, that intrigued me. If the methanogens were taxonomically grouped on the basis of common methanogenic biochemistry, then the resulting taxon contained a wide variety of morphologies (including Gram stain differences) and growth conditions. The alternative morphologically based grouping would cast methanogenic biochemistry to the taxonomic winds. This last is how the seventh edition of *Bergey's Manual* had done it (2); but in the eighth edition (methanogenic) biochemistry had been used as the taxonomic desideratum (22). (A similar conundrum was posed by bacterial photosynthesis, in which case molecular analysis would resolve the issue differently—in favor of a reticulated biochemistry spread across the bacterial organismal phylogenetic tree.) We had the technology here and now to settle the phylogenetic issue methanogenesis raised. Methanogens went to the top of the to-do list.

The only difficulty was that, at the time—which at the latest was spring 1974 (the semester before George Fox joined the lab as my post doc)—a technology that would allow growing methanogens readily and safely in radioactive medium had not yet been developed. It would soon be (1). William (Bill) Balch,

a graduate student in Wolfe's lab, was about to develop such a method for other reasons: he would grow methanogens in pressurized serum bottles, a technique that was not only efficient and adaptable, but, what was important to us, safe for working with radioactively labeled cultures. By 1976 it had become possible for the two labs to collaborate.

When the first methanogen 16S rRNA oligonucleotide catalog rolled off the production line (June 1976), I was stunned by what we had; and thereafter our entire focus was to be on methanogens and the new group of organisms, the archaebacteria, that they would come to represent. We had discovered a "third form of life," a new "urkingdom"! Methanogens (and their yet undiscovered relatives) stood genealogically completely apart from the other bacteria we had so far characterized (which we came to call "eubacteria") and from the (small but representative) cluster of eucaryotes whose rRNA catalogs we had done. We were in for the ride of our scientific lives!

HITCHING UP THE TEAM

Like everyone else in those days, I had tacitly accepted that all bacteria were "prokaryotes." That is, they were all of a kind; all had stemmed from some common procaryotic ancestor (29, 31, 32). Suddenly confronted with a procaryote that didn't seem to be a procaryote, I had to ask myself why I had believed all bacteria to be procaryotes in the first place? No good reason, it turned out! When analyzed scientifically the "procaryote" didn't wash; there was no hard evidence, not even sound scientific reasoning to back it up (31, 32). We had hard evidence in hand and were in the process of gathering a lot more.

Within six months we had the rRNA catalogs of another five or so methanogens, with others in production, and *all* of them possessed the same anomalous type of rRNA (9, 11). That was it! We had shown methanogenesis to be monophyletically distributed. The exceptional variability in the group's morphologies and habitats was what needed explaining.

I had shared my "eureka moment" (upon assembling the first methanogen rRNA catalog) with my then post doc, George Fox, who, along with Bill Balch, had gotten the project actually up and running. George and I became the first to see the phylogenetic "light," to realize that there were actually *three* primary lines of descent (urkingdoms) on this planet, *not two*!

Convincing ourselves was not the problem. Convincing others was. It would be a hard sell. For reasons I could not understand at the time, literally all biologists believed in "*the prokaryote*." And it was not your typical scientific belief—always open to question. This was dogma, unshakable doctrine! Some biologists would take years to overcome their procaryote prejudice. Some have yet to do so. Most of those who came to accept the three urkingdom notion were in fact not even genuine "converts"; while they accepted eubacteria and archaebacteria to be genealogically distinct (representing two separate origins of bacteria), they held firm to the notion that both of them still had the *same* generic cellular organization (31, 32, 41).

So many times I have heard that archaea are "still just bacteria (procaryotes)" or words to that effect—they have the same *cellular organization*! What does this mean? In a purely scientific sense, it means nothing! The resemblances between the two are in essence negative—features not found in eucaryotic cells (viewed under a light microscope). It means nothing that archaebacteria and eubacteria both show simple rod, coccoid, or spiral shapes; it means nothing that both are typically much smaller than eucaryotic cells, that neither has any microscopically visible intracellular inclusions. All that these facts tell us is that neither type possesses certain features characteristic of eucaryotes. Nothing has been (or could be) said about the "common organization" the two supposedly share!

Consider an analogous situation on the metazoan level. Three kinds of eyes exist among animals. Their similarity is defined only functionally. Structurally the three are worlds apart. Might not the same situation exist with regard to archaebacterial and eubacterial organizations (41)? After all, there are no facts to refute the idea, and the extremely gross traits (cited above) that all "procaryotes" were supposed to share could easily be rationalized in general terms having to do with lifestyle more than anything else. Its organization is the most complex attribute the cell has. No biologist would accept that a trait this complex could have arisen more than once. Nor would any biologist accept that two independently evolved "procaryotic" cell types would be so organizationally similar that their differences would be trivial, uninteresting! The above metazoan example makes perfectly clear that if evolution builds two complex traits that resemble one another in some overall functional sense, it cannot build them so alike on the underlying (complex) structural level that they will fail to show major (nonhomologic) differences.

In sum, the notion that eubacteria and archaebacteria could have evolved independently yet have arrived at so similar a cellular organization as not to be of interest to the biologist, is absurd; it *contradicts all that we know about evolution!* Microbiologists were simply having their cake and eating it too. More has to be said about the "procaryote" simplification

and will be below, for it is pivotal to microbiology's development (or lack thereof) in the twentieth century.

MAPPING OUT THE TERRITORY

Having discovered what we thought to be a third primary line of descent (represented at the time solely by methanogens), we needed to explore its evolutionary implications. The theory of genealogical descent would *per force* be put to its greatest test yet! Evolution as we know it demanded that the new "urkingdom" display two fundamental characteristics:

1. It should comprise a number of major organismal groups very different from one another in their overall phenotypes. (My colleague Norman Pace used to refer to this as "kingdom level" diversity.)

2. The phenotypic features integral to the warp and woof of the basic archaeal cell design should strongly distinguish it from the eubacterial cell design (and vice versa)—features at least as striking as those that distinguish plants from animals, for example.

Fortunately, a small cadre of scientists had quickly come around to the three-urkingdom perspective (in addition to our collaborators Wolfe and Balch). In particular, I would mention the Germans Otto Kandler, Wolfram Zillig, and Karl Stetter. Kandler, whose studies at the time centered on (bacterial) cell walls, had visited the University of Illinois in early 1977 (prior to the publication of our work). He was one of the first outsiders whom we told about the three-kingdom concept, and the only one to understand and accept it upon first hearing! Wolfram Zillig, a group leader at the Martinsried Max Planck Institute, was an expert in DNA-dependent RNA-polymerases. Both he and Kandler had encountered certain anomalies in their own research that could be neatly explained on the basis of a three-urkingdom hypothesis. Karl Stetter, who had trained with both Kandler and then Zillig, also appreciated the significance of the archaebacteria—and proceeded to fashion a career as the greatest of the "swashbuckling" archaea hunters of the twentieth century; I wouldn't be surprised if most of the isolated archaeal species today have come from his laboratory. These three individuals were initially the most effective of all in promoting the study of the archaebacteria, especially among Europeans. They were a strong second front, as it were, in part because Europeans had been less affected by the procaryote propaganda that so dominated (micro)biology in North America.

WHERE ARE THE COUSINS?

Methanogens offered no clues as to their cousins. The unusual coenzymes that methanogenic biochemistry utilizes seemed to occur almost nowhere else. With Kandler's visit to Urbana, however, there came a break. Kandler already knew (as did others working in the cell wall field) that the walls of extremely halophilic bacteria did not contain peptidoglycan, which at the time had been taken as a defining characteristic of the "procaryote" (31). Moreover, he had just determined that the wall of one methanogenic species was also aberrant (though not of a halobacterial type) (15).

The walls of the would-be archaeabacteria were not the most important clue, however, for they would turn out to be nonhomologous among themselves. But the walls were pointing us in the right direction. The critical clue was to be atypical lipids possessed by the halophiles—ether-linked and branched chain (not ester-linked and straight chain, as the lipids of bacteria and eucaryotes are). These strange lipids would ultimately be found in all the archaebacteria characterized. They were the first of the universal phenotypic characteristics found exclusively in the archaebacteria!

Ether-linked lipids had been discovered by Morris Kates in the early 1960s (16, 20). Kates took them, as did everyone else at the time, to be adaptive, arising independently in certain organisms that grow in extreme environments. Shortly thereafter, the microbiologist Thomas Brock isolated two strange new types of bacteria that inhabited thermophilic niches, *Sulfolobus* and *Thermoplasma* (the former discovered also in Italy, and named *Caldariella*) (3, 7, 8). Another Tom, Thomas Langworthy, went on to show that the lipids of Brock's new isolates were of the ether-linked type but somewhat different from the extreme halophile lipids: *Thermoplasma* and *Sulfolobus* had "tetra-ether" lipids, which were back-to-back covalently linked versions of the halophile (diether) lipids (20, 21).

The pieces of the archaebacterial puzzle now began to fall into place. Not only were the phenotypically diverged cousins of the methanogens beginning to show up, but so were the traits common to all archaebacteria. The rRNA sequences that defined the archaebacteria, though phylogenetically very telling, should not be considered traits in the true (informative) phenotypic, organismal sense. These sequences (in effect gene sequences) tell you nothing about the overall phenotype, either of the group as a whole, or of the subordinate subgroupings therein. A tree based on rRNA sequence comparisons is basically just an abstract *genealogical* tree. Look at it as only a road map interconnecting towns. The contours of the map begin to appear, the towns bustle with people, only when the phenotypically informative traits become known. Those features common (and unique) to all archaebacteria then reveal the overall landscape of the archaebacterial world. The first common features to appear were:

1. The unusual ether-linked, branched-chain lipids.

2. The characteristic subunit structures of their DNA-dependent RNA polymerases (more similar to their eucaryotic than their [eu]bacterial counterparts) (48).

3. An unusual and characteristic modified version of the so-called common arm sequence in tRNAs, the archaeal version of which uniquely had the sequence Ψ-Ψ-C-G, rather than the customary eubacterial and eucaryotic T-Ψ-C-G (12). Also, the so-called D-loop of archaebacterial tRNAs contains no "D," dihydrouridine (12).

4. Cell walls that lack peptidoglycan, a negative trait, equivalent, say, to our metaphorical landscape's having no lakes (15).

5. A largely unique spectrum of antibiotic sensitivities vis-à-vis the eubacteria (4).

Now, the rRNA analyses began to tell us even more: the new archaebacterial urkingdom was larger than we first thought. While the halobacteria and *Thermoplasma* lay within the phylogenetic confines defined by the methanogens (44), the rRNAs of the *Sulfolobales* showed these organisms to be only a sister group to the others. (The in-group species would be called the *Euryarchaeota*; the *Sulfolobales* in turn would be the *Crenarchaeota* [43].) And, a notably deep (and still not understood) phylogenetic divide separated the two groups (39, 44).

REACTIONARY SCIENCE AT ITS BEST

To this point my narrative may have felt like emerging from a dark forest (ignorance) to find before you a beautiful panoramic vista, with the clear sky of a bright future above you—a sky clear except for one small cloud on the horizon (a small cloud called the "procaryote"). As it came nearer, however, that small cloud would darken and broaden into a violent storm (as we will now see).

Things had gone well with the archaebacteria to begin with—*before* their formal presentation to the scientific community and the public at large. By the fall of 1977 the time for a formal debut had come; now the world would see the new urkingdom, the hitherto unknown "third form of life." I had alerted both the NSF and NASA, my funding agencies, that what we were about to publish might be of particular interest to them, might make a little splash—the storm had yet to materialize. The two agencies were indeed interested and decided to make a joint statement to the press on the day the first of our two articles appeared in print. That would be the 3rd of November 1977. I realized something bigger than anticipated was afoot when one or two days before the formal press release the phone started ringing and a reporter from *The New York Times* showed up in my office.

November 3rd. There they were! The archaebacteria! Right there on the front page of *The New York Times* (and a lot of other newspapers around the globe).

But unbeknownst to anyone at the time, a telling coincidence had occurred, one that set the stage for an epic drama, a struggle for the soul of biology—a drama in which today we are all actors. November 3rd just so happened to be the date chosen by the then president of the U.S. National Academy of Sciences, Philip Handler, to release an official statement heralding the dawn of the cloning era (signaled by the recent cloning in bacteria of the gene for the growth hormone somatotropin). As it can now be appreciated, that fortuitous coincidence was a foretaste of the first skirmish in the ideological struggle between what would become the biomedical-industrial complex and resurgent evolution. Our "third form of life," which touched on one of the deepest chords in human nature (i.e., where we came from), completely wiped the press release announcing the era of "Man-the-medical-miracle" off the front pages of the papers. I was overjoyed at the public's appreciation of our work (9, 42)!

The celebratory mood, the joy at the public and general scientific reaction, was short-lived, however. As alluded to above, resistance to the concept had already surfaced to some extent in the microbiological community. Now the storm hit full force! On the day *The New York Times* announced our discovery of a "third form of life" on its front page, my colleague Ralph Wolfe received a telephone call from his friend, the Nobel Laureate Salvador Luria (whom he did not initially identify), an upset Salvador Luria. According to Wolfe, Luria told him in no uncertain terms to publically dissociate himself from this scientific fakery or face the ruination of his career. In a recent recounting of the episode (47), Ralph said he was so humiliated he "wanted to crawl under something and

hide," but he did tell Luria that supporting evidence for the claim had been published. When informed that the scientific journal was the *Proceedings of the National Academy of Sciences* (a fact that had been mentioned in *The New York Times* article), Luria, a bit befuddled, had blurted out that the October issue had just arrived but he hadn't looked at it yet. Fortunately, Ralph left for a planned out-of-town family gathering the next day (47) and thereby escaped any further humiliation.

As you might expect, I saw the episode and its overall significance differently. How could this Luria fellow have the temerity to excoriate his friend and my colleague like that? What pulpit was *he* preaching from? It appears that he had blustered at Ralph something to the effect that: "Everybody knows that all bacteria are procaryotes; there can't be any such thing as a 'third form of life!'" Irony of ironies! As time (and the diligence of a particular scientific historian) have shown, the fakery lay not in *our* work, but in the procaryote concept itself (26): it is now clear that the "procaryote" was mere guesswork (more on this below.) But in the heyday of the procaryote, which this was, the true believers were out to pillory us for our heresy: how dare we slander their *Procaryote!*

That day in November 1977 had laid bare a fundamental structural problem that affected all of biology, and the condition of microbiology was only the most obvious symptom of it. The name of that problem was *molecular biology*—under whose hegemony twentieth-century biology had developed. Molecular biology's world view was that of classical physics—a world view that is inimical to biology. Without doubt spectacular progress occurred under molecular biology's hegemony, but the result has been a superhighway to a biological dead end—unless one looks at bioengineering as the ultimate goal of biology! Our encounter with microbiologists over the archaebacteria was by no means the parochial taxonomic squabble one might think. The issue was not confined to microbiology, nor even to biology as a whole. The ultimate issue is the relationship between biology and society—their conjoined future. The archaebacteria were telling twentieth-century biology (microbiology in particular) that it had forsaken its roots and had no more chance of becoming mankind's ultimate view of biology than an oak tree has of growing on a flat rock. Such advice ran afoul of biology's conventional wisdom, and biology, microbiology in particular, was reacting accordingly.

As I stated earlier, one of my goals in establishing the program of phylogenetic analysis in the first place had been to restore a sadly lacking evolutionary perspective/spirit to biology. I had long cautioned that biology was on the verge of being taken over by "science mongers and technological adventurists." The great fear was that biology would cease being a basic scientific discipline, becoming merely an olio of separate subdisciplines, whose common focus was applied problems—the complaints of the society, as it were. This is acceptable and proper up to a point, of course, but beyond that point the framework of basic biology crumbles and the basic science turns into engineering. And *this must not happen*! The "procaryote" was both the symbol and the linchpin of biology's structural problem. (Try to imagine how different twentieth-century biology might have been had microbiologists stayed true to their birthright, i.e., the world of microorganisms—rather than turning their discipline into a theme park for molecular gawking.)

The discovery of the archaea had provided one of those special moments when things are put into grand perspective: it was a wake-up call to microbiology and provided a time for biology as a whole to reassess its head-long dash into reductionism. Introspective opportunities such as this are rare, ephemeral, and transient. In contrast the grip of convention is strong, persistent, and compelling. The discovery of the archaea was a moment that should have been grasped. Some of us tried hard to bring this about. If only microbiology would change back to the kind of discipline that Beijerinck knew and was trying to develop, there would be hope. But there were so many road blocks, so many citadels to conquer. The immediate road block, of course, was the mighty "prokaryote," which seemed to deny everything that biology is—organism, ecology, evolution. From the late 1970s onward I percieved defanging the procaryote as absolutely essential if microbiology (and biology) were to be set right again, and I devoted much of my energy to it.

DECONSTRUCTING THE PROCARYOTE

The prokaryote-eukaryote dichotomy of the 1960s was not an astonishing formulation. It neither took scientists by surprise nor opened up new avenues of research. It was a rhetorical discovery, one which involved summoning lost words from a foreign scientist in an obscure journal and synthesizing contemporary data based on molecular biology and electron microscopy. It was greeted with accolades not because of its novelty, but because it seemed to close, once and for all, long-simmering issues. It confirmed and clarified the differences between bacteria and blue-green algae on the one hand, and viruses and the cells of protists, fungi, plant and animal, on the other. The belief in the monophyly of bacteria, moved by historical inertia, and strengthened by molecular biology's model organism [*E. coli*], resulted in a crowning achievement: the legitimizing of the new kingdom, Monera, or of the superkingdom, Prokaryota. The prokaryote-eukaryote dichotomy thus marks a signal moment in the development of biology (26).

And, I will add, a moment that brought microbiology's conceptual development to a halt! Stanier and van Niel's "procaryote" paper of 1962 (32) intended to, and did, accomplish three things for the field. (i) It stated outright the problem that all microbiologists intuitively knew to exist from the outset: ". . . the abiding intellectual scandal of bacteriology has been the absence of a clear concept of a bacterium . . . the problem of defining these organisms as a group in terms of their biological organization is clearly still of great importance, and remains unsolved" (32)—thereby exposing the discipline's roots in *organism* (biological organization). (ii) It severed these roots by declaring the old (phylogenetic) approach to the problem unworkable and unnecessary—false assertions both. (iii) It bunched together what remained and stuffed it into the narrow vase of reductionism—having left out those parts that wouldn't fit. As a result, we have before us today a conceptual bouquet that is thoroughly wilted—unified in name only, the "discipline of microbiology."

Microbiology's historical development and the climate in which it occurred underlie and define the whole issue. That history effectively began in the last half of the nineteenth century, when microbiology was first attempting to gather itself into a coherent biological discipline. The principals whose works best epitomize this phase in microbiology's history are Ferdinand Cohn, Louis Pasteur, Martinus Beijerinck, and Serge Winogradski.

It was self-evident to Cohn that bacteria stood apart from other forms of life. He distinguished the "Schizophytes" as the ". . . first and simplest division of living beings . . ." (5, p. 201). Beijerinck also subscribed to this general view, but from a remarkably enlightened perspective. For Beijerinck the microbial world was ". . . that part of nature which deals with the lowest limits of the organic world, and which constantly keeps before our minds the profound problem of the origin of life itself" (quoted in reference 34). Beijerinck has been pigeonholed historically as the father of microbial ecology (34), but that narrowing of him hardly does the man or his scientific perspective justice. In his day Beijerinck (unlike most of his successors) was a complete microbiologist/biologist. He did not concern himself solely with culturing individual bacterial species and studying them physiologically (although he had made revolutionary contributions to the area by his introduction of enrichment culturing methodology). By his own account, his "... approach can be concisely stated as the study of microbial ecology, i.e., of the relation between environmental conditions and the special forms of life corresponding to them" (34). Beijerinck stood for a microbiology that was biology in the full sense, which could synthesize from its various roots a greater whole: its medical origins (Koch's world); its foundation in practical ecological concerns (fermentation, Pasteur's somewhat reductionist medical world); Beijerinck's own pioneering work on geomicrobiology and virology; and the rich and tangled involvement of microbes with the whole of Nature (so well captured by the image of the Winogradski column—or the hot spring colorations at, say, Yellowstone Park). For Beijerinck microbial ecology was definitely not a secondary pursuit, derivative of biochemistry. His microbiology was synthetic and holistic, not reductionistic, mechanistic, or derivative.

Beijerinck's perspective might well have been the broadest and most coherent foundation that microbiology could ever have known. But circumstances all worked against the development of his holistic viewpoint: The methodologies required to usefully realize it would not be in place for roughly half a century; there was no phylogenetic framework within which to structure and develop it, and (unlike developmental biology) microbiology had no conceptual sea wall to staunch the surge of reductionism—a reductionism that denied holism and, so, evolution per se. Finally, there was the overarching "linear" perspective that had worked so well in the world of classical physics, and with which classical physics (through molecular biology) was now infecting biology (40). The immediate success of molecular biology's confident reductionism gained it a string of converts among conceptually struggling microbiologists.

It comes as no surprise, then, that Beijerinck's appointed successor, Kluyver, who was trained as a (natural products) chemist, would take microbiology in a doggedly reductionist direction. The "unity of biochemistry" became microbiology's new leit motif, and that plus the perceived need to elucidate all the nuanced variations on the central biochemical themes, constituted the territory to be conquered (17, 18, 36). The new approach to microbiology that Kluyver's "Delft School" pioneered was known as "comparative biochemistry" (17). The biology, however, lay mainly in the name. (A holistic biologist would be fully justified in calling this a "piñata paradigm"—for at the end of the day the "piñata" (the organism) is gone, but there remains plenty of [molecular] candy for the kids!)

Every scientific discipline reflects a particular scientific mythology, which is its axiomatic foundation.

In microbiology's case that mythology needed to be grounded in "organism," in biological organization. This had not happened. By the mid-twentieth century the piñata paradigm, then in full force, had yielded bacterial biochemistry galore—all manner of new biochemicals and biochemical pathways had been uncovered and the underlying unity of biochemistry seemed assured. Still, one of the best-recognized microbiologists of his day, Roger Stanier, found a need to denounce the state of microbiology (quoted above) as an "abiding scandal" (32). He had trenchantly sensed that microbiology's scientific mythology had yet to be properly established.

Unfortunely, Stanier and van Niel's purpose in writing *The Concept of a Bacterium* (32) was not to *address* the long-standing foundational issue. It was, as said, to get *rid* of it, to cover it up—and in so doing (appear to) provide microbiology with its badly needed (organismal) touchstone. The formal definition of "prokaryote" read as follows: "The distinctive property of bacteria and blue-green algae is the procaryotic nature of their cells" (32), from which the statement "... we can therefore safely infer a common [prokaryotic] origin for the whole group in the remote evolutionary past" must follow (31). No more, no less!

In the term's obscure past (26), the prefix "pro-" had obviously expressed evolutionary relationship of some sort between bacteria and the "higher forms" of life. Both Cohn and Beijerinck had seen bacteria as primitive and perhaps predating the "higher forms"— as did many other microbiologists then and later. Evolutionary relationship was *not* intended, however, by the "new" procaryote; it was not a part of its new wardrobe. An evolutionary perspective had effectively been proscribed! Thus the "pro-" in "procaryote" would now remain unexplained in the term's revived/revised usage (32). Note that "procaryote" is for all intents and purposes absent from the literature before the 1960s (26). The word "procaryote" fails even to appear in what at the time was microbiology's primary teaching instrument, the first (1957) edition of *The Microbial World* (30). Yet the term and concept were the centerpiece in the second (1963) edition (31). Within a decade of its announcement "procaryote" was on every biologist's lips; it had become conventional wisdom (29). What was this bizarre turn of events all about?

It was all about "The King is dead; long live the King!" It was about regime change, conceptual transition. Microbiology's now moribund attempts to pull itself together as an organismal discipline (in keeping with Beijerinck's vision) had altogether failed. Microbiologists had been unable to develop the needed framework of a "natural classification." Microbial ecology had been stopped in its tracks by its inability to define niches in organismal terms—to representatively sample, isolate, or phylogenetically characterize the bacteria therein. Bacterial evolution (phylogeny) had never gotten beyond the hopeful speculation stage—although that in itself had given microbiology a certain sense of direction. Yet the whole enterprise was now declared null and void—a waste of time (26, 29, 30).

Ironically, this cynical, dismissive attitude that "procaryote" embodied developed just when the molecular methodology for solving the problem of bacterial relationships was coming onto the scene, and the coming of a new field of "protein taxonomy" had already been proclaimed (6). Organismal thinking, however, now went against reductionist microbiology's grain.

Microbiologists of an earlier day had struggled to define bacteria holistically—as their botanist and zoologist colleagues had done so successfully with plants and animals. However, microbiology's new reductionist outlook increasingly trivialized the *emergence* that was "organism" (biological organization). And by midcentury, the time had come to eliminate organismal thinking entirely and move on (26).

The slight of hand by which this transition was accomplished is remarkable in its simplicity. Bacteria were flat-out *declared* all to be of a kind—no scientific discussion needed! Placing them all into a newly named highest level taxon, *Procaryotae* (22), served reflexively to justify the *coup*. Classification of bacterial species, which henceforth would serve only utilitarian purposes, was relegated to the dull pedantry of determinative classification (31, 35).

The key biological problem that bacteria presented, their biological organization, was simply filed away under "unimaginable complexity." No one knew, or could *possibly* know anything about it! Nevertheless, that organization was declared to be the same, of the same genre, in *all* bacteria. What a convenient conceptual "black hole"! The organism, the biology, evolution, disappeared into the impenetrable sink of *organizational* sameness! All of the *biological* concerns disappeared, like peas in a shell game. In this way microbiology could "know itself" to be an organismal discipline in camera, as it were, without having to admit it (through its actions) to the reductionist world around it!

Microbiologists will some day have to admit that in accepting the procaryote, they were rejoicing in the Emperor's New Clothes. And the archaea, like the child in the nursery tale, had shouted the obvious.

The famed physicist Erwin Schroedinger condemns what he calls "guesswork solutions" to scientific questions. Science must reject these because such

a ". . . fake removes the urge to seek after a tenable answer." Adding, "So efficiently may attention be diverted that the answer is missed even when, by good luck, it comes close at hand" (28). There you have it: *the procaryote*!

Nevertheless, only by seeing the "procaryote" for what it *really* is—scientifically valid *or not*—can microbiology get back on its naturally intended course. *The whole intent of the "procaryote" was to supply bacteriology with the semblance of a hungered-for mythology, the axiomatic foundation the discipline required to know and present itself as a coherent organismal discipline.* Among other things, this dynamic would explain the dogged loyalty microbiologists have shown to the concept, their rapid and unquestioning acceptance of it, as well as their immediate rejection (a decade and a half later) of the three-urkingdom concept (which had exposed their exposed "Emperor"); it would explain microbiologists' consistent adherence to the modes of thought and academic organization inherent in the simplistic "procaryote/eucaryote" dichotomy—not to mention microbiology's persistent blindness with regard to evolution in general.

RECAPITULATION AND CONCLUSION

Where Are We?

This narrative has followed in both fact and feeling the discovery of the archaea. It began on a note of joy, the feeling of power, potential, and flat-out awe that accompanies any major discovery. The archaea represented a radical departure in the way we looked at life on this planet. Biologists had thought that all life on earth arose originally from one of two primary lines of descent, either the eucaryotic or the so-called procaryotic. The archaea were here to set that right. There were three, not two, primary lines of descent, and the procaryote was a false concept—which never had any scientific evidence to support it in the first place, and had never been put to proper scientific test.

Throughout the history of microbiology bacteria had been seen as a grouping of some sort. But that grouping, which customarily went under the names Schizomycetes or Monera, was in effect gradistically defined: bacteria were those "simple" organisms that were not eucaryotic in cellular type. The eucaryote may have been phylogenetically defined, but not the procaryote. Phylogenetic relationships among the bacteria had been impossible to determine (until the advent of molecular sequencing), and this made for a chronic foundational problem in the discipline. As Roger Stanier pointed out in the 1960s, microbiology did not know itself, for it had no biological concept of the organisms it studied (32). Conceptually ostracized by molecular reductionism, microbiologists had sought to resolve this foundational problem and move on—that is, join the reductionist ranks. There was no need to resolve the matter scientifically in this climate; it could simply be done away with rhetorically, by declaration!

Thus, by fiat, all bacteria shared some hypothetical common "procaryotic" cellular organization and, therefore, stemmed from a common procaryotic ancestry (31). The purpose of the procaryote guesswork had merely been to make microbiology's foundational problem go away, by effectively taking the organism out of consideration.

The discovery of the archaea would settle this foundational issue scientifically, not rhetorically. Molecular sequencing evidence showed that the bacteria were indeed only a *gradistically* defined group. Phylogenetically, "bacteria" comprised *two* distinct groupings, not one (41). Microbiology's foundational issue clearly had not been settled; the nature of the *organism* must remain a central concern for the discipline.

The message of the archaea was only superficially one of organismal discovery. There was a deeper, more important message embedded therein: namely, that in adopting reductionist fundamentalism (40)—which denied the basic nature of both organisms and evolution—microbiology (and biology as a whole) had been going down the wrong road for at least half a century, and it was time to stop and reassess the matter.

Though reluctant at first, microbiology did eventually come around to accepting the archaea. However, if one were to consider this acceptance a victory for the archaeal perspective, it had to be a pyrrhic one. All that microbiology had done was to add a new high-level grouping to its taxonomy. Microbiology's world view had not changed one iota in the process! There was still *no* recognition that microbiology is a true evolutionary, organismal discipline! The creative energy inherent in the archaeal discovery had been siphoned off—their call for biological reformation (microbiology in particular) drained of its spirit. For me the archaea (at this point) were a failure. They had effectively changed *nothing*!

As a discipline with no self-image, no proper axiomatic foundation, microbiology could have been only a passive bystander in the titanic struggle that was twentieth-century biology. For most of the twentieth century the discipline simply drifted with the conceptual tides. After several decades of cajoling I finally gave up on trying to reform microbiology (the discipline as academically constituted, that is). The problem of the microbial world clearly remains, but the inflexible and ineffective formal discipline of mi-

crobiology that emerged from the twentieth century seems incapable of dealing with it.

The problem of the microbial world has long ago transcended microbiology and is on its way to becoming a forefront for biology as a whole in the twenty-first century. Microorganisms dominate the planet's biosphere. They make up most of the planet's living biomass. They are the metabolic engines of the planet. They were essential to the evolution of all macroscopic life on earth (both the mitochondrion and the chloroplast originated as bacterial endosymbionts). Without bacteria all life on this planet would disappear in short order—the same cannot be said for macroscopic life! We need to assimilate this message fully and study the microbial world accordingly.

Biology today is in conceptual crisis. I do not consider microbiology to be the primary problem, however. Molecular reductionism is. Microbiology—a properly constituted microbiology—is the basis for a solution. Beijerinck showed us the beginnings of such a microbiology a long (and long forgotten) time ago. Here was (could have been) microbiology in the full sense—not that pale residue of it one sees later in the reductionist piñata paradigm. For Beijerinck (see above) the organism was no mere sum of its parts; first and foremost the study of bacteria was ecological, one in which the organismal *community* and its environment were a paramount concern. At the end of the day microbiology (as microbial ecology) was a study in ". . . the profound problem of the origin of life itself" (34)—and this would include the evolution of the organism-environment dichotomy in the first place.

Beijerinck's microbiology was ultimately a study in the emergence of biological organization. Yet *emergence* was something proscribed by the new sum-of-the-parts world of molecular biology. No reductionist, Beijerinck seemed to view that which emerges from a given level of biological organization as just as important, just as fundamentally biological as what underlay and gave rise to the given level in the first place.

If there is to be a basic biology in the future (instead of an applied, engineering discipline) then biology must be freed from the shackles of reductionism. (This is especially important in microbiology's case, where many of its current practitioners still view biochemistry, metabolism, as the *essence* of a bacterium—to them bacteria are still bags of enzymes!)

If you were to ask a thorough-going molecular reductionist today what *fundamental* problems in biology remain to be solved, the answer you are most likely to get would be "none." How could there be any such when *The Gene* (in the molecularist's sense) had answered the great question of biology, "What is Life?" (27). To the molecular coterie of the mid-twentieth century, molecular biology seemed "the biology to end all biology" (14)!

Going into the twentieth century, biologists considered evolution to be the foundational concept of biology. Molecular biology was quick to replace evolution with *The Gene*, however. For these reductionists, evolution, far from being significant to any fundamental understanding of biology, was no more than an epiphenomenon (14)! As it now stands, however, it is not that evolution is epiphenomenological, it is that a reductionist view of biology is *incomplete and misleading.*

The twentieth century gave rise to two conceptual revolutions in science, one in biology—the effects of which all biologists working today can feel in their scientific bones—and (a more subtle and philosophically profound) one in physics. Both concerned the reductionist world view. The *biological* revolution promulgated reductionism (40). The one in *physics* brought it into question. Incredible! Just as molecular biology was rebuilding biology along reductionistic lines, the discipline of physics (which had given rise to [classical] scientific reductionism in the first place) was in the process of dismantling it. This is surely one of the great ironies in biology's history—a grand twentieth-century biological cathedral erected on an old and crumbling nineteenth-century foundation!

In replacing evolution (descent with variation) as the fount and focus of biology by the gene, molecular biology had replaced a (nascent) process perspective by a mechanistic materialistic one. If evolution (biology) is anything, it is a quintessentially nonlinear process—the very aspect of reality that could not be encompassed by the linear mathematics that had *defined and delimited* the world view of classical physics. Thus, the molecularist cannot even *begin* to recognize evolution, much less accord it the scientific respectability that is its due. And that is why the molecular paradigm now finds itself aground on the shoals of (biological) complexity, biology's true nature.

Almost needless to say, evolution, though antithetical to twentieth-century molecular reductionism, is in complete harmony with the complex dynamic perspective in physics today. Both are studies in emergent organization. Biologists are going to see major changes in their world as the synthesis of the studies of biology and complex systems—which will become the biology of the twenty-first century—comes into being.

The Dawning of a New Biology

Have you noticed that nonlinear mathematics (pictorially presented) has an innate aesthetic appeal (I am thinking here of those delightful books of fractal representations one can see on so many bookshelves these days)? These types of pictures are not new; only the underlying mathematics is. Cultural images such as these go back to the dawn of mankind. Nonlinear mathematics seems somehow to resonate deeply in the human psyche.

To me the juxtaposition of aesthetic feeling and scientific understanding here means one thing! In the (near) future, when evolution is couched in a nonlinear mathematical framework, representations of evolution will be not only spectacular in their simplicity, but compelling in their beauty. There is a deep (yet to be appreciated) connection between evolution and our study thereof.

Although the discovery of the archaea had not been transformative of microbiology, a seed had been planted and it quietly began to grow. Interest in isolating novel and interesting bacteria was revived. Bacterial taxonomists began to see rRNA sequence as the gold standard for species identification and classification. It seemed possible for the first time that the day would come when all the bacterial species could be brought together in one grand, comprehensive classification—all those that could be cultivated, that is. Microbiologists had long thought that the reserve of undetected bacterial novelty had been pretty well exhausted, that most of the high-level bacterial taxa had been discovered. The archaea, of course, had shaken that belief somewhat, but even with their discovery nothing spectacular, world shaking, was expected to come from further studies in bacterial diversity. How wrong we were!

While I was visiting my colleague Norman Pace in Denver back in the early 1980s, he suggested a novel idea for how to identify bacteria (24). Since bacteria could now be identified on the basis of their rRNAs alone, why bother growing the organism in the laboratory? Why not take the rRNAs (or their genes) directly from the setting in which the organisms naturally grow? That way, you would not be at the mercy of the vagaries of laboratory cultivation and so could, in principle (by isolating enough of the same type of gene directly from a given environment), have an *unbiased* representation of all the bacterial types in that environment.

Such a powerful methodology changes even the way organism/environment questions are initially framed—just as Beijerinck's enrichment culturing had done in its day (34). The difference between the new and the old methodology would be like shopping with or without access to a shopping catalog. If you had such a catalog, you might even decide you wanted to look for something other than what you originally intended (or in addition to it). Very powerful! What's more, Pace told me, all of this could be done simply by adapting some of the currently available molecular genetic technology.

Pace's idea was no innocent extension of the bacterial characterizations we and others had been doing (based on cultured strains). It was a whole new way of exploring the organism-environment relationship. Pace had questioned what all microbiologists had previously taken for granted—namely, that to identify and know anything about the relationship of a particular bacterial species to other bacteria, that species had first to be grown in the laboratory. Culturing methods imposed crippling limitations on what organisms could be isolated. There was no way one could study organisms in their natural settings in any comprehensive manner (24). Pace told me he planned to go up to Yellowstone National Park as soon as possible to get some samples to test out his idea. I tagged along. Unbeknownst to any of us, Pace's methodology (and the extensions of it to come) were opening up the future of twenty-first-century biology.

By the early 1980s my laboratory and others had characterized several hundred bacterial strains by the rRNA method. In 1987 I would review the field (39) and conclude that there are about a dozen major bacterial groupings (phyla) that we know of, and that when all was said and done several more might surface. How wrong that projection was! To date Pace's methodology has produced in the range of 100,000 unique rRNA species from various environments, and these represent on the order of 50 new bacterial phyla. *But* there is no sign of letup—almost all of the separate environmental rRNA isolates have no 100% sequence matches anywhere else in the rRNA sequence database. In other words, even this large a number of rRNA reads has been a remarkably sparse sampling of the total number of "species" that are out there in Nature. So large a number of unique species of rRNA with no end in sight is starting to question the very concept of bacterial species itself! The advent of environmental genome sequencing has done nothing to resolve the situation. It has only deepened the conundrum and brought it more sharply into focus.

The Microbiology Yet To Come

Given the context of this chapter I will be relatively brief. Molecular reductionism is now spent as a conceptual force and has settled into being a most useful body of *technology*. Microbial biology in the meanwhile has undergone a conceptual and methodological revolution of its own, freeing itself from its self-inflicted intellectual confinement.

The future of biology lies in microbial ecology. This is not the *faux* ecology that microbiologists knew throughout most of the twentieth century. That was a "real world" defined by microbial biochemistry, in which microbial ecology was merely a divertissement. Neither is it the "traditional," academically constituted, (macro) ecology, which never grounded itself in the reality of basic twentieth-century biology and so ignored the sleeping giant of microbial ecology, and still does so. It is Beijerinck's microbial ecology: fundamental in its own right; evolutionary to its core. What most distinguishes Beijerinck from later-day microbiologists is the feeling for the emergence of the whole. And that is what will distinguish twenty-first-century biology from the biology of the molecular era.

It is remarkable to see the confluence that is bringing about the new biology. Molecular biology, despite its aversion to evolution, is moving in an evolutionary direction—compelled by its own technology (principally the capacity to sequence genes). Microbiology, after focusing itself for the better part of a century on mechanism (to the near exclusion of microbial ecology and a total disregard of evolution) is also being compelled by developments in these "secondary" areas to reorient itself precisely in their direction. A synthesis of all biological disciplines around an evolutionary focus is in the making.

The difference between what twentieth-century biology was and what twenty-first-century biology will be is subtle (abstract) but profound. The twentieth-century biologist's goal was to see biology as mechanism, to understand it in terms of the static, material aspect of Reality. Form was everything! In this view evolution becomes an epiphemonenon—a property, an incidental characteristic of (mechanistic) biological systems. The twenty-first-century biologist has a different goal: to see biology as arising out of complex dynamic process. In this view biological form becomes phenomenological—a characteristic of a Universal Evolutionary Process. To put it in the vernacular, "Evolution came first!"

The question "What is life?" is precisely the question "What is evolution?"

Acknowledgments. The author's studies are supported by grants from the Department of Energy and the National Aeronautics and Space Administration.

REFERENCES

1. **Balch, W. E., and R. S. Wolfe.** 1976. New approach to the cultivation of methanogenic bacteria: 2-mercaptoethanesulfonic acid (HS-CoM)-dependent growth of Methanobacterium ruminantium in a pressurized atmosphere. *Appl. Environ. Microbiol.* **32:**781–791.
2. **Breed, R. S., E. G. D. Murray, and N. R. Smith (ed.).** 1957. *Bergey's Manual of Determinative Bacteriology*, 7th ed. Williams & Wilkins, Baltimore, Md.

3. Brock, T. D., K. M. Brock, R. T. Belly, and R. L. Weiss. 1972. Sulfolobus: a new genus of sulfur-oxidizing bacteria living at low pH and high temperature. *Arch. Mikrobiol.* **84**:54–68.

4. Cammarano, P., A. Teichner, P. Londei, M. Acca, B. Nicolaus, J. L. Sanz, and R. Amils. 1985. Insensitivity of archaebacterial ribosomes to protein synthesis inhibitors. Evolutionary implications. *EMBO J.* **4**:811–816.

5. Cohn, F. 1875. Üntersuchungenüber Bacterien II. *Beitr. Biol. Pflanz* **3**:141–208.

6. Crick, F. H. C. 1958. The biological replication of macromolecules. *Symp. Soc. Exp. Biol.* **12**:138–163.

7. Darland, G., T. D. Brock, W. Samsonoff, and S. F. Conti. 1970. A thermophilic, acidophilic mycoplasma isolated from a coal refuse pile. *Science* **170**:1416–1418.

8. De Rosa, M., A. Gambacorta, G. Millonig, and J. D. Bu'Lock. 1974. Convergent characters of extremely thermophilic acidophilic bacteria. *Experientia* **30**:866–868.

9. Fox, G. E., L. J. Magrum, W. Balch, R. S. Wolfe, and C. R. Woese. 1977. Classification of methanogenic bacteria by 16S ribosomal RNA characterization. *Proc. Natl. Acad. Sci. USA* **74**:4537–4541.

10. Fox, G. E., K. R. Pechman, and C. R. Woese. 1977. Comparative cataloging of 16S ribosomal RNA: molecular approach to procaryotic systematics. *Int. J. Syst. Bacteriol.* **27**:44–57.

11. Fox, G. E., E. Stackebrandt, R. B. Hespell, J. Gibson, J. Maniloff, T. A. Dyer, R. S. Wolfe, W. E. Balch, R. Tanner, L. J. Magrum, L. B. Zablen, R. Blakemore, R. Gupta, L. Bonen, B. J. Lewis, D. A. Stahl, K. R. Luehrsen, K. N. Chen, and D. R. Woese. 1980. The phylogeny of prokaryotes. *Science* **209**:457–463.

12. Gupta, R. 1984. *Halobacterium volcanii* tRNAs. Identification of 41 tRNAs covering all amino acids, and the sequences of 33 class I tRNAs. *J. Biol. Chem.* **259**:9461–9471.

13. Jones. D., and P. H. Sneath. 1970. Genetic transfer and bacterial taxonomy. *Bacteriol. Rev.* **34**:40–81.

14. Judson, H. F. 1996. *The Eighth Day of Creation: Makers of the Revolution in Biology.* Cold Spring Harbor Laboratory Press, Plainview, N.Y.

15. Kandler, O., and Hippe, H. 1977. Lack of peptidoglycan in the cell walls of *Methanosarcina barkeri. Arch. Microbiol.* **113**:57–60.

16. Kates, M. 1972. Ether-linked lipids in extremely halophilic bacteria, p. 351–398. *In* F. Snyder (ed.), *Ether Lipids, Chemistry and Biology.* Academic Press, New York, N.Y.

17. Kluyver, A. J. 1931. *The Chemical Activities of Microorganisms.* University Press, London, United Kingdom.

18. Kluyver, A. J., and H. J. L. Donker. 1926. Die Einheit in der Biochemie. *Chem. Zell. Gew.* **13**:134–190.

19. Lamour, V., S. Quevillon, S. Diriong, V. C. N'Guyen, M. Lipinski, and M. Mirande. 1994. Evolution of the Glx-tRNA synthetase family: the glutaminyl enzyme as a case of horizontal gene transfer. *Proc. Natl. Acad. Sci. USA* **91**:8670–8674.

20. Langworthy, T. A., M. E. Smith, and W. R. Mayberry. 1972. Long-chain glycerol diether and polyol dialkyl glycerol triether lipids of Sulfolobus acidocaldarius. *J. Bacteriol.* **112**:1193–1200.

21. Langworthy, T. A., M. E. Smith, and W. R. Mayberry. 1974. A new class of lipopolysaccharide from Thermoplasma acidophilum. *J. Bacteriol.* **119**:106–116.

22. Murray, R. G. E. 1974. A place for bacteria in the living world, p. 4–9. *In* R. E. Buchanan and N. E. Gibbons (ed.), *Bergey's Manual of Determinative Bacteriology,* 8th ed. Williams & Wilkins, Baltimore, Md.

23. Nagel, G. M., and R. F. Doolittle. 1991. Evolution and relatedness in two aminoacyl-tRNA synthetase families. *Proc. Natl. Acad. Sci. USA* **88**:8121–8125.

24. Olsen, G. J., D. J. Lane, S. J. Giovannoni, N. R. Pace, and D. A. Stahl. 1986. Microbial ecology and evolution: a ribosomal RNA approach. *Annu. Rev. Microbiol.* **40**:337–365.

25. Sanger, F., G. G. Brownlee, and B. G. Barrell. 1965. A two-dimensional fractionation procedure for radioactive nucleotides. *J. Mol. Biol.* **13**:373–398.

26. Sapp, J. 2005. The prokaryote-eukaryote dichotomy: meanings and mythology. *Microbiol. Mol. Biol. Rev.* **69**:292–305.

27. Schroedinger, E. 1944. *What is Life?* Cambridge University Press, Cambridge, United Kingdom.

28. Schroedinger, E. 1954. *Nature and the Greeks.* Cambridge University Press, Cambridge, United Kingdom.

29. Stanier, R. Y. 1970. Organization and control in prokaryotic and eukaryotic cells, p. 39–54. *In* H. P. Charles and B. C. J. G. Knight (ed.), The Society for General Microbiology, Symposium 20. Cambridge University Press, Cambridge, United Kingdom.

30. Stanier, R. Y., M. Doudoroff, and E. A. Adelberg. 1957. *The Microbial World.* Prentice-Hall, Inc., Engelwood Cliffs, N.J.

31. Stanier, R. Y., M. Doudoroff, and E. A. Adelberg. 1963. *The Microbial World,* 2nd ed., Prentice-Hall, Inc., Engelwood Cliffs, N.J.

32. Stanier, R. Y., and C. B. van Niel. 1962. The concept of a bacterium. *Archiv. Mikrobiol.* **42**:17–35.

33. Uchida, T., L. Bonen, H. W. Schaup, B. J. Lewis, L. Zablen, and C. Woese. 1974. The use of ribonuclease U2 in RNA sequence determination. Some corrections in the catalog of oligomers produced by ribonuclease T1 digestion of Escherichia coli 16S ribosomal RNA. *J. Mol. Evol.* **3**:63–77.

34. van Iterson, G., Jr., L. E. den Dooren de Jong, and A. J. Kluyver. 1983. *Martinus Beijerinck: His Life and Work.* Science Tech Inc., Madison, Wisc.

35. van Niel, C. B. 1946. The classification and natural relationships of bacteria. *Cold Spring Harbor Symp. Quant. Biol.* **11**:285–301.

36. van Niel, C. B. 1949. The "Delft school" and the rise of general microbiology. *Bacteriol. Rev.* **13**:161–174.

37. Vishwanath, P., P. Favaretto, H. Hartman, S. C. Mohr, and T. F. Smith. 2004. Ribosomal protein-sequence block structure suggests complex prokaryotic evolution with implications for the origin of eukaryotes. *Mol. Phylogenet. Evol.* **33**:615–625.

38. Woese, C. R. 1970. The genetic code in prokaryotes and eukaryotes. *In* H. P. Charles and B. C. J. G. Knight (ed.), *Organization and Control in Prokaryotic and Eukaryotic Cells,* p. 39–54. The Society for General Microbiology, Symposium 20. Cambridge University Press, United Kingdom.

39. Woese, C. R. 1987. Bacterial evolution. *Microbiol. Rev.* **51**:221–271.

40. Woese, C. R. 2004. A new biology for a new century. *Microbiol. Mol. Biol. Rev.* **68**:173–186.

41. Woese, C. R., and G. E. Fox. 1977. The concept of cellular evolution. *J. Mol. Evol.* **10**:1–6.

42. Woese, C. R., and G. E. Fox. 1977. The phylogenetic structure of the procaryotic domain: the primary kingdoms. *Proc. Natl. Acad. Sci. USA* **74**:5088–5090.

43. Woese, C. R., O. Kandler, and M. L. Wheelis. 1990. Towards a natural system of organisms: proposal for the domains Archaea, Bacteria, and Eucarya. *Proc. Natl. Acad. Sci. USA* **87**:4576–4579.

44. Woese, C. R., and G. J. Olsen. 1986. Archaebacterial phylogeny: perspectives on the urkingdoms. *Syst. Appl. Microbiol.* **7**:161–177.

45. Woese, C. R., M. L. Sogin, D. A. Stahl, B. J. Lewis, and L. Bonen. 1976. A comparison of the 16S ribosomal RNAs from mesophilic and thermophilic bacilli. *J. Mol. Evol.* **7**:197–213.

46. **Woese, C. R., M. L. Sogin, and L. A. Sutton.** 1974. Procaryote phylogeny. I. Concerning the relatedness of Aerobacter aerogenes to Escherichia coli. *J. Mol. Evol.* 3:293–299.

47. **Wolfe, R. S.** 2001. The Archaea: a personal overview of the formative years. *In* M. Dworkin et al. (ed.), *The Prokaryotes: An Evolving Electronic Resource for the Microbiological Community*, 3rd ed., release 3.7, November 2, 2001. Springer-Verlag, New York, N.Y.

48. **Zillig, W., K. O. Stetter, and D. Janekovic.** 1979. DNA-dependent RNA polymerase from the archaebacterium *Sulfolobus acidocaldarius. Eur. J. Biochem.* **96:**597–604.

49. **Zuckerkandl, E., and L. Pauling.** 1965. Molecules as documents of evolutionary history. *J. Theoret. Biol.* **8:**357–366.

Archaea: Molecular and Cellular Biology
Edited by Ricardo Cavicchioli
© 2007 ASM Press, Washington, D.C.

Chapter 2

General Characteristics and Important Model Organisms

Arnulf Kletzin

INTRODUCTION

The studies conducted by Carl Woese and coworkers in the 1970s opened up new perspectives in biological taxonomy and the origin of species (14, 15, 96, 434, 435, 443) (see Chapter 1). Ribosomal 16S (and 18S) RNA (rRNA) and DNA (rDNA) became the most important molecule for constructing phylogenetic dendrograms that spanned the entire world of living organisms (e.g., 357). The studies obliterated the dichotomous distinction between noncomposite "prokaryotes" and composite "higher eukaryotes," and allowed a third "urkingdom" to be identified: the *Archaea* (Archaebacteria) (96, 434). The concept of a third domain of life explained puzzling biochemical observations, such as the unusual composition of cell walls, membrane lipids, and RNA polymerase that were present in some microorganisms (96). The third domain was initially poorly accepted and vigorously challenged in the scientific community (50, 220, 221). However, the concept of the *Archaea* was advanced through seminal studies performed by Otto Kandler, Ralph S. Wolfe, Karl O. Stetter, and Wolfram Zillig (e.g., 436), which included, in particular, the isolation and characterization of numerous archaeal (and bacterial) extremophiles by Karl Stetter and Wolfram Zillig (see Chapter 1).

Initial studies on the *Archaea* defined three physiologically different groups: methanogens, haloarchaea, and sulfur-dependent thermophiles (96). The understanding of the physiological and phylogenetic diversity of *Archaea* has improved as the number of isolates has increased. All cultivated archaea up to the year 2005 (212) may be regarded as extremophiles (with respect to their growth temperature, osmolarity, or pH) or methanogens (many of which are also extremophiles). Natural habitats for extremophilic archaea include terrestrial hot springs of volcanic origin, hydrothermal vents on the ocean floor, and a diverse range of highly saline, acidic, alkaline, and anaerobic environments. In contrast, mesophilic and psychrophilic archaea are ubiquitous and present in most terrestrial and aquatic environments that have been analyzed (51, 353). As a consequence, *Archaea* exhibit a diverse range of cellular adaptations. For example, energy conservation by sulfur oxidation or reduction is a hallmark of the (hyper-) thermophilic archaea, an adaptation that links to the abundance of sulfur in volcanic environments where they proliferate (159). In these largely light-independent ecosystems, sulfur-dependent and chemolithoautotrophic archaea are among the most important primary producers. In contrast, mesophilic and psychrophilic archaea are present in rivers, ice, and lakes (51, 353). Methanogens are ubiquitous in freshwater sediments, wetlands, and animals. Members of the *Crenarchaeota* are present in cold and temperate soils worldwide. Molecular ecology studies indicate that *Archaea* fulfill important ecological roles in most ecosystems. In contrast to the abundance of molecular ecological data, very few of these archaea (with the possible exception of the methanogens), have been isolated and cultivated in the laboratory (51, 353).

This chapter provides an overview of the *Archaea* and some of their morphological, physiological, biochemical, and molecular properties, providing a platform for the more specific chapters on archaeal molecular cell biology. This chapter also introduces model organisms and systems that have been used to study fundamental properties and principles of archaeal biology, in addition to those that have served as models for understanding the biology of more complex eucaryal cells.

Arnulf Kletzin • Institute of Microbiology and Genetics, Darmstadt University of Technology, Schnittspahnstrasse 10, D-64287 Darmstadt, Germany.

PHYLOGENY OF *ARCHAEA* AND THE ORIGIN OF LIFE

Due to their cellular organization being similar to *Bacteria*, *Archaea* were misclassified until comparison of their molecular traits revealed that they form a fundamentally distinct domain of life (see Chapter 1). The tripartite tree based on 16S rDNA sequences is supported by biochemical data and is now accepted by the majority of scientists as being the best representation of the evolution of the three domains of life. Nevertheless, the three-domain concept has been repeatedly challenged. The "eocyte tree" separated the domain *Archaea* into several kingdoms (220) and eventually led to the proposal of the "ring of life"

(369). A different proposal classified *Bacteria* as the only "Prokaryotes" and divided them in two sub-kingdoms, and placed the *Archaea* with the *Eucarya* in a kingdom of "Neomura" (50).

The coherence of *Archaea* as an evolutionary group, and the distinction of the *Euryarchaeota/Crenarchaeota* kingdoms are evident at the molecular level (Fig. 1). Numerous common biochemical features and the results of comparative genomics support the overall architecture of the *Archaea* tree (40, 41, 95, 328) (see Chapter 19). It is supported by many studies using amino acid sequences of individual proteins or combined sets of functional protein classes for tree construction, such as ribosomal or translation proteins (e.g., elongation factors or RNA polymerases)

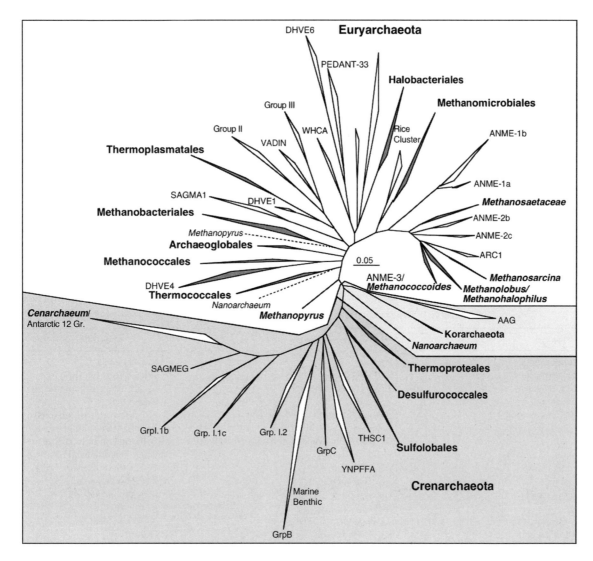

Figure 1. 16S rDNA phylogenetic dendrogram of the *Archaea*. Branches representing cultivated strains (boldface lettering and dark polygons); branches representing sequences determined from molecular ecology studies (white polygons). Alternative branch points of *Nanoarchaeum* and *Methanopyrus* (dashed lines) (40, 42). Reproduced with modifications from *Nature Reviews Microbiology* (353) with permission of the publisher.

(40, 41, 72, 95, 369) (see Chapters 8 and 19). The validity of these trees, and hence the interpretations about evolution, is based on the expectation that genes from central information-processing pathways are less likely to be successfully exchanged with foreign counterparts (even homologous proteins) than genes for metabolic enzymes, which may more easily be replaced by functional analogs. Most archaeal proteins required for replication, transcription, and translation have homologs with the highest level of sequence identity to proteins within the *Archaea*. In addition, these proteins share more similarity with eucaryal proteins than with bacterial proteins. In contrast, archaeal proteins involved in metabolic pathways and regulatory functions are generally more similar to *Bacteria* than to *Eucarya*.

Crenarchaeota and *Euryarchaeota* are not only defined by branching orders of dendrograms, but also by distinct and nontrivial differences in the replicative functions within members of these two kingdoms (95, 187) (see Chapters 3 and 19). Some genes are absent in one kingdom and not in the other kingdom (Table 1). For example, *Euryarchaeota* possess heterodimeric DNA polymerases of the D family, the ssDNA-binding protein RPA, a homotrimeric sliding clamp protein PCNA, histones of the eucaryal type, and the bacterial-like cell division protein, FtsZ. In contrast, the *Crenarchaeota* harbor the ribosomal protein S30, the ssDNA-binding protein SSB, and a heterotrimeric PCNA.

Despite the overall consistency of the archaeal 16S tree, there are still many puzzling features. Most prominent is the seemingly illogical branching order of the *Euryarchaeota*. The anaerobic *Archaeoglobales* and the essentially aerobic *Thermoplasmatales* and *Halobacteriales* disrupt the methanogens into several taxonomic groups, a result that is supported by most protein-based comparisons (Fig. 1) (41). In contrast, all methanogens, *Archaeoglobales* and *Halobacteriales* possess an RNA polymerase B subunit that is encoded by two subunit genes, while the *Thermoplasmatales* do not (Fig. 2) (40). Other archaea possess a single large subunit. It is highly improbable that this split occurred twice in evolution as the separation always occurs at the same position in the gene. There are several possible explanations: (i) the dendrograms do not give the historically correct branching order; (ii) the potential for methanogenesis, which requires an extraordinary number of genes for the enzymes and for the cofactor biosynthesis, has been either acquired or lost repeatedly in archaeal evolution. Support for this latter possibility comes from the observation that some bacterial methylotrophs oxidize methane using reverse methanogenesis, and the genes are very likely to have been acquired from *Archaea* (55); (iii) a split in the RNA polymerase B subunit genes could have been rejoined in the *Thermoplasmatales*. At present it is neither clear which (if any) of these alternatives is correct nor whether the methanogens are monophyletic.

The difficulty in inferring the phylogeny of methanogens is due in part to the placement of the "difficult" and possibly fast-evolving species, *Methanopyrus kandleri*, which has a tendency to move its branch point in phylogenetic trees depending on the dataset analyzed (Fig. 1) (40, 371) (see Chapter 19).

Table 1. Replication proteins in *Crenarchaeota* and *Euryarchaeota*[a]

Feature	Crenarchaeota	Euryarchaeota
Replication origin(s)	*Sulfolobus*: 2–3	*Pyrococcus*: 1
Origin recognition	Cdc6/ORC	Cdc6/ORC
Replicative helicase[b]	MCM (single homolog)	MCM (single or multiple homologs)
Helicase loader	Cdc6/ORC	Cdc6/ORC
ssDNA-binding protein[c]	SSB (one subunit)	RPA (one or three subunits)
Primase[c]	Primase (two subunits)	Primase (two subunits)
Polymerase[b]	PolB (one or multiple homologs)	PolB (one or multiple homologs)
		PolD (heterodimer)
DNA sliding clamp[c]	PCNA (heterotrimer)	PCNA (homotrimer)
Clamp loader	RFC (two subunits)	RFC (two subunits)
Removal of primers	Flap endonuclease 1	Flap endonuclease 1
	RNase H	RNase H
Lagging strand maturation	ATP-dependent DNA ligase	ATP-dependent DNA ligase
Cell division protein	Not yet identified	FtsZ
Ribosomal protein S30[d]	+	−

[a]Data compiled from Kelman and White (187). MCM, minichromosome maintenance; ORC, origin recognition complex; PCNA, proliferating cell nuclear antigen; REC, replication factor C; RPA, replication protein A (*Eucarya*-like); SSB, ssDNA-binding protein (*Bacteria*-like).
[b]Single gene or multiple paralogous genes in genomes.
[c]Subunits in the mature proteins and corresponding genes in genomes.
[d]Gene present (+) or absent (−) in genomes.

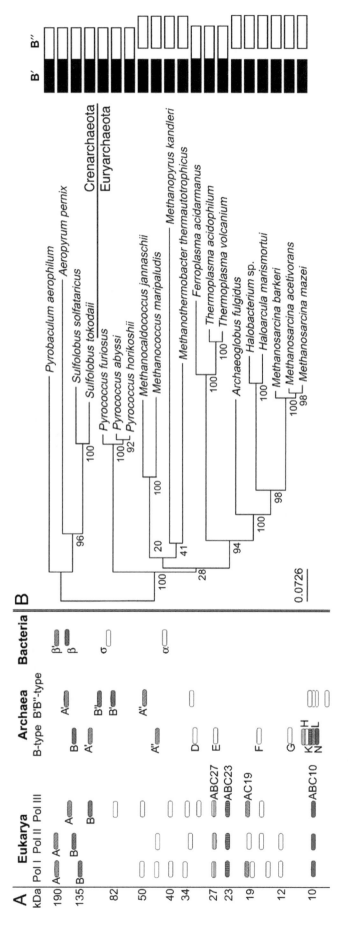

Figure 2. Subunit composition and phylogenetic dendrogram of RNA polymerase amino acid sequences. (A) Virtual SDS-polyacrylamide gel showing the subunit sizes of the subunits of the different RNA polymerases. Homologous subunits are shown in identical shading. Reproduced with modifications from *Proceedings of the National Academy of Sciences of the United States of America* (225) with permission of the publisher. (B) Phylogenetic dendrogram of concatenated RNA polymerase amino acid sequences and presence of a single subunit or two-subunit B homologs. Reproduced with minor modifications from *Genome Biology* (40) with permission of the publisher.

Methanopyrus is always placed as one of the most deeply rooting *Archaea* in 16S rDNA dendrograms (for example, deeper than *Archaeoglobus*) separate from other methanogens. It branches in different positions when comparing concatenated sets of all transcription versus translation proteins from archaeal genomes, where it groups with the *Methanobacteriales* (Fig. 2) (40). In addition, it has the split of RNA polymerase B subunit gene. The "difficult" phylogeny may relate to the approximately fourfold higher "indel" (insertions and deletions in otherwise conserved genes) frequency in the *Methanopyrus* genome compared with other *Archaea*. This correlates with the long branches in dendrograms and is consistent with a much faster rate of evolution than other *Archaea*. If this interpretation is correct, it indicates that the evolutionary rate is not constant for (micro-) organisms. This would be consistent with recent constructions of the archaeal tree, which report that the 16S rDNA does not give reliable branching points for many higher taxa (328).

A fast rate of evolution has also been suggested for *Nanoarchaeum equitans* (see Chapter 19). *N. equitans* is a symbiotic, hyperthermophilic archaeon that has the smallest known genome of cellular organisms (481 kbp) and lives in close association with a sulfur-dependent anaerobic hyperthermophile, *Ignicoccus* (see Chapter 14). *N. equitans* was proposed to be a member of a new, deeply branching kingdom, the *Nanoarchaeota* (152), which diverged before the *Crenarchaeota/Euryarchaeota* split. However, rDNA and protein phylogenies give conflicting outcomes (42). It has been proposed that genome reduction led to a deletion of many presumed "vital" genes, leading to a parasitic rather than a symbiotic relationship with *Ignicoccus*. The genome displays extensive rearrangements including the virtual absence of operons, which are otherwise a feature of many archaeal genomes (255) (see Chapters 6 and 8). *N. equitans* possesses many genes in common with members of the *Euryarchaeota* and more than half of the protein-encoding genes have best BLAST hits with euryarchaeal species. *N. equitans* may represent a highly derived and fast-evolving euryarchaeal lineage related to the *Thermococcales* (Fig. 1) (42, 255, 328). The distinctiveness of the rRNA sequence of *N. equitans* may have arisen as a result of the combined effects of adaptation to hyperthermophily, symbiotic lifestyle, genome reduction, and rapid evolutionary change (42, 429).

Members of the fourth, deep-branching lineage of the *Archaea*, the *Korarchaeota*, have been enriched from hot environments, but pure isolates have not been cultivated (19). Most inference about members of the *Korarchaeota* has been derived from metage-nomic studies (277). A sequencing project of a *Korarchaeota*-containing community is under way, and this will provide a great deal more information about this kingdom in the near future (http://www.jgi.doe.gov/sequencing/why/CSP2006/korarchaeota.html).

Many hyperthermophiles grow chemolithoautotrophically with inorganic energy sources using CO_2 as the sole carbon source. In addition, most hyperthermophiles (*Archaea* and *Bacteria*) have deep branching points in phylogenetic dendrograms, and most have short branch lengths (Fig. 1). These findings have led to a hypothesis that the last universal common ancestor(s) (LUCA) was a hyperthermophilic chemolithotroph, and that life originated in hot environments, such as submarine hydrothermal vents (e.g., 386, 419). This popular but also controversial hypothesis has excited other scientists interested in the origin of life, as it has important implications for prebiotic evolution and for the properties of the first cellular organisms; the debate about the hypothesis is heated and ongoing. Support for a "hyperthermophilic origin of life" came from hypotheses that invoked metabolism evolving on iron sulfide (FeS) surfaces during prebiotic times (autocatalytic anabolist hypothesis), and from experiments showing that some of these reactions are chemically possible (151, 183, 420). However, many reasons have been used to argue against the "hyperthermophilic origin of life" hypothesis (360), including the following:

- The G+C content of rRNA from hyperthermophiles is high (average, 60%), irrespective of their average genome G+C content (e.g., many *Sulfolobales* have ~35%). The high G+C bias in rRNA genes of hyperthermophiles leads to long-branch attraction artifacts (and other effects) in dendrogram calculations that may affect the reliability of the tree (433).
- Protein phylogenies do not always support the 16S rDNA trees that place hyperthermophiles at the root of the tree (40, 42, 202).
- To cope with temperature effects, hyperthermophiles have developed several other modifications besides 16S rRNA composition that help to protect against thermal denaturation. For example, the degree of tRNA modification increases with increasing optimal growth temperature (T_{opt}) (283). Similarly, the degree of cyclization of membrane lipids increases with temperature (see "Membranes" below) (107). Many hyperthermophiles also contain significant amounts of thermoprotective secondary metabolites, such as cyclic bisphosphoglycerate (in some methanogens) (235, 364) and dimyo-

inositol phosphate (in *Pyrococcus* spp.) (359). All of these modifications require specific biosynthetic pathways. It seems improbable that these modifications were present at early stages of cellular evolution, or in LUCA, and were subsequently lost during adaptation to lower-temperature environments. It may be expected that these mechanisms of modification developed over evolutionary time in response to adaptation to hyperthermophily.

- The half-life of some important biomolecules, including NAD(P)(H) and ATP, decreases dramatically with increasing temperature and decreasing pH, thereby decreasing the likelihood that life could have first evolved under these conditions (63).
- High temperatures may cause reactions to proceed in a less controlled fashion than lower temperatures, thereby making it more difficult for microorganisms to evolve at hot rather than cold temperatures.

A recent, alternative hypothesis is that life evolved in stratified seafloor hydrothermal mounds, where microcompartments the size of bacterial or archaeal cells were formed from FeS precipitates (256), and which may have served as scaffolds for cell formation prior to the presence of lipid vesicles. According to this hypothesis, organic precursors were synthesized abiotically at high temperatures, whereas cellular life forms evolved in inorganic FeS microcompartments in warm (not hot) zones of submarine hydrothermal vents (Fig. 3). This hypothesis takes into account that (i) the structure of the minerals in the long-living "Lost City" hydrothermal vent field are percolated by warm (40 to 90°C), gradually cooling vent fluids, and (ii) FeS precipitates with a similar structure were detected at ancient hydrothermal sites (337), and their formation was simulated in the laboratory (394). These 1- to 100-μm-wide microcompartments would provide a confined space with catalytically active surfaces (nickel/iron), which eliminates the need for lipid membranes to enclose the cellular precursors. It is hypothesized that temperature, pH, mineral, and redox gradients within the larger structures would create the dynamic disequilibria required for metabolic reactions, and populations of viruslike RNA molecules, encoding small numbers of proteins, would be the agents of variation and selection (213). The sorting of genetic elements among different compartments could result in the proliferation of increasingly complex molecular ensembles and the selection of entities that had achieved replicative advantages. According to this hypothesis, LUCA was an inorganically housed assemblage of expressed and exchangeable genetic elements. Escape from these confinements was possible only after the evolution of DNA replication and membrane biosynthesis. *Archaea* and *Bacteria* (and probably other, now extinct lineages) could have escaped separately, which would explain some of the fundamental differences between both domains (213). Many aspects of this hypothesis are compatible with the "autocatalytic anabolist" and the "RNA world" hypotheses. It is difficult to reconstruct this environment and test it experimentally, and as a result the field will remain open for debate and speculation.

GENERAL CHARACTERISTICS OF *ARCHAEA*

The overall cellular architecture of *Archaea* is similar to *Bacteria*. The number of known morphological types of *Archaea* is less than in *Bacteria*. For example, there are no multicellular stages with cellular differentiation, no mycelia-forming archaea similar to streptomycetes, or structures similar to the periplasmic flagella and the flexible body structure of spirochetes. On the other hand, members of the *Archaea* do possess morphological features with no counterpart in the *Bacteria*. These include the amoebalike *Thermoplasma* and *Ferroplasma* cells (115), flat, irregular-shaped *Haloferax* spp., and rectangular haloarchaea reminiscent of a postage stamp (*Haloquadratum* or "Walsby's square bacterium," Fig. 4) (421). The metabolic diversity of *Archaea* appears similar to that of *Bacteria*, although the present understanding is incomplete due to incomplete sampling of cultivatable isolates. With the notable exception of methanogenesis, almost all metabolic pathways discovered also exist among *Bacteria*. Photosynthetic bacteriorhodopsin (BR), which was thought to be present only in haloarchaea, has recently been detected in planktonic bacteria (22, 23), whereas "classical" photosynthesis with any type of chlorophyll has not been found in any archaea.

A number of unique molecular properties that characterize *Archaea* were recognized in seminal studies from several laboratories in the 1970s and 1980s (see Chapter 1). These properties helped to substantiate that *Archaea* are fundamentally different from *Bacteria* despite their similar cellular organization (reviewed in reference 436); they include:

- the presence of phytanyl ether instead of fatty ester lipids in the membranes.
- the absence of canonical peptidoglycan and a frequent use of proteinaceous S layers.

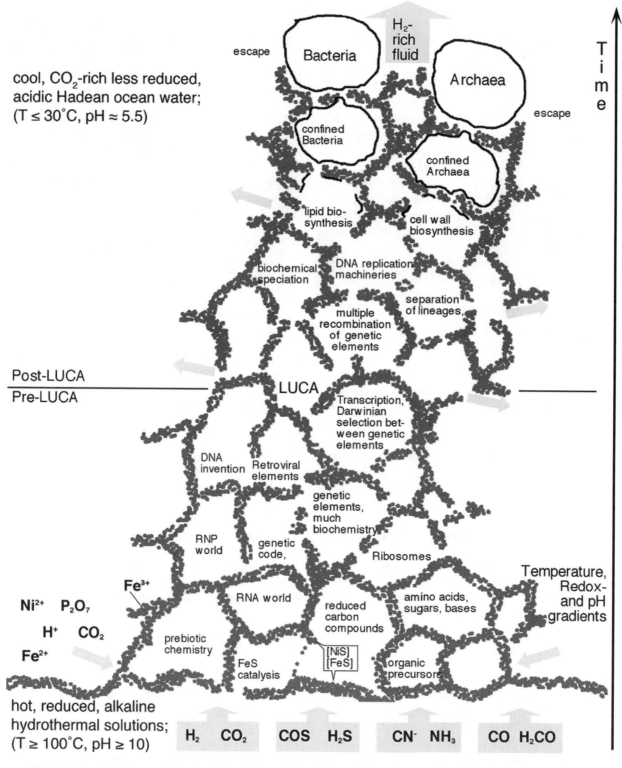

Figure 3. Hypothetical scenario for the origin of living cells from (in)organic precursors. Archaeal and bacterial cells are shown escaping from within naturally formed inorganic metal sulfide-based compartments in a 3.8-Ga-old hydrothermal vent. The compartments (1 to 100 mm in diameter) and vent structures are schematic and not drawn to scale. LUCA, last universal common ancestor. Reproduced with modifications from *Trends in Genetics* (213) with permission of the publisher.

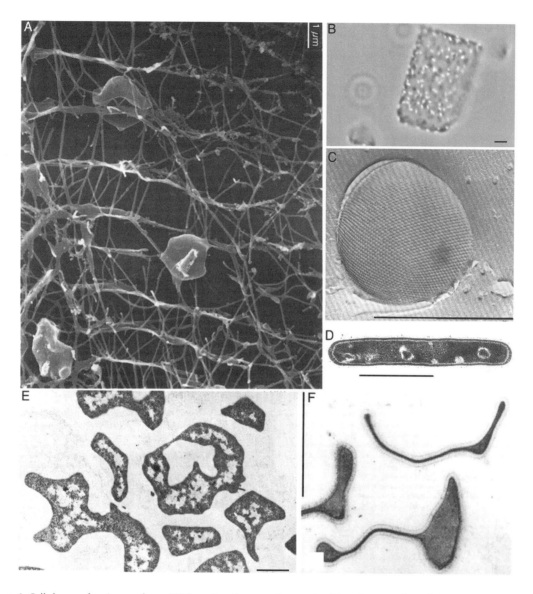

Figure 4. Cell shapes of various archaea. (A) Scanning electron micrograph of *Pyrodictium* cells within a network of extracellular cannulae (388). (B) Phase-contrast micrograph of *Haloquadratum walsbyi* with gas vesicles. Photograph courtesy of T. Hechler, Darmstadt. (C) Electron micrograph of a freeze-etched *Metallosphaera sedula* cell showing the hexagonal lattice of the S layer (157). (D) Ultrathin section of a *Methanothermus fervidus* cell (387). (E) Ultrathin section of *Methanogenium carici* cells (330). (F) Ultrathin section of *Pyrodictium abyssi* cells (95). Bars, 1 µm. Panels A and C to E reproduced from *Bergey's Manual of Systematic Bacteriology* with permission of the publisher. Panel F reproduced from *Theoretical Population Biology* (95) with permission of the publisher.

- the complexity of the RNA polymerase.
- the observation that elongation factor G is ADP-ribosylated by diphtheria toxin.
- the use of unmodified initiator methionine in translation, similar to *Eucarya*.

A general overview of the characteristic properties of archaeal cells is provided below.

Membranes

The presence of phytanyl ether instead of fatty acid ester lipids in the cytoplasmic membranes is one of the most characteristic features that is present in all *Archaea* (see Chapter 15). The ability of archaeal membranes to function effectively under severe environmental stress, e.g., gradients in excess of five pH units [see "(Thermo-)acidophilic archaea: *Thermoplas-*

matales and *Sulfolobales*," later in this chapter] (250), has prompted intensive efforts to characterize the nature and biogenesis of membrane lipids. *Archaea* have 2,3-di-O-alkyl-*sn*-glycerol (diether lipids, C_{20} or C_{25}) made of isoprenoid units, which are analogs of conventional diacylglycerides, but with the opposite stereochemistry to *Bacteria* (1,2-*sn*-glycerol) (107, 181, 292). In addition, many archaea have a wide spectrum of glycerol or nonitol dialkyl (C_{40}) tetraether lipids, which form a monolayer instead of a bilayer membrane, and have no ester lipid counterpart (107). The tetraether monolayers exhibit unparalleled resistance to proton transfer and maintain rigidity at high temperatures (250). However, the distribution of di- and tetraether lipids in *Archaea* does not simply correlate with any particular growth characteristic (e.g., temperature) (reviewed in reference 250). Most, but not all hyperthermophiles contain tetraether lipids. The "hyperacidophiles" (pH optimum, 2.5 or lower) comprising all *Thermoplasmatales* and some of the facultative autotrophic *Sulfolobales*, are the only group with an almost exclusive use of this lipid type, whereas heterotrophic *Sulfolobales* (pH optimum, 2.5 to 3.5) (125, 250) contain predominantly diether lipids (227). The isoprenoid chains in *Archaea* can be unsaturated or cyclized and contribute to changes in membrane fluidity. The average cyclization number of hydrocarbon chains can increase with growth temperature (107, 181).

The isoprenoid biosynthesis utilizes the mevalonate pathway, similar to animals, whereas the recently discovered nonmevalonate pathway is essential in plants, apicomplexan parasites, and many bacteria (39, 80). The hydroxymethylglutaryl-CoA reductase (HMGR) is the rate-limiting enzyme in the mevalonate pathway. There are two paralogous classes of HMGR, a eucaryal/archaeal and a bacterial class. Whereas most archaea contain the eucaryal/archaeal class, the *Archaeoglobales* and *Thermoplasmatales* use the bacterial class, an observation that is interpreted as an example of lateral gene transfer (see Chapter 15) (38). The competitive HMGR inhibitor mevinolin or its precursor Lovastatin is used as selection marker in haloarchaeal transformation systems (see Chapter 21) (222, 223).

Biological membranes need to be permeable to water to enable the buildup of osmotic pressure within the cell, and to maintain the water balance. Water permeability is mediated by aquaporins, ubiquitous transmembrane proteins belonging to the family of major intrinsic proteins (MIPs) (see Chapter 16). MIPs facilitate the passive transport not only of water but also small neutral solutes (e.g., glycerol) across cell membranes (82). Only one archaeal aquaporin, AqpM from *Methanothermobacter marburgensis*,

has been described and structurally characterized (216, 233). It has sequence and structural similarity to eucaryal and bacterial aquaporins and exhibits high stability against denaturation. The aquaporin promotes water but not glycerol permeability, suggesting that it might keep neutral solutes inside the cell. Homologs are found in most but not all archaeal genomes. It is not clear how water transport is mediated in *Archaea* that do not contain this porin.

Cell Walls and Extracellular Structures

The absence of murein in the cell walls (96), and, with the exception of *Ignicoccus* (see "*Desulfurococcales*" and Fig. 19, later in this chapter), the absence of a periplasmic space are features that distinguish *Archaea* from *Bacteria*. Only two specialized taxa of *Archaea* have a polysaccharide cell wall, whereas the majority have proteinaceous S layers (176) (see Chapter 15). The *Methanobacteriales*, one of orders of methanogens (see Chapter 13), stain gram positive and they are the only *Archaea* with cell wall components similar to peptidoglycan (see Chapter 15). The cell wall consists of pseudomurein, a β-1,3-linked polymer of N-acetylglucosamine and N-talosaminglucoronic acid cross-linked with oligopeptides (176). Among the haloarchaea, *Halococcus* species possess a thick cell wall made of a complex, highly sulfated heteropolysaccharide, which contains the unusual components N-acetyl gulosaminuronic acid and N-glycyl-substituted glucosamine moieties. *Halococcus* cells do not require high salt concentrations in the media to maintain cellular integrity in contrast to other haloarchaea (351, 383). This might explain the endurance of *Halococcus* cells in atypical, low-salt habitats (313). The related *Natronococcus occultus* has a cell wall with a significantly different structure consisting of repeating units of a glycoconjugate with poly-(L-glutamine) (279).

S layers of glycoproteins can represent the sole cell wall component in *Archaea* and can provide a high level of rigidity to the envelope (79, 261) (see Chapter 14). Planctomycetales are the only *Bacteria* known to lack peptidoglycan and to contain a proteinaceous S layer similar to *Archaea* (239). *Thermoplasma* and *Ferroplasma* lack a cell wall and have a cytoplasmic membrane with embedded glycoproteins and lipoglycan (31, 88, 226). *Thermoplasma* grows in hypotonic media. Large amounts of glycoproteins within the membrane are indicative of a network that maintains cellular integrity and may represent an evolutionary product of an ancestral S layer.

Archaeal S layers are made of large proteins anchored in the cytoplasmic membrane and form regular, symmetrical arrays when viewed by electron microscopy (64) (see Chapter 14). S-layer proteins assume a hexagonal symmetry in most cases (Fig. 4), although several examples of tetragonal symmetries are known (302, 303, 354). The hyperthermophilic member of the *Crenarchaeota*, *Staphylothermus marinus*, provides an interesting example of the general architecture of archaeal S layers, although it is an unusual type. The scaffold with a tetragonal symmetry is formed by an extended $\alpha_4\beta_4$-glycoprotein termed tetrabrachion (Fig. 5), which forms a long stalk (α-subunit, coiled coil) and a canopy-like network ($\alpha\beta$-subunits) enclosing a "quasi-periplasmic space" (302, 303). The subunits are generated by proteolytic cleavage of a single gene product. A subtilisin-type protease is attached in the middle of the stalks, consistent with the growth of *S. marinus* at the expense of peptides and proteins.

The S layers of most *Archaea* enclose a "quasi-periplasmic space" with variable thickness and with pores of varying sizes that provide room for hydrolytic enzymes (e.g., *Staphylothermus*; Fig. 5), substrate-binding proteins, electron transport components, and glycosyl residues of membrane proteins. It has been speculated that the space might be packed

with proteins and provide a specific extracellular environment close to the membrane. This could be particularly important for the "hyperacidophiles," which maintain steep pH gradients across the membrane, and for the protection of proteins with metallic components, such as iron-sulfur clusters or molybdenum cofactors. Thus, the "quasi-periplasmic space" may provide a pH gradient buffering capacity to the otherwise harsh external conditions.

Many archaea, including the cell wall-less *Thermoplasma* and *Ferroplasma*, are flagellated (Fig. 6). The flagella seem unrelated to their bacterial counterparts in terms of amino acid composition, structure, and morphogenesis, although their mechanics and overall function are similar (polar or peritrichous flagellation, formation of bundles, and reversible rotation) (see Chapter 18). Archaeal flagellar filaments share some features with bacterial type IV pili but are unique in other aspects (18). Archaeal flagella have a smaller diameter than bacterial flagella and a smooth surface, but the diameter is larger than type IV pili. In contrast to type IV pili, they are synthesized with a signal peptide. They are glycosylated and sulfated in a similar way to S-layer proteins. In *Methanococcus*

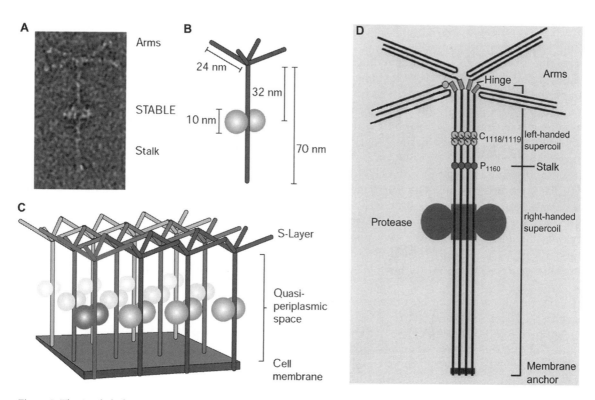

Figure 5. The *Staphylothermus marinus* S-layer protein tetrabrachion. (A) Electron micrograph of the negatively stained tetrabrachion-protease complex. (B) Schematic model of the complex with dimensions. (C) Schematic model of the cell surface of *Staphylothermus*. (D) Proposed folding topology with cysteine residues and the unique proline residue separating left- and right-handed supercoils. Figure compiled and reproduced from *Current Biology* (259) and the *Journal of Molecular Biology* (302) with permission of the publishers.

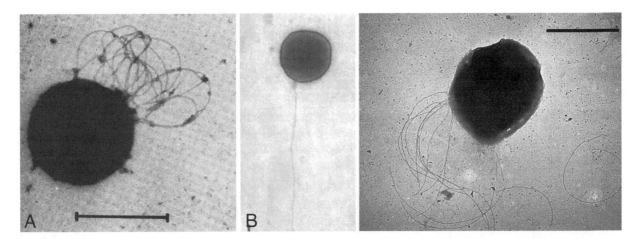

Figure 6. Flagellated archaea. (A) *Thermococcus celer.* Reproduced from *Bergey's Manual of Systematic Bacteriology* (208) with permission of the publisher. (B) *Thermoplasma acidophilum.* Reproduced from the *Journal of Bacteriology* (31) with permission of the publisher. (C) *Halobacterium salinarum* PHH4. Photograph courtesy of T. Hechler, Darmstadt.

voltae the flagellum is composed of four proteins: FlaA, FlaB1, FlaB2, and FlaB3. The attached glycan is a novel N-linked trisaccharide composed of modified mannose and glucose residues linked to asparagine residues. The same trisaccharide was identified on a tryptic peptide of the S-layer protein from *M. voltae*, implicating a common glycosylation pathway in flagella and S-layer modification (417).

Flagella biosynthesis, motor structures, and switching of flagellar rotation has been studied in detail in *Halobacterium salinarum* (see Chapter 18). Similar to *Bacteria*, sensory transducers trigger flagellar switching (see Chapter 11). A novel transducer (MpcT) that responds to membrane potential ($\Delta\Psi$) has been characterized in *H. salinarum* (209) (see "Bacteriorhodopsins and light-driven ATP synthesis" below). The structure and function of the flagella motor of gram-negative bacteria has been well studied, in contrast to the archaeal motor where anchoring in the cytoplasmic membrane and S layer is poorly understood. Fla gene clusters have been identified in several archaeal genome sequences, but homologs of motor proteins have not (18).

Pili or piluslike structures were observed in many archaea (e.g., 155), and conjugative plasmids have been identified (332, 352) (see Chapter 5). The structure and molecular characteristics of the pilus from the SM1 euryarchaeon, referred to as "hami," was analyzed in detail. It is one of the most exciting new extracellular structures recently identified in *Archaea* (Fig. 7) (265–267). SM1 grows in cold sulfurous springs in southern Germany in assemblages reminiscent of "strings of pearls" (see Table 7 below). The inner part of each pearl consists of $\sim 10^7$ cells of SM1,

while the exterior and the filament connectors are formed either by a *Thiothrix* or an ε-proteobacterium (IMB1) (335). SM1 can be grown in situ on polyethylene fabrics to form high-density biofilms but has not been cultivated in the laboratory. The archaeal cells maintain a distance of approximately 2 to 4 µm from each other and are connected by a thick web of appendages (Fig. 7). High-resolution electron microscopic pictures showed that the hami are made of a regular structure with a defined base and tip. The central region, up to 2 µm in length, is made of up to 60 repeating units of a 46 × 4 nm, elongated coiled trimer that consists of a 40-kDa filamentous protein that is 4 nm in diameter. The trimer is stable against physical and chemical denaturation, and the amino acid sequence of the protein is not similar to any sequences in GenBank. The protein trimer sitting at the tip of the filament is partially uncoiled and bent back toward the cell forming a structure reminiscent of a three-pointed grappling or fishing hook.

Chromosomes, DNA Structure, and Replication

Archaeal chromosomes are circular, similar to those of most *Bacteria*, with sizes typically between ~1.5 and 6 Mbp (e.g., 5.75 Mbp in *Methanosarcina acetivorans*). *Methanosarcina* species have relatively large genomes and are the only *Archaea* thought to have acquired large numbers of genes from other species (gene gatherers), similar to the bacterial streptomycetes and rhizobia that have genomes up to 10 Mbp in size; the latter appear to have acquired numerous genes for transporters, two-component systems, polymer hydrolases, and oxygenases for the

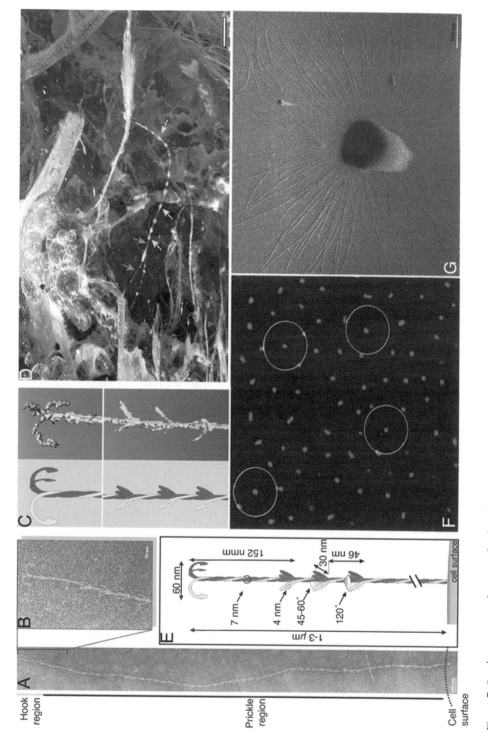

Figure 7. *See the separate color insert for the color version of this illustration.*) The euryarchaeon SM1 and its extracellular appendages ("hami"). (A) Electron micrograph of a "hamus." (B) Enlargement of the hook region. (C) Simplified model of a hamus with the three filaments shown in different colors and 3D reconstruction from cryoelectron microscopy. (D) "String of pearls," archaeal/bacterial community in cold, sulfurous spring water. (E) Hamus model with dimensions. (F) Natural biofilm hybridized with an SM1-specific fluorescent probe; circle diameter, 4 μm. (G) Pt-shadowed electron micrograph of a single SM1 cell with appendages. Figure compiled, modified, and reproduced from *Biospektrum* (264) and *Molecular Microbiology* (265) with permission of the publishers.

degradation of unknown substrates (106) (see Chapters 13 and 19). In contrast, most chemolithotrophs have small genomes between 1.5 and 2 Mbp. The smallest genome (481 kbp) was found in the symbiotic archaeon, *N. equitans* (see "Phylogeny of *Archaea* and the origin of life," above). Gene organization (e.g., clustering in putative operons) and coding density (75 to 90% gene encoding) in *Archaea* are similar to the *Bacteria* (e.g., 105). Some archaea occasionally have self-splicing group II introns (138, 198) but group I introns have not been found (see Chapter 7).

Chromatin proteins are used to organize DNA in the cell and to prevent unspecific condensation (see Chapter 4). Distinct families of small, basic, and abundant chromatin proteins are present in all three domains of life (reviewed recently in reference 340). Archaeal chromatin proteins fall into two families, histones and Alba proteins. Both have homologs in *Eucarya*. Histones have been well characterized in *Euryarchaeota* (340), and have more recently been identified in *Crenarchaeota* (e.g., Sargasso sea environmental genome scaffolds and the genome of *Cenarchaeum symbiosum*) (60). Similar to those from *Euryarchaeota*, histones from *Crenarchaeota* form dimers in solution. They oligomerize further in contact with DNA and form compact structures that resemble eucaryal nucleosomes. In contrast, the role of the Alba proteins is less well defined as they bind both double-stranded DNA (dsDNA) and single-stranded RNA (ssRNA). Unlike *Eucarya*, there is no consensus model for archaeal chromatin as each organism has a characteristic set of chromatin proteins with unique regulatory features. Nevertheless, a good understanding is emerging of how chromatin proteins maintain a compact and structured chromosome that is still accessible to gene expression complexes (340). Histones may also contribute to the stabilization of DNA against thermally induced strand separation in the hyperthermophilic archaea (274) (see Chapter 19). The assumption that the G+C content is a determinant of DNA thermostability in hyperthermophiles is wrong. Hyperthermophilic archaea have a G+C content between ~35% (e.g., *Sulfolobus*) and ≥60% (e.g., *Pyrodictium*), which is similar to the range of G+C content in mesophiles. The only and notable exceptions are rRNA genes (274, 348).

Insight into mechanisms of DNA replication in *Archaea* has progressed very rapidly in the past few years (see Chapter 3). Most archaeal replication proteins are more similar to those found in *Eucarya* than to the analogous proteins in *Bacteria* (187). For example, *Archaea* use ATP-dependent DNA ligases, similar to *Eucarya*, as opposed to NAD-dependent and phylogenetically unrelated bacterial enzymes

(206). However, *Archaea* only possess a subset of the eucaryal machinery and possess features of replication that are not found in other organisms. *Pyrococcus abyssi* was the first archaeon for which a single origin of replication was predicted by skew analysis and subsequently confirmed experimentally (273) (see Chapters 3 and 19). In silico analyses suggested that other archaeal species might contain more than one origin, and this was recently demonstrated for *Sulfolobus solfataricus* and *Sulfolobus acidocaldarius* (246, 329). Most of the replication proteins have been biochemically examined (Table 1).

Transcription

The first indications that the transcription machinery and most of the information-processing machinery in *Archaea* are more closely related to *Eucarya* than to *Bacteria* came from in vitro inhibitor studies (see Chapter 6). The *Sulfolobus* and *Halobacterium* DNA-dependent RNA polymerases (RNAPs) were found to be resistant to rifampicin, streptolydigin, and α-amanitin, similar to eucaryal RNAPs (453, 454). The sequences of archaeal RNAPs resemble eucaryal RNAP II (and III) and consist of up to 13 different subunits (Fig. 2) (25, 449, 453). The B subunit is homologous to the B (second largest) subunit of the eucaryal nuclear polymerases I, II, and III and to the bacterial β-subunit (Fig. 2). The combined A and C subunits in *Archaea* correspond to the largest A subunits of the eucaryal RNAPs and to the bacterial β'. Because the RNAP is composed of highly conserved and variable regions it has been useful for phylogenetic analyses (see Chapter 19). Moreover, the combined size of the RNAP genes (~8,500 bp) provides a large dataset (~2,000 conserved amino acids), which may provide a better estimate of cellular evolution than rDNA sequences (203).

Similar to *Eucarya*, the typical archaeal promoter contains an AT-rich TATA box ~22 to 28 bp upstream of the transcription initiation site (322, 323) (see Chapter 6). In addition, a TATA box-binding factor (TBP) and transcription factor B (TFB) participate in promoter recognition and direct the RNAP promoter binding (21, 25, 26, 320, 323). These transcription factors are sufficient for initiating transcription, whereas additional factors are required in *Eucarya* (21). In contrast, most of the regulatory proteins and the mechanism of gene regulation are more similar to *Bacteria* (24). Transcription termination seems to involve pyrimidine or T-rich stretches; for example, 3' termini have been mapped in vivo to a location immediately downstream of a TTTTTYT sequence that was part of a pyrimidine-rich region of 16 to 19 nt (321, 431). In *Methanothermobacter thermautotrophicus*, RNAP was found to terminate in vitro at intergenic sequences similar to bacterial terminators, and stable RNA hairpins did not appear to be important (341).

Posttranscriptional Gene Silencing

The mechanism of RNA interference (RNAi), also termed posttranscriptional gene silencing, was one of the most exciting discoveries of the past decade in the field of gene regulation in *Eucarya* (241). Many of the proteins responsible for RNAi have recently been identified in *Archaea* and *Bacteria* and have become excellent models for studying the structure, function, and specificity of RNAi proteins (247, 295, 296, 374, 442). In RNAi, an RNase III-like enzyme (Dicer) cleaves dsRNA into short (21 to 23 nt) RNA duplexes, referred to as guide RNAs or siRNAs, which have 2-nt overhangs at the 3′ ends and 5′-phosphates (134). One of the strands is incorporated into the RNA-induced silencing complex (RISC), which then targets a complementary mRNA. A minimal RISC consists of the guide RNA plus a single protein called argonaute (Ago). The protein mediates duplex formation and subsequently cleaves the complementary target mRNA via an endonuclease domain, termed slicer or PIWI. The mRNA is further degraded by exonucleases, and gene expression is thus silenced. Silencing is usually long lasting and stable, which makes RNAi an ideal tool for studying gene function in *Eucarya*. A related pathway utilizes micro RNAs (miRNAs) that are known to regulate translation of an estimated 20% of all human genes. Micro RNAs are transcribed as precursors with stable hairpin structures that are processed by the Dicer enzymes into short imperfect duplex structures resembling siRNA (432). Processed miRNAs are then integrated into RISCs that target the 3′-UTR of many mRNA transcripts and interfere with translation initiation.

Dicer and Ago are both multidomain proteins (Fig. 8) with multiple functions (241). It has been difficult to dissect their structure and activities. Sequences with similarity to the PIWI domain from Ago are present in many archaea and in bacteria. Full-length Ago is also present in several *Euryarchaeota*, but the degree of sequence conservation is low even within this archaeal kingdom. For example, Ago from *Pyrococcus furiosus* and *Methanocaldococcus jannaschii* share only 27% amino acid identity. The recently solved X-ray structures of Ago from *Archaeoglobus fulgidus*, *P. furiosus*, and the bacterium *Aquifex aeolicus* revealed that the PIWI domain has structural similarity to RNase H (Fig. 8) (247, 295, 296, 374, 442). Mutations within the RNase H domain of human Ago 2 inactivate RISC and demonstrate that this domain is responsible for the nuclease activity (242, 327). Guided by the structure of the *P. furiosus* enzyme (374), it has also been demonstrated that the human Ago 2 enzyme possesses the Slicer activity of RISC.

The mammalian Dicer is a large protein with, among other domains, a DEAD box RNA helicase, a PAZ, and two RNase domains (Fig. 8) (241). The helicase and RNase domains have homologs in bacterial and archaeal genomes: the *P. furiosus* Hef helicase and *Aq. aeolicus* double strand-specific RNase III, respectively, have low sequence similarity (A. Kletzin, unpublished results). The two archaeal enzymes are presently the only members of their respective families that have three-dimensional (3D) structures available (108, 210, 280). Thus, the eucaryal RNAi enzymes dicer and slicer seem to be assemblies of several bacterial and archaeal precursor domains, which, in combination in the eucaryal enzymes, allows recognition and targeting of specific mRNAs for degradation.

The biological importance of cellular posttranscriptional gene silencing, in particular, in *Bacteria* and *Archaea*, is not fully understood. It could function as an evolutionarily conserved endogenous defense to the presence of double-stranded RNA, and thus a defense mechanism against invading viral DNA, RNA, (retro-) transposons, and other heterologous double-stranded genetic elements that may potentially harm the genetic integrity of the host (56). Gene silencing may also play a role as a regulatory mechanism involving the synthesis of antisense RNA, a phenomenon that has only been shown to occur in very few archaeal systems. For example, RNA-mediated gene regulation has been demonstrated for the immunity transcript of the haloarchaeal phage ΦH (392, 393). The processing activity is not sequence specific but depends on the presence of an RNA duplex. RNase activity was demonstrated in this system, but the enzyme responsible was not identified. These observations and the growing number of small noncoding RNAs (see Chapter 7) point to a potentially widespread role of RNAi-like mechanisms in *Archaea*.

Translation

Superficially, the archaeal translation machinery is similar to the bacterial machinery. Common features are 70S ribosomes with 30S and 50S subunits, similar-length ribosomal RNAs, transcriptional and translational coupling, the presence of polycistronic messages, and a frequent but nonexclusive use of Shine-Dalgarno sequences as ribosomal binding sites (reviewed in reference 244) (see Chapter 8). However, the archaeal translation system includes numerous

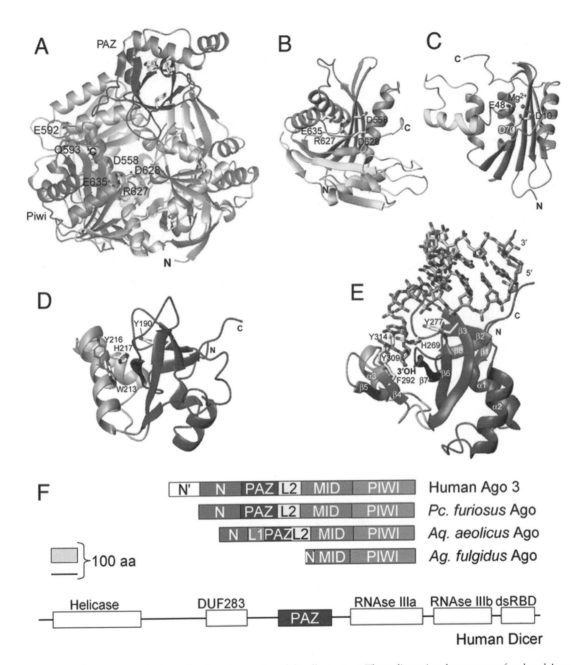

Figure 8. (*See the separate color insert for the color version of this illustration.*) Three-dimensional structures of archaeal Argonaute proteins. (A) 3D structure of *P. furiosus* Ago with the PAZ domain (blue) and the PIWI domain (green/yellow) (PDB code 1U04). (B, C) Similarity of the *P. furiosus* PIWI domain (B) with the catalytic core of the *E. coli* RNase H1 (C) (PDB code 1RDD) with the catalytic DDE triad and bound Mg^{2+} ion highlighted. The *P. furiosus* PIWI domain has a putative, similar catalytic DDE triad and a conserved Arg (position 627). (D, E) 3D structure of the *P. furiosus* PAZ domain (D) and comparison with the homologous domain of human Ago1 bound to an siRNA mimic (E) (PDB code 1SI3). (F) Domain structures of Ago proteins, including N-terminal, linker (L1 and L2), PAZ, Mid, and PIWI domains and of human Dicer comprising a DEXH helicase, a PAZ, two RNase III, and dsRBD domains and a conserved domain of unknown function (DUF). Panels A to E reproduced with modifications from *Current Opinion in Structural Biology* (241) with permission of the publisher.

specific features not present in *Bacteria*, some of which are specific to *Archaea* and others that are similar to *Eucarya*. For example, almost all antibiotics inhibiting bacterial translation are ineffective in the *Archaea* (and the *Eucarya*) and vice versa, regardless

of whether they affect the 30S or 50S subunits. Bacteria use N-formyl-methionyl-tRNA for translational start codons, while *Archaea* and *Eucarya* use unmodified methionine. Approximately 21 genes encoding putative translation initiation factors have been identified in archaeal genomes, most of which have homologs in the *Eucarya*.

A feature that is common to *Archaea* and *Eucarya* and that is therefore diagnostic for discriminating *Archaea* from *Bacteria* is the finding that that elongation factor 2 (EF-2) is ADP-ribosylated by diphtheria toxin (193) (see Chapter 8). The archaeal and eucaryal EF-2s contain diphthamide, a posttranslationally modified histidine, which is the target of the toxin (Fig. 9) (175) (see Chapter 11). A short region that contains a histidine residue in EF-2 proteins is absent in the homologous EF-G protein from *Bacteria*, thereby rendering *Bacteria* resistant to the toxin (232). The in vivo role of diphthamide is not clear. *Sulfolobus* and yeast EF-2 overproduced in *Escherichia coli* contain the unmodified histidine, and these recombinant proteins had indistinguishable stability and in vitro translation activity from the diphthamide-containing native enzymes (68). This is an area open for investigation, because not all of the genes required for biosynthesis of diphthamide have been identified in *Eucarya* and only one of them, DHP2, has been identified in archaeal genomes (243). Moreover, diphthamide biosynthesis has not been experimentally examined in *Archaea*.

The translation initiation factor IF-5D contains hypusine in *Crenarchaeota* and deoxyhypusine in *Euryarchaeota* (Fig. 9) (20, 362) (see Chapters 8, 9, and 11). The lysine derivative hypusine (N^ε-[4-amino-2-hydroxybutyl]-l-lysine) is synthesized in a two-step reaction catalyzed by the enzymes deoxyhypusine synthase and deoxyhypusine hydroxylase

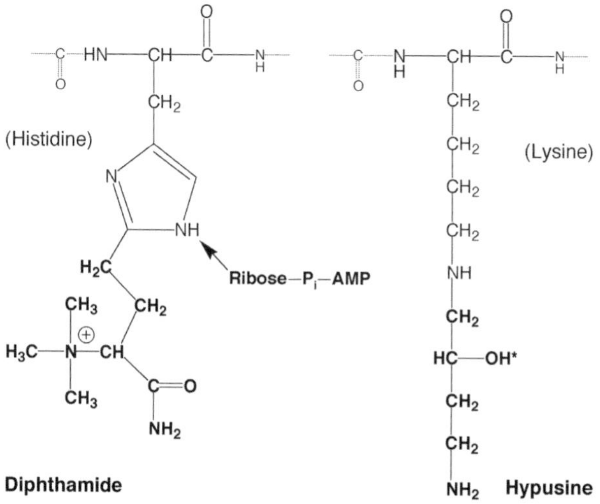

(294). The synthase but not the hydroxylase gene has been identified in the *Eucarya* and *Archaea* (43). Crystal structures of IF-5D have been obtained from three hyperthermophilic archaea, but equivalent structures are not available from *Eucarya* (196, 298, 439). The structures reveal that the hypusine is positioned at the tip of an extended and exposed loop of conserved amino acid residues, which appears to mimic the anticodon loop of a tRNA molecule. Elongation factor P (EF-P) from *Bacteria* has low sequence but high structural similarity to archaeal IF-5D and contains a conserved and unmodified lysine residue at the equivalent position (43, 439). The EF-P appears to stimulate peptidyltransferase activity, whereas the role of IF-5D has not been clearly defined. Conditional IF-5D mutants in yeast exhibit cell cycle arrest (293), and a similar effect was observed in *Archaea* by using the deoxyhypusine synthase inhibitor N^1-guanyl-1,7-diaminoheptane (166). These data suggest that the initiation factor might be important for the translation of cell cycle proteins.

One of the outstanding advances in structural biology and in translation research was obtaining the 3D structure of the 70S ribosome. Ribosomes of two species, the archaeon *Haloarcula marismortui* and the thermophilic bacterium *Thermus thermophilus* were resolved to 2.4 and ~3 Å, respectively (Fig. 10). In addition, functional complexes of tRNA molecules, initiation and elongation factors have been analyzed by cryogenic microscopy reconstruction studies guided by the 3D model from X-ray crystallography (282, 384, 441). Not only the initiation and elongation reactions have been studied in *Archaea*, but also cytosolic factors required for the exit of the nascent polypeptide chain from the ribosome. The 3D structure of a homodimeric "nascent polypeptide-associated complex" previously identified in *Eucarya* and conserved in *Bacteria* and *Archaea* has recently been solved from *Methanothermobacter marburgensis* (377). The protein is associated with ribosomes and seems to have a role in cellular protein quality control.

mRNA recognition and initiation of translation in *Archaea* appear to occur by more than one mechanism (244) (see Chapter 8). Ribosome-binding sites (RBSs) are present upstream of start codons of many genes (244). However, other mRNAs have an extremely short or absent 5'-untranslated leader, preventing the use of an RBS for ORF recognition (29, 131, 372). In vivo studies using reporter genes in *Haloferax volcanii* have shown that the leaderless transcripts often show the strongest translation levels, whereas leader sequences seem to act in downregulation of the translation activities and fine tuning of gene expression (345). Contrasting results were obtained with a *Sulfolobus* cell-free in vitro translation

Figure 9. Chemical structures of diphthamide and hypusine. Modifications (bold type) of the amino acids histidine and lysine to form diphthamide and hypusine, respectively. ADP ribosylation site catalyzed by diphtheria toxin (arrow).

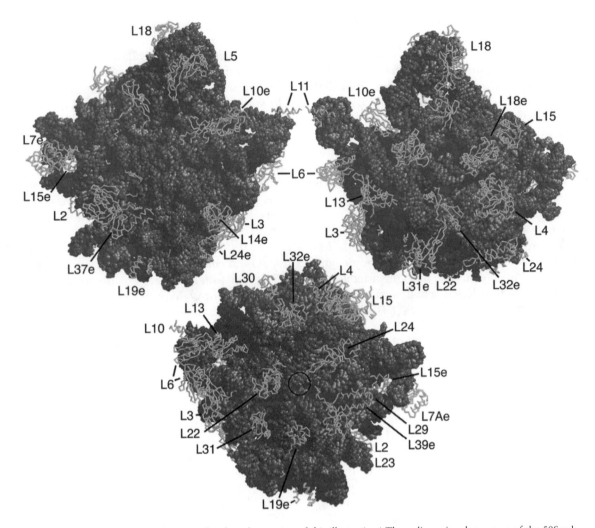

Figure 10. (*See the separate color insert for the color version of this illustration.*) Three-dimensional structure of the 50S subunit of the *Haloarcula marismortui* ribosome. The ribosome arm around ribosomal protein L1 was omitted (for a more complete picture see reference 200). Figure drawn from the coordinates from PDB entry 1QVF (200); ribosomal RNAs are displayed in red (backbone) and gray (bases), proteins are displayed as yellow backbone ribbons. Top left, crown view; top right, back view; bottom, bottom view; the circle indicates the position of the polypeptide exit tunnel.

system, which resulted in decreased translation levels with leaderless transcripts (57). The percentage of leaderless transcripts seems to vary within the *Archaea*. Ten transcripts were mapped in *Pyrobaculum aerophilum*, and all were leaderless, and it was proposed that all transcripts in this organism should be leaderless (372). Similar bioinformatic predictions were made for other archaea (424), but few experimental data are available (131).

Archaeal aminoacyl-tRNA synthesis displays a significant level of divergence from the *E. coli* model system (see Chapter 9). Novel archaeal mechanisms have been described for the aminoacylation of tRNAs with Asp, Cys, Glu, and Lys, some of which have been subsequently found in the *Bacteria* (310). *Ar-*

chaea incorporate the 21st amino acid, selenocysteine, in response to UGA stop codons and a *cis*-acting selenocysteine insertion sequence element (SECIS) in the mRNA, which forms a short stem-loop structure in the RNA. The 22nd amino acid, pyrrolysine, was detected recently in five methanogenic archaea and in one bacterium. Pyrrolysine is inserted in response to UAG stop codons and requires a specific, but as yet undefined, sequence element (PYLIS) (444) (see Chapter 9).

A novel method of recoding nonsense codons for the incorporation of rare amino acids into foreign proteins has been described for *E. coli*, yeast, and mammalian cells (437), that relies on the expression of archaeal aminoacyl-tRNA synthetases in host cells.

In addition, many archaeal tRNAs cannot be charged by host synthetases. The tRNA and corresponding synthetase from *M. jannaschii* was expressed and coupled to the in vivo incorporation of the amino acid O-methyl-L-tyrosine into the dihydrofolate reductase in *E. coli* encoded by an amber nonsense codon (425). To overcome the restrictions of the triplet code, recoding was successful using a *Pyrococcus horikoshii* type I lysyl-tRNA synthetase, which allowed the site-specific incorporation of L-homoglutamine in response to a quadruplet codon, AGGA. The novel codon was resistant to frame-shift suppression. This system also permitted the simultaneous incorporation of a second unnatural amino acid at distinct positions in myoglobin (8). More than 60 unnatural amino acids have been translationally incorporated into proteins. Many provide side chains with functional groups poised for covalent linkage with dyes and other markers, such as spin labels. In the future, selenocysteine may also be translationally incorporated into proteins by site-directed mutagenesis without the need for extended RNA secondary structures. This would greatly expand the feasibility of performing spectroscopic studies on cysteine-liganded metal centers.

ARCHAEAL MODEL ORGANISMS AND IMPORTANT TAXA

An ideal model organism may be considered as one that can reproduce and grow quickly, be easily handled in the laboratory, and possess characteristics analogous to those of many other organisms. No such organism is found in the microbial world because microorganisms exist in markedly different habitats and exhibit enormous physiological diversity. At present, there is not a single archaeon for which the genetic, physiological, and biochemical tools have been developed to an extent that is similar to *E. coli*. Even if such an organism does emerge, it can only be a model for a limited range of physiological traits and represent a distinct component of archaeal diversity.

Rather than one model organism, a broad range of archaea have proven useful for studying morphology, physiology, molecular mechanisms of adaptation, and so forth. These include *Methanothermobacter thermautotrophicus*, *M. marburgensis*, and *Methanosarcina* spp. for methanogenesis; *Thermoplasma* for proteolysis; *Halobacterium* for light-driven proton translocation, gene regulation, chemotaxis, and gas vesicles synthesis; *Archaeoglobus* for sulfate reduction; *Pyrococcus* and *Acidianus ambivalens* for inorganic sulfur metabolism; *Sulfolobus*, *A. ambivalens*, *Pyrococcus*, and the methanogens for electron transport chains; *Sulfolobus* and *Pyrococcus* for

DNA replication and transcription; *Halobacterium*, *Haloarcula*, and *Sulfolobus* for translation, and the list continues.

There are no model organisms representing archaeal pathogens for the simple reason that there is not a single pathogen known, despite the fact that *Archaea* are natural components of the microbiota of many, if not all, multicellular animals, especially herbivores. It would be premature to assume that archaeal pathogens do not exist because they have not been identified (52). *Archaea* are known to excrete proteases and cellulases, and some can colonize metazoan hosts. Therefore, there is no fundamental barrier for the pathogenic potential toward animals or plants. However, even with the use of modern molecular biology techniques (i.e., not just relying on cultivation), no archaeon has been clearly linked as the causative agent of disease. One recent report describes *Methanobrevibacter oralis* as a potential cause of endodontic infections, but others were unable to find a correlation between this methanogen and the disease (370, 415). Even the growth of members of the *Halobacteriales* on dried salted fish may present only aesthetic issues as the *Archaea* do not cause illness. As a consequence, *Archaea* can be regarded as nonpathogenic. An interesting consequence of this is that pathogen-related biosafety issues do not hinder laboratory work.

In the remaining sections of this chapter, the properties of major archaeal taxa are described according to their ecology and molecular similarity. Important characteristics of some of the key organisms are included.

Halophilic Archaea

Halophilic microorganisms and their habitats

Halophilic microorganisms and, in particular, halophilic archaea have developed mechanisms of osmoprotection that allow them to thrive in habitats containing high levels of various ions (up to saturation, e.g., 5.2 M NaCl) (291). Hypersaline habitats are widespread around the world and can be differentiated due to their genesis. Thalassohaline waters are dominated by NaCl and typically arise from seawater evaporation. These include natural or manmade environments such as crystallizer ponds of solar salterns, salt mine drainage waters, coastal splash zones, tide pools, and brine springs from underground salt deposits. Athalassohaline waters originate from inland surface water that leaves behind high concentrations of divalent cations such as Mg^{2+} (Dead Sea, Israel) or carbonate (122). Hypersaline environments vary in pH from near neutral (e.g., the

Dead Sea, Israel; pH $\sim$ 6) to alkaline, carbonate-rich waters (e.g., Lake Magadi, Kenya, and Big Soda Lake, Nevada; pH $\sim$ 9 to 10) (122, 289).

A large diversity of microorganisms of all three domains occupy both thalassohaline and athalassohaline habitats. These ecosystems can be very productive in terms of CO_2 fixation and cell densities. The most important primary producers are green algae (*Dunaniella*) and anoxygenic phototrophic bacteria such as *Ectothiorhodospira* and *Halorhodospira* (300). In contrast, extremely halophilic archaea of the order *Halobacteriales* are chiefly aerobic chemoorganoheterotrophs, growing with amino acids, peptides, organic acids, and carbohydrates. A high percentage of these grow facultatively anaerobic, either by respiration of nitrate, DMSO, and trimethylamine, and/or by fermentation (291). Novel strains were recently isolated that could grow on substrates other than complex organic nutrients, chemolithoautotrophically, or by anaerobic respiration with terminal electron acceptors such as sulfate, fumarate, or sulfur (83, 375). *H. salinarum* is well known for its capacity to grow photoorganoheterotrophically under low oxygen tension using the retinal protein BR as a light-driven proton pump (286). This property is restricted to a small number of *Halobacteriales* species.

Many types of bacterial metabolism have not been found in extreme halophiles. This might be due to incomplete sampling or difficulties in cultivation. However, bioenergetic constraints may also limit certain types of chemolithotrophy and anaerobic respiration to habitats with lower salt concentrations. For example, free-energy calculations show that chemolithoautotrophic growth of halophilic methanogens on CO_2/H_2, or heterotrophic growth on acetate should not be possible at salinities exceeding 1 M (291). Extremely halophilic methanogens have been isolated on substrates including trimethylamine (e.g., *Methanohalobium evestigatum*) (446). Neither *Bacteria* nor *Archaea* have been cultivated that (i) oxidize their substrates completely, (ii) grow at salinities above 15%, and (iii) gain cellular energy by autotrophic ammonia or nitrite oxidation, by the anammox reaction, or by sulfate reduction (291).

Adaptation to hyperosmolarity

Maintaining an osmotic balance and protection of proteins from high salt concentrations are two aspects important for cellular life in high-salt environments. Cells typically have to maintain a lower water activity of the cytoplasm than the surrounding brine to maintain sufficient turgor pressure (123). Halophilic archaea are exceptional in keeping the internal salt concentrations isoosmotic to the environment.

Two strategies exist within the microbial world to cope with the osmotic stress:

1. A strategy based on compatible solutes requires the accumulation of small organic molecules such as glycine-betaine, ectoine, polyols, sugars, or amino acids to balance the external salt concentration, while maintaining a low internal salt concentration. In this case, salt adaptation of cytoplasmic proteins is not required. Compatible solutes are used by extremely halophilic methanogenic archaea and most halophilic bacteria, whereas it is uncommon among *Halobacteriales*. A single report documents that some haloalkaliphilic strains accumulate 2-sulfotrehalose as a compatible solute when grown under low-nutrient conditions (67).

2. The archaea in the order *Halobacteriales*, the halophilic anaerobic bacteria of the order *Haloanaerobiales*, and *Salinibacter* use a "salt-in" strategy (269, 292). KCl and NaCl are accumulated in the cytoplasm to a concentration osmotically equivalent to the external medium (e.g., 4.2 M KCl plus 1 M NaCl in *H. salinarum*) (123). Cytoplasmic proteins need to be protected from precipitation or denaturation due to dehydration. This is accomplished by an adaptive change in the amino acid composition. Most proteins from haloarchaea (and also from *Salinibacter*) contain a small proportion of hydrophobic residues and an excess of acidic amino acids. The acidic amino acids are located predominantly on the surface of haloarchaeal proteins attracting cations in sufficient amounts to preserve a hydrated shell and to enhance solubility. As a consequence, most haloarchaeal proteins have an acidic pI (4 to 5) and depend on high-salt concentrations for function, in vivo and in vitro, although the salt can be replaced for in vitro experiments by neutral osmoprotectants such as sugar alcohols or glycerol. The ion balance is maintained by Na^+ transport from the cytoplasm by an Na^+/H^+ antiporter and a K^+ symporter, which are both driven by the proton motive force (123).

The dependence of haloarchaeal proteins on high salt concentrations in general renders protein structure studies difficult; however, two structures of general importance have been determined. The ribosome of *H. marismortui* was specifically selected for structural studies because the salt masks negative charges of the ribosomal RNA. The structure has been refined to 2.4 Å resolution, and many cocrystallizations with bound antibiotics have been successful (Fig. 10). The other extremely well studied haloarchaeal protein, for which there is an X-ray structure, is bacteriorhodopsin and its relatives, the halorhodopsins and sensory rhodopsins (see "Bacteriorhodopsins and light-driven ATP synthesis" and Fig. 13 below). In contrast, only eight other X-ray and 2 NMR structures have been solved from the *Halobacteriales*. These structures are for malate dehydrogenase, dihydrofolate reductase, dodecin (a dodecameric protein of unknown function), DpsA (an icosatetrameric protein involved in iron homeostasis), a nucleotide diphosphate kinase, a catalase peroxidase, an α-amylase, and a [2Fe-2S] ferredoxin (structures compiled from the structures browser at http://www.ncbi.nih.gov). To put this number in perspective, structures of more than 50 different proteins have been generated from *Pc. furiosus*, many of which did not come from the structural genomics project (see Chapter 20).

Halobacteriales

Typical habitats of *Halobacteriales* are salt lakes such as the Dead Sea, the salt lake of the East African rift valley, and man-made habitats, including solar salterns, which consist of a chain of crystallizer ponds with successively increasing salt concentrations. While cell counts of bacterial and eucaryal species decrease with increasing salt concentration, haloarchaea become more predominant and densities may exceed 1×10^9 ml^{-1}. The fascinating reddish color of solar salterns results from orange and red bacterioruberins (primarily C_{50} carotenoids) in the haloarchaeal membranes, whose primary function is the protection of the cells against photo-oxic damage caused by exposure to the intense sunlight of the Earth's lower latitudes (Fig. 11). The pink or violet retinal-based pigments bacteriorhodopsin and halorhodopsin contribute to the coloration. Colonies of most *Halobacteriales* on agar plates are red or reddish in color. Haloarchaea are also found, inoculated with the added salt in hides, soy sauce, and dried fish (e.g., *H. salinarum*).

An exciting discovery was the detection of viable haloarchaea in fossil halite (NaCl) deposits (reviewed in reference 382), some of which have been cultivated (Table 2) (381). Massive sedimentation of halite and other salt deposits occurred during several periods in the Earth's history, including the late Permian and early Triassic, ~245 to 280 million years ago (445). Some of the rock salts contain pink crystals from haloarchaea (Fig. 11). It is assumed that haloarchaea were entrapped in the halite upon evaporation, most likely in microinclusions of saturated brines in the salt (123). The mechanisms of persistence remain speculative, and it has been debated whether the haloarchaea were entrapped at the time of the rock salt formation or whether they invaded the salt in more recent times through the action of percolating ground water (260).

Figure 11. (*See the separate color insert for the color version of this illustration.*) Haloarchaea in liquid cultures and within salt crystals. (A) Cultures of *Haloferax* and *Halorubrum*: first flask (front), *H. volcanii* WFD11 wild type; second flask, *H. volcanii* WFD11 gas vesicle ΔD mutant (see Fig. 14); third flask, *H. volcanii* WFD11 gas vesicle ΔD mutant complemented with the *gvpD* gene; fourth flask, *Halorubrum vacuolatum* wild type. (B) Himalayan rock salt ("Eubiona"; Claus, GmbH, Baden-Baden, Germany). (C, D) Crystals formed from dried *Halobacterium* cultures (cells trapped within). Bars, 1 cm. Crystals courtesy of F. Pfeifer, Darmstadt, Germany. Photographs by F. Pfeifer, Darmstadt, Germany (panel A), and A. Kletzin (panels B to D).

It is well known that haloarchaea can be trapped in salt crystals and remain viable (Fig. 11) (123). While this is not a method commonly used by culture collections, many isolates have been preserved this way (292).

Archaea tend to predominate in the oxic zones of the water column of hypersaline environments while *Bacteria* are more abundant in anoxic sediments (270). The majority of the archaeal fraction of anoxic zones is *Halobacteriales* despite their predominantly aerobic metabolism. Considerable numbers of extremely halophilic methanogens are also found that grow on trimethylamine and related compounds; isolates include *Methanohalophilus mahii* and *Methanohalobium evestigatum* with optimal salt concentrations of 12% and 24%, respectively (297, 446). 16S rDNA signatures of soil *Crenarchaeota* have also been detected from the microbial mats at the bottom of salterns and salt lakes, although none have been cultivated (270).

Twenty-two genera are described that belong to the *Halobacteriales* (Table 2). They have a high degree of morphological variation (Fig. 12). Many, such

Table 2. Characteristics of halophilic *Halobacteriales* (*Euryarchaeota*)[a]

Species	Morphology	Source habitat	Temperature optimum (°C)	pH optimum	NaCl optimum (M) (range)	Anaerobic growth[b]	No. of species in genus	Remarks[c]
Halobacterium salinarum NRC-1 and DSM 670	Rods	Salted cow hide	35–50	5.2–8	4–5 (3–5.2)	Arginine fermentation	3	G, photoheterotrophic with bacteriorhodopsin, gas vesicles
Halobacterium noricense	Rods	Permian rock salt, Austria	45	5.2–7.7	3 (2.2–5.2)	–		Requires 0.6–0.9 M Mg^{2+}
Halalkalicoccus tibetensis	Coccoid	Lake Zabuye, Tibet, China	40	9.5–10	3.4 (1.4–5.2)	–	1	
Haloarcula marismortui	Pleomorphic flat	Dead Sea, Israel	40–50	Neutral	3.4–3.9 (1.7–5.1)	Nitrate	10	G*, crystal structure of ribosomes; genome composed of one chromosome and 8 (mega-) plasmids
Halobaculum gomorrense	Rod	Dead Sea, Israel	40	6–7	1.5–2.5 (1.0–2.5)	Nitrate	1	G*, requires 0.6–1.0 M Mg^{2+}
Halobiforma haloterrestris	Pleomorphic	Hypersaline soil, Aswan, Egypt	42	7.5	3.4 (2.2–5.2)	–	3	Polyhydroxybutyric acid granules
Halococcus morrhuae	Coccus	Dead Sea, Israel	30–45	Neutral	3.5–4.5 (2.5–5.2)	–	5	Sulfated heteropolysaccharide cell wall
Haloferax mediterranei	Pleomorphic flat	Saltern, Spain	40	6.5	2.9 (2.2–5)	Nitrate	9	Gas vesicles
Haloferax volcanii	Pleomorphic flat	Dead Sea, Israel	40	Neutral	1.7–2.5	Nitrate, sulfur (weak)		G
Haloferax sulfurifontis	Pleomorphic flat	Zodletone spring, Oklahoma, USA	32–37	6.4–6.8	2.2–2.6 (1.1–5.2)	Sulfur		First species with sulfur respiration
Halogeometricum borinquense	Pleomorphic flat	Saltern, Puerto Rico	40	7	3.4–4.3 (1.4–5.2)	Nitrate	1	Gas vesicles
Halomicrobium mukohataei	Rod	Salt flats, Argentina	40	Neutral	3–3.5 (2.5–4.5)	Nitrate	1	
Haloquadratum walsbyi	Pleomorphic flat, square	Salterns, Sinai, Egypt	40	Neutral	3.3	n.r.	1	G, giant cells observed (≤40 × 40 μm); polyhydroxybutyric acid granules; requires pyruvate; high Mg^{2+} tolerance (>2 M)

Species	Morphology	Site of isolation	Temp (°C)	pH	NaCl, M (range)	Anaerobic growth[b]	No.	Comments
Halorhabdus utahensis	Pleomorphic	Great Salt Lake, Utah, USA	50	6.7–7.1	4.7 (1.6–5.1)	Glucose fermentation, sulfur respiration	1	Requires sugars, no growth on peptides/amino acids
Halorubrum saccharovorum	Rods	Saltern, California, USA	50	Neutral	3.5–4.5 (1.5–5.2)	–	14	
Halorubrum vacuolatum	Rods	Lake Magadi, Kenya	35–40	Neutral	3.5 (2.6–5.1)	–		Gas vesicles
Halorubrum lacusprofundi	Rods	Deep Lake, Antarctica	33–40	Neutral	3.5 (2.6–5.1)	–		Gas vesicles
Halosimplex carlsbadense	Rods	Permian rock salt, New Mexico, USA	37–40	7–8	4.4	–	1	Growth only on defined media: acetate + glycerol, glycerol + pyruvate, or pyruvate; three nonidentical 16S rDNA genes
Haloterrigena thermotolerans	Rods	Salterns, Cabo Rojo, Puerto Rico	50	7–7.5	3–3.5 (2.2–5)	–	4	Growth also at 60°C
Natrialba asiatica	Rods	Beach sand, Japan	35–40	6.6–7.8	3.5–4 (2.0–5.2)	–	7	G*, no colony pigmentation
Natrialba magadii	Rods	Lake Magadi, Kenya	37–40	9.5	3.5 (2–5.2)	–	6	Red colony pigmentation
Natrinema pallidum	Rods	Salted cod	37–40	7.2–7.6	3.4–4.3	–	3	
Natronobacterium gregoryi	Rods	Lake Magadi, Kenya	37	9.5	3 (2–5.2)	–	5	
Natronococcus occultus	Coccus	Lake Magadi, Kenya	35–40	9.5	3.4–3.8 (1.4–5.2)	–	2	
Natronolimnobius baerhuensis	Rods	Soda lakes, Inner Mongolia, China	37–45	9.0–9.5	3.4 (2.7–5)	Thiosulfate		
Natronomonas pharaonis	Rods	Wadi Natrun, Egypt	45	9.5–10	3.5 (2–5.2)	–	1	G
Natronorubrum tibetense	Pleomorphic	Alkaline salt lake, Tibet, China	45	9	3.4	–	4	
"*Natronorubrum thiooxidans*"	Pleomorphic	Kulunda steppe, Altai, Russia	30–32	7.5–8	3.5 (2.5–5)	–	4	Mixotrophic growth: acetate + thiosulfate oxidation in symbiosis with tetrathionate-oxidizing proteobacterium

[a]Data are compiled from Oren (292), and from original descriptions accessed from the taxonomy browser (http://www.ncbi.nih.gov). All genera of the *Halobacteriales* are represented. All strains grow aerobically by respiration, and colonies are usually red.

[b]n.r., not reported; –, no anaerobic growth.

[c]G, genome sequence available; G*, draft genome sequence (120).

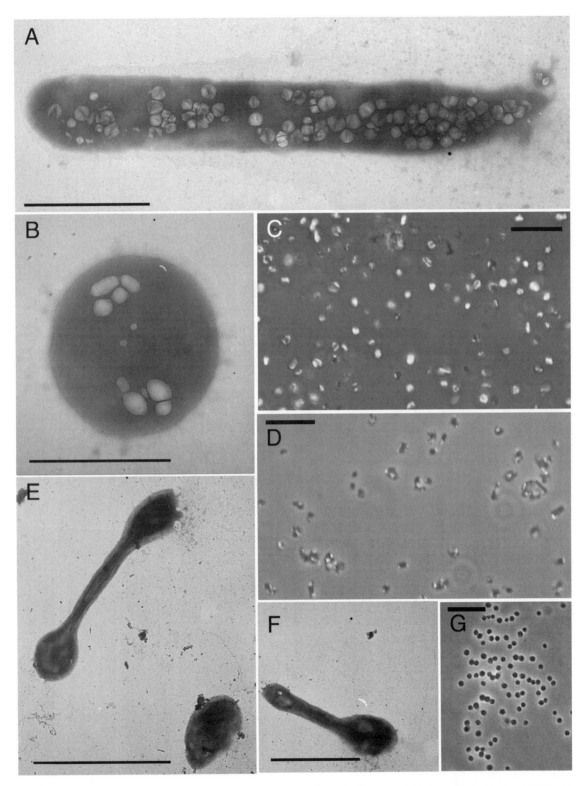

Figure 12. Morphology of haloarchaea. (A) Negative stain image of aerobically grown *Halobacterium salinarum* PHH1 with gas vesicles. (B) Negative stain image of aerobically grown *H. mediterranei* with gas vesicles. (C) Phase-contrast micrograph of *H. volcanii* WFD11 gas vesicle ΔD mutant (see Fig. 14). (D) Phase-contrast micrograph of *Halorubrum vacuolatum* with gas vesicles. (E, F) Negative stain electron micrographs of *H. mediterranei* cells showing pleomorphic shape. (G) Phase-contrast micrograph of *H. volcanii* WFD11 wild type. Photographs courtesy of T. Hechler, Darmstadt.

as *Halobacterium*, are rods, and others, such as *Natronococcus*, are cocci. Many are flagellated and motile. The most characteristic morphology for members of the *Halobacteriales* (e.g., *Haloferax*) are flattened cells that look like pleomorphic discs when viewed from the top and thin rods when viewed from the edge (Fig. 12). This morphology is otherwise uncommon in the *Archaea*, with the exception of *Pyrodictium* spp. (see "*Pyrodictiaceae,*" below). Some haloarchaea have a square shape that resembles a postage stamp (also referred to as "Walsby's square bacterium"; Fig. 4) (423). The length of the sides of square cells is typically ≤10 μm, with some occasionally ≤40 μm, and the thickness of cells is typically 0.25 μm. The thickness apparently remains constant with cell growth and cell division (192, 390). Cells with this morphology are stable only when the cytoplasm is isoosmotic with the environment. Square archaea account for up to 30% of the population in many solar salterns and salt lakes worldwide. Despite being observed for many years, it was not until 2004 that isolates were first propagated in the laboratory and the species *Haloquadratum walsbyi* was described (33, 48). The isolation was successful using conventional methods and considerable persistence. *H. walsbyi* cells contain gas vesicles, similar to many other haloarchaea (Fig. 4) (see "Gas vesicles and transformation systems," below). The genome for *H. walsbyi* is currently being sequenced (http://www.biochem.mpg.de/oesterhelt/).

New strains have recently been isolated from salt-marsh sediments from the River Colne, Essex, United Kingdom, that grow at substantially lower salt concentrations than other haloarchaea (2.5 to 5% NaCl) and appear to represent a new genus within the *Halobacteriaceae* (313). Presumably, other haloarchaea adapted to low-salt concentrations will be found in the future and further expand knowledge of *Archaea* that are adapted to less extreme habitats.

The hypersaline Deep Lake (Vestfold Hills, Antarctica) is interesting because it has a salt concentration near saturation (3.6 to 4.8 M) and remains ice-free throughout the year (51). The water temperature fluctuates between +10°C and −15°C. In situ microbial productivity is low, and only two isolates have been obtained, a *Dunaniella* species (*Eucarya*) and *Halorubrum lacusprofundi* (51, 99). The latter is a heterotrophic haloarchaeon with a T_{opt} in the mesophilic range (Tables 2 and 7) (99). Cold adaptation has been linked to the presence of unsaturated diether lipid in cells grown at 12°C (111).

Hypersaline waters accumulate in depressions in the Red Sea and in a range of other, mostly warm, marine zones worldwide. The temperatures can occasionally exceed 60°C. Thermophilic and halophilic microorganisms have not yet been cultivated from the brines. Molecular analysis revealed the presence of novel and previously uncultivated archaea and bacteria (74) and the retrieval of potentially valuable and novel hydrolases. An exciting finding was an esterase that showed remarkable activity in response to temperature, salt, and pressure (89).

Bacteriorhodopsins and light-driven ATP synthesis

Some species, such as *H. salinarum*, synthesize the membrane protein BR, primarily under conditions of light and low aeration. It contains covalently bound retinal as the single chromophor inserted between seven transmembrane helices (Fig. 13). BR is a light-driven proton pump generating an electrochemical gradient used for ATP synthesis by a proton-translocating, membrane-bound F_0F_1 ATP synthase. BR has a purple coloration with a marked absorption maximum at about 570 nm (reviewed, for example, in references 75, 229). The molecules are not evenly distributed in the cytoplasmic membranes but localized in patches with the appearance and properties of two-dimensional (2D) crystals. This purple membrane can easily be isolated with sucrose density gradient centrifugation and separated from the reddish bacterioruberin-containing cytoplasmic membrane (see Chapter 22).

The resolution of the 3D structure of the integral membrane domains of BR was the first example of the use of the regular array of 2D crystals to enhance electron microscopic images to high resolution (142). Additional spectroscopic studies revealed that the retinal in BR is an all-*trans* conformation of the C_{20} chain in the ground state (Fig. 13). It becomes excited upon light absorption and converts to the *cis* conformation that is coupled to the transfer of a proton to the outside surface of the membrane. The retinal returns to the ground state with the uptake of a proton from the cytoplasmic side of the membrane (75, 229). The electrochemical gradient can also drive Na^+ export and K^+ and nutrient import, in addition to ATP synthesis. The Na^+/H^+ antiporter also drives the uptake of amino acids indirectly, which is mediated by a Na^+ symport system.

In addition to BR, *H. salinarum* synthesizes at least three different retinal proteins with other functions. Chloride is accumulated as a counter ion to K^+ in the cytoplasm by a light-driven chloride pump, named halorhodopsin. The retinal binds Cl^- and transfers it into the cell. Sensory rhodopsins control phototaxis together with two-component regulators and affect flagellar rotation (378) (see Chapters 11 and 18). These retinal proteins were thought to only be present in archaea until the related proteorhodopsin

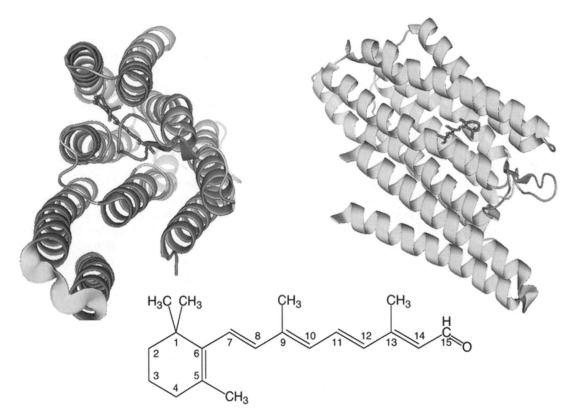

Figure 13. Three-dimensional model of the *Natronomonas pharaonis* sensory rhodopsin highlighting α-helices and retinal (black) and chemical structure of the all-*trans* form of retinal (PDB code 1H2S).

genes were detected in uncultured marine bacteria, and xanthorhodopsin genes in the extremely halophilic bacterium, *Salinibacter ruber* (13). In *S. ruber*, one of the genes resembles bacterial proteorhodopsins, while the others are of the haloarchaeal type. The function of proteorhodopsins appears to be similar to BR.

Gas vesicles and transformation systems

Many haloarchaea synthesize gas vesicles that are easily recognized as bright refractive bodies under phase-contrast microscopy (Figs. 4 and 12). These flotation devices are filled with gas in a watertight proteinaceous shell (422). Gas vesicles are used to adjust the buoyant density of the cells, enabling them to float at the water surface (Fig. 11) or, more commonly, to adjust the mean density of the cell to float in zones of the water column that are optimal for the growth of the organism (e.g., oxygen tension). Gas vesicles have mainly been studied in halophilic archaea and in cyanobacteria, but they are not restricted to these groups and occur in many freshwater bacteria such as *Ancalomicrobium* and *Magnetospirillum* (380, 422). Surprisingly, gas vesicle genes have also been found in soil bacteria, including *Streptomyces* spp. and *Bacil-*

lus megaterium (238, 414), and the heterologous expression of the *B. megaterium* gene cluster led to gas vesicle production in *E. coli*. Gas vesicles are also present in *Methanosarcina* and *Methanothrix* (188), and gas vesicle genes are present in genome sequence of *Methanosarcina barkeri* (Fig. 14).

Haloarchaeal gas vesicles are spindle- or cylinder-shaped structures with conical ends, ~200 nm in diameter and 400 to 1,500 nm in length (287, 422). The vesicles consist of more than 90% of the small hydrophobic protein, GvpA, that forms semicrystalline, 4.5-nm-wide helical ribs that are perpendicular to the gas vesicle's longitudinal axis. The second structural protein, GvpC, contains internal amino acid repeats that are reported to bind to each of two different ribs (422), and it is thought to stabilize the vesicles in cyanobacteria. The deletion of the *gvpC* gene in recombinant haloarchaea resulted in pleomorphic gas vesicles with variable diameter and rounded ends, instead of regularly arranged vesicles (287). These results suggest that GvpC plays a role in the determination of the shape of gas vesicles (287).

Deletion and overexpression analysis showed that some of the proteins encoded in the haloarchaeal gene cluster (e.g., GvpD and GvpE) have regulatory

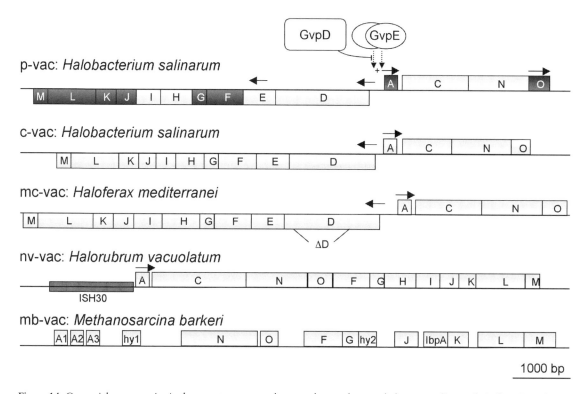

Figure 14. Gas vesicle operons in *Archaea*. *gvp* genes are shown as boxes above or below according to their direction of transcription, and promoter regions are denoted by arrows; *p-vac*, plasmid-encoded *vac* region; *c-vac*, chromosomally encoded *vac* region; *mc-vac*, chromosomally encoded *vac* region of *H. mediterranei*; ΔD, in-frame deletion mutant of the *mc-vac* region, leading to gas vesicle-overproducing *H. volcanii* transformants; *nv-vac*, chromosomally encoded *vac* region of *Halorubrum vacuolatum* (formerly *Natronobacterium vacuolatum*); ISH30, insertion element; *mb-vac*, chromosomally encoded *vac* region from the genome of *Methanosarcina barkeri*; GvpE and GvpD, transcriptional activator and repressor, respectively. Modified from a figure provided by F. Pfeifer, Darmstadt, with reference to several sources (287, 305, 306) and the *M. barkeri* genome sequence (GenBank NC_007355).

functions and are not required for gas vesicle synthesis. Other genes encoded in the cluster (e.g., GvpI and GvpH) play a role in determining the length, stability, or number of vesicles per cell. The minimal region required to synthesize small vesicles comprises eight of the fourteen *gvp* genes that are present in most haloarchaeal *gvp* clusters (Fig. 14) (287). The *Methanosarcina barkeri gvp* cluster contains the same conserved genes that have been described as essential. The role of most of the genes has not been well defined. It has been suggested that some genes (F,G,J,L,M) might encode minor structural proteins, since the proteins are detectable in immunochemical analysis of whole vesicles (368). The amino acid sequences of GvpJ and GvpM are similar to the main structural protein GvpA (305) and may participate in gas vesicle formation by serving as a scaffold or nucleation core.

Haloarchaea, such as *Haloferax* and *Halorubrum*, contain a single *gvp* cluster in the genome. Multiplications of the *gvpA* gene are present in *Methanosarcina* (Fig. 14) and are common in bacteria. A haloarchaeal

exception is *H. salinarum* PHH1, which contains two complete and similar *gvp* clusters. One cluster (c-vac) is encoded on a large megaplasmid and the other (p-vac) on a 150-kbp plasmid, pHH1. A deletion strain PHH4, which contains only the c-vac region, produces long, cylinder-shaped vesicles during stationary phase (304). The expression of the p-vac region is constitutive and produces small, spindle-shaped vesicles. GvpD and GvpE are involved in regulation of the *gvp* genes (288). GvpE is a dimer-forming, DNA-binding, coiled-coil, leucine-zipper protein that is required for transcriptional activation (217) (see Chapter 6). GvpD is a repressor that interacts with GvpE through protein-protein contacts (457). An in-frame deletion of large parts of the *Haloferax mediterranei gvpD* gene (mc *gvpD*) results in a gas vesicle-overproducing phenotype in *H. volcanii* recombinants (Figs. 11, 12, and 14) (87). The cells that are normally flat, pleomorphic discs, become coccoid and resemble inflated balls.

The analysis of gas vesicle synthesis and regulation has required the use of a *gvp*-negative *H. volcanii* strain and transformation systems based on three different mutually compatible haloarchaeal shuttle vectors (32) (see Chapter 21). The plasmids for *H. volcanii* include pWL101 and pMS20, which are general cloning vectors that utilize native promoters of cloned genes for their expression, and pJAS35, which carries the strong constitutive ferredoxin promoter for overexpression of introduced genes. Quantitative measurements of gene expression levels can also be determined using a haloarchaeal β-galactosidase (*bgaH*) reporter gene (124). In addition, it is possible to generate chromosomal gene knockouts (299). Haloarchaea are the most easily genetically manipulated members of the *Archaea* (299).

Archaeocins

Bacteriocins are excreted peptides that are ribosomally translated or synthesized by peptidyl transferases, which have bactericidal activity toward related, non-bacteriocin-producing bacterial strains. Many members of the *Halobacteriales* produce Archaeocins, proteinaceous archaeal antimicrobials. At least ten different archaeocins (also termed halocins) have been characterized (reviewed in reference 285). *Sulfolobus islandicus* also produces a 20-kDa archaeocin (sulfolobicin) that is attached to S-layer-derived protein particles (311). Sulfolobicin only has activity against its own species. In contrast, halocins have activity against many different species (285). Archaeocin activity is usually detectable in cultures at the beginning of stationary phase. Halocins can be differentiated into two classes. Most consist of proteins of 30- to 35-kDa molecular mass, while the microhalocins are small peptides of 3 to 4 kDa. The large proteinaceous halocins are easily inactivated by dialysis against low-salt buffers. The peptide microhalocins are more robust and can withstand extensive heat treatment and desalting. Both size classes of halocins are sensitive to proteases. The 36 amino acids of the microhalocin HalS8 are excised from the interior of a 311-amino-acid precursor protein by an unknown protease activity. Microhalocins exhibit cross-kingdom toxicity as they are active against *Sulfolobus* species (*Crenarchaeota*) in addition to a broad spectrum of antihaloarchaeal activity (*Euryarchaeota*).

The mechanisms of halocin immunity was recently studied with the 76-amino-acid microhalocin, HalC8, which is excreted by *Halobacterium* strain AS7092 (395). This halocin has high stability to denaturing agents and has a wide spectrum of antiarchaeal and antibacterial activity that includes most haloarchaea and some halophilic bacteria. HalC8 is the C-terminal proteolysis product of a 283-amino-acid precursor protein. HalI and HalC8 are localized to the cytoplasmic membrane. The unprocessed precursor and the N-terminal 207-amino-acid domain (HalI) are able to block halocin activity of mature HalC8 in vitro. Heterologous expression of the *halI* gene in the halocin-sensitive strain, *Haloarcula hispanica*, induced a remarkable HalC8 resistance indicating that the HalI domain might cause immunity. In vitro assays confirmed that a strong interaction between HalI and HalC8 exists. The precursor protein appears not only to be required to transfer the halocin across the cytoplasmic membrane but also the proteolysis product(s) seem(s) to play an important role in immunity (395).

Hyperthermophilic Archaea

Hyperthermophilic microorganisms and their habitats

Hyperthermophiles are defined as microorganisms having a T_{opt} of 80°C or higher, and thermophiles as having a T_{opt} of 50 to 80°C (386). Hyperthermophiles are found in most hot-water environments worldwide. One hyperthermophilic archaeon has been reported to grow at temperatures up to 121°C (179). Most hyperthermophiles are *Archaea* (two hyperthermophilic genera of *Bacteria* versus >20 of *Archaea*; 159). Hyperthermophilic (*) and thermophilic species are found in many orders and genera of the *Archaea*. Among the methanogens, all species of the genera *Methanopyrus**, *Methanothermus**, *Methanothermobacter*, *Methanocaldococcus**, and *Methanotorris** are thermophiles or hyperthermophiles. Hyperthermophiles or thermophiles are also represented by members of the *Archaeoglobales**, *Picrophilus*, *Thermoplasma*, *Thermococcales**, *Nanoarchaeum**, *Desulfurococcales**, *Thermoproteales**, and *Sulfolobales**.

Typical habitats are terrestrial hot springs and solfataric areas in geologically active zones, volcanic hot spots, geological plate-spreading or subduction zones, and fault lines (Fig. 15). Other hot-water environments that have not been well studied include the deep-subsurface aquifers of the Great Artesian Basin in Australia (≤100°C), which cover vast areas of the continent and contain significant microbial activity (197), and the deep-sea brines (≤70°C) at the bottom of deep-sea clefts in the Mediterranean Sea and Red Sea (see "*Halobacteriales*," above) (74). More studies have been performed on the spectacular "black smokers" submarine structures emitting superheated (≤350 to 400°C) H_2S-reduced fluids that form dark clouds when released into the oxidized, cold (2 to 4°C) seawater (135). The more common "white smokers" (20 to 350°C) also harbor many hy-

Figure 15. (*See the separate color insert for the color version of this illustration.*) Solfatara and Pisciarelli fumaroles. (Left) Fumaroles in the Solfatara caldera (Pozzuoli near Naples, Italy) with deposition of sulfur, mercury, and arsenic salts. (Right) Fumarole-heated hole with boiling water, typical of habitats for *Sulfolobales* (Pisciarelli, near Naples, Italy). Photos taken by A. Kletzin.

perthermophiles. Cold seawater is the source of venting, which seeps into the fissures of ridge systems, reaching heated rocks and/or magma. It is extruded after becoming enriched with many minerals from passage through rock (e.g., Si, Fe, Mn, Zn, Cu) while remaining depleted of others (Mg, Mo). Insoluble metal sulfides, sulfates, and/or silica precipitate as soon as the vent discharge mixes with seawater, and this leads to the formation of porous and impressive vent structures (215, 418). The pores are densely populated with hyperthermophilic microorganisms, including methanogens, fermentative *Thermococcales*, and heterotrophic or autotrophic sulfate and sulfur reducers (324, 399).

Submarine vents are not restricted to deep-sea settings. There are many examples of shallow hot-water vents close to islands, such as those at a depth of ~3 m near Vulcano Island beach, Italy. These were the sources of several well-studied hyperthermophiles, including *P. furiosus*/*P. woesei*, *Thermococcus celer*, and *Pyrodictium occultum* (Tables 3 to 5) (91, 389, 448). Similarly, *Hyperthermus butylicus* originates from a similar site at a depth of 8 m near São Miguel, Açores, Portugal (447).

Microorganisms are thermally stratified within smoker walls according to their optimal growth conditions, suggesting that the steep gradients of temperature, pH, redox potential, and substrates provide numerous niches for specialized species of anaerobic and aerobic bacteria and archaea (136, 399). Microbial ecosystems are also thought to exist beneath hy-

drothermal vent systems and to be dominated by hyperthermophilic, chemolithotrophic methanogens and heterotrophic sulfur-reducing *Thermococcales* ("HyperSLiME": *hyper*thermophilic *s*ubvent *l*ithoautotrophic *m*icrobial *e*cosystem) (398). An interesting example of an established thermophilic hydrothermal environment is the "Lost City" that is located ~15 km away from the Mid-Atlantic-Ridge (184) and consists of carbonate towers up to 30 m in height with an age of >30,000 years. The alkaline vent fluids contain large amounts of H_2 and methane but essentially no CO_2. Methane-producing *Methanosarcinales* are abundant in hotter regions of the "Lost City" vents (361), and anaerobic methane-oxidizing archaea from the ANME-1 group have been detected in cooler regions of the vents [see "(Anaerobic) methane oxidation by methanotrophic archaea," below] (185).

The metabolism of sulfur and inorganic sulfur compounds (ISCs) is one of the hallmarks of hyperthermophilic archaea, and this is reflected in the number of species that use ISCs as electron acceptor or donor (Tables 3 to 9). This type of metabolism correlates with the abundance of S^0 and ISCs in volcanic settings. ISCs can account for $\leq 10\%$ of the dry volume of the gaseous emissions from terrestrial sources or $\leq 1\%$ of fluid volume emitted from hydrothermal vents (Fig. 15). The proportion of SO_2, SO_3, H_2S, and S^0 can vary (391). For example, H_2S dominates in geologically older terrestrial hydrothermal fields and submarine vents. In contrast, SO_2 and SO_3 are more abundant in younger or more active volcanic sites.

Table 3. Characteristics of hyperthermophilic *Thermococcales* (*Euryarchaeota*) and *Nanoarchaeum* growing at neutral or alkaline pH[a]

Species	Source/habitat	Temperature optimum (range) (°C)	pH optimum (range)	Metabolism	Substrates	Electron acceptors	No. of species in genus	Remarks[b]
Pyrococcus furiosus/woesei[c]	Shallow vent, Vulcano Island, Italy	100–103 (70–105)	7 (5–9)	Heterotrophic/ mixotrophic	Amino acids, maltose, complex organics	S^0, H^+ organics	5	G, structural genomic project
Pyrococcus horikoshii	Okinawa Trough vents, Northeastern Pacific Ocean	98 (80–102)	7 (5–8)	Heterotrophic/ mixotrophic	Complex organics	S^0, organics		G
Pyrococcus abyssi	Deep-sea vent, North Fiji Basin, Pacific Ocean	96 (67–100)	6.8 (4–8.5)	Heterotrophic/ mixotrophic	Amino acids, maltose, starch, pyruvate, complex organics	S^0, organics		G
Thermococcus kodakaraensis	Shallow vent, Kodakara Island, Kagoshima, Japan.	96 (67–100)	6.9 (4–8.5)	Heterotrophic/ mixotrophic	Amino acids, maltose, complex organics	S^0, organics	29	G
Thermococcus litoralis	Shallow vent, Bay of Lucrino, Naples, Italy	88 (65–95)	7.2 (6.2–8.5)	Heterotrophic/ mixotrophic	Peptides, pyruvate	S^0, organics		
Thermococcus celer	Shallow vent, Vulcano Island, Italy	87 (–93)	5.8 (4–7)	Heterotrophic/ mixotrophic	Amino acids, sugars, peptides	S^0, organics		High G+C species
Thermococcus alkaliphilus	Shallow vent, Vulcano Island, Italy	85 (54–91)	9 (6.5–10)	Heterotrophic/ mixotrophic	Amino acids, peptides	S^0, organics		
Palaeococcus ferrophilus	Deep-sea vent, North Fiji Basin, Pacific Ocean	83 (60–88)	6.0 (4–8)	Heterotrophic/ mixotrophic	Complex organics	S^0, organics	2	
Nanoarchaeum equitans	Kolbeinsey Ridge, Iceland	90 (70–98)	6 (4–8.5)	Unclear	H_2/CO_2	S^0	1	Grows exclusively with *Igniococcus* strain KIN/4 in symbiosis; organism with smallest genome to date

[a]Data compiled from Bertoldo and Anthranikian (28) and from original descriptions from the taxonomy browser (http://www.ncbi.nih.gov). All *Thermococcales* strains grow anaerobically by fermentation. Cells are coccoid. All genera of the *Thermococcales* are represented.
[b]G, genome sequence available.
[c]Very closely related and probably strains of a single species (133).

Table 4. Characteristics of anaerobic, hyperthermophilic sulfate reducers from the *Euryarchaeota* (*Archaeoglobus*, *Ferroglobus*, and *Geoglobus*) and *Crenarchaeota* (*Caldivirga*) kingdoms[a]

Species	Morphology	Source/habitat	Temperature optimum (range) (°C)	pH range	Metabolism	Substrates	Electron acceptors	No. of species in genus	Remarks[b]
Archaeoglobus fulgidus	Irregular coccoid	Shallow vents, Vulcano Island, Italy; submarine vents, oil wells	83 (50–90)	5.5–8	Facultatively chemolitho-autotrophic	H_2, lactate, pyruvate, formate	Sulfate, sulfite, thiosulfate	4	G
Ferroglobus placidus	Irregular coccoid	Shallow vents, Vulcano Island, Italy	85 (65–95)	6–8.5	Facultatively chemolitho-autotrophic	Fe^{2+}, sulfide, H_2	Nitrate, thiosulfate	1	Anaerobic degradation of acetate and aromatic hydrocarbons
Geoglobus ahangari	Pleomorphic	Submarine hydrothermal vents, Guaymas	88 (65–90)	6–7.5	Facultatively chemolitho-autotrophic	H_2, organic acids, amino acids, complex organics	Fe^{3+}, sulfide, H_2	1	
Caldivirga maquilingensis	Straight or curved rods	Mt. Maquiling, Laguna, Philippines	85 (60–92)	2.3–6.4	Heterotrophic	Glycogen, gelatin, amino acids, complex organics	Sulfate, S^0, thiosulfate	1	

[a]Data compiled from several sources (129, 137, 162, 180).
[b]G, genome sequence available.

43

Table 5. Characteristics of hyperthermophilic *Desulfurococcales* (*Crenarchaeota*) growing at approximately neutral pH[a]

Species	Morphology	Source/habitat	Temperature optimum (range) (°C)	pH optimum (range)	Aerobic/ anaerobic	Metabolism	Substrates	Electron acceptors	No. of species in genus	Remarks[b]
Desulfurococcus mucosus	Cocci	Terrestrial hot spring areas, Iceland	76–93	6 (4.5–7)	Anaerobic	Heterotrophic	Complex organics	S^0	5	Fermentation, sulfur respiration
Aeropyrum pernix	Cocci	Shallow marine vent, Kodakara Island, Japan	90–95 (70–100)	7 (5–9)	Aerobic	Heterotrophic	Complex organics	O_2	2	G; thiosulfate stimulatory
Ignicoccus hospitalis KIN/4	Cocci	Marine vents, Kolbeinsey Ridge, Iceland	90 (70–98)	6	Anaerobic	Chemolitho-autotroph	H_2	S^0	3	S^0/H_2 autotrophy
Staphylothermus marinus	Cocci	Shallow marine vent, Vulcano Island, Italy	65–98	4.5–8.5	Anaerobic	Heterotrophic	Peptides, complex organics	S^0	2	
Stetteria hydrogenophila	Cocci	Shallow vent, Milos, Greece	95 (68–102)	6 (4.5–7)	Anaerobic	Mixotrophic	H_2, complex organics	S^0, thiosulfate	1	
Sulfophobococcus zilligii	Cocci	Terrestrial hot spring areas, Iceland	85 (70–95)	7.5 (6.5–8.5)	Anaerobic	Heterotrophic	Yeast extract	Organics	1	Fermentation
Thermodiscus maritimus	Disc-shaped cocci	Shallow vent, Vulcano, Island, Italy	90 (75–98)	5.5 (5–7)	Anaerobic	Heterotrophic	Complex organics	S^0, organics	1	Fermentation, sulfur respiration
Thermosphaera aggregans	Cocci	Terrestrial hot spring areas, Yellowstone, WY, USA	85 (65–90)	6.5 (5–7)	Anaerobic	Heterotrophic	Complex organics	Organics	1	Fermentation
Pyrodictium occultum	Disc-shaped, flat	Marine shallow vents, Vulcano Island, Italy	105 (85–110)	5.5 (4.5–7.2)	Anaerobic	Chemolitho-autotroph	H_2	S^0, thiosulfate	3	S^0/H_2 autotrophy
Pyrodictium abyssi	Discs	Deep-sea vent, TAG site, Mid-Atlantic Ridge	97 (80–110)	5.5 (4.7–7.1)	Anaerobic	Chemolitho-autotroph	H_2, organics	S^0		S^0/H_2 autotrophy; fermentation
Hyperthermus butylicus	Cocci	Marine shallow vent, São Miguel, Açores	95–106 (72–108)	7	Anaerobic	Heterotrophic	Peptides	Organics	1	G, fermentation of peptides
Pyrolobus fumarii	Irregular cocci	Deep-sea vent, TAG site, Mid-Atlantic Ridge	106 (90–113)	5.5 (4–6.5)	Anaerobic/ micro-aerophilic	Chemolitho-autotroph	H_2	Nitrate, thiosulfate, O_2	1	"Hottest" validly described organism
Strain 121	Cocci	Deep-sea Mothra vent field, Juan de Fuca Ridge, east Pacific	106 (80–121)	n.r.[c]	Anaerobic	Heterotrophic	Formate	Fe^{3+}	-	"Hottest organism"
Acidilobus aceticus	Cocci	Terrestrial acidic hot springs, Kamchatka, Russia	85 (60–92)	3.8 (2–6)	Anaerobic	Heterotrophic	Starch, complex organics	S^0, organics	1	
"*Caldococcus noboribetus*"	Cocci	Terrestrial acidic hot springs, Japan	92 (70–92)	3 (1.5–4)	Anaerobic	Heterotrophic	Complex organics	S^0, organics	1	
Ignisphaera aggregans	Cocci	Terrestrial solfataric fields, Rotorua, New Zealand	92–95 (72–108)	6.4 (5.4–7)	Anaerobic	Heterotrophic	Sugars, carbohydrates, complex organics	Organics	1	Strictly fermentative, consortia with *Pyrobaculum sp.*

[a]Data compiled from Huber and Stetter (155), and from original descriptions obtained from the taxonomy browser (http://www.ncbi.nih.gov). All genera of the *Desulfurococcales* are represented.
[b]G, genome sequence available.
[c]n.r., not reported.

Table 6. Characteristics of hyperthermophilic *Thermoproteales* (*Crenarchaeota*) growing at acidic to neutral pH[a]

Species	Source/habitat	Temperature optimum (range) (°C)	pH optimum (range)	Aerobic/anaerobic	Metabolism	Substrates	Electron acceptors	No. of species in genus	Remarks[b]
Thermoproteus tenax	Solfataric field, Krafla, Iceland	88 (70–98)	5.5 (2.5–6)	Anaerobic (moderately aerotolerant)	Facultatively lithoautotrophic	H₂, amino acids, sugars, complex organics	S⁰	3	G
Thermoproteus neutrophilus	Solfataric field, Iceland	85 (–97)	6.5 (5–7.5)	Anaerobic	Facultatively lithoautotrophic	H₂, amino acids, sugars	S⁰		
Pyrobaculum aerophilum	Maronti Beach, Ischia Island, Italy	100 (75–104)	7 (5.8–9)	Facultatively anaerobic	Facultatively lithoautotrophic	H₂, thiosulfate, complex organics	O₂, nitrate, nitrite, Fe³⁺, arsenate	7	G, this represents the only marine species of *Pyrobaculum*
Pyrobaculum islandicum	Geothermal power plant, Iceland	100 (74–102)	6 (5–7)	Anaerobic	Facultatively lithoautotrophic	H₂, complex organics	S⁰, thiosulfate, sulfite, Fe³⁺, cysteine, oxid. glutathione		
Thermocladium modestus	Acidic hot spring areas, Japan	75 (45–82)	4.2 (2.6–5.9)	Anaerobic (moderately aerotolerant)	Heterotrophic	Complex organics	S⁰, thiosulfate, sulfate, cysteine	1	
Caldivirga maquilingensis	Acidic hot springs, Philippines	85 (60–92)	3.7–4.2 (2.3–6.4)	Anaerobic (moderately aerotolerant)	Heterotrophic	Complex organics	S⁰, thiosulfate, sulfate	1	
Vulcanisaeta distributa	Hot spring areas, eastern Japan	85–90	4–4.5	Anaerobic	Heterotrophic	Complex organics	S⁰, thiosulfate,	2	
Thermofilum pendens	Solfataric fields, Iceland	88 (70–95)	5.5 (4–6.5)	Anaerobic	Heterotrophic	Peptides	S⁰	2	Requires extract of other archaea

[a]Data compiled from Huber et al. (153), and from original descriptions from the taxonomy browser (http://www.ncbi.nih.gov). Cells are stiff rods (*Thermoproteaceae*) or very thin long rods (*Thermofilaceae*). All genera of the *Thermoproteales* are represented.
[b]G, genome sequence available.

Table 7. Characteristics of the cold-adapted *Nitrosopumilus*, *Cenarchaeum* (*Crenarchaeota*), and SM1 (*Euryarchaeota*)[a]

Species	Morphology	Source/habitat	Temperature (°C)	pH	Aerobic/ anaerobic growth	Type of metabolism	Electron acceptors	Electron donors	Remarks[b]
Nitrosopumilus maritimus	Short, thin rods	Aquarium seawater tank, Seattle, WA, USA	28	7–7.2	Aerobic	Chemolithotrophic	O_2	NH_3	
Cenarchaeum	Short, thin rods	Pacific coast off California, USA	8–18	n.r.	Aerobic	n.r.	O_2	n.r.	G, endosymbiont in the marine sponge *Axinella mexicana*
Euryarchaeon SM1	Cocci	Cold sulfidic springs, Germany	10–11	7.2–7.3	n.r.	n.r.	n.r.	n.r.	Cultivation only in situ, consortia with *Thiothrix*-like bacteria

[a]The cold-adapted haloarchaeon *H. lacusprofundi* is described in Table 2, and the methanogens *M. burtonii* and *M. frigidum* are described in Table 10. Data compiled from several sources (51, 144, 212, 335). n.r., not reported.
[b]G, genome sequence available.

46

Table 8. Characteristics of thermoacidophilic *Thermoplasmatales* (*Euryarchaeota*)[a]

Species	Morphology	Source/habitat	Temperature optimum (°C)	pH optimum (range)	Aerobic/ anaerobic growth	Substrates	Electron acceptors	No. of species in genus	Remarks[b]
Thermoplasma acidophilum	Pleomorphic	Smoldering coal refuse piles, solfataric fields	59	1–2 (0.5–4)	f	Peptides (carbohydrates)	O_2, S^0	2	G
Thermoplasma volcanium	Pleomorphic	Solfataric fields	59	2 (1–4)	f	Peptides (carbohydrates)	O_2, S^0		G
Picrophilus torridus	Irregular cocci	Geothermally heated acid soil, Kawayu Onsen, Japan	60	0.7 (0–3.5)	Aerobic	Yeast extract	O_2	2	G
Ferroplasma acidiphilum	Pleomorphic	Metal-leaching plant	35	1.7 (1.3–2.2)	Aerobic	Fe_2, pyrite	O_2, S^0	3	G, autotrophic
Ferroplasma acidarmanus	Pleomorphic	Acid mine drainage, Iron Mountain, SD, USA	37	1.2 (0–2.5)	Aerobic	Fe_2, pyrite, yeast extract	O_2, S^0	2	G, autotrophic, salt and transition-metal tolerant

[a]Data compiled from Huber and Stetter (156), and from original descriptions from the taxonomy browser (http://www.ncbi.nih.gov). All genera of the *Thermoplasmatales* are represented.
[b]G, genome sequence available.

Table 9. Characteristics of thermoacidophilic *Sulfolobales* (*Crenarchaeota*)[a]

Species	Source/habitat	Temperature optimum (range) (°C)	PH optimum (range)	Metabolism	Aerobic/ anaerobic growth	Substrates	Electron acceptors	No. of species in genus	Remarks[b]
Sulfolobus acidocaldarius	Solfataric fields, Yellowstone, WY, USA	70–75 (55–85)	2–3 (1–6)	Aerobic	Heterotrophic/ mixotrophic	Sugars, complex organics, H_2	O_2	10	G
Sulfolobus solfataricus	Solfataric fields, Solfatara (Naples), Italy	78–85 (50–88)	3–4 (2–5.5)	Aerobic	Heterotrophic/ mixotrophic	Sugars, starch, complex organics, H_2	O_2		G
Sulfolobus tokodaii	Solfataric fields, Japan	80 (70–85)	2.5–3 (2–5)	Aerobic	Heterotrophic/ chemolithoautotrophic	Complex organics, amino acids, S^0	O_2		G, sulfur oxidation
Sulfolobus shibatae	Solfataric fields, Japan	81 (–86)	3	Aerobic	Heterotrophic/ mixotrophic	Complex organics, sugars, H_2	O_2		Lysogenic host of virus SSV1
Sulfolobus metallicus	Solfataric fields, Iceland	65 (50–75)	2–3 (1–4.5)	Aerobic	Obligatory chemolithoautotrophic	S^0, sulfidic ores (pyrite, sphalerite, and chalcopyrite)	O_2		Ore bioleaching
Acidianus ambivalens	Solfataric fields, Leirhnukur fissure, Iceland	80 (–87)	2.5 (0.8–3.5)	Facultatively anaerobic	Obligatory chemolithoautotrophic	H_2, S^0	O_2, S^0	7	S^0/H_2 autotrophy, sulfur oxidation
Acidianus infernus	Solfataric fields, Solfatara (Naples), Italy	85–90 (55–95)	2 (1–5.5)	Facultatively anaerobic	Obligatory chemolithoautotrophic	H_2, S^0, sulfidic ores (pyrite, sphalerite, and chalcopyrite)	O_2, S^0		S^0/H_2 autotrophy, sulfur oxidation, Knallgas reaction, ore bioleaching
Acidianus brierleyi	Solfataric fields, Yellowstone, WY, USA	70 (40–75)	1.5–2 (1–6)	Facultatively anaerobic	Facultatively chemolithoautotrophic	Complex organics, H_2, S^0, sulfidic ores (pyrite, sphalerite, and chalcopyrite)	O_2, S^0		S^0/H_2 autotrophy, sulfur oxidation, Knallgas reaction, ore bioleaching
Metallosphaera sedula	Solfataric fields, Solfatara (Naples), Italy; smoldering slag heaps	75 (50–80)	2–3 (1–4.5)	Aerobic	Facultatively chemolithoautotrophic	Complex organics, H_2, S^0, sulfidic ores (pyrite, sphalerite, and chalcopyrite)	O_2	3	Sulfur oxidation, Knallgas reaction, ore bioleaching
Stygiolobus azoricus	Solfataric fields, Caldeira Velha, São Miguel, Açores, Portugal	80 (57–89)	2.5–3 (1–5.5)	Anaerobic	Obligatory chemolithoautotrophic	H_2	S^0	1	S^0/H_2 autotrophy
Sulfurisphaera ohwakuensis	Solfataric fields, Japan	84 (63–92)	2 (1–5)	Facultatively anaerobic	Heterotrophic/ mixotrophic	Complex organics, H_2, S^0	O_2, S^0	1	
Sulfurococcus mirabilis	Smoldering coal refuse piles, solfataric fields	70–75	2–3	Aerobic	Facultatively chemolithoautotrophic	Complex organics, S^0, sulfidic ores (pyrite, sphalerite, and chalcopyrite)	O_2,	2	

[a]Data compiled from Huber and Prangishvili (154), and from original descriptions from the taxonomy browser (http://www.ncbi.nih.gov). Cells of all species are irregular cocci, and sometimes lobed cocci. All genera of the *Sulfolobales* are represented.
[b]G, genome sequence available.

48

Elemental sulfur is deposited from H_2S by comproportionation (i.e., the opposite of disproportionation) with SO_2 or oxidation with air. In seawater, sulfate is the predominant sulfur species (and the second most abundant anion) providing the main substrate for sulfate reducers in submarine vents (see "Hyperthermophilic sulfate-reducing *Euryarchaeota*: the *Archaeoglobales*," below) (263). Oxidation and reduction of ISCs form an important energy source for hyperthermophilic archaea but there are also numerous species that are able to use other inorganic or organic substrates for energy metabolism (e.g. nitrate, ferric iron, etc; Tables 3 to 8), and it is likely that most of the metabolic pathways represented in mesophiles will also be found in hyperthermophiles.

Thermal adaptation

The large number and variety of (hyper-) thermophilic archaea attest to the ability of cells to adapt to growth at high temperatures. The fluidity of membranes is adjusted by varying the proportion of tetraether versus diether lipids and/or by the degree of cyclization of the lipids (see "Membranes" above). The degree of modification in stable RNAs, in particular, tRNAs, tends to increase with growth temperature (283). The histones in *Archaea*, which complex and organize DNA, may also protect DNA against thermal denaturation (see "Chromosomes, DNA structure and replication," above). Most small molecules, such as (d)NTPs, NAD(P)(H), and vitamins, are reasonably stable at high temperatures at neutral pH but not at acidic pH (10). The internal pH of the thermoacidophile *S. acidocaldarius* is 6.5, whereas it is 4.6 for *Picrophilus oshimae*. The difference in their internal pH suggests that the half-lives of thermal denaturation of small molecules may indeed limit growth at high temperatures for *P. oshimae* but not for *S. acidocaldarius* (268) (see "*Thermoplasmatales*," below). Some hyperthermophiles accumulate small molecules in the cytoplasm, such as cyclic 2,3-bisphosphoglycerate (*Methanopyrus, Methanothermus*) or di-myoinositol phosphate (*Pyrococcus*), which may provide thermal protection (359, 364).

The thermostability, -lability, and -activity of enzymes broadly correlates with the growth temperature of the organism, and the optimal catalytic activities of thermostable enzymes tend to exceed the T_{opt} of the host (416). For example, the amylopullanase from *Thermococcus litoralis* is optimally active at 117°C, which is 29°C above the organism's T_{opt} (45). The Arrhenius plots of orthologous enzymes from mesophiles and thermophiles are typically linear, suggesting that the enzymes' functional conformations remain unchanged in their respective temperature

ranges (416). The overall sequence and structural similarity between homologous enzymes from different thermal classes that possess identical catalytic mechanisms is often very high, making it difficult to identify thermostability determinants of comparative protein sets. The successful expression of protein-encoding genes from hyperthermophiles in mesophilic hosts such as *E. coli* demonstrates that proteins from hyperthermophiles tend to be intrinsically thermostable and fold properly far below the growth temperatures of their native hosts. This is also true for proteins from hyperthermophiles that synthesize high concentrations of cyclic 2,3-bisphosphoglycerate or di-myoinositol phosphate (359, 364). In general, thermostability appears to arise from a few highly specific alterations within individual proteins. Stability against melting seems to arise from changes in protein rigidity, which is affected by a combination of ion pairs and ion-pair networks, hydrogen bonding, and hydrophobic and Van der Waals interactions. In addition, intersubunit interactions, shortened or more rigidly anchored loops, stabilization of the N- and C-terminal ends of proteins, and less solvent-exposed hydrophobic interactions can all lead to a decrease in the ability of protein domains to move and therefore contribute to the protein's stability (416, 427).

A large amount of TF55 chaperonin-like protein is synthesized by *Pyrodictium brockii* when it is grown at 108°C, suggesting that chaperonins in general may be important under conditions close to the maximum growth temperature (307) (see Chapter 10). Only one gene that is common to all hyperthermophiles—encoding a reverse gyrase—has been identified in genomewide comparisons (see Chapter 19). The enzyme induces positive supercoiling into DNA via an ATP-dependent topoisomerase I activity, and it was thought to play a major role in thermostabilization of DNA (94, 194). However, a reverse gyrase gene knockout in *Thermococcus kodakaraensis* does not change its phenotype or growth kinetics (11). Although this does not rule out that reverse gyrase can be part of a DNA thermoadaptation mechanism, it shows that it is not essential for this organism and further highlights the difficulties in attempting to define general rules for microbial adaptation to high temperatures.

Heterotrophic Hyperthermophilic *Euryarchaeota*: The *Thermococcales*

Heterotrophic *Thermococcales* and especially *Thermococcus* spp. are easily enriched and isolated. They all grow heterotrophically under anaerobic conditions, usually by fermentation of carbohydrates and/or peptides. Elemental sulfur is either essential or stimulatory for growth (Table 3). Three genera have been described, *Pyrococcus* (six species), *Thermococcus* (30 species), and *Palaeococcus* (two species) (28). *Thermococcales* species are abundant organisms in marine environments, shallow and deep-sea hydrothermal vents, and continental and offshore oil wells, whereas only a minority has been isolated from terrestrial samples (e.g., 331). In general, the cells are regular or irregular cocci, 1 to 2.5 μm in diameter, have an S layer with hexagonal symmetry, and are flagellated (Fig. 16) (28).

Two of the three genera of the *Thermococcales* have been well defined by their 16S rDNA phylogeny and other taxonomic features. The genus *Pyrococcus* was originally used to describe marine organisms with a low G+C content (38 to 48%) growing optimally at ≥100°C. *Pyrococcus* species form a coherent phylogenetic group that clusters with the chitin-degrading, *T. chitinophilus* (28). The two *Palaeococcus* species are characterized by growth at ≤90°C, pH 4 to 8, and moderate salinity, and branch more deeply in the order *Thermococcales* (6, 28, 400). *P. ferrophilus* is barophilic ($P_{opt} \sim 30$ MPa). Two species are distinguished by their G+C content (*P. ferrophilus*, 53%; *P. helgensonii*, 42%). The genus *Thermococcus* encompasses at least 30 species that differ in their physiology and are phylogenetically separated into several clades that include isolates with a broad range of G+C content (40 to 58%). The two species *T. alcaliphilus* and *T. acidoaminovorans* are thermoalkaliphiles, while others are mostly neutrophilic. For the purposes of this chapter, neutrophiles, acidophiles, and alkaliphiles are referred to as microorganisms that grow optimally in the range pH 5 to 8, 0.7 to 4, and 9 to 11, respectively.

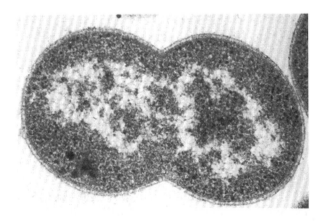

Figure 16. Electron micrograph of *Thermococcus celer*. Reproduced from *Bergey's Manual of Systematic Bacteriology* (208) with permission of the publisher.

Thermococcales grow quickly, with doubling times between 25 and 70 min. Many species are sufficiently aerotolerant to be handled without requiring strict anaerobic conditions. They can be easily grown in anaerobic jars on solid media (Gelrite or Phytagel). *Thermococcales* are the most common source of commercially available proofreading DNA polymerases for PCR applications (254) (see Chapter 22). Many other enzymes of potential interest have also been described from *Thermococcales*, including proteases, esterases, and ATP-dependent DNA ligases (78, 206, 258). Genome sequences were determined and published for *P. furiosus*, *P. abyssi*, *P. horikoshii*, and *T. kodakaraensis* (http://www.ncbi.nih.gov), and the genome sequence of *T. gammatolerans* (France) has not yet been published (see 41, 42) (see Chapter 19). *Pyrococcus furiosus* has been selected as a model organism for a comprehensive transcriptomic/proteomic/structural genomic project, and as of January 2006, 30 X-ray structures were available in the Protein Data Bank, with many more in progress (http://www.secsg.org) (see Chapter 20). An in vitro transcription system has been developed for *P. furiosus*, and this has been instrumental in studying the features of archaeal transcription that are similar to the *Eucarya* (see Chapter 6).

T. kodakaraensis is a promising archaeal model organism for future research, although few publications are available to date. A versatile gene knockout system based on auxotrophy markers has been developed. *T. kodakaraensis* is naturally competent and efficient in recombination, and this allows specific genes to be deleted without producing polar effects (346, 347). Single, double, and triple mutations of various biosynthesis genes (Δ*pyrF*, Δ*trpE*, Δ*hisD*) have been constructed. The deletion of the reverse gyrase gene was performed with this system, demonstrating that it is not essential for growth at high temperatures (11) (see "Thermal adaptation," above).

A remarkable feature of several of the *Thermococcales* is their high resistance to ionizing (γ) radiation. At least three *Thermococcus* isolates have been obtained by applying a dose of 30 kGy from a ^{60}Co source in enrichment cultures (171, 172). The level of radiation resistance is the same order of magnitude as for the bacterium *Deinococcus radiodurans*, which exhibits a D_{10} survival dose (10% colony-forming units after irradiation) of ~16 kGy (e.g., compared with a D_{10} dose for *E. coli* of ~0.7 kGy) (290). Stationary-phase *P. abyssi* cultures are able to completely repair their fully fragmented chromosomes (double-strand breaks) within 2 h of irradiation (173). *P. abyssi* seems to respond to DNA damage by uncoupling DNA repair from DNA synthesis and thus prevents the accumulation of genetic errors by exporting damaged DNA. Several typical repair proteins, including RadA, replication protein A, and replication factor C remain chromatin bound before and after irradiation, and active chromatin-bound repair and replication complexes are therefore thought to remain available to counteract DNA damage (173).

The three *Pyrococcus* genome sequences have been used for the analysis of genome evolution and for determining their origins of replication (273, 458) (see Chapter 19). *P. horikoshii* and *P. abyssi* have very similar genomes (Fig. 17). It was concluded from the syntenic fragments that one reversion and one trans-

position event is sufficient to explain the differences in the *P. abyssi* genome after an inferred recent divergence from *P. horikoshii* (458). In *P. horikoshii* and *P. abyssi*, most genes are predicted to be transcribed in the same direction as DNA replication within each of the two replichores (two chromosome halves defined by replication origin and terminus). In contrast, the *P. furiosus* genome appears to be more scrambled and only short regions of synteny exist with the other two species. *P. furiosus* contains numerous recombined segments, many of which correlate with boundaries of insertion elements, suggesting that transposi-

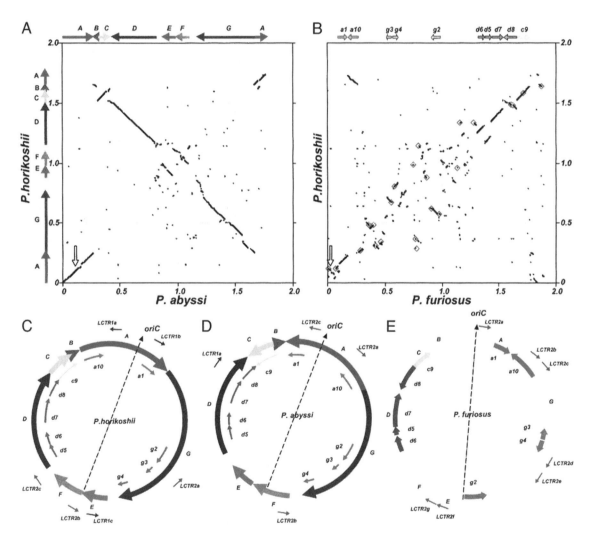

Figure 17. Genome rearrangements in *Pyrococcus* species. (A, B) Pairwise genome dot plots: (A) solid arrows with capital letters indicate large synteny segments conserved between *P. horikoshii* and *P. abyssi*; (B) shaded arrows represent the location and relative orientation of synteny segments conserved in all three species; small squares denote position of IS elements in the *P. furiosus* genome. (C to E) Genome organization and synteny segments in the three *Pyrococcus* species. Location and orientation of the major syntenic genome fragments conserved between *P. horikoshii* and *P. abyssi* are marked as above; 10 smaller syntenic fragments conserved also with *P. furiosus* are marked with medium arrows. Dashed arrows represent the two replichores defined by the origin of replication (*oriC*) and the putative terminus of replication at the bottom of the arrows; LCTR1 and 2, two families of long conserve tandem repeats occurring in all species (458). Reproduced with modifications from *Nucleic Acids Research* (458) with permission of the publisher.

tion events have caused many disruptions in the genome. In *P. furiosus*, colinearity of genome transcription and replication only occurs for highly transcribed genes.

Metalloproteins, sugar, sulfur, and hydrogen metabolism in *P. furiosus*

The reduction of S^0 or protons coupled to the oxidation of carbohydrate and peptides to CO_2 and small organic compounds (e.g., acetate or alanine), is one of the most common pathways of energy metabolism in hyperthermophilic archaea, in particular, for members of the *Thermococcales* (7, 385) (see Chapters 12 and 20). *P. furiosus* has served as a model archaeon for studying glucose, maltose, and amino acid metabolism and sulfur reduction, although several important aspects remain to be determined. Peptides are the optimal substrate for fermentation in *P. furiosus*, and S^0 is required, a property which is also true for other *Pyrococcus* species (3). In contrast, lower growth rates are observed during fermentation of maltose. If S^0 is included in maltose growth medium, however, significant H_2S formation occurs despite S^0 not being required for growth or increasing growth rate.

P. furiosus catabolizes sugars using a modified Embden-Meyerhof-Parnas glycolytic pathway (EMP). Subsequently, pyruvate:ferredoxin oxidoreductase converts pyruvate into acetyl-CoA and CO_2. The pathway involves most steps known from bacterial glycolysis but utilizes a few unusual enzymes. The modified EMP in *Pyrococcus* includes the ADP-dependent hexokinase and phosphofructokinase instead of ATP-dependent enzymes (189, 190). The most peculiar step in the glycolytic pathway of many hyperthermophilic archaea is the single-step oxidation of glyceraldehyde 3-phosphate (GAP) to 3-phosphoglycerate. The reaction is catalyzed by a tungsten-containing GAP:ferredoxin oxidoreductase that is related to other aldehyde:ferredoxin oxidoreductases (AOR, Fig. 18) instead of a NAD-dependent GAP dehydrogenase (334).

Three different soluble proteins with S-reducing activity have been purified from *P. furiosus*. Two of these enzymes are hydrogenases with additional S^0 or polysulfide reductase activity using H_2 or NADPH as electron donors (sulfhydrogenases) (248). The third enzyme is a ferredoxin:NADP oxidoreductase with a broad substrate range, which oxidizes and/or reduces many different substrates, including FAD, polysulfides, NAD(P)(H), and O_2 (249). A membrane-bound hydrogenase (MBH) purified from *P. furiosus* did not reduce S^0 or polysulfides (343). MBH activities increased in the presence of maltose and in the absence of S^0 (3). The enzyme was part of a membrane-bound complex, which appears to couple maltose fermentation to proton translocation across the membrane, driven by proton reduction with ferredoxin as electron donor. It constitutes a novel and very simple anaerobic electron transport chain. Identifying this protein complex helped to resolve questions concerning energy conservation in *Pyrococcus* cells growing in the absence of S^0 (342). However, it has not been determined whether S^0 reduction is coupled to electron transport (3, 342). Inducible transcripts (microarrays) and proteins (2D gels) were identified in sulfur-grown, compared with sulfur-free, cultures. Two genes, *sipA* and *sipB*, which are divergently transcribed from a common promoter region, had the highest increase in mRNA levels (363). The gene products are membrane-associated proteins with unknown function (3, 342). It remains to be determined whether the so far elusive energy-converting sulfur reductase will be found in *P. furiosus*.

Related to its capacity for sulfur metabolism, *Pyrococcus* has been an important source for metalloproteins (in addition to some methanogens and *Sulfolobus*), which have served as models for understanding redox processes in the *Archaea*, *Bacteria*, and *Eucarya*. Important findings from research into *Pyrococcus* and *Thermococcus* metalloproteins include the structure determination of a family of tungsten-containing AORs (Fig. 18) (334), characterization of the superoxide reductase/rubrerythrin oxygen protection mechanism (1, 440), and analysis of membrane-bound, proton-translocating, and proton-reducing hydrogenases (140, 342). These achievements provided an important foundation for the *Pyrococcus* structural genomics project, part of which specifically focuses on the metalloproteome (http://www.secsg.org).

Hyperthermophilic Sulfate-Reducing *Euryarchaeota*: The *Archaeoglobales*

Archaeoglobus fulgidus was the first archaeon to be isolated that grows by dissimilatory sulfate reduction (2). *Archaeoglobus* strains are often the first to grow when enriching sulfate reducers from samples of natural or man-made sulfate-rich biotopes. The number of archaea with this growth property on sulfate is limited, with the only other cultivatable isolate being *Caldivirga* (*Thermoproteales*, *Crenarchaeota*). *A. fulgidus* uses organic acids such as lactate and/or H_2 as electron donors to reduce sulfate, thiosulfate, or sulfite to H_2S. In contrast, the phylogenetically related archaeon, *Ferroglobus placidus*, mainly grows autotrophically using nitrate or thiosulfate as electron acceptors and Fe^{2+}, H_2, and H_2S as electron donors (Table 4) (129). *Ferroglobus* is one of the few archaea

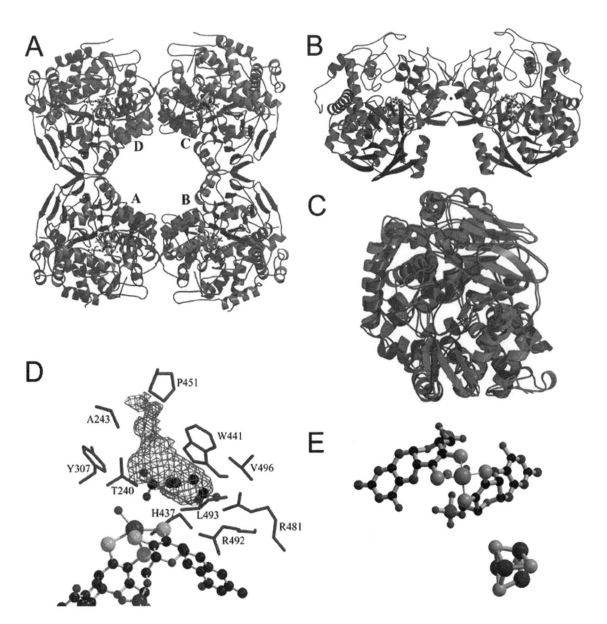

Figure 18. (*See the separate color insert for the color version of this illustration.*) Three-dimensional structures of tungsten-containing aldehyde:ferredoxin oxidoreductases from *Pyrococcus furiosus*. (A) Cartoon of the formaldehyde:ferredoxin oxidoreductase (FOR), homotetrameric holoenzyme (150). (B) Cartoon of the aldehyde:ferredoxin oxidoreductase (AOR) homodimeric holoenzyme (53). (C) Peptide chains of AOR (cyan) superimposed on FOR (magenta) showing close structural similarity (150). (D) Active-site cavity of the FOR with surrounding residues and glutarate shown (150). (E) [4Fe-4S] cluster and the W-(bis-tungstopterin) cofactor of the AOR (53). FOR images reproduced from the *Journal of Molecular Biology* with permission of the publisher (150); AOR images reproduced from *Science* (53) with permission of the publisher.

reported to oxidize acetate anaerobically; others include the phylogenetically related *Geoglobus* and many methanogens (180, 403). *Ferroglobus* is also the only archaeon reported to oxidize aromatic hydrocarbons anaerobically (404). *Archaeoglobus* and relatives have been repeatedly isolated from oil fields, where they grow at the oil/water interface, and are partially responsible for well souring and H$_2$S production. *A. fulgidus* was one of the first members of

the *Archaea* to have its genome sequenced (201), revealing a large number of genes and operon-like structures potentially involved in the degradation of numerous biopolymers and small compounds, including the β-oxidation of fatty acids.

The biochemistry of dissimilatory sulfate reduction is essentially the same as in the assimilatory pathway (3'-phosphoadenylylsulfate, has not yet been found in *Archaea*). Three enzymes are required (equations 1 to 3; see Fig. 25 below). They are localized in the cytoplasm, and all have been purified from *A. fulgidus* (62, 376).

$$\text{ATP:sulfate adenylyltransferases}$$
$$\text{ATP} + \text{SO}_4^{2-} \leftrightarrow \text{APS} + \text{PP}_i \qquad (1)$$

$$\text{Adenylylsulfate (APS) reductase}$$
$$\text{APS} + 2\,\text{e}^- + \text{H}^+ \leftrightarrow \text{AMP} + \text{HSO}_3^- \qquad (2)$$

$$\text{Sulfite reductase}$$
$$\text{HSO}_3^- + 6\,\text{e}^- + 6\,\text{H}^+ \leftrightarrow \text{HS}^- + 3\,\text{H}_2\text{O} \qquad (3)$$

ATP:sulfate adenylyltransferases (equation 1) are key enzymes in sulfate reduction. The homooligomeric enzymes are found in all organisms capable of dissimilatory sulfate reduction and consist of 41 to 69 kDa subunits (376). The distinction between different types of the enzyme does not correlate with the organisms' phylogeny but rather with growth temperature. The enzymes from thermophiles, but not from mesophiles, contain a zinc-binding site, which may contribute to the thermal stabilization of the protein (396).

The crystal structure of the *A. fulgidus* dissimilatory APS reductase (equation 2) serves as a model for this class of enzymes. Structural and spectroscopic studies facilitated the determination of the reaction mechanism (103, 104). APS reductases are heterodimeric, FAD-containing, and iron-sulfur (FeS) cluster-containing enzymes (62). The flavoprotein subunit is paralogous to the large subunit of succinate dehydrogenases (SDH). The smaller subunit contains two FeS clusters in the APS reductase and does not have paralogs or orthologs in other enzymes. The flavin is located at the bottom of a pocket of the large subunit (104) in proximity to the first [4Fe-4S] cluster. The appearance of a broad radical signal in EPR spectra following incubation with AMP and sulfite indicates that flavosemiquinone may interact with cluster I. These results suggest that an FAD N(5)-sulfite adduct is the intermediate during both the oxidative and the reductive reactions.

The dissimilatory, siroheme-containing sulfite reductases (DSRs; equation 3) catalyze the six-electron reduction from sulfite to sulfide, and vice versa. They are present in all anaerobic sulfate or sulfite reducers and in some sulfide-oxidizing bacteria, and they have served as specific, amplifiable gene markers in molecular ecology studies (178). The heterodimeric enzymes may have arisen from a gene duplication event. The *dsrA* and *dsrB* gene products have a moderate level of similarity (~30% amino acid identity between *A. fulgidus* DsrA and DsrB). Both subunits are strictly conserved between the *Archaea* and *Bacteria* (*A. fulgidus* [archaeon] and *Desulfovibrio vulgaris* [bacterium] have ~60% identity) (62). The siroheme

in the DSRs is a complex cofactor consisting of a heme molecule electrically coupled to a [4Fe-4S] cluster.

A DSR with an unusual domain composition, cofactor specificity, and function (sulfite detoxification) was isolated from *M. jannaschii* (170). *M. jannaschii* and *Methanothermobacter thermautotrophicus* grew with normally toxic levels of sulfite, utilizing it as the sole source of sulfur. A protein with an N-terminal F_{420} dehydrogenase and a C-terminal DsrA domain was identified that had F_{420}-dependent sulfite reductase activity and contained siroheme. Homologs were found exclusively in the genome sequences of other strictly hydrogenotrophic and sulfite-resistant thermophilic methanogens. This type of sulfite resistance may provide an evolutionary selective advantage.

The *Archaeoglobales* are phylogenetically similar to methanogenic archaea and use several of the coenzymes of methanogenesis to oxidize organic substrates (141). For example, lactate is converted via pyruvate to acetyl-CoA, and CO_2 is coupled to the reduction of ferredoxins. An acetylsynthase/carbon monoxide dehydrogenase complex, similar to that found in methanogens, cleaves the C–C bond in acetyl-CoA generating both an enzyme-bound methyl group and bound CO. CO and the methyl-group are oxidized to CO_2 by reactions with coenzymes that are characteristic of the methanogens (see Chapter 13). However, it is not clear how electrons are transferred to the enzymes of the sulfate reduction pathway.

Hyperthermophilic and Predominantly Neutrophilic *Crenarchaeota*: *Thermoproteales* and *Desulfurococcales*

Cultivated crenarchaeota have mainly been isolated from volcanically heated terrestrial habitats, although some have been isolated from marine environments. A hallmark of the hyperthermophilic *Crenarchaeota* is chemolithoautotrophic growth, which makes them important primary producers in these environments. Many isolates require elemental sulfur, prompting their description as "sulfur-dependent and thermophilic Archaebacteria." However, several heterotrophs and autotrophs have been identified that can grow at lower temperatures, indicating that this branch of the *Crenarchaeota* represents diverse physiologies.

The thermophilic/hyperthermophilic *Crenarchaeota* are divided into three orders: *Thermoproteales*, *Desulfurococcales*, and *Sulfolobales*. The thermoacidophilic *Sulfolobales* form a phylogenetically coherent group that is distinct from the *Thermoproteales* and *Desulfurococcales* (see "*Sulfolobus*, *Acidianus*, and relatives," below). In contrast, the description of the *Thermoproteales* and *Desulfurococcales* is not clear. The *Pyrodictiaceae* have been described as a separate order, "*Pyrodictiales*," that is distinct from the *Desulfurococcales* (e.g., 386). However, more recent descriptions group the genera *Pyrodictium*, *Hyperthermus*, and *Pyrolobus* into one family within the *Desulfurococcales* (155).

Most of the *Desulfurococcales* and *Thermoproteales* are anaerobes, although both taxa include aerobes, such as *Aeropyrum* and *Pyrobaculum* (Table 5). Anaerobic chemolithoautotrophic growth with H_2 as electron donor and S^0 as electron acceptor ("S^0/H_2 autotrophy") (92) is characteristic of many hyper/thermophilic members of the *Crenarchaeota* and is probably the most important energy-yielding reaction for CO_2 fixation in volcanic hot springs and submarine vents. Other electron acceptors include thiosulfate, nitrate, nitrite, ferric iron, and sulfate. Many species grow with organic substrates, either by fermentation (e.g., *Hyperthermus*) or by anaerobic respiration coupled to inorganic electron acceptors. Organic nutrients are often oxidized completely to CO_2, even under anaerobic conditions (e.g., *Thermoproteus*).

Desulfurococcales

All *Desulfurococcales* are hyperthermophiles and include all isolates over the past 25 years that are able to grow at the highest temperatures: *Pyrodictium*, *Pyrolobus*, and strain 121. Most strains grow anaerobically, and a few grow aerobically or facultatively anaerobically. Energy is gained from the oxidation of hydrogen under autotrophic conditions using elemental sulfur, thiosulfate, nitrate, or nitrite as electron acceptors, and CO_2 as a carbon source. Alternatively, organotrophic growth can occur by aerobic or anaerobic respiration, or by fermentation of organic substrates. The culturable *Desulfurococcales* isolates are divided into two families: the *Desulfurococcaceae* include a large number of genera and diverse physiological types, while the *Pyrodictiaceae* contain only three genera (155). *Desulfurococcales* cells are all coccoid or disc shaped and occur singly or in aggregates. The diameter of cells is typically 0.5 to 4 μm.

Desulfurococcaceae. The *Desulfurococcus* genus includes five formally described species and several incompletely described strains that were isolated from terrestrial volcanic samples (155). They grow anaerobically with organic substrates (peptides and/or carbohydrates), either by fermentation or sulfur respiration (Table 5) (155). Similar conditions are required for *Sulphophobococcus zilligii* and *Thermosphaera aggregans*, which grow anaerobically by fermentation

of complex organic nutrients, when sulfur is absent (Table 5). *Thermosphaera aggregans* was one of the first microorganisms to be predicted by environmental 16S rDNA analysis to exist, prior to being isolated from the same environment (158). *Acidilobus aceticus* (pH_{opt}, 3.8) and the related *Caldococcus noboribetus* (pH_{opt}, 3.0) require complex nutrients for growth, and growth is stimulated by the addition of sulfur (155). Acetate is one of the main products of fermentation. The two isolates cluster together phylogenetically and represent a rare example of thermoacidophilic anaerobic crenarchaeota, which are not members of the *Sulfolobales* order (Tables 5 and 8).

One of the first archaeal introns was identified in the 23S rRNA gene of *Desulfurococcus mobilis* (230) (see Chapter 7). After excision, the intron ligates to form a stable circular RNA that encodes a site-specific endonuclease. Cleavage and exon-splicing reactions resembled those of type II tRNA introns in *Eucarya* (199, 428).

The marine *Desulfurococcaceae* comprise the genera *Aeropyrum*, *Ignicoccus*, *Staphylothermus*, *Stetteria*, and *Thermodiscus*. *Staphylothermus*, *Thermodiscus*, and *Stetteria* require complex nutrients in addition to S^0 (155). *Staphylothermus* degrades proteins and peptides, and possesses a characteristic S-layer structure with long stalks and 4-fold symmetry (Fig. 5) (see Chapter 14). *Stetteria* grows mixotrophically with organic nutrients, H_2, and S^0 or thiosulfate.

The two *Aeropyrum* species differ from other members of the *Desulfurococcales* by being strictly aerobic and growing on complex organic nutrients. Thiosulfate is likely to be oxidized to sulfate and stimulates the growth of *Aeropyrum pernix*. The availability of the genome sequence of *A. pernix* (the first published for a member of the *Crenarchaeota*) (182) underpinned efforts to determine 3D protein structures. One of the most interesting outcomes was the resolution of a voltage-gated potassium channel (KvAP) and the determination of channel's opening mechanism in response to membrane polarization (Fig. 19). The channel contains a K^+ pore that is surrounded by "sensors" (consisting of helices) near the pore, and "voltage-sensor paddles" on the outer perimeter of the pore. In the crystal structure of the closed channel, the paddles are located near the intracellular side of the channel. In response to membrane depolarization, the paddles appear to move across the membrane toward the outside and open the channel (168, 169, 234).

In contrast to the members of *Desulfurococcaceae* described above, the three *Ignicoccus* species grow by S^0/H_2-autotrophy. *I. islandicus* and *I. pacificus* are formerly described (Table 5), whereas *Igni-*

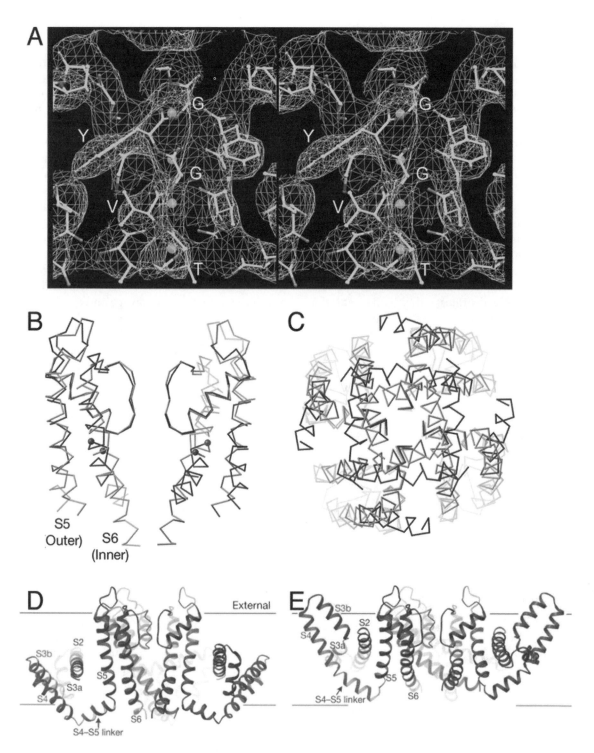

Figure 19. (*See the separate color insert for the color version of this illustration.*) Model of the *Aeropyrum* voltage-gated K$^+$-channel KvAP and comparison with the *Streptomyces lividans* KcsA K$^+$ channel. (A) Stereo view of the KvAP pore with electron density map contoured at 1.0 Δ carbon (yellow), nitrogen (blue), oxygen (red), potassium (green). (B, C) α-Carbon traces of the KvAP pore (blue) and the *Streptomyces lividans* KcsA K$^+$ channel (green) shown as a side view (B) and end-on from the intracellular side (C); S5, S6, outer and inner helices; glycine-gating hinges (red spheres). (D, E) Models of the closed (D) and open (E) KvAP structures based on the positions of the paddles (red), the pore and the S5 and S6 helices of KcsA. Reproduced with modifications from *Nature* (168, 169) with permission of the publisher.

coccus sp. KIN4/I is not. Strain KIN4/I may be related to *I. islandicus* as it was sourced from the same geographical region. *Ignicoccus* is the only member of the *Archaea* that possesses a true periplasmic space formed by an outer membrane, rather than an S layer (Fig. 20) (see Chapter 14). The periplasm is much wider than in gram-negative bacteria (20 to 400 nm) and contains membrane-bound vesicles. The outer membrane contains lipids and regularly arranged proteins, in addition to pores (24 nm diameter) that are surrounded by a ring of regularly arranged particles (130 nm diameter). The function of all these membrane features is unknown (275, 315).

Unusually small symbiotic (or parasitic) cells of *N. equitans* were discovered attached to the outer membrane of *Ignicoccus* sp. KIN4/I (Fig. 20) (152). Whereas the *Ignicoccus* strain grew successfully in monoculture, it has not been possible to grow *N. equitans* without the host. The circular genome of *N. equitans* (481 kbp) encodes 585 genes and is the smallest known genome for a free-living cell (429). Seventy-three percent of the genes have homologs in GenBank. In contrast, *Acanthamoeba polyphaga mimivirus* has the largest viral genome (linear, 1.19 Mbp) and encodes 911 genes, for which ~10% have homologs in GenBank (110). The *Nanoarchaeum* 16S rDNA sequence is significantly different from other archaea, and the genus has been placed into a novel kingdom, the *Nanoarchaeota*. The phylogenetic placement is unclear (see "Phylogeny of *Archaea* and the origin of life," above), and it has been argued that it might have evolved rapidly and be derived from the

Euryarchaeota (see Chapter 19). Because of the divergence of the 16S rDNA of *N. equitans*, the organism was unable to be detected in environmental samples with "universal" 16S *Archaea* primers, but is now able to be detected using specific primers.

The genome of *N. equitans* has revealed a number of interesting evolutionary features. For example, several of the tRNA genes are split and the two gene segments are physically separated in the genome (see Chapter 7) contributing to the debate concerning whether modern tRNAs evolved from split precursors (317). It is not clear what benefits the two species, *Ignicoccus* and *Nanoarchaeum*, derive from their association. For example, it is not known what metabolites are transferred between them. The lipid composition in the membranes of both species changes in parallel when growth temperature is changed, suggesting that *Nanoarchaeum* may utilize the lipid pool from the *Ignicoccus* host (164). The discovery of *Nanoarchaeum/Ignicoccus* has promoted numerous avenues for new research into the *Archaea*.

Pyrodictiaceae. The *Pyrodictiaceae* comprise three genera with a total of five to six species that have all been isolated from marine environments (155). They grow optimally above 100°C and include the species that grow at the highest temperatures (Table 5). *Pyrodictium occultum*, *P. brockii*, and *P. abyssi* were the first microorganisms found to grow optimally above 100°C (T_{max}, 110°C) (155, 389). *Pyrodictium* cells are disc shaped, flat (0.1 to 0.2 μm), and pleomorphic, similar to *Haloferax* and

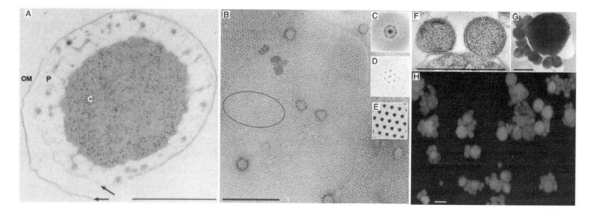

Figure 20. (*See the separate color insert for the color version of this illustration.*) Electron micrograph and fluorescence images of *Ignicoccus* and *Nanoarchaeum*. (A) Transmission electron micrograph of thin-sectioned *Ignicoccus* cell with broad periplasmic space (P) and budded vesicles; OM, outer membrane, C, cytoplasm, bar, 1 μm. (B) Negative stained *Ignicoccus* outer membrane, highlighting power spectra of image field (C to E) (275). Panels A to E reproduced from *Biochemical Society Symposia* (275) with permission of the publisher. (F) Ultrathin section of *Nanoarchaeum* cells attached to the outer membrane of *Ignicoccus* sp. KIN/4. (G) Platinum shadowing of *Ignicoccus* cell with several *Nanoarchaeum* cells attached (left side of photograph). (H) Confocal laser-scanning micrograph using *Nanoarchaeum* (red) and *Ignicoccus*-specific probes (green). Panels F to H reproduced from *Nature* (152) with permission of the publisher.

Haloquadratum (Fig. 4). *Pyrodictium* grows in hypotonic media, in contrast to the extreme halophiles that have a cytoplasm that is isosmotic with the medium. It is not clear how *Pyrodictium* cells maintain their shape with a positive turgor pressure inside the cell, because they only appear to have a proteinaceous S layer to physically support the cell. Cells are covered with a hexagonal S layer anchored to the membrane with stalks (see Chapter 14). The resulting pseudoperiplasmic space has a constant width of ~35 nm. The cells are embedded in a network of hollow tubules (cannulae), which have an outer diameter of 25 nm. The cannulae comprise three similar and probably homologous glycoproteins (278). Daughter cells remain connected to the mother cell by the cannulae, which are anchored in the S layer. Growing cultures form macroscopically visible flocks and biofilms.

Pyrodictium species are anaerobes growing by S^0/H_2 autotrophy. S^0-reducing and H_2-oxidizing chemolithoautotrophs require at least two membrane-bound enzymes for energy conversion: a hydrogenase, and sulfur reductase (SR) or polysulfide reductase (PSR; see Fig. 26) that are coupled in a short electron transport chain. Both enzymes have been characterized from *P. brockii* (308, 309), and in more detail from *P. abyssi* (71). A 520-kDa membrane-bound multienzyme complex purified from *P. abyssi* is composed of nine subunits and has hydrogenase and SR activities (71). It contains a variety of transition metals, heme *b* and heme *c*, but no Mo or W that are typically present in SR or PSR. The single complex contains all the constituents necessary for the entire electron transport from H_2 to S^0.

S^0 undergoes a transition from crystalline rhombic α-S_8, or monoclinic β-S_8, to the biologically inaccessible polymeric (liquid) μ-sulfur at 113 and 119°C, respectively. As a result, microorganisms that require available S^0 for growth are limited to temperatures below these values. Consistent with this, the two microorganisms with the highest T_{max} use other electron acceptors. *Pyrolobus fumarii* is a formally described species that has a T_{max} of 113°C and a T_{opt} of 106°C; it is unable to grow at temperatures $\leq$90°C (155). *Pyrodictium* and *Pyrolobus* species have been shown to survive autoclaving at 121°C for one hour. *Pyrolobus* grows chemolithoautotrophically with H_2 as electron donor and thiosulfate, nitrate, or a low partial pressure of oxygen as terminal acceptors, with H_2S, ammonia, and water as products, respectively. One published report exists for strain 121 (a formal description is not available) that describes its growth on formate with Fe^{3+} as electron acceptor at a T_{max} of 121°C and a T_{opt} of 106°C (179). The 16S rDNA sequence is most similar to other species of *Pyrodictiaceae* (95

to 96%), suggesting that it also belongs to the same family.

The only representative of the third genus of the *Pyrodictiaceae*, *Hyperthermus butylicus*, differs from *Pyrodictium* and *Pyrolobus* by being entirely dependent on complex organic substrates for growth and being unable to grow chemolithoautotrophically (447). It has a T_{opt} of 103°C and performs mixed butyric acid fermentation from amino acids, producing large amounts of C_4 and C_5 organic acids by an unknown pathway. The annotated genome sequence of *H. butylicus* is being finalized. Presently, it is the microorganism with the highest T_{opt} for which genome sequence data are available. Its capacity to grow at such high temperatures should reveal thermostable enzymes of commercial interest (see Chapter 22). In addition, *H. butylicus* is a heterotroph with the ability to degrade various organic polymers and therefore synthesizes numerous dehydrogenases and esterases, which may also be of commercial value (58).

Thermoproteales

Members of the *Thermoproteales* have been isolated from volcanically heated terrestrial, acidic, and neutral springs, soil biotopes, and water and mud holes, in addition to *Pyrobaculum aerophilum*, which is the sole marine isolate from a vent in shallow water off the coast of Ischia Island, Italy (Table 6) (153). In contrast to the *Desulfurococcales*, which are coccoid or pleomorphic, the *Thermoproteales* form stiff rods (Fig. 21). Many strains produce branched and/or "golf club"-like structures (stiff rods with a coccoid extension or daughter cell at their ends). Most members of the *Thermoproteales* are anaerobic hyperthermophiles, with the exception of *Thermocladium modestus*, which is a thermophile (T_{opt}, 75°C). The two different families of *Thermoproteales*, the *Thermoproteaceae* and *Thermofilaceae*, can be distinguished by morphology. The *Thermoproteaceae* form stiff rods, $\geq$0.4 μm in diameter. In contrast, the two *Thermofilaceae* species form thin rods (0.15 to 0.35 μm in diameter).

Thermoproteales grow heterotrophically and/or chemolithoautotrophically with a T_{opt} of 75 to 100°C (153). *Thermoproteus neutrophilus* and *Pyrobaculum* species have a neutral pH_{opt} (6 to 7), whereas *Caldivirga*, *Vulcanisaeta*, and *Thermocladium* are acidophiles (pH_{opt}, 3.7 to 4.5), and several others are moderate acidophiles (pH_{opt}, 5 to 6). Growth is usually best in media of low ionic strength. Most *Thermoproteales* grow anaerobically, whereas *Pyrobaculum oguniense* grows aerobically, and several other species, including *P. aerophilum*, *T. Modestus*, and *Caldivirga maquilingensis*, are able to grow at low

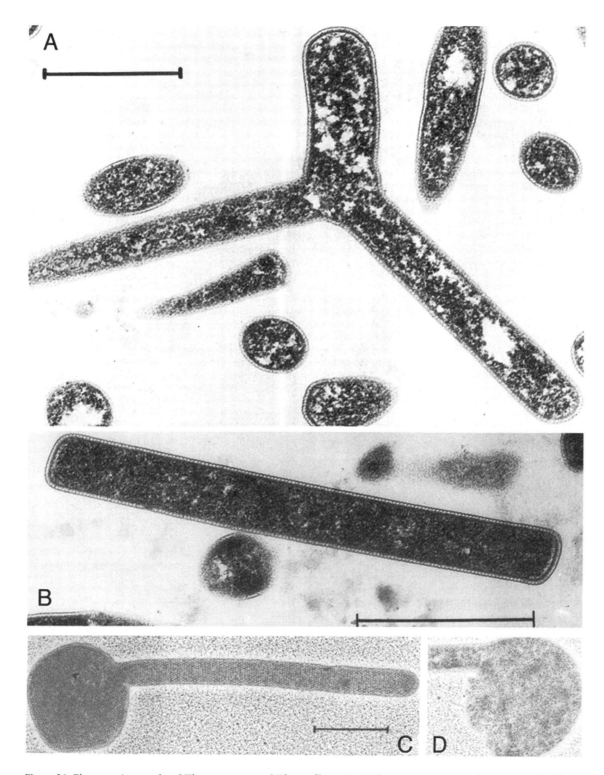

Figure 21. Electron micrographs of *Thermoproteus* and *Thermofilum*. (A, B) *Thermoproteus tenax* cells with branched form. (C, D) *Thermofilum pendens* with golf club structure. Reproduced from *Bergey's Manual of Systematic Bacteriology* (451, 452) with permission of the publisher.

oxygen tension. *Caldivirga* is the only known sulfate-reducing member of the *Crenarchaeota*.

Thermoproteus tenax and *P. aerophilum* have become crenarchaeal model organisms, and their genomes have been sequenced. *T. tenax* grows by chemolithoautotrophic oxidation of H_2, or chemoorganoheterotrophically with S^0 as electron acceptor. It is sufficiently oxygen tolerant to permit cultivation without strict anaerobic handling. The enzymology and regulation of sugar breakdown, and gluconeogenesis have been studied extensively in *T. tenax* (see Chapter 12). The related species, *Thermoproteus neutrophilus*, has been used to study enzymes and mechanisms of autotrophic CO_2 fixation via the reductive citric acid cycle. The same pathway seems to function in *Pyrobaculum islandicum*. In contrast, *Pyrodictium* species appear to utilize ribulose bisphosphate carboxylase (160). Similar to *T. tenax*, *P. aerophilum* grows facultatively chemolithoautotrophically with H_2 and organic nutrients as electron donors. However, it differs from *T. tenax* by using the electron acceptors, nitrate, oxygen, thiosulfate, arsenate, or selenate. Sulfur inhibits growth of *P. aerophilum* but not the growth of other species of this genus (153).

The only known archaeal nitrate reducers are *P. aerophilum*, *P. fumarii* from the *Crenarchaeota*, and *F. placidus* and some species of *Haloferax* from the *Euryarchaeota* (4, 129, 257). Several metalloenzymes have been isolated from *Pyrobaculum* spp. The membrane-bound dissimilatory nitrate reductase contains molybdenum, several FeS clusters, and heme *b* and consists of three subunits similar to nitrate reductases from bacterial mesophiles (4). A nitric oxide reductase displays menaquinol:NO oxidoreductase activity and contains heme and nonheme iron in a 2:1 ratio (70). Additional *Pyrobaculum* enzymes that have been purified include a tungsten aldehyde oxidoreductase that is similar to the enzymes from *Pyrococcus* and a siroheme sulfite reductase similar to an *Archaeoglobus* enzyme (61, 130).

Cenarchaeum, *Nitrosopumilus*, and Uncultured Mesophilic *Crenarchaeota*

Almost all cultivated members of the *Crenarchaeota* are hyperthermophiles. However, members of the *Crenarchaeota* have been detected in molecular ecology studies from numerous cold or temperate environments (reviewed in references 51, 353). Despite their ecological relevance, *Archaea* from cold environments have been significantly understudied. The available cultivatable isolates are restricted to six formally characterized methanogens, one haloarchaeon, and one "cold crenarchaeon" (51). In addition, *Cenarchaeum symbiosum* (*Crenarchaeota*) and "euryarchaeon SM1" (Fig. 7) have been studied but are unable to be cultivated as a monoculture (Table 7) (see "Cell walls and extracellular structures," above) (51).

Symbiotic populations of *C. symbiosum* are present in the marine sponge *Axinella mexicana*, and related phylotypes have been recovered from other sponges in different oceanic regions, illustrating that this specific metazoan-archaeal association is widespread (51). It is difficult to separate the archaeal cells from the sponge tissue and bacterial endosymbiots. However, sufficient archaeal biomass has been obtained to enable the construction of a genomic fosmid library and to fulfill the requirements for a genome-sequencing project of *C. symbiosum* (http://web.mit.edu/esi/html/researchsub/organize.html). It was an interesting discovery that *Cenarchaeum* is not monoclonal in *Axinella* and is present as a population of strains with nucleotide sequence differences ≤10% and with additional rearrangements within coding regions. *C. symbiosum* was also the first cold-adapted member of the *Crenarchaeota* to have a functional gene (DNA polymerase) expressed in *E. coli* (356).

It took considerable effort and patience before the first cultivated, free-living member of the *Crenarchaeota* was obtained from a nonthermophilic environment (212). From the sequence analysis of large genomic DNA fragments cloned directly from the environment, it was predicted that some members of the *Crenarchaeota* may be able to utilize inorganic nitrogen compounds for energy metabolism (405). Despite this prediction, it was a revelation when "*Nitrosopumilus maritimus*" was isolated and found to grow chemolithoautotrophically by nitrification (212). It appears to be phylogenetically positioned as a novel order of the *Crenarchaeota* that is related to a number of uncultivated organisms from environmental samples. Little else has been characterized about its cellular properties.

(Thermo-) Acidophilic Archaea: *Thermoplasmatales* and *Sulfolobales*

Solfataric fields, named after the Solfatara caldera near Naples, Italy (Fig. 17), are typical habitats of archaeal thermoacidophiles (Tables 8 and 9). Many oxidize sulfur and ISCs for energy conservation and are largely responsible for the low pH values observed in these environments (reviewed in references 205, 207). Solfataras are typically small terrestrial, steam-heated pools of surface water or mud, with cell densities sometimes in excess of 1×10^8 ml^{-1} and a pH of 1 to 4. Two phylogenetically unrelated groups of archaea dominate solfataras. The majority of microorganisms are lobed, irregular cocci of 1 to 2 μm in diameter, that grow aerobically and/or anaerobically at ambient boiling temperature in their natural environments and belong to six different genera of the obligatory acidophilic *Sulfolobales* (Table 9) (154).

In contrast, *Thermoplasma* and *Picrophilus* (members of the *Euryarchaeota*) are found in lower-temperature pools, and very acidic soils with a low water activity, in and around solfataric fields (Table 8) (156). *Picrophilus* spp. are the most acidophilic organisms with a pH_{opt} of 0.7 at 59°C, and growth is possible at pH 0 (354, 355). A comparison of the genome sequences of *Sulfolobales* (three species) and *Thermoplasmatales* (four species) revealed that a significant fraction of the protein-encoding genes from *Sulfolobus* species had their best match with homologs in the genomes of *Thermoplasmatales* species (with the exception of homologs from the other strains of *Sulfolobus*) (336). This indicates that a high level of gene exchange is occurring under conditions that may be expected to rapidly degrade extracellular DNA and indicates that genome composition can be strongly influenced by the ecosystem in a way that is largely independent of an organism's phylogeny (see Chapter 5).

Ferroplasma acidophilum and *F. acidarmanus* belong to a third genus of the *Thermoplasmatales* and were isolated from acid mine drainage where they were found to oxidize a variety of metals for energy conservation. They were the first mesophilic archaea to be isolated that do not belong to the methanogens or haloarchaea (114, 115).

The prototypes of thermoacidophilic archaea are the *Thermoplasmatales* and the *Sulfolobales*. This phenotypic class is slowly expanding and includes isolates from other archaeal lineages, such as anaerobic *Acidilobus* (*Desulfurococcales*), and *Vulcanisaeta*, *Thermocladium*, and *Caldivirga* (*Thermoproteales*) (312). However, there are few reports that describe biochemical or molecular properties of these latter thermoacidophiles.

Sulfolobus solfataricus and *S. acidocaldarius* have proven to be model organisms for studies on biochemical pathways, enzymology, cell biology, gene regulation, DNA replication, and the developments of genetic tools in *Archaea*. *S. acidocaldarius* and *A. ambivalens* have developed into model organisms for studying heterotrophic and chemolithoautotrophic energy conservation, respectively. Studies on *Thermoplasma* have advanced the understanding of proteolytic and peptidolytic enzymes (e.g., proteasome). *Picrophilus* provides a model for acidophily, although it has recently been reported that intracellular proteins from *Ferroplasma* are more adapted to low intracellular pH than *Picrophilus* (90) (see "*Thermoplasmatales*," next paragraph).

Thermoplasmatales

Thermoplasma acidophilum was isolated from smoldering, self-igniting coal refuse piles and gained considerable attention due to its unusual habitat and growth conditions (Table 8) and the fact that it is devoid of a cell wall (156). Studies of its membrane composition, and subsequent studies of *S. acidocaldarius*, led to the discovery of phytanyl ether lipids in thermophilic archaea (227, 228). *Thermoplasma* species have been detected worldwide in moderately heated, but always very acidic, pools and soils of solfataric fields. They thrive aerobically between 40 and 68°C at pH 0 to 4 on organic, preferentially proteinaceous nutrients and can grow anaerobically provided that S^0 is present to function as a terminal electron acceptor (156).

In contrast to *Thermoplasma*, *Picrophilus* possesses a cell wall (Fig. 22), consisting of a proteinaceous S layer with tetragonal symmetry. *Picrophilus* grows aerobically at pH 0–2 on yeast extract (T_{opt}, 60°C), similar to *Thermoplasma*, but it is unable to grow anaerobically. The cytoplasmic pH for *Picrophilus* is 4.5 (pH 5.6 for *Ferroplasma*), which is the lowest measured for a living cell (113, 413). Some low-molecular-mass compounds are very unstable under these conditions and rapidly degrade. For example, the in vitro half-life for NADPH is only 2.4 min at 60°C and 1.7 min at 65°C (10). Because of the pH of the environment and cytoplasm, *Picrophilus* enzymes may be expected to be adapted to low pH. However, unlike the secreted proteins, the pH_{opt} of heterologously produced enzymes is near neutrality for most of the cytoplasmic enzymes, although it should be noted that only a limited number of *Picrophilus* enzymes have been studied.

The genus *Ferroplasma* comprises slow-growing, iron-oxidizing archaea that are devoid of a cell wall (Fig. 22) and require large amounts of iron for energy conservation when growing chemolithoautotrophically (12 mg of protein g^{-1} Fe^{2+}) (114). *F. acidophilum* was isolated from a metal-bioleaching pilot plant, and *F. acidarmanus* was isolated from Richmond Mine at Iron Mountain, Calif. (Table 8) (156). *Ferroplasma* species have a pH_{opt} of 1.3 to 1.7 and a T_{opt} of 35 to 42°C and grow also anaerobically with organic nutrients, provided that Fe^{3+} is supplied as electron acceptor. *Ferroplasma* is highly tolerant to potentially toxic (transition) metals, consistent with the presence of these compounds in bioleaching plants and acid mine effluents; the concentration of Fe is 28 to 111 g $liter^{-1}$, and the concentrations of Zn, Cu, Cd, and As ions exceed 2.5 g $liter^{-1}$, 380 mg $liter^{-1}$, 250 mg $liter^{-1}$ and 53 mg $liter^{-1}$, respectively, at Iron Mountain (76, 77).

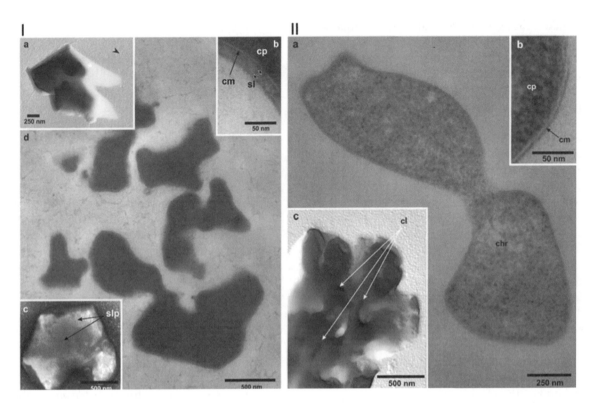

Figure 22. Morphology and ultrastructure of *Picrophilus torridus* and *Ferroplasma acidiphilum*. (I) *Picrophilus torridus*: (a) Pt-shadowed *P. torridus* showing the outline of the cell shape; (b) ultrathin section of cytoplasmic membrane (cm) and S layer (sl); (c) negatively stained individual cell; (d) ultrathin section showing pleomorphic morphology. (II) *Ferroplasma acidiphilum*: (a) ultrathin section of *F. acidiphilum* showing irregular cell shape; chr, translucent chromatoid region; (b) enlargement of the cell envelope with cytoplasmic membrane (cm) and no cell wall; (c) Pt-shadowed cells showing cytoplasmic lobes (cl) protruding from the surface. Reproduced from *Environmental Microbiology* (115) with permission from the publisher.

Genome sequence data of *Ferroplasma* spp. were highly represented in the first environmental genome project (metagenome) that was conducted on DNA sourced from a biofilm taken from acid mine drainage effluents at the Richmond Mine (408). From the 76 Mb of sequence data, the genomes of the dominant bacterium, a *Leptospirillum* group II species, and the dominant archaeon, *Ferroplasma* sp. type II, were largely reconstructed. Large scaffolds were also obtained for another *Ferroplasma* strain, which is closely related to *F. acidarmanus* Fer 1 (type I), an uncultivated archaeon (G-plasma) (408) that belongs to a novel genus within the *Thermoplasmatales*, and other bacteria, including another *Leptospirillum* (group III) and a *Sulfobacillus* species. The reconstruction of cellular metabolisms found a considerable degree of specialization. For example, *Leptospirillum* group III appeared to be the only organism group in the community that possessed nitrogen fixation (*nif*) genes.

Recently, a novel membrane-bound α-glucosidase from *Ferroplasma acidophilum* was found to have an acidic pH$_{opt}$ and required iron for activity (90). Three other cytoplasmic α-glucosidases and a carboxylesterase had similar properties (113). The acidic pH$_{opt}$ of enzymes compared with a more neutral pH (5.6) of the cytoplasm indicates that low-pH cellular compartments may exist in the organism. The iron dependence of enzymes may indicate that iron is not fully excluded from the cytoplasm and the enzymes have evolved a requirement for metal ions. Despite its relatively slow growth rate and the low fermentation yields, *Ferroplasma* represents a useful model for studying adaptation to acidophily.

Thermoplasma acidophilum has served as an important model organism for studying proteolysis and the consecutive action of proteases and peptidases (258). The proteasome was discovered in *T. acidophilum* and found to be similar to the eucaryal counterpart (see Chapter 10). The proteasome plays a central role in the degradation of misfolded and inactive (ubiquitinylated in the case of *Eucarya*) cellular proteins. The architecture of the proteasome was elucidated by X-ray crystallography using the 20S proteasome from *T. acidophilum* and subsequently from yeast (127, 245). The cylinder-shaped core enzyme is composed of 28 protein subunits that are arranged in four stacked rings of seven subunits each, which form an elongated cylinder with three large cavities connected by narrow constrictions (Fig. 23) (126). The four rings of the *Thermoplasma* enzyme are made from two subunits in an $\alpha_7\beta_7\beta_7\alpha_7$ stoichiometry, whereas the eucaryal enzymes consist of up to 14 different subunits in the same three-dimensional arrangement. The active site of the 700-kDa particle lies in the central chamber of the interior. The proteasome associates with AAA "unfoldases" (proteasome-activating nucleotidases) and catalyzes an energy-dependent degradation of proteins into fragments of 3 to 30 amino acids in length. The actual substrate for the archaeal proteasomes is not known, but there are indications that phosphorylated proteins might be the targets (no ubiquitin homolog has been unequivocally identified in *Archaea*) (30, 258).

The peptide products of proteasomes are further degraded by another large complex. This complex consists of several interacting enzymes whose framework is provided by the "tricorn peptidase" (named for its tricornlike shape) and was also purified and crystallized from *T. acidophilum*. It consists of multiple copies of a single, 120-kDa polypeptide chain (Fig. 23). Six subunits form hexamers that further

assemble into an icosahedral capsid with a molecular mass of 14.6 MDa. The crystal structure of the 720-kDa hexamer shows that each monomer consists of five separate domains, which may function by coordinating the steps of substrate channeling to the active site (126). The peptidase digests oligomeric peptides to tri- and dipeptides. These are then sequentially degraded to free amino acids by peptidases (tricorn interacting factors F1, F2, and F3) that are small enough to fill the open spaces in the tricorn assembly; the partial structures of these have been generated (219). Thus, protein unfoldases, proteasomes, the tricorn protease, and its interacting factors form a supramolecular protein degradation machinery that processes proteins from their folded state through to single amino acids that can be reused for cellular metabolism (37, 126).

The tricorn protease is not widely distributed in *Archaea*, and different peptidases are present for peptide breakdown. A tetrahedral-shaped, dodecameric 480-kDa aminopeptidase (TET) has been crystallized from *P. horikoshii*, and TET has also been purified from *H. marismortui* (37, 338). This self-compartmentalizing complex degrades some proteins and most peptides down to amino acids. TET contains a binuclear zinc active center, and proteolysis occurs independently of NTP hydrolysis. The pores for substrate access have a maximal diameter of 10 Å, which allows only small peptides to enter the active site. This protease has homologs in many archaea and bacteria (see Chapter 10).

Sulfolobus, Acidianus, and relatives

Sulfolobus acidocaldarius was the first hyperthermophile isolated (44). Cells are irregular, lobed, and sometimes motile cocci occur singly with cell diameters of 1 to 2 μm (Fig. 24). It grows aerobically either by the chemolithoautotrophic oxidation of S^0, thiosulfate, sulfidic ores, or H_2, or heterotrophically with various organic substrates. The cell morphology of the five to six genera of *Sulfolobales* is relatively uniform (Fig. 24; Table 9), and most cells in samples of acidic solfataras (as well as isolates) are microscopically indistinguishable (154).

The *Sulfolobales* genera tend to be distinguished by physiological properties rather than by phylogenetic differences in their 16S rDNA sequences (reviewed in reference 154). All *Sulfolobales* have a T_{opt} of 65 to 95°C and pH_{opt} of 2 to 4. The species of the genus *Sulfolobus* are characterized by aerobic growth and a high metabolic versatility. Most *Sulfolobus* strains, including the model organisms *S. acidocaldarius*, *S. solfataricus*, and *S. tokodaii*, can be routinely grown with peptides and/or sugars as carbon and

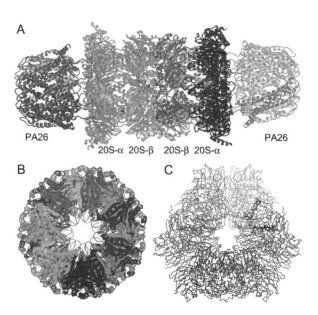

A

PA26   PA26

20S-α   20S-β   20S-β   20S-α

B   C

Figure 23. (*See the separate color insert for the color version of this illustration.*) Three-dimensional structures of the *T. acidophilum* proteasome and tricorn protease. (A) Side view of the 26S proteasome/activator particle with the two sets of seven terminal PA26 subunits and the two $\alpha_7\beta_7\beta_7\alpha_7$ rings (PDB code 1YA7) (93). (B) Top view of the 20S proteasome core particle showing the sevenfold symmetry (PDB code 1PMA) (245). (C) Top view of the homohexameric tricorn protease complexed with a tridecameric peptide derivative (PDB code 1N6E) (195).

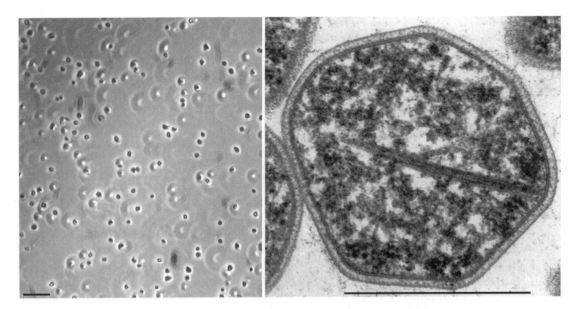

Figure 24. Electron micrographs of *Acidianus ambivalens*. (Left) Phase-contrast micrograph; bar 10 μm. Photograph courtesy of K. Lauber, Darmstadt, Germany. (Right) Transmission electron micrograph of thin section; bar 1 μm. Photograph courtesy of W. Zillig, Martinsried, Germany.

energy sources at pH 3 and 75 to 85°C (Table 9). *S. metallicus* is a facultatively chemolithoautotrophic, sulfur and metal sulfide-oxidizing aerobe that is frequently found in metal-rich, self-heating mine heaps and bioleaching operations. These growth properties are atypical for the genus, but not more generally for the *Sulfolobales* order.

Stygiolobus species are anaerobic chemolithoautotrophic sulfur reducers. The genus *Acidianus* (Fig. 24) comprises several chemolithoautotrophic and facultatively anaerobic and physiologically versatile species. The genus was initially characterized as growing aerobically by S^0 oxidation to sulfuric acid, or anaerobically by H_2 oxidation with S^0 as the electron acceptor. In addition to these metabolic properties, *Acidianus* species grow by H_2 oxidation with O_2 (Knallgas reaction), anaerobic respiration with various electron acceptors (e.g., molybdate, Fe^{3+}, or arsenate), or oxidation of tetrathionate and metal sulfides (e.g., pyrite, chalcopyrite, and sphalerite), which causes bioleaching (339). Some *Acidianus* strains are facultative heterotrophs. The bioleaching species of *Acidianus*, *Sulfolobus metallicus*, and *Metallosphaera* contribute to base and precious metal extraction and to the formation of acidic drainage downstream of mines and slag heaps (154). *Metallosphaera* spp. are strict aerobes and metal leachers, and they grow at more moderate temperatures (T_{opt}, 65 to 72°C) similar to *Acidianus brierleyi* and *Sulfolobus thuringiensis*. The genus *Sulfurisphaera* comprises facultatively anaerobic strains that are also facultative heterotrophs.

A sixth genus, *Sulfurococcus*, has been described, although the cultures were never deposited in a culture collection and are probably no longer available.

Many *Sulfolobales* isolates were termed "*Sulfolobus*," although subsequent nucleic acid hybridization and 16S rDNA sequencing suggested that they were distantly related to each other and should be phylogenetically reclassified (Fig. 25) (154). For example, *Sulfurisphaera ohwakuensis*, *Sulfolobus tokodaii*, and *S. yangmingiensis* have high 16S rDNA nucleotide identity (≥99%) and should probably be considered a single genus or species (154), while the similarity between *S. tokodaii* and *S. solfataricus* is only ~90%, suggesting that they might belong to separate genera (Fig. 25). Several type strains (e.g., *S. acidocaldarius* and *S. solfataricus*) were unable to oxidize S^0 a number of years after they were isolated (284), a property that had been demonstrated (367) and used to originally define *Sulfolobus* (44, 455). Most of the early isolates were purified from natural samples by successive rounds of serial dilution and not by plating (44, 455) and were likely to have included cocultures of heterotrophs and autotrophs. This is no longer an issue for pure cultures that have been purified by using improved plating techniques for hyperthermophiles (e.g., 450).

Compared with many archaea, the heterotrophic *Sulfolobales* strains are easy to handle and cultivate and have led to the generation of systems for genetic manipulation (e.g., 5) and made them model organisms for studying many cellular processes. Numerous

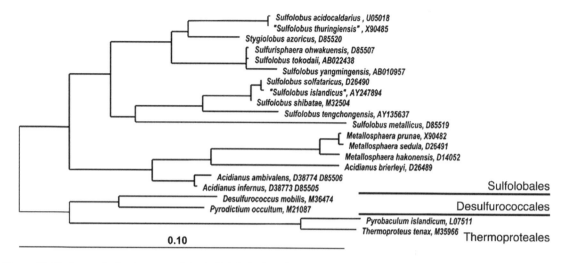

Figure 25. Phylogenetic dendrogram of the *Sulfolobales* based on 16S rDNA sequences. Scale bar, 10 estimated exchanges within 100 nucleotides. Reproduced with minor modifications from *The Prokaryotes* (154) with permission of the publisher.

3D structures for proteins from *Sulfolobales* are in the Protein Data Bank. Chemolithoautotrophic *Sulfolobales* are more difficult to grow on plates because they grow slowly and S^0 needs to be precipitated on solid media, or colloidal sulfur used (450). Complete genome sequences are available for *S. solfataricus*, *S. tokodaii*, and *S. acidocaldarius*.

Sulfur metabolism in *A. ambivalens*

Oxidation and reduction of sulfur and inorganic sulfur compounds is a characteristic often associated with the *Archaea*. The reduction of S^0 is a common physiological property (Fig. 26) (7). However, sulfur oxidation is primarily a feature of acidophiles as the oxidation product causes a marked decrease in the pH of the environment. *Acidianus* species can do both; depending on the culture conditions, they reduce S^0 to hydrogen sulfide or oxidize S^0 to sulfuric acid.

Sulfur reduction in *A. ambivalens*. Sulfur reduction in anaerobically grown *A. ambivalens* requires sulfur reductase (SR) and hydrogenase activities comparable to those of *Pyrodictium* spp. (231) (see "*Pyrodictiaceae*" above). The two enzymes have been purified. The hydrogenase gene is present in a multigene cluster that includes genes for a NiFe and an FeS subunit. The subunits are not very similar to other hydrogenases (highest pairwise similarity, 40%) (231). The SR operon consists of a five-gene cluster, *sreABCDE*. The deduced amino acid sequences show similarity to molybdoenzymes of the dimethyl sulfoxide/nitrate reductase family (231), and molybdenum

was found in solubilized membrane fractions, suggesting that the SR is a molybdoprotein. The molecular composition of the *A. ambivalens* SR is similar to the *Wolinella succinogenes* polysulfide reductase. Both enzymes consist of homologous catalytic and electron transfer subunits presumably oriented toward the (pseudo-) periplasm, and nonhomologous membrane anchors (231). It is assumed that an electrochemical gradient is generated via a Q-cycle mechanism, similar to *bc1* complexes of the respiratory chain ("Q cycle" describes a mechanism where reduction of the quinone leads to proton uptake in the cytoplasm, and reoxidation of the quinone leads to release of protons into the periplasm) (406) (see "Energy metabolism: aerobic electron transport chains" below).

A gene cluster that is similar to *A. ambivalens* *sreABCDE* is present in the genome of *S. solfataricus* (but not in *S. acidocaldarius* or *S. tokodaii*), suggesting that it should have the capacity to grow by heterotrophic anaerobic S^0 respiration, but attempts to demonstrate this have been unsuccessful so far (A. Kletzin, unpublished results). None of the three *Sulfolobus* genomes contain hydrogenase genes, indicating that they are not expected to be able to grow lithotrophically with H_2-like *A. ambivalens*.

Sulfur oxidation in *A. ambivalens*. The ability to aerobically or anaerobically oxidize S^0 and ISCs is widespread in the microbial world (Fig. 26). Bacterial sulfur oxidizers are physiologically and phylogenetically diverse. In contrast, only a few archaeal sulfur oxidizers are known, and these are members of the *Sulfolobales*. The biochemistry and electron transport chains of S^0 oxidation was determined primarily using

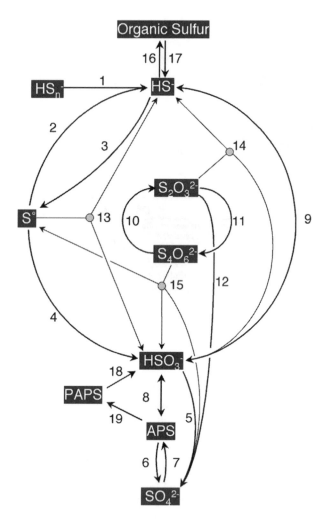

Figure 26. The bioinorganic sulfur cycle. Sulfur cycle depicted with enzyme reactions involving sulfur compounds. Enzymes: 1, polysulfide reductase; 2, sulfur reductase; 3, sulfide:quinone oxidoreductase or sulfide:cytochrome *c* oxidoreductase; 4, sulfur oxygenase; 5, sulfite:acceptor oxidoreductase (the acceptor is cytochrome *c* in most bacteria) or sulfite oxidase; 6, ATP sulfurylase or adenylylsulfate:phosphate adenylyltransferase (APAT); 7, ATP sulfurylase; 8, adenylylsulfate (APS) reductase; 9, sulfite reductase; 10, tetrathionate reductase; 11, thiosulfate:acceptor oxidoreductase; 12, Sox complex; 13, sulfur oxygenase reductase; 14, thiosulfate reductase; 15, tetrathionate hydrolase (there is also a trithionate hydrolase in addition); 16, *O*-acetylserin or *O*-phosphoserine sulfhydrolases; 17, cysteine desulfurase; 18, APS kinase; 19, 3'-phosphoadenylylsulfate (PAPS) reductase. Gray circles denote disproportionation reactions. Reactions of cysteine breakdown are omitted. Compiled from several sources (205, 214, 401) with permission of the publishers.

Acidianus species (205, 207) and found to differ significantly from the bacterial models.

The oxidation of S^0 to sulfuric acid proceeds in at least two steps and involves intermediates such as sulfite, thiosulfate, and tetrathionate (Fig. 26) (e.g., reviewed in references 100, 101, 186, 401). (Thermo-) acidophilic archaea oxidize S^0 with a soluble sulfur

oxygenase reductase (86, 139, 204). Thiosulfate and sulfite are oxidized by membrane-bound oxidoreductases (271, 456). In contrast, neutrophilic bacteria oxidize S^0 and ISCs with a periplasmic Sox multienzyme complex (100, 101). Sox genes are present in genomes of numerous mesophilic and thermophilic bacteria, but not in archaea (100).

The sulfur oxygenase reductase (SOR), which performs the initial step in the S^0 oxidation pathway of aerobic archaea, is unique. It produces sulfite, thiosulfate, and hydrogen sulfide in an oxygen-dependent sulfur disproportionation reaction when incubated at elevated temperatures with S^0. Oxygen, but no other cofactors, is required for activity (86, 139, 204):

$$4\ S^0 + O_2 + 4\ H_2O \rightarrow 2\ HSO_3^- \\ + 2\ H_2S + 2\ H^+ \tag{4}$$

Thiosulfate is probably a nonenzymic product of sulfite condensation with excess S^0. The SOR was purified from *A. ambivalens,* and the *sor* gene was expressed in *E. coli* to produce an active enzyme (409). Other SORs have been purified from *A. brierleyi* and *A. tengchongensis* (86, 139). Other *sor* genes are present in the genomes of *S. tokodaii, F. acidarmanus, Picrophilus acidarmanus,* and the hyperthermophilic bacterium, *Aquifex aeolicus* (409), but absent from *S. solfataricus* and *S. acidocaldarius.*

Three conserved cysteine residues are present in SOR enzymes. One of the Cys residues (Cys31 in *A. ambivalens*) was shown by site-directed mutagenesis to be essential, while mutagenesis of the other two Cys residues reduced specific activity (54, 412). EPR spectroscopy and redox titration showed that the SOR contains a mononuclear nonheme iron center in the high-spin Fe^{3+} state and with a low midpoint potential (E_o', -268 mV). The reduction potential was more than 300 mV lower than typically found for this type of iron center, and low enough to explain the S^0-reducing activity of the enzyme (E_o' [H_2S/S^0] = -270 mV) (409).

The SOR crystal structure determined to 1.7 Å resolution shows that the enzyme is a spherical homoicosatetramer (i.e., 24 subunits) with an external diameter of 150 Å (410, 411). It surrounds an empty cavity with a diameter of 71 to 107 Å (Fig. 27). A bidentate glutamate, two histidine ligands, and two water molecules coordinate the iron center in a "2-His 1-Carboxylate facial triad" structural motif (59, 411). Mutation of any of the three iron ligands resulted in loss of activity and iron-binding capabilities (412). Residue Cys31 has a persulfide modification. The iron center and the three Cys residues are buried in a pocket in the interior of each monomer and are accessible only from the interior cavity. The core

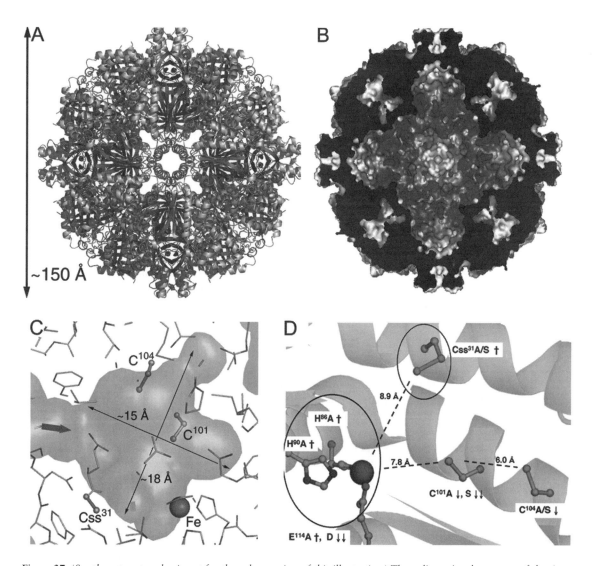

Figure 27. (*See the separate color insert for the color version of this illustration.*) Three-dimensional structure of the *A. ambivalens* sulfur oxygenase reductase. (A) The SOR holoenzyme. Cartoon representation viewed along the crystallographic four-fold axis; cyan, α-helices; purple, β-sheets; red spheres, Fe ions. (B) Molecular accessible surface representation in the same orientation of inner surface of the sphere, color-coded according to the calculated electrostatic potentials: red, $\leq -10 \pm 1$ κT/e; white, neutral; blue, $\geq +10 \pm 1$ κT/e. (C) Cavity surface representation of the catalytic pocket, with conserved cysteines and iron highlighted; gray arrow, cavity entrance. (D) Effect of mutants on SOR activity; †, zero activity; ↓, reduced activity; ⇓, strongly reduced activity. The core active site composed of the Fe site and the persulfide-modified Css31 is highlighted within ellipsoids. Reproduced with minor modifications from *Science* (411) with permission from the publisher.

active site is composed of the iron site and the modified Cys31 (Fig. 27). Substrate entry proceeds through the hydrophobic channels along the 4-fold axes of the sphere. The SOR thus provides an enclosed reaction compartment separated from the cytoplasm. The presence of a persulfide shows that S^0 is likely to be covalently bound to Cys31. The linear sulfur chain is aligned to the iron site and replaces the water ligands, poising the iron site for dioxygen binding and activation (411).

H_2S, sulfite, and thiosulfate oxidation. The oxidation of sulfite, sulfide, and thiosulfate in *A. ambivalens* requires membrane-bound, proton-pumping oxidoreductases since the SOR does not couple S^0 oxidation to electron transport or substrate-level phosphorylation. Homologs of the *sqr* gene are present in many archaeal genomes; however, H_2S oxidoreductase activity (sulfide:quinone oxidoreductase) (402) has not been experimentally detected yet, despite considerable effort (A. Kletzin, unpublished results).

A tetrathionate-forming, membrane-bound thiosulfate:quinone oxidoreductase (TQO) was purified from *A. ambivalens*. Oxygen electrode measurements showed that electrons were transported from thiosulfate to molecular oxygen via the terminal heme copper quinol:oxygen oxidoreductase. The 102-kDa holoenzyme consists of two subunits in an $\alpha_2\beta_2$ stoichiometry (DoxDA). The fate of the tetrathionate is not clear. There is a possibility, however, that a thiosulfate/tetrathionate cycle exists (271). Tetrathionate is unstable in the presence of strong reductants and is reduced to thiosulfate in vitro at high temperatures. H_2S and sulfite might re-reduce tetrathionate formed by the TQO and thus feed electrons indirectly from the S^0 disproportionation reaction catalyzed by the SOR into the quinone pool.

Energy metabolism: aerobic electron transport chains

Aerobic electron transport chains have been studied primarily in *Crenarchaeota* (*S. acidocaldarius* and *A. ambivalens* and, more recently, *A. pernix*), to a lesser extent in *Euryarchaeota* (various *Halobacteriales*, *T. acidophilum* (301). The prototypical aerobic electron transfer chains from mitochondria and bacteria (e.g., *Paracoccus denitrificans*) consist of complexes I to IV (Fig. 28). In contrast, electron transport chains in most other bacteria and archaea are composed of multiple heterologous and often redundant complexes, which are expressed under growth-dependent conditions.

The components of aerobic electron transport chains of *Archaea* are described below according to their similarity to the canonical complexes I to IV and their associated functions. The components of electron transport for oxygen reduction and chemolithoautotrophic aerobic sulfur oxidation are linked in *A. ambivalens* (see "Sulfur oxidation in *A. ambivalens*," above). ATP synthases are not described and have been reviewed recently (128, 272).

I. The rotenone-sensitive NADH:quinone oxidoreductase (NQO or complex I) accepts electrons from NADH and reduces quinone, thereby pumping protons (e.g., 406). Complex I consists of up to 42 subunits in *Eucarya* and 14 in *Bacteria* (termed either NuoA-N or Nqo1-14) (344). It is L-shaped in electron micrographs with a large cytoplasmic and transmembrane domains. Nine iron-sulfur clusters are present in the cytoplasmic domain (Fig. 28) (148). NADH-oxidizing versions of complex I with identical subunit composition and cofactor specificity to bacterial or eucaryal complexes have not been found in *Archaea*. However, the proton-pumping $F_{420}H_2$ dehydrogenases from *M. barkeri* and *A. fulgidus* are functionally very similar (see Chapter 13). The archaeal enzymes are

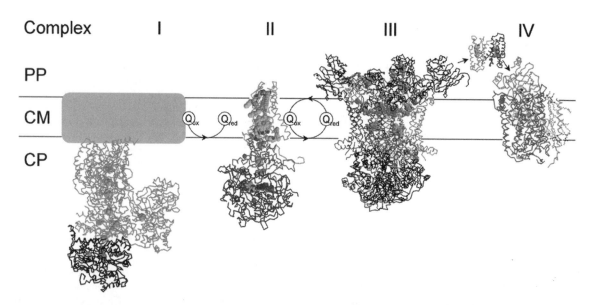

Figure 28. (*See the separate color insert for the color version of this illustration.*) Canonical respiratory chain in bacteria and mitochondria. Scheme based on 3D structures with the exception of the membrane domain of complex I, for which a structure is not available. Domains that have not been identified in *Archaea* are shown in black. PP, periplasm; CM, cytoplasmic membrane; CP, cytoplasm; Q, quinols/quinones. The figure was prepared from the coordinates of PDB entries 1FUG (complex I, *Thermus thermophilus*), 1NEK (complex II, *E. coli*), 1KYO (complex III, *Saccharomyces cerevisiae*), 1EHK (complex IV, *Thermus thermophilus*), and 2CCY (cytochrome *c*, *Rhodospirillum molischianum*).

composed of 11 subunits that are homologous to subunits from the bacterial NQO (Nqo 4 to 14; Fig. 29). Two additional subunits, FpoO and FpoF, are required to oxidize the substrate F_{420} and to replace the bacterial equivalent, NADH-accepting subunits (Nqo 1 to 3) that are missing in the archaeal enzymes. The Ech hydrogenases from several methanogens also pump protons. However, while they are homologous subunits and are considered to be precursors of complex I, they serve a different function in methanogens (140). Other aerobic archaea, including *P. aerophilum, A. pernix,* and the *Thermoplasmatales* encode the same 11 subunits as *Methanosarcina*. However, these complexes have not been studied and the electron donors are not known.

The alternative, "rotenone-insensitive" type II NADH dehydrogenase (NDH-2) that is present in many bacteria and eucarya was purified from the membrane fraction of *A. ambivalens* (117). The gene is present in genome sequences from the *Sulfolobales, Thermoplasmatales,* and *H. salinarum*. The single-sub-unit flavoproteins have the same enzymatic activity as complex I but are unable to pump protons. Transmembrane helices or extensive hydrophobic regions are not predicted from the sequence of the archaeal NDH-2, and it is speculated to be embedded at the membrane surface by amphipathic helices rather than being an integral membrane protein (16). Consistent with its expected role as an electron transport chain, the *A. ambivalens* type II NADH dehydrogenase and the cytochrome aa_3 terminal oxidase were reconstituted with caldariella quinone in liposomes and showed NADH-dependent oxygen consumption (117).

II. Succinate:quinone oxidoreductases (SQOs; Fig. 28) are membrane-bound enzymes that catalyze the oxidation of succinate to fumarate with quinone as the electron acceptor while the related, paralogous fumarate reductases (FRDs) perform the reverse reaction (224). All SQOs and FRDs have a flavin-containing, large catalytic subunit (SqoA/FrdA) and a smaller iron-sulfur protein (SqoB/FrdB). Most of the enzymes contain one 4Fe-4S, 3Fe-4S, and 2Fe-2S

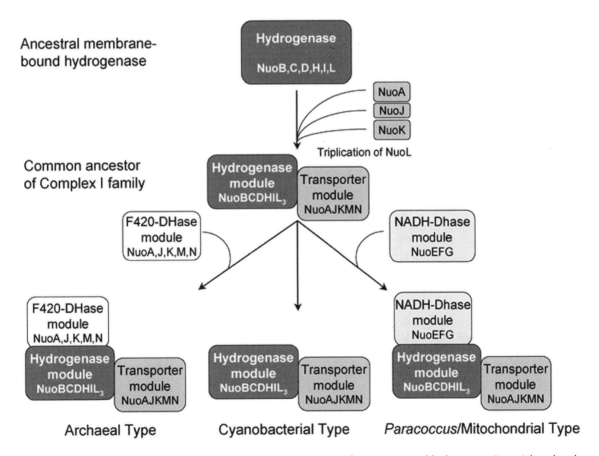

Figure 29. Hypothetical scheme of the modular evolution of complex I from an ancestral hydrogenase. Bacterial, archaeal, and cyanobacterial complexes emerged by acquisition of specific modules. Dark gray, hydrogenase module; light gray, transporter module. Reproduced and modified from the *Journal of Bioenergetics and Biomembranes* (102) with permission of the publisher.

cluster each in the SqoB subunit. SQOs fall into five families (A to E) differing from each other in the number, hydrophobicity, similarity, and heme content of the remaining one to two subunit(s) that serve as membrane anchors and quinone-binding site. For example, in the mitochondrial family C, SQRs, SqoC, and D are transmembrane proteins with one bound heme b.

Many archaeal genomes contain single or multiple copies of genes encoding SQO/FRD family proteins that have significant similarity to bacterial and eucaryal counterparts (349). The *Sulfolobales* are the only archaea to harbor family E enzymes, which are significantly different from standard SQOs. The four-subunit proteins were purified from membrane fraction, but the operons lack a gene encoding an obvious transmembrane subunit. Instead, they contain hydrophilic, putative anchor proteins with amphipathic helices (SqoE and F), suggesting that the complex "swims" in the membrane (236). A cysteine-rich, putative FeS subunit (SqoE) that is related to heterodisulfide reductases has paralogs in many microorganisms (163). SqoE may be an electron transfer protein with multiple functions. Unlike other SqoB proteins in other families, SqoB in family E has two 4Fe clusters instead of one 4Fe and one 3Fe (119, 165). Only SqoA is similar in all SQO families.

III. The *bc1* complex (Ubiquinole:cytochrome c oxidoreductase) couples electron transfer to proton pumping by a Q-cycle mechanism (Fig. 28) (344). The *bc1* complexes consist of several subunits, including an integral membrane protein with two b-type hemes, and two membrane-anchored proteins with one heme $c1$ moiety and a Rieske iron-sulfur cluster, respectively. The brightly colored Rieske proteins contain an unusual 2Fe-2S cluster, in which one iron ion is coordinated to the protein by two nitrogen ligands from histidine residues and the other by two sulfur ligands from cysteine residues.

Canonical *bc1* complexes have not been found in *Archaea*, with the exception of a haloarchaeal analog that has been reported to be enriched but not yet purified (379). Analogous complexes containing multiple hemes but lacking the *c1* component were found, for example, in whole cell EPR studies of *S metallicus* (118) and in other *Sulfolobales*. This was inferred from findings that CbsAB from membranes of *S. acidocaldarius* is a highly glycosylated cytochrome b (145) and that two Rieske proteins were present, SoxF and SoxL (147, 358). It was subsequently found that SoxL and CbsAB were encoded in the same pentacistronic operon (*cbsAB-soxLN-odsN*) containing two b-type cytochromes (CbsA and SoxN), one Rieske protein (SoxL), and two proteins of unknown function. Homologous genes are present

in *Metallosphaera* and *A. ambivalens*. The functions of this novel complex and its electron acceptors are not known. A distinct *S. acidocaldarius* complex that is analogous to *bc1* complexes is described below (complex IV).

IV. Oxygen reductases (terminal oxidases) catalyze the reduction of dioxygen to water with a variety of electron donors that include quinoles, reduced small electron transfer proteins (e.g., cytochrome c), copper-containing azurin, and high-potential, iron-sulfur proteins (Fig. 28) (301, 344). The prototypes of oxygen reductases are the cytochrome aa_3 complexes of the respiratory chain of mitochondria and bacteria. They are heme copper enzymes, characterized by the presence of a binuclear center at the active site consisting of a heme iron (a_3) and a copper ion (CuB), which together bind and reduce dioxygen coupled to proton uptake from the cytoplasm (69). A second heme a present in most oxygen reductases is involved in electron transfer toward the active site, and a second, binuclear copper site (CuA) is present in the cytochrome c oxidases. Terminal oxidases pump protons via a proton-conducting channel that is distinct from the proton uptake channel, resulting in the electrochemical gradient being generated by two separate mechanisms.

Two different oxygen reductase complexes have been isolated from *S. acidocaldarius*. Both pump protons, and both act as quinol oxidases. One is composed of four subunits encoded by genes in the *soxABCD* operon (112). Its composition is similar to that of bacterial quinol oxidases. The SoxB subunit contains the typical cytochrome aa_3 diheme/copper active site. SoxA is homologous to the copper A-containing subunit of cytochrome c oxidases but lacks the copper site consistent with its function as a quinol oxidase. SoxC is a diheme cytochrome a subunit with an unknown function.

The second oxygen reductase (SOX-M) forms a larger complex and is encoded by the *soxMHGFEI* operon (211). SOX-M activity is blocked by cyanide and strongly inhibited by aurachin C and D derivatives, suggesting that the complex has two proton-pumping sites (211, 301). The complex is unusual because it combines an oxygen reductase with a *bc1* analogue. SOX-M consists of two subcomplexes. The oxygen reductase subcomplex is a bb_3-type terminal oxidase containing CuA and CuB. The oxygen-reducing subunit SoxM contains CuB and one b and one b_3-type heme. The CuA-containing SoxH subunit is similar to mitochondrial cytochrome c oxidases. The *bc1*-like subcomplex consists of a Rieske FeS-protein (SoxF) and a homolog of cytochrome b that hosts two a-type hemes (SoxG). The blue copper protein sulfocyanin (SoxE) functionally links the two subcomplexes (211). With these properties, SOX-M differs markedly from the SoxABCD complex (301).A high-resolution 3D structure (1.1 Å) is available of the soluble domain of the SoxF Rieske protein (35).

The proton-pumping, cytochrome $aa3$ quinol oxidase extracted from membranes of aerobically grown *A. ambivalens* has little similarity to the *S. acidocaldarius* complexes (116). The enzyme was composed of five subunits encoded in two separate operons (*doxBCEF* and *doxDA*) (314). DoxDA was later identified as the organism's thiosulfate:quinone oxidoreductase (see "H_2S, sulfite, and thiosulfate oxidation," above) (271). The *doxB* gene encodes the *aa3*-type heme-containing active subunit, while *doxC* encodes subunit II, which does not possess a CuA center.

Other respiratory chain components have been isolated from a variety of archaea, including *b*-type cytochromes from *H. salinarum* (132), two oxygen reductases from *A. pernix* (*aa3* and *ba* types) (161), and cytochromes and another type II NADH oxidase from *S. metallicus* (17, 118). The respiratory chain of the haloalkaliphilic archaeon *Natronobacterium pharaonis* has also been studied (350). A small peripheral blue copper membrane protein, halocyanin, appears to have a similar function to plastocyanin. A small, two-subunit *bc* cytochrome that lacks a Rieske FeS center may be analogous to the *bc1* complex, and a large heterodimer carries two different *ba3*-type hemes and copper and may be the terminal oxidase.

The archaeal respiratory chain complexes are diverse and differ from the canonical respiratory chains of mitochondria, *P. denitrificans*, and other complexes found in *Bacteria*. Thus archaeal complexes highlight the evolutionary diversity of respiratory chains and, in particular, expand the knowledge of aerobic respiration in extreme habitats.

Methanogenic Archaea

Numerous members of the *Euryarchaeota* produce methane as the major end product of their energy metabolism (see Chapter 13). Methane and biological methane production were discovered in the eighteenth century when the Italian physicist Alessandro Volta collected gas from anaerobic sediments of swamps and marshes and showed that it was flammable (combustible air) (253). Methane formation is the last step in the anoxic biodegradation of biologically derived organic compounds (and often xenobiotics) where electron acceptors other than CO_2 are limiting. Methane is important in the global carbon cycle and is a greenhouse gas. Approximately 80 to 85% of the global annual production (600 to 1,200 Tg) of methane is biogenic (141, 319). Biogenic methane may be produced exclusively by methanogenic archaea, although recently it was suggested that plants may be a source of methane (≤10 to 20%). Plant methane is reported to be liberated by nonenzymatic and presently unknown reaction(s) (191). Bacteria also produce small amounts of methane from the demethylation of methyl group carriers (e.g., *S*-adenosylmethionine) (325).

Natural habitats for methanogens include anoxic freshwater swamps, ocean and lake sediments, hydrothermal vents, animal digestive tracts (e.g., the rumen, termite hindgut, and humans), and within anaerobic protozoa. Important man-made habitats include rice paddies, landfills, and anaerobic sludge digesters of sewage treatment plants. The typical growth substrates of methanogens are H_2 and CO_2, or short-chain (C_1 to C_5) organic compounds (reviewed in references 34, 109, 141, 188). In some environments, methanogens are out-competed for these substrates by bacteria (e.g., homoacetogenic and sulfate-reducing bacteria in the termite hindgut and marine sediments, respectively). Approximately two thirds of methane that is produced does not reach the atmosphere as it is reoxidized, either aerobically by methylotrophic bacteria or anaerobically by a nonculturable consortia of methylotrophic archaea and sulfate-reducing bacteria (319, 365).

The Black Sea benthos and marine sediments near to methane hydrate deposits are examples of habitats where "methanotrophic methanogens" reside (methanotrophic archaea closely related to methanogenic archaea of the order *Methanosarcinales*) (365). These archaea perform an important process in the global carbon cycle by performing anaerobic oxidation of methane (AOM) to CO_2, coupled to the reduction of sulfate to H_2S, and they are present in sufficient abundance in the Black Sea benthos to be studied in situ.

Cultivated methanogens belong exclusively to the *Euryarchaeota*. They are phylogenetically very diverse and are divided into four to five orders (Fig. 1) (see "Phylogeny of *Archaea* and the origin of life," above).

Physiology of methanogenesis

Small carbon compounds are converted to methane by pure cultures of methanogens. The most abundant substrates are $H_2 + CO_2$, formate, and acetate. Additional C-1 substrates include methanol, mono-, di-, or trimethylamine, and dimethyl sulfide, and some higher order alcohols such as ethanol, 2-propanol, 2-butanol, and cyclopentanol. These substrates are converted stoichiometrically to methane and CO_2. Methanogens do not tend to use typical bacterial growth substrates, such as sugars, amino acids, or complex nutrient mixtures (e.g., yeast extract). However, some of these compounds can stimulate growth, and many methanogens require more complex nutrients or acetate for carbon assimilation (34, 109, 141, 188). Cultures of methanogens and anaerobic bacteria can degrade complex carbon sources in a cooperative and symbiotic manner. Cocultivation of H_2-producing and sugar-utilizing clostridia with H_2-consuming methanogens keeps the partial pressure of H_2 low and facilitates growth of both types of microorganisms. Substrate-level phosphorylation (typical for fermentative bacteria), does not seem to occur in methanogens. Chemolithoautotrophic growth on CO_2 plus H_2 is a feature of methanogens that is not restricted by phylogeny. In contrast, acetoclastic methanogens that disproportionate acetate to CH_4 and CO_2 are confined to the *Methanosarcinales* (188).

All methanogens have a blue-green fluorescence under the microscope that results from a specific archaeal flavin derivative, F_{420}. Methanogens utilize several highly specific coenzymes for carbon reduction, coenzymes that are found in only a few specialized bacteria. Some cofactors are C_1 carriers (methanofurane and methanopterin), while others are redox active; HS-HTP (also known as 7-mercaptoheptanoylthreonine phosphate or coenzyme B), F_{420}, nickel-containing F_{430}, and in some species, methanophenazine. All pathways of methanogenesis converge at the cofactor, coenzyme M (CoM). CoM and HS-HTP form an energy-rich mixed disulfide. The reduction of this disulfide is the energy-yielding step in methanogenesis (65, 141).

H_2 is a typical electron donor for CO_2 reduction, and electrons can also be derived from formate, CO, or specific alcohols. Methanol and methylamines are converted by a methylotrophic pathway to CH_4 and CO_2 in a 3:1 ratio. The biochemistry of methanogenesis has been extensively reviewed (65, 66, 141) (see Chapter 13), and key steps are summarized below.

CO₂ reduction: CO_2 is activated by a molybdenum- or tungsten-containing formylmethanofuran dehydrogenase, which is an ion pump catalyzing the reversible formylation of methanofuran coupled to the reduction to the formyl state with electrons derived from H_2 via F_{420}. A sodium motive force drives this endergonic reaction.

Formyl transfers and reduction: The formyl group is transferred to a methanopterin-containing enzyme, dehydrated, and reduced to the methyl level in two steps. The methyl group is then transferred to CoM. This reaction is exergonic and catalyzed by a Na^+-pumping methyltransferase enzyme complex, thus creating a sodium motive force.

Methyl-CoM reduction: The F_{430}-containing methyl reductase accepts the methyl group from CoM, forming a Ni-methyl intermediate. Methane is liberated following reduction by electrons derived from the formation of a mixed disulfide consisting of coenzymes M and B. The exergonic reaction is catalyzed by the soluble methylcoenzyme M reductase.

Heterodisulfide reduction: The mixed disulfide is the terminal electron acceptor. It is reduced to the individual components by the heterodisulfide reductases, which are central enzymes in methanogens. The electron donor is H_2.

Acetate cleavage: Acetate is activated by a kinase in species such as *Methanosarcina thermophila* and *Methanosarcina barkeri* when growing on acetate. The key steps of the pathway are catalyzed by the acetyl-CoA decarbonylase/synthase (ACDS), a complex nickel enzyme that cleaves the C–C and C–S bonds in acetyl-CoA and oxidizes CO to CO_2. The methyl group is transferred to tetrahydrosarcinapterin.

Methanobacteriales and *Methanopyrus*

The *Methanobacteriales* are predominantly rod-shaped and form chains or filaments. Most species stain gram positive because of pseudomurein in cell walls. *Methanobacteriales* isolates have been obtained from all kinds of strictly anaerobic habitats, including sewage digestors, hydrothermal vents and terrestrial volcanic hot springs, freshwater sediments, rice paddies, and gastrointestinal tracts of animals. They tend to grow optimally at low-salt concentrations with a T_{opt} of 15 to 97°C (34).

The order *Methanobacteriales* consists of two families, the *Methanobacteriaceae* and *Methanothermaceae*. The *Methanobacteriaceae* are divided into the four, widely distributed genera, *Methanobacterium*, *Methanobrevibacter*, *Methanosphaera*, and *Methanothermobacter*. All grow by H_2 oxidation, with CO_2 as terminal electron acceptor, although some species can use formate, alcohols, and/or CO as alternative electron donors (34). *Methanobacterium formicicum* is frequently isolated from sewage sludge and has been an intensively studied strain. *Methanobrevibacter* species are typical inhabitants of ruminants, termites, and other animals, including humans (240). *Methanobrevibacter oralis* has been isolated from human subgingival plaque (237), while various *Methanobrevibacter smithii* strains appear to be the most abundant archaeal species in the mouth and in the intestines (73, 237). These two species of *Methanobrevibacter* have been implicated as possible causative agents of periodontitis in humans, although other research groups have not been able to confirm the results (237, 370, 415). *Methanosphaera* species differ from other members of this family by reducing methanol instead of CO_2 as a terminal electron acceptor. The genome sequence of *Methanosphaera stadtmanae* was recently released (GenBank: NC_007681) (34).

The genus *Methanothermobacter* was recently proposed, with members being derived from the genus *Methanobacterium* (426). *Methanothermobacter thermautotrophicus* and *M. marburgensis* were previously described as *Methanobacterium thermautotrophicum* strain Delta-H and strain Marburg, respectively. The genome sequence of *Methanothermobacter thermautotrophicum* was one of the first archaeal genomes to be published (373). *Methanothermobacter* species are H_2/CO_2 utilizing thermophiles with a T_{opt} of 55 to 65°C (34). Some strains can also use formate to reduce CO_2. Most strains were isolated from anaerobic sludge digestors, despite the fact that the T_{opt} of all species exceeds the typical digester temperature (37°C). Relatives of *Methanothermobacter* are found worldwide in hot springs.

In contrast to the *Methanobacteriaceae*, the *Methanothermaceae* are represented by only one genus, *Methanothermus*. The two species of this genus, *M. fervidus* and *M. sociabilis*, were isolated from near-neutral Icelandic solfataras and grow by H_2 oxidation coupled to CO_2 reduction with a T_{max} of 97°C (34).

Methanopyrus kandleri was isolated from a deep hydrothermal vent in the Gulf of California and has a similar physiology to members of the *Methanothermaceae* (47). It is capable of growing at a temperature that is hotter than any other methanogen and has a growth temperature range of 80 to 110°C utilizing H_2/CO_2. *M. kandleri* has the highest intracellular concentrations of cyclic 2,3-bisphosphoglycerate (1.1 M), and this may help to stabilize proteins against thermal denaturation (366). Consistent with the high concentration of this intracellular solute, most *Methanopyrus* proteins exhibit optimal activity at high-salt concentrations (>1 M). *M. kandleri* has an unusual membrane composition consisting of terpenoid lipids. The phylogeny of *Methanopyrus* has been debated, and it appears either to be a member of a distinct order, the *Methanopyrales*, or a fast-evolved *Methanobacteriales* species (see "Phylogeny of *Archaea* and the origin of life," above) (see Chapter 19). •

Methanococcales

The *Methanococcales* form a phylogenetically and physiologically coherent order of methanogens. For example, the similarity of the 16S rDNAs of the mesophilic *Methanococcus maripaludis* and the hyperthermophilic *Methanocaldococcus jannaschii* is 88% (430). All members are fast-growing (0.5 to 2 h doubling time), irregular cocci that require salt. They all utilize CO_2/H_2, while only a few species are able to use formate (but no other substrate) as electron donor. Similar to *M. maripaludis*, some *Methanococcales* fix atmospheric nitrogen. The cell wall is composed of a proteinaceous S layer, and most strains are motile. The greatest phenotypic diversity in this order relates to their T_{opt}.

The *Methanococcales* form two families with two genera in each (430). The *Methanocaldococcaceae* comprise the hyperthermophilic genera *Methanocaldococcus* and *Methanotorris*, while the mesophiles and moderate thermophiles are in the family *Methanococcaceae*. The genus *Methanococcus* includes mesophilic species, with thermophilic species in the genus *Methanothermococcus*. All isolates of *Methanococcales* have come from marine environment, such as *Methanococcus vannielii* from the San Francisco Bay and *M. voltae* from estuarine sediments in Florida. The moderately thermophilic *Methanothermococcaceae* were isolated from geothermally heated coastal sediments and from North Sea oil fields. The hyperthermophilic species within the genera *Methanocaldococcus* and *Methanotorris* are widespread in submarine hydrothermal systems.

Methanocaldococcus jannaschii, previously named *Methanococcus jannaschii*, has the distinction of being the first hyperthermophile to be isolated from deep-sea vents, and the first cultivated hyperthermophilic methanogen (174). It was also the first archaeon, and only the second microorganism, to have its complete genome sequenced (46). As a result, it has become a model organism, in particular, for structural biology. Numerous crystal structures have been determined for proteins with known and unknown functions, including the cell division protein FtsZ, the FLAP and splicing endonucleases, topoisomerases, translation initiation factors, DNA-binding proteins, ribosomal proteins, and many more (http://www.ncbi.nlm.nih.gov/Structure/). *M. jannaschii* was isolated from deep-sea vents with in situ hydrostatic pressure of 20 to 30 MPa and, as a result, has served as a model organism for studying the effects of pressure on growth, metabolism, and enzyme activities. Several enzymes (e.g., hydrogenases) have significantly increased half-lives at elevated pressure (97). Pressure has also been shown to exert a lipid-ordering effect (177).

Genetic systems have been developed for *Methanococcus voltae* and *M. maripaludis*, including methods for transformation, shuttle and expression vectors, antibiotic resistance markers, and reporter genes (reviewed in reference 407) (see Chapter 21). A phage transduction system has been described for *M. voltae* (81). The genome sequence is available from *M. maripaludis* (143). Mechanisms of gene regulation of motility, nitrogen fixation, and hydrogenases have been extensively studied in *M. voltae* (430). Four hydrogenase operons are present; two encode selenium-dependent enzymes containing selenocysteine residues at the active site NiFe cluster, and the others encode selenium-free enzymes with the canonical set of four cysteine residues coordinating a Ni ion. Both pairs include coenzyme F_{420}-reducing and F_{420}-non-reducing enzymes. The selenium-free hydrogenases are encoded by two operons divergently transcribed from a 453-bp intergenic region. Gene expression was only observed when selenium was absent in the growth medium and seems to be negatively and positively regulated from a silencer region and a binding site for a transcriptional activator (281).

Methanomicrobiales

Methanomicrobiales have a diverse morphology that includes rods, irregular cocci, and disc-shaped cells (109). They are also distinguished from other methanogens by their cell wall and lipid composition. All utilize H_2/CO_2, and some species may use alcohols and/or formate. *Methanomicrobiales* are generally distinguished from the *Methanobacteriales* by their proteinaceous S-layer cell wall, and from the *Methanosarcinales* by their inability to use acetate or organic C_1 compounds (other than formate) as substrates for methanogenesis. However, some species require acetate as an organic carbon source for assimilation. At least twenty-four species grouped into nine genera and three families have been described, highlighting the cultivatable diversity of this taxon. The *Methanomicrobiales* have not been as well studied as other methanogens. For example, only a single 3D structure is available (the luciferase-like, F_{420}-dependent, secondary alcohol dehydrogenase from *Methanoculleus thermophilicus*) (12). Genome sequence data are available for the psychrophile *Methanogenium frigidum* (348), and recently from *Methanospirillum hungatei* (GenBank: NC_007796).

Methanomicrobiales species have been isolated from sewage sludge digestors, marine and freshwater sediments, oil field reservoirs, and the rumen (109). Molecular ecology studies have identified *Methanomicrobium mobile* as a predominant methanogen in a wood-fermenting bioreactor, representing ~90% of the methanogen population (252). This predominance has not been observed in other types of digestors, which may reflect its requirement for specific wood substrates. *Methanomicrobiales* and *Methanosarcinales* have been found to represent the majority of methanogens in a study of 21 different digestors (318). *Methanomicrobiales* are the second most abundant methanogens typically found in ruminants, *Methanosarcinales* being the most abundant (e.g., 397). In the bovine rumen, *Methanomicrobium mobile* and *Methanobrevibacter* species were found in abundance (167). The psychrophile *Methanogenicum frigidum* was isolated from the cold hypolimnion of a seawater-derived Antarctic lake and grows between 1 and 17°C (Table 10) (51, 98). As a psychrophile, *M. frigidum* expands the temperature range for cultivated methanogens (98). In contrast, the phylogenetically related thermophile, *Methanoculleus thermophilicum*, was isolated from sediments of high-temperature effluent channels of a nuclear power plant (326).

A few species of *Methanomicrobiales* are endosymbionts of anaerobic protozoa that are associated with the hosts' hydrogenosome or cytoplasm. Hydrogenosome association suggests a hydrogen transfer to the methanogen (27, 84, 85). For example, the anaerobic protozoa *Metopus*, *Trimyema*, and *Pelomyxa* harbor endosymbionts related to *Methanocorpusculum* and *Methanoplanus* that are phylogenetically different from their free-living relatives (84).

Cells of *Methanospirillum hungatei* are rod shaped and show a unique ultrastructure when viewed with the electron microscope. Similar to other methanomicrobiales, they contain an S layer (109). However, individual *M. hungatei* cells often form chains that are further enclosed by a proteinaceous paracrystalline sheath. The *M. hungatei* sheath is resistant to denaturants, salts, or proteases, although it can be dissociated by reducing agents, suggesting that disulfide bridges may function to stabilize the structure. Twenty percent of the sheath's total mass consists of phenol-soluble proteins. A regular array of 2.8-nm-wide particles form the sheath (reviewed in reference 109) (see Chapter 14). The sheath can withstand pressures of 30 to 40 MPa (measured by atomic force microscopy). It has been speculated that the sheath may function as a pressure regulator, opening in response to high intracellular CH_4 pressures and allowing a gas exchange with H_2 and CO_2 (438). Within the sheath, highly permeable spacer plugs separate the cells at their poles, indicating that they might play a role in cell division (109).

Methanosarcinales

Methanosarcinales species are widespread in all types of anaerobic environments. Cells are either sheathed rods or coccoid and often in groups or clus-

Table 10. Characteristics of selected species of methanogens[a]

Order and species	Morphology	Source/habitat	Temperature optimum (°C)	pH optimum	Substrates for methanogenesis	Total no. of species in genus	Remarks[b]
Methanobacteriales and							
Methanobacterium							
Methanobacterium formicicum	Long rods	Anaerobic sludge digestors	30–45	7–7.5	H_2/CO_2; formate	13	
Methanobacterium bryantii	Long rods	Coculture with *Methanobacillus omelianski*	30–45	6.5–7.5	H_2/CO_2		
Methanobrevibacter ruminantium	Short rods	Bovine rumen	37–39	6.3–6.8	H_2/CO_2; formate	12	
Methanobrevibacter smithii	Coccoid, chains	Intestines of many animals, including humans	37	6.9–7.4	H_2/CO_2; formate		
Methanobrevibacter oralis	Short rods	Bovine rumen	35–38	6.3–6.8	H_2/CO_2		
Methanosphaera stadtmanae	Coccoid	Human feces	30–40	6.5–6.9	Methanol and H_2		G
Methanothermobacter marburgensis	Long rods	Anaerobic sludge digestors	65	7.2–7.6	H_2/CO_2	6	G
Methanothermobacter thermautotrophicus	Long rods	Anaerobic sludge digestors	65–70	6.7–7	H_2/CO_2; some strains on formate		Growth stimulated by tungstate
Methanothermobacter wolfei	Rods, sometimes coccoid	Anaerobic sludge digestors	n.r.[c]	n.r.	H_2/CO_2; formate		Presence of a periplasm
Methanothermus fervidus	Rods	Icelandic hot spring	80–85	6.5	H_2/CO_2	2	Temperature range 80–110°C, requires salt
Methanopyrus kandleri	Rods	Abyssal hot vents, Guaymas, and Kolbeinsey Ridge	95–98	≈6.5	H_2/CO_2	1	
Methanococcales							
Methanococcus vannielii	Cocci	Marine sediments	35–40	6.5	H_2/CO_2; formate	4	Motile, halophilic (≤5% salt)
Methanococcus voltae	Cocci	Marine sediments	35–40	6.5	H_2/CO_2; formate		Motile, halophilic (≤5% salt)
Methanococcus maripaludis	Cocci	Marine sediments	35–40	6.5	H_2/CO_2; formate		Motile, halophilic (≤5% salt), N_2 fixation
Methanothermococcus thermolithotrophicus	Cocci	Submarine vents	60–70	6.5–7.5	H_2/CO_2; formate	2	Motile, halophilic (≤9% salt, N_2 fixation (some)
Methanocaldococcus jannaschii	Cocci	White smoker East Pacific Rise	80–85	6–6.5	H_2/CO_2	5	G; motile, halophilic (≤5% salt)
Methanotorris igneus	Cocci	Kolbeinsey Ridge, Iceland	85–88	5.5–6	H_2/CO_2	2	Nonmotile, halophilic (≤7% salt)
Methanomicrobiales							
Methanocorpusculum parvum	Irregular cocci	Sour whey digestor	35–40	6.7–7.5	H_2/CO_2; formate; 2-propanol + CO_2	5	Motile, requires tungstate
Methanoculleus thermophilus	Irregular cocci	Sludge digestors, marine sediments	55–60	6.7–7.2	H_2/CO_2; formate	6	Motile
Methanofollis liminatans	Irregular cocci	Industrial wastewater digestor	40	7	H_2/CO_2; formate; CO_2 + 2-propanol or 2-butanol or cyclopentanol	5	Motile, tungstate stimulatory
Methanogenium frigidum	Irregular cocci	Ace Lake, Antarctica	15	7.5–7.9	H_2/CO_2; formate	3	G*, temperature 0–17°C, slightly halophilic

(Continued)

Table 10. *Continued*

Order and species	Morphology	Source/habitat	Temperature optimum (°C)	pH optimum	Substrates for methanogenesis	Total no. of species in genus	Remarks[b]
Methanolacinia paynteri	Irregular cocci	Marine sediments	40	7	H_2/CO_2; CO_2 + 2-propanol, 2-butanol, cyclopentanol	1	Highly irregular morphology
Methanomicrobium mobile	Slightly curved, short rods	Rumen fluid	40	6.1–6.7	H_2/CO_2; formate	1	
Methanoplanus endosymbiosus	Slightly curved, short rods	Marine ciliate *Metopus contortus*	32	6.8–7.3	H_2/CO_2; formate	3	Endosymbiont
Methanospirillum hungatei	Long curved rods	Various	30–37	6.6–7.4	H_2/CO_2; formate; CO_2 + 2-propanol, 2-butanol	1	G, flagella and sheaths
Methanocalculus halotolerans	Coccoid	Oil-producing well	38	7.6	H_2/CO_2; formate	4	Halophile, salt concentration 0–12%
Methanosarcinales							
Methanosaeta concilii	Straight rods	Anaerobic sludge digestors	35–40	7.1–7.4	Acetate	2	
Methanomicrococcus blatticola	Flat polygonal spheroids	Cockroach hindgut	39	7.2–7.7	H_2 + methanol or methylamines	1	
Methanothrix thermophila	Coccoid, aggregates	Khlorid Lake, Kamchatka, Russia	50	7.5	Acetate, methanol, methylamines	1	
Methanococcoides burtonii	Irregular cocci, aggregates	Ace Lake, Antarctica	23	7.7	Methanol, methylamines	3	G*, motile
Methanohalobium evestigatum	Flat polygonal spheroids, aggregates	Salt lagoons, Sivash, Ukraine	50	7–7.5	Methylamines	1	Extremely halophilic: optimal NaCl concentration 4.2 M
Methanohalophilus mahii	Irregular cocci, aggregates	Great Salt Lake, Utah, USA	35	7.5	Methanol, methylamines	4	Halophilic: optimal NaCl concentration 2 M
Methanolobus oregonensis	Irregular cocci, aggregates	Alkaline, saline aquifer from Oregon, USA	35	8.6	Methanol, methyl sulfides, methylamines	5	
Methanosalsum zhilinae	Irregular cocci, aggregates	Wadi el Natrun, Egypt	45	9.2	Methanol, methylsulfides, methylamines	1	Motile, moderately halophilic: optimal NaCl concentration 0.7 M
Methanosarcina acetivorans	Cocci, aggregates	Marine sediments	35–40	6.5–7	Acetate; methanol, methylamines	9	
Methanosarcina barkeri	Cocci, aggregates	Anaerobic sludge digestors	45	7.0	H_2/CO_2; acetate; methylamines, methanol		G, gas vesicles
Methanosarcina mazei	Cocci, aggregates	Anaerobic sludge digestors	40–42	6.8–7.2	H_2/CO_2; acetate; methylamines, methanol		G
Methanosarcina vacuolata	Cocci, aggregates	Anaerobic sludge digestors	40	7.5	Acetate		Gas vesicles
Methanomethylovorans hollandica	Cocci, aggregates	Eutrophic lake, The Netherlands	37	6.5–7.0	Methanol, dimethylsulfide, methylamines, methane thiol		Gas vesicles

76

[a]Data compiled from several sources (34, 109, 188, 430), including the taxonomy browser (http://www.ncbi.nih.gov) and the German culture collection (http://www.dsmz.de). All currently recognized genera are represented but not all species.
[b]G, genome sequence available; G*, draft genome sequence.
[c]n.r., not reported.

ters. They usually have a proteinaceous cell wall, while some are additionally surrounded by a sheath or acidic heteropolysaccharide; pseudomurein is not present (reviewed in reference 188). All acetoclastic methanogens are species of *Methanosarcinales* (65). This order represents methanogens, which are capable of utilizing the widest range of substrates (CO_2/H_2, disproportionation of C_1 compounds or methylated amines, chloroform, methyl sulfides, and acetate), and some individual strains can use most of these substrates (188).

The two *Methanosarcinales* families, the *Methanosarcinaceae* and the *Methanosaetaceae*, are distinguished by morphological and physiological properties. The *Methanosarcinaceae* have a G+C content of 35 to 46% and form coccoid or pseudosarcinal cells; many are motile and can use a wide range of substrates. The *Methanosarcinaceae* include halophilic genera isolated from hypersaline environments. *Methanosaeta concilii* and *Methanosaeta thermophila*, the only two species of the *Methanosaetaceae*, are nonmotile sheathed rods. They use only acetate for methanogenesis. Their G+C content is 49 to 54%. *Methanosaeta* strains often outcompete *Methanosarcina* spp. in environments where acetate concentrations are low due to their transport systems possessing a low k_m for the substrate. In contrast, *Methanosarcina* strains often predominate at low-pH values and high-acetate concentrations (188). The nine cultivatable *Methanosarcina* species are slightly halotolerant and/or slightly halophilic and are nonmotile. They metabolize acetate, methylamines, methanol, and CO. Some strains also grow on CO_2/H_2 (188), and several can dechlorinate chloroform and other highly chlorinated small alkans (9, 49, 149). Some species, such as *Msc. vacuolatum* and a number of *Methanosarcina mazei* strains, form gas vesicles (36). Gas vesicle synthesis genes are present in the genome sequence of *Methanosarcina barkeri* (Fig. 14) but not the *M. mazei* strains. At least four species were isolated from anaerobic digestors, where they are particularly common (188). Knowledge of the metabolic diversity of the *Methanosarcinales* was recently improved, and *Methanosarcina acetivorans* strain C2A was found to form acetate and formate, rather than methane, as the major metabolic end products for energy conservation when growing on CO (333). Consistent with an inhibitory effect of CO, methane production was found to decrease with increasing CO concentration.

Methanosarcina barkeri, *M. mazei*, and *M. acetivorans* are the most well studied methanogens and are important model organisms for studies on acetoclastic methanogenesis, transcription, and chaperonins (65, 146, 251). Genome sequences are available

from *M. barkeri* (GenBank: NC_007355), *M. acetivorans* C2A (GenBank: NC_003552), and *M. mazei* Gö-1 (GenBank: NC_003901).

Other genera of the *Methanosarcinaceae* are moderately halophilic, growing on organic C_1 compounds at the salinity of seawater (188). *Methanolobus* and *Methanococcoides* species were isolated from sea sediments or from terrestrial salt lakes. Similar to *Methanogenium frigidum*, *Methanococcoides burtonii* was isolated from the anoxic hypolimnion of Ace Lake, Antarctica, and is a psychrophile. The genome sequence was recently completed, and the organism serves as a model for cold adaptation (51, 121, 348). In contrast, *Methanohalophilus* and *Methanosalsum* species have been isolated from hypersaline lakes or soda lakes, respectively (188). They grow at optimal salt concentrations of 0.7 to 1.5 M (Table 10). *Methanosalsum zhiliniae* and *Methanolobus orogensis* are the only *Methanosarcinales* isolates with an alkaline pH optimum (9.2 and 8.6, respectively), and other members of this family are neutrophiles. In contrast, *Methanohalobium evestigatum* is a moderately thermophilic halophile with an optimal salt concentration of 4.3 M, a T_{opt} of 50°C that is found in hypersaline, NaCl-saturated environments. *M. evestigatum* grows on methylamines, while other, more moderately halophilic, methanogens (e.g., *Methanococcoides burtonii*) can also use methanol.

(Anaerobic) methane oxidation by methanotrophic archaea

Numerous aerobic bacteria oxidize methane using a methyl oxygenase for the primary activation of the substrate. Alternatively, several cultivated methylotrophic bacteria, such as *Methylococcus capsulatus*, have a reversed methanogenesis pathway for methane oxidation (55). A third means of methane oxidation is AOM, which is carried out anaerobically by uncultivated methylotrophic archaea that are closely related to the *Methanosarcinales* (reviewed in reference 365). Several phylogenetically distinct lineages that are associated with AOM have been identified (Fig. 1). They appear in situ to be in a syntrophic association with sulfate-reducing δ-proteobacteria. The association is required to support metabolism by an extracellular electron transfer and to keep the free-energy change below the limits of -10 kJ mol^{-1} for methanogenic archaea and -19 kJ mol^{-1} for sulfate-reducing bacteria.

The AOM pathway involves enzymes from methanogenesis, including a homolog of the methyl CoM reductase (365). The Black Sea microbial mats are composed of >50% of archaea from the ANME 1 cluster. In contrast to the open ocean, the water body of the Black Sea is anoxic at shallow depth. Large amounts of methane are liberated in the sediments, which allows the development of large, slow-growing pillars that consist of microbial mats of AOM archaea and sulfate-reducing proteobacteria (262, 276). In the mats, a homolog of methyl-CoM reductase represents up to 7% of total protein (218). The reductase purified directly from mat samples primarily catalyzed the reverse reaction (oxidation of methane to methyl-CoM). The enzyme had comparable biochemical properties to the canonical methanogen enzymes, with exception that the Ni-containing F_{430} cofactor of the enzyme contained an unknown chemical modification that increased its molecular mass from 905 Da to 951 Da. The Black Sea mats performed anaerobic, sulfate-dependent AOM in the laboratory, and methanogenesis with different substrates did not occur. These results are strong evidence that the mats carry out AOM rather than methanogenesis (218, 365).

Very recently, AOM was found to occur in anaerobic freshwater ecosystems within methane-producing sediments (316). A consortium consisting of methane-oxidizing archaea and novel bacteria was retrieved from a canal in the Netherlands and was cultivated successfully in the laboratory by coupling AOM to denitrification. The *Bacteria* belong to a phylogenetic branch, which had not been isolated before and was known solely from molecular ecological studies. These results suggest that AOM is much more widespread than anticipated and that it could be coupled to a variety of anaerobic respiration processes.

PERSPECTIVE: THE NEXT FIVE YEARS

Research on *Archaea* has generated impressive results in the almost 30 years that have passed since their recognition as a fundamentally different domain of life. Research has focused on the discovery and descriptions of novel organisms, the search for *Archaea* in habitats other than hot springs, salt lakes, and anaerobic sediments, the determination of their fundamental cellular processes, and the determination of their major physiological pathways. An increasing number of *Archaea* have become amenable to genetic techniques, although a great scope exists for this to expand. Commercial applications have also surfaced, the most obvious being the proofreading DNA polymerases used for PCR.

Areas where significant progress may be expected in the next five years include:

Isolation. It is likely that an increasing number of nonextremophilic and/or nonmethanogenic isolates will be obtained. The novel strains will expand understanding of the range of archaeal metabolic diversity and biochemical targets for enzymological studies. The isolates should begin to replace the uncultivated species that have been identified from molecular ecology, and particularly large-scale metagenomic studies, and add culturable *Archaea* to the "orphan branches" of the phylogenetic tree (Fig. 1).

Consortia. In addition to attempting to isolate individual species of *Archaea*, focus is required on studying symbiotic and other mutualistic consortia of microorganisms that can only be grown in coculture or are not amenable to any form of laboratory-based cultivation. The Iron Mountain acid mine ecosystem (408) and the consortia of denitrifying anaerobic methane oxidizers (316) are examples of recently identified consortia that contain novel groups of *Archaea*. Realization of the importance of these systems is likely to stimulate future research of other types of archaeal consortia.

Metagenomics. Metagenomic approaches have been responsible for dramatically expanding knowledge of microbial diversity and producing the highly diverged phylogenetic tree that, in particular, highlights the number of uncultivated species (Fig. 1) (353). The future will add an increasing number of metagenomes, which should also aid in the isolation of culturable strains of *Archaea* and *Bacteria*. Metagenomics is likely to enable complete archaeal genomes to be constructed without having isolated, or even recognized, the existence of the microorganism in the environment. Fluorescence staining, stable isotope probing, and optical tweezers techniques could make it possible to identify and isolate previously unknown species from environmental samples. One consequence of these studies will be the generation of an expanded database of commercially and scientifically valuable enzymes that will be able to be identified using increasingly powerful bioinformatic tools.

Structural biology. The availability of archaeal genome sequences has triggered a large increase in protein structural data (see Chapter 20). Individual research groups and coordinated structural genomics programs have both played important roles. The structural genomics programs have passed the inevitable early-lag phase and are presently generating large volumes of data, and this is set to increase significantly in the near future. In particular, hyper/thermophilic archaea (and bacteria) are well suited as sources of proteins for structural studies. With improvements in high-throughput robots for crystallization using extremely small volumes of proteins, archaeal targets will be able to be screened at increasingly higher speeds. The result of this archaeal research will be to significantly improve the fundamental understanding of cellular processes not only in the *Archaea*, but also in all other forms of life.

Modeling. One of the current trends in systems biology is the attempt to model and predict metabolic fluxes within microorganisms and in ecosystems, signal transduction networks in large regulons and whole cells, and in vivo protein-protein interactions. These types of approaches will be enormously helpful in devising hypotheses about all aspects of cell biology and ecology, thereby enabling predictions to be tested experimentally. *Archaea* have been and will continue to be useful models for a broad variety of systems biology studies.

Cell biology. Central processes of replication, transcription, and translation have been elucidated to a greater or lesser degree in all organisms. However, important details of these processes remain to be determined (see Chapters 3, 6, and 8), and a great deal remains to be discovered about other cellular processes. It is likely that the speed in which knowledge is gained will continue to increase. The *Archaea* will remain model organisms for studying the more complicated eucaryal cellular systems. Molecular interaction studies will help to resolve the protein-protein interaction networks that occur throughout the archaeal cell cycle, and cryotomography will be a valuable tool for reconstructing high-resolution images of the living cell.

Physiology. As physiological studies of the *Archaea* expand to the same degree as has occurred with *Bacteria*, the extent of metabolic capacity and the pathways of metabolism will begin to be realized (see Chapter 12), and will illustrate the diversity of physiological types that exist within the archaeal world.

Acknowledgments. Special thanks are due to Christa Schleper and Melanie Jonuscheit (Bergen, Norway), who gave me the project and without whom this chapter would not have been possible. Special thanks are also due to Rick Cavicchioli for his excellent editing of the chapter and for his patience. Further thanks are due to Felicitas Pfeifer (Darmstadt, Germany) for critically reading the manuscript and for the some of the *Halobacteriales* figures and to Torsten Hechler (Darmstadt) for the phase-contrast and EM pictures of the *Halobacteriales*. This work was supported by funds from the Deutsche Forschungsgemeinschaft (Kl-885/3-3) and the German Bundesministerium für Bildung und Forschung (Förderkennzeichen 0313106 and 6408415).

REFERENCES

1. Abreu, I. A., A. V. Xavier, J. LeGall, D. E. Cabelli, and M. Teixeira. 2002. Superoxide scavenging by neelaredoxin: dismutation and reduction activities in anaerobes. *J. Biol. Inorg. Chem.* 7:668–674.

2. Achenbach-Richter, L., K. O. Stetter, and C. R. Woese. 1987. A possible biochemical missing link among archaebacteria. *Nature* 327:348–349.

3. Adams, M. W., J. F. Holden, A. L. Menon, G. J. Schut, A. M. Grunden, C. Hou, A. M. Hutchins, F. E. Jenney, Jr., C. Kim, K. Ma, G. Pan, R. Roy, R. Sapra, S. V. Story, and M. F. Verhagen. 2001. Key role for sulfur in peptide metabolism and in regulation of three hydrogenases in the hyperthermophilic archaeon *Pyrococcus furiosus*. *J. Bacteriol.* 183:716–724.

4. Afshar, S., E. Johnson, S. de Vries, and I. Schroder. 2001. Properties of a thermostable nitrate reductase from the hyperthermophilic archaeon *Pyrobaculum aerophilum*. *J. Bacteriol.* 183:5491–5495.

5. Albers, S. V., M. Jonuscheit, S. Dinkelaker, T. Urich, A. Kletzin, R. Tampe, A. J. Driessen, and C. Schleper. 2006. Production of recombinant and tagged proteins in the hyperthermophilic archaeon *Sulfolobus solfataricus*. *Appl. Environ. Microbiol.* 72:102–111.

6. Amend, J. P., D. R. Meyer-Dombard, S. N. Sheth, N. Zolotova, and A. C. Amend. 2003. *Palaeococcus helgesonii* sp. nov., a facultatively anaerobic, hyperthermophilic archaeon from a geothermal well on Vulcano Island, Italy. *Arch. Microbiol.* 179:394–401.

7. Amend, J. P., and E. L. Shock. 2001. Energetics of overall metabolic reactions of thermophilic and hyperthermophilic Archaea and bacteria. *FEMS Microbiol. Rev.* 25:175–243.

8. Anderson, J. C., N. Wu, S. W. Santoro, V. Lakshman, D. S. King, and P. G. Schultz. 2004. An expanded genetic code with a functional quadruplet codon. *Proc. Natl. Acad. Sci. USA* 101:7566–7571.

9. Andrews, E. J., and P. J. Novak. 2001. Influence of ferrous iron and ph on carbon tetrachloride degradation by *Methanosarcina thermophila*. *Water Res.* 35:2307–2313.

10. Angelov, A., O. Futterer, O. Valerius, G. H. Braus, and W. Liebl. 2005. Properties of the recombinant glucose/galactose dehydrogenase from the extreme thermoacidophile, *Picrophilus torridus*. *FEBS J.* 272:1054–1062.

11. Atomi, H., R. Matsumi, and T. Imanaka. 2004. Reverse gyrase is not a prerequisite for hyperthermophilic life. *J. Bacteriol.* 186:4829–4833.

12. Aufhammer, S. W., E. Warkentin, H. Berk, S. Shima, R. K. Thauer, and U. Ermler. 2004. Coenzyme binding in F_{420}-dependent secondary alcohol dehydrogenase, a member of the bacterial luciferase family. *Structure* 12:361–370.

13. Balashov, S. P., E. S. Imasheva, V. A. Boichenko, J. Anton, J. M. Wang, and J. K. Lanyi. 2005. Xanthorhodopsin: a proton pump with a light-harvesting carotenoid antenna. *Science* 309:2061–2064.

14. Balch, W. E., G. E. Fox, L. J. Magrum, C. R. Woese, and R. S. Wolfe. 1979. Methanogens: Reevaluation of a unique phylogenetic group. *Microbiol. Rev.* 43:260–296.

15. Balch, W. E., L. J. Magrum, G. E. Fox, R. S. Wolfe, and C. R. Woese. 1977. An ancient divergence among the bacteria. *J. Mol. Evol.* 9:305–311.

16. Bandeiras, T. M., C. Salgueiro, A. Kletzin, C. M. Gomes, and M. Teixeira. 2002. *Acidianus ambivalens* type-II NADH dehydrogenase: genetic characterisation and identification of the flavin moiety as FMN. *FEBS Lett.* 531:273–277.

17. Bandeiras, T. M., C. A. Salgueiro, H. Huber, C. M. Gomes, and M. Teixeira. 2003. The respiratory chain of the ther-

mophilic archaeon *Sulfolobus metallicus*: studies on the type-II NADH dehydrogenase. *Biochim. Biophys. Acta* **1557:** 13–19.

18. Bardy, S. L., S. Y. Ng, and K. F. Jarrell. 2004. Recent advances in the structure and assembly of the archaeal flagellum. *J. Mol. Microbiol. Biotechnol.* **7:**41–51.

19. Barns, S. M., C. F. Delwiche, J. D. Palmer, and N. R. Pace. 1996. Perspectives on archaeal diversity, thermophily and monophyly from environmental rRNA sequences. *Proc. Natl. Acad. Sci. USA* **93:**9188–9193.

20. Bartig, D., H. Schümann, and F. Klink. 1990. The unique post-translational modification leading to deoxyhypusine or hypusine is a general feature of the archaebacterial kingdom. *Syst. Appl. Microbiol.* **13:**112–116.

21. Bartlett, M. S. 2005. Determinants of transcription initiation by archaeal RNA polymerase. *Curr. Opin. Microbiol.* **8:** 677–684.

22. Beja, O., L. Aravind, E. V. Koonin, M. T. Suzuki, A. Hadd, L. P. Nguyen, S. B. Jovanovich, C. M. Gates, R. A. Feldman, J. L. Spudich, E. N. Spudich, and E. F. DeLong. 2000. Bacterial rhodopsin: evidence for a new type of phototrophy in the sea. *Science* **289:**1902–196.

23. Beja, O., E. N. Spudich, J. L. Spudich, M. Leclerc, and E. F. DeLong. 2001. Proteorhodopsin phototrophy in the ocean. *Nature* **411:**786–789.

24. Bell, S. D. 2005. Archaeal transcriptional regulation—variation on a bacterial theme? *Trends Microbiol.* **13:**262–265.

25. Bell, S. D., and S. P. Jackson. 1998. Transcription and translation in Archaea: a mosaic of eukaryal and bacterial features. *Trends Microbiol.* **6:**222–228.

26. Bell, S. D., C. P. Magill, and S. P. Jackson. 2001. Basal and regulated transcription in Archaea. *Biochem. Soc. Trans.* **29:**392–395.

27. Berger, J., and D. H. Lynn. 1992. Hydrogenosome-methanogen assemblages in the echinoid endocommensal plagiopylid ciliates, *Lechriopyla mystax* Lynch, 1930 and *Plagiopyla minuta* Powers, 1933. *J. Protozool.* **39:**4–8.

28. Bertoldo, C., and G. Anthranikian. 2004. The order Thermococcales. *In* M. Dworkin (ed.), *The Prokaryotes: An Evolving Electronic Resource for the Microbiological Community*, 3rd ed., release 3.15 ed. Springer-Verlag, New York, N.Y. [Online.] http://link.springer-ny.com/link/service/books/10125/.

29. Betlach, M., J. Friedman, H. W. Boyer, and F. Pfeifer. 1984. Characterization of a halobacterial gene affecting bacterioopsin gene expression. *Nucleic Acids Res.* **12:**7949–7959.

30. Bienkowska, J. R., H. Hartman, and T. F. Smith. 2003. A search method for homologs of small proteins. Ubiquitin-like proteins in prokaryotic cells? *Protein Eng.* **16:**897–904.

31. Black, F. T., E. A. Freundt, O. Vinther, and Christiansen. 1979. Flagellation and swimming motility of *Thermoplasma acidophilum*. *J. Bacteriol.* **137:**456–460.

32. Blaseio, U., and F. Pfeifer. 1990. Transformation of *Halobacterium halobium*: development of vectors and investigation of gas vesicle synthesis. *Proc. Natl. Acad. Sci. USA* **87:** 6772–6776.

33. Bolhuis, H., E. M. Poele, and F. Rodriguez-Valera. 2004. Isolation and cultivation of Walsby's square archaeon. *Environ. Microbiol.* **6:**1287–1291.

34. Bonin, A. S., and D. R. Boone. 2004. The order Methanobacteriales. *In* M. Dworkin (ed.), *The Prokaryotes: An Evolving Electronic Resource for the Microbiological Community*, 3rd ed., release 3.16 ed. Springer-Verlag, New York, N.Y. [Online.] http://141.150.157.117:8080/prokPUB/index.htm.

35. Bonisch, H., C. L. Schmidt, G. Schafer, and R. Ladenstein. 2002. The structure of the soluble domain of an archaeal Rieske iron-sulfur protein at 1.1 A resolution. *J. Mol. Biol.* **319:**791–805.

36. Boone, D. R., and R. A. Mah. 2001. *Methanosarcina*, p. 269–276. *In* D. R. Boone and R. W. Castenholz (ed.), *Bergey's Manual of Systematic Bacteriology*, 2nd ed. vol. 1. Springer, New York, N.Y.

37. Borissenko, L., and M. Groll. 2005. Crystal structure of TET protease reveals complementary protein degradation pathways in prokaryotes. *J. Mol. Biol.* **346:**1207–1219.

38. Boucher, Y., H. Huber, S. L'Haridon, K. O. Stetter, and W. F. Doolittle. 2001. Bacterial origin for the isoprenoid biosynthesis enzyme HMG-CoA reductase of the archaeal orders Thermoplasmatales and Archaeoglobales. *Mol. Biol. Evol.* **18:** 1378–1388.

39. Boucher, Y., M. Kamekura, and W. F. Doolittle. 2004. Origins and evolution of isoprenoid lipid biosynthesis in archaea. *Mol. Microbiol.* **52:**515–527.

40. Brochier, C., P. Forterre, and S. Gribaldo. 2004. Archaeal phylogeny based on proteins of the transcription and translation machineries: tackling the *Methanopyrus kandleri* paradox. *Genome Biol.* **5:**R17.

41. Brochier, C., P. Forterre, and S. Gribaldo. 2005. An emerging phylogenetic core of Archaea: phylogenies of transcription and translation machineries converge following addition of new genome sequences. *BMC Evol. Biol.* **5:**36.

42. Brochier, C., S. Gribaldo, Y. Zivanovic, F. Confalonieri, and P. Forterre. 2005. Nanoarchaea: representatives of a novel archaeal phylum or a fast-evolving euryarchaeal lineage related to Thermococcales? *Genome Biol.* **6:**R42.

43. Brochier, C., P. Lopez-Garcia, and D. Moreira. 2004. Horizontal gene transfer and archaeal origin of deoxyhypusine synthase homologous genes in bacteria. *Gene* **330:**169–176.

44. Brock, T. D., K. M. Brock, R. T. Belly, and R. L. Weiss. 1972. *Sulfolobus*: A new genus of sulfur- oxidizing bacteria living at low pH and high temperature. *Arch. Microbiol.* **84:**54–68.

45. Brown, S. H., and R. M. Kelly. 1993. Characterization of amylolytic enzymes, having both alpha-1,4 and alpha-1,6 hydrolytic activity, from the thermophilic archaea *Pyrococcus furiosus* and *Thermococcus litoralis*. *Appl. Environ. Microbiol.* **59:**2614–2621.

46. Bult, C. J., O. White, G. J. Olsen, L. Zhou, R. D. Fleischmann, G. G. Sutton, J. A. Blake, L. M. FitzGerald, R. A. Clayton, J. D. Gocayne, A. R. Kerlavage, B. A. Dougherty, J. F. Tomb, M. D. Adams, C. I. Reich, R. Overbeek, E. F. Kirkness, K. G. Weinstock, J. M. Merrick, A. Glodek, J. L. Scott, N. S. M. Geoghagen, and J. C. Venter. 1996. Complete genome sequence of the methanogenic archaeon, *Methanococcus jannaschii*. *Science* **273:**1058–1073.

47. Burggraf, S., K. O. Stetter, P. Rouviere, and C. R. Woese. 1991. *Methanopyrus kandleri*—an archael methanogen unrelated to all other known methanogens. *Syst. Appl. Microbiol.* **14:**346–351.

48. Burns, D. G., H. M. Camakaris, P. H. Janssen, and M. L. Dyall-Smith. 2004. Cultivation of Walsby's square haloarchaeon. *FEMS Microbiol. Lett.* **238:**469–473.

49. Cabirol, N., R. Villemur, J. Perrier, F. Jacob, B. Fouillet, and P. Chambon. 1998. Isolation of a methanogenic bacterium, *Methanosarcina* sp. strain FR, for its ability to degrade high concentration of perchloroethylene. *Can. J. Microbiol.* **44:** 1142–1147.

50. Cavalier-Smith, T. 2002. The neomuran origin of archaebacteria, the negibacterial root of the universal tree and bacterial megaclassification. *Int. J. Syst. Evol. Microbiol.* **52:**7–76.

51. Cavicchioli, R. 2006. Cold-adapted Archaea. *Nat. Rev. Microbiol.* **4:**331–343.

52. Cavicchioli, R., P. M. Curmi, N. Saunders, and T. Thomas. 2003. Pathogenic archaea: do they exist? *Bioessays* **25**:1119–1128.

53. Chan, M. K., S. Mukund, A. Kletzin, M. W. W. Adams, and D. C. Rees. 1995. The 2.3 Å resolution structure of the tungstoprotein aldehyde ferredoxin oxidoreductase from the hyperthermophilic archaeon *Pyrococcus furiosus*. *Science* **267**:1463–1469.

54. Chen, Z. W., C. Y. Jiang, Q. She, S. J. Liu, and P. J. Zhou. 2005. Key role of cysteine residues in catalysis and subcellular localization of sulfur oxygenase-reductase of *Acidianus tengchongensis*. *Appl. Environ. Microbiol.* **71**:621–628.

55. Chistoserdova, L., J. A. Vorholt, and M. E. Lidstrom. 2005. A genomic view of methane oxidation by aerobic bacteria and anaerobic archaea. *Genome Biol.* **6**:208.

56. Collins, R. E., and X. Cheng. 2005. Structural domains in RNAi. *FEBS Lett.* **579**:5841–5849.

57. Condo, I., A. Ciammaruconi, D. Benelli, D. Ruggero, and P. Londei. 1999. Cis-acting signals controlling translational initiation in the thermophilic archaeon *Sulfolobus solfataricus*. *Mol. Microbiol.* **34**:377–384.

58. Corre, E. 2000. Approaches moléculairs de la diversité microbienne de deux environnements extrêmes: les sources hydrothermales profondes et les réservoirs pétroliers. Doctoral thesis, p. 86–87. Université de Bretange Occidentale, Brest.

59. Costas, M., M. P. Mehn, M. P. Jensen, and L. Que, Jr. 2004. Dioxygen activation at mononuclear nonheme iron active sites: enzymes, models, and intermediates. *Chem. Rev.* **104**:939–986.

60. Cubonova, L., K. Sandman, S. J. Hallam, E. F. Delong, and J. N. Reeve. 2005. Histones in crenarchaea. *J. Bacteriol.* **187**:5482–5485.

61. Dahl, C., M. Molitor, and H. G. Trüper. 2001. Siroheme-sulfite reductase-type protein from *Pyrobaculum islandicum*. *Methods Enzymol.* **331**:410–419.

62. Dahl, C., and H. G. Trüper. 2001. Sulfite reductase and APS reductase from *Archaeoglobus fulgidus*. *Methods Enzymol.* **331**:427–441.

63. Daniel, R. M., and D. A. Cowan. 2000. Biomolecular stability and life at high temperatures. *Cell. Mol. Life Sci.* **57**:250–264.

64. Deatherage, J. F., K. A. Taylor, and L. A. Amos. 1983. Three-dimensional arrangement of the cell wall protein of *Sulfolobus acidocaldarius*. *J. Mol. Biol.* **167**:823–848.

65. Deppenmeier, U. 2004. The membrane-bound electron transport system of *Methanosarcina* species. *J. Bioenerg. Biomembr.* **36**:55–64.

66. de Poorter, L. M., W. G. Geerts, A. P. Theuvenet, and J. T. Keltjens. 2003. Bioenergetics of the formyl-methanofuran dehydrogenase and heterodisulfide reductase reactions in *Methanothermobacter thermautotrophicus*. *Eur. J. Biochem.* **270**:66–75.

67. Desmarais, D., P. E. Jablonski, N. S. Fedarko, and M. F. Roberts. 1997. 2-Sulfotrehalose, a novel osmolyte in haloalkaliphilic archaea. *J. Bacteriol.* **179**:3146–3153.

68. de Vendittis, E., M. R. Amatruda, G. Raimo, and V. Bocchini. 1997. Heterologous expression in *Escherichia coli* of the gene encoding an archaeal thermoacidophilic elongation factor 2. Properties of the recombinant protein. *Biochimie* **79**:303–308.

69. de Vries, S., and I. Schroder. 2002. Comparison between the nitric oxide reductase family and its aerobic relatives, the cytochrome oxidases. *Biochem. Soc. Trans.* **30**:662–667.

70. de Vries, S., M. J. Strampraad, S. Lu, P. Moenne-Loccoz, and I. Schroder. 2003. Purification and characterization of the MQH2:NO oxidoreductase from the hyperthermophilic archaeon *Pyrobaculum aerophilum*. *J. Biol. Chem.* **278**:35861–35868.

71. Dirmeier, R., M. Keller, G. Frey, H. Huber, and K. O. Stetter. 1998. Purification and properties of an extremely thermostable membrane-bound sulfur-reducing complex from the hyperthermophilic *Pyrodictium abyssi*. *Eur. J. Biochem.* **252**:486–491.

72. Doolittle, W. F. 1999. Phylogenetic classification and the universal tree. *Science* **284**:2124–2129.

73. Eckburg, P. B., E. M. Bik, C. N. Bernstein, E. Purdom, L. Dethlefsen, M. Sargent, S. R. Gill, K. E. Nelson, and D. A. Relman. 2005. Diversity of the human intestinal microbial flora. *Science* **308**:1635–1638.

74. Eder, W., W. Ludwig, and R. Huber. 1999. Novel 16S rRNA gene sequences retrieved from highly saline brine sediments of kebrit deep, red Sea. *Arch. Microbiol.* **172**:213–218.

75. Edmonds, B. W., and H. Luecke. 2004. Atomic resolution structures and the mechanism of ion pumping in bacteriorhodopsin. *Front. Biosci.* **9**:1556–1566.

76. Edwards, K. J., P. L. Bond, T. M. Gihring, and J. F. Banfield. 2000. An archaeal iron-oxidizing extreme acidophile important in acid mine drainage. *Science* **287**:1796–1799.

77. Edwards, K. J., M. O. Schrenk, R. Hamers, and J. F. Banfield. 1998. Microbial oxidation of pyrite: Experiments using microorganisms from an extreme acidic enviroment. *Am. Mineral.* **83**:1444–1453.

78. Egorova, K., and G. Antranikian. 2005. Industrial relevance of thermophilic Archaea. *Curr. Opin. Microbiol.* **8**:649–655.

79. Eichler, J. 2003. Facing extremes: archaeal surface-layer (glyco)proteins. *Microbiology* **149**:3347–3351.

80. Eisenreich, W., A. Bacher, D. Arigoni, and F. Rohdich. 2004. Biosynthesis of isoprenoids via the non-mevalonate pathway. *Cell. Mol. Life Sci.* **61**:1401–1426.

81. Eiserling, F., A. Pushkin, M. Gingery, and G. Bertani. 1999. Bacteriophage-like particles associated with the gene transfer agent of *Methanococcus voltae* PS. *J. Gen. Virol.* **80**:3305–3308.

82. El Karkouri, K., H. Gueune, and C. Delamarche. 2005. MIPDB: a relational database dedicated to MIP family proteins. *Biol. Cell.* **97**:535–543.

83. Elshahed, M. S., K. N. Savage, A. Oren, M. C. Gutierrez, A. Ventosa, and L. R. Krumholz. 2004. *Haloferax sulfurifontis* sp. nov., a halophilic archaeon isolated from a sulfide- and sulfur-rich spring. *Int. J. Syst. Evol. Microbiol.* **54**:2275–2279.

84. Embley, T. M., and B. J. Finlay. 1993. Systematic and morphological diversity of endosymbiotic methanogens in anaerobic ciliates. *Antonie Leeuwenhoek* **64**:261–271.

85. Embley, T. M., B. J. Finlay, and S. Brown. 1992. RNA sequence analysis shows that the symbionts in the ciliate *Metopus contortus* are polymorphs of a single methanogen species. *FEMS Microbiol. Lett.* **97**:57–62.

86. Emmel, T., W. Sand, W. A. König, and E. Bock. 1986. Evidence for the existence of a sulfur oxygenase in *Sulfolobus brierleyi*. *J. Gen. Microbiol.* **132**:3415–3420.

87. Englert, C., G. Wanner, and F. Pfeifer. 1992. Functional analysis of the gas vesicle gene cluster of the halophilic archaeon *Haloferax mediterranei* defines the vac-region boundary and suggests a regulatory role for the *gvpD* gene or its product. *Mol. Microbiol.* **6**:3543–3550.

88. Faguy, D. M., D. P. Bayley, A. S. Kostyukova, N. A. Thomas, and K. F. Jarrell. 1996. Isolation and characterization of flagella and flagellin proteins from the Thermoacidophilic archaea *Thermoplasma volcanium* and *Sulfolobus shibatae*. *J. Bacteriol.* **178**:902–905.

89. Ferrer, M., O. V. Golyshina, T. N. Chernikova, A. N. Khachane, V. A. Martins Dos Santos, M. M. Yakimov, K. N. Timmis, and P. N. Golyshin. 2005. Microbial enzymes mined from the Urania deep-sea hypersaline anoxic basin. *Chem. Biol.* 12:895–904.

90. Ferrer, M., O. V. Golyshina, F. J. Plou, K. N. Timmis, and P. N. Golyshin. 2005. A novel alpha-glucosidase from the acidophilic archaeon *Ferroplasma acidiphilum* strain Y with high transglycosylation activity and an unusual catalytic nucleophile. *Biochem. J.* 391:269–276.

91. Fiala, G., and K. O. Stetter. 1986. *Pyrococcus furiosus sp. nov.* represents a novel genus of marine heterotrophic archaebacteria growing optimally at 100°C. *Arch. Microbiol.* 145:56–61.

92. Fischer, F., W. Zillig, K. O. Stetter, and G. Schreiber. 1983. Chemolithoautotropic metabolism of anaerobic extremely thermophilic archaebacteria. *Nature* 301:511–513.

93. Forster, A., E. I. Masters, F. G. Whitby, H. Robinson, and C. P. Hill. 2005. The 1.9 A structure of a proteasome-11S activator complex and implications for proteasome-PAN/PA700 interactions. *Mol. Cell* 18:589–99.

94. Forterre, P. 2002. A hot story from comparative genomics: reverse gyrase is the only hyperthermophile-specific protein. *Trends Genet.* 18:236–237.

95. Forterre, P., C. Brochier, and H. Philippe. 2002. Evolution of the Archaea. *Theor. Popul. Biol.* 61:409–422.

96. Fox, G. E., E. Stackebrandt, R. B. Hespell, J. Gibson, J. Maniloff, T. A. Dyer, R. S. Wolfe, W. E. Balch, R. S. Tanner, L. J. Magrum, L. B. Zablen, R. Blakemore, R. Gupta, L. Bonen, B. J. Lewis, D. A. Stahl, K. R. Luehrsen, K. N. Chen, and C. R. Woese. 1980. The phylogeny of procaryotes. *Science* 209:457–463.

97. Frankenberg, R. J., M. Andersson, and D. S. Clark. 2003. Effect of temperature and pressure on the proteolytic specificity of the recombinant 20S proteasome from *Methanococcus jannaschii*. *Extremophiles* 7:353–360.

98. Franzmann, P. D., Y. Liu, D. L. Balkwill, H. C. Aldrich, E. Conway de Macario, and D. R. Boone. 1997. *Methanogenium frigidum* sp. nov., a psychrophilic, H2-using methanogen from Ace Lake, Antarctica. *Int. J. Syst. Bacteriol.* 47:1068–1072.

99. Franzmann, P. D., E. Stackebrandt, K. Sanderson, J. K. Volkman, C. D. E, P. L. Stevenson, T. A. Mcmeekin, and H. R. Burton. 1988. *Halobacterium lacusprofundi sp. nov.*, a halophilic bacterium isolated from Deep Lake, Antarctica. *Syst. Appl. Microbiol.* 11:20–27.

100. Friedrich, C. G., F. Bardischewsky, D. Rother, A. Quentmeier, and J. Fischer. 2005. Prokaryotic sulfur oxidation. *Curr. Opin. Microbiol.* 8:253–259.

101. Friedrich, C. G., D. Rother, F. Bardischewsky, A. Quentmeier, and J. Fischer. 2001. Oxidation of reduced inorganic sulfur compounds by bacteria: emergence of a common mechanism? *Appl. Environ. Microbiol.* 67:2873–2882.

102. Friedrich, T. 2001. Complex I: a chimaera of a redox and conformation-driven proton pump? *J. Bioenerg. Biomembr.* 33:169–177.

103. Fritz, G., T. Buchert, and P. M. Kroneck. 2002. The function of the [4Fe-4S] clusters and FAD in bacterial and archaeal adenylylsulfate reductases. Evidence for flavin-catalyzed reduction of adenosine 5″-phosphosulfate. *J. Biol. Chem.* 277:26066–26073.

104. Fritz, G., A. Roth, A. Schiffer, T. Buchert, G. Bourenkov, H. D. Bartunik, H. Huber, K. O. Stetter, P. M. Kroneck, and U. Ermler. 2002. Structure of adenylylsulfate reductase from the hyperthermophilic *Archaeoglobus fulgidus* at 1.6-A resolution. *Proc. Natl. Acad. Sci. USA* 99:1836–1841.

105. Fütterer, O., A. Angelov, H. Liesegang, G. Gottschalk, C. Schleper, B. Schepers, C. Dock, G. Antranikian, and W. Liebl. 2004. Genome sequence of *Picrophilus torridus* and its implications for life around pH 0. *Proc. Natl. Acad. Sci. USA* 101:9091–9096.

106. Galagan, J. E., C. Nusbaum, A. Roy, M. G. Endrizzi, P. Macdonald, W. FitzHugh, S. Calvo, R. Engels, S. Smirnov, D. Atnoor, A. Brown, N. Allen, J. Naylor, N. Stange-Thomann, K. DeArellano, R. Johnson, L. Linton, P. McEwan, K. McKernan, J. Talamas, A. Tirrell, W. Ye, A. Zimmer, R. D. Barber, I. Cann, D. E. Graham, D. A. Grahame, A. M. Guss, R. Hedderich, C. Ingram-Smith, H. C. Kuettner, J. A. Krzycki, J. A. Leigh, W. Li, J. Liu, B. Mukhopadhyay, J. N. Reeve, K. Smith, T. A. Springer, L. A. Umayam, O. White, R. H. White, E. Conway de Macario, J. G. Ferry, K. F. Jarrell, H. Jing, A. J. Macario, I. Paulsen, M. Pritchett, K. R. Sowers, R. V. Swanson, S. H. Zinder, E. Lander, W. W. Metcalf, and B. Birren. 2002. The genome of *M. acetivorans* reveals extensive metabolic and physiological diversity. *Genome Res.* 12:532–542.

107. Gambacorta, A., A. Trincone, B. Nicolaus, L. Lama, and M. De Rosa. 1993. Unique features of lipids of Archaea. *Syst. Appl. Microbiol.* 16:518–527.

108. Gan, J., J. E. Tropea, B. P. Austin, D. L. Court, D. S. Waugh, and X. Ji. 2005. Intermediate states of ribonuclease III in complex with double-stranded RNA. *Structure (Cambridge)* 13:1435–1442.

109. Garcia, J.-L., B. Ollivier, and W. B. Whitman. 2001. The order Methanomicrobiales. *In* M. Dworkin (ed.), *The Prokaryotes: An Evolving Electronic Resource for the Microbiological Community*, 3rd ed., release 3.6 ed. Springer-Verlag; New York, N.Y. [Online.] http://141.150.157.117:8080/prokPUB/index.htm.

110. Ghedin, E., and J. M. Claverie. 2005. Mimivirus relatives in the Sargasso sea. *Virol. J.* 2:62.

111. Gibson, J. A., M. R. Miller, N. W. Davies, G. P. Neill, D. S. Nichols, and J. K. Volkman. 2005. Unsaturated diether lipids in the psychrotrophic archaeon *Halorubrum lacusprofundi*. *Syst. Appl. Microbiol.* 28:19–26.

112. Gleissner, M., U. Kaiser, E. Antonopoulos, and G. Schafer. 1997. The archaeal SoxABCD complex is a proton pump in *Sulfolobus acidocaldarius*. *J. Biol. Chem.* 272:8417–8426.

113. Golyshina, O. V., P. N. Golyshin, K. N. Timmis, and M. Ferrer. 2006. The 'pH optimum anomaly' of intracellular enzymes of *Ferroplasma acidiphilum*. *Environ. Microbiol.* 8:416–425.

114. Golyshina, O. V., T. A. Pivovarova, G. I. Karavaiko, T. F. Kondrateva, E. R. Moore, W. R. Abraham, H. Lunsdorf, K. N. Timmis, M. M. Yakimov, and P. N. Golyshin. 2000. *Ferroplasma acidiphilum gen. nov., sp. nov.*, an acidophilic, autotrophic, ferrous-iron-oxidizing, cell-wall-lacking, mesophilic member of the *Ferroplasmaceae fam. nov.*, comprising a distinct lineage of the Archaea. *Int. J. Syst. Evol. Microbiol.* 50:997–1006.

115. Golyshina, O. V., and K. N. Timmis. 2005. *Ferroplasma* and relatives, recently discovered cell wall-lacking archaea making a living in extremely acid, heavy metal-rich environments. *Environ. Microbiol.* 7:1277–1288.

116. Gomes, C. M., C. Backgren, M. Teixeira, A. Puustinen, M. L. Verkhovskaya, M. Wikstrom, and M. I. Verkhovsky. 2001. Heme-copper oxidases with modified D- and K-pathways are yet efficient proton pumps. *FEBS Lett.* 497:159–164.

117. Gomes, C. M., T. M. Bandeiras, and M. Teixeira. 2001. A new type-II NADH dehydrogenase from the archaeon *Acidianus ambivalens*: characterization and in vitro reconstitution of the respiratory chain. *J. Bioenerg. Biomembr.* 33:1–8.

118. Gomes, C. M., H. Huber, K. O. Stetter, and M. Teixeira. 1998. Evidence for a novel type of iron cluster in the respiratory chain of the archaeon *Sulfolobus metallicus*. *FEBS Lett.* **432**:99–102.

119. Gomes, C. M., R. S. Lemos, M. Teixeira, A. Kletzin, H. Huber, K. O. Stetter, G. Schäfer, and S. Anemüller. 1999. The unusual iron sulfur composition of the *Acidianus ambivalens* succinate dehydrogenase complex. *Biochim. Biophys. Acta* **1411**:134–141.

120. Goo, Y. A., J. Roach, G. Glusman, N. S. Baliga, K. Deutsch, M. Pan, S. Kennedy, S. DasSarma, W. V. Ng, and L. Hood. 2004. Low-pass sequencing for microbial comparative genomics. *BMC Genomics* **5**:3.

121. Goodchild, A., M. Raftery, N. F. Saunders, M. Guilhaus, and R. Cavicchioli. 2005. Cold adaptation of the Antarctic archaeon, *Methanococcoides burtonii* assessed by proteomics using ICAT. *J. Proteome Res.* **4**:473–480.

122. Grant, W. D. 2004. Half a lifetime in soda lakes, p. 17–31. *In* A. Ventosa (ed.), *Halophilic Microorganisms*. Springer, New York, N.Y.

123. Grant, W. D. 2004. Life at low water activity. *Philos. Trans. R. Soc. Lond. B Biol. Sci.* **359**:1249–1266; discussion, 1266–1267.

124. Gregor, D., and F. Pfeifer. 2001. Use of a halobacterial *bgaH* reporter gene to analyse the regulation of gene expression in halophilic archaea. *Microbiology* **147**:1745–1754.

125. Grogan, D. W. 1989. Phenotypic characterization of the archaebacterial genus Sulfolobus: comparison of five wild-type strains. *J. Bacteriol.* **171**:6710–6719.

126. Groll, M., M. Bochtler, H. Brandstetter, T. Clausen, and R. Huber. 2005. Molecular machines for protein degradation. *Chembiochem* **6**:222–256.

127. Groll, M., L. Ditzel, J. Lowe, D. Stock, M. Bochtler, H. D. Bartunik, and R. Huber. 1997. Structure of 20S proteasome from yeast at 2.4 Å resolution. *Nature* **386**:463–471.

128. Gruber, G., H. Wieczorek, W. R. Harvey, and V. Muller. 2001. Structure-function relationships of A-, F- and V-ATPases. *J. Exp. Biol.* **204**:2597–2605.

129. Hafenbradl, D., M. Keller, R. Dirmeier, R. Rachel, P. Rossnagel, S. Burggraf, H. Huber, and K. O. Stetter. 1996. *Ferroglobus placidus* gen. nov., sp. nov., A novel hyperthermophilic archaeum that oxidizes Fe^{2+} at neutral pH under anoxic conditions. *Arch. Microbiol.* **166**:308–314.

130. Hagedoorn, P. L., T. Chen, I. Schroder, S. R. Piersma, S. de Vries, and W. R. Hagen. 2005. Purification and characterization of the tungsten enzyme aldehyde:ferredoxin oxidoreductase from the hyperthermophilic denitrifier *Pyrobaculum aerophilum*. *J. Biol. Inorg. Chem.* **10**:259–269.

131. Hain, J., W. D. Reiter, U. Hudepohl, and W. Zillig. 1992. Elements of an archaeal promoter defined by mutational analysis. *Nucleic Acids Res.* **20**:5423–5428.

132. Hallberg Gradin, C., and A. Colmsjo. 1989. Four different b-type cytochromes in the halophilic archaebacterium, *Halobacterium halobium*. *Arch. Biochem. Biophys.* **272**:130–136.

133. Hamilton-Brehm, S. D., G. J. Schut, and M. W. Adams. 2005. Metabolic and evolutionary relationships among *Pyrococcus* species: genetic exchange within a hydrothermal vent environment. *J. Bacteriol.* **187**:7492–7429.

134. Hammond, S. M. 2005. Dicing and slicing: the core machinery of the RNA interference pathway. *FEBS Lett.* **579**:5822–5829.

135. Hannington, M. D., I. R. Jonasson, P. M. Herzig, and S. Peterson. 1995. Physical and chemical processes of seafloor mineralization at mid-ocean ridge, p. 115–157. *In* S. E. Humphris, R. A. Zierenberg, L. S. Mullineaux, and R. E. Thomson (ed.), *Seafloor Hydrothermal Systems: Physical, Chemical, Biolog-ical, and Geological Interactions*, vol. 91. American Geophysical Union, Washington, D.C.

136. Harmsen, H., D. Prieur, and C. Jeanthon. 1997. Distribution of microorganisms in deep-sea hydrothermal vent chimneys investigated by whole-cell hybridization and enrichment culture of thermophilic subpopulations. *Appl. Environ. Microbiol.* **63**:2876–2883.

137. Hartzell, P., and D. Reed. 2002. The genus *Archaeoglobus*. *In* M. Dworkin (ed.), *The Prokaryotes: An Evolving Electronic Resource for the Microbiological Community*, 3rd ed., release 3.8 ed. Springer-Verlag, New York, N.Y. [Online.] http://141.150.157.117:8080/prokPUB/index.htm.

138. Haugen, P., D. M. Simon, and D. Bhattacharya. 2005. The natural history of group I introns. *Trends Genet.* **21**:111–119.

139. He, Z., Y. Li, P. Zhou, and S. Liu. 2000. Cloning and heterologous expression of a sulfur oxygenase/reductase gene from the thermoacidophilic archaeon *Acidianus* sp. S5 in *Escherichia coli*. *FEMS Microbiol. Lett.* **193**:217–221.

140. Hedderich, R. 2004. Energy-converting [NiFe] hydrogenases from archaea and extremophiles: ancestors of complex I. *J. Bioenerg. Biomembr.* **36**:65–75.

141. Hedderich, R. C., and W. B. Whitman. 2005. Physiology and biochemistry of the methane-producing Archaea. *In* M. Dworkin (ed.), *The Prokaryotes: An Evolving Electronic Resource for the Microbiological Community*, 3rd ed., release 3.20 ed. Springer-Verlag, New York, N.Y. [Online.] http://141.150.157.117:8080/prokPUB/index.htm.

142. Henderson, R., and P. N. Unwin. 1975. Three-dimensional model of purple membrane obtained by electron microscopy. *Nature* **257**:28–32.

143. Hendrickson, E. L., R. Kaul, Y. Zhou, D. Bovee, P. Chapman, J. Chung, E. Conway de Macario, J. A. Dodsworth, W. Gillett, D. E. Graham, M. Hackett, A. K. Haydock, A. Kang, M. L. Land, R. Levy, T. J. Lie, T. A. Major, B. C. Moore, I. Porat, A. Palmeiri, G. Rouse, C. Saenphimmachak, D. Soll, S. Van Dien, T. Wang, W. B. Whitman, Q. Xia, Y. Zhang, F. W. Larimer, M. V. Olson, and J. A. Leigh. 2004. Complete genome sequence of the genetically tractable hydrogenotrophic methanogen *Methanococcus maripaludis*. *J. Bacteriol.* **186**:6956–6969.

144. Henneberger, R., C. Moissl, T. Amann, C. Rudolph, and R. Huber. 2006. New insights into the lifestyle of the cold-loving SM1 euryarchaeon: natural growth as a monospecies biofilm in the subsurface. *Appl. Environ. Microbiol.* **72**:192–199.

145. Hettmann, T., C. L. Schmidt, S. Anemuller, U. Zahringer, H. Moll, A. Petersen, and G. Schafer. 1998. Cytochrome *b558/566* from the archaeon *Sulfolobus acidocaldarius*. A novel highly glycosylated, membrane-bound b-type hemoprotein. *J. Biol. Chem.* **273**:12032–12040.

146. Hickey, A. J., E. Conway de Macario, and A. J. Macario. 2002. Transcription in the archaea: basal factors, regulation, and stress-gene expression. *Crit. Rev. Biochem. Mol. Biol.* **37**:537–599.

147. Hiller, A., T. Henninger, G. Schäfer, and C. L. Schmidt. 2003. New genes encoding subunits of a cytochrome *bc1*-analogous complex in the respiratory chain of the hyperthermoacidophilic crenarchaeon *Sulfolobus acidocaldarius*. *J. Bioenerg. Biomembr.* **35**:121–131.

148. Hinchliffe, P., and L. A. Sazanov. 2005. Organization of iron-sulfur clusters in respiratory complex I. *Science* **309**:771–774.

149. Holliger, C., G. Schraa, A. J. Stams, and A. J. Zehnder. 1990. Reductive dechlorination of 1,2-dichloroethane and chloroethane by cell suspensions of methanogenic bacteria. *Biodegradation* **1**:253–261.

150. Hu, Y., S. Faham, R. Roy, M. W. W. Adams, and D. C. Rees. 1999. Formaldehyde ferredoxin oxidoreductase from *Pyro-*

coccus furiosus: the 1.85 Å resolution crystal structure and its mechanistic implications. *J. Mol. Biol.* **286**:899–914.

151. Huber, C., W. Eisenreich, S. Hecht, and G. Wachtershauser. 2003. A possible primordial peptide cycle. *Science* **301**:938–940.

152. Huber, H., M. J. Hohn, R. Rachel, T. Fuchs, V. C. Wimmer, and K. O. Stetter. 2002. A new phylum of Archaea represented by a nanosized hyperthermophilic symbiont. *Nature* **417**:63–67.

153. Huber, H., R. Huber, and K. Stetter. 2002. The order Thermoproteales. *In* M. Dworkin (ed.), *The Prokaryotes: An Evolving Electronic Resource for the Microbiological Community*, 3rd ed., release 3.8 ed. Springer-Verlag, New York, N.Y. [Online.] http://141.150.157.117:8080/prokPUB/index.htm.

154. Huber, H., and D. Prangishvili. 2005. The order Sulfolobales. *In* M. Dworkin (ed.), *The Prokaryotes: An Evolving Electronic Resource for the Microbiological Community*, 3rd ed., release 3.19 ed. Springer-Verlag, New York, N.Y. [Online.] http://link.springer-ny.com/link/service/books/10125/.

155. Huber, H., and K. Stetter. 2002. Desulfurococcales. *In* M. Dworkin (ed.), *The Prokaryotes: An Evolving Electronic Resource for the Microbiological Community*, 3rd ed., release 3.11 ed. Springer-Verlag, New York, N.Y. [Online.] http://141.150.157.117:8080/prokPUB/index.htm.

156. Huber, H., and K. Stetter. 2002. The order Thermoplasmatales. *In* M. Dworkin (ed.), *The Prokaryotes: An Evolving Electronic Resource for the Microbiological Community*, 3rd ed., release 3.8 ed. Springer-Verlag, New York, N.Y. [Online.] http://141.150.157.117:8080/prokPUB/index.htm.

157. Huber, H., and K. O. Stetter. 2001. *Metallosphaera*, p. 204–206. *In* D. R. Boone and R. W. Castenholz (ed.), *Bergey's Manual of Systematic Bacteriology*, 2nd ed., vol. 1. Springer, New York, N.Y.

158. Huber, R., D. Dyba, H. Huber, S. Burggraf, and R. Rachel. 1998. Sulfur-inhibited *Thermosphaera aggregans sp. nov.*, a new genus of hyperthermophilic archaea isolated after its prediction from environmentally derived 16S rRNA sequences. *Int. J. Syst. Bacteriol.* **48**(Pt 1):31–38.

159. Huber, R., H. Huber, and K. O. Stetter. 2000. Towards the ecology of hyperthermophiles: biotopes, new isolation strategies and novel metabolic properties. *FEMS Microbiol. Rev.* **24**:615–623.

160. Hugler, M., H. Huber, K. O. Stetter, and G. Fuchs. 2003. Autotrophic CO_2 fixation pathways in archaea (Crenarchaeota). *Arch. Microbiol.* **179**:160–173.

161. Ishikawa, R., Y. Ishido, A. Tachikawa, H. Kawasaki, H. Matsuzawa, and T. Wakagi. 2002. *Aeropyrum pernix* K1, a strictly aerobic and hyperthermophilic archaeon, has two terminal oxidases, cytochrome *ba3* and cytochrome *aa3*. *Arch. Microbiol.* **179**:42–49.

162. Itoh, T., K. Suzuki, P. C. Sanchez, and T. Nakase. 1999. Caldivirga maquilingensis gen. nov., sp. nov., a new genus of rod-shaped crenarchaeote isolated from a hot spring in the Philippines. *Int. J. Syst. Bacteriol.* **49**(Pt 3):1157–1163.

163. Iwasaki, T., A. Kounosu, M. Aoshima, D. Ohmori, T. Imai, A. Urushiyama, N. J. Cosper, and R. A. Scott. 2002. Novel [2Fe-2S]-type redox center C in SdhC of archaeal respiratory complex II from *Sulfolobus tokodaii* strain 7. *J. Biol. Chem.* **277**:39642–39648.

164. Jahn, U., R. Summons, H. Sturt, E. Grosjean, and H. Huber. 2004. Composition of the lipids of *Nanoarchaeum equitans* and their origin from its host *Ignicoccus* sp. strain KIN4/I. *Arch. Microbiol.* **182**:404–413.

165. Janssen, S., G. Schäfer, S. Anemüller, and R. Moll. 1997. A succinate dehydrogenase with novel structure and properties from the hyperthermophilic archaeon *Sulfolobus acidocaldarius*: Genetic and biophysical characterization. *J. Bacteriol.* **179**:5560–5569.

166. Jansson, B. P., L. Malandrin, and H. E. Johansson. 2000. Cell cycle arrest in archaea by the hypusination inhibitor N(1)-guanyl-1,7-diaminoheptane. *J. Bacteriol.* **182**:1158–1161.

167. Jarvis, G. N., C. Strompl, D. M. Burgess, L. C. Skillman, E. R. Moore, and K. N. Joblin. 2000. Isolation and identification of ruminal methanogens from grazing cattle. *Curr. Microbiol.* **40**:327–332.

168. Jiang, Y., A. Lee, J. Chen, V. Ruta, M. Cadene, B. T. Chait, and R. MacKinnon. 2003. X-ray structure of a voltage-dependent K^+ channel. *Nature* **423**:33–41.

169. Jiang, Y., V. Ruta, J. Chen, A. Lee, and R. MacKinnon. 2003. The principle of gating charge movement in a voltage-dependent K+ channel. *Nature* **423**:42–48.

170. Johnson, E. F., and B. Mukhopadhyay. 2005. A new type of sulfite reductase—a novel coenzyme F420-dependent enzyme from the methanarchaeon *Methanocaldococcus jannaschii*. *J. Biol.Chem.* **280**:39776–38786.

171. Jolivet, E., E. Corre, S. L'Haridon, P. Forterre, and D. Prieur. 2004. *Thermococcus marinus sp. nov.* and *Thermococcus radiotolerans sp. nov.*, two hyperthermophilic archaea from deep-sea hydrothermal vents that resist ionizing radiation. *Extremophiles* **8**:219–227.

172. Jolivet, E., S. L'Haridon, E. Corre, P. Forterre, and D. Prieur. 2003. *Thermococcus gammatolerans sp. nov.*, a hyperthermophilic archaeon from a deep-sea hydrothermal vent that resists ionizing radiation. *Int. J. Syst. Evol. Microbiol.* **53**:847–851.

173. Jolivet, E., F. Matsunaga, Y. Ishino, P. Forterre, D. Prieur, and H. Myllykallio. 2003. Physiological responses of the hyperthermophilic archaeon "*Pyrococcus abyssi*" to DNA damage caused by ionizing radiation. *J. Bacteriol.* **185**:3958–3961.

174. Jones, W. J., J. A. Leigh, F. Mayer, C. R. Woese, and R. S. Wolfe. 1983. *Methanococcus jannaschii sp. nov.*, an extremely thermophilic methanogen from a submarine hydrothermal vent. *Arch. Microbiol.* **136**:254–261.

175. Jorgensen, R., A. R. Merrill, and G. R. Andersen. 2006. The life and death of translation elongation factor 2. *Biochem. Soc. Trans.* **34**:1–6.

176. Kandler, O., and H. Konig. 1998. Cell wall polymers in Archaea (Archaebacteria). *Cell. Mol.. Life Sci.* **54**:305–308.

177. Kaneshiro, S. M., and D. S. Clark. 1995. Pressure effects on the composition and thermal behavior of lipids from the deep-sea thermophile *Methanococcus jannaschii. J. Bacteriol.* **177**:3668–3672.

178. Kappler, U., and C. Dahl. 2001. Enzymology and molecular biology of prokaryotic sulfite oxidation. *FEMS Microbiol. Lett.* **203**:1–9.

179. Kashefi, K., and D. R. Lovley. 2003. Extending the upper temperature limit for life. *Science* **301**:934.

180. Kashefi, K., J. M. Tor, D. E. Holmes, C. V. Gaw Van Praagh, A. L. Reysenbach, and D. R. Lovley. 2002. *Geoglobus ahangari gen. nov., sp. nov.*, a novel hyperthermophilic archaeon capable of oxidizing organic acids and growing autotrophically on hydrogen with Fe(III) serving as the sole electron acceptor. *Int. J. Syst. Evol. Microbiol.* **52**:719–728.

181. Kates, M. 1992. Archaebacterial lipids: structure, biosynthesis and function. *Biochem. Soc. Symp.* **58**:51–72.

182. Kawarabayasi, Y., Y. Hino, H. Horikawa, S. Yamazaki, Y. Haikawa, K. Jin-no, M. Takahashi, M. Sekine, S. Baba, A. Ankai, H. Kosugi, A. Hosoyama, S. Fukui, Y. Nagai, K. Nishijima, H. Nakazawa, M. Takamiya, S. Masuda, T. Funahashi, T. Tanaka, Y. Kudoh, J. Yamazaki, N. Kushida, A. Oguchi, and H. Kikuchi. 1999. Complete genome

sequence of an aerobic hyper-thermophilic crenarchaeon, *Aeropyrum pernix* K1. *DNA Res.* 6:83–101, 145–152.

183. Keller, M., E. Blochl, G. Wachtershauser, and K. O. Stetter. 1994. Formation of amide bonds without a condensation agent and implications for origin of life. *Nature* 368:836–838.

184. Kelley, D. S., J. A. Karson, D. K. Blackman, G. L. Fruh-Green, D. A. Butterfield, M. D. Lilley, E. J. Olson, M. O. Schrenk, K. K. Roe, G. T. Lebon, and P. Rivizzigno. 2001. An off-axis hydrothermal vent field near the Mid-Atlantic Ridge at 30 degrees N. *Nature* 412:145–149.

185. Kelley, D. S., J. A. Karson, G. L. Fruh-Green, D. R. Yoerger, T. M. Shank, D. A. Butterfield, J. M. Hayes, M. O. Schrenk, E. J. Olson, G. Proskurowski, M. Jakuba, A. Bradley, B. Larson, K. Ludwig, D. Glickson, K. Buckman, A. S. Bradley, W. J. Brazelton, K. Roe, M. J. Elend, A. Delacour, S. M. Bernasconi, M. D. Lilley, J. A. Baross, R. E. Summons, and S. P. Sylva. 2005. A serpentinite-hosted ecosystem: the Lost City hydrothermal field. *Science* 307:1428–1434.

186. Kelly, D. P., J. K. Shergill, W. P. Lu, and A. P. Wood. 1997. Oxidative metabolism of inorganic sulfur compounds by bacteria. *Antonie Van Leeuwenhoek* 71:95–107.

187. Kelman, Z., and M. F. White. 2005. Archaeal DNA replication and repair. *Curr. Opin. Microbiol.* 8:669–676.

188. Kendall, E. M., and D. R. Boone. 2004. The order Methanosarcinales. *In* M. Dworkin (ed.), *The Prokaryotes: An Evolving Electronic Resource for the Microbiological Community*, 3rd ed., release 3.17 ed. Springer-Verlag, New York, N.Y. [Online.] http://141.150.157.117:8080/prokPUB/index.htm.

189. Kengen, S. W., J. E. Tuininga, C. H. Verhees, J. van der Oost, A. J. Stams, and W. M. de Vos. 2001. ADP-dependent glucokinase and phosphofructokinase from *Pyrococcus furiosus*. *Methods Enzymol.* 331:41–53.

190. Kengen, S. W. M., and A. J. M. Stams. 1994. Growth and energy conservation in batch cultures of *Pyrococcus furiosus*. *FEMS Microbiol. Lett.* 117:305–309.

191. Keppler, F., J. T. Hamilton, M. Brass, and T. Rockmann. 2006. Methane emissions from terrestrial plants under aerobic conditions. *Nature* 439:187–191.

192. Kessel, M., and Y. Cohen. 1982. Ultrastructure of square bacteria from a brine pool in Southern Sinai. *J. Bacteriol.* 150:851–860.

193. Kessel, M., and F. Klink. 1980. Archaebacterial elongation factor is ADP-ribosylated by diphtheria toxin. *Nature* 287:250–251.

194. Kikuchi, A., and K. Asai. 1984. Reverse gyrase—a topoisomerase which introduces positive superhelical turns into DNA. *Nature* 309:677–681.

195. Kim, J. S., M. Groll, H. J. Musiol, R. Behrendt, M. Kaiser, L. Moroder, R. Huber, and H. Brandstetter. 2002. Navigation inside a protease: substrate selection and product exit in the tricorn protease from *Thermoplasma acidophilum*. *J. Mol. Biol.* 324:1041–1050.

196. Kim, K. K., L. W. Hung, H. Yokota, R. Kim, and S. H. Kim. 1998. Crystal structures of eukaryotic translation initiation factor 5A from *Methanococcus jannaschii* at 1.8 Å resolution. *Proc. Natl. Acad. Sci. USA* 95:10419–10424.

197. Kimura, H., M. Sugihara, H. Yamamoto, B. K. Patel, K. Kato, and S. Hanada. 2005. Microbial community in a geothermal aquifer associated with the subsurface of the Great Artesian Basin, Australia. *Extremophiles* 9:407–414.

198. Kjems, J., and R. A. Garrett. 1985. An intron in the 23S ribosomal RNA gene of the archaebacterium *Desulfurococcus mobilis*. *Nature* 318:675–677.

199. Kjems, J., and R. A. Garrett. 1988. Novel splicing mechanism for the ribosomal RNA intron in the archaebacterium *Desulfurococcus mobilis*. *Cell* 54:693–703.

200. Klein, D. J., P. B. Moore, and T. A. Steitz. 2004. The roles of ribosomal proteins in the structure assembly, and evolution of the large ribosomal subunit. *J. Mol. Biol.* 340:141–177.

201. Klenk, H. P., R. A. Clayton, J. F. Tomb, O. White, K. E. Nelson, K. A. Ketchum, R. J. Dodson, M. Gwinn, E. K. Hickey, J. D. Peterson, D. L. Richardson, A. R. Kerlavage, D. E. Graham, N. C. Kyrpides, R. D. Fleischmann, J. Quackenbush, N. H. Lee, G. G. Sutton, S. Gill, E. F. Kirkness, B. A. Dougherty, K. McKenney, M. D. Adams, B. Loftus, and J. C. Venter. 1997. The complete genome sequence of the hyperthermophilic, sulphate- reducing archaeon *Archaeoglobus fulgidus*. *Nature* 390:364–370.

202. Klenk, H. P., T. D. Meier, P. Durovic, V. Schwass, F. Lottspeich, P. P. Dennis, and W. Zillig. 1999. RNA polymerase of *Aquifex pyrophilus*: implications for the evolution of the bacterial *rpoBC* operon and extremely thermophilic bacteria. *J. Mol. Evol.* 48:528–451.

203. Klenk, H. P., and W. Zillig. 1994. DNA-dependent RNA polymerase subunit B as a tool for phylogenetic reconstruction: branching topology of the archaeal domain. *J. Mol. Evol.* 38:420–432.

204. Kletzin, A. 1989. Coupled enzymatic production of sulfite, thiosulfate, and hydrogen sulfide from sulfur: purification and properties of a sulfur oxygenase reductase from the facultatively anaerobic archaebacterium *Desulfurolobus ambivalens*. *J. Bacteriol.* 171:1638–1643.

205. Kletzin, A. 2006. Metabolism of inorganic sulfur compounds in Archaea, p. 261–274. *In* R. A. Garrett and H.-P. Klenk (ed.), *Archaea Evolution, Physiology, and Molecular Biology*. Blackwell Publishing, Oxford, United Kingdom.

206. Kletzin, A. 1992. Molecular characterization of a DNA ligase gene of the extremely thermophilic archaeon *Desulfurolobus ambivalens* shows close phylogenetic relationship to eukaryotic ligases. *Nucleic Acids Research* 20:5389–5396.

207. Kletzin, A., T. Urich, F. Müller, T. M. Bandeiras, and C. M. Gomes. 2004. Dissimilatory oxidation and reduction of elemental sulfur in thermophilic archaea. *J. Bioenerg. Biomembr.* 36:77–91.

208. Kobayashi, T. 2001. *Thermococcus*, p. 342–346. *In* D. R. Boone and R. W. Castenholz (ed.), *Bergey's Manual of Systematic Bacteriology*, 2nd ed., vol. 1. Springer, New York, N.Y.

209. Koch, M. K., and D. Oesterhelt. 2005. MpcT is the transducer for membrane potential changes in *Halobacterium salinarum*. *Mol. Microbiol.* 55:1681–1694.

210. Komori, K., R. Fujikane, H. Shinagawa, and Y. Ishino. 2002. Novel endonuclease in Archaea cleaving DNA with various branched structure. *Genes Genet. Syst.* 77:227–241.

211. Komorowski, L., W. Verheyen, and G. Schafer. 2002. The archaeal respiratory supercomplex SoxM from *S. acidocaldarius* combines features of quinole and cytochrome *c* oxidases. *Biol. Chem.* 383:1791–1799.

212. Konneke, M., A. E. Bernhard, J. R. de la Torre, C. B. Walker, J. B. Waterbury, and D. A. Stahl. 2005. Isolation of an autotrophic ammonia-oxidizing marine archaeon. *Nature* 437:543–546.

213. Koonin, E. V., and W. Martin. 2005. On the origin of genomes and cells within inorganic compartments. *Trends Genet.* 21:647–654.

214. Kopriva, S., and A. Koprivova. 2004. Plant adenosine 5′-phosphosulphate reductase: the past, the present, and the future. *J. Exp. Bot.* 55:1775–1783.

215. Koski, R. A., I. A. Jonasson, D. C. Kadko, V. K. Smith, and F. L. Wong. 1994. Compositions, growth mechanisms, and temporal relations of hydrothermal sulfide-sulfate-silica chimmneys at the northern Cleft segment, Juan de Fuca Ridge. *J. Geophys. Res. Solid Earth* 99:4813–4832.

216. **Kozono, D., X. Ding, I. Iwasaki, X. Meng, Y. Kamagata, P. Agre, and Y. Kitagawa.** 2003. Functional expression and characterization of an archaeal aquaporin. AqpM from *Methanothermobacter marburgensis. J. Biol. Chem.* **278:**10649–10656.

217. **Kruger, K., T. Hermann, V. Armbruster, and F. Pfeifer.** 1998. The transcriptional activator GvpE for the halobacterial gas vesicle genes resembles a basic region leucine-zipper regulatory protein. *J. Mol. Biol.* **279:**761–771.

218. **Kruger, M., A. Meyerdierks, F. O. Glockner, R. Amann, F. Widdel, M. Kube, R. Reinhardt, J. Kahnt, R. Bocher, R. K. Thauer, and S. Shima.** 2003. A conspicuous nickel protein in microbial mats that oxidize methane anaerobically. *Nature* **426:**878–881.

219. **Kyrieleis, O. J., P. Goettig, R. Kiefersauer, R. Huber, and H. Brandstetter.** 2005. Crystal structures of the tricorn interacting factor F3 from *Thermoplasma acidophilum,* a zinc aminopeptidase in three different conformations. *J. Mol. Biol.* **349:**787–800.

220. **Lake, J. A.** 1987. Prokaryotes and archaebacteria are not monophyletic: rate invariant analysis of rRNA genes indicates that eukaryotes and eocytes form a monophyletic taxon. *Cold Spring Harbor Symp. Quant. Biol.* **52:**839–46.

221. **Lake, J. A.** 1983. Ribosome evolution: the structural bases of protein synthesis in archaebacteria, eubacteria, and eukaryotes. *Prog. Nucleic Acid Res. Mol. Biol.* **30:**163–194.

222. **Lam, W. L., and W. F. Doolittle.** 1992. Mevinolin-resistant mutations identify a promoter and the gene for a eukaryote-like 3-hydroxy-3-methylglutaryl-coenzyme—a reductase in the archaebacterium *Haloferax volcanii. J. Biol. Chem.* **267:**5829–5834.

223. **Lam, W. L., and W. F. Doolittle.** 1989. Shuttle vectors for the archaebacterium *Halobacterium volcanii. Proc. Natl. Acad. Sci. USA* **86:**5478–5482.

224. **Lancaster, C. R.** 2002. Succinate:quinone oxidoreductases: an overview. *Biochim. Biophys. Acta* **1553:**1–6.

225. **Langer, D., J. Hain, P. Thuriaux, and W. Zillig.** 1995. Transcription in archaea: similarity to that in eucarya. *Proc. Natl. Acad. Sci. USA* **92:**5768–5772.

226. **Langworthy, T.** 1985. Lipids of archaebacteria, p. 459–497. *In* C. R. Woese and R. S. Wolfe (ed.), *The Bacteria,* vol. VIII. *A Treatise on Structure and Function: Archaebacteria.* Academic Press, Orlando, Fla.

227. **Langworthy, T. A.** 1977. Comparative Lipid-Composition of Heterotrophically and Autotrophically Grown *Sulfolobus acidocaldarius. J. Bacteriol.* **130:**1326–1332.

228. **Langworthy, T. A., P. F. Smith, and W. R. Mayberry.** 1972. Lipids of *Thermoplasma acidophilum. J. Bacteriol.* **112:**1193–1200.

229. **Lanyi, J. K.** 2004. Bacteriorhodopsin. *Annu. Rev. Physiol.* **66:**665–688.

230. **Larsen, N., H. Leffers, J. Kjems, and R. A. Garrett.** 1986. Evolutionary divergence between the ribosomal RNA operons of *Halococcus morrhuae* and *Desulfurococcus mobilis. Syst. Appl. Microbiol.* **7:**49–57.

231. **Laska, S., F. Lottspeich, and A. Kletzin.** 2003. Membrane-bound hydrogenase and sulfur reductase of the hyperthermophilic and acidophilic archaeon *Acidianus ambivalens. Microbiology* **149:**2357–2371.

232. **Lechner, K., G. Heller, and A. Bock.** 1988. Gene for the diphtheria toxin-susceptible elongation factor 2 from *Methanococcus vannielii. Nucleic Acids Res.* **16:**7817–7826.

233. **Lee, J. K., D. Kozono, J. Remis, Y. Kitagawa, P. Agre, and R. M. Stroud.** 2005. Structural basis for conductance by the archaeal aquaporin AqpM at 1.68 A. *Proc. Natl. Acad. Sci. USA* **102:**18932–18937.

234. **Lee, S. Y., A. Lee, J. Chen, and R. MacKinnon.** 2005. Structure of the KvAP voltage-dependent K$^+$ channel and its de-

235. **Lehmacher, A., A. B. Vogt, and R. Hensel.** 1990. Biosynthesis of cyclic 2,3-diphosphoglycerate. Isolation and characterization of 2-phosphoglycerate kinase and cyclic 2,3-diphosphoglycerate synthetase from *Methanothermus fervidus. FEBS Lett.* **272:**94–98.

236. **Lemos, R. S., A. S. Fernandes, M. M. Pereira, C. M. Gomes, and M. Teixeira.** 2002. Quinol:fumarate oxidoreductases and succinate:quinone oxidoreductases: phylogenetic relationships, metal centres and membrane attachment. *Biochim. Biophys. Acta* **1553:**158–170.

237. **Lepp, P. W., M. M. Brinig, C. C. Ouverney, K. Palm, G. C. Armitage, and D. A. Relman.** 2004. Methanogenic Archaea and human periodontal disease. *Proc. Natl. Acad. Sci. USA* **101:**6176–6181.

238. **Li, N., and M. C. Cannon.** 1998. Gas vesicle genes identified in *Bacillus megaterium* and functional expression in *Escherichia coli. J. Bacteriol.* **180:**2450–2458.

239. **Liesack, W., H. König, H. Schlesner, and P. Hirsch.** 1986. Chemical composition of the peptidoglycan-free cell envelopes of budding bacteria of the *Pirella Planctomyces* group. *Arch. Microbiol.* **145:**361–366.

240. **Lin, C., and T. L. Miller.** 1998. Phylogenetic analysis of *Methanobrevibacter* isolated from feces of humans and other animals. *Arch. Microbiol.* **169:**397–403.

241. **Lingel, A., and M. Sattler.** 2005. Novel modes of protein-RNA recognition in the RNAi pathway. *Curr. Opin. Struct. Biol.* **15:**107–115.

242. **Liu, J., M. A. Carmell, F. V. Rivas, C. G. Marsden, J. M. Thomson, J. J. Song, S. M. Hammond, L. Joshua-Tor, and G. J. Hannon.** 2004. Argonaute2 is the catalytic engine of mammalian RNAi. *Science* **305:**1437–1441.

243. **Liu, S., G. T. Milne, J. G. Kuremsky, G. R. Fink, and S. H. Leppla.** 2004. Identification of the proteins required for biosynthesis of diphthamide, the target of bacterial ADP-ribosylating toxins on translation elongation factor 2. *Mol. Cell. Biol.* **24:**9487–9497.

244. **Londei, P.** 2005. Evolution of translational initiation: new insights from the archaea. *FEMS Microbiol. Rev.* **29:**185–200.

245. **Lowe, J., D. Stock, B. Jap, P. Zwickl, W. Baumeister, and R. Huber.** 1995. Crystal structure of the 20S proteasome from the archaeon *T. acidophilum* at 3.4 A resolution. *Science* **268:**533–539.

246. **Lundgren, M., A. Andersson, L. Chen, P. Nilsson, and R. Bernander.** 2004. Three replication origins in *Sulfolobus* species: synchronous initiation of chromosome replication and asynchronous termination. *Proc. Natl. Acad. Sci. USA* **101:**7046–7051.

247. **Ma, J. B., Y. R. Yuan, G. Meister, Y. Pei, T. Tuschl, and D. J. Patel.** 2005. Structural basis for 5′-end-specific recognition of guide RNA by the *A. fulgidus* Piwi protein. *Nature* **434:**666–670.

248. **Ma, K., and M. W. Adams.** 2001. Hydrogenases I and II from *Pyrococcus furiosus. Methods Enzymol.* **331:**208–216.

249. **Ma, K., and M. W. Adams.** 1994. Sulfide dehydrogenase from the hyperthermophilic archaeon *Pyrococcus furiosus:* a new multifunctional enzyme involved in the reduction of elemental sulfur. *J. Bacteriol.* **176:**6509–6517.

250. **Macalady, J. L., M. M. Vestling, D. Baumler, N. Boekelheide, C. W. Kaspar, and J. F. Banfield.** 2004. Tetraether-linked membrane monolayers in *Ferroplasma* spp: a key to survival in acid. *Extremophiles* **8:**411–419.

251. **Macario, A. J., M. Malz, and E. Conway de Macario.** 2004. Evolution of assisted protein folding: the distribution of the

main chaperoning systems within the phylogenetic domain archaea. *Front. Biosci.* **9:**1318–1332.

252. **Macario, A. J. L., M. W. Peck, E. C. d. Macario, and D. P. Chynoweth.** 1991. Unusual methanogenic flora of a wood-fermenting anaerobic bioreactor. *J. Appl. Bacteriol.* **71:** 31–37.

253. **Madigan, M., and J. Martinko.** 2006. *Brock Biology of Microorganisms,* 11th ed, p. 426. Pearson Prentice Hall, Upper Saddle River, N.J.

254. **Majernik, A. I., E. R. Jenkinson, and J. P. Chong.** 2004. DNA replication in thermophiles. *Biochem. Soc. Trans.* **32:**236–239.

255. **Makarova, K. S., and E. V. Koonin.** 2005. Evolutionary and functional genomics of the Archaea. *Curr. Opin. Microbiol.* **8:**586–594.

256. **Martin, W., and M. J. Russell.** 2003. On the origins of cells: a hypothesis for the evolutionary transitions from abiotic geochemistry to chemoautotrophic prokaryotes, and from prokaryotes to nucleated cells. *Philos. Trans. R. Soc. Lond. B Biol. Sci.* **358:**59–83, discussion 83–85.

257. **Martinez-Espinosa, R. M., D. J. Richardson, J. N. Butt, and M. J. Bonete.** 2006. Respiratory nitrate and nitrite pathway in the denitrifier haloarchaeon *Haloferax mediterranei.* *Biochem. Soc. Trans.* **34:**115–117.

258. **Maupin-Furlow, J. A., M. A. Gil, M. A. Humbard, P. A. Kirkland, W. Li, C. J. Reuter, and A. J. Wright.** 2005. Archaeal proteasomes and other regulatory proteases. *Curr. Opin. Microbiol.* **8:**720–728.

259. **Mayr, J., A. Lupas, J. Kellermann, C. Eckerskorn, W. Baumeister, and J. Peters.** 1996. A hyperthermostable protease of the subtilisin family bound to the surface layer of the archaeon *Staphylothermus marinus.* *Curr. Biol.* **6:**739–749.

260. **McGenity, T. J., R. T. Gemmell, W. D. Grant, and H. Stan-Lotter.** 2000. Origins of halophilic microorganisms in ancient salt deposits. *Environ. Microbiol.***2:**243–250.

261. **Messner, P., G. Allmaier, C. Schaffer, T. Wugeditsch, S. Lortal, H. Konig, R. Niemetz, and M. Dorner.** 1997. Biochemistry of S-layers. *FEMS Microbiol. Rev.* **20:**25–46.

262. **Michaelis, W., R. Seifert, K. Nauhaus, T. Treude, V. Thiel, M. Blumenberg, K. Knittel, A. Gieseke, K. Peterknecht, T. Pape, A. Boetius, R. Amann, B. B. Jorgensen, F. Widdel, J. Peckmann, N. V. Pimenov, and M. B. Gulin.** 2002. Microbial reefs in the Black Sea fueled by anaerobic oxidation of methane. *Science* **297:**1013–1015.

263. **Middelburg, J. J.** 2000. The geochemical sulfur cycle, p. 33–46. *In* P. N. L. Lens and P. Hulshoff (ed.), *Environmental Technologies to Treat Sulfur Pollution.* IWA Publishing, London, United Kingdom.

264. **Moissl, C., A. Briegel, H. Engelhardt, and R. Huber.** 2005. Enterhaken und Stacheldraht: Verblüffende Strukturen aus der archaeelen Nanowelt. *Biospektrum* **11:**732–733.

265. **Moissl, C., R. Rachel, A. Briegel, H. Engelhardt, and R. Huber.** 2005. The unique structure of archaeal 'hami,' highly complex cell appendages with nano-grappling hooks. *Mol. Microbiol.* **56:**361–70.

266. **Moissl, C., C. Rudolph, and R. Huber.** 2002. Natural communities of novel archaea and bacteria with a string-of-pearls-like morphology: molecular analysis of the bacterial partners. *Appl. Environ. Microbiol.* **68:**933–937.

267. **Moissl, C., C. Rudolph, R. Rachel, M. Koch, and R. Huber.** 2003. In situ growth of the novel SM1 euryarchaeon from a string-of-pearls-like microbial community in its cold biotope, its physical separation and insights into its structure and physiology. *Arch. Microbiol.* **180:**211–217.

268. **Moll, R., and G. Schäfer.** 1988. Chemiosmotic H⁺ cycling across the plasma membrane of the thermoacidophilic ar-

chaebacterium *Sulfolobus acidocaldarius.* *FEBS Lett.* **232:** 359–363.

269. **Mongodin, E. F., K. E. Nelson, S. Daugherty, R. T. Deboy, J. Wister, H. Khouri, J. Weidman, D. A. Walsh, R. T. Papke, G. Sanchez Perez, A. K. Sharma, C. L. Nesbo, D. MacLeod, E. Bapteste, W. F. Doolittle, R. L. Charlebois, B. Legault, and F. Rodriguez-Valera.** 2005. The genome of *Salinibacter ruber:* convergence and gene exchange among hyperhalophilic bacteria and archaea. *Proc. Natl. Acad. Sci. USA* **102:**18147–18152.

270. **Moune, S., P. Caumette, R. Matheron, and J. C. Willison.** 2003. Molecular sequence analysis of prokaryotic diversity in the anoxic sediments underlying cyanobacterial mats of two hypersaline ponds in Mediterranean salterns. *FEMS Microbiol. Ecol.* **44:**117–130.

271. **Müller, F. H., T. M. Bandeiras, T. Urich, M. Teixeira, C. M. Gomes, and A. Kletzin.** 2004. Coupling of the pathway of sulphur oxidation to dioxygen reduction: characterization of a novel membrane-bound thiosulphate:quinone oxidoreductase. *Mol. Microbiol.* **53:**1147–1160.

272. **Müller, V.** 2004. An exceptional variability in the motor of archael A1A0 ATPases: from multimeric to monomeric rotors comprising 6–13 ion binding sites. *J. Bioenerg. Biomembr.* **36:**115–125.

273. **Myllykallio, H., P. Lopez, P. Lopez-Garcia, R. Heilig, W. Saurin, Y. Zivanovic, H. Philippe, and P. Forterre.** 2000. Bacterial mode of replication with eukaryotic-like machinery in a hyperthermophilic archaeon. *Science* **288:**2212–2215.

274. **Nakashima, H., S. Fukuchi, and K. Nishikawa.** 2003. Compositional changes in RNA, DNA and proteins for bacterial adaptation to higher and lower temperatures. *J. Biochem. (Tokyo)* **133:**507–513.

275. **Nather, D. J., and R. Rachel.** 2004. The outer membrane of the hyperthermophilic archaeon *Ignicoccus:* dynamics, ultrastructure and composition. *Biochem. Soc. Trans.* **32:**199–203.

276. **Nauhaus, K., T. Treude, A. Boetius, and M. Kruger.** 2005. Environmental regulation of the anaerobic oxidation of methane: a comparison of ANME-I and ANME-II communities. *Environ. Microbiol.* **7:**98–106.

277. **Nercessian, O., A. L. Reysenbach, D. Prieur, and C. Jeanthon.** 2003. Archaeal diversity associated with in situ samplers deployed on hydrothermal vents on the East Pacific Rise (13 degrees N). *Environ. Microbiol.* **5:**492–502.

278. **Nickell, S., R. Hegerl, W. Baumeister, and R. Rachel.** 2003. *Pyrodictium cannulae* enter the periplasmic space but do not enter the cytoplasm, as revealed by cryo-electron tomography. *J. Struct. Biol.* **141:**34–42.

279. **Niemetz, R., U. Karcher, O. Kandler, B. J. Tindall, and H. Konig.** 1997. The cell wall polymer of the extremely halophilic archaeon *Natronococcus occultus.* *Eur. J. Biochem.* **249:**905–911.

280. **Nishino, T., K. Komori, D. Tsuchiya, Y. Ishino, and K. Morikawa.** 2005. Crystal structure and functional implications of *Pyrococcus furiosus* hef helicase domain involved in branched DNA processing. *Structure (Cambridge)* **13:**143–153.

281. **Noll, I., S. Muller, and A. Klein.** 1999. Transcriptional regulation of genes encoding the selenium-free [NiFe]-hydrogenases in the archaeon *Methanococcus voltae* involves positive and negative control elements. *Genetics* **152:**1335–1341.

282. **Noller, H. F.** 2005. RNA structure: reading the ribosome. *Science* **309:**1508–1514.

283. **Noon, K. R., R. Guymon, P. F. Crain, J. A. McCloskey, M. Thomm, J. Lim, and R. Cavicchioli.** 2003. Influence of temperature on tRNA modification in archaea: *Methanococcoides burtonii* (optimum growth temperature [T_{opt}], 23 de-

grees C) and *Stetteria hydrogenophila* (T$_{opt}$, 95 degrees C). *J. Bacteriol.* 185:5483–5490.

284. **Norris, P. R., and D. B. Johnson.** 1998. Acidophilic Microorganisms, p. 133–154. *In* K. Horikoshi and W. D. Grant (ed.), *Extremophiles: Microbial Life in Extreme Environments.* John Wiley, New York, N.Y.

285. **O'Connor, E. M., and R. F. Shand.** 2002. Halocins and sulfolobicins: the emerging story of archaeal protein and peptide antibiotics. *J. Ind. Microbiol. Biotechnol.* 28:23–31.

286. **Oesterhelt, D., and W. Stoeckenius.** 1973. Functions of a new photoreceptor membrane. *Proc. Natl. Acad. Sci. USA* 70:2853–2857.

287. **Offner, S., A. Hofacker, G. Wanner, and F. Pfeifer.** 2000. Eight of fourteen *gvp* genes are sufficient for formation of gas vesicles in halophilic archaea. *J. Bacteriol.* 182:4328–4336.

288. **Offner, S., and F. Pfeifer.** 1995. Complementation studies with the gas vesicle-encoding p-vac region of *Halobacterium salinarium* PHH1 reveal a regulatory role for the *p-gvpDE* genes. *Mol. Microbiol.* 16:9–19.

289. **Ollivier, B., P. Caumette, J. L. Garcia, and R. A. Mah.** 1994. Anaerobic bacteria from hypersaline environments. *Microbiol. Rev.* 58:27–38.

290. **Omelchenko, M. V., Y. I. Wolf, E. K. Gaidamakova, V. Y. Matrosova, A. Vasilenko, M. Zhai, M. J. Daly, E. V. Koonin, and K. S. Makarova.** 2005. Comparative genomics of *Thermus thermophilus* and *Deinococcus radiodurans*: divergent routes of adaptation to thermophily and radiation resistance. *BMC Evol. Biol.* 5:57.

291. **Oren, A.** 1999. Bioenergetic aspects of halophilism. *Microbiol. Mol. Biol. Rev.* 63:334–348.

292. **Oren, A.** 2000. The order Halobacteriales. *In* M. Dworkin (ed.), *The Prokaryotes: An Evolving Electronic Resource for the Microbiological Community.* 3rd ed., release 3.2. Springer-Verlag, New York, N.Y. [Online.] http://141.150.157.117:8080/prokPUB/index.htm.

293. **Park, M. H., Y. A. Joe, and K. R. Kang.** 1998. Deoxyhypusine synthase activity is essential for cell viability in the yeast *Saccharomyces cerevisiae. J. Biol. Chem.* 273:1677–1683.

294. **Park, M. H., and E. C. Wolff.** 1988. Cell-free synthesis of deoxyhypusine. Separation of protein substrate and enzyme and identification of 1,3-diaminopropane as a product of spermidine cleavage. *J. Biol. Chem.* 263:15264–15269.

295. **Parker, J. S., S. M. Roe, and D. Barford.** 2004. Crystal structure of a PIWI protein suggests mechanisms for siRNA recognition and slicer activity. *EMBO J.* 23:4727–4737.

296. **Parker, J. S., S. M. Roe, and D. Barford.** 2005. Structural insights into mRNA recognition from a PIWI domain-siRNA guide complex. *Nature* 434:663–666.

297. **Paterek, J. R., and P. H. Smith.** 1988. *Methanohalophilus mahii* gen. nov., sp. nov., a methylotrophic halophilic methanogen. *Int. J. Syst. Bacteriol.* 83:122–123.

298. **Peat, T. S., J. Newman, G. S. Waldo, J. Berendzen, and T. C. Terwilliger.** 1998. Structure of translation initiation factor 5A from *Pyrobaculum aerophilum* at 1.75 A resolution. *Structure* 6:1207–1214.

299. **Peck, R. F., S. Dassarma, and M. P. Krebs.** 2000. Homologous gene knockout in the archaeon *Halobacterium salinarum* with *ura3* as a counterselectable marker. *Mol. Microbiol.* 35:667–676.

300. **Perdrós-Alió, C.** 2004. Trophic ecology of solar salterns, p. 17–31. *In* A. Ventosa (ed.), *Halophilic Microorganisms.* Springer, New York, N.Y.

301. **Pereira, M. M., T. M. Bandeiras, A. S. Fernandes, R. S. Lemos, A. M. Melo, and M. Teixeira.** 2004. Respiratory chains from aerobic thermophilic prokaryotes. *J. Bioenerg. Biomembr.* 36:93–105.

302. **Peters, J., W. Baumeister, and A. Lupas.** 1996. Hyperthermostable surface layer protein tetrabrachion from the archaebacterium *Staphylothermus marinus*: evidence for the presence of a right-handed coiled coil derived from the primary structure. *J. Mol. Biol.* 257:1031–1041.

303. **Peters, J., M. Nitsch, B. Kuhlmorgen, R. Golbik, A. Lupas, J. Kellermann, H. Engelhardt, J. P. Pfander, S. Muller, K. Goldie, et al.** 1995. Tetrabrachion: a filamentous archaebacterial surface protein assembly of unusual structure and extreme stability. *J. Mol. Biol.* 245:385–401.

304. **Pfeifer, F., U. Blaseio, P. Ghahraman, et al.** 1988. Dynamic plasmid populations in *Halobacterium halobium. J. Bacteriol.* 170:3718–3724.

305. **Pfeifer, F., D. Gregor, A. Hofacker, P. Plosser, and P. Zimmermann.** 2002. Regulation of gas vesicle formation in halophilic archaea. *J. Mol. Microbiol. Biotechnol.* 4:175–181.

306. **Pfeifer, F., P. Zimmermann, S. Scheuch, and S. Sartorius-Neef.** 2005. Gene regulation and initiation of translation in halophilic Archaea, p. 201–216. *In* N. Gunde-Cimerman, A. Oren, and A. Plemenitas (ed.), *Adaptation to Life in High Salt Concentrations in Archaea, Bacteria, and Eukarya,* vol. 9. Springer, Dordrecht, The Netherlands.

307. **Phipps, B. M., A. Hoffmann, K. O. Stetter, and W. Baumeister.** 1991. A novel ATPase complex selectively accumulated upon heat shock is a major cellular component of thermophilic archaebacteria. *EMBO J.* 10:1711–1722.

308. **Pihl, T. D., L. K. Black, B. A. Schulman, and R. J. Maier.** 1992. Hydrogen-oxidizing electron transport components in the hyperthermophilic archaebacterium *Pyrodictium brockii. J. Bacteriol.* 174:137–143.

309. **Pihl, T. D., and R. J. Maier.** 1991. Purification and characterization of the hydrogen uptake hydrogenase from the hyperthermophilic archaebacterium *Pyrodictium brockii. J. Bacteriol.* 173:1839–1844.

310. **Praetorius-Ibba, M., and M. Ibba.** 2003. Aminoacyl-tRNA synthesis in archaea: different but not unique. *Mol. Microbiol.* 48:631–637.

311. **Prangishvili, D., I. Holz, E. Stieger, S. Nickell, J. K. Kristjansson, and W. Zillig.** 2000. Sulfolobicins, specific proteinaceous toxins produced by strains of the extremely thermophilic archaeal genus *Sulfolobus. J. Bacteriol.* 182:2985–2988.

312. **Prokofeva, M. I., I. V. Kublanov, O. Nercessian, T. P. Tourova, T. V. Kolganova, A. V. Lebedinsky, E. A. Bonch-Osmolovskaya, S. Spring, and C. Jeanthon.** 2005. Cultivated anaerobic acidophilic/acidotolerant thermophiles from terrestrial and deep-sea hydrothermal habitats. *Extremophiles* 9:437–448.

313. **Purdy, K. J., T. D. Cresswell-Maynard, D. B. Nedwell, T. J. McGenity, W. D. Grant, K. N. Timmis, and T. M. Embley.** 2004. Isolation of haloarchaea that grow at low salinities. *Environ. Microbiol.* 6:591–595.

314. **Purschke, W. G., C. L. Schmidt, A. Petersen, and G. Schäfer.** 1997. The terminal quinol oxidase of the hyperthermophilic archaeon *Acidianus ambivalens* exhibits a novel subunit structure and gene organization. *J. Bacteriol.* 179:1344–1353.

315. **Rachel, R., I. Wyschkony, S. Riehl, and H. Huber.** 2002. The ultrastructure of *Ignicoccus*: evidence for a novel outer membrane and for intracellular vesicle budding in an archaeon. *Archaea* 1:9–18.

316. **Raghoebarsing, A. A., A. Pol, K. T. van de Pas-Schoonen, A. J. Smolders, K. F. Ettwig, W. I. Rijpstra, S. Schouten, J. S. Damste, H. J. Op den Camp, M. S. Jetten, and M. Strous.** 2006. A microbial consortium couples anaerobic methane oxidation to denitrification. *Nature* 440:918–921.

317. Randau, L., M. Pearson, and D. Soll. 2005. The complete set of tRNA species in *Nanoarchaeum equitans. FEBS Lett.* 579:2945–2497.

318. Raskin, L., D. D. Zheng, M. E. Griffin, P. G. Stroot, and P. Misra. 1995. Characterization of microbial communities in anaerobic bioreactors using molecular probes. *Antonie Leeuwenhoek* 68:297–308.

319. Reeburgh, W. S. 1996. "Soft spots" in the global methane budget, p. 334–352. *In* M. E. Lidstrom and F. R. Tabita (ed.), *Microbial Growth on C1 Compounds.* Kluwer Academic Publishers, Dordrecht, The Netherlands.

320. Reeve, J. N. 2003. Archaeal chromatin and transcription. *Mol. Microbiol.* 48:587–598.

321. Reiter, W.-D., P. Palm, and W. Zillig. 1988. Transcription termination in the archaebacterium *Sulfolobus*: signal structures and linkage to transcription initiation. *Nucleic Acids Res.* 16:2445–2459.

322. Reiter, W. D., U. Hudepohl, and W. Zillig. 1990. Mutational analysis of an archaebacterial promoter: essential role of a TATA box for transcription efficiency and start-site selection in vitro. *Proc. Natl. Acad. Sci. USA* 87:9509–9513.

323. Reiter, W. D., P. Palm, and W. Zillig. 1988. Analysis of transcription in the archaebacterium *Sulfolobus* indicates that archaebacterial promoters are homologous to eukaryotic pol II promoters. *Nucleic Acids Res.* 16:1–19.

324. Reysenbach, A. L., and S. L. Cady. 2001. Microbiology of ancient and modern hydrothermal systems. *Trends Microbiol.* 9:79–86.

325. Rimbault, A., P. Niel, H. Virelizier, J. C. Darbord, and G. Leluan. 1988. l-Methionine, a Precursor of Trace Methane in Some Proteolytic Clostridia. *Appl. Environ. Microbiol.* 54:1581–1586.

326. Rivard, C. J., and P. H. Smith. 1082. Isolation and characterization of a thermophilic marine methanogenic bacterium, *Methanogenium thermophilicum* sp. nov. *Int. J. Syst. Bacteriol.* 32:430–436.

327. Rivas, F. V., N. H. Tolia, J. J. Song, J. P. Aragon, J. Liu, G. J. Hannon, and L. Joshua-Tor. 2005. Purified Argonaute2 and an siRNA form recombinant human RISC. *Nat. Struct. Mol. Biol.* 12:340–349.

328. Robertson, C. E., J. K. Harris, J. R. Spear, and N. R. Pace. 2005. Phylogenetic diversity and ecology of environmental Archaea. *Curr. Opin. Microbiol.* 8:638–642.

329. Robinson, N. P., I. Dionne, M. Lundgren, V. L. Marsh, R. Bernander, and S. D. Bell. 2004. Identification of two origins of replication in the single chromosome of the archaeon *Sulfolobus solfataricus. Cell* 116:25–38.

330. Romesser, J. A. 2001. *Methanogenium*, p. 256–258. *In* D. R. Boone and R. W. Castenholz (ed.), *Bergey's Manual of Systematic Bacteriology*, 2nd ed., vol. 1. Springer, New York, N.Y.

331. Ronimus, R. S., A. Reysenbach, D. R. Musgrave, and H. W. Morgan. 1997. The phylogenetic position of the *Thermococcus* isolate AN1 based on 16S rRNA gene sequence analysis: a proposal that AN1 represents a new species, *Thermococcus zilligii* sp. nov. *Arch. Microbiol.* 168:245–248.

332. Rosenshine, I., R. Tchelet, and M. Mevarech. 1989. The mechanism of DNA transfer in the mating system of an archaebacterium. *Science* 245:1387–1389.

333. Rother, M., and W. W. Metcalf. 2004. Anaerobic growth of *Methanosarcina acetivorans* C2A on carbon monoxide: an unusual way of life for a methanogenic archaeon. *Proc. Natl. Acad. Sci. USA* 101:16929–16934.

334. Roy, R., and M. W. Adams. 2002. Tungsten-dependent aldehyde oxidoreductase: a new family of enzymes containing the pterin cofactor. *Met. Ions Biol. Syst.* 39:673–697.

335. Rudolph, C., C. Moissl, R. Henneberger, and R. Huber. 2004. Ecology and microbial structures of archaeal/bacterial strings-of-pearls communities and archaeal relatives thriving in cold sulfidic springs. *FEMS Microbiol. Ecol.* 50:1–11.

336. Ruepp, A., W. Graml, M. L. Santos-Martinez, K. K. Koretke, C. Volker, H. W. Mewes, D. Frishman, S. Stocker, A. N. Lupas, and W. Baumeister. 2000. The genome sequence of the thermoacidophilic scavenger *Thermoplasma acidophilum. Nature* 407:508–513.

337. Russell, M., A. Hall, A. Cairns-Smith, and P. Braterman. 1988. Submarine hot springs and the origin of life. *Nature* 336:117.

338. Russo, S., and U. Baumann. 2004. Crystal structure of a dodecameric tetrahedral-shaped aminopeptidase. *J. Biol. Chem.* 279:51275–51281.

339. Sand, W., T. Gehrke, P. G. Jozsa, and A. Schippers. 2001. (Bio) chemistry of bacterial leaching − direct vs. indirect bioleaching. *Hydrometallurgy* 59:159–175.

340. Sandman, K., and J. N. Reeve. 2005. Archaeal chromatin proteins: different structures but common function? *Curr. Opin. Microbiol.* 8:656–661.

341. Santangelo, T. J., and J. N. Reeve. 2006. Archaeal RNA polymerase is sensitive to intrinsic termination directed by transcribed and remote sequences. *J. Mol. Biol.* 355:196–210.

342. Sapra, R., K. Bagramyan, and M. W. Adams. 2003. A simple energy-conserving system: proton reduction coupled to proton translocation. *Proc. Natl. Acad. Sci. USA* 100:7545–7450.

343. Sapra, R., M. F. Verhagen, and M. W. Adams. 2000. Purification and characterization of a membrane-bound hydrogenase from the hyperthermophilic archaeon *Pyrococcus furiosus. J. Bacteriol.* 182:3423–3428.

344. Saraste, M. 1999. Oxidative phosphorylation at the fin de siecle. *Science* 283:1488–1493.

345. Sartorius-Neef, S., and F. Pfeifer. 2004. *In vivo* studies on putative Shine-Dalgarno sequences of the halophilic archaeon *Halobacterium salinarum. Mol. Microbiol.* 51:579–588.

346. Sato, T., T. Fukui, H. Atomi, and T. Imanaka. 2005. Improved and versatile transformation system allowing multiple genetic manipulations of the hyperthermophilic archaeon *Thermococcus kodakaraensis. Appl. Environ. Microbiol.* 71:3889–3899.

347. Sato, T., T. Fukui, H. Atomi, and T. Imanaka. 2003. Targeted gene disruption by homologous recombination in the hyperthermophilic archaeon *Thermococcus kodakaraensis* KOD1. *J. Bacteriol.* 185:210–220.

348. Saunders, N. F., T. Thomas, P. M. Curmi, J. S. Mattick, E. Kuczek, R. Slade, J. Davis, P. D. Franzmann, D. Boone, K. Rusterholtz, R. Feldman, C. Gates, S. Bench, K. Sowers, K. Kadner, A. Aerts, P. Dehal, C. Detter, T. Glavina, S. Lucas, P. Richardson, F. Larimer, L. Hauser, M. Land, and R. Cavicchioli. 2003. Mechanisms of thermal adaptation revealed from the genomes of the Antarctic Archaea *Methanogenium frigidum* and *Methanococcoides burtonii. Genome Res.* 13:1580–1588.

349. Schäfer, G., S. Anemüller, and R. Moll. 2002. Archaeal complex II: 'classical' and 'non-classical' succinate:quinone reductases with unusual features. *Biochim. Biophys. Acta* 1553:57–73.

350. Scharf, B., R. Wittenberg, and M. Engelhard. 1997. Electron transfer proteins from the haloalkaliphilic archaeon *Natronobacterium pharaonis*: possible components of the respiratory chain include cytochrome bc and a terminal oxidase cytochrome ba3. *Biochemistry* 36:4471–4479.

351. Schleifer, K., J. Sther, and H. Mayer. 1982. Chemical composition and structure of teh cell wall of *Halococcus morrhuae. Zentbl. Bacteriol. Hyg. I Abt. Orig.* C3:171–178.

352. Schleper, C., I. Holz, D. Janekovic, J. Murphy, and W. Zillig. 1995. A multicopy plasmid of the extremely thermophilic archaeon *Sulfolobus* effects its transfer to recipients by mating. *J. Bacteriol.* **177:**4417–4426.

353. Schleper, C., G. Jurgens, and M. Jonuscheit. 2005. Genomic studies of uncultivated archaea. *Nat. Rev. Microbiol.* **3:**479–488.

354. Schleper, C., G. Puehler, I. Holz, A. Gambacorta, D. Janekovic, U. Santarius, H. P. Klenk, and W. Zillig. 1995. *Picrophilus gen. nov., fam. nov.*: a novel aerobic, heterotrophic, thermoacidophilic genus and family comprising archaea capable of growth around pH 0. *J. Bacteriol.* **177:**7050–7059.

355. Schleper, C., G. Puhler, B. Kuhlmorgen, and W. Zillig. 1995. Life at extremely low pH. *Nature* **375:**741–742.

356. Schleper, C., R. V. Swanson, E. J. Mathur, and E. F. DeLong. 1997. Characterization of a DNA polymerase from the uncultivated psychrophilic archaeon *Cenarchaeum symbiosum*. *J. Bacteriol.* **179:**7803–7811.

357. Schloss, P. D., and J. Handelsman. 2004. Status of the microbial census. *Microbiol. Mol. Biol. Rev.* **68:**686–691.

358. Schmidt, C. L., S. Anemuller, and G. Schafer. 1996. Two different respiratory Rieske proteins are expressed in the extreme thermoacidophilic crenarchaeon *Sulfolobus acidocaldarius*: cloning and sequencing of their genes. *FEBS Lett.* **388:**43–46.

359. Scholz, S., J. Sonnenbichler, W. Schafer, and R. Hensel. 1992. Di-myo-inositol-1,1′-phosphate: a new inositol phosphate isolated from *Pyrococcus woesei*. *FEBS Lett.* **306:**239–242.

360. Schoonen, M. A., Y. Xu, and J. Bebie. 1999. Energetics and kinetics of the prebiotic synthesis of simple organic acids and amino acids with the FeS-H_2S/FeS$_2$ redox couple as reductant. *Orig. Life Evol. Biosph.* **29:**5–32.

361. Schrenk, M. O., D. S. Kelley, S. A. Bolton, and J. A. Baross. 2004. Low archaeal diversity linked to subseafloor geochemical processes at the Lost City Hydrothermal Field, Mid-Atlantic Ridge. *Environ. Microbiol.* **6:**1086–1095.

362. Schümann, H., and F. Klink. 1989. Archaebacterial protein contains hypusine, a unique amino acid characteristic for eukaryotic translation initiation factor 4D *Syst. Appl. Microbiol.* **11:**103–107.

363. Schut, G., J. Zhou, and M. Adams. 2001. DNA microarray analysis of the hyperthermophilic archaeon *Pyrococcus furiosus*: evidence for a new type of sulfur-reducing enzyme complex. *J. Bacteriol.* **183:**7027–7036.

364. Shima, S., D. A. Herault, A. Berkessel, and R. K. Thauer. 1998. Activation and thermostabilization effects of cyclic 2,3-diphosphoglycerate on enzymes from the hyperthermophilic *Methanopyrus kandleri*. *Arch. Microbiol.* **170:**469–472.

365. Shima, S., and R. K. Thauer. 2005. Methyl-coenzyme M reductase and the anaerobic oxidation of methane in methanotrophic Archaea. *Curr. Opin. Microbiol.* **8:**643–648.

366. Shima, S., R. K. Thauer, and U. Ermler. 2004. Hyperthermophilic and salt-dependent formyltransferase from *Methanopyrus kandleri*. *Biochem. Soc. Trans.* **32:**269–272.

367. Shivvers, D. G., and T. D. Brock. 1973. Oxidation of elemental sulfur by *Sulfolobus acidocaldarius*. *J. Bacteriol.* **114:**706–710.

368. Shukla, H. D., and S. DasSarma. 2004. Complexity of gas vesicle biogenesis in *Halobacterium* sp. strain NRC-1: identification of five new proteins. *J. Bacteriol.* **186:**3182–3186.

369. Simonson, A. B., J. A. Servin, R. G. Skophammer, C. W. Herbold, M. C. Rivera, and J. A. Lake. 2005. Decoding the genomic tree of life. *Proc. Natl. Acad. Sci. USA* **102**(Suppl. 1):6608–6613.

370. Siqueira, J. F., Jr., I. N. Rocas, J. C. Baumgartner, and T. Xia. 2005. Searching for Archaea in infections of endodontic origin. *J. Endod.* **31:**719–722.

371. Slesarev, A. I., K. V. Mezhevaya, K. S. Makarova, N. N. Polushin, O. V. Shcherbinina, V. V. Shakhova, G. I. Belova, L. Aravind, D. A. Natale, I. B. Rogozin, R. L. Tatusov, Y. I. Wolf, K. O. Stetter, A. G. Malykh, E. V. Koonin, and S. A. Kozyavkin. 2002. The complete genome of hyperthermophile *Methanopyrus kandleri* AV19 and monophyly of archaeal methanogens. *Proc. Natl. Acad. Sci. USA* **99:**4644–4649.

372. Slupska, M. M., A. G. King, S. Fitz-Gibbon, J. Besemer, M. Borodovsky, and J. H. Miller. 2001. Leaderless transcripts of the crenarchaeal hyperthermophile *Pyrobaculum aerophilum*. *J. Mol. Biol.* **309:**347–360.

373. Smith, D. R., L. A. Doucette-Stamm, C. Deloughery, H. Lee, J. Dubois, T. Aldredge, R. Bashirzadeh, D. Blakely, R. Cook, K. Gilbert, D. Harrison, L. Hoang, P. Keagle, W. Lumm, B. Pothier, D. Qiu, R. Spadafora, R. Vicaire, Y. Wang, J. Wierzbowski, R. Gibson, N. Jiwani, A. Caruso, D. Bush, J. N. Reeve, et al. 1997. Complete genome sequence of *Methanobacterium thermoautotrophicum* deltaH: functional analysis and comparative genomics. *J. Bacteriol.* **179:**7135–7155.

374. Song, J. J., S. K. Smith, G. J. Hannon, and L. Joshua-Tor. 2004. Crystal structure of Argonaute and its implications for RISC slicer activity. *Science* **305:**1434–1437.

375. Sorokin, D. Y., T. P. Tourova, and G. Muyzer. 2005. Oxidation of thiosulfate to tetrathionate by an haloarchaeon isolated from hypersaline habitat. *Extremophiles* **9:**501–504.

376. Sperling, D., U. Kappler, H. G. Trüper, and C. Dahl. 2001. Dissimilatory ATP sulfurylase from *Archaeoglobus fulgidus*. *Methods Enzymol.* **331:**419–427.

377. Spreter, T., M. Pech, and B. Beatrix. 2005. The crystal structure of archaeal nascent polypeptide-associated complex (NAC) reveals a unique fold and the presence of a ubiquitin-associated domain. *J. Biol. Chem.* **280:**15849–15854.

378. Spudich, J. L., and H. Luecke. 2002. Sensory rhodopsin II: functional insights from structure. *Curr. Opin. Struct. Biol.* **12:**540–546.

379. Sreeramulu, K., C. L. Schmidt, G. Schafer, and S. Anemuller. 1998. Studies of the electron transport chain of the euryarcheon *Halobacterium salinarum*: indications for a type II NADH dehydrogenase and a complex III analog. *J. Bioenerg. Biomembr.* **30:**443–453.

380. Staley, J. T. 1968. *Prosthecomicrobium* and *Ancalomicrobium*: new prosthecate freshwater bacteria. *J. Bacteriol.* **95:**1921–1942.

381. Stan-Lotter, H., T. J. McGenity, A. Legat, E. B. Denner, K. Glaser, K. O. Stetter, and G. Wanner. 1999. Very similar strains of *Halococcus salifodinae* are found in geographically separated permo-triassic salt deposits. *Microbiology* **145** (Pt 12):3565–3574.

382. Stan-Lotter, H., C. Radax, T. J. McGenity, A. Legat, M. Pfaffenhuemer, H. Wieland, C. Gruber, and E. B. M. Denner. 2004. From intraterrestrials to extraterrestrials—viable haloarchaea in ancient salt deposits, p. 17–31. *In* A. Ventosa (ed.), *Halophilic Microorganisms*. Springer, New York, N.Y.

383. Steensland, H., and H. Larsen. 1969. A study of the cell envelope of the halobacteria. *J. Gen. Microbiol.* **55:**325–336.

384. Steitz, T. A. 2005. On the structural basis of peptide-bond formation and antibiotic resistance from atomic structures of the large ribosomal subunit. *FEBS Lett.* **579:**955–958.

385. Stetter, K. O. 1999. Extremophiles and their adaptation to hot environments. *FEBS Lett.* **452:**22–25.

386. Stetter, K. O. 1996. Hyperthermophilic procaryotes. *FEMS Microbiol. Rev.* **18:**149–158.

387. Stetter, K. O. 2001. *Methanothermus*, p. 233–235. *In* D. R. Boone and R. W. Castenholz (ed.), *Bergey's Manual of Systematic Bacteriology*, 2nd ed., vol. 1. Springer, New York, N.Y.

388. Stetter, K. O. 2001. *Pyrodictium*, p. 192–195. *In* D. R. Boone and R. W. Castenholz (ed.), *Bergey's Manual of Systematic Bacteriology*, 2nd ed., vol. 1. Springer, New York, N.Y.

389. Stetter, K. O. 1982. Ultrathin mycelia-forming organisms from submarine volcanic areas having an optimum growth temperature of 105°C. *Nature* **300:**258–260.

390. Stoeckenius, W. 1981. Walsby's square bacterium: fine structure of an orthogonal procaryote. *J. Bacteriol.* **148:**352–360.

391. Stoiber, R. E. 1995. Volcanic gases from subaerial volcanoes on Earth, p. 308–319. *In* T. J. Ahrens (ed.), *Global Earth Physics: a Handbook Of Physical Constants*, vol. 1. American Geophysical Union, Washington, D.C.

392. Stolt, P., and W. Zillig. 1993. Antisense RNA mediates transcriptional processing in an archaebacterium, indicating a novel kind of RNase activity. *Mol. Microbiol.* **7:**875–882.

393. Stolt, P., and W. Zillig. 1994. Transcription of the halophage phi H repressor gene is abolished by transcription from an inversely oriented lytic promoter. *FEBS Lett.* **344:**125–128.

394. Stone, D. A., and R. E. Goldstein. 2004. Tubular precipitation and redox gradients on a bubbling template. *Proc. Natl. Acad. Sci. USA* **101:**11537–11541.

395. Sun, C., Y. Li, S. Mei, Q. Lu, L. Zhou, and H. Xiang. 2005. A single gene directs both production and immunity of halocin C8 in a haloarchaeal strain AS7092. *Mol. Microbiol.* **57:**537–549.

396. Taguchi, Y., M. Sugishima, and K. Fukuyama. 2004. Crystal structure of a novel zinc-binding ATP sulfurylase from *Thermus thermophilus* HB8. *Biochemistry* **43:**4111–4118.

397. Tajima, K., T. Nagamine, H. Matsui, M. Nakamura, and R. I. Aminov. 2001. Phylogenetic analysis of archaeal 16S rRNA libraries from the rumen suggests the existence of a novel group of archaea not associated with known methanogens. *FEMS Microbiol. Lett.* **200:**67–72.

398. Takai, K., T. Gamo, U. Tsunogai, N. Nakayama, H. Hirayama, K. H. Nealson, and K. Horikoshi. 2004. Geochemical and microbiological evidence for a hydrogen-based, hyperthermophilic subsurface lithoautotrophic microbial ecosystem (HyperSLiME) beneath an active deep-sea hydrothermal field. *Extremophiles* **8:**269–282.

399. Takai, K., T. Komatsu, F. Inagaki, and K. Horikoshi. 2001. Distribution of archaea in a black smoker chimney structure. *Appl. Environ. Microbiol.* **67:**3618–3629.

400. Takai, K., A. Sugai, T. Itoh, and K. Horikoshi. 2000. *Palaeococcus ferrophilus* gen. nov., sp. nov., a barophilic, hyperthermophilic archaeon from a deep-sea hydrothermal vent chimney. *Int. J. Syst. Evol. Microbiol.* **50**(Pt 2)**:**489–500.

401. Takakuwa, S. 1992. Biochemical aspects of microbial oxidation of sulfur compounds, p. 1–43. *In* S. Oae (ed.), *Organic Sulfur Chemistry: Biochemical Aspects*. CRC Press, Boca Raton, Fla.

402. Theissen, U., M. Hoffmeister, M. Grieshaber, and W. Martin. 2003. Single eubacterial origin of eukaryotic sulfide: quinone oxidoreductase, a mitochondrial enzyme conserved from the early evolution of eukaryotes during anoxic and sulfidic times. *Mol. Biol. Evol.* **20:**1564–1574.

403. Tor, J. M., K. Kashefi, and D. R. Lovley. 2001. Acetate oxidation coupled to Fe(iii) reduction in hyperthermophilic microorganisms. *Appl. Environ. Microbiol.* **67:**1363–1365.

404. Tor, J. M., and D. R. Lovley. 2001. Anaerobic degradation of aromatic compounds coupled to Fe(III) reduction by *Ferroglobus placidus*. *Environ. Microbiol.* **3:**281–287.

405. Treusch, A. H., S. Leininger, A. Kletzin, S. C. Schuster, H. P. Klenk, and C. Schleper. 2005. Novel genes for nitrite reductase and Amo-related proteins indicate a role of uncultivated mesophilic crenarchaeota in nitrogen cycling. *Environ. Microbiol.* **7:**1985–1995.

406. Trumpower, B. L., and R. B. Gennis. 1994. Energy transduction by cytochrome complexes in mitochondrial and bacterial respiration: the enzymology of coupling electron transfer reactions to transmembrane proton translocation. *Annu. Rev. Biochem.* **63:**675–716.

407. Tumbula, D. L., and W. B. Whitman. 1999. Genetics of *Methanococcus*: possibilities for functional genomics in Archaea. *Mol. Microbiol.* **33:**1–7.

408. Tyson, G. W., J. Chapman, P. Hugenholtz, E. E. Allen, R. J. Ram, P. M. Richardson, V. V. Solovyev, E. M. Rubin, D. S. Rokhsar, and J. F. Banfield. 2004. Community structure and metabolism through reconstruction of microbial genomes from the environment. *Nature* **428:**37–43.

409. Urich, T., T. M. Bandeiras, S. S. Leal, R. Rachel, T. Albrecht, P. Zimmermann, C. Scholz, M. Teixeira, C. M. Gomes, and A. Kletzin. 2004. The sulphur oxygenase reductase from *Acidianus ambivalens* is a multimeric protein containing a low-potential mononuclear non-haem iron centre. *Biochem. J.* **381:**137–146.

410. Urich, T., R. Coelho, A. Kletzin, and C. Frazao. 2005. The sulfur oxygenase reductase from *Acidianus ambivalens* is an icosatetramer as shown by crystallization and Patterson analysis. *Biochim. Biophys. Acta* **1747:**267–270.

411. Urich, T., C. M. Gomes, A. Kletzin, and C. Frazao. 2006. X-ray structure of a self-compartmentalizing sulfur cycle metalloenzyme. *Science* **311:**996–1000.

412. Urich, T., A. Kroke, C. Bauer, K. Seyfarth, M. Reuff, and A. Kletzin. 2005. Identification of core active site residues of the sulfur oxygenase reductase from *Acidianus ambivalens* by site-directed mutagenesis. *FEMS Microbiol. Lett.* **248:**171–176.

413. van de Vossenberg, J. L., A. J. Driessen, W. Zillig, and W. N. Konings. 1998. Bioenergetics and cytoplasmic membrane stability of the extremely acidophilic, thermophilic archaeon Picrophilus oshimae. *Extremophiles* **2:**67–74.

414. van Keulen, G., D. A. Hopwood, L. Dijkhuizen, and R. G. Sawers. 2005. Gas vesicles in actinomycetes: old buoys in novel habitats? *Trends Microbiol.* **13:**350–354.

415. Vianna, M. E., G. Conrads, B. P. Gomes, and H. P. Horz. 2006. Identification and quantification of archaea involved in primary endodontic infections. *J. Clin. Microbiol.* **44:**1274–1282.

416. Vieille, C., and G. J. Zeikus. 2001. Hyperthermophilic enzymes: sources, uses, and molecular mechanisms for thermostability. *Microbiol. Mol. Biol. Rev.* **65:**1–43.

417. Voisin, S., R. S. Houliston, J. Kelly, J. R. Brisson, D. Watson, S. L. Bardy, K. F. Jarrell, and S. M. Logan. 2005. Identification and characterization of the unique N-linked glycan common to the flagellins and S-layer glycoprotein of *Methanococcus voltae*. *J. Biol. Chem.* **280:**16586–16593.

418. von Damm, K. L. 1990. Seafloor hydrothermal activity: black smoker chemistry and chimmneys. *Annu. Rev. Earth Planet. Sci.* **18:**173–204.

419. Wachtershauser, G. 1990. Evolution of the first metabolic cycles. *Proc. Natl. Acad. Sci. USA* **87:**200–204.

420. Wächtershäuser, G. 2002. Origin of life: RNA world versus autocatalytic anabolism. *In* M. Dworkin (ed.), *The Prokaryotes: An Evolving Electronic Resource for the Microbiological Community*, 3rd ed., release 3.8 ed. Springer-Verlag, New York, N.Y. [Online.] http://141.150.157.117:8080/prokPUB/index.htm.

421. Walsby, A. E. 2005. Archaea with square cells. *Trends Microbiol.* **13:**193–195.

422. Walsby, A. E. 1994. Gas vesicles. *Microbiol. Rev.* **58:**94–144.

423. Walsby, A. E. 1980. A square bacterium. *Nature* **283:**69–71.

424. Wan, X. F., S. M. Bridges, and J. A. Boyle. 2004. Revealing gene transcription and translation initiation patterns in archaea, using an interactive clustering model. *Extremophiles* **8:**291–299.

425. Wang, L., A. Brock, B. Herberich, and P. G. Schultz. 2001. Expanding the genetic code of *Escherichia coli*. *Science* **292:**498–500.

426. Wasserfallen, A., J. Nolling, P. Pfister, J. Reeve, and E. Conway de Macario. 2000. Phylogenetic analysis of 18 thermophilic *Methanobacterium* isolates supports the proposals to create a new genus, Methanothermobacter gen. nov., and to reclassify several isolates in three species, *Methanothermobacter thermautotrophicus comb. nov.*, *Methanothermobacter wolfeii comb. nov.*, and *Methanothermobacter marburgensis sp. nov. Int. J. Syst. Evol. Microbiol.* 50(Pt 1):43–53.

427. Watanabe, K., Y. Hata, H. Kizaki, Y. Katsube, and Y. Suzuki. 1997. The refined crystal structure of Bacillus cereus oligo-1,6-glucosidase at 2.0 A resolution: structural characterization of proline-substitution sites for protein thermostabilization. *J. Mol. Biol.* 269:142–153.

428. Watanabe, Y., S. Yokobori, T. Inaba, A. Yamagishi, T. Oshima, Y. Kawarabayasi, H. Kikuchi, and K. Kita. 2002. Introns in protein-coding genes in Archaea. *FEBS Lett.* 510:27–30.

429. Waters, E., M. J. Hohn, I. Ahel, D. E. Graham, M. D. Adams, M. Barnstead, K. Y. Beeson, L. Bibbs, R. Bolanos, M. Keller, K. Kretz, X. Lin, E. Mathur, J. Ni, M. Podar, T. Richardson, G. G. Sutton, M. Simon, D. Soll, K. O. Stetter, J. M. Short, and M. Noordewier. 2003. The genome of *Nanoarchaeum equitans*: insights into early archaeal evolution and derived parasitism. *Proc. Natl. Acad. Sci. USA* 100:12984–12988.

430. Whitman, W. B., and C. Jeanthon. 2002. Methanococcales. *In* M. Dworkin (ed.), *The Prokaryotes: An Evolving Electronic Resource for the Microbiological Community*, 3rd ed., release 3.9 ed. Springer-Verlag, New York, N.Y. [Online.] http://141.150.157.117:8080/prokPUB/index.htm.

431. Wich, G., H. Hummel, M. Jarsch, U. Bär, and A. Böck. 1986. Transcription signals for stable RNA genes in *Methanococcus. Nucleic Acids Res.* 6:2459–2479.

432. Wienholds, E., and R. H. Plasterk. 2005. MicroRNA function in animal development. *FEBS Lett.* 579:5911–5922.

433. Woese, C. R., L. Achenbach, P. Rouviere, and L. Mandelco. 1991. Archaeal phylogeny: Reexamination of the phylogenetic position in light of certain composition-induced artifacts. *Syst. Appl. Microbiol.* 14:364–371.

434. Woese, C. R., and G. E. Fox. 1977. Phylogenetic structure of the prokaryotic domain: the primary kingdoms. *Proc. Natl. Acad. Sci. USA* 74:5088–5090.

435. Woese, C. R., O. Kandler, and M. L. Wheelis. 1990. Toward a natural system of organisms: Proposal for the domains Archaea, Bacteria, and Eukarya. *Proc. Natl. Acad. Sci. USA* 87:4576–4579.

436. Wolfe, R. 2002. The Archaea: a personal overview of the formative years. *In* M. Dworkin (ed.), *The Prokaryotes: An Evolving Electronic Resource for the Microbiological Community.* 3rd ed., release 3.11. Springer-Verlag, New York, N.Y. [Online.] http://141.150.157.117:8080/prokPUB/index.htm.

437. Xie, J., and P. G. Schultz. 2005. An expanding genetic code. *Methods* 36:227–238.

438. Xu, W., P. J. Mulhern, B. L. Blackford, M. H. Jericho, M. Firtel, and T. J. Beveridge. 1996. Modeling and measuring the elastic properties of an archaeal surface, the sheath of *Methanospirillum hungatei*, and the implication of methane production. *J. Bacteriol.* 178:3106–3112.

439. Yao, M., A. Ohsawa, S. Kikukawa, I. Tanaka, and M. Kimura. 2003. Crystal structure of hyperthermophilic archaeal initiation factor 5A: a homologue of eukaryotic initiation factor 5A (eIF-5A). *J. Biochem. (Tokyo)* 133:75–81.

440. Yeh, A. P., Y. Hu, F. E. Jenney, Jr., M. W. Adams, and D. C. Rees. 2000. Structures of the superoxide reductase from *Pyrococcus furiosus* in the oxidized and reduced states. *Biochemistry* 39:2499–2508.

441. Yonath, A. 2005. Antibiotics targeting ribosomes: resistance, selectivity, synergism and cellular regulation. *Annu. Rev. Biochem.* 74:649–679.

442. Yuan, Y. R., Y. Pei, J. B. Ma, V. Kuryavyi, M. Zhadina, G. Meister, H. Y. Chen, Z. Dauter, T. Tuschl, and D. J. Patel. 2005. Crystal structure of *A. aeolicus* argonaute, a site-specific DNA-guided endoribonuclease, provides insights into RISC-mediated mRNA cleavage. *Mol. Cell.* 19:405–419.

443. Zablen, L. B., M. S. Kissil, C. R. Woese, and D. E. Buetow. 1975. Phylogenetic origin of the chloroplast and prokaryotic nature of its ribosomal RNA. *Proc. Natl. Acad. Sci. USA* 72:2418–2422.

444. Zhang, Y., P. V. Baranov, J. F. Atkins, and V. N. Gladyshev. 2005. Pyrrolysine and selenocysteine use dissimilar decoding strategies. *J. Biol. Chem.* 280:20740–20751.

445. Zharkov, M. A. 1981. *History of Palaeozoic Salt Accumulation.* Springer, New York, N.Y.

446. Zhilina, T. N., and G. A. Zavarzin. 1987. *Methanohalobium evestigatus, n. gen., n. sp.*—an extremely halophilic methanogenic archaebacterium. *Dokl. Akad. Nauk SSSR* 293:464–468.

447. Zillig, W., I. Holz, D. Janekovic, H. P. Klenk, E. Imsel, J. Trent, S. Wunderl, V. H. Forjaz, R. Coutinho, and T. Ferreira. 1990. *Hyperthermus butylicus*, a hyperthermophilic sulfur-reducing archaebacterium that ferments peptides. *J. Bacteriol.* 172:3959–65.

448. Zillig, W., I. Holz, H.-P. Klenk, J. Trent, S. Wunderl, D. Janekovic, E. Imsel, and B. Haas. 1987. *Pyrococcus wosei*, sp. nov., an ultra-thermophilic marine archaebacterium, representing a novel order, *Thermococcales. Syst. Appl. Microbiol.* 9:62–70.

449. Zillig, W., H. P. Klenk, P. Palm, G. Puhler, F. Gropp, R. A. Garrett, and H. Leffers. 1989. The phylogenetic relations of DNA-dependent RNA polymerases of archaebacteria, eukaryotes, and eubacteria. *Can. J. Microbiol.* 35:73–80.

450. Zillig, W., A. Kletzin, C. Schleper, I. Holz, D. Janekovic, J. Hain, M. Lanzendörfer, and J. K. Kristjansson. 1994. Screening for *Sulfolobales*, their plasmids and their viruses in icelandic solfataras. *Syst. Appl. Microbiol.* 16:609–628.

451. Zillig, W., and A. L. Reysenbach. 2001. *Thermofilum*, p. 178–179. *In* D. R. Boone and R. W. Castenholz (ed.), *Bergey's Manual of Systematic Bacteriology*, 2nd ed., vol. 1. Springer, New York, N.Y.

452. Zillig, W., and A. L. Reysenbach. 2001. *Thermoproteus*, p. 171–173. *In* D. R. Boone and R. W. Castenholz (ed.), *Bergey's Manual of Systematic Bacteriology*, 2nd ed., vol. 1. Springer, New York, N.Y.

453. Zillig, W., K. O. Stetter, and D. Janekovic. 1979. DNA-dependent RNA polymerase from the archaebacterium *Sulfolobus acidocaldarius. Eur. J. Biochem.* 96:597–604.

454. Zillig, W., K. O. Stetter, and M. Tobien. 1978. DNA-dependent RNA polymerase from *Halobacterium halobium. Eur. J. Biochem.* 91:193–199.

455. Zillig, W., K. O. Stetter, S. Wunderl, W. Schulz, H. Priess, and I. Scholz. 1980. The *Sulfolobus-*"Caldariella" group: taxonomy on the basis of the structure of DNA-dependent RNA polymerases. *Arch. Microbiol.* 125:259–269.

456. Zimmermann, P., S. Laska, and A. Kletzin. 1999. Two modes of sulfite oxidation in the extremely thermophilic and acidophilic archaeon *Acidianus ambivalens. Arch. Microbiol.* 172:76–82.

457. Zimmermann, P., and F. Pfeifer. 2003. Regulation of the expression of gas vesicle genes in *Haloferax mediterranei*: interaction of the two regulatory proteins GvpD and GvpE. *Mol. Microbiol.* 49:783–794.

458. Zivanovic, Y., P. Lopez, H. Philippe, and P. Forterre. 2002. *Pyrococcus* genome comparison evidences chromosome shuffling-driven evolution. *Nucleic Acids Res.* 30:1902–1910.

Archaea: Molecular and Cellular Biology
Edited by Ricardo Cavicchioli
© 2007 ASM Press, Washington, D.C.

Chapter 3

DNA Replication and Cell Cycle

Si-houy Lao-Sirieix, Victoria L. Marsh, and Stephen D. Bell

INTRODUCTION

Replication of DNA is of fundamental importance to all living organisms. Inappropriate or untimely replication can have severe consequences at the cellular and organismal level. Replication must also be highly accurate to ensure faithful propagation of the genetic information, yet errors must be tolerated to permit the generation of diversity upon which evolutionary selective processes can act. This chapter describes the recent advances that have been made in understanding the biochemical players that facilitate the complex macromolecular process that mediates faithful replication of archaeal chromosomes. Furthermore, it is clear that once DNA replication is initiated it must progress to completion. Thus, cells have dedicated periods within the cell cycle to devote energy and metabolites toward the highly costly task of replicating the genome. The current state of knowledge of the machineries that drive the archaeal cell cycle is discussed.

Conceptually and mechanistically, DNA replication can be split into several stages. First, a site at which replication initiates, an origin of replication, must be defined at the molecular level (Fig. 1A and B). This leads to the recruitment of a DNA helicase, an enzyme that utilizes energy in the form of ATP, to unwind the double helix of DNA at the origin, and to expose the single-stranded DNA template for synthesis of new DNA (Fig. 1C). The single-stranded DNA is stabilized by specialized single-strand-binding proteins, and DNA synthesis is initiated by the generation of a short oligoribonucleotide primer (Fig. 1D). This primer is then extended by cellular DNA polymerases (Fig. 1E). An additional complexity arises at this point due to innate chemical asymmetry of DNA. All known primases and DNA polymerases can only

synthesize DNA in a 5′ to 3′ polarity. To overcome this potential problem, as the two strands of DNA are copied, one strand, the leading strand, is synthesized continuously, while the other strand is synthesized in short segments (Okazaki fragments), which require processing and joining to form covalently intact DNA molecules (Fig. 1F and G). Thus, there are very different requirements imposed on leading- and lagging-strand DNA polymerases. More specifically, the polymerase on the leading strand must synthesize DNA highly processively, remaining attached to the template for potentially many megabases. In contrast, the lagging-strand DNA polymerase must constantly undergo a repetitive cycle of recruitment, synthesis of a short DNA molecule followed by release of the template. These highly differing properties are imposed, at least in part, by the regulated association of DNA polymerase with an accessory factor, the sliding clamp. The sliding clamp also serves as a central nexus for the coordination of the proteins involved in processing and joining Okazaki fragments (Fig. 1G).

Table 1 summarizes the proteins that catalyze this complex cascade of events. It is readily apparent from this list that the machineries employed by the *Archaea* and *Eucarya* to perform these multiple tasks are closely related. Furthermore, although analogous activities are found in bacterial cells, the proteins that catalyze the events are nonorthologous with their archaeo-eucaryal counterparts. The relationship between the archaeal and eucaryal proteins, coupled with the fact that the archaeal machinery is generally perceived as a simplified version of the eucaryal machinery, has generated considerable interest in the *Archaea* as an experimentally tractable model for the fundamentally related yet massively more complex eucaryal replication machinery (5, 32, 59).

Si-houy Lao-Sirieix, Victoria L. Marsh, and Stephen D. Bell • MRC Cancer Cell Unit, Hutchison MRC Research Centre, Hills Road, Cambridge CB2 2XZ, United Kingdom.

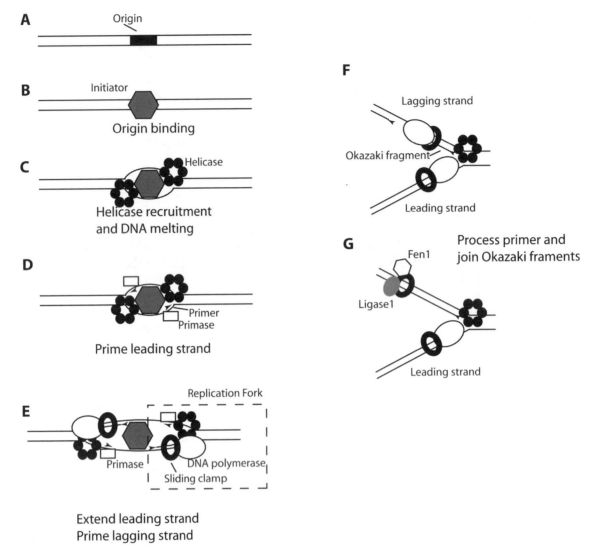

Figure 1. Cartoon of the steps involved in DNA replication. Panels A to E describe the assembly of replication fork components at an origin of replication. SSB has been omitted for visual clarity. Panels F and G show detail at a single replication fork.

ORIGINS OF ORIGINS

Despite the similarity between the DNA replication proteins in *Archaea* and *Eucarya*, there is a fundamental difference in the way that genomic DNA is organized in these two domains of life. In eucarya, chromosomes contain linear DNA molecules; in contrast, archaea and most bacteria have circular genomes. In addition, the DNA content of eucaryal cells can be three to four orders of magnitude greater than that in bacteria and archaea.

This difference in organization and abundance of DNA between *Eucarya* and *Bacteria* is mirrored by the fundamentally different way in which they organize DNA replication. Clearly, eucaryal cells do not take 10,000 times as long as bacterial cells to replicate their DNA. Rather, eucarya exploit a strategy of utilizing multiple initiation sites along their chromosomes that are typically 10 to 300 kb apart (4, 96). The way in which these origins are defined and regulated is beyond the scope of this chapter but has been reviewed elsewhere recently (82, 96). In contrast, bacteria use a single origin of replication per chromosome (53). Given the organizational similarity between bacterial and archaeal chromosomes, it might be anticipated that archaea would also have a single origin per chromosome. This supposition was confirmed by the pioneering work of the Forterre laboratory, which provided strong evidence that there was a single origin of replication (*oriC*), in the *Pyrococcus* genome (87). Initial bioinformatics and labeling studies predicted a single origin, and this was confirmed by a two-dimensional (2D) gel-mapping approach that allows the identification of replication intermedi-

Table 1. Identity of the factors that catalyze the various stages of DNA replication described in Fig. 1

Factor	Archaea[a]	Eucarya[a]	Bacteria (E. coli)
Initiator	Orc1/Cdc6	ORC	DnaA
Helicase loader	Orc1/Cdc6	Cdc6 + Cdt1	DnaC
Helicase	MCM	MCM complex	DnaB
Single-strand binding	Various[b]	RPA	SSB
Primase	Primase (2)	Primosome (4)	DnaG
Replicative DNA polymerase	B-type (C+E)[c] D-type (E)	Pol δ and ε	DNA Pol III
Sliding clamp	PCNA	PCNA	β-Clamp
Clamp loader	RF-C	RF-C	γ-Complex
Ligase	DNA lig 1	DNA lig 1	Ligase
Processing	RNaseH/Fen1	RNaseH/Dna2/Fen1	RNaseH

[a]The eucaryal ORC complex has six subunits, Orc1 to Orc6. Archaea possess proteins that are homologous to both Orc1 and Cdc6 and are termed Orc1/Cdc6; see the main text for discussion.
[b]Distinct archaeal species have unique single strand binding proteins, hence the designation of "various."
[c]There is a bifurcation in the distribution of polymerases in the *Crenarchaeota* (C) and *Euryarchaeota* (E). Crenarchaea possess only B-type replicative DNA polymerases, whereas euryarchaea have both B- and D-family enzymes.

ates (77). Subsequently, the precise initiation site was mapped using high-resolution methodologies (78). The *Pyrococcus* origin lies immediately upstream of the gene encoding the *Pyrococcus* homolog of the eucaryal DNA replication factors Orc1 and Cdc6 (see footnote to Table 1). This organization was immediately reminiscent of that seen in many bacteria, where the gene for the DnaA initiator is adjacent to the origin of replication (83). Thus, *Pyrococcus* appears to use a bacterial-like mode of DNA replication with eucaryal-like machinery (87).

At this time, however, bioinformatics studies, based on the poorly understood observation that leading and lagging strands often possess distinct sequence properties, were beginning to give hints that other archaeal species may possess more than one origin of replication (114). In particular, *Halobacterium* was proposed to have two origins of replication, both adjacent to genes for Orc1/Cdc6 homologs. However, a targeted genetic screen for autonomous replication function only found evidence for one of these predicted origins (upstream of the *orc7* gene for an Orc1/Cdc6 homolog). As this screen tested a limited number of targets, it is currently unclear if additional origins may exist in *Halobacterium* (7).

The first experimental evidence for multiple origins of replication in an archaeon came from studies of the crenarchaeon *Sulfolobus solfataricus*. A 2D gel-mapping analysis, employing a targeted approach, identified origins of replication (*oriC1* and *oriC2*) in noncoding regions upstream of two of the three *Sulfolobus* homologs of Orc1/Cdc6 (*cdc6-1* and *cdc6-3*) (97). No origin activity could be found within 20 kb of the final Orc1/Cdc6 homolog, *cdc6-2*. The targeted nature of this screen meant that additional origins may exist in the *Sulfolobus* chromosome, and a whole-genome marker-frequency analysis by Bernander and colleagues provided strong evidence for a

third origin of replication, about 80 kb removed from *cdc6-2* (71). The marker-frequency analysis also indicated that, following release from a cell cycle block, replication initiated from all three origins simultaneously, suggesting a mechanism for coordinated control at these three sites. Whether this is due to colocalization of the origins in a single controlling "factory" structure or due to the presence of a *trans*-acting "start" signal is currently unknown.

Thus far, all archaeal origins characterized share some common features. They have stretches that are highly rich in A and T bases, indicative of readily meltable DNA (7, 87, 97). Indeed, the precise nucleotides at which replication initiates have been mapped in *Pyrococcus* and *Sulfolobus* and are adjacent to, or within, these candidate duplex unwinding elements (DUEs). In addition, all origins possess a number of repeated sequence elements. Biochemical experiments using purified *Sulfolobus* Orc1/Cdc6 homologs revealed that these sequence motifs are bound by differing subsets of the Orc1/Cdc6 homologs (97). Indeed, extensive inverted repeat elements, termed origin recognition boxes (ORBs), were bound by the Cdc6-1 protein (Fig. 2). Candidate ORB elements were also identified in the *Pyrococcus* and *Halobacterium* origins and were demonstrated to be capable of binding *Sulfolobus* Cdc6-1 in vitro. Thus, it appears that ORB consensus elements are a feature of several archaeal origins of replication. However, *oriC2* in *Sulfolobus* does not have full ORB elements, but rather has a shorter motif that corresponds to a central conserved core of the ORB element (termed mini-ORB or mORB). These mORBs bound Cdc6-1 with reduced affinity compared with full ORBs. Cdc6-3 protein stimulated binding of Cdc6-1 to mORBs (97).

It may be significant that in *Methanothermobacter thermautotrophicus*, an organism that lacks a

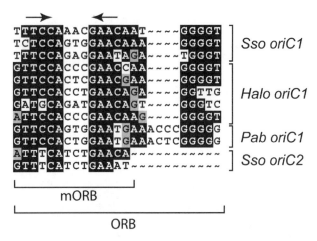

Figure 2. Sequence conservation of ORM and mini-ORB (m-ORB) elements at archaeal origins of replication. These serve as binding sites for orthologs of the *Sulfolobus* Cdc6-1 protein. The arrows indicate an imperfect inverted repeat found in the elements. Sso, *S. solfataricus*; Halo, *Halobacterium* NRC1; Pab, *P. abyssii*).

Cdc6-3 homolog, the predicted origin has no full ORB elements but, instead, has multiple mORB elements (18). It is possible that cooperative interactions between the *M. thermautotrophicus* homologs of *Sulfolobus* Cdc6-1 bound to the mORBs may contribute to high-affinity binding to this origin. In this light, note that the bacterial initiator protein, DnaA, can bind cooperatively to noncanonical DnaA-binding sites (84).

The studies on *Sulfolobus* Orc1/Cdc6 proteins also revealed that the Cdc6-2 protein bound to multiple repeats at the origins (97). Furthermore, these repeats overlapped ORB elements or Cdc6-3-binding sites. This organization suggested that Cdc6-2 might compete for binding to DNA with Cdc6-1 and/or Cdc6-3. Analyses of the protein levels of Orc1/Cdc6s over the cell cycle revealed that Cdc6-1 and Cdc6-3 are highest in prereplicative and replicating cells and that Cdc6-2 peaks in postreplicative cells. This tight temporal partitioning of protein levels, coupled with the overlapping nature of the binding sites for Cdc61/3 and Cdc6-2, suggests distinct roles for the proteins. Possibly the simplest hypothesis is that Cdc6-1 and Cdc6-3 act to promote replication and the Cdc6-2 serves as an inhibitor of re-replication events.

Thus, it appears that while some archaea possess single origins of replication, other species possess multiple origins. Why do some archaea, with genomes smaller than many bacteria, have multiple origins? One possibility may lie in the slow rate of genome replication seen in *Sulfolobus* species; this organism is estimated to replicate DNA at about 6 kb min^{-1} (71). To ensure sufficiently rapid replication of the genome, *Sulfolobus* may therefore have evolved multiple ori-

gins. In contrast, *Pyrococcus* uses a single *oriC*, and it appears that the fork progresses at 20 kb min^{-1} (87). It is also possible that species with multiple origins exploit differential usage of the origins to modulate the growth rate of the cells. In this regard, it will be of considerable interest to determine whether there is differential usage of origins in *Sulfolobus*.

MELTING THE ORIGIN

In bacteria, the initiator protein, DnaA binds the origin and facilitates an initial melting of the duplex, resulting in the replicative helicase, DnaB, being recruited to a melted bubble of DNA (83, 84). In contrast, there is no evidence in either eucarya or archaea that recognition of the origin(s), by origin recognition complex (ORC) or Orc1/Cdc6, respectively, melts DNA. It is, therefore, currently assumed that minichromosome maintenance (MCM), the presumptive replicative helicase, is initially recruited onto double-stranded DNA. In both bacteria and eucarya, the initiator proteins require additional cofactors to facilitate recruitment and loading of the replicative helicase. Bacteria utilize DnaC to facilitate this reaction (2, 12, 26, 33, 38), and eucarya use Cdt1 and Cdc6 (4). There are no clear primary sequence homologs of Cdt1 in archaeal genomes. Furthermore, as discussed in the footnote to Table 1, archaeal Orc1/Cdc6s share similarity to both Orc1 and Cdc6, and it has been proposed that these proteins may play joint roles in marking origins and loading MCM. However, to date no system for MCM loading in the *Archaea* has been published and, thus, protein requirements for the reaction remain undefined.

Nevertheless, strong indications have emerged from the work of Kelman and colleagues that there is indeed a direct physical and functional interaction between archaeal Orc1/Cdc6 homologs and MCM complex (103). Several laboratories (19, 22, 42, 62, 80, 101) have demonstrated that the homomultimeric archaeal MCM (a double hexamer in *M. thermautotrophicus* and single hexamer in *Sulfolobus* and *Archaeoglobus*) possesses processive helicase activity and moves in a 3′ to 5′ direction. It had previously been observed that the helicase activity of bacterial DnaB was inhibited by the helicase loader, DnaC. In a conceptually analogous series of experiments, Kelman and colleagues demonstrated that the *M. thermautotrophicus* Orc1/Cdc6 proteins could inhibit the helicase activity of MCM (103). Furthermore, they revealed that a direct physical interaction could be detected between Orc1/Cdc6s and MCM. Thus, while additional, as yet unidentified factors may play roles in the MCM-loading process, it is likely that the

Orc1/Cdc6 proteins will play an active role in the reaction.

While the basis of MCM recruitment to origins of replication remains poorly defined at the molecular level, there have been a number of studies of the structure and function of the MCM complex. Most of the studies have been on the *M. thermautotrophicus* MCM, although some recent biochemical studies of *Sulfolobus* MCM have revealed insight into the mechanism by which the helicase affects movement along DNA.

MCM

Archaeal MCMs are large homomultimeric machines that harness the energy released by ATP hydrolysis to bring about melting of double-stranded DNA. The archaeal MCM monomer is approximately 70 kDa and has the domain organization cartooned in Fig. 3, with an approximately 30-kDa N-terminal region, followed by an AAA$^+$ ATPase domain and a poorly conserved helix-turn-helix at the C-terminal end of the protein. Electron microscopic studies have revealed that *M. thermautotrophicus* MCM forms ring shaped structures (22, 89, 113). Diverse studies have described a range of different stoichiometries for the *M. thermautotrophicus* MCM; with single hexamer, double hexamer, and heptameric forms of the protein described. In addition, filamentous forms have been seen (21). This degree of structural heterogeneity appears to be common to a number of phylogenetically diverse ring-shaped helicases. For example, the gp4 helicase of bacteriophage T7 has been observed in both hexameric and heptameric forms (105, 107).

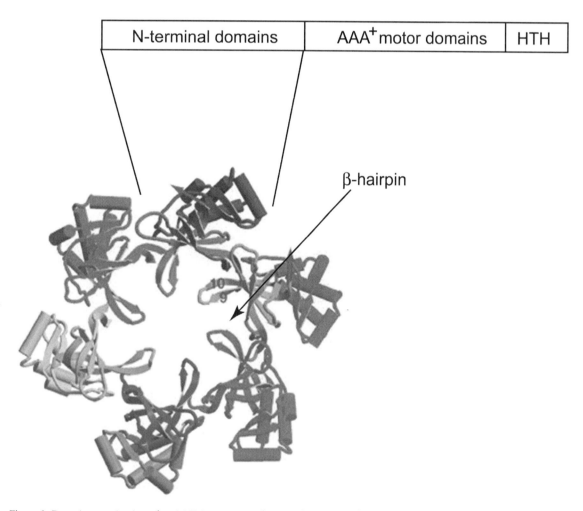

Figure 3. Domain organization of an MCM monomer. The crystal structure of the N-terminal region of *M. thermautotrophicus* has been solved. This region forms a double hexamer; for simplicity, only one hexamer is shown. The DNA-binding β-hairpin structures are indicated. HTH, helix-turn-helix domain. Modified from *Nature Structural Biology* (34) with permission of the publishers.

In MCM, the central hole in the ring is of sufficient width to accommodate a double helix of DNA. The crystal structure of the first 286 amino acids of *M. thermautotrophicus* MCM has been solved and shows that this N-terminal section of the molecule has a sixfold symmetric structure that forms a double hexamer, with individual component hexamers aligning in a head-to-head manner (106). In the structure, a zinc-binding motif appears to be involved in mediating hexamer-hexamer interactions. Mutational analysis of this motif demonstrated little effect on the multimeric status of the protein, but the mutations did affect MCM binding of single-stranded DNA (91).

The structure also revealed that each subunit in the component hexamers contributes a β-hairpin motif that points in toward the center of the ring (Fig. 3). Mutation of highly conserved basic residues at the tips of the hairpins reduces the ability of MCM to interact with DNA (34). Interestingly, the mutations have distinct effects depending on the presence of the C-terminal domain of the protein. When these mutations were in a truncated (amino acids 1 to 286) *M. thermautotrophicus* MCM, DNA binding was abrogated. In contrast, similar mutations in full-length *Sulfolobus* MCM resulted in a quantitative reduction, but not loss, of DNA-binding affinity and helicase activity. These data suggest additional DNA-binding sites are present in the C-terminal domain(s) of the protein (80).

Inspection of the sequence of the AAA⁺ domain of MCM revealed a striking similarity with helicases of the phylogenetically distinct Superfamily 3 (exemplified by the large T antigen, Tag, of eucaryal simian virus 40 [SV40]). Tag has been shown to possess a β-hairpin insertion in the AAA⁺ domain (39, 68). Furthermore, this hairpin has been observed to move during the ATP-binding and hydrolysis cycle of Tag, leading to the proposal that it brings about the power stroke of the helicase along DNA (39). A similar sequence insertion has been noted in MCM, and mutational analyses have revealed that mutation of absolutely conserved basic residues in the proposed tip of this hairpin have only a modest effect on the DNA-binding activity of the enzyme but completely abolish the helicase activity of MCM (80). It appears, therefore, that MCM has two sets of DNA-binding β-hairpins, one in the N-terminal domain and one in the AAA⁺ domain. Mutation of either alone only modestly reduces DNA binding, but when both hairpins are mutated, the helicase can no longer bind DNA. By analogy with SV40 Tag, it has been proposed that the C-terminal hairpin drives MCM along DNA, with the N-terminal hairpin playing a minor role (if any) in this process.

It is currently unclear what the precise mechanism is that enables MCM to mediate strand separa-

tion during the helicase reaction. Kelman and colleagues have elegantly demonstrated that *M. thermautotrophicus* MCM can translocate along either single- or double-stranded DNA (104). They have also showed that the presence of a flap, in the form of a protruding DNA 5′ end, allows MCM to peel this strand off the DNA (Fig. 4A). This could suggest that MCM acts as a molecular bulldozer, stripping the strands apart ahead of the main body of the enzyme. In a related model, Onesti and colleagues in their electron microscopic studies (89) proposed that pores in walls of the ring-shaped helicase may serve as exit pores for the displaced strand (Fig. 4B). Finally, it is possible that, in vivo, MCM is acting as a double pump, with individual hexamers bound to double-stranded DNA and pumping against each other, thereby forcing DNA into the center of the helicase and leading to unwinding (Fig. 4C). Regardless of the mechanism by which MCM unwinds DNA, the result is the generation of single-stranded DNA that can act as a template for initiation and elongation of DNA synthesis.

SINGLE-STRANDED DNA-BINDING PROTEINS

Single-stranded (ss) DNA-binding proteins (SSBs) do not possess any enzymatic activity but have a structural role. Their main function is to stabilize ssDNA by preventing the formation of secondary structures and

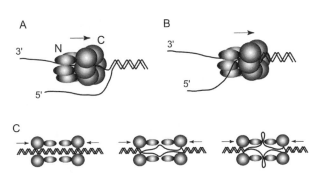

Figure 4. Models for the mechanism of DNA unwinding by the MCM helicase. (A) Single hexamer of MCM translocating along single-stranded DNA in a 3′ to 5′ direction with the C-terminal AAA⁺ domain leading. As it translocates, it unzips DNA ahead of it. A similar situation is shown in B, the difference being that the displaced strand is passed out through an exit channel in the MCM hexamer. An implication of this model is that the motor domain of MCM would bind to double-stranded DNA and the N-terminal domains would bind to single-stranded DNA. (C) A cutaway model of a double hexamer of MCM, with only two of the subunits of each hexamer shown. In this model, the two hexamers are held together by the N-terminal domains, and rather than hexamers moving on DNA, DNA is pumped into the central cavity of the double hexamer. Single-stranded loops of DNA are generated, and these are extruded from the body of the enzyme.

to protect it from chemical modifications by coating the unwound DNA. The cellular functions of SSBs have been best studied in *Eucarya*, where they are involved in various stages of DNA replication, recombination, and repair, during which ssDNA occurs (112).

SSBs bind DNA via a particular structure called an oligonucleotide/oligosaccharide-binding fold or OB fold (86). In bacteria, the functional SSB is a homotetramer that wraps 65 nucleotides. Each monomer possesses one OB fold and an acidic C-terminal tail involved in protein-protein interactions (92, 93). In eucarya, the SSB is called Replication Protein A, RPA. It consists of a heterotrimer formed by RPA70, RPA32, and RPA14, each subunit being named after its molecular mass (13). RPA70 possesses four OB folds and a C-terminal zinc finger motif whose function is still unclear. One OB fold is found in each RPA32 and RPA14, but not all of these are necessary for DNA binding (8, 9). In archaea, the sequence arrangements of SSBs and their multimerization level vary between bacterial SSB-like and eucaryal RPA-like arrangements (Table 2).

Euryarchaeal SSBs range from a monomeric SSB in *Methanocaldococcus jannaschii*, which has four OB folds and a zinc finger motif near its C terminus, reminiscent of that in RPA70 (58), to the heterotrimeric RPA from *Pyrococcus furiosus*, composed of RPA41, RPA32, and RPA14 (64). The largest subunit, RPA41, consists of an OB fold and a C-terminal zinc finger motif. RPA41 was coimmunoprecipitated with RadA, a RecA/Rad51 family protein, and a Holliday junction resolvase, suggesting a role in strand exchange during homologous recombination. RPA41 also coimmunoprecipitated with DNA replication enzymes (64). More recently, three SSBs from the mesophilic archaeon *Methanosarcina acetivorans* were characterized: RPA1, RPA2, and RPA3 (95). Surprisingly, these three polypeptides did not form a heterotrimer but homomultimerized. Moreover, RPA1 possesses four OB folds but no zinc finger motif, whereas RPA2 and RPA3 have two OB folds and a zinc finger motif. Taken together, these observations suggested that MacRPAs function independently of each other (95).

By far the most studied archaeal SSB is that of the crenarchaeon *S. solfataricus*. It is a 16-kDa monomer with a sequence arrangement similar to that of bacteria, as it consists of a single OB fold and an acidic C-terminal tail (43, 110). However, the crystal structure of *S. solfataricus* SSB showed that its structural architecture is more similar to the DNA-binding domain of RPA than to the bacterial SSB (63). The flexible C-terminal tail of *S. solfataricus* SSB is not involved in DNA binding but is believed to play a role in protein-protein interactions (110). Recently, this acidic tail was shown to interact with RNA polymerase (Richard et al., 2004) and to overcome transcriptionally repressive effects of the chromatin protein Alba (3). Furthermore, *S. solfataricus* SSB increased the RecA-mediated DNA strand exchange in *Escherichia coli* (43) and activated the in vitro activity of *S. shibatae* reverse gyrase, even when DNA was coated with the chromatin protein Sul7d (88).

DNA PRIMASES

Because replicative DNA polymerases cannot synthesise DNA de novo, they require the presence of specialized enzymes to complete DNA replication. These enzymes, DNA primases, are by definition DNA-dependent RNA polymerases. They initiate DNA replication once only on the leading strand but several times on the lagging strand by synthesizing RNA primers de novo, which are then elongated by DNA polymerases to form the Okazaki fragments.

Table 2. Composition of SSBs in *Archaea*, *Eucarya*, and *Bacteria*, highlighting differences in subunit composition and architecture

Organism	Form in solution	Name of each subunit	No. of OB folds per monomer	Motif at C terminus	Reference
E. coli	Homotetramer		1	Acidic tail	93
Saccharomyces cerevisiae	Heterotrimer	RPA70	4	Zinc finger	13
		RPA32	1		
		RPA14	1		
M. jannaschii	Monomer		4	Zinc finger	58
P. furiosus	Heterotrimer	RPA41	1	Zinc finger	64
		RPA32	1		
		RPA14	1		
M. acetivorans	Homotetramer/homodimer	RPA1	4		95
	Homodimer	RPA2	2	Zinc finger	
	Homodimer	RPA3	2	Zinc finger	
S. solfataricus	Monomer or homotetramer		1	Acidic tail	43, 110

Bacterial primases, as exemplified by *E. coli* DnaG, are often physically associated with the replicative helicase. Indeed in some bacteriophage, such as T7, the primase and helicase are found in a single polypeptide (36).

Archaeal primases are heterodimers consisting of a catalytic and a regulatory subunit. Initial studies of the biochemical properties of the *P. furiosus* small catalytic subunit on its own gave the very surprising finding that the enzyme was capable of synthesizing 2.4 kb long DNA, not RNA, products in vitro (10). However, subsequent work revealed that, when the regulatory subunit was present, the dimeric primase was then capable of synthesizing both DNA and RNA (70). Additionally, the DNA products were shorter (0.7 kb). Subsequent in vitro studies of *P. horikoshii* and *S. solfataricus* enzymes also showed that the primase can synthesize long RNA and DNA products de novo, suggesting that this property exists in all archaeal primases (65, 75). In eucarya, the core primase heterodimer corresponds to the archaeal heterodimeric primase. However, in eucarya the core primase further associates with DNA polymerase α and the B subunit to form the polα-primase complex (reviewed in reference 67). Since archaeal primases are not stably associated with polα, it is tempting to suggest that their dual RNA and DNA synthesis capability allows archaeal primases, first, to initiate the formation of RNA primers and, second, to extend these primers to form RNA-DNA hybrids.

Until recently, only the crystal structure of the *Pyrococcus* primase catalytic subunit on its own was available (1, 50). The crystallographic analysis showed that the catalytic subunit consists of two domains: a larger α/β-domain, which includes the catalytic site (the prim domain), and a smaller α-helical domain of unknown function. Surprisingly, a zinc-binding motif was also found not far from the catalytic site. Recently, the crystal structure of the heterodimeric primase of *S. solfataricus* offered the first insight into the architectural organization of the primase large regulatory subunit PriL (66). This subunit consists of an entirely α-helical domain forming the bulk of PriL and a smaller α/β-domain responsible for the interaction with the catalytic subunit PriS (Fig. 5). In PriS, the α-helical domain is much smaller than that of the *Pyrococcus* enzyme and the zinc-binding motif comparatively longer. However, the organization of the prim domain is conserved in the *Pyrococcus* and the *S. solfataricus* primases.

When the *Pyrococcus* primase catalytic subunit structure was solved, it was found that the arrangement of the triad of catalytic aspartate residues was very similar to that of the DNA pol X family member, eucaryal pol β. Note, however, that the topology of

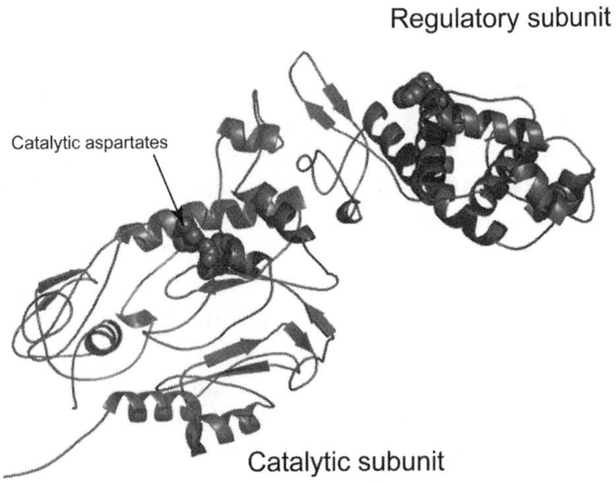

Figure 5. Crystal structure of the heterodimeric core primase of *S. solfataricus*. The two subunits are indicated, as are the catalytic aspartate residues. As the regulatory subunit is spatially removed from the catalytic site, it is proposed that this subunit exerts its effect by modulating primer length (see reference 66).

the fold that contains the aspartate residues is unrelated in primase and pol β, which is suggestive of convergent evolution of the protein structure (1, 67).

DNA pol β is a member of the DNA pol X family that includes pol λ, pol μ, pol σ and terminal deoxynucleotidyltransferase (TdT). These members of the pol X family play roles in DNA replication, repair, and recombination processes (49, 94). Surprisingly, the primase from *S. solfataricus* possesses a TdT activity (28, 65). Taken together with the fact that pol λ, and pol μ can initiate RNA and DNA synthesis de novo (94), it is likely that the particular arrangement of catalytic residues in primase and pol X has been selected for its enzymatic flexibility. Because most archaea do not possess a pol X family DNA repair polymerase, these data also suggest that it is possible that primase plays a role in DNA damage repair processes in archaea.

Finally, it appears that the archaeo-eucaryal primase catalytic fold may be derived from an evolutionarily ancient source (51, 67). Recent work has found distant relatives of this fold in diverse molecules, including the ORF904 product of the pRN1 plasmid from *S. islandicus (69)*, the primase-helicase of bacteriophage phBC6A51 (79), and the primase-like molecule, LigD, involved in NHEJ (29).

REPLICATIVE DNA POLYMERASES

DNA polymerases act to extend the primers generated by primase complexes (49). These enzymes are divided into six families based on phylogenetic relationships: families A, B, C, D, X, and Y. Not all of these are replicative polymerases, however; some are utilized solely for DNA repair. The *E. coli* replication fork is built around the DNA pol III holoenzyme, a large assembly containing two copies of the family C DNA polymerase subunit, subunit α, which is encoded by the *dnaE* gene. One of these polymerase subunits is thought to replicate the lagging strand and one the leading strand (81). This was thought to be the case for all bacteria until two polymerases were identified in *Bacillus subtilus;* one of these encoded by a gene homologous to *E. coli dnaE*, *dnaE*$_{BS}$ and the other by *polC*. Pol C and DnaE$_{BS}$ are thought to be responsible for leading-strand and lagging-strand DNA synthesis, respectively (30).

Eucaryal DNA synthesis requires three B-family polymerases: Pol α, Pol δ, and Pol ε. Pol δ and Pol ε are the major replicative polymerases, with Pol δ acting on the lagging strand and both acting on the leading strand of DNA replication. Pol α is involved in lagging-strand Okazaki fragment synthesis by extending primers, which are subsequently transferred to Pol δ (37).

Similar to eucarya, the replicative polymerases utilized in the *Archaea* were initially thought to be solely members of the family B polymerases based on predictions from complete genome sequences; for example, *S. solfataricus* was predicted to contain three family B members: Pol B1, Pol B2, and Pol B3. It was later established that although multiple B-family polymerases are a feature of several crenarchaea, such as *Aeropyrum pernix*, which has had two activities identified from cell extracts (14), euryarchaeal DNA replication requires a B-family polymerase and another, unique heterodimeric polymerase, Pol D. Pol D is composed of two subunits, DP1 and DP2, DP2 being the polymerase domain and DP1 the 3'-5'-exonuclease domain (15). The archaeal B-family polymerases were found to uniquely possess a "read ahead" recognition pocket for uracil bases (23) that is produced as a result of cytosine deamination. This results in replication halting four bases from the primer/template junction and prevents the mutation of C-G base pairing to A-T pairing (35). This feature may be particularly important for archaea that inhabit high-temperature environments, as deamination is more prevalent under these conditions, and, if undetected and uncorrected, genetic integrity would be rapidly lost by these organisms.

Crystal structures have been determined for several of the archaeal B-family polymerases. These have revealed the classical "right hand" organization of the C-terminal domain, with catalytic residues present in the palm domain. The N-terminal domain has two subdomains, the first being the editing 3'-5'-exonuclease domain. The other N-terminal subdomain forms a fold related to an RNA-binding motif and is the site of the uracil recognition pocket.

The relationship between Pol B and Pol D in the euryarchaea has been investigated (44). It has been established that both are replicative polymerases, but they possess different biochemical properties and have been hypothesized to perform specific roles at the replication fork. Pol D is capable of primer extension of both RNA and DNA primers, although not to full length. However, the addition of proliferating cell nuclear antigen (PCNA) (see below) stimulates this extension. Pol B, on the other hand, is only capable of extending DNA primers; these are extended to the full length of the template in the absence of PCNA. The inability of Pol B to extend RNA primers cannot be overridden by the addition of PCNA. Pol D readily achieves strand displacement of DNA primers. However, displacement of RNA primers requires the additional presence of PCNA. In contrast, Pol B is incapable of achieving strand displacement unless PCNA is present; even then, only DNA primers can be displaced. These data have given rise to a model whereby Pol B is the leading-strand polymerase, after initiation by Pol D, and Pol D is also the lagging-strand polymerase (44).

As the crenarchaea contain more than one B-family homolog, it is tempting to speculate that, as is seen in the *Eucarya*, the *Euryarchaeota* kingdom, and some members of the *Bacteria*, these homologs may also have specific and distinct roles at the replication fork.

SLIDING CLAMPS

Sliding clamps are proteins with no known enzymatic activity, whose principal function is to increase the processivity or activity of proteins with which they interact by tethering them to duplex DNA. Although the amino acid sequences of sliding clamps are very different in the three domains of life, their three-dimensional structure is very similar: they all form a pseudohexameric ring-shaped structure that can accommodate a double-stranded DNA in its center (54) (Fig. 5). *E. coli* sliding clamp, the β-clamp, consists of a homodimer with each polypeptide arranged in three domains to form a quasi-hexameric structure. In eucarya, the processivity clamp is called PCNA and is a homotrimer. Each monomer is formed of two domains that allow the formation of the pseudohexameric ring-shaped structure. Sliding clamps are well known for their role in DNA replication, but they also interact with factors involved in other cellular processes, such as DNA repair and recombination, and cell cycle regulators (reviewed in references 60, 109, 111). The interaction of PCNA with other proteins occurs principally via a conserved motif, the PCNA-interacting protein box or PIP motif, which is usually present at the N- or the C-terminal end of the proteins (111). Recently, the structural basis for the interaction of the flap endonuclease FEN1 and PCNA was uncovered in *Archaeoglobus fulgidus* (20) and in humans (98).

PCNA is well conserved between *Eucarya* and *Archaea*, and there appears to be at least one PCNA homolog in each archaeal species. In the euryarchaea *P. furiosus*, *M. thermautotrophicus*, and *Thermococcus fumicolans*, only one homolog of PCNA was found that forms a homotrimer that activates the activity of various DNA polymerases (16, 46, 61). The elucidation of the crystal structure of *P. furiosus* PCNA suggested that the thermostability of this protein compared with that of eucaryal sliding clamps may be due to the increased number of ion pairs, and thus electrostatic interactions, within the protein (76). The crenarchaea appear to be slightly more complex, as they possess up to three PCNA homologs. In *A. pernix*, the PCNA subunits appear capable of forming both homotrimeric and heteromultimeric complexes (25). Although each PCNA is capable of increasing the activity of DNA polymerases, PCNA3 appears to be the most efficient. In another member of the *Crenarchaeota*, *S. solfataricus*, PCNA exists as a heterotrimer of three distinct subunits. No homomultimerization could be detected for individual subunits. The formation of the heterotrimer occurs in an obligate order: PCNA1 and PCNA2 interact first before PCNA3 can complete the ring (31).

PCNAs are ring-shaped molecules and, to load onto DNA, require a ring-opening and -shutting reaction. In eucarya, the clamp loader, RF-C, mediates this reaction. Although archaea encode RF-C, many archaeal PCNAs are capable of self-loading onto DNA without the help of a clamp loader; the presence of RF-C does, nevertheless, stimulate the process (16, 25, 61).

CLAMP LOADER

Generally, in all three domains of life, the clamp loader is a heteropentamer consisting of one large subunit and four smaller ones. Each of these monomers belongs to the AAA$^+$ ATPase family (27, 52), and although ATP hydrolysis is not required to load PCNA onto DNA, the presence of one or more ATPs is necessary.

In *E. coli*, the clamp loader is the γ-complex and consists of three different subunits, with a stoichiometry of $\gamma_3\delta\delta'$, that form the pentamer (54–57). The

eucaryal clamp loader is composed of five different subunits called RF-C1 to RF-C5 (24). Within these subunits, seven segments of amino acids (boxes II to VIII) are highly conserved. These include the Walker A and B motifs responsible for NTP-binding and ATPase activity, which are located in boxes III and V, respectively. The large subunit RF-C1 possesses an additional conserved domain in its N-terminal region (24). In comparison with the two other domains of life, the archaeal clamp loader is simpler because only two genes homologous to RF-C have been identified in fully sequenced genomes. This means that all archaeal RF-Cs are composed of only two subunits: RF-C$_L$ (large subunit) and RF-C$_S$ (small subunit).

To date, a eucaryal-like pentameric organization of the clamp loader (4 RF-C$_S$, 1 RF-C$_L$) has been identified in only two archaea, *S. solfataricus* and *A. fulgidus* (90, 99). In both cases, the RF-C complex increased the activity of DNA pol B and DNA pol D, in the presence of PCNA. In *M. thermautotrophicus*, the RF-C appears to be a hexamer capable of loading PCNA onto singly nicked circular DNA and of activating DNA synthesis by pol B in the presence of PCNA (61). In *P. furiosus* and in *P. abyssi*, the composition of the RF-C complex is still unclear. It was proposed to be three to four RF-C$_S$s and one to two RF-C$_L$s for *P. furiosus* (17) and either a hexamer (two RF-C$_S$s and four RF-C$_L$s) or a trimer (two RF-C$_S$s and one RF-C$_L$s) for *P. abyssi* (46). Both PfuRF-C and PabRF-C were capable of activating the activity of DNA pol D in the presence of PCNA (17, 45).

For a long time, the loading of PCNA onto DNA by RF-C was believed to require ATP hydrolysis. However, studies on PabRF-C and AfuRF-C demonstrated that ATP binding to the clamp loader, but not hydrolysis, is necessary to load the processivity clamp onto DNA (45, 99, 100). More recently the importance of ATP binding and hydrolysis by AfuRF-C for the loading of PCNA was studied in detail (100). The authors suggested that the initial binding of two molecules of ATP by RF-C is necessary for its interaction with PCNA. This interaction then leads to the binding of two additional ATP molecules and allows the loading of PCNA onto DNA. The subsequent hydrolysis of three ATP molecules results in the release of RF-C from the PCNA-DNA complex, which allows the sliding clamp to bind other proteins (e.g., DNA polymerase) and to fulfill its processivity function. According to this model, the pentameric AfuRF-C can only bind four ATP molecules like the eucaryotic RF-C (40, 41).

Recent structural snapshots have been obtained of the yeast (11) and archaeal (85) RF-C PCNA complexes by crystallography and electron microscopy, respectively (108). In the archaeal structure, the horseshoe-shaped RF-C stacks on top of the ring-shaped PCNA. The large subunit of RF-C contacts one PCNA subunit, and two or three of the small subunits contact the remaining two subunits of PCNA (85). The high-resolution yeast structure shows a similar arrangement. However, in this case the replication factor C (RFC) small subunits, rather than lying flat on PCNA, spiral upward from the ring (11). It is possible, therefore, that RF-C mediates opening of the PCNA ring by a sideways pull on one or more subunits, resulting in opening of the ring into a structure resembling a lock washer (Fig. 6).

ARCHAEAL CELL CYCLE

While an ever-growing body of data has yielded considerable insight into the form and function of the archaeal DNA replication machinery, much less is known about the details of the archaeal cell cycle and its control. Indeed, what little is known appears to be suggesting that diverse mechanisms may be employed to regulate chromosome copy number, to coordinate DNA replication and cell division, and even to mediate the process of cell division itself.

This latter point is emphasized by the finding that, while euryarchaeal genomes encode clear homologs of the bacterial FtsZ cell division protein, the crenarchaeal genomes do not. FtsZ is homologous to the key eucaryal cytoskeleton component, tubulin. FtsZ plays an essential role in the septation process in bacteria and presumably those archaea in which it is found. Bacterial FtsZ forms the so-called Z ring that forms at the junction of the two, soon-to-be daughter cells. The ring then contracts and thereby drives septation. FtsZ rings have also been observed in the archaeon *Haloferax mediterranei*, providing evidence for an analogous role for this protein in archaea. It is intriguing that no clear FtsZ or tubulin homologs are detected in the available crenarchaeal genomes and the identity of the proteins involved in cell division in these organisms remains unknown.

Flow cytometric analyses have been performed on several archaeal species within the *Euryarchaeota* and *Crenarchaeota*. The crenarchaeal *Sulfolobus* species are among the best studied and have been shown to have a cell cycle that varies between one copy of the chromosome in newborn cells and two copies in a postreplicative period (perhaps analogous to eucaryal G2 phase of the cell cycle). The single chromosome (G1-like) phase of the cell cycle is very short and the G2 period is the dominant feature of the *Sulfolobus* cell cycle. Furthermore, cells in stationary phase have two fully replicated copies of the chromosome (6, 48). This makes intuitive sense; an organism

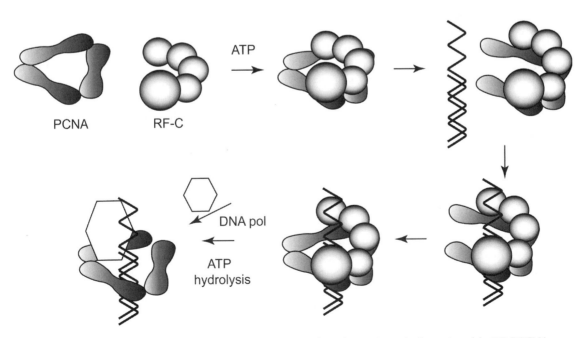

Figure 6. Model for loading of the PCNA clamp by RF-C. Binding of ATP by RF-C permits formation of the RF-C-PCNA complex. This leads to PCNA opening, presumably by repositioning of the RF-C subunits. DNA is loaded and the clamp resealed. None of these steps require ATP hydrolysis. ATP hydrolysis is, however, required for the next stage, recruitment of DNA polymerase (DNA pol) and release of RF-C.

living aerobically in a high-temperature environment will experience high levels of DNA damage. By biasing a one genome-two genome cell cycle to the phase with two copies of the chromosome, the organism enhances its chances of repairing DNA damage by the error-free method of homologous recombination.

The euryarchaeon *A. fulgidus*, a high-temperature anaerobe, has a cell cycle distribution similar to *Sulfolobus*, with a short prereplicative phase and prolonged postreplicative phase. However, in contrast to the tight control between one and two chromosome copies in *Sulfolobus*, *Archaeoglobus* cells can possess one, two, three, or four copies of the chromosome. Furthermore, conditions were observed in which the majority of *Archaeoglobus* cells in stationary phase had a single chromosome (72).

Studies of *Methanococcus jannaschii* revealed that it had a complex number and pattern of copies of its chromosome, with possibly as many as 15 copies present during exponential growth, and between one and five copies in stationary-phase cells (74).

Recent work has established that *M. thermautotrophicus* possesses two, four, or eight copies of the genome (73). Although some four chromosome cells were present in stationary phase, the majority of cells had two copies of the chromosome. However, microscopy revealed that the two copies of the chromosome present in stationary phase were present in discrete nucleoids. This contrasts with stationary-phase *Sulfolobus* cells, in which the two chromosomes appear to be in a single nucleoid structure. In *M. thermautotrophicus* the number of nucleoids observed correlated well with the number of genomes present, leading to the proposal that, in this organism, chromosome segregation occurred rapidly after, or even concomitant with, DNA replication. This suggests that in contrast to eucarya, where, during G2, sister chromatids are paired postreplicatively, *M. thermautotrophicus* may segregate its chromosomes in a manner akin to that seen in bacteria (73, 102). Whether this bacterial-like paradigm for chromosome segregation can be extended to other archaea is not yet known.

While some important basic cell cycle parameters have now been established for a range of archaeal species, there is still very little information available about the mechanisms that drive cell cycle progression and about the variation of protein levels during the cycle. This is due in part to the relative paucity of genetic tools for archaeal species and also due to the technical challenge of achieving good methods of cell cycle synchronization for archaeal cells. It has been possible to achieve partial synchronization of *Sulfolobus acidocaldarius* cultures by using treatment with acetic acid to arrest the cycle (by an as yet unknown mechanism) followed by release into fresh medium (71). A degree of synchrony has also been achieved for *Halobacterium* by using release from a block imposed by the DNA synthesis inhibitor, aphidicolin (47).

The latter study, performed by Herrmann and Soppa, examined the expression profile of two homologs of structural maintenance of chromosome (SMC) proteins (*sph1* and *hp24*), which could play a role in DNA segregation, DNA repair, and/or cell division (47). This study found evidence for cell cycle regulation of the abundance of both genes' transcripts. In agreement with the transcript profile, protein levels of Sph1 also showed modulation. This study also examined nucleoid distribution during the cell cycle, and the results suggested that chromosome segregation was concomitant with DNA replication, as was proposed for *M. thermautotrophicus* (see above), in a mode akin to that employed by bacteria.

Finally, as mentioned above, studies of the levels of *Sulfolobus* Orc1/Cdc6 homologs revealed that Cdc6-1 and Cdc6-3 are highest in G1- and S-phase cells, while Cdc6-2 peaks in G2 cells (97). In bacteria, the DnaA replication initiator protein plays dual roles as a replication and transcription factor; whether this is the case for the Orc1/Cdc6s is currently unknown. However, it is tempting to speculate that the fact that origins of replication lie immediately upstream of the *cdc6-1* and *cdc6-3* may provide a mechanism for coordinating expression of the initiator factors with origin activity. The scheme described in Fig. 7 illustrates a simple binary oscillator that could coordinately modulate the activities of the origins and expression of initiator and regulator proteins. If the Orc1/Cdc6 proteins also modulate the expression of additional downstream genes, then this simple loop could lie at the heart of the *Sulfolobus* cell cycle. Given the availability of purified Orc1/Cdc6 proteins and a defined in vitro transcription system for *Sulfolobus*, it should be possible to test the predictions of the model.

PERSPECTIVE: THE NEXT FIVE YEARS

Tremendous advances have taken place in our knowledge of the function of individual archaeal DNA replication proteins over the past few years. Clear goals for the near future lie in the integration of the individual components into coordinated in vitro systems designed to address the complex macromolecular interactions and transactions that mediate processes such as MCM loading and helicase-linked replication fork progression.

G1/S phase

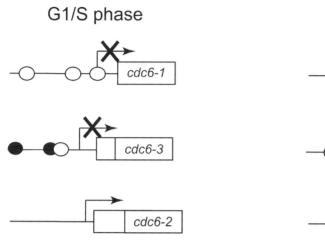

Origins active and *cdc6-1* and *cdc6-3* repressed

cdc6-2 expressed and gradually accumulating

G2 phase

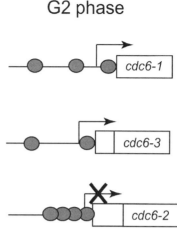

cdc6-1 and *cdc6-3* de-repressed or activated

cdc6-2 repressed

Key

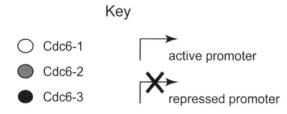

Figure 7. A model for the interplay between origin activity and transcriptional regulation of *cdc6-1* and *cdc6-3* genes. The model postulates that upon binding to the origins adjacent to their own genes, Cdc6-1 and/or Cdc6-3 exert negative regulation of their own expression. It is further proposed that the *cdc6-2* gene becomes active at late S phase. This could conceivably be by Cdc6-1 and/or Cdc6-3 activating *cdc6-2* expression (not shown). In G2, remaining Cdc6-1 and Cdc6-3 protein levels decay and Cdc6-2 levels rise. Cdc6-2 could act as an activator of *cdc6-1* and *cdc6-3* and as repressor of its own expression, thereby reducing its own levels, elevating Cdc6-1 and Cdc6-3, and preparing cells for another round of replication following division.

Beyond the challenges of understanding the mechanistic architecture of the replication apparatus lies the necessity of improving our understanding of the molecular basis of the archaeal cell cycle. How is progression of the cycle driven; will posttranslational modifications and targeted degradation of key proteins lie at the heart of the cycle, or will it be powered by comparatively simple transcriptional feedback loops? Given that the crenarchaeal and euryarchaeal cell division systems appear to have some fundamental differences, not the least being the presence or absence of FtsZ, it may be an oversimplification to talk of an "archaeal" cell cycle. It is possible that distinct mechanisms for controlling cell cycle progression exist in the two phyla. With ever-improving systems for the genetic manipulation for an increasing range of archaeal species and the power of transcriptomic and proteomic approaches, it can

be anticipated that progress on these issues will be rapid and exciting.

REFERENCES

1. **Augustin, M. A., R. Huber, and J. T. Kaiser.** 2001. Crystal structure of a DNA-dependent RNA polymerase (DNA primase). *Nat. Struct. Biol.* **8:**57–61.
2. **Baker, T. A., B. E. Funnell, and A. Kornberg.** 1987. Helicase action of Dnab protein during replication from the *Escherichia coli* chromosomal origin *in vitro. J. Biol. Chem.* **262:**6877–6885.
3. **Bell, S. D., C. H. Botting, B. N. Wardleworth, S. P. Jackson, and M. F. White.** 2002. The interaction of Alba, a conserved archaeal, chromatin protein, with Sir2 and its regulation by acetylation. *Science* **296:**148–151.
4. **Bell, S. P., and A. Dutta.** 2002. DNA replication in eukaryotic cells. *Ann. Rev. Biochem.* **71:**333–374.
5. **Bernander, R.** 2000. Chromosome replication, nucleoid segregation and cell division in Archaea. *Trends Microbiol.* **8:**278–283.

6. Bernander, R., and A. Poplawski. 1997. Cell cycle characteristics of thermophilic archaea. *J. Bacteriol.* **179:**4963–4969.

7. Berquist, B. R., and S. DasSarma. 2003. An archaeal chromosomal autonomously replicating sequence element from an extreme halophile, Halobacterium sp. strain NRC-1. *J. Bacteriol.* **185:**5959–5966.

8. Bochkarev, A., and E. Bochkareva. 2004. From RPA to BRCA2: lessons from single-stranded DNA binding by the OB-fold. *Curr. Opin. Struct. Biol.* **14:**36–42.

9. Bochkarev, A., E. Bochkareva, L. Frappier, and A. M. Edwards. 1999. The crystal structure of the complex of replication protein A subunits RPA32 and RPA14 reveals a mechanism for single- stranded DNA binding. *EMBO J.* **18:**4498–4504.

10. Bocquier, A. A., L. D. Liu, I. K. O. Cann, K. Komori, D. Kohda, and Y. Ishino. 2001. Archaeal primase: bridging the gap between RNA and DNA polymerases. *Curr. Biol.* **11:**452–456.

11. Bowman, G. D., M. O'Donnell, and J. Kuriyan. 2004. Structural analysis of a eukaryotic sliding DNA clamp-clamp loader complex. *Nature* **429:**724–730.

12. Bramhill, D., and A. Kornberg. 1988. A model for initiation at origins of DNA-replication. *Cell* **54:**915–918.

13. Brill, S. J., and B. Stillman. 1991. Replication factor-a from *Saccharomyces cerevisiae* is encoded by 3 essential genes co-ordinately expressed at S-ohase. *Genes Dev.* **5:**1589–1600.

14. Cann, I. K., S. Ishino, N. Nomura, Y. Sako, and Y. Ishino. 1999. Two family B DNA polymerases from *Aeropyrum pernix*, an aerobic hyperthermophilic crenarchaeote. *J. Bacteriol.* **181:**5984–5892.

15. Cann, I. K., K. Komori, H. Toh, S. Kanai, and Y. Ishino. 1998. A heterodimeric DNA polymerase: evidence that members of Euryarchaeota possess a distinct DNA polymerase. *Proc. Natl. Acad. Sci. USA* **95:**14250–14255.

16. Cann, I. K. O., S. Ishino, I. Hayashi, K. Komori, H. Toh, K. Morikawa, and Y. Ishino. 1999. Functional interactions of a homolog of proliferating cell nuclear antigen with DNA polymerases in Archaea. *J. Bacteriol.* **181:**6591–6599.

17. Cann, I. K. O., S. Ishino, M. Yuasa, H. Daiyasu, H. Toh, and Y. Ishino. 2001. Biochemical analysis of replication factor C from the hyperthermophilic archaeon *Pyrococcus furiosus*. *J. Bacteriol.* **183:**2614–2623.

18. Capaldi, S. A., and J. M. Berger. 2004. Biochemical characterization of Cdc6/Orc1 binding to the replication origin of the euryarchaeon *Methanothermobacter thermoautotrophicus*. *Nucleic Acids Res.* **32:**4821–4832.

19. Carpentieri, F., M. De Felice, M. De Falco, M. Rossi, and F. M. Pisani. 2002. Physical and functional interaction between the mini-chromosome maintenance-like DNA helicase and the single-stranded DNA binding protein from the crenarchaeon *Sulfolobus solfataricus*. *J. Biol. Chem.* **277:**12118–12127.

20. Chapados, B. R., D. J. Hosfield, S. Han, J. Z. Qiu, B. Yelent, B. H. Shen, and J. A. Tainer. 2004. Structural basis for FEN-1 substrate specificity and PCNA- mediated activation in DNA replication and repair. *Cell* **116:**39–50.

21. Chen, Y. J., X. O. Yu, R. Kasiviswanathan, J. H. Shin, Z. Kelman, and E. H. Egelman. 2005. Structural polymorphism of *Methanothermobacter thermautotrophicus* MCM. *J. Mol. Biol.* **346:**389–394.

22. Chong, J. P. J., M. K. Hayashi, M. N. Simon, R. M. Xu, and B. Stillman. 2000. A double-hexamer archaeal minichromosome maintenance protein is an ATP-dependent DNA helicase. *Proc. Natl. Acad. Sci. USA* **97:**1530–1535.

23. Connolly, B. A., M. J. Fogg, G. Shuttleworth, and B. T. Wilson. 2003. Uracil recognition by archaeal family B DNA polymerases. *Biochem. Soc. Trans.* **31:**699–702.

24. Cullmann, G., K. Fien, R. Kobayashi, and B. Stillman. 1995. Characterization of the 5 Replication Factor-C Genes of *Saccharomyces cerevisiae*. *Mol. Cell. Biol.* **15:**4661–4671.

25. Daimon, K., Y. Kawarabayasi, H. Kikuchi, Y. Sako, and Y. Ishino. 2002. Three proliferating cell nuclear antigen-like proteins found in the hyperthermophilic archaeon *Aeropyrum pernix*: Interactions with the two DNA polymerases. *J. Bacteriol.* **184:**687–694.

26. Davey, M. J., L. H. Fang, P. McInerney, R. E. Georgescu, and M. O'Donnell. 2002. The DnaC helicase loader is a dual ATP/ADP switch protein. *EMBO J.* **21:**3148–3159.

27. Davey, M. J., D. Jeruzalmi, J. Kuriyan, and M. O'Donnell. 2002. Motors and switches: AAA+ machines within the replisome. *Nat. Rev. Mol. Cell. Biol.* **3:**826–835.

28. De Falco, M., A. Fusco, M. De Felice, M. Rossi, and F. M. Pisani. 2004. The DNA primase of *Sulfolobus solfataricus* is activated by substrates containing a thymine-rich bubble and has a 3′-terminal nucleotidyl-transferase activity. *Nucleic Acids Res.* **32:**5223–5230.

29. Della, M., P. L. Palmbos, H. M. Tseng, L. M. Tonkin, J. M. Daley, L. M. Topper, R. S. Pitcher, A. E. Tomkinson, T. E. Wilson, and A. J. Doherty. 2004. Mycobacterial Ku and ligase proteins constitute a two-component NHEJ repair machine. *Science* **306:**683–685.

30. Dervyn, E., C. Suski, R. Daniel, C. Bruand, J. Chapuis, J. Errington, L. Janniere, and S. D. Ehrlich. 2001. Two essential DNA polymerases at the bacterial replication fork. *Science* **294:**1716–1719.

31. Dionne, I., R. K. Nookala, S. P. Jackson, D. A. J., and S. D. Bell. 2003. A heterotrimeric PCNA in the hyperthermophilic archaeon *Sulfolobus solfataricus*. *Mol. Cell* **11:**275–282.

32. Dionne, I., N. P. Robinson, A. T. McGeoch, V. L. Marsh, A. Reddish, and S. D. Bell. 2003. DNA replication in the hyperthermophilic archaeon *Sulfolobus solfataricus*. *Biochem. Soc. Trans.* **31**(Pt 3):674–676.

33. Fang, L. H., M. J. Davey, and M. O'Donnell. 1999. Replisome assembly at oriC, the replication origin of *E. coli*, reveals an explanation for initiation sites outside an origin. *Mol. Cell* **4:**541–553.

34. Fletcher, R. J., B. E. Bishop, R. P. Leon, R. A. Sclafani, C. M. Ogata, and X. S. Chen. 2003. The structure and function of MCM from archaeal *M. thermoautotrophicum*. *Nat. Struct. Biol.* **10:**160–167.

35. Fogg, M. J., L. H. Pearl, and B. A. Connolly. 2002. Structural basis for uracil recognition by archaeal family B DNA polymerases. *Nat. Struct. Biol.* **9:**922–927.

36. Frick, D. N., and C. C. Richardson. 2001. DNA primases. *Annu. Rev. Biochem.* **70:**39–80.

37. Fukui, T., K. Yamauchi, T. Muroya, M. Akiyama, H. Maki, A. Sugino, and S. Waga. 2004. Distinct roles of DNA polymerases delta and epsilon at the replication fork in *Xenopus* egg extracts. *Genes Cells* **9:**179–191.

38. Funnell, B. E., T. A. Baker, and A. Kornberg. 1987. *In vitro* assembly of a prepriming complex at the origin of the *Escherichia coli* chromosome. *J. Biol. Chem.* **262:**10327–10334.

39. Gai, D. H., R. Zhao, D. W. Li, C. V. Finkielstein, and X. S. Chen. 2004. Mechanisms of conformational change for a replicative hexameric helicase of SV40 large tumor antigen. *Cell* **119:**47–60.

40. Gomes, X. V., and P. M. J. Burgers. 2001. ATP utilization by yeast replication factor C I. ATP-mediated interaction with DNA and with proliferating cell nuclear antigen. *J. Biol. Chem.* **276:**34768–34775.

41. Gomes, X. V., S. L. G. Schmidt, and P. M. J. Burgers. 2001. ATP utilization by yeast replication factor C II. Multiple step-

wise ATP binding events are required to load proliferating cell nuclear antigen onto primed DNA. *J. Biol. Chem.* **276:** 34776–34783.

42. Grainge, I., S. Scaife, and D. Wigley. 2003. Biochemical analysis of components of the pre-replication complex of *Archaeoglobus fulgidus*. *Nucleic Acids. Res.* **31:**4888–4898.

43. Haseltine, C. A., and S. C. Kowalczykowski. 2002. A distinctive single-stranded DNA-binding protein from the archaeon *Sulfolobus solfataricus*. *Mol. Microbiol.* **43:**1505–1515.

44. Henneke, G., D. Flament, U. Hubscher, J. Querellou, and J. P. Raffin. 2005. The hyperthermophilic euryarchaeota *Pyrococcus abyssi* likely requires the two DNA polymerases D and B for DNA replication. *J. Mol. Biol.* **350:**53–64.

45. Henneke, G., Y. Gueguen, D. Flament, P. Azam, J. Querellou, J. Dietrich, U. Hubscher, and J. P. Raffin. 2002. Replication factor C from the hyperthermophilic archaeon *Pyrococcus abyssi* does not need ATP hydrolysis for clamp- loading and contains a functionally conserved RFC PCNA-binding domain. *J. Mol. Biol.* **323:**795–810.

46. Henneke, G., J. P. Raffin, E. Ferrari, Z. O. Jonsson, J. Dietrich, and U. Hubscher. 2000. The PCNA from *Thermococcus fumicolans* functionally interacts with DNA polymerase delta. *Biochem. Biophys. Res. Commun.* **276:**600–606.

47. Herrmann, U., and J. Soppa. 2002. Cell cycle-dependent expression of an essential SMC-like protein and dynamic chromosome localization in the archaeon *Halobacterium salinarum*. *Mol. Microbiol.* **46:**395–409.

48. Hjort, K., and R. Bernander. 2001. Cell cycle regulation in the hyperthermophilic crenarchaeon *Sulfolobus acidocaldarius*. *Mol. Microbiol.* **40:**225–234.

49. Hubscher, U., G. Maga, and S. Spadari. 2002. Eukaryotic DNA polymerases. *Annu. Rev. Biochem.* **71:**133–163.

50. Ito, N., O. Nureki, M. Shirouzu, S. Yokoyama, and F. Hanaoka. 2003. Crystal structure of the *Pyrococcus horikoshii* DNA primase-UTP complex: implications for the mechanism of primer synthesis. *Genes Cells* **8:**913–923.

51. Iyer, L. M., E. V. Koonin, D. D. Leipe, and L. Aravind. 2005. Origin and evolution of the archaeo-eukaryotic primase superfamily and related palm-domain proteins: structural insights and new members. *Nucleic Acids. Res.* **33:** 3875–3896.

52. Iyer, L. M., D. D. Leipe, E. V. Koonin, and L. Aravind. 2004. Evolutionary history and higher order classification of AAA plus ATPases. *J. Struct. Biol.* **146:**11–31.

53. Jacob, F., S. Brenner, and F. Kuzin. 1963. On the regulation of DNA replication in bacteria. *Cold Spring Harbor Symp. Quant. Biol.* **28:**329–348.

54. Jeruzalmi, D., M. O'Donnell, and J. Kuriyan. 2002. Clamp loaders and sliding clamps. *Curr. Opin. Struct. Biol.* **12:** 217–224.

55. Jeruzalmi, D., M. O'Donnell, and J. Kuriyan. 2001. Crystal structure of the processivity clamp loader gamma (gamma) complex of *E. coli* DNA polymerase III. *Cell* **106:**429–441.

56. Jeruzalmi, D., M. O'Donnell, and J. Kuriyan. 2001. Structural studies of DNA polymerase processivity clamp loading. *Mol. Biol. Cell* **12:**2191.

57. Jeruzalmi, D., O. Yurieva, Y. X. Zhao, M. Young, J. Stewart, M. Hingorani, M. O'Donnell, and J. Kuriyan. 2001. Mechanism of processivity clamp opening by the delta subunit wrench of the clamp loader complex of E-coli DNA polymerase III. *Cell* **106:**417–428.

58. Kelly, T. J., P. Simancek, and G. S. Brush. 1998. Identification and characterization of a single-stranded DNA- binding protein from the archaeon *Methanococcus jannaschii*. *Proc. Natl. Acad. Sci. USA* **95:**14634–14639.

59. Kelman, L. M., and Z. Kelman. 2003. Archaea: an archetype for replication initiation studies? *Mol. Microbiol.* **48:**605–616.

60. Kelman, Z., and J. Hurwitz. 1998. Protein-PCNA interactions: a DNA-scanning mechanism? *Trends Biochem. Sci.* **23:**236–238.

61. Kelman, Z., and J. Hurwitz. 2000. A unique organization of the protein subunits of the DNA polymerase clamp loader in the archaeon *Methanobacterium thermoautotrophicum* Delta H. *J. Biol. Chem.* **275:**7327–7336.

62. Kelman, Z., J. K. Lee, and J. Hurwitz. 1999. The single minichromosome maintenance protein of *Methanobacterium thermoautotrophicum* Delta H contains DNA helicase activity. *Proc. Natl. Acad. Sci. USA* **96:**14783–14788.

63. Kerr, I. D., R. I. M. Wadsworth, L. Cubeddu, W. Blankenfeldt, J. H. Naismith, and M. F. White. 2003. Insights into ssDNA recognition by the OB fold from a structural and thermodynamic study of *Sulfolobus* SSB protein. *EMBO J.* **22:**2561–2570.

64. Komori, K., and Y. Ishino. 2001. Replication protein A in *Pyrococcus furiosus* is involved in homologous DNA recombination. *J. Biol. Chem.* **276:**25654–25660.

65. Lao-Sirieix, S. H., and S. D. Bell. 2004. The heterodimeric primase of the hyperthermophilic archaeon *Sulfolobus solfataricus* possesses DNA and RNA primase, polymerase and 3'-terminal nucleotidyl transferase activities. *J. Mol. Biol.* **344:**1251–1263.

66. Lao-Sirieix, S. H., R. K. Nookala, P. Roversi, S. D. Bell, and L. Pellegrini. 2005. Structure of the heterodimeric core primase. *Nat. Struct. Mol. Biol.* **12:**1137–1144.

67. Lao-Sirieix, S. H., L. Pellegrini, and S. D. Bell. 2005. The promiscuous primase. *Trends Genet.* **21:**568–572.

68. Li, D. W., R. Zhao, W. Lilyestrom, D. H. Gai, R. G. Zhang, J. A. DeCaprio, E. Fanning, A. Jochimiak, G. Szakonyi, and X. J. S. Chen. 2003. Structure of the replicative helicase of the oncoprotein SV40 large tumour antigen. *Nature* **423:**512–518.

69. Lipps, G., S. Rother, C. Hart, and G. Krauss. 2003. A novel type of replicative enzyme harbouring ATPase, primase and DNA polymerase activity. *EMBO J.* **22:**2516–2525.

70. Liu, L. D., K. Komori, S. Ishino, A. A. Bocquier, I. K. O. Cann, D. Kohda, and Y. Ishino. 2001. The archaeal DNA primase—biochemical characterization of the p41-p46 complex from *Pyrococcus furiosus*. *J. Biol. Chem.* **276:**45484–45490.

71. Lundgren, M., A. Andersson, L. M. Chen, P. Nilsson, and R. Bernander. 2004. Three replication origins in *Sulfolobus* species: Synchronous initiation of chromosome replication and asynchronous termination. *Proc. Natl. Acad. Sci. USA* **101:**7046–7051.

72. Maisnier-Patin, S., L. Malandrin, N. K. Birkeland, and R. Bernander. 2002. Chromosome replication patterns in the hyperthermophilic euryarchaea *Archaeoglobus fulgidus* and *Methanocaldococcus (Methanococcus) jannaschii*. *Mol. Microbiol.* **45:**1443–1450.

73. Majernik, A. I., M. Lundgren, P. McDermott, R. Bernander, and J. P. J. Chong. 2005. DNA content and nucleoid distribution in *Methanothermobacter thermautotrophicus*. *J. Bacteriol.* **187:**1856–1858.

74. Malandrin, L., H. Huber, and R. Bernander. 1999. Nucleoid structure and partition in *Methanococcus jannaschii*: an Archaeon with multiple copies of the chromosome. *Genetics* **152:**1315–1323.

75. Matsui, E., M. Nishio, H. Yokoyama, K. Harata, S. Darnis, and I. Matsui. 2003. Distinct domain functions regulating de novo DNA synthesis of thermostable DNA primase from hyperthermophile *Pyrococcus horikoshii*. *Biochemistry* **42:** 14968–14976.

76. **Matsumiya, S., Y. Ishino, and K. Morikawa. 2001.** Crystal structure of an archaeal DNA sliding clamp: proliferating cell nuclear antigen from *Pyrococcus furiosus*. *Protein Sci.* **10:** 17–23.

77. **Matsunaga, F., P. Forterre, Y. Ishino, and H. Myllykallio. 2001.** In vivo interactions of archaeal Cdc6/Orc1 and minichromosome maintenance proteins with the replication origin. *Proc. Natl. Acad. Sci. USA* **98:**11152–11157.

78. **Matsunaga, F., C. Norais, P. Forterre, and H. Myllykallio. 2003.** Identification of short 'eukaryotic' Okazaki fragments synthesized from a prokaryotic replication origin. *EMBO Rep.* **4:**154–158.

79. **McGeoch, A. T., and S. D. Bell. 2005.** Eukaryotic/Archaeal primase and MCM proteins encoded in a bacteriophage genome. *Cell* **120:**167–168.

80. **McGeoch, A. T., M. A. Trakselis, R. A. Laskey, and S. D. Bell. 2005.** Organization of the archaeal MCM complex on DNA and implications for helicase mechanism. *Nat. Struct. Mol. Biol.* **12:**756–762

81. **McHenry, C. S. 2003.** Chromosomal replicases as asymmetric dimers: studies of subunit arrangement and functional consequences. *Mol. Microbiol.* **49:**1157–1165.

82. **Mechali, M. 2001.** DNA replication origins: from sequence specificity to epigenetics. *Nat. Rev. Genet.* **2:**640–645.

83. **Messer, W. 2002.** The bacterial replication initiator DnaA. DnaA and *oriC*, the bacterial mode to initiate DNA replication. *FEMS Microbiol. Rev.* **26:**355–374.

84. **Messer, W., F. Blaesing, D. Jakimowicz, M. Krause, J. Majka, J. Nardmann, S. Schaper, H. Seitz, C. Speck, C. Weigel, G. Wegrzyn, M. Welzeck, and J. Zakrzewska-Czerwinska. 2001.** Bacterial replication initiator DnaA. Rules for DnaA binding and roles of DnaA in origin unwinding and helicase loading. *Biochimie* **83:**5–12.

85. **Miyata, T., T. Oyama, K. Mayanagi, S. Ishino, Y. Ishino, and K. Morikawa. 2004.** The clamp-loading complex for processive DNA replication. *Nat. Struct. Mol. Biol.* **11:**632–636.

86. **Murzin, A. G. 1993.** Ob(oligonucleotide oligosaccharide binding)-fold—common structural and functional solution for nonhomologous sequences. *EMBO J.* **12:**861–867.

87. **Myllykallio, H., P. Lopez, P. Lopez-Garcia, R. Heilig, W. Saurin, Y. Zivanovic, H. Philippe, and P. Forterre. 2000.** Bacterial mode of replication with eukaryotic-like machinery in a hyperthermophilic archaeon. *Science* **288:**2212–2215.

88. **Napoli, A., A. Valenti, V. Salerno, M. Nadal, F. Garnier, M. Rossi, and M. Ciaramella. 2005.** Functional interaction of reverse gyrase with single-strand binding protein of the archaeon *Sulfolobus*. *Nucleic Acids Res.* **33:**564–576.

89. **Pape, T., H. Meka, S. X. Chen, G. Vicentini, M. van Heel, and S. Onesti. 2003.** Hexameric ring structure of the full-length archaeal MCM protein complex. *EMBO Rep.* **4:**1079–1083.

90. **Pisani, F. M., M. De Felice, F. Carpentieri, and M. Rossi. 2000.** Biochemical characterization of a clamp-loader complex homologous to eukaryotic replication factor C from the hyperthermophilic archaeon *Sulfolobus solfataricus*. *J. Mol. Biol.* **301:**61–73.

91. **Poplawski, A., B. Grabowski, S. F. Long, and Z. Kelman. 2001.** The zinc finger domain of the archaeal minichromosome maintenance protein is required for helicase activity. *J. Biol. Chem.* **276:**49371–49377.

92. **Raghunathan, S., A. G. Kozlov, T. M. Lohman, and G. Waksman. 2000.** Structure of the DNA binding domain of E-coli SSB bound to ssDNA. *Nat. Struct. Biol.* **7:**648–652.

93. **Raghunathan, S., C. S. Ricard, T. M. Lohman, and G. Waksman. 1997.** Crystal structure of the homo-tetrameric

DNA binding domain of *Escherichia coli* single-stranded DNA-binding protein determined by multiwavelength x-ray diffraction on the selenomethionyl protein at 2.9-angstrom resolution. *Proc. Natl. Acad. Sci. USA* **94:**6652–6657.

94. **Ramadan, K., I. V. Shevelev, G. Maga, and U. Hubscher. 2004.** De novo DNA synthesis by human DNA polymerase lambda, DNA polymerase mu and terminal deoxyribonucleotidyl transferase. *J. Mol. Biol.* **339:**395–404.

95. **Robbins, J. B., M. C. Murphy, B. A. White, R. I. Mackie, T. Ha, and I. K. O. Cann. 2004.** Functional analysis of multiple single-stranded DNA-binding proteins from Methanosarcina acetivorans and their effects on DNA synthesis by DNA polymerase BI. *J. Biol. Chem.* **279:**6315–6326.

96. **Robinson, N. P., and S. D. Bell. 2005.** Origins of DNA replication in the three domains of life. *FEBS J.* **272:**3757–3766.

97. **Robinson, N. P., I. Dionne, M. Lundgren, V. L. Marsh, R. Bernander, and S. D. Bell. 2004.** Identification of two origins of replication in the single chromosome of the archaeon *Sulfolobus solfataricus*. *Cell* **116:**25–38.

98. **Sakurai, S., K. Kitano, H. Yamaguchi, K. Hamada, K. Okada, K. Fukuda, M. Uchida, E. Ohtsuka, H. Morioka, and T. Hakoshima. 2005.** Structural basis for recruitment of human flap endonuclease 1 to PCNA. *EMBO J.* **24:**683–693.

99. **Seybert, A., D. J. Scott, S. Scaife, M. R. Singleton, and D. B. Wigley. 2002.** Biochemical characterisation of the clamp/clamp loader proteins from the euryarchaeon *Archaeoglobus fulgidus*. *Nucleic Acids Res.* **30:**4329–4338.

100. **Seybert, A., and D. B. Wigley. 2004.** Distinct roles for ATP binding and hydrolysis at individual subunits of an archaeal clamp loader. *EMBO J.* **23:**1360–1371.

101. **Shechter, D. F., C. Y. Ying, and J. Gautier. 2000.** The intrinsic DNA helicase activity of *Methanobacterium thermoautotrophicum* Delta H minichromosome maintenance protein. *J. Biol. Chem.* **275:**15049–15059.

102. **Sherratt, D. J. 2003.** Bacterial chromosome dynamics. *Science* **301:**780–785.

103. **Shin, J. H., B. Grabowski, R. Kasiviswanathan, S. D. Bell, and Z. Kelman. 2003.** Regulation of minichromosome maintenance helicase activity by Cdc6. *J. Biol. Chem.* **278:**38059–38067.

104. **Shin, J. H., Y. Jiang, B. Grabowski, J. Hurwitz, and Z. Kelman. 2003.** Substrate requirements for duplex DNA translocation by the eukaryal and archaeal minichromosome maintenance helicases. *J. Biol. Chem.* **278:**49053–49062.

105. **Singleton, M. R., M. R. Sawaya, T. Ellenberger, and D. B. Wigley. 2000.** Crystal structure of T7 gene 4 ring helicase indicates a mechanism for sequential hydrolysis of nucleotides. *Cell* **101:**589–600.

106. **Tenney, D. J., A. K. Sheaffer, W. W. Hurlburt, M. Bifano, and R. K. Hamatake. 1995.** Sequence-dependent primer synthesis by the Herpes Simplex Virus helicase-primase complex. *J. Biol. Chem.* **270:**9129–9136.

107. **Toth, E. A., Y. Li, M. R. Sawaya, Y. F. Cheng, and T. Ellenberger. 2003.** The crystal structure of the bifunctional primase-helicase of bacteriophage T7. *Mol. Cell* **12:**1113–1123.

108. **Trakselis, M. A., and S. D. Bell. 2004.** The loader of the rings. *Nature* **429:**708–709.

109. **Vivona, J. B., and Z. Kelman. 2003.** The diverse spectrum of sliding clamp interacting proteins. *FEBS Lett.* **546:**167–172.

110. **Wadsworth, R. I. M., and M. F. White. 2001.** Identification and properties of the crenarchaeal single-stranded DNA binding protein from *Sulfolobus solfataricus*. *Nucleic Acids Res.* **29:**914–920.

111. **Warbrick, E. 2000.** The puzzle of PCNA's many partners. *Bioessays* **22:**997–1006.

112. **Wold, M. S.** 1997. Replication protein A: a heterotrimeric, single-stranded DNA-binding protein required for eukaryotic DNA metabolism. *Annu. Rev. Biochem.* **66:**61–92.

113. **Yu, X., M. S. VanLoock, A. Poplawski, Z. Kelman, T. Xiang, B. K. Tye, and E. A. Egelman.** 2002. The *Methanobacterium thermoautotrophicum* MCM protein can form heptameric rings. *EMBO Rep.* **3:**792–797.

114. **Zhang, R., and C. Zhang.** 2003. Multiple replication origins of the archaeon *Halobacterium* species NRC-1. *Biochem. Biophys. Res. Commun.* **302:**728–734.

Archaea: Molecular and Cellular Biology
Edited by Ricardo Cavicchioli
© 2007 ASM Press, Washington, D.C.

Chapter 4

DNA-Binding Proteins and Chromatin

RACHEL SAMSON AND JOHN N. REEVE

INTRODUCTION

The genomes of all organisms have to be compacted to fit within the confines of a cell or nuclear compartment. In higher eucarya, two meters of DNA are condensed to fit within a nucleus that is ~10 μm in diameter, and bacteria and archaea must similarly compact their genomes to fit within a cell that is a thousand times shorter (25, 33, 53, 85). The very high concentrations of proteins and nucleic acids that are present in vivo aid DNA condensation by disrupting DNA-solvent interactions, and cations also neutralize many repulsive negative charges along the DNA backbone (53). Given such conditions of macromolecular crowding and ionic environment, the issue in vivo is not *how* to collapse a large chromosome structure, but how to do so in a manner that preserves flexibility and access to the DNA molecule for replication and gene expression. In the three biological domains, several structurally unrelated families of chromatin proteins exist that have apparently evolved independently to facilitate genome condensation, prevent DNA aggregation, and allow expression. As described, some of these chromatin proteins are present predominantly in one biological domain, but there are overlaps arguing for ancient common ancestries or lateral gene transfer.

BACTERIAL CHROMATIN PROTEINS AND NUCLEOIDS

Procaryotic genomic DNA and associated proteins together form an irregularly shaped structure, designated the nucleoid (33, 53). In *Escherichia coli*, this has been shown to have a central mass from which long supercoiled loops of transcriptionally active DNA extend (4). In many bacteria, including *E. coli*, members of the HU family are the most abundant chromatin proteins. The length of the DNA bound by different HU homologs is reported to range from ~9 to 35 bp per dimer, with binding causing a sharp kink and minor groove widening and introducing negative superhelicity (25, 37, 77). Cocrystals with DNA have revealed that HU dimers have two long flexible β-ribbon arms that extend from a central globular body and interact with the phosphodiester backbone within the minor groove of the DNA helix (77). *E. coli* mutants lacking HU are viable, indicating that other chromatin proteins can provide compensatory functions (25). HU-family members are also present in *Giardia lamblia* and dinoflagellates that could be remnants of endosymbiotic events (81, 86), as is almost certainly the case for HU in plant plastids (69). Alternatively, these eucaryal HU-family members may have been acquired by horizontal gene transfer from bacteria, as was most likely the origin of the archaeal HU-family members present in the *Thermoplasma* lineage (22).

The *E. coli* nucleoid contains several other chromatin proteins, the most abundant being H-NS and FIS, which also function as transcription activators and repressors (3, 25). H-NS binds with high affinity to bent DNA structures and can constrain supercoils in DNA. FIS binds both nonspecifically and in a sequence-dependent manner to DNA, generating severe bends in the DNA helix. It is frequently reported that bacterial chromatin proteins change in abundance with growth phase (3, 24, 66), and in this regard, H-NS increases 5- to 10-fold, whereas FIS decreases in abundance in *E. coli* cells as cultures enter the stationary phase. Such changes are likely to be related to, if not responsible for, many of the changes in gene expression that occur in different growth phases, but this has not yet been definitively documented. Changes in the chromatin proteins present also occur

Rachel Samson and John N. Reeve • Department of Microbiology, Ohio State University, Columbus, OH 43210.

when bacteria differentiate. For example, *Bacillus subtilis* spores contain DNA-binding proteins, designated small acid-soluble proteins (SASPs), that are not present in vegetative cells (71). The SASPs establish the structure of the spore nucleoid, protect the spore DNA from radiation and desiccation damage, and silence gene expression. Similarly, the nucleoid of the metabolically inert, infectious form of *Chlamydia* is compacted into a tight spherical structure by Hc1 and Hc2 (60), two proteins related to eucaryal histone H1 that are not present in vegetative cells.

EUCARYAL HISTONES AND CHROMATIN

In contrast to the range of different chromatin proteins identified in bacteria, almost all eucaryal genomes are compacted into nucleosomes, chromatin, and chromosomes by essentially the same four proteins, histones H2A, H2B, H3, and H4. A nucleosome core consists of a histone $(H3+H4)_2$ tetramer flanked on either side by a histone (H2A+H2B) heterodimer. This histone octamer binds and wraps ~146 bp of DNA in 1.65 negative superhelical turns to form a nucleosome unit (46). All four nucleosome core histones form the same structural motif known as the histone fold (HF) with three α-helices (α1, α2, α3) separated by two short β-strand loops (L1, L2) (46, 68). N- and C-terminal sequences extend from the HF that participate in higher-order nucleosome polymerization. They also provide the targets for posttranslation acetylation, methylation, phosphorylation, and ubiquitinylation events that regulate eucaryal chromatin structure and gene expression (18, 55, 74). Acetylation of histone-tail lysine residues reduces histone-DNA affinity and interactions between neighboring nucleosomes, and this facilitates transcription factor access to the DNA and so relieves repression of gene expression imposed by chromatin structure. Histone-modifying enzymes, such as histone deacetylases and histone acetyltransferases, are therefore frequently identified as activators or repressors of eucaryal gene expression (52, 82). The DNA between eucaryal nucleosomes is bound by proteins, designated linker histones H1 or H5, although these are structurally unrelated to the nucleosome core histones and do not have HFs (38).

ARCHAEAL CHROMATIN PROTEINS

The following sections describe several different families of archaeal chromatin proteins with unrelated structures, but with the common properties of abundance, small size, positive charge, and ability to bind to DNA with little or no sequence specificity (64, 85). It is presumed that these proteins compact archaeal genomic DNA in vivo and prevent DNA aggregation, and that they probably also participate in DNA replication, repair, and gene expression. These roles in vivo have not, however, yet been established experimentally and, in the one published report, the introduction of mutations into archaeal chromatin genes was not lethal (34).

Archaeal Histones

Archaeal histones were first discovered in methanogens (67), and then later in other members of the *Euryarchaeota* but, until very recently, were noticeably absent from all *Crenarchaeota* (64, 68). This argued that histones evolved after the divergence of the two major archaeal lineages, consistent hypotheses that posit that the eucaryal nucleus originated from an euryarchaeon (50). However, the *Nanoarchaeum equitans* genome was found to contain two histone genes, demonstrating that histones are present in this third very deep branching archaeal lineage (13), and histone-encoding genes have now also been identified in genomic DNA that most likely originated from a marine crenarchaeon, and also in the genome of the crenarchaeal marine sponge symbiont, *Cenarchaeum symbiosum* (19). Histones do therefore exist in mesophilic crenarchaeota but have apparently been lost, and presumably functionally replaced, in hyperthermophilic crenarchaeota by the proteins, such as Sul7, Sul10a, and Sul10b (Alba), described in sections below.

Most archaeal histones are just HFs without N- or C-terminal extensions (Fig. 1). The HF is stable only in a dimer configuration, and archaeal histones form both homodimers and heterodimers in vitro, and presumably also in vivo, in species that have multiple histones (64, 68). In this regard, *Methanosarcina* species have the largest archaeal genomes but have only one histone gene, whereas the much smaller genomes of *Methanocaldococcus jannaschii* and *Methanosphaera stadtmanae* encode six and seven different archaeal histones, respectively. The most thoroughly characterized archaeal histone is HMfB from *Methanothermus fervidus* (21), and this is the archetype of the most prevalent family of archaeal histones, proteins with 65 to 69 residues that form one HF. The HF is stabilized by an intramolecular salt bridge between an arginine in L2 (R52 in HMfB) and an aspartate in α3 (D59 in HMfB) and by intermolecular hydrophobic interactions between α2 and α3 residues within the core of a dimer (Fig. 2A). In the presence of DNA, archaeal histone dimers associate to form tetramers that bind and wrap ~90 bp of DNA (5, 47, 80, 87). The archaeal nucleosomes generated (61) re-

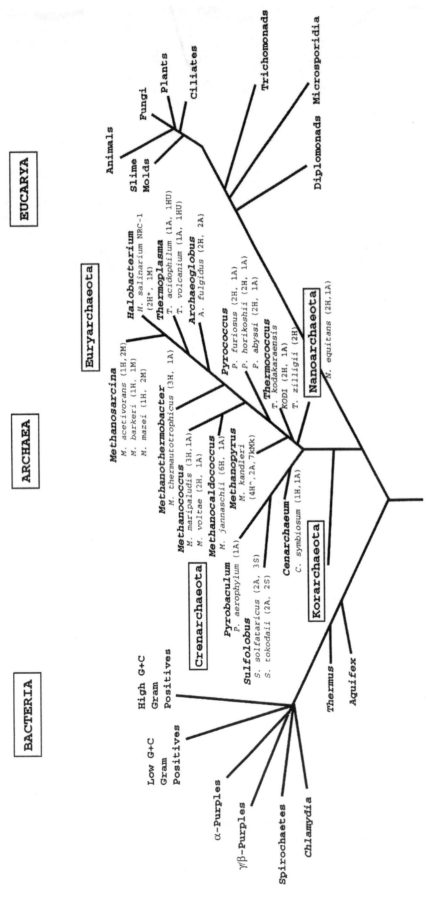

Figure 1. Phylogenetic tree based on rDNA sequence alignments of selected organisms. Branch lengths do not reflect evolutionary distances, but the branching orders are correctly represented. The numbers of histones (H), Sul7d (S), Alba (A), MC1 (M), 7kMk and HU-family members encoded in the genomes of representative archaea are denoted in parentheses. The *H. salinarium* NRC-1 histone has two HFs in one polypeptide (*). *M. kandleri* has two members of the HMfB family of archaeal histones with one HF (^), and also HMk, an archaeal histone with two HFs.

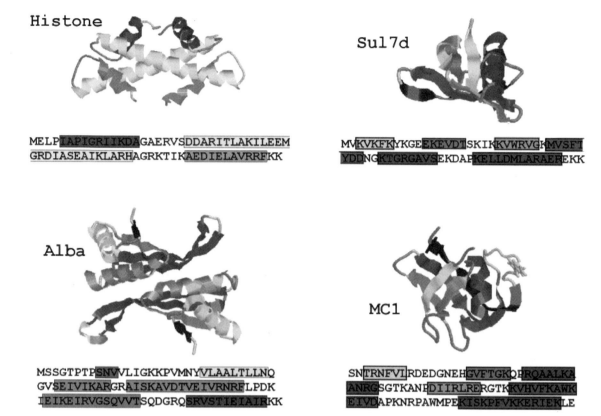

Figure 2. (*See the separate color insert for the color version of this illustration.*) Sequences and structures of representative archaeal chromatin proteins. Primary sequences of HMfB from *M. fervidus* (A), Sul7d (Sac7d) from *S. acidocaldarius* (B), Alba (Sso10b1) from *S. solfataricus* (C), and MC1 from *Methanosarcina* sp. CHTI55 (D) are shown below the corresponding protein structure. The figure was constructed using structures available from the Protein Data Bank (11). Regions with α-helical and β-strand structures are colored identically in the sequence and in the corresponding structure.

tain the flexibility to wrap DNA in either a positive or negative supercoil (47), and this flexibility may help accommodate the activities of DNA polymerases, RNA polymerase, and topoisomerases that generate local DNA supercoiling. Histones bind to the sugar-phosphate backbone of DNA (46) and so can incorporate any DNA sequence into a nucleosome, although DNA sequences that readily distort to accept alternating major and minor groove compressions are preferentially bound. Consistent with genome sequences having evolved to facilitate their own packaging, dinucleotide repeats that facilitate nucleosome assembly are overrepresented in the genomes of histone-containing species, including archaea (6, 70). In *M. fervidus*, the archaeal histones HMfA and HMfB constitute ~4% of total soluble protein, sufficient for one histone tetramer per ~67 bp of genomic DNA, assuming a genome size of ~1.7 Mbp. Chromatin immunoprecipitation studies indicate that most genomic sequences in this hyperthermophilic archaeon are associated in vivo with histones (61, 62).

Eucaryal histone (H3+H4) heterodimers are asymmetric (46). This asymmetry positions surface-located DNA-binding residues appropriately and also prevents higher-order histone dimer-dimer oligomerization. In this regard, two archaeal histones, MkaH in *Methanopyrus kandleri* and HHb in *Halobacterium salinarum* NRC1, have inherently asymmetric HF "dimer" structures with two nonidentical HFs tandemly linked in one polypeptide chain (Fig. 1; reference 64). This guarantees HF heterodimer formation and an asymmetry that could direct nucleosome assembly (27, 58). How archaeal histone homodimer polymerization is limited in *Methanosarcina* species that have only one archaeal histone gene is an intriguing question.

Archaeal histones that do have C-terminal extensions have been identified in methanococcal species (64). The extensions are ~25 residues in length and are predicted to fold into α-helices, but they share no sequence identity with eucaryal histone tails. Deletion of the C-terminal extension reduced the thermostability of such a histone (MJ1647) from *M. jannaschii*, but increased its ability to bind DNA (43). Apparently, in vivo, the C-terminal extension of MJ1647 obstructs its homopolymerization and so helps direct the assembly of archaeal nucleosomes in *M. jannaschii* with asymmetric heterotetramer histone cores.

The only amino acid within the HF of a eucaryal nucleosome core histone that has been shown to undergo posttranslation regulatory modification is K69 in the expanded L1 region of histone H3 (12). Intriguingly, NEQ0288, one of the two histones in *N. equitans* (13), has a related sequence including a lysine residue present at the same location within L1. It will be interesting to determine whether this lysine is a target for regulatory methylation in NEQ0288 as is K69 in H3 (12).

Sul7d

Sul7d is the generic name given to a family of very abundant, highly conserved ~7-kDa proteins (1, 44). Individually, they are designated Sac7d in *Sulfolobus acidocaldarius*, Sso7d in *Sulfolobus solfataricus*, and Sto7d in *Sulfolobus todakaii* (Fig. 1). These hyperthermophilic crenarchaeotes do not have histones. Members of the Sul7d family constitute up to ~5% of total cellular protein. They bind DNA noncooperatively with one monomer contacting ~4 bp of DNA (15, 29, 44, 51). Binding unwinds the DNA and introduces negative superhelicity which could compensate for the overwinding of DNA that occurs inherently at hyperthermophilic growth temperatures (see Chapter 19). When bound to closed DNA molecules, Sul7d unwinding of the duplex increases DNA writhe and compacts positively supercoiled and relaxed DNAs, but decompacts negatively supercoiled DNA molecules (44, 54).

NMR solution and X-ray crystal structures have been solved for Sul7d-DNA complexes. The barrel-shaped protein consists of a double-stranded β-sheet packed against a triple-stranded β-sheet, followed by a C-terminal amphiphilic α-helix (29, 40) (Fig. 2B). Residues in the triple-stranded β-sheet contact four bases in the minor groove of DNA, with the hydrophobic side chains of V26 and M29 intercalating between bases and causing a ~66° bend in the DNA (1, 63, 76). The protein binds largely through non-polar interactions with deoxyribose units in the minor groove, although electrostatic interactions, salt bridges, and hydrogen bonds also occur that together provide strong DNA binding with a low degree of sequence specificity (8).

Although the mechanism and biological function(s) are unknown, 5 of the 14 lysine residues in Sul7d are specifically monomethylated in vivo. This methylation does not change the DNA affinity or

thermal stability of the protein (76). In addition to its likely role in chromosome compaction, Sul7d has a surprising number of additional activities. It facilitates the annealing of complementary single-stranded DNAs and has RNase, ATP-dependent protein chaperone, and protein renaturation activities (30, 31, 72).

Sul10b/Alba

Hyperthermophilic *Sulfolobus* species contain Sul10a and Sul10b, in addition to Sul7d, and whereas Sul7d and Sul10a have limited phylogenetic distribution, Sul10b is widely distributed throughout the *Archaea*. Individual proteins have been designated Sso10b in *S. solfataricus*, Ssh10b in *S. shibatae*, Sac10b in *S. acidocaldarius*, Afu10b in *Archaeoglobus fulgidus*, and Mja10b in *M. jannaschii* (Fig.1; references 20, 32, 83, 84, and 89). Sso10b has also been designated Alba, based on the observation that *acetylation lowers DNA-binding affinity* (10). Crystal structures of Sso10b/Alba dimers from *S. solfataricus*, *A. fulgidus*, and *M. jannaschii* have revealed a β1-α1-β2-α2-β3-β4 monomer secondary structure with an overall topology similar to that of the C-terminal domain of *E. coli* initiation factor IF3, and the N-terminal DNA-binding domain of DNase I (83, 84, 89). The Alba monomers are associated in an antiparallel orientation, via residues in the α2-helix and the β3- and β4-strands. The dimer has two β-hairpin structures extending from a globular central body (Fig. 2C). Alba dimers bind DNA cooperatively and can constrain DNA into negative supercoils (10, 35, 88). In one model proposed for DNA binding, the central body of the dimer is positioned across the major DNA groove, and the β-hairpins extend to make contacts within the flanking minor grooves (84).

A very intriguing observation is that Alba (Sso10b) from *S. solfataricus* is acetylated at K16 in vivo, and acetylated Alba has a 3-fold lower affinity for DNA than nonacetylated Alba. The enzymes responsible for Alba acetylation and deacetylation in *S. solfataricus* have been characterized, and acetylation and deacetylation in vitro has been established (10, 48, 49). When deacetylated Alba binds to the template DNA, transcription in vitro is repressed and this repression is relieved by acetylation (10). This is reminiscent of the acetylation and deacetylation of lysine residues in eucaryal histone tails regulating eucaryal chromatin structure and gene expression (18, 52, 74, 82). K16 acetylation may not directly reduce Alba affinity for DNA, but rather may modulate Alba dimer-dimer interactions required for Alba polymerization and so binding along a DNA duplex. Regardless of the molecular details, this indicates that chromatin structure, and thereby gene expression, might be regulated in archaea by chromatin protein modifications. To date, Alba acetylation has only been documented in *S. solfataricus*, and not all Alba family members have a lysine at the site of K16 in Sso10b (89).

A bioinformatics analysis concluded that Alba is a member of a superfamily of proteins, most of which are involved in RNA metabolism, for example, subunits of RNase P, RNase MRP, and Mdp2 (2). Together with the very wide distribution of Alba proteins in archaea, this has led to the hypothesis that Alba may be an RNA-binding protein in some archaea that has been recruited in other lineages to become a chromatin protein (2). Alba does bind to both DNA and RNA in *Sulfolobus* species (32, 49), but chromatin immunoprecipitation experiments argue convincingly that Alba is bound to genomic DNA and functions as a chromatin protein in *S. solfataricus* (49).

As is often the case with chromatin proteins, some archaea have more than one Alba (Alba1 = Sso10b1; Alba2 = Sso10b2 [17, 35]). In *A. fulgidus* and *M. kandleri*, the two Alba proteins have very similar sequences, consistent with recent gene duplication events, but in *S. solfataricus*, *S. tokodaii*, *S. acidocaldarius*, and *Aeropyrum pernix*, the Alba1 and Alba2 sequences are only 30 to 40% identical. In *S. solfataricus*, in stationary phase, Alba1 is 20-fold more abundant than Alba2, suggesting that Alba2 may have a regulatory rather than structural function. In vitro, (Alba1+Alba2) heterodimers bind simultaneously to two DNA duplexes at a protein to DNA ratio of 1 dimer/12 bp, which results in DNA compaction (35). DNA is similarly compacted by binding of Alba1 homodimers to two DNA duplexes at this ratio, but when increased to 1 dimer/6 bp, Alba1 homodimers coat the DNA to form nucleoprotein filaments that result in little or no DNA compaction. To form such a filament, Alba1 homodimers assemble end to end (84, 89). The residues at this dimer-dimer interface are not conserved in Alba2, and Alba2 homodimers and (Alba1+Alba2) heterodimers do not form nucleoprotein filaments (17). The biological function of Alba2 may therefore be to form (Alba1+Alba2) heterodimers that limit nucleoprotein filament formation by Alba1 homodimers. Alba2 limiting Alba1 oligomerization could be another mechanism of regulation, based on chromatin structure, added to the regulation based on K16 acetylation-deacetylation of Alba1 in *S. solfataricus* (85).

Sul10a

Sul10a is the generic name of an abundant ~11 kDa DNA-binding protein investigated from *S. acidocaldarius* (Sac10a) and *S. solfataricus* (Sso10a) (16, 26, 36). X-ray crystal and NMR solution structures have established that Sso10a dimers are highly elongated, with a two-stranded antiparallel coiled-coil rod central region separating two globular winged-helix DNA-binding domains. This structure is similar to structures established for MotA, CueR, and ModE bacterial transcription regulators, and for a DNA replication terminator protein, RTP. Consistent with Sac10a being a sequence-specific DNA-binding protein, it exhibits a pronounced binding preference for poly(dA/dT) over poly(dG/dC) or *E. coli* genomic DNA. However, winged-helix DNA-binding motifs are not only present in sequence-specific transcription factors (see Chapter 6). This structure is also present in the eucaryal H1 linker histone (38). Electron microscopy has revealed that Sac10a binds and coats double-stranded DNA at high protein to DNA ratios, and under such conditions Sac10a binding does introduce supercoiling into closed circular DNA molecules.

MC1

Methanogen chromosomal protein 1 (MC1) is an abundant ~93-amino-acid residue chromosomal protein characterized from *Methanosarcina* that, based on genome sequencing, is also present in *Halobacterium* species (Fig. 1). An NMR solution structure has been established for MC1 from *Methanosarcina* sp. CHTI55 (56). This has a barrel-shaped protein fold similar to that of Sul7d, with a double-stranded β-sheet linked by an α-helix to a triple-stranded β-sheet (Fig. 2D). Photochemical cross-linking of MC1-DNA complexes has established that the DNA-binding region is located between K69 and K87. This region contains part of a conserved loop located between the second and third β-strands of the triple β-sheet and three conserved lysine residues and one tryptophan residue that likely interact directly with DNA (56). MC1 monomers bind noncooperatively to double-stranded DNA kinking the DNA to an angle of 116° (14, 23, 42). Although MC1 binding to DNA appears to be sequence independent, preferential binding has been observed with some DNAs that most likely have readily deformable sequences (57). MC1 has strong affinity for bent, negatively supercoiled and four-way junction DNAs (78, 79). DNase I protection results suggest that the DNA wraps around MC1, and, although no structure has yet been reported for an MC1-DNA complex, models have been proposed in which MC1 binds to both the inside and outside of a curved DNA molecule (56).

7kMk

In addition to the MkaH histone, *M. kandleri* also contains three other chromatin proteins with molecular masses of ~7, 10, and 30 kDa (59). The 7-kDa protein, designated 7kMk, has been characterized and shown to be a member of the ribbon-helix-helix family. It is homodimer in solution and likely folds into a structure with an N-terminal β-strand followed by four α-helices (59). 7kMk binds to double-stranded DNA without any apparent sequence specificity, forming looplike structures and introducing negative supercoils. Two binding models have been proposed in which the DNA is either wrapped around a preformed 7kMk protein core, or 7kMk binds to the DNA cooperatively, gradually bending the DNA into a left-handed superhelical loop (59).

HU-Related Proteins

Members of the *Thermoplasma* lineage are euryarchaeota that do not have histones, but rather have HU-related chromatin proteins (65). HTa from *Thermoplasma acidophilum* was, in fact, the first archaeal chromatin protein to be studied in detail (22). The sequence of this 89-amino-acid residue protein shows a common ancestry with bacterial HU proteins. HTa binding to DNA reduces the contour length of DNA and stabilizes the DNA against heat denaturation. Nucleoprotein particles were calculated to have an HTa tetramer core that bound and circularized 40 bp of DNA (21), but this structure now seems very unlikely, given the extreme DNA distortion that circularizing 40 bp would require. It seems most likely that HTa binds to DNA and forms complexes similar to those formed by bacterial HU proteins (25, 77), but this remains to be determined.

Participation of Chromatin Proteins in Transcription and DNA Metabolism

It is apparent that many different chromatin proteins have evolved, all of which must bind and compact DNA into complexes that are readily disassembled, or that are inherently compatible with DNA replication and transcription machineries. It is now understood that chromatin proteins that were originally considered to have only architectural functions, such as HU in bacteria and the eucaryal nucleosome core histones, actually participate in regulating gene expression (3, 18, 25, 55, 74), and it has been proposed that differences in bacterial versus eucaryal chromatin structure result in fundamentally different mechanisms of regulating gene expression (75). In eucarya, histone-compacted chromatin is considered to be generically repressive. Gene expression therefore requires transcription activators, for example, histone acetylases that help disassemble chromatin and so allow transcription factor access to the DNA (18, 52,

55, 74, 82). In contrast, the nucleoid structure in bacteria is argued not to prevent transcription factor access, and therefore promoter-specific repressors are needed to prevent inappropriate gene expression (75). The validity and extensions of these arguments to archaea remain to be determined. The archaeal transcription initiation machinery does most closely resemble the eucaryal RNA polymerase (RNAP) II system (see Chapter 6), but most of the transcription regulation documented to date involves repressors rather than activators (7, 9, 64). Archaeal transcription initiation in vitro is inhibited by archaeal histone assembly of the promoter into an archaeal nucleosome, but downstream archaeal nucleosomes do not block archaeal RNAP elongation after initiation (87).

As most archaeal histones lack N- and C-terminal tails, and given the absence of genomic evidence for archaeal homologs of eucaryal chromatin-remodeling complexes, and the direct evidence that archaeal histones isolated from *M. jannaschii* do not have post-translation modifications (28), it seems unlikely that gene expression in archaea is regulated by histone modification. Archaeal histones could, however, still participate in regulating gene expression as different archaeal histones in the same species have different affinities for different DNA sequences (6, 47). So by controlling their relative abundances, and possibly also the extent of homodimer versus heterodimer formation, archaeal nucleosome assembly could be specifically positioned in vivo to regulate gene expression (68). By using chromatin immunoprecipitation (62), after in vivo histone-to-DNA crosslinking, it should be possible to establish where nucleosomes are located throughout an archaeal genome under different growth conditions, and therefore, how such positioning relates to gene expression.

As described, Alba (Sso10b1) binding to DNA is modulated by posttranslation acetylation in *S. solfataricus* and, in this archaeon, gene expression may well be regulated by modifying chromatin structure (10, 35, 884). It is clearly important to determine whether such regulation extends beyond *Sulfolobus* species, in particular, to species where the residue in Alba at the position homologous to K16 in Alba1/Sso10b1 is not a lysine (87). Some archaeal genomes encode homologs of the Sir2 deacetylase that deacetylates Alba1/Sso10b1 in *S. solfataricus* that do not encode Alba proteins.

Few archaeal DNA replication/repair investigations to date have studied the participation of archaeal chromatin proteins (39, 41, 63), although Sul7d has been shown to modulate the primer extension and excision activities of the *S. solfataricus* proofreading DNA polymerase B1 (PolB1). Sul7d apparently binds to and stabilizes double-stranded

DNA, inhibiting primer cleavage by PolB1 and encouraging primer extension (45). The presence of Sul7d did not affect cleavage of single-stranded DNA or the proofreading ability of the polymerase, suggesting that Sul7d may play a role in balancing the polymerization versus exonuclease activities of this DNA polymerase in favor of polymerization. Binding Sul7d to the DNA did not inhibit DNA unwinding by the *S. solfataricus* replicative MCM helicase. However, DNA unwinding by this helicase was inhibited by Alba1, and this inhibition was reduced by acetylation of Sso10b/Alba, consistent with acetylation reducing the affinity of Alba1 for DNA (48). Studies with the *Methanothermobacter thermautotrophicus* MCM helicase have established that this enzyme can unwind DNA bound by an archaeal histone into an archaeal nucleosome, or bound into a transcription preinitiation complex, but this enzyme could not unwind DNA held in a transcription elongation complex (73).

PERSPECTIVE: THE NEXT FIVE YEARS

With the accumulation of genome sequences, it is now apparent that most archaea have the capacity to synthesize several different chromatin proteins. Some of these proteins (e.g., histones, Alba) have wide distribution, whereas others (e.g., Sul7d, 7kMk, MC1) have restricted phylogenetic distributions (Fig. 1). Based on bacterial precedents (25), it seems likely that many of these proteins will have compensating and overlapping functions in chromatin organization, but probably very specific and noncompensating functions in regulating gene expression. Also based on bacterial precedents, archaeal chromatin proteins are probably not individually essential for viability, but by mutational inactivation, their roles in specific DNA metabolic events may be identified. Currently, the consequences of adding purified archaeal chromatin proteins to in vitro DNA replication, repair, and transcription systems are being determined, and these experiments will identify specific reactions that are inhibited or stimulated by chromatin proteins binding to the template DNA. Crosslinking chromatin proteins to genomic DNA in vivo, followed by the isolation and identification of the DNA bound by microarray hybridizations, will identify where chromatin proteins are located in vivo, and this will (or will not) be correlated with specific gene expression. But to prove the role of a chromatin protein in vivo in any complex process will require substantially more sophisticated genetics than is currently available for archaea. However, the development of such genetic technology is on the horizon (see Chapter 21), and

with such developments there will be rapid progress in understanding and dissecting the detailed structural and regulatory roles of archaeal chromatin proteins.

Acknowledgments. Our research on chromatin proteins and gene expression in archaea at Ohio State University is supported by grants from the National Institutes of Health and the U.S. Department of Energy.

REFERENCES

1. **Agback, P., H. Baumann, S. Knapp, R. Ladenstein, and T. Hard.** 1998. Architecture of nonspecific protein-DNA interactions in the Sso7d-DNA complex. *Nat. Struct. Biol.* **5:** 579–584.
2. **Aravind, L., L. M. Iyer, and V. Anantharaman.** 2003. The two faces of Alba: the evolutionary connection between proteins participating in chromatin structure and RNA metabolism. *Genome Biol.* **4:**R64.
3. **Azam, T. A., A. Iwata, A. Nishimura, S. Ueda, and A. Ishihama.** 1999. Growth phase-dependent variation in protein composition of the *Escherichia coli* nucleoid. *J. Bacteriol.* **181:** 6361–6370.
4. **Azam, T. A., S. Hiraga, and A. Ishihama.** 2000. Two types of localization of the DNA-binding proteins within the *Escherichia coli* nucleoid. *Genes Cells* **5:**613–626.
5. **Bailey, K. A., C. S. Chow, and J. N. Reeve.** 1999. Histone stoichiometry and DNA circularization in archaeal nucleosomes. *Nucleic Acids Res.* **27:**532–536.
6. **Bailey, K. A., S. L. Pereira, J. Widom, and J. N. Reeve.** 2000. Archaeal histone selection of nucleosome positioning sequences and the procaryotic origin of histone-dependent genome evolution. *J. Mol. Biol.* **303:**25–34.
7. **Bartlett, M. S.** 2005. Determinants of transcription initiation by archaeal RNA polymerase. *Curr. Opin. Microbiol.* **8:** 677–684.
8. **Bedell, J. L., S. P. Edmondson, and J. W. Shriver.** 2005 Role of a surface tryptophan in defining the structure, stability, and DNA binding of the hyperthermophile protein Sac7d. *Biochemistry* **44:**915–925.
9. **Bell, S. D.** 2005. Archaeal transcriptional regulation—variation on a bacterial theme? *Trends Microbiol.* **13:**262–265.
10. **Bell, S. D., C. H. Botting, B. N. Wardleworth, S. P. Jackson, and M. F. White.** 2002. The interaction of Alba, a conserved archaeal chromatin protein, with Sir2 and its regulation by acetylation. *Science* **296:**148–151.
11. **Berman, H. M., J. Westbrook, Z. Feng, G. Gilliland, T. N. Bhat, H. Weissig, I. N. Shindyalov, and P. E. Bourne.** 2000. The Protein Data Bank. *Nucleic Acids Res.* **28:**235–242.
12. **Briggs, S. D., T. Xiao, Z.-W. Sun, J. A. Caldwell, J. Shabanowitz, D. F. Hunt, C. D. Allis, and B. D. Strahl.** 2002. Trans-histone regulatory pathway in chromatin. *Nature* **418:**498.
13. **Brochier, C., S. Gribaldo, Y. Zivanovic, F. Confalonieri, and P. Forterre.** 2005. Nanoarchaea: representatives of a novel archaeal phylum or a fast-evolving euryarchaeal lineage related to *Thermococcales*? *Genome Biol.* **6:**R42.
14. **Cam, E. L., F. Culard, E. Larquet, E. Delain, and J. A. Cognet.** 1999. DNA bending induced by the archaebacterial histone-like protein MC1. *J Mol Biol* **285:**1011–21.
15. **Chen, C. Y., T. P. Ko, T. W.Lin, C. C. Chou, C. J. Chen, and A. H. Wang.** 2005 Probing the DNA kink structure induced by the hyperthermophilic chromosomal protein Sac7d. *Nucleic Acids Res.* **33:**430–438.

16. Chen, L., L. R. Chen, X. E. Zhou, Y. Wang, M. A. Kahsai, A. T. Clark, S. P. Edmondson, Z. J. Liu, J. P. Rose, B. C. Wang, E. J. Meehan, and J. W. Shriver. 2004. The hyperthermophile protein Sso10a is a dimer of winged helix DNA-binding domains linked by an antiparallel coiled coil rod. *J. Mol. Biol.* **341**:73–91.

17. Chou, C-C., T-W. Lin, C-Y. Chen, and A.H-J. Wang. 2003. Crystal structure of the hyperthermophilic archaeal DNA-binding protein Sso10b2 at a resolution of 1.85 angstroms. *J. Bacteriol.* **185**:4066–4073.

18. Cosgrove, M. S., J. D. Boeke, and C. Wolberger. 2004. Regulated nucleosome mobility and the histone code. *Nat. Struct. Mol. Biol.* **11**:1037–1043.

19. Cubonova, L., K. Sandman, S. J. Hallam, E. F. Delong, and J. N. Reeve. 2005. Histones in *Crenarchaea*. *J. Bacteriol.* **187**:5482–5485.

20. Cui, Q., Y. Tong, H. Xue, L.Huang, Y. Feng, Y., and J. Wang. 2003. Two conformations of archaeal Ssh10b. *J. Biol. Chem.* **278**:51015–51022.

21. Decanniere, K., A. M. Babu, K. Sandman, J. N. Reeve, and U. Heinemann. 2000. Crystal structures of recombinant histones HMfA and HMfB from the hyperthermophilic archaeon *Methanothermus fervidus*. *J. Mol. Biol.* **303**:35–47.

22. DeLange, R. J., G. R. Green, and D. G. Searcy. 1981. A histone-like protein (HTa) from *Thermoplasma acidophilum*. I. Purification and properties. *J. Biol. Chem.* **256**:900–904.

23. De Vuyst, G., S. Aci, D. Genest, and F. Culard. 2005. Atypical recognition of particular DNA sequences by the archaeal chromosomal MC1 protein. *Biochemistry* **44**:10369–10377.

24. Dinger, M. E., G. J. Baillie, and D. R. Musgrave. 2000. Growth phase-dependent expression and degradation of histones in the thermophilic archaeon *Thermococcus zilligii*. *Mol. Microbiol.* **36**:876–885.

25. Drlica, K., and J. Rouviere-Yaniv. 1987. Histone-like proteins of bacteria. *Microbiol. Rev.* **51**:301–319.

26. Edmondson, S. P., M. A. Kahsai, R. Gupta, and J. W. Shriver. 2004. Characterization of Sac10a, a hyperthermophile DNA-binding protein from *Sulfolobus acidocaldarius*. *Biochemistry* **43**:13026–13036.

27. Fahrner, R. L., D. Cascio, J. A. Lake, and A. Slesarev. 2001. An ancestral nuclear protein assembly: crystal structure of the *Methanopyrus kandleri* histone. *Protein Sci.* **10**:2002–2007.

28. Forbes, A. J., S. M. Patrie, G. K. Taylor, Y. B. Kim, L. Jiang, and N. L. Kelleher. 2004. Targeted analysis and discovery of posttranslational modifications in proteins from methanogenic archaea by top-down MS. *Proc. Natl. Acad. Sci. USA* **101**:2678–2683.

29. Gao, Y. G., S. Y. Su, H. Robinson, S. Padmanabhan, L. Lim, B. S. McCrary, S. P. Edmondson, J. W. Shriver, and A. H. Wang. 1998. The crystal structure of the hyperthermophile chromosomal protein Sso7d bound to DNA. *Nat. Struct. Biol.* **5**:782–786.

30. Guagliardi, A., L. Cerchia, and M. Rossi. 2002. The Sso7d protein of *Sulfolobus solfataricus*: in vitro relationship among different activities. *Archaea* **1**:87–93.

31. Guagliardi, A., L. Mancusi, and M. Rossi. 2004. Reversion of protein aggregation mediated by Sso7d in cell extracts of *Sulfolobus solfataricus*. *Biochem. J.* **381**:249–255.

32. Guo, R., H. Xue, and L. Huang. 2003. Ssh10b, a conserved thermophilic archaeal protein, binds RNA *in vivo*. *Mol. Microbiol.* **50**:1605–1615.

33. Hayat, M. A., and D. A. Mancarella. 1995. Nucleoid proteins. *Micron* **26**:461–480.

34. Heinicke, I., J. Muller, M. Pittelkow, and A. Klein. 2004. Mutational analysis of genes encoding chromatin proteins in the archaeon *Methanococcus voltae* indicates their involvement in the regulation of gene expression. *Mol. Genet. Genomics* **272**:76–87.

35. Jelinska, C., M. J. Conroy, C. J. Craven, A. M. Hounslow, P. A. Bullough, J. P. Waltho, G. L. Taylor, and M. F. White. 2005. Obligate heterodimerization of the archaeal Alba2 protein with Alba1 provides a mechanism for control of DNA packaging. *Structure* **13**:963–971.

36. Kahsai, M. A., B. Vogler, A. T. Clark, S. P. Edmondson, and J. W. Shriver. 2005. Solution structure, stability, and flexibility of Sso10a: a hyperthermophile coiled-coil DNA-binding protein. *Biochemistry* **44**:2822–2832.

37. Kamau, E., N. D. Tsihlis, L. A. Simmons, and A. Grove. 2005. Surface salt bridges modulate the DNA site size of bacterial histone-like HU proteins. *Biochem. J.* **390**:49–55.

38. Kasinsky, H. E., J. D. Lewis, J. B. Dacks and J. Ausió. 2001. Origin of H1 linker histones. *FASEB J.* **15**:34–42

39. Kelman, Z., and M. F. White. 2005. Archaeal DNA replication and repair. *Curr. Opin. Microbiol.* **8**:669–676.

40. Ko, T. P., H. M. Chu, C. Y Chen, C. C. Chou, and A. H. Wang. 2004. Structures of the hyperthermophilic chromosomal protein Sac7d in complex with DNA decamers. *Acta Crystallogr. D Biol. Crystallogr.* **60**:1381–1387.

41. Kvaratskhelia, M., B. N. Wardleworth, C. S. Bond, J. M. Fogg, D. M. Lilley, and M. F. White. 2002. Holliday junction resolution is modulated by archaeal chromatin components *in vitro*. *J. Biol. Chem.* **277**:2992–2996.

42. Le Cam, E., F. Culard, E. Larquet, E. Delain, and J. A. H. Cognet. 1999. DNA bending induced by the archaebacterial histone-like protein MC1. *J. Mol. Biol.* **285**:1011–1021.

43. Li, W. T., K. Sandman, S. L. Pereira, and J. N. Reeve. 2000. MJ1647, an open reading frame in the genome of the hyperthermophile *Methanococcus jannaschii*, encodes a very thermostable archaeal histone with a C-terminal extension. *Extremophiles* **4**:43–51.

44. Lopez-Garcia, P., S. Knapp, R. Ladenstein, and P. Forterre. 1998. In vitro DNA binding of the archaeal protein Sso7d induces negative supercoiling at temperatures typical for thermophilic growth. *Nucleic Acids Res.* **26**:2322–2328.

45. Lou, H., Z. Duan, X. Huo, and L. Huang. 2004. Modulation of hyperthermophilic DNA polymerase activity by archaeal chromatin proteins. *J Biol Chem* **279**:127-32.

46. Luger, K., A. W. Mader, R. K. Richmond, D. F. Sargent, and T. J. Richmond. 1997. Crystal structure of the nucleosome core particle at 2.8 Å resolution. *Nature* **389**:251–260.

47. Marc, F., K. Sandman, R. Lurz, and J. N. Reeve. 2002. Archaeal histone tetramerization determines DNA affinity and the direction of DNA supercoiling. *J. Biol. Chem.* **277**:30879–30886.

48. Marsh, V. L., A. T. McGeoch, and S. D. Bell. 2006. Influence of chromatin and single strand binding proteins of the activity of an archaeal MCM. *J. Mol. Biol.* **357**:1345–1350.

49. Marsh, V. L., S. Y. Peak-Chew, and S. D. Bell. 2005. Sir2 and the acetyltransferase, Pat, regulate the archaeal chromatin protein, Alba. *J. Biol. Chem.* **280**:21122–21128.

50. Martin, W. 2005. Archaebacteria (Archaea) and the origin of the eukaryotic nucleus. *Curr. Opin. Microbiol.* **8**:630–637.

51. McAfee, J. G., S. P. Edmondson, I. Zegar, and J. W. Shriver. 1996. Equilibrium DNA binding of Sac7d protein from the hyperthermophile *Sulfolobus acidocaldarius*: fluorescence and circular dichroism studies. *Biochemistry* **35**:4034–4045.

52. Mellor, J. 2005. The dynamics of chromatin remodeling at promoters. *Mol. Cell* **19**:147–157.

53. Minsky, A. 2004. Information content and complexity in the high-order organization of DNA. *Annu. Rev. Biophys. Biomol. Struct.* **33**:317–342.

54. Napoli, A., Y. Zivanovic, C. Bocs, C. Buhler, M. Rossi, P. Forterre, and M. Ciaramella. 2002. DNA bending, compaction and negative supercoiling by the architectural protein Sso7d of *Sulfolobus solfataricus*. *Nucleic Acids Res.* **30**:2656–2662.

55. Narliker, G. J., H.-Y. Fan, and R. E. Kingston. 2002. Cooperation between complexes that regulate chromatin structure and transcription. *Cell* **108**:475–487.

56. Paquet, F., F. Culard, F. Barbault, J. C. Maurizot, and G. Lancelot. 2004. NMR solution structure of the archaebacterial chromosomal protein MC1 reveals a new protein fold. *Biochemistry* **43**:14971–14978.

57. Paradinas, C., A. Gervais, J. C. Maurizot, and F. Culard. 1998. Structure-specific binding recognition of a methanogen chromosomal protein. *Eur. J. Biochem.* **257**:372–379.

58. Pavlov, N. A., D. I. Cherny, T. M. Jovin, and A. I. Slesarev. 2002. Nucleosome-like complex of the histone from the hyperthermophile *Methanopyrus kandleri* (MkaH) with linear DNA. *J. Biomol. Struct. Dyn.* **20**:207–214.

59. Pavlov, N. A., D. I. Cherny, I. V. Nazimov, A. I. Slesarev, and V. Subramaniam. 2002. Identification, cloning and characterization of a new DNA-binding protein from the hyperthermophilic methanogen *Methanopyrus kandleri*. *Nucleic Acids Res.* **30**:685–694.

60. Pedersen, L. B., S. Birkelund, and G. Christiansen.1996. Purification of recombinant *Chlamydia trachomatis* histone H1-like protein Hc2, and comparative functional analysis of Hc2 and Hc1. *Mol. Microbiol.* **20**:295–311.

61. Pereira, S. L., R. A. Grayling, R. Lurz, and J. N. Reeve. 1997. Archaeal nucleosomes. *Proc. Natl. Acad. Sci. USA* **94**:12633–12637.

62. Pereira, S. L., and J. N. Reeve. 1999 Archaeal nucleosome positioning sequence from *Methanothermus fervidus*. *J. Mol. Biol.* **289**:675–681.

63. Peters, W. B., S. P. Edmondson, and J. W. Shriver. 2005. Effect of mutation of the Sac7d intercalating residues on the temperature dependence of DNA distortion and binding thermodynamics. *Biochemistry* **44**:4794–4804.

64. Reeve, J. N. 2003. Archaeal chromatin and transcription. *Mol. Microbiol.* **48**:587–598.

65. Ruepp, A., W. Graml, M. L. Santos-Martinez, K. K. Koretke, C. Volker, H. W. Mewes, D. Frishman, S. Stocker, A. N. Lupas, and W. Baumeister. 2000. The genome sequence of the thermoacidophilic scavenger *Thermoplasma acidophilum*. *Nature* **407**:508–513.

66. Sandman, K., R. A. Grayling, B. Dobrinski, R. Lurz, and J. N. Reeve. 1994. Growth-phase-dependent synthesis of histones in the archaeon *Methanothermus fervidus*. *Proc. Natl. Acad. Sci. USA* **91**:12624–12628.

67. Sandman, K., J. A. Krzycki, B. Dobrinski, R. Lurz, and J. N. Reeve. 1990. DNA binding protein HMf, from the hyperthermophilic archaebacterium *Methanothermus fervidus*, is most closely related to histones. *Proc. Natl. Acad. Sci. USA* **87**:5788–5791.

68. Sandman, K., and J. N. Reeve. 2000. Structure and functional relationships of archaeal and eucaryal histones and nucleosomes. *Arch. Microbiol.* **173**:165–169.

69. Sato, N., K. Terasawa, K. Miyajima, and Y. Kabeya. 2003. Organization, developmental dynamics, and evolution of plastid nucleoids. *Int. Rev. Cytol.* **232**:217–262.

70. Schieg, P., and H. Herzel, H. 2004. Periodicities of 10-11 bp as indicators of the supercoiled state of genomic DNA. *J. Mol. Biol.* **343**:891–901.

71. Setlow, P. 1995. Mechanisms for the prevention of damage to DNA in spores of *Bacillus* species. *Annu. Rev. Microbiol.* **49**:29–54.

72. Shehi, E., S. Serina, G. Fumagalli, M. Vanoni, R. Consonni, and L. Zetta. 2001. The Sso7d DNA-binding protein from *Sulfolobus solfataricus* has ribonuclease activity. *FEBS Lett.* **497**:131–136.

73. Shin, J. H., T. J. Santangelo, Y. Xie, J. N. Reeve, and Z. Kelman. Archaeal MCM helicase can unwind DNA bound by archaeal histones and transcription factors. *J. Biol. Chem.*, in press.

74. Strahl, B. D., and C. D. Allis. 2000. The language of covalent histone modifications. *Nature* **403**:41–45.

75. Struhl, K. 1999. Fundamentally different logic of gene regulation in eukaryotes and prokaryotes. *Cell* **98**:1–4.

76. Su, S., Y.-G. Gao, H. Robinson, Y-C. Liaw, S. P. Edmondson, J. W. Shriver, and A. H.-J. Wang. 2000. Crystal structures of the chromosomal proteins Sso7d/Sac7d bound to DNA containing T-G mismatched base pairs. *J. Mol. Biol.* **303**:395–403.

77. Swinger, K. K., K. M. Lemberg, Y. Zhang, and P. A. Rice. 2003. Flexible DNA bending in HU-DNA cocrystal structures. *EMBO J.* **22**:3749–3760.

78. Teyssier, C., B. Laine, A. Gervais, J. C. Maurizot, and F. Culard. 1994. Archaebacterial histone-like protein MC1 can exhibit a sequence-specific binding to DNA. *Biochem. J.* **303**:567–573.

79. Teyssier, C., F. Toulme, J. P. Touzel, A. Gervais, J. C. Maurizot, and F. Culard. 1996. Preferential binding of the archaebacterial histone-like MC1 protein to negatively supercoiled DNA minicircles. *Biochemistry* **35**:7954–7958.

80. Tomschik, M., M. A. Karymov, J. Zlatanova, and S. H. Leuba, S.H. 2001. The archaeal histone-fold protein HMf organizes DNA in *bona fide* chromatin fibres. *Structure* **9**:1201–1211.

81. Triana, O., N. Galanti, N. Olea, U. Hellman, C. Wernstedt, H. Lujan, C. Medina, and G. C. Toro. 2001. Chromatin and histones from *Giardia lamblia*: a new puzzle in primitive eukaryotes. *J. Cell. Biochem.* **82**:573–582.

82. Tsukiyama, T. 2002. The in vivo functions of ATP-dependent chromatin-remodelling factors. *Nat. Rev. Mol. Cell. Biol.* **3**:422–429.

83. Wang, G., R. Guo, M. Bartlam, H. Yang, H. Xue, Y. Liu, L. Huang, and Z. Rao. 2003. Crystal structure of a DNA binding protein from the hyperthermophilic euryarchaeon *Methanococcus jannaschii*. *Protein Sci.* **12**:2815–2822.

84. Wardleworth, B. N., R. J. Russell, S. D. Bell, G. L. Taylor, and M. F. White. 2002. Structure of Alba: an archaeal chromatin protein modulated by acetylation. *EMBO J.* **21**:4654–4662.

85. White, M. F., and S. D. Bell. 2002. Holding it together: chromatin in the Archaea. *Trends Genet.* **18**:621–626.

86. Wong, J. T., D. C. New, J. C. Wong, and V. K. Hung. 2003. Histone-like proteins of the dinoflagellate *Crypthecodinium cohnii* have homologies to bacterial DNA-binding proteins. *Eukaryot. Cell* **2**:646–650.

87. Xie, Y., and J. N. Reeve. 2004. Transcription by an archaeal RNA polymerase is slowed but not blocked by an archaeal nucleosome. *J. Bacteriol.* **186**:3492–3498.

88. Xue, H., R. Guo, Y. Wen, D. Liu, and L. Huang. 2000. An abundant DNA binding protein from the hyperthermophilic archaeon *Sulfolobus shibatae* affects DNA supercoiling in a temperature-dependent fashion. *J. Bacteriol.* **182**:3929–3933.

89. Zhao, K., X. Chai, and R. Marmorstein. 2003. Structure of a Sir2 substrate, Alba, reveals a mechanism for deacetylation-induced enhancement of DNA binding. *J. Biol. Chem.* **278**:26071–26077.

Archaea: Molecular and Cellular Biology
Edited by Ricardo Cavicchioli
© 2007 ASM Press, Washington, D.C.

Chapter 5

Mechanisms of Genome Stability and Evolution†

DENNIS W. GROGAN

INTRODUCTION: THE SIGNIFICANCE OF GENETIC PROCESSES IN ARCHAEA

When researchers describe the scientific motivations to study archaea, certain themes tend to recur. One is the evolutionary centrality of the *Archaea*, reflecting the very early origin of this domain (122). This centrality implies that reconstructions of early cellular evolution must account for the observed molecular properties of archaea, and, conversely, that individual molecular properties of archaea should be interpreted with respect to the wider evolutionary framework. Another theme emphasizes the uniqueness of archaea, in particular, those from "extreme" environments. Many of these archaea have proven to be exceptional organisms, and a number of them define physiological limits of life on earth. Archaea have been a rich source of phenomena that have revised certain rules of biology and could not be discovered in the well-established bacterial and eucaryal model species.

Both the evolutionary centrality and the uniqueness of archaea have proven their validity and heuristic value over nearly three decades of research, but in juxtaposition these two ideas also create certain interesting tensions. For example, the uniqueness of archaea would seem, logically, to extend to their evolution. Unless evolutionary forces and mechanisms are intrinsically immutable across biology, archaeal evolution should exhibit certain features not evident in bacteria or eucarya, reflecting the extreme molecular and cellular divergence among the three domains. Does archaeal evolution in fact deviate from the picture(s) evident from research on bacteria and unicellular eucarya?

This broad, almost philosophical, question can be reduced to more concrete terms if one considers the roles of genetic processes in the molecular evolution of microorganisms. According to modern Darwinian theory, a fundamental "separation of powers" distinguishes processes that produce genetic variation in natural populations (mutation, migration, and recombination) from processes that filter it out (drift and selection); the resulting interplay determines the rate and course of change (Fig. 1). Properties of mutation and recombination can be measured under laboratory conditions and are expected to contribute more or less directly to the "nearly neutral" variation that dominates evolution at the molecular level (88). Even in cases where selection dictates the successful phenotype, properties of the genetic processes may still determine the route taken to reach it. In addition, the five primary processes depicted in Fig. 1 are themselves affected by aspects of the organism's biochemistry, physiology, and ecology and can be expected, conversely, to shape the genetic properties of a microbial lineage over evolutionary time.

As detailed throughout this volume, archaea have many molecular and cellular features not seen in bacteria or eucarya, and many archaeal lineages have apparently been living for a very long time in environments considered (even by microbiological standards) to be "extreme," "harsh," or "unusual." It thus seems legitimate and significant to ask whether the third form of life employs a third form of genetics in its daily survival and evolutionary diversification. This question also serves to unify and organize the admittedly limited data on basic molecular-genetic processes in archaea, and to identify for future attention topics of particular significance. Accordingly, this chapter examines the "natural genetics" of methanogenic, halophilic, and thermophilic archaea, progressing from the molecular scale to cells and finally to populations. Related topics addressed in other chapters include DNA replication (see Chapter 3), the evolutionary aspects of archaeal genomes revealed by

---

†In memory of Wolfram Zillig.

Dennis W. Grogan • Department of Biological Sciences, University of Cincinnati, Cincinnati, OH 45221-0006.

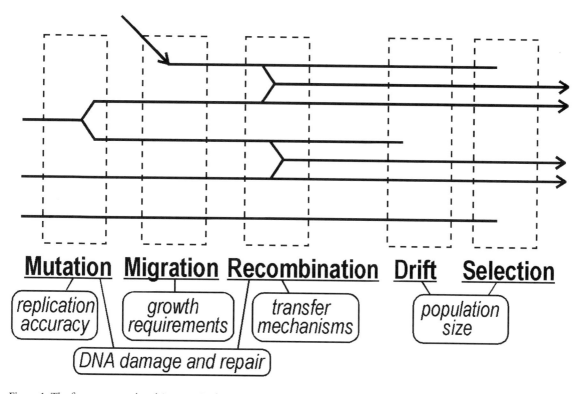

Figure 1. The five processes that drive genetic change in natural populations. Each horizontal line represents a distinct genotype (i.e., genome sequence) present in a hypothetical population of a microbial species as a function of time. New genotypes arise in this population by mutation, immigration from other populations, and recombination. Conversely, genotypes are eliminated randomly by drift, or according to functional properties determined by particular alleles, by selection. As a result of these natural processes, the genetic composition of the population changes irreversibly over time. For a comprehensive discussion, see Ridley (88).

sequencing (see Chapter 19), and experimental techniques used to manipulate archaea genetically (see Chapter 21).

SPONTANEOUS MUTATION

All DNA replication systems make mistakes, and the possible mistakes outnumber the correct base at each position in the nucleotide sequence. Thus, the error rate and the molecular nature of the errors made represent two fundamental genetic properties that can vary among microorganisms and affect genome stability and evolution. The isolation of "mutator" strains provides a practical example of this. In haploid microorganisms, a single mutation can inactivate an important accuracy-enforcing mechanism (such as DNA mismatch repair), yielding a strain with an elevated rate of spontaneous mutation and an altered mutational spectrum (13, 52). These mutator strains seldom exhibit serious growth defects; in fact, certain selections and growth regimens can actually give them a selective advantage, which facilitates their iso-

lation (64). However, the fact that nearly all microorganisms isolated from nature have multiple accuracy-enforcing mechanisms that are not essential for basic cell viability implies that success in nature requires a higher level of genetic fidelity than does growth of pure cultures under laboratory conditions.

The importance of genetic accuracy for evolutionary fitness becomes more obvious when mutation rates are compared for microorganisms ranging from bacteriophage to bacteria to fungi. Replication of these genomes involves widely varying error rates per base pair (10^{-7} to 10^{-11} events per generation), but remarkably constant rates per genome (about 0.003 events per replication) (15). This conservation of a rate per genome at the expense of the rate per base pair implies a strong selective force in nature that rewards this genomic rate as generally optimal. Accordingly, this genomic rate provides a biologically meaningful reference for evaluating the replication fidelity of a microorganism. Does this selective force extend to archaea, and, if so, can archaea growing in harsh, potentially mutagenic conditions attain the prescribed level of fidelity?

The technical demands for accurate mutation-rate measurement have so far been satisfied only in the thermoacidophilic crenarchaeon *Sulfolobus acidocaldarius*. Essentially all mutations conferring resistance to 5-fluoro-orotic acid (FOA) inactivate the *pyrE* or *pyrF* gene of this species and can be detected with high efficiency by plating. For the greatest accuracy, genic rates calculated from fluctuation tests using maximum-likelihood methods (47) are corrected for the effects, if any, of phenotypic lag and relative fitness of mutant versus parent. In *S. acidocaldarius*, this approach showed that *pyrE* and *pyrF* mutations occurred at a combined rate of 3.3×10^{-7} events per cell division. This corresponds to about 2.6×10^{-7} phenotypically detected mutational events per kilobase pair (kbp), which matches the average value calculated from several metabolic genes of *Escherichia coli* (42). Thus, replication of these two target genes in *S. acidocaldarius* at 75 to 80°C exhibits fidelity comparable to that of similar genes in *E. coli* growing at 37°C.

Calculating the mutation rate of the entire genome requires two additional parameters: (i) the fraction of mutations that go undetected, and (ii) the genome size. Ideally, parameter (i) is determined empirically from the spectrum of spontaneous mutation. To ensure the *S. acidocaldarius* genomic rate was not underestimated, the *pyrF* gene (site of relatively few mutations) was excluded as a target, and parameter (i) was estimated more generously for the *pyrE* gene than indicated from the actual spectrum (36). Parameter (ii) was estimated from published flow cytometry data to within one percent of the value later determined by sequencing (9, 36). The resulting genomic rate, 0.0018 mutational events per replication, falls slightly below the highly conserved value of 0.003 obtained for diverse mesophiles (15). This result confirms that a strong, but as yet ill-defined, selection for a common error rate per genome applies to unicellular life across three domains and contrasting ecological niches. It also demonstrates that hyperthermophilic archaea can meet this optimal level of accuracy despite the genotoxicity of their growth conditions and their peculiar situation with respect to DNA repair genes (see "DNA repair," below). In fact, despite incorporating assumptions that would tend to overestimate the genomic rate, the calculations depict *S. acidocaldarius* as having the most stable genome of any organism known so far. To the extent that molecular evolution consists primarily in the accumulation of nearly neutral mutations (88), this result suggests a relatively slow molecular clock for lineages of *Sulfolobus* and perhaps other archaea that share the genetic properties of *S. acidocaldarius* (see "Transposable genetic elements," below).

One qualification of these conclusions has yet to be confirmed experimentally, namely, the extent to which the *pyrE* gene is representative of the *S. acidocaldarius* genome. However, it is possible to estimate the likelihood of this by assuming that the overall error rate of replicating a typical *S. acidocaldarius* gene replication is determined primarily by the length of its longest mononucleotide tract, which is consistent with the mutational spectrum of *pyrE* (described below). By this criterion, only *S. acidocaldarius* genes containing tracts of eight or more A (T), or tracts of six or more G (C), are expected to mutate more frequently than *pyrE*. Analysis of the genome sequence identifies only 27 A (T) tracts and only 151 G (C) tracts meeting these criteria (D. Grogan, unpublished results). This predicts a higher mutational load from a maximum of about 190 of the 2,200 genes predicted in the *S. acidocaldarius* genome. The quantitative effect on the genomic mutation rates is more difficult to estimate, but is likely to be modest. For example, assuming an average tenfold elevation over the *pyrE* rate in the 190 genes, counterbalanced by a tenfold reduction in the approximately 700 genes having mononucleotide tracts shorter than those in *pyrE*, yields less than a twofold increase in genomic mutation rate.

A spectrum of spontaneous mutation provides additional information about genetic fidelity, including (i) which molecular errors occur during DNA replication in vivo and escape correction, (ii) the relative frequencies of these various uncorrected errors, and (iii) any association of particular errors with particular DNA sequences. Spectra of loss-of-function mutations have been determined for genes of *E. coli*, *Saccharomyces cerevisiae*, and other well-studied microorganisms, enabling *S. acidocaldarius* to be compared with known systems in molecular terms. Although each of the target genes listed in Table 1 may be subject to some idiosyncrasy, trends emerge that suggest functional differences in the accuracy of DNA replication and repair. In *S. acidocaldarius*, for example, expansions and contractions of mononucleotide and triplet repeats account for the vast majority of spontaneous mutations. This is an obvious contrast to the *S. cerevisiae* spectrum, but it also differs from the *E. coli* spectrum, in that the *pyrE* mutations occur at several sites, as opposed to a single "hot spot" in the *lacI* gene (38). Expansions and contractions of short repeated sequences like these are attributed to slippage of the nascent and template strands during DNA synthesis; migration of the resulting mismatch back along the nascent strand helps these events escape correction by polymerase proofreading (32).

Conversely, two types of mutations appear to be less frequent in *S. acidocaldarius* than in *E. coli* and

Table 1. Comparison of mutational spectra across the three domains[a]

Mutation	S. acidocaldarius pyrE	E. coli lacI	S. cerevisiae URA3
A:T to G:C	1.0	0.7	5.5
G:C to A:T	7.9	6.2	22.2
Total transitions	**8.9**	**6.9**	**27.8**
A:T to T:A	2.0	1.0	21.1
A:T to C:G	<1	1.0	5.5
G:C to T:A	<1	1.8	14.4
G:C to C:G	1.0	0.4	14.4
Total transversions	**3.0**	**4.1**	**55.5**
−1	61.4	4.0	8.9
+1	17.8	1.1	1.1
±block[b]	3.0	72.0	1.1
Total frameshifts	**82.2**	**77.1**	**11.1**
Other duplications	5.0	1.0	4.4
Other deletions	1.0	9.9	1.1
TE insertions[c]	<1	1.1	<1

[a]Data are percentages of independent, spontaneous mutants, representing 101 *pyrE* mutants of *S. acidocaldarius* (36), 729 *lacI* mutants of *E. coli* (38), and 90 *URA3* mutants of *S. cerevisiae* (57, 98).
[b]Includes all mutations in which a short (2- to 5-bp) unit is added to, or subtracted from, a naturally occurring repeated sequence, even when this does not change the reading frame.
[c]TE, transposable genetic element.

S. cerevisiae. The first represents insertions by transposable genetic elements (TEs), which are common in *E. coli* (38) and less common in *S. cerevisiae* (91). Absence of TE insertions in the *S. acidocaldarius* spectrum reflects a dearth of complete insertion sequences in this species and apparent inactivity of the few that do occur (9). Also underrepresented in the *S. acidocaldarius* spectrum are deletions. As demonstrated by analysis of a larger set of mutants, deletions account for only about 0.4% of all spontaneous mutations in *pyrE* and do not exhibit any statistical association with direct repeats or inverted repeats at their end points (37). In contrast, bacterial and eucaryal systems produce deletions at much higher rates, and the majority of these occur between short direct repeats and inverted repeats (13, 29, 38). Thus, the common modes of deletion formation in bacteria and eucarya seem to be inefficient in *S. acidocaldarius*, and this may enhance genome stability. In particular, the genomes of *Sulfolobus* species and other archaea have clusters of short, regularly spaced repeats (SRSRs) (78). In *S. acidocaldarius*, these SRSR clusters consist of 4 to 133 copies of 24-bp direct repeats separated by short, unique sequences (9). This length (24 bp) is enough to encourage the deletion of tandem duplications in *S. acidocaldarius* (37). It will thus be of interest to measure the genetic stability of SRSR clusters (which have been proposed to play roles in chromosome segregation) in *S. acidocaldarius* or other archaea.

Transposable Genetic Elements

Mutations caused by TEs do not reflect molecular properties of genome replication and repair, but they occur frequently in many archaea and can have a major impact on genome evolution. The most abundant TEs in archaeal genomes are insertion sequences (ISs), which represent the smallest TEs capable of independent transposition (7). Most ISs consist of a transposase-encoding gene flanked by short inverted repeats (IRs), which provide the recognition and cleavage sites for the transposase. In the genome, these sequences are typically flanked by even shorter direct repeats (DRs) that represent duplication of the target site upon transposition (62). Due to the high informational density of the DNA in most unicellular hosts, ISs often inactivate genes when they propagate, and, unlike antibiotic-resistance transposons, they encode no extra genes that can benefit the host directly. Thus, ISs seem to represent a classic example of "selfish DNA."

The mesophilic extreme halophile *Halobacterium salinarum* (which encompasses strains previously designated *Halobacterium halobium*, *Halobacterium cutirubrum*, and *Halobacterium salinarum*) contains several families of insertion sequences, each present in multiple copies (81). Some of these ISs transpose with high frequency among the various plasmids and lower G+C-content chromosomal DNA of this species, which together constitute about 30% of the total genomic DNA (80, 93, 94). As a result, the function of some environmentally important genes, notably those for bacterio-opsin (*bop*), and gas vacuoles (*vac*) are inactivated at very high frequencies (80). At least two of these IS families, designated ISH23 and ISH27, have also been shown to generate deletions at high frequencies (82).

In methanogens, analysis of genomic DNA sequences has identified a diversity of ISs, and only one species, *Methanothermobacter thermautotrophicus* (formerly called *Methanobacterium thermautotrophicum*), has so far been found to lack them completely (7). In addition to these native elements, TEs from eucarya have been used as genetic tools in methanogens (see Chapter 21).

Multiple, diverse ISs are also the rule in the genomes of hyperthermophiles. In particular, the genome of *Sulfolobus solfataricus* contains more ISs than any other bacterial or archaeal genome reported so far. About 10% of the 2.99 Mb *S. solfataricus* P2 genome consists of ISs representing at least 10 families present in an average of 25 copies each. Most of these ISs copies appear to be intact (7) and to transpose actively, as indicated by the frequency of spontaneous rearrangements detected in the course of genome sequencing (105). Genetic assays confirm that transposition of several different ISs cause the vast majority of spontaneous *pyrE* and *pyrF* mutations in the related strain *S. solfataricus* strain P1. As a result, the rate of forming *pyrE* and *pyrF* mutants is 10 to 100 times higher in *S. solfataricus* than in *S. acidocaldarius* (65). Additional ISs have been found by examination of other *Sulfolobus* isolates. For example, six of seven ISs found to transpose in *Sulfolobus* "*islandicus*" were distinct from the ISs detected previously in *S. solfataricus* strains by sequencing or genetic assay (5).

The observed phylogenetic diversity of ISs in *Sulfolobus* is accompanied by diversity of functional properties. Transposition frequencies of seven ISs in various *S.* "*islandicus*" isolates varied over a 250-fold range. The ISs also differed in target-site specificity, although the region between the TATA box and the transcription start site of *pyrE* seemed to represent a preferred region for transposition in both *S. solfataricus* and *S.* "*islandicus*" (5, 65). A more striking difference among the ISs in *S.* "*islandicus*" related to precise excision. This is a significant form of TE excision, because it represents the primary route to restoring gene function when a TE has transposed into a gene and has been shown to be mediated by DNA-metabolizing systems of the host, rather than of the TE (61, 85). Only one of the seven *S* "*islandicus*" ISs evaluated was lost precisely from *pyrE*, thereby

restoring growth without uracil; ironically, this was also the only IS having no IRs or DRs to facilitate deletion (5). Precise removal of this IS (provisionally called ISC1926) is consistent with the fact that it belongs to a family of TEs which transpose via precise excision of a circular intermediate that subsequently integrates at a new site (62). Conversely, the fact that none of the other ISs were deleted precisely at any detectable frequency (i.e., about 5×10^{-10}) remains consistent with the observation that short, hyphenated (i.e., nontandem) IRs and DRs do not promote deletion in another *Sulfolobus* species (37).

The observed failure of typical *Sulfolobus* ISs to excise precisely even under strong selection implies that when they insert into a host gene, the loss of gene function is normally irreversible. This has important implications for genome evolution (see Chapter 19), because in the absence of other mechanisms (such as transfer of an intact copy of the gene from another cell), this property effectively shields IS insertions from selection for removal. This does not mean that IS insertions cannot be lost, but it does predict that the loss will normally be incremental and slow and will not restore gene function. BLAST analyses show that certain ISs are represented only by fragmentary relics in the sequenced *Sulfolobus* genomes (5), and that other ISs present as multiple, full-length copies are also accompanied by numbers of nonfunctional copies bearing distinct mutations (44, 105). This pattern seems consistent with slow removal of ISs by incremental mutation causing accumulation of inactive copies, although others have postulated a mechanism that specifically inactivates ISs when they become too numerous in a genome (7). Another pattern seen in *S. solfataricus* is the congregation of ISs in certain regions of the chromosome and insertion into each other (7). This may be explained by the abundance and activity of ISs in this species, which results in the largest and least deleterious target sites for IS transposition being provided by other ISs.

In addition to full-length ISs and inactive IS fragments, archaea harbor short elements that resemble miniature inverted-repeat transposable elements, or MITEs (84). This type of TE was first discovered as abundant, 50- to 200-bp repeated sequences in eucaryal chromosomes (75). The mobility of MITEs is thought to reflect complementation by ISs, since MITEs possess IRs matching those of full-length, functional ISs present elsewhere in the genome (75). Two classes of MITEs are distinguished by the nature of the interior sequence: Type I MITEs retain sequences of the corresponding helper IS, as though they were derived from a related IS by partial deletion of the transposase gene. Type II MITEs have sequences between the IRs that are unrelated to those of the helper IS, and a basis for their formation remains unclear. Like the corresponding ISs, at least some of the MITEs found in *Sulfolobus* species can transpose and inactivate genes, as demonstrated experimentally in *S. "islandicus"* (5).

DNA REPAIR

All cells have multiple, distinct systems of specialized enzymes that cooperate to keep the genome intact and ready for replication. Figure 2 summarizes these systems according to the biochemical strategy they employ to deal with DNA damage (for comprehensive review, see reference 26). Certain forms of damage (alkylation and UV-induced dimerization) are simply reversed, whereas oxidized or fragmented bases, various helix-distorting adducts, and mismatched bases must be excised and replaced. Should the DNA damage escape repair by these mechanisms, two additional strategies allow replication to proceed anyway: *translesion synthesis* (TLS) by specialized DNA polymerases, which are often inaccurate and thus mutagenic, and homologous recombination (HR). These last two strategies do not actually remove the DNA lesion and are thus more accurately termed "tolerance pathways." The components of these seven pathways have been defined by experimental analysis of bacteria and eucarya (26), and homologs of the corresponding genes can be found in all three domains (Table 2).

According to computational analysis of complete genome sequences, putative genes for alkyltransferase (AT), base-excision repair (BER), and HR occur in essentially all archaeal genera; representative examples are summarized in Table 2. In many cases, the corresponding biochemical function has also been detected. For example, alkyltransferase and DNA *N*-glycosylase activities have been detected in cell-free extracts of thermophilic archaea (50, 111) and in proteins expressed from archaeal genes in *E. coli* (10, 56, 60, 95). HR has been detected by genetic assays in several archaea (see "Genetic recombination," below), and has been shown to require *radA*, the archaeal homolog of eucaryal *RAD51* and bacterial *recA* genes, in *Haloferax volcanii* (124). The RadA proteins of other archaea catalyze nucleoprotein filament formation and strand-exchange reactions in vitro (69, 102), and the *radA* gene of *H. volcanii* was also shown to contribute to survival of DNA damage (124). Highly efficient HR in *Pyrococcus* and related archaea is indicated by survival of extremely high doses of ionizing radiation and reconstruction of a complete chromosome from the resulting small DNA fragments (14).

Putative photoreactivation (PR), TLS, MMR, and NER genes present a more complex distribution in which the number of DNA repair systems in archaea seems to decline with increasing growth temperature. For example, genes encoding DNA photolyases are evident in only about half of the archaeal genera included in Table 2. The validity of this indication from genome sequences is reinforced by the fact that PR has been demonstrated by experiment in each of the three archaea encoding putative photolyases (20, 45, 67, 68, 123). The DNA photolyase of *M. thermautotrophicus* was also purified to homogeneity and characterized in vitro (46). However, lack of photoreactivation has not been confirmed in the archaea that lack putative photolyase genes. Genes encoding error-prone family-Y DNA polymerases show a similar pattern, being evident primarily in mesophilic archaea and *Sulfolobus* spp. (Table 2). Mutagenesis of *S. acidocaldarius* by a wide range of DNA-damaging agents provides indirect evidence in vivo of error-prone TLS (86, 123), and assays in vitro have confirmed the bypass properties of family-Y DNA polymerase from *Sulfolobus* species (6, 48). X-ray crystallographic structures of two such enzymes have been resolved and suggest a structural basis for frameshifting during TLS (59, 108). However, the archaea that lack these family-Y polymerases have apparently not been examined for a lack of damage-induced mutagenesis.

Proteins responsible for MMR in bacteria and eucarya show similarity to the corresponding *E. coli* proteins MutS and MutL, whereas NER systems are of two distinct types. Bacterial NER requires only three proteins (UvrABC), whereas eucaryal NER requires at least seven (83). The bacterial and eucaryal NER proteins have low similarity to each other, but within their respective domains, they are well conserved (18). Among archaea, homologs of bacterial and eucaryal NER genes co-occur, and the three repair systems (MMR, bacterial NER, and eucaryal NER) seem to be distributed according to the optimal growth temperature of the species (Table 2). Thus, most archaea growing optimally below about 55°C encode genes representing a complete MMR system, a complete bacterial NER system, and a partial eucaryal NER system missing *RAD14/XPA*, *RAD 4/XPC*, *RAD 23/hR23b*, and *RAD10/ERCC1* homologs. With one exception so far (*M. thermautotrophicus*), archaea growing optimally above about 55°C lack MMR genes, bacterial NER genes (*uvrABC* homologs), and at least one of the eucaryal NER-gene homologs found in mesophilic archaea, without gaining homologs of any additional eucaryal NER genes.

A few of the thermophilic archaea, primarily *Sulfolobus* species, have been examined for their ability to repair DNA damage in vivo, with no obvious de-

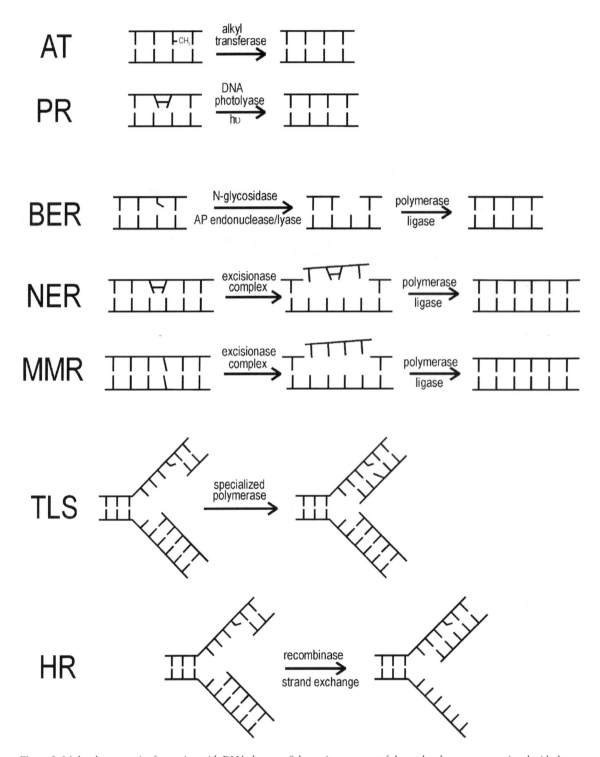

Figure 2. Molecular strategies for coping with DNA damage. Schematic summary of the molecular events associated with damage reversal, damage excision, and damage tolerance. Abbreviations: AT, alkyl transfer; PR, photoreactivation; BER, base excision repair; NER, nucleotide excision repair; MMR, mismatch repair; TLS, *trans*-lesion synthesis; HR, homologous recombination. Some of the processes (HR, in particular) are shown greatly simplified; for comprehensive review, see Friedberg et al. (26).

Table 2. Representation of major DNA repair pathways in microbial genomes[a]

Genus	T_{opt}[b]	AT	PR	BER	NER[c]	MMR	TLS	HR[d]
Archaea								
Methanosarcina	37	+	+	+	Uvr	+	+	RadA
Halobacterium	50	+	+	+	Uvr	+	+	RadA
Thermoplasma	59	?	No	+	No	No	No	RadA
Methanothermobacter	65	+	+	+	Uvr	No	No	RadA
Sulfolobus[e]	80	+	+	+	No	No	+	RadA
Archaeoglobus	83	+	No	+	No	No	No	RadA
Pyrobaculum[e]	98	+	No	+	No	No	No	RadA
Pyrococcus	98	+	No	+	No	No	No	RadA
Bacteria								
Escherichia	37	+	+	+	Uvr	+	+	RecA
Thermotoga	80	+	No	+	Uvr	+	No	RecA
Eucarya								
Saccharomyces	30	+	+	+	Rad	+	+	Rad51

[a]Compiled from The Institute for Genomic Research Comprehensive Microbial Resource (TIGR CMR, version 16.0; http://pathema.tigr.org/tigr-scripts/CMR/CmrHomePage.cgi). In the table, "+" indicates that sufficient relevant COGs (gene families) are represented in the genome to form a functional pathway; "?" indicates a match of marginal significance ($P > 10^{-7}$), and "No" indicates that the necessary homologs are not present. For pathway abbreviations, see the legend to Fig. 2.

[b]Temperature (°C) yielding fastest growth (approximate).

[c]"Uvr" indicates a complete pathway of the bacterial type; "Rad" indicates a complete pathway of the eucaryal type.

[d]Family of recombinase found; see Sandler et al. (92) for molecular differences among the families.

[e]Members of the *Crenarchaeota*; all other archaea listed are members of the *Euryarchaeota* (122).

fects detected (see "An alternative form of NER?" below). The simplest reconciliation of these two observations (i.e., absence of genes required for MMR and NER on one hand, and absence of obvious DNA repair defects on the other), would seem to be that the thermophilic and hyperthermophilic archaea have alternative molecular strategies that assume the function of classical MMR and NER systems but do not involve homologous proteins. Meanwhile, specific experimental evidence supporting or contradicting the idea of alternative MMR or NER strategies in hyperthermophilic archaea remains limited and remarkably balanced, as summarized in the next paragraph.

An Alternative Form of MMR?

The primary evidence for postreplicational MMR in hyperthermophilic archaea comes from the low rate of spontaneous mutation measured in *S. acidocaldarius*, which is about 0.1% of mutation rates of MMR-deficient bacteria (42, 64). There is so far no evidence that other *Sulfolobus* species replicate their DNA less accurately than *S. acidocaldarius* does, since the higher mutation rates reported for them can be attributed to IS transposi-

tion. However, a distantly related crenarchaeon, *Pyrobaculum aerophilum*, has been claimed to be a natural mutator, based on the length heterogeneity of mononucleotide tracts in its genomic DNA (24). Nearly all (16 of 18) of the longest mononucleotide tracts of this organism were recovered in multiple lengths by cloning, and most of the variable sites were located between, not within, predicted genes. The available data neither exclude nor prove fundamental genetic differences between *S. acidocaldarius* and *P. aerophilum*. Differences are suggested, for example, by failure to observe length heterogeneity in the mononucleotide tracts of *S. acidocaldarius* during genome sequencing (L. Chen, personal communication), despite the fact that several of these are as long as the variable tracts in *P. aerophilum*. On the other hand, the *Pyrobaculum* DNA used for cloning came from a large population of cells (23), suggesting the possibility of accumulated spontaneous mutations. In addition, the spontaneous mutation spectrum of *S. acidocaldarius* is dominated by the types of slipped-strand events that characterize MMR-deficient mutants of bacteria and eucarya (32, 36). Note that *S. acidocaldarius* could conceivably achieve a low overall mutation rate in the absence of MMR, provided (i) the genome has relatively few long mononucleotide tracts and (ii) extremely accurate proofreading occurs within the replicative polymerase complex. As described above (Spontaneous mutation), preliminary examination of the *S. acidocaldarius* genome sequence supports (i). With respect to (ii), the observed error rate of replicating *pyrE*, 7.8×10^{-10} per bp, is only about 15-fold lower than that of bacteriophage T_4 replication, which does not involve postreplicational MMR (16). Thus, a replicative complex with 15-fold more efficient proofreading than the T_4 replicase could, in theory, explain the mutational properties of *S. acidocaldarius* (J. Drake, personal communication).

An Alternative Form of NER?

The broad conservation of NER systems among bacteria and eucarya argues that this form of repair, which typically operates on large, helix-distorting DNA lesions, is fundamental to evolutionary fitness (18). In addition, the conservation of the latter portion of the eucaryal pathway in all archaea suggests that these "downstream" enzymes, primarily helicases and structure-specific endonucleases, have a function that is being maintained by selection. In particular, the only NER protein homologs found in all eucarya and consistently absent from hyperthermophilic archaea are those that bind to the DNA lesion and thus initiate the process of excising it as an oligonucleotide. Accordingly, the only "novel" DNA repair proteins that may need to be hypothesized for many archaea represent alternatives to the eucaryal damage-recognition proteins (Table 2). However, this argument based on gene conservation leads to a very different conclusion when applied to the mesophilic and moderately thermophilic archaea. These archaea encode complete, apparently functional NER systems of the bacterial type (Table 2) (67, 68, 74). Thus if the eucaryal-NER homologs indeed have a cellular function in these species, it would seem to be distinct from that of the bacterial NER homologs, which these archaea also encode.

Experimental evidence for NER in hyperthermophilic archaea remains limited. Survival curves indicate that *S. acidocaldarius* has about half of the dark-repair capacity for UV photoproducts of Uvr$^+$ *E. coli* (123), which would be a significant level of NER. In addition, UV photoproducts are converted in the dark to apparently recombinogenic lesions that are removed at physiological temperature but not at room temperature (101). However, the *S. acidocaldarius* genome has recently been shown to encode a UVDE homolog (9), and the predicted function of this protein is to initiate removal of UV photoproducts by an alternative pathway (127). Thus, this gene

product could explain the observed UV responses of *S. acidocaldarius*. Biochemical activities of the eucaryal-NER homologs encoded by *S. solfataricus* have been confirmed to be consistent with "downstream" (i.e., late) events of NER (89), and T_4 endonucleaseV-sensitive sites disappear from the DNA of UV-irradiated *S. solfataricus* cells incubated in the dark at physiological temperature (73). Since the genomes of *S. solfataricus* and most other archaea encode no UVDE homolog (9), the latter result provides perhaps the strongest experimental evidence to date for some alternative form of NER in hyperthermophilic archaea. It remains unclear, however, whether this system is functionally equivalent to conventional NER. Like UVDE, for example, it may use an alternative mechanism, or it may lack the broad substrate range of conventional NER.

Uracil-Sensitive DNA Synthesis

As a rule, archaea encode multiple DNA polymerases of the B family, at least one of which is presumed to replicate the genome. Because of technological interest, most attention has been focused on the DNA polymerases of hyperthermophilic archaea, and these were first shown by Lasken et al. (55) to have the unusual property of stalling at uracil residues in the template strand. This is now known to result from a uracil-specific binding pocket in the polymerase that scans the template strand 4 to 6 nt ahead of the catalytic site (31). The biological function commonly proposed for this feature is avoidance of transition mutations initiated by dC deamination. However, as the molecular properties of archaeal DNA replication and BER become clear (see Chapter 3), a more complex rationale for this phenomenon seems increasingly necessary.

For example, systematic measurements of binding affinity reveal the striking molecular specificity of the phenomenon. The special pocket found in archaeal polymerases binds only uracil, not structurally similar derivatives, and only when incorporated into ssDNAs longer than about 15 nt (107). Thus, stalling is neither induced by other potentially mutagenic pyrimidine bases, nor by uracil in a context distinct from a replication fork. The fact that the complex stalls with the uracil residue in ssDNA protected by the enzyme (25) would seem to pose serious obstacles to BER. Repair strategies that involve dissociation of the polymerase must initially protect the uracil from BER, because it is within ssDNA. Regression of the replication fork would allow the uracil-containing strand to reanneal to its original partner so that BER could repair the lesion, although this may be complicated by the proximity of the template uracil residue to the nascent 3′ end. Alternatively, displacement of the replicative polymerase by a specialized TLS polymerase would, in principle, allow for accurate replication past the uracil, in a manner analogous to UV photoproducts (6, 117). Another possibility is that a uracil DNA glycosylase breaks the replication fork by excising the uracil from the ssDNA, thereby triggering double-strand-break repair (via HR) and reassembly of the replication fork (51). In any case, assays in vitro indicate that, without some form of rescue, the stalled complex should eventually insert an A opposite the U and continue (25).

These various mechanistic possibilities nevertheless fail to explain why uracil-induced stalling should occur only in family-B DNA polymerases of archaea. The need for BER at high temperatures remains obvious, but there is no evidence that hyperthermophilic archaea do not excel at this form of DNA repair. These archaea typically encode multiple, diverse DNA N-glycosylases which exhibit high activity in cell extracts and seem to be part of a sophisticated, versatile BER system (50, 60, 95). In particular, no reason has emerged explaining why hyperthermophilic bacteria, which tend to have higher G+C content in their DNA than their archaeal counterparts, should not also need this form of "read-ahead" proofreading. The observation that archaeal family-B DNA polymerases fail to discriminate effectively against deaminated dNTPs (40, 55) raises the possibility that the primary source of dU in archaeal DNA is incorporation by polymerase, rather than spontaneous deamination of dC residues. Thus, the more serious fate avoided by polymerase stalling may be the double-strand breaks from BER of dU residues placed close to each other on opposite strands (35), or some other problem yet to be identified.

GENETIC RECOMBINATION

In its most general sense, genetic recombination means the creation of new DNA sequences from existing sequences by processes involving strand exchange rather than error-prone synthesis. This definition accommodates recent progress in bacterial and yeast systems that has blurred historical distinctions among replication, repair, and recombination (51, 112), and encompasses a diversity of molecular events that all have important effects on the stability and evolution of microbial genomes. Under this broad definition, four types of recombination can be distinguished, according to the role played by the DNA sequences involved (Table 3). General, or homologous, recombination (HR) requires significant intervals (≥50 nt) of identical sequence to effect strand exchange, but places few other constraints on the sequence. This reflects its mechanism, in which partner recognition is mediated primarily by the DNA sequences. In contrast, site-specific recombination requires a specialized enzyme to bind a specific, short DNA sequence in both partners and catalyze strand exchange between these short sequences. Transposition resembles site-specific recombination in requiring protein recognition of one DNA partner in the exchange (i.e., the ends of transposable element), but the other partner (the target sequence) is typically chosen with little or no sequence specificity. Finally, "illegitimate" recombination breaks and joins DNA sequences with negligible influence of the sequences involved. Some situations make it difficult to distinguish illegitimate and homologous recombination, as recombinase-independent break-and-join events may be facilitated by very short matching sequences (microhomologies) (12).

Table 3. Distinct types of genetic recombination

Type, subtype	Defining properties	Genomic consequences	Archaeal examples[a]
Homologous	Requires extensive sequence identity		
Reciprocal	All markers preserved in the products	Crossovers	Targeted integration of plasmid constructs
Nonreciprocal	Marker(s) eliminated	Gene conversion	Intragenic recombination in S. acidocaldarius (hypothesized)
Site-specific	Integrase/excisionase operates on a specific sequence within both partner DNAs	Virus/plasmid integration and excision	SSV1, conjugal plasmids of Sulfolobus
Transposition[b]	Transposase operates on a specific DNA sequence at the TE boundaries	Relocation of TE	Insertion sequences, MITEs
Illegitimate	DNAs broken and rejoined essentially at random	Large rearrangements, error-prone repair of double-strand breaks	None identified

[a]See text for evidence and relevant literature.
[b]Also classified as "nonconservative site-specific" recombination.

Although HR has genetic consequences, in particular, in conjunction with DNA transfer (discussed below in "DNA transfer"), it appears to serve primarily as part of a system that reassembles replication forks that have stalled or fallen apart (51, 112). However, even a strict conservation of this role across all three domains leaves room for significant differences in the genetic properties of HR. In addition, archaeal recombinases (RadA proteins) differ structurally from bacterial RecA proteins, and less so from eucaryal Rad51 proteins (92). Furthermore, some archaea encode RadA paralogs, represented by the RadB proteins of *Haloferax*, *Thermococcus*, and *Pyrococcus* species (1, 49). The *Pyrococcus* RadB protein has been reported to interact with a DNA polymerase subunit, a putative Holliday junction resolvase, and RadA, depending on parameters such as ATP concentration (49). *H. volcanii* RadB is not essential for HR (1) and thus appears to play a yet-undefined role in DNA metabolism.

Consistent with the apparent ubiquity of RadA orthologs in archaea, HR has been documented in diverse archaeal species. For example, the generation of selectable phenotypes from genetically marked strains (auxotrophs and resistance mutants) has been demonstrated with *H. volcanii*, *M. voltae*, *M. thermautotrophicus*, and *S. acidocaldarius* (4, 34, 71, 125). Since the corresponding strains lack episomes, these results provide strong presumptive evidence for marker incorporation by homologous recombination. Other studies have incorporated artificial DNA constructs into archaeal genomes via homologous recombination (see Chapter 21).

Functional properties capable of distinguishing archaeal HR from HR in classic bacterial and eucaryal systems include (i) the way in which substrate (homology) length affects recombination efficiency, (ii) the relative frequency of reciprocal versus nonreciprocal events, and (iii) the effect of occasional mismatches on recombination frequency. With respect to (i), the efficiency of reciprocal recombination (i.e., the yield of crossovers) relates in a characteristic way to the distance between the markers. Below a certain threshold, called the "minimal efficient processing segment," or MEPS, recombination is extremely rare and often independent of RecA/Rad51 function (106). At larger distances, the frequency of recombinants increases in proportion to the effective cross section of the detector (i.e., the interval between the two markers). At still greater distances, multiple crossovers between markers become increasingly frequent, diminishing the increase in recombinants and leading eventually to a plateau. This frequency/distance pattern of reciprocal HR occurs in diverse systems, although it exhibits quantitative differences

among them. Thus, MEPSs observed in bacteriophage, bacteria, and eucaryal cells are about 40, 70, and 250 nt, respectively, whereas the midpoint of the proportional-increase region corresponds to roughly 2,000, 20,000, and 30,000 nt, respectively (97, 106, 110).

In contrast, HR between *pyrE* mutations in the *S. acidocaldarius* chromosome, initiated by conjugation, does not fit this consensus behavior. Frequency-versus-distance data have provided no statistical support for an MEPS, and the region of proportionality extends to only about 50 nt, beyond which recombination was found to be fairly constant as a function of marker separation (39). Although an effect of the (yet-undefined) DNA transfer mechanism cannot be excluded, the same frequency-versus-distance behavior was observed with over 50 combinations of mutations located at various points in *pyrE* and was not altered by eightfold stimulation of recombination by prior UV irradiation (39); it thus seems to be a robust property of the mode of HR that accompanies conjugation in this species. The most similar frequency-versus-distance behavior identified so far in other genetic systems is that of a strand-annealing pathway of double-strand break repair in *S. cerevisiae* (114). Based on this and other information, the *S. acidocaldarius* results seem easiest to reconcile with nonreciprocal recombination in which base-pairing allows relatively short patches of single-stranded donor DNA to simply replace corresponding ssDNA segments of the recipient (39). A similar mechanism may also explain the relative efficiency with which linear DNAs transfer markers to the *S. acidocaldarius* chromosome and the fact that this occurs even when the homologous sequence is very short (53).

Reciprocal recombination has been demonstrated in other archaea by the integration of nonreplicating circular DNAs into the chromosome, usually in the course of allele replacement (see Chapter 21). In *Methanococcus voltae*, the heterozygotic intermediate formed by the first crossover was readily detected in transformant clones and found to be stable in the absence of selection for at least 40 generations following initial selection (27). Efforts to detect such intermediates were not reported in similar manipulations of two hyperthermophilic archaea (96, 126), and were unsuccessful in *S. acidocaldarius*, despite the use of PCR after only six generations of growth (53). Thus, it remains unclear whether outcomes of recombination between circular DNAs and host genomes differ among the major groups of archaea.

Examples of site-specific recombination in archaea include viruses that integrate into the host chromosome at specific sites. The best-studied example, SSV1, integrates into the *Sulfolobus shibatae* genome by site-specific recombination at a tRNA[arg] gene. This interrupts the integrase gene of the virus, which presumably stabilizes the integrated provirus (87). Purified SSV1 integrase catalyzes both forward and reverse recombination reactions in vitro (72). Under these conditions, the site-specific recombination requires a minimal sequence of only 19 bp, corresponding to the distal portion of the anticodon stem and loop of the tRNA (103). Similar features have been found in integration of certain bacteriophage (87), and more recently, in conjugative plasmids integrated into *Sulfolobus* genomes. Unlike the integration of SSV1, these latter events do not disrupt the integrase gene itself (104)

Several examples of IS transposition have also been documented in archaea. Most of these IS have relatives in bacteria and eucarya, and the corresponding IS families represent diverse molecular strategies of transposition (62). According to the data available, ISs in halophilic archaea seem unusual with regard to the high frequency with which they transpose and delete adjacent sequences (81, 82, 93). On the other hand, the archaeal ISs that have been characterized appear to resemble their bacterial counterparts in transposing preferentially into the 5′-untranslated (i.e., promoter) regions of target genes (5, 65, 81).

Illegitimate recombination (as suggested by its name), is rare and difficult to detect specifically by genetic assays; as a result, it has not been analyzed systematically in any archaeon. In eucarya and bacteria, illegitimate recombination often results from the activity of type II DNA topoisomerases and from nonhomologous end-joining of double-strand breaks (17, 41). In *S. acidocaldarius*, analysis of sequences at the end points of large spontaneous deletions could not support or exclude either possibility (37).

DNA Transfer

The mechanisms that enable cells to receive DNA from the environment, defective virons, or other cells are in most respects independent of homologous recombination and arguably less important for maintaining structural integrity of the genome. However, in combination with HR or other mechanisms that allow the sequence to persist in the recipient, DNA transfer provides microorganisms with the opportunity to acquire ecologically useful alleles and genes. According to theoretical models, this source of genetic variation greatly accelerates the evolutionary adaptation of those lineages, relative to lineages that lack DNA transfer (8). The three modes of DNA transfer known in bacteria (transformation, transaction, and conjugation) also operate in archaea. However, similar to bacteria, the events are rare and detecting them experimentally requires genetic selections,

which are not available for many archaea. Natural transformation (i.e., uptake of naked DNA under conditions that would be considered physiologically normal) has so far been reported only for *M. voltae* and occurred at low frequency (4). Documented examples of transduction and conjugation are not dramatically more numerous, but they occur in diverse archaeal clades and exhibit certain interesting differences from the classical systems.

Transduction

Two rather different examples of DNA transfer by virion-like particles have been observed in methanogens. In the first example, a lytic virus (phage), ψM1, of *M. thermautotrophicus* has been shown to transduce chromosomal genetic markers of its host at frequencies of 10^{-6} to 10^{-4} recombinants per particle (70). In the second example, cells of *M. voltae* shed small icosahedral particles that resemble bacteriophage and contain small fragments of host DNA (19). Incubation of these particles with suitably marked recipients generates recombinants at higher frequencies than achieved by transformation of this species. This "voltae transfer agent" (VTA) loses its activity relatively rapidly, which has complicated its analysis (3). Most features, including the absence of any viral DNA or cell killing associated with the particles, argue that VTA is not a classical transducing phage (virus) (3), although similar transfer phenomena have been described in a few bacteria, most notably *Rhodobacter capsulatus* (54).

Conjugation

Haloferax volcanii was the first archaeon to demonstrate cell-to-cell transfer of DNA, by generating prototrophic recombinants from mixtures of two stable, genetically distinct auxotrophs. The transfer of chromosomal DNA required some stabilization of cell-cell contacts (71), which is also seen with some bacterial and eucaryal conjugation systems (121). Unlike bacterial systems, however, it did not require genetically distinct donor and recipient strains (71). A point of origin and direction of transfer have not been tested in this or other archaeal conjugation systems, because of the difficulty of producing multiply marked strains and other experimental limitations. *H. volcanii* can also transmit certain plasmids to *Haloferax mediterranei*, but transfer of the same plasmids in the reverse direction, or transfer from *H. volcanii* to other species, could not be detected (116). It thus remains unclear whether successful transfer requires certain functions missing, in particular, host/plasmid combinations or is constrained by other aspects of the donor and recipient.

Conjugational transfer of a plasmid between *Sulfolobus* species was first observed for a large plasmid, pNOB8, by coculture of its native host with an excess of recipient cells having a distinct genomic restriction pattern (100). A number of such self-transmissible plasmids have since been found in isolates of *S. solfataricus*, *S.* "*islandicus*," and related species. The mode of transmission appears to be analogous to that of classical bacterial conjugation, in the sense that it involves a plasmid-containing donor and a plasmid-free recipient (8). The *Sulfolobus* conjugative plasmids are rather diverse, and only a few genes can be implicated in plasmid replication and cell-to-cell transfer by the criterion of conservation among these plasmids. This suggests that these archaeal plasmids use simpler transmission mechanisms than many bacterial conjugative plasmids do (33).

The remaining example of archaeal conjugation, found in *S. acidocaldarius*, resembles the *H. volcanii* system in that chromosomal markers are transferred and the process involves no free plasmid and no genetic distinction between donor and recipient strain, aside from the chromosomal markers used to select recombinants (34). The effects of various manipulations on the efficiency of the *S. acidocaldarius* system indicate that conjugation can initiate rapidly upon cell mixing and can occur efficiently in liquid medium (i.e., without stabilization of cell-cell contact by adsorption to a surface) (28). Genetic properties suggest that much of the donor DNA is ultimately incorporated into the recipient as relatively small fragments. Specifically, (i) unselected markers exhibited negligible genetic linkage to selected markers in three-factor crosses, even when separated by only 500 to 600 bp, and (ii) when one parental strain was forced to serve as the donor by moderate gamma irradiation, strains bearing large deletions were very poor recipients (39).

Gene Flow in Natural Populations

Historically, genetic analysis of unicellular organisms has involved designating a particular isolate (clone) as "wild type" and using it to represent the species. This focus on a specific clone provides for defined genotypes, which are very useful in experimental genetics, but it does not address the extensive genetic variation that occurs in natural populations of microorganisms. Conversely, the historic focus of classical population genetics on plant and animal systems does not address the dynamics unique to haploid, unicellular organisms, or their phenotypically cryptic variation (58). Fortunately, study of microbial populations has gained both sophistication and momentum with the advent of large-scale DNA sequencing, and some of the recent analyses of microbial populations have examined archaea.

Animal and plant speciation has long been understood in terms of geographic separation of populations, leading to restricted gene (allele) flow between populations, reproductive isolation, and relatively unrestrained genetic divergence (88). Geographical isolation has been considered irrelevant for microorganisms because their large numbers and small size favor efficient, long-range dispersal and globally homogenous populations (i.e., "everything is everywhere, the environment selects") (22). This view is reinforced for bacteria and archaea by many studies that employ only low-resolution phylotypes (e.g., 16S rRNA sequences), which typically fail to reveal geographic patterning. Higher-resolution genetic analysis also indicates a lack of population structure, or "biogeography," of some bacterial species that have cosmopolitan habitats and facile dispersal (90). Thus, the archaea (and bacteria) that lack specific survival forms and populate "extreme" environments provide a strategic test of the generality of the "no biogeography" assumption. This reflects the fact that the anthropocentric designation "extremophile" identifies microorganisms for which the expansive world of "normal" habitats cannot support growth and reproduction. Accordingly, the dispersal of viable cells (migration), and the resulting sharing of alleles among populations should be inefficient and potentially outpaced by genetic changes unique to each locale (i.e., the creation of new alleles through mutation and the loss of alleles through selection and genetic drift) (Fig. 1). If so, each local population will accumulate its own unique alleles, reflecting its own unique history, and the overall genetic divergence between populations will parallel the physical distance between them, reflecting the dominant role played by the geographical separation as a barrier to migration.

This predicted genetic impact of restricted migration has been confirmed by multilocus sequence typing (MLST) for populations of *S.* "*islandicus*" distributed throughout the Northern Hemisphere. Seventy-eight individuals (clones) isolated from acidic hot springs and solfataras were first determined to be conspecific by the criterion of greater than about 99% nucleotide identity of nine protein-encoding, "housekeeping" genes (119). These nucleotide differences were then analyzed phylogenetically and found to define five major clades within this species. Each clade corresponded to a distinct geographical region: the central Kamchatka peninsula, the southern Kamchatka peninsula, Iceland, Lassen Volcanic National Park (California), and Yellowstone National Park (Wyoming). The total genetic divergence between pairs of individuals increased with the distance between their sites of origin and did not correlate with differences of temperature, pH, or regional geology. Thus, the genetic divergence of local *Sulfolobus* populations was attributed primarily to restricted dispersal, rather than locale-specific selection (119). This situation may extend to other microorganisms that require special conditions for growth and survival. A similar analysis has, for example, indicated genetic isolation of at least one population of *Pyrococcus furiosus* (21).

Limited migration elevates the importance of mutation and recombination in the evolution of natural populations (Fig. 1). The relative rates of mutation and recombination can be assessed by statistical analysis of sequence differences (polymorphisms) among closely related individuals; this has been done in at least three divergent archaea, with generally congruent results. In the first study, the natural abundance of a *Ferroplasma* species in an acid mine biofilm enabled its population structure to be characterized by sequencing of DNAs cloned directly from the biofilm (118). The genomic segments revealed a relatively limited number of polymorphisms within the population, but many different combinations of these polymorphisms. The resulting genotypic diversity of the *Ferroplasma* species greatly exceeded that of the dominant bacterium in the community, a *Leptospirillum* species (118). In a second study, isolates of various *Halorubrum* species cultivated from Spanish salterns exhibited low levels of linkage disequilibrium, as assessed by MLST, indicating frequent recombination among lineages (76). In a third study, detailed MLST analysis of *S. "islandicus"* isolates identified 16 distinct six-locus genotypes among only 60 isolates recovered from one geothermal site (120). The phylogenies (evolutionary histories) of the six loci disagreed, and for any two genotypes, differences were much more common between different loci (genes) than within a locus. According to three different statistical measures, a typical allele in this population was several times more likely to have been created by a recombination event than by a mutational event (120).

All three studies indicate extensive shuffling of DNA sequences among closely related archaeal lineages. Such prevalence of recombination over mutation is not unique to archaea; certain bacteria, notably naturally competent pathogens, exhibit a similar population structure, for example (113). However, most bacteria exhibit a highly clonal population structure in which such recombination is rare, and this has also been reported for bacteria in "extreme" biotopes, such as geothermal springs (79). In contrast, no archaea have so far been confirmed to exhibit clonal population structures, despite considerable phyloge-

netic and ecological diversity of the few species that have been examined.

PERSPECTIVE: THE NEXT FIVE YEARS

Measuring the Functional Genetic Properties of Archaea

The biological differences separating the *Archaea* from the *Bacteria* on the one hand, and from the *Eucarya* on the other hand, are too great to allow the nature of genetic processes in archaea to be predicted by analogy to either of the other domains. In fact, it seems likely that major groups of archaea differ among themselves with respect to genetic properties, as suggested by the unequal distribution of putative DNA repair genes. As a result, genetic properties of archaea must continue to be measured experimentally, which demands the development and validation of appropriate quantitative assays. This effort has only begun, making comparisons among archaea, bacteria, and eucarya highly asymmetric and, thus, tenuous. In this context, the hypothesis that archaea, or major archaeal groups, have their own distinctive set of genetic properties is important, not because of its existing support, but because it focuses attention on an area of strategic importance for elucidating and exploiting the molecular and cellular biology of archaea. Testing this hypothesis will involve measuring molecular processes central to the survival, reproduction, and evolution of archaea and to the development of experimental tools for establishing gene function at the molecular level. It thus seems appropriate to identify for special scrutiny some of the characteristics that seem to be emerging as distinctive properties of archaea, even though the experimental data remain limited. These properties include (i) genetic conditions or states of archaea, (ii) genetic mechanisms that lead to such states, and (iii) unusual features of certain archaeal enzymes presumed to have genetic impacts.

How Do Hyperthermophilic Archaea Avoid Mutation?

Sulfolobus spp. have well-defined molecular mechanisms, in the form of error-prone DNA polymerases, for converting DNA lesions into mutations (6). The low error rates these archaea achieve during normal growth therefore imply other, highly accurate DNA replication and repair mechanisms that compete effectively with the error-prone processes, and thereby convert lesions into intact DNA. The latter mechanisms remain mysterious, however, because of the lack of identifiable damage-recognition proteins that initiate MMR or NER. The availability of complete genome sequences and sensitive mass-spectrometric techniques may allow proteins that bind specifically to DNA mismatches or UV photoproducts to be unambiguously identified in these archaea. It will also be important to develop assays in vivo that investigate the functional roles of the Rad1/XPF and Rad2/XPG homologs of archaea. Do these proteins complete the process of NER in hyperthermophilic archaea initiated by other proteins? Does their role differ in the archaea that encode a complete set of UvrABC homologs?

Archaeal DNA synthesis exhibits interesting molecular features (see Chapter 3), including some relevant to genome stability. For example, hyperthermophilic archaea synthesize very short Okazaki fragments (66), implying a highly discontinuous lagging strand, whereas discontinuity of the leading strand is suggested by the active uracil N-glycosylases of these archaea, combined with poor discrimination of their family-B DNA polymerases against dUTP (55). (The latter property, in combination with "read-ahead uracil proofreading," limits the effectiveness of archaeal polymerases in PCR [40]). Are both leading and lagging strands discontinuous in these archaea? If so, does this discontinuity mark newly synthesized DNA for important molecular processes? Though speculative, this question is significant because it represents a potential basis for the strand discrimination required by postreplicational MMR, whether of a conventional or novel type. Discontinuity has been proposed to be the primary mechanism by which conventional MMR discriminates daughter from template strands in eucarya and most bacteria (11), and alternative MMR systems can be imagined which could also exploit daughter-strand discontinuity. For example, if two helicases of opposite polarity were recruited to a mismatch (in a manner analogous to eucaryal NER), translation along the DNA in both directions would ultimately displace the erroneous (daughter) strand from the correct (template) strand and prepare the region for accurate resynthesis (35). It may soon be possible to test certain predictions of these hypotheses, either through genetic manipulation of uracil DNA glycosylase and dUTPase levels in vivo, or through assays in vitro with cell extracts and model substrates.

How Does Homologous Recombination Affect Genome Stability?

HR has been used to disrupt specific genes in many archaea, but it has rarely been analyzed in archaea as a genetic process. It thus remains unclear how the recombination mediated in vivo by archaeal RadA proteins compares with that initiated by RecA and Rad51 proteins in bacteria and eucarya, respectively. Furthermore, the limited data on properties of archaeal HR seem to differ even within a single genus. Recombination of exogenous DNA with the host genome is reported to be virtually inactive in *S. solfataricus* strain P1, moderately active in a related *Sulfolobus* isolate, and highly active in *S. acidocaldarius* (39, 43, 126). HR between chromosomal mutations in *S. acidocaldarius* conjugation assays is surprisingly efficient over very short distances, which seems to demand a nonreciprocal mechanism (39). It will be important to determine whether this mode of recombination is also important over longer distances and in contexts outside of conjugation.

A mode of HR that operates on short regions of sequence identity may offer advantages for repairing damaged genomes (14), but it also threatens genomic stability during normal growth by promoting inappropriate recombination between nearly identical sequences in different regions of the genome. The deleterious consequences of such "ectopic" recombination, including gene disruption and gross rearrangements, provide a rationale for the second major function of known MMR proteins, namely, inhibition of HR between sequences containing a few nucleotide differences (115). Many of the hyperthermophilic archaea that lack identifiable MMR proteins also have low G+C genomes and high purine content in the nontranscribed strand of genes (77). These compositional biases limit DNA sequence complexity and thereby tend to create substrates for ectopic recombination. It will thus be important to determine how HR in hyperthermophilic archaea responds to minor sequence divergence and how it compares with HR in mesophilic archaea. If differences are observed, it will also be important to investigate the role of the "conventional" MMR homologs encoded in the mesophilic archaea in these differences.

Are Archaea Adapted for Adaptation?

Compared with bacteria and unicellular eucarya, archaea that have been analyzed in genetic terms have rather low rates of neutral mutation (which ignores the activity of TEs) and high rates of recombination. This combination of properties would seem ideal for evolutionary adaptation. Recombination of existing, functional sequences is predicted to be a much more efficient strategy for improving fitness than mutation de novo, reflecting the fact that very few mutations are beneficial and many are disadvantageous. Superiority of the recombinational strategy also has experimental support (30). It will be important, however, to test this pattern of low-mutation and high-recombination rates across a wider range of archaeal taxa, environments, and life-history traits than has been possible in the past. Because the relative rates of mutation and recombination can be inferred from statistical analysis of sequence polymorphisms, sequencing DNA of abundant but uncultivated archaea cloned directly from "moderate" habitats, such as marine sponges, pelagic plankton, or the rhizosphere of terrestrial plants (2, 99, 109) may fill in important gaps in our knowledge about the relative roles of mutation and recombination in archaea. Ultimately, combining computational genetic analyses of natural populations with experimental analyses of cultured archaea will help clarify how molecular mechanisms of archaea determine genetic properties and how genetic properties of archaea affect genome stability and evolution.

Acknowledgments. Work in the author's laboratory was sponsored by grants from the U.S. Office of Naval Research and the National Science Foundation.

REFERENCES

1. Allers, T., and M. Mevarech. 2005. Archaeal genetics—the third way. *Nat. Rev. Genet.* **6:**58–73.

2. Beja, O., E. V. Koonin, L. Aravind, L. T. Taylor, H. Seitz, J. L. Stein, D. C. Bensen, R. A. Feldman, R. V. Swanson, and E. F. DeLong. 2002. Comparative genomic analysis of archaeal genotypic variants in a single population and in two different oceanic provinces. *Appl. Environ. Microbiol.* **68:**335–345.

3. Bertani, G. 1999. Transduction-like gene transfer in the methanogen *Methanococcus voltae*. *J. Bacteriol.* **181:**2992–3002.

4. Bertani, G., and L. Baresi. 1987. Genetic transformation in the methanogen *Methanococcus voltae* PS. *J. Bacteriol.* **169:**2730–2738.

5. Blount, Z. D., and D. W. Grogan. 2005. New insertion sequences of *Sulfolobus*: functional properties and implications for genome evolution in hyperthermophilic archaea. *Mol. Microbiol.* **55:**312–325.

6. Boudsocq, F., S. Iwai, F. Hanaoka, and R. Woodgate. 2001. *Sulfolobus solfataricus* P2 DNA polymerase IV (Dpo4): an archaeal DinB-like DNA polymerase with lesion-bypass properties akin to eukaryotic pol Eta. *Nucleic Acids Res.* **29:**4607–4616.

7. Brugger, K., P. Redder, Q. She, F. Confalonieri, Y. Zivanovic, and R. A. Garrett. 2002. Mobile elements in archaeal genomes. *FEMS Microbiol. Lett.* **206:**131–141.

8. Burger, R. 1999. Evolution of genetic variability and the advantage of sex and recombination in changing environments. *Genetics* **153:**1055–1069.

9. Chen, L., K. Brugger, M. Skovgaard, P. Redder, Q. She, E. Torarinsson, B. Greve, M. Awayez, A. Zibat, H. P. Klenk, and R. A. Garrett. 2005. The genome of *Sulfolobus acidocaldarius*, a model organism of the Crenarchaeota. *J. Bacteriol.* **187:**4992–4999.

10. Chung, J. H., M. J. Suh, Y. I. Park, J. A. Tainer, and Y. S. Han. 2001. Repair activities of 8-oxoguanine DNA glycosylase from *Archaeoglobus fulgidus*, a hyperthermophilic archaeon. *Mutat. Res.* **486:**99–111.

11. Claverys, J. P., and S. A. Lacks. 1986. Heteroduplex deoxyribonucleic acid base mismatch repair in bacteria. *Microbiol. Rev.* **50:**133–165.

12. Conway, C., C. Proudfoot, P. Burton, J. D. Barry, and R. McCulloch. 2002. Two pathways of homologous recombination in *Trypanosoma brucei*. *Mol. Microbiol.* **45:**1687–1700.

13. Dillon, D., and D. Stadler. 1994. Spontaneous mutation at the mtr locus in *Neurospora*: the molecular spectrum in wild-type and a mutator strain. *Genetics* **138:**61–74.

14. DiRuggiero, J., N. Santangelo, Z. Nackerdien, J. Ravel, and F. T. Robb. 1997. Repair of extensive ionizing-radiation DNA damage at 95 degrees C in the hyperthermophilic archaeon *Pyrococcus furiosus*. *J. Bacteriol.* **179:**4643–4645.

15. Drake, J. W., B. Charlesworth, D. Charlesworth, and J. F. Crow. 1998. Rates of spontaneous mutation. *Genetics* **148:**1667–1686.

16. Drake, J. W., and L. S. Ripley. 1994. Mutagenesis, p. 98–124. *In* J. D. Karam, J. W. Drake, K. N. Kreuzer, G. Mosig, D. H. Hall, F. A. Eiserling, L. W. Black, E. K. Spicer, E. Kutter, K. Carlson, and E. S. Miller (ed.), *Molecular Biology of Phage T₄*. ASM Press, Washington, D.C.

17. Ehrlich, S. D., H. Bierne, E. d'Alencon, D. Vilette, M. Petranovic, P. Noirot, and B. Michel. 1993. Mechanisms of illegitimate recombination. *Gene* **135:**161–166.

18. Eisen, J. A., and P. C. Hanawalt. 1999. A phylogenomic study of DNA repair genes, proteins, and processes. *Mutat. Res.* **435:**171–213.

19. Eiserling, F., A. Pushkin, M. Gingery, and G. Bertani. 1999. Bacteriophage-like particles associated with the gene transfer agent of *Methanococcus voltae* PS. *J. Gen. Virol.* **80(Pt 12):**3305–3308.

20. Eker, A. P. M., L. Formenoy, and L. E. A. De Wit. 1991. Photoreactivation in the extreme halophilic archaebacterium *Halobacterium cutirubrum*. *Photochem. Photobiol.* **53:**643–645.

21. Escobar-Paramo, P., S. Ghosh, and J. Diruggiero. 2005. Evidence for genetic drift in the diversification of a geographically isolated population of the hyperthermophilic archaeon *Pyrococcus*. *Mol. Biol. Evol.* **22:**2297–2303.

22. Finlay, B. J. 2002. Global dispersal of free-living microbial eukaryote species. *Science* **296:**1061–1063.

23. Fitz-Gibbon, S., A. J. Choi, J. H. Miller, K. O. Stetter, M. I. Simon, R. Swanson, and U. J. Kim. 1997. A fosmid-based genomic map and identification of 474 genes of the hyperthermophilic archaeon *Pyrobaculum aerophilum*. *Extremophiles* **1:**36–51.

24. Fitz-Gibbon, S. T., H. Ladner, U. J. Kim, K. O. Stetter, M. I. Simon, and J. H. Miller. 2002. Genome sequence of the hyperthermophilic crenarchaeon *Pyrobaculum aerophilum*. *Proc. Natl. Acad. Sci. USA* **99:**984–989.

25. Fogg, M. J., L. H. Pearl, and B. A. Connolly. 2002. Structural basis for uracil recognition by archaeal family B DNA polymerases. *Nat. Struct. Biol.* **9:**922–927.

26. Friedberg, E. C., G. C. Walker, and W. Siede. 1995. *DNA Repair and Mutagenesis*. ASM Press, Washington, D.C.

27. Gernhardt, P., O. Possot, M. Foglino, L. Sibold, and A. Klein. 1990. Construction of an integration vector for use in the archaebacterium *Methanococcus voltae* and expression of a eubacterial resistance gene. *Mol. Gen. Genet.* **221:**273–279.

28. Ghane, F., and D. W. Grogan. 1998. Chromosomal marker exchange in the thermophilic archaeon *Sulfolobus acidocaldarius*: physiological and cellular aspects. *Microbiology* **144:**1649–1657.

29. Glickman, B. W., and L. S. Ripley. 1984. Structural intermediates of deletion mutagenesis: a role for palindromic DNA. *Proc. Natl. Acad. Sci. USA* **81:**512–516.

30. Goddard, M. R., H. C. Godfray, and A. Burt. 2005. Sex increases the efficacy of natural selection in experimental yeast populations. *Nature* **434:**636–640.

31. Greagg, M. A., M. J. Fogg, G. Panayotou, S. J. Evans, B. A. Connolly, and L. H. Pearl. 1999. A read-ahead function in archaeal DNA polymerases detects promutagenic template-strand uracil. *Proc. Natl. Acad. Sci. USA* **96:**9045–9050.

32. Greene, C. N., and S. Jinks-Robertson. 1997. Frameshift intermediates in homopolymer runs are removed efficiently by yeast mismatch repair proteins. *Mol. Cell. Biol.* **17:**2844–2850.

33. Greve, B., S. Jensen, K. Bruegger, W. Zillig, and R. Garrett. 2004. Genomic comparison of archaeal conjugative plasmids from *Sulfolobus. Archaea* **1:**231–239.

34. Grogan, D. W. 1996. Exchange of genetic markers at extremely high temperatures in the archaeon *Sulfolobus acidocaldarius. J. Bacteriol.* **178:**3207–3211.

35. Grogan, D. W. 2004. Stability and repair of DNA in hyperthermophilic Archaea. *Curr. Issues Mol. Biol.* **6:**137–144.

36. Grogan, D. W., G. T. Carver, and J. W. Drake. 2001. Genetic fidelity under harsh conditions: analysis of spontaneous mutation in the thermoacidophilic archaeon *Sulfolobus acidocaldarius. Proc. Natl. Acad. Sci. USA* **98:**7928–7933.

37. Grogan, D. W., and J. E. Hansen. 2003. Molecular characteristics of spontaneous deletions in the hyperthermophilic archaeon *Sulfolobus acidocaldarius. J. Bacteriol.* **185:**1266–1272.

38. Halliday, J. A., and B. W. Glickman. 1991. Mechanisms of spontaneous mutation in DNA repair-proficient *Escherichia coli. Mutat. Res.* **250:**55–71.

39. Hansen, J. E., A. C. Dill, and D. W. Grogan. 2005. Conjugational genetic exchange in the hyperthermophilic archaeon *Sulfolobus acidocaldarius*: intragenic recombination with minimal dependence on marker separation. *J. Bacteriol.* **187:**805–809.

40. Hogrefe, H. H., C. J. Hansen, B. R. Scott, and K. B. Nielson. 2002. Archaeal dUTPase enhances PCR amplifications with archaeal DNA polymerases by preventing dUTP incorporation. *Proc. Natl. Acad. Sci. USA* **99:**596–601.

41. Ikeda, H., K. Shiraishi, and Y. Ogata. 2004. Illegitimate recombination mediated by double-strand break and end-joining in *Escherichia coli. Adv. Biophys.* **38:**3–20.

42. Jacobs, K. L., and D. W. Grogan. 1997. Rates of spontaneous mutation in an archaeon from geothermal environments. *J. Bacteriol.* **179:**3298–3303.

43. Jonuscheit, M., E. Martusewitsch, K. M. Stedman, and C. Schleper. 2003. A reporter gene system for the hyperthermophilic archaeon *Sulfolobus solfataricus* based on a selectable and integrative shuttle vector. *Mol. Microbiol.* **48:**1241–1252.

44. Kawarabayasi, Y., Y. Hino, H. Horikawa, K. Jin-no, M. Takahashi, M. Sekine, S. Baba, A. Ankai, H. Kosugi, A. Hosoyama, S. Fukui, Y. Nagai, K. Nishijima, R. Otsuka, H. Nakazawa, M. Takamiya, Y. Kato, T. Yoshizawa, T. Tanaka, Y. Kudoh, J. Yamazaki, N. Kushida, A. Oguchi, K. Aoki, S. Masuda, M. Yanagii, M. Nishimura, A. Yamagishi, T. Oshima, and H. Kikuchi. 2001. Complete genome sequence of an aerobic thermoacidophilic crenarchaeon, Sulfolobus tokodaii strain7. *DNA Res.* **8:**123–140.

45. Kiener, A., R. Gall, T. Rechsteiner, and T. Leisinger. 1985. Photoreactivation in *Methanobacterium thermoautotrophicum. Arch. Microbiol.* **143:**147–150.

46. Kiener, A., I. Husain, A. Sancar, and C. Walsh. 1989. Purification and properties of *Methanobacterium thermoautotrophicum* DNA photolyase. *J. Biol. Chem.* **264:**13880–13887.

47. Koch, A. L. 1982. Mutation and growth rates from Luria-Delbruck fluctuation tests. *Mutat. Res.* **95:**129–143.

48. Kokoska, R. J., K. Bebenek, F. Boudsocq, R. Woodgate, and T. A. Kunkel. 2002. Low fidelity DNA synthesis by a Y family DNA polymerase due to misalignment in the active site. *J. Biol. Chem.* **277:**19633–19638.

49. Komori, K., T. Miyata, J. DiRuggiero, R. Holley-Shanks, I. Hayashi, I. K. Cann, K. Mayanagi, H. Shinagawa, and Y. Ishino. 2000. Both RadA and RadB are involved in homologous recombination in *Pyrococcus furiosus. J. Biol. Chem.* **275:**33782–33790.

50. Koulis, A., D. A. Cowan, L. H. Pearl, and R. Savva. 1996. Uracil-DNA glycosylase activities in hyperthermophilic micro-organisms. *FEMS Microbiol. Lett.* **143:**267–271.

51. Kreuzer, K. N. 2005. Interplay between DNA replication and recombination in prokaryotes. *Annu. Rev. Microbiol.* **59:**43–67.

52. Kunz, B. A., X. L. Kang, and L. Kohalmi. 1991. The yeast rad18 mutator specifically increases G.C—T.A transversions without reducing correction of G-A or C-T mismatches to G.C pairs. *Mol. Cell. Biol.* **11:**218–225.

53. Kurosawa, N., and D.W. Grogan. 2005. Homologous recombination of exogenous DNA with the *Sulfolobus acidocaldarius* genome: properties and uses. *FEMS Lett.* **253:**141–149.

54. Lang, A. S., and J. T. Beatty. 2001. The gene transfer agent of *Rhodobacter capsulatus* and "constitutive transduction" in prokaryotes. *Arch. Microbiol.* **175:**241–249.

55. Lasken, R. S., D. M. Schuster, and A. Rashtchian. 1996. Archaebacterial DNA polymerases tightly bind uracil-containing DNA. *J. Biol. Chem.* **271:**17692–17696.

56. Leclere, M. M., M. Nishioka, T. Yuasa, S. Fujiwara, M. Takagi, and T. Imanaka. 1998. The O^6-methylguanine-DNA methyltransferase from the hyperthermophilic archaeon *Pyrococcus* sp. KOD1: a thermostable repair enzyme. *Mol. Gen. Genet.* **258:**69–77.

57. Lee, G. S., E. A. Savage, R. G. Ritzel, and R. C. von Borstel. 1988. The base-alteration spectrum of spontaneous and ultraviolet radiation-induced forward mutations in the URA3 locus of *Saccharomyces cerevisiae. Mol. Gen. Genet.* **214:**396–404.

58. Levin, B. R., and C. T. Bergstrom. 2000. Bacteria are different: observations, interpretations, speculations, and opinions about the mechanisms of adaptive evolution in prokaryotes. *Proc. Natl. Acad. Sci. USA* **97:**6981–6985.

59. Ling, H., F. Boudsocq, R. Woodgate, and W. Yang. 2001. Crystal structure of a Y-family DNA polymerase in action: a mechanism for error-prone and lesion-bypass replication. *Cell* **107:**91–102.

60. Liu, J., B. He, H. Qing, and Y. W. Kow. 2000. A deoxyinosine specific endonuclease from hyperthermophile, *Archaeoglobus fulgidus*: a homolog of Escherichia coli endonuclease V. *Mutat. Res.* **461:**169–177.

61. Lundblad, V., A. F. Taylor, G. R. Smith, and N. Kleckner. 1984. Unusual alleles of recB and recC stimulate excision of inverted repeat transposons Tn10 and Tn5. *Proc. Natl. Acad. Sci. USA* **81:**824–828.

62. Mahillon, J., and M. Chandler. 1998. Insertion sequences. *Microbiol. Mol. Biol. Rev.* **62:**725–774.

63. Reference deleted.

64. Mao, E. F., L. Lane, J. Lee, and J. H. Miller. 1997. Proliferation of mutators in A cell population. *J. Bacteriol.* **179:**417–422.

65. Martusewitsch, E., C. W. Sensen, and C. Schleper. 2000. High spontaneous mutation rate in the hyperthermophilic archaeon *Sulfolobus solfataricus* is mediated by transposable elements. *J. Bacteriol.* **182:**2574–2581.

66. Matsunaga, F., C. Norais, P. Forterre, and H. Myllykallio. 2003. Identification of short 'eukaryotic' Okazaki fragments synthesized from a prokaryotic replication origin. *EMBO Rep.* **4:**154–158.

67. McCready, S. 1996. The repair of ultraviolet light-induced DNA damage in the halophilic archaebacteria, *Halobacterium*

cutirubrum, Halobacterium halobium and *Haloferax vol-canii. Mutat. Res.* 364:25–32.

68. McCready, S., and L. Marcello. 2003. Repair of UV damage in *Halobacterium salinarum. Biochem. Soc. Trans.* 31:694–698.

69. McIlwraith, M. J., D. R. Hall, A. Z. Stasiak, A. Stasiak, D. B. Wigley, and S. C. West. 2001. RadA protein from Archaeoglobus fulgidus forms rings, nucleoprotein filaments and catalyses homologous recombination. *Nucleic Acids Res.* 29:4509–4517.

70. Meile, L., P. Abendschein, and T. Leisinger. 1990. Transduction in the archaebacterium *Methanobacterium thermoautotrophicum* Marburg. *J. Bacteriol.* 172:3507–3508.

71. Mevarech, M., and R. Werczberger. 1985. Genetic transfer in *Halobacterium volcanii. J. Bacteriol.* 162:461–462.

72. Muskhelishvili, G., P. Palm, and W. Zillig. 1993. SSV1-encoded site-specific recombination system in *Sulfolobus shibatae. Mol. Gen. Genet.* 237:334–342.

73. Napoli, A., A. Valenti, V. Salerno, M. Nadal, F. Garnier, M. Rossi, and M. Ciaramella. 2004. Reverse gyrase recruitment to DNA after UV light irradiation in *Sulfolobus solfataricus. J. Biol. Chem.* 279:33192–33198.

74. Ogrunc, M., D. F. Becker, S. W. Ragsdale, and A. Sancar. 1998. Nucleotide excision repair in the third kingdom. *J. Bacteriol.* 180:5796–5798.

75. Oosumi, T., B. Garlick, and W. R. Belknap. 1996. Identification of putative nonautonomous transposable elements associated with several transposon families in *Caenorhabditis elegans. J. Mol. Evol.* 43:11–18.

76. Papke, R. T., J. E. Koenig, F. Rodriguez-Valera, and W. F. Doolittle. 2004. Frequent recombination in a saltern population of *Halorubrum. Science* 306:1928–1929.

77. Paz, A., D. Mester, I. Baca, E. Nevo, and A. Korol. 2004. Adaptive role of increased frequency of polypurine tracts in mRNA sequences of thermophilic prokaryotes. *Proc. Natl. Acad. Sci. USA* 101:2951–2956.

78. Peng, X., K. Brugger, B. Shen, L. Chen, Q. She, and R. A. Garrett. 2003. Genus-specific protein binding to the large clusters of DNA repeats (short regularly spaced repeats) present in *Sulfolobus* genomes. *J. Bacteriol.* 185:2410–2417.

79. Petursdottir, S. K., G. O. Hreggvidsson, M. S. Da Costa, and J. K. Kristjansson. 2000. Genetic diversity analysis of *Rhodothermus* reflects geographical origin of the isolates. *Extremophiles* 4:267–274.

80. Pfeifer, F., and M. Betlach. 1985. Genome organization in *Halobacterium halobium*: a 70 kb island of more (AT) rich DNA in the chromosome. *Mol. Gen. Genet.* 198:449–455.

81. Pfeifer, F., M. Betlach, R. Martienssen, J. Friedman, and H. W. Boyer. 1983. Transposable elements of *Halobacterium halobium. Mol. Gen. Genet.* 191:182–188.

82. Pfeifer, F., and U. Blaseio. 1989. Insertion elements and deletion formation in a halophilic archaebacterium. *J. Bacteriol.* 171:5135–5140.

83. Prakash, S., and L. Prakash. 2000. Nucleotide excision repair in yeast. *Mutat. Res.* 451:13–24.

84. Redder, P., Q. She, and R. A. Garrett. 2001. Non-autonomous mobile elements in the crenarchaeon *Sulfolobus solfataricus. J. Mol. Biol.* 306:1–6.

85. Reddy, M., and J. Gowrishankar. 1997. Identification and characterization of ssb and uup mutants with increased frequency of precise excision of transposon Tn10 derivatives: nucleotide sequence of *uup* in *Escherichia coli. J. Bacteriol.* 179:2892–2899.

86. Reilly, M. S., and D. W. Grogan. 2002. Biological effects of DNA damage in the hyperthermophilic archaeon *Sulfolobus acidocaldarius. FEMS Microbiol. Lett.* 208:29–34.

87. Reiter, W. D., P. Palm, and S. Yeats. 1989. Transfer RNA genes frequently serve as integration sites for prokaryotic genetic elements. *Nucleic Acids Res.* 17:1907–1914.

88. Ridley, M. 2004. Natural selection and genetic drift in molecular evolution, p. 156–193. *In Evolution*, 3rd ed. Blackwell Science Ltd., Malden, Mass.

89. Roberts, J., and M. F. White. 2005. An archaeal endonuclease displays key properties of both eukaryal XPF-ERCC1 and Mus81. *J. Biol. Chem.* 280:5924–5928.

90. Roberts, M. S., and F. M. Cohan. 1995. Recombination and migration rates in natural populations of *Bacillus subtilis* and *Bacillus mojavensis. Evolution* 49:1081–1094.

91. Rose, M., and F. Winston. 1984. Identification of a Ty insertion within the coding sequence of the *S. cerevisiae URA3* gene. *Mol. Gen. Genet.* 193:557–560.

92. Sandler, S. J., P. Hugenholtz, C. Schleper, E. F. DeLong, N. R. Pace, and A. J. Clark. 1999. Diversity of *radA* genes from cultured and uncultured archaea: comparative analysis of putative RadA proteins and their use as a phylogenetic marker. *J. Bacteriol.* 181:907–915.

93. Sapienza, C., and W. F. Doolittle. 1982. Unusual physical organization of the *Halobacterium* genome. *Nature* 295:384–389.

94. Sapienza, C., M. R. Rose, and W. F. Doolittle. 1982. High-frequency genomic rearrangements involving archaebacterial repeat sequence elements. *Nature* 299:182–185.

95. Sartori, A. A., and J. Jiricny. 2003. Enzymology of base excision repair in the hyperthermophilic archaeon *Pyrobaculum aerophilum. J. Biol. Chem.* 278:24563–24576.

96. Sato, T., T. Fukui, H. Atomi, and T. Imanaka. 2003. Targeted gene disruption by homologous recombination in the hyperthermophilic archaeon *Thermococcus kodakaraensis* KOD1. *J. Bacteriol.* 185:210–220.

97. Scheerer, J. B., and G. M. Adair. 1994. Homology dependence of targeted recombination at the Chinese hamster APRT locus. *Mol. Cell. Biol.* 14:6663–6673.

98. Scheller, J., A. Schurer, C. Rudolph, S. Hettwer, and W. Kramer. 2000. MPH1, a yeast gene encoding a DEAH protein, plays a role in protection of the genome from spontaneous and chemically induced damage. *Genetics* 155:1069–1081.

99. Schleper, C., E. F. DeLong, C. M. Preston, R. A. Feldman, K. Y. Wu, and R. V. Swanson. 1998. Genomic analysis reveals chromosomal variation in natural populations of the uncultured psychrophilic archaeon *Cenarchaeum symbiosum. J. Bacteriol.* 180:5003–5009.

100. Schleper, C., I. Holz, D. Janekovic, J. Murphy, and W. Zillig. 1995. A multicopy plasmid of the extremely thermophilic archaeon *Sulfolobus* effects its transfer to recipients by mating. *J. Bacteriol.* 177:4417–4426.

101. Schmidt, K. J., K. E. Beck, and D. W. Grogan. 1999. UV stimulation of chromosomal marker exchange in *Sulfolobus acidocaldarius*: implications for DNA repair, conjugation and homologous recombination at extremely high temperatures. *Genetics* 152:1407–1415.

102. Seitz, E. M., and S. C. Kowalczykowski. 2000. The DNA binding and pairing preferences of the archaeal RadA protein demonstrate a universal characteristic of DNA strand exchange proteins. *Mol. Microbiol.* 37:555–560.

103. Serre, M. C., C. Letzelter, J. R. Garel, and M. Duguet. 2002. Cleavage properties of an archaeal site-specific recombinase, the SSV1 integrase. *J. Biol. Chem.* 277:16758–16767.

104. She, Q., B. Shen, and L. Chen. 2004. Archaeal integrases and mechanisms of gene capture. *Biochem. Soc. Trans.* 32:222–226.

105. She, Q., R. K. Singh, F. Confalonieri, Y. Zivanovic, G. Allard, M. J. Awayez, C. C. Chan-Weiher, I. G. Clausen, B. A. Curtis, A. De Moors, G. Erauso, C. Fletcher, P. M. Gordon,

I. Heikamp-de Jong, A. C. Jeffries, C. J. Kozera, N. Medina, X. Peng, H. P. Thi-Ngoc, P. Redder, M. E. Schenk, C. Theriault, N. Tolstrup, R. L. Charlebois, W. F. Doolittle, M. Duguet, T. Gaasterland, R. A. Garrett, M. A. Ragan, C. W. Sensen, and J. Van der Oost. 2001. The complete genome of the crenarchaeon Sulfolobus solfataricus P2. *Proc. Natl. Acad. Sci. USA* **98**:7835–7840.

106. Shen, P., and H. V. Huang. 1986. Homologous recombination in *Escherichia coli*: dependence on substrate length and homology. *Genetics* **112**:441–457.

107. Shuttleworth, G., M. J. Fogg, M. R. Kurpiewski, L. Jen-Jacobson, and B. A. Connolly. 2004. Recognition of the pro-mutagenic base uracil by family B DNA polymerases from archaea. *J. Mol. Biol.* **337**:621–634.

108. Silvian, L. F., E. A. Toth, P. Pham, M. F. Goodman, and T. Ellenberger. 2001. Crystal structure of a DinB family error-prone DNA polymerase from *Sulfolobus solfataricus*. *Nat. Struct. Biol.* **8**:984–989.

109. Simon, H. M., C. E. Jahn, L. T. Bergerud, M. K. Sliwinski, P. J. Weimer, D. K. Willis, and R. M. Goodman. 2005. Cultivation of mesophilic soil crenarchaeotes in enrichment cultures from plant roots. *Appl. Environ. Microbiol.* **71**:4751–4760.

110. Singer, B. S., L. Gold, P. Gauss, and D. H. Doherty. 1982. Determination of the amount of homology required for recombination in bacteriophage T₄. *Cell* **31**:25–33.

111. Skorvaga, M., N. D. Raven, and G. P. Margison. 1998. Thermostable archaeal O^6-alkylguanine-DNA alkyltransferases. *Proc. Natl. Acad. Sci. USA* **95**:6711–6715.

112. Smith, G. R. 2001. Homologous recombination near and far from DNA breaks: alternative roles and contrasting views. *Annu. Rev. Genet.* **35**:243–274.

113. Smith, J. M., E. J. Feil, and N. H. Smith. 2000. Population structure and evolutionary dynamics of pathogenic bacteria. *Bioessays* **22**:1115–1122.

114. Sugawara, N., G. Ira, and J. E. Haber. 2000. DNA length dependence of the single-strand annealing pathway and the role of *Saccharomyces cerevisiae* RAD59 in double-strand break repair. *Mol. Cell. Biol.* **20**:5300–5309.

115. Surtees, J. A., J. L. Argueso, and E. Alani. 2004. Mismatch repair proteins: key regulators of genetic recombination. *Cytogenet. Genome Res.* **107**:146–159.

116. Tchelet, R., and M. Mevarech. 1994. Interspecies genetic transfer in halophilic archaebacteria. *Syst. Appl. Microbiol.* **16**:578–581.

117. Tippin, B., P. Pham, and M. F. Goodman. 2004. Error-prone replication for better or worse. *Trends Microbiol.* **12**:288–295.

118. Tyson, G. W., J. Chapman, P. Hugenholtz, E. E. Allen, R. J. Ram, P. M. Richardson, V. V. Solovyev, E. M. Rubin, D. S. Rokhsar, and J. F. Banfield. 2004. Community structure and metabolism through reconstruction of microbial genomes from the environment. *Nature* **428**:37–43.

119. Whitaker, R. J., D. W. Grogan, and J. W. Taylor. 2003. Geographic barriers isolate endemic populations of hyperthermophilic archaea. *Science* **301**:976–978.

120. Whitaker, R. J., D. W. Grogan, and J. W. Taylor. 2005. Recombination shapes the natural population structure of the hyperthermophilic archaeon *Sulfolobus* '*islandicus.*' *Mol. Biol. Evol.* **22**:2354–2361.

121. Wilkins, B. M. 1995. Gene transfer by bacterial conjugation: diversity of systems and functional specializations, p. 59–88. *In* S. Baumberg, J. P. W. Young, E. M. H. Wellington, and J. R. Saunders (ed.), *Population Genetics of Bacteria*, vol. 1. Cambridge University Press, Cambridge, United Kingdom.

122. Woese, C. R., O. Kandler, and M. L. Wheelis. 1990. Towards a natural system of organisms: proposal for the domains Archaea, Bacteria, and Eucarya. *Proc. Natl. Acad. Sci. USA* **87**:4576–4579.

123. Wood, E. R., F. Ghane, and D. W. Grogan. 1997. Genetic responses of the thermophilic archaeon *Sulfolobus acidocaldarius* to short-wavelength UV light. *J. Bacteriol.* **179**:5693–5698.

124. Woods, W. G., and M. L. Dyall-Smith. 1997. Construction and analysis of a recombination-deficient (radA) mutant of *Haloferax volcanii*. *Mol. Microbiol.* **23**:791–797.

125. Worrell, V. E., D. P. Nagle, Jr., D. McCarthy, and A. Eisenbraun. 1988. Genetic transformation system in the archaebacterium *Methanobacterium thermoautotrophicum* Marburg. *J. Bacteriol.* **170**:653–656.

126. Worthington, P., V. Hoang, F. Perez-Pomares, and P. Blum. 2003. Targeted disruption of the alpha-amylase gene in the hyperthermophilic archaeon *Sulfolobus solfataricus*. *J. Bacteriol.* **185**:482–488.

127. Yonemasu, R., S. J. McCready, J. M. Murray, F. Osman, M. Takao, K. Yamamoto, A. R. Lehmann, and A. Yasui. 1997. Characterization of the alternative excision repair pathway of UV-damaged DNA in *Schizosaccharomyces pombe*. *Nucleic Acids Res.* **25**:1553–1558.

Archaea: Molecular and Cellular Biology
Edited by Ricardo Cavicchioli
© 2007 ASM Press, Washington, D.C.

Chapter 6

Transcription: Mechanism and Regulation

MICHAEL THOMM

INTRODUCTION

The biochemical machinery involved in the processes of DNA replication (see Chapter 3), transcription, and translation (see Chapter 8) shows a striking similarity and phylogenetic relationship to the equivalent machinery in eucarya (12, 38, 72, 99, 102). In particular, RNA polymerase (RNAP) and the basal transcriptional machinery of archaea share many properties with the eucaryal RNA polymerase II (RNAP II) transcription apparatus (12, 83, 99).

The first step in the initiation process is the recognition of an AT-rich promoter element (TATA box) by the archaeal TATA-binding protein (TBP) (Fig. 1; references 87, 98). The archaeal TATA box is located ~25 to 30 bp upstream of the transcription start site, (37, 45, 90). TBP is highly conserved in sequence and function (12, 94), and human and yeast TBPs can functionally replace archaeal TBP in a mesophilic archaeal cell-free transcription system (107). The TBP-promoter complex is bound by TFB, the second archaeal factor, which is closely related in structure and function to eucaryal TFIIB (Fig. 1; references 12, 47). Transcription polarity is governed by the interaction of promoter-bound TFB with a conserved motif immediately upstream of the TATA box, the B recognition element (BRE) (14, 17). This ternary complex recruits the archaeal RNAP, which is bound to the DNA region downstream of the TATA box. The DNA-binding site of the RNAP extends 3′ from the TATA box to position +18 (Fig. 1; reference 96). All complete archaeal genome sequences contain a homolog of the α-subunit of the eucaryal general transcription factor TFIIE. Archaeal TFE is not absolutely required for cell-free transcription but stimulates transcription of some promoters under certain conditions (9, 41). Homologs of the TATA-box-associated factors (TAFs) and eucaryal basal transcription factors TFIIA, TFIIH, and TFIIF have not been detected in archaeal genomes, although the existence of a TBP interaction protein TIP26 (71) raises the possibility that the machinery directing transcription in archaea is more complex. In general, the archaeal transcriptional machinery can be considered a simplified version, and the evolutionary precursor of the more complex eucaryal machinery. Thus, the archaeal system is a useful model for investigating the mechanism and structure of the eucaryal transcriptional machinery.

The archaeal RNAP has a structure resembling eucaryal RNA polymerases. The subunit complexity and sequence of individual subunits is highly conserved between eucarya and archaea. The small subunits E, H, K, L, and N have homologs in eucaryal enzymes but not in the *Escherichia coli* RNAP (13, 60, 105). The RNAP from *Methanocaldococcus jannaschii* has been assembled in vitro from individual subunits. The complex formed by the subunits D, L, N, and P serves as a scaffold for the association of the subunits A′, A″ and B′, B″ forming the active site of the enzyme (105).

In contrast to the basal machinery, most putative transcriptional regulators identified in archaeal genomes (1, 36, 56) are homologs of bacterial proteins carrying helix-turn-helix motifs (11, 20). Regulators of archaeal transcription repress initiation by preventing TFB/TBP access to the TATA-box region (11, 74) or RNAP recruitment to the transcription start site (103). Two regulators, TrmB and TrpY (see "The sugar transport regulator, TrmB," and "Regulation of the tryptophan operon by TrpY," below), block access of TBP/TFB or RNA polymerase in a promoter-dependent manner (61, 62, 110). The most widely represented archaeal regulators in archaeal genomes are members of the Lrp/AsnC family (20). Similar to the global regulator Lrp (leucine-responsive-regulatory

Michael Thomm • Universität Regensburg, Lehrstuhl für Mikrobiologie, Universitätsstrasse 31, D-93053 Regensburg, Germany.

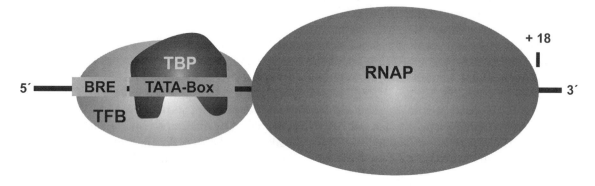

Figure 1. Initiation of transcription in archaea. The first step of promoter recognition is binding of TBP to the archaeal TATA box. This complex is stabilized by the association of TFB. Bound TFB interacts with the purine-rich BRE sequence 5' of the TATA box. This complex recruits the RNA polymerase that binds to the DNA region downstream of the TATA box and covers the transcription start site and the DNA downstream region to position +18.

protein) of *E. coli*, they appear to function as repressors (19, 32, 74) and activators of transcription (77, 79). Several archaeal regulators with no homologs to bacterial or eucaryal regulators, including GvpE (54), Phr (103), and TrmB (61), have been identified and characterized in some detail.

Due to the lack of tractable genetic systems for most archaeal genera (see Chapter 21), the physiological function of many transcriptional regulators remains obscure. The development of whole-genome microarrays (92, 93) coupled with cell-free transcription experiments using fragmented chromosomal DNA as template may provide a useful tool for the elucidation of the set of genes regulated by global archaeal regulators such as Phr, TrmB, and archaeal homologs of the LrpA/AsnC family.

THE TRANSCRIPTIONAL MACHINERY

Transcription Signals

A TATA box at position −25 was identified as the first structural determinant of archaeal promoters both in the *Euryarchaeota* (98) and the *Crenarchaeota* (87). Mutational analyses of the significance of this sequence in cell-free transcription systems for *Methanococcus* (33) and *Sulfolobus* (49) confirmed the significance of this sequence as a promoter signal (40, 42, 86). In *Haloferax volcanii* the function of the TATA box as a major promoter signal was demonstrated in vivo (82). Further work revealed some variation of the TATA-box sequence among various genera of archaea (29, 94, 101). However, the involvement of the TATA box as a major determinant of archaeal promoters, and the general mechanism of initiation (see "Transcription factors" and "The mechanism of transcription," below), seems to be conserved in the *Archaea*. A second conserved promoter signal immediately upstream of the archaeal TATA box, the BRE, is the principal determinant for the orientation of transcription (14). This purine-rich sequence (consensus RNWAAW; R = purine; W = A or T; N = any base) interacts with promoter-bound TFB, and this interaction defines transcriptional polarity.

Much less is known about the signals directing termination of transcription. Oligo-dT sequences downstream of archaeal genes were proposed as candidate sequences for archaeal terminators (21, 87, 108). One study using mutagenesis of the sequence 5′-TTTTAATT-3′ provided evidence for its significance as a terminator signal in the *Methanococcus* tRNA^{Val} gene (97). However, in contrast to the eucaryal polII system, where four consecutive T residues act as a terminator signal (35), five T residues were not sufficient for efficient termination of the tRNA^{Val} gene, and sequences encoding the tRNA were clearly required for efficient termination in vitro. In contrast to the *Methanococcus* cell-free system, the *Sulfolobus* and *Pyrococcus* cell-free systems are unable to terminate transcription accurately, suggesting that additional components are required for efficient termination. The process of termination of transcription is not well understood. Further work is required to determine the specific sequences involved in termination and the factors catalyzing the termination process.

Transcription Factors

The inability of purified archaeal RNAP to initiate transcription accurately in vitro provided the first evidence for the existence of archaeal transcription factors. The first reconstituted transcription systems were strictly dependent on the addition of protein fractions devoid of RNAP activity (33, 49). A dimeric transcription factor from *Methanococcus* was purified to homogeneity (44). This factor could be replaced in cell-free transcription experiments by eucaryal TBPs (107) and was shown to bind to the archaeal TATA box (37). Sequence analysis of archaeal genomes revealed the presence of homologs of eucaryal TBP and TFB (69, 77, 90), and heterologous expression of these proteins and analysis of their function in *Pyrococcus* and *Sulfolobus* transcription systems revealed that these genes encoded homologs of the eucaryal transcription factors TBP and TFIIB (47, 84). Thus, a minimal archaeal transcription system is defined by the factors TBP, TFB, and RNAP (Fig. 1).

Archaeal TBPs consist of two directly repeated protein domains. These two copies of the repeat show

a high degree (~40%) of amino acid identity. Archaeal TBPs lack the N-terminal domain characteristic of eucaryal TBPs and possess a highly acidic C-terminal tail (reviewed in reference 12). Analysis of the crystal structure of *Pyrococcus* TBP also revealed striking similarity to eucaryal counterparts. *Pyrococcus* TBP is a saddle-shaped molecule. The DNA-binding part of the molecule is located on the underside of the saddle (30).

The structural organization of TFB is similar to that of eucaryal TFIIB. TFB consists of three domains, an N-terminal domain harboring a zinc-ribbon and a B finger (22, 81), and a C-terminal region comprising two direct repeats of about 90 amino acids, including a helix-turn-helix motif close to the C terminus. The structure of the N-terminal domain has been solved (112). The metal-binding motif is highly conserved throughout the *Archaea* and *Eucarya* (95). Deletion of the N-terminal domain did not prevent formation of the TBP-TFB promoter complex because N-terminal truncated versions of TFB were used for structural analyses of promoter-bound TFB/TBP, both in the eucaryal and the archaeal system (53, 75). Mutation of a conserved amino acid close to the zinc ribbon inactivated TFB-dependent transcription at some promoters. This TFB mutant was still able to form a TBP-TFB-promoter complex but had lost the ability to recruit the RNAP (14). Two-hybrid analyses suggest an interaction of the N-terminal domain with subunit K of RNAP (68). The finding that reconstituted RNAP from *M. jannaschii* lacking subunit K is still able to transcribe promoters in vitro (105) indicates that subunit K is not absolutely required for RNAP recruitment. More components of the RNA polymerase, and probably also other domains of TFB, seem to be involved in RNAP-TFB interaction. Structural analyses of the yeast TFIIB-RNAPII complex revealed that the N-terminal domain of TFIIB lies in the RNAPII channel harboring the template strand of the transcription bubble and the DNA-RNA hybrid that forms at early stages of transcription. Site-specific photochemical cross-linking has shown that TFB cross-links to DNA upstream and downstream of the TATA box (7, 8, 88). In addition, TFB cross-links to a DNA segment at the transcription start site that is part of the open complex. By using recruitment-independent nonspecific transcription assays, a stimulatory postrecruitment function of *M. jannaschii* TFB was shown. The B-finger domain of TFB was essential for the observed stimulation of abortive initiation (106).

All the available structural evidence is consistent with the archaeal TFB interacting with DNA and RNA polymerase in a manner similar to eucaryal TFIIB (34, 77). Taken together, the structural and cross-linking data suggest that TFB has the following functions (Fig. 2). The C-terminal core domain binds to TBP and recognizes the BRE. The N-terminal Zn ribbon interacts with the dock domain of RNAP. The contacts between promoter DNA and RNAP are weak in the closed complex. During open complex formation, the template strand is inserted into the active center of RNAP. The B finger stabilizes the template strand in the active site of the enzyme and facilitates a DNA/rNTP configuration that favors the catalytic activity of the RNAP (Fig. 2; 34, 106).

Eucaryal TFIIE is a heterotetramer ($\alpha_2\beta_2$). It recruits TFIIH and is involved in the RNAP phosphorylating activity of TFIIH. Archaeal genomes encode a homolog of the N-terminal half of the α-subunit of TFE. The crystal structure of an N-terminal fragment of *Sulfolobus* TFE has been solved (73). It contains a winged helix-turn-helix structure. A modest stimulatory effect of archaeal TFE on in vitro transcription of some weak promoters was shown (9, 41). The moderate level of stimulation of transcription by TFE was confirmed using a completely purified system, consisting of recombinant TBP and TFB and a RNAP reconstituted from bacterially produced subunits (80). This finding excludes the possibility that strong stimulatory effects of TFE were underestimated owing to the presence of minor quantities of TFE in RNAP preparations. Werner and Weinzierl (106) recently showed that mutants of the Zn-ribbon and B-finger domain of TFB, which are unable to recruit RNAP, can be complemented in vitro by TFE. Furthermore, experiments using heteroduplex templates located upstream and downstream of the transcription start site suggest a role of TFE in promoting DNA melting and/or template loading. Taken together these data suggest that both factors have an active role in influencing the catalytic properties of RNAP and act synergistically during initiation of transcription.

Archaea contain a cleavage induction factor, TFS. This protein is homologous to the C-terminal part of eucaryal elongation factor TFIIS and in addition to the small subunits B12, 2, A12, 6, and C11 of eucaryal RNAPs I to III (43). This protein does not purify with archaeal RNAP. Experiments with paused elongation complexes showed that TFIIS induces a cleavage activity in the archaeal RNAP. TFS confers a proofreading activity to RNAP by inducing the release of dinucleotides from the 3' end of nascent RNA (59). It would be valuable to examine the role that TFS plays in enabling RNAP to overcome barriers to RNA elongation (e.g., histones or other DNA-associated proteins), and its possible role in transcription termination.

SSB

The single-stranded DNA (ssDNA)-binding proteins of crenarchaeota have a domain organization similar to *E. coli* SSB. This includes a single domain with an oligonucleotide-binding fold for ssDNA binding which is separated by a flexible spacer from an acidic C-terminal tail. The archaeal SSB of *Sulfolobus* interacts with RNAP via the C-terminal acidic tail (89). Under TBP-limiting conditions SSB stimulates transcription, suggesting that it acts at the level of RNAP recruitment or initiation. SSB also appears to be able to replace TBP in transcription assays. A contamination of RNAP preparations with trace amounts of TBP is unlikely but cannot be excluded.

The repression of transcription observed in the presence of the major chromatin component of hyperthermophiles, the ALBA (Sso 10b) protein, is relieved by SSB (89). The specific interaction of the acidic tail of SSB with RNA polymerase is critical for this stimulation of transcription. Localized melting of DNA in the AT-rich TATA-box region mediated by

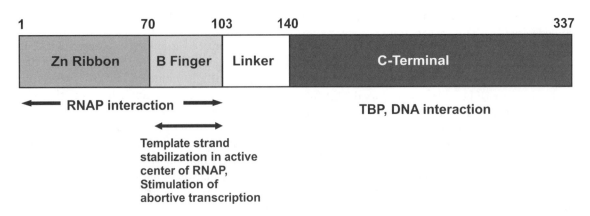

Figure 2. Domain structure of TFB. The major structural features of TFB and their interactions with other components of the transcriptional machinery.

SSB may overcome the requirement for TBP. Analyses of DNA templates harboring AT-rich sequences located within structural genes will reveal whether the ability of SSB to replace TBP is promoter specific, or whether SSB-directed initiation can also occur within AT-rich open reading frames.

RNA Polymerase

The similarity of the multisubunit archaeal RNAP to eucaryal RNAP was one of the first recognized eucaryal features of the *Archaea* (39, 113). Archaeal RNAPs consist of 11 subunits, B, A′, A′, D, E, F, L, H, N, K, and P (*Crenarchaeota, Pyrococcus;* Fig. 3), or 12 subunits. The largest subunit, B, is split into two subunits in methanogens (91), yielding 12 subunits to give the subunit composition, A′, B′, B′, A′, D, E, F, L, H, N, K, and P. The homology of RNAP subunits in the three domains of life is shown in Fig. 3. In general, the larger subunits are paralogs, although the sequence similarities are much more pronounced between *Archaea* and *Eucarya* than be-

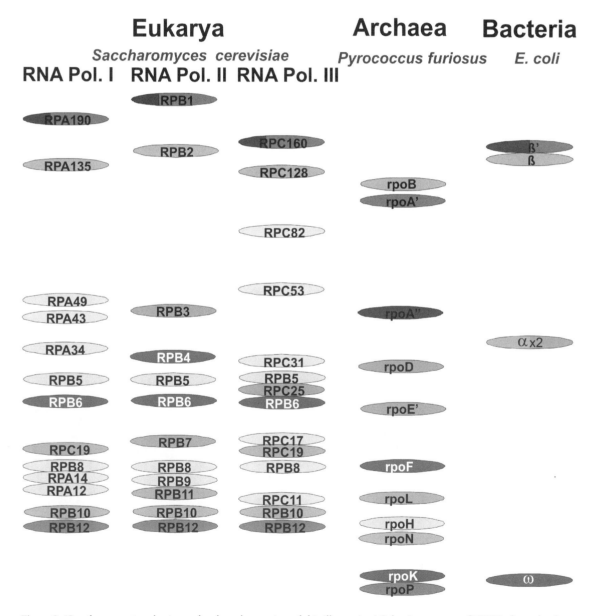

Figure 3. (*See the separate color insert for the color version of this illustration.*) Subunit structure of RNAPs from the three domains of life. The largest subunit in the *Eucarya* and β′ in the *Bacteria* is split into two subunits, A1 and A2, in the *Archaea*. In methanogens, subunit B is also split into two polypeptides, B′ and B′. Different parts of bacterial subunit α are encoded by the genes for the archaeal subunits D and L. Subunits E1, F, H, N, and P are only shared between the *Archaea* and *Eucarya*. The pattern shown is based on separation of subunits by polyacrylamide gel electrophoresis under denaturing conditions. The numbers in the subunits of the eucaryal RNAP A (I), B (II), and C (III) indicate the molecular mass.

tween *Bacteria* and *Archaea* or *Bacteria* and *Eucarya*. The archaeal paralog of the bacterial subunit β′ is split into two polypeptides in archaea: A′ and A″. A′ is related to the C-proximal half of Rpb1 and β′, and A″ is related to the N-terminal half of Rpb1 and β′. The subunits D and L are related to Rpb3 and a part of α, and L is related to Rpb11 and to a different segment of α (Fig. 3).

The RNAP was reconstituted from the methanogen *M. jannaschii* from 12 recombinant subunits (105). The α_2 dimer of bacterial RNAP serves as a platform for assembly of the larger subunits (50). Studies of Rpb3 and Rpb11 subunits suggest that the archaeal D-L heterodimeric complex is the structural and functional equivalent of the bacterial α-homodimer (31, 104). The D-L-N-P subcomplex was used as a platform for reconstitution of an active *M. jannaschii* RNAP. For reconstitution of *M. jannaschii* RNAP, the subunits D-L and E-F, which are not soluble when expressed separately, were coexpressed as soluble heterodimers. The *Pyrococcus* RNAP that was reconstituted from 11 separately expressed subunits exhibited wild-type levels of promoter-specific activity (Naji et al., unpublished). This finding indicates that preassembly of the D-L-N-P complex is not a prerequisite for the assembly of an archaeal RNAP.

The availability of the recombinant enzyme was used to investigate the function of essential domains of the enzyme and the contributions of small subunits to RNAP function, and to identify the minimal subunit configuration for specific RNA synthesis. The complex A′-A′-B′-B′-D-L was soluble but devoid of activity. Activity was restored by the addition of subunits N and P (105). Subunit K, the paralog of bacterial ω and eucaryal RpB6, was not required for activity but did enhance the activity of the minimal active assembly (A′-A′-B′-B′-D-L-N-P) twofold. Two hybrid analyses have shown that *Sulfolobus* K interacts specifically with the N-terminal domain of TFB (68). Hence, interaction of K with the zinc ribbon of TFB seems to participate in recruitment of RNAP by the TBP-TFB-promoter complex (Fig. 4). However, it is not essential for that step since omitting K from reconstitution reactions does not completely abolish promoter-directed transcription.

The zinc ribbon of eucaryal TFIIB does not bind to RpB6 but binds to a surface pocket of the dock, wall, and clamp domain of RNAP II, which is formed by the two large subunits Rpb1 and Rpb2 (24). This finding suggests an interesting difference in the molecular architecture of eucaryal and archaeal RNAP. Subunit H is the homolog of eucaryal Rpb5. Rpb5

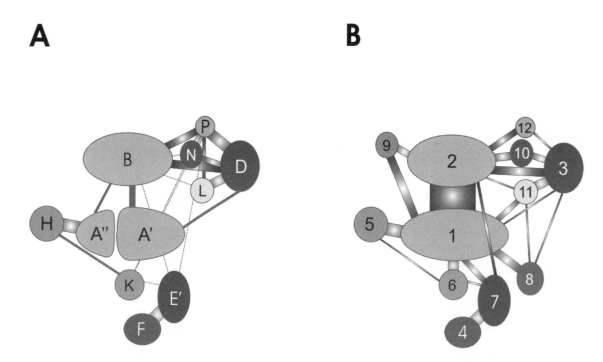

Figure 4. (*See the separate color insert for the color version of this illustration.*) Structural similarity of *Pyrococcus* RNAP (A) and yeast RNAPII (B). Comparison of interactions of an archaeal RNAP inferred from Far-Western analysis with interactions of yeast RNAPII observed in the crystal structure of the enzyme. The width of the lines connecting subunits is a measure of the intensity of the interaction. Modified from *Science* (27) with additional data from *Proceedings of the National Academy of Sciences USA* (2).

and Rpb1 form the lower jaw of RNAP II and are likely to be involved in contacting the template DNA. Subunit H stimulated the promoter-specific activity of the archaeal enzyme up to 10-fold, indicating that it contributes considerably to specific activity of the reconstituted enzyme. The eucaryal homologs of subunits E/F, Rpb4/7, are essential (34), but while E/F are easily incorporated into the recombinant archaeal RNAP, they are not essential for its activity (105). The large subunits of archaeal RNAP A′ and B′ contain the metal A and metal B motifs of RNAP II (27) involved in Mg^{2+} chelating near the catalytic center of RNAP. Site-specific mutagenesis of conserved amino acids in these two motifs completely abolished or substantially reduced specific activity of the archaeal enzyme, indicating a high degree of conservation of the basic mechanism of transcription between these subunits from the *Archaea* and *Eucarya*.

The interaction of recombinant subunits of the *Pyrococcus* RNAP have been extensively studied to investigate the molecular structure of archaeal RNAP (Naji et al., unpublished). Far Western blot analysis revealed strong interactions of DLN and P, the proposed platform for assembly and interactions with subunit B (Fig. 4A). The B-P-N-L-D complex is connected to the rest of the enzyme via B-A′, N-A′, and P-E′ interactions. Subunits E-F also show strong contacts and can be purified with the main complex by gel filtration (36a). In summary, the structural similarity of *Pyrococcus* RNAP to eucaryal RNAP II is striking (Fig. 4). However, determination of the crystal structure of the archaeal enzyme is required to confirm and refine the similarities and differences inferred from biochemical analyses.

THE MECHANISM OF TRANSCRIPTION

A comprehensive and consistent view of the mechanism of transcription by archaeal RNAP has been gained from the recently solved structures of the yeast RNAPII (2, 26) and the TFIIB-RNAPII complex (22), protein-protein cross-linking analyses in the eucaryal system (24, 25), protein-DNA cross-linking analyses in the archaeal system (7, 8, 88), analyses of open complex formation and paused transcription complexes (16, 46, 96), and analyses of reconstituted RNAP and archaeal TFB variants (105, 106) (Figs. 4 and 5). As a first step, the saddle-shaped archaeal TBP binds to the minor groove of an 8-bp TATA box. This leads to bending of DNA by approximately 65° (53). The process of TBP binding is stabilized and stimulated by TFE. TFB associates with the TBP-promoter complex whereby the C-terminal domain makes contacts with TBP and the BRE of the archaeal promoter

and defines the polarity of archaeal transcription (17). The RNAP is recruited by interactions of the TFB zinc ribbon with the dock domain of RNA polymerase and with subunit K. The contacts of RNAP with DNA in the closed complex are weak (Fig. 5A). In the open complex, the template strand comes into contact with the active site and is stabilized by the B finger of TFB (Fig. 5B). TFE stabilizes this complex by closing the mobile clamp of RNAP via subunits E/F. During promoter clearance, RNAP loses contact to transcription factors (Fig. 5), and TFB and TFE are probably released. On strong promoters TBP remains bound in complexes containing transcripts of 4 to 24 nucleotides (109). On weak promoters TBP dissociates after promoter clearance (34). The nascent RNA is directed toward subunits E/F (Fig. 5C) in the elongation complex.

The position of *Pyrococcus* RNAP on the DNA, the extension and location of the transcription bubble, and the length of the RNA-DNA hybrid in transcription complexes paused at various positions has been analyzed by exonuclease III and potassium permanganate footprinting (96). In complexes stalled at position +5, RNAP binds downstream to transcription factors and the binding site of the enzyme extends to position +18 (Fig. 6A). The RNAP is in close contact with promoter-bound transcription factors and therefore an upstream end of the RNAP DNA-binding site cannot be defined by this technique. The transcription bubble comprises 12 bp and extends from −7 to +5 (similar to the bubble found in the open complex) (12, 46). The first major transition during early transcription is observed in complexes stalled at position +6/+7 (Fig. 6B). An upstream end of RNAP can now be defined, indicating a conformational change of RNAP, but the downstream end of the RNAP is still located at position +18. This finding indicates that 7 nucleotides of RNA can be synthesized without any translocation of RNAP along the template. The transcription bubble of this complex extends from −7 to +9, indicating that reclosure of the DNA of the open complex has not yet occurred at this stage of transcription. The second major transition occurs at position +10/+11 (Fig. 6C). In this complex, translocation of the downstream edge of RNAP to position +24 occurs and reclosure of the DNA in the region upstream of the transcription initiation site begins, indicative of promoter clearance. The transcription bubble extends over 17 bp and the RNA-DNA hybrid over 9 bp, similar to the extent of coverage that occurs in subsequent elongation complexes. The distance of the active center to the downstream edge of RNAP is ~12 nucleotides.

The archaeal enzyme appears to be committed to elongation at register +10. This transition occurs in

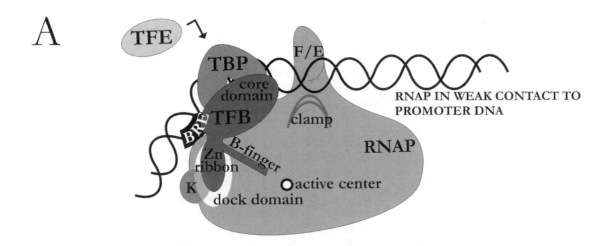

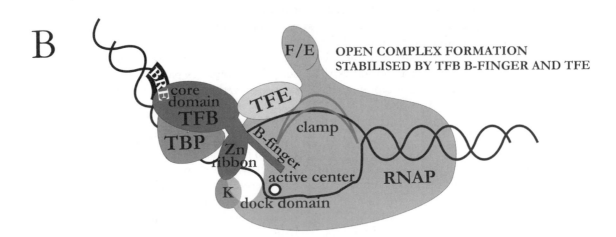

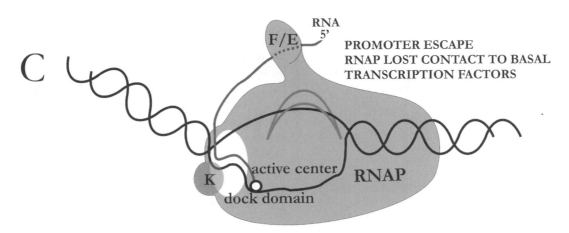

Figure 5. Mechanism of transcription by an archaeal RNAP. (A) TFE facilitates binding of the TFB zinc ribbon domain to the core domain of RNAP. (B) After open complex formation, the B finger of TFB stabilizes the template strand in the active center of RNAP. TFE provides additional stability to this complex by closing the clamp of RNAP. (C) After synthesis of a transcript longer than 10 nucleotides, RNAP reaches the elongation committed state. RNAP moves synchronously with RNA synthesis from this point.

all multisubunit RNAPs in members of all three domains of life. One characteristic of the archaeal system is that the distance of the catalytic center of RNAP to the front edge of the footprint is 12 bp; this is shorter than in bacterial and eucaryal RNAPs, where this distance is approximately 18 bp. Between positions +11 and +20, RNAP and the transcription bubble move synchronously with RNA synthesis. There is no evidence for discontinuous translocation of the archaeal RNAP in later stages of transcription.

REGULATION OF TRANSCRIPTION

Considering the eucaryal nature of the basal transcriptional machinery, the detection of bacterial-like regulators in the genomes of archaea (1, 55, 56) was a surprise. A mosaic archaeal transcriptional machinery was proposed, implying the regulation of a eucaryal basal machinery by bacterial regulators. However, many regulators characteristic of the *Archaea* have been identified and characterized. These findings demonstrate that unique pathways of regulation of transcription exist in the *Archaea*, and these deserve special attention.

The Lrp Family Contains Paralogs of Bacterial Proteins with Repressor and Activator Activities

Bacterial Lrp homologs have been identified in 94% of the analyzed archaeal genomes (20). Members of the Lrp family are small DNA-binding proteins. In *E. coli*, members of the Lrp family regulate the transcription of 10% of genes. The *E. coli* regulon includes genes involved in amino acid metabolism and in pili synthesis. Binding of bacterial Lrp can repress or activate transcription, and binding of its modulator, leucine, can either stimulate or reduce the repressor or activator activity of Lrp at specific promoters.

Archaeal paralogs of Lrp were among the first biochemically characterized archaeal transcriptional regulators (15, 19, 28). The crystal structure of LrpA from *Pyrococcus* has been solved (63). The N-terminal part of the protein contains a helix-turn-helix motif involved in binding to dyad symmetry elements in the DNA (78). The N-terminal part is connected with a hinge to the C-terminal domain. In bacterial homologs this region has been shown to be involved in the response to leucine and the activation of transcription. Similar to bacterial Lrp, archaeal homologs tend to form homooligomers. Free LrpA from *Pyrococcus* exists mainly as a dimer, and DNA-bound LrpA exists mainly as a tetramer (19).

LrpA and Lrs14 from *Sulfolobus* inhibit transcription of their own genes in vitro. The DNA-binding site of LrpA overlaps the RNAP-binding site, and DNA-bound LrpA inhibits transcription by blocking RNA polymerase recruitment (28). *Sulfolobus* Lrs14 binds to the DNA region harboring the TATA box. It acts on an earlier step of preinitiation complex assembly by preventing binding of TBP/TFB (15, 74). Aside from autoregulation of transcription in vitro, no information is available on the physiological function of *Sulfolobus* Lrs14 and *Pyrococcus* LrpA. In contrast, a Lrp homolog of *M. jannaschii*, Ptr2, was found to affect transcription of the ferredoxin A and rubredoxin 2 genes, suggesting that it is involved in regulation of redox reactions. However, unlike LrpA and Lrs14, which are repressors of their own promoters, Ptr2 is an activator of these two promoters (77).

Ptr2 conveys a stimulatory effect on transcription by facilitating recruitment of TBP to the promoter. Ptr2 binds to two adjacent palindromic sites. Mutational analysis of this bipartite upstream-activating sequence (UAS) showed that the promoter proximal site is sufficient for Ptr2-dependent promoter activation of the rubredoxin and ruberythrin genes (79). The UAS differs significantly from the UAS of the bacterial *ilvH* promoter, which is bound by bacterial Lrp. In *M. jannaschii* six binding sites within a DNA segment of 200 bp are involved in binding two Lrp octamers, forming a compact "UASome" (51). The assembly of archaeal Lrps into octamers and helical arrays in crystals (52, 63) led to the proposal of an UAS structure in which DNA is wrapped around arrays of laterally acting protein dimers. However, despite the potential of archaeal Lrp-like molecules to form higher-ordered structures, analysis of the UAS in a *M. jannaschii* cell-free system has shown that a single *cis*-acting site is sufficient for activation of transcription. Although a single UAS is reminiscent of simple bacterial promoters, analysis of Ptr2-mediated activation revealed a clear difference from bacterial activators. Ptr2 activates transcription by recruiting a eucaryal transcription factor and not by interacting with subunits (α and σ) of RNAP. Future analyses of transcriptional regulators will reveal whether archaeal activators have evolved that directly modulate RNAP binding. The subunits D and L, which are related to a different part of bacterial α, would be the candidate binding partners for such a regulator (σ factors do not exist in the *Archaea*).

A second Lrp protein family, represented by LysM, also appears to be involved in activating transcription. It is a part of a gene cluster encoding enzymes for lysine biosynthesis. Transcription of part of the *lys* gene cluster (*lysWXJK*) is induced upon lysine starvation in vivo (18). LysM binds to the promoter of *lysW*. Lysine can modulate the DNA-binding properties of LysM. In the absence of lysine, LysM binding to DNA is enhanced. LysM has been suggested to possibly play a role as an activator of transcription of genes for lysine biosynthesis. However, LysM has not been reported to affect cell-free transcription of the *lys* operon. This may be due to the absence of a hitherto unknown coactivator in the cell-free system.

Negative Regulation: MDR1, NrpR, TrpY

MDR1, a metal-dependent repressor

The function of MDR1, an *Archaeoglobus fulgidus* homolog of the bacterial metal-dependent repressor DtxR, was studied in the *Sulfolobus* cell-free transcription system and in *Archaeoglobus* cells (11). The gene encoding MDR1 is located upstream of three genes encoding an iron-importing ABC transporter, and all four genes are cotranscribed as a polycistronic transcription unit. In vivo expression of the MDR-1 gene was strictly dependent on metal ion availability. In assays based on chromatin immunoprecipitation, binding of MDR1 to operator DNA in *Archaeoglobus* cells was found to depend on the presence of bivalent cations. MDR1 binds cooperatively in a metal-dependent manner to three operator sequences located between positions -18 and $+67$ relative to the transcription start site. MDR1 represses transcription from its own promoter in vitro by preventing RNAP recruitment. This repression also depends on bivalent cations. DNA binding of the homolog of MDR1 from *Corynebacterium* Dxtr is also metal dependent. MDR1 was the first archaeal repressor whose binding to operators was shown to be influenced by an inducer in vitro and in vivo.

The regulator of nitrogen fixation in methanogens, NrpR

Some methanogens utilize dinitrogen as a nitrogen source. A regulator of *nif* gene expression, NrpR, has been characterized in *Methanococcus maripaludis*. NrpRs from *M. maripaludis*, *M. thermautotrophicus*, and *M. jannaschii* are tetrameric DNA-binding proteins. The archaeal NrpR contains an N-terminal winged helix-turn-helix domain and two conserved domains that may function in dimerization or multimerization. Homologs lacking one or more of the three domains are present in other methanogens and *Archaeoglobus* and were not found in crenarchaeota, suggesting that NrpR represents a novel family of regulators unique to euryarchaeota (64).

M. maripaludis can utilize ammonia, alanine, and dinitrogen as sources of nitrogen. NrpR controls the transcription of the *nif* operon by binding cooperatively to two tandem operator sequences, OR_1 and OR_2, located downstream of the transcription start site. The stronger and promoter proximal NrpR-binding site (OR_1) can mediate repression of *nif* transcription during growth on ammonia. Both OR_1 and OR_2 are required for intermediate repression during growth on alanine. 2-Oxoglutarate is an intracellular indicator of nitrogen deficiency and binds to NrpR and lowers its binding affinity to the operators (65). Hence, 2-oxoglutarate acts as inducer of *nif* gene expression in archaea and this induction is brought about by NrpR. This is the first archaeal system where the roles of a repressor, inducer, and two operators have been investigated both in vivo and in vitro. However, the mode of interaction of NrpR with the transcriptional machinery is unexplored. The occurrence of NrpR is restricted to euryarchaeota, but cooperative binding of two repressor dimers to tandem operators resembles bacterial repressor systems. NrpR binds downstream of the transcription start site and is likely to inhibit RNAP recruitment.

Regulation of the tryptophan operon by TrpY

Tryptophan synthesis is energetically very expensive, and as a consequence expression of the *trp* genes is tightly regulated. TrpY is a regulator that contains an N-terminal helix-turn-helix-DNA-binding motif and a C-terminal domain that binds tryptophan as an allosteric effector. The *M. thermautotrophicus* *trpY* gene is transcribed divergently from a promoter overlapping the promoter of the *trpEGCFBAD* operon (110). The TATA boxes of these promoters are separated by one helical turn and are therefore located on different faces of the DNA helix. TrpY binds to four TRP boxes (consensus TGTACA) located in the overlapping promoter region between *trpY* and *trpE*. In cell-free transcription experiments, TrpY has been found to autorepress its promoter in the absence of tryptophan while expression of the tryptophan operon and a separately encoded gene of Trp pathway, *trpB2*, is only inhibited in the presence of tryptophan. TrpY is a dimer in solution and blocks transcription at the *trpY* operators 1 and 2, probably by inhibiting RNAP recruitment. Binding of tryptophan to TrpY appears to induce a conformational change leading to increased affinity of TrpY to the nonconsensus TRP boxes 3 and 4 downstream of the TATA box of the *trpE* gene. In this conformational state, TrpY retains its potential to bind to the box 1 and box 2 consensus operator sequences, thereby regulating expression of the *trpY* gene. The box 3 and box 4 TrpY binding sites overlap with the TATA box of the *trpE* and *trpY* promoter, respectively. The

TrpY-tryptophan, repressor-ligand complex acts by a different mechanism by competing with TBP binding to the TATA box. Due to the close spacing of the divergent promoters, simultaneous transcription initiation does not seem possible. The mechanism controlling expression of *trp* genes and/or *trpY* is not yet understood, although regulation at the level of translation has been proposed (110).

In silico analyses revealed that TrpY paralogs are present in other euryarchaeota but not in crenarchaeota (36). TrpY acts superficially like the tryptophan-sensitive bacterial regulator, TrpR (111). However, there is no evidence for a common ancestry of these proteins (110).

The Production of Specialized Organelles in *Haloarchaea*: Further Examples of Positive Regulation

In *Haloarchaea*, the genes involved in formation of gas vesicles and the purple membrane are under positive gene regulatory control. Fourteen genes (*gvp* genes) are involved in gas vesicle synthesis in *Halobacterium salinarum* (48). GvpE is an activator of GvpA, the major structural protein of gas vesicles (54). GvpE contains a leucine zipper motif similar to that found in the eucaryal activator GCN4. However, aside from this motif it has no sequence similarity to eucaryal transcriptional regulators. GvpE therefore appears to represent a type of activator that is unique to the *Archaea*. A site upstream of the BRE sequence of one GvpE-activated promoter is involved in transcriptional regulation (48). Cell-free transcription systems are not available for haloarchaea, and the exact mechanism of GvpE-mediated activation is unclear. However, its activity is inhibited by the GvpD repressor. GvpE and GvpD interact in vitro, and an interaction of these proteins in vivo may be responsible for GvpD-mediated inhibition of GvpE.

The transcriptional regulator Bat regulates expression of genes responsible for the synthesis of purple membranes in *Halobacterium* (6). Synthesis of the purple membrane is highly induced in response to light intensity and low-oxygen tension. Bat contains a photoresponsive cGMP-binding domain (GAF), a bacterial AraC-type helix-turn-helix domain, and a PAS/PAC domain involved in sensing the redox status of the cell. The *bop* gene directs the expression of the major structural protein of the purple membrane. Its promoter was thoroughly investigated using mutagenesis (3, 4). The DNA gyrase inhibitor novobiocine blocks *bop* gene induction, suggesting that supercoiling of DNA is stimulating *bop* transcription. A DNA-supercoiling sensitivity site and an UAS upstream of BRE were identified as specific features of this promoter. The sensitivity of the bop gene to supercoiling was correlated to the presence of an alternating purine-pyrimidine sequence (RY box) overlapping the TATA box of the *bop* promoter by 4 nt. A *bop*-like UAS, the putative binding site for Bat, is conserved upstream of three additional Bat-regulated genes in *Halobacterium* which are presumably involved in retinal and carotenoid biosynthesis. A family of regulators related to Bat exists in *Halobacterium*, suggesting that regulatory networks responding to environmental changes in light and oxygen availability have evolved in halophilic archaea (6).

Regulation of the Heat Shock Response

Even though the hyperthermophile *Pyrococcus* grows at temperatures higher than 100°C, it has evolved a heat shock response to cope with higher-temperature fluxes that occur in its natural environment (58). The major chaperone classes Hsp100, Hsp90/Hsp83, and Hsp70 (DnaK) are absent from the genomes of hyperthermophilic archaea, although they are present in some mesophilic and psychrophilic species (58, 67). The major heat shock proteins predicted by a bioinformatics analysis (36), a Hsp60-like chaperonin (thermosome), two other chaperones belonging to the AAA$^+$ family, and a Hsp20-like small heat shock protein, were found to be highly induced upon heat shock (57, 93). The *AAA*$^+$ and *hsp20* promoters were investigated in cell-free transcription experiments. A palindromic sequence overlapping the transcription start site (TTT. .T. .C. . .G. .A. .AAA) was identified as a characteristic feature of *Pyrococcus* heat shock promoters (Fig. 6). A putative regulator of the heat shock response, Phr, was found to selectively inhibit transcription of these templates (103) by binding to a conserved, inverted-repeat heat shock element. The binding site of Phr overlaps the transcription start site, and operator-bound Phr inhibits RNAP recruitment to the promoter (Fig. 7).

The crystal structure of Phr has been solved. It is a winged helix-turn-helix protein that contains four helices in the N-terminal domain with a C-terminal domain that is involved in dimer formation and possibly in effector binding. Mutational analyses revealed that amino acids in three helices and the wing region are involved in operator recognition, suggesting a novel mode of DNA-protein interaction. Phr is conserved among *Euryarchaeota* (36). At present it is unknown whether an environmental factor (e.g., heat), an effector molecule, and/or an additional regulator modulates DNA binding of Phr.

In the genome of *Halobacterium* strain NRC-1, six copies of TBP and seven of TFB (5) have been identified. This indicates that alternative TBP-TFB-

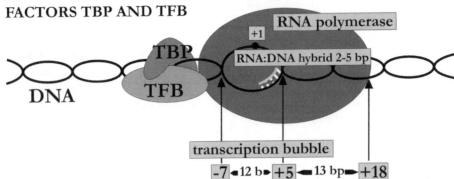

A TRANSCRIPTION AT POSITION +5: RNAP IN CLOSE CONTACT TO TRANSCRIPTION FACTORS TBP AND TFB

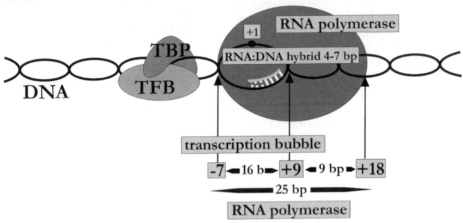

B FIRST TRANSITION AT POSITIONS +6/+7: MOVEMENT OF UPSTREAM END OF RNAP

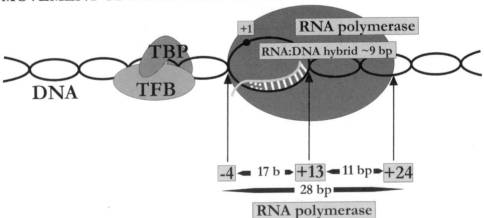

C SECOND TRANSITION AT POSITIONS +10/+11: MOVEMENT OF BOTH ENDS OF RNAP

Figure 6. The two major transitions in archaeal transcription initiation. (A) In the preinitiation complex and during synthesis of the first five nucleotides, RNAP is in close contact with transcription factors and the transcription bubble extends from position -7 to $+5$. (B) After synthesis of 6/7 nucleotides, the upstream edge of RNAP loses contact with transcription factors but the downstream edge is unchanged. (C) At position $+10/+11$, promoter clearance occurs and RNAP moves continuously to enable RNA synthesis.

Pyrococcus heat shock promoter

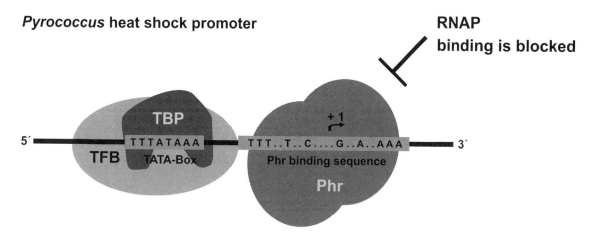

Figure 7. Interaction of a *Pyrococcus* heat shock regulator (Phr) with heat shock promoters. Phr binds specifically to a conserved palindromic sequence of archaeal heat shock promoters overlapping the transcription start site. When bound to the DNA, Phr blocks RNAP recruitment. The factors modulating the DNA-binding properties of Phr are unknown.

RNAP complexes may form and provides an explanation for the diversity of halophilic promoters and the regulation of heat shock response. Multiple copies of TBP and TFB also exist in other haloarchaea, and two TFB copies are present in *Pyrococcus*: TFB1 (identical to TFB), the factor used for cell-free transcription experiments, and TFB2, not investigated thus far. Heat shock-induced upregulation of some TFB genes from haloarchaea and of TFB2 from *Pyrococcus* have been reported (93, 100). *Pyrococcus* TFB2 can replace TFB1 in cell-free transcription experiments, but *Pyrococcus* heat shock promoters did not show any specificity for TFB2, suggesting that TFB2 is not involved specifically in transcription of heat shock promoters (M. Micorescu, A. Franke, M. Thomm, and M. Bartlett, manuscript in preparation).

In bacteria, the 5'-untranslated region (5'-UTR) of some cold shock-induced genes contains a conserved 11-nucleotide-sequence element, referred to as the cold box. A 113-nucleotide 5'-untranslated region was found upstream of a DEAD-box RNA helicase of the Antarctic methanogen *Methanococcoides burtonii* (66). This 5'-UTR contains a sequence closely matching a bacterial cold-box element. Therefore, *Bacteria*-like regulatory elements in 5'-UTRs seem to be involved in cold adaptation in *Archaea*. In the bacterium *Bradyrhizobium*, at least five heat shock genes are under the control of a conserved 100-nucleotide DNA segment (ROSE) positioned precisely in the 5'-UTR between the transcriptional and translational start sites (76). This *cis*-acting element confers temperature control by preventing translation at physiological growth temperatures. Although ROSE is involved mainly in the regulation of small heat shock proteins that also play a major role in *Archaea* (57, 58), there is no evidence that a similar mechanism relying on a secondary structure of RNA in the 5'-UTR is operating in the regulation of heat shock response in archaeal cells.

The Sugar Transport Regulator, TrmB: a Molecule Responding to Different Ligands in a Promoter-Dependent Manner

In *Pyrococcus*, two distinct ABC transporter systems exist for the uptake of maltose/trehalose (*mal* genes) and maltodextrins (*mdx* genes). The expression of the operon encoding these ABC transporter systems is regulated by TrmB, a global regulator. The interaction of TrmB with the *malE* promoter of the maltose ABC transporter system and with the *mdxE* promoter of the maltodextrin ABC transporter system has been studied (61, 62). At the *malE* promoter, TrmB binds to a sequence overlapping the TATA box and inhibits cell-free transcription. This is likely to occur by inhibiting TBP/TFB binding to the promoter. The transcriptional inhibition of *malE* is reversed by maltose or trehalose, which binds to the repressor and causes a change in conformation leading to dissociation of TrmB from the operator (Fig. 8). When TrmB is bound at the *mdxE* promoter, a different situation is encountered. (i) The DNA recognition site is different in sequence and location and overlaps the transcription start site (Fig. 8). (ii) The addition of maltose or trehalose does not release TrmB from the promoter, suggesting that binding to DNA induces a conformational change in the protein. In contrast, in cell-free transcription reactions, the addition of the substrate (maltodextrins) of this transporter system causes TrmB to dissociate from the promoter and relieves inhibition of RNA synthesis (Fig. 8). TrmB recognizes two different sequences. Its property to respond in a

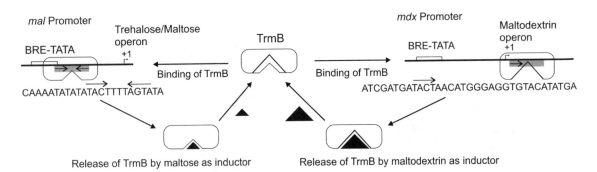

Figure 8. The archaeal regulator TrmB responds to different ligands when bound at different promoters. In the absence of any ligands, TrmB binds to its operator sequences at the maltose (*mal*) and maltodextrin (*mdx*) promoter. At the *mal* promoter, TrmB binding is influenced by maltose as inducer, but not by maltodextrins. At the *mdx* promoter maltose has no effect on TrmB binding but maltodextrins lower its affinity for the operator. The TrmB-binding sites differ substantially at both promoters, and the TrmB-binding site overlaps the transcription start site at the *mal* promoter, and the BRE/TATA box sequence at the *mdx* promoter. The smaller triangle represents maltose, and the larger triangle, maltodextrins. The TATA box is indicated; the DNA-binding sequence of TrmB is represented by a shaded box and shown on both promoters, and the binding sequence is shown below TrmB. The transcription start site is indicated by +1. The binding site of TrmB contains a palindrome at the *mal* promoter and is represented by two horizontal arrows. Only one half of it is conserved in the *mdx* promoter.

different manner to ligands depending on its association with a specific DNA sequence is unique for transcriptional regulators and reveals that the *Archaea* have evolved regulatory mechanisms that do not appear to be present in the *Bacteria* and *Eucarya*.

THE ROMA APPROACH

The lack of genetic systems in many archaea hampers analysis of transcriptional regulation in vivo. For example, 9 and 5 copies of Lrp-like genes are present in the genomes of *Pyrococcus furiosus* and *Sulfolobus solfataricus*, respectively (20). However, the only function assigned to *Pyrococcus* Lrp-A is inhibition of its own promoter in cell-free transcription assays. Analysis of the first complete-genome DNA microarray for a hyperthermophilic archaeon revealed a high degree of coordinated regulation in cells of *Pyrococcus* (92). Of the 2,065 open reading frames (ORFs) annotated in the genome of *P. furiosus*, the expression of 125 of them differed by more than five-fold between cultures grown with peptides or maltose as a primary carbon source (see Chapter 20). However it is difficult to infer from these global analyses the effects of particular regulators on the modulation of transcription.

A novel approach to identify the targets of regulators uses fragmented chromosomal DNA as a template for cell-free transcription reactions (Fig. 9). The transcripts from this template can be labeled and hybridized to microarrays. A comparison of the microarray hybridization patterns of transcripts obtained in the absence and presence of a given regulator will identify the genes affected by this regulator. Initial results (79) indicate that chromosomal DNA from archaea can be used successfully as template in cell-free transcription reactions. This in vitro transcriptomic approach, ROMA (runoff transcription/macroarray analysis; 23, 79), may be a useful tool to identify the target genes of archaeal regulators.

EVOLUTIONARY IMPLICATIONS

The striking similarity of the archaeal and eucaryal genetic machinery described in this chapter and the chapters on translation (Chapter 8) and replication (Chapter 3) sheds new light on the evolution of the eucaryal cell. The commonly accepted theory is that the eucaryal cell developed independently after separation of three major phylogenetic lineages. The lineage leading to *Archaea* and *Eucarya* was separated later in evolution, between 1.9 to 1.5 billion years ago, whereas the lineage leading to *Bacteria* branched earlier (3.5 to 1.9 billion years ago). This implies that *Eucarya* and *Archaea* might have a common evolutionary history of possibly two billion years; this may account for the similarity of their genetic machinery. But this does not explain how the eucaryal cell was generated. Also, the endosymbiosis hypothesis provides only an explanation of how modern eucaryal cells containing organelles derived from free-living *Bacteria* have evolved from preexisting eucaryal cells. But how was the first eucaryal cell containing an archaeal transcriptional machinery generated in evolution? Did the last common ancestor of *Archaea* and *Eucarya* already contain the basic

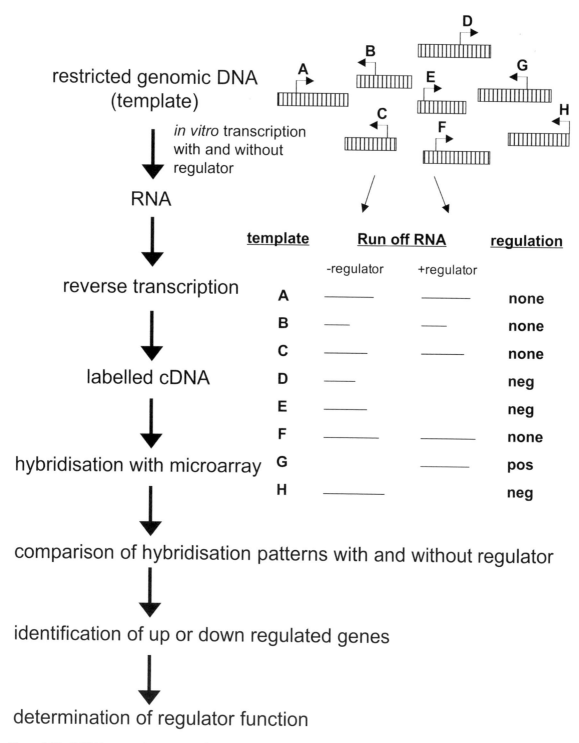

Figure 9. The ROMA approach. Fragmented genomic DNA is transcribed in a cell-free transcription system in the presence and absence of a regulator. Each DNA fragment harbors on average one promoter. Some promoters are unaffected (no regulation) by a given regulator, some are upregulated (by activators), and some are downregulated (by repressors). The in vitro RNA is converted to labeled cDNA and hybridized with a whole-genome microarray. By comparison of the hybridization patterns of RNAs synthesized in the presence and absence of a regulator the genes modulated by the regulatory protein can be identified.

archaeal machinery that was later evolved in *Eucarya* to the more complicated machinery of modern *Eucarya*, including major components such as the TATA box containing promoters TBP, TFB, and RNAP? The other possibility is that the eucaryal cell was generated by a fusion of an archaeal cell, providing the basis for the eucaryal nucleus and cytoplasm containing the genetic machinery, and a bacterial cell carrying the energy-generating electron transport system. An elegant theory proposes that the eucaryal cell has arisen through symbiotic association of an anaerobic, strictly hydrogen-dependent archaeon (possibly a methanogen) with a hydrogen- and CO_2-producing facultative anaerobe able to respire (possibly a proteobacterium) (70). The hydrogen dependence of the latter host (the archaeal cell) provides a strong selective force in evolution for irreversible association and subsequent incorporation of the hydrogen-producing proteobacterium (the symbiont). The resulting primitive eucaryal cell would have a cytoplasm with an archaeal genetic machinery and autotrophic metabolism and a heterotrophic symbiont generating ATP from organic compounds either by anaerobic (and possibly later also aerobic) respiration or via fermentation. This hypothesis provides an ingenious explanation for the archaeal nature of the eucaryal transcriptional machinery, as well as a new endosymbiosis hypothesis. The host for endosymbiosis of the bacterial cell was not a differentiated eucaryal cell but a hydrogen-dependent archaeon. If this hypothesis is true, then archaeal machinery is not eucarya-like; rather, the eucaryal machinery is archaeal as it is derived from *Archaea*.

PERSPECTIVE: THE NEXT FIVE YEARS

The archaeal transcriptional machinery is the evolutionary ancestor of the more complex eucaryal machinery, and recent reports describing the post-recruitment function of TFB and TFE suggest that it serves as a useful model to address questions that may be difficult to address in eucaryal systems. Although many regulators resemble bacterial repressors, unique regulatory mechanisms have been elucidated. This highlights that investigation of regulatory pathways, an area still in its infancy, will reveal novel insight into the mechanism and the evolution of regulatory principles in the *Archaea*. Future analyses of archaeal transcription need to take into account the role played by archaeal histones (85) and chromosome-associated proteins, such as Alba (10) (see Chapter 4).

Acknowledgments. I thank Gudrun Vierke, Patrizia Spitalny, and Annette Keese for artwork.

REFERENCES

1. Aravind, L., and E. V. Koonin. 1999. DNA-binding proteins and evolution of transcription regulation in the archaea. *Nucleic Acids Res.* 27:4658–4670.
2. Armache, K. J., H. Kettenberger, and P. Cramer. 2003. Architecture of initiation-competent 12-subunit RNA polymerase II. *Proc. Natl. Acad. Sci. USA* 100:6964–6968.
3. Baliga, N. S., and S. DasSarma. 1999. Saturation mutagenesis of the TATA box upstream activator sequence in the haloarchaeal *bop* gene promoter. *J. Bacteriol.* 181:2513–2518.
4. Baliga, N. S., and S. DasSarma. 2000. Saturation mutagenesis of the haloarchaeal *bop* gene promoter: identification of DNA supersoiling sensitivity sites and absence of TFB recognition element and UAS enhancer activity. *Mol. Microbiol.* 36:1175–1183.
5. Baliga, N. S., Y. A. Goo, W. V. Ng, L. Hood, C.J. Daniels, and S. DasSarma. 2000. Is gene expression in *Halobacterium NRC-1* regulated by multiple TBP and TFB transcription factors? *Mol. Microbiol.* 35:1184–1185.
6. Baliga, N. S., S. P. Kennedy, W. V. Ng, L. Hood, and S. DasSarma. 2001. Genomic and genetic dissection of an archaeal regulon. *Proc. Natl. Acad. Sci. USA* 98:2521–2525.
7. Bartlett, M. S., M. Thomm, and E. P. Geiduschek. 2000. The orientation of DNA in an archaeal transcription initiation complex. *Nat. Struct. Biol.* 7:782–785.
8. Bartlett, M. S., M. Thomm, and E. P. Geiduschek. 2004. Topography of the euryarchaeal transcription initiation complex. *J. Biol. Chem.* 279:5894–5903.
9. Bell, S. D., A. B. Brinkman, J. van der Oost, and S. P. Jackson. 2001. The archaeal TFIIEalpha homologue facilitates transcription initiation by enhancing TATA-box recognition. *EMBO Rep.* 2:133–138.
10. Bell, S. D., C. H. Botting, B. N. Wardleworth, S.P. Jackson, and M. F. White. 2002. The interaction of Alba, a conserved archaeal chromatin protein, with Sir2 and its regulation by acetylation. *Science* 296:148–151.
11. Bell, S. D., S. S. Cairns, R.L. Robson, and S. P. Jackson. 1999. Transcriptional regulation of an archaeal operon *in vivo* and *in vitro*. *Mol. Cell* 4:971–982.
12. Bell. S. D., and S. P. Jackson. 1998. Transcription and translation in Archaea: a mosaic of eukaryal and bacterial features. *Trends Microbiol.* 6:222–228.
13. Bell, S. D., and S. P. Jackson. 2000. Charting a course through RNA polymerase. *Nature Struct. Biol.* 7:703–705.
14. Bell, S. D., and S. P. Jackson. 2000. The role of transcription factor B in transcription initiation and promoter clearance in the archaeon Sulfolobus acidocaldarius. *J. Biol. Chem.* 275:12934–12940.
15. Bell, S. D., and S. P. Jackson. 2000. Mechanism of autoregulation by an archaeal transcriptional repressor. *J. Biol. Chem.* 275:31624–31629.
16. Bell, S. D., C. Jaxel, M. Nadal, P. F. Kosa, and S. P. Jackson. 1998. Temperature, template topology, and factor requirements of archaeal transcription. *Proc. Natl. Acad. Sci. USA* 95:15218–15222.
17. Bell, S. D., P. L. Kosa, P. D. Sigler, and S. P. Jackson. 1999. Orientation of the transcription preinitiation complex in archaea. *Proc. Natl. Acad. Sci. USA* 96:13662–13667.
18. Brinkman, A. B., S. D. Bell, R. J. Lebbink, W. M. de Vos, and J. van der Oost. 2002. The *Sulfolobus solfataricus* Lrp-like protein LysM regulates lysine biosynthesis in response to lysine availability. *J. Biol. Chem.* 277:29537–29549.
19. Brinkman, A. B., I. Dahlke, J. E. Tuininga, T. Lammers, V. Dumay, and E. de Heus. 2000. An Lrp-like transcriptional

regulator from the archaeon *Pyrococcus furiosus* is negatively autoregulated. *J. Biol. Chem.* **275:**38160–38169.

20. Brinkman, A. B., T. J. Ettema, W. M. de Vos, and J. van der Oost. 2003. The Lrp family of transcriptional regulators. *Mol. Microbiol.* **48:**287–294

21. Brown, J., M. Thomm, G. Beckler, G. Frey, K. O. Stetter, and J. N. Reeve. 1988. Archaebacterial RNA polymerase binding site and transcription of the *hisA* gene of *Methanococcus vannielii. Nucleic Acids Res.* **16:**135–164.

22. Bushnell, D. A., K. D. Westover, R. E. Davis, and R. D. Kornberg. 2004. Structural basis of transcription: an RNA polymerase II-TFIIB cocrystal at 4.5 Ångströms. *Science* **303:**983–988.

23. Cao, M., P. A. Kobel, M. M. Morshedi, M. F. Wu, C. Paddon, and J. D. Heimann. 2002. Defining the *Bacillus subtilis* sigma (W) regulon: a comparative analysis of promoter consensus search, run-off transcription/macroarray analysis (ROMA), and transcriptional profiling approaches. *J. Mol. Biol.* **316:**443–457.

24. Chen, H. T., and S. Hahn. 2003. Binding of TFIIB to RNA polymerase II: mapping the binding site for the TFIIB zinc ribbon domain within the preinitiation complex. *Mol. Cell* **12:**437–447.

25. Chen, H. T., and S. Hahn. 2004. Mapping the location of TFIIB within the RNA polymerase II transcription preinitiation complex: a model for the structure of the PIC. *Mol. Cell.* **119:**169–180.

26. Cramer, P., D. A. Bushnell, J. Fu, A. L. Gnatt, B. Maier-Davis, N. E. Thompson, R. R. Burgess, A. M. Edwards, P. R. David, and R. D. Kornberg. 2000. Architecture of RNA polymerase II and implications for the transcription mechanism. *Science* **288:**640–649.

27. Cramer, P., D. A. Bushnell, and R. D. Kornberg. 2001. Structural basis of transcription: RNA polymerase II at 2.8 Å resolution. *Science* **292:**1863–1876

28. Dahlke, I., and M. Thomm. 2002. A *Pyrococcus* homologue of the Leucine-responsive regulatory protein, LrpA, inhibits transcription by abrogating RNA polymerase recruitment. *Nucleic Acids Res.* **30:**701–710.

29. Danner, S., and J. Soppa. 1996. Characterization of the distal promoter element of halobacteria *in vivo* using saturation mutagenesis and selection. *Mol. Microbiol.* **19:**1265–1276.

30. DeDecker, B., R. O'Brien, P. J. Fleming, J. H. Geiger, S. Jackson, and P. B. Sigler. 1996. The crystal structure of a hyperthermophilic archaeal TATA-box binding protein. *J. Mol. Biol.* **264:**1072–1084.

31. Eloranta, J. J., A. Kato, M. Teng, and R. O. Weinzierl. 1998. *In vitro* assembly of an archaeal D-L-N RNA polymerase subunit complex reveals a eukaryote-like structural arrangement. *Nucleic Acids Res.* **26:**5562–5567.

32. Florentino, G., R. Cannio, M. Rossi, and S. Bartolucci. 2003. Transcriptional regulation of the gene encoding an alcohol dehydrogenase in the archaeon *Sulfolobus solfataricus* involves multiple factors and control elements. *J. Bacteriol.* **185:**3926–3934.

33. Frey, G., M. Thomm, B. Brüdigam, H. Gohl, and W. Hausner. 1990. An archaebacterial cell-free transcription system. The expression of tRNA genes from *Methanococcus vannielii* is mediated by a transcription factor. *Nucleic Acids Res.* **18:**1361–1367.

34. Geiduschek, E. P., and M. Ouhammouch. 2005. Archaeal transcription and its regulators. *Mol. Microbiol.* **56:**1397–1407.

35. Geiduschek, E. P., and G. P. Tocchini-Valentini. 1988. Transcription by RNA polymerase III. *Annu. Rev. Biochem.* **57:**873–914.

36. Gelfand, M. S., E. V. Koonin, and A. A. Mironov. 2000. Prediction of transcription regulatory sites in Archaea by a comparative genomic approach. *Nucleic Acids Res.* **28:**695–705.

36a. Goede, B., S. Naji, O. von Kampen, K. Ilg, and M. Thomm. 2006. Protein-protein interactions in the archaeal transcriptional machinery: binding studies of isolated RNA polymerase subunits and transcription factors. *J. Biol. Chem.* **281:**30581–30592.

37. Gohl, H. P., B. Gröndahl, and M. Thomm. 1995. Promoter recognition in archaea is mediated by transcription factors: Identification of a TFB from *Methanococcus thermolithotrophicus* as archaeal TATA-binding protein. *Nucleic Acids Res.* **23:**3837–3841.

38. Grabowski, B., and Z. Kelman. 2003. Archaeal DNA replication: Eukaryal proteins in a bacterial context. *Annu. Rev. Microbiol.* **57:**487–516.

39. Gropp, F., W. D. Reiter, A. Sentenac, W. Zillig, R. Schnabel, M. Thomm, and K. O. Stetter. 1986. Homologies of components of DNA-dependent RNA polymerases of Archaebacteria, Eukaryotes and Eubacteria. *Syst. Appl. Microbiol.* **7:**95–101.

40. Hain, J., W. D. Reiter, U. Hüdepohl, and W. Zillig. 1992. Elements of an archaeal promoter defined by mutational analysis. *Nucleic Acids Res.* **20:**5423–5428.

41. Hanzelka, B. L., T. J. Darcy, and J. N. Reeve. 2001. TFE, an archaeal transcription factor in *Methanobacterium thermoautotrophicum* related to eucaryal transcription factor TFIIFalpha. *J. Bacteriol.* **183:**1813–1818.

42. Hausner, W., G. Frey, and M. Thomm. 1991. Control regions of an archaeal gene. A TATA box and initiator element promote cell-free transcription of the RNA^{Val} gene of *Methanococcus vannielii. J. Mol. Biol.* **222:**495–508.

43. Hausner, W., U. Lange, and M. Musfeldt. 2000. Transcription factor S, a cleavage induction factor of the archaeal RNA polymerase. *J. Biol. Chem.* **275:**12393–12399.

44. Hausner, W., and M Thomm. 1993. Purification and characterization of a general transcription factor, aTFB from the archaeon *Methanococcus thermolithotrophicus. J. Biol. Chem.* **268:**24047–24052

45. Hausner, W., and M. Thomm. 1995. The translation product of the presumptive *Thermococcus celer* TATA-binding protein sequence is a transcription factor related in structure and function to *Methanococcus* Factor B. *J. Biol. Chem.* **270:**17649–17651.

46. Hausner, W., and M. Thomm. 2001. Events during initiation of archaeal transcription: Open complex formation and DNA-protein interactions. *J. Bacteriol.* **183:**3025–3031.

47. Hausner, W., J. Wettach, C. Hethke, and M. Thomm. 1996. Two transcription factors related with the eucaryal transcription factors TATA-binding protein and transcription factor IIB direct promoter recognition by an archaeal RNA polymerase. *J. Biol. Chem.* **271:**30144–30148.

48. Hofacker, A, K. M. Schmitz, A. Cichonczyk, S. Sartorius-Neef, and F. Pfeifer. 2004. GvpE- and GvpD-mediated transcription regulation of the p-gvp genes encoding gas vesicles in Halobacterium salinarum. *Microbiology* **150:**1829–1838.

49. Hüdepohl, U., W. D. Reiter, and W. Zillig. 1990. *In vitro* transcription of two rRNA genes of the archaebacterium *Sulfolobus* sp. B 12 indicates a factor requirement for specific initiation. *Proc. Natl. Acad. Sci. USA* **87:**5851–5855.

50. Ishihama, A. 1981. Subunit assembly of E. *coli* RNA polymerase. *Adv. Biophys.* **14:**1–35.

51. Jafri, S., S. Chen, and J. M. Calvo. 2002. *ilvIH* operon expression in *Escherichia coli* requires Lrp binding to two distinct regions of DNA. *J. Bacteriol.* **184:**5293–5300.

52. Koike, H., S. A. Ishijima, L. Clowney, and M. Suzuki. 2004. The archaeal feast/famine regulatory protein: Potential roles

of its assembly forms for regulating transcription. *Proc. Natl. Acad. Sci. USA* **101**:2840–2845.

53. Kosa, P. F., G. Ghosh, B. S. DeDecker, and P. B. Sigler. 1997. The 2.1-Å crystal structure of an archaeal preinitiation complex: TATA-box-binding protein/transcription factor (II) B core/TATA-box. *Proc. Natl. Acad. Sci. USA* **94**:6042–6047.

54. Krüger, K., T. Hermann, V. Armbruster, and F. Pfeifer. 1998. The transcriptional activator GvpE for the halobacterial gas vesicle genes resembles a basic region leucine-zipper regulatory protein. *J. Mol. Biol.* **279**:761–771.

55. Kyrpides, N. C., and C. A. Ouzounis. 1995. The eubacterial transcriptional activator Lrp is present in the archaeon *Pyrococcus furiosus. Trends Biochem. Sci.* **20**:140–141.

56. Kyrpides, N. C., and C. A. Ouzounis. 1999. Transcription in archaea. *Proc. Natl. Acad. Sci. USA* **96**:8545–50.

57. Laksanalamai, P., D. L. Maeder, and F. T. Robb. 2001. Regulation and mechanism of action of the small heat shock protein from the hyperthermophilic archaeon *Pyrococcus furiosus. J. Bacteriol.* **183**:5198–5202.

58. Laksanalamai, P., T. A. Whitehead, and F. T. Robb. 2004. Minimal protein-folding systems in hyperthermophilic archaea. *Nat. Rev. Microbiol.* **2**:315–324.

59. Lange, U., and W. Hausner. 2004. Transcriptional fidelity and proofreading in Archaea and implications for the mechanism of TFS-induced RNA cleavage. *Mol. Microbiol.* **52**:1133–1143.

60. Langer, D., J. Hain, P. Thuriaux, and W. Zillig. 1995. Transcription in Archaea: Similarity to that in Eukarya. *Proc. Natl. Acad. Sci. USA* **92**:5768–5772.

61. Lee, S. J., A. Engelmann, R. Horlacher, Q. Qu, G. Vierke, C. Hebbeln, M. Thomm, and W. Boos. 2003. TrmB, a sugar-specific transcriptional regulator of the trehalose/maltose ABC transporter from the hyperthermophilic archaeon *Thermococcus litoralis. J. Biol. Chem.* **278**:983–990.

62. Lee, S. J., C. Moulakakis, W. Hausner, S. M. Koning, M. Thomm, and W Boos. 2005. TrmB, a sugar sensing regulator of ABC transporter genes in *Pyrococcus furiosus* exhibits dual promoter specificity and is controlled by different inducers. *Mol. Microbiol.* **57**:1797–1807.

63. Leonard, P. M., S. H. J. Smits, S. E. Sedelnikova, A. B. Brinkman, W. M. de Vos, J. Van der Oost, D. W. Rice, and J. B. Rafferty. 2001. Crystal structure of the Lrp-like transcription regulator from the archaeon *Pyrococcus furiosus. EMBO J.* **20**:990–997.

64. Lie, T. J., and J. A. Leigh. 2003. A novel repressor of *nif* and *ginA* expression in the methanogenic archaeon *Methanococcus maripaludis. Mol. Microbiol.* **47**:235–246.

65. Lie, T. J., G. E. Wood, and J. A. Leigh. 2005. Regulation of *nif* expression in *Methanococcus maripaludis. J. Biol. Chem.* **200**:5236–5241.

66. Lim, J., T. Thomas, and R. Cavicchioli. 2000. Low temperature regulated DEAD-box RNA helicase from the Antarctic archaea, *Methanococcoides burtonii. J. Mol. Biol.* **297**:553–567.

67. Macario, A. J. L., and E. Conway de Macario. 2001. The molecular chaperone system and other anti-stress mechanisms in archaea. *Front. Biosci.* **6**:262–283.

68. Magilli, C. P., S. P. Jackson, and S. D. Bell. 2001. Identification of a conserved archaeal RNA polymerase subunit contacted by the basal transcription factor TFB. *J. Biol. Chem.* **276**:46693–46696.

69. Marsh, T. L., C. I. Reich, R. B. Whitelock, and G. J. Olsen. 1994. Transcription factor IID in the Archaea: Sequences in the *Thermococcus celer* genome would encode a product closely related to the TATA-binding protein of eukaryotes. *Proc. Natl. Acad. Sci. USA* **91**:4180–4184.

70. Martin, W., and M. Müller. 1998. The hydrogen hypothesis for the first eukaryote. *Nature* **392**:37–41.

71. Matsuda, T., M. Fujikawa, S. Ezaki, T. Imanaka, M. Morikawa, and S. Kanaya. 2001. Interaction of TIP26 from a hyperthermophilic archaeon with TFB/TBP/DNA ternary complex. *Extremophiles* **5**:177–182.

72. Matsunaga, F., P. Forterre, Y. Ishino, and H. Myllykallio. 2001. *In vivo* interactions of archaeal Cdc6/Orc1 and minichromosome maintenance proteins with the replication origin. *Proc. Natl. Acad. Sci. USA* **98**:11152–11157.

73. Meinhart, A., J. Blobel, and P. Cramer. 2003. An extended winged helix domain in general transcription factor E/IIE alpha. *J. Biol. Chem.* **278**:48267–48274.

74. Napoli, A., J. van der Oost, C. W. Sensen, R. L. Charlebois, M. Rossi, and M. Ciaramella. 1999. An Lrp-like protein of the hyperthermophilic archaeon *Sulfolobus solfataricus* which binds to its own promoter. *J. Bacteriol.* **181**:1474–1480.

75. Nikolov, D. B., H. Chen, E. D. Halay, A. A. Usheva, K. Hisatake, and D. K. Lee. 1995. Crystal structure of a TFIIB-TBP-TATA-element ternary complex. *Nature* **377**:119–128.

76. Nocker, A., T. Hausherr, S. Balsiger, N.-P. Krstulovic, H. Hennecke, and F. Narberhaus. 2001. A mRNA-based thermosensor controls expression of rhizobial heat shock genes. *Nucleic Acids Res.* **29**:4800–4807.

77. Ouhammouch, M., R. E. Dewhurst, W. Hausner, M. Thomm, and E. P. Geiduschek. 2003. Activation of archaeal transcription by recruitment of the TATA-binding protein. *Proc. Natl. Acad. Sci. USA* **100**:5097–5102.

78. Ouhammouch, M., and E. P. Geiduschek. 2001. A thermostable platform for transcriptional regulation: the DNA-binding properties of two Lrp homologs from the hyperthermophilic archaeon *Methanococcus jannaschii. EMBO J.* **20**:146–156.

79. Ouhammouch, M., G. E. Langham, W. Hausner, A. J. Simpson, N. M. A. El-Sayed, and E. P. Geiduschek. 2005. Promoter architecture and response to a positive regulator of archaeal transcription. *Mol. Microbiol.* **56**:625–637.

80. Ouhammouch, M., F. Werner, R. O. Weinzierl, and E. P. Geiduschek. 2004. A fully recombinant system for activator-dependent archaeal transcription. *J. Biol. Chem.* **279**:51719–51721.

81. Ouzounis, C., and C. Sander. 1992. TFIIB, an evolutionary link between the transcription machineries of archaebacteria and eukaryotes. *Cell* **71**:189–190.

82. Palmer, J. R., and C. J. Daniels. 1995. *In vivo* definition of an archaeal promoter. *J. Bacteriol.* **177**:1844–1849.

83. Pühler, G., H. Leffers, F. Gropp, P. Palm, H. P. Klenk, F. Lottspeich, R. A. Garrett, and W. Zillig. 1989. Archaebacterial DNA-dependent RNA polymerases testify to the evolution of the eukaryotic nuclear genome. *Proc. Natl. Acad. Sci. USA* **86**:4569–4573.

84. Qureshi, S. A., S. D. Bell, and S. P. Jackson. 1997 Factor requirements for transcription in the archaeon *Sulfolobus shibatae. EMBO J.* **16**:2927–2936.

85. Reeve, J. N. 2003. Archaeal chromatin and transcription. *Mol. Microbiol.* **48**:587–598.

86. Reiter, W. D., U. Hüdepohl, and W. Zillig. 1990. Mutational analysis of an archaebacterial promoter: Essential role of a TATA box for transcription efficiency and start-site selection. *Proc. Natl. Acad. Sci. USA* **87**:9509–9513.

87. Reiter, W. D., P. Palm, and W. Zillig. 1988. Analysis of transcription in the archaebacterium *Sulfolobus* indicates that archaebacterial promoters are homologous to eukaryotic pol II promoters. *Nucleic Acids Res.* **16**:11–19.

88. Renfrow, M. B., N. Naryshkin, L. M. Lewis, H. T. Chen, R. M. Ebright, and R. A. Scott. 2004. Transcription factor B

contacts promoter DNA near the transcription start site of the archaeal transcription initiation complex. *J. Biol. Chem.* 279:2825–2831.

89. Richard, D. J., S. D. Bell, and M. F. White. 2004. Physical and functional interaction of the archaeal single-stranded DNA-binding protein SSB with RNA polymerase. *Nucleic Acids Res.* 32:1065–1074.

90. Rowlands, T., P. Baumann, and S. P. Jackson. 1994. The TATA-binding protein: A general transcription factor in Eukaryotes and Archaebacteria. *Science* 264:1326–1329.

91. Schnabel, R., M. Thomm, R. Gerardy-Schahn, W. Zillig, K. O. Stetter, and J. Huet. 1983. Structural homology between different archaebacterial DNA-dependent RNA polymerases analyzed by immunological comparison of their components. *EMBO J.* 2:751–755.

92. Schut, G. J., S. D. Brehm, S. Datta, and M. W. Adams. 2003. Whole-genome DNA microarray analysis of a hyperthermophile and an archaeon: *Pyrococcus furiosus* grown on carbohydrates or peptides. *J. Bacteriol.* 185:3935–3947.

93. Shockley, K. R., D. E. Ward, S. R. Chhabra, S. B. Conners, C. I. Montero, and R. M. Kelly. 2003. Heat shock response by the hyperthermophilic archaeon *Pyrococcus furiosus*. *Appl. Environ. Microbiol.* 69:2365–2371.

94. Soppa, J. 1999. Normalized nucleotide frequencies allow the definition of archaeal promoter elements for different archaeal groups and reveal base-specific TFB contacts upstream of the TATA box. *Mol. Microbiol.* 31:1589–1601.

95. Soppa, J. 1999. Transcription initiation in Archaea: facts, factors and future aspects. *Mol. Microbiol.* 31:1295–1305.

96. Spitalny, P., and M. Thomm. 2003. Analysis of the open region and of DNA-protein contacts of archaeal RNA polymerase transcription complexes during transition from initiation to elongation. *J. Biol. Chem.* 278:30497–30505.

97. Thomm, M., W. Hausner, and C. Hethke. 1994. Transcription factors and termination of transcription in *Methanococcus*. *Syst. Appl. Microbiol.* 16:148–155.

98. Thomm, M., and G. Wich. 1988. An archaebacterial promoter element for stable RNA genes with homology to the TATA box of higher eukaryotes. *Nucleic Acids Res.* 16:151–163.

99. Thomm, M. 1996. Archaeal transcription factors and their role in transcription initiation. *FEMS Microbiol. Rev.* 18:159–171.

100. Thomson, D. K., J. R. Palmer, and C. J. Daniels. 1999. Expression and heat-responsive regulation of a TFIB homologue from the archaeon *Haloferax volcanii. Mol. Microbiol.* 33:1081–1092.

101. Torarinsson, E., H. P. Klenk, and R. A. Garrett. 2005. Divergent transcriptional and translational signals in Archaea. *Environ. Microbiol.* 7:47–54.

102. Trakselis, M. A., and S. D. Bell. 2004. The loader of the rings. *Nature* 429:708–709.

103. Vierke, G., A. Engelmann, C. Hebbeln, and M. Thomm. 2003. A novel archaeal transcriptional regulator of heat shock response. *J. Biol. Chem.* 278:18–26.

104. Werner, F., J. J. Eloranta, and R. O. Weinzierl. 2000. Archaeal RNA polymerase subunits F and P are *bona fide* homologs of eukaryotic RPB4 and RPB12. *Nucleic Acids Res.* 28:4299–305.

105. Werner, F., and R. O. Weinzierl. 2002. A recombinant RNA polymerase II-like enzyme capable of promoter-specific transcription. *Mol. Cell* 10:635–646.

106. Werner, F., and R. O. Weinzierl. 2005. Direct modulation of RNA polymerase core functions by basal transcription factors. *Mol. Cell. Biol.* 25:8344–8355.

107. Wettach, J., H. P. Gohl, H. Tschochner, and M. Thomm. 1995. Functional interaction of yeast and human TATA-binding proteins with an archaeal RNA polymerase and promoter. *Proc. Natl. Acad. Sci. USA* 92:472–476.

108. Wich, G., H. Hummel, M. Jarsch, U. Bär, and A. Böck. 1986. Transcription signals for stable RNA genes in *Methanococcus. Nucleic Acids Res.* 14:2447–2479.

109. Xie, Y., and J. N. Reeve. 2004. Transcription by Methanothermobacter thermoautotrophicus RNA polymerase in vitro releases archaeal transcription factor B but not TATA-box binding protein from the template DNA. *J. Bacteriol.* 186:6306–6310.

110. Xie, Y., and J. N. Reeve. 2005. Regulation of tryptophan operon expression in the archaeon *Methanothermobacter thermautotrophicus. J. Bacteriol.* 187:6419–6429.

111. Yanofsky, C. 2003. Using studies of tryptophan metabolism to answer basic biology questions. *J. Biol. Chem.* 278:10859–10878.

112. Zhu, W., Q. Zeng, C. M. Colangelo, M. Lewis, M. F. Summers, and R. A. Scott. 1996. The N-terminal domain of TFIIB from *Pyrococcus furiosus* forms a zinc ribbon. *Nat. Struct. Biol.* 3:122–124.

113. Zillig, W., K. O. Stetter, R. Schnabel, J. Madon, and A. Gierl. 1982. Transcription in Archaebacteria. *Zbl. Bakt. Hyg., I. Abt. Orig. C* 3:218–227.

Archaea: Molecular and Cellular Biology
Edited by Ricardo Cavicchioli
© 2007 ASM Press, Washington, D.C.

Chapter 7

RNA Processing

GABRIELE KLUG, ELENA EVGUENIEVA-HACKENBERG, ARINA D. OMER,
PATRICK P. DENNIS, AND ANITA MARCHFELDER

INTRODUCTION

Archaea represent a distinct phylogenetic lineage that is separate and distinct from *Bacteria* and *Eucarya*. Although they lack a nucleus and their gene organization, genome structure, and basic central metabolism are bacterial-like, their DNA replication, transcription, and translation machineries are eucaryal-like. This review primarily deals with the production and function of the archaeal translational machinery and the mechanisms used for posttranscriptional regulation of gene expression. The core components of the translational apparatus are represented by the three major classes of RNA: mRNA, which contains a copy of the genetic information carried in the DNA that is decoded into the amino acid sequence of proteins during translation; tRNA, which decodes the information carried on the mRNA and aligns amino acids for polymerization; and rRNA, which is the core component of the ribosome, the ribonucleoprotein machine that coordinates the decoding process and catalyzes the formation of the peptide bonds during amino acid polymerization (see Chapter 8). In addition to the three major classes of RNA, *Archaea*, similar to *Bacteria* and *Eucarya*, contain a plethora of other types of small RNAs that function in the processing, modification, and assembly of the translational apparatus or in the control of gene expression at the translational level.

MATURATION OF TRANSFER RNA ENDS

Most organisms contain between 40 and 50 different tRNA molecules that read one or more of the 61 or 62 different sense codons through specific codon-anticodon interactions and position the appropriate amino acid for insertion into the growing polypeptide chain. Recent data indicate that tRNAs are dynamic and have coevolved with the ribosome to ensure speed and accuracy in the translation process (37, 42, 123). The archaeal genes that encode the different tRNAs are either individually transcribed, cotranscribed with other tRNA genes, or cotranscribed with other types of genes (20). In all cases the tRNAs are contained within a longer precursor (pre-tRNA), which requires nuclease processing at the 5′ and 3′ ends of the mature tRNA sequence. In general, the 3′-terminal CCA sequence present in all tRNAs is not encoded in the tRNA gene; after numerous nucleotide modifications are introduced into the tRNA, the CCA trinucleotide is added by a terminal transferase to produce the mature tRNA molecules (70) (Fig. 1). Additional complexity in the tRNA processing and modification pathway occurs in cases where archaeal tRNA genes contain an intron (usually in the anticodon loop of the tRNA). Excision of the intron, and exon ligation, is closely linked to modification and processing (36).

Maturation of the tRNA 5′ End by RNase P

In all organisms the 5′-leader sequence of the pre-tRNA is removed by a universally conserved endonuclease, RNase P (EC 3.1.26.5; reviewed in reference 55). RNase P is a ribonucleoprotein complex composed of a single RNA and one (in *Bacteria*) or more than one (in *Archaea* and *Eucarya*) protein subunit. Most bacterial and archaeal RNase P RNAs are structurally similar (type A); the exceptions are *Bacil-*

Gabriele Klug and Elena Evguenieva-Hackenberg • Institut für Mikrobiologie und Molekularbiologie, Universität Giessen, Heinrich-Buff-Ring 26-32, 35392 Giessen, Germany.   Arina D. Omer • Department of Biochemistry and Molecular Biology, University of British Columbia, Vancouver, BC, Canada, V6T 1Z3.   Patrick P. Dennis • Division of Molecular and Cellular Biosciences, National Science Foundation, 4201 Wilson Blvd., Arlington, VA 22230.   Anita Marchfelder • Molekulare Botanik, Universität Ulm, Albert-Einstein-Allee 11, 89069 Ulm, Germany.

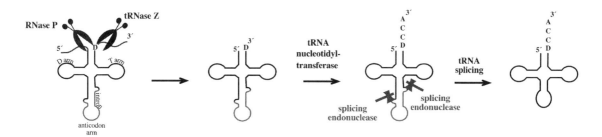

Figure 1. Maturation of precursor tRNA in *Archaea*. Precursor tRNAs are transcribed with 5'-leader and 3'-trailer sequences that have to be removed to yield a functional tRNA. Processing at the 5' end is performed by the ubiquitous enzyme RNase P. The tRNA 3'-end maturation is catalyzed by the endonuclease tRNase Z; after this the 3'-terminal CCA sequence is added by the tRNA nucleotidyltransferase. Some archaeal tRNA precursors contain introns that are removed by the splicing endonuclease, which recognizes the bulge-helix-bulge (BHB) structural motif that forms between the exon/intron and intron/exon boundaries. The two halves of the tRNA that are generated by intron excision are joined by a tRNA ligase activity that has not yet been identified or characterized. In addition to nucleolytic processing, numerous nucleosides in the tRNA are subjected to modification. After all these processing and modification steps, the tRNA is ready for aminoacylation. The order of processing events is not known, and the scheme depicted is not necessarily what occurs in vivo.

lus and relatives (type B), *Thermomicrobium* (type C), and *Methanocaldococcus* and relatives (type M). Eucaryal RNase P RNAs are distinct (type E) from those of bacteria and archaea (63).

Under elevated ionic conditions, the RNAs from bacteria are catalytically active in vitro in the absence of protein (62). Some archaeal type A RNAs exhibit similar, albeit weak, ribozyme activity and can be reconstituted with the *Bacillus subtilis* protein to create catalytically proficient chimeric holoenzymes (106). In contrast, the eucaryal RNAs require the protein components for in vitro catalytic activity. In all organisms, the genes encoding both the RNA and protein components of RNase P are essential for viability (105).

The bacterial RNase P has a single protein subunit of rather small size, whereas eucaryal RNase P enzymes contain up to ten protein subunits, none of which are related to the single bacterial protein (28, 152). In the genome of *Methanothermobacter thermautotrophicus* four open reading frames (ORFs) encode proteins with similarity to the yeast nuclear RNase P protein subunits Pop4 (MTH11), Pop5 (MTH687), Rpp1p (MTH688), and Rpr2p (MTH1618). All four *M. thermautotrophicus* ORFs have obvious homologs in other archaeal genomes (63). Thus, the protein subunits of archaeal RNase P enzymes are clearly homologous to eucaryal RNase P subunits rather than those of bacteria. Based on the homology of protein subunits and the similarity of the protein-protein contacts in the yeast and archaeal RNase P complexes, it seems likely that the structures of the holoenzymes from the *Eucarya* and *Archaea* are remarkably similar (64). The endonuclease activity that processes the 5' end of tRNAs has been recovered by reconstituting recombinant protein subunits and in vitro transcribed RNA from *M. thermautotrophicus* and *Pyrococcus horikoshii*; all four archaeal protein subunits were re-

quired for activity (17, 84). Taken together, these observations imply that the archaeal RNase P RNP is a chimera of bacterial and eucaryal parts: the RNA subunit of the ribonucleoprotein RNase P more closely resembles the bacterial homolog, whereas the protein components are related to those found in the eucaryal complex.

Processing at the tRNA 3′ End by tRNase Z

In *Bacteria*, two modes of tRNA 3′-end maturation exist; either the 3′-trailer sequence is digested by a combination of exonucleases (49) or the endonuclease tRNase Z (EC 3.1.26.11, reviewed in reference 144) cleaves the precursor to remove the trailer (96, 107). In *Eucarya* the tRNA 3′-trailer sequence is removed by the endonuclease tRNase Z, which exists in two forms. The short form (which is the one found in *Bacteria*) is about 250 to 350 amino acids in length, and the long form is 750 to 900 amino acids in length. All archaeal organisms analyzed so far use the short form of tRNase Z endonuclease to remove the tRNA 3′ trailer (126, 127). The enzyme cleaves immediately 3′ to the discriminator base (the discriminator is the first unpaired base, extending on the 3′ end from the tRNA acceptor stem) leaving a 3′-hydroxyl group to allow immediate addition of the CCA trinucleotide by the tRNA terminal transferase enzyme. tRNase Z enzymes belong to the metallo-β-lactamase family, which is characterized by a specific structural fold, two central β-sheets flanked on each side by α-helices (6). Members of this family require at least one manganese, iron, or zinc ion for activity. The *Escherichia coli* tRNase Z requires zinc. The bacterial, archaeal, and the two paralogous eucaryal (the long and short form) tRNase Z enzymes in general exhibit high sequence similarity to each other.

The tRNase Z enzymes from four archaea have been isolated: *Pyrococcus furiosus, Methanococcus jannaschii, Haloferax volcanii,* and *Pyrobaculum aerophilum* (96, 127; S. Schubert and A. Marchfelder, in preparation). All four have been expressed in *E. coli* and have tRNA-processing activity in vitro. The recombinant halophilic enzyme is inhibited in vitro by KCl concentrations higher than 100 mM (Schubert and Marchfelder, in preparation). Similarly, extracts from *H. volcanii* that were dialyzed with molecular weight cutoff filters of 30 kDa are also inactive in 3′-tRNA end processing at high-salt concentrations (126). Thus, the tRNase Z is one of a few enzymes from a halophilic archaeon that is not active in vitro under high salt concentrations.

Addition of the 3′-Terminal CCA Sequence

Mature tRNAs from all three domains of life carry the CCA sequence at their 3′ termini (Fig. 1). The CCA terminal sequence is critical for the base-pairing interaction between the 3′ end of the tRNA and the A (loop nucleotides around position 2552; numbering based on *E. coli* 23S rRNA) and P sites (loop nucleotides around position 2251) within the large subunit rRNA (123). Some bacteria and archaea encode tRNA genes that include the CCA sequence, whereas most of the bacterial and archaeal and all of the eucaryal tRNAs require addition of the 3′-terminal CCA sequence. The enzyme responsible for synthesis and regeneration of the CCA sequence is the ATP (CTP):tRNA nucleotidyltransferase (EC 2.7.7.25; also referred to as tRNA terminal transferase); the gene encoding this enzyme has been identified in all three domains. The tRNA nucleotidyltransferases belong to the superfamily of nucleotidyltransferases that also includes poly(A) polymerase and DNA polymerase β. The ability of the tRNA nucleotidyltransferase to add specific nucleotides in the absence of a nucleic acid template makes it an intriguing polymerase. In organisms where many or most tRNA genes do not encode the CCA, the tRNA nucleotidyltransferase gene is essential (35). The first identification of the archaeal homolog was difficult (155), because of the low level of sequence similarity to the bacterial/eucaryal proteins and the low abundance of the protein in cell extracts (39, 155). Mutational analysis of the *Sulfolobus* enzyme showed that there is only a single active site for addition of the CCA nucleotides (35, 156). The archaeal tRNA nucleotidyltransferase from *Archaeoglobus fulgidus* has been cocrystallized with different substrates (tRNA-C, tRNA-CC, and tRNA-CCA) (153). The structures show that the tRNA acceptor stem remains fixed on the enzyme as C_{75} and A_{76} are added (it is not known

yet whether this is also true for addition of C_{74}), while the growing 3′ end refolds to reposition the new 3′-hydroxyl group relative to the incoming nucleotide and the catalytic nucleotidyltransferase motif.

The tRNA nucleotidyltransferases are divided into two classes (class I and class II) that are distinguished by their amino acid sequences (155). Class I tRNA nucleotidyltransferases are present in *Archaea,* whereas class II tRNA nucleotidyltransferases are found in *Bacteria* and *Eucarya.* In certain deep-rooting bacteria like *Aquifex aeolicus* and *Deinococcus radiodurans,* CCA-adding is the joint responsibility of related class II CC- and A-adding enzymes (138). The CC is added by one activity (the CC-adding enzyme) and the A is added by a second, closely related activity (the A-adding enzyme).

INTRONS IN ARCHAEAL TRANSCRIPTS

Introns that disrupt the exon-coding regions of genes have been found in all three domains of life. Intronic sequences are transcribed by RNA polymerases and are removed from the transcript by endonuclease excision. The functional RNA is formed by ligation or splicing of the flanking exons. Three distinct splicing mechanisms for (i) group I introns, (ii) group II, group III, and spliceosomal introns, and (iii) archaeal introns and eucaryal nuclear tRNA introns are known (27, 91, 95, 98, 108). Group I and group III introns have not been detected in archaea, whereas group II introns have been found only in the *Methanosarcinaceae,* members of the *Euryarchaeota* (40, 139). In contrast, several archaeal rRNA and tRNA genes and a single protein-coding gene (149) have been found to contain short intron sequences ranging from 14 to 106 nt in length that are related to the eucaryal nuclear tRNA introns.

Archaeal Group II Introns

The *Methanosarcinaceae* are *Euryarchaeota* with large genomes in the range of 4.1 to 5.75 Mb (48, 58). They are the only archaea that contain group II introns in their genome sequences (40, 139). These introns are similar to bacterial class D group II introns and chloroplast-like class 1 introns. In *M. acetivorans,* 21 group II introns were detected, including seven that lack an internal ORF (40). The other 14 introns that contain an internal ORF appear to encode reverse transcriptases. The internal ORF is, in general, considered to be a canonical feature of all group II introns (40). The ORF-less archaeal group II introns, together with ORF-less group II introns from the cyanobacterium *Thermosynechococcus elongatus*

BP-1, are the only known exceptions to this highly conserved feature.

Introns in tRNA, rRNA, and mRNA Genes

The introns present in tRNA and rRNA genes are variable in sequence but contain a highly conserved structural motif (the bulge-helix-bulge or BHB motif) that forms between the 5′ exon/intron and 3′ intron/exon boundaries in the transcribed RNA (136). Similar introns ranging in size from 11 to 60 nt have been found in eucaryal tRNAs, but these lack the BHB structural motif. Eucaryal tRNA introns are always located in the anticodon loop between nucleotides 37 and 38. Archaeal tRNA introns are most often located at the same position. They can also exist at other locations in the tRNA, however, including the anticodon, 5′ to the anticodon, the amino acid acceptor stem, the D loop, the T loop, the anticodon stem, and the variable arm (93). In a few cases two introns have been found in a single tRNA gene (93).

Many other intron-related anomalies have been observed in the genomes of some archaeal species. In *Nanoarchaeum equitans*, several of the tRNA genes are split into 5′- and 3′-half genes that are well separated on the chromosome and separately transcribed. The trailer of the 5′-half gene and the leader of the 3′-half gene are partially complementary and form an extended helix that includes the anticodon stem of the mature tRNA (see Chapters 8 and 9). The imperfect helix contains a BHB motif and is believed to be a substrate for the splicing endonuclease that initiates a *trans*-splicing reaction generating the mature tRNAs (115). In *Crenarchaeota*, some of the BHB-associated introns that occur in the 23S rRNA genes (23, 41, 77) are large and contain ORFs that encode a homing endonuclease with a conserved LAGLIDADG motif (91). These introns are also excised by the splicing endonuclease (78). In three species of thermophilic and hyperthermophilic crenarchaeota, short introns that generate the BHB motif at the exon/intron boundaries have been detected in the gene encoding the aCbf5 protein. This protein is a component of archaeal H/ACA pseudouridylation guide RNPs (see "Modification of tRNA and rRNA nucleosides," below). A reverse transcriptase (RT) PCR assay was used to demonstrate that the introns are removed from the mRNAs in vivo (149).

The BHB motif that forms at the exon/intron/exon junctions of intron-containing archaeal tRNAs is necessary and suffcient for cleavage by the archaeal splicing endonuclease. In contrast, the eucaryal enzyme uses tRNA structure to recognize the positions of intron excision (108, 141). The archaeal introns are removed by a two-step mechanism. The splicing endonuclease cleaves the intron boundaries at both ends, leaving the 5′ exon with a 2′,3′ cyclic phosphate end group and the 3′ exon with a 5′-OH end group. The tRNA ligase subsequently joins the tRNA exon halves to yield the complete tRNA while circularizing the intron. At the present time no ligase protein has been identified, and it has been suggested that ligation may be RNA catalyzed and mediated by the BHB motif (45, 115). The eucaryal tRNA-splicing pathway has one more step than the archaeal pathway and uses a protein ligase. The tRNA-splicing endonuclease performs a similar reaction, leaving the same end groups, whereas the protein ligase joins the tRNA exon halves by a complex series of reactions requiring ATP and GTP and leaves a 2′-phosphate at the splice site junction. A third enzyme, the 2′-phosphotransferase, is required to remove the 2′-phosphate.

The Splicing Endonuclease

Both the archaeal and eucaryal splicing endonucleases have been identified (EC 3.1.27.9). Within the *Archaea*, three different forms of the splicing endonucleases have been identified. In *Euryarchaeota* the splicing endonuclease (SE) is either an α_2-homodimer (e.g., two 37-kDa subunits in *H. volcanii*) or an α_4-homotetramer (e.g., four 20-kDa subunits in *M. jannaschii*). In *Crenarchaeota* (e.g., in *Sulfolobus solfataricus*) and *Nanoarchaeota* the splicing endonuclease is a heterotetramer consisting of paralogous $\alpha_2\beta_2$-subunits (137). The euryarchaeal SE is more stringent, recognizing only canonical BHB motifs, whereas the crenarchaeal SE also recognizes noncanonical BHB motifs (24). In the *Eucarya* the enzyme is an $\alpha\beta\gamma\delta$-heterotetramer consisting of a 54-kDa, a 44-kDa, a 34-kDa, and a 15-kDa subunit (141). The eucaryal splicing endonuclease cleaves tRNA introns at position 37/38 within a correctly folded tRNA molecule. The 44- and 34-kDa subunits are paralogs to each other and homologs to the archaeal splicing endonuclease α-subunit.

The archaeal splicing endonuclease can not splice eucaryal tRNA introns, but the eucaryal enzyme can recognize and cleave a synthetic RNA substrate containing an archaeal BHB motif in vitro and in vivo (50, 54, 57). The splicing endonuclease may have evolved from the more primitive homotetrameric α_4-type enzyme (93). In addition to the homotetrameric α_4-form, after divergence of *Crenarchaeota* and *Euryarchaeota*, a homodimeric form [$(\alpha-\alpha)_2$] of tRNA endonuclease may have arisen in the *Euryarchaeota* as a result of a duplication that fused two α-like sequences together within a single gene to form the larger protein. In *Crenarchaeota*, the heterotetramer $\alpha_2\beta_2$-form may have appeared following a duplication of the α-gene and divergence of one of the genes to encode the paralogous β-subunit. In the *Eucarya*, a heterotetramer αβγδ appears to have evolved (141) through the retention of the αβ-subunits and the addition of two new protein subunits.

The tRNA Ligase

In addition to the splicing endonuclease, a tRNA ligase (EC 6.5.1.3) is involved in tRNA intron splicing. In the *Eucarya*, at least three distinct ligase activities from yeast, plants, and animals have been observed (52, 108). In yeast, the enzyme is a monomeric 92-kDa protein containing three intrinsic activities: an N-terminal adenylyltransferase domain that resembles T4 RNA ligase 1, a central domain that resembles T4 polynucleotide kinase (without 3'-phosphate activity), and a C-terminal cyclic phosphodiesterase (CPDase) domain (5, 109, 125). Yeast tRNA ligase is also responsible for nonspliceosomal splicing of mRNA in the unfolded protein response pathway (129). In plants it is an enzyme of 125 kDa (*Arabidopsis thaliana*), which contains motifs for three intrinsic enzyme activities: an adenylytransferase/ligase domain, a polynucleotide kinase domain, and a cyclic phosphodiesterase domain (52). Comparison of the yeast and plant ligase sequences reveals no similarities, although the physical order of intrinsic enzymatic activities has been conserved (52). The broad substrate range of plant tRNA ligases clearly suggests that the action of this enzyme is not limited to pre-tRNA splicing or tRNA repair (52). The archaeal ligase reaction appears to be somewhat similar to the animal type, because in both cases, the junction phosphate is derived from the precursor (159). Neither archaeal nor animal enzymes have been sufficiently purified for detailed characterization.

MATURATION OF RIBOSOMAL RNA

The genes for ribosomal RNA (rrn) are located in operons in the archaeal genome and are transcribed to produce multicistronic precursor RNAs (Fig. 2). The number of rRNA operons per genome varies between one (extreme thermophiles) and four (*Methanococcus vannielii*) operons per genome (60) (see Chapter 8). In general, bacterial rRNA operons contain the 16S rRNA gene, one or more tRNA genes in the internal transcribed spacer (ITS), the 23S rRNA gene, the 5S rRNA gene, and one or more distal tRNA genes (111). A similar organization is generally observed in *Euryarchaeota*, where the rRNA operon contains the 16S, 23S, and 5S rRNA genes with a tRNA[Ala] gene located in the ITS and a tRNA[Cys] gene

in the distal position (Fig. 2A). In contrast, in the *Crenarchaeota* (e.g., *Thermoproteus tenax*, *Desulfurococcus mobilis*, *Sulfolobus acidocaldarius*) the 5S and tRNA genes are positioned elsewhere in the genome and are not part of the rRNA operons (80, 111). In euryarchaeal halophiles, the rrn operons are preceded by a series of up to ten tandemly arranged promoters that serve as the sites for transcription initiation and generate transcripts with 5'-external-transcribed spacer (ETS) sequences of variable length (44). The bacterial and archaeal pre-rRNA transcripts in general contain inverted repeats surrounding the 16S and 23S RNA sequences that form extended helical structures and contain the sites for the initial endonucleolytic cleavage and excision of pre-16S and pre-23S from the primary transcript (Fig. 2B). In *E. coli* the endonuclease responsible for the pre-rRNA excision is the helix-specific RNase III. No RNase III-like enzyme has been identified in the *Archaea* (38). The inverted repeats surrounding the large rRNAs in the *Archaea* contain a BHB motif that is recognized by the splicing or BHB endonuclease (29). Extensive conservation of sequence within the 5'-ETS in halophiles suggests that these sequences play an active role in processing of the rRNA or assembly of ribosomal subunits (46).

Two early studies used nuclease protection and primer extension assays to define the intermediates generated during processing of the primary rRNA transcript from two canonical rrn operons (typical euryarchaeotal rrn operons as described above): the single-rrn operon in *Halobacterium salinarum* and the canonical rrnA operon in *Haloarcula marismortui* (Fig. 2B) (29, 47). Those studies demonstrated (i) that the BHB motif within the processing stem surrounding the 16S sequence is a site required for the excision of pre-16S from the primary transcript; (ii) that excision of pre-16S is a prerequisite for RNase P-mediated cleavage at the 5' end of the spacer tRNA[Ala]; (iii) that processing at the 3' end of the spacer tRNA[Ala] is endonucleolytic, is temporarily unordered, and can occur at any point in the processing pathway; (iv) that maturation of the spacer tRNA[Ala] can be disrupted by endo- or exonuclease cleavages within the mature tRNA sequence; (v) that the maturation of the 23S rRNA can occur directly (albeit at low efficiency) without endonuclease excision at the BHB site in the processing stem; and (vi) that the 5S sequence is excised as precursor and trimmed at the 5' and 3' ends. Moreover, similar characterization of the noncanonical rrnB operon of *H. marismortui* (which differs from the canonical rrn operon shown in Fig. 2A since it does not encode tRNA[Ala] and it lacks a BHB motif in the 16S rRNA-processing stem) demonstrates that there is no precursor cleavage in the 5'-ETS preceding

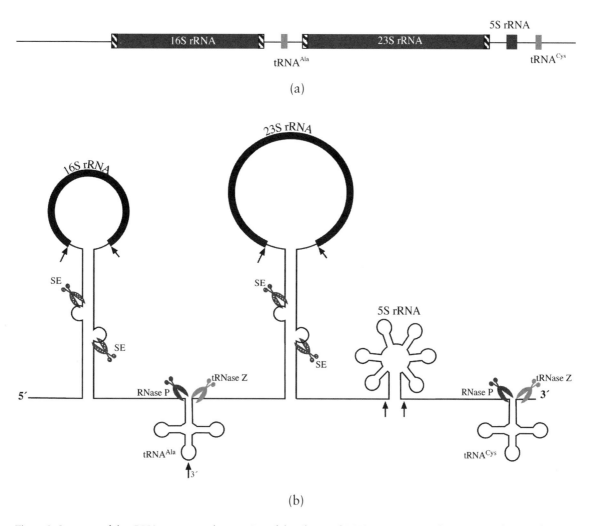

Figure 2. Structure of the rRNA operons and processing of the ribosomal RNA precursor. (a) The structure of a typical rRNA operon from *Euryarchaeota* is shown. 16S, 23S, 5S rRNA, tRNA^Ala, and tRNA^Cys genes are represented by solid black boxes; inverted repeats flanking the 16S and 23S genes are indicated by hatched boxes. Sequences are not drawn to scale. (b) The rRNA operon is transcribed as a multicistronic precursor molecule and is cleaved at numerous sites by a variety of endonucleases (indicated by black arrows and scissors). The splicing endonuclease (SE) cleaves at the BHB motifs to excise the precursor 16S and precursor 23S rRNA sequences, and RNase P and tRNase Z remove tRNA^Ala from the internal transcribed spacer region (47). The activities responsible for maturation at the 5′ and 3′ ends of the 16S and 23S rRNAs have not been identified or characterized. There is some evidence to suggest that the 5S sequence is excised from the primary transcript as a precursor (indicated by black arrows upstream and downstream from the 5S rRNA) and trimmed by a few nucleotides at both the 5′ and 3′ ends to generate the mature 5S sequence (47). The cleavage site in the anticodon loop of the tRNA^Ala (indicated by a black arrow and 3′) was mapped as a 3′-end site and there was no corresponding 5′-end site at the same position (47). Thus this particular 3′ end is either generated by exonucleolytic trimming from the 3′ end of the tRNA, or it is generated by an endonucleolytic cleavage in the anticodon loop and the other product containing the tRNA 3′ half is degraded. The processing sites for the tRNA^Cys have not been mapped experimentally, but tRNA^Cys 5′ and 3′ processing is very likely to be performed by RNase P and tRNase Z, respectively. The 5′-ETS is located upstream of the 16S rRNA, ITS1 is located between the 16S and the 23S rRNA, ITS2 is located between the 23S and the 5S rRNA, and the 3′-ETS is downstream of the 5S rRNA.

the 16S sequence and that excision occurs by endonuclease cleavage at the 5′ end of mature 16S rRNA. To summarize, in vivo analysis of rRNA-processing intermediates indicates that when BHB motifs are present, they are efficiently used for the excision of pre-16S or pre-23S rRNA from the rRNA operon primary transcript.

Cleavage of the BHB-processing sites within pre-rRNA seems to be identical to the cleavage of archaeal introns. Recent studies employing cloning of small noncoding RNAs and PCR reactions across cleavage junctions suggest that the BHB cleavage products can be ligated to each other, yielding circularized pre-16S or pre-23S rRNAs and a ligated leader-spacer-trailer molecule (135). These spacer sequences often contain an RNA kink turn (K-turn) structural motif, which introduces a sharp turn in the backbone structure of the RNA. The aL7Ae protein binds to these K-turn motifs that are formed by box C- and box D-like sequences (45, 135, 157). These box sequences were first recognized as conserved features of eucaryal and archaeal C/D box small RNAs that guide 2'-O-ribose methylation in rRNA. The circular pre-rRNA intermediates are rapidly processed to yield the mature rRNAs.

Introns are sometimes found in archaeal rRNAs: in the 16S rRNA gene of *P. aerophilum* (23), the 23S rRNA genes of *D. mobilis* (77), and the 23S rRNA gene of *Staphylothermus marinus* (79). Due to the presence of the BHB motif that forms between the exon/intron and intron/exon junctions, it is expected that the ribosomal RNA introns are removed by the same splicing endonuclease that removes introns from tRNAs and that has been implicated in the excision of pre-16S and pre-23S from the precursor rRNA transcript (see "Maturation of transfer RNA ends," above). The large free introns derived from pre-rRNAs have been observed as stable and abundant circular RNAs in certain crenarchaeota. In addition tRNA introns have also been found as circular molecules (91, 122).

MODIFICATION OF tRNA AND rRNA NUCLEOSIDES

Almost one hundred different modified nucleosides have been characterized in rRNA, tRNA, mRNA, and other RNAs (118), and virtually all of these are introduced after transcription as modifications of the normal adenosine (A), guanosine (G), cytidine (C), and uridine (U) residues. The proportion of modified nucleotides in tRNA can approach 50% or more (reviewed in reference 14). Modifications in RNA offer an important mechanism for stabilizing the structure of the RNA across the entire temperature range of natural habitats for microorganisms (85, 99). Some RNA modifications are present in all three domains, suggesting that these modifications were present in a progenitor, whereas others are specific for each domain, suggesting that these modifications have evolved after the three domains of life diverged. The two most frequent modifications in RNA are ribose methylation and the pseudouridylation. In *Archaea* and *Eucarya* ribonucleoprotein complexes that use base complementarity to direct modification to specific locations in the target RNA sometimes mediate these modifications. The box C/D RNPs guide 2'-O-ribose methylation (76, 103) and box H/ACA RNPs guide the isomerization of uridine to pseudouridine (9, 59, 134).

Box C/D Methylation Guide RNPs

There are 67 sites of 2'-O-ribose methylation in the rRNA of *S. acidocaldarius* (94). This number is roughly equivalent to the number found in eucarya, such as yeast or human, and an order of magnitude greater that the number found in bacteria such as *E. coli*. Both *Archaea* and *Eucarya* use box C/D RNPs to guide the majority of these methylations. In addition, archaea also use the RNP guide complexes to guide methylation to many positions in tRNAs, whereas in eucarya, all of these modifications are believed to be mediated by proteins that recognize sequence and structural features of the RNA substrate, without a guide RNA. The archaeal RNP complex consists of the C/D box sRNA guide and three proteins: the core RNA-binding protein aL7Ae, the aNop5 (also referred to as aNop56), and the methyltransferase, aFib. The aL7Ae binds directly to box C/D RNAs at the K-turn motifs (81) formed by conserved box C and D, and box C' and D' sequences (32, 86, 116, 151), and nucleates the further addition of two copies of the aNop5 and aFib proteins. The aNop5 protein recognizes features in the RNA K turn that are stabilized by the binding of the aL7Ae protein and serves as a bridge between the sRNA and the catalytic aFib subunit. The aFib protein component of C/D box RNPs has an *S*-adenosylmethionine-binding domain and has been implicated in the methyltransferase activity of the complex (1, 45, 104, 116, 140). Base pairing of the guide RNA with the target RNA positions the substrate nucleotide for 2'-O-methylation by fibrillarin (26, 76). The methylation guide RNP complexes from *Sulfolobus*, *M. jannaschii*, and *A. fulgidus* have been reconstituted from purified recombinant components and are active in in vitro site-specific methylation (104, 116, 140, 158).

In one unusual instance a C/D box methylation guide sequence was identified in the intron of the tRNA^Trp gene from several euryarchaeal species (36). Subsequently it was shown that the guide sequence within the intron was used for methylation of C34 and U39 of tRNA^Trp from *H. volcanii*, and that methylation at both positions precedes removal of the intron (18, 36, 130).

Only a single protein 2'-O-methyltransferase, which catalyzes the archaeal specific methylation of tRNA C56 to Cm56, has been identified (118). It is present in all archaeal genome sequences, except for the crenarchaeon, *P. aerophilum*, which employs an sRNP to modify C56 (118).

Box H/ACA Pseudouridylation Guide RNPs

In *Eucarya* there are dozens of pseudouridine modifications in rRNA; H/ACA guide RNPs mediate their incorporation. In contrast, pseudouridine modifications in rRNA are rare in *Bacteria* and *Archaea*. *Bacteria* use proteins to introduce these modifications, whereas in at least some instances, *Archaea* use the guide H/ACA RNP machines. Four protein components of the archaeal H/ACA RNPs have been identified: aCbf5, aGar1, aL7Ae, and aNop10 (19, 51, 68, 89, 121, 148, 150). Active H/ACA RNPs have been reconstituted from recombinant components from *P. furiosus* and *Pyrococcus abyssi* (8, 31). The evidence indicates that aCbf5 is the pseudouridine synthase which catalyzes the modification (8, 148). The aL7Ae protein is present to stabilize a K-turn motif; this stabilized kink is required for optimal activity of the complex. The other two components are believed to be involved in binding and release of the substrate RNA target and are required for optimal activity (8, 31).

Pseudouridine is a common modification in other types of RNAs; most or all of these modifications appear to be introduced by psuedouridine synthetase enzymes that recognize localized sequence and structure in the substrate RNA and do not involve guide RNPs (30, 82, 101, 102, 154).

The aL7Ae Binds to Many Small Noncoding RNAs

The aL7Ae protein is a component of both C/D box and H/ACA RNPs. The protein plays a critical role early in the assembly pathway of the box C/D box RNP complex (104), but not in box H/ACA RNP complexes (8). The aL7Ae protein recognizes the K-turn motif that is found in many different RNAs from all three domains of life (45, 81, 86). In the *Archaea*, K-turn motifs have been shown to occur in rRNA, C/D box sRNAs, and H/ACA sno-like RNAs (121), and it has been demonstrated that aL7Ae specifically interacts with the K-turn motif in all three of these RNA classes (10, 86, 121). By employing immunoaffinity precipitation with antibodies against aL7Ae, Omer and colleagues showed that there is a large and diverse class of RNAs in *S. solfataricus* that interact with this protein (157). In addition to the C/D box and H/ACA box guide RNAs, the following small RNA fragments were recovered from a library screen: (i) sense strand mRNA sequences originating from within or overlapping with ORFs, (ii) RNAs from intergenic regions, (iii) antisense RNAs from within or overlapping with sense strand ORFs or C/D box sRNAs, (iv) internal regions of 7S RNA, and (v) fragments of rRNAs and tRNAs. Hüttenhoffer and col-

leagues (134) have constructed a similar library of nonselected, small, noncoding RNAs from *S. solfataricus*. Many of the clones from this library overlap with the Omer library (157), and many contain C- and D-box sequences and are able to bind the aL7Ae protein. The antisense noncoding RNAs (ncRNAs) from these complementary libraries may be important posttranscriptional regulators of gene expression, although the mechanism of action has not been investigated (157).

PROCESSING OF mRNA

The processing of mRNA in *Bacteria* and *Eucarya* is highly ordered and contributes to the regulation of gene expression. Half-lives can vary considerably among individual mRNA species. The stabilities of several bacterial transcripts depend on external factors such as temperature or oxygen tension (reviewed in reference 133), whereas the stabilities of eucaryal transcripts vary in response to cellular stimuli and stage of differentiation (reviewed in reference 120). In bacteria the half-lives of individual segments of polycistronic transcripts can vary considerably and contribute to the nonstoichiometric production of operon-encoded proteins (61, 117, 131).

The organizational structure of bacterial and eucaryal mRNAs differ significantly (Fig. 3). Eucaryal mRNAs are mostly monocistronic and may contain introns that disrupt coding regions and have to be removed by splicing. The 5′ end of the transcript is modified by the addition of a methylated guanosine cap, and the 3′ end contains a long nontemplated poly(A) tail that stabilizes the transcript. In contrast, bacterial mRNAs are often polycistronic, have a triphosphate at the 5′ end, and occasionally contain a short nontemplated poly(A) tail at the 3′ end. The 5′-triphosphate provides protection against the endoribonuclease RNase E, whereas processing products with 5′-monophosphates are attacked by RNase E unless hidden within a region of stable secondary structure (92). The short poly(A) tail at the 3′ end of bacterial mRNAs contributes to degradation by providing a platform for the loading of exoribonucleases that can then overcome stable secondary structures that normally protect the RNA from degradation (87).

Similar to bacterial mRNAs, archaeal mRNAs are often polycistronic, and no specialized cap has been found at the 5′ end (21, 22). The mechanisms of mRNA decay in the *Archaea* are presently poorly understood, but the existing data suggest that *Archaea* combine bacterial and eucaryal features of mRNA processing and degradation, and these mech-

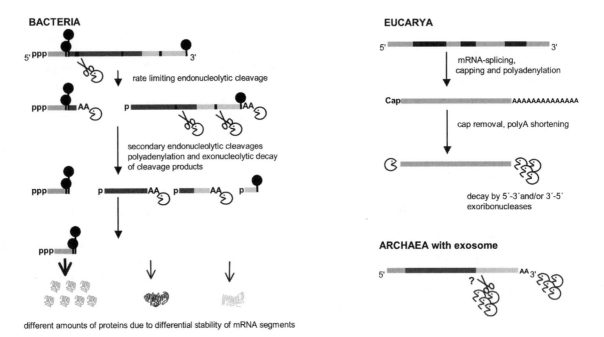

Figure 3. Principles of mRNA decay in *Bacteria* and *Eucarya* and model for mRNA decay by exosomes in *Archaea*. The addition of poly(A) destabilizes transcripts in *Bacteria*, and in the nucleus of *Eucarya*. Exonucleolytic activity of the exosome has been demonstrated, but the involvement of endonucleases in mRNA turnover is still uncertain. Differential stabilities of mRNA segments can control protein levels in *Bacteria* and *Archaea*. Differently shaded bars indicate different ORFs in *Bacteria* and *Archaea*, and introns and exons in *Eucarya*. Short black bars indicate endonucleolytic cleavage sites in bacterial mRNA. Endoribonucleases are symbolized by scissors, exoribonucleases by "pacmen," the exosome by multiple "pacmen." The black hair-pin structures represent stabilizing mRNA secondary structures.

anisms differ significantly among the members within the domain.

Stability of mRNA

In general, the half-lives of bacterial mRNAs are much shorter than those of eucaryal mRNAs. Recent studies have used microarray technology to determine the respective half-lives of all expressed mRNAs in *E. coli* (12, 128), *B. subtilis* (65), and *Saccharomyces cerevisiae* (147). In all these organisms a wide range of stabilities was found for individual mRNAs. More than 80% of the *E. coli* and *B. subtilis* mRNAs exhibited half-lives in the range of 3 to 8 min, whereas in *S. cerevisiae* a mean half-life of 23 min was calculated (147).

Little is known about the stability of mRNAs in *Archaea*. In an early study, the growth of *M. vannielii* was inhibited by addition of bromoethanesulfonate or by the removal of hydrogen, and the decay of several mRNA species was quantified by Northern blot analysis (67). Under these conditions, the half-lives of the *secY, mcr, mva,* and *argG* transcripts varied from 7 to 57 min.

Two, more recent studies have used actinomycin D to block transcription in *S. solfataricus* or *Halo-*

ferax mediterranei (13, 73). In *S. solfataricus*, the following mRNA half-lives were determined by Northern blot analysis: *tfb1* (transcription factor TFIIB paralog), 2 h; *sod* (superoxide dismutase), 2 h; *dgh1* (glucose dehydrogenase), 37 min; *malA* (α-glucosidase), 25 min; *tfb2* (transcription factor TFIIB), 13.5 min; *gln1* (one of three glutamine synthetases), 6.3 min (13). These data suggest that some archaeal mRNAs have significantly longer average half-life than most mRNAs from *E. coli* and *B. subtilis*. The sequences or structural features that contribute to these apparent differences in half-life and the pathway for degradation have not been identified.

In *H. mediterranei*, the divergently transcribed *gvpACNO* and *gvpDEFGHIJKLM* operons encode proteins required for gas vesicle formation, the intercellular structures that are used to regulate the buoyancy of cells in highly saline environments. All of the full-length transcripts and detectable transcript fragments derived from the two operons use either the *gvpA* or *gvpD* promoters, and all share the common 5'-promoter proximal sequences. Full-length operon transcripts are rare, and shorter transcripts with 3' ends more proximal in the operons are always more abundant than longer transcripts with 3' ends located more distal in the operons (119). It is unclear if the mRNA fragments from the two operons are the result of premature transcription termination within the operons or from endonuclease cleavage and differential stability (as shown for a bacterial polycistronic transcript in Fig. 3). In an attempt to distinguish between these two alternatives, actinomycin D was used to inhibit transcription and Northern hybridization was used to follow the decay of the various mRNA fragments. The half-lives of the *gvp* transcript fragments varied from 4 min to 80 min under different conditions of growth. There was a strong correlation between fragment length and stability: the shorter the fragment, the longer the half-life. This correlation was used to argue that the transcript fragments are generated by endonucleotic cleavage rather than premature transcription termination, and that the distal sequences released following cleavage are selectively degraded. This results in a directional 3′ to 5′ degradation of the mRNAs. The most stable and most abundant transcript fragment was the *gvpA* mRNA encoding the major gas vesicle structural protein. Overall, this study suggests that selective transcript stability contributes to the regulation of gas vesicle gene expression. At the present time there is no indication of what the signals for endonuclease cleavage within the mRNA transcripts might be and how the proximal fragment might be stabilized and the distal fragment destabilized.

In both the *S. solfataricus* and the *H. mediterranei* studies, the interpretation of the data is complicated by the use of actinomycin D as an inhibitor of transcription. Actinomycin D inhibits transcription and DNA replication fork progression by intercalating into DNA. If intercalation occurs at random, it means that longer transcripts are more susceptible to inhibition than shorter transcripts. Moreover, uridine incorporation was used to monitor the efficacy of inhibition. It is well known from studies in bacteria that uridine incorporation is efficient only when rRNA synthesis is high and there is a net accumulation of RNA in the culture. Since rRNAs are derived from the transcription of very large operons (~5,000 nt), rRNA synthesis is extremely sensitive to inhibition by actinomycin D. Indeed, there may be synthesis of shorter promoter proximal mRNA fragments even in the absence of detectable uridine incorporation. At this point the data should be taken as suggestive; a more comprehensive understanding will require new technological advances.

Bacterial Type mRNA-Degrading Enzymes

In *E. coli*, and probably other gram-negative bacteria, the endoribonuclease RNase E plays a very important role in mRNA turnover by catalyzing the initial endonuclease cleavage of mRNA molecules (87). An RNase E-like RNA degrading activity has been reported in halophilic *Archaea* (56), but the corresponding protein has not identified. Archaeal genomes encode proteins with very limited sequence similarity to the bacterial endoribonuclease RNase E. One of these proteins, FAU-1 from *P. furiosus*, binds to an AU-rich RNA sequence (75); whether FAU-1 or other RNase E homologs exhibit RNase activity remains to be determined. Some archaeal genomes encode RNase R homologs. RNase R is a 3′ to 5′ exoribonuclease that is involved in bacterial rRNA and tmRNA (an RNA that functions as both mRNA and tRNA during the process of translation) processing and mRNA turnover (e.g., 33, 34, 69). The role of RNase R in mRNA decay in members of the *Archaea* is currently under investigation.

An Exosome-like RNA-Degrading Complex in *Archaea*

In *Bacteria* and *Eucarya*, large multiprotein complexes are involved in mRNA turnover. The degradosome consists of endoribonuclease, exoribonuclease, and helicase activities and has been characterized in *E. coli* (25), *Rhodobacter capsulatus* (74), and a psychrophilic strain of *Pseudomonas syringae* (112) (Fig. 4). The central component of the complex is the endoribonuclease RNase E, which is responsible for the cleavage that initiates the degradation of most mRNA molecules. In *E. coli* PNPase (polynucleotide phosphorylase), a major 3′- to 5′-exoribonuclease is associated with the degradosome, whereas in *P. syringae* RNase E interacts with the hydrolytic 3′- to 5′-exoribonuclease RNase R. These activities act on the natural mRNA 3′ end or the ends generated by endonuclease cleavage by RNase E. The helicase component presumably functions to unwind mRNA secondary structure and allow processive 3′-exonuclease degradation. These RNA-degrading complexes have however been identified only in gram-negative bacteria. Additional enzyme components found in degradosome complexes are enolase and polyphosphate kinase in *E. coli* (16, 25), and the transcription termination factor, Rho, in *R. capsulatus* (74).

The eucaryal exosome has been implicated in rRNA processing and mRNA degradation (2, 3, 114). The eucaryal exosome consists of ten essential proteins known or predicted to possess 3′- to 5′-exonuclease activity (43, 97, 142) (Fig. 4). Six of these proteins (Rrp41 to 43, Rrp45 to 46, and Mtr3) contain an RNase PH domain (RPD) typical for phosphorolytic exoribonucleases. The hydrolytic exoribonucleases Rrp4 and Rrp40 are similar to each other and to another exosome component, the RNA-binding protein Csl4. The hydrolytic exoribonuclease Rrp44

bacterial degradosome complexes

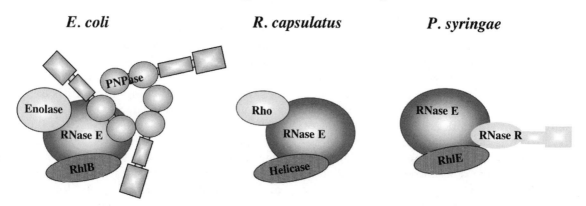

E. coli   *R. capsulatus*   *P. syringae*

core exosome complexes

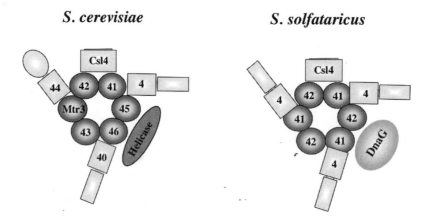

S. cerevisiae   *S. solfataricus*

Figure 4. RNA-degrading protein complexes in members of the three domains of life. The degradosome of gram-negative bacteria is organized around endoribonuclease E. Association of RNase E with an exoribonuclease and one or more helicases seems to be conserved. The exoribonuclease PNPase consists of a trimer, each of which contains two RNase PH domains, one KH- and one S1 RNA-binding domain. A similar hexameric structure was found for the central ring of the exosome in *Archaea* and *Eucarya*, which is composed of six subunits with RNase PH domain. The exosome subunits Rrp4 and Rrp40 show the typical KH- and S1 RNA-binding domains of hydrolytic exoribonucleases. Rrp44 and Csl4, and RNase R comprise the S1 DNA-binding domain. In the *Sulfolobus* exosome core, Rrp41 is catalytically active while Rrp42 is not but contributes to the structuring of the Rrp41 active site.

resembles the bacterial RNase R. This core exosome interacts with additional proteins, including helicases in both its cytoplasmic or nuclear form. The bacterial PNPase is a trimer of subunits, each containing two RPD domains that form a hexameric ring surrounding a central pore (132). It was proposed that the six RPD domains of the eucaryal exosome are arranged in a similar hexameric structure (4, 113).

An analysis of archaeal genomes revealed the presence of orthologs of exosome proteins in all se-

quenced species, with the exception of halophiles and *M. jannaschii* (83). The existence of a eucaryal-like exosome was demonstrated for *S. solfataricus* (53). Four of the *Sulfolobus* exosome proteins have counterparts in the eucaryal exosome (Rrp4, Rrp41, Rrp42, Csl4). An additional protein with sequence similarity to the bacterial DnaG primase was identified as a subunit of the *Sulfolobus* exosome. Based on the subunit composition, a hexameric ring of three Rrp41 and three Rrp42 proteins was suggested. This conformation of the *S. solfataricus* complex was recently confirmed by reconstitution experiments (146a) and by resolving the crystal structure at 2.8-Å resolution; the complex is active in the degradation of mRNA (90).

It has not been determined whether the archaeal exosome has only exoribonucleolytic activity or whether it also serves to organize proteins with endonucleolytic activity in vivo (Fig. 4). Structure-guided mutagenesis studies revealed that the activity of the complex resides within the active sites of the Rrp41 subunits, all three of which face the same side of the hexameric structure. The Rrp42 subunit is inactive but contributes to the structuring of the Rrp41 active site (90).

Polyadenylation of Archaeal mRNAs

Most eucaryal transcripts carry poly(A) tails of ~200 nt. The decay of eucaryal polyadenylated mRNA is initiated by a shortening of the poly(A) to about 15 nt and is followed by cap removal and 5'-3'-exonucleolytic decay (11, 72). More than 20% of *E. coli* mRNAs carry poly(A) tails of 10 to 40 nt; this feature contributes to the destabilization of these mRNAs (15, 100, 124). Poly(A) tails are also added to mRNA fragments that result from RNase E cleavage of full-length mRNAs (66). There is evidence that PNPase and RNase E are involved in poly(A) removal (71, 146). RNase E-catalyzed poly(A) removal is an endonucleolytic process regulated by the phosphorylation status at the 5' end of the mRNA transcript. The rate of poly(A) removal by RNase E is much higher when the 5' end carries a monophospate group, compared with substrates with triphosphate at the 5' end (146).

Some archaeal mRNAs were reported to carry short poly(A) tails (21, 145). The role of these tails in mRNA decay has not been studied. The eucaryal poly(A) polymerases are related to the archaeal CCA-adding enzymes (7), suggesting that these archaeal enzymes possibly have a second function as poly(A) polymerases. More recent analyses have demonstrated the presence of poly(A) tails up to 30 nt in length in *S. solfataricus* mRNAs, but no poly(A) tails have been detected in *H. volcanii* mRNAs (110). The same study revealed that the exosome-like complex of *S. solfataricus* constitutes the major polyadenylating activity in this organism. Polyadenylation is catalyzed by the same Rrp41 active sites that are used for degradation (90). Ongoing studies are addressing whether polyadenylation exists in other branches of the *Archaea*, and which enzymes catalyze polyadenylation and degradation.

Recent studies revealed that polyadenylation of transcripts by the TRAMP complex in the nucleus stimulates degradation by the exosome (88, 143).

This suggests that the ancestral role of poly(A) addition was stimulation of exonucleolytic degradation, and this function is maintained in modern *Bacteria* and *Archaea*.

PERSPECTIVE: THE NEXT FIVE YEARS

In *Archaea* (as well as *Bacteria* and *Eucarya*) our understanding of the processing, modification, and maturation of the major RNA components of the translation machinery (i.e., tRNA and rRNA) is, in many ways, still rudimentary. Nevertheless, the information that is known for rRNA and tRNA is serving as a foundation for understanding the processing, maturation, and degradation of other, less well characterized classes of RNA. In recent years, there has been a revolutionary realization that organisms contain a very large number of small ncRNAs that play an essential role in the regulation of gene expression at various levels. Many of these ncRNA molecules are likely to be synthesized as precursors and subjected to processing and maturation. While still poorly understood, recent evidence suggests that they may control the function, stability, and turnover, including RNA-quality control, of stable RNAs and mRNAs. There is a high likelihood that many of these issues will be resolved in the near future and that many of these processes will be defined within the context of the three-dimensional structure of the cell. It will be interesting to see whether *Archaea* contain subcellular structures that are the progenitor to eucaryal RNA-processing and assembly structures, such as the nucleolus, cajal bodies, and P bodies.

Acknowledgments. We apologize to all colleagues who contributed to the knowledge on RNA processing in *Archaea* and were not cited here due to the lack of space. We thank Hildburg Beier, James W. Brown, Roger Garrett, and Hong Li for critical reading of the manuscript and helpful discussion. Work in the laboratory of A.M. is funded by FCI, VolkswagenStiftung, DFG, and Fritz-Thyssen-Stiftung. Work on mRNA processing in the laboratory of G.K. was funded by DFG, VolkswagenStiftung, and FCI. Work in the laboratory of A.D.O. and P.P.D. was supported by the Canadian Institutes for Health Research, the National Science Foundation, and the Natural Sciences and Engineering Research Council of Canada. Any opinions, findings, and conclusions expressed in this paper are those of the authors and do not necessarily reflect the views of the National Science Foundation.

REFERENCES

1. Aittaleb, M., R. Rashid, Q. Chen, J. R. Palmer, C. J. Daniels, and H. Li. 2003. Structure and function of archaeal box C/D sRNP core proteins. *Nat. Struct. Biol.* 10:256–263.
2. Allmang, C., J. Kufel, G. Chanfreau, P. Mitchell, E. Petfalski, and D. Tollervey. 1999. Functions of the exosome in

rRNA, snoRNA and snRNA synthesis. *EMBO J.* **18:**5399–5410.

3. Allmang, C., P. Mitchell, E. Petfalski, and D. Tollervey. 2000. Degradation of ribosomal RNA precursors by the exosome. *Nucleic Acids Res.* **28:**1684–1691.

4. Aloy, P., F. D. Ciccarelli, C. Leutwein, A. C. Gavin, G. Superti-Furga, P. Bork, B. Bottcher, and R. B. Russell. 2002. A complex prediction: three-dimensional model of the yeast exosome. *EMBO Rep.* **3:**628–635.

5. Apostol, B. L., S. K. Westaway, J. Abelson, and C. L. Greer. 1991. Deletion analysis of a multifunctional yeast tRNA ligase polypeptide. Identification of essential and dispensable functional domains. *J. Biol. Chem.* **266:**7445–7455.

6. Aravind, L. 1999. An evolutionary classification of the metallo-beta-lactamase fold proteins. *In Silico Biol.* **1:**69–91.

7. Aravind, L., and E. V. Koonin. 1999. DNA polymerase beta-like nucleotidyltransferase superfamily: identification of three new families, classification and evolutionary history. *Nucleic Acids Res.* **27:**1609–1618.

8. Baker, D. L., O. A. Youssef, M. I. Chastkofsky, D. A. Dy, R. M. Terns, and M. P. Terns. 2005. RNA-guided RNA modification: functional organization of the archaeal H/ACA RNP. *Genes Dev.* **19:**1238–1248.

9. Balakin, A. G., L. Smith, and M. J. Fournier. 1996. The RNA world of the nucleolus: two major families of small RNAs defined by different box elements with related functions. *Cell* **86:**823–834.

10. Ban, N., P. Nissen, J. Hansen, P. B. Moore, and T. A. Steitz. 2000. The complete atomic structure of the large ribosomal subunit at 2.4 A resolution. *Science* **289:**905–920.

11. Beelman, C. A., and R. Parker. 1995. Degradation of mRNA in eukaryotes. *Cell* **81:**179–183.

12. Bernstein, J. A., A. B. Khodursky, P. H. Lin, S. Lin-Chao, and S. N. Cohen. 2002. Global analysis of mRNA decay and abundance in *Escherichia coli* at single-gene resolution using two-color fluorescent DNA microarrays. *Proc. Natl. Acad. Sci. USA* **99:**9697–9702.

13. Bini, E., V. Dikshit, K. Dirksen, M. Drozda, and P. Blum. 2002. Stability of mRNA in the hyperthermophilic archaeon *Sulfolobus solfataricus*. *RNA* **8:**1129–1136.

14. Björk, G. R. 1995. Biosynthesis and function of modified nucleosides, p. 165–205. *In* D. Söll and U. RajBhandary (ed.), *tRNA: Structure, Biosynthesis, and Function*. ASM Press, Washington, D.C.

15. Blum, E., A. J. Carpousis, and C. F. Higgins. 1999. Polyadenylation promotes degradation of 3'-structured RNA by the *Escherichia coli* mRNA degradosome in vitro. *J. Biol. Chem.* **274:**4009–4016.

16. Blum, E., B. Py, A. J. Carpousis, and C. F. Higgins. 1997. Polyphosphate kinase is a component of the *Escherichia coli* RNA degradosome. *Mol. Microbiol.* **26:**387–398.

17. Boomershine, W. P., C. A. McElroy, H. Y. Tsai, R. C. Wilson, V. Gopalan, and M. P. Foster. (2003) Structure of Mth11/Mth Rpp29, an essential protein subunit of archaeal and eukaryotic RNase P. *Proc. Natl. Acad. Sci. USA* **100:**15398–15403.

18. Bortolin, M. L., J. P. Bachellerie, and B. Clouet-d'Orval. 2003. In vitro RNP assembly and methylation guide activity of an unusual box C/D RNA, cis-acting archaeal pre-tRNA(Trp). *Nucleic Acids Res.* **31:**6524–6535.

19. Bousquet-Antonelli, C., Y. Henry, J. P. G'Elugne, M. Caizergues-Ferrer, and T. Kiss. 1997. A small nucleolar RNP protein is required for pseudouridylation of eukaryotic ribosomal RNAs. *EMBO J.* **16:**4770–4776.

20. Brown, J. W., C. J. Daniels, and J. N. Reeve. 1989. Gene structure, organization, and expression in archaebacteria. *Crit. Rev. Microbiol.* **16:**287–338.

21. Brown, J. W., and J. N. Reeve. 1985. Polyadenylated, noncapped RNA from the archaebacterium *Methanococcus vannielii*. *J. Bacteriol.* **162:**909–917.

22. Brown, J. W., and J. N. Reeve. 1986. Polyadenylated RNA isolated from the archaebacterium *Halobacterium halobium*. *J. Bacteriol.* **166:**686–688.

23. Burggraf, S., N. Larsen, C. R. Woese, and K. O. Stetter. 1993. An intron within the 16S ribosomal RNA gene of the archaeon *Pyrobaculum aerophilum*. *Proc. Natl. Acad. Sci. USA* **90:**2547–2550.

24. Calvin, K., M. D. Hall, F. Xu, X. Song, and H. Li. 2005. Biochemical and structural characterization of the catalytic subunit of a novel RNA splicing endonuclease. *J. Mol. Biol.* **353:**952–960.

25. Carpousis, A. J. 2002. The Escherichia coli RNA degradosome: structure, function and relationship in other ribonucleolytic multienzyme complexes. *Biochem. Soc. Trans.* **30:**150–155.

26. Cavaille, J., M. Nicoloso, and J. P. Bachellerie. 1996. Targeted ribose methylation of RNA in vivo directed by tailored antisense RNA guides. *Nature* **383:**732–735.

27. Cech, T. R. 1990. Self-splicing of group I introns. *Annu. Rev. Biochem.* **59:**543–568.

28. Chamberlain, J. R., Y. Lee, W. S. Lane, and D. R. Engelke. 1998. Purification and characterization of the nuclear RNase P holoenzyme complex reveals extensive subunit overlap with RNase MRP. *Genes Dev.* **12:**1678–1690.

29. Chant, J., and P. Dennis 1986. Archaebacteria: transcription and processing of ribosomal RNA sequences in *Halobacterium cutirubrum*. *EMBO J.* **5:**1091–1097.

30. Charette, M., and M. W. Gray. 2000. Pseudouridine in RNA: what, where, how, and why. *IUBMB Life* **49:**341–351.

31. Charpentier, B., S. Muller, and C. Branlant. 2005. Reconstitution of archaeal H/ACA small ribonucleoprotein complexes active in pseudouridylation. *Nucleic Acids Res.* **33:**3133–3144.

32. Charron, C., X. Manival, A. Clery, V. Senty-Segault, B. Charpentier, N. Marmier-Gourrier, C. Branlant, and A. Aubry. 2004. The archaeal sRNA binding protein L7Ae has a 3D structure very similar to that of its eukaryal counterpart while having a broader RNA-binding specificity. *J. Mol. Biol.* **342:**757–773.

33. Cheng, Z. F., and M. P. Deutscher. 2002. Purification and characterization of the *Escherichia coli* exoribonuclease RNase R. Comparison with RNase II. *J. Biol. Chem.* **277:**21624–21629.

34. Cheng, Z. F., and M. P. Deutscher. 2005. An important role for RNase R in mRNA decay. *Mol. Cell* **17:**313–318.

35. Cho, H. D., and A. M. Weiner. 2004. A single catalytically active subunit in the multimeric *Sulfolobus shibatae* CCA-adding enzyme can carry out all three steps of CCA addition. *J. Biol. Chem.* **279:**40130–40136.

36. Clouet d'Orval, B., M. L. Bortolin, C. Gaspin, and J. P. Bachellerie. 2001. Box C/D RNA guides for the ribose methylation of archaeal tRNAs. The tRNATrp intron guides the formation of two ribose-methylated nucleosides in the mature tRNATrp. *Nucleic Acids Res.* **29:**4518–4529.

37. Cochella, L., and R. Green. 2005. An active role for tRNA in decoding beyond codon:anticodon pairing. *Science* **308:**1178–1180.

38. Condon, C., and H. Putzer. 2002. The phylogenetic distribution of bacterial ribonucleases. *Nucleic Acids Res.* **30:**5339–5346.

39. Cudny, H., J. R. Lupski, G. N. Godson, and M. P. Deutscher. 1986. Cloning, sequencing, and species relatedness of the *Escherichia coli* cca gene encoding the enzyme tRNA nucleotidyltransferase. *J. Biol. Chem.* **261:**6444–6449.

40. Dai, L., and S. Zimmerly. 2003. ORF-less and reverse-transcriptase-encoding group II introns in archaebacteria, with a pattern of homing into related group II intron ORFs. *RNA* 9:14–19.

41. Dalgaard, J. Z., and R. A. Garrett. 1992. Protein-coding introns from the 23S rRNA-encoding gene form stable circles in the hyperthermophilic archaeon *Pyrobaculum organotrophum*. *Gene* 121:103–110.

42. Daviter, T., F. V. T. Murphy, and V. Ramakrishnan. 2005. Molecular biology. A renewed focus on transfer RNA. *Science* 308:1123–1124.

43. Decker, C. J. 1998. The exosome: a versatile RNA processing machine. *Curr. Biol.* 8:R238–R240.

44. Dennis, P. P. 1985. Multiple promoters for the transcription of the ribosomal RNA gene cluster in *Halobacterium cutirubrum*. *J. Mol. Biol.* 186:457–461.

45. Dennis, P. P., and A. Omer. 2005. Small non-coding RNAs in Archaea. *Curr. Opin. Microbiol.* 8:685–694.

46. Dennis, P. P., A. G. Russell, and M. Moniz De Sa. 1997. Formation of the 5′ end pseudoknot in small subunit ribosomal RNA: involvement of U3-like sequences. *RNA* 3:337–343.

47. Dennis, P.P., S. Ziesche, and S. Mylvaganam. 1998. Transcription analysis of two disparate rRNA operons in the halophilic archaeon *Haloarcula marismortui*. *J. Bacteriol.* 180:4804–4813.

48. Deppenmeier, U., A. Johann, T. Hartsch, R. Merkl, R. A. Schmitz, R. Martinez-Arias, A. Henne, A. Wiezer, S. Baumer, C. Jacobi, H. Bruggemann, T. Lienard, A. Christmann, M. Bomeke, S. Steckel, A. Bhattacharyya, A. Lykidis, R. Overbeek, H. P. Klenk, R. P. Gunsalus, H. J. Fritz, and G. Gottschalk. 2002. The genome of *Methanosarcina mazei*: evidence for lateral gene transfer between bacteria and archaea. *J. Mol. Microbiol. Biotechnol.* 4:453–461.

49. Deutscher, M. P. (1995) tRNA processing nucleases, p. 51–65. *In* D. Söll and U. RajBhandary (ed.), *tRNA: Structure, Biosynthesis, and Function*. ASM Press, Washington, D.C.

50. Di Segni, G., L. Borghese, S. Sebastiani, and G. P. Tocchini-Valentini. 2005. A pre-tRNA carrying intron features typical of Archaea is spliced in yeast. *RNA* 11:70–76.

51. Dragon, F., V. Pogacic, and W. Filipowicz. 2000. In vitro assembly of human H/ACA small nucleolar RNPs reveals unique features of U17 and telomerase RNAs. *Mol. Cell. Biol.* 20, 3037–3048.

52. Englert, M., and H. Beier. 2005. Plant tRNA ligases are multifunctional enzymes that have diverged in sequence and substrate specificity from RNA ligases of other phylogenetic origins. *Nucleic Acids Res.* 33:388–399.

53. Evguenieva-Hackenberg, E., P. Walter, E. Hochleitner, F. Lottspeich, and G. Klug. 2003. An exosome-like complex in *Sulfolobus solfataricus*. *EMBO Rep.* 4:889–893.

54. Fabbri, S., P. Fruscoloni, E. Bufardeci, E. Di Nicola Negri, M. I. Baldi, D. G. Attardi, E. Mattoccia, and G. P. Tocchini-Valentini. 1998. Conservation of substrate recognition mechanisms by tRNA splicing endonucleases. *Science* 280:284–286.

55. Frank, D. N., and N. R. Pace. 1998. Ribonuclease P: unity and diversity in a tRNA processing ribozyme. *Annu. Rev. Biochem.* 67:153–180.

56. Franzetti, B., B. Sohlberg, G. Zaccai, and A. von Gabain. 1997. Biochemical and serological evidence for an RNase E-like activity in halophilic Archaea. *J. Bacteriol.* 179:1180–1185.

57. Fruscoloni, P., M. I. Baldi, and G. P. Tocchini-Valentini. 2001. Cleavage of non-tRNA substrates by eukaryal tRNA splicing endonucleases. *EMBO Rep.* 2:217–221.

58. Galagan, J. E., C. Nusbaum, A. Roy, M. G. Endrizzi, P. Macdonald, W. FitzHugh, S. Calvo, R. Engels, S. Smirnov, D. Atnoor, A. Brown, N. Allen, J. Naylor, N. Stange-Thomann, K. DeArellano, R. Johnson, L. Linton, P. McEwan, K. McKernan, J. Talamas, A. Tirrell, W. Ye, A. Zimmer, R. D. Barber, I. Cann, D. E. Graham, D. A. Grahame, A. M. Guss, R. Hedderich, C. Ingram-Smith, H. C. Kuettner, J. A. Krzycki, J. A. Leigh, W. Li, J. Liu, B. Mukhopadhyay, J. N. Reeve, K. Smith, T. A. Springer, L. A. Umayam, O. White, R. H. White, E. Conway de Macario, J. G. Ferry, K. F. Jarrell, H. Jing, A. J. Macario, I. Paulsen, M. Pritchett, K. R. Sowers, R. V. Swanson, S. H. Zinder, E. Lander, W. W. Metcalf, and B. Birren. 2002. The genome of *M. acetivorans* reveals extensive metabolic and physiological diversity. *Genome Res.* 12:532–542.

59. Ganot, P., M. L. Bortolin, and T. Kiss. 1997. Site-specific pseudouridine formation in preribosomal RNA is guided by small nucleolar RNAs. *Cell* 89:799–809.

60. Garrett, R. A., J. Dalgaard, N. Larsen, J. Kjems, and A. S. Mankin. 1991. Archaeal rRNA operons. *Trends Biochem. Sci.* 16:22–26.

61. Grunberg-Manago, M. 1999. Messenger RNA stability and its role in control of gene expression in bacteria and phages. *Annu. Rev. Genet.* 33:193–227.

62. Guerrier-Takada, C., K. Gardiner, T. Marsh, N. Pace, and S. Altman. 1983. The RNA moiety of ribonuclease P is the catalytic subunit of the enzyme. *Cell* 35:849–857.

63. Hall, T.A., and J. W. Brown. 2002. Archaeal RNase P has multiple protein subunits homologous to eukaryotic nuclear RNase P proteins. *RNA* 8:296–306.

64. Hall, T. A., and J. W. Brown. 2004. Interactions between RNase P protein subunits in archaea. *Archaea* 1:247–254.

65. Hambraeus, G., C. von Wachenfeldt, and L. Hederstedt. 2003. Genome-wide survey of mRNA half-lives in *Bacillus subtilis* identifies extremely stable mRNAs. *Mol. Genet. Genomics* 269:706–714.

66. Haugel-Nielsen, J., E. Hajnsdorf, and P. Regnier. 1996. The rpsO mRNA of *Escherichia coli* is polyadenylated at multiple sites resulting from endonucleolytic processing and exonucleolytic degradation. *EMBO J.* 15:3144–3152.

67. Hennigan, A. N., and J. N. Reeve. 1994. mRNAs in the methanogenic archaeon *Methanococcus vannielii*: numbers, half-lives and processing. *Mol. Microbiol.* 11:655–670.

68. Henras, A., Y. Henry, C. Bousquet-Antonelli, J. Noaillac-Depeyre, J. P. Gelugne, and M. Caizergues-Ferrer. 1998. Nhp2p and Nop10p are essential for the function of H/ACA snoRNPs. *EMBO J.* 17:7078–7090.

69. Hong, S. J., Q. A. Tran, and K. C. Keiler. 2005. Cell cycle-regulated degradation of tmRNA is controlled by RNase R and SmpB. *Mol. Microbiol.* 57:565–575.

70. Hopper, A. K. and E. M. Phizicky. 2003. tRNA transfers to the limelight. *Genes Dev.* 17:162–180.

71. Huang, H., J. Liao, and S. N. Cohen. 1998. Poly(A)- and poly(U)-specific RNA 3′ tail shortening by *E. coli* ribonuclease E. *Nature* 391:99–102.

72. Jacobson, A., and S. W. Peltz. 1996. Interrelationships of the pathways of mRNA decay and translation in eukaryotic cells. *Annu. Rev. Biochem.* 65:693–739.

73. Jäger, A., R. Samorski, F. Pfeifer, and G. Klug. 2002. Individual gvp transcript segments in *Haloferax mediterranei* exhibit varying half-lives, which are differentially affected by salt concentration and growth phase. *Nucleic Acids Res.* 30:5436–5443.

74. Jäger, S., O. Fuhrmann, C. Heck, M. Hebermehl, E. Schiltz, R. Rauhut, and G. Klug. 2001. An mRNA degrading complex in *Rhodobacter capsulatus*. *Nucleic Acids Res.* 29:4581–4588.

75. Kanai, A., H. Oida, N. Matsuura, and H. Doi. 2003. Expression cloning and characterization of a novel gene that encodes

the RNA-binding protein FAU-1 from *Pyrococcus furiosus*. *Biochem. J.* 372:253–261.

76. **Kiss-Laszlo, Z., Y. Henry, J. P. Bachellerie, M. Caizergues-Ferrer, and T. Kiss.** 1996. Site-specific ribose methylation of preribosomal RNA: a novel function for small nucleolar RNAs. *Cell* 85:1077–1088.

77. **Kjems, J., and R. A. Garrett.** 1985. An intron in the 23S ribosomal RNA gene of the archaebacterium *Desulfurococcus mobilis*. *Nature* 318:675–677.

78. **Kjems, J., and R. A. Garrett.** 1988. Novel splicing mechanism for the ribosomal RNA intron in the archaebacterium *Desulfurococcus mobilis*. *Cell* 54:693–703.

79. **Kjems, J., and R. A. Garrett.** 1991. Ribosomal RNA introns in archaea and evidence for RNA conformational changes associated with splicing. *Proc. Natl. Acad. Sci. USA* 88:439–443.

80. **Kjems, J., H. Leffers, R. A. Garrett, G. Wich, W. Leinfelder, and A. Bock.** 1987. Gene organization, transcription signals and processing of the single ribosomal RNA operon of the archaebacterium *Thermoproteus tenax*. *Nucleic Acids Res.* 15:4821–4835.

81. **Klein, D. J., T. M. Schmeing, P. B. Moore, and T. A. Steitz.** 2001. The kink-turn: a new RNA secondary structure motif. *EMBO J.* 20:4214–4221.

82. **Koonin, E. V.** 1996. Pseudouridine synthases: four families of enzymes containing a putative uridine-binding motif also conserved in dUTPases and dCTP deaminases. *Nucleic Acids Res.* 24:2411–2415.

83. **Koonin, E. V., Y. I. Wolf, and L. Aravind.** 2001. Prediction of the archaeal exosome and its connections with the proteasome and the translation and transcription machineries by a comparative-genomic approach. *Genome Res.* 11:240–252.

84. **Kouzuma, Y., M. Mizoguchi, H. Takagi, H. Fukuhara, M. Tsukamoto, T. Numata, and M. Kimura.** 2003. Reconstitution of archaeal ribonuclease P from RNA and four protein components. *Biochem. Biophys. Res. Commun.* 306:666–673.

85. **Kowalak, J. A., J. J. Dalluge, J. A. McCloskey, and K. O. Stetter.** 1994. The role of posttranscriptional modification in stabilization of transfer RNA from hyperthermophiles. *Biochemistry* 33:7869–7876.

86. **Kuhn, J. F., E. J. Tran, and E. S. Maxwell.** 2002. Archaeal ribosomal protein L7 is a functional homolog of the eukaryotic 15.5kD/Snu13p snoRNP core protein. *Nucleic Acids Res.* 30:931–941.

87. **Kushner, S. R.** 2002. mRNA decay in Escherichia coli comes of age. *J. Bacteriol.* 184:4658–4665; discussion, 4657.

88. **LaCava, J., J. Houseley, C. Saveanu, E. Petfalski, E. Thompson, A. Jacquier, and D. Tollervey.** 2005. RNA degradation by the exosome is promoted by a nuclear polyadenylation complex. *Cell* 121:713–724.

89. **Lafontaine, D. L., C. Bousquet-Antonelli, Y. Henry, M. Caizergues-Ferrer, and D. Tollervey.** 1998. The box H + ACA snoRNAs carry Cbf5p, the putative rRNA pseudouridine synthase. *Genes Dev.* 12:527–537.

90. **Lorentzen, E., P. Walter, S. Fribourg, E. Evgenieva-Hackenberg, G. Klug, and E. Conti.** 2005. The archaeal exosome core is a hexameric ring structure with three catalytic subunits. *Nat. Struct. Mol. Biol.* 12:575–581.

91. **Lykke-Andersen, J., C. Aagaard, M. Semionenkov, and R. A. Garrett.** 1997. Archaeal introns: splicing, intercellular mobility and evolution. *Trends Biochem. Sci.* 22:326–331.

92. **Mackie, G. A.** 1998. Ribonuclease E is a 5′-end-dependent endonuclease. *Nature* 395:720–723.

93. **Marck, C., and H. Grosjean.** 2003. Identification of BHB splicing motifs in intron-containing tRNAs from 18 archaea: evolutionary implications. *RNA* 9:1516–1531.

94. **McCloskey, J. A., and J. Rozenski.** 2005. The Small Subunit rRNA Modification Database. *Nucleic Acids Res.* 33:D135–138.

95. **Michel, F., and J. L. Ferat.** 1995. Structure and activities of group II introns. *Annu. Rev. Biochem.* 64:435–461.

96. **Minagawa, A., H. Takaku, M. Takagi, and M. Nashimoto.** 2004. A novel endonucleolytic mechanism to generate the CCA 3′ termini of tRNA molecules in *Thermotoga maritima*. *J. Biol. Chem.* 279:15688–15697.

97. **Mitchell, P., E. Petfalski, A. Shevchenko, M. Mann, and D. Tollervey.** 1997. The exosome: a conserved eukaryotic RNA processing complex containing multiple 3′×5′ exoribonucleases. *Cell* 91:457–466.

98. **Moore, M. J., C. C. Query, and P. A. Sharp.** 1993. Splicing of precursors to mRNA by the spliceosome, p. 303–357. *In* R. F. Gesteland and J. F. Atkins (ed.), *The RNA World*. Cold Spring Harbor Laboratory Press, Plainview, N.Y.

99. **Noon, K. R., R. Guymon, P. F. Crain, J. A. McCloskey, M. Thomm, J. Lim, and R. Cavicchioli.** 2003. Influence of temperature on tRNA modification in archaea: *Methanococcoides burtonii* (optimum growth temperature [T_{opt}], 23 degrees C) and *Stetteria hydrogenophila* (T_{opt}, 95 degrees C). *J. Bacteriol.* 185:5483–5490.

100. **O'Hara, E. B., J. A. Chekanova, C. A. Ingle, Z. R. Kushner, E. Peters, and S. R. Kushner.** 1995. Polyadenylylation helps regulate mRNA decay in Escherichia coli. *Proc. Natl. Acad. Sci. USA* 92:1807–1811.

101. **Ofengand, J., M. Del Campo, and Y. Kaya.** 2001. Mapping pseudouridines in RNA molecules. *Methods* 25:365–373.

102. **Ofengand, J., and M. Fournier.** 1998. The pseudouridine residues of rRNA: number, location, biosynthesis and function, p. 229–253. *In* H. Grosjean and R. Benne (ed.), *Modification and Editing of RNA*. AMS Press, Washington, D.C.

103. **Omer, A. D., T. M. Lowe, A. G. Russell, H. Ebhardt, S. R. Eddy, and P. P. Dennis.** (2000) Homologs of small nucleolar RNAs in Archaea. *Science* 288:517–522.

104. **Omer, A. D., S. Ziesche, H. Ebhardt, and P. P. Dennis.** 2002. In vitro reconstitution and activity of a C/D box methylation guide ribonucleoprotein complex. *Proc. Natl. Acad. Sci. USA* 99:5289–5294.

105. **Pace, N. R., and J. W. Brown.** 1995. Evolutionary perspective on the structure and function of ribonuclease P, a ribozyme. *J. Bacteriol.* 177:1919–1928.

106. **Pannucci, J. A., E. S. Haas, T. A. Hall, J. K. Harris, and J. W. Brown.** 1999. RNase P RNAs from some Archaea are catalytically active. *Proc. Natl. Acad. Sci. USA* 96:7803–7808.

107. **Pellegrini, O., J. Nezzar, A. Marchfelder, H. Putzer, and C. Condon.** 2003. Endonucleolytic processing of CCA-less tRNA precursors by RNase Z in *Bacillus subtilis*. *EMBO J.* 22:4534–4543.

108. **Phizicky, E. M., and C. L. Greer.** 1993. Pre-tRNA splicing: variation on a theme or exception to the rule? *Trends Biochem. Sci.* 18:31–34.

109. **Phizicky, E. M., R. C. Schwartz, and J. Abelson.** 1986. *Saccharomyces cerevisiae* tRNA ligase. Purification of the protein and isolation of the structural gene. *J. Biol. Chem.* 261:2978–2986.

110. **Portnoy, V., E. Evgenieva-Hackenberg, F. Klein, P. Walter, G. Klug, and G. Schuster.** 2005. RNA polyadenylation in Archaea: not observed in *Haloferax* while the exosome polyadenylates RNA in *Sulfolobus*. *EMBO Rep.* 6:1188–1193.

111. **Potter, S., P. Durovic, A. Russell, X. Wang, D. de Jong-Wong, and P. P. Dennis.** 1995. Preribosomal RNA processing in archaea: characterization of the RNP endonuclease mediated processing of precursor 16S rRNA in the thermoacidophile *Sulfolobus acidocaldarius*. *Biochem. Cell. Biol.* 73:813–823.

112. Purusharth, R. I., F. Klein, S. Sulthana, S. Jager, M. V. Jagannadham, E. Evguenieva-Hackenberg, M. K. Ray, and G. Klug. 2005. Exoribonuclease R interacts with endoribonuclease E and an RNA helicase in the psychrotrophic bacterium *Pseudomonas syringae* Lz4W. *J. Biol. Chem.* **280:**14572–14578.

113. Raijmakers, R., W. V. Egberts, W. J. van Venrooij, and G. J. Pruijn 2002. Protein-protein interactions between human exosome components support the assembly of RNase PH-type subunits into a six-membered PNPase-like ring. *J. Mol. Biol.* **323:**653–663.

114. Raijmakers, R., G. Schilders, and G. J. Pruijn. 2004. The exosome, a molecular machine for controlled RNA degradation in both nucleus and cytoplasm. *Eur. J. Cell Biol.* **83:** 175–183.

115. Randau, L., R. Munch, M. J. Hohn, D. Jahn, and D. Söll. 2005. *Nanoarchaeum equitans* creates functional tRNAs from separate genes for their 5′- and 3′-halves. *Nature* **433:**537–541.

116. Rashid, R., M. Aittaleb, Q. Chen, K. Spiegel, B. Demeler, and H. Li. 2003. Functional requirement for symmetric assembly of archaeal box C/D small ribonucleoprotein particles. *J. Mol. Biol.* **333:**295–306.

117. Rauhut, R., and G. Klug. 1999. mRNA degradation in bacteria. *FEMS Microbiol. Rev.* **23:**353–370.

118. Renalier, M. H., N. Joseph, C. Gaspin, P. Thebault, and A. Mougin. 2005. The Cm56 tRNA modification in archaea is catalyzed either by a specific 2′-O-methylase, or a C/D sRNP. *RNA* **11:**1051–1063.

119. Röder, R., and F. Pfeifer. 1996 Influence of salt on the transcription of the gas-vesicle genes of *Haloferax mediterranei* and identification of the endogenous transcriptional activator gene. *Microbiology* **142**(Pt 7):1715–1723.

120. Ross, J. 1995. mRNA stability in mammalian cells. *Microbiol. Rev.* **59:**423–450.

121. Rozhdestvensky, T. S., T. H. Tang, I. V. Tchirkova, J. Brosius, J. P. Bachellerie, and A. Hüttenhofer. 2003. Binding of L7Ae protein to the K-turn of archaeal snoRNAs: a shared RNA binding motif for C/D and H/ACA box snoRNAs in Archaea. *Nucleic Acids Res.* **31:**869–877.

122. Salgia, S. R., S. K. Singh, P. Gurha, and R. Gupta. 2003. Two reactions of *Haloferax volcanii* RNA splicing enzymes: joining of exons and circularization of introns. *RNA* **9:**319–330.

123. Samaha, R. R., R. Green, and H. F. Noller. 1995. A base pair between tRNA and 23S rRNA in the peptidyl transferase centre of the ribosome. *Nature* **377:**309–314.

124. Sarkar, N. 1997. Polyadenylation of mRNA in prokaryotes. *Annu. Rev. Biochem.* **66:**173–197.

125. Sawaya, R., B. Schwer, and S. Shuman. 2003. Genetic and biochemical analysis of the functional domains of yeast tRNA ligase. *J. Biol. Chem.* **278:**43928–43938.

126. Schierling, K., S. Rösch, R. Rupprecht, S. Schiffer, and A. Marchfelder. 2002. tRNA 3′ end maturation in archaea has eukaryotic features: the RNase Z from *Haloferax volcanii. J. Mol. Biol.* **316:**895–902.

127. Schiffer, S., S. Rösch, and A. Marchfelder. 2002. Assigning a function to a conserved group of proteins: the tRNA 3′-processing enzymes. *EMBO J.* **21:**2769–2777.

128. Selinger, D. W., R. M. Saxena, K. J. Cheung, G. M. Church, and C. Rosenow. 2003. Global RNA half-life analysis in *Escherichia coli* reveals positional patterns of transcript degradation. *Genome Res.* **13:**216–223.

129. Sidrauski, C., R. Chapman, and P. Walter. 1998. The unfolded protein response: an intracellular signalling pathway with many surprising features. *Trends Cell. Biol.* **8:**245–249.

130. Singh, S. K., P. Gurha, E. J. Tran, E. S. Maxwell, and R. Gupta. 2004 Sequential 2′-O-methylation of archaeal pre-tRNATrp nucleotides is guided by the intron-encoded but trans-acting box C/D ribonucleoprotein of pre-tRNA. *J. Biol. Chem.* **279:** 47661–47671.

131. Steege, D. A. 2000. Emerging features of mRNA decay in bacteria. *RNA* **6:**1079–1090.

132. Symmons, M. F., M. G. Williams, B. F. Luisi, G. H. Jones, and A. J. Carpousis. 2002. Running rings around RNA: a superfamily of phosphate-dependent RNases. *Trends Biochem. Sci.* **27:**11–18.

133. Takayama, K., and S. Kjelleberg. 2000. The role of RNA stability during bacterial stress responses and starvation. *Environ. Microbiol.* **2:**355–365.

134. Tang, T. H., N. Polacek, M. Zywicki, H. Huber, K. Brugger, R. Garrett, J. P. Bachellerie, and A. Hüttenhofer. 2005. Identification of novel non-coding RNAs as potential antisense regulators in the archaeon *Sulfolobus solfataricus. Mol. Microbiol.* **55:**469–481.

135. Tang, T. H., T. S. Rozhdestvensky, B. C. d'Orval, M. L. Bortolin, H. Huber, B. Charpentier, C. Branlant, J. P. Bachellerie, J. Brosius, and A. Hüttenhofer. 2002. RNomics in Archaea reveals a further link between splicing of archaeal introns and rRNA processing. *Nucleic Acids Res.* **30:**921–930.

136. Thompson, L. D., and C. J. Daniels. 1990. Recognition of exon-intron boundaries by the *Halobacterium volcanii* tRNA intron endonuclease. *J. Biol. Chem.* **265:**18104–18111.

137. Tocchini-Valentini, G. D., P. Fruscoloni, and G. P. Tocchini-Valentini. 2005. Structure, function, and evolution of the tRNA endonucleases of Archaea: an example of subfunctionalization. *Proc. Natl. Acad. Sci. USA* **102:**8933–8938.

138. Tomita, K., and A. M. Weiner. 2001. Collaboration between CC- and A-adding enzymes to build and repair the 3′-terminal CCA of tRNA in *Aquifex aeolicus. Science* **294:**1334–1336.

139. Toro, N. 2003. Bacteria and Archaea Group II introns: additional mobile genetic elements in the environment. *Environ. Microbiol.* **5:**143–151.

140. Tran, E. J., X. Zhang, and E. S. Maxwell. 2003. Efficient RNA 2′-O-methylation requires juxtaposed and symmetrically assembled archaeal box C/D and C′/D′ RNPs. *EMBO J.* **22:**3930–3940.

141. Trotta, C. R., F. Miao, E. A. Arn, S. W. Stevens, C. K. Ho, R. Rauhut, and J. N. Abelson. 1997. The yeast tRNA splicing endonuclease: a tetrameric enzyme with two active site subunits homologous to the archaeal tRNA endonucleases. *Cell* **89:**849–858.

142. van Hoof, A., and R. Parker. 1999. The exosome: a proteasome for RNA? *Cell* **99:**347–350.

143. Vanacova, S., J. Wolf, G. Martin, D. Blank, S. Dettwiler, A. Friedlein, H. Langen, G. Keith, and W. Keller. 2005. A new yeast poly(A) polymerase complex involved in RNA quality control. *PLoS Biol.* **3:**e189.

144. Vogel, A., O. Schilling, B. Späth, and A. Marchfelder. 2005. The tRNase Z family of proteins. Physiological functions, substrate specificity and structural properties. *Biol. Chem.* **386:**1253–1264.

145. Volkl, P., P. Markiewicz, C. Baikalov, S. Fitz-Gibbon, K. O. Stetter, and J. H. Miller. 1996. Genomic and cDNA sequence tags of the hyperthermophilic archaeon *Pyrobaculum aerophilum. Nucleic Acids Res.* **24:**4373–4378.

146. Walsh, A. P., M. R. Tock, M. H. Mallen, V. R. Kaberdin, A. von Gabain, and K. J. McDowall. 2001. Cleavage of poly(A) tails on the 3′-end of RNA by ribonuclease E of *Escherichia coli. Nucleic Acids Res.* **29:**1864–1871.

146a. Walter, P., F. Klein, E. Lorentzen, A. Ilchmann, G. Klug, and E. Evguenieva-Hackenberg. 2006. Characterization of native and reconstituted exosome complexes from the hyperthermophilic archaeon *Sulfolobus solfataricus. Mol. Microbiol.* **62:**1076–1089.

147. Wang, Y., C. L. Liu, J. D. Storey, R. J. Tibshirani, D. Herschlag, and P. O. Brown. 2002. Precision and functional specificity in mRNA decay. *Proc. Natl. Acad. Sci. USA* **99**:5860–5865.

148. Watanabe, Y., and M. W. Gray. 2000. Evolutionary appearance of genes encoding proteins associated with box H/ACA snoRNAs: cbf5p in *Euglena gracilis*, an early diverging eukaryote, and candidate Gar1p and Nop10p homologs in archaebacteria. *Nucleic Acids Res.* **28**:2342–2352.

149. Watanabe, Y., S. Yokobori, T. Inaba, A. Yamagishi, T. Oshima, Y. Kawarabayasi, H. Kikuchi, and K. Kita. 2002. Introns in protein-coding genes in Archaea. *FEBS Lett.* **510**:27–30.

150. Watkins, N. J., A. Gottschalk, G. Neubauer, B. Kastner, P. Fabrizio, M. Mann, and R. Lührmann. 1998. Cbf5p, a potential pseudouridine synthase, and Nhp2p, a putative RNA-binding protein, are present together with Gar1p in all H BOX/ACA-motif snoRNPs and constitute a common bipartite structure. *RNA* **4**:1549–1568.

151. Watkins, N. J., V. Segault, B. Charpentier, S. Nottrott, P. Fabrizio, A. Bachi, M. Wilm, M. Rosbash, C. Branlant, and R. Lührmann. 2000. A common core RNP structure shared between the small nucleoar box C/D RNPs and the spliceosomal U4 snRNP. *Cell* **103**:457–466.

152. Xiao, S., F. Scott, C. A. Fierke, and D. R. Engelke. 2002. Eukaryotic ribonuclease P: a plurality of ribonucleoprotein enzymes. *Annu. Rev. Biochem.* **71**:165–189.

153. Xiong, Y., and T. A. Steitz. 2004. Mechanism of transfer RNA maturation by CCA-adding enzyme without using an oligonucleotide template. *Nature* **430**:640–645.

154. Yu, Y. T., R. M. Terns, and M. P. Terns. 2005. Mechanisms and functions of RNA-guided RNA modification, p. 223–262. *In* Grosjean, H. (ed.), *Fine-Tuning of RNA Functions by Modification and Editing*, vol. 12. *Topics in Current Genetics*. Springer Verlag, New York, N.Y.

155. Yue, D., N. Maizels, and A. M. Weiner. 1996. CCA-adding enzymes and poly(A) polymerases are all members of the same nucleotidyltransferase superfamily: characterization of the CCA-adding enzyme from the archaeal hyperthermophile Sulfolobus shibatae. *RNA* **2**:895–908.

156. Yue, D., A. M. Weiner, and N. Maizels. 1998. The CCA-adding enzyme has a single active site. *J. Biol. Chem.* **273**:29693–29700.

157. Zago, M. A., P. P. Dennis, and A. D. Omer. 2005. The expanding world of small RNAs in the hyperthermophilic archaeon Sulfolobus solfataricus. *Mol. Microbiol.* **55**:1812–1828.

158. Ziesche, S. M., A. D. Omer, and P. P. Dennis. 2004. RNA-guided nucleotide modification of ribosomal and non-ribosomal RNAs in Archaea. *Mol. Microbiol.* **54**:980–993.

159. Zofallova, L., Y. Guo, and R. Gupta. 2000. Junction phosphate is derived from the precursor in the tRNA spliced by the archaeon *Haloferax volcanii* cell extract. *RNA* **6**:1019–1030.

Archaea: Molecular and Cellular Biology
Edited by Ricardo Cavicchioli
© 2007 ASM Press, Washington, D.C.

Chapter 8

Translation

PAOLA LONDEI

INTRODUCTION

Translation is the last step of the gene expression pathway, whereby the genetic message carried from the DNA in the form of messenger RNA (mRNA) is converted into the final product: a polypeptide chain. The synthesis of a protein is an extraordinarily complex process, which requires the participation of a host of small and large molecules that are centered around what is perhaps the cell's most wonderful macromolecular machine, the ribosome. The importance of the translational apparatus in the cellular economy is perhaps best underscored by the fact that up to one-fourth of the overall metabolic energy in an actively growing cell is devoted to the synthesis of translational components.

Being a fundamental cellular process, translation is well conserved across the three domains of life. It is now generally accepted that the translation apparatus must have already attained a fair degree of complexity in the Last Universal Common Ancestor (LUCA) of extant cells. The LUCA was likely to have been endowed with ribosomes that are very similar in structure and function to present-day ribosomes, in addition to having tRNAs, aminoacyl-tRNA synthetases, and some translation factors that are also similar to those of extant cells (4, 18, 57, 61, 81). However, there are also important differences in composition and complexity of the protein synthesis machinery of present-day cells in the three domains of life. Until recently, conventional wisdom stated that there were two different versions of the translation apparatus: a simple, streamlined form present in the *Bacteria* and a more complex form found in *Eucarya*. This dichotomy seemed to be the logical result of the different organization and lifestyles of "eukaryotic" and "prokaryotic" cells (see Chapter 1). According to this logic, the "simpler" bacteria, whose basic evolu-

tionary strategy consisted of maximizing the velocity of growth and multiplication, had gene expression machinery made of fewer and often smaller components. In contrast, the complexity of *Eucarya* (e.g., metazoans) naturally led to the expectation that they would have a sophisticated and complicated gene expression apparatus.

The discovery of the third domain of life, the *Archaea*, has challenged in many ways the classical textbook dichotomy between "prokaryotes" and "eukaryotes" (119). As far as cellular organization is concerned, all known archaea are unicellular prokaryotes. However, a host of phylogenetic, molecular, biochemical, and genomic studies have now shown that the *Archaea* are clearly distinct from the other prokaryotic order, the *Bacteria*, and are instead specifically related to the *Eucarya*. The affinity between the *Archaea* and the *Eucarya* is particularly evident in the structure of the translation apparatus, to the extent that, far from being "simple prokaryotes," the *Archaea* show an unexpected complexity in some aspects of translation, and this is currently one of the more puzzling and challenging topics in need of experimental investigation (11, 30). Due to this, the study of the translational machinery in archaea is not only interesting in its own right but also provides opportunities for gaining new insight into the evolutionary history of the protein synthesis mechanism.

The importance of translation to cellular function can not be overstated. The emergence of translation as a process was key to the evolution of modern cellular life (118). Primitive "life" based on self-replicating nucleic acids without translation is conceivable. For example, the hypothesis of an "RNA world" posits the existence of RNA-based ancestral entities, entirely without a proteome (20, 47, 68, 88). However, successful self-propagating cells were created only with the emergence of translation,

Paola Londei • Department of Medical Biochemistry, Biology and Physics (DIBIFIM), Università degli Studi di Bari (Policlinico), Bari, Italy.

which established a stable link between a nucleic acid-based genotype and a protein-based phenotype. It is very unlikely that we will ever completely understand how translation came into existence. However, the discovery of the *Archaea* has rekindled interest in the study of early cellular evolution and has made it possible to tackle the problem of the evolution of translation from a novel vantage point (118). This chapter describes what is known about the translational apparatus and the protein-synthesis mechanism in archaea. It also highlights how this understanding increased and will further lead to a greater understanding of the evolution of translation in all forms of life.

COMPOSITION AND STRUCTURE OF THE ARCHAEAL TRANSLATION APPARATUS

General Overview

The composition of the translation apparatus is basically the same in all cells and cell organelles. The ribosomes are large ribonucleoprotein complexes composed of two unequal subunits. In archaea, the small ribosomal subunits sediment at about 30S and contain a single RNA molecule of about 1,500 nt (small-subunit [SSU] rRNA) and 25 to 28 proteins, depending on the organism. The large ribosomal subunits sediment at about 50S and contain a large RNA molecule of about 3,000 nt (large-subunit [LSU] rRNA), a smaller one of about 120 nt (5S RNA), and 35 to 40 proteins, depending on the organism.

Other essential components of the protein synthesis machinery that are found in all cells are specific sets of proteins known as translation factors. These are necessary to assist the different stages of translation, i.e., initiation, elongation, and termination. To date, computational analyses of about 20 complete archaeal genome sequences have identified six putative initiation factors (IFs), two elongation factors (EFs), and one release or termination factor (RF). It is possible, however, that other factors will be identified in the future through the progress of biochemical and genetic studies of archaeal translation.

Comparative genomic analyses have also revealed that the three domains of life possess a complex set of tRNAs and aminoacyl-tRNA synthetases. Archaeal tRNAs and their charging enzymes are covered in Chapters 7 and 9. Other proteins that participate in the synthesis of the translation apparatus or in the posttranslational processing of proteins are described in Chapter 7 (rRNA processing and modifying enzymes) and Chapter 10 (chaperones and proteasomes).

Archaeal Ribosomes

Historical overview

The basic architecture of the ribosome is evolutionarily conserved. Nevertheless, it has been known for many years that domain-specific differences exist between bacterial and eucaryal ribosomes. The latter are bigger and compositionally more complex than the former; their component rRNAs are larger by several hundred nucleotides, and they typically contain more and somewhat larger proteins. The structural differences between bacterial and eucaryal ribosomes are also underscored by their differential sensitivity to ribosome-targeted antibiotics.

Biochemical and structural analyses of archaeal ribosomes started in the eighties, soon after the discovery of the third domain of life. Note that the discovery of the *Archaea* can be attributed to the field of ribosomal studies, since it stemmed from the recognition that the sequences of archaeal SSU rRNAs formed a coherent cluster sharply separated from the equally coherent clusters formed by the bacterial and eucaryal SSU rRNA sequences (120) (see Chapters 1 and 2).

The gross chemical composition of archaeal ribosomes from different species of *Crenarchaeota* and *Euryarchaeota* was initially analyzed by many techniques such as gel electrophoresis and density gradient sedimentation (112). The general finding was that the ribosomes and ribosomal subunits of most archaea had sedimentation coefficients of 30S, 50S, and 70S and included 16S, 23S, and 5S rRNAs, similar to their bacterial counterparts. However, the ribosomes of certain archaea contain more proteins than bacterial ribosomes; this was the case, in particular, for the small subunit (75). A larger number of proteins was a characteristic of sulfur-dependent thermophiles, while halophiles and methanogens had ribosomes more similar in composition to their bacterial counterparts (3). Observations made by electron microscopy also showed that the ribosomes of sulfur-dependent thermophiles had morphological characteristics similar to those of eucaryal ribosomes (66–67). This was particularly the case for the small ribosomal subunits, which possessed a "bill" on the head and "lobes" of the body, similar to those observed in eucaryal, but not bacterial, counterparts. The fact that these features were absent in the ribosomes of halophiles and some methanogens led Lake and coworkers to propose that the sulfur-dependent archaea constituted a separate order (referred to as the "eocytes") specifically related to eucarya (96). Although the "eocyte" hypothesis proved untenable, data accumulated since the beginning of the genome-sequencing era have essentially confirmed that archaeal ribosomes have a composition that is more similar to eucaryal ribosomes, despite the overall size of the archaeal ribosomes being more similar to bacterial ribosomes.

Archaeal rRNAs

A survey of the rRNAs present in 21 fully sequenced archaeal genomes is shown in Table 1. The sizes of archaeal SSU RNA range from about 1430 to ~1,500 nt, while LSU rRNAs are ~2,900 to 3,100 nt in size. However, "extra-large" rRNAs are observed

Table 1. Size[a] and number[b] of genes for ribosomal RNAs in *Archaea*

Organism	23S	16S	5S	5.8S
Crenarchaeota				
A. pernix	4,413[c]	1,423[c]	119	167
			132	
S. solfataricus	3,049	1,496	121	
S. tokodaii	3,012	1,445	125	
P. aerophylum	3,024	2,210[d]	130	
Euryarchaeota				
Archeoglobales				
A. fulgidus	2,931	1,491	123	
Halobacteriales				
H. marismortui	2,921 (3)	1,471 (3)	121 (3)	
Halobacterium sp.				
NRC-1	2,905 (3)	1,472 (3)	122 (3)	
Methanobacteriales				
M. thermauto-				
trophicus	3,028	1,478 (2)	126	
	3,034		128	
Methanococcales				
M. jannaschii	2,889	1,474	119 (2)	
	2,948	1,477	119	
M. maripaludis	2,956 (3)	1,391 (3)	114 (4)	
Methanosarcinales				
M. mazei	2,892 (3)	1,473 (3)	134 (2)	
			132 (1)	
M. acetivorans	2,831	1,429 (3)	134 (2)	
	2,848		132 (1)	
	2,948			
Methanopyrales				
M. kandleri	3,097	1,511	132	
Thermococcales				
P. furiosus	3,048	1,446	125	
			121	
P. horikoshii	3,857	1,494	121 (2)	
P. abyssi	3,017	1,502	121 (2)	
T. kodarakaensis	3,028	1,497	125 (2)	
Thermoplasmata				
T. acidophilum	3,044	1,470	122	
T. volcanium	2,906	1,469	122	
P. torridus	2,895	1,468	120	
Nanoarchaeota				
N. equitans	2,861	1,344	122	

[a]Number of nucleotides in each rRNA gene.
[b]Number of genes, if more than one, is indicated in parentheses, except when the genes are of different length, in which case their size is shown in full.
[c]Gene containing two introns.
[d]Gene containing one intron.

in some archaea. The aberrantly large sizes are due to the presence of introns, which are found, although infrequently, in both the SSU and LSU rRNA genes. For example, the 2,210-nt 16S RNA gene of *Pyrobaculum aerophylum* includes a 713-nt intron. Likewise, the LSU rRNA gene of *Aeropyrum pernix*, which measures 4,413 nt, is endowed with a 202- and a 575-nt intron. *A. pernix* strain K1 also harbors an intron in the SSU rRNA, making it the one species that possesses introns in both the LSU and the SSU rRNAs (84). Other archaeal genera with rRNA introns are *Desulfurococcus* and *Thermoproteus* (45, 56). Intron splicing is carried out by an archaeal splicing endonuclease that recognizes a characteristic secondary structure, described as the helix-bulge-helix motif (77).

The small 5S rRNA is rather constant in size in all archaea, ranging from a minimal length of 119 nt to a maximum of 134 nt. Notably, *A. pernix* is unique in apparently containing a 5.8S RNA homolog of 167 nt in addition to two different 5S RNA genes of 119 and 132 nt.

The primary and secondary structures of the rRNAs are extremely well conserved throughout evolution. The secondary structures of archaeal LSU and SSU rRNAs are almost superimposable on their bacterial counterparts. Only very subtle differences are evident, and these consist of the occasional addition, lengthening, or shortening of hairpin elements of helical segments (Fig. 1).

Ribosomal Proteins

The availability of genome sequences has enabled a detailed picture of the protein composition of archaeal ribosomes to be derived (69). Sixty-eight r-proteins (of a known total of 102) are represented in *Archaea*; 28 belong to the SSU and 40 to the LSU. Thirty four (15 in the SSU and 19 in the LSU) are universal proteins present in the ribosomes of all three domains of life. Another 33 r-proteins (13 SSU and 20 LSU) are shared by *Archaea* and *Eucarya* but not by *Bacteria*. No r-proteins are shared by *Bacteria* and *Archaea* but absent in *Eucarya*. Only one r-protein (LXa in the LSU) is unique to the *Archaea* (69).

The total number of r-proteins in *Bacteria, Archaea*, and *Eucarya* is 57, 68, and 78, respectively. The ribosomal protein complement of archaeal ribosomes is entirely represented in the *Eucarya*. The similarity between the protein composition of archaeal and eucaryal ribosomes is further illustrated by the fact that, among the universal r-proteins, the archaeal and eucaryal homologs have the highest level of sequence identity. On the other hand, the substantial divergence of the bacterial ribosome from the archaeal/

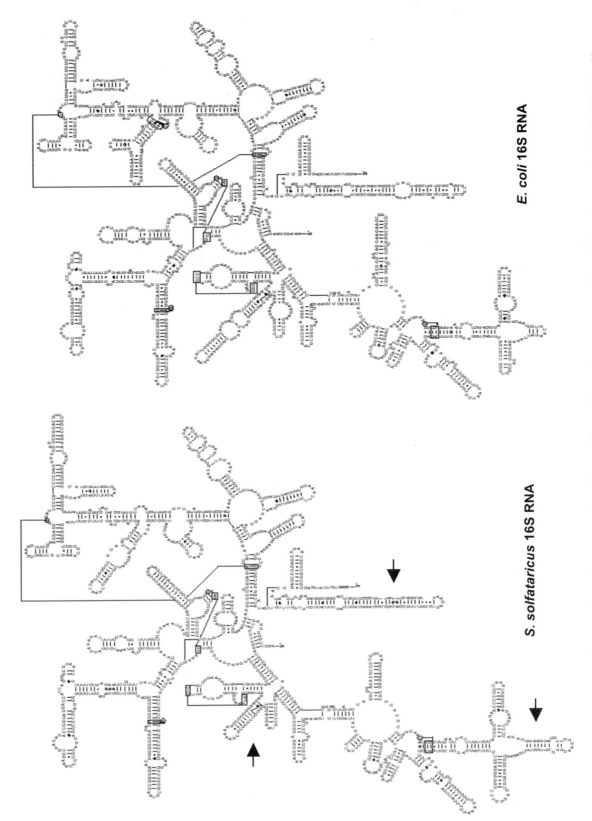

Figure 1. Comparison of 16S RNA secondary structure in archaea and bacteria. Secondary structure models are shown for one archaeal 16S rRNA (*S. solfataricus*) and one bacterial (*E. coli*) 16S RNA. The regions where the structures differ are indicated by arrows. The black lines represent identified tertiary interactions between nucleotides. Data taken from the http://www.rna.icmb.utexas.edu/. The details of the database are described in reference 17.

S. solfataricus 16S RNA

E. coli 16S RNA

eucaryal type is highlighted by the fact that 23 r-proteins are exclusive to bacteria (69).

The protein composition of archaeal ribosomes is not the same in all archaea and reflects the phylogeny of the host species (69). The branching order of several archaeal genera in the "universal tree of life" is shown in Fig. 2. On the whole, ribosomes of the early-branching *Crenarchaeota* (e.g., *Sulfolobus*) appear to have more ribosomal proteins than the *Euryarchaeota* (e.g., *Thermoplasma*). At one extreme, the crenarchaeon *A. pernix* is endowed with the full complement of 68 archaeal r-proteins, while the ribosomes of the late-branching halobacteria and thermoplasmales lack as many as 10 proteins, approaching the number found in bacteria (Table 2). All of the missing proteins are those that are shared by the *Archaea* and *Eucarya*. Moreover, protein loss seems to have evolved gradually in the 50S subunit, while being stepwise in the 30S particle, the 30S having either 28 r-proteins in all *Crenarchaeota* or 25 in all *Euryarchaeota* (Table 2).

This progressive reduction of the number of ribosomal proteins is not observed in bacterial and eucaryal ribosomes, whose composition remains essentially the same in most species, with only some exceptions in parasitic ones such as *Giardia lamblia* (69).

The fact that the archaeal species with "heavy" ribosomes (larger numbers of r-proteins) are deep-branching suggests that they were also present in the ancestor of the *Archaea* and *Eucarya*. The eucaryal ribosomes may have gained more proteins in the course of evolution, while the bacteria appeared to undergo a marked reduction in their complement of ribosomal proteins. It may be inferred that the 34 universal proteins were present in the LUCA. In support of this, a recent study of the composition of a "minimal" or "ancestral" ribosome reached the conclusion that it should contain 35 to 40 proteins (81).

Ribosome Architecture

The general architecture of the ribosome is conserved in all cells. However, both the small and the large ribosomal subunits have domain-specific RNA and protein composition. As noted above ("Historical overview"), the shape of eucaryal ribosomes is somewhat different from that of bacterial ribosomes (3). In recent years, spectacular progress has been made using X-ray crystallography and cryo-electron microscopy to analyze ribosome architecture. It is somewhat paradoxical that, although most data on ribosome function has been obtained with *Escherichia coli*

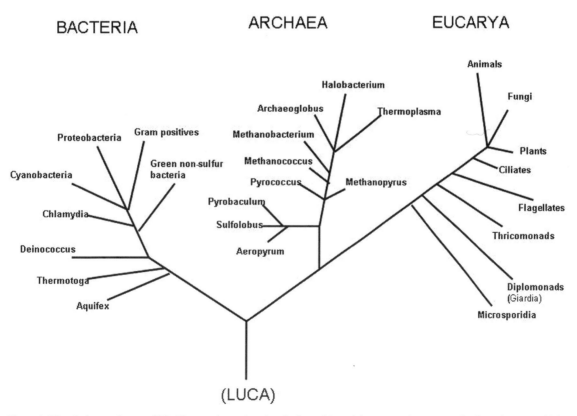

Figure 2. The phylogenetic tree of life. Unrooted tree showing the branching of the principal species in the three domains of life. Adapted from Lecompte et al. (69).

Table 2. Differential protein composition of archaeal ribosomes

Organism	Missing SSU r-proteins (of 28 total)	Missing LSU r-proteins[a] (of 39 total)
Crenarchaeota		
A. pernix	–[b]	
P. aerophylum	–	L35ae
S. solfataricus	–	L35ae; L38e
S. tokodaii	–	=[c]
Euryarchaeota		
P. furiosus	S25e; S26e; S30e	L13e; L38e
P. horikoshii	=	=
P. abyssi	=	=
T. kodakaraensis	=	=
M. kandleri	=	L13e; L38e
M. jannaschii	=	L13e; L38e; L35ae
M. thermauto-		
trophicus	=	=
M. mazei	=	L13e; L14e; L34e; L35ae; L38e
A. fulgidus	=	=
T. acidophilum	=	L13e; L14e; L30e; L34e; L35ae; L38e; Lxa
T. volcanii	=	=
Halobacterium		
sp. NRC-1	=	=
H. marismortui	=	=

[a]Protein L41 is not included in the table because its presence/absence is uncertain in several species.
[b]–, no proteins missing.
[c]=, same missing proteins as above.

ribosomes, the most detailed structural studies have been performed on the ribosomes of other bacteria and archaea. *E. coli* ribosomes have proven to be particularly resistant to good crystallization. The best crystallographic analyses of ribosome structure have been performed using components from extremophilic bacteria (e.g., *Thermus thermophilus* and *Deinococcus radiodurans*) (16, 42) or archaea.

Currently, the best model of the large ribosomal subunit has been obtained from crystals of the *Haloarcula marismortui* LSU (6). *H. marismortui* has a relatively small ribosome (having a number of r-proteins similar to bacteria). Therefore, the *H. marismortui* 50S structure is not a good model for studying ribosomes that are composed of the full array of archaeal r-proteins. For example, the archaeal-specific protein, LXa, is not present in *H. marismortui* (or other halophiles). Despite this limitation, some interesting observations can be made; one relates to protein L7ae, which is present in the *Eucarya* and the *Archaea* but not present in *Bacteria*. L7ae was originally identified as a ribosomal protein. However, it was subsequently found to function also as a compo-

nent of the machinery for rRNA posttranscriptional modification (87). As described in more detail below (Ribosome biosynthesis), L7ae has homology with the eucaryal protein snu13p, which is the RNA-binding element of the snoRNPs involved in posttranscriptional modification of the rRNA transcripts (60). L7ae is clearly identifiable in the three-dimensional structure of the *H. marismortui* 50S subunit, showing that it is a bona fide ribosomal protein. However, consistent with its multifunctional character, it is located at the periphery of the subunit and is one of the few r-proteins that makes contact with only one rRNA domain. Its function in the ribosome is unclear (6). Future studies examining the structures of whole ribosomal subunits from other archaea will help to identify the architectural features that may be unique to archaeal ribosomes.

Ribosome biosynthesis

The only information that is available on the mechanism of ribosome biosynthesis in archaea relates to the processing and maturation of the rRNA transcripts. With the exception of the thermoplasmales, the LSU and SSU rRNA genes are located adjacent to one another in archaeal genomes and are cotranscribed as a single large precursor molecule, similar to bacteria and eucarya.

An important step of rRNA maturation that differentiates the three domains of life is the posttranscriptional chemical modification of certain nucleotides. In bacteria, relatively few types of rRNA modifications occur, and these are mainly limited to the bases: the most common are A and G methylations and a particular kind of isomerization that transforms uracil into pseudo-uracil. Among the *Archaea*, the *Euryarchaeota* display a pattern of rRNA posttranscriptional modification that is similar to *Bacteria*. In contrast, the rRNAs of the *Crenarchaeota* exhibit a diverse and abundant pattern of modifications similar to *Eucarya*. In addition to base methylation and pseudo-uridilations, numerous examples of methylation on the 2'-OH group of ribose have been observed (79).

Remarkably, the *Crenarchaeota* and *Eucarya* employ the same elaborate mechanism for inserting ribose methylations in their rRNAs. In both types of cells, the rRNA-methylating enzymes are ribonucleoprotein (RNP) complexes containing small RNA molecules (5, 51, 52, 113). In these complexes, the proteins provide the catalytic activity, while the RNA molecule acts as the guide for target recognition. The small RNAs found in the RNPs involved in 2'-OH ribose methylation are called the C/D box sRNAs (snoRNAs in the *Eucarya*). All C/D box sRNAs contain two conserved sequences, the C box (RUGAUGA, where R stands for purine) and the D box (CUGA) located near the 5′ and 3′ end, respectively. In all archaeal and some eucaryal C/D box snoRNAs, a second set of conserved sequences, C′ and D′, is located in the central region of the molecule between the C and D boxes (55, 86). The C/D box snoRNAs guide site-specific modification by base-pairing with the pre-rRNA at the target sites. Specifically, C/D box snoRNAs contain one, or sometimes two, sequences of 10 to 21 nt located immediately upstream of D and D′ boxes and complementary to the target RNA. Methylation takes place on the ribose of the nucleotide located exactly five bp upstream of box D or D′ (19, 86).

In eucarya, four core proteins associate with all C/D box snoRNAs: NOP56, NOP58, fibrillarin, and a 15.5-KDa protein (termed Snu13p in yeast). Fibrillarin probably acts as the methylase enzyme (82, 117), while the RNA-binding component of the complex is the 15.5-kDa protein. All archaeal genomes sequenced to date contain genes encoding proteins homologous to the components of eucaryal snoRNA complexes: fibrillarin, Nop56/58 (a single protein for NOP56 and NOP58), and L7ae. As described above (see "Ribosome architecture"), the latter is also a protein of the large ribosomal subunit (6). It has been experimentally demonstrated that the components of the archaeal C/D box sRNAs do associate and form complexes that carry out site-specific ribose methylation in vitro (86, 87, 126, 129). On the basis of these data, it has been proposed that the sRNP-based machinery for site-specific ribose methylation originated in thermophilic archaea (probably because of the need to stabilize the rRNAs against thermal denaturation) and was subsequently inherited by the *Eucarya*.

During ribosome synthesis, rRNA processing and modification proceeds in parallel with the assembly of the ribosomal proteins on the rRNA in a highly integrated and regulated fashion. In vivo studies of ribosome assembly have not been performed in archaea. However, in vitro studies have been performed for almost two decades. The ribosomal subunits of the crenarchaeon, *Sulfolobus solfataricus* (76), and of the euryarchaeon, *Haloferax mediterranei* (101, 102), can be functionally reconstituted in vitro from individual RNA and protein components. Assembly of thermophilic and halophilic ribosomes in vitro requires high temperature and high salt, respectively, conditions that reflect the organism's natural environment. Aside from these basic requirements, assembly is spontaneous, albeit rather slow. In this respect, both the "large" (*S. solfataricus*) and "small" (*H. mediterranei*) types of archaeal ribosomes resemble their bacterial counterparts, which also spontaneously assem-

ble in vitro (83). In contrast, functional reconstitution of eucaryal ribosomes in vitro has not been achieved, despite the efforts of many investigators for more than three decades. Even though the archaeal (and bacterial) ribosomes can spontaneously assemble in vitro from the separate rRNA and protein components, this does not exclude the need for chaperones or other accessory factors in vivo, in particular, to enhance the rate of ribosome biosynthesis and to regulate ribosome assembly. In support of this view, in bacteria the chaperone protein DnaK seems to be involved in ribosome biogenesis (1). Similar to many aspects of archaeal protein synthesis, ribosome biogenesis requires a great deal more study.

Archaeal mRNAs

Few experimental studies have been performed on the structure of archaeal gene transcripts. However, it is clear that the translation of polycistronic mRNAs into multiple polypeptides is common in archaea (3). Moreover, it has been known for many years that archaeal mRNA, similar to bacterial mRNA, contains ribosome-binding sites similar to the Shine-Dalgarno (SD) sequences in bacteria; accordingly, archaeal SSU ribosomal RNAs have corresponding anti-SD sequences (28).

Recent computational analyses of genome sequences that analyzed the position of transcription start sites, initiation codons, and potential ribosome-binding motifs (SD sequences) generated interesting findings about unique aspects of mRNA structure in *Archaea*. The putative structure of archaeal transcripts have been identified from the genome sequences of the hyperthermophilic archaea, *S. solfataricus* and *P. aerophilum* (105, 109). Based on the predicted location of promoters and the start sites of transcription, a large proportion of mRNAs from these two organisms appear to lack a 5′-untranslated region (5′-UTR; leaderless mRNAs). This has been experimentally verified for *P. aerophylum* (109). The absence of a 5′-UTR means that a SD motif is unavailable for ribosome binding. However, the downstream genes of leaderless polycistronic transcripts usually contain identifiable SD motifs upstream of their initiation codons. On the basis of these data, it was proposed that two distinct mechanisms for translation initiation exist in *Archaea*: one of a bacterial type, based on SD motifs and operating mainly on the internal cistrons of polycistronic mRNAs, and another of an unknown nature employed for initiating translation of leaderless cistrons (12, 114).

A more recent survey of archaeal genomes predicted the presence of two distinct types of archaeal transcripts (115). Group A genomes include several (but nor all) *Crenarchaeota*, and the *Euryarchaeota*, thermoplasmales, halobacteria, and *Nanoarchaeum equitans*. These are predicted to produce a high proportion (~50%) of leaderless transcripts. In some of these genomes the genes located internally in polycistronic transcripts are preceded by clearly identifiable SD motifs. However, in other group A genomes, such as those of *N. equitans* and *P. aerophilum*, the internal genes in cistrons often lack identifiable SD-like sequences.

Group B genomes are predicted to mainly produce transcripts which possess SD motifs ahead of the initiation codons of both the first and the internal genes in operons (or of genes in monocistronic transcripts), and only produce a few leaderless transcripts. Group B genomes include a diverse array of species that includes methanogens, as well as the *Crenarchaeota*, *A. pernix* and *Hyperthermus butylicus*, and the pyrococcales.

Although this study confirmed that leaderless transcripts appeared to be produced in several archaea, it also showed the marked heterogeneity that exists within the *Archaea*. An analysis of the phylogenetic distribution of group A and group B genomes indicates that leaderless transcripts are primarily found in late-branching organisms (see Fig. 2), suggesting a more recent evolutionary origin. However, this conclusion contrasts with studies showing that leaderless mRNAs are universally translatable by all kinds of ribosomes, a finding that argues in favor of their primitive status (36). Further studies are necessary to shed more light on this very interesting aspect of archaeal translation.

GENOME ORGANIZATION OF TRANSLATION COMPONENTS

The availability of several complete genome sequences of diverse archaeal species allows a meaningful comparison of the distribution and organization of the various RNA- and protein-encoding genes whose products constitute the translation apparatus. These consist of the numerous genes that encode the components of the ribosome: the three ribosomal RNAs (16S, 23S, and 5S) and the 68 possible r-proteins. In addition, there are genes encoding tRNAs and the accessory proteins that function in translation initiation, elongation, and termination.

The size and number of rRNA genes found in 21 complete archaeal genomes are summarized in Table 1. Most species have a single copy of the LSU and SSU RNA genes. The exceptions are exclusively among the halophiles and the methanogens, most of which have two or three LSU and SSU rRNA genes. Some

of the late-branching euryarchaeota have mutiple SSU and LSU rRNA genes, although members of the thermoplasmales, *Thermoplasma acidophilum* and *Thermoplasma volcanii*, have single copies. All *Crenarchaeota*, the thermococcales and the archeoglobales have single SSU and LSU genes.

There is a greater tendency toward redundancy of the 5S RNA genes. Greater than half of the sequenced genomes contain two to four 5S rRNA genes (Table 1). Remarkably, the *A. pernix* genome seems to have an eucarya-like arrangement of rRNA genes, since it includes a 5.8S RNA gene located about 300 bp away from the 23S RNA gene and possibly cotranscribed with it.

The LSU and SSU RNA genes are located close to one another in most archaeal species, and the available experimental data (77) indicate that they tend to be cotranscribed, similar to bacteria and eucarya. The exceptions are the thermoplasmales and *N. equitans*, whose LSU and SSU rRNA genes are located in distinct regions of the genome and are transcribed independently. In contrast, the 5S RNA genes are generally unlinked from the LSU and SSU RNA genes, except in halophiles and most methanogens. In the *Methanococcus maripaludis* genome, three 5S RNA genes are arranged in three individual LSU-SSU rRNA gene clusters, and the fourth has an independent location, whereas in *Methanopyrus kandleri* the single 5S RNA gene is unlinked from the large rRNA genes.

The organization of r-protein genes in archaeal genomes is interesting. In bacteria, about half of the r-protein genes are clustered in the two large operons (spectinomycin [spc] and S10), and their composition is conserved in most species (23). In the *Archaea*, over one-third of the r-protein genes are included in a few large clusters that closely resemble the bacterial spc and S10 operons in the type and order of genes. The r-protein genes that are present in these clusters are particularly interesting as most of them are conserved in all three domains of life.

Figure 3 shows the organization of the S10-like and spc-like r-protein gene clusters in 19 complete archaeal genomes. These clusters represent 22 of the 68 r-protein genes. The organization of the same genes in *E. coli* is shown for comparison in the last row of Fig. 3. Gene order tends to be conserved in the *Archaea* and resembles that of the *Bacteria*. No experimental information is available about the transcription patterns of these clusters, making it difficult to know to what extent they are organized into functional operons. Close clustering (genes located less than 50 bp apart) may be indicative of an operon structure; if so, the pyrococci appear to have only two large operons of r-proteins (Fig. 3). In contrast, the genes are arranged in smaller, operon-like clusters in other archaea, in-

SSO	L3	L4	L23	L2	S19	S22	S3	L29			S17	L14	L24	S4	L5	S14	S8	L6	L32	L19	L18	S5	L30	L15
STO	L3	L4	L23	L2	S19	S22	S3	L29			S17	L14	L24	S4	L5	S14	S8	L6	L32	L19	L18	S5	L30	L15
AFU	L3	L4	L23	L2	S19	L22	S3	L29		HY	S17	L14	L24	S4	L5	S14	S8	L6	L32	L19	L18	S5	L30	L15
APE	L3	L4	L23	L2	S19	L22	S3	L29		HY	S17	L14	L24	S4	L5	S14	S8	L6	L32	L19	L18	S5	L30	L15
PFU	L3	L4	L23	L2	S19	L22	S3	L29	SUI	HY	S17	L14	L24	S4	L5	S14	S8	L6	L32	L19	L18	S5	L30	L15
PHO	L3	L4	L23	L2	S19	L22	S3	L29	SUI	HY	S17	L14	L24	S4	L5	S14	S8	L6	L32	L19	L18	S5	L30	L15
PAB	L3	L4	L23	L2	S19	L22	S3	L29	SUI	HY	S17	L14	L24	S4	L5	S14	S8	L6	L32	L19	L18	S5	L30	L15
TKO	L3	L4	L23	L2	S19	L22	S3	L29	SUI	HY	S17	L14	L24	S4	L5	S14	S8	L6	L32	L19	L18	S5	L30	L15
PAE	L3	L4	L23				S3	L29	SUI			L14	L24		L5	S14	S8	L6	L32	L19	L18	S5	L30	
MKA	L3	L4	L23	L2		L22	S3	L29	SUI		S17	L14	L24	S4	L5	S14	S8	L6	L32					L15
MMA	L3	L4	L23	L2	S19	L22	S3	L29			S17	L14	L24	S4	L5	S14	S8	L6	L32	L19	L18	S5	L30	L15
MAC	L3	L4	L23	L2	S19	L22	S3	L29			S17	L14	L24	S4	L5	S14	S8	L6	L32	L19	L18	S5	L30	L15
MTH	L3	L4	L23	L2	S19	L22	S3	L29	SUI	HY	S17	L14	L24	S4	L5	S14	S8	L6	L32	L19	L18	S5	L30	L15
MJA	L3	L4	L23	L2	S19	L22	S3	L29			S17	L14	L24	S4	L5	S14	S8	L6	L32	L19	L18	S5	L30	L15
MMP	L3	L4	L23	L2	S19	L22	S3	L29			S17	L14	L24	S4	L5		S8	L6	L32	L19	L18	S5	L30	L15
HMA	L3	L4	L23	L2	S19	L22	S3	L29			S17	L14	L24	S4	L5	S14	S8	L6		L19	L18	S5	L30	L15
H-sp	L3	L4	L23	L2	S19	L22	S3	L29		HY	S17	L14	L24	S4	L5	S14	S8	L6	L32	L19	L18	S5	L30	L15
TAC	L3	L4	L23	L2	S19	L22	S3	L29	SUI	HY	S17	L14	L24	S4	L5	S14	S8	L6	L32	L19	L18	S5	L30	L15
TVO	L3	L4	L23	L2	S19	L22	S3	L29	SUI	HY	S17	L14	L24	S4	L5	S14	S8	L6	L32	L19	L18	S5	L30	L15
ECO	S10	L3	L4	L23	L2	S19	L22	S3	L16	L29	S17	L14	L24	S4	L5	S14	S8	L6			L18	S5	L30	L15

Figure 3. (*See the separate color insert for the color version of this illustration.*) Organization of the main ribosomal protein gene clusters in archaeal genomes. SSO, *Sulfolobus solfataricus*; STO, *Sulfolobus tokodaii*; AFU, *Archaeoglobus fulgidus*; APE, *Aeropyrum pernix*; PFU, *Pyrococcus furiosus*; PHO, *Pyrococcus horikoshii*; PAB, *Pyrococcus abyssi*; TKO, *Thermococcus kodakaraensis*; PAE, *Pyrobaculum aerophylum*; MKA, *Methanopyrus kandleri*; MMA, *Methanosarcina mazei*; MAC, *Methanosarcina acetivorans*; MTH, *Methanothermobacter thermautotrophicus*; MJA, *Methanococcus jannaschii*; MMP, *Methanococcus maripaludis*; HMA, *Haloarcula marismortui*; H-sp, *Halobacterium* sp. NRC1; TAC, *Thermoplasma acidophilum*; TVO, *Thermoplasma volcanii*. The last line (ECO) shows for comparison the organization of the same genes in *E. coli* that is also present in most bacteria. Genes that are within 50 bp of each other, and may therefore be cotranscribed, are indicated in the same color. Domain-specific genes are underlined.

dicating that they have smaller functional units. The distinct functional organization of these r-protein genes between different members of the *Archaea* is indicative of extensive gene rearrangements. Moreover, while in most species such units tend to remain located in the same region of the genome, in a few cases they are located far apart from one another. This is the case for *M. kandleri*, in which the r-proteins form three hypothetical operons (Fig. 3) separated by about half a million base pairs from one another. For *P. aerophilum*, some genes have been separated from the conserved clusters and are located elsewhere in the genome. For example, the L6 gene is missing from the cluster that commences with S17 and is found, on its own, elsewhere in the genome. Similarly, the L22 gene is separated from its neighbor S3 by a few hundred base pairs (Fig. 2).

On the whole, the similarity of the clustering order of the r-protein genes in archaea and bacteria (Fig. 3) strongly suggests that it is an ancestral feature that was present in the LUCA and predates the radiation of the three primary domains. The alternative hypothesis is that similar gene clustering is due to convergent evolution. This could be explained by positive selection due to a functional advantage that occurred independently in the *Archaea* and *Bacteria*. However, it is not apparent what this advantage might have been as the gene clusters shown in Fig. 3 demonstrate that they can exist in a "broken" form in extant bacteria and archaea.

In some archaea, the r-protein genes have a greater tendency to group together than in others. Moreover, in some species (e.g., *Methanosarcina mazei*), the clustered genes tend to be very close to each other, lacking spacer tracts between them and often even overlapping by a few base pairs, while in other species (e.g., *Methanocaldococcus jannaschii*), they are generally separated by at least a few base pairs. No correlation is apparent between these patterns of gene organization and the phylogenetic position of the archaeal species in which they occur. For example, there is a strong clustering of the r-protein genes in late-branching species (*Thermoplasmales*) and early-branching organisms (pyrococci and *A. pernix*), as well as broken clusters for early- and late-branching organisms, *P. aerophylum* and *H. marismortui*, respectively (see Fig. 2). Research examining the expression and regulation of r-protein genes in archaea is required before an understanding can be obtained of the functional significance, if any, of the differences in gene organization among the various species.

Ribosomal protein genes (other than those shown in Fig. 3), in general, are found in individual genomic locations or in small groups of a few genes, sometimes intermingled with other translational genes such as those encoding elongation and initiation factors. Most of the r-protein genes that are shared by the *Archaea* and the *Eucarya* have this kind of organization. In all archaea, the one archaeal-specific r-protein gene, LXa, is found in a cluster that contains other genes encoding translational components that are present in the *Archaea* and the *Eucarya* but not the *Bacteria* (e.g., r-protein L30, the putative initiation factor eIF6, and the chaperone-like protein, prefoldin).

Genes encoding translation factors are more or less scattered throughout all archaeal genomes. The four genes encoding the universal initiation factors YciH/SUI1, IF1/IF1A, IF2/IF5B, and EFP/IF5A (see "Translation initiation factors," below, and Table 3) tend to be unlinked from other translational genes and are likely to be individually transcribed. The genes encoding the α-, β-, and γ-subunits of the archaeal/eucaryal initiation factor a/eIF2 occupy separate locations in all known archaeal genomes and must therefore be transcribed independently. Moreover, the a/eIF2-α, -β, and -γ genes are seldom clustered with

Table 3. Translation initiation factors in *Archaea*

Factor name	*Eucarya*	*Bacteria*	Structure	Function in *Archaea*	Function in other domains
aIF1A	eIF1A	IF1	Solved	Unknown	*Bacteria*, stimulates aIF2 *Eucarya*, assists scanning
eIF5B	IF2		Solved	Unknown	*Bacteria*, binds fmet-tRNA; *Eucarya*, subunit joining
aSUI1	eIF1/SUI1	YciH (some phyla)	Solved	Unknown	*Bacteria*, unknown; *Eucarya*, fidelity factor
a/eIF2 (αβγ)	eIF2 (αβγ)	–	Solved (subunits)	Binds met-tRNAi	Binds met-tRNAi
aIF6	eIF6	–	Solved	Unknown	Inhibits subunit association
aIF5A[a]	eIF5A	EFP	Solved	Unknown	*Bacteria*: stimulates *first* peptide bond
aIF2B (α,β,δ)[b]	eIF2B (α,β,γ,δ,ε)	–	Solved	Unknown	eIF2 recycling

[a]Behaves as a specialized elongation factor.
[b]Involvement in translation initiation uncertain (see text). The nomenclature for the archaeal proteins varies in the literature, and the form chosen for this table is based on that used in the author's laboratory (74).

other translational genes, with the exception of a/eIF2-α gene, which in most genomes is located in the vicinity of r-proteins S27 and L44. Note that the gene for the a/eIF2 β-subunit is clustered and possibly cotranscribed with several putative cell-division genes in some genomes. The other archaeal/eucaryal IF, aIF6, tends to be clustered with a few archaeal/eucaryal type r-protein genes and the archaeal specific r-protein, LXa. The aIF6 gene cluster contains the gene encoding the putative chaperone protein prefoldin, which is predicted to be highly expressed in many archaea (46). The closeness of the genes in the aIF6 cluster indicates they may be arranged in an operon and cotranscribed. However, no experimental data are available to confirm this.

The genes for the two translation elongation factors EF1A and EF2 are unlinked in all archaeal genomes, although in several cases they are clustered with other translation genes. Finally, the gene encoding the putative translation termination factor aRF1 is in general not clustered with other genes encoding components of the protein synthesis apparatus.

MECHANISM OF TRANSLATION

General Overview

The process of protein synthesis develops along an orderly series of steps and is conserved in all three domains. The first step is initiation, whereby the ribosomes interact with the translation start region and guide the initiator tRNA onto the initiation codon, thereby setting the correct reading frame for decoding. Initiation controls to a large extent the speed and efficiency of protein synthesis and has therefore evolved to be the target for most mechanisms of translational control. Several IFs are required to assist initiation. The next stage is elongation, which consists of three distinct steps: (i) adaptation, during which aminoacyl-tRNAs, in a ternary complex with EF1A and GTP, recognize the proper codon located in the ribosomal A site; (ii) transpeptidation, during which the peptide bond is formed; (iii) translocation, during which the ribosome slides by one codon toward the 3' terminus of the mRNA with the aid of EF2. Finally, termination occurs when a stop codon enters the A site and is recognized by the proper RF, which catalyzes the release of the completed polypeptide chain from the ribosome.

The mechanism of protein synthesis has been elucidated in fine detail in bacteria and, to a lesser extent, in eucarya. In contrast, studies on the mechanism of translation in archaea are still in their infancy, and very little experimental information is available;

the current extent of knowledge is summarized in the following sections.

Translation Initiation

It has been recognized for many years that the initiation step of translation has diverged greatly in the domains of life. More than three decades of research have revealed that the mechanism and the cellular machinery of translation initiation in the *Bacteria* and the *Eucarya* have little similarity. Eucaryal ribosomes normally initiate translation by means of a "scanning" mechanism, whereby the 40S subunits, aided by many protein factors and carrying the specific initiator tRNA charged with unmodified methionine (met-tRNAi), slide along the message until the initiation codon is found (58). In contrast, bacterial ribosomes bind directly to the mRNAs with the aid of the RNA/RNA interaction between the SD sequence on the mRNA and the anti-SD sequence on the SSU rRNA. To assist the initiation reactions, bacteria have a minimal set of protein factors, only three proteins, compared with more than a dozen in eucarya (37). Most of the initiation factors are domain specific. These differences have been attributed to the different complexity of eucaryal and prokaryotic cells and to the need of the former to evolve more sophisticated mechanisms for translational regulation. The *Archaea* have proven to be ideal for investigating the divergent evolution of the translation initiation step. As a result, considerable effort has recently been dedicated to the understanding of the features of translation initiation in archaea.

Ribosome/mRNA interaction

Currently, the best understood aspect of translation initiation in archaea is the mechanism of ribosome/mRNA interaction. Studies carried out in the 1980s identified the mechanism to be very similar to that employed by the *Bacteria*, i.e., based on the use of the SD/anti-SD interaction to promote ribosomal recognition of the translation initiation sites (43, 94). However, it was recognized that some archaeal mRNAs failed to fit this simple model, i.e., the leaderless mRNAs that had been detected in halophiles (13). At the time of this discovery, the leaderless mRNAs were regarded as an oddity that did not challenge the essence of the SD-mediated scheme of mRNA/ribosome interaction. It was argued that leaderless mRNAs were also found, although infrequently, in bacteria and bacteriophages, and the problem of their interaction with the ribosomes had apparently been settled by a study carried out on the transcript of the lambda repressor gene (107). The leaderless mRNA contained a special sequence downstream of the initiation codon, termed the "downstream box," which mediated mRNA/ribosome interaction by pairing with a tract of the 16S rRNA located at nt 1469 to 1483 (numbering relative to the *E. coli* 16S rRNA) (33, 107). Leaderless mRNAs were generally thought to possess either a "downstream box" or a "cryptic" SD sequence located a few base pairs downstream of the initiation codon that were recognized by the SSU with the customary RNA/RNA interaction mechanism (13). Subsequent studies challenged these conclusions, however. One concern was that recognition of leaderless mRNAs was widespread in some archaea (105, 109). Another concern was that the putative "anti-downstream box sequence" is not conserved in archaeal SSU RNAs. The third issue related to the demonstration that, even in bacteria, disruption of the putative antidownstream boxes did not inhibit the translation of the leaderless mRNAs (64, 85). As a result, the view emerged that the *Archaea* employ two distinct mechanisms of mRNA/ribosome interaction: one that was poorly understood and operated on the leaderless mRNAs, and the other based on the SD/anti-SD interaction that was specific for the leadered messages with SD motifs (114).

The first set of experimental data supporting the existence of two distinct translation mechanisms in *Archaea* was obtained using *S. solfataricus* as the model organism. With the aid of a cell-free translation system, it was demonstrated that, when present, SD motifs are essential for translational initiation (25). The disruption of SD motifs by site-directed mutagenesis was found to completely inhibit in vitro translation, a result that was more pronounced than that found in bacteria where disruption of SD sequences usually reduces, but does not abolish, protein production. However, other archaea, such as *Halobacterium salinarum*, may have a less stringent requirement for SD motifs, as their disruption leads to a reduction of translational efficiency but not to a total inhibition of protein synthesis (103).

A most remarkable finding was that the in vitro translation of the mRNAs whose SD motifs had been disrupted could be rescued by entirely deleting the 5'-UTR, i.e., by rendering the mRNA leaderless (25). These results indicate that an individual mRNA can be translated by two different routes, depending on whether it possesses a 5'-UTR endowed with a SD motif.

The mechanism for leaderless mRNA/ribosome interaction is poorly understood. Some experimental information was obtained from in vitro studies carried out with purified translational components of *S. solfataricus* (12). In the presence of SD motifs, *S. solfataricus* SSU interacted directly and strongly with leadered mRNAs in the absence of any other translation components. Unlike their bacterial counterparts, the binary 30S/mRNA complexes that formed were very stable. In contrast, leaderless mRNAs were unable to form binary complexes with 30S subunits. A 30S subunit/leaderless mRNA interaction could be detected only in the presence of met-tRNAi (12), supporting the idea that codon-anticodon pairing was required for initiation site recognition, as previously observed for leaderless mRNA translation in *E. coli* (36). This mechanism has some similarity to eucarya, where ribosomes identify the translation start site by also requiring the presence of met-tRNAi in the P site for the 40S subunit to land on the correct AUG initiation codon.

Gene function does not correlate with mRNA being "leaderless" or "leadered." However, the (still scarce) experimental data available allow the general conclusion that leaderless mRNAs are less efficiently translated than leadered mRNA (e.g., reference 25). This contention is consistent with a genomic study that showed that highly expressed genes in archaea tended to have SD sequences upstream of their initiation codons (48).

An important question raised by the existence of two initiation mechanisms in archaea is whether they evolved at the same time, or whether one mechanism preceded the other in the course of evolution. Several observations argue that "leaderless" initiation is the ancestral mechanism. The most compelling evidence is that leaderless mRNAs are universally translatable (at least in vitro) by archaeal, bacterial, and eucaryal ribosomes (36). Since "normal" eucaryal mRNAs are poorly, if at all, translated in bacterial systems (and vice versa), this is a remarkable fact that argues for a common conserved mechanism underlying leaderless translation. The recent observation that most mRNAs are leaderless in the protozoan *G. lamblia* (73) is interesting. If *G. lamblia* is a primitive eucarya, this provides support for leaderless mRNAs being ancestral. *Giardia* occupy a deep branch in the eucaryal evolutionary tree (see Fig. 2). However, it is still unclear whether this is a true indication of being primitive or is an artifact of evolutionary analysis due to an abnormally fast evolutionary rate, a frequent occurrence in parasitic organisms such as *G. lamblia*.

Additional evidence supporting the ancestral nature of leaderless mRNAs is that, at least in bacteria, their translation seems to have no stringent requirement for initiation factors, especially if performed by 70S ribosomes (116). As a small set of IFs are common to all three domains (see "Translation initiation factors," below), the minimal number of accessory factors required for translation initiation may be indicative of what was present in primitive cells.

There are also data arguing against leaderless mRNAs being the primitive form. First, leaderless mRNAs are especially abundant in late-branching archaeal species, while early-branching species tend to have leadered mRNAs with SD motifs (115). Based on these data, the latter would be predicted to be the "ancestral" mRNA structure. If so, the prevalence of leaderless mRNAs in later-evolved, and especially in extremely thermophilic, archaeal species may reflect a physiological requirement that is presently not understood. Second, it has been argued that the polycistronic arrangement of genes, and the presence of SD motifs, is likely to predate the branching of the three domains (74). The r-protein genes are often clustered in a similar order (and sometimes cotranscribed) in both bacteria and archaea (Fig. 3). It is very unlikely, albeit not impossible, that this has arisen from convergent evolution. In early-branching archaea, such as *A. pernix* and *H. butylicus* (Fig. 2), most genes with SD motifs use AUG, GUG, and UUG as initiation codons in roughly the same proportion, while in most other species AUG is by far the prevalent initiation signal (115). This suggests that a "primitive" function of SD motifs may have been to ensure correct ribosome positioning on the translation initiation site, in a way not strictly dependent on the presence of an optimal initiation codon. In support of this view, in *S. solfataricus* the 30S ribosomes can form stable binary complexes with mRNAs endowed with SD motifs, even if these are not followed by a proper initiation codon (12).

Translation Initiation Factors

It is well known from a wealth of detailed genetic and biochemical studies that bacteria and eucarya greatly differ in the type and usage of the proteins that assist and modulate translation initiation. Only three IFs exist in bacteria. The principal factor, IF2, is an RNA-binding G-protein of ~90 kDa that performs the essential task of promoting the correct binding of the initiator tRNA (f-met-tRNA) to the ribosomal P site. IF1 and IF3 assist initiation by hindering premature subunit association, and IF3 assists by discouraging recognition of nonoptimal initiation codons (37).

In contrast, *Eucarya* require a much more elaborate complement of IFs (91). The cap-binding factor, eIF4F (absent in bacteria), is necessary to unwind the mRNA to allow its interaction with the ribosome. To successfully perform "scanning" for the initiator AUG codon, the eucaryal 40S subunits must carry the initiator tRNA (met-tRNAi) and the proteins eIF1, eIF1A, and eIF3 (92). eIF1 and eIF1A are both required for the correct identification of the start codon, while one of the roles of eIF3 is to connect the ribosome with the scanning factor eIF4F. Met-tRNAi binding to the 40S subunits is promoted by the G-protein eIF2, a heterotrimeric complex, whose subunits are not homologous to the bacterial factor, IF2 (50, 62). Adaptation of met-tRNAi in the P site is accompanied by the hydrolysis of the eIF2-bound GTP, whereupon the factor dissociates from the ribosome. However, eIF2 does not have spontaneous GTPase activity and requires a GTPase activator factor, eIF5, to trigger GTP hydrolysis. Moreover, the reactivation of eIF2-GDP has an obligatory requirement for a GTP/GDP exchange factor, the pentameric protein, eIF5B (50). After the establishment of the codon-anticodon interaction, eIF5B (a G-protein) stimulates subunit joining, leading to the formation of the monomeric ribosome 80S (93). Finally, eIF5A is required to trigger the formation of the first peptide bond (80).

The availability of complete genome sequences from several archaea has made it possible to identify homologs of translation initiation factors from bacteria and eucarya and thereby formulate hypotheses about their role in archaea. The original surveys proved surprising because, although archaea were expected to have a restricted number of initiation factors, similar to bacteria, archaeal genomes contained a host of genes homologous to eucaryal factors, the only exception being the lack of those involved in cap recognition (11, 30) (Table 3).

The evolution of IFs is similar to ribosomal proteins in several ways: (i) the primary sequences of the putative archaeal initiation factors are always more similar to eucaryal proteins; (ii) a (small) set of factors is shared by all three domains of life; (iii) some IFs are shared by the *Archaea* and the *Eucarya* and not by the *Bacteria*; (iv) no IF is shared by the *Archaea* and the *Bacteria* and absent in the *Eucarya*. In addition, the *Eucarya* and the *Bacteria*, but not the *Archaea*, possess unique factors not present in the other domains. This latter observation may not be surprising as the archaeal IFs have been identified by sequence comparisons with IFs from the other two domains. A detailed biochemical and genetic analysis of archaeal translational initiation may well lead to the discovery of IFs unique to the third domain.

In contrast to the evolutionary considerations that continue to highlight the similarities between archaeal and eucaryal translation, comparatively little is known about the function of the putative IFs in archaea. However, X-ray or NMR structures are available for all of the putative archaeal IFs listed in Table 3; these have been obtained with a view to gaining insight into the function of the eucaryal homologs, rather than the function in archaea per se.

The one archaeal IF for which a function has been experimentally defined is the trimeric protein homologous to the eucaryal factor eIF2. In the literature, this factor is referred to as aIF2 or a/eIF2; here, the second designation is employed, while using aIF2 for the monomeric protein homologous to bacterial IF2 and eucaryal eIF5B.

Eucaryal IF2 is an important translation initiation factor, as it specifically interacts with the initiator tRNA (met-tRNAi) and carries it to the 40S ribosomal subunit (50). In bacteria, the same essential function is carried out by a monomeric protein, also called IF2 but unrelated in sequence to any of the eIF2 subunits (37). Until recently, the prevalent rationale for the divergence of the tRNAi-binding factors in bacteria and eucarya was that the latter had to evolve a more complex protein to achieve a more sophisticated regulation of translational initiation. In fact, eIF2 is a main target for eucaryal translational regulation. However, the fact that the archaea resemble the eucarya in having both eIF2-like and IF2-like factors shows that cellular complexity has probably nothing to do with the usage of these translation initiation factors.

Similar to eIF2, a/eIF2 is composed of three subunits that associate to form a heterotrimeric complex (90, 123, 124). This complex is smaller than its eucaryal counterpart, due to the much reduced length of its β-subunit (15 kDa instead of ~50 kDa). In all archaea, the γ-subunit is the largest protein (~45 kDa) followed by the α-subunit (~30 kDa). The archaeal β-polypeptide is smaller than its eucaryal counterpart due to the lack of the domains that, in the eucaryal subunit, interact with the guanine nucleotide exchange factor, eIF2B, and with the GTPase activator, eIF5. This is consistent with the observation that all archaea lack a homolog of eIF5, as well as three of the five subunits of eucaryal eIF2B. Archaeal genomes do include homologs of the α-, β-, and δ-subunits of eIF2B but lack counterparts of the γ- and ε-subunits that catalyze guanine nucleotide exchange on eIF2. Therefore, it is probable that the archaeal homologs of the eIF2B α-, β-, and δ-proteins have a function unrelated to guanine nucleotide exchange (62).

X-ray or NMR structures are available for all three subunits of a/eIF2. The largest subunit, γ, has a striking resemblance to the EF-1A (EF-Tu in bacteria) (97, 104), consistent with the fact that it contains the guanine nucleotide-binding domain and is principally involved in the interaction with met-tRNAi. The small β-subunit contains a zinc-finger motif (39). The α-subunit is composed of three distinct domains, two of which have RNA-binding properties (124).

The function of a/eIF2 from *Pyrococcus abyssi* (123) and *S. solfataricus* (90) has been explored by in vitro biochemical assays using the factor reconstituted from the cloned recombinant subunits. These experiments revealed that, similar to its eucaryal counterpart, a/eIF2 specifically binds met-tRNAi and carries it to the ribosome. However, several features functionally differentiate the archaeal and the eucaryal proteins. One feature relates to the nature of the tRNA-binding site. For a/eIF2, an αγ-dimer is necessary and sufficient to achieve a stable interaction with met-tRNAi, while met-tRNAi binding seems to involve mainly the γ- and β-subunits of eIF2 (29, 32). The eucaryal α-polypeptide participates mainly in the regulation of eIF2 activity; its phosphorylation, triggered by various metabolic cues, inhibits the eIF2B-catalyzed exchange of GDP with GTP, thereby blocking the protein in its inactive form (24, 121). Another functional difference is that a/eIF2 has a similar affinity for GDP and GTP and therefore does not require a guanine nucleotide exchange factor to be reactivated (90). This finding is consistent with the lack of a complete homolog of eIF2B in archaeal genomes (62). Thus, eucaryal-type functional regulation of a/eIF2 (phosphorylation of the α-subunit and inhibition of guanine nucleotide exchange) should not exist in archaea. However, it has recently been reported that the α-subunit of *Pyrococcus horikoshii* a/eIF2 is phosphorylated by a specific protein kinase (111), although no data are yet available on the possible functional significance of this remarkable finding.

Another difference between eIF2 and a/eIF2 is the probable lack of a GTPase activator protein for the latter. GTP hydrolysis of eucaryal eIF2-phosphate is triggered by the helper factor eIF5, and no recognizable homolog is present in the genome sequences of archaea. It is therefore likely that a/eIF2 has an intrinsic, ribosome-triggered GTPase activity, although this has not yet been demonstrated experimentally. Alternatively, GTP hydrolysis of a/eIF2 may be facilitated by an unidentified GTPase activator.

The function of the remaining IFs listed in Table 3 has not been determined, although functions have been speculated for some of them. A particularly interesting protein is aIF2 (aIF5B), which, with aIF1A (homolog to IF1 in bacteria and eIF1A in eucarya), is one of the two IFs present in all domains (63). Prior to this discovery (63), IF2 was thought to be strictly a bacterial protein, whose function in eucarya (specific recognition of the initiator tRNA) was fulfilled by an unrelated and more complex factor, eIF2. The function of the eucaryal IF2 homolog, eIF5B, has recently been clarified. Although nonessential, the factor is important to promote the joining of the ribosomal subunits after the 40S particle has correctly recognized the initiation codon (93). Similar to IF2, eIF5B has an intrinsic ribosome-dependent GTPase activity; GTP hydrolysis accompanies the ejection of the factor from the ribosome after its task has been accomplished (71).

There are few published experimental data addressing the function of the archaeal IF2-like factor. The only study performed in vivo showed that the *M. jannaschii* IF2 homolog can partially rescue yeast mutants lacking eIF5B (70), thus demonstrating that aIF2 is to some extent functionally analogous to eIF5B. Preliminary in vitro data also suggest that *S. solfataricus* aIF2 promotes the binding of met-tR-NAi to the ribosome (74).

In contrast to the paucity of functional data, there is much detailed information about the structure of aIF2. Crystallographic studies on the *Methanothermobacter thermautotrophicus* aIF2 (98) show that it is characteristically shaped as a chalice (Fig. 4). The globular "cup" of the chalice (N-terminal region) includes three domains, the first of which is the guanine nucleotide-binding domain. The "stem" of the chalice is a long α-helix, while the globular "base"

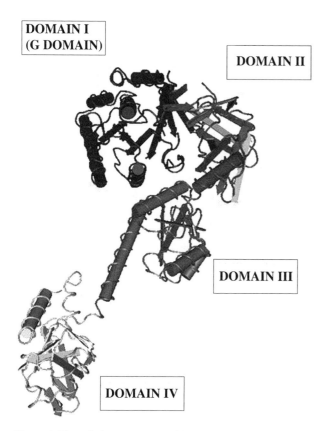

DOMAIN I (G DOMAIN)

DOMAIN II

DOMAIN III

DOMAIN IV

Figure 4. The "chalice" structure of the archaeal IF2-like translation initiation factor. The crystal structure of the archaeal translation initiation factor aIF2, homologous to eucaryal eIF5B and bacterial IF2, is shown. The four protein domains are indicated. Data taken from the NCBI structure data bank: PDB: 1G7T viewed with Cn3D 4.1.

(domain IV) corresponds to the C-terminal domain known to bind f-met-tRNA in bacterial IF2 (38). The aIF2 proteins are smaller, in general, than their eucaryal and bacterial homologs, as they lack the long, poorly conserved and functionally uncharacterized N-terminal region that is present in both IF2 and eIF5B. The structures of aIF2-GDP and aIF2-GTP show some differences, suggesting that the protein undergoes a conformational change upon GTP hydrolysis that modifies its affinity for the ribosome (98).

The function of the second universal IF in archaea (aIF1A) remains undetermined, although the X-ray structures of the factors from all three domains have been solved (10, 106). Recent experimental data suggest that aIF1A forms a relatively stable complex with aIF2 (74). A similar type of interaction occurs between eIF5B and eIF1A. Bacterial IF2 and IF1 do not form a complex in solution, although they may interact on the surface of the ribosome, as suggested by earlier cross-linking data (15) and by a recent cryo-electron microscopy study (2). On the basis of these data, it has been proposed that a complex of the IF1-like and IF2-like factors is a universal and ancestral feature of translational initiation, whose conserved function would be to enhance the affinity of the P site for the initiator tRNA and to stimulate subunit joining (61).

The small protein, aSUI (aIF1) has homologs in all eucarya (SUI1 in yeast; eIF1 in vertebrates) and in a limited number of bacterial species (YCiH in *E. coli*) (26). Phylogenetic analysis shows that SUI1 is probably an ancestral factor that has been subsequently lost by most bacteria, possibly because its function has been replaced by another protein (61). In archaea, aSUI1 interacts with the 30S subunits, although its precise function in translation initiation has not been determined. In eucarya, SUI1/eIF1 is an essential protein that controls the fidelity of initiation codon recognition and probably also elongation (27).

A very interesting initiation factor present only in the *Archaea* and the *Eucarya* is the 25-kDa protein, aIF6 (eIF6). The function of this factor in the eucarya has been studied in some detail but remains somewhat enigmatic. In yeast, eIF6 is an essential protein that is found in the nucleolus and in the cytoplasm, where it associates with the 60S ribosomal subunit (8, 108). The main phenotype observed in conditional mutants lacking the factor is a defect in the synthesis of 60S ribosomes, specifically a block in the processing of the rRNA 26S precursor (9). However, the cytoplasmic, 60S-bound eIF6 behaves as a ribosome antiassociating factor, preventing the formation of 80S particles and thereby inhibiting protein synthesis (108). According to a recent report, the dissociation of eIF6 from mammalian ribosomes requires the phosphorylation of the factor, which is accomplished when certain environmental cues activate a specific kinase (21). Thus, eIF6 resembles eIF2 in being a general regulator of protein synthesis. It remains to be determined whether eIF6 can serve two functions in translation. Clearly, functional studies of aIF6 will be of fundamental value in advancing understanding of the cellular role of this interesting protein.

The universal protein a(e)IF5A (EFP in bacteria) is usually classed as a translation initiation factor. However, since this protein does little to help the selection of the translation start site and functions as a specialized elongation factor, it is described in the following section.

Elongation

Elongation is the core process of protein synthesis and as a result is extremely well conserved in the three domains. The three canonical phases of elongation (adaptation, transpeptidation, and translocation) proceed with similar modalities in all cells. Likewise, in all three domains of life, elongation is assisted by the two highly conserved, paralogous elongation factors, EF1A and EF2 (EFTu and EFG in bacteria, respectively). Both EFs are G-proteins that interact with the ribosome in the GTP-bound form and are released following GTP hydrolysis as inactive EF-GDP complexes. EF1A has a much higher affinity for GDP than for GTP and requires the guanine nucleotide exchange factor, EF1B (EFTs in bacteria), to be recycled.

Similar to most components of the translational apparatus, the archaeal elongation factors are closer in sequence to their eucaryal counterparts. Elongation factor-based evolutionary trees were used to infer the root of the universal tree, identifying the archaea and the eucarya as sister domains (46).

Interesting new insight about the archaea was obtained from analysis of elongation, including the incorporation of the rare amino acids, selenocysteine and pyrrolysine, respectively dubbed the 21st and 22nd amino acids of the genetic code (see Chapter 9).

Selenocysteine (sec) is responsible for the presence of the trace element selenium in proteins. Selenocysteine is incorporated during elongation in response to an internal UGA stop codon that precedes a specific RNA hairpin termed SECIS (selenocysteine insertion structure) (see Chapter 9). Incorporation requires the action of the specialized elongation factor SELB, which recognizes a specialized tRNA, sec-tRNAsec (34, 95, 100). This pathway is common to all three domains of life, although specific differences exist among them. In bacteria, the SECIS is adjacent to the UGA codon and is bound directly by SELB,

which carries sec-tRNAsec, thereby leading to the insertion of sec in the appropriate position (95). In the *Eucarya*, the SECIS is instead placed in the 3′-UTR of the mRNA and is indirectly recognized by SELB with the aid of the adaptor protein SBP2 (34). The different SECIS-binding abilities of bacterial and eucaryal SELB are due to structural differences in the C-terminal domain of the protein (domain IV). In bacteria, the SELB domain IV is longer and can interact directly with the SECIS, while in the eucarya it is shorter and requires the help of SBP2 to contact the SECIS.

In the archaea, SELB homologs have only been identified in methanococci (100). Sequence analyses revealed that these proteins have a C-terminal domain IV that is even shorter than that of their eucaryal homolog, suggesting that its interaction with the SECIS may also be mediated by an adaptor. In support of this, the SECIS elements of archaeal mRNAs encoding selenoproteins are in the 3′-UTR, similar to eucarya (99). However, a SBP2 homolog has not been identified in archaeal genomes.

Very recently, the structure of *M. maripaludis* SELB was solved by X-ray crystallography, yielding interesting insights into the function of the protein (72). Remarkably, archaeal SELB has a shape resembling a chalice that is very similar to that of the initiation factor aIF2 (aIF5B) (Fig. 3). Since bacterial IF2 is an RNA-binding protein that binds f-met-tRNA, it has been suggested that, despite the reduced size of its domain IV, the archaeal SELB may interact directly with the SECIS, similar to its bacterial counterpart. A computer model of the SELB/ribosome complex shows that domain IV points toward the 3′-mRNA entrance where the SECIS is likely to be located and an interaction established (125). These findings would explain the apparent lack of a SBP2 homolog in archaea and raise the interesting possibility that aIF2 (aIF5B) contacts the mRNA as a part of its function.

While the synthesis of selenoproteins was established in bacterial systems, the discovery of pyrrolysine was based in the *Archaea*. Pyrrolysine is a lysine derivative that was originally identified in methylamine methyltransferase genes of *Methanosarcina barkeri* and subsequently found in other methanogens, where it may play an exclusive role (44). Like sec, pyrrolysine is encoded by an UAG codon, which is decoded by an unusual tRNA with a CUA anticodon. The latter is encoded by *pylT*, located near a methyltransferase gene cluster. The adjacent *pylS* gene encodes a class II aminoacyl-tRNA synthetase that charges the *pylT*-derived tRNA with lysine; this tRNA synthetase is not closely related to known lysyl-tRNA synthetases (128). However, unlike sec, the insertion of pyrrolysine into the growing polypeptide chain does not appear to require the action of a specialized EF, and the pyrrolysine tRNA is probably recognized by EF-1A. It is also unlikely that pyrrolysine insertion requires a *cis*-acting mRNA element similar to the SECIS. Recent data suggest that UAG is rarely used as a termination codon in the species containing pyrrolysine and may therefore be used as a normal sense codon (128).

Another interesting feature of translation elongation concerns the formation of the first peptide bond, which seems to require a specialized factor acting at the interface between initiation and elongation. The factor is a universal protein known as EFP in *Bacteria* and a(e)IF5A in *Archaea* and *Eucarya*. The factors from the three domains are clearly homologous, although the archaeal and eucaryal polypeptides have the highest level of sequence identity (63). Biochemical studies, mainly carried out on the bacterial protein, have established that it functions to catalyze the formation of the first peptide bond (35). Considerably more experimentation is required to determine its role in translation.

In contrast, structural information is available for bacterial EFP and archaeal aIF5A. The bacterial factor is composed of three beta-barrel domains. It has an L-shaped structure reminiscent of a tRNA, suggesting that it might interact with the ribosomal tRNA binding site(s) by virtue of "molecular mimicry." It appears to bind both ribosomal subunits and to stimulate the peptidyltransferase center on the 50S particle (40). The archaeal factor (structures are available for *M. jannaschii*, *P. aerophylum*, and *P. horikoshii* aIF5A) is somewhat shorter than its bacterial homolog and includes only two, instead of three, beta-barrel domains (49, 89, 122). Due to the structural difference, it has a rodlike shape rather than an L shape and may therefore interact preferentially with the large ribosomal subunit, although no experimental data are available to address this. A remarkable feature of aIF5A, which is shared with its eucaryal homolog, is the presence of a uniquely modified lysine, known as hypusine (N-ε-(4-aminobutyl-2-hydroxy) lysine). A conserved lysine is present in the corresponding position in the bacterial protein, and it is not posttranslationally modified to hypusine. The functional role of hypusine is poorly understood, although it is known that inactivation of the hypusine-forming enzyme in yeast is lethal. Note that the presence of hypusine in archaea was one of the first "eucaryal" features to be identified in the *Archaea* (7).

EFP is one of the few bacterial translation factors that is larger than its archaeal/eucaryal counterparts. e/aIF5A may have evolved from an EFP-like ancestor following the loss of one of its three protein domains. Moreover, in archaea and eucarya, an unknown protein may fulfill the function of the missing domain (40).

Termination

Translation terminates when a stop codon enters the ribosomal A site. In *Archaea*, *Bacteria*, and *Eucarya*, the recognition of the stop codon and the release of the completed protein from the ribosome is carried out by a set of proteins termed release or termination factors (RFs). The mechanism of termination is well understood in bacteria, while several aspects are still to be elucidated in eucarya. In archaea, very few experimental studies have been performed that address this step of protein synthesis. The most significant information derives from in silico analysis of bacterial and eucaryal homologs of the termination machinery in archaeal genomes. The analyses have revealed that, similar to initiation and elongation, archaeal termination involves eucaryal-like proteins.

In bacteria and eucarya, class 1 termination factors recognize stop codons and release the completed polypeptide by promoting the hydrolysis of the ester bond that is anchored to the tRNA in the P site. *Bacteria* possess two class 1 termination factors: RF1 recognizes UAA and UAG, and RF2 recognizes UAA and UGA. In contrast, the *Eucarya* appear to employ a single factor, eRF1, to recognize all three stop codons (54). All archaeal genomes include genes encoding a polypeptide homologous to eRF1 (referred to as aRF1), while no bacterial RF2 homologs have been detected. Therefore, the *Archaea* appear to resemble the *Eucarya* in the use of a single factor for stop codon recognition. Consistent with this, the *M. jannaschii* aRF1 promotes termination on eucaryal ribosomes (31). Despite intensive computational analyses, no meaningful similarity has been detected between the bacterial and the archaeal/eucaryal class 1 RF, which indicates two distinct protein families have evolved (54). The structure of human eRF1 resembles a tRNA by having an elongated shape with two "arms," similar to the acceptor and anticodon stems of a tRNA (110). This "molecular mimicry" may reflect the fact that both RF1 and tRNA bind in the ribosomal A site. However, the precise mechanism for stop codon recognition by the RF1 proteins has not been determined.

Given the analagous function of aRF1 and eRF1, the archaeal proteins may be expected to have a shape similar to their eucaryal counterparts. However, a unique feature of aRF1 proteins is that they are extensively truncated at their C-termini in comparison to both bacterial and eucaryal RF1 (54). The C-terminal domain of RF1 proteins is involved in interacting with class 2 RF, and these appear to be absent in archaea.

In addition to the class 1 RF, the *Bacteria* and the *Eucarya*, but not the *Archaea*, possess a class 2 RF, referred to as RF3 in *Bacteria* and eRF3 in *Eucarya*.

Class 2 RF proteins are G-proteins that do not participate in the peptide release reaction itself. The function of RF3 has been analyzed in some detail, and its main task seems to be to accelerate the recycling of class 1 RF proteins after translation termination (53, 127). RF3 interacts with the ribosomes in a GDP-bound form, and the ribosome itself then functions as a guanine nucleotide-exchange factor (GEF) that promotes the release of GDP and the binding of GTP to the factor. RF3-GTP then catalyzes the release of RF1/RF2 from the ribosome and detaches following GTP hydrolysis. RF3 is not essential in bacteria (53), while its eucaryal ortholog eRF3 is an essential factor that binds to eRF1. However, the cellular role of eRF3 remains enigmatic (54).

In addition to class 1 and class 2 RFs, the *Bacteria* possess an additional essential termination factor, referred to as ribosome recycling factor (RRF). This factor appears to be only in bacteria, as no homologs have been identified in the *Archaea* or the *Eucarya*.

Information to date indicates that the *Archaea* are endowed with a simplified version of the eucaryal translation termination mechanism that utilizes a single class 1 RF without a RF3 or RRF protein. Detailed experimental studies are required to determine whether the *Archaea* possess additional, unique termination factors that may perform the role of the RF3 and/or RRF proteins from the other two domains.

Translational Regulation

Few studies have examined translational regulation in archaea, and these have mainly addressed autoregulation, i.e., regulation by a protein of its own mRNA. Autoregulation of r-protein operons of methanogenic archaea was first demonstrated approximately ten years ago (41). Little has appeared in the literature since this time; progress may have been hampered by unavailability of suitable experimental tools (see Chapter 21). The mechanism of translational autoregulation in methanogens appears to be essentially the same as in bacteria. A well-studied case is the *M. vannielii* L1 ribosomal protein operon (encoding the r-proteins L1, L10, and L12), which is transcribed as a single polycistronic mRNA (41, 59, 78). The regulatory protein L1 interacts preferentially with its binding site on the 23S rRNA and, when in excess, also binds to the regulatory target site of its mRNA, thereby inhibiting translation of all three cistrons in the operon. The regulatory mRNA site, a structural mimic of the rRNA-binding site for L1, is located within the L1 gene about 30 nt downstream of the ATG initiation codon (59). A similar regulatory mechanism also exists in *M. thermolithotrophicum*

and *M. jannaschii*. However, it is doubtful that this mechanism exists in nonmethanogenic archaea (59).

An interesting example of translational regulation that was recently identified in archaea involves cotranslational frameshifting, whereby a contiguous polypeptide is synthesized from two, out-of-frame ORFs (split genes). In genes with frameshifts, the gene is interrupted by an in-frame stop codon, and a full-length protein is produced by the ribosome-changing register and continuing translation. The presence of several split genes in archaea has recently been documented (22). In at least one case, an uninterrupted polypeptide was shown to be synthesized following a programmed -1 frameshift (22), a mechanism that operates in bacteria. However, the details of this archaeal mechanism for frame-shifting have not been examined.

PERSPECTIVE: THE NEXT FIVE YEARS

The study of translation in archaea is still in its infancy; much more work is required before a comprehensive understanding of its similarities and differences with the other two domains of life are determined. Topics that are likely to progress rapidly in the near future are the following:

Mechanism for Leaderless mRNA Translation

Leaderless transcripts have attracted considerable attention recently because they are produced in abundance by some archaea, they are decoded by a novel and poorly understood mechanism, and they have anticipated evolutionary importance. Understanding leaderless translation will first require extended structural analysis of the structure of the putative leaderless transcripts, many of which have been identified solely on the basis of in silico studies. It will be important to compare the translational efficiency of true leaderless mRNAs that have their 5′ ends coinciding with the initiation codon, versus quasi-leaderless mRNAs that have a few nucleotides upstream of the initiation codon. It should be determined how the length of the short 5′-UTRs influence translation and whether there are any preferred base compositions or sequences that may affect translational efficiency. Another open question is whether an initiator AUG is required for optimal translation of leaderless mRNAs. The biogenesis of the leaderless mRNAs also needs to be explored, as it is possible that some may be generated by posttranscriptional processing of longer transcripts. A very important task will be to determine which initiation factors are required for leaderless mRNA translation.

Function of the Translation Initiation Factors

The unexpected similarity between the archaeal and eucaryal IFs has attracted much attention and has prompted many structural studies of the archaeal proteins in the past few years. However, except for a/eIF2, the function of the putative archaeal translation initiation factors remains undetermined. It is important to establish which factors are required for the decoding of leadered and leaderless mRNAs, what is the precise function of the universal IFs in archaea (in particular, the IF2-like protein that has somewhat different roles in bacteria and eucarya), and whether archaeal-specific IFs exist. An important prerequisite for enabling rapid advancement is to improve the experimental tools for the molecular analysis of translation. For example, the available techniques for performing molecular genetics in archaea are still far from being well standardized and of easy general use (see Chapter 21). However, such techniques are essential to enable the exploration of the function of the translation components by means of mutational and gene knockout studies. For example, in vivo studies are needed to determine which of the archaeal IFs are encoded by essential genes.

Tools for in vitro analysis also need to be developed. To date, cell-free systems for the translation of natural archaeal mRNAs have been described only for the crenarchaeon *S. solfataricus* (25). Clearly, in the archaeal field there is a general lack of highly resolved systems (similar to those available for the *Bacteria* and the *Eucarya*) for the biochemical dissection of translational events. Fortunately, this issue is beginning to be seriously addressed, and meaningful advancements can be expected in the next few years.

Translational Regulation

Possibly the most interesting topic to be tackled in the next few years relates to the presence in archaea of translation regulation mechanisms similar to those in eucarya. This is indicated by the presence in the *Archaea* of two IFs (a/eIF2 and aIF6), the homologs of which are important in regulating eucaryal protein synthesis. The activity of both proteins in eucarya is modulated by phosphorylation. eIF2 is functionally inactivated through the phosphorylation of its α-subunit. Recently, the α-subunit of a/eIF2 from *P. horikoshii* was reported to be phosphorylated in vitro by a specific kinase (111). This finding needs to be confirmed and substantiated by functional studies. The biological significance of a/eIF2 phosphorylation is not clear, since the protein has a similar affinity for GDP and GTP (90) and cannot be regulated by

G-nucleotide exchange similar to its eucaryal counterpart. A possibility is that α-subunit phosphorylation regulates the interaction between the α- and γ-subunits, or the binding of the factor to the ribosome. The development of this line of research may elucidate why the common ancestor of the *Archaea* and the *Eucarya* adopted a trimeric factor for carrying met-tRNAi to the ribosome.

Another translation factor that seems to have a general regulatory role in eucaryal translation is eIF6. However, the function of this protein is poorly understood compared with eIF2, as it appears to participate both in regulating 80S ribosome formation and in ribosome synthesis (9, 21, 108). Determining the function of archaeal IF6 will shed light on the role of the eucaryal homolog. In general, the study of a/eIF2 and aIF6 may lead to exciting new insights into the evolutionary history of not only the archaeal, but also the eucaryal mechanisms of translational regulation.

REFERENCES

1. **Alix, J. H., and K. H. Nierhaus.** 2003. DnaK-facilitated ribosome assembly in *Escherichia coli* revisited. *RNA* 9:787–793.
2. **Allen, G. S., A. Zavialov, R. Gursky, M. Ehrenberg, and J. Frank.** 2005. The cryo-EM structure of a translation initiation complex from *Escherichia coli. Cell* 121:703–712.
3. **Amils, R., P. Cammarano, and P. Londei.** 1993. Translation in Archaea, p. 393–437. *In* M. Kates, D. Kushner, and A. Matheson (ed.), *Biochemistry of Archaea.* New Comprehensive Biochemistry Series. Elsevier, Amsterdam, The Netherlands.
4. **Anantharaman, V., E. V. Koonin, and L. Aravind.** 2002. Comparative genomics and evolution of proteins involved in RNA metabolism. *Nucleic Acids Res.* 30:1427–1464.
5. **Bachellerie, J. P., J. Cavaille, and A. Huttenhofer.** 2002. The expanding snoRNA world. *Biochimie* 84:775–790.
6. **Ban, N., P. Nissen, J. Hansen, P. B. Moore, and T. A. Steitz.** 2000. The complete atomic structure of the large ribosomal subunit at 2.4 A resolution. *Science* 289:905–920.
7. **Bartig. D., K. Lemkemeier, J. Frank, F. Lottspeich, and F. Klink.** 1992. The archaebacterial hypusine-containing protein. Structural features suggest common ancestry with eukaryotic translation initiation factor 5A. *Eur. J. Biochem.* 204:751–758.
8. **Basu, U., K. Si, H. Deng, and U. Maitra.** 2003. Phosphorylation of mammalian eukaryotic translation initiation factor 6 and its *Saccharomyces cerevisiae* homologue Tif6p: evidence that phosphorylation of Tif6p regulates its nucleocytoplasmic distribution and is required for yeast cell growth. *Mol. Cell. Biol.* 23:6187–6199.
9. **Basu, U., K. Si, J. R. Warner, and U. Maitra.** 2001. The *Saccharomyces cerevisiae* TIF6 gene encoding translation initiation factor 6 is required for 60S ribosomal subunit biogenesis. *Mol. Cell. Biol.* 21:1453–1462.
10. **Battiste, J. L., T. V. Pestova, C. U. Hellen, and G. Wagner.** 2000. The eIF1A solution structure reveals a large RNA-binding surface important for scanning function. *Mol. Cell* 5:109–119.
11. **Bell. S. D., and S. P. Jackson.** 1998. Transcription and translation in Archaea: a mosaic of eukaryal and bacterial features. *Trends Microbiol.* 6:222–228.

12. Benelli, D., E. Maone, and P. Londei. 2003. Two different mechanisms for ribosome/mRNA interaction in archaeal translation initiation. *Mol. Microbiol.* 50:635–643.

13. Betlach, M., J. Friedman, H. W. Boyer, and F. Pfeifer. 1984. Characterization of a halobacterial gene affecting bacterioopsin gene expression. *Nucleic Acids Res.* 12:7949–7959.

14. Bini, E., V. Dikshit, K. Dirksen, M. Drozda, and P. Blum. 2002. Stability of mRNA in the hyperthermophilic archaeon *Sulfolobus solfataricus. RNA* 8:1129–1136.

15. Boileau, G., P. Butler, J. W. Hershey, and R. R. Traut. 1983. Direct cross-links between initiation factors 1, 2, and 3 and ribosomal proteins promoted by 2-iminothiolane. *Biochemistry* 22:3162–3170.

16. Brodersen, D. E., W. M. Clemons, Jr., A. P. Carter, B. T. Wimberly, and V. Ramakrishnan. 2002. Crystal structure of the 30S ribosomal subunit from *Thermus thermophilus*: structure of the proteins and their interactions with 16S RNA. *J. Mol. Biol.* 316:725–768.

17. Cannone. J. J., S. Subramanian, M. N. Schnare, J. R. Collett, L. M. D'Souza, Y. Du, B. Feng, N. Lin, L. V. Madabusi, K. M. Muller, N. Pande, Z. Shang, N. Yu, and R. R. Gutell. 2002. The Comparative RNA Web (CRW) site: an online database of comparative sequence and structure information for ribosomal, intron, and other RNAs. *BioMed Central Bioinform.* 3:15.

18. Castresana, J. 2001. Comparative genomics and bioenergetics. *Biochim. Biophys. Acta* 1506:147–162.

19. Cavaille, J., M. Nicoloso, and J P. Bachellerie. 1996. Targeted ribose methylation of RNA in vivo directed by tailored antisense RNA guides. *Nature* 383:732–735.

20. Cech, T. R. 1993. The efficiency and versatility of catalytic RNA: implications for an RNA world. *Gene* 135:33–36.

21. Ceci, M., C. Gaviraghi, C. Gorrini, L. A. Sala, N. Offenhauser, et al. 2003. Release of eIF6 (p27BBP) from the 60S subunit allows 80S ribosome assembly. *Nature* 426:579–584.

22. Cobucci-Ponzano, B., M. Rossi, and M. Moracci. 2005. Recoding in archaea. *Mol. Microbiol.* 55:339–348.

23. Coenye, T., and P. Vandamme. 2005. Organisation of the S10, spc and *alpha* ribosomal protein gene clusters in prokaryotic genomes. *FEMS Microbiol. Lett.* 242:117–126.

24. Colthurst, D. R., D. G. Campbell, and C. G. Proud. 1987. Structure and regulation of eukaryotic initiation factor eIF-2. Sequence of the site in the alpha subunit phosphorylated by the haem-controlled repressor and by the double-stranded RNA-activated inhibitor. *Eur. J. Biochem.* 166:357–363.

25. Condo, I., A. Ciammaruconi, D. Benelli, D. Ruggero, and P. Londei. 1999. Cis-acting signals controlling translational initiation in the thermophilic archaeon *Sulfolobus solfataricus. Mol. Microbiol.* 34:377–384.

26. Cort, J. R., E. V. Koonin, P. A. Bash, and M. A. Kennedy. 1999. A phylogenetic approach to target selection for structural genomics: solution structure of YciH. *Nucleic Acids Res.* 27:4018–4027.

27. Cui, Y., J. D. Dinman, T. G. Kinzy, and S. W. Peltz. 1998. The Mof2/Sui1 protein is a general monitor of translational accuracy. *Mol. Cell Biol.* 18:1506–1516.

28. Daalgard, J. Z., and R. A. Garrett. 1993. Archaeal hyperthermophile genes, p. 535–563. *In* M. Kates, D. Kushner, and A. Matheson (ed.), *Biochemistry of Archaea.* New Comprehensive Biochemistry Series. Elsevier, Amsterdam. The Netherlands.

29. Das, A., M. K. Bagchi, P. Ghosh-Dastidar, and N. K. Gupta. 1982. Protein synthesis in rabbit reticulocytes. A study of peptide chain initiation using native and beta-subunit-depleted eukaryotic initiation factor 2. *J. Biol. Chem.* 257:1282–1288.

30. Dennis, P. P. 1997. Ancient ciphers: translation in Archaea. *Cell* 89:1007–1010.

31. Dontsova, M., L. Frolova, J. Vassilieva, W. Piendl, L. Kisselev, and M. Garber. 2000. Translation termination factor aRF1 from the archaeon *Methanococcus jannaschii* is active with eukaryotic ribosomes. *FEBS Lett.* 472:213–216.

32. Erickson, F. L., and E. M. Hannig. 1996. Ligand interactions with eukaryotic translation initiation factor 2: role of the gamma-subunit. *EMBO J.* 15:6311–6320.

33. Etchegaray, J. P., and M. Inouye. 1999. Translational enhancement by an element downstream of the initiation codon in Escherichia coli. *J. Biol. Chem.* 274:10079–10085.

34. Fagegaltier, D., N. Hubert, K. Yamada, T. Mizutani, P. Carbon, and A. Krol. 2000. Characterization of mSelB, a novel mammalian elongation factor for selenoprotein translation. *EMBO J.* 19:4796–4805.

35. Glick, B. R., S. Chladek, and M. C. Ganoza. 1979. Peptide bond formation stimulated by protein synthesis factor EF-P depends on the aminoacyl moiety of the acceptor. *Eur. J. Biochem.* 97:23–28.

36. Grill, S., C. O. Gualerzi, P. Londei, and U. Blasi. 2000. Selective stimulation of translation of leaderless mRNA by initiation factor 2: evolutionary implications for translation. *EMBO J.* 19:4101–4110.

37. Gualerzi, C., and C. L. Pon. 1990. Initiation of mRNA translation in prokaryotes. *Biochemistry* 29:5881–5889.

38. Guenneugues, M., E. Caserta, L. Brandi, R. Spurio, S. Meunier, et al. 2000. Mapping the fMet-tRNA(f)(Met) binding site of initiation factor IF2. *EMBO J.* 19:5233–5240.

39. Gutierrez, P., M. J. Osborne, N. Siddiqui, J. F. Trempe, C. Arrowsmith, and K. Gehring. 2004. Structure of the archaeal translation initiation factor aIF2 beta from *Methanobacterium thermoautotrophicum*: implications for translation initiation. *Protein Sci.* 13:659–667.

40. Hanawa-Suetsugu, K. S. Sekine, H. Sakai, C. Hori-Takemoto, T. Terada, et al. 2004. Crystal structure of elongation factor P from *Thermus thermophilus* HB8. *Proc. Natl. Acad. Sci. USA* 101:9595–9600.

41. Hanner,, M., C. Mayer, C. Kohrer, G. Golderer, P. Grobner, and W. Piendl. 1994. Autogenous translational regulation of the ribosomal MvaL1 operon in the archaebacterium *Methanococcus vannielii. J. Bacteriol.* 176:409–418.

42. Harms, J., F. Schluenzen, R. Zarivach, A. Bashan, S. Gat, et al. 2001. High resolution structure of the large ribosomal subunit from a mesophilic eubacterium. *Cell* 107:679–688.

43. Hennigan, A. N., and J. N. Reeve. 1994. mRNAs in the methanogenic archaeon *Methanococcus vannielii*: numbers, half-lives and processing. *Mol. Microbiol.* 11:655–670.

44. Ibba, M., and D. Soll. 2002. Genetic code: introducing pyrrolysine. *Curr. Biol.* 12:R464–R466.

45. Itoh, T., K. Suzuki, and T. Nakase. 1998. Occurrence of introns in the 16S rRNA genes of members of the genus *Thermoproteus. Arch. Microbiol.* 170:155–161.

46. Iwabe, N., K. Kuma, M. Hasegawa, S. Osawa, and T. Miyata. 1989. Evolutionary relationship of archaebacteria, eubacteria, and eukaryotes inferred from phylogenetic trees of duplicated genes. *Proc. Natl. Acad. Sci. USA* 86:9355–9359.

47. Joyce, G. F. 2002. The antiquity of RNA-based evolution. *Nature* 418:214–221.

48. Karlin, S., J. Mrazek, J. Ma, and L. Brocchieri. 2005. Predicted highly expressed genes in archaeal genomes. *Proc. Natl. Acad. Sci. USA* 102:7303–7308.

49. Kim, K. K., L. W. Hung, H. Yokota, R. Kim, and S. H. Kim. 1998. Crystal structures of eukaryotic translation initiation factor 5A from *Methanococcus jannaschii* at 1.8 A resolution. *Proc. Natl. Acad. Sci. USA* 95:10419–10424.

50. Kimball, S. R. 1999. Eukaryotic initiation factor eIF2. *Int. J. Biochem. Cell. Biol.* 31:25–29.

51. Kiss, T. 2001. Small nucleolar RNA-guided post-transcriptional modification of cellular RNAs. *EMBO J.* **20:**3617–3622.

52. Kiss, T. 2002. Small nucleolar RNAs: an abundant group of noncoding RNAs with diverse cellular functions. *Cell* **109:**145–148.

53. Kisselev, L., M. Ehrenberg, and L. Frolova. 2003. Termination of translation: interplay of mRNA, rRNAs and release factors? *EMBO J.* **22:**175–182.

54. Kisselev, L. L., and R. H. Buckingham. 2000. Translational termination comes of age. *Trends Biochem. Sci.* **25:**561–566.

55. Kiss-Laszlo, Z., Y. Henry, and T. Kiss. 1998. Sequence and structural elements of methylation guide snoRNAs essential for site-specific ribose methylation of pre-rRNA. *EMBO J.* **17:**797–807.

56. Kjems, J., and R. A. Garrett. 1988. Novel splicing mechanism for the ribosomal RNA intron in the archaebacterium *Desulfurococcus mobilis*. *Cell* **54:**693–703.

57. Koonin, E. V. 2003. Comparative genomics, minimal genesets and the last universal common ancestor. *Nat. Rev. Microbiol.* **1:**27–36.

58. Kozak, M. 1999. Initiation of translation in prokaryotes and eukaryotes. *Gene* **234:**187–208.

59. Kraft, A., C. Lutz, A. Lingenhel, P. Grobner, and W. Piendl. 1999. Control of ribosomal protein L1 synthesis in mesophilic and thermophilic archaea. *Genetics* **152:**1363–1372.

60. Kuhn, J. F., E. J. Tran, and E. S. Maxwell. 2002. Archaeal ribosomal protein L7 is a functional homolog of the eukaryotic 15.5kD/Snu13p snoRNP core protein. *Nucleic Acids Res.* **30:**931–941.

61. Kyrpides, N., R. Overbeek, and C. Ouzounis. 1999. Universal protein families and the functional content of the last universal common ancestor. *J. Mol. Evol.* **49:**413–423.

62. Kyrpides, N. C., and C. R. Woese. 1998. Archaeal translation initiation revisited: the initiation factor 2 and eukaryotic initiation factor 2B alpha-beta-delta subunit families. *Proc. Natl. Acad. Sci. USA* **95:**3726–3730.

63. Kyrpides, N. C., and C. R. Woese. 1998. Universally conserved translation initiation factors. *Proc. Natl. Acad. Sci. USA* **95:**224–228.

64. La Teana, A., A. Brandi, M. O'Connor, S. Freddi, and C. L. Pon. 2000. Translation during cold adaptation does not involve mRNA-rRNA base pairing through the downstream box. *RNA* **6:**1393–1402.

65. Lake, J. A. 1983. Evolving ribosome structure: domains in archaebacteria, eubacteria, and eucaryotes. *Cell* **33:**318–319.

66. Lake, J. A. 1985. Evolving ribosome structure: domains in archaebacteria, eubacteria, eocytes and eukaryotes. *Annu. Rev. Biochem.* **54:**507–530.

67. Lake, J. A., E. Henderson, M. Oakes, and M. W. Clark. 1984. Eocytes: a new ribosome structure indicates a kingdom with a close relationship to eukaryotes. *Proc. Natl. Acad. Sci. USA* **81:**3786–3790.

68. Lamond, A. I., and T. J. Gibson. 1990. Catalytic RNA and the origin of genetic systems. *Trends Genet.* **6:**145–149.

69. Lecompte, O., R. Ripp, J. C. Thierry, D. Moras, and O. Poch. 2002. Comparative analysis of ribosomal proteins in complete genomes: an example of reductive evolution at the domain scale. *Nucleic Acids Res.* **30:**5382–5390.

70. Lee, J. H., S. K. Choi, A. Roll-Mecak, S. K. Burley, and T. E. Dever. 1999. Universal conservation in translation initiation revealed by human and archaeal homologs of bacterial translation initiation factor IF2. *Proc. Natl. Acad. Sci. USA* **96:**4342–4347.

71. Lee, J. H., T. V. Pestova, B. S. Shin, C. Cao, S. K. Choi, and T. E. Dever. 2002. Initiation factor eIF5B catalyzes second GTP-dependent step in eukaryotic translation initiation. *Proc. Natl. Acad. Sci. USA* **99:**16689–16694.

72. Leibundgut, M., C. Frick, M. Thanbichler, A. Bock, and N. Ban. 2005. Selenocysteine tRNA-specific elongation factor SelB is a structural chimaera of elongation and initiation factors. *EMBO J.* **24:**11–22.

73. Li, L., and C. C. Wang. 2004. Capped mRNA with a single nucleotide leader is optimally translated in a primitive eukaryote, *Giardia lamblia*. *J. Biol. Chem.* **279:**14656–14664.

74. Londei, P. 2005. Evolution of translational initiation: new insights from the archaea. *FEMS Microbiol. Rev.* **29:**185–200.

75. Londei, P., A. Teichner, P. Cammarano, M. De Rosa, and A. Gambacorta. 1983. Particle weights and protein composition of the ribosomal subunits of the extremely thermoacidophilic archaebacterium *Caldariella acidophila*. *Biochem. J.* **209:**461–470.

76. Londei, P., J. Teixido, M. Acca, P. Cammarano, and R. Amils. 1986. Total reconstitution of active large ribosomal subunits of the thermoacidophilic archaebacterium *Sulfolobus solfataricus*. *Nucleic Acids Res.* **14:**2269–2285.

77. Lykke-Andersen, J., C. Aagaard, M. Semionenkov, and R. A. Garrett. 1997. Archaeal introns: splicing, intercellular mobility and evolution. *Trends Biochem. Sci.* **22:**326–331.

78. Mayer, C., C. Kohrer, P. Grobner, and W. Piendl. 1998. MvaL1 autoregulates the synthesis of the three ribosomal proteins encoded on the MvaL1 operon of the archaeon *Methanococcus vannielii* by inhibiting its own translation before or at the formation of the first peptide bond. *Mol. Microbiol.* **27:**455–468.

79. McCloskey, J. A., and J. Rozenski. 2005. The Small Subunit rRNA Modification Database. *Nucleic Acids Res.* **33:**D135–D138.

80. Merrick, W. C. 1992. Mechanism and regulation of eukaryotic protein synthesis. *Microbiol. Rev.* **56:**291–315.

81. Mushegian, A. 2005. Protein content of minimal and ancestral ribosome. *RNA* **11:**1400–1404.

82. Niewmierzycka, A., and S. Clarke. 1999. S-Adenosylmethionine-dependent methylation in *Saccharomyces cerevisiae*. Identification of a novel protein arginine methyltransferase. *J. Biol. Chem.* **274:**814–824.

83. Nomura, M. 1973. Assembly of bacterial ribosomes. *Science* **179:**864–873.

84. Nomura, N., Y. Sako, and A. Uchida. 1998. Molecular characterization and postsplicing fate of three introns within the single rRNA operon of the hyperthermophilic archaeon *Aeropyrum pernix* K1. *J. Bacteriol.* **180:**3635–3643.

85. O'Connor, M., T. Asai, C. L. Squires, and A. E. Dahlberg. 1999. Enhancement of translation by the downstream box does not involve base pairing of mRNA with the penultimate stem sequence of 16S rRNA. *Proc. Natl. Acad. Sci. USA* **96:**8973–8978.

86. Omer, A. D., T. M. Lowe, A. G. Russell, H. Ebhardt, S. R. Eddy, and P. P. Dennis. 2000. Homologs of small nucleolar RNAs in Archaea. *Science* **288:**517–522.

87. Omer, A. D., S. Ziesche, H. Ebhardt, and P. P. Dennis. 2002. In vitro reconstitution and activity of a C/D box methylation guide ribonucleoprotein complex. *Proc. Natl. Acad. Sci. USA* **99:**5289–5294.

88. Orgel, L. E. 2004. Prebiotic chemistry and the origin of the RNA world. *Crit. Rev. Biochem. Mol. Biol.* **39:**99–123.

89. Peat, T. S., J. Newman, G. S. Waldo, J. Berendzen, and T. C. Terwilliger. 1998. Structure of translation initiation factor 5A from *Pyrobaculum aerophilum* at 1.75 A resolution. *Structure* **6:**1207–1214.

90. Pedulla, N., R. Palermo, D. Hasenohrl, U. Blasi, P. Cammarano, and P. Londei. 2005. The archaeal eIF2 homologue:

functional properties of an ancient translation initiation factor. *Nucleic Acids Res.* **33**:1804–1812.

91. Pestova, T. V., and C. U. Hellen. 2001. Functions of eukaryotic factors in initiation of translation. *Cold Spring Harbor Symp. Quant. Biol.* **66**:389–396.

92. Pestova, T. V., and V. G. Kolupaeva. 2002. The roles of individual eukaryotic translation initiation factors in ribosomal scanning and initiation codon selection. *Genes Dev.* **16**:2906–2922.

93. Pestova, T. V., I. B. Lomakin, J. H. Lee, S. K. Choi, T. E. Dever, and C. U. Hellen. 2000. The joining of ribosomal subunits in eukaryotes requires eIF5B. *Nature* **403**:332–335.

94. Ramirez, C., A. K. E. Kopke, D.-C. Yang, T. Boeckh, and A. T. Matheson. 1993. The structure, function and evolution of archaeal ribosomes, p. 439–466. *In* M. Kates, D. Kushner, and A. Matheson (ed.), *Biochemistry of Archaea.* New Comprehensive Biochemistry Series, Elsevier, Amsterdam, The Netherlands.

95. Ringquist, S., D. Schneider, T. Gibson, C. Baron, A. Bock, and L. Gold. 1994. Recognition of the mRNA selenocysteine insertion sequence by the specialized translational elongation factor SELB. *Genes Dev.* **8**:376–385.

96. Rivera, M. C., and J. A. Lake. 1992. Evidence that eukaryotes and eocyte prokaryotes are immediate relatives. *Science* **257**:74–76.

97. Roll-Mecak, A., P. Alone, C. Cao, T. E. Dever, and S. K. Burley. 2004. X-ray structure of translation initiation factor eIF2gamma: implications for tRNA and eIF2alpha binding. *J. Biol. Chem* **279**:10634–10642.

98. Roll-Mecak, A., C. Cao, T. E. Dever, and S. K. Burley. 2000. X-Ray structures of the universal translation initiation factor IF2/eIF5B: conformational changes on GDP and GTP binding. *Cell* **103**:781–792.

99. Rother, M., A. Resch, W. L. Gardner, W. B. Whitman, and A. Bock. 2001. Heterologous expression of archaeal selenoprotein genes directed by the SECIS element located in the 3′ non-translated region. *Mol. Microbiol.* **40**:900–908.

100. Rother, M., R. Wilting, S. Commans, and A. Bock. 2000. Identification and characterisation of the selenocysteine-specific translation factor SelB from the archaeon *Methanococcus jannaschii*. *J. Mol. Biol.* **299**:351–358.

101. Sanchez, M. E., P. Londei, and R. Amils. 1996. Total reconstitution of active small ribosomal subunits of the extreme halophilic archaeon *Haloferax mediterranei*. *Biochim. Biophys. Acta* **1292**:140–144.

102. Sanchez, M. E., D. Urena, R. Amils, and P. Londei. 1990. In vitro reassembly of active large ribosomal subunits of the halophilic archaebacterium *Haloferax mediterranei*. *Biochemistry* **29**:9256–9261.

103. Sartorius-Neef, S., and F. Pfeifer. 2004. In vivo studies on putative Shine-Dalgarno sequences of the halophilic archaeon *Halobacterium salinarum*. *Mol. Microbiol.* **51**:579–588.

104. Schmitt, E., S. Blanquet, and Y. Mechulam. 2002. The large subunit of initiation factor aIF2 is a close structural homologue of elongation factors. *EMBO J.* **21**:1821–1832.

105. Sensen, C. W., H. P. Klenk, R. K. Singh, G. Allard, C. C. Chan, et al. 1996. Organizational characteristics and information content of an archaeal genome: 156 kb of sequence from *Sulfolobus solfataricus* P2. *Mol. Microbiol.* **22**:175–191.

106. Sette, M., P. van Tilborg, R. Spurio, R. Kaptein, M. Paci, et al. 1997. The structure of the translational initiation factor IF1 from *E. coli* contains an oligomer-binding motif. *EMBO J.* **16**:1436–1643.

107. Shean, C. S., and M. E. Gottesman. 1992. Translation of the prophage lambda cI transcript. *Cell* **70**:513–522.

108. Si, K., and U. Maitra. 1999. The *Saccharomyces cerevisiae* homologue of mammalian translation initiation factor 6 does not function as a translation initiation factor. *Mol. Cell. Biol.* **19**:1416–1426.

109. Slupska, M. M., A. G. King, S. Fitz-Gibbon, J. Besemer, M. Borodovsky, and J. H. Miller. 2001. Leaderless transcripts of the crenarchaeal hyperthermophile *Pyrobaculum aerophilum*. *J. Mol. Biol.* **309**:347–360.

110. Song, H., P. Mugnier, A. K. Das, H. M. Webb, D. R. Evans, et al. 2000. The crystal structure of human eukaryotic release factor eRF1—mechanism of stop codon recognition and peptidyl-tRNA hydrolysis. *Cell* **100**:311–321.

111. Tahara, M., A. Ohsawa, S. Saito, and M. Kimura. 2004. In vitro phosphorylation of initiation factor 2 alpha (aIF2 alpha) from hyperthermophilic archaeon *Pyrococcus horikoshii* OT3. *J. Biochem. (Tokyo)* **135**:479–485.

112. Teichner, A., P. Londei, and P. Cammarano. 1986. Intralineage diversity of archaebacterial ribosomes. A dichotomy of ribosome features separates methanococcaceae and sulfur-dependent thermophiles from other archaebacterial taxa. *J. Mol. Evol.* **23**:343–353.

113. Tollervey, D. 1996. Small nucleolar RNAs guide ribosomal RNA methylation. *Science* **273**:1056–1057.

114. Tolstrup, N., C. W. Sensen, R. A. Garrett, and I. G. Clausen. 2000. Two different and highly organized mechanisms of translation initiation in the archaeon *Sulfolobus solfataricus*. *Extremophiles* **4**:175–179.

115. Torarinsson, E., H. P. Klenk, and R. A. Garrett. 2005. Divergent transcriptional and translational signals in Archaea. *Environ. Microbiol.* **7**:47–54.

116. Udagawa, T., Y. Shimizu, and T. Ueda. 2004. Evidence for the translation initiation of leaderless mRNAs by the intact 70 S ribosome without its dissociation into subunits in eubacteria. *J. Biol. Chem.* **279**:8539–8546.

117. Wang, H., D. Boisvert, K. K. Kim, R. Kim, and S. H. Kim. 2000. Crystal structure of a fibrillarin homologue from *Methanococcus jannaschii*, a hyperthermophile, at 1.6 A resolution. *EMBO J.* **19**:317–323.

118. Woese, C. R. 2001. Translation: in retrospect and prospect. *RNA* **7**:1055–1167.

119. Woese, C. R., O. Kandler, and M. L. Wheelis. 1990. Towards a natural system of organisms: proposal for the domains Archaea, Bacteria, and Eucarya. *Proc. Natl. Acad. Sci. USA* **87**:4576–4579.

120. Woese, C. R., L. J. Magrum, and G. E. Fox. 1978. Archaebacteria. *J. Mol. Evol.* **11**:245–251.

121. Yang, W., and A. G. Hinnebusch. 1996. Identification of a regulatory subcomplex in the guanine nucleotide exchange factor eIF2B that mediates inhibition by phosphorylated eIF2. *Mol. Cell. Biol.* **16**:6603–6616.

122. Yao, M., A. Ohsawa, S. Kikukawa, I. Tanaka, and M. Kimura. 2003. Crystal structure of hyperthermophilic archaeal initiation factor 5A: a homologue of eukaryotic initiation factor 5A (eIF-5A). *J. Biochem. (Tokyo)* **133**:75–81.

123. Yatime, L., E. Schmitt, S. Blanquet, and Y. Mechulam. 2004. Functional molecular mapping of archaeal translation initiation factor 2. *J. Biol. Chem.* **279**:15984–15993.

124. Yatime, L., E. Schmitt, S. Blanquet, and Y. Mechulam. 2005. Structure-function relationships of the intact aIF2alpha subunit from the archaeon *Pyrococcus abyssi*. *Biochemistry* **44**:8749–8756.

125. Yoshizawa, S., L. Rasubala, T. Ose, D. Kohda, D. Fourmy, and K. Maenaka. 2005. Structural basis for mRNA recognition by elongation factor SelB. *Nat. Struct. Mol. Biol.* **12**:198–203.

126. Zago, M. A., P. P. Dennis, and A. D. Omer. 2005. The expanding world of small RNAs in the hyperthermophilic archaeon *Sulfolobus solfataricus. Mol. Microbiol.* **55**:1812–1828.

127. Zavialov, A. V., L. Mora, R. H. Buckingham, and M. Ehrenberg. 2002. Release of peptide promoted by the GGQ motif of class 1 release factors regulates the GTPase activity of RF3. *Mol. Cell* **10**:789–798.

128. Zhang, Y., P. V. Baranov, J. F. Atkins, and V. N. Gladyshev. 2005. Pyrrolysine and selenocysteine use dissimilar decoding strategies. *J. Biol. Chem.* **280**:20740–20751.

129. Ziesche, S. M., A. D. Omer, and P. P. Dennis. 2004. RNA-guided nucleotide modification of ribosomal and non-ribosomal RNAs in Archaea. *Mol. Microbiol.* **54**:980–993.

Archaea: Molecular and Cellular Biology
Edited by Ricardo Cavicchioli
© 2007 ASM Press, Washington, D.C.

Chapter 9

Features of Aminoacyl-tRNA Synthesis Unique to *Archaea*

CARLA POLYCARPO, KELLY SHEPPARD, LENNART RANDAU, ALEXANDRE AMBROGELLY,
ALEXANDER MACHADO CARDOSO, SHUYA FUKAI, STEPHANIE HERRING, MICHAEL HOHN,
YUKO NAKAMURA, HIROYUKI OSHIKANE, SOTIRIA PALIOURA, JUAN CARLOS SALAZAR,
JING YUAN, OSAMU NUREKI, AND DIETER SÖLL

INTRODUCTION

Aminoacyl-tRNAs (aa-tRNAs) are the essential substrates for ribosomal protein synthesis and are central in ensuring accurate translation of the genetic message. Two processes are involved in the formation of aa-tRNA: (i) the transcription of tRNA genes and processing of the transcript to mature tRNA and (ii) the aminoacylation of the individual tRNA species with their proper (cognate) amino acid. These processes have been well studied in bacteria and eucarya; these studies led to the acceptance of a unified view of the mechanism of aa-tRNA synthesis. Thus, it was unexpected when studies of aa-tRNA formation in archaea revealed exceptions to this scheme. In this chapter, four unique aspects of archaeal aa-tRNA formation that led to a much deeper understanding of this process not only in the *Archaea*, but in all domains of life, are discussed. The topics are: (i) processing of half-tRNA genes to mature tRNA in *Nanoarchaeum equitans*, (ii) RNA-dependent cysteine synthesis in methanogens, (iii) pyrrolysyl-tRNA formation in the *Methanosarcinaceae*, and (iv) glutaminyl-tRNA synthesis in archaea.

tRNA BIOSYNTHESIS

tRNA molecules are formed by transcription of tRNA genes into longer precursor tRNA molecules that are converted to mature tRNAs by nuclease processing, enzymatic addition of the 3′-terminal CCA sequence, and formation of the many modified nucleosides present in these ancient RNA molecules (30).

A closer look at the properties of tRNAs of archaea displays a variety of features that are either shared only with *Eucarya* or only with *Bacteria*, placing them "tRNA-wise" in the middle of these two domains (34). The genomic distribution and promoter organization of tRNA genes serves as an evident example of this characteristic. Similar to the situation in *Eucarya*, archaeal tRNA genes are not clustered in operons but possess individual promoters. However, the promoters themselves display features that they share with *Bacteria* (21). *Archaea* possess a single complex RNA polymerase transcribing all genes, which, unlike the eucaryal transcription machinery, does not rely on conserved sequence elements within the tRNA gene. All bacterial and archaeal genomes instead display external promoters upstream of the tRNA genes. Comparisons of these upstream sequences revealed a conserved TATA-like box A element (see Chapter 6), located approximately 25 base pairs upstream of the transcription start site (41). A second element, termed box B, contains the transcription start site; the RNA usually starts with a purine residue. To generate functional tRNA molecules, the 5′ end and the 3′ end of the primary tRNA transcript (pre-tRNA) need to be processed.

At the 5′ end, this task is performed by RNase P in a single endonucleolytic cleavage of the pre-tRNA (18). RNase P is an essential ribonucleoprotein that is ubiquitously present in all tRNA-synthesizing cells and cellular compartments. The only known exceptions to this are three genomes in which no definable RNase P RNA sequence could be identified. Apart from the bacterium *Aquifex aeolicus*, these organisms include *Nanoarchaeum equitans* and *Pyrobaculum*

Carla Polycarpo, Kelly Sheppard, Lennart Randau, Alexandre Ambrogelly, Alexander Machado Cardoso, Stephanie Herring, Michael Hohn, Sotiria Palioura, Juan Carlos Salazar, and Jing Yuan • Departments of Molecular Biophysics and Biochemistry, Yale University, New Haven, CT 06520-8114. Dieter Söll • Departments of Molecular Biophysics and Biochemistry and of Chemistry, Yale University, New Haven, CT 06520-8114. Shuya Fukai, Yuko Nakamura, Hiroyuki Oshikane, and Osamu Nureki • Department of Biological Information, Graduate School of Bioscience and Biotechnology, Tokyo Institute of Technology, Yokohama-shi, Kanagawa 226-8501, Japan.

aerophilum (33). RNase P is thought to recognize the tRNA structure of the precursor tRNA by a set of interactions between the catalytic RNA subunit and the tRNA's T stem and acceptor stem, together with important residues in the 5′-leader sequence and the CCA 3′ terminus. The 5′ end of the histidine tRNA exhibits a further domain-specific tRNA feature; it contains one additional guanosine compared with all other tRNA species, which was determined to be crucial for aminoacylation by histidyl-tRNA synthetase (10). In *Eucarya* this important additional residue is added posttranscriptionally by the enzyme tRNA^His guanylyltransferase (11, 20), while it is already encoded in the bacterial and archaeal tRNA^His genes whose transcripts are processed at position −1 by RNase P (8).

Much progress in understanding 3′-end processing of pre-tRNA has been made recently (30). The RNase Z from *Haloferax volcanii* was shown to cleave tRNA precursors 3′ to the discriminator base, which resembles eucaryal 3′-tRNA processing (50). The cleavage efficiency of RNase Z was shown to be influenced by the length of the tRNA's acceptor stem, whereas the 3′-terminal CCA sequence was not required for activity (51). This CCA sequence, which is required for amino acid attachment to tRNA, is added to the 3′ end processed pre-tRNA by the CCA-adding enzyme, as most archaeal (and all eucaryal) tRNA genes do not encode the 3′-terminal CCA sequence.

Special attention should be drawn to introns that interrupt the continuity of many tRNA genes (Fig. 1). These sequences are removed during tRNA splicing to generate full-length tRNAs (1). While bacteria contain self-splicing tRNAs, eucarya and archaea use a classical enzymatic process. A splicing endonuclease excises the intron from the pre-tRNA to yield two tRNA half-molecules, a 5′-half-exon with a 2′,3′-cyclic phosphate and a 3′-half-exon with 5′-hydroxyl terminus (32). These tRNA half-molecules are then joined by an RNA ligase. The homotetrameric archaeal splicing endonuclease from *Methanocaldococcus jannaschii* is well characterized; its crystal structure is known, and the enzyme recognizes a consensus bulge-helix-bulge structure (3 nt bulge, 4 bp, 3 nt bulge) found at the intron-exon junctions of intron-containing tRNAs, rRNA, and even mRNA (58, 62, 66). The location of eucaryal tRNA introns is conserved; they are found between nucleotides 37 and 38 of the tRNA, one nucleotide 3′ of the anticodon. However, many archaeal tRNA genes contain introns with unique features; they occur at different positions, such as the tRNA acceptor stem, D loop, D stem, T loop, variable loop, and anticodon stem. Furthermore, the BHB (bulge-helix-bulge) motifs postulated to form at the intron-exon junctions of archaeal tRNAs show divergence from the canonical structure (35). Introns with most deviation from the canonical BHB structure and position are found in the *Crenarchaeota*

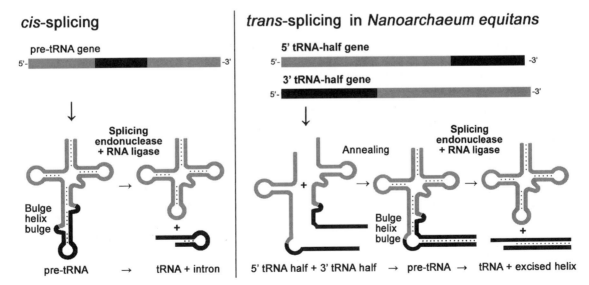

Figure 1. Schematic representation of *cis-* and *trans*-splicing. Conventional *cis*-splicing involves a splicing endonuclease that recognizes and cleaves a bulge-helix-bulge RNA motif in the pre-tRNA leading to the excision of the intron (black). An RNA ligase generates the mature tRNA (gray). The unique *trans*-splicing of tRNA observed in *N. equitans* requires the annealing of intervening reverse complementary sequences (black) found in the primary transcripts of a 5′-tRNA half-gene and a 3′-tRNA half-gene. Noncanonical splicing motifs are accommodated at the junctions and recognized by the *N. equitans* splicing endonuclease.

and in *N. equitans*. Their occurrence correlates with the presence of a novel heteromeric ($\alpha_2\beta_2$) splicing endonuclease that was shown to recognize more divergent splicing motifs (9, 62). A tRNA ligase that joins the two tRNA exons has been partially characterized in *H. volcanii* (71).

SPLIT tRNA GENES IN *N. EQUITANS*

Bioinformatic analysis of the *N. equitans* genome sequence uncovered a phenomenon unique to this archaeon (45, 46). In addition to 34 normal tRNA genes, four (tRNA^{Ile}, tRNA^{Met}, tRNA^{Trp}, and tRNA^{Tyr}) were discovered to have an intron. They were expected to be processed by a normal *cis*-splicing process (Fig. 1). However, the genes for tRNA^{Gln}, tRNA^{Glu}, tRNA^{His}, tRNA$_i^{Met}$, and tRNA^{Lys} were encoded as half-genes scattered throughout the genome. Biochemical analysis showed that *N. equitans* utilizes a unique *trans*-splicing mechanism to obtain full-length tRNA species from these half-genes; this requires the assembly of individually expressed RNA precursors encoding the 5′- and 3′-tRNA halves (44–46). The tRNA halves form a 12- to 14-nucleotide GC-rich RNA duplex between the end of the 5′-tRNA half and the beginning of the 3′-tRNA half. This structure probably provides the primary nucleation site that facilitates folding of the tRNA body. Noncanonical BHB motifs form at the junction between this RNA duplex and the tRNA; these RNA molecules could be cleaved in vitro by the heteromeric splicing endonuclease of *N. equitans* (44) or *Sulfolobus solfataricus* (63) but not by the canonical archaeal enzyme from *Archaeoglobus fulgidus* (44). Why is *N. equitans* the only known organism with split tRNA genes? Since site-specific integration of archaeal viruses or conjugative plasmids occurs exclusively at tRNA genes (53), one may speculate that there might be an adaptive value for such a small genome in providing resistance to the integration of mobile DNA elements by tRNA half-genes.

AMINOACYL-tRNA FORMATION

Once mature tRNA has been generated each tRNA species needs to be acylated (charged) with the correct amino acid. This occurs in the process of aminoacylation (reviewed in detail in reference 26) which "matches" the amino acid carried by the tRNA with the corresponding anticodon (see Chapter 8). This is primarily achieved by the direct attachment of an amino acid to the corresponding tRNA by an aminoacyl-tRNA synthetase. However, since many

organisms lack the complete set of 20 aminoacyl-tRNA synthetases (aaRSs), many biochemical, genetic, and genomic studies revealed the existence of an essential indirect two-step pathway that also provides correctly charged aa-tRNA.

Direct Aminoacylation of tRNA

The aminoacyl-tRNA synthetases are an ancient family of enzymes that esterify an amino acid to the 3′ end of the cognate tRNA species. By using ATP to activate the amino acid, they display remarkable specificity solely for the cognate substrates. A further quality control step of many synthetases is an editing activity that hydrolyzes incorrectly activated amino acids or mischarged aa-tRNAs (24).

tRNA-Dependent Amino Acid Conversion of Misacylated Aminoacyl-tRNA

Several essential aa-tRNA species are created by this indirect route (Fig. 2). This pathway is mainly used for the synthesis of Gln-tRNA and Asn-tRNA (in *Bacteria* and *Archaea*) (64), of Sec-tRNA (in all three domains) (5), and for Cys-tRNA (in methanogens) (49; see below). The first step in this indirect pathway relies on a nondiscriminating aaRS with relaxed tRNA specificity to generate a misacylated tRNA. For instance, the Gln-tRNA biosynthesis by this route takes advantage of a nondiscriminating GluRS to synthesize Glu-tRNA^{Gln} (Fig. 2B). Then a tRNA-dependent heterotrimeric amidotransferase (GatCAB) converts the misacylated aa-tRNA by amidation to the cognate Gln-tRNA^{Gln} (12). A similar pathway exists to form Asn-tRNA^{Asn} using a nondiscriminating AspRS and the same tRNA-dependent amidotransferase (GatCAB) (13, 38). However, in *Archaea*, Gln-tRNA^{Gln} is made by the archaea-specific heterodimeric GatDE enzyme. The current understanding of the *Methanothermobacter thermautotrophicus* Glu-tRNA^{Gln} amidotransferase GatDE is discussed below (see "The mechanism of Gln-tRNA formation").

The biosynthesis of Sec-tRNA^{Sec} is known in bacteria (5). It involves the formation of Ser-tRNA^{Sec} by seryl-tRNA synthetase and then the Ser → Sec conversion by selenocysteine synthase. Currently, it is unclear how Sec is formed in eucarya or archaea, but it is believed not to follow the bacterial route.

Cys-tRNA^{Cys} FORMATION IN ORGANISMS LACKING A CANONICAL CysRS

Cysteine is part of the twenty canonical amino acid repertoire ubiquitously used for protein synthesis.

A

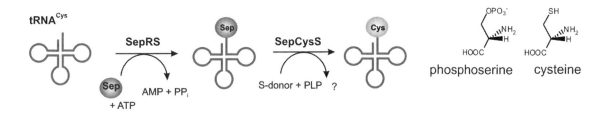

B

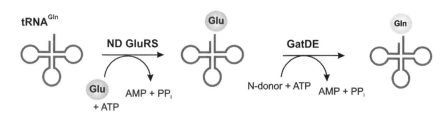

Figure 2. Schematic representations of indirect pathways for aminoacyl-tRNA formation. (A) Route of Cys-tRNA formation in *M. jannaschii*, and structures of phosphoserine and cysteine. (B) The indirect route for Gln-tRNA^{Gln} formation.

Cysteinyl-tRNA synthetase is a class I aaRS that produces Cys-tRNA^{Cys}, an essential substrate for insertion of cysteine into proteins. While cysteine is present in *M. jannaschii*, *M. thermautotrophicus*, and *Methanopyrus kandleri* proteins to a similar extent as found in other organisms, the genomes of these organisms do not encode a canonical CysRS. Despite intensive efforts by different groups, the question as to how Cys-tRNA^{Cys} might be synthesized in these organisms remained unanswered for more than a decade (2). A novel approach using the combination of anaerobic biochemical purification and proteomic analysis of the chromatographic fractions finally allowed the resolution of this intriguing puzzle (49). Starting from a cell-free *M. jannaschii* extract and employing a rigorous identification of aa-tRNA by acid gel electrophoresis and Northern blot, two proteins and two low-molecular-weight factors were discovered to be essential for Cys-tRNA^{Cys} formation (Fig. 2A). The first protein, an unusual aaRS homologous to the PheRS α-subunit, selectively acylates tRNA^{Cys} with phosphoserine (Sep) (Fig. 2A). The second enzyme subsequently converts Sep-tRNA^{Cys} to Cys-tRNA^{Cys} in the presence of pyridoxal phosphate and a still unidentified sulfur donor. The two enzymes were named phosphoseryl-tRNA synthetase (SepRS, encoded by *sepS*) and Sep-tRNA:Cys-tRNA synthase (SepCysS), respectively. Searches in genomic databases revealed the presence of homologs of these proteins not only in *M. jannaschii*, *M. kandleri*, and *M. thermautotrophicus* but also in genomes of other archaea such as *M. maripaludis*, *Methanococcoides burtonii*, *Methanospirillum hungatei*, *A. fulgidus*, and the *Methanosarcinaceae*, organisms that already possess the canonical CysRS (49).

An indication of the physiological and evolutionary significance of such duplication in these organisms was provided by genetic experiments in *M. maripaludis*. While the inactivation of the *cysS* gene encoding the canonical CysRS had no effect on cell growth in different media (54), a *M. maripaludis* strain with a *sepS* deletion displayed cysteine auxotrophy (49). The dispensability of the canonical CysRS together with the auxotrophic nature of the *M. maripaludis sepS* knockout strain demonstrates that (i) in the presence of exogenous cysteine the SepRS/SepCysS pathway and CysRS are functionally equivalent in Cys-tRNA synthesis and (ii) the indirect route to Cys-tRNA is the sole source of free cellular cysteine in this organism (49). In *M. maripaludis* the SepRS/SepCysS pathway provides the cell with both Cys-tRNA^{Cys} and free cysteine. Most of the other archaea have genes homologous to genes involved in cysteine biosynthesis in bacteria and eucarya. In these organisms, the physiological significance for the coexistence of the SepRS/SepCysS pathway with a tRNA-independent cysteine biosynthesis pathway and a CysRS is not yet understood. Possibly the

SepRS/SepCysS pathway is still involved in free cysteine biosynthesis under specific growth conditions. It is also possible that the coexistence in *M. maripaludis* of both a direct and an indirect route to Cys-tRNA may correspond to a transient evolutionary stage that will eventually result in the displacement of the indirect pathway.

Analyses of the phylogenetic distribution have shown that the direct and indirect pathways are of equally ancient origin (40). In *M. jannaschii, M. kandleri,* and *M. thermautotrophicus* the SepRS/SepCysS pathway may be strictly restricted to Cys-tRNA^{Cys} formation as free cysteine might not only be absent from the cytoplasm but also possibly toxic for the cell. Indeed, no homologs of tRNA-independent cysteine biosynthesis genes can be identified in the genome sequence of these organisms. Consequently, these organisms might alleviate the lack of free cysteine by using inorganic sulfur source for various cellular metabolic needs. While *M. jannaschii* tRNA molecules were shown to contain thio-nucleotides (e.g., 2-thiouridine [36]), the cysteine desulfurase enzymes that activate the sulfur atom from cysteine for tRNA modification appear to be lacking. However, the metal-cluster proteins that act as a sulfur carrier between the cysteine desulfurase and the tRNA modification enzymes are widespread in these organisms (67), strongly suggesting that sulfur atoms required for tRNA modification originate from an inorganic source. Fixation of sulfur from inorganic sulfide or sulfite was also shown to allow synthesis of other metabolites in *M. jannaschii* such as homocysteine and phosphosulfolactate, key intermediates in methionine and coenzyme M biosynthesis, respectively (19, 67). Finally, *M. jannaschii* ProRS was shown to efficiently mischarge in vitro cysteine onto tRNA^{Pro}, thus potentially compromising translational fidelity and subsequently cell viability (2, 57). While organisms from the bacterial and eucaryal domains have dealt with the danger of a promiscuous ProRS by acquiring an editing mechanism able to clear mischarged Cys-tRNA^{Pro} (3, 48), no such mechanism can be identified in archaea and in *M. jannaschii* in particular. The absence of selective pressure to acquire a Cys-tRNA^{Pro} editing mechanism could be explained by the absence of formation of the mischarged tRNA species due to a low ratio of cysteine:proline concentrations in the cell. Maintaining a low cellular cysteine concentration is then crucial.

Phylogenetic analyses suggest that class I CysRS emerged in bacteria and spread in archaea through multiple horizontal gene transfer events (69). The SepRS/SepCysS pathway appears to be the ancestral archaeal/eucaryal specific tRNA cysteinylation and cysteine biosynthesis system. By separating the amino-acylation and amino acid biosynthetic functions, the acquisition of both a tRNA-independent cysteine biosynthetic route and a CysRS may have provided a significant evolutionary advantage for the organisms, thus explaining the progressive displacement of the ancestral SepRS/SepCysS pathway. A still ongoing evolutionary process for the establishment of modern day tRNA cysteinylation metabolism is in good accord with the idea that cysteine may have been a late addition to the genetic code (7). Discovery of the tRNA-dependent cysteine biosynthetic route in *M. jannaschii* may have implications that reach far beyond the only problem of the formation of Cys-tRNA^{Cys} in three methanogenic archaea. Indeed, the fact that cysteine biosynthesis in *M. jannaschii* proceeds via Sep-tRNA^{Cys} and a subsequent Sep → Cys conversion catalyzed by SepCysS suggests that the same (or a similar) enzyme may also carry out the Sep → Sec transformation, the missing step in Sec formation in archaea and eucarya (49).

PYRROLYSINE: AN UNEXPECTED AMINO ACID DISCOVERED IN *METHANOSARCINA*

The *Methanosarcinaceae* are an exception among the methanogens, as they are able to use compounds like methanol, methylated thiols, and methylamines as energy sources (6). Methylation of coenzyme M (CoM) initiates methanogenesis; the different pathways of trimethylamine-, dimethylamine-, or mono-methylamine-specific CoM methyl transfer require different methyltransferases responsible for the demethylation of these energy sources and subsequent methylation of a different corrinoid protein that is subsequently demethylated by a methylcobamide:CoM methyltransferase.

When the *Methanosarcina barkeri* methylamine methyltransferase genes were investigated, each of them was shown to contain an in-frame amber (UAG) codon that does not act as a translation stop during synthesis of the methyltransferase. Initial attempts to characterize by mass spectrometry the UAG-encoded residue in the *M. barkeri* monomethylamine methyltransferase MtmB protein, identified lysine (29). However, the crystal structure of this enzyme revealed the UAG-encoded residue to be a new amino acid (22). The structure (Fig. 2) showed that lysine was modified with a C4-substituted pyrroline derivative. This 22nd cotranslationally inserted amino acid was named pyrrolysine (22). The identity of the C4-substituent was initially not established, but a later crystal structure of MtmB combined with chemical synthesis (23) of the amino acid characterized the Pyl structure as having a methyl group. This was confirmed very recently, as mass spectra of native preparations of the three different methyltransferases agreed with the expected mass number of proteins containing one pyrrolysine residue (55).

Concurrent with the detection of pyrrolysine was another important finding: the discovery of a UAG suppressor tRNA in *M. barkeri* (56) which might be the tRNA that will be charged with Pyl and that decodes (on the ribosome) the UAG codon with Pyl. This tRNA[Pyl] (encoded by *pylT)* has an unusual tRNA structure (56) and has only two modified nucleosides (43). The *pylT* gene is in an operon arrangement with three other open reading frames, *pylS, pylB,* and *pylC.* PylS appeared to encode a class II aminoacyl-tRNA synthetase-like protein. The other three genes are thought to be involved in pyrrolysine biosynthesis (56). Based on the presence of *pylS* and *pylT* and the knowledge of many genomes, the machinery for Pyl insertion appears to be present in the *Methanosarcinaceae* family and in the bacterium D*esulfitobacterium hafniense* (56). In addition, a proteomic analysis showed the presence of the *pyl* genes in the Antarctic archaeon *Methanococcoides burtonii* (17).

Originally it was thought (43, 56) that the mechanism of Pyl-tRNA[Pyl] synthesis involved an initial formation of Lys-tRNA[Pyl] and subsequent Lys → Pyl conversion in a tRNA-dependent manner as is known for Sec-tRNA[Sec] (5) and for Cys-tRNA[Cys] synthesis in *M. jannaschii* (49). This idea was supported by the report that PylS was able to charge tRNA[Pyl] with Lys (56) and by the unexpected finding that the combination of *M. barkeri* class I LysRS (25) and *M. barkeri* class II LysRS was required to form Lys-tRNA[Pyl] (43). It is pertinent to mention that, in archaea, only the *Methanosarcinaceae* contain both classes of lysyl-tRNA synthetase. A ternary complex of two aaRSs and one tRNA has been modeled (59) and may have played a role in the evolution of the two classes of tRNA syn-

thetases (47). Although these aminoacylation results suggested tRNA-dependent amino acid modification as the path to Pyl-tRNA[Pyl], one could not dismiss the possibility that Pyl was present as a metabolite ready to be charged directly onto tRNA[Pyl] by a special aaRS. Analysis of the *M. barkeri* genome revealed only *pylS* as the remaining unannotated synthetase-related gene. Using chemically synthesized Pyl it was demonstrated that PylS forms Pyl-tRNA[Pyl] but not Lys-tRNA[Pyl] (Fig. 3) (42). Although these in vitro data were in disagreement with the earlier PylS study (56), they were fully supported by an independent investigation (4). Addition of Pyl to the growth medium of *Escherichia coli* expressing the *M. barkeri mtmB* gene gave further in vivo support to the notion that PylS specifically charges Pyl to the cognate tRNA[Pyl] (4). Therefore PylS was called pyrrolysyl-tRNA synthetase (PylRS).

How does Pyl get inserted into proteins? Is the particular UAG codon reassigned (recoded) and specifically identified to the ribosome during translation? Or is Pyl inserted at random UAG codons in an inefficient process by the tRNA[Pyl] amber suppressor? To consider this it is important to recall how UGA is recoded for Sec; this is achieved by an essential EF-Tu-like protein (SelB) and a particular hairpin structure present in the mRNA (SECIS element) of the selenoprotein (5).

The bioinformatic investigation of the *Methanosarcinaceae* genomes revealed no genus-specific open reading frames similar to EF-Tu except for significantly truncated SelB paralogs. Thus, Pyl-tRNA[Pyl] may be delivered to the ribosome by the normal elongation factor(s). Such a possibility is supported by the fact that Pyl-tRNA[Pyl] in *E. coli* behaved like a normal amber suppressor (4); consistent with this is the observation that an anticodon mutated Lys-tRNA[Pyl] is recognized by *Thermus thermophilus* EF-Tu (60, 70). However, analyses with RNA-folding programs pre-

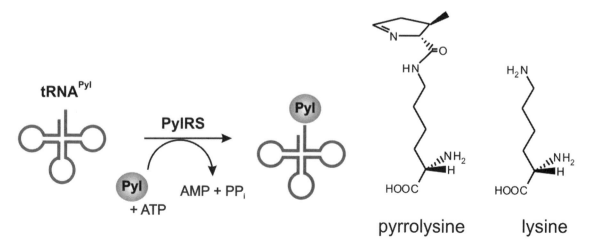

Figure 3. Schematic representation of the pyrrolysyl-tRNA[Pyl] formation by PylRS and structures of pyrrolysine and lysine.

dicted hairpin-like structures (named PYLIS) six nucleotides downstream of the "pyrrolysine" UAG codon in the *mtmB* genes of the *Methanosarcinaceae* (39) and the *D. hafniense mttB* gene (28). A synthetic "PYLIS element" based on the *M. barkeri mtmB1* sequence was shown by chemical and enzymatic probing experiments to fold in the predicted way (61). A global bioinformatic survey of the genes encoding selenoproteins and expected Pyl-containing proteins notes that the mechanism of UGA/Sec recoding may differ from that of the UAG/Pyl reassignment, and strongly suggests that Pyl-tRNA^{Pyl} may act as a normal suppressor tRNA (70).

THE MECHANISM OF Gln-tRNA FORMATION

The tRNA-dependent transamidation pathways are the primordial route of Gln-tRNA and Asn-tRNA formation (Fig. 2B) (27, 69). The resulting tRNA aa-tRNA amidotransferases are exciting enzymes as they are the result of a recruitment of amino-acid-metabolizing enzymes (amidases and asparaginases) to become components of enzymes crucial in protein synthesis. The best-characterized enzyme to date is the heterodimeric archaea-specific Glu-tRNA^{Gln} amidotransferase, GatDE (Fig. 4) (14, 54, 64). GatDE, like GatCAB, catalyzes three unique but coupled reactions. (i) The enzyme is a kinase that forms the activated intermediate γ-phosphoryl-Glu-tRNA^{Gln} (P-Glu-tRNA^{Gln}) by ATP hydrolysis (68). (ii) The enzyme is a glutaminase that generates ammonia by glutamine or asparagine hydrolysis (14). (iii) The enzyme is an amidotransferase that, using the ammonia liberated in step 2, amidates P-Glu-tRNA^{Gln} to Gln-tRNA^{Gln}.

Structural and sequence homology along with biochemical results implies that GatD belongs to the family of L-asparaginases and that it is the enzyme subunit responsible for liberating ammonia (14, 54, 64). L-Asparaginases generate ammonia by hydrolyzing Gln → Glu or Asn → Asp. Mutational studies of *M. thermautotrophicus* GatD have confirmed the importance of four residues conserved in asparaginase active sites (T101, T177, D178, and K254 in *M. thermautotrophicus* GatD) for the glutaminase activity of GatD (14). In addition, GatD dimerizes in a fashion similar to other type I asparaginases; the structure of the *Pyrococcus abyssi* GatD homodimer is superimposable with the L-asparaginase from the same organism (54).

GatE along with GatB belongs to an isolated protein family (54, 64). An insertion domain of about 190 aa present in GatE is the major difference between the two proteins (61). Structurally the domain resembles a domain found in the bacterial AspRS enzymes and may explain why GatDE is only able to use Glu-tRNA^{Gln} as a substrate (54). GatE has been implicated in the kinase activity of the holoenzyme, responsible for the formation of P-Glu-tRNA^{Gln} (14). This activated intermediate has not yet been experimentally trapped with pure GatDE, as had been accomplished using crude *Bacillus subtilis* extract (68). A recent structure of the *M. thermautotrophicus* GatDE complexed to tRNA^{Gln} shows GatE binds tRNA in a cradle domain lined by amino acids conserved in the GatE/GatB protein family. Those residues are involved in making up the kinase and transamidase active sites of GatE (Fig. 4) (40a, 54).

The mode of interaction of the two enzyme subunits is now known (Fig. 4). A tunnel, made up by a series of very conserved aa residues in the two subunits, connects the asparaginase active site in GatD to the amidotransferase active site on GatE and allows the delivery of ammonia for the final amidation step. It appears that in the *P. abyssi* apoenzyme structure (54) the tunnel is closed and the catalytically important Thr in GatD (14) is 7 Å away from the GatD active site (54). This suggests that, upon Glu-tRNA binding, conformational changes occur in GatDE which move the Thr into a position enabling catalysis and opening the tunnel to channel the resulting ammonia for amidation of Glu-tRNA bound to GatE (54). Such conformational changes would explain the biochemical data that GatDE is only able to function as a glutaminase in the presence of Glu-tRNA^{Gln} (14), implying a tight coupling between the three activities of Glu-tRNA^{Gln} amidotransferase.

Why do archaea have a unique enzyme for Gln-tRNA synthesis that was not replaced during evolution by lateral gene transfer with GlnRS (69)? In archaeal proteins, Asn and Gln are underrepresented relative to their levels in proteins from mesophilic bacteria and eucarya (37). Thus, the presence of the "more efficient" GlnRS would not provide the same selective advantage it might give to some bacteria and to eucarya. Differences in tRNA identity elements between archaeal and eucaryal or bacterial tRNA^{Gln} sequences may also explain why *glnS* did not transfer to archaea (64); *E. coli* and yeast GlnRS enzymes are unable to aminoacylate *M. thermautotrophicus* tRNA^{Gln} unless this tRNA species was given the *E. coli* tRNA^{Gln} identity elements (64). Differences in amino acid metabolism might be another reason why GlnRS is not found in archaea and might explain why GatDE developed uniquely, since GatD em-

M. thermautotrophicus GatDE:tRNA^{Gln} Complex

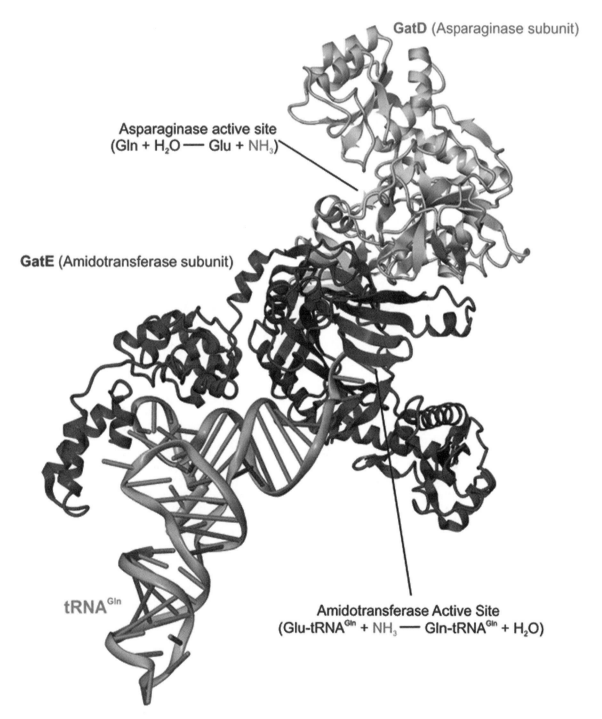

GatD (Asparaginase subunit)

Asparaginase active site
(Gln + H$_2$O —— Glu + NH$_3$)

GatE (Amidotransferase subunit)

tRNA^{Gln}

Amidotransferase Active Site
(Glu-tRNA^{Gln} + NH$_3$ —— Gln-tRNA^{Gln} + H$_2$O)

Figure 4. (*See the separate color insert for the color version of this illustration.*) Crystal structure of the *M. thermautotrophicus* GatDE complexed with tRNA^{Gln}. The dimer of the heterodimeric GatDE (thus forming a heterotetramer) binds two tRNA molecules. The asparaginase active site of GatD and the kinase/amidotransferase active site of GatE are distantly separated, connected with a "molecular tunnel."

ploys a different catalytic strategy in generating ammonia compared with GatA (64). Much remains to be done to elucidate the role of GatDE in archaeal protein synthesis and cell metabolism.

PERSPECTIVE: THE NEXT FIVE YEARS

Studies on aaRSs in archaea have led to an exciting expansion of our knowledge of how the Genetic Code is translated and expanded. The use of designed orthogonal aminoacyl-tRNA synthetase/tRNA pairs (15) led to exiting achievements of incorporation of unusual amino acids into proteins (summarized in reference 65). While such studies have relied on the in vitro design of novel tRNA and aaRS pairs, it is likely that the "naturally evolved" PylRS:tRNA^{Pyl} or SepRS:tRNA^{Cys} pairs may prove far more efficient. Further studies on PylRS or SepRS mediated recoding in archaea may be beneficial for devising more efficient systems for incorporation of nonnatural amino acids in other systems (31).

Acknowledgments. Work in the authors' laboratory was supported by grants from the National Institute of General Medical Sciences and the Department of Energy.

REFERENCES

1. **Abelson, J., C. R. Trotta, and H. Li.** 1998. tRNA splicing. *J. Biol. Chem.* 273:12685–12688.

2. **Ambrogelly, A., S. Kamtekar, A. Sauerwald, B. Ruan, D. Tumbula-Hansen, D. Kennedy, I. Ahel, and D. Söll.** 2004. CystRNA^{Cys} formation and cysteine biosynthesis in *Methanocaldococcus jannaschii*: two faces of the same problem? *Cell. Mol. Life Sci.* 61:2437–2445.

3. **An, S., and K. Musier-Forsyth.** 2004. Trans-editing of CystRNA^{Pro} by *Haemophilus influenzae* YbaK protein. *J. Biol. Chem.* 279:42359–42362.

4. **Blight, S. K., R. C. Larue, A. Mahapatra, D. G. Longstaff, E. Chang, G. Zhao, P. T. Kang, K. B. Greenchurch, M. K. Chan, and J. A. Krzycki.** 2004. Direct charging of tRNA$_{CUA}$ with pyrrolysine *in vitro* and *in vivo*. *Nature* 431:333–335.

5. **Böck, A., M. Thanbichler, M. Rother, and A. Resch.** 2005. Selenocysteine, p. 320–327. *In* M. Ibba, C. S. Francklyn, S. Cusack (ed.), *Aminoacyl-tRNA Synthetases*. Landes Bioscience, Georgetown, Tex.

6. **Boone, D. R., W. B. Whitman, and P. Rouvière.** 1993. Diversity and taxonomy of methanogens, p. 35–80. *In* J. G. Ferry (ed.), *Methanogenesis*. Chapman & Hall, New York, N.Y.

7. **Brooks, D. J., and J. R. Fresco.** 2002. Increased frequency of cysteine, tyrosine and phenylalanine residues since the last universal ancestor. *Mol. Cell. Proteomics* 1:125–131.

8. **Burkard, U., I. Willis, and D. Söll.** 1988. Processing of histidine transfer RNA precursors: abnormal cleavage site for RNase P. *J. Biol. Chem.* 263:2447–2451.

9. **Calvin, K., M. D. Hall, F. Xu, S. Xue, and H. Li.** 2005. Structural characterization of the catalytic subunit of a novel RNA splicing endonuclease. *J. Mol. Biol.* 353:952–960.

10. **Connolly, S. A., A. E. Rosen, K. Musier-Forsyth, and C. S. Francklyn.** 2004. G-1:C73 recognition by an arginine cluster in the active site of *Escherichia coli* histidyl-tRNA synthetase. *Biochemistry* 43:962–969.

11. **Cooley, L., B. Appel, and D. Söll, D.** 1982. Post-transcriptional nucleotide addition is responsible for the maturation of the 5′-terminus of histidine tRNA. *Proc. Natl. Acad. Sci. USA* 79:6475–6479.

12. **Curnow, A. W., K. Hong, R. Yuan, S Martins, W. Winkler, T. M. Henkin, and D. Söll.** 1997. Glu-tRNA^{Gln} amidotransferase: a novel heterotrimeric enzyme required for correct decoding of glutamine codons during translation. *Proc. Natl. Acad. Sci. USA* 94:11819–11826.

13. **Curnow, A. W., M. Ibba, and D. Söll.** 1996. tRNA-dependent asparagine formation. *Nature* 382:589–590.

14. **Feng, L., K. Sheppard, D. Tumbula-Hansen, and D. Söll.** 2005. Gln-tRNA^{Gln} formation from Glu-tRNA^{Gln} requires cooperation of an asparaginase and a Glu-tRNA^{Gln} kinase. *J. Biol. Chem.* 280:8150–8155.

15. **Furter, R.** 1998. Expansion of the genetic code: site-directed p-fluoro-phenylalanine incorporation in *Escherichia coli*. *Protein Sci.* 7:419–426.

16. **Galagan, J. E., C. Nusbaum, A. Roy, M. G. Endrizzi, P. Macdonald, W. FitzHugh, S. Calvo, R. Engels, S. Smirnov, D. Atnoor, A. Brown, N. Allen, J. Naylor, N. Stange-Thomann, K. DeArellano, R. Johnson, L. Linton, P. McEwan, K. McKernan, J. Talamas, A. Tirrell, W. Ye, A. Zimmer, R. D. Barber I. Cann, D. E. Graham, D. A. Grahame, A. M. Guss, R. Hedderich, C. Ingram-Smith, H. C. Kuettner, J. A. Krzycki, J. A. Leigh, W. Li, J. Liu, B. Mukhopadhyay, J. N. Reeve, K. Smith, T. A. Springer, L. A. Umayam, O. White, R. H. White, E. Conway de Macario, J. G. Ferry, K. F. Jarrell, H. Jing, A. J. Macario, I. Paulsen, M. Pritchett, K. R. Sowers, R. V. Swanson, S. H. Zinder, E. Lander, W. W Metcalf, and B. Birren.** 2002. The genome of *M. acetivorans* reveals extensive metabolic and physiological diversity. *Genome Res.* 12:532–542.

17. **Goodchild, A., N. F. Saunders, H. Ertan, M. Raftery, M. Guilhaus, P. M. Curmi, and R. Cavicchioli.** 2004. A proteomic determination of cold adaptation in the antarctic archaeon, *Methanococcoides burtonii*. *Mol. Microbiol.* 53:309–321.

18. **Gopalan, V., A. Vioque, and S. Altman.** 2002. RNase P: variations and uses. *J. Biol. Chem.* 277:6759–6762.

19. **Graham, D. E., H. Xu, and R. H. White.** 2002. Identification of coenzyme M biosynthetic phosphosulfolactate synthase. A new family of sulfonate-biosynthesizing enzymes. *J. Biol. Chem.* 277:13421–13429.

20. **Gu, W., J. E. Jackman, A. J. Lohan, M. W. Gray, and E. M. Phizicky.** 2003. tRNA^{His} maturation: an essential yeast protein catalyzes addition of a guanine nucleotide to the 5- end of tRNA^{His}. *Genes Dev.* 17:2889–2901.

21. **Hain, J., W. D. Reiter, U. Hudepohl, and W. Zillig.** 1992. Elements of an archaeal promoter defined by mutational analysis. *Nucleic Acids Res.* 20:5423–5428.

22. **Hao, B., W. Gong, T. K. Ferguson, C. M. James, J. A. Krzycki, and M. K. Chan.** 2002. A new UAG-encoded residue in the structure of a methanogen methyltransferase. *Science* 296:1462–1466.

23. **Hao, B., G. Zhao, P. T. Kang, J. A. Soares, T. K. Ferguson, J. Gallucci, J. A. Krzycki, and M. K. Chan.** 2004. Reactivity and chemical synthesis of L-pyrrolysine-the 22nd genetically encoded amino acid. *Chem. Biol.* 11:1317–1324.

24. **Hendrickson, T. L., and P. Schimmel.** 2003. Transfer RNA-dependent amino acid discrimination by aminoacyl-tRNA synthetases, p. 34–69. *In* J. Lapointe and L. Brakier-Gingras (eds.), *Translation Mechanisms*. Kluwer Academic Plenum Publishers, Dordrecht, The Netherlands.

25. **Ibba, M., S. Morgan, A. W. Curnow, D. R. Pridmore, U. C. Vothknecht, W. Gardner, W. Lin, C. R. Woese, and D. Söll.**

1997. A euryarchaeal lysyl-tRNA synthetase: resemblance to class I synthetases. *Science* 278:1119–1122.

26. Ibba, M., and D. Söll. 2000. Aminoacyl-tRNA synthesis. *Annu. Rev. Biochem.* 69:617–650.

27. Ibba, M., and D. Söll. 2001. The renaissance of aminoacyl-tRNA synthesis. *EMBO Rep.* 2:382–387.

28. Ibba, M., and D. Söll. 2004. Aminoacyl-tRNAs: setting the limits of the genetic code. *Genes Dev.* 18:731–738.

29. James, C. M., T. K. Ferguson, J. F. Leykam, and J. A. Krzycki. 2001. The amber codon in the gene encoding the monomethylamine methyltransferase isolated from *Methanosarcina barkeri* is translated as a sense codon. *J. Biol. Chem.* 276:34252–34258.

30. Klug, G., E. Evguenieva-Hackenberg, A. D. Omer, P. P. Dennis, and A. Marchfelder. 2007. RNA processing. p. 158–174. *In* R. Cavicchioli (ed.), *Archaea: Molecular and Cellular Biology.* ASM Press, Washington, D.C.

31. Köhrer, C., E. L. Sullivan, and U. L RajBhandary. 2004. Complete set of orthogonal 21st aminoacyl-tRNA synthetase-amber, ochre and opal suppressor tRNA pairs: concomitant suppression of three different termination codons in an mRNA in mammalian cells. *Nucleic Acids Res.* 32:6200–6211.

32. Li, H., C. R. Trotta, and J. Abelson. 1998. Crystal structure and evolution of a transfer RNA splicing enzyme. *Science* 280:279–284.

33. Li, Y., and S. Altman. 2004. In search of RNase P RNA from microbial genomes. *RNA* 10:1533–1540.

34. Marck, C., and H. Grosjean. 2002. tRNomics: analysis of tRNA genes from 50 genomes of Eukarya, Archaea, and Bacteria reveals anticodon-sparing strategies and domain-specific features. *RNA* 8:1189–1232.

35. Marck, C., and H. Grosjean. 2003. Identification of BHB splicing motifs in intron-containing tRNAs from 18 archaea: evolutionary implications. *RNA* 9:1516–1531.

36. McCloskey, J. A., D. E. Graham, S. Zhou, P. F. Crain, M. Ibba, J. Konisky, D. Söll, and G. J. Olsen. 2001. Post-transcriptional modification in archaeal tRNAs: identities and phylogenetic relations of nucleotides from mesophilic and hyperthermophilic *Methanococcales.* *Nucleic Acids Res.* 29:4699–4706.

37. Michelitsch, M. D., and J. S. Weissman. 2000. A census of glutamine/asparagine-rich regions: implications for their conserved function and the prediction of novel prions. *Proc. Natl. Acad. Sci. USA* 97:11910–11915.

38. Min, B., J. T. Pelaschier, D. E. Graham, D. Tumbula-Hansen, and D. Söll. 2002. Transfer RNA-dependent amino acid biosynthesis: an essential route to asparagine formation. *Proc. Natl. Acad. Sci. USA* 99:2678–2683.

39. Namy, O., J.-P. Rousset, S. Napthine, and I. Brierley. 2004. Reprogrammed genetic decoding in cellular gene expression. *Mol. Cell* 13:157–168.

40. O'Donoghue, P., A. Sethi, C. R. Woese, and Z. A. Luthey-Schulten. 2005. The evolutionary history of Cys-tRNA^{Cys} formation. *Proc. Natl. Acad. Sci. USA* 102:19003–19008.

40a. Oshikane, H., K. Sheppard, S. Fukai, Y. Nakamura, R. Ishitani, T. Numata, R. L. Sherrer, L. Feng, E. Schmitt, M. Panvert, S. Blanquet, Y. Mechulam, D. Söll, and O. Nureki. 2006. Structural basis of RNA-dependent recruitment of glutamine to the genetic code. *Science* 312:1950–1954.

41. Palmer, J. R., and C. J. Daniels. 1995. *In vivo* definition of an archaeal promoter. *J. Bacteriol.* 177:1844–1849.

42. Polycarpo, C., A. Ambrogelly, A. Bérubé, S. M. Winbush, J. A. McCloskey, P. F. Crain, J. L. Wood, and D. Söll. 2004. An aminoacyl-tRNA synthetase that specifically activates pyrrolysine. *Proc. Natl. Acad. Sci. USA* 101:12450–12454.

43. Polycarpo, C., A. Ambrogelly, and B. Ruan, D. Tumbula-Hansen, S. F. Ataide, R. Ishitani, S. Yokoyama, O. Nureki,

M. Ibba, and D. Söll. 2003. Activation of the pyrrolysine suppressor tRNA requires formation of a ternary complex with class I and class II lysyl-tRNA synthetases. *Mol. Cell* 12:287–294.

44. Randau, L., K. Calvin, M. Hall, J. Yuan, H. Li, and D. Söll. 2005. A heteromeric splicing endonuclease of *Nanoarchaeum equitans* cleaves non-canonical bulge-helix-bulge motifs of joined tRNA halves. *Proc. Natl. Acad. Sci. USA* 102:17934–17939.

45. Randau, L., M. Pearson, and D. Söll. 2005. The complete set of RNA species in *Nanoarchaeum equitans.* *FEBS Lett.* 579:2945–2947.

46. Randau, L., R. Münch, M. J. Hohn, D. Jahn, and D. Söll. 2005. *Nanoarchaeum equitans* creates functional tRNAs from separate genes for their 5′-and 3′-halves. *Nature* 433:537–541.

47. Ribas de Pouplana, L., and P. Schimmel. 2001. Two classes of tRNA synthetases suggested by sterically compatible dockings on tRNA acceptor stem. *Cell* 104:191–193.

48. Ruan, B., and D. Söll. 2005. The bacterial YbaK protein is a Cys-tRNA^{Pro} and Cys-tRNA^{Cys} deacylase. *J. Biol. Chem.* 280:25887–25891.

49. Sauerwald, A., W. Zhu, T. A. Major, H. Roy, S. Palioura, D. Jahn, W. B. Whitman, J. R. Yates III, M. Ibba, and D. Söll. 2005. RNA-dependent cysteine biosynthesis in archaea. *Science* 307:1969–1972.

50. Schierling, K, S. Rösch, R. Rupprecht, S. Schiffer, and A. Marchfelder. 2002. tRNA 3-end maturation in archaea has eukaryotic features: the RNase Z from *Haloferax volcanii.* *J. Mol. Biol.* 316:895–902.

51. Schiffer, S., S. Rösch, and A. Marchfelder. 2003. Recombinant RNase Z does not recognize CCA as part of the tRNA and its cleavage efficiency is influenced by acceptor stem length. *Biol. Chem.* 384:333–342.

52. Reference deleted.

53. She, Q., B. Shen, and L. Chen. 2004. Archaeal integrases and mechanisms of gene capture. *Biochem. Soc. Trans.* 32:222–226.

54. Schmitt, E., M. Panvert, S. Blanquet, and Y. Mechulam. 2005. Structural basis for tRNA-dependent amidotransferase function. *Structure* 13:1421–1433.

55. Soares, J. A., L. Zhang, R. L. Pitsch, N. M. Kleinholz, R. B. Jones, J. J. Wolff, J. Amster, K. B. Green-Church, and J. A. Krzycki. 2005. The residue mass of L-pyrrolysine in three distinct methylamine methyltransferases. *J. Biol. Chem.* 280:36962–36969.

56. Srinivasan, G., C. M. James, and J. A. Krzycki. 2002. Pyrrolysine encoded by UAG in archaea: charging of a UAG-decoding specialized tRNA. *Science* 296:1459–1462.

57. Stathopoulos, C., W. Kim, T. Li, I. Anderson, B. Deutsch, S. Palioura, W. Whitman, and D. Söll. 2001. Cysteinyl-tRNA synthetase is not essential for viability of the archaeon *Methanococcus maripaludis.* *Proc. Natl. Acad. Sci. USA* 98:14292–14297.

58. Tang, T. H., T. S. Rozhdestvensky, B. C. d-Orval, M. L. Bortolin. H. Huber, B. Charpentier, C. Branlant. J. P. Bachellerie, J. Brosius, and A. Hüttenhofer. 2002. RNomics in Archaea reveals a further link between splicing of archaeal introns and rRNA processing. *Nucleic Acids Res.* 30:921–930.

59. Terada, T., O. Nureki, R. Ishitani, A. Ambrogelly, M. Ibba, D. Söll, and S. Yokoyama. 2002. Functional convergence of two lysyl-tRNA synthetases with unrelated topologies. *Nat. Struct. Biol.* 9:257–262.

60. Theobald-Dietrich, A., M. Frugier, R. Giegé, and J. Rudinger-Thirion. 2004. Atypical archaeal tRNA pyrrolysine transcript behaves towards EF-Tu as a typical elongator tRNA. *Nucleic Acids Res.* 32:1091–1096.

61. Theobald-Dietrich, A., R. Giegé, and J. Rudinger-Thirion. 2005. Evidence for the existence in mRNAs of a hairpin element responsible for ribosome dependent pyrrolysine insertion into proteins. *Biochimie* 87:813–817.

62. Thompson, L. D., and C. J. Daniels. 1990. Recognition of exon-intron boundaries by the *Halobacterium volcanii* tRNA intron endonuclease. *J. Biol. Chem.* **265:**18104–18111.

63. Tocchini-Valentini, G. D., P. Fruscoloni, and G. P. Tocchini-Valentini. 2005. Structure, function, and evolution of the tRNA endonucleases of Archaea: an example of subfunctionalization. *Proc. Natl. Acad. Sci. USA* **102:**8933–8938.

64. Tumbula, D. L., H. D. Becker, W.-Z. Chang, and D. Söll. 2000. Domain-specific recruitment of amide amino acids for protein synthesis. *Nature* **407:**106–110.

65. Wang, L., and P. G. Schultz. 2004. Expanding the genetic code. *Angew. Chem. Int. Ed. Engl.* **44:**34–66.

66. Watanabe, Y., S. Yokobori, T. Inaba, A. Yamagishi. T. Oshima, Y. Kawarabayasi, H. Kikuchi, and K. Kita. 2002. Introns in protein-coding genes in Archaea. *FEBS Lett.* **510:**27–30.

67. White, R. H. 2003. The biosynthesis of cysteine and homocysteine in *Methanococcus jannaschii. Biochim. Biophys. Acta.* **1624:**46–53.

68. Wilcox, M. 1969. Gamma-glutamyl phosphate attached to glutamine-specific tRNA. A precursor of glutaminyl-tRNA in *Bacillus subtilis. Eur. J. Biochem.* **11:**405–12.

69. Woese, C. R., G. J. Olsen, M. Ibba, and D. Söll. 2000. Aminoacyl-tRNA synthetases, the genetic code, and the evolutionary process. *Microbiol. Mol. Biol. Rev.* **64:**202–236.

70. Zhang, Y., P. V. Baranov, J. F. Atkins, and V. N. Gladyshev. 2005. Pyrrolysine and selenocysteine use dissimilar decoding strategies. *J. Biol. Chem.* **280:**20740–20751.

71. Zofallova, L., Y. Guo, and R. Gupta, R. 2000. Junction phosphate is derived from the precursor in the tRNA spliced by the archaeon *Haloferax volcanii* cell extract. *RNA* **6:**1019–1030.

Archaea: Molecular and Cellular Biology
Edited by Ricardo Cavicchioli
© 2007 ASM Press, Washington, D.C.

Chapter 10

Protein-Folding Systems

FRANK T. ROBB, RYO IZUKA, AND MASAFUMI YOHDA

INTRODUCTION

All organisms face a protein-folding problem, which is the requirement to convert their proteins from a random coil conformation emerging from the ribosome into homogeneous, precisely folded states. For survival and efficient growth under normal conditions, cells must be able to maintain the majority of their proteins in native conformations and to recover proteins damaged by exposure to stressors. Protein folding in the *Archaea* is a fascinating topic because many archaeal species are able to grow, and seemingly fold and salvage their proteins, in extreme conditions that might be expected to preclude proper folding.

Archaeal hyperthermophiles grow at temperatures up to 113°C (6), and possibly as high as 121°C (6, 38), archaeal psychrophiles as low as −2°C (11). Thermoacidophiles grow in solfataric environments at pH 0 to 1, and extreme halophiles in evaporative salt deposits supersaturated with Na^+, K^+, or Ca^{2+} (10, 73) (see Chapter 2). Adaptations to environmental extremes include the synthesis of heat-stable and impermeable phytanyl ether-linked membrane lipids (20) (see Chapters 2 and 15) and exceptionally stable outer surface glycoproteins (14) (see Chapter 14) that protect the interior milieu of the cells from the harsh external environment. However, microbial cells cannot be thermally insulated, and all the components of cellular metabolism, including the protein-folding pathways, must be adapted to function under the prevailing temperature regimes.

The compositional and conformational adaptations of proteins that result in intrinsic stability under moderate or extreme conditions are fairly well understood (80). Protein adaptations include highly charged exterior surfaces, rigid structures maintained by multiple ion pair networks, tight hydrophobic core packing, and overall compact protein structures achieved by increased packing density to minimize internal voids (96). Adaptive amino acid substitutions can be detected through comparisons of protein sequences and compositional biases in large data sets. High-temperature adaptation is associated with a high content of the charged amino acids (lysine, arginine, glutamate, and aspartate) that will promote increased surface charge and ion pair formation. Other compositional changes include a higher residue volume and a decrease in charged nonpolar amino acids on the surface of proteins (26) (see Chapter 16). In extremely stable proteins, high intrinsic stability of the proteins affects folding processes as additional desolvation energy is required to localize the charged residues within the hydrophobic interior of the protein (15, 16). The maintenance of functional proteomes in hyperthermophiles thus raises questions relating to the energetics of folding at very high temperatures, a topic that is poorly understood in terms of energetics and the pathways of chaperone activity. This chapter describes the known members of archaeal protein-folding pathways, including not only the heat-shock-regulated members, but also the non-heat-shock-regulated protein chaperones.

GENOME SIZE AND COMPLEXITY OF THE REPERTOIRES OF PROTEIN CHAPERONES

The published genomes of archaeal species span a tenfold size range, from the 0.49-Mbp genome of *Nanoarchaeum equitans* to the 5.75-Mbp genome of *Methanosarcina acetivorans* (see Chapters 2 and 19).

Frank T. Robb • Center of Marine Biotechnology, University of Maryland Biotechnology Institute, 701 E. Pratt Street, Baltimore, MD 21202. **Ryo Izuka** • Laboratory of Bio-Analytical Chemistry, Graduate School of Pharmaceutical Sciences, The University of Tokyo, 7-3-1 Hongo, Bunkyo-ku, Tokyo 113-0033, Japan. **Masafumi Yohda** • Department of Biotechnology and Life Science, Tokyo University of Agriculture and Technology, 2-24-16 Naka-cho, Koganei, Tokyo 184-8588, Japan.

Coding density in these circular genomes is high, with *N. equitans*, a parasitic hyperthermophile, having the smallest cellular genome and providing the highest coding density recorded to date (99). The hyperthermophiles, with optimal growth temperatures (T_{opt}) >80°C, tend to have smaller genomes than mesophilic species such as *Halobacterium* strain NRC1 and the *Methanosarcina* species. As a result, protein-folding strategies are relatively simple or more complex depending on genome size and the number of homologs and paralogs that they encode.

Inventories of chaperones found in the genomes of *Archaea* include representatives of protein families characteristic of *Eucarya*, including the prefoldins, small heat shock proteins (sHsp), and ATP-dependent chaperonins (Table 1). Two major classes of eucaryal chaperones (Hsp100 and Hsp90/Hsp83) are absent in archaeal genome sequences. The chaperones that are shared with bacteria, including the "Chaperone Machine" (60), which is composed of Hsp70 (DnaK), Hsp40 (DnaJ), and GrpE, only occur in the larger complete genome sequences of mesophilic archaea (58, 61), as well as the psychrophilic methanogen, *Methanococcoides burtonii* (21; R. Cavicchiolli, personal communication). Hyperthermophiles represented by *Pyrococcus* spp, *Sulfolobus* spp, *Pyrobaculum aerophilum*, *Methanocaldococcus jannaschii*, *Methanopyrus kandleri*, *Archaeoglobus fulgidus*, *N. equitans*, and *Picrophilus torridus* (78) do not have Hsp90, DnaK, DnaJ, GrpE, Hsp33, and Hsp10 homologs (Table 1). The smaller archaeal genomes lack the highest molecular weight chaperones found in eucarya. The Hsp100/Clp protein family contains the largest ATP-dependent heat shock proteins so far characterized.

In *Escherichia coli*, degradation of denatured proteins is mediated by the cooperative functions of the ClpA and ClpP proteins. ClpA has protein remodeling functions in addition to "protein repair"

functions. In *E. coli*, ClpA alone can reactivate replication initiator protein, RepA, from an inactive RepA dimer to an active RepA monomer (90). The major chaperone classes, Hsp100 and Hsp90/Hsp83, are absent from the genomes of the hyperthermophilic archaea, although they are present in several mesophilic and thermophilic archaea (Table 1). For example, the thermophilic methanogen, *Methanothermobacter thermautotrophicus* contains a ClpA/B homolog, which could have been acquired by lateral gene transfer from bacteria.

In *Saccharomyces cerevisiae*, Hsp100 and Hsp-104 have important roles in acquired thermotolerance (57). A characteristic feature of Hsp100/Clp, Hsp104, and HslU/V (the high-molecular-weight chaperones) is the presence of conserved AAA$^+$ domains. These AAA$^+$ heat shock proteins share sequence similarity with the CDC48 and NSF proteins in the *S. cerevisiae* and human genomes, respectively. Clp/Hsp100 proteins can also function cooperatively with other chaperones, such as Hsp70, Hsp40, or GrpE. However, the mesophilic methanogen *Methanosarcina mazei* has a bacterial-type GroEL/GroES system that can substitute functionally for the *E. coli* components in vitro (53).

CHAPERONES AND THERMOTOLERANCE

Organisms exposed to a brief sublethal heat shock develop tolerance to otherwise lethal temperatures. This phenomenom is referred to as acquired thermotolerance. It has been well established in a diverse range of organisms (e.g., *Drosophila*, yeast, *E. coli*) that Hsp (heat shock protein) induction is responsible for acquired thermotolerance. In the *Archaea*, evidence for an adaptive thermotolerance response linked to chaperone expression was first discovered in the hyperthermophilic archaeal species, *Sulfolobus shibatae* (92, 93). Acquired thermotolerance was achieved following heat shock at 88°C for 60 min, which enabled the cells to survive a normally lethal exposure at 95°C for 40 min (92). Acquired thermotolerance was accompanied by the synthesis of high levels of the chaperone Hsp60.

Other environmental stressors, including treatment of yeast cells with 2.5% NaCl, 4% ethanol, or exposure to reduced pH (e.g., pH 5) induce thermotolerance (35). Cells exposed to sublethal levels of these types of stressors display increased synthesis of Hsps compared with nonstressed cells, demonstrating that Hsps can be involved in cellular responses to a variety of stressors in addition to heat shock.

The gene for a putative AAA$^+$ homolog (NP_579611) from the hyperthermophile, *Pyrococcus fu-*

Table 1. Occurrence of different classes of HSPs in the three domains

HSP	Occurrence of HSPs in:		
	Bacteria	*Eucarya*	*Archaea*
HSP100s	ClpA, ClpB, HslU	HSP100	Absent[a]
HSP90s	HtpG	Hsp90, Hsp83	Absent
HSP70s	DnaK	Hsp70, Hsc70	Hsp70[b]
HSP60s	GroEL/ES	TCP-1	TF55, Thermosome, Cpn60
sHSPs	IbpA, IbpB (*E. coli*)	sHSp, α-crystallin	sHSP

[a]ClpA/B homologs are found in *M. thermoautotrophicum*.
[b]Hsp70s are absent in most thermophiles and hyperthermophiles except *A. pernix*, in which the putative mitochondrial HSP70 has been identified from the complete genome sequence.

riosus, is up-regulated following heat shock at 105°C by induction of a repressor protein, Phr (*heat shock regulator protein*) (97) (see Chapter 6). A single *phr* gene is present in all three *Pyrococcus* genome sequences (*P. furiosus*, *P. abyssi*, and *P. horikoshii*), and encodes a 24-kDa basic protein. In *P. furiosus*, the promoters of the heat shock inducible *hsp20* (Pfu-*shsp*) and *aaa+* ATPase genes have highly conserved dyad operator sites (51). The expression of the *phr* gene was not induced by heat shock, suggesting that the Phr protein may be required at both normal growth and heat shock temperatures. Repression is relieved by an unknown mechanism during heat shock, and a *cis*-acting regulatory sequence has been described that may be important for heat shock regulation (97). Aligning the upstream regions of the AAA+-encoding genes from *P. furiosus* and *P. abyssi* enabled conserved regions to be identified which may be Phr-binding sites in both organisms (53). The promoter region of the AAA+ gene from *P. abyssi* also has phr recognition motifs similar to the promoter of the heat-inducible, small heat shock protein (*shsp*) gene from *P. furiosus*, indicating that these two species may be using a common heat shock regulatory mechanism (45).

Since the Clp/Hsp100 genes are absent in archaea (with the exception of the *Methanosarcina* spp.), it is possible that the AAA+ proteins from hyperthermophilic archaea may fulfill a similar functional role by combining with as-yet unidentified cochaperones. In this regard, the Lon protease from the thermophile, *Thermoplasma* spp., is an example of a naturally occurring fusion protein that contains a AAA+ domain (4). The heat-inducible AAA+ proteins (and unidentified functional partner proteins) may be functionally analogous to the Clp proteases, GroEL/GroES (bacterial chaperonin), or DnaJ/K (bacterial chaperones).

HtpX is a putative membrane-bound metalloprotease in bacteria and is ubiquitous in archaea, although it is annotated in many archaeal genomes as a conserved hypothetical protein. One copy of the *htpX* gene is present in the genomes of *P. furiosus* and *P. abyssi*, and two copies are present in each of the genomes of *P. horikoshii* and *S. solfataricus*. Similar to AAA+ protein genes, *htpX* is heat inducible in *M. jannaschii* (7), *Archaeoglobus fulgidus* (17, 81), and *P. furiosus* (85) (P. Laksanalamai, J. DiRuggiero, F. Robb, and T. Lowe, manuscript in preparation). It is possible that htpX may have a cellular function that is similar to, or complements, the heat shock-inducible AAA+ proteins.

ORIGINS OF NSF AND HslUV

The heat shock-inducible AAA+ proteins in the archaea are very distantly related to chaperone-associated AAA+ modules in eucarya, but are homologous to the yeast CDC48 proteins. CDC48 proteins are molecular chaperones that are crucial for correct cell division in eucarya, and regulate spindle disassembly following mitosis. The AAA+ proteins from *Sulfolobus* species have high sequence identity (40 to 55%) to the yeast CDC48 homolog and to the related human NSF proteins; NSF proteins modify the conformations of specific integral membrane proteins. It seems likely that this has resulted from nonorthologous gene displacement (53). The AAA+ domain present in a major class of heat shock-regulated chaperones in archaea may have originated from a membrane-localized bifunctional chaperone, which was an ancestor of the modern NSF proteins. NSF and HslU are structurally similar proteases, and it is plausible that the archaeal homologs of NSF may have features that allow them to carry out HslU-like proteolytic functions (55). Either the heat inducible AAA+ proteins, or the proteasome-associated AAA+ subunit in archaea, could resemble the progenitor of the NSF function in eucarya, and might have been one of the core functions present in the "proto-eukaryote" (4). These chaperone-associated AAA+ modules in eucarya could have arisen by the acquisition of another AAA+ module, introducing ATP hydrolysis into an ancestral folding system that lacked ATPases.

PREFOLDINS

Prefoldins are universally present in eucarya and archaea, with similar structures, but are absent in bacteria. The prefoldins are "holdase" chaperones whose crystal structure was first resolved from the archaeon *Methanothermobacter thermautotrophicus* (86, 91). The chaperone has been likened to a jellyfish in shape, with a globular "body" with six canonical, antiparallel coiled coils (the "tentacles") with their N and C domains oriented outwardly from an oligomerization domain (Fig. 1). The coiled-coil "tentacles" form a cavity lined with hydrophobic patches that secure nonnative target proteins (59). In archaea, with one exception, prefoldins are hexamers consisting of two α-subunits and four β-subunits, which act as generalized holding chaperones. The archaeal prefoldins bind to a wide range of non-native proteins in vitro, although their intracellular substrates are not known. Although similar in overall structure, the eucaryal prefoldins consist of six nonidentical subunits (two α-class and four β-class subunits) and, in contrast to archaeal prefoldins, bind specifically to the ribosome-nascent forms of actins and tubulins (56).

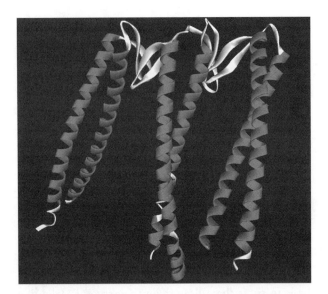

Figure 1. Structure of the prefoldin from *M. thermautotrophicus*. The globular body and coiled coils extensions in the "jellyfish" model form an adjustable cage that accommodates and binds proteins by a clamp mechanism (59). Reproduced from *Nature Reviews Microbiology* (53) with permission of the publisher.

Several recent lines of evidence indicate that prefoldins can act cooperatively with chaperonins, such as HSP60, and load nonnative proteins into their cavity. The prefoldin tentacles are capable of flexing outward to accommodate both small (14 kDa, lysozyme) and large (62 kDa, firefly luciferase) proteins in the cavity formed by the "tentacles" to prevent their aggregation (56, 59). The holding-and-release mechanism of the archaeal prefoldins has recently been elucidated (72, 107). In addition, the transfer of nonnative substrates to chaperonins has been well characterized using surface plasmon resonance and takes place between the prefoldins and chaperonins from one species of *Pyrococcus*, but not when chaperones from different *Pyrococcus* species are used (72). The hyperthermophilic methanogen, *M. jannaschii*, encodes genes for the α- and β-subunits of prefoldin. However, a unique third prefoldin subunit is encoded by the *pfdγ* gene and is heat shock regulated, unlike the α- and β-subunits (7, 28, 34, 68). This system raises new questions regarding the functional assignments of this heat shock-inducible prefoldin and the sHSPs, since they have overlapping chaperone activities in vitro.

SMALL HEAT SHOCK PROTEINS

Putative *shsp* genes are present in all archaeal genome sequences including *N. equitans*. Both the sHSPs and vertebrate α-crystallins are holdase-type molecular chaperones (28, 34, 68). The sHSPs and α-crystallins have a monomeric molecular weight range of 15 to 40 kDa and typically form polydisperse multimeric complexes in vivo. However, in the *Archaea* biochemical characterization is limited to thermophilic and hyperthermophilic organisms.

Only two crystal structures of sHSPs from unrelated organisms, *M. jannaschii* (42) and *Triticum aestivum* (wheat) (95), have been reported. The sHSPs share amino acid sequence similarity with the central core of vertebrate eye lens α-crystallin proteins, which are conserved in this family of proteins through all domains of life. The sHSP proteins have relatively low amino acid sequence similarity, and their quaternary structures are dissimilar, but the monomeric structures of these proteins are almost identical. Their specific functional mechanisms may be determined by their individual quaternary structures and their cognate target proteins and chaperone partners. The archaeal sHSPs can prevent denatured proteins from aggregating under strong denaturing conditions, and in some cases, are able to refold denatured proteins (43, 51, 82, 94). The sequences of the N- and C-terminal domains of archaeal sHSPs differ, and this variability is responsible for the great variety of multisubunit structures that form. The N-terminal domain of the *M. jannaschii* sHSP16.5 is disordered in the crystal structure, but low-resolution features have been resolved by cryoelectron microscopy. This domain is essential for proper holdase function in sHSP16.5 (41).

The copy number of sHSP-encoding genes is variable among archaeal species. The thermophilic and hyperthermophilic archaea contain one, two, or three *shsp* homologs. Hyperthermophilic species growing optimally near 100°C have one *shsp* gene, with the exception of *Pyrobaculum aerophilum*, which has two homologs (51). *Thermoplasma acidophilum* and all the *Sulfolobus* spp. represented by genome sequences each have three *shsp* homologs. However, one of the sHSPs in *T. acidophilum* appears to have domains that are similar to the two ATPase domains of ArsA from *E. coli* (83). *Sulfolobus solfataricus* and *S. tokodaii* have one 14- to 15-kDa and two 20- to 21-kDa sHSPs each. The mesophilic methanogens *M. acetivorans* and *M. mazei* GoE1 contain three and four *shsp* homologs, respectively. However, one of the two sHSPs from *M. acetivorans* (NP_619401) does not appear to belong to the α-crystalline-type HSPs. The genome sequence of *Halobacterium* NRC-1 has the highest paralogy, encoding five sHSPs that all clearly belong to the α-crystallin family. It seems likely that the multiple sHSPs encoded in a single species perform a range of potentially overlapping cellular functions; however, this has not been assessed experimentally.

The role of sHSPs in protein folding is still a topic of active investigation in both archaea and eucarya. They can maintain solubility of nonnative proteins under physiological conditions indefinitely, for example, in the eye lens, displaying a remarkable capacity for binding nonnative target proteins present in greater concentration than the chaperones. The binding capacity of eucaryal α,β-crystallins for nonnative proteins is greatly stimulated by serine phosphorylation of the sHSP, and the dynamic reordering of sHSP complexes is required for solubilization of nonnative proteins (84). Although archaeal systems for protein phosphorylation have been described (see Chapter 11), it is unknown whether the archaeal chaperones are phosphorylated.

Recently, reconstitution of a protein-refolding pathway in vitro was described (52). Denatured *Taq* polymerase was reactivated cooperatively at 100°C by a mixture of sHSP or prefoldin with HSP60 from *P. furiosus* in an ATP-dependent folding pathway. The cooperative protein salvage pathway depends on the presence of HSP60 and ATP for full activity (Fig. 2). The sHSPs (Fig. 2A) and prefoldins (Fig. 2B) appear to fulfill similar roles in this system, namely to transfer denatured proteins to the HSP60 chaperonin, although the differences in their structures suggest that they transfer nonnative proteins to chaperonins by different mechanisms. The rate of refolding of *Taq* polymerase was minimal when just the holdase chaperones were present and was greatly increased when HSP60 and ATP were present.

NAC PROTEINS

The nascent polypeptide associated complex (NAC) was first isolated from bovine brain cytosol and recognized as a molecular chaperone. Multiple subunits of NAC were first characterized in yeast, and formation of NAC complexes with ribosomes appears to be critical for folding and export of eucaryal proteins (76). The mechanism of eucaryal NAC complexes is not well understood; several hypotheses exist, and studies are ongoing. The hypothesis that NAC proteins prevent inappropriate interaction between newly synthesized polypeptide chains and other cellular factors appears to be well supported. NAC functions in archaea may be similar to bacterial trigger factor (TF), since TF homologs are absent from all archaeal genomes. NAC proteins have been shown to be involved in translational control and localization of Oskar mRNA. Unlike eucarya, archaea do not have multiple NAC subunits. The recent characterization of a NAC protein from *S. solfataricus* (89) revealed that it is a homodimer. The monomers

have two domains, formed by the N- and C-terminal regions of the protein. The N-terminal domain is homologous to NAC α-subunits in eucarya. All archaeal NAC proteins in archaea have a C-terminal ubiquitin associated (UBA) domain. At present, the hypothesis that the complex interacts with ubiquitin is speculative (89). Putative ubiquitin homologs occur in several archaeal genomes but are missing from several others, and consequently the UBA domain, which is strongly conserved in all archaeal genomes, may have functions unrelated to ubiquitin binding.

THE GROUP II CHAPERONINS IN ARCHAEA: MECHANISTIC INSIGHTS FROM MINIMALISTIC CHAPERONINS

The chaperonins are ubiquitous molecular chaperones that form double-ring assemblies of subunits with a molecular mass of 60 to 70 kDa. The resulting structures have a large central cavity where nonnative proteins can undergo productive folding in an ATP-dependent manner (9, 27). The paradigm for chaperonin-assisted protein folding has been the group I GroE system from *E. coli*, consisting of the GroEL chaperonin and associated GroES cochaperonin, which are generally found in bacteria and eucaryal organelles of bacterial origin (27, 87).

STRUCTURE AND SUBUNIT COMPOSITION OF ARCHAEAL GROUP II CHAPERONINS

The archaeal group II chaperonins form toroidal double rings with an eight- or ninefold symmetry, consisting of homologous subunits (25). The archaeal chaperonins are composed of up to five sequence-related subunits. *Sulfolobus* species (2, 37), *Haloferax volcanii* (54), *M. mazei* (45), and *M. burtonii* (21; R. Cavicchioli, personal communication) contain three chaperonin genes. Table 2 lists the number of subunits per genome and subunit composition of chaperonins from characterized members of the *Archaea*. Recently, it was found that there are five chaperonin subunits (Hsp60-1, -2, -3, -4, and -5) in *M. acetivorans*. Among them, Hsp60-1, Hsp60-2, and Hsp60-3 have orthologs in *Methanosarcinaceae*, but others, Hsp60-4 and Hsp60-5, occur only in *M. acetivorans*. The HSP60-4 and Hsp60-5 paralogs may represent the third class of chaperonin that may be ancestral to two widely distributed group I and II orthologs (62).

The subunit composition of the chaperonin complexes in several archaea changes with growth temperature (32, 37, 102). The chaperonin from the hyperthermophilic archaeon *Thermococcus* sp. strain KS-1

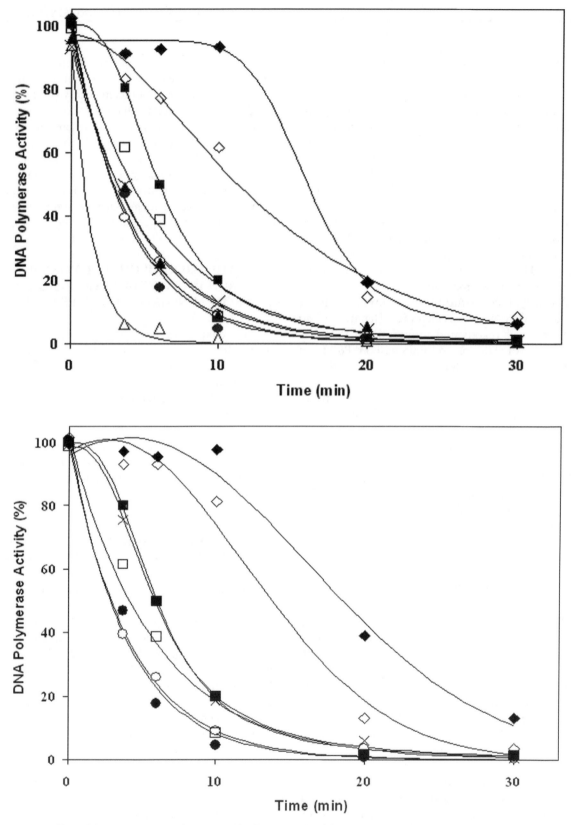

Figure 2. Effect of chaperones sHsp and Hsp60 on the thermostability of *Taq* DNA polymerase in the presence of *P. furiosus* molecular chaperones. (A) Inactivation of *Taq* polymerase in the presence of individual subunits of sHsp (△), Hsp60 (□), Hsp60-Mg^{2+}-ATP (■), sHsp and HSP60 (◇), and sHsp and Hsp60-Mg^{2+}-ATP (◆). The controls are reactions without the addition of chaperones (○) and with the addition of Mg^{2+} and ATP (●). (B) Inactivation of *Taq* polymerase in the presence of individual subunits of prefoldin, prefoldin α (△) and β (▲), prefoldin complex (✖), Hsp60 (□), Hsp60-Mg^{2+}-ATP (■), prefoldin and HSP60 (◇), and prefoldin and Hsp60-Mg^{2+}-ATP (◆). The controls are reactions without the addition of chaperones (○) and with the addition of Mg^{2+} and ATP (●). Reproduced from *Biotechnology and Bioengineering* (52) with permission of the publisher.

Table 2. Archaeal chaperonin: number of subunits encoded per genome

Organisms	Subunit species[a]	Rotational symmetry[a]	Reference
Crenarchaeota			
Aeropyrum pernix	2 (α, β)	NR	39
Pyrobaculum aerophilum	2 (α, β)	NR	67, 75
Pyrodictium occultum	2 (α, β)	8	36
Sulfolobus acidocaldarius	3 (α, β, γ)	NR	2
Sulfolobus shibatae	3 (α, β, γ)	9	2, 75
Sulfolobus solfataricus	3 (α, β, γ)	9	37, 46
Sulfolobus tokodaii	3 (α, β, γ)	NR	2
Euryarchaeota			
Archaeoglobus fulgidus	2 (α, β)	8	17, 75
Halobacterium sp. NRC-1	2 (α, β)	NR	70
Haloferax volcanii	3 (CC, T1, 2, 3)	NR	54
Methanocaldococcus jannaschii	1	NR	47
Methanococcus thermolithotrophicus	1	8	18
Methanococcus maripaludis	1	NR	50
Methanopyrus kandleri	1	8	1, 67
	I: 1	NR	62
	II: 5 (Hsp60-1, -2, -3, -4, -5)	NR	
Methanosarcina acetivorans	I: 1	NR	62
Methanosarcina barkeri	II: 3 (1, 2, 3)	NR	
Methanosarcina mazei	I: 1	NR	45
	II: 3 (α, β, γ)	8?	
Methanothermobacter thermautotrophicus	2 (α, β)	NR	88
Picrophilus torridus	2	NR	19
Pyrococcus abyssi	1	NR	www.genoscope.cns.fr/Pab/
Pyrococcus furiosus	1	NR	79
Pyrococcus horikoshii	1	NR	72
Thermoplasma acidophilum	2 (α, β)	8	71, 98
Thermoplasma volcanium	2 (α, β)	NR	40
Thermococcus kodakaraensis	2 (α, β)	NR	33, 101
Thermococcus sp. strain KS-1	2 (α, β)	8	104, 106

[a]*Methanosarcina acetivorans*, *M. barkeri*, and *M. mazei* contain both group I (refer to "I") and group II (refer to "II") chaperonins. NR, not reported (72).

is composed of two highly sequence-related subunits, α and β (106), that form a heterooligomer with variable subunit composition in vivo (102). Expression of α- and β-subunits is regulated differently, and only the α-subunit is thermally inducible (102). The proportion of the α-subunit in *Thermococcus* KS-1 chaperonin increases with temperature, and the β-subunit-rich chaperonin is more thermostable than the α-subunit-rich chaperonin (103). The hyperthermoacidophilic archaeon, *S. shibatae*, contains group II chaperonins composed of up to three different subunits (α, β, and γ). Expression of the α- and β-subunits is increased by heat shock and decreased by cold shock (37). On the other hand, expression of the γ-subunit gene is undetectable at heat shock temperatures and low at normal growth conditions, but induced by cold shock (22, 37). A cold-adaptation response in *M. burtonii* has also been studied by using ICAT proteomic profiling (22). The halophilic archaeon *H. volcanii* has three group II chaperonins genes, *cct1*, *cct2*, and *cct3*, which are all expressed but to differing levels (54). Deletion of *cct3* has no effect on the activity of the chaperonin complex, but

loss of *cct1* leads to ~50% reduction in the purified chaperonin ATPase activity (58). The precise functional properties and physiological significance of the heterologous subunit composition of archaeal group II chaperonin subunits is still the subject of active investigation.

The prototype crystal structure of the group II chaperonin is shown in Fig. 3. This structure from the thermoacidophilic archaeon, *T. acidophilum*, has shown that the subunit architectures are very similar to group I chaperonins, except for differences in the helical protrusion region (8, 13, 44). The helical protrusion in group II chaperonins may provide a functional role equivalent to the GroES subunit of group I chaperonins by sealing off the central cavity of the chaperonin complex (29, 64) (Fig. 3).

PROTEIN-FOLDING MECHANISM OF ARCHAEAL GROUP II CHAPERONINS

Although nucleotide and amino acid sequences of many archaeal chaperonins have been reported,

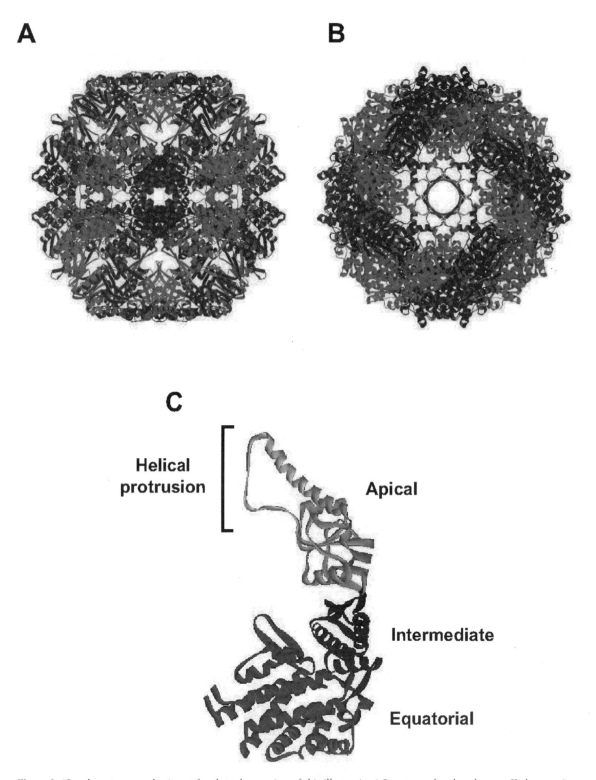

Figure 3. (*See the separate color insert for the color version of this illustration.*) Structure of archaeal group II chaperonin from *Thermoplasma acidophilum*. (*A* and *B*) The side view and top view of the crystal structure of *T. acidophilum* chaperonin, respectively. The α subunits are shown in dark green, and the β subunits are shown in dark blue. The hexadecameric structure was drawn using MOLSCRIPT (48). (C) The subunit structure of *T. acidophilum* chaperonin. Apical, intermediate, and equatorial domains are represented by green, blue, and red, respectively. The helical protrusion is highlighted by yellow. The figure was drawn with the Viewer Light 5.0 software (Accelrys).

there are comparatively few reports on their functional characterization; these include the native chaperonin from *S. solfataricus* (23, 24) and *Thermococcus* KS-1 (18, 102), and recombinant chaperonins from *Methanococcus thermolithotrophicus* (18), *Pyrococcus horikoshii* (72), and *Methanococcus maripaludis* (5, 49, 50), *T. acidophilum* (5), and *Thermococcus* KS-1 (106). The group II chaperonin from *Thermococcus* KS-1 has been studied in most detail. The *Thermococcus* KS-1, α- and β-subunits coassemble to form double-ring homooligomers (α-chaperonin and β-chaperonins, respectively), and are able to capture denatured proteins and fold them in an ATP-dependent manner in vitro (105, 106). Taking advantage of this, significant progress has been made in defining functional mechanisms (29–31, 105) (see "Role of the helical protrusion" and "Functional asymmetry," below). The known properties, arrest and ATPase activity, and structural characteristics of archaeal chaperonins appear in Table 3.

ROLE OF THE HELICAL PROTRUSION

The helical protrusion is strictly conserved among group II chaperonins (25). From crystallographic studies (13, 44), the region was thought to function as a substitute for the cochaperonin groES of the group I chaperonins and to be important for binding to unfolded proteins. To elucidate the exact role of the helical protrusion of a group II chaperonin in its molecular chaperone function, three deletion mutants of *Thermococcus* KS-1 α-chaperonin were constructed, lacking one third, two thirds, and the whole of the helical protrusion, respectively. Protease sensitivity assays and small-angle X-ray scattering (SAXS) experiments were performed to examine the conformational changes of the wild-type and mutant proteins. While the binding of ATP to the wild-type protein induced a structural transition corresponding to the closure of the built-in lid, it did not cause significant structural changes in the mutant proteins. Although

Table 3. Structural and functional characteristics of archaeal group II chaperonins

Organism	Subunit species	Native or recombinant	Rotational symmetry	ATPase activity	Arrest activity[a]	Folding activity	Reference
Crenarchaeota							
Sulfolobus shibatae	3	Native	9	Trace	+	NR	37, 93
Sulfolobus solfataricus	3	Native	9	Trace	+	+	2, 23, 24, 46, 63
Sulfolobus tokodaii	3	Native	NR	Trace	+	−	2, 69
		Recombinant (α, β)[b]	NR	Trace	+	−	
Pyrodictium brockii	NR[i]	Native	8	NR	NR	NR	75
Pyrodictium occultum	2	Native	8	+	NR	NR	66, 75
		Recombinant (α, β)	8	+	+	NR	
		Recombinant (α+β)[c]	8	+	+	NR	
Euryarchaeota							
Archaeoglobus fulgidus	2	Native	8	NR	NR	NR	17, 75
Haloferax volcanii	3	Native	NR	+	NR	NR	54
Methanococcus jannaschii	1	Native	NR	+	+	−[d]	47
Methanococcus	1	Recombinant[e]	8	+	+	+	18
Methanococcus maripaludis	1	Recombinant	NR	+	+	+	50
Methanopyrus kandleri	1	Native	8	+	NR	NR	1, 3
		Recombinant	8	+	+	NR	
(5)	(5)[3]	Native	8 (α: β: γ = 2:1:1)	NR	NR	NR	45
		Recombinant	8 (αγ)[f]	+	+ (αβγ)[g]	−(αβγ)	
Pyrococcus horikoshii	1	Recombinant	NR	+	+	+	72
		Native	8	Trace	+	NR	5, 71, 98
Thermoplasma acidophilum	2	Recombinant (α, β)	8	+	NR	NR	
		Recombinant (α+β)	8	+	Trace	Trace	
Thermococcus kodakaraensis	2	Recombinant (α, β)	NR	+	+(β)[b]	NR	33, 101
Themococcus sp. strain KS-1	2	Native	NR	+	+	+	104, 106
		Recombinant (α, β)	8	+	+	+	

[a] "Arrest activity" means the binding activity to nonnative proteins.
[b] α- and β-subunits are separately expressed in *E. coli* and purified.
[c] α- and β-subunits are coexpressed in *E. coli* and purified.
[d] The measurement is carried out at 30°C.
[e] Reconstituted complex of purified subunit.
[f] Reconstituted complex of α- and γ-subunits.
[g] Reconstituted complex of α-, β-, and γ-subunits.
[h] Purified β-subunits prevent thermal inactivation of yeast alcohol dehydrogenase.
[i] NR, not reported.

the mutants effectively protected proteins from thermal aggregation, the ATP-dependent protein-folding ability was remarkably diminished. The results indicate that the helical protrusion is not necessarily important for binding to unfolded proteins, but its ATP-dependent conformational change mediates folding of captured unfolded proteins (29, 31).

FUNCTIONAL ASYMMETRY

Although many studies have been carried out on the lid conformation, the steps of the ATPase cycle remained obscure until recently. The pathway was revealed by determining the effects of ADP-beryllium fluoride (BeF$_x$) complex formation on the *Thermococcus* KS-1 α chaperonin (30). Biochemical assays, electron microscopic observations, and SAXS measurements demonstrate that α-chaperonin incubated with ADP and BeF$_x$ exists in an asymmetric conformation; one ring is open, and the other is closed. The binding of ADP under conditions of inhibition of ATPase activity by BeF$_x$ resulted in freezing the alternation of conformational changes in the complex. The result indicates that α-chaperonin shares the inherent functional asymmetry of bacterial (9, 27) and eucaryal cytosolic chaperonins (64, 65).

Even though there is a difference between archaeal and eucaryal group II chaperonins, ATP binding is sufficient to close the lid of archaeal chaperonins (30), while the lid closure of CCT is triggered by the transition state of ATP hydrolysis (64).

Addition of ADP and BeF induced the *Thermococcus* KS-1 α-chaperonin to encapsulate unfolded proteins in the closed ring but did not trigger their folding. Moreover, the α-chaperonin incubated with ATP and BeF$_x$ adopted a symmetric closed conformation, and its functional turnover was inhibited. The existing evidence indicates that asymmetric and symmetric molecules are present in the functional ATPase cycle of archaeal group II chaperonins (30).

A schematic model for the functional mechanism of archaeal group II chaperonins is depicted in Fig. 4 (30). In the absence of nucleotides, the chaperonin is maintained in the open conformation and captures nonnative polypeptides. Although the exact location of the substrate-binding site has not been determined experimentally, it is likely that nonnative substrates interact via the exposed hydrophobic surface of the apical domain. This is supported by the finding that the helical protrusion region is not required for the recognition and binding of the substrate protein. Subsequently, the binding of ATP leads to a conformational change to the asymmetric structure, in a similar manner to ATP and GroES binding with the substrate-bound ring of bacterial GroEL. In contrast to the GroEL/ES model, protein folding is not induced by the closing of the lid, and further conformational change seems to be required. The asymmetric conformation changes to the symmetric closed conformation when ATP binds to the other ring. The substrate is released from the cavity wall into the hydrophilic central cavity, where productive folding occurs (Fig. 4). Consequently, the release of the γ-phosphate generated by ATP hydrolysis in the folding active ring triggers the opening of the lid and release of the substrate (Fig. 4). The two rings of group II chaperonins may alternate as a folding chamber, similar to GroEL, and

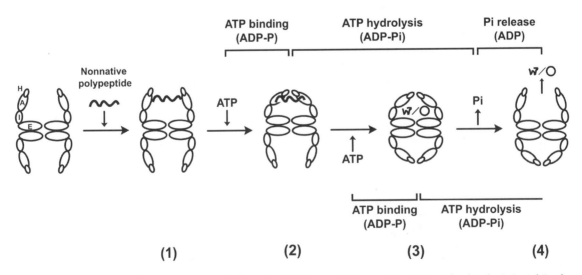

Figure 4. Schematic model for the reaction mechanism of archaeal group II chaperonins. See text for details. A, I, and E refer to the apical, intermediate, and equatorial domains, respectively. H represents the helical protrusion. Reproduced from *Journal of Biological Chemistry* (104) with permission of the publisher.

the archaeal group II chaperonins may have a molecular mechanism similar to GroEL and its partner GroES. The cyclic switching of chaperonin chambers may be compared with the compression/decompression cycles of a two-stroke gasoline engine.

OCCURRENCE OF GROUP I CHAPERONINS IN *METHANOSARCINA* SPECIES

The genome sequences of several *Methanosarcina* species have been completed, for example *M. mazei* (12) and *M. acetivorans* (21), and have revealed that these species contain both group I and II chaperonin genes (45, 61). In *M. mazei*, both chaperonin complexes are formed from chaperonin oligomers that are expressed from genes that are moderately induced by heat stress (45). How group I and II chaperonins share the protein-folding duties in the cell is a very interesting question that is open to speculation. In *Methanosarcina* species, it is proposed that 20 to 35% of genes were acquired horizontally from bacteria, including group I chaperonin genes. The coexistence of both groups of chaperonins in the same cytosol in the *Methanosarcina* species provides a useful model system for studying the differential substrate specificities of the group I and II chaperonins, and for elucidating how newly synthesized proteins are sorted from the ribosome to the appropriate chaperonin for folding (45, 62).

PERSPECTIVE: THE NEXT FIVE YEARS

Molecular chaperones are diverse and eclectic, and every cell carries a unique repertoire of different independent or cooperative protein-folding machines. In the *Archaea*, chaperones that are stress inducible have received the most attention thus far. Perhaps the most understandable cellular functions of chaperones occur during cell stress by salvaging nonnative proteins and recruiting them to join the pool of stable proteins, to prevent their demise as intracellular aggregates. In the eucarya and bacteria, chaperones are also known to participate in many fundamental cellular processes in nonstressed cells including DNA replication, regulation of gene expression, cell division, membrane translocation, protein folding, and protein remodeling (100). For example, the ClpA proteins in bacteria can mediate protein folding, unfolding, assembly, and disassembly without themselves being part of the final complex (74). Open questions remain regarding the archaeal protein-folding systems as to how posttranslational modeling functions are partitioned between the known

chaperones and chaperones or cochaperones that have not yet been discovered.

In the case of chaperonin and prefoldin, the compact nature of the protein-folding pathways in normal and stressed cells has been accomplished by double-duty assignments of these chaperones to normal, as well as stress-responsive protein-folding pathways (45). The mechanisms of protein folding in more complex eucaryal pathways have become accessible through the analysis of chaperones from archaea, due to their simple architecture and exceptional stability. New insights into posttranslational processing and protein salvage will very likely emerge in the next five years as new developments, such as tractable genetic systems (see Chapter 21), are more widely applied in the archaea.

One aspect of protein trafficking in archaea that has received little attention is the decisive mechanism that sends irreversibly misfolded proteins to their proteolytic fates. The proteasome has been characterized with respect to the PAN (*proteasome-activating nucleotidase*) system. PAN serves as a sensor for unfolded proteins. Binding activates ATP hydrolysis, which stimulates substrate unfolding, gate opening in the 20S complex, and protein translocation (3). Pan A and Pan B proteins prime entry into the 20S complex and the PAN regulatory complex, a homolog of the eucaryal 19S ATPase (77). However, the mechanisms of PAN action in archaea are still the subject of active research. The complex function of protein entry into the proteasome is controlled in eucarya by ubiquitinylation, a prerequisite for proteolysis by the proteasome. While all archaea have putative ubiquitin-binding domains in the NAC proteins (89), the factors controlling the gating of proteins into the proteasome and the restraints exerted on native proteins, preventing their entry into the protein turnover process, have not been determined. This enigma in archaeal molecular biology is very likely to be addressed and solved within the next five years.

Acknowledgments. This is contribution number 06-141 from the Center of Marine Biotechnology.

F.T.R. was supported by National Science Foundation grant 9809352.

REFERENCES

1. Andra, S., G. Frey, M. Nitsch, W. Baumeister, and K. O. Stetter. 1996. Purification and structural characterization of the thermosome from the hyperthermophilic archaeum *Methanopyrus kandleri*. *FEBS Lett.* **379:**127–131.
2. Archibald, J. M., J. M. Logsdon, and W. F. Doolittle. 1999. Recurrent paralogy in the evolution of archaeal chaperonins. *Curr. Biol.* **9:**1053–1056.
3. Benaroudj, N., P. Zwickl, E. Seemuller, W. Baumeister, and A. L. Goldberg. 2003. ATP hydrolysis by the proteasome regulatory complex PAN serves multiple functions in protein degradation. *Mol. Cell* **11:**69–78.

4. Besche, H. T., N. Tamura, and P. F. Zwickl. 2004. Mutational analysis of conserved AAA+ residues in the archaeal Lon protease from Thermoplasma acidophilum. *FEBS Lett.* 574:161–166.

5. Bigotti, M. G., and A. R. Clarke. 2005. Cooperativity in the thermosome. *J. Mol. Biol.* 348:13–26.

6. Blochl, E., R. Rachel, S. Burggraf, D. Hafenbradl, H. W. Jannasch, and K. O. Stetter. 1997. *Pyrolobus fumarii,* gen. and sp. nov., represents a novel group of archaea, extending the upper temperature limit for life to 113 degrees C. *Extremophiles* 1:14–21.

7. Boonyaratanakornkit, B. B., A. J. Simpson, T. A. Whitehead, C. M. Fraser, N. M. El-Sayed, and D. S. Clark. 2005. Transcriptional profiling of the hyperthermophilic methanarchaeon *Methanococcus jannaschii* in response to lethal heat and non-lethal cold shock. *Environ. Microbiol.* 7:789–797.

8. Braig, K. 1998. Chaperonins. *Curr. Opin. Struct. Biol.* 8:159–165.

9. Bukau, B., and A. L. Horwich. 1998. The Hsp70 and Hsp60 chaperone machines. *Cell* 92:351–366.

10. Burston, S. G., and A. R. Clarke. 1995. Molecular chaperones: physical and mechanistic properties. *Essays Biochem.* 29:125–136.

11. Cavicchioli, R. 2006. Cold-adapted archaea. *Nat. Rev. Microbiol.* 4:331–343.

12. Deppenmeier, U., A. Johann, T. Hartsch, R. Merkl, R. A. Schmitz, R. Martinez-Arias, A. Henne, A. Wiezer, S. Baumer, C. Jacobi, H. Bruggemann, T. Lienard, A. Christmann, M. Bomeke, S. Steckel, A. Bhattacharyya, A. Lykidis, R. Overbeek, H. P. Klenk, R. P. Gunsalus, H. J. Fritz, and G. Gottschalk. 2002. The genome of *Methanosarcina mazei:* evidence for lateral gene transfer between bacteria and archaea. *J. Mol. Microbiol. Biotechnol.* 4:453–461.

13. Ditzel, L., J. Lowe, D. Stock, K. O. Stetter, H. Huber, R. Huber, and S. Steinbacher. 1998. Crystal structure of the thermosome, the archaeal chaperonin and homolog of CCT. *Cell* 93:125–138.

14. Eichler, J. 2003. Facing extremes: archaeal surface-layer (glyco)proteins. *Microbiology* 149:3347–3351.

15. Elcock, A. 1998. The stability of salt bridges at high temperatures: implications for hyperthermophilic proteins. *J. Mol. Biol.* 284:489–502.

16. Elcock, A. H. 2001. Prediction of functionally important residues based solely on the computed energetics of protein structure. *J. Mol. Biol.* 312:885–896.

17. Emmerhoff, O. J., H. P. Klenk, and N. K. Birkeland. 1998. Characterization and sequence comparison of temperature-regulated chaperonins from the hyperthermophilic archaeon *Archaeoglobus fulgidus. Gene* 215:431–438.

18. Furutani, M., T. Iida, T. Yoshida, and T. Maruyama. 1998. Group II chaperonin in a thermophilic methanogen, *Methanococcus thermolithotrophicus.* Chaperone activity and filament-forming ability. *J. Biol. Chem.* 273:28399–28407.

19. Futterer O. A. A., H. Liesegang, G. Gottschalk, C. Schleper, B. Schepers, C. Dock, G. Antranikian, and W. Liebl. 2004. Genome sequence of *Picrophilus torridus* and its implications for life around pH 0. *Proc. Natl. Acad. Sci. USA* 101:9091–9096.

20. Gabriel, J. L., and P. L.-G.. Chong. 2000. Molecular modeling of archaebacterial bipolar tetraether lipid membranes. *Chem. Phys. Lipids* 105:193–200.

21. Galagan, J. E., C. Nusbaum, A. Roy, M. G. Endrizzi, P. Macdonald, W. FitzHugh, S. Calvo, R. Engels, S. Smirnov, D. Atnoor, A. Brown, N. Allen, J. Naylor, N. Stange-Thomann, K. DeArellano, R. Johnson, L. Linton, P. McEwan, K. McKernan, J. Talamas, A. Tirrell, W. Ye, A. Zimmer, R. D. Bar-

ber, I. Cann, D. E. Graham, D. A. Grahame, A. M. Guss, R. Hedderich, C. Ingram-Smith, H. C. Kuettner, J. A. Krzycki, J. A. Leigh, W. Li, J. Liu, B. Mukhopadhyay, J. N. Reeve, K. Smith, T. A. Springer, L. A. Umayam, O. White, R. H. White, E. Conway de Macario, J. G. Ferry, K. F. Jarrell, H. Jing, A. J. Macario, I. Paulsen, M. Pritchett, K. R. Sowers, R. V. Swanson, S. H. Zinder, E. Lander, W. W. Metcalf, and B. Birren. 2002. The genome of *M. acetivorans* reveals extensive metabolic and physiological diversity. *Genome Res.* 12:532–542.

22. Goodchild A. R. M., N. F. Saunders, M. Guilhaus, and R. Cavicchioli. 2005. Cold adaptation of the Antarctic archaeon, Methanococcoides burtonii assessed by proteomics using ICAT. *J. Proteome Res.* 4:473–480.

23. Guagliardi, A., L. Cerchia, S. Bartolucci, and M. Rossi. 1994. The chaperonin from the archaeon *Sulfolobus solfataricus* promotes correct refolding and prevents thermal denaturation in vitro. *Protein Sci.* 3:1436–1443.

24. Guagliardi, A., L. Cerchia, and M. Rossi. 1995. Prevention of in vitro protein thermal aggregation by the *Sulfolobus solfataricus* chaperonin. Evidence for nonequivalent binding surfaces on the chaperonin molecule. *J. Biol. Chem.* 270:28126–28132.

25. Gutsche, I., L. O. Essen, and W. Baumeister. 1999. Group II chaperonins: new TRiC(k)s and turns of a protein folding machine. *J. Mol. Biol.* 293:295–312.

26. Haney, P. J., J. H. Badger, G. L. Buldak, C. I. Reich, C. R. Woese, and G. J. Olsen. 1999. Thermal adaptation analyzed by comparison of protein sequences from mesophilic and extremely thermophilic *Methanococcus* species. *Proc. Natl. Acad. Sci. USA* 96:3578–3583.

27. Hartl, F. U., and M. Hayer-Hartl. 2002. Molecular chaperones in the cytosol: from nascent chain to folded protein. *Science* 295:1852–1858.

28. Horwitz, J. 1992. Alpha-crystallin can function as a molecular chaperone. *Proc. Natl. Acad. Sci. USA* 89:10449–10453.

29. Iizuka, R., S. So, T. Inobe, T. Yoshida, T. Zako, K. Kuwajima, and M. Yohda. 2004. Role of the helical protrusion in the conformational change and molecular chaperone activity of the archaeal group II chaperonin. *J. Biol. Chem.* 279:18834–18839.

30. Iizuka, R., T. Yoshida, N. Ishii, T. Zako, K. Takahashi, K. Maki, T. Inobe, K. Kuwajima, and M. Yohda. 2005. Characterization of archeal group II chaperonin-ADP-metal fluoride complexes: implications that group II chaperonins operate as a "two-stroke engine." *J. Biol. Chem.* 280:40375–40383.

31. Iizuka, R., T. Yoshida, Y. Shomura, K. Miki, T. Maruyama, M. Odaka, and M. Yohda. 2003. ATP binding is critical for the conformational change from an open to closed state in archaeal group II chaperonin. *J. Biol. Chem.* 278:44959–44965.

32. Izumi, M., S. Fujiwara, M. Takagi, K. Fukui, and T. Imanaka. 2001. Two kinds of archaeal chaperonin with different temperature dependency from a hyperthermophile. *Biochem. Biophys. Res. Commun.* 280:581–587.

33. Izumi, M., S. Fujiwara, M. Takagi, S. Kanaya, and T. Imanaka. 1999. Isolation and characterization of a second subunit of molecular chaperonin from Pyrococcus kodakaraensis KOD1: analysis of an ATPase-deficient mutant enzyme. *Appl. Environ. Microbiol.* 65:1801–1805.

34. Jacob, U., M. Gaestel, E. Katrin, and J. Buchner. 1993. Small heat shock proteins are molecular chaperones. *J. Biol. Chem.* 268:1517–1520.

35. Jeffries, T. W., and Y.-S. Jin. 2000. Ethanol and thermotolerance in the bioconversion of xylose by yeasts. *Adv. Appl. Microbiol.* 47:221–268.

36. Kagawa, H. K., J. Osipiuk, N. Maltsev, R. Overbeek, E. Quaite-Randall, A. Joachimiak, and J. D. Trent. 1995. The

60 kDa heat shock proteins in the hyperthermophilic archaeon *Sulfolobus shibatae*. *J. Mol. Biol.* **253**:712–725.

37. Kagawa, H. K., T. Yaoi, L. Brocchieri, R. A. McMillan, T. Alton, and J. D. Trent. 2003. The composition, structure and stability of a group II chaperonin are temperature regulated in a hyperthermophilic archaeon. *Mol. Microbiol.* **48**:143–156.

38. Kashefi, K. L., and D. R. Lovley. 2003. Extending the upper temperature limit for life. *Science* **301**:934.

39. Kawarabayasi, Y., Y. Hino, H. Horikawa, S. Yamazaki, Y. Haikawa, K. Jin-no, M. Takahashi, M. Sekine, S. Baba, A. Ankai, H. Kosugi, A. Hosoyama, S. Fukui, Y. Nagai, K. Nishijima, H. Nakazawa, M. Takamiya, S. Masuda, T. Funahashi, T. Tanaka, Y. Kudoh, J. Yamazaki, N. Kushida, A. Oguchi, K. Aoki, K. Kubota, Y. Nakamura, N. Nomura, Y. Sako, and H. Kikuchi. 1999. Complete genome sequence of an aerobic hyper-thermophilic crenarchaeon, *Aeropyrum pernix* K1. *DNA Res.* **6**:83–101.

40. Kawashima, T., Y. Yamamoto, H. Aramaki, T. Nunoshiba, T. Kawamoto, K. Watanabe, M. Yamazaki, K. Kanehori, N. Amano, Y. Ohya, K. Makino, and M. Suzuki. 1999. Determination of the complete genomic DNA sequence of *Thermoplasma volcanium* GSS1. *Proc. Jpn. Acad.* **75**:213–218.

41. Kim, D. R., I. Lee, S. C. Ha, and K. K. Kim. 2003. Activation mechanism of HSP16.5 from *Methanococcus jannaschii*. *Biochem. Biophys. Res. Commun.* **307**:991–998.

42. Kim, K. K., R. Kim, and S. H. Kim. 1998. Crystal structure of a small heat-shock protein. *Nature* **394**:595–599.

43. Kim, R., K. K. Kim, H. Yokota, and S. H. Kim. 1998. Small heat shock protein of *Methanococcus jannaschii*, a hyperthermophile. *Proc. Natl. Acad. Sci. USA* **95**:9129–9133.

44. Klumpp, M., and W. Baumeister. 1998. The thermosome: archetype of group II chaperonins. *FEBS Lett.* **430**:73–77.

45. Klunker, D., B. Haas, A. Hirtreiter, L. Figueiredo, D. J. Naylor, G. Pfeifer, V. Muller, U. Deppenmeier, G. Gottschalk, F. U. Hartl, and M. Hayer-Hartl. 2003. Coexistence of group I and group II chaperonins in the archaeon *Methanosarcina mazei*. *J. Biol. Chem.* **278**:33256–33267.

46. Knapp, S., I. Schmidt-Krey, H. Hebert, T. Bergman, H. Jornvall, and R. Ladenstein. 1994. The molecular chaperonin TF55 from the Thermophilic archaeon *Sulfolobus solfataricus*. A biochemical and structural characterization. *J. Mol. Biol.* **242**:397–407.

47. Kowalski, J. M., R. M. Kelly, J. Konisky, D. S. Clark, and K. D. Wittrup. 1998. Purification and functional characterization of a chaperone from *Methanococcus jannaschii*. *Syst. Appl. Microbiol.* **21**:173–178.

48. Kraulis, P. 1991. MOLSCRIPT: A program to produce both detailed and schematic plots of protein structures. *J. Appl. Crystallogr.* **24**:946.

49. Kusmierczyk, A. R., and J. Martin. 2001. Chaperonins—keeping a lid on folding proteins. *FEBS Lett.* **505**:343–347.

50. Kusmierczyk, A. R., and J. Martin. 2003. Nucleotide-dependent protein folding in the type II chaperonin from the mesophilic archaeon *Methanococcus maripaludis*. *Biochem. J.* **371**:669–673.

51. Laksanalamai, P., D. L. Maeder, and F. T. Robb. 2001. Regulation and mechanism of action of the small heat shock protein from the hyperthermophilic archaeon *Pyrococcus furiosus*. *J. Bacteriol.* **183**:5198–5202.

52. Laksanalamai, P., A. R. Pavlov, A. I. Slesarev, and F. T. Robb. 2006. Stabilization of Taq DNA polymerase at high temperature by protein folding pathways from a hyperthermophilic archaeon, *Pyrococcus furiosus*. *Biotechnol. Bioeng.* **93**:1–5.

53. Laksanalamai, P., T. A. Whitehead, and F. T. Robb. 2004. Minimal protein-folding systems in hyperthermophilic archaea. *Nat. Rev. Microbiol.* **2**:315–324.

54. Large, A. T., E. Kovacs, and P. A. Lund. 2002. Properties of the chaperonin complex from the halophilic archaeon *Haloferax volcanii*. *FEBS Lett.* **532**:309–312.

55. Lenzen, C. U., D. Steinmann, S. W. Whiteheart, and W. I. Weis. 1998. Crystal structure of the hexamerization domain of N-ethylmaleimide-sensitive fusion protein. *Cell* **94**:525–536.

56. Leroux, M. R., M. Fandrich, D. Klunker, K. Siegers, A. N. Lupas, J. R. Brown, E. Schiebel, C. M. Dobson, and F. U. Hartl. 1999. MtGimC, a novel archaeal chaperone related to the eukaryotic chaperonin cofactor GimC/prefoldin. *EMBO J.* **18**:6730–6743.

57. Lindquist, S. 1992. Heat-shock proteins and stress tolerance in microorganisms. *Curr. Opin. Genet. Dev.* **2**:748–755.

58. Lund, P. A., A. T. Large, and G. Kapatai. 2003. The chaperonins: perspectives from the Archaea. *Biochem. Soc. Trans.* **31**:681–685.

59. Lundin, V. F., P. C. Stirling, J. Gomez-Reino, J. C. Mwenifumbo, J. M. Obst, J. M. Valpuesta, and M. R. Leroux. 2004. Molecular clamp mechanism of substrate binding by hydrophobic coiled-coil residues of the archaeal chaperone prefoldin. *Proc. Natl. Acad. Sci. USA* **101**:4367–4372.

60. Macario, A. J., and E. Conway De Macario. 2001. The molecular chaperone system and other anti-stress mechanisms in archaea. *Front. Biosci.* **6**:D262–D283.

61. Macario, E., D. L. Maeder, and A. J. L. Macario. 2003. Breaking the mould: Archaea with all four chaperoning systems. *Biochem. Biophys. Res. Commun.* **301**:811–812.

62. Maeder, D. L., A. J. Macario, and E. C. de Macario. 2005. Novel chaperonins in a prokaryote. *J. Mol. Evol.* **60**:409–416.

63. Marco, S., D. Urena, J. L. Carrascosa, T. Waldmann, J. Peters, R. Hegerl, G. Pfeifer, H. Sack-Kongehl, and W. Baumeister. 1994. The molecular chaperone TF55. Assessment of symmetry. *FEBS Lett.* **341**:152–155.

64. Meyer, A. S., J. R. Gillespie, D. Walther, I. S. Millet, S. Doniach, J. Frydman. 2003. Closing the folding chamber of the eukaryotic chaperonin requires the transition state of ATP hydrolysis. *Cell* **113**:369–381.

65. Miller, E. J., A. S. Meyer, and J. Frydman. 2006. Modeling of possible subunit arrangements in the eukaryotic chaperonin TRiC. *Protein Sci.* **15**:1422–1526.

66. Minuth, T., G. Frey, P. Lindner, R. Rachel, K. O. Stetter, and R. Jaenicke. 1998. Recombinant homo- and hetero-oligomers of an ultrastable chaperonin from the archaeon Pyrodictium occultum show chaperone activity in vitro. *Eur. J. Biochem.* **258**:837–845.

67. Minuth, T., M. Henn, K. Rutkat, S. Andra, G. Frey, R. Rachel, K. O. Stetter, and R. Jaenicke. 1999. The recombinant thermosome from the hyperthermophilic archaeon *Methanopyrus kandleri*: in vitro analysis of its chaperone activity. *Biol. Chem.* **380**:55–62.

68. Muchowski, P. J., L. G. Hays, J. R. Yates III, and J. I. Clark. 1999. ATP and the core "alpha-Crystallin" domain of the small heat-shock protein alphaB-crystallin. *J. Biol. Chem.* **274**:30190–30195.

69. Nakamura, N., H. Taguchi, N. Ishii, M. Yoshida, M. Suzuki, I. Endo, K. Miura, and M. Yohda. 1997. Purification and molecular cloning of the group II chaperonin from the acidothermophilic archaeon, Sulfolobus sp. strain 7. *Biochem. Biophys. Res. Commun.* **236**:727–732.

70. Ng, W. V., S. P. Kennedy, G. G. Mahairas, B. Berquist, M. Pan, H. D. Shukla, S. R. Lasky, N. Baliga, V. Thorsson, J. Sbrogna, S. Swartzell, D. Weir, J. Hall, T. A. Dahl, R. Welti, Y. A. Goo, B. Leithauser, K. Keller, R. Cruz, M. J. Danson, D. W. Hough, D. G. Maddocks, P. E. Jablonski, M. P. Krebs, C. M. Angevine, H. Dale, T. A. Isenbarger, R. F. Peck, M.

Pohlschrod, J. L. Spudich, K.-H. Jung, M. Alam, T. Freitas, S. Hou, C. J. Daniels, P. P. Dennis, A. D. Omer, H. Ebhardt, T. M. Lowe, P. Liang, M. Riley, L. Hood, and S. DasSarma. 2000. Genome sequence of *Halobacterium* species NRC-1. *Proc. Natl. Acad. Sci. USA* 97:12176–12181.

71. Nitsch, M., M. Klummp, A. Lupas, and W. Baumeister. 1997. The thermosome: alternating alpha and beta-subunits within the chaperonin of the archaeon *Thermoplasma acidophilum*. *J. Mol. Biol.* 267:142–149.

72. Okochi, M., H. Matsuzaki, T. Nomura, N. Ishii, and M. Yohda. 2005. Molecular characterization of the group II chaperonin from the hyperthermophilic archaeum *Pyrococcus horikoshii* OT3. *Extremophiles* 9:127–134.

73. Oren, A. 2002. Diversity of halophilic microorganisms: environments, phylogeny, physiology, and applications. *J. Ind. Microbiol. Biotechnol.* 28:56–63.

74. Pak M., and S. Wickner. 1997. Mechanism of protein remodeling by ClpA chaperone. *Proc. Natl. Acad. Sci. USA* 94:4901–496.

75. Phipps, B. M., A. Hoffmann, K. O. Stetter, and W. Baumeister. 1991. A novel ATPase complex selectively accumulated upon heat shock is a major cellular component of thermophilic archaebacteria. *EMBO J.* 10:1711–1722.

76. Reimann, B., J. Bradsher, J. Franke, E. Hartmann, M. Wiedmann, S. Prehn, and B. Wiedmann. 1999. Initial characterization of the nascent polypeptide-associated complex in yeast. *Yeast* 15:397–407.

77. Reuter, C. J., S. J. Kaczowka, and J. A. Maupin-Furlow. 2004. Differential regulation of the PanA and PanB proteasome-activating nucleotidase and 20S proteasomal proteins of the haloarchaeon *Haloferax volcanii*. *J. Bacteriol.* 186:7763–7772.

78. Robb, F. T. 2003. Chapter 10, Genomics of thermophiles. *In* C. M. Fraser, T. D. Read, and K. F. Nelson (ed.), *Microbial Genomics*. Humana Press, Totowa, N.J.

79. Robb, F. T., D. L. Maeder, J. R. Brown, J. DiRuggiero, M. D. Stump, R. K. Yeh, R. B. Weiss, and D. M. Dunn. 2001. Genomic sequence of hyperthermophile, *Pyrococcus furiosus*: Implications for physiology and enzymology. *Methods Enzymol.* 330:134–157.

80. Robb, F. T., and D. L. Maeder. 1998. Novel evolutionary histories and adaptive features of proteins from hyperthermophiles. *Curr. Opin. Biotechnol.* 9:288–291.

81. Rohlin, L., J. D. Trent, K. Salmon, U. Kim, R. P. Gunsalus, and J. C. Liao. 2005. Heat shock response of *Archaeoglobus fulgidus*. *J. Bacteriol.* 187:6046–6057.

82. Roy, S. K., T. Hiyama, and H. Nakamoto. 1999. Purification and characterization of the 16-kDa heat-shock-responsive protein from the thermophilic cyanobacterium *Synechococcus vulcanus*, which is an alpha-crystallin-related, small heat shock protein. *Eur. J. Biochem.* 262:406–416.

83. Ruepp, A., B. Rockel, I. Gutsche, W. Baumeister, and A. N. Lupas. 2001. The chaperones of the archaeon *Thermoplasma acidophilum*. *J. Struct. Biol.* 135:126–138.

84. Shashidharamurthy, R., H. A. Koteiche, J. Dong, and H. S. Mchaourab. 2005. Mechanism of chaperone function in small heat shock proteins: dissociation of the HSP27 oligomer is required for recognition and binding of destabilized T4 lysozyme. *J. Biol. Chem.* 280:5281–5289.

85. Shockley, K., D. E. Ward, S. R. Chhabra, S. B. Conners, C. I. Montero, and R. M. Kelly. 2003. Heat shock response by the hyperthermophilic archaeon *Pyrococcus furiosus*. *Appl. Environ. Microbiol.* 69:2365–2371.

86. Siegert, R., M. R. Leroux, C. Scheufler, F. U. Hartl, and I. Moarefi. 2000. Structure of the molecular chaperone prefoldin: unique interaction of multiple coiled coil tentacles with unfolded proteins. *Cell* 103:621–632.

87. Sigler, P. B. 1995. Unliganded GroEL at 2.8 A: structure and functional implications. *Philos. Trans. R. Soc. Lond. B Biol. Sci.* 348:113–119.

88. Smith, D. R., L. A. Doucette-Stamm, C. Deloughery, H.-M. Lee, J. Dubois, T. Aldredge, R. Bashirzadeh, D. Blakely, R. Cook, K. Gilbert, D. Harrison, L. Hoang, P. Keagle, W. Lumm, B. Pothier, D. Qiu, R. Spadafora, R. Vicare, Y. Wang, J. Wierzbowski, R. Gibson, N. Jiwani, A. Caruso, D. Bush, H. Safer, D. Patwell, S. Prabhakar, S. McDougall, G. Shimer, A. Goyal, S. Pietrovski, G. M. Church, C. J. Daniels, J.-I. Mao, P. Rice, J. Nolling, and J. N. Reeve. 1997. Complete genome sequence of *Methanobacterium thermoautotrophicum* deltaH: functional analysis and comparative genomics. *J. Bacteriol.* 179:7135–7155.

89. Spreter, T., M. Pech, and B. Beatrix. 2005. The crystal structure of archaeal nascent polypeptide-associated complex (NAC) reveals a unique fold and the presence of a ubiquitin-associated domain. *J. Biol. Chem.* 280:15849–15854.

90. Squires, C. L., S. Pedersen, B. M. Ross, and C. Squires. 1991. ClpB is the *Escherichia coli* heat shock protein F84.1. *J. Bacteriol.* 173:4254–4262.

91. Stirling, P. C., V. F. Lundin, and M. R. Leroux. 2003. Getting a grip on non-native proteins. *EMBO Rep.* 4:565–570.

92. Trent, J. D., M. Gabrielsen, B. Jensen, J. Neuhard, and J. Olsen. 1994. Acquired thermotolerance and heat shock proteins in thermophiles from the three phylogenetic domains. *J. Bacteriol.* 176:6148–6152.

93. Trent, J. D., E. Nimmesgern, J. S. Wall, F. U. Hartl, and A. L. Horwich. 1991. A molecular chaperone from a thermophilic archaebacterium is related to the eukaryotic protein t-complex polypeptide-1. *Nature* 354:490–493.

94. Usui, K., T. Yoshida, T. Maruyama, and M. Yohda. 2001. Small heat shock protein of a hyperthermophilic archaeum, *Thermococcus* sp. strain KS-1, exists as a spherical 24 mer and its expression is highly induced under heat-stress conditions. *J. Biosci. Bioeng.* 92:161–166.

95. van Montfort, R. L., E. Basha, K. L. Friedrich, C. Slingsby, and E. Vierling. 2001. Crystal structure and assembly of a eukaryotic small heat shock protein. *Nat. Struct. Biol.* 8:1025–1030.

96. Vetriani, C., D. L. Maeder, N. Tolliday, K.-S. Yip, T. J. Stillman, K. L. Britton, D. Rice, H. H. Klump, and F. T. Robb. 1998. Protein thermostability above 100°C: A key role for ionic interactions. *Proc. Natl. Acad. Sci. USA* 95:12300–12305.

97. Vierke, G., A. Engelmann, C. Hebbeln, and M. Thomm. 2003. A novel archaeal transcriptional regulator of heat shock response. *J. Biol. Chem.* 278:18–26.

98. Waldmann, T., A. Lupas, J. Kellermann, J. Peters, and W. Baumeister. 1995. Primary structure of the thermosome from *Thermoplasma acidophilum*. *Biol. Chem. Hoppe Seyler* 376:119–126.

99. Waters, E., M. J. Hohn, I. Ahel, D. E. Graham, M. D. Adams, M. Barnstead, K. Y. Beeson, L. Bibbs, R. Bolanos, M. Keller, K. Kretz, X. Lin, E. Mathur, J. Ni, M. Podar, T. Richardson, G. G. Sutton, M. Simon, D. Soll, K. O. Stetter, J. M. Short, and M. Noordewier. 2003. The genome of *Nanoarchaeum equitans*: insights into early archaeal evolution and derived parasitism. *Proc. Natl. Acad. Sci. USA* 100:12984–12988.

100. Wickner, S., S. Gottesman, D. Skowyra, J. Hoskins, K. McKenney, and M. R. Maurizi. 1994. A molecular chaperone, ClpA, functions like DnaK and DnaJ. *Proc. Natl. Acad. Sci. USA* 91:12218–12222.

101. Yan, Z., S. Fujiwara, K. Kohda, M. Takagi, and T. Imanaka. 1997. In vitro stabilization and in vivo solubilization of foreign proteins by the beta subunit of a chaperonin from the hy-

perthermophilic archaeon *Pyrococcus* sp. strain KOD1. *Appl. Environ. Microbiol.* **63:**785–789.

102. Yoshida, T., A. Ideno, S. Hiyamuta, M. Yohda, and T. Maruyama. 2001. Natural chaperonin of the hyperthermophilic archaeum, *Thermococcus* strain KS-1: a hetero-oligomeric chaperonin with variable subunit composition. *Mol. Microbiol.* **39:**1406–1413.

103. Yoshida, T., A. Ideno, R. Suzuki, M. Yohda, and T. Maruyama. 2002. Two kinds of archaeal group II chaperonin subunits with different thermostability in *Thermococcus* strain KS-1. *Mol. Microbiol.* **44:**761–769.

104. Yoshida, T., R. Kawaguchi, and T. Maruyama. 2002. Nucleotide specificity of an archaeal group II chaperonin from *Thermococcus* strain KS-1 with reference to the ATP-dependent protein folding cycle. *FEBS Lett.* **514:**269–274.

105. Yoshida, T., R. Kawaguchi, H. Taguchi, M. Yoshida, T. Yasunaga, T. Wakabayashi, M. Yohda, and T. Maruyama. 2002. Archaeal group II chaperonin mediates protein folding in the cis-cavity without a detachable GroES-like co-chaperonin. *J. Mol. Biol.* **315:**73–85.

106. Yoshida, T., M. Yohda, T. Iida, T. Maruyama, H. Taguchi, K. Yazaki, T. Ohta, M. Odaka, I. Endo, and Y. Kagawa. 1997. Structural and functional characterization of homo-oligomeric complexes of alpha and beta chaperonin subunits from the hyperthermophilic archaeum *Thermococcus* strain KS-1. *J. Mol. Biol.* **273:**635–645.

107. Zako, T., R. Iizuka, M. Okochi, T. Nomura, T. Ueno, H. Tadakuma, M. Yohda, and T. Funatsu. 2005. Facilitated release of substrate protein from prefoldin by chaperonin. *FEBS Lett.* **579:**3718–3724.

Archaea: Molecular and Cellular Biology
Edited by Ricardo Cavicchioli
© 2007 ASM Press, Washington, D.C.

Chapter 11

Sensing, Signal Transduction, and Posttranslational Modification

PETER J. KENNELLY

INTRODUCTION

The ability to coordinate molecular functions and modulate them in response to changes in the status of both its internal milieu and external environment is essential to the continuation of any and all life forms (Fig. 1). Without recourse to regulatory controls, adaptations, and checkpoints an organism will succumb, sooner or later, to the exhaustion of the raw materials needed to sustain existence or an ill-timed leap into a resource-intensive and demandingly complex process such as cell division. Hence, all contemporary life forms, no matter how "simple," "primitive," or "ancient" they may appear, must contain a basic suite of sensor-response machinery (Table 1).

The extremophilic organisms that dominate the domain *Archaea* have challenged long-established concepts regarding the parameters that define a habitable environment. How do the *Archaea* sense and respond to the extremes of acidity, temperature, salinity, pressure, etc. that characterize their distinctive environmental niches? Have the *Archaea* developed unique mechanisms for regulating the many novel metabolic pathways to which they are host? Do archaea communicate with one another or with other organisms?

At this point in time, we know comparatively little regarding molecular regulatory mechanisms in the *Archaea*. The mechanisms that control archaeal gene transcription and chemotaxis are described in Chapters 6 and 18, respectively. The present chapter focuses on other modalities of signal transduction, including intracellular second messengers, feedback regulation, and posttranslational modifications such as the phosphorylation-dephosphorylation of proteins (Fig. 2).

RECEPTORS

How "Sensitive" Are the *Archaea?*

Do the *Archaea* possess the extensive sensor-response machinery typical of free-living *Bacteria* and *Eucarya?* Given the unique nature of the habitats occupied by the first recognized members of the third domain of life, one's first instinct might be to say no. Wouldn't the hostility of these extreme environments toward conventional life forms suggest that their archaeal inhabitants were effectively quarantined from the rest of the biosphere? Moreover, the unequivocal nature of the descriptor "extreme" implies an inherent constancy. When viewed from our own terrestrial frame of reference, these environments in general appear to be invariantly extreme in their acidity, salinity, temperature, etc. The sensor-response needs of an organism living in such physically monotonous and biologically sterile niches therefore would be expected to be minimal in extent and primitive in nature.

However, the *Archaea* are not ecological hermits. With the development of sensitive molecular probes such as the polymerase chain reaction, it has become apparent that archaea permeate the biosphere. Conversely, we now realize that a phylogenetically diverse range of organisms populate the extreme habitats formerly regarded as homogenous archaeal ghettoes. It is therefore not surprising to find that the *Archaea* respond to a wide range of environmental factors, including the type and availability of sources of carbon, energy, and essential elements, as well as variations in the pH, salinity, and temperature of their surroundings (Table 2). However, while it is clear that archaea are "sensitive," many fundamental questions remain. What types of sensor-response mechanisms have these phylogenetically distinct, and oftentimes

Peter J. Kennelly • Department of Biochemistry, Mail Stop 0308, Virginia Polytechnic Institute and State University, Blacksburg, VA 24061.

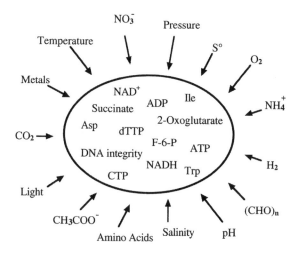

Figure 1. Environmental variables and internal cues known or likely to be monitored by members of the *Archaea*.

metabolically unique, organisms developed to cope with life at terrestrial extremes? What can the *Archaea* tell us about the origins and evolution of molecular sensor-response pathways across the phylogenetic spectrum?

Do Archaea Communicate with Their Neighbors?

Little is known regarding cell-cell communication systems in the *Archaea*. Many archaea populate syntrophic microbial consortia (215, 221), implying the existence of both intraspecies and interspecies signaling mechanisms. Empirical studies of archaeal-archaeal and archaeal-bacterial communication have been few in number and preliminary in nature (131, 193, 222). Inspection of archaeal genomes has revealed them to be devoid of homologs of the prototypic bacterial quorum-sensing proteins LuxS and LuxR (286).

Archaeal Receptors

While it is clear that the *Archaea* are responsive to changes in their surroundings, our knowledge of the specific environmental cues that are monitored and the receptors that recognize them remains limited. Most of the archaeal receptors characterized to date fall into five broad functional classes: sensors for amino acids, sensors for light energy, sensors for diatomic gases, sensors for osmotic stress, and sensors of electrical potential. In addition, bioinformatic analyses have identified several proteins whose topography suggests that they function as transmembrane receptors.

Amino acids and betaines

The haloarchaeon, *Halobacterium salinarium*, employs three different types of receptors to detect amino acids and derivatives thereof. HtrII is a classic transmembrane receptor consisting of an extracellular serine-binding domain fused to an intracellular transmitter domain, specifically a methyl-accepting chemotaxis protein (MCP) (117). By contrast, BasB, a receptor for branched-chain and sulfur-containing amino acids, and CosB, which binds the modified amino acid betaine, also known as *N,N,N*-trimethylglycine, lack either membrane domains or fused signaling domains. It has been postulated that these proteins are tethered to the cell surface by lipid anchors, where they interact with the extracellular domains of their cognate MCPs, BasT and CosT (154). The haloarchaeal arginine receptor Car, on the other hand, is a soluble MCP-fusion protein that resides within the cytoplasm. Car relies on the activity of the arginine:ornithine antiporter to deliver ligands to its vicinity (285). Each of these receptors triggers a two-component signaling cascade that regulates the activity of flagellar motor proteins (see "Two-component system" below, and Chapter 18). Examination of archaeal genome sequences has revealed that several additional archaea contain deduced MCP-associated chemoreceptors, including *Archaeoglobus fulgidus*, *Haloarcula marismortui*, *Methanococcus maripaludis*, *Methanosarcina acetivorans*, *Methanosarcina mazei*, *Pyrococcus abyssi*, and *Pyrococcus horikoshii* (95, 289).

Light

Many haloarchaea contain a brace of sensory rhodopsins, designated SRI and SRII, each of which is configured to sense a specific color of light via the photo-induced isomerization of retinal (115) (see Chapter 14). SRI is activated by orange light, while SRII is sensitive to blue-green wavelengths. Both sensory rhodopsins are coupled to two-component signaling cascades via their cognate MCPs, HtrI and HtrII, respectively. Intriguingly, HtrII also serves as the chemoreceptor for the amino acid serine (see "Amino acids and betaines," above).

Oxygen and other gases

The *Archaea* employ a variety of heme proteins as receptors for oxygen and, potentially, other diatomic gases such as NO and CO. The aerotactic behavior of *H. salinarium,* for example, is mediated by two heme-containing proteins designated HtrVIII and HemAT-Hs. HtrVIII is an integral membrane protein in which a heme-containing cytochrome oxidase-like

Table 1. Definitions of terms and abbreviations used in Chapter 11

Allosteric regulation. The modulation of the functional properties of a protein via the binding of a molecule (**allosteric effector**) to a site distinct from that at which the function takes place. The association of an allosteric effector with its protein target is noncovalent and, hence, reversible. As the gross concentration of an allosteric effector decreases, it will dissociate from its cognate target protein.

ATCase. Aspartate transcarbamolyase.

CACHE domain. A deduced extracellular domain conserved among certain <u>ca</u>lcium channels and <u>che</u>motaxis receptors.

cAMP. The second messenger 3′,5′-cyclic AMP.

c-diGMP. The second messenger bis (3′,5′-cyclic diguanylic acid).

cGMP. The second messenger 3′,5′-cyclic GMP.

CHASE domain. A general designation for a series of deduced *c*yclase/*h*istidine kinase-*a*ssociated *s*ensing *e*xtracellular domains. Numbers are used to designate the various classes of CHASE domains, i.e., CHASE1, CHASE2, etc.

Covalent modification. Any structural change involving the formation and/or rupture of a covalent bond. Examples include phosphorylation, proteolysis, glycosylation, and disulfide formation.

Downstream and **upstream.** Terms used to define the relative positions of the molecular components within a signal transduction cascade. Downstream refers to a frame of reference proceeding from the signal to the target, and upstream refers to the converse.

EF-2. Elongation factor 2.

ePK. The "eukaryote-like" protein kinases. The most prolific family of protein-serine/threonine/tyrosine kinases in nature. The designator derives from the fact that the members of this family were thought, for many years, to be confined exclusively to members of the *Eucarya* ("eukaryotes").

Feed-forward regulation. Feed-forward activation is that form of feedback regulation in which an increase in the level of the indicator metabolite leads to the stimulation of the activity of a target enzyme.

Feedback inhibition. Feedback inhibition is that form of feedback regulation in which an increase in the level of the indicator metabolite triggers a decrease in the activity of a target enzyme.

Feedback regulation. Regulation of an enzyme catalyzing one of the early committed steps in a pathway or process in which an end product of the pathway, or some closely related pathway, serves as an indicator metabolite.

GAF domain. A cyclic nucleotide-binding domain. The acronym GAF is derived from the names of some of the proteins in which it appears: c<u>G</u>MP-stimulated phosphodiesterase, *Anabaena* <u>a</u>denylate cyclase, and bacterial transcription factor <u>F</u>hlA.

GAPN. Glyceraldehyde-3-phosphate dehydrogenase.

GDH. Glutamate dehydrogenase.

Hierarchical regulation. Numerous proteins are targeted by multiple transmission domains, covalent modifications, second messengers, and/or indicator metabolites. In those cases where these regulatory modalities act in something other than a simple, additive manner, the overall regulatory mechanism is termed hierarchal.

Histidine kinase (domain). A protein kinase that autophosphorylates on a conserved histidine residue.

HPr. Histidine-rich protein.

Hpt domain. Histidine phosphotransfer domain. A signal transmission unit sometimes encountered in extended two-component signal transduction cascades. Hpt domains are phosphorylated on a conserved histidine residue. The predominant function of Hpt domains is to shuttle phosphoryl groups between response regulator domains.

Hybrid histidine kinase. A polypeptide that contains, in addition to a histidine kinase domain, one or more response regulator and/or Hpt domains.

IF-5A. Initiation factor 5A.

Indicator metabolite. A metabolite that also serves as an allosteric effector.

KTN domain. KTN stand for <u>K</u>+ transport, <u>n</u>ucleotide-<u>b</u>inding domain. KTN domains are found within the potassium transport proteins of several bacteria, where they presumably bind nucleotides that serve as allosteric regulators of channel activity.

MCP. Methyl-accepting chemotaxis protein. A conserved family of signal transmission domains that generally are fused to a chemotactic or other type of sensor domain. The name derives from the fact that these proteins/domains are subject to covalent modification by methylation of glutamic acid side chains. Methylation serves to modulate the sentivity of the sensor. MCPs generally act on an associated or fused histidine kinase.

ORF. Open reading frame.

Continued on following page

Table 1. *Continued*

PAS domain. PAS is an acronym formed from the names of three members of the protein family, <u>P</u>er, the period clock protein from *Drosophila melanogaster*; <u>A</u>rnt, an aryl hydrocarbon nuclear transporter from vertebrates; and <u>Sim</u>, "single-minded" protein from *Drosophila*. PAS domains provide a scaffold for a variety of different prosthetic groups, can be mounted to serve as sensors for light, gases, etc.

PPM. One of the two most abundant families of protein-serine/threonine phosphatases. The membership of the PPM family includes protein phosphatase-2C and SpoIIE.

PPP. One of the two most abundant families of protein-serine/threonine phosphatases. The membership of the PPP family includes protein phosphatase-1, protein phosphatase-2A, and protein phosphatase-2B (calcineurin).

Protein kinase (PK). Any enzyme that catalyzes the transfer of a phosphoryl group from a donor substrate such as ATP to an amino acid side chain of a protein. The amino acid side chains most frequently targeted by this covalent modification include the hydroxyl moieties of serine, threonine, and tyrosine; the carboxylate of aspartic acid; and the nitrogen atoms within the imidazole ring of histidine. If a PK catalyzes its own modification by phosphorylation, it is said to be **autophosphorylated.** Autophosphorylation may occur in *cis* or *trans*. Five major superfamilies of protein kinases have been identified to date, the "eukaryote-like" protein kinases, histidine kinases, myosin heavy chain/eIF-2 or α-kinases, the isocitrate dehydrogenase kinase/phosphatases, and the HPr kinases.

Protein phosphatase. Any enzyme that catalyzes the hydrolysis of the covalent bond linking a phosphoryl group to an amino acid side chain on a protein.

PTP. A general term for protein-tyrosine phosphatases. Specific families of PTPs include the conventional PTPs (cPTPs) and low-molecular-weight PTPs (LMW PTPs).

Receptor. Any protein containing a functionally competent sensor domain.

Redox regulation. Redox stands for reduction-oxidation. Redox regulation refers to a change in functional status effected by the addition or removal of one or more electrons. This may (e.g., disulfide formation) or may not (e.g., transition ferrous to ferric iron) be accompanied by the formation and/or rupture of a covalent bond.

Response regulator domain. A domain that autophosphorylates on a conserved aspartic acid residue using an autophosphorylated histidine kinase as substrate. Response regulator domains are commonly fused to or associated with transcription factors or modulators of flagellar motor proteins.

Second messenger. Second messengers, such as cAMP or Ca^{2+}, are dedicated allosteric effectors that are introduced into the interior of a cell in response to an extracellular signal (the **first messenger**). The allosteric effector serves as a surrogate, or second, messenger for the external signal.

Sensor. The molecular species with which the signal directly interacts. Common sensors include ligand-binding sites, individual amino acid side chains, and prosthetic groups such as heme.

Sensor protein/domain. The minimum macromolecular unit that can serve as a functionally competent sensor.

Sensor-response. The selective alteration of one or more cellular processes as a consequence of the appearance, disappearance, or shift of a specific internal or external signal.

Sensor-response pathway, sometimes referred to as a **signal transduction cascade**. The molecules responsible for effecting selective alterations in cellular processes in response to changes in a specific internal or external signal:

$$\text{Signal} \rightarrow \text{signal-response machinery} \rightarrow \text{target} \rightarrow \text{cellular effect.}$$

Signal. Any species that acts via a sensor-response pathway to effect a specific cellular response. Signals may be biological (e.g., other cells and organisms), chemical (e.g., toxins, nutrients, metabolites, pH), or physical (e.g., surfaces, light, temperature, pressure) in nature.

Signal transduction. The process of sensing and responding to a signal that originates from a source external to the cell, such as the surrounding environment or other cells. This subset of general sensor-response events is defined by the requirement to relay or "transduce" the signal, either directly or indirectly, across the barrier of the cell membrane.

Signal transmission protein/domain (also referred to as a **transmitter** or **signaling domain**). A protein or domain thereof whose function is to regulate another protein. Transmitters may act by producing a second messenger, by binding to or dissociating from another protein, or by catalyzing the covalent modification of another protein. It is not uncommon for the multiple transmission proteins to be linked together to form a **signal transduction cascade**.

Target. A protein(s) (e.g., transcription factors, enzymes, cytoskeletal components) whose functional properties must ultimately be altered (regulated) to effect the desired cellular response.

Transmembrane receptor. A receptor that spans the cell membrane. In general, transmembrane receptors are configured with their sensor domain facing the exterior of the cell with their transmitter domain or associated transmitter protein facing the interior.

Two-component system (sometimes referred to as the **His-Asp phosphorelay system**). A signal transmission unit whose basic core consists of a histidine kinase domain and a response regulator domain. Two-component systems are found in most bacteria, as well as some archaea and eucarya. In the *Archaea*, the two-component paradigm is employed almost exclusively for the control of chemo- and phototaxis (see Chapter 18) and gene transcription (see Chapter 6).

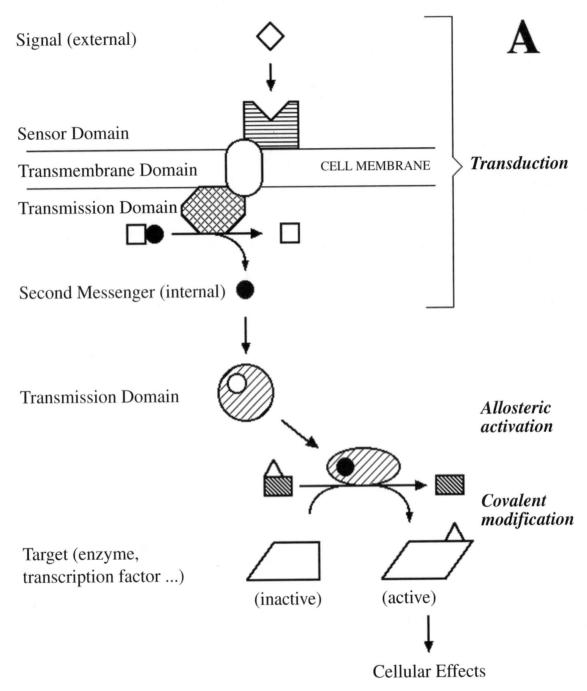

Figure 2. Basic elements of biological sensor-response pathways. (A) Hypothetical multistep signal transduction pathway in which an external signal (open diamond) interacts with a transmembrane receptor complex to activate a target protein. The sensor-response pathway comprises two steps. In the first, the transmission domain of the receptor complex produces a second messenger (filled circle) that, in turn, serves as an allosteric activator for a second transmission domain (hatched circle) that catalyzes the covalent modification (open triangle) of the target (open quadrilateral). In this example, binding of the allosteric ligand and covalent modification both activate their respective target proteins by altering their conformation. (B) Hypothetical multistep biosynthetic pathway that is subject to feedback inhibition by one of the products of the final enzyme in the pathway (filled diamond). In this case, the indicator metabolite binds to and allosterically activates a sensor-transmitter fusion protein that subsequently binds to and inhibits the activity of the first enzyme in the pathway (target). The second and third enzymes in the pathway are denoted by diagonal hatching and cross hatching, respectively. See Table 1 for definitions of terms used.

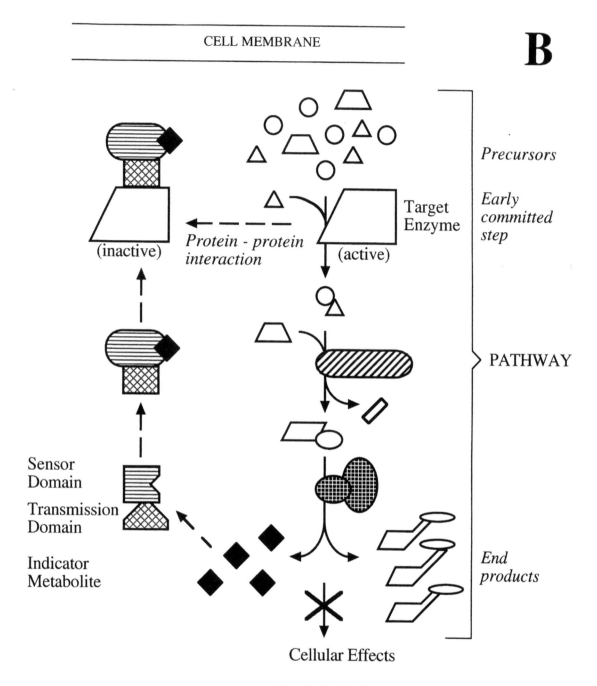

Figure 2. *Continued*

receptor domain is fused to an MCP-family signal transmitter domain (42). HemAT-Hs is a soluble protein in which a myoglobin-like domain is fused to an MCP (118). In addition to HemAT-Hs, simple globins lacking fusion domains have been characterized from *Aeropyrum pernix* and *M. acetivorans* (89). These archaeal "hemoglobins" bind NO and CO as well as oxygen in vitro. It has been suggested that redox-mediated interconversion of the heme iron between the Fe(II) and Fe(III) state (the latter of which does not

bind gases) may occur in these heme proteins, thereby enabling them to sense changes in both oxygen levels and redox state (43).

The transcription factor Bat from *Halobacterium* sp. NRC-1 (106) is an example of the third, and potentially largest and most diverse type of heme-based oxygen sensor in the *Archaea*. Bat is a soluble protein containing a PAS domain, a putative cGMP-binding domain, and a helix-turn-helix domain (22). The acronym PAS is derived from *Per*, the period clock

Table 2. Responses of the *Archaea* to environmental cues

Archaeon	Environmental cue	Response monitored	References
Archaeoglobus fulgidus	pH, temperature, metals, oxygen, xenobiotics	Biofilm production	166
	Salinity, temperature	Osmolyte levels	204
	Heat shock	Gene transcription	244
Ferroplasma acidarmanus	Copper	Gene transcription and protein expression	21
Haloarcula marismortui	Glucose and acetate	Enzyme activity	39
Halobacterium NRC-1	Light	Transcriptome and proteome	22
Halobacterium halobium	Glucose, acetate, benzoate, $NiSO_4$, histidine, asparagines, leucine, methionine, quinine, phenol	Chemotaxis	261, 278
	Light	Phototaxis	112, 278
	Light	Protein phosphorylation	277, 279
Halobacterium salinarium	Betaine, choline, carnitine	Chemotaxis	154
	Diesel oil	Proteomics	174
Haloferax mediterranei	Growth phase, salinity	mRNA stability	124
	Growth phase, salinity, light, oxygen	Gene transcription	230
Haloferax volcanii	Growth phase	Gene transcription	239
	Heat shock	Gene transcription	159, 295, 296
	Salinity	Gene transcription	28
	Salinity	Proteome	210
	Amino acids	Transporter activity	133
Halomonas elongata	Salinity	Proteome	210
Halorubrum sp.	Osmotic shock	Membrane composition	172
Metallosphaera sedula	Heterotrophy vs. chemolithotrophy	Gene transcription	134
	Heat shock, nutrient levels	Proteome, respiration	226
Methanothermobacter thermautotrophicus	Amino acids	Enzyme activity	97
	Temperature	Osmolyte levels	111
Methanococcoides burtonii	Cold adaptation	Proteome	99, 100
Methanococcus igneus	Salinity, temperature	Osmolyte levels	53
Methanococcus jannaschii	Salinity	Gene transcription	231
	Heat shock, cold shock	Gene transcription	35
Methanococcus maripaludis	Salinity	Gene transcription	231
	Alanine	Gene transcription and enzyme activity	182
	Ammonia	Enzyme activity	140
	Ammonia	Gene transcription	56
	Hydrogen	Gene transcription	321
Methanococcus thermolithotrophicus	Salinity	Osmolyte synthesis	202
Methanococcus voltae	Acetate, isoleucine, leucine	Chemotaxis	271
Methanohalophilus portucalensis	Salinity	Osmolyte levels	243
Methanosarcina acetivorans	Salinity	Gene transcription	231
	Methanol	Gene expression and proteome	247
Methanosarcina barkeri	Salinity	Gene transcription	231
	Ammonia	Enzyme activity	185
	Salinity	Osmolyte levels	38
Methanosarcina mazei	Salinity	Gene transcription	231
	Heat shock	Transcription factor level and/or activity	63
	Heat shock	Gene transcription	165
	Carbon and nitrogen sources	Gene transcription	304
Methanosarcina thermophila	Salinity	Osmolyte level	275
Methanothermus fervidus	Temperature	Osmolyte levels	111
Methanothermus sociabilis	Temperature	Osmolyte levels	111
Natronobacterium pharaonis	Light	Phototaxis	259
Pyrococcus strain ES4	Heat shock	Proteome	116
Pyrococcus furiosus	Cold shock	Transcriptome	313
	Heat shock	Transcriptome	269
	Salinity, temperature	Osmolyte levels	203
	Elemental sulfur	Enzyme activity	1
	Carbon source	Gene transcription	68

Continued on following page

Table 2. *Continued*

Archaeon	Environmental cue	Response monitored	References
	Carbon source	Enzyme activity and gene transcription	302
	Carbon source	Transcriptome	263
Pyrodictium occultum	Heat shock	Protein level	232
Sulfolobus acidocaldarius	UV light	Genetic reversion and recombination frequency	320
Sulfolobus shibatae	Heat shock	Protein level	298
Sulfolobus solfataricus	Carbon source	Gene transcription	86, 109, 123
	UV irradiation and actinomycin D	Gene transcription	257
	Lysine	Gene transcription	40
Thermococcus strain ES-1	Elemental sulfur	Enzyme activity	194
Thermococcus barophilus	Pressure	Proteome	201
Thermococcus litoralis	Carbon source	Gene transcription	68
Thermoproteus tenax	Autotrophy vs. heterotrophy	Gene transcription and enzyme activity	45

protein from *Drosophila melanogaster;* Arnt, an aryl hydrocarbon receptor nuclear transporter from vertebrates; and Sim, "single-minded" protein from *Drosophila* (293). PAS domains provide a scaffold on which several different prosthetic groups can be mounted, including heme, FAD, or the chromophore 4-hydroxycinnamic acid. The nature of the prosthetic group determines whether a particular PAS domain senses oxygen, redox state, or light. PAS domains are found in a variety of archaea, including *A. fulgidus, Haloarcula marismortuii, Halobacterium* sp. NRC-1, *H. salinarium, Methanococcoides burtonii, Methanospirillum hungatei, M. thermautotrophicus, M. acetivorans, Methanosarcina barkeri, M. mazei, Natronomonas pharaonis,* and *P. abyssi,* where they are oftentimes fused with either an MCP domain or a two-component histidine kinase (9, 19, 94, 293). However, empirical evidence verifying the functional properties of these receptorlike proteins is lacking.

Osmolarity

Archaea such as *Haloferax volcanii* (172), *M. jannaschii* (147), *Thermoplasma acidophilum* (148), and *Thermoplasma volcanium* (148) sense and adapt to osmotic stress through the action of mechanosensitive ion channels. Rather than sampling solute concentration directly, these channels respond to stress-induced changes in the physical properties of the cell membrane. These channels are widely dispersed throughout the phylogenetic spectrum, implying early evolutionary origins (149).

Temperature

The farnesyl diphosphate/geranylgeranyl diphosphate synthase, *Tk*-IsdA, from the extreme thermophile

Thermococcus kodakaraensis catalyzes the condensation of C_5 isoprenoid units with allylic diphosphate to produce C_{15}- and C_{20}-diphosphates. These polyisoprenoids serve as precursors for the synthesis of squalene and its derivatives or membrane di- and tetraether lipids, respectively. Product composition is controlled by temperature, with increased temperature favoring synthesis of C_{15} over C_{20} products (90). The mechanism of regulation involves a temperature-mediated conformational change that alters the position of a gating tyrosine residue, Tyr-81.

Electrical potential

The activity of the flagellar motor proteins that regulate the swimming behavior of the haloarchaeon *H. salinarium* is sensitive to changes in proton motive force across the cell membrane. Gene knockout studies have implicated a membrane-bound MCP, MpcT, as the relevant sensor (152). As McpT is a small protein lacking a recognizable extracellular domain, it is proposed that McpT responds directly to changes in membrane potential, rather than sensing shifts in the level of some chemical species such as H^+.

In general, the translocation of potassium across the membranes of archaeal cells is mediated by homologs of the channels responsible for "gating" ion currents during neurotransmission in eucarya. The archaeal proteins KvAP from *A. pernix* (253, 254) and MVP from *M. jannaschii* (265) display all the features characteristic of the voltage-gated ion channels of the *Eucarya*. Recombinant versions of the channels are selective for potassium and responsive to changes in membrane polarization. In addition, both KvAp and MVP can be inactivated by the spider- and scorpion-derived toxins that target their eucaryal counterparts. It has been postulated that the voltage-gated ion channels of the *Archaea* play a role in environmental adaptation and metabolic changes (158).

Cryptic receptors

A variety of putative receptors have been identified based on circumstantial evidence in the form of (i) potential transmembrane topography; (ii) fusion to a likely signal-propagating domain such as an MCP, histidine kinase, protein-serine/threonine/tyrosine kinase, or protease; and (iii) conservation across a large number of organisms or association with multiple classes of signaling domains (96, 300). However, the ligands for these putative receptors have yet to be identified. A handful of these domains have been encountered in archaeal genomes, including CACHE, CHASE4, and CHASE6.

CACHE refers to an extracellular domain conserved among certain *ca*lcium channels and *che*motaxis receptors (7). Based on its fusion with some bacterial MCPs, it has been suggested that CACHE domains bind small ligands such as citrate or amino acids. Archaea encoding potential CACHE-containing receptor proteins include *Halobacterium* sp. NRC-1, *M. burtonii*, *M. acetivorans*, *M. mazei*, *M. hungatei*, and *P. horikoshii* (7, 14, 19, 292). Members of the genus *Methanosarcina* encode polypeptides in which both CACHE and MCP domains are fused to a potential ligand-gated ion channel similar to the acetylcholine receptors of the higher *Eucarya* (292).

CHASE stands for *cyclase/histidine kinase-associated sensing extracellular* (8, 214). Six classes of CHASE domains have been deduced using computational methods, two of which are found in the *Archaea* (CHASE4 and CHASE6). CHASE4 domains form the predicted extracellular domains of histidine kinases in *A. fulgidus*, *M. acetivorans*, *M. barkeri*, and *M. mazei* (19, 327). A CHASE6 domain is fused to a domain of unknown function in *Halobacterium* sp. NRC-1 (327).

In addition, a variety of deduced transmembrane and lipid-anchored secreted proteins have been identified whose sequence indicates that they possess ligand-binding potential. However, these proteins lack endogenous intracellular signal transmission domains (Table 1). Thus, they would require an associated MCP, adenylate cyclase, protein kinase, membrane channel, phosphodiesterase, or other partner to form a functionally competent signal transduction unit. Potential ligands include metals (31, 208), di- and oligonucleotides (3, 31), amino acids and dipeptides (31), phosphate (31), and sugars and sugar phosphates (3, 31). Other extracellular proteins contain domains implicated in cell-cell adhesion events in the *Eucarya* such as β-helix domains, β-propeller domains, and polycystic kidney disease domains (130, 233).

SECOND MESSENGERS

The term "second messenger" was coined to describe a set of specialized intracellular signaling molecules that are synthesized or released in response to the binding of ligands, or first messengers, to extracellular receptors. The best-known example of a second messenger is 3′,5′-cyclic AMP (cAMP), which is produced by the enzyme adenylate cyclase when the hormone epinephrine binds to the β-adrenergic receptor. Second messengers in general elicit cellular effects by acting as allosteric modulators of key proteins (Fig. 2). Oftentimes allosteric regulators bind to and modulate the activity of other signal transmission proteins, such as protein kinases or protein phosphatases, which in turn act on "downstream" target proteins. The resulting multistep signal transduction cascade provides both a means for constructing branches for reaching multiple proteins and, perhaps more importantly, a source of connective nodes within cellular sensor-response networks that facilitate the integration of multiple inputs. In other cases, a second messenger will bind to and directly modulate the enzymes and other proteins that form the ultimate targets of a sensor-response pathway.

Cyclic Nucleotides

Cyclic nucleotides such as cAMP, 3′,5′-cyclic GMP (cGMP), and bis(3′,5′-cyclic diguanylic) acid (c-diGMP) are universally employed as regulators of gene expression and metabolic activity in bacteria and eucarya. However, although the presence of cAMP in the *Archaea* was reported nearly twenty years ago (175), very little is known about the role of cyclic nucleotides in these organisms (30). A sizable majority, 80%, of the *Archaea* encode potential class 2 adenylate cyclases within their genomes (95). However, stereotypical cyclic nucleotide phosphodiesterases appear to be lacking (95).

c-diGMP is a novel second messenger that was recently discovered in the *Bacteria*. Examination of archaeal genome sequences for potential diguanylate cycles has yielded equivocal results. Pei and Grishin (227) reported that several proteins from the *Archaea*, *A. fulgidus*, *M. jannaschii*, *M. thermautotrophicus*, and *P. horikoshii*, contain the GGDEF motif that is characteristic of bacterial diguanylate cyclases. However, Galperin (95) argues that any perceived resemblance is too faint to constitute a reliable predictor of physiologic function.

While few archaeal proteins display the classic cyclic nucleotide-binding sites commonly found in the *Eucarya* and *Bacteria* (233), several possess a more recently recognized class of cyclic nucleotide-binding motif, the GAF domain (9). GAF is an acronym assembled from the names of the proteins in which it was first identified, i.e., the cGMP-stimulated phosphodiesterases, *Anabaena* adenylate cyclases, and the bacterial transcription factor FhlA (16, 113). Since most archaea contain only a single recognizable cyclase within their genomes, these archaeal GAFs are likely to bind cAMP instead of cGMP. Many of these GAF domains are fused to two-component histidine kinases (96), suggesting that archaea may utilize cAMP-mediated phosphorylation cascades. The triggers for these implied cascades remain to be elucidated, as none of the deduced archaeal adenylate cyclases identified to date contain a transmembrane domain (95), nor do they possess other recognizable functional domains beyond those implicated in catalysis (267).

Calcium

In the *Eucarya*, calcium has long vied with cAMP for the title of most prolific second messenger. While it has been speculated that calcium plays a prominent role as a second messenger in bacteria, little evidence has emerged to support this supposition (70, 209, 272). Although a protein capable of activating a calmodulin-dependent phosphodiesterase in a calcium-dependent manner was isolated from *H. salinarium* (246), examinations of archaeal genome sequences failed to reveal the presence of the calmodulin/parvalbumin family's characteristic calcium-binding motif, the EF hand (233, 242). Nonetheless, many archaea possess the basic prerequisites for modulating the level of free calcium in their cytosol: channels for the admittance of calcium into their cytoplasm (49) and the ATP-driven pumps required to maintain a low basal intracellular calcium concentration (65, 303).

Several secreted hydrolases from the *Archaea* (98, 145, 236, 248, 258) bind to and are stabilized by calcium. However, it is difficult to envision a regulatory or signaling function for calcium involving these extracellular enzymes. The handful of intracellular proteins that have been reported to reversibly bind calcium include an ATP-dependent DNA ligase from *Sulfolobus shibatae* (163), as well as several proteins from *M. thermautotrophicus*, namely elongation factor-1ß, MTH1880 (173), and a calcium-gated potassium channel (129). Only the calcium-gated potassium channel exhibited calcium-dependent effects on its activity or structure that were characteristic of the binding of an allosteric ligand.

Inositol Phosphates

In the *Eucarya*, multiply phosphorylated forms of the hexose inositol serve as second messengers in signal transduction (121). While many thermophilic archaea are capable of synthesizing phosphorylated forms of inositol, these compounds are used as compatible solutes to combat osmotic stress and appear to have no role in cellular regulation (216).

ALLOSTERIC REGULATION BY METABOLITES

Regulation via the binding of dissociable effectors is not confined to dedicated second messengers such as cAMP. A variety of metabolites also serve as allosteric regulators of key enzymes. In the most basic form of allosteric regulation (feedback inhibition), a biosynthetic end product inhibits the activity of an enzyme responsible for catalyzing an early committed step in its biosynthesis (224). Early intervention is more efficient as it avoids the accumulation of unneeded pathway intermediates. Other "indicator metabolites" are employed to coordinate flux through related pathways, such as those producing the purine and pyrimidine nucleotide building blocks of DNA, or to intervene when changes in global cellular parameters, such as energy status or redox state, dictate that local control mechanisms be overridden.

Acetolactate Synthase

Acetolactate synthase catalyzes the first committed steps in the synthesis of the branched-chain amino acids valine, leucine, and isoleucine: the condensation of pyruvate with either a second molecule of pyruvate or 2-ketobutyrate to yield carbon dioxide and either acetolactate or acetohydroxybutyrate, respectively. The acetolactate synthase activities in extracts from three species of *Methanococcus*, *M. aeolicus*, *M. maripaludis*, and *M. voltae*, are sensitive to feedback inhibition by branched-chain amino acids (323). However, the pattern of inhibition observed was species specific. The acetolactate synthase activity in extracts from *M. aeolicus* was sensitive solely to valine, while the activity in extracts from *M. voltae* could be inhibited by either valine or isoleucine, suggesting that the enzyme in *M. voltae* was sensitive to both amino acids. While the acetolactate synthase activity in extracts of *M. maripaludis* also was sensitive to valine and isoleucine, the maximum degree of inhibition obtained with either amino acid was only about 50%. Such behavior is suggestive of the existence of either two isoforms of the enzyme, each of which was sensitive to a single feedback regulator, or possibly, additive inhibition of a single enzyme.

Aspartate Transcarbamoylase

The enzyme aspartate transcarbamoylase (AT-Case) catalyzes the initial step in de novo pyrimidine biosynthesis. The form of ATCase found in *Escherichia coli* is used, literally, as a textbook example for two important phenomena: cooperativity and allosteric regulation (184). The enterobacterial enzyme is a dodecamer comprising equal numbers of catalytic and regulatory subunits. Binding of the substrate, aspartate, results in homotypic positive cooperativity. *E. coli* ATCase is subject to feedback inhibition by the pathway end product CTP, whose efficacy can be potentiated by a second pathway end product, UTP. ATP activates the enzyme via a mechanism that largely overrides the inhibitory effects of CTP, thus ensuring that pyrimidine nucleotide biosynthesis keeps pace with purine nucleotide biosynthesis during periods of vigorous biosynthetic activity.

Although the evolutionary history of the carbamoyltransferases is a complex one, the archaeal ATCases appear to constitute a monophyletic group (160, 161). Nevertheless, those archaeal ATCases studied to date exhibit significant diversity in their kinetic and regulatory properties. Those from *Halobacterium cutirubrum* (219) and *P. abyssi* (235, 301) displayed the type of cooperative kinetic behavior and pattern of allosteric regulation characteristic of the iconic ATCase from *E. coli*. On the other hand, the enzyme from *M. jannaschii* displays little evidence of either cooperativity or allosteric regulation by nucleotides (107). Intermediate between these extremes is the ATCase from *S. acidocaldarius*, which displays cooperative kinetics and activation by ATP, but is activated rather than inhibited by CTP and UTP (71).

Biosynthesis of Aromatic Amino Acids

The aromatic amino acids phenylalanine, tyrosine, and tryptophan are synthesized via a branched pathway whose last common step is the synthesis of chorismate. In many bacteria, total flux through the pathway and distribution of that flux among its branches are regulated by an elegant network of allosteric feedback events. While genomic and biochemical analyses indicate that flux through the early, common portion of the pathway responsible for the de novo synthesis of aromatic amino acids is not subject to allosteric feedback inhibition in the *Archaea* (262), considerable evidence has appeared for allosteric control of end-product distribution.

The first branch point in the biosynthetic pathway for the aromatic amino acids is formed by anthranilate synthase and prephenate synthase, which direct flux from chorismate toward tryptophan and phenylalanine plus tyrosine, respectively. The deduced amino acid sequence of the anthranilate synthase from *H. volcanii* reportedly contains the conserved residues involved in feedback inhibition of its bacterial homologs (164). Examination of anthranilate synthases isolated from *A. fulgidus* (47), *M. thermautotrophicus* (97), *S. solfataricus* (299), and *Thermococcus kodakaraensis* (291) revealed that each is inhibited by tryptophan.

The second branch of the pathway bifurcates immediately after the synthesis of prephenate from chorismate. Prephenate dehydrogenase catalyzes the first committed step in the synthesis of tyrosine, while prephenate dehydratase catalyzes the first committed step in the synthesis of phenylalanine from prephenate. Like its bacterial counterpart, the prephenate dehydrogenase from the haloarchaeon *Methanohalophilus mahii* is subject to feedback inhibition by its cognate end product, tyrosine (87). Likewise, the prephenate dehydratase from this organism is subject to multivalent feedback control (87). Phenylalanine inhibits the enzyme in vitro, while amino acids such as tyrosine, methionine, leucine, and isoleucine activate it. Similarly, the prephenate dehydrogenase from *Halobacterium vallismortis* is inhibited by phenylalanine and activated by isoleucine (127). While several other amino acids were also observed to exert allosteric effects in vitro, the concentrations required were supraphysiological. Feedback regulation is apparently not ubiquitous among the *Archaea*. It has recently been reported that both the prephenate dehydratase and the prephenate dehydrogenase activities in *M. maripaludis* are insensitive to aromatic amino acids (234).

dCTP Deaminase/dUTP Diphosphatase

dCTP deaminase/dUTP diphosphatase represents a uniquely archaeal fusion, within a single protein domain, of the active-site components responsible for catalyzing two reactions in the synthesis of uridine and deoxyuridine nucleotides from cytosine and deoxycytosine nucleotides (120). The deoxyuridine nucleotides produced by this enzyme also serve as the precursors to deoxythymine nucleotides, such as dTTP, which acts as a feedback inhibitor of this bifunctional enzyme (180).

Flagellar Motors

Switching the direction of rotation of flagellar motors in response to chemo-, aero-, and phototactic signals in *H. salinarium* requires the presence of the tricarboxylic acid (TCA) cycle intermediate fumarate (206). Fumarate serves as a switching or potentiating factor whose binding to flagellar motors renders the proteins responsive to CheY, the terminal phosphoprotein of the two-component signaling cascade (see "Two-component system," below) that becomes activated upon stimulation of the haloarchaeon's chemo-, aero-, and photoreceptors (212). The switch itself is under the control of rhodopsin-containing photoreceptors, whose activation triggers the release of a pool of membrane-bound fumarate molecules (211). While not a second messenger in the strictest definition of the term, its sequestration from central metabolism (see Chapter 12) renders this membrane-bound pool of fumarate the functional equivalent of a "classic" second messenger such as calcium.

Glyceraldehyde 3-Phosphate Dehydrogenase

The glycolytic enzymes of the *Archaea* display little evidence of the type of multinodal allosteric control so frequently observed in the *Bacteria* and *Eucarya* (305). A notable exception to this pattern is the unusual NAD^+-dependent, nonphosphorylating glyceraldehyde 3-phosphate dehydrogenase (GAPN) from the archaeon *Thermoproteus tenax*. The catalytic activity of this enzyme is sensitive to several metabolites (44, 162). NADP(H), NADH, and ATP inhibit GAPN by reducing its affinity for NAD^+, while AMP, ADP, glucose 1-phosphate, and fructose 6-phosphate act in a reciprocal fashion. When the enzyme was incubated with equivalent concentrations of the most potent inhibitor (NADPH) and activator (glucose 1-phosphate), a net twofold activation of the enzyme was observed despite the fact that the K_d of the former was the lower of the two: 0.3 versus 1.0 μM (44). Thus, binding of allosteric activators may override the effects of allosteric inhibitors for affecting the catalytic efficiency of the enzyme.

KtrA Potassium Transporter

The genome of the methanoarchaeon *M. jannaschii* encodes two proteins whose sequences suggest that they mediate the active transport of potassium ions. One of these, KtrA, contains a cytoplasmic KTN (K^+ transport, *nucleotide binding*) domain that includes a Rossmann nucleotide-binding fold (158), and binds NAD^+ or NADH in vitro (245). A comparison of X-ray crystal structures indicates that the NADH and NAD^+ act in a reciprocal manner on the channel. NAD^+ impedes opening of the channel by binding in a manner that restricts movement of the hinge region of its protein subunits (245). However, when NADH is bound these hinge regions remain flexible, thereby permitting potassium transport to proceed (245). It therefore appears that importation

of potassium is regulated by the energy status of the organism, as reflected by the ratio of the reduced versus oxidized form of the indicator metabolite NAD(H).

Nitrogen Metabolism

Glutamate dehydrogenase plays a central role in nitrogen metabolism in all living organisms (48). The haloarchaeon, *H. salinarium*, contains two forms of the enzyme, one $NADP^+$ dependent (NADP-GDH) and another NAD^+ dependent (NAD-GDH). A comparison of the K_m values of the two glutamate dehydrogenases for ammonia indicates that NADP-GDH, which exhibits a very low K_m for ammonia, acts to funnel nitrogen into anabolic pathways (34). On the other hand, NAD-GDH is thought to participate in amino acid catabolism (33). During catabolism, the oxidative deamination of glutamate yields NADH and 2-oxoglutarate; the latter is further oxidized via the Krebs cycle and ultimately yields ATP. NAD-GDH is activated by a variety of amino acids and inhibited by several Krebs cycle intermediates, including fumarate, oxaloacetate, succinate, and malate (32). It has not been firmly established whether these inhibitory dicarboxylates bind to the active site or to a distinct, allosteric regulatory site. However, each was observed to be kinetically noncompetitive with respect to both glutamate and NAD^+, consistent with an allosteric mode of action.

Several archaea are capable of fixing atmospheric nitrogen when other sources of this key element are lacking (176). In diazotrophic bacteria, 2-oxoglutarate (a central intermediate in nitrogen metabolism) serves as a critical indicator of intracellular nitrogen levels (218). Therefore, it is not surprising that in certain archaea, 2-oxoglutarate stimulates the activity of key enzymes in nitrogen metabolism: nitrogenase from *M. maripaludis* (69) and glutamine synthetase from *M. mazei* (72). Both of these enzymes associate with the archaeal homolog of the PII protein found in photosynthetic plants and bacteria. In these latter organisms, binding of 2-oxoglutarate to PII negates this protein's ability to inhibit the activity of the nitrogen-metabolizing enzymes with which it associates. It appears likely that archaeal PII proteins also mediate the action of this indicator metabolite.

Phosphoenolpyruvate Carboxylase

Phosphoenolpyruvate carboxylase (PEPC) catalyzes the synthesis of oxaloacetate, a key intermediate in the Krebs cycle. The PEPCs from *S. acidocaldarius* (256) and *S. solfataricus* (79) are inhibited by malate and the amino acid aspartate, a behavior exhibited by bacterial and eucaryal PEPCs. By contrast, PEPC from *Methanococcus sociabilis* (255) is insensitive to these compounds.

Ribonucleotide Reductase

Ribonucleotide reductase catalyzes the synthesis of deoxyribonucleotides from corresponding ribonucleotides. Thus far, only two archaeal ribonucleotide reductases, members of the class II family of coenzyme B_{12}-dependent enzymes, have been characterized (75, 240). Feedback inhibition by dATP, an indicator of intracellular dNTP pools, is commonly observed among the ribonucleotide reductases. It is therefore somewhat surprising that the ribonucleotide reductase from the archaeon *T. acidophilum* displays little sensitivity to dATP. However, the enzyme does exhibit a complex pattern of feedback regulation in which other nucleotides modulate both its catalytic efficiency and substrate specificity (75). For example, dTTP markedly stimulates activity toward GDP, while inhibiting activity toward ADP and CDP by about one-half. dGTP, on the other hand, stimulates activity toward ADP and inhibits activity toward CDP and GDP. dCTP stimulates activity toward GDP with little effect on activity toward either ADP or CDP. While stimulation of activity toward GDP by dCTP appears logical because the guanine and cytosine form one of the base pairs in double-stranded DNA, the effects of the other feedback regulators are less easily rationalized. The enzyme from *P. furiosus*

behaves more conventionally, as it reportedly is "most strongly" inhibited by dATP (240).

Uracil Phosphoribosyltransferase

Uracil phosphoribosyltransferase catalyzes condensation of uracil with phosphoribosylpyrophosphate to form UMP and pyrophosphate, a key step in the pyrimidine salvage pathway. The uracil phosphoribosyltransferase from the archaeon *S. solfataricus* is dramatically stimulated by GTP. When GTP is saturating, k_{cat} increases nearly 20-fold while the K_m values for uracil and PRPP decrease 10- and 2-fold, respectively (17, 126). UMP inhibits the enzyme in a hierarchal, CTP-dependent manner. While feedback regulation of the de novo biosynthesis of pyrimidines by pathway end products and feed-forward stimulation by purine nucleotides are common among bacteria and archaea (see "Aspartate transcarbamoylase," above), this behavior is atypical for pyrimidine salvage pathways. Its unique status renders the sophisticated allosteric regulatory mechanism of the uridine phosphoribosyltransferase from *S. solfataricus* even more intriguing.

Other Potential Binding Domains for Allosteric Ligands

Examination of archaeal genomes has revealed the presence of several domains proposed to bind small molecule ligands (Table 3). In many cases, these

Table 3. Intracellular small molecule-binding domains in the *Archaea*[a]

Domain name	Acronym	Hypothetical ligand(s)
Aspartokinase, chorismate mutase, TyrA	ACT	Amino acids, purines
ATP cone	N/A	ATP
Cystathione β-synthase	CBS	S-Adenosylmethionine
Double-stranded β-helix	DSBH	Carbohydrates, oxalate, amino acids
Ferredoxins	Fx	Redox state
cGMP phosphodiesterase, adenylate cyclase, FhlA	GAF	Cyclic nucleotides, tetrapyrroles, formate
Heavy metal-associated	HMA	Copper and other metals
Nitrogen fixation protein X	NIFX	Molybdate, iron
Per-Arndt-Sim	PAS	Oxygen, light, redox state
Periplasmic binding protein type II	PBP-II	N-Acetylserine, thiosulfate, amino acids, sugars
Regulation of amino acid metabolism	RAM	Amino acids
Transcriptional regulators, cation-transporting ATPases, and hydrogenases	TRASH	Heavy metals
Transporter-associated OB domain	T-OB	Sulfate, molybdate, sugars
TrkA protein	TRKA	NAD(H)
Universal stress protein A	USPA	NTPs
Vinyl 4 reductase	4VHR	Hydrocarbons

[a]Deduced from analyses of genome sequences. For further information, see references 8, 15, 78, and 233.

putative ligand-binding domains are fused to additional domains that may function as targets for allosteric regulation. These deduced functional domains include membrane transporters, metabolic enzymes, DNA-binding proteins, proteases, and esterases (9, 15). Several examples also have been encountered in which individual polypeptide chains contain two or more classes of small molecule-binding domains, suggesting a potential role as signal integrators.

PROTEIN-SERINE/THREONINE/TYROSINE PHOSPHORYLATION

The covalent modification of proteins via phosphorylation of the hydroxyl groups of serine, threonine, and tyrosine represents Nature's most versatile and prolific mechanism for regulating cellular processes (322). The intrinsic ability of the phosphoryl group to dramatically alter local charge and hydrophilicity renders it a potent agent for perturbing the phyiscochemical characteristics of a protein on both a local and a global scale. While regulation by allosteric ligands requires the addition of a large, three-dimensionally complex binding domain to a protein, the basic structural requirement for attaching a phosphoryl group consists of a suitably positioned amino acid side chain containing a nucleophilic hydroxyl, carboxyl, amino, or imino group. Consequently, covalent phosphorylation is a regulatory mechanism compatible with and adaptable to an extraordinarily diverse spectrum of proteins. Equally important, a phosphorylated protein (phosphoprotein) can be restored to its original, dephosphorylated state through the intervention of a protein phosphatase (136). Evidence from both genomics and biochemistry indicate that the covalent modification of proteins via phosphorylation and dephosphorylation is a nearly universal attribute of the free-living members of the domain *Archaea*.

Phosphoproteins

Approximately a dozen proteins have been identified in the *Archaea* that are subject to phosphorylation on the hydroxyl groups of serine, threonine, and/or tyrosine (Table 4).

Protein-serine/threonine/tyrosine kinases

Three protein kinases exhibiting the capacity to catalyze their self- or autophosphorylation, a common behavior for enzymes of this type, have been identified in the *Archaea*: Rio1 (169) and Rio2 (167, 168) from *A. fulgidus* and SsoPK2 from *S. solfataricus* (190). While autophosphorylation often constitutes a prerequisite for the efficient phosphorylation of exogenous substrate proteins, this would appear not to be the case for either Rio1 or SsoPK2. Versions of these proteins that had been rendered incapable of autophosphorylation by mutagenic alterations displayed full activity toward exogenous proteins in vitro. Inspection of the X-ray structures of the phos-

Table 4. Archaeal phosphoproteins[a,b]

Protein	Archaeon	Phosphoamino acid	Function	Reference
Methyltransferase activation protein	*M. barkeri*	n.d.[c]	Autophosphorylated form activates methyltransferase	59
90-kDa aminopeptidase	*S. solfataricus*	n.d.	ND	57
Cdc6	*M. thermautotrophicus*	P-Ser	DNA-stimulated autophosphorylation	102
Cdc6	*S. acidocaldarius*	n.d.	ND	64
Glycogen synthase	*S. acidocaldarius*	Acid stable	ND	50
Phenylalanyl-tRNA synthetase	*T. kodakaraensis*	P-Tyr	ND	128
Phosphomannomutase	*T. kodakaraensis*	P-Tyr	ND	128
RNA terminal phosphate cyclase	*T. kodakaraensis*	P-Tyr	Nb	128
SsoPK2	*S. solfataricus*	P-Ser	Autophosphorylation	190
SsoPK3	*S. solfataricus*	P-Thr	ND	191
Initiation factor 2-α	*P. horikoshii*	Ser-48	ND	290
D-Gluconate dehydratase	*S. solfataricus*	n.d.	Activates	145
Phosphohexomutase	*S. solfataricus*	Ser-309	Inhibits	237
Rio1	*A. fulgidus*	Ser-108	Autophosphorylation	169
Rio2	*A. fulgidus*	Ser-128	Autophosphorylation	168

[a]Included are proteins for which the nature of protein-phosphate bond has been determined, as well as polypeptides that can be radiolabeled with [^{32}P]phosphate or which are immunoreactive with antibodies against phosphotyrosine.
[b]Components of the two-component system are not included (see "Two-component system," below).
[c]n.d., not determined.

pho and dephospho forms of Rio2 suggest that autophosphorylation influences ATP binding by this protein kinase, but this has not been tested experimentally (168). By contrast, SsoPK3, a protein-serine kinase from the membrane fraction of *S. solfataricus*, does not undergo autophosphorylation in vitro (191). However, SsoPK3 is phosphorylated in situ by a second, threonine-preferring protein kinase that resides in the membrane of *S. solfataricus* (191).

CDC6 and methyltransferase-activating protein

Several other archaeal proteins known to bind ATP have been reported to undergo autophosphorylation, including Cdc6 from both *M. thermautotrophicus* (102) and *S. acidocaldarius* (64), and the methyltransferase-activating protein from *M. barkeri* (59). While the nature and function of the phosphorylation of Cdc6 remains controversial, autophosphorylation of the methyltransferase-activating protein from *M. barkeri* renders it capable of stimulating the activity of the MT-1 methyltransferase (59). Activation requires stoichiometric levels of the phosphoprotein, even in the presence of excess ATP. Hence, the autophosphorylated protein would appear to activate its cognate methyltransferase by binding to it.

Phosphohexosemutases

In the case of the phosphohexosemutase from *S. solfataricus*, mapping the site of modification on the X-ray structure of a bacterial homolog suggested in what manner, and by what mechanism, phosphorylation affects the enzyme's catalytic activity (237).

These models placed the site of phosphorylation, Ser-309, squarely within the substrate-binding site. Here, a phosphoryl group would be expected to inhibit catalysis by erecting an electrosteric barrier to substrate binding (Fig. 3). Introduction of a negative charge into the active site by mutagenically altering the phosphoacceptor serine to aspartate verified this prediction. A phosphohexosemutase from another thermophile, *T. kodakaraensis*, is phosphorylated on tyrosine (128). However, neither the location of the phosphoacceptor tyrosine residue(s) nor the effect of phosphorylation on catalytic activity was reported.

Phenylalanyl-tRNA synthetase and RTCB

The tyrosine-phosphorylated phosphomannomutase from *T. kodakaraensis* was isolated by affinity chromatography using an inactive form of the protein-tyrosine phosphatase *Tk*-PTP (128). Two other proteins also adhered to this column and, following elution, displayed immunoreactivity toward antibodies against phosphotyrosine. The first was the β-chain of a deduced phenylalanyl-tRNA synthetase. The second was a polypeptide, RTCB, of unspecified function that is encoded within the operon for RNA terminal phosphate cyclase (128).

Initiation factor 2

It is highly likely that the phosphorylation of the α-subunit of initiation factor 2 from *P. horikoshii* inhibits this factor in a manner similar to that observed in the *Eucarya* (290). Not only is the site of phosphorylation on the archaeal protein similar to that of

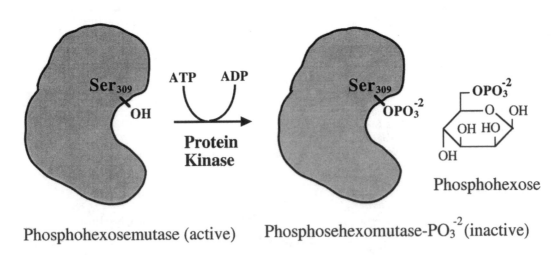

Figure 3. Phosphorylation of phosphohexosemutase from *S. solfataricus*. Ser-309 on the phosphohexosemutase inhibits catalysis by electrosterically interfering with the binding of substrate phosphohexoses.

its eucaryal homolog, but it also can be phosphorylated by an RNA-dependent protein kinase (PKR) from *Homo sapiens*. PKR is one of several protein kinases that are known to phosphorylate eucaryal initiation factor 2 in vivo (54). It is therefore particularly noteworthy that *P. horikoshii* contains a homolog of PKR, PH0512, that phosphorylates archaeal initiation factor 2 in vitro (290).

D-Gluconate dehydratase

Phosphorylation of D-gluconate dehydratase from *S. solfataricus* appears to activate the enzyme, as incubation of the isolated phosphoprotein with alkaline or acid phosphatase leads to a concomitant loss of phosphate and activity (144). Neither the impact nor the specific site of modification by phosphorylation appears to have been determined for any of the remaining proteins listed in Table 4.

Protein-Serine/Threonine/Tyrosine Kinases

At least five distinct molecular paradigms have evolved for catalyzing the phosphorylation of proteins on the side-chain hydroxyl groups of serine, threonine, and tyrosine: the eucaryal protein kinases, the myosin heavy-chain kinase/elongation factor-2 kinases, the isocitrate dehydrogenase kinase/phosphatase, the histidine-rich protein (HPr) kinases, and the Rsb/Spo kinases (137, 138). Among this quintet, the eucaryal protein kinases (ePKs) have emerged as the numerically dominant and phylogenetically cosmopolitan species (179, 196, 197, 233). Deduced members of the extended ePK superfamily are found throughout the *Eucarya* (196) and most of the *Bacteria* (95, 137). It is therefore perhaps not surprising that a majority of archaeal genome sequences encode one or more (typically 3 to 5) potential ePKs (95, 139), or that ePKs constitute the only apparent source of protein-serine/threonine/tyrosine kinase activity in the *Archaea* (138).

Whereas the vast majority of eucaryal and bacterial ePKs resemble the catalytic subunit of the prototypic cAMP-dependent protein kinase (197), most archaeal ePKs belong to the RIO and piD261 subfamilies (179). These subfamilies deviate from the prototypic ePKs in two respects. First, they lack a basic amino acid residue in subdomain VIb, the catalytic loop, which distinguishes the serine/threonine- and tyrosine-specific forms of classic ePKs. Second, their C-terminal protein-substrate-binding domain is shorter and lacks several of the structural features conserved in classic ePKs (167–169).

Phylogenetic analyses indicate that the RIO/piD261 subfamilies formed the progenitors of the classic ePKs (10, 81, 179). It is thus somewhat ironic that the former have been dubbed "atypical" ePKs (197). It has been hypothesized that the classic ePK paradigm originated in the *Eucarya*, wherein it underwent explosive expansion and from which it was acquired by the *Bacteria* by lateral gene transfer (179). Note that the *Archaea* are not completely devoid of proteins resembling classic ePKs, as was initially believed. *S. solfataricus*, for example, contains at least two open reading frames (ORFs) encoding protein kinases that resemble classic ePKs (138).

Rio1 and Rio2

The ORFs encoding five archaeal ePKs have been cloned and their protein products characterized: Rio1 (169) and Rio2 (167, 168) from *A. fulgidus*, PH0512 from *P. horikoshii* (290), and SsoPK2 (190) and SsoPK3 (191) from *S. solfataricus*. Both Rio1 and Rio2 catalyze their own phosphorylation, as well as that of several exogenous proteins such as histone H1 and myelin basic protein. In addition, Rio1 transphosphorylates itself (169). X-ray crystallographic analysis of Rio2 revealed the presence of an N-terminal winged helix-turn-helix motif of the type typically associated with DNA-binding proteins (167). No such domain is present in Rio1. It is tempting to speculate that the winged helix-turn-helix domain of Rio2 targets this enzyme to the components of a protein-oligonucleotide complex in vivo (170, 171).

SsoPK2 and SsoPK3

Similar to Rio1 and Rio2, SsoPK2 from *S. solfataricus* catalyzes its autophosphorylation as well as the phosphorylation of exogenous proteins such as casein, myelin basic protein, mixed histones, and a chemically modified form of lysozyme in vitro (190). In every instance, the enzyme targeted serine residues for modification. The rate at which these proteins were phosphorylated was very low, which raises questions as to whether proteins serve as the physiologic target of this phosphotransferase. SsoPK2 contains a large, 300-residue, N-terminal domain of unknown function. SsoPK3, which was isolated as a phosphoprotein present in membrane extracts of *S. solfataricus*, also possesses a large (180 residue) N-terminal domain of undetermined function, as well as a potential leucine zipper near the C terminus (191). While SsoPK3 displayed protein-serine/threonine kinase activity toward exogenous proteins such as casein, bovine serum albumin, myelin basic protein, and chemically modified lysozyme, it exhibited no propensity to autophosphorylate in vitro.

PH0512

PH0512 from *P. horikoshii* was identified as a potential homolog of the double-stranded, RNA-activated protein kinase that phosphorylates eucaryal initiation factor 2-α in the *Eucarya* (290). The recombinant protein product phosphorylated archaeal initiation factor 2-α in vitro. The site of phosphorylation, Ser-48, corresponds to the site of phosphorylation of its eucaryal homolog, Ser-51. PH0152 appears to be the first archaeal ePK for which a cognate substrate has been identified.

Regulation of archaeal protein kinases

Little is known about the mechanisms by which archaeal protein kinases are regulated. However, most archaea contain at least one deduced ePK that contains a predicted transmembrane domain, implicating them as components of transmembrane receptors (95). In *H. volcanii*, expression of mRNA encoding a deduced ePK is responsive to changes in environmental salinity (28). SsoPK3 from *S. solfataricus* is phosphorylated by a second membrane-associated protein kinase (191), most likely a glycosylated, threonine-preferring protein kinase (188, 189). It has not been determined whether phosphorylation of SsoPK3 affects its catalytic properties.

Protein-Serine/Threonine/Tyrosine Phosphatases

ORFs representing four distinct families of protein-serine/threonine/tyrosine phosphatases have been discerned in archaeal genomes. They are the PPP-family protein phosphatases, the PPM-family protein phosphatases, the conventional protein-tyrosine phosphatases (cPTP), and the low-molecular-weight protein-tyrosine phosphatases (LMW-PTP) (136, 138).

PPP family

The eucaryal representatives of the PPP family, which include protein phosphatase-1, protein phosphatase-2A, and calcineurin, are serine/threonine-specific metalloenzymes (23). By contrast, the bacterial members of the PPP family are highly promiscuous, hydrolyzing protein-bound phosphoserine, phosphothreonine, phosphotyrosine, and phosphohistidine (136, 138). Most, but not all bacterial PPPs, require the presence of exogenous divalent metal ions for activity (136). Three members of the PPP family have been characterized from the archaeal organisms: PP1-arch1 from *S. solfataricus* (178), PP1-arch2 from *Methanosarcina thermophila* TM-1 (274), and Py-PP1 from *M. abyssi* (195). Each shares 25 to 30% se-

quence identity with eucaryal PPPs, and 15 to 19% identity with bacterial PPPs. Similar to their eucaryal counterparts, they are serine/threonine specific. However, all three require the addition of a divalent metal ion, usually Mn^{2+}, for activity.

PPM family

The PPM family of protein phosphatases is found in about half of the *Bacteria* and all members of the *Eucarya* (23, 136). All forms of the enzyme characterized to date require the presence of an exogenous divalent metal ion for activity. The vast majority are serine/threonine specific. However, forms that also hydrolyze protein-bound phosphotyrosine in vitro have recently been discovered in the cyanobacteria (181, 252). Among the nearly two-dozen archaeal genome sequences released to date, only one ORF encoding a member of the PPM family of divalent metal ion-dependent protein-serine/threonine phosphatases has been detected: TVN0703 from *T. volcanium* (138). The gene encoding this protein phosphatase was presumably acquired via lateral transfer from a bacterium. The protein product of this gene has not been characterized.

Protein-tyrosine phosphatases

Both the cPTPs and LMW-PTPs are found throughout the *Eucarya* and in many of the *Bacteria* (137, 268). All known LMW-PTPs are tyrosine specific. On the other hand, the cPTPs have diverged into several subfamilies. Some, most notably the VH1 family, are dual specific, i.e., they hydrolyze protein-bound phosphoserine/phosphothreonine and phosphotyrosine residues (5). Only one archaeal PTP, *Tk*-PTP from *T. kodakaraensis*, has been characterized (128). *Tk*-PTP is a member of the VH1-like branch of the cPTPs (5) and exhibits dual-specific phosphatase activity in vitro (128).

Distribution "pattern"

The heterogeneous distribution of these various protein phosphatases complicates the assignment of specific roles to particular enzymes or enzyme families. While at least one ORF encoding a deduced protein-serine/threonine/tyrosine phosphatase is present in every archaeon that possesses an ePK, no one family is found in all archaea (138). The most widely distributed families are the PPPs and cPTPs, each of which appears in roughly one-half of the *Archaea*. Phylogenetic analysis indicates that the PPPs (136) and, possibly, cPTPs (5) were probably present in the last universal common ancestor. If this is the case, then some archaea lost PPP and/or cPTP genes through the course of evolution.

Archaeal genomes display a striking imbalance between the number of deduced ePKs and countervailing protein phosphatases (138). In particular, in the archaea that contain four or more ePKs, the number of recognizable protein phosphatases is in general one-half, or less, of the number of deduced protein kinases. *S. solfataricus*, for example, may contain as many as eight ePKs but possesses only two protein phosphatases: a PPP and a cPTP. The ratio in *P. horikoshii* is four potential ePKs to one cPTP. This apparent imbalance suggests either that (i) the *Archaea* contain one or more families of yet-to-be-identified protein phosphatases, (ii) archaeal protein phosphatases serve as a largely static pool of protein phosphatase activity, deferring modulation of the phosphorylation state of individual proteins to the protein kinase, or (iii) the substrate specificity of archaeal protein phosphatases is modulated via control of their spatial distribution rather than their gross catalytic activity. This latter mode of regulation is employed by many eucaryal members of the PPP-family (55).

TWO-COMPONENT SYSTEM

The term two-component, or His-Asp phosphorelay, system refers to a set of conserved phosphotransfer domains that are employed in modular fashion to construct signal transmission cascades linking a wide range of sensors to their intracellular targets (Fig. 4) (80, 229, 284). The ubiquitous core modules for these cascades consist a pair of partner phosphotransferases known as histidine kinases (103) and response regulators (52, 283). Upon stimulation of an associated receptor, a histidine kinase autophosphorylates on a conserved histidine residue. The resulting phosphoprotein serves as the specific phosphodonor substrate for its partner-response regulator, which autophosphorylates on a conserved aspartate residue. The autophosphorylated response regulator, in turn, triggers an appropriate cellular response either through interaction with a target protein, such as a flagellar motor (37), or through activation of a fused "output" domain, usually a DNA-binding domain or enzyme (283, 284). Many response regulators can employ acetyl phosphate as an alternative phosphodonor substrate in vitro. Consequently, it has been postulated that some organisms utilize acetyl phosphatase as an indicator metabolite, dubbed an "acetate switch," that acts, at least in part, by fueling the autophosphorylation of select response regulator proteins (319).

Two-component systems differ in several fundamental respects from protein-serine/threonine/tyrosine phosphorylation cascades. First, autophosohorylation is the predominant mechanism of phosphorylation in the two-component system, whereas protein-serine/threonine/tyrosine phosphorylation cascades rely primarily on phosphotransfer reactions catalyzed by protein kinases that are distinct from the phosphoacceptor protein. This reliance of autophosphorylation imposes greater constraints on the proteins that ultimately can be targeted for regulation by this mechanism than does phosphorylation by an exogenous catalyst. Second, the chemical nature of the phosphoryl moieties formed during two-component signaling differs significantly from that of protein-serine/threonine/tyrosine phosphorylation. The phosphoramide and mixed acid anhydride bonds of phosphohistidine and phosphoaspartate are significantly higher in energy than the phosphoester bonds of phosphoserine, phosphothreonine, and phosphotyrosine (46, 52, 91, 315). Therefore, the thermodynamic barrier to the "reverse' reaction, i.e., the transfer of the phosphoryl group from the phosphoaspartate of a response regulator to the histidine of a partner phosphodonor protein, is relatively low.

The facility with which phosphoryl groups can be transferred in either direction between histidine and aspartate residues has been exploited to expand on the binary architecture of archetypical two-component signal transduction systems. These more extensive cascades often employ a third conserved module called a histidine phosphotransfer, or Hpt, domain. Hpt domains shuttle phosphoryl groups between response regulator domains, thus facilitating the construction of extended, oftentimes branched, two-component cascades (11, 114). Although it has been speculated that Hpt domains catalyze the phosphotransfer reactions in which they participate, they exhibit neither kinase nor phosphatase activity in isolation (229, 284). It therefore seems likely that phosphotransfer to and from the phosphoacceptor histidine of Hpt proteins is catalyzed by the autophosphorylation activity of their cognate response regulator domains. In this scenario, the phosphohistidine-containing Hpt domain functions as phosphodonor substrate in a manner analogous to an autophosphorylated histidine kinase. Phosphorylation of Hpt domains by autophosphorylated response regulators would take place via the simple reversal of this reaction.

Distribution of Two-Component Domains among the *Archaea*

The two-component system appears to have originated in the *Bacteria*, from which it radiated into members of the other phylogenetic domains (142, 156). An examination of the current library of archaeal genome

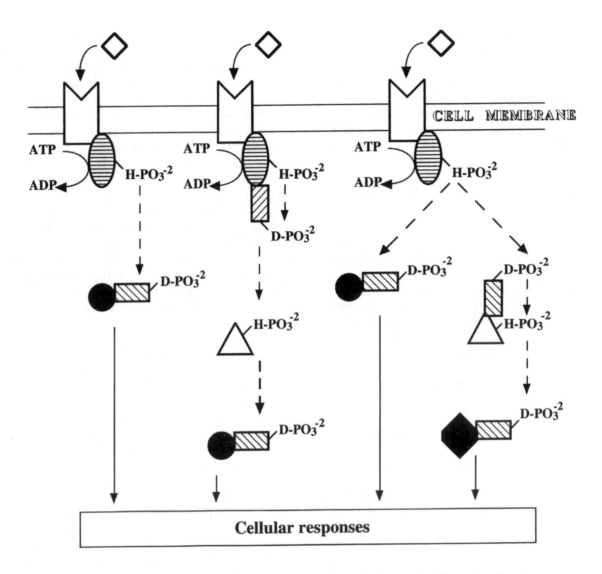

Figure 4. Examples of typical architectures of two-component signal transduction cascades. Shown are schematic representations of three hypothetical two-component signal transduction cascades. For each example, an external signal (open diamond) activates a histidine kinase (hatched oval) by binding to its transmembrane receptor domain (open pentagon). Response regulator domains are represented as diagonally striped rectangles. Output domains (filled circles and filled hexagons); Hpt domains (open triangles); phosphoryl transfer events (hatched arrows); conserved histidine (H) and conserved aspartate (D) residues within each two-component domain. A basic two-component signaling cascade (left); an extended two-component cascade employing a hybrid histidine kinase, i.e., one that is fused to a response regulator domain, and phosphoryl shuttle via an Hpt domain (middle); a branched two-component cascade whose right-hand branch includes an response regulator domain-Hpt domain fusion protein that serves as a phosphoryl group shuttle bridging the histidine kinase to a downstream target response regulator protein (right).

sequences reveals that more than half contain open reading frames encoding deduced histidine kinase and response regulator domains (Table 5). However, the distribution is strikingly skewed. While the majority of the *Euryarchaeota* possess deduced two-component systems, both the *Nanoarchaeota* and *Crenarchaeota* are devoid of histidine kinase and response regulator domains (18, 19),

Speculation as to the cause(s) of this polarized distribution pattern, such as transfer to an ancient eu-

ryarchaeon (19), may be premature, as both the *Crenarchaeota* and *Nanoarchaeota* are poorly represented in the current collection of genome sequences. In addition, it may be noteworthy that the four crenarchaeal and sole nanoarchaeal exemplars are all thermophiles, as the possession of two-component systems within the *Euryarchaeota* displays a rough correlation with growth temperature (Table 5). Specifically, nine of the ten mesophiles and the sole psychrophile possess multiple deduced two-component systems. By

Table 5. Distribution of ORFs encoding potential two-component signal transduction domains within genomes of *Archaea*

Archaeon	Phylum	T[a]	HK[b] (reference)	RR (reference)	Hpt
Aeropyrum pernix	C	95	None (2, 3, 5)	None (3–5)	None (3)
Archaeoglobus fulgidus	E	83	14 (1, 3, 5), 15 (2), 23 (6), 21 (7)	11 (1, 3, 4, 7)	1 (3), none (6)
Ferroplasma acidarmanus	E	40	None (7)	None (7)	n.d.
Halobacterium NRC-1	E	37	10 (6), 12 (7), 14 (3, 8)	4 (4), 6 (8), 7 (7)	1 (3), none (6)
Halobacterium salinarium	E	37	13 (5)	≥2 (R & O 1996)	n.d.
Haloarcula marismortui	E	37	59 (5), 61 (7)	43 (7)	n.d.
Methanothermobacter thermautotrophicus	E	65	16 (1, 5), 15 (2, 7)	8 (3), 9 (4), 10 (1), 11 (7)	None (3)
Methanococcoides burtonii	E	23	35 (7)	15 (7)	n.d.
Methanococcus jannaschii	E	85	None (1, 2, 3, 5, 7)	None (1, 3, 4, 7)	None (3)
Methanococcus maripaludis	E	37	3 (5, 7)	3 (7)	n.d.
Methanosarcina acetivorans	E	37	64 (6), 52 (7)	19 (7), 20 (4)	1 (6)
Methanosarcina barkeri	E	37	26 (7)	13 (4, 7)	n.d.
Methanosarcina mazei	E	37	28 (6), 31 (7), 33 (5)	17 (4, 7)	None (6)
Methanopyrus kandleri	E	110	None (7)	None (4, 7)	n.d.
Methanospirillum hungatei	E	37	39 (7)	78 (7)	n.d.
Nanoarchaeum equitans	N	90	None (5, 7)	None (7)	n.d.
Natronomonas pharaonis	E	20	33 (7)	20 (7)	n.d.
Picrophilus torridus	E	50	None (5)		n.d.
Pyrobaculum aerophilum	C	100	None (5)	None (4)	n.d.
Pyrococcus abyssi	E	96	1 (3, 5, 7)	1 (3), 2 (4, 7)	None (3)
Pyrococcus furiosus	E	96	None (94)	None (4)	n.d.
Pyrococcus horikoshii	E	98	1 (1, 2, 3, 5, 7)	1 (1, 3), 2 (4, 7)	1 (3)
Sulfolobus solfataricus	C	80	None (5)	None (4)	n.d.
Sulfolobus tokodaii	C	80	None (5)	None (4)	n.d.
Thermococcus kodakaraensis	E	102	1 (7)	2 (7)	n.d.
Thermoplasma acidophilum	E	59	None (1, 3, 5, 7)	None (1, 3, 4, 7)	None (3)
Thermoplasma volcanium	E	60	None (5, 7)	None (4, 7)	n.d.

[a]Temperature optimum in degrees Celsius as reported in reference 19 with the exception of *M. burtonii*, which is 23°C (100).

[b]Abbreviations used: C, *Crenarchaeota*; E, *Euryarchaeota*; N, *Nanoarchaeota*; HK, histidine kinase domain; RR, response regulator domain; Hpt, histidine phosphotransfer domain; n.d., not determined.

contrast, only five of the eleven thermophilic members of the *Euryarchaeota* possess deduced two-component domains. Thus, two-component systems may yet be found among undiscovered mesophilic members of the *Crenarchaeota* and *Nanoarchaeota*.

Only a handful of deduced Hpt domains have been identified in the *Archaea* (Table 5). In every instance, they are found in *Euryarchaeota* that possess deduced histidine kinase and response regulator domains.

Physiological Roles of Archaeal Two-Component Systems

One prominent function of archaeal two-component regulatory systems is to transmit the signals that guide chemo-, photo-, and/or aerotactic behavior (see Chapter 18). Homologs of the proteins that comprise the core of the bacterial signal transduction cascades responsible for guiding chemotaxis are encoded in the genomes of thirteen members of the *Euryarchaeota* (19, 95). These core units include MCPs, the CheA

histidine kinase, and two response regulators: CheY, which modulates flagellar motor function, and CheB, whose methylesterase domain modulates the sensitivity of the MCPs (20, 289). The predicted functional properties of each of these archaeal homologs have been verified using the genetically malleable halo-archaeon *H. salinarium* (see Chapter 21) as a model (249–251, 287).

Probable output domains have been identified in a handful of other deduced archaeal response regulators. These include DNA-binding domains suggestive of transcriptional regulators in *M. burtonii* (100) and *H. marismortui* (19), and glycosyltransferase domains in *M. thermautotrophicus* (19). However, the vast majority of archaeal response regulator domains fall into the "one-domain" category; i.e., they lack a large, fused output domain (18, 19). Such one-domain regulators are presumed to act on cellular targets via protein-protein interactions, as is the case with the prototypic one-domain regulator CheY, or to serve as intermediates that shuttle phosphoryl groups in an extended two-component cascade (19).

Dephosphorylation of Response Regulator Domains

While no archaeal protein has been directly demonstrated to exhibit protein-histidine or protein-aspartate phosphatase activity, archaeal genomes encode several polypeptides analogous to those implicated as two-component phosphatases in the *Bacteria*. These include the phosphohistidine phosphatase SixA (207) and the chemotaxis-specific phosphoaspartate phosphatase CheC (288). Since SixA has been found only in a few crenarchaea, which themselves are devoid of two-component systems, the archaeal version of this enzyme presumably plays a role in metabolism rather than in sensor-response processes. On the other hand, open reading frames encoding deduced CheC phosphatases have been identified in many of the same euryarchaea in which two-component systems reside (e.g., *A. fulgidus, Halobacterium* NRC-1, *H. salinarium, M. acetivorans, M. mazei, M. maripaludis, N. pharaonis, P. abyssi, P. horikoshii,* and *T. kodakaraensis* [146]), but not in other euryarchaea or the *Nanoarchaeota* and *Crenarchaeota*. This correlation strongly implicates CheC as a prime catalyst for the dephosphorylation of those response regulator domains that participate in aero-, chemo-, and phototactic sensor-response processes. No archaeal homologs of the regulator aspartyl-phosphate, RAP, phosphatases (238) have been identified.

For those two-component systems not involved in mediating archaeal taxis, the most likely sources of protein phosphatase activity are the histidine kinase and response regulator domains themselves. Many bacterial response regulators possess intrinsic autophosphatase activity that often is enhanced through their association with other proteins (284). Notably, one of the best known of these enhancers, the CheZ "phosphatase" (225) of the β- and γ-proteobacteria (326), is absent from the *Archaea* despite the pervasiveness of other Che proteins (156). Several histidine kinases also have been revealed to be bifunctional; i.e., they can hydrolyze the phosphoaspartyl moieties of their cognate-response regulators (284). Deletion of the autophosphorylated histidine residue of the bacterial EnvZ histidine kinase abrogates phosphotransfer but not protein-aspartate phosphatase activity, indicating that dephosphorylation takes place via a direct hydrolytic mechanism rather than via a "reverse kinase" mechanism (119, 270).

OTHER COVALENT MODIFICATIONS

Archaeal proteins are the targets for a wide range of posttranslational modifications (74). Several of these appear to be structural in nature (e.g., glycosylation, fatty acylation, isoprenylation, and prolyl *cis-trans* isomerization) and are not discussed in this chapter. In addition, certain archaeal proteins contain the atypical amino acids selenocysteine (132) or pyrrolysine (108, 273) (see Chapter 9). Selenocysteine is commonly found in archaeal redox enzymes (317), while pyrrolysine is restricted to methylamine methyltransferases, where it is speculated to serve as a catalytic electrophile (273). While the scarcity of these amino acids imbues them with a functional resemblance to the modified amino acids produced via posttranslational modification, both selenocysteine (317) and pyrrolysine (157) are incorporated into the nascent polypeptide chain during translation from charged tRNA precursors. They are thus frequently referred to as the 21st and 22nd genetically encoded amino acids, respectively.

The posttranslational modifications discussed below were highlighted because they have been demonstrated, at least in some instances, to modulate the function of one or more target proteins from the *Archaea* or other organisms.

Acetylation

Two forms of protein acetylation have been reported in the *Archaea*. The first is acetylation of the amino terminus of ribosomal protein HmaS7 from the haloarchaeon *H. marismortui*, which presumably occurs as a structural processing event (150). The second is the acetylation of the side-chain amino group of Alba (26), a broad-specificity DNA-binding protein found in thermophilic archaea and some eucarya (316). Alba and its relatives are thought to perform roles analogous to histones in archaeal chromatin (see Chapters 3 and 4). Acetylation of Alba decreases its affinity for DNA ~40-fold, blunting its ability to repress transcription in a reconstituted system (26).

Two distinct mechanisms have been proposed to account for the effects of lysine acetylation upon Alba's affinity for DNA. Using the three-dimensional structure of Alba from *S. solfataricus*, models of Alba-DNA complexes were constructed and showed that the modified lysine residue, Lys-16, resided in an area of direct protein-DNA contact (312). Acetylation would be expected to directly interfere with protein-DNA binding by introducing steric hindrance and eliminating the potential for protonation, which would introduce a positive charge. Zhao et al. (325), on the other hand, observed that mutation of the acetyl-acceptor lysine in Alba from *A. fulgidus* disrupts homooligomerization, a prerequisite for high-affinity binding to DNA.

The modulation of DNA-protein interaction by the acetylation of lysine residues is reminiscent of the situation in the *Eucarya*, where acetylation is one of a cadre of reversible covalent modifications that modulate chromatin structure by targeting broad-specificity DNA-binding proteins, such as histones (58, 104). This parallel was further illustrated by the observation that the enzyme responsible for catalyzing the deacetylation of acetyl-Alba is a homolog of the eucaryal histone deacetylase, Sir2 (26). However, several archaeal genomes that encode Alba lack a recognizable Sir2 deacetylase, suggesting that the "histone code" is not employed by every archaeon (88, 316). Moreover, the recently identified Alba acetyltransferase, Pat, is a homolog of a bacterial enzyme and not a eucaryal histone acetylase (200). Thus, despite employing some common components, the archaeal and eucaryal chromatin acetylation-deacetylation systems may have developed independently (200).

Deamidation

In certain MCPs, sites of regulatory methylation (see "Methylation," below) are created posttranslationally by the enzymatic deamidation of glutamine side chains (141). In the bacterium *Bacillus subtilis*, the source of this glutamine deamidase activity has been traced to the product of the gene *cheD* (141). Homologs of CheD have been identified in the genome sequences of several archaea, including *A. fulgidus*, *H. salinarium*, *P. abyssi*, and *P. horikoshii* (146).

Diphthamidation

In common with their eucaryal homologs, a conserved histidine in elongation factor 2 (EF-2) is covalently modified to form 2-(3-carboxyamide-3-(trimethylammonio)propyl) histidine, commonly referred to as diphthamide (223). This covalent modification appears to be universal among the *Archaea*, as all versions of EF-2 tested to date are sensitive to the ADP-ribosylating activity of the diphtheria toxin, which specifically targets the conserved diphthamide residue (139). The physiologic role of the diphthamide residue in archaeal EF-2 is not known. An unmodified protein was functionally competent when assayed in vitro (66), and archaeal EF-2 does not appear to be ADP ribosylated inside the cell.

Disulfide Formation

Reversible changes in protein structure and function induced by the formation and reduction of disulfide bonds provide a logical means for exerting redox-mediated control of cellular processes. Behavior suggestive of such a control mechanism has been reported for four archaeal proteins: coenzyme F_{390} hydrolase from *M. thermautotrophicus* (307), A_1 ATPase from *M. mazei* (177), inositol monophosphatase/fructose bisphosphatase from *A. fulgidus* (282), and ferredoxin from *P. furiosus* (266). In each of the first three proteins, reduction of the critical disulfide bond markedly enhances its catalytic activity. However, no functional role for the reducible disulfide bond in the archaeal ferredoxin has been elucidated. The activity of coenzyme F_{390} synthetase, whose activity opposes that of the coenzyme F_{390} hydrolase, is also subject to redox regulation. However, in the case of the synthetase, regulation is mediated through changes in the redox state of one of its substrates, coenzyme F_{420}, and not via modification of the protein catalyst. Specifically, the reduced form of coenzyme F_{420}, 1,5-dihydro- of coenzyme F_{420}, is a potent competitive inhibitor of the synthetase (306). Thus, under reducing conditions where coenzyme F_{390} hydrolase is activated, the accumulation of the reduced coenzyme F_{420} results in the concomitant inhibition of the hydrolase (Fig. 5).

Hypusination

In both the *Archaea* and *Eucarya*, a conserved lysine residue in initiation factor-5A (IF-5A) undergoes a unique covalent modification to form N-ε-(4-aminobutyl-2-hydroxy) lysine, commonly referred to as hypusine (24). Modification takes place in two steps that are catalyzed by the enzymes deoxyhypusine synthase and deoxyhypusine hydroxylase (41). Hypusination is essential in the *Archaea*, as inhibition of deoxyhypusine synthase leads to growth arrest in *H. halobium*, *Haloferax mediterranei*, *S. acidocaldarius*, and *S. solfataricus* (125). It has been speculated, based on an inspection of the three-dimensional structure of archaeal IF-5A, that hypusine directly participates in the binding of mRNA (324).

Methylation

A wide range of archaeal proteins is subject to covalent modification by methylation (Table 6). Prominent among them are MCPs associated with many of the receptors that mediate aero-, chemo-, and phototaxis via two-component signaling cascades. The methylation of γ-carboxylate group of glutamate in MCPs, forming the corresponding methyl ester, is a dynamic process modulated through the opposing catalytic actions of methyltransferases and methylesterases (37, 289). The activity of these enzymes is modulated by the MCP's cognate histidine kinase in such a manner that stimulation of the signal transduction cascade generally promotes methylation (43, 153, 183, 228, 260, 280). MCP methyla-

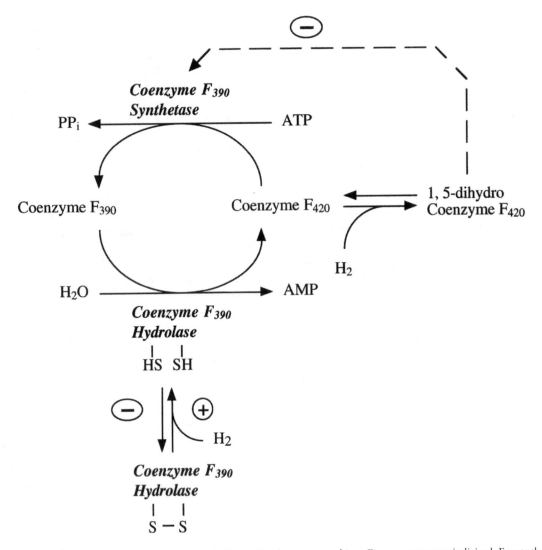

Figure 5. Redox regulation of coenzyme F_{390} metabolism in *M. thermautotrophicus*. Enzyme names are italicized. Events that stimulate (plus sign) or inhibit (minus sign) enzymatic activity; cysteine sulfhydryl groups (SH); cystine disulfides (S—S).

tion modulates the sensitivity of the receptor complex to stimuli, thereby enabling cells to recalibrate their sensor-response machinery as they progress along a gradient of attractant or repellant (2, 250, 287). While some MCPs contain a single site of methylation, others extend the dynamic range of this adaptive mechanism through the utilization of multiple methylation sites whose modification shifts sensitivity in a progressive manner (205).

Other archaeal methylation events target the ε-amino group of lysine (Table 6). Subjects of lysine methylation in the archaeon *S. solfataricus* include the DNA-binding proteins Sso7c (220) and Sso7d (25). In addition, a lysine methyltransferase with activity toward the putative chromatin protein MC1-α has recently been identified in *M. mazei* (198). Methylation of archaeal DNA-binding proteins may per-

form a similar role in modulating chromatin structure in the *Archaea* as does histone methylation in the *Eucarya*. However, only minor differences were reported for the thermodynamics of DNA binding by methylated and nonmethylated versions of Sso7d (192). Alternatively, as the Sso7d protein has a chromodomain-fold that contains a methyllysine recognition motif, methylation may facilitate homooligomerization (316).

It has been speculated that N-methylation of lysine may enhance the thermostability of proteins in hyperthermophilic archaea (199). This certainly appears to be the case for the β-glycosidase from *S. solfataricus* (85). On the other hand, despite the fact that the degree of methylation of lysine residues in Sso7d increases in concert with growth temperature (25), methylation does not noticeably improve the thermal

Table 6. Methylated proteins from the *Archaea*[a]

Protein	Archaeon	Amino acid(s)	Reference
MCPs			
BasT	*H. salinarium*	O-γ-methylglutamate	154
Car	*H. salinarium*	O-γ-methylglutamate	285
HtrI and HtrII	*H. salinarium*	O-γ-methylglutamate	228
HtrVII and HtrXI	*H. salinarium*	O-γ-methylglutamate	43
HtrVIII	*H salinarium*	O-γ-methylglutamate	42
MpcT/Htr14	*H. salinarium*	O-γ-methylglutamate	152
Other proteins			
50S ribosomal proteins HL3, HL, HL10, HL11, HL14	*H. cutirubrum*	ND[b]	6
Methyl-CoM reductase (α subunit)	*M. thermautotrophicus*	2-(S)-methylglutamine, 5-(S)-methylarginine, 1-N-methylhistidine, S-methylcysteine	264
Methyl-CoM reductase (α subunit)	*M. barkeri*	5-(S)-methylarginine, 1-N-methylhistidine, S-methylcysteine	76
P_2 ribonuclease	*S. solfataricus*	N-ε-methyllysine	92
P_3 ribonuclease	*S. solfataricus*	N-ε-methyllysine	93
Glutamate dehydrogenase	*S. solfataricus*	N-ε-methyllysine	199
β-Glycosidase	*S. solfataricus*	N-ε-methyllysine	85
Ferredoxin	*Sulfolobus* sp. strain 7	N-ε-methyllysine	310
Sso7c	*S. solfataricus*	N-ε-methyllysine	220
Sso7d	*S. solfataricus*	N-ε-methyllysine	25

[a]Included are proteins for which the nature of methyl-protein bond has been directly defined as well as proteins observed to incorporate radioactive methyl groups.
[b]ND, not determined.

stability of this protein in vitro (151). Similarly, methylation did not significantly alter either the thermal stability or catalytic properties of the P2 (92) or P3 (93) ribonucleases from *S. solfataricus*.

The methyl-coenzyme M reductases from *M. barkeri* and *M. thermautotrophicus* contain a multiplicity of methylated amino acids, including 1-*N*-methylhistidine, *S*-methylcysteine, 5-(*S*)-methylarginine, and 2-(*S*)-methylglutamine (101). These methylated amino acids, along with another unusual modified amino acid, thioglycine, cluster around the enzyme's active site. Their proximity to the site of catalysis suggests that these modifications play an important structural and/or functional role(s) in the enzyme. 2-(*S*)-Methylglutamine is unusual as the addition of the methyl group involves the formation of a C—C bond, rather than the C—O, C—N, and C—S bonds encountered in other methylated amino acids.

Another form of protein methylation encountered in the *Archaea* is associated with the repair of damaged proteins. The aberrant amino acid β- or isoaspartate is formed when the side chains of aspartate and asparagine undergo nucleophilic attack by the amido nitrogen of the adjacent peptide bond. Hydrolysis of the resulting cyclic succinimide can take place with release of either the side-chain carboxylate,

yielding L-aspartate, or the α-carboxylate, yielding isoaspartate. Methylation of the α-carboxylate by a repair methyltransferase and *S*-adenosylmethionine provides a thermodynamically accessible route for reforming the cyclic succinimide and, subsequently, L-aspartate. By contrast to the repair methyltransferases from the *Eucarya* and *Bacteria*, the isoaspartate methyltransferase from the hyperthermophilic archaeon *P. furiosus* targets D-aspartate as well as isoaspartate in peptide substrates (105, 294). Formation, and hence repair, of the latter also proceeds via formation of the same cyclic succinimide intermediate as isoaspartate.

Poly(ADP-ribosyl)ation

In *Eucarya*, poly(ADP-ribose) polymerases use the dinucleotide NAD$^+$ as a precursor for the synthesis of protein-bound polymers of ADP-ribose that can reach 200 units in length. This posttranslational modification predominantly targets nuclear proteins such as histones, as well as the poly(ADP-ribose) polymerase itself (60, 67). Poly(ADP-ribose) polymerase activity has been detected in the hyperthermophilic archaeon *S. solfataricus* (82). The protein substrates of this activity include the poly(ADP-ribose) polymerase itself (60) as well as a small basic DNA-binding protein (82). In common with its eucaryal counterparts, the poly(ADP-ribose) polymerase from *S. solfataricus* interacts with DNA (83). Single-stranded DNA moderately stimulates the activity of the archaeal enzyme, while double stranded DNA had little or no effect (84).

It is widely presumed that (poly)ADP-ribosylation regulates the functional properties of proteins, as is the case with other covalent modifications such as protein phosphorylation-dephosphorylation (60, 67). However, many fundamental questions concerning the regulation of proteins by poly(ADP-ribosyl)ation remain unanswered, such as whether this modification is reversible in vivo. It is thus of considerable interest that the genome of *A. fulgidus* encodes a protein, AF1521, containing a MACRO domain (4, 135). Computational analysis indicates that MACRO domains form the catalytic core of ADP-ribose phosphoesterases (13). Consistent with this prediction, AF1521 binds ADP-ribose with nanomolar affinity (4, 135) and displays detectable phosphohydrolase activity toward this nucleotide in vitro (135).

Regulated Proteolysis

Targeted, controlled proteolytic events can play important roles in determining when and where specific proteins manifest their functional properties (73). Examination of archaeal genomes provides intriguing, albeit circumstantial, evidence for regulated proteolysis in the *Archaea*. For example, ORFs encoding deduced transmembrane proteases of the rhomboid family have been identified in several archaea (155). In addition to receptor-like proteases, some archaea encode putative metalloproteases containing CBS domains, which have been suggested to act as potential allosteric binding sites for small molecules (9). Another potential avenue for controlling protease activity is provided by archaeal serpins (122, 242), members of a family of highly target-specific inhibitors of serine- and cysteine-proteases.

Several examples of archaeal zymogens have been reported. Not surprisingly, many of these involve proteolytic enzymes such as pyrolysin from *P. furiosus* (309), a chymotrypsinogen B-like enzyme from *N. pharaonis* (281), and the β-subunits of the archaeal 20S proteasome (328). Activation of the latter is deferred until assembly of the proteasome (see Chapter 10), at which time a *trans*-autoproteolytic event unmasks latent catalytic activity (110).

Zymogen activation has been observed for some nonproteases as well. Two pyruvoyl-dependent decarboxylases from *M. jannaschii*, S-adenosylmethionine decarboxylase (143) and arginine decarboxylase

(297), autocatalytically cleave larger precursor proteins into separate alpha and beta subunits. The cleavage event generates the pyruvoyl moiety from the N-terminal serine of the alpha chain, which forms a Schiff's base with substrates during catalysis.

Ubiquitination

In the *Eucarya*, modification by the covalent attachment of the small, 76-residue protein, ubiquitin, serves a variety of regulatory functions; the best known function is tagging proteins destined for degradation by the 26S proteasome (314). In 1993, a protein from the archaeon *T. acidophilum* was isolated whose N-terminal sequence resembled that of ubiquitin (318). More recently, ORFs encoding ubiquitin-like proteins have been detected in the genomes of *A. fulgidus*, *Aquifex aeolicus*, *M. thermautotrophicus*, and *T. volcanium* (29). However, while proteins reactive with antibodies against ubiquitin have been detected in the haloarchaeon *Natronococcus occultus* (217), it is unclear how this modification is effected in the *Archaea* as no homologs of the eucaryal ubiquitin-conjugating system have been reported. It has been postulated that the nascent polypeptide-associated complex from *Methanothermobacter marburgensis*, which contains an ubiquitin-associated domain, may play some role in protein ubiquitination in the *Archaea* (276).

PERSPECTIVE: THE NEXT FIVE YEARS

Are the *Archaea* "Introspective"?

The number of recognizable transmembrane receptors in most archaea appears to be exceedingly small (95), in particular, if those receptors associated with chemotactic and phototactic sensor-response cascades are discounted. Moreover, the genomes of nonchemotactic *Archaea* encode a limited number and range of deduced signal transmission proteins of the types normally associated with transmembrane signal transduction, one or perhaps two adenylate cyclases, and a handful of protein-serine/threonine/tyrosine kinases and phosphatases (95, 138). In contrast, the number and variety of deduced cytoplasmic sensors appear to be much greater (95). In addition to a fairly wide range of allosteric domains that monitor the levels of key metabolites, the *Archaea* contain a significant number of proteins possessing cytoplasmic receptor-like domains such as PAS domains, globins, and CBS domains. The balance between extra- and intracellular sensors thus appears, at first glance, to be heavily biased toward internal sensors. This implies that many archaea are, as Galperin (95) has coined, "introspective," i.e., they respond to changes in their surroundings if and when they impact the status of some key intracellular molecules.

Does "Introspective" Mean Simple?

While internal sensors play an important role in cellular regulation, reliance on derivative, internal cues as the means of viewing and responding to changes in the surrounding environment would appear to entail several significant drawbacks. These include a limited ability to discriminate between different environmental variables and a significant time lag between the onset of an environmental change and its detection. The former imposes a corresponding limit on the range of available responses. Thus, the impact of a predominantly reactive strategy on an organism's prospects for survival, at first glance, seems clear.

So how and why have these "introspective" archaea managed to survive for eons? One possible explanation is that the rudimentary nature of the sensor-response machinery found in many archaea is, in fact, appropriately scaled to survival in a specialized environmental niche. A basic and (relatively) sluggish suite of receptors might be sufficient to monitor and respond to environments characterized by changes that are limited in number and rate. It therefore may be noteworthy that archaea dwelling in less extreme and more cosmopolitan environments (e.g., the halo-archaea and certain methanogens) have acquired (or retained?) extensive suites of sensor-response machinery, presumably to cope with the demands of their more dynamic and competitive habitats.

However, the bias toward internal sensing may reflect an adaptive response to the harsh nature of the surrounding environment. Extremes of heat, pH, and the combination thereof severely stress the chemical and conformational stability of proteins. Whenever practicable under these circumstances, sheltering sensors inside the protection of the cell membrane may afford significant advantages. The moderate and carefully buffered pH of the cytoplasm minimizes the rate of deleterious hydrolytic reactions while providing unfettered access to the cell's internal protein maintenance and repair services. Species that readily diffuse through the membrane, such as oxygen or carbon dioxide, are particularly well suited for monitoring by internal receptors, a fact reflected in the large number of internal PAS and globin domains. Other nutrients that are routinely transported into the cytoplasm for assimilation can also be monitored internally at little additional cost.

However, it must be borne in mind that the ability to gauge the degree to which the *Archaea* are "introspective" is predicated on the ability to identify internal and external sensors from genome sequence information. A consistent leitmotif of genomic analyses has been the recognition that the structure and function of a large proportion of the proteins encoded in living organisms remain unidentified. Presumably these include numerous archaeal sensors. The unconventional nature of many archaeal habitats and the unique features of their metabolism certainly would be expected to stimulate the development of novel, and hence difficult to recognize, receptors. Moreover, the *Archaea* possess ample raw material for this purpose, as one in five archaeal proteins features a deduced transmembrane domain (12, 311). Lipid-anchored secreted proteins, many of which exhibit potential ligand binding domains, provide an additional reservoir of potential external sensors (31). While future discoveries of new receptors may show the *Archaea* to be more outward looking than currently estimated, the perception that they are generally introspective is likely to be enduring.

Insights into the Evolution of Protein Regulation

It appears likely that allosteric feedback regulation represents nature's most ancient mechanism for modulating protein function. Its elegant simplicity and apparent economy (by utilizing preexisting metabolites and protein domains) constitute compelling arguments for a seminal role in the pantheon of macromolecular regulation. Existing metabolites serve as indicators of status, while allosteric "receptor" domains are fabricated via the duplication or fusion of extant genes. The discovery of riboswitches suggests that allosteric modulation of macromolecular function may predate the emergence of proteins (308).

When did other regulatory mechanisms appear? What did the sensor-response network of early archaea look like? Galperin (95) observes that the only signal transmission proteins that approach universal distribution among the *Archaea* are adenylate cyclase and members of the extended ePK family. It would therefore appear that cAMP represents a universal and extremely ancient second messenger (95). Protein-serine/threonine/tyrosine phosphorylation also appears to predate the divergence of the three domains of life (138). Phylogenetic analyses suggest that the ePKs (233), PPP-family protein phosphatases (27, 136), and, potentially, the cPTPs (5) and LMW PTPs (27) trace their lineage back to the last universal common ancestor.

The fact that protein phosphorylation (and dephosphorylation) requires the development of a separate and specialized enzyme species, the protein kinase (and phosphatase), would appear at first glance to argue that regulation by diffusible allosteric ligands must predate this covalent modification by a considerable margin. By employing materials already on hand, allosterism would appear to impose minimal overhead costs. However, two points should be considered in determining when these resultant mechanisms originated and evolved. First, the RNA-world model (51) indicates that phosphoester and phosphoramide chemistry predates the emergence of proteins as the primary source of catalytic activity in cells. Thus, it is likely that phosphotransferases and phosphoesterases were among the first protein catalysts to emerge. Second, it is relatively easy to propagate this mechanism once the specialized catalyst is at hand (321).

If one accepts the premise that the phosphorylation and dephosphorylation of proteins is an ancient process, two questions emerge. First, given that virtually every archaeon possesses one or more ePKs, why is there so little consistency in the distribution of counterpart protein phosphatases among the Archaea? Second, why do so few archaea possess prototypic ePKs, as the preponderance of the prototypic form of ePK in the Eucarya indicates that it is superior to the RIO and piD231 ePKs that dominate in the Archaea?

Do Unique Sensor-Response Mechanisms Remain To Be Discovered in the Archaea?

As the sensor-response machinery within the Archaea is delved into in more depth, novel variations on established themes and completely unique mechanisms will undoubtedly be discovered. Many "orphan" ligand-binding domains will be united with their signal transmission partners, and vice versa. Of particular interest will be the sensor-response mechanisms utilized by the Archaea for cell-cell communication. Do the Archaea utilize secreted messengers (as occurs in bacterial quorum sensing) to modulate growth and coordinate activities with their neighbors in biofilms and other microbial consortia? Do they engage in contact-mediated detection? The presence of deduced extracellular proteins containing homologs of domains found in some eucaryal cell surface proteins, e.g., polycystic kidney disease, β-propeller, and β-helix domains (130, 233), is suggestive of a role for contact-mediated communication.

One potential explanation for the apparent imbalance between the number of protein-serine/threonine/tyrosine kinases and protein-serine/threonine/tyrosine phosphatases that is typical of most archaea is the existence of one or more unrecognized families of protein-serine/threonine/tyrosine phosphatases in the Archaea. Likewise, the paucity of stereotypical second messengers in the Archaea suggests the pres-

ence of novel species as well. Recently, both a novel family of protein-tyrosine phosphatases (213) and a new second messenger, cyclic di-GMP (61), were discovered in the Bacteria. Given that the members of the bacterial domain have been the subject of decades of intensive study, it appears highly likely that the Archaea will not only be found to contain new sensor-response mechanisms and molecules, but that they will provide new insights into this vital process in other organisms as well.

Acknowledgments. I apologize for any and all errors or omissions committed during the preparation of this chapter. I gratefully acknowledge the continued support of the National Science Foundation, specifically Grant MCB-0315122, for my laboratory's research on the Archaea.

REFERENCES

1. **Adams, M. W. W., J. F. Holden, A. L. Menon, G. J. Schut, A. M. Grunden, C. Hu, A. M. Hutchins, F. E. Jenney, Jr., C. Kim, K. Ma, G. Pan, R. Roy, R. Sapra, S. V. Story, and M. F. J. M. Verhagen.** 2001. Key role for sulfur in peptide metabolism and in regulation of three hydrogenases in the hyperthermophilic archaeon Pyrococcus furiosus. J. Bacteriol. 183:716–724.

2. **Alam, M., M. Lebert, D. Oesterhelt, and G. L. Hazelbauer.** 1989. Methyl-accepting chemotaxis proteins in Halobacterium halobium. EMBO J. 8:631–639.

3. **Albers, S. V. and A. J. M. Driessen.** 2002. Signal peptides of secreted proteins of the archaeon Sulfolobus solfataricus: a genomic survey. Arch. Microbiol. 177:209–216.

4. **Allen, M. D., A. M. Buckle, S. C. Cordell, J. Lowe, and M. Bycroft.** 2003. The crystal structure of AF1521, a protein from Archaeoglobus fulgidus with homology to the non-histone domain of MacroH2A, J. Mol. Biol. 330:503–511.

5. **Alonso, A., S. Burkhalter, J. Sasin, L. Tautz, J. Bogets, H. Huynh, M. C.D. Bremer, L. J. Holsinger, A. Godzik, and T. Mustelin.** 2004. The minimal essential core of a cysteine-based protein-tyrosine phosphatase revealed by a novel 16-kDa VH1-like phosphatase, VHZ. J. Biol. Chem. 279:35768–35774.

6. **Amaro, A. M., and C. A. Jerez.** 1984. Methylation of ribosomal proteins in bacteria: Evidence of conserved modification of the eubacterial 50S subunit. J. Bacteriol. 158:84–93.

7. **Anantharaman, V., and L. Aravind.** 2000. Cache—a signaling domain common to animal Ca^{2+}-channel subunits and a class of prokaryotic chemotaxis receptors. Trends Biochem. Sci. 25:535–537.

8. **Anantharaman, V., and L. Aravind.** 2001. The CHASE domain: a predicted ligand-binding module in plant cytokinin receptors and other eukaryotic and bacterial receptors. Trends Biochem. Sci. 26:579–591.

9. **Anantharaman, V., E. V. Koonin, and L. Aravind.** 2001. Regulatory potential, phyletic distribution and evolution of ancient, intracellular small-molecule-binding domains. J. Mol. Biol. 307:1271–1292.

10. **Angermayr, M., A. Roidl, and W. Brandlow.** 2002. Yeast Rio1p is the founding member of a novel subfamily of protein serine kinases involved in control of cell cycle progression. Mol. Microbiol. 44:309–324.

11. **Appleby, J. L., J. S. Parkinson, and R. B. Bourret.** 1996. Signal transduction via the multi-step phosphorelay: not necessarily a road less traveled. Cell 86:845–848.

12. Arai, M., M. Ikeda, and T. Shimizu. 2003. Comprehensive analysis of transmembrane topologies in prokaryotic genomes. *Gene* 304:77–86.

13. Aravind, L. 2001. The WWE domain: a common interaction module in protein ubiquitination and ADP ribosylation. *Trends Biochem. Sci.* 26:273–275.

14. Aravind, L., V. Anantharaman, and L. M. Iyer. 2003. Evolutionary connections between bacterial and eukaryotic signaling systems: a genomic perspective. *Curr. Opin. Microbiol.* 6:490–497.

15. Aravind, L. and E. V. Koonin. 1999. DNA-binding proteins and evolution of transcription regulation in the *Archaea*. *Nucleic Acids Res.* 27:4658–4670.

16. Aravind, L., and C. P. Ponting. 1997. The GAF domain: An evolutionary link between diverse phototransducing proteins. *Trends Biochem. Sci.* 22:458–459.

17. Arent, S., P. Harris, K. F. Jensen, and S. Larsen. 2005. Allosteric regulation and communication between subunits in uracil phosphoribosyltransferase from *Sulfolobus solfataricus*. *Biochemistry* 44:883–892.

18. Ashby, M. K. 2004. Survey of the number of two-component response regulator genes in the complete and annotated genome sequences of prokaryotes. *FEMS Microbiol. Lett.* 231:277–281.

19. Ashby, M. K. 2006. Distribution, structure, and diversity of "bacterial" genes encoding two-component proteins in the Euryarchaeota. *Archaea* 2:11–30.

20. Baker, M. D., P. M. Wolanin, and J. B. Stock. 2005. Signal transduction in bacterial chemotaxis. *BioEssays* 28:9–22.

21. Baker-Austin, C., M. Dopson, M. Wexler, R. G. Sawers, and P. L. Bond. 2005. Molecular insight into extreme copper resistance in the extremophilic archaeon 'Ferroplasma acidarmanus' Fer1. *Microbiology* 151:2637–2646.

22. Baliga, N. S., S. P. Kennedy, W. V. Ng, L. Hood, and S. DasSarma. 2001. Genomic and genetic dissection of an archaeal regulon. *Proc. Natl. Acad. Sci. USA* 98:2521–2525.

23. Barford, D. 1996. Molecular mechanisms of the protein serine/threonine phosphatases. *Trends Biochem. Sci.* 21:407–412.

24. Bartig, D., K. Lemkemeier, J. Frank, F. Lottspeich, and F. Klink. 1992. The archaebacterial hypusine-containng protein. Strucutral features suggest common ancestry with eukaryotic translation initiation factor 5A. *Eur. J. Biochem.* 204:751–758.

25. Baumann, H., T. Lundback, R. Ladenstein, and T. Hard. 1994. Solution structure and DNA-binding properties of a thermostable protein from the archaeon *Sulfolobus solfataricus*. *Nat. Struct. Biol.* 1:808–819.

26. Bell, S. D., C. H. Botting, B. N. Wardelworth, S. P. Jackson, and M. F. White. 2002. The interaction of Alba, a conserved archaeal chromatin protein, with Sir2 and its regulation by acetylation. *Science* 296:148–151.

27. Bhaduri, A., and R. Sowdhamini. 2005. Genome-wide survey of prokaryotic O-phosphatases. *J. Mol. Biol.* 352:736–752.

28. Bidle, K. A. 2003. Differential expression of genes influenced by changing salinity using RNA arbitrarily primed PCR in the archaeal halophile *Haloferax volcanii*. *Extremophiles* 7:1–7.

29. Bienkowska, J. R., H. Hartman, and T. F. Smith. 2004. A search method for homologs of small proteins. Ubiquitin-like proteins in prokaryotic cells? *Protein Eng.* 16:897–904.

30. Bini, E. and P. Blum. 2001. Archaeal catabolite repression: a gene regulatory paradigm. *Adv. Appl. Microbiol.* 50:39–367.

31. Bolhuis, A. 2002. Protein transport in the halophilic archaeon *Halobacterium* sp. NRC-1: A major role for the twin-arginine translocation pathway. *Microbiology* 148:3335–3346.

32. Bonete, M. J., F. Perez-Pomares, J. Ferrer, and M. L. Camacho. 1996. NAD-glutamate dehydrogenase from *Halobacterium halobium*: Inhibition and activation by TCA intermediates and amino acids. *Biochim. Biophys. Acta* 1289:14–24.

33. Bonete, M. J., M. L. Camacho, and E. Cadenas. 1989. Kinetic mechanism of *Halobacterium halobium* NAD +-glutamate dehydrogenase. *Biochim. Biophys. Acta* 24:150–155.

34. Bonete, M. J., M. L. Camacho, and E. Cadenas. 1990. Analysis of the kinetic mechanism of halophilic NADP-dependent glutamate dehydrogenase. *Biochim. Biophys. Acta* 5:305–310.

35. Boonyaratanakornkit, B. B., A. J. Simpson, T. A. Whitehead, C. M. Fraser, N. M. A. El-Sayed, and D. S. Clark. 2005. Transcriptional profiling of the hyperthermophilic methanarchaeon *Methanococcus jannaschii* in response to lethal heat and non-lethal cold shock. *Environ. Microbiol.* 7:789–797.

36. Botsford, J. L, and J. G. Harman. 1992. Cyclic AMP in prokaryotes. *Microbiol. Rev.* 56:100–122.

37. Bourret, R. B., and A. M. Stock. 2002. Molecular information processing: Lessons from bacterial chemtaxis. *J. Biol. Chem.* 277:9625–9628.

38. Brabbban A. D., E. N. Orcutt, and S. H. Zinder. 1999. Interactions between nitrogen fixation and osmoregulation in the methanogenic archaeon *Methanosarcina barkeri* 227. *Appl. Environ. Microbiol.* 65:1222–1227.

39. Brasen, C., and P. Schonheit. 2004. Regulation of acetate and acetyl-CoA converting enzymes during growth on acetate and/or glucose in the halophilic archaeon *Haloarcula marismortui*. *FEMS Microbiol. Lett.* 241:21–26.

40. Brinkman, A. B., S. D. Bell, R. J. Lebbink, W. M. deVos, and J. van der Oost. 2002. The *Sulfolobus solfataricus* Lrp-like protein LysM regulates lysine biosynthesis in response to lysine availability. *J. Biol. Chem.* 277:29537–29549.

41. Brochier, C., P. Lopez-Garcia, and D. Moreira. 2004. Horizontal gene transfer and archaeal origin of deoxyhypusine synthase homologous genes in bacteria. *Gene* 330:169–176.

42. Brooun, A., J. Bell, T. Freitas, R. W. Larsen, and M. Alam. 1998. An archaeal aeotaxis transducer combines subunit I core structures of eukaryotic cytochrome *c* oxidase and eubacterial methyl-accepting chemotaxis proteins. *J. Bacteriol.* 180:1642–1646.

43. Brooun, A., W. Zhang, and M. Alam. 1997. Primary structure and functional analysis of the soluble transducer protein HtrXI in the archaeon *Halobacterium salinarium*. *J. Bacteriol.* 179:2963–2968.

44. Brunner, N. A., H. Brinkmann, B. Siebers, and R. Hensel. 1998. NAD$^+$-dependent glyceraldehyde-3-phosphate dehydrogenase from *Thermoproteus tenax*. *J. Biol. Chem.* 273:6149–6156.

45. Brunner, N. A., B. Siebers, and R. Hensel. 2001. Role of two different glyceraldehydes-3-phosphate dehydrogenases in controlling the reversible Embden-Meyerhof-Parnas pathway in *Thermoproteus tenax*: Regulation on protein and transcript level. *Extremophiles* 5:101–109.

46. Buss, J. E., and J. T. Stull. 1983. Measurement of chemical phosphate in proteins. *Methods Enzymol.* 99:7–14.

47. Byrnes, W. M., and V. L. Vilker. 2004. Extrinsic factors potassium chloride and glycerol induce thermostability in recombinant anthranilate synthase from *Archaeoglobus fulgidus*. *Extremophiles* 8:455–462.

48. Cabello, P., M. D. Roldan, and C. Moreno-Vivian. 2004. Nitrate reduction and the nitrogen cycle in archaea. *Microbiology* 150:3527–3546.

49. Cai, X., and J. Lytton. 2004. The cation/Ca^{2+} exchanger superfamily: phylogenetic analysis and structural implications. *Mol. Biol. Evol.* 21:1692–1703.

50. Cardona, S., F. Remonsellez, N. Guiliani, and C. A. Jerez. 2001. The glycogen-bound polyphosphate kinase from *Sulfolobus acidocaldarius* is actually a glycogen synthase. *Appl. Environ. Microbiol.* 67:4773–4780.

51. Cech, T. R. 1993. The efficiency and versatility of catalytic RNA: Implications for an RNA world. *Gene* **135**:33–36.

52. Cho, H. S., J. G. Pelton, D. Yan, S. Kustu, and D. E. Wemmer. 2001. Phosphoaspartates in bacterial signal transduction. *Curr. Opin. Struct. Biol.* **11**:679–684.

53. Ciulla, R. A., S. Burggraf, K. O. Stetter, and M. F. Roberts. 1994. Occurrence and role of di-myo-inositol-1,1 (prm1)-phosphate in *Methanococcus igneus*. *Appl. Environ. Microbiol.* **60**:3660–3664.

54. Clemens, M. J. 2004. Targets and mechanisms for the regulation of translation in malignant transformation. *Oncogene* **23**:3180–3188.

55. Cohen, P. T. W. 2002. Protein phosphatase 1—targeted in many directions. *J. Cell Sci.* **115**:241–256.

56. Cohen-Kupiec, R., C. Blank, and J. A. Leigh. 1997. Transcriptional regulation in *Archaea: In vivo* demonstration of a repressor binding site in a methanogen. *Proc. Natl. Acad. Sci. USA* **94**:1316–1320.

57. Condo, L., D. Ruggero, R. Reinhardt, and P. Londel. 1998. A novel amino-peptidase associated with the 60 kDa chaperonin in the thermophilic archaeon *Sulfolobus solfataricus*. *Mol. Microbiol.* **29**:775–785.

58. Cosgrove, M. S., and C. Wolberger. 2005. How does the histone code work? *Biochem. Cell Biol.* **83**:468–476.

59. Daas, P. J. H., R. W. Wassenaar, P. Willemsen, R. J. Theunissen, J. T. Keltjens, C. van der Drift, and G. D. Vogels. 1996. Purification and properties of an enzyme involved in the ATP-dependent activation of the methanol:2-mercaptoethanesulfonic acid methyltransferase reaction in *Methanosarcina barkeri*. *J. Biol. Chem.* **271**:22339–22345.

60. D'Amours, D., S. Desnoyers, I. D'Silva, and G. G. Poirier. 1999. Poly(ADP-ribosyl)ation reactions in the regulation of nuclear functions. *Biochem. J.* **342**:249-268.

61. D'Argenio, D. A., and S. I. Miller. 2004. Cyclic di-GMP as a bacterial second messenger. *Microbiology* **150**:2497–2502.

62. DasSarma, S., S. P. Kennedy, B. Berquist, W. V. Ng, N. S. Baliga, J. L. Spudich, M. P. Krebs, J. A. Eisen, C. H. Johnson, and L. Hood. 2001. Genomic perspective on the photobiology of *Halobacterium* species NRC-1, a phototrophic, phototactic, and UV-tolerant haloarchaeon. *Photosynth. Res.* **70**:3–17.

63. De Biase, A., A. J. L. Macario, and E. C. De Macario. 2002. Effect of heat stress on promoter binding by transcription factors in the cytosol of the archaeon *Methanosarcina mazeii*. *Gene* **282**:189–197.

64. De Felice, M., L. Esposito, B. Pucci, F. Carpentieri, M. De Falco, M. Rossi, and F. M. Pisani. 2003. Biochemical characterization of a CDC6-like protein from the crenarchaeon *Sulfolobus solfataricus*. *J. Biol. Chem.* **278**:46424–46431.

65. De Hertogh, B., A. C. Lantin, P. V. Baret, and A. Goffeau. 2004. The archaeal P-Type ATPases. *J. Bioenerg. Biomemb.* **36**:135–142.

66. De Vendittis, E., M. R. Amatruda, G. Raimo, and V. Bocchini. 1997. Heterologous expression in *Escherichia coli* of the gene encoding an archaeal thermoacidophilic elongation factor 2. Properties of the recombinant protein. *Biochimie* **79**:303–308.

67. Diefenbach, J., and A. Burkle. 2005. Introduction to poly (ADP-ribose) metabolism. *Cell. Mol. Life Sci.* **62**:721–730.

68. DiRuggiero, J., D. Dunn, D. L. Maeder, R. Holley-Shanks, J. Chatard, R. Horlacher, R. T. Robb, W. Boos, and R. B. Weiss. 2000. Evidence of recent lateral gene transfer among hyperthermophilic *Archaea*. *Mol. Microbiol.* **38**:684–693.

69. Dodsworth, J. A., N. C. Cady, and J. A. Leigh. 2005. 2-Oxoglutarate and the PII homologs $Nifl_1$ and $Nifl_2$ regulate nitrogenase activity in cell extracts of *Methanaococcus maripaludis*, *Mol. Microbiol.* **56**:1527–1538.

70. Dominguez, D. C. 2004. Calcium signaling in bacteria. *Mol. Microbiol.* **54**:291–297.

71. Durbecq, V., T. L. Thia-Toong, D. Charlier, V. Villeret, M. Roovers, R. Wattiez, C. Legrain, and N. Glansdorff. 1999. Aspartate carbamoyltransferase from the thermoacidophilic archaeon *Sulfolobus acidocaldarius*. *Eur. J. Biochem.* **264**:233–241.

72. Ehlers, C., K. Weidenbach, K. Veit, K. Forchhammer, and R. A. Schmitz. 2005. Unique mechanistic features of posttranslational regulation of glutamine synthetase activity in *Methanosarcina mazei* strain Gö1 in response to nitrogen availability. *Mol. Microbiol.* **55**:1841–1854.

73. Ehrmann, M., and T. Clausen. 2004. Proteolysis as a regulatory mechanism. *Annu. Rev. Genet.* **38**:709–724.

74. Eichler, J., and M. W. W. Adams. 2005. Posttranslational protein modification in *Archaea*. *Microbiol. Mol. Biol. Rev.* **69**:393–425.

75. Eliasson, R., E. Pontis, A. Jordan, and P. Reichard. 1999. Allosteric control of three B_{12}-dependent (class II) ribonucleotide reductases. *J. Biol. Chem.* **274**:7182–7189.

76. Ermler, U., W. Grabarse, S. Shima, M. Goubeaud, and R. K. Thaure.1997. Crystal structure of methyl-coenzyme M reductase: The key enzyme of biological methane formation. *Science* **278**:1457–1462.

77. Ettema, T. J. G., A. B. Brinkman, T. H. Tani, J. B. Rafferty, and J. van der Oost. 2002. A novel ligand-binding domain involved in regulation of amino acid metabolism in prokaryotes. *J. Biol. Chem.* **277**:37464–37468.

78. Ettema, T. J. G., M. A. Hyunen, W. M. de Vos, and J. van der Oost. 2003. TRASH: A novel metal-binding domain predicted to be involved in heavy-metal sensing, trafficking, and resistance. *Trends Biochem. Sci.* **28**:170–173.

79. Ettema, T. J. G., K. S. Makarova, G. L. Jellema, H. J. Gierman, E. V. Koonin, M. A. Huynen, W. M. de Vos, and J. van der Oost. 2004. Identification and functional verification of archaeal-type phosphoenolpyruvate carboxylase, a missing link in archaeal central carbohydrate metabolism. *J. Bacteriol.* **186**:7754–7762.

80. Fabret, C., V. A. Feher, and J. A. Hoch. 1999. Two-component signal transduction in *Bacillus subtilis*: How one organism sees its world. *J. Bacteriol.* **181**:1975–1983.

81. Facchin, S., S. Sarno, O. Marin, R. Lopreiato, G. Sartori, and L. A. Pinna. 2002. Acidophilic nature of yeast PID261/BUD32, a putative ancestor of eukaryotic protein kinases. *Biochem. Biophys. Res. Commun.* **296**:1366–1371.

82. Faraone-Mennella, M. R., and B. Farina. 1995. In the thermophilic archaeon *Sulfolobus solfataricus* a DNA-binding protein is *in vitro* (ADPribosyl)ated. *Biochem. Biophys. Res. Commun.* **208**:55–62.

83. Faraone-Mennella, M. R., A. Gambacorta, B. Nicolaus, and B. Farina. 1998. Purification and biochemical characterization of a poly(ADP-ribose) polymerase-like enzyme from the thermophilic archaeon *Sulfolobus solfataricus*. *Biochem. J.* **335**:441–447.

84. Faraone-Mennella, M. R., G. Piccialli, P. De Luca, S. Castellano, A. Giordano, D. Rigano, L. De Napoli, and B. Farina. 2002. Interaction of the ADP-ribosylating enzyme from the hyperthermophilic archaeon *S. solfataricus* with DNA and ss-oligo deoxy ribonucleotides. *J. Cell Biochem.* **85**:146–157.

85. Febbraio, F., A. Andolfo, F. Tanfani, R. Briante, F. Gentile, S. Formisano, C. Vaccaro, A. Scire, E. Bertoli, P. Pucci, and R. Nucci. 2004. Thermal stability and aggregation of *Sulfolobus solfataricus* β-glycosidase are dependent upon the N-ε-methylation of specific lysyl residues. *J. Biol. Chem.* **279**:10185–10194.

86. Fiorentino, G., R. Cannio, M. Rossi, and S. Bartolucci. 2003. Transcriptional regulation of the gene encoding and alcohol

dehydrogenase in the archaeon *Sulfolobus solfataricus* involves multiple factors and control elements. *J. Bacteriol.* **185:**3926–3934.

87. Fischer, R. S., C. A. Bonner, D. R. Boone, and R. A. Jensen. 1993. Clues from a halophilic methanogen about aromatic amino-acid biosynthesis in Archaebacteria. *Arch. Microbiol.* **160:**440–446.

88. Forbes, A. J., S. M. Patrie, G. K. Taylor, Y. B. Kim, L. Jiang, and N. L. Kelleher. 2004. Targeted analysis and discovery of posttranslational modifications in proteins from methanogenic *Archaea* by top-down MS. *Proc. Natl. Acad. Sci. USA* **2:**2678–2683.

89. Freitas, T. A. K., S. Hou, E. M. Dioum, J. A. Saito, J. Newhouse, G. Gonzalez, M. A. Gilles-Gonzalez, and M. Alam. 2004. Ancestral hemoglobins in *Archaea*. *Proc. Natl. Acad. Sci. USA* **101:**6675–6690.

90. Fujiwara, S., A. Yamanaka, K. Hirooka, A. Kobayashi, T. Imanaka, and E. Fukusaki. 2004. Temperature-dependent modulation of farnesyl diphosphate/geranylgeranyl diphosphate synthase from hyperthermophilic *Archaea*. *Biochem. Biophys. Res. Commun.* **325:**1066–1074.

91. Fujitaki, J. M., and R. A. Smith. 1984. Techniques in the detection and characterization of phosphoramidate-containing proteins. *Methods Enzymol.* **107:**23–36.

92. Fusi, P., M. Grisa, E. Mombelli, R. Consonni, P. Tortora, and M. Vanoni. 1995. Expression of s synthetic gene encoding P2 ribonuclease from the extreme thermoacidophilic archaebacterium *Sulfolobus solfactaricus* in mesophylic host. *Gene* **154:**99–103.

93. Fusi, P., M. Grisa, G. Tedeschi, A. Megri, A. Guerritore, and P. Tortora. 1995. An 8.5-kDa ribonuclease from the extreme thermophilic archaebacterium *Sulfolobus solfataricus*. *FEBS Lett.* **360:**187–190.

94. Galperin, M. Y. 2004. Bacterial signal transduction network in a genomic prespective. *Environ. Microbiol.* **6:**552–567.

95. Galperin, M. Y. 2005. A census of membrane-bound and intracellular signal transduction proteins in bacteria: bacterial IQ, extroverts and introverts. *BMC Microbiol.* **5:**35.

96. Galperin, M. Y., A. N. Nikolskaya, and E. V. Koonin. 2001. Novel domains of the prokaryotic two-component signal transduction system. *FEMS Microbiol. Lett.* **203:**11–21.

97. Gast, D. A., U. Jenal, A. Wasserfallen, and T. Leisinger. 1994. Regulation of tryptophan biosynthesis in *Methanobacterium thermoautotrophicum* Marburg. *J. Bacteriol.* **176:**4590–4596.

98. Goldman, S., K. Hecht, H. Eisenberg, and M. Mevarech. 1990. Extracellular Ca^{2+}-dependent inducible alkaline phosphatase from the extremely halophilic archaebacterium *Haloarcula marismortui*. *J. Bacteriol.* **172:**7065–7070.

99. Goodchild, A., M. Raftery, N. F. W. Saunders, M. Guilhaus, and R. Cavicchioli. 2005. Cold adaptation of the Antarctic archaeon, *Methanococcus burtonii* assessed by proteomics using ICAT. *J. Proteome Res.* **4:**473–480.

100. Goodchild, A., N. F. W. Saunders, H. Ertan, M. Raftery, M. Guilhaus, P. M. G. Curmi, and R. Cavicchioli. 2004. A proteomic determination of cold adaptation in the Antarctic archaeon, *Methanococcus burtonii*. *Mol. Microbiol.* **53:**309–321.

101. Grabarse, W., F. Mahlert, S. Shima, R. K. Thauer, and U. Ermler. 2000. Comparison of three methyl-coenzyme M reductases from phylogenetically distant organisms: Unusual amino acid modification, conservation and adaptation. *J. Mol. Biol.* **303:**329–344.

102. Grabowski, B., and Z. Kelman. 2001. Autophosphorylation of archaeal Cdc6 homologs is regulated by DNA. *J. Bacteriol.* **183:**5459–5464.

103. Grebe, T. W., and J. B. Stock. 1999. The histidine protein kinase superfamily. *Adv. Microbial Physiol.* **41:**139–227.

104. Gregory, P. D., K. Wagner, and W. Horz. 2001. Histone acetylation and chromatin remodeling. *Exp. Cell Res.* **265:**195–202.

105. Griffith, S. C., M. R. Sawaya, D. R. Boutz, N. Thapar, J. E. Katz, C. Clarke, and T. O. Yeates. 2001. Crystal structure of protein repair methyltransferase from *Pyrococcus furiosus* with its ʟ-isoaspartyl peptide substrate. *J. Mol. Biol.* **313:**1103–1116.

106. Gropp, F., and M. C. Betlach. 1994. The *bat* gene of *Halobacterium halobium* encodes a trans-acting oxygen inducibility factor. *Proc. Natl. Acad. Sci. USA* **91:**5474–5479.

107. Hack, E. S., T. Vorobyova, J. B. Sakash, J. M. West, C. P. Macol, G. Herve, M. K. Williams, and E. R. Kantrowitz. 2000. Characterization of the aspartate transcarbamoylase from *Methanococcus jannaschii*. *J. Biol. Chem.* **275:**15820–15827.

108. Hao, B., W. Gong, T. K. Ferguson, C. M. James, J. A. Krzycki, and M. K. Chan. 2002. A new UAG-encoded residue in the structure of a methanogen methyltransferase. *Science* **296:**1462–1466.

109. Haseltine, C., R. Montalvo-Rodriguex, E. Bini, A. Carl, and P. Blum. 1999. Coordinate transcriptional control in the hyperthermophilic archaeon *Sulfolobus solfataricus*. *J. Bacteriol.* **181:**3920–3927.

110. Heinemeyer, W., M. Fischer, T. Krimmer, U. Stachon, and D. H. Wolf. 1997. The active sites of the eukaryotic 20 S proteasome and their involvement in subunit precursor processing. *J. Biol. Chem.* **272:**25200–25209.

111. Hensel, R., and H. Konig. 1988. Thermoadaption of methanogenic bacteria by intracellular ion concentration. *FEMS Microbiol. Lett.* **49:**75–79.

112. Hildebrand, E., and N. Dencher. 1975. Two photosystems controlling behavioural responses of *Halobacterium halobium*. *Nature* **257:**46–48.

113. Ho, Y. S. J., L. M. Burden, and J. H. Hurley. 2000. Structure of AGF domain, a ubiquitous signaling motif and a new class of cyclic GMP receptor. *EMBO J.* **19:**5288–5299.

114. Hoch, J. A. 2000. Two-component and phosphorelay signal transduction. *Curr. Opin. Microbiol.* **3:**165–170.

115. Hoff, W. D., K. H. Jung, and J. L. Spudich. 1997. Molecular mechanism of photosignaling by archaeal sensory rhodopsins. *Annu. Rev. Biophys. Biomol. Struct.* **26:**223–258.

116. Holden, J. F., and J. A. Baross. 1993. Enhanced thermotolerance and temperature-induced changes in protein composition in the hyperthermophilic archaeon ES4. *J. Bacteriol.* **175:**2839–2843.

117. Hou, S., A. Brooun, H. S. Yu, T. Freitas, and M. Alam. 1998. Sensory rhodopsin II transducer HrtII is also responsible for serine chemotaxis in the archaeon *Halobacterium salinarum*. *J. Bacteriol.* **180:**1600–1602.

118. Hou, S., R. W. Larsen, D. Boudko, C. W. Riley, E. Karatan, M. Zimmer, G. W. Ordal, and M. Alam. 2000. Myoglobin-like aerotaxis transducers in *Archaea* and *Bacteria*. *Nature* **403:**540–544.

119. Hsing, W., and T. J. Silhavy. 1997. Function of conserved histidine-243 in phosphatase activity of EnvZ, the sensor for porin osmoregulation in *Escherichia coli*. *J. Bacteriol.* **179:**3729–3735.

120. Huffman, J. L., H. Li, R. H. White, and J. A. Tainer. 2003. Structural basis for recognition and catalysis by the bifunctional dCTP deaminase and dUTPase from *Methanococcus jannaschii*. *J. Mol. Biol.* **331:**885–896.

121. Irvine, R. F. 2003. 20 years of Ins(1,4,5)P₃, and 40 years before. *Nat. Rev. Mol. Cell Biol.* **4:**586–590.

122. Irving, J. A., P. J. M. Steenbakkers, A. M. Lesk, H. J. M. Op den Camp, R. N. Pike, and J. C. Whisstock. 2002. Serpins in prokaryotes. *Mol. Biol. Evol.* **19:**1881–1890.

123. Izzo, V., E. Notomista, A. Picardi, F. Pennacchio, and A. Di Donato. 2005. The thermophilic archaeon *Sulfolobus sol-*

fataricus is able to grow on phenol. *Res. Microbiol.* **156:** 677–689.

124. Jager, A., R. Samorski, F. Pfeifer, and G. Klug. 2002. Individual *gvp* transcript segments in *Haloferax mediterranei* exhibit varying half-lives, which are differentially affected by salt concentration and growth phase. *Nucleic Acids Res.* **30:**5436–5443.

125. Jansson, B. P. M., L. Malandrin, and H. E. Johansson. 2000. Cell cycle arrest in *Archaea* by the hypusination inhibitor N¹-guanyl-1,7-diaminoheptane. *J. Bacteriol.* **182:**1158–1161.

126. Jensen, K. F., S. Arent, S. Larsen, and L. Schack. 2005. Allosteric properties of the GTP activated and CTP inhibited uracil phosphoibosyltransferase from the thermoacidophilic archaeon *Sulfolobus solfataricus*. *FEBS J.* **272:**1440–1453.

127. Jensen, R. A., T. A. d'Amato, and L. I. Hochstein. 1988. An extreme-halophilic archaebacterium possesses the interlock type of prephenate dehydratase characteristic of Gram-positive eubacteria. *Arch. Microbiol.* **148:**365–371.

128. Jeon, S.-J., S. Fujiwara, M. Takagi, M. Tanaka, and T. Imanaka. 2002. *Tk*-PTP, a protein-tyrosine phosphatase from the hyperthermophilic archaeon *Thermococcus kodakaraensis* KOD1: enzymatic characteristics and identification of its substrate proteins. *Biochem. Biophys. Res. Commun.* **295:**508–514.

129. Jiang, Y., A. Lee, J. Chen, M. Cadene, B. T. Chait, and R. MacKinnon. 2002. Crystal structure and mechanism of a calcium-gated potassium channel. *Nature* **30:**516–522.

130. Jing, H., J. Takagi, J. Liu, S. Lindgren, R. Zhang, A. Joachimiak, J. Wang, and T. A. Springer. 2002. Archaeal surface layer proteins contain β propeller, PKD, and β helix domains and are related to metazoan cell surface proteins. *Structure* **10:**1453–1464.

131. Johnson, M. R., C. I. Montero, S. B. Conners, K. R. Shockley, S. L. Bridger, and R. M. Kelly. 2005. Population density-dependent regulation of exopolysaccharide formation in the hyperthermophilic bacterium *Thermotoga maritime*. *Mol. Microbiol.* **55:**664–674.

132. Jones, J. B., G. L. Dilworth, and T. C. Stadtman. 1979. Occurrence of selenocysteine in the selenium-dependent formate dehydrogenase of *Methanococcus vannielii*. *Arch. Biochem. Biophys.* **195:**255–260.

133. Kaidoh, K., S. Miyauchi, A. Abe, S. Tanabu, T. Nara, and N. Kamo. 1996. Rhodamine 123 efflux transporter in *Haloferax volcanii* is induced when cultured under 'metabolic stress' by amino acids: the efflux system resembles that in a doxorubicin-resistant mutant. *Biochem. J.* **314:**355–359.

134. Kappler, U., I. Lindsay, and A. G. McEwan. 2005. Respiratory gene clusters of *Metallosphaera sedula*—differential expression and transcriptional organization. *Microbiology.* **151:**35–43.

135. Karras, G. I., G. Kustatscher, H. R. Buhecha, M. D. Allen, C. Pugieux, F. Sait, M. Bycroft, and A. G. Ladurner. 2005. The macro domain is an ADP-ribose binding module. *EMBO J.* **24:**1911–1920.

136. Kennelly, P. J. 2001. Protein phosphatases: a phylogenetic perspective. *Chem. Rev.* **101:**2291–2312.

137. Kennelly, P. J. 2002. Protein kinases and protein phosphatases in prokaryotes: A genomic perspective. *FEMS Microbiol. Lett.* **206:**1–8.

138. Kennelly, P. J. 2003. Archaeal protein kinases and protein phosphatases—insights from genomics and biochemistry. *Biochem. J.* **370:**373–389.

139. Kessel, M., and F. Klink. 1980. Archaebacterial elongation factor is ADP-ribosylated by diphtheria toxin. *Nature* **287:** 250–251.

140. Kessler, P. S., and J. A. Leigh. 1999. Genetics of nitrogen regulation in *Methanococcus maripaludis*. *Genetics* **152:** 1343–1351.

141. Khristich, C. J., and G. W. Ordal. *Bacillus subtilis* CheD is a chemorecpetoor modification enzyme required for chemotaxiis. *J. Biol. Chem.* **277:**25356–25362.

142. Kim, D.-J., and S. Forst. 2001. Genomic analysis of the histidine kinase superfamily in bacteria and archaea. *Microbiology* **147:**1197–1212.

143. Kim, A. D., D. E. Graham, S. H. Seeholzer, and G. D. Markham. 2000. S-Adenosylmethionine decarboxylase from the archaeon *Methanococcus jannaschii*: Identification of a novel family of pyruvoyl enzymes. *J. Bacteriol.* **182:**6667–6672.

144. Kim, S., and S. B. Lee. 2005. Identification and characterization of *Sulfolobus solfataricus* D-gluconate dehydratase: a key enzyme in the non-phosphorylated Entner-Doudoroff pathway. *Biochem. J.* **387:**271–280.

145. Kim, J. W., H. A. Terc, L. O. Flowers, M. Whiteley, and T. L. Peeples. 2001. Novel, thermostable family-13-like glycoside hydrolase from *Methanococcus jannaschii*. *Folia Microbiol.* **46:**475–481.

146. Kirby, J. R., C. J. Khristich, M. M. Saulmon, M. A. Zimmer, L. F. Garriity, I. B. Zhulin, and G. W. Ordal. 2001. CheC is related to the family of flagellar switch proteins and acts independently from CheD to control chemotaxis in *Bacillus subtilis*. *Mol. Microbiol.* **42:**573–585.

147. Kloda, A., and B. Martinac. 2001. Structural and functional differences between two homologous mechanosensitive channels of *Methanococcus jannaschii*. *EMBO J.* **20:**1888–1896.

148. Kloda, A., and B. Martinac. 2001. Mechanosensitive channel of *Thermoplasma*, the cell wall-less *Archaea*: Cloning and molecular characterization. *Cell Biochem. Biophys.* **34:** 321–347.

149. Kloda, A., and B. Martinac. 2002. Common evolutionary origins of mechanosensitive ion channels in *Archaea*, *Bacteria* and cell-walled *Eukarya*. *Archaea* **1:**35–44.

150. Klussmann, S., P. Franke, U. Bergmann, S. Kostka, and B. Wittmann-Liebold. 1993. N-Terminal modification and amino-acid sequence of the ribosomal protein HmaS7 from *Haloarcula marismortui* and homology studies to other ribosomal proteins. *Biol. Chem. Hoppe-Seyler* **374:**305–312.

151. Knapp, S., A. Karshikoff, K. D. Berndt, P. Christova, B. Atanasov, and R. Ladenstein. 1996. Thermal unfolding of the DNA-binding protein Sso7d from the hyperthermophile *Sulfolobus solfataricus*. *J. Mol. Biol.* **264:**1132–1144.

152. Koch, M. K., and D. Oesterhelt. 2005. MpcT is the transducer for membrane potential changes in *Halobacterium salinarium*. *Mol. Microbiol.* **55:**1681–1694.

153. Kokoeva, M. V., and D. Oesterhelt. 2000. BasT, a membrane-bound transducer protein for amino acid detection in *Halobacterium salinarium*. *Mol. Microbiol.* **35:**647–656.

154. Kokoeva, M. V., K. F. Storch, C. Klein, and D. Oesterhelt. 2002. A novel mode of sensory transduction in *Archaea*: binding protein-mediated chemotaxis towards osmoprotectants and amino acids. *EMBO J.* **21:**2312–2322.

155. Koonin, E. V., K. S. Makarova, I. B. Rogozin, L. Davidovic, M. C. Letellier, and L. Pellegrini. 2003. The rhomboids: a nearly ubiquitous family of intramembrane serine proteases that probably evolved by multiple ancient horizontal gene transfers. *Genome Biol.* **4:**R19.1–R19.12.

156. Koretke, K. K., A. N. Lupas, P. V. Warren, M. Rosenberg, and J. R. Brown. 2000. Evolution of the two-component regulatory system. *Mol. Biol. Evol.* **17:**1956–1970.

157. Krzycki, J. A. 2005. The direct genetic encoding of pyrrolysine. *Curr. Opin. Microbiol.* **8:**706–712.

158. Kuo, M. M.-C., W. J. Haynes, S. H. Loukin, C. Kung, and Y. Saimi. 2005. Prokaryotic K⁺ channels: from crystal structures to diversity. *FEMS Microbiol. Rev.* **29:**961–985.

159. Kuo, Y. P., D. K. Thompson, A. St. Jean, R. L. Charlebois, and C. J. Daniels. 1977. Characterization of two heat shock

genes from *Haloferax volcanii*: A model system for transcription regulation in the *Archaea*. *J. Bacteriol.* 179:6318–6324.

160. **Labedan, B., A. Boyen, M. Baetens, D. Charlier, P. Chen, R. Cunin, V. Durbeco, N. Glansdorff, G. Herv, C. Legrain, Z. Liang, C. Purcarea, M. Roovers, R. Sanchez, T. L. Toong, M. Van de Casteel, F. van Vliet, Y. Xu, and Y. F. Zhang.** 1999. The evolutionary history of carbamoyltransferases: a complex set of paralogous genes was already present in the last universal common ancestor. *J Mol Evol.* 49:461–473.

161. **Labedan, B., Y. Xu, D. G. Naumoff, and N. Glansdorff.** 2003. Using quaternary structures to assess the evolutionary history of proteins: the case of the aspartate carbamoyltransferase. *Mol. Biol. Evol.* 21:364–373.

162. **Lai, M. C., D. R. Yang, and M. J. Chuang.** 1999. Regulatory factors associated with synthesis of the osmolyte glycine betaine in the halophilic methanoarchaeon *Methanohalophilus portucalensis*. *Appl. Environ. Microbiol.* 65:828–833.

163. **Lai, X., H. Shao, F. Hao, and L. Haung.** 2002. Biochemical characterization of an ATP-dependent DNA ligase from the hyperthermophilic crenarchaeon *Sulfolobus shibatae*. *Extremophiles* 6:469–477.

164. **Lam, W. L., S. M. Logan, and W. F. Doolittle.** 1992. Genes for tryptophan Biosynthesis in the halophilic archaebacterium *Haloferax volcanii*: The *trpDFEG* cluster. *J. Bacteriol.* 174:1694–1697.

165. **Lange, M., T. Tolker-Nielsen, S. Molin, and B. K. Ahring.** 2000. *In situ* reverse transcription-PCR for monitoring gene expression in individual *Methanosarcina mazei* S-6 cells. *Appl. Environ. Microbiol.* 66:1796–1800.

166. **LaPaglia, C., and P. L. Hartzell.** 1997. Stress-induced production of biofilm in the hyperthermophile *Archaeoglobus fulgidus*. *Appl. Environ. Microbiol.* 63:3158–3163.

167. **LaRonde-LaBlanc, N., and A. Wlodawer.** 2004. Crystal structure of *A. fulgidus* Rio2 defines a new family of serine protein kinases. *Structure* 12:1585–1594.

168. **LaRonde-LeBlanc, N., T. Guszczynski, T. Copeland, and A. Wlodawer.** 2005. Autophosphorylation of *Archaeoglobus fulgidus* Rio2 and crystal structures of its nucleotide-metal ion complexes. *FEBS J.* 272:2800–2810.

169. **LaRonde-LeBlanc, N., T. Guszczynski, T. Copeland, and A. Wlodawer.** 2005. Structure and activity of the atypical serine kinase Rio1. *FEBS J.* 272:3698–3713.

170. **LaRonde-LeBlanc, N., and A. Wlodawer.** 2005. A family portrait of the RIO kinases. *J. Biol. Chem.* 280:37297–37300.

171. **LaRonde-LeBlanc, N., and A. Wlodawer.** 2005. The RIO kinases: An atypical protein kinase family required for ribosome biogenesis and cell cylce progression. *Biochim. Biophys. Acta* 1754:14–24.

172. **Le Dain, A. C., N. Saint, A. Kloda, A. Ghazi, and B. Martinac.** 1998. Mechanosensitive ion channels of the archaeon *Haloferax volcanii*. *J. Biol. Chem.* 273:12116–12119.

173. **Lee, C. H., J. W. Jung, A. Yee, C. H. Arrowsmith, and W. Lee.** 2004. Solution structure of a novel calcium binding protein, MTH1880, from *Methanobacterium thermoautotrophicum*. *Protein Sci.* 13:1148–1154.

174. **Lee, M. S., W. A. Joo, and C. W. Kim.** 2004. Identification of a novel protein D3UPCA from *Halobacterium salinarum* and prediction of its function, *Proteomics* 4:3622–3631.

175. **Leichtling, B. H., H. V. Rickenberg, R. J. Seely, D. E. Fahrney, and N. R. Pace.** 1986. The occurrence of cyclic AMP in *Archaebacteria*. *Biochem. Biophys. Res. Commun.* 136:1078–1082.

176. **Leigh, J. A.** 2000. Nitrogen fixation in methanogens: The archaeal perspective. *Curr. Issues Mol. Biol.* 2:125–131.

177. **Lemker, T., G. Gruber, R. Schmid, and V. Muller.** 2003. Defined subcomplexes of the A$_1$ ATPase from the archaeon

Methanosarcina mazei Gö1: biochemical properties and redox regulation. *FEBS Lett.* 544:206–209.

178. **Leng, J., A. J. Cameron, S. Buckel, and P. J. Kennelly.** 1995. Isolation and cloning of a protein-serine/threonine phosphatase from an archaeon. *J. Bacteriol.* 177:6510–6517.

179. **Leonard, C. J., L. Aravind, and E. V. Koonin.** 1998. Novel families of putative protein kinases in *Bacteria* and *Archaea*: evolution of the 'eukaryotic' protein kinase superfamily. *Genome Res.* 8:1038–1047.

180. **Li, H., H. Xu, D. E. Graham, and R. H. White.** 2003. The *Methanococcus jannaschii* dCTP deaminase is a bifunctional deaminase and diphosphatase. *J. Biol. Chem.* 278:11100–11106.

181. **Li, R., M. B. Potters, L. Shi, and P. J. Kennelly.** 2005. The protein phosphatases of *Synechocystis* sp. strain PCC6803: Open reading frames *sll1033* and *sll1387* encode enzymes that exhibit both protein-serine and protein-tyrosine phosphatase activity *in vitro*. *J. Bacteriol.* 187:5877–5884.

182. **Lie, T. J., and J. A. Leigh.** 2002. Regulatory response of *Methanococcus maripaludis* to alanine, an intermediate nitrogen source. *J. Bacteriol.* 184:5301–5306.

183. **Lindbeck, J. C., E. A. Goulbourne, Jr., M. S. Johnson, and B. L. Taylor.** 1995. Aerotaxis in *Halobacterium salinarium* is methylation-dependent. *Microbiology* 141:2945–2953.

184. **Lipscomb, W. N.** 1994. Aspartate transcarbamylase from *Escherichia coli*: activity and regulation. *Adv. Enzymol. Relat. Areas Mol. Biol.* 68:67–151.

185. **Lobo, A. L., and S. H. Zinder.** 1990. Nitrogenase in the archaebacterium *Methanosarcina barkeri* 227. *J. Bacteriol.* 172:6789–6796.

186. **Lopalco, P., S. Lobasso, F. Babudri, and A. Corcelli.** 2004. Osmotic shock stimulates *de novo* synthesis of two cardiolipins in an extreme halophilic archaeon. *J. Lipid Res.* 45:194–201.

187. **Lorentzen, E., R. Hensel, T. Knura, H. Ahmed, and E. Pohl.** 2004. Structural basis of allosteric regulation and substrate specificity of the non-phosphorylating glyceraldehyde 3-phosphate dehydrogenase from *Thermoproteus tenax*. *J. Mol. Biol.* 341:815–828.

188. **Lower, B. H., K. M. Bischoff, and P. J. Kennelly.** 2000. The archaeon *Sulfolobus solfataricus* contains a membrane-associated protein kinase that preferentially phosphorylates threonine residues. *J. Bacteriol.* 182:3452–3459.

189. **Lower, B. H., and P. J. Kennelly.** 2002. The membrane-associated protein-serine/threonine kinase from *Sulfolobus solfataricus* is a glycoprotein. *J. Bacteriol.* 184:2614–2619.

190. **Lower, B. H., and P. J. Kennelly.** 2003. Open reading frame *sso2387* from the archaeon *Sulfolobus solfataricus* encodes a polypeptide with protein-serine kinase activity. *J. Bacteriol.* 185:3436–3445.

191. **Lower, B. H., M. B. Potters, and P. J. Kennelly.** 2004. A phosphoprotein from the archaeon *Sulfolobus solfataricus* with protein-serine kinase activity. *J. Bacteriol.* 186:463–472.

192. **Lundback, T., H. Hansson, S. Knapp, R. Ladenstein, and T. Hard.** 1998. Thermodynamic characterization of non-sequence-specific DNA-binding by the Sso7d protein from *Sulfolobus solfataricus*. *J. Mol. Biol.* 276:775–786.

193. **Luo, H.W., H. Zhang, T. Suzuki, S. Hattori, and Y. Kamagata.** 2002. Differential expression of methanogenesis genes of *Methanothermobacter thermoautotrophicus* (formerly *Methanobacterium thermoautotrophicum*) in pure culture and in cocultures with fatty acid-oxidizing syntrophs. *Appl. Environ. Microbiol.* 68:1173–1179.

194. **Ma, K., H. Loessner, J. Heider, M. K. Johnson, and M. W. W. Adams.** 1995. Effects of elemental sulfur on the metabolism of the deep-sea hyperthermophilic archaeon *Thermococcus* strain ES-1: Characterization of a sulfur-regulated, non-heme iron alcohol dehydrogenase. *J. Bacteriol.* 177:4748–4756.

195. Mai, B., G. Frey, R. V. Swanson, E. J. Mathur, and K. O. Stetter. 1998. Molecular cloning and functional expression of a protein-serine/threonine phosphatase from the hyperthermophilic archaeon *Pyrodictium abyssi* TAG11. *J. Bacteriol.* **180:**4030–4035.

196. Manning, G., G. D. Plowman, T. Hunter, and S. Sudarsanam. 2002. Evolution of protein kinase signaling from yeast to man. *Trends Biochem. Sci.* **27:**514–520.

197. Manning, G., D. B. Whyte, R. Martinez, T. Hunter, and S. Sudarsanam. 2002. The protein kinase complement of the human kinase. *Science* **298:**1912–1934.

198. Manzur, K. L., and M. M. Zhou. 2005. An archaeal SET domain protein exhibits distinct lysine methyltransferase activity towards DNA-associated protein MC1-α. *FEBS Lett.* **579:** 3859–3865.

199. Maras, B., V. Consalvi, R. Chiaraluce, L. Politi, M. De Rosa, F. Bossa, R. Scandurra, and D. Barra. 1992. The protein sequence of glutamate dehydrogenase from *Sulfolobus solfataricus*, a thermoacidophilic archaebacterium. *Eur. J. Biochem.* **203:**81–87.

200. Marsh, V. L., S. Y. Peak-Chew, and S. D. Bell. 2005. Sir2 and the acetyltransferase, Pat, regulate the archaeal chromatin protein, Alba. *J. Biol. Chem.* **280:**21122–21128.

201. Marteinsson, V. T., A. L. Reysenbach, J. L. Birrien, and D. Prieur. 1999. A stress protein is induced in the deep-sea barophilic hyperthermophile *Thermoccocus barophilus* when grown under atmospheric pressure. *Extremophiles* **3:**277–282.

202. Martin, D. D., R. A. Ciulla, P. M. Robinson, and M. R. Roberts. 2000. Switching osmolyte strategies: response of *Methanococcus thermolithotrophicus* to changes in external NaC1. *Biochim. Biophys. Acta* **1524:**1–10.

203. Martins, L. O., and H. Santos. 1995. Accumulation of mannosylglycerate and di-myo-inositol-phosphate by *Pyrococcus furiosus* in response to salinity and temperature. *Appl. Environ. Microbiol.* **61:**3299–3303.

204. Martins, L. O., R. Huber, H. Huber, K. O. Stetter, M. S. Da Costa, and H. Santos. 1997. Organic solutes in hyperthermophilic *Archaea*. *Appl. Environ. Microbiol.* **63:**896–902.

205. Marwan, W., S. I. Bibikov, M. Montrone, and D. Oesterhelt. 1995. Mechanism of photosensory adaptation in *Halobacterium salinarium*. *J. Mol. Biol.* **246:**493–499.

206. Marwan, W., W. Schafer, and D. Oesterhelt. 1990. Signal transduction in *Halobacterium* depends on fumarate. *EMBO J.* **9:**355–362.

207. Matsubara, M., and T. Mizuno. 2000. The SixA phospho-histidine phosphatase modulates the ArcB phosphorelay signal transduction in *Escherichia coli*. *FEBS Lett.* **470:**118–124.

208. Mattar, S., B. Scharf, S. B. H. Kent, K. Rodewald, D. Oesterhelt, and M. Engelhard. 1994. The primary structure of halocyanin, and archaeal blue copper protein, predicts a lipid anchor for membrane fixation. *J. Biol. Chem.* **269:**14939–14945.

209. Michiels, J., C. Xi, J. Verhaert, and J. Vanderleyden. 2002. The Functions of Ca^{2+} in bacteria: a role for EF-hand proteins? *Trends Microbiol.* **10:**87–93.

210. Mojica, F. J. M., E. Cisneros, C. Ferrer, F. Rodriguez-Valera, and G. Juez. 1997. Osmotically induced response in representatives of halophilic prokaryotes: The bacterium *Halomonas elongata* and the archaeon *Haloferax volcanii*. *J. Bacteriol.* **179:**5471–5481.

211. Montrone, M., W. Marwan, H. Grunberg, S. Mubeleck, C. Starostzik, and D. Oesterhelt. 1993. Sensory rhodopsin-controlled release of the switch factor fumarate in *Halobacterium salinarium*. *Mol. Microbiol.* **10:**1077–1085.

212. Montrone, M., D. Oesterhelt, and W. Marwan. 1996. Phosphorylation-independent bacterial chemoresponses correlate

213. Morona, J. K., R. Morona, D. C. Miller, and J. C. Paton. 2002. *Streptococcus pneumoniae* capsule biosynthesis protein CpsB is a novel manganese-dependent phosphotyrosine-protein phosphatase. *J. Bacteriol.* **184:**577–583.

214. Mougel, C., and I. B. Zhulin. 2001. CHASE: an extracellular sensing domain common to transmembrane receptors from prokaryotes, lower eukaryotes and plants. *Trends Biochem. Sci.* **26:**582–584.

215. Nauhaus, K., T. Treude, A. Boetius, and M. Kruger. 2005. Environmental regulation of the anaerobic oxidation of methane: A comparison of ANME-I and ANME-II communities. *Environ. Microbiol.* **7:**98–106.

216. Neelon, K., Y. Wang, B. Stec, and M. F. Roberts. 2005. Probing the mechanism of the *Archaeoglobus fulgidus* inositol-1-phospate synthase, *J. Biol. Chem.* **280:**11475–11482.

217. Nercessian, D., R. E. De Castro, and R. D. Conde. 2002. Ubiquitin-like proteins in halobacteria. *J. Basic Microbiol.* **42:**277–283.

218. Ninfa, A. J., and P. Jiang. 2005. PII signal transduction proteins: sensors of α-ketoglutarate that regulate nitrogen metabolism. *Curr. Opin. Microbiol.* **8:**168–173.

219. Norberg, P., J. G. Kaplan, and D. J. Kushner. 1973. Kinetics and regulation of the salt-dependent aspartate transcarbamylase of *Halobacterium cutirubrum*, *J. Bacteriol.* **113:**680–686.

220. Oppermann, U. C. T., S. Knapp, V. Bonetto, R. Ladenstein, and H. Jornvall. 1998. Isolation and structure of repressor-like proteins from the archaeon *Sulfolobus solfataricus*. *FEBS Lett.* **432:**141–144.

221. Overmann, J., and K. Schubert. 2002. Phototrophic consortia: Model systems for symbiotic interrelations between prokaryotes. *Arch. Microbiol.* **177:**201–208.

222. Paggi, R. A., C. B. Martone, C. Fuqua, and R. E. De Castro. 2003. Detection of quorum sensing signals in the haloalkaliphilic archaeon *Natronococcus occultus*. *FEMS Microbiol. Lett.* **221:**49–52.

223. Pappenheimer, A. M. Jr., P. C. Dunlop, K. W. Adolph, and J. W. Bodley. 1983. Occurrence of diphthamide in archaebacteria. *J. Bacteriol.* **153:**1342–1347.

224. Pardee, A. B., and G. P.-V. Reddy. 2003. Beginning of feedback inhibition, allostery, and multi-protein complexes. *Gene* **321:**17–23.

225. Parkinson, J. S. 2003. Bacterial chemotaxis: a new player in response regulator dephosphorylation. *J. Bacteriol.* **185:** 1492–1494.

226. Peeples, T. L., and R. M. Kelly. 1995. Bioenergetic response of the extreme thermoacidophile *Metallosphaera sedula* to thermal and nutritional stresses. *Appl. Environ. Microbiol.* **61:** 2314–2321.

227. Pei, J., and N. V. Grishin. 2001. GGDEF domain is homologous to adenylyl cyclase. *Proteins: Struct. Funct. Genet.* **42:** 210–216.

228. Perazzona, B., and J. L. Spudich. 1999. Identification of methylation sites and effects of phototaxis stimuli on transducer methylation in *Halobacterium salinarium*. *J. Bacteriol.* **181:**5676–5683.

229. Perraud, A.-L., V. Weiss, and R. Gross. 1999. Signballing pathways in two-component phosphorelay systems. *Trends Microbiol.* **7:**115–120.

230. Pfeifer, F., D. Gregor, A. Hofacker, P. Plosser, and P. Zimmermann. 2002. Regulation of gas vesicle formation in halophilic *Archaea*. *J. Mol. Microbiol. Biotechnol.* **4:**175–181.

231. Pfluger, K., S. Baumann, G. Gottschalk, W. Lin, H. Santos, and V. Muller. 2003. Lysine-2,3-aminomutase and β-lysine acetyltransferase genes of methanogenic *Archaea* are salt in-

with changes in the cytoplasmic level of fumarate. *J. Bacteriol.* **178:**6882–6887.

duced and are essential for the biosynthesis of N$^\varepsilon$-acetyl-β-lysine and growth at high salinity. *Appl. Environ. Microbiol.* 69:6047–6055.

232. Phipps, B. M., A. Hoffmann, K. O. Stetter, and W. Baumeister. 1991. A novel ATPase complex selectively accumulated upon heat shock is a major cellular component of thermophilic archaebacteria. *EMBO J.* 10:1711–1722.

233. Ponting, C. P., L. Aravind, J. Schultz, P. Bork, and E. V. Koonin. 1999. Eukaryotic signaling domain homologues in *Archaea* and *Bacteria*. Ancient ancestry and horizontal gene transfer. *J. Mol. Biol.* 289:729–745.

234. Porat, I., B. W. Waters, Q. Teng, and W. B. Whitman. 2004. Two biosynthetic pathways for aromatic amino acids in the archaeon *Methanococcus maripaludis*. *J. Bacteriol.* 186:4940–4950.

235. Purcarea, C., G. Herve, M. M. Ladjimi, and R. Cunin. 1997. Aspartate transcarbamylase from the deep-sea hyperthermophilic archaeon *Pyrococcus abyssi*: genetic organization, structure, and expression in *Escherichia coli*. *J. Bacteriol.* 179:4143–4157.

236. Rashid, N., J. Cornista, S. Ezaki, T. Fukui, H. Atomi, and T. Imanaka. 2002. Characterization of an archaeal cyclodextrin glucanotransferase with a novel C-terminal domain. *J. Bacteriol.* 184:777–784.

237. Ray, W. K., S. M. Keith, A. M. DeSantis, J. P. Hunt, T. J. Larson, R. F. Helm, and P. J. Kennelly. 2005. A phosphohexomutase from the archaeon *Sulfolobus solfataricus* is covalently modified by phosphorylation on serine. *J. Bacteriol.* 187:4270–4275.

238. Reizer, J., A. Reizer, M. Perego, and M. H. Saier, Jr. 1997. Characterization of a family of bacterial response regulator aspartyl-phosphate (RAP) phosphatases. *Microb. Comp. Genomics* 2:103–111.

239. Reuter, C. J., S. J. Kaczowka, and J. A. Maupin-Furlow. 2004. Differential regulation of the PanA and PanB proteasome-activating nucleotidase and 20S proteasomal proteins of the haloarchaeon *Haloferax volcanii*. *J. Bacteriol.* 186:7763–7772.

240. Riera, J., F. T. Robb, R. Weiss, and M. Fontecave. 1997. Ribonucleotide reductase in the archaeon *Pyrococcus furiosus*: A critical enzyme in the evolution of DNA genomes? *Proc. Natl. Acad. Sci. USA* 94:475–478.

241. Rigden, K. J. and M. Y. Galperin. 2004. The DxDxDG motif for calcium binding: Multiple structural contexts and implications for evolution. *J. Mol. Biol.* 343:971–984.

242. Roberts, T. H., J. Hejgaard, N. F. W. Saunders, R. Cavicchioli, and P. J. G. Curmi. 2004. Serpins in unicellular *Eukarya*, *Archaea*, and *Bacteria*: sequence analysis and evolution. *J. Mol. Evol.* 59:437–447.

243. Robertson, D. E., M. Lai, R. P. Gunsalus, and M. F. Roberts. 1992. Composition, variation, and dynamics of major osmotic solutes in *Methanohalophilus* strain FDF1. *Appl. Environ. Microbiol.* 58:2438–2443.

244. Rohlin, L., J. D. Trent, K. Salmon, U. Kim, R. P. Gunsalus, and J. C. Liao. 2005. Heat shock response of *Archaeoglobus fulgidus*. *J. Bacteriol.* 187:6046–6057.

245. Roosild, T. P., S. Miller, I. R. Booth, and S. Choe. 2002. A mechanism of regulating transmembrane potassium flux through a ligand-mediated conformational switch. *Cell* 109:781–791.

246. Rotharmel, T., and G. Wagner. 1995. Isolation and characterization of a calmodulin-like protein from *Halobacterium salinarium*. *J. Bacteriol.* 177:864–866.

247. Rother, M., P. Boccazzi, A. Bose, M. A. Pritchett, and W. W. Metcalf. 2005. Methanol-dependent gene expression demonstrates that methyl-coenzyme M reductase is essential for

Methanosarcina acetivorans C2A and allows isolation of mutants with defects in regulation of the methanol utilization pathway. *J. Bacteriol.* 187:5552–5559.

248. Rudiger, A., P. L. Jorgensen, and G. Antranikian. 1995. Isolation and characterization of a heat-stable pullulanase from the hyperthermophilic archaeon *Pyrococcus woesei* after cloning and expression of its gene in *Escherichia coli*. *Appl. Environ. Microbiol.* 61:567–575.

249. Rudolph, J., and D. Oesterhelt. 1995. Chemotaxis and phototaxis require a CheA histidine kinase in the archaeon *Halobacterium salinarium*. *EMBO J.* 14:667–673.

250. Rudolph, J., and D. Oesterhelt. 1996. Deletion analysis of the *che* operon in the archaeon *Halobacterium salinarium*. *J. Mol. Biol.* 258:548–554.

251. Rudolph, J., N. Tolliday, C. Schmitt, S. C. Schuster, and D. Oesterhelt. 1995. Phosphorylation in halobacterial signal transduction. *EMBO J.* 14:4249–4257.

252. Ruppert, U., A. Irmler, N. Kloft, and K. Forchhammer. 2002. The novel protein phosphatase PphA from *Synechocystis* PCC 6803 controls dephosphorylation of the signaling protein P$_{II}$. *Mol. Microbiol.* 44:855–864.

253. Ruta, V., and R. MacKinnon. 2004. Localization of the voltage-sensor toxin receptor on KvAP. *Biochemistry* 43:10071–10079.

254. Ruta, V., Y. Jiang, A. Lee, J. Chen, and R. MacKinnon. 2003. Functional analysis of an archaebacterial voltage-dependent K$^+$ channel. *Nature* 422:180–185.

255. Sako, Y., K. Takai, A. Uchida, and Y. Ishida. 1996. Purification and characterization of phosphoenolpyruvate carboxylase from the hyperthermophilic archaeon *Methanothermus sociabilis*. *FEBS Lett.* 392:148–152.

256. Sako, Y., K. Takai, T. Nishizaka, and Y. Ishida. 1997. Biochemical relationship of phosphoenolpyruvate carboxylases (PEPCs) from the thermophilic *Archaea*. *FEMS Microbiol. Lett.* 153:159–165.

257. Salerno, V., A. Napoli, M. F. White, M. Rossi, and M. Ciaramella. 2003. Transcriptional response to DNA damage in the archaeon *Sulfolobus solfataricus*. *Nucleic Acids Res.* 31:6127–6138.

258. Savchenko, A., C. Vieille, S. Kang, and J. G. Zeikus. 2002. *Pyrococcus furiosus* α-amylase is stabilized by calcium and zinc. *Biochemistry* 41:6193–6201.

259. Scharf, B., and E. K. Wolff. 1994. Phototactic behavior of the archaebacterial *Natronobacterium pharaonis*. *FEBS Lett.* 340:114–116.

260. Schmiz, A. 1981. Methylation of membrane proteins is involved in chemosensory and photosensory behavior in *Halobacterium halobium*. *FEBS Lett.* 125:205–207.

261. Schmiz, A., and E. Hildebrand. 1979. Chemosensory responses of *Halobacterium halobium*. *J. Bacteriol.* 140:749–753.

262. Schofield, L. R., M. L. Patchett, and E. B. Parker. 2003. Expression, purification, and characterization of 3-deoxy-D-*arabino*-heptulosonate 7-phosphate synthase from *Pyrococcus furiosus*. *Protein Expr. Purif.* 34:17–27.

263. Schut, G. J., S. D. Brehm, S. Datta, and M. W. W. Adams. 2003. Whole-genome DNA microarray analysis of a hyperthermophile and an archaeon: *Pyrococcus furiosus* grown on carbohydrates or peptides. *J. Bacteriol.* 185:3935–3947.

264. Selmer, T., J. Kahnt, M. Goubeaud, S. Shima, W. Grabarse, U. Ermler, and R. K. Thauer. 2000. The biosynthesis of methylated amino acids in the active site region of methyl-coenzyme M reductase. *J. Biol. Chem.* 275:3755–3760.

265. Sesti, F., S. Rajan, R. Gonzalez-Colsao, N. Nikolaeva, and S. A. N. Goldstein. Hyperpolarization moves S$ sensors inward to open MVP, a methanococcal voltage-gated potassium channel. *Nat. Neurosci.* 6:353–361.

266. Sham, S., L. Calzolai, P. L. Wang, K. Bren, H. Haarklau, P. S. Brereton, M. W. W. Adams, and G. N. La Mar. 2002. A solution NMR molecular model for the aspartate-ligated, cubane cluster containing ferredoxin from the hyperthermophilic archeaon *Pyrococcus furiosus*. *Biochemistry* **41:**12498–12508.

267. Shenoy, A. R., and S. S. Visweswariah. 2004. Class III nucleotide cyclases in bacteria and archaebacteria: Lineage-specific expansion of adenylyl cyclases and a dearth of guanylyl cyclases. *FEBS Lett.* **561:**11–21.

268. Shi, L., M. Potts, and P. J. Kennelly. 1998. The serine, threonine, and/or tyrosine-specific protein kinases and protein phosphatases of prokaryotic organisms: a family portrait. *FEMS Microbiol. Rev.* **22:**229–253.

269. Shockley, K. R., D. E. Ward, S. R. Chhabra, S. B. Conners, C. I. Montero, and R. M. Kelly. 2003. Heat shock response by the hyperthermophilic archaeon *Pyrococcus furiosus*. *Appl. Environ. Microbiol.* **69:**2365–2371.

270. Skarphol, K., J. Waukau, and S. A. Forst. 1997. Role of His243 in thhe phosphatase activity of EnvZ in *Escherichia coli*. *J. Bacteriol.* **179:**1413–1416.

271. Sment, K. A., and J. Konisky. 1989. Chemotaxis in the archaebacterium *Methanococcus voltae*. *J. Bacteriol.* **171:**2870–2872.

272. Smith, R. J. 1995. Calcium and bacteria. *Adv. Microb. Physiol.* **37:**83–133.

273. Soares, J. A., L. Zhang, R. L. Pitsch, N. M. Kleinholz, R. B. Jones, J. J. Wolff, J. Amster, K. B. Green-Church, and J. A. Krzycki. 2005. The residue mass of L-pyrrolysine in three distinct methylamine methyltransferases. *J. Biol. Chem.* **280:**36962–36969.

274. Solow, B., J. C. Young, and P. J. Kennelly. 1997. Gene cloning and expression and characterization of a toxin-sensitive protein phosphatase from the methanogenic archaeon *Methanosarcina thermophila* TM-1. *J. Bacteriol.* **179:**5072–5075.

275. Sowers, K. R., and R. P. Gunsalus. 1995. Halotolerance in *Methanosarcina* spp.: role of N^ε-acetyl-β-lysine, α-glutamate, glycine betaine, and K^+ as compatible solutes for osmotic adaptation. *Appl. Environ. Microbiol.* **61:**4382–4388.

276. Spreter, T., M. Pech, and B. Beatrix. 2005. The crystal structure of archaeal nascent polypeptide-associated complex (NAC) reveals a unique fold and the presence of a ubiquitin-associated domain. *J. Biol. Chem.* **280:**15849–15854.

277. Spudich, E. N., and J. L. Spudich. 1981. Photosensitive phosphoproteins in halobacteria: Regulatory coupling of transmembrane proton flux and protein dephosphorylation. *J. Cell Biol.* **91:**895–900.

278. Spudich, J. L., and W. Stoeckenius. 1979. Photosensory and chemosensory behavior of *Halobacterium halobium*. *Photobiochem. Photobiophys.* **1:**43–53.

279. Spudich, J. L., and W. Stoeckenius. 1980. Light-mediated retinal-dependent reversible phosphorylation of *Halobacterium* proteins. *J. Biol. Chem.* **255:**5501–5503.

280. Spudich, E. N., T. Takahashi, and J. L. Spudich. 1989. Sensory rhodopsins I and II modulate the methylation/demethylation system in *Halobacterium halobium* phototaxis. *Proc. Natl. Acad. Sci. USA* **86:**7746–7750.

281. Stan-Lotter, H., E. Doppler, M. Jarosch, C. Radax, C. Gruber, and K. Inatomi. 1999. Isolation of a chymotrypsinogen B-like enzyme from the archaeon *Natronomonas pharaonis* and other halobacteria. *Extremophiles* **3:**153–161.

282. Stieglitz, K. A., B. A. Seaton, J. F. Head, B. Stec, and M. F. Roberts. 2003. Unexpected similarity in regulation between an archaeal inositol monophosphatase/fructose bisphosphatase and chloroplast fructose bisphosphatase. *Protein Sci.* **12:**760–767.

283. Stock, J., and S. De Re. 2000. Signal transduction: Response regulators on and off. *Curr. Biol.* **10:**R420–R424.

284. Stock, A. M., V. L. Robinson, and P. N. Goudreau. 2000. Two-component signal transduction. *Annu. Rev. Biochem.* **69:**182–215.

285. Storch, K. F., J. Rudolph, and D. Oesterhelt. 1999. Car: a cytoplasmic sensor responsible for arginine chemotaxis in the archaeon *Halobacterium salinarum*. *EMBO J.* **18:**1146–1158.

286. Sun, J., R. Daniel, I. Wagner-Dobler, and A. P. Zeng. 2004. Is autoinducer-2 a universal signal for interspecies communication: A comparative genomic and pathlogenetic analysis of the synthesis and signal transduction pathways. *BMC Evol. Biol.* **4:**36–47.

287. Sundberg, S. A., M. Alam, M. Lebert, J. L. Spudich, D. Oesterhelt, and G. L. Hazelbauer. 1990. Characterization of *Halobacterium halobium* mutants defective in taxis. *J. Bacteriol.* **172:**2328–2335.

288. Szurmant, H., T. J. Muff, and G. W. Ordal. 2004. *Bacillus subtilis* CheC and FliY are members of a novel class of CheY-P-hydrolyzing proteins in the chemotactic signal transduction cascade. *J. Biol. Chem.* **279:**21787–21792.

289. Szurmant, H., and G. W. Ordal. 2004. Diversity of chemotaxis mechanisms among the *Bacteria* and *Archaea*. *Microbiol. Mol. Biol. Rev.* **68:**301–319.

290. Tahara, M., A. Ohsawa, S. Saito, and M. Kimura. 2004. *In vitro* phosphorylation of initiation factor 2α (aIF2α) from hyperthermophilic archaeon *Pyrococcus horikoshii* OT3. *J. Biochem. (Tokyo)* **135:**479–485.

291. Tang, X. F., S. Ezaki, H. Atomi, and T. Imanaka. 2001. Anthranilate synthase without an LLES motif from a hyperthermophilic archaeon is inhibited by tryptophan. *Biochem. Biophys. Res. Commun.* **281:**858–865.

292. Tanseem, A., L. M. Iyer, E. Jacobsson, and L. Aravind. 2004. Identification of the prokaryotic ligand-gated ion channels and their implications for the mechanisms and origins of animal Cys-loop ion channels. *Genome Biol.* **6:**R4.

293. Taylor, B. L., and I. B. Zhulin. 1999. PAS domains. Internal sensors of oxygen, redox potential, and light. *Microbiol. Mol. Biol. Rev.* **63:**479–506.

294. Thapar, N., S. C. Griffith, T. O. Yeates, and S. Clarke. 2002. Protein repair methyltransferase from the hyperthermophilic archaeon *Pyrococcus furiosus*. *J. Biol. Chem.* **227:**1058–1065.

295. Thompson, D. K., and C. J. Daniels. 1998. Heat shock inducibility of an archaeal TATA-like promoter is controlled by adjacent sequence elements. *Mol. Microbiol.* **27:**541–551.

296. Thompson, D. K., J. R. Palmer, and C. J. Daniels. 1999. Expression and heat-responsive regulation of a TFIIB homologue from the archaeon *Haloferax volcanii*. *Mol. Microbiol.* **33:**1081–1092.

297. Tolbert, W. D., D. E. Graham, R. H. White, and St. E. Ealick. 2003. Pyruvoyl-dependent arginine decarboxylase from *Methanococcus jannaschii*: crystal structures of the self-cleaved and S53A proenzyme forms. *Structure* **11:**285–294.

298. Trent, J. D., J. Osipiuk, and T. Pinkau. 1990. Acquired thermotolerance and heat shock in the extremely thermophilic archaebacterium *Sulfolobus* sp. strain B12. *J. Bacteriol.* **172:**1478–1484.

299. Tutino, M. L., A. Tosco, G. Marino, and G. Sannia. 1997. Expression of *Sulfolobus solfactaricus trpE* and *trpG* genes in *E. coli*. *Biochem. Biophys. Res. Commun.* **230:**306–310.

300. Ulrich, L. E., E. V. Koonin, and I. B. Zhulin. 2005. One-component systems dominate signal transduction in prokaryotes. *Trends Microbiol.* **13:**52–56.

301. Van Boxstael, S., D. Maes, and R. Cunin. 2005. Aspartate transcarbamylase from the hyperthermophilic archaeon *Py-*

rococcus abyssi. Insights into cooperative and allosteric mechanisms. *FEBS J.* 272:2670–2683.

302. van der Oost, J., G. Schut, S. W. M. Kengen, W. R. Hagen, M. Thomm, and W. M. de Vos. 1998. The ferredoxin-dependent conversion of glyceraldehyde-3-phosphate in the hyperthermophilic archaeon *Pyrococcus furiosus* represents a novel site of glycolytic regulation. *J. Biol. Chem.* 273:28149–28154.

303. Varecka, L., P. Smigan, and M. Greksak. 1996. The presence of H⁺ and Na⁺-linked Ca²⁺ extruding systems in *Methanobacterium thermoautotrophicum*. *FEBS Lett.* 399:171–174.

304. Veit, K. C. Ehlers, and R. A. Schmitz. 2005. Effects of nitrogen and carbon sources on transcription of soluble methyltransferases in *Methanosarcina mazei* Stain Go1. *J. Bacteriol.* 187:6147–6154.

305. Verhees, C. H., S. W. M. Kengen, J. E. Tuininga, G. J. Schut, M. W. W. Adams, W. M. de Vos, and J. van der Oost. 2003. The unique features of glycolytic pathways in *Archaea*. *Biochem. J.* 375:231–246.

306. Vermeij, P., F. J. Detmers, F. J. Broers, J. T. Keltjens, and C. Van der Drift. 1994. Purification and characterization of coenzyme F390 synthetase from *Methanobacterium thermoautotrophicum* (strain delta H). *Eur. J. Biochem.* 226:185–191.

307. Vermeij, P., E. Vinke, J. T. Keltjens, and C. Van Der Drift. 1995. Purification and properties of coenzyme F390 hydrolase from *Methanobacterium thermoautotrophicum* (strain Marburg). *Eur. J. Biochem.* 234:592–597.

308. Vitreschak, A. G., D. A. Rodionov, A. A. Mironov, and M. S. Gelfand. 2004. Riboswitches: The oldest mechanism for the regulation of gene expression? *Trends Genet.* 20:44–50.

309. Voorhorst, W. G.B., R. I. L. Eggen, A. C. M. Geerling, C. Platteeuw, R. J. Siezen, and W. M. de Vos. 1996. Isolation and characterization of the hyperthermostable serine protease, pyrolysin, and its gene from the hyperthermophilic archaeon *Pyrococcus furiosus*. *J. Biol. Chem.* 271:20426–20431.

310. Wakagi, T., T. Fujii, and T. Oshima. 1996. Molecular cloning, sequencing, and heterologous expression of a novel zinc-containing ferredoxin gene from a thermoacidophilic archaeon *Sulfolobus* sp. strain 7. *Biochem. Biophys. Res. Commun.* 225:489–493.

311. Wallen, E., and G. Von Heijne. 1998. Genome-wide analysis of integral membrane proteins from eubacterial, archaean, and eukaryotic organisms. *Protein Sci.* 7:1029–1038.

312. Wardleworth, B. N., R. J. M. Russell, S. D. Bell, G. L. Taylor, and M. F. White. 2002. Structure of Alba: An archaeal chromatin protein modulated by acetylation. *EMBO J.* 21:8654–8662.

313. Weinbert, M. V., G. J. Schut, S. Brehm, S. Datta, and M. W. W. Adams. 2005. Cold shock of a hyperthermophilic ar-

chaeon: *Pyrococcus furiosus* exhibits multiple responses to a suboptimal growth temperature with a key role for membrane-bound glycoproteins, *J. Bacteriol.* 187:336–348.

314. Welchman, R. L., C. Gordon, and R. J. Mayer. 2005. Ugiquitin and ubiquitin-like proteins as multifunctional signals. *Nat. Rev. Mol. Cell Biol.* 6:599–609.

315. Westheimer, F. H. 1987. Why nature chose phosphates. *Science* 235:1173–1178.

316. White, M. F., and S. D. Bell. 2002. Holding it together: Chromatin in the *Archaea*. *Trends Genet.* 18:621–626.

317. Wilting, R., S. Schorling, B. C. Persson, and A. Bock. 1997. Selenoprotein synthesis in archaea: Identification of an mRNA element of *Methanococcus jannaschii* probably directing selenocysteine insertion. *J. Mol. Biol.* 266:637–641.

318. Wolf, S., F. Lottspeich, and W. Baumeister. 1993. Ubiquitin found in the archaebacterium *Thermoplasma acidophilum*. *FEBS Lett.* 326:42–44.

319. Wolfe, A. J. 2005. The acetate switch. *Microbiol. Mol. Biol. Rev.* 69:12–50.

320. Wood, E. R., F. Ghane, and D. W. Grogan. 1997. Genetic responses of the thermophilic archaeon *Sulfolobus acidocaldarius* to short-wavelength UV light. *J. Bacteriol.* 179:5693–5698.

321. Wood, G. E., A. K. Haydock, and J. A. Leigh. 2003. Function and regulation of the formate dehydrogenase genes of the methanogenic archaeon *Methanoccus maripaludis*. *J. Bacteriol.* 185:2548–2554.

322. Wurgler-Murphy, S. M., D. M. King, and P. J. Kennelly. 2004. The Phosphorylation Site Database: A guide to the serine-, threonine-, and/or tyrosine-phosphorylated proteins in prokaryotic organisms. *Proteomics* 4:1562–1570.

323. Xing, R., and W. B. Whitman. 1987. Sulfometuron methyl-sensitive and –resistant acetolactate synthases of the archaebacteria *Methanococcus* spp. *J. Bacteriol.* 169:4486–4492.

324. Yao, M., A. Ohsawa, S. Kikukawa, I. Tanaka, and M. Kimura. 2003. Crystal structure of hyperthermophilic archaeal initiation factor 5A: A homologue of eukaryotic initiation factor 5A (eIF-5A). *J. Biochem. (Tokyo)* 133:75–81.

325. Zhao, K., X. Chai, and R. Marmorstein. 2003. Structure of a Sir2 substrate, Alba, reveals a mechanism for deacetylation-induced enhancement of DNA binding. *J. Biol. Chem.* 278:26071–26077.

326. Zhao, R., E. J. Collin, R. B. Bourret, and R. E. Silversmith. 2002. Structure and catalytic mechanism of the *E coli* chemotaxis phosphatase CheZ. *Nature Struct. Biol.* 9:570–575.

327. Zhulin, I. B., A. N. Nikolskaya, and M. Y. Galperin. 2003. Common extracellular sensory domains in transmembrane receptors for diverse signal transduction pathways in bacteria and archaea. *J. Bacteriol.* 185:285–294.

328. Zwickl, P., J. Kleinz, and W. Baumeister. 1994. Critical elements in proteasome assembly. *Nat. Struct Biol.* 1:765–770.

Archaea: Molecular and Cellular Biology
Edited by Ricardo Cavicchioli
© 2007 ASM Press, Washington, D.C.

Chapter 12

Central Metabolism

MICHAEL J. DANSON, HENRY J. LAMBLE, AND DAVID W. HOUGH

INTRODUCTION

Organisms utilizing organic nutrients do so to supply the precursors of all their cellular components and to generate energy for biosynthesis and other endergonic processes. The degradative pathways by which these nutrients are metabolized are known as "catabolic" routes, whereas the biosynthetic pathways are referred to as "anabolic" routes. The metabolic link between these two processes is provided by the pathways of *central metabolism*, the reactions of which also serve as the main energy-generating routes. Organisms growing autotrophically also have these same pathways of central metabolism, although the primary function in these cells is to provide biosynthetic precursors, while energy is produced photosynthetically or via chemolithotrophic reactions.

The pathways of central metabolism are therefore at the heart of an organism's total metabolic capacity, and their wide conservation suggests they were an early evolutionary invention. Consistent with this view are common themes that are found spanning the *Archaea*, *Bacteria*, and *Eucarya*, although variations are observed that reflect not only phylogeny but also particular lifestyles and requirements. With this in mind, the principal aim of this chapter is to describe the central metabolic pathways of the *Archaea* and to identify the unique or unusual features of archaeal metabolism.

Several reviews on archaeal central metabolism have been written previously. However, they have necessarily relied mainly on enzymological data, and what makes a new review important at this stage is that we are now able to integrate these data with information from genome sequences. While taking this line, it will become clear that the currently-fashionable "Systems Biology" approach depends on data input from both biochemistry and molecular biology; indeed, in various areas of central metabolism, gene

sequences, at best, predict enzymic activities, whereas biochemical studies actually define them.

THE PATHWAYS OF CENTRAL METABOLISM IN THE *ARCHAEA*

The chapter is structured in four main sections. The first two form the centerpiece of central metabolism: the conversion of sugars to pyruvate, and then the metabolic fate of pyruvate, either to organic end products or to CO_2 by complete oxidation via the citric acid cycle. It is directly into these two sets of pathways that all nutrients essentially feed and from which biosynthesis commences. However, it is necessary and instructive to add two further sections. First, growth on acetate is discussed as this may involve an additional cyclic pathway, the glyoxylate cycle. Second, the catabolism of amino acids is included; while these do feed into the citric acid cycle, catabolism of branched-chain amino acids in particular deserves a special mention as it is in these reactions that the presence of a family of multienzyme complexes was discovered, which were until recently thought to be absent from all archaea.

To fulfil the objective of integrating biochemical and genomic data, discussions and analyses are concentrated on those archaea whose genomes have been sequenced. However, inevitably there will be crucial data from archaeal organisms for which the complete DNA sequence is not known, and reference will be made to these where necessary.

The Metabolism of Monosaccharides to Pyruvate

Catabolism of glucose

Virtually all organisms from all three domains of life have the ability to metabolize glucose to pyruvate.

Michael J. Danson, Henry J. Lamble, and David W. Hough • Centre for Extremophile Research, Department of Biology and Biochemistry, University of Bath, Bath BA2 7AY, United Kingdom.

This applies regardless of whether an organism grows anaerobically or aerobically, heterotrophically or autotrophically, or whether it employs fermentative metabolism. The conventional Embden-Meyerhof (EM) pathway is perhaps the most widely recognized and thoroughly investigated central metabolic pathway (48), and this conservation has led to a perceived fundamental and sacrosanct nature of the pathway. However, in reality there is considerable variability in the route of glucose catabolism in different organisms, and in 1952 a fundamentally distinct pathway for the breakdown of glucose to pyruvate was reported in *Pseudomonas saccharophila* (41). This Entner-Doudoroff (ED) pathway was revealed by a characteristic difference to the EM pathway in the labeling pattern of pyruvate when 1-^{14}C-glucose is metabolized (Fig. 1). It is now apparent that the ED pathway of glucose metabolism has a considerably wider distribution than was originally appreciated (23, 49, 86). Organisms from all three domains of life have been shown to possess ED-type pathways, often alongside the EM pathway. In the *Archaea*, EM and ED pathways, and variations and combinations thereof, represent the predominant routes for glucose catabolism. In certain bacteria and lower eucarya, other pathways, most notably the oxidative pentose phosphate pathway and the phosphoketolase pathway, can also contribute to glycolytic flux (24).

The classical EM and ED pathways (Fig. 2) both begin with the phosphorylation of glucose to glucose 6-phosphate, which may be performed by a specific glucokinase or a broad specificity hexokinase. Alternatively, the phosphoenolpyruvate-dependent phosphotransferase system, which is employed by many bacteria for sugar uptake, phosphorylates glucose as it is transported into the cell. In the EM pathway, phosphoglucose isomerase then converts glucose 6-phosphate to fructose 6-phosphate, which is further phosphorylated by phosphofructokinase to produce fructose 1,6-bisphosphate. Fructose-1,6-bisphosphate aldolase then catalyzes the cleavage of the C6 sugar, fructose 1,6-bisphosphate, into two C3 compounds, glyceraldehyde 3-phosphate and dihydroxyacetone phosphate. A second molecule of glyceraldehyde 3-phosphate is produced from dihydroxyacetone phosphate via the action of triose-phosphate isomerase.

In the classical ED pathway, the glucose 6-phosphate is oxidized by glucose-6-phosphate dehydrogenase to produce 6-phosphogluconate, which is then dehydrated by 6-phosphogluconate dehydratase to

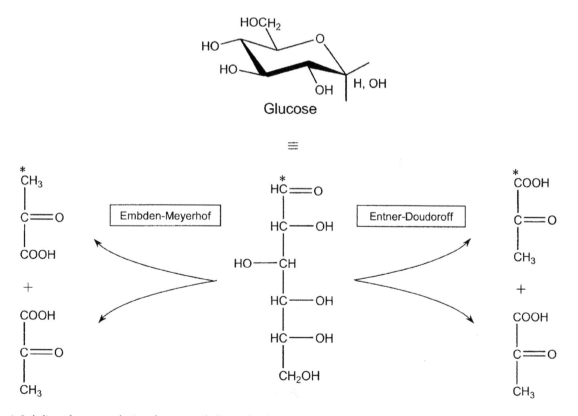

Figure 1. Labeling of pyruvate during glucose catabolism. The characteristic labeling pattern of pyruvate resulting from glucose catabolism by the Embden-Meyerhof and Entner-Doudoroff pathways.

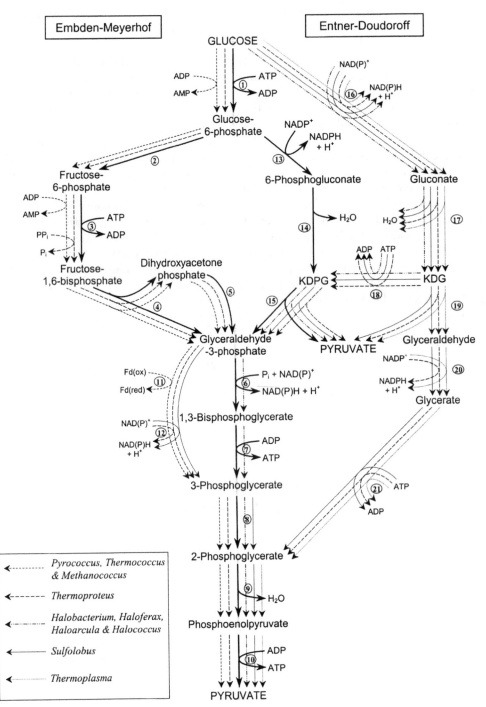

Figure 2. Pathways of glucose metabolism. The classical Embden-Meyerhof and Entner-Doudoroff pathways of bacteria and eucarya are shown in bold with each step shown connected by large full arrows, while the alternative pathways of selected archaeal genera are displayed with various small arrows (see key). Unless specified, the cofactor usage is as shown for the classical pathways. Enzymes are denoted by numbers: 1 = glucokinase, 2 = phosphoglucose isomerase, 3 = phosphofructokinase, 4 = fructose-1,6-bisphosphate aldolase, 5 = triose-phosphate isomerase, 6 = glyceraldehyde-3-phosphate dehydrogenase, 7 = phosphoglycerate kinase, 8 = phosphoglycerate mutase, 9 = enolase, 10 = pyruvate kinase, 11 = glyceraldehyde-3-phosphate ferredoxin oxidoreductase, 12 = nonphosphorylating glyceraldehyde-3-phosphate dehydrogenase, 13 = glucose-6-phosphate dehydrogenase, 14 = 6-phosphogluconate dehydratase, 15 = KDPG aldolase, 16 = glucose dehydrogenase, 17 = gluconate dehydratase, 18 = KDG kinase, 19 = KDG aldolase, 20 = glyceraldehyde dehydrogenase, 21 = glycerate kinase. In *Sulfolobus* species it is not yet clear whether the conversion of glyceraldehyde to glycerate is catalyzed by glyceraldehyde dehydrogenase (20) or glyceraldehyde oxidoreductase; see text for details. The reactions involved in the conversion of glucose, or other C6 sugars, to C3 intermediates make up the upper pathway, whereas the lower pathway refers to the conversion of C3 intermediates to pyruvate.

give 2-keto-3-deoxy-6-phosphogluconate (KDPG). KDPG aldolase then catalyzes the cleavage of the C6 compound, KDPG, to produce two C3 compounds, glyceraldehyde 3-phosphate and pyruvate. In both EM and ED pathways, glyceraldehyde 3-phosphate is converted to pyruvate by a five-step reaction sequence. It is first phosphorylated and oxidized by glyceraldehyde-3-phosphate dehydrogenase to give 1,3-bisphosphoglycerate. This is then converted to 3-phosphoglycerate by phosphoglycerate kinase, coupled to the synthesis of ATP from ADP. 3-Phosphoglycerate is rearranged by phosphoglycerate mutase to 2-phosphoglycerate, which is dehydrated by enolase to phosphoenolpyruvate. Finally, pyruvate kinase converts phosphoenolpyruvate to pyruvate, with the synthesis of another molecule of ATP from ADP. For each molecule of glucose, the classical EM pathway has an overall ATP yield of two, while the classical ED pathway has an overall yield of one.

The nature of glucose catabolism in the *Archaea* is of considerable interest because of its fundamental importance to understanding their unique biochemistry, and because of the evolutionary significance of the highly conserved pathways and enzymes. As such it has been the subject of several reviews both before (27, 30) and after (126, 166) the publication of genome sequences. The ensuing text will attempt to provide an up-to-date report of the novel pathways, enzymes, and cofactors found in archaeal glucose metabolism, and compare and contrast them to the classical pathways of eucarya and bacteria. In addition, differences in the metabolism of other monosaccharides, the regulation of the pathways, and gluconeogenesis will be considered. Inevitably the literature represents a considerable bias toward research into the metabolism of particular sugars (especially glucose) in certain "model" organisms, in particular, those employing fermentative metabolism. Fortuitously, the species that have been the focus of most of the research into archaeal metabolism represent a diverse range of archaeal genera and growth environments. The combination of genome sequence and biochemical data from these organisms has thus allowed a complete picture of glucose metabolism in the *Archaea* to emerge.

Halobacterium, *Haloarcula*, *Halococcus*, and *Haloferax*.

Some important early work on archaeal glucose metabolism was performed with the extreme halophile, *Halobacterium saccharovorum*, isolated from a saltern in San Francisco Bay, California (159). Enzymological studies on cell extracts revealed that glucose metabolism occurs by a part-phosphorylative variant of the ED pathway (160). In this pathway, instead of being phosphorylated, glucose is first oxidized to gluconate by glucose dehydrogenase. A dehydratase enzyme then converts gluconate to KDG, which is phosphorylated to KDPG by KDG kinase. KDPG is then metabolized as in the classical ED pathway to yield two molecules of pyruvate (Fig. 2). As with the classical ED pathway, this variant has an overall yield of one ATP per molecule of glucose. This part-phosphorylative pathway had previously been observed in *Rhodopseudomonas sphaeroides* (157) and several other bacterial species, including *Clostridium aceticum*, where it is employed for the metabolism of gluconate (5).

Subsequent enzymological studies have shown that the part-phosphorylative ED pathway also accounts for glucose metabolism in other halophilic archaeal genera, including *Haloferax*, *Haloarcula*, and *Halococcus* (119, 144, 145). In most cases, the ED enzymes have been shown to be inducibly expressed during growth on glucose, and not produced constitutively (72, 119, 154). A ^{13}C-glucose labeling study in *Halococcus saccharolyticus* has confirmed that glucose metabolism occurs exclusively via the part-phosphorylative ED pathway in this organism (72). The first enzyme of the pathway, glucose dehydrogenase, has been purified and characterized from *Haloferax mediterranei* and shown to be a Zn^{2+}-containing enzyme with a dual specificity for NAD^+ and $NADP^+$ (11). A complete glycolytic EM pathway has not been detected in any halophilic archaeon, and it therefore appears that the different halophilic genera perform glucose catabolism exclusively via the part-phosphorylative ED pathway.

It has become clear that most halophiles can grow on carbohydrate energy sources, despite initial predictions to the contrary (119). Even species that are reported to be unable to use carbohydrates, such as *Halobacterium* sp. NRC-1, have been shown to possess the genes of the part-phosphorylative ED pathway. In the published genome sequence of this organism (105), genes have been annotated for all enzymes of the part-phosphorylative ED pathway and, interestingly, genes for the upper pathway enzymes, glucose dehydrogenase, gluconate dehydratase, and KDPG aldolase, appear in a cluster, without the KDG kinase gene. There is a near-full complement of genes for the EM sequence, although no gene for 6-phosphofructokinase has been annotated (166). A similar complement of genes was found subsequently to be present in the genome of *Haloarcula marismortui*, although in this organism there is no cluster of ED genes (8). Genes for glucose dehydrogenase, gluconate dehydratase, and KDG kinase from the two organisms have a high level of amino acid sequence identity, although there is no clear KDPG aldolase ortholog in *H. marismortui*. Additionally, the predicted

KDPG aldolase gene from *Haloferax alicantei* (65) does not have a clear ortholog in either genome, implying a diverse evolutionary history to this gene within the halophilic archaea. As yet, it is unclear whether halophilic organisms with a highly-restricted metabolic capability, such as *Halosimplex carlsbadense* (170), also retain the genes and enzymes for the ED pathway.

The route of glucose metabolism in halophilic archaea appears to contrast with the situation in halophilic bacteria. A recent enzymological study on the extremely halophilic bacterium *Salinibacter ruber* concluded that glucose is metabolized via the classical ED pathway, and not the part-phosphorylative variant, despite the physiological closeness between this organism and halophilic archaea of the family *Halobacteriaceae* (108). This observation requires further investigation and may have interesting evolutionary implications pertaining to the lateral transfer of metabolic genes between the *Bacteria* and *Archaea*.

Sulfolobus. *Sulfolobus* species exhibit considerable metabolic diversity and versatility and are commonly considered to be opportunistic heterotrophs, capable of utilizing a wide range of carbohydrate energy sources (57, 131). Early work on *Sulfolobus solfataricus* showed that glucose was metabolized via a nonphosphorylative variant of the ED pathway (36). Glucose-labeling studies have suggested that a similar variant pathway accounts for metabolism in the closely-related organism *S. acidocaldarius* (142) and other *Sulfolobus* species (175). In the nonphosphorylative ED pathway (Fig. 2), glucose is metabolized to KDG as described for the part-phosphorylative variant, although, in this case, KDG is cleaved directly to glyceraldehyde and pyruvate by KDG aldolase. Glyceraldehyde dehydrogenase or glyceraldehyde oxidoreductase is then thought to oxidize glyceraldehyde to glycerate, which is phosphorylated by glycerate kinase to give 2-phosphoglycerate. A second molecule of pyruvate can then be produced from this by the actions of enolase and pyruvate kinase, as occurs in the classical pathways. A similar nonphosphorylative pathway had previously been shown to be responsible for the metabolism of gluconate in *Aspergillus niger* (40). Note that the nonphosphorylative ED pathway proceeds with no net yield of ATP. This is in contrast to the classical and part-phosphorylative routes described previously, both of which have an ATP yield of one mole per mole of glucose.

Glucose dehydrogenase from *S. solfataricus* has been purified and characterized and found to be a highly-expressed and active enzyme, with specificity for NAD^+ and $NADP^+$, as was observed with the enzyme from *H. mediterranei* (98). Gluconate dehydratase and KDG aldolase, which both possess (β/α) structures, have also been purified and characterized biochemically (16, 99). Enzymes of the lower pathway, from glyceraldehyde to 2-phosphoglycerate, have not been fully characterized in *Sulfolobus* spp., although a molybdenum-containing glyceraldehyde oxidoreductase has been discovered in *S. acidocaldarius* (78), and putative glyceraldehyde dehydrogenase genes have been annotated in *S. solfataricus* (166). The enzyme activities constituting a complete EM pathway have not been detected in any *Sulfolobus* species, with a notable absence of phosphofructokinase activity.

Unexpectedly, the genome sequence of *S. solfataricus* (146) revealed that the genes encoding gluconate dehydratase and KDG aldolase were in an ED cluster, upstream of putative genes encoding KDG kinase and nonphosphorylating glyceraldehyde-3-phosphate dehydrogenase. Subsequent biochemical analysis of the natural and recombinant enzymes confirmed these activities and also revealed that the KDG aldolase displays KDPG aldolase activity (1). This work provides convincing evidence that, in fact, *S. solfataricus* metabolizes glucose via the part-phosphorylative ED pathway, which may occur alone or in parallel with the nonphosphorylative pathway (Fig. 2).

The NADP-dependent, nonphosphorylating glyceraldehyde-3-phosphate dehydrogenase represents a bypass of the conventional glyceraldehyde-3-phosphate dehydrogenase and phosphoglycerate kinase, and is thought to operate only in the catabolic direction. An orthologous gene is also present in the genome of *Halobacterium* sp. NRC-1 (166). Although the corresponding enzyme activity has not been confirmed, it is possible that halophilic archaea also employ this enzyme in the catabolic direction. The result of this enzyme bypass is that the modified part-phosphorylative pathway that occurs in *S. solfataricus* has a net ATP yield of zero.

It is not clear what advantage the organism would gain by having parallel non- and part-phosphorylative pathways, or what regulates which pathway is favored. For twenty years or more it has been known that microorganisms frequently employ a variety of pathways for glucose metabolism, depending on growth conditions and substrates (110). A powerful example is *Thiobacillus* A2, in which ED, EM, and oxidative pentose-phosphate pathways constitute varying proportions of glycolytic flux, depending on culture conditions (174). One could speculate that the presence of two variant ED pathways may give *S. solfataricus* more metabolic versatility, providing a greater opportunity to generate intermediates for biosynthesis. For example, during glycolytic growth conditions there may be a requirement for glyceraldehyde

3-phosphate, which is used in the production of deoxyribose 5-phosphate for nucleoside biosynthesis (116). It should also be considered that the nonphosphorylative pathway permits the conversion of glucose to pyruvate in only three steps, without any ATP input, which may permit the supply of carbon to the energy-generating citric acid cycle under conditions of low ATP.

In contrast to halophilic archaea, the ED enzymes in *S. solfataricus* appear to be constitutively expressed. Specific activities of glucose dehydrogenase, gluconate dehydratase, and KDG aldolase have been shown to be unaffected by culture age or growth substrate (glucose or yeast extract) (36). *Sulfolobus* species have been shown to maintain an energy store in the form of glycogen (94), which may explain the presence of the ED pathway enzymes even during growth on noncarbohydrate food sources.

Thermoplasma. *Thermoplasma* species are thermoacidophilic, facultatively anaerobic, obligate heterotrophs with growth requirements of 33 to 67°C and pH 0.5 to 4 (141). Enzymological studies on one model organism, *Thermoplasma acidophilum* (32), have revealed that glucose metabolism occurs via the nonphosphorylative ED pathway (17) (Fig. 2), as reported in *Sulfolobus* species. However, in *T. acidophilum* the activities of the lower pathway have been confirmed enzymologically. The first enzyme of the pathway, glucose dehydrogenase, has been comprehensively characterized and was found to exhibit dual cofactor specificity for NAD and NADP (151). The complete genome sequence of *T. acidophilum* (129) revealed the expected complement of genes for the nonphosphorylative ED pathway, while key activities and genes of the EM pathway, notably phosphofructokinase, have not been found. A similar distribution of homologous genes has also been found in the genome of the closely related organism *Thermoplasma volcanium* (81). No ortholog of the *S. solfataricus* KDG kinase gene can be identified in the genomes of *Thermoplasma* species, so that the presence of the part-phosphorylative ED pathway in *Thermoplasma* remains an open question.

Recently, the genome sequence of *Picrophilus torridus* has revealed a similar profile of metabolic genes to *Thermoplasma* species (54). This organism, and the closely related species *Picrophilus oshimae*, exhibits a close physiological relationship to *Thermoplasma*, and it is likely that they metabolize glucose in the same way. The first enzyme of the nonphosphorylative ED pathway, glucose dehydrogenase, has been characterized from *P. torridus*, and it exhibits properties similar to the enzyme from *T. acidophilum* (6).

Pyrococcus, Thermococcus, Methanococcus, and Other Hyperthermophiles. *Pyrococcus* species are hyperthermophilic members of the *Euryarchaeota*, with typical growth requirements of 70 to 100°C, 0.5 to 5% NaCl and pH 5 to 9. They are strictly anaerobic heterotrophs and can be cultured on minimal salt media with a variety of carbon sources, such as maltose and starch. The most researched model organism is *Pyrococcus furiosus* (45). Extensive investigation by ^{13}C-labeling and enzymological studies has revealed glucose metabolism to proceed by a modified EM pathway (Fig. 2), with several critical differences to the classical pathway (83, 135). Although virtually the same sequence of chemical transformations is performed to convert glucose to pyruvate, there are notable differences in the cofactor usage in this organism, and many of the enzymes are not evolutionarily related to known bacterial or eucaryal enzymes. The first step is catalyzed by an ADP-dependent (AMP-forming) glucokinase, which is induced by growth on sugar substrates (85). An unusual phosphoglucose isomerase, a member of the cupin superfamily, then catalyzes the isomerization of glucose 6-phosphate to fructose 6-phosphate (58, 165). Fructose 6-phosphate phosphorylation is catalyzed by an ADP-dependent (AMP-forming) phosphofructokinase (161), which was found to be a member of a novel, evolutionarily distinct family of phosphofructokinases (family C) (125). As observed in the part-phosphorylative ED pathway of *S. solfataricus*, glyceraldehyde 3-phosphate is converted directly to 3-phosphoglycerate, bypassing the activities of glyceraldehyde-3-phosphate dehydrogenase and phosphoglycerate kinase. However, in *P. furiosus* this reaction is catalyzed by a unique, inducible, ferredoxin-dependent glyceraldehyde-3-phosphate oxidoreductase (164). Phosphoglycerate mutase has also been characterized and was again found to represent a new protein family, unrelated to bacterial and eucaryal enzymes (163). It has been shown that the glycolytic route in *P. furiosus* has a net yield of one mole of ATP per mole of glucose (83).

The complete genome sequence of *P. furiosus* has revealed genes for all the expected EM enzymes and a notable absence of ED enzyme genes (100). In addition, the genomes of *Pyrococcus horikoshii* (79) and *Pyrococcus abyssi* (22) have been sequenced and show a similar distribution of orthologous glycolytic enzyme genes. This provides good evidence that the sole route of glucose catabolism in the genus *Pyrococcus* is via the same modified EM pathway.

In addition to *Pyrococcus*, some research has also been performed on glucose catabolism in other organisms of the order *Thermococcales*. Notably, there is good evidence that species of *Thermococcus* employ a glycolytic pathway very similar to that of *Pyrococcus*. For example, the complete pathway has been characterized in *Thermococcus celer* and *Thermococcus litoralis* by enzyme assays and labeling studies (142). In addition, an ADP-dependent glucokinase has been characterized in *T. litoralis* (92), and an ADP-dependent phosphofructokinase has been characterized in *T. zilligii* (124), with both enzymes possessing similar properties to the enzymes from *P. furiosus*. The genome sequence of *Thermococcus kodakaraensis* (53) revealed orthologous sequences for all the expected enzymes of the modified EM pathway.

Species of *Methanococcus* also possess an EM glycolytic pathway similar to that in *Pyrococcus*. The most researched model organism is the strictly-anaerobic, hyperthermophilic autotroph, *Methanococcus jannaschii* (75). An ADP-dependent phosphofructokinase has been characterized in this organism and was also found to possess glucokinase activity (130, 165). Additionally, highly divergent phosphoglucose isomerase (127) and phosphoglycerate mutase (56) genes have been discovered. The published genomic sequence revealed all the expected orthologs of EM pathway enzymes (18) and an absence of ED pathway genes. A comprehensive study of glycolytic enzyme activities has been performed on the closely related organism *Methanococcus maripaludis* (177). Enzymes of the EM pathway were detected, while enzymes of the ED and oxidative pentose phosphate pathways could not be found.

Given the strictly autotrophic nature of *Methanococcus* spp., it may seem surprising that they possess a pathway for glucose catabolism. However, several methanogenic genera are known to contain glycogen stores (93) that are metabolized by the modified EM pathway. A comprehensive investigation of phosphofructokinase in methanogenic archaea has revealed that a number of glycogen-containing organisms, both mesophiles and thermophiles from the genera *Methanococcus* and *Methanosarcina*, possess the activity (168). Organisms that do not possess glycogen stores, such as *Methanothermobacter thermautotrophicus*, were found not to possess any ADP- or ATP-dependent phosphofructokinase activity, which suggests they do not contain a complete glycolytic pathway. The inability to find key glycolytic genes in the genome sequence of *M. thermautotrophicus* provides further support for this assertion (150, 166). However, it has been documented that 1-^{13}C-glucose and 6-^{13}C-glucose are metabolized by cells of the organism to form exclusively 3-^{13}C-2,3-cyclopyrophosphoglycerate, convincingly demonstrating that an EM glycolytic pathway *is* present (43, 44). This discrepancy could be explained by a novel path-

way of glucose metabolism, or evolutionarily-distant enzymes of the EM pathway, possibly with unusual cofactor dependence.

Several other hyperthermophilic genera have a similar glycolytic route to those described above, with only minor variations. For example, the hyperthermophilic, sulfate-reducer *Archaeoglobus fulgidus* strain 7324 has been shown by enzymological studies to possess all the required activities of the modified EM pathway (96), and ADP-dependent kinases for glucose (97) and fructose 6-phosphate (60) have been characterized. However, orthologous genes could not be found in the genome sequence of strain VC16, and so it is unclear whether the EM pathway is employed universally in this genus (89). The hyperthermophilic anaerobe *Desulfurococcus amylolyticus* has been shown by labeling and enzymological studies to use a similar EM pathway for glucose catabolism (142), although, in this organism, ATP-dependent kinases were found for the phosphorylation of glucose and fructose 6-phosphate. The aerobic, hyperthermophilic crenarchaeon *Aeropyrum pernix* also appears to employ a modified EM pathway with ATP-dependent kinases. The ATP-dependent glucose phosphorylation is catalyzed by a broad-specificity hexokinase, with similarity to the ROK group of bacterial sugar kinases (59). The conversion of glucose 6-phosphate to fructose 6-phosphate is performed by an unusual bifunctional phosphoglucose/phosphomannose isomerase (61). The ATP-dependent phosphofructokinase is a family B enzyme, otherwise only found in certain enterobacteria, which contrasts with the family C ADP-dependent enzymes of most hyperthermophilic archaea (123). The published genome sequence of this organism revealed the expected EM pathway genes and suggested that a nonphosphorylating glyceraldehyde-3-phosphate dehydrogenase is employed instead of a ferredoxin-dependent enzyme (80).

An important comparison should be made between metabolism in hyperthermophilic archaea and the pathways present in hyperthermophilic bacteria. Metabolism has been investigated in the hyperthermophilic bacterium *Thermotoga maritima*, and under the conditions used, glucose was metabolized by both the classical EM pathway and the classical ED pathway (142). None of the archaeal modifications to the EM pathway was found, such as a nonphosphorylating glyceraldehyde-3-phosphate dehydrogenase or kinases with unusual cofactor specificity. Furthermore, no non- or part-phosphorylative variants of the ED pathway have been detected. One interesting novel feature is the presence of a fusion enzyme with phosphoglycerate kinase and triose-phosphate isomerase activities (138). The published genome sequence supports these observations as it contains the expected profile of genes that encode the enzymes of the classical glycolytic pathways (104).

Thermoproteus. One other model organism that should be considered to give a complete overview of archaeal glucose metabolism is the anaerobic, hyperthermophilic crenarchaeaon *Thermoproteus tenax*, which is capable of autotrophic and heterotrophic growth. The pathways and enzymes present have been pieced together by a combination of genomic, enzymological, and microbiological techniques (149). Remarkably, this organism appears to contain all three variants of glucose metabolic pathways found in archaea; it can use the nonphosphorylative and part-phosphorylative ED pathways and its own variant of the EM pathway (Fig. 2). Notably, EM metabolism in *T. tenax* involves a number of distinct enzymes to those in *P. furiosus*, including a broad-specificity ATP-dependent hexokinase of the ROK group (38), a family A pyrophosphate-dependent phosphofructokinase (148), and an NAD-dependent, nonphosphorylating glyceraldehyde-3-phosphate dehydrogenase (14). A novel class I fructose-1,6-bisphosphate aldolase was also discovered in this organism (147), and orthologous sequences were subsequently identified in virtually all archaeal genomes (166). This enzyme contrasts with the class II enzyme found to be present in thermophilic bacteria, such as *Thermus aquaticus* (34). As in *Sulfolobus*, a single aldolase has been shown to be responsible for the cleavage of KDPG and KDG, and it has been suggested that the part-phosphorylative and nonphosphorylative ED pathways may function in parallel (149). It has been reported that cells grown with glucose and yeast extract metabolize 80 to 90% of glucose via the modified EM pathway and the remainder via the modified ED pathway(s) (142). The physiological implications of the parallel EM and ED pathways in *T. tenax* are not clear but are likely to be affected by the culture conditions employed.

Catabolism of other sugars

The vast majority of work on archaeal metabolism has focused on investigating the pathways of glucose metabolism, and to date there has been little research into how other hexose or pentose sugars enter central metabolism. This is a pertinent question given the ability of many saccharolytic archaea to metabolize alternative carbohydrates.

One example that *is* well documented is the metabolism of fructose in halophilic archaea. Work on *Haloarcula vallismortis* (3) and *Halococcus saccharolyticus* (72) has shown that this sugar enters the EM pathway via ketohexokinase-catalyzed, ATP-dependent phosphorylation to form fructose 1-phosphate; subsequent phosphorylation by 1-phosphofructokinase produces fructose 1,6-bisphosphate. The presence of class I and II fructose-1,6-bisphosphate aldolases has been documented in halophilic archaea, and they catalyze the next step in fructose metabolism (37). *Haloarcula vallismortis* and *H. mediterranei* have also been shown to catabolize sucrose and mannitol via this route after initial conversion to fructose (4). It is possible that this route is used for fructose catabolism in other archaea, such as *Sulfolobus* species, which have been shown to grow on sucrose as the sole carbon and energy source (57). A similar route of fructose metabolism via the EM pathway exists in many bacteria, although in this case, the initial phosphorylation of exogenous fructose accompanies uptake by the phosphotransferase transporter (24).

In *S. solfataricus*, it has been discovered that galactose, the C-4 epimer of glucose, is metabolized by the same nonphosphorylative ED pathway enzymes that perform glucose catabolism (98, 99). The three enzymes of the upper pathway, glucose dehydrogenase, gluconate dehydratase, and KDG aldolase, were found to have the necessary substrate promiscuity to permit their activity with substrates displaying either configuration at C-4. Similarly, enzymes of the part-phosphorylative ED pathway, which may occur in parallel with the nonphosphorylative variant in *S. solfataricus*, have also recently been found to have the same substrate promiscuity for the metabolism of both sugars (H. J. Lamble, D. W. Hough, and M. J. Danson, unpublished observations).

The catabolism of galactose by both part-phosphorylative (33) and nonphosphorylative (39) variants of the ED pathway has been documented in other organisms, although in these cases the galactose catabolic pathway consists of separate, inducible enzymes. It has been suggested that the 'metabolic pathway promiscuity' observed in *Sulfolobus* may also exist in other archaeal genera such as *Thermoplasma*, *Picrophilus*, *Haloferax*, and *Thermoproteus* (54, 98). However, it has recently been found that specific ED enzymes are induced in *Haloferax volcanii* when the organism is cultured in galactose-containing media (Lamble, Hough and Danson, unpublished), implying that a separate ED pathway is used for galactose catabolism in this organism. Additionally, it is reported that the gluconate dehydratase of the ED pathway in *Thermoproteus tenax* does not have activity with galactonate (1), although it is still possible that other enzymes of the pathway are employed for the metabolism of both sugars in this organism.

In *P. furiosus* the first step of galactose catabolism appears to be phosphorylation at C-1, and a specific galactokinase has been characterized (167). This suggests that galactose is metabolized via the Leloir pathway, in which galactose 1-phosphate is converted to glucose 1-phosphate by means of uridine nucleotide intermediates (64). It is not yet clear whether any archaea use the tagatose 6-phosphate pathway for galactose catabolism; this pathway performs the same series of chemical transformations as the EM pathway, but with enzymes that are specific for the alternative configuration at C-4.

There may be other examples of "promiscuous" central metabolic pathways in the *Archaea*. For example, the modified EM pathways reported in *A. pernix*, *Pyrobaculum aerophilum*, and *T. tenax* seem to be promiscuous for the metabolism of mannose, the C-2 epimer of glucose. The broad-specificity hexokinases found in these organisms have activity with glucose and mannose, phosphorylating both sugars at C-6. Each of these organisms also contains a bifunctional phosphoglucose/phosphomannose isomerase, which converts both compounds to fructose 6-phosphate. This situation contrasts with bacteria and eucarya, which employ a specific phosphomannose isomerase (61).

A question remains over the entry point of several other sugars into central metabolism in the *Archaea*, and will no doubt be addressed by future research. One important area to investigate is how pentose sugars can enter central metabolism. Some pentoses, such as D-xylose and L-arabinose, are known to support the growth of certain archaea, although to date there has been little investigation into how they are metabolized. One recent report has described how a specific D-xylose dehydrogenase is induced in *H. marismortui* during growth on this sugar (71), and it therefore seems likely that the oxidation of D-xylose to D-xylonate is the first step in its metabolism. The glucose dehydrogenase from both *T. acidophilum* and *S. solfataricus* has been found to use D-xylose as a substrate (Lamble, Hough and Danson, unpublished). This may imply that the glucose dehydrogenase from the "promiscuous" ED pathway in these organisms also acts as the first step in D-xylose catabolism. This proposed reaction in pentose catabolism contrasts with the situation in bacteria, where D-xylose, L-arabinose, and D-ribose are commonly metabolized via the pentose phosphate pathway.

The pentose phosphate pathway

The text above has catalogued the different variations to the classical EM and ED pathways that have been documented in model archaeal organisms. To date there is little substantive evidence that any other pathways are involved in glucose catabolism in the *Archaea*. This situation contrasts with glucose metabolism in bacteria and lower eucarya, where the pentose phosphate pathway (115) can also contribute to the glycolytic flux. The pentose phosphate pathway has an oxidative part whereby glucose 6-phosphate is converted to ribulose 5-phosphate with the release of CO_2. It also has a nonoxidative part involving transaldolase and transketolase, into which the EM pathway intermediates glyceraldehyde 3-phosphate and fructose 6-phosphate can enter. An extension to the pathway, the phosphoketolase pathway, involves the action of phosphoketolases on intermediates of the nonoxidative part and is found in some lactic acid bacteria (76). When employed in glucose catabolism, the different variations result in characteristic labeling patterns, as described for the EM and ED pathways (Fig. 1), which permit their relative contributions to be assessed. In reality, these pathways rarely contribute significantly to glycolytic flux and are most commonly employed for tetrose and pentose biosynthesis (24).

A comprehensive phylogenetic analysis of the genes for enzymes of the pentose phosphate pathway in the *Archaea* has been performed (153). *M. jannaschii*, *T. acidophilum*, and *T. volcanium* are reported to possess a complete nonoxidative pentose phosphate pathway that is employed for the synthesis of ribose 5-phosphate. Other organisms are predicted to use the ribulose monophosphate pathway for pentose biosynthesis. There is no evidence for a complete oxidative pathway in any sequenced archaeal genome, and genes for glucose-6-phosphate dehydrogenase or 6-phosphogluconate dehydrogenase have not been found. However, it should be considered that the sequences may be too distantly related to known enzymes to be detected by comparative genomics. Certain species that do not have a complete nonoxidative pentose phosphate pathway, such as *A. pernix* and *S. solfataricus*, do still possess a gene for transketolase, which is predicted to function in the synthesis of erythrose 4-phosphate for the biosynthesis of aromatic amino acids.

A novel pathway of glucose metabolism may operate alongside the EM route in *Thermococcus zilligii* (176). A labeling pattern was observed that is consistent with the pentose phosphoketolase pathway, and it is proposed that glucose is linked to this pathway by first being converted to formate and pentose phosphate. This novel pathway requires confirmation by enzyme characterization but represents the best demonstration so far of an alternative to the EM or ED pathway, in any archaeon. Another suggestion that an alternative pathway may operate in archaea came from the labeling pattern resulting from glucose metabolism by autotrophically grown strains of *Sulfolobus* (175), and this work should be reconsidered in the light of the novel pathway proposed in *T. zilligii*. It is possible that, as a wider range of organisms is investigated under a variety of growth conditions, a greater contribution of alternative and/or novel glycolytic pathways may be revealed.

Gluconeogenesis

Gluconeogeneis is essential for the generation of biosynthetic intermediates and polysaccharide energy stores and is commonly active during growth on energy sources other than hexoses. The pathway of gluconeogenesis is ubiquitous in all three phylogenetic domains and occurs by a reversal of the chemical transformations of the classical EM pathway (Fig. 3). Two critical irreversible enzymatic steps of this pathway involve alternative enzymes in archaeal gluconeogenesis—the reactions catalyzed by pyruvate kinase and phosphofructokinase. In addition, those archaea that use a nonphosphorylating glyceraldehyde-3-phosphate dehydrogenase or oxidoreductase employ the traditional enzyme reactions of phosphoglycerate kinase and glyceraldehyde-3-phosphate dehydrogenase in the gluconeogenic direction.

To bypass the pyruvate kinase reaction during gluconeogenesis, a number of possible enzymes have been reported. Most importantly, phosphoenolpyruvate (PEP) synthase homologs are present in all sequenced archaeal genomes and catalyze PEP formation from pyruvate, coupled with the conversion of ATP to AMP + P_i. In addition, several archaea have been found to contain a predicted gene for PEP carboxykinase; this enzyme has been characterized in *Thermococcus kodakaraensis* (51) and catalyzes the GTP-dependent conversion of oxaloacetate to PEP, coupled to the release of CO_2. A gene for malic enzyme has also been characterized in *T. kodakaraensis* (52), and orthologous genes are present in several other archaea. This enzyme catalyzes the conversion of malate into pyruvate, coupled to the reduction of $NADP^+$ and the release of CO_2. In *T. tenax* a further alternative has been assayed; the enzyme pyruvate phosphate dikinase catalyzes the reversible interconversion of pyruvate and PEP, coupled to the formation of AMP + PP_i from P_i + ATP. It is suggested that this enzyme may operate in both glycolytic and gluconeogenic pathways (149), although its physiological significance and distribution have not yet been established. Which enzyme is employed as the entry point of gluconeogenesis depends on the particular growth substrate an organism is utilizing, and is intrinsically connected to how the citric acid cycle is functioning (see "The citric acid cycle," below).

The other step that requires an alternative enzyme in archaeal gluconeogenesis is the reaction cat-

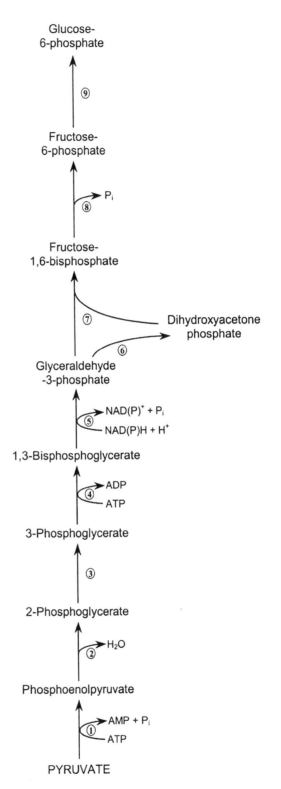

Figure 3. Gluconeogenesis. The reactions of the gluconeogenic pathway. Enzymes are denoted by numbers: 1 = phosphoenolpyruvate synthase, 2 = enolase, 3 = phosphoglycerate mutase, 4 = phosphoglycerate kinase, 5 = glyceraldehyde-3-phosphate dehydrogenase, 6 = triose-phosphate isomerase, 7 = fructose-1,6-bisphosphate aldolase, 8 = fructose-1,6-bisphosphatase, 9 = phosphoglucose isomerase.

alyzed by phosphofructokinase, the reverse reaction being performed by fructose-1,6-bisphosphatase. Of the sequenced archaeal genomes, only that of *Halobacterium* NRC-1 was found to contain a canonical fructose-1,6-bisphosphatase (type I). However, a novel bifunctional inositol-1-phosphatase/fructose-1,6-bisphosphatase (type IV) was subsequently found in *M. jannaschii* (155), and orthologs of this are present in several other archaeal genomes. Yet another fructose-1,6-bisphosphatase (type V) has been characterized in the genome of *T. kodakaraensis*, with orthologs in virtually all archaeal genome sequences (117).

The other enzyme steps of the glycolytic pathway are readily reversible and permit archaea to form glycolytic intermediates for biosynthesis. The pathway of gluconeogenesis has been characterized by enzymological and labeling studies in several organisms. For example, in the autotroph *Methanothermobacter thermautotrophicus*, it has been demonstrated that labeled carbon from $^{14}CO_2$ is incorporated into fructose phosphate and glucose phosphate (69), and in *P. furiosus*, enzymological studies have confirmed all the required enzyme activities for conversion of pyruvate to glucose 6-phosphate (133). The genes required for gluconeogenesis have been annotated in all the sequenced archaeal genomes, although a gene for phosphoglucose isomerase could not be identified in the genomes of *M. thermautotrophicus* or *A. fulgidus* (166). However, the latter analysis contradicts biochemical studies, which suggest that phosphoglucose isomerase activity is in fact present in strains of both organisms (43, 96). It therefore seems that a complete gluconeogenic pathway to glucose 6-phosphate is present in all archaeal genera. In certain organisms, glyceraldehyde 3-phosphate and fructose 6-phosphate may be used directly for biosynthesis via the nonoxidative pentose phosphate pathway. In others, such as *Sulfolobus*, *Thermococcus*, and *Methanococcus*, glucose 6-phosphate may be used for the synthesis of glycogen.

Conventionally, glycogen is synthesized by conversion of glucose 6-phosphate to glucose 1-phosphate and its subsequent conversion to UDP-glucose. Glycogen synthase then condenses this compound to the growing polymer chain, with the release of UDP. There has been little characterization of the enzymes of glycogen metabolism in archaea, although glycogen synthase has been characterized in *S. acidocaldarius* (19, 94). The putative genes of glycogen synthesis are found in a cluster in the *S. solfataricus* genome, alongside putative genes for glycogen-degrading enzymes. A similar cluster has also been described in the genome sequence of *T. tenax* (149). Further biochemical characterization is required to establish the enzymes and regulation of glycogen synthesis and breakdown in the *Archaea*.

The ubiquity of the gluconeogenic pathway attaches possible evolutionary significance to this pathway. While many alternative glycolytic pathways have been documented, the gluconeogenic route is conserved throughout all three domains of life. In addition, it is striking that the enzymes of the lower pathway (pyruvate to glyceraldehyde 3-phosphate) are significantly more conserved than the upper pathway enzymes, which have been found to belong to several different sequence families. Indeed, a detailed analysis of the distribution and phylogenies of glycolytic and gluconeogenic enzymes has led to the proposal that the gluconeogenic pathway is evolutionarily more ancient (126).

Regulation of glucose metabolism

Given the fundamental metabolic role of the interconversion of glucose and pyruvate in all three domains of life, the regulation of the enzymes and pathways involved is of critical importance. There has been little investigation into the precise nature of the mechanisms that underpin the regulation of glycolysis and gluconeogenesis in the *Archaea*. However, recent work on *T. tenax* and *P. furiosus* has begun to elucidate the control of the pathways in these organisms. The fact that a nonphosphorylating glyceraldehyde-3-phosphate dehydrogenase or oxidoreductase is employed solely in the glycolytic direction in many archaea provides a novel regulatory control point. In *P. furiosus*, regulation occurs primarily at the transcript level, and a strong upregulation of glyceraldehyde-3-phosphate oxidoreductase is found during growth on cellobiose compared with growth on pyruvate (164). In *T. tenax*, the nonphosphorylating glyceraldehyde-3-phosphate dehydrogenase operates under allosteric control; the enzyme is inhibited by NADPH, NADP+, NADH, and ATP and is activated by AMP, glucose 1-phosphate, fructose 6-phosphate, ADP, and ribose 5-phosphate (15). This allosteric control ensures that the enzyme is most active when intracellular conditions require a higher level of glycolytic flux. The presence of orthologs of the *T. tenax* enzyme in several other archaea suggests that this strategy may also be employed in other organisms. Taken together with evidence of transcript level regulation, it appears that the interconversion of glyceraldehyde 3-phosphate and 3-phosphoglycerate is the critical control point in archaeal glycolysis/gluconeogenesis. This contrasts with the situation in bacteria and eucarya, where phosphofructokinase and pyruvate kinase are the critical regulatory control points, operating under strict transcript and allosteric control.

In *P. furiosus*, a recent whole-genome microarray analysis (139) has revealed that the glycolytic enzymes phosphoglucose isomerase, phosphofructokinase, and triose phosphate isomerase are also upregulated during growth on carbohydrates. In addition, the gluconeogenic enzymes glyceraldehyde-3-phosphate dehydrogenase, phosphoglycerate kinase, and fructose-1,6-bisphosphate aldolase are upregulated during growth on peptides. A similar study with *H. volcanii* revealed an upregulation of several ED pathway genes when growth substrate was switched from amino acids to glucose (178). In *T. tenax*, there is evidence that pyruvate kinase is upregulated at the transcript level during heterotrophic growth (137). There is no evidence for allosteric control of this enzyme in the *Archaea*, other than positive cooperativity with its substrates PEP and ADP (70).

In certain organisms, such as *Thermoplasma* spp., glycolysis and gluconeogenesis are performed by separate pathways. This may simplify the regulatory requirements in these organisms, as fewer enzymes are required to function in both catabolic and anabolic directions. Furthermore, the existence of "promiscuous" metabolic pathways in certain archaea may simplify the regulatory requirements during growth on different energy substrates. Conventionally, growth on an alternative sugar requires an organism to activate the transcription and expression of new enzymes and pathways, via complex regulatory mechanisms. If the same pathway of enzymes is used for more than one sugar, then this is not required and may permit an organism to adjust more efficiently to alternative energy sources.

Another process that may play a role in the regulation of archaeal glycolysis/gluconeogenesis is phosphorylation, and it has recently been suggested that certain metabolic enzymes in *S. solfataricus* are subject to control by a phosphorylation-dephosphorylation mechanism (88, 120) (see Chapter 11). However, the precise nature of this mechanism, and its contribution to the regulation of archaeal carbohydrate metabolism, remains to be elucidated.

The Metabolic Fate of Pyruvate

Oxidation to acetyl-CoA via pyruvate oxidoreductase

In contrast to the diversity of catabolic pathways leading from hexoses to pyruvate, there is a distinct unity in the manner in which the *Archaea* convert pyruvate to acetyl-coenzyme A (acetyl-CoA), namely via a pyruvate ferredoxin (Fd) oxidoreductase (POR) (reviewed in reference 140). This enzyme is active in all archaea, whether aerobic or anaerobic, and catalyzes the oxidative decarboxylation of pyruvate:

$$\text{Pyruvate} + \text{CoASH} + \text{Fd}_{ox} \rightarrow \text{acetyl-CoA} + CO_2 + \text{Fd}_{red}$$

The decarboxylation reaction is a thiamine pyrophosphate (TPP)-dependent process, and the acetyl group is handed directly to CoA. FeS centers in the enzyme serve to direct electron flow to the electron acceptor, which is ferredoxin in all archaea so far investigated. The cell ultimately disposes of the electrons as H_2, H_2S, or an organic acid.

Many archaeal PORs (e.g., from *P. furiosus*, *A. fulgidus*, and *M. thermautotrophicus*) are octameric in nature ($\alpha_2\beta_2\gamma_2\delta_2$), with whole M_r values at about 240,000 (90, 95, 158). The POR from *Halobacterium halobium* is an $\alpha_2\beta_2$-tetramer (113), whereas heterodimeric enzymes have been characterized from *S. solfataricus* (179) and *A. pernix* (106). Sequence analyses of the four-subunit PORs indicate that the β-subunit contains a TPP-binding motif and four conserved cysteines that might bind an [4Fe-4S] cluster, and that the δ-subunit contains two conserved [4Fe-4S] cluster-binding motifs (140). By combining these data with electron paramagnetic resonance studies on the *P. furiosus* holoenzyme and the δ-component, a mechanistic model is proposed whereby the oxidative decarboxylation of pyruvate to acetyl-CoA is catalyzed by the β-subunit, and electron flow from the 2-oxo acid to ferredoxin is via the δ-protein (Fig. 4).

In the *Archaea*, the POR is one member of a family of enzymes that also includes 2-oxoglutarate (α-ketoglutarate; KGOR) and branched-chain 2-oxoacid (BCOR) oxidoreductases; all three are shown to be homologous by sequence comparisons (140). Furthermore, in *P. furiosus*, for example, the δ-, α-, and β-genes of POR lie in a cluster close to a similar cluster for the BCOR, with the γ-subunit of the two enzymes being encoded by a single upstream gene. Not all OR enzymes are substrate specific. For example, *A. pernix* has two sets of genes encoding ORs that both utilize pyruvate, but only one of which can efficiently accept 2-oxoglutarate as a substrate; interestingly, neither enzyme can use the branched-chain 2-oxoacids (106). *S. solfataricus* also has an OR that oxidatively decarboxylates both pyruvate and 2-oxoglutarate (50).

Finally, note that the oxidative decarboxylation of pyruvate to acetyl-CoA can be reversed using reduced ferredoxin and the oxidoreductase. In combination with PEP synthase, this is important for gluconeogensis to function during growth on acetate and for those archaea that can fix CO_2 via a reductive citric acid cycle; both these processes are considered in more detail below (see "The citric acid cycle" and "Growth on acetate and the glyoxylate cycle").

The production of acetate

Following the decarboxylation of pyruvate, many archaea are able to convert acetyl-CoA to ac-

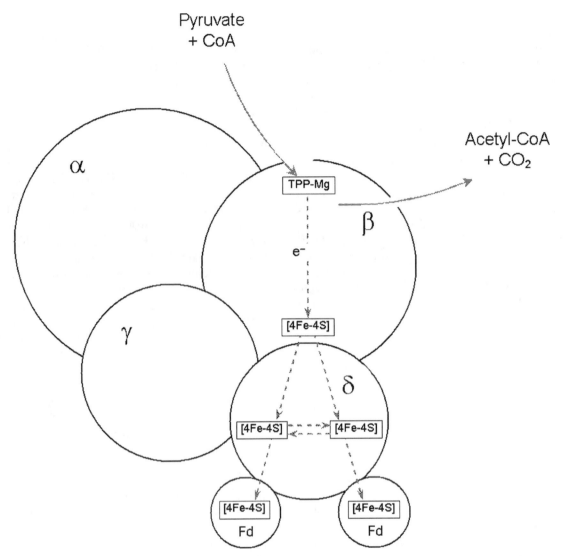

Figure 4. Pyruvate ferredoxin oxidoreductase. Schematic representation of the four-subunit ($\alpha\beta\gamma\delta$) pyruvate ferredoxin oxidoreductase and the proposed pathway of electron flow (adapted from reference 140). Ferredoxin (Fd) is the electron acceptor. CoA, coenzyme A; TPP, thiamine pyrophosphate; [4Fe-4S], iron sulfur cluster.

etate with the concomitant production of ATP. Indeed, in anaerobic hyperthermophilic archaea such as *P. furiosus*, this reaction represents the major energy-conserving step in the fermentation of sugars and pyruvate (136). A unique feature of the *Archaea* is that they possess an ADP-dependent acetyl-CoA synthetase (ACD) to catalyze this production of ATP in a one-step conversion:

$$\text{Acetyl-CoA} + \text{ADP} + \text{Pi} \rightleftarrows \text{acetate} + \text{CoA} + \text{ATP}$$

This contrasts with the situation in all bacteria, which use a two-step mechanism involving phosphate acetyltransferase and acetate kinase, with acetylphosphate being generated as a metabolic intermediate.

Archaeal ACD was first detected in *T. acidophilum* (26) and was subsequently characterized in a variety of halophilic, hyperthermophilic, and methanogenic organisms (134). The oligomeric nature of these ACDs varies between the $\alpha_2\beta_2$-type in *Pyrococcus* (55, 101) and *Pyrobaculum* (12), and homologous homodimers representing gene and polypeptide $\alpha\beta$-fusions in *Haloarcula* (12), *Archaeoglobus*, and *Methanococcus* species (103). The different enzymes have differing substrate specificities, with the ACD from *P. aerophilum*, for example, utilizing acetyl-CoA, isobutyryl-CoA, and phenylacetyl-CoA. Thus ACDs play a role in the metabolism of both aliphatic and aromatic amino acids, as well as energy generation from the catabolism of sugars.

The production of alanine

Many anaerobic archaea produce alanine in addition to acetate while fermenting sugars and other nutrients. Kengen and Stams (84) found that *P. furiosus* produced substantial amounts of L-alanine during batch growth on cellobiose, maltose, or pyruvate; ratios of alanine to acetate, which was also produced, varied from 0.07 to 0.8, depending on the redox potential of the terminal electron acceptor. Alanine formation from pyruvate was shown to occur via an alanine aminotransferase, and it was suggested that this enzyme might operate in conjunction with glutamate dehydrogenase and a ferredoxin:NADP$^+$ oxidoreductase to recycle the electron acceptors involved in catabolism. In support of this, the genes encoding the aminotransferase and the glutamate dehydrogenase appear to be coregulated at the transcriptional level, the expression of both being induced when the cells were grown on pyruvate (171).

Similarly, *Thermococcus profundus* excretes L-alanine into the medium (91). High activity of alanine aminotransferase was present in the cells, but no alanine dehydrogenase activity could be detected. It was suggested that alanine formation may be initiated by ammonia incorporation by glutamate dehydrogenase, followed by amino transfer from glutamate to pyruvate by the aminotransferase. Another *Thermococcus* species, *T. kodakaraensis*, has also been shown to produce alanine and acetate when grown on starch or pyruvate (53).

The finding that members of the *Thermotogales*, one of the deepest branching genera within the *Bacteria*, also produce L-alanine during glucose fermentation has led to the view that this may be an ancestral metabolic characteristic (118).

The citric acid cycle

The citric acid cycle was discovered in aerobic organisms as the pathway through which all nutrients can be completely oxidized to CO_2 and H_2O, with a considerably higher yield of energy (ATP) than is gained from either Embden-Meyerhof (EM) or the various Entner-Doudoroff (ED) pathways. Following the conversion of hexoses to pyruvate by the EM or ED pathways, and its subsequent decarboxylation to acetyl-CoA, the oxidative cycle "begins" with the condensation of acetyl-CoA (C_2) with oxaloacetate (C_4) to form the C_6-compound citrate (Fig. 5). The pathway then comprises a series of chemical transformations to permit the loss of two carbons as CO_2 and the removal of hydrogen atoms by the cofactors NAD(P)$^+$ and FAD, the reoxidation of which with molecular oxygen yields ATP and H_2O. From the re-

maining 4-carbon compound, oxaloacetate is regenerated to complete the cycle. However, intermediates of the cycle are removed for biosynthesis, which necessitates the replenishment of oxaloacetate by the so-called anaplerotic reactions. For example, pyruvate generated from sugar catabolism can be partitioned between its oxidative decarboxylation to acetyl-CoA and its ATP-dependent carboxylation to oxaloacetate by pyruvate carboxylase, the anaplerotic route. Clearly, this partitioning is highly regulated in most organisms.

Unlike the different routes that have evolved for the catabolism of glucose to pyruvate, the metabolic intermediates of the citric acid cycle are remarkably consistent throughout the *Archaea*, *Bacteria*, and *Eucarya*, making it one of the least variant of the central metabolic pathways in terms of its chemical transformations (27). However, considerable variation is seen, in particular in archaea and bacteria, with respect to the "completeness" of the cycle and the use to which it is put (67). In turn, these variations clearly reflect the lifestyle of each organism, in terms of their aerobic/anaerobic and autotrophic/heterotrophic modes of growth.

The oxidative citric acid cycle. In the aerobic archaea, acetyl-CoA generated from pyruvate can be oxidized to CO_2 and H_2O via an oxidative citric acid cycle (Fig. 5), and energy may be generated via oxidative phosphorylation. The constituent enzymes have been assayed in a variety of halophilic archaea, and in *T. acidophilum* and *S. acidocaldarius* (2, 28), and the genes have been identified in the genome sequences of these and of *Halobacterium* NRC-1, *Picrophilus torridus*, *T. volcanium*, *S. solfataricus*, *S. tokadaii*, *P. aerophilum*, *Ferroplasma acidarmanus*, and *A. pernix*. This is the same set of enzymes as those in the citric acid cycle of the aerobic bacteria and eucarya, although it is thought that it is the 2-oxoglutarate oxidoreductase, and not the dehydrogenase complex, that converts 2-oxoglutarate to succinyl-CoA in the *Archaea*.

The reductive citric acid cycle. In some archaea, the citric acid cycle operates in the reductive mode to fix CO_2 during autotrophic growth (Fig. 6). Two key enzymes are required to reverse the cycle: 2-oxoglutarate oxidoreductase, which reductively carboxylates succinyl-CoA to 2-oxoglutarate using reduced ferredoxin, and ATP-citrate lyase to drive the formation of acetyl-CoA from citrate. The acetyl-CoA can then be reductively carboxylated to pyruvate via pyruvate oxidoreductase, again using reduced ferredoxin. Note that the two reductive reactions using reduced ferredoxin catalyzed by the oxidoreductases

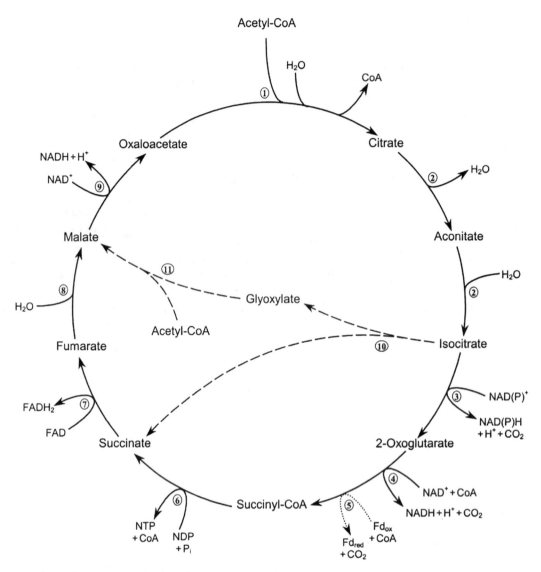

Figure 5. The oxidative citric acid cycle and the glyoxylate cycle. The reactions of the citric acid cycle are denoted by solid arrows, and the reactions unique to the glyoxylate cycle are shown with dotted lines. Enzymes are denoted by numbers: 1 = citrate synthase, 2 = aconitase, 3 = isocitrate dehydrogenase, 4 = 2-oxoglutarate dehydrogenase complex (aerobic bacteria and eucarya), 5 = 2-oxoglutarate ferredoxin oxidoreductase (archaea), 6 = succinate thiokinase, 7 = succinate dehydrogenase, 8 = fumarase, 9 = malate dehydrogenase, 10 = isocitrate lyase, 11 = malate synthase.

are not possible using NADH and the corresponding 2-oxoacid dehydrogenase complexes. The operation of this pathway in *Thermoproteus neutrophilus* has been supported by enzymic assays and radiolabeling studies (132), and the required enzymes have also been found in *Pyrobaculum islandicum* (66) and *T. tenax* (149). Also note that, while autotrophic CO_2 fixation is common in the *Archaea*, not all members use the reverse citric acid cycle. A reductive acetyl-CoA pathway is used in the euryarchaea *Archaeoglobus*, *Ferroplasma*, and the methanogens (66) (see Chapter 13). On the other hand, members of the *Crenarchaeota*, including *Metallosphaera sedula*,

Acidianus ambivalens, *Acidianus brierleyi*, and *Sulfolobus* species contain key enzymes of the 3-hydroxypropionate cycle for CO_2 fixation (66).

A number of questions still remain on the modes of CO_2 fixation in the *Archaea*. For example, ribulose-1,5-bisphosphate carboxylase/oxygenase (RubisCO) is present in halophiles, methanogens, and several thermophiles (46). The enzyme from these archaea is catalytically active in a ribulose bisphosphate-dependent CO_2-fixation reaction, but its role in a reductive pentose phosphate pathway is uncertain because a gene for phosphoribulokinase, which generates the substrate (ribulose bisphosphate) for the

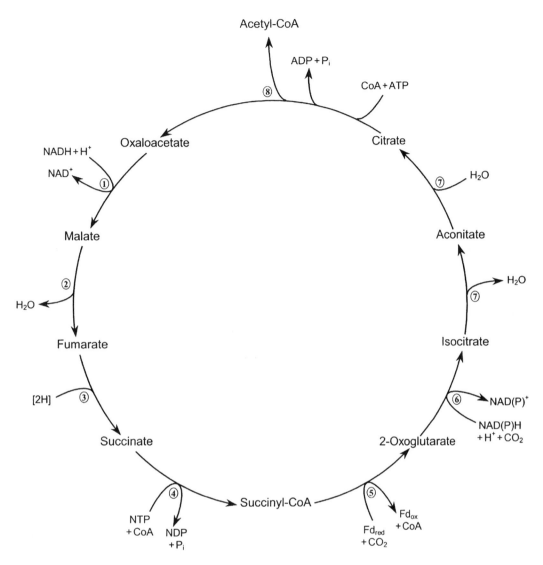

Figure 6. The reductive citric acid cycle. Enzymes are denoted by numbers: 1 = malate dehydrogenase, 2 = fumarase, 3 = fumarate reductase, 4 = succinate thiokinase, 5 = 2-oxoglutarate ferredoxin oxidoreductase, 6 = isocitrate dehydrogenase, 7 = aconitase, 8 = ATP citrate lyase.

RubisCO reaction, has not been identified. Enzymic assays performed with alternative substrates have provided evidence for a previously uncharacterized pathway for the synthesis of ribulose bisphosphate from 5-phospho-D-ribose-1-pyrophosphate in *M. jannaschii* and other methanogenic archaea (47). The enzyme responsible has been purified and the gene identified from the N-terminal protein sequence; interestingly, there are good homologs of the gene in methanogenic, thermophilic, and halophilic archaea. However, the quantitative significance of RubisCo in CO_2 fixation in these autotrophic archaea has not yet been assessed.

The cycle in anaerobic archaea. In contrast to aerobic archaea, anaerobic members utilize the citric acid cycle primarily for biosynthetic purposes, and therefore may not be expected to have the complete oxidative cycle. Identifying genes in the citric acid cycle is complicated by their sequence similarity to genes that are involved in other pathways. For example, isocitrate dehydrogenase (citric acid cycle) and 3-isopropylmalate dehydrogenase (leucine metabolism) share significant sequence identity, as do the various members of the 2-oxoacid Fd oxidoreductases. As a result, it is difficult to make assignments about substrate specificity.

Careful analysis of the genome sequence of the anaerobe, *T. tenax*, revealed the presence of all the genes of an oxidative cycle, suggesting that it might therefore be functional under heterotrophic growth conditions (149). The genes for a complete cycle are present in *A. fulgidus* (89) and some of the enzymes have been characterized (156). In *P. furiosus*, all the genes, except for malate dehydrogenase, are present in the genome sequence; however, a putative malate oxidoreductase gene has been annotated that may produce an enzyme with equivalent function (121). *P. abyssi* and *P. horikoshii* are quite different from *P. furiosus* as they lack the first three enzymes of the cycle, namely citrate synthase, aconitase, and isocitrate dehydrogenase (67, 100).

Under anaerobic conditions, the complete citric acid cycle requires terminal electron acceptors other than oxygen; these include sulfur, sulfate, thiosulfate, and nitrate (136).

Other anaerobic archaea appear to have partial citric acid cycles. For example, in the methanogens partial versions of the cycle may exist that fulfil an anabolic function (see Chapter 13). In *Methanosarcina barkerii*, enzyme assays show that 2-oxoglutarate can be synthesized via citrate, aconitate, and isocitrate, whereas in *M. thermautotrophicus* the enzymes for an incomplete reductive cycle have been assayed to enable 2-oxoglutarate synthesis via succinate (10). Genome sequences confirm the presence of the necessary genes for these partial cycles (150), although homologs for some of the other citric acid cycle enzymes can also be detected. Indeed, with the prediction of a novel aconitase in the *Archaea* (102), *M. thermautotrophicus* might have a complete set of genes for the citric acid cycle enzymes, although experimental verification of enzymic activities is required. *T. kodakaraensis* appears to lack the genes for several key cycle enzymes, but confirmatory enzymic analyses have not yet been performed (53).

Regulation of the citric acid cycle. The citric acid cycle is a multifunctional pathway, serving to oxidize pyruvate to CO_2 and H_2O, providing a variety of metabolites for biosynthetic reactions, and being the main generator of energy under aerobic conditions. Consequently, in bacteria and eucarya, the flux through the cycle is tightly regulated, with allosteric feedback inhibition and covalent control via phosphorylation-dephosphorylation being the two most common regulatory mechanisms (reviewed in reference 82) (see Chapter 11). However, not all the enzymes of the citric acid cycle are controlled in these organisms; regulation has only been consistently observed for the pyruvate dehydrogenase complex, citrate synthase, isocitrate dehydrogenase and the 2-oxoglutarate dehydrogenase complex. A variety of metabolites serve to effect the control of these enzymes, depending on the organism and its nutrient source.

In contrast to the large volume of literature describing the regulation of the citric acid cycle enzymes from bacteria and eucarya, there is little information for the archaea. The pyruvate and 2-oxoglutarate dehydrogenase complexes are replaced by the equivalent Fd oxidoreductases in archaea, for which no control mechanisms have been reported. Archaeal citrate synthases are isosterically inhibited by ATP with K_i values similar to those reported for citrate synthases from gram-positive bacteria and eucarya (28); however, the physiological significance of this inhibition has been questioned (172). Finally, no regulation of an archaeal isocitrate dehydrogenase has been reported.

In *P. furiosus* there is evidence of a strong upregulation in the transcription of citrate synthase, aconitase and isocitrate dehydrogenase during growth on maltose compared with growth on peptides (139). In contrast, there was no upregulation of these transcripts when *H. volcanii* was grown on glucose after growth on casamino acids (178).

Growth on Acetate and the Glyoxylate Cycle

Several archaea can grow on acetate. The first metabolic step for acetate utilization is the conversion to acetyl-CoA catalyzed by AMP-forming acetyl-CoA synthetase:

$$Acetate + ATP + CoA \rightarrow acetyl\text{-}CoA + AMP + PP_i$$

In the aerobic archaea, acetyl-CoA enters the citric acid cycle for energy production. However, from an anabolic viewpoint, this poses a problem as the two carbon atoms will be lost as CO_2, and the extraction of any cycle intermediate for biosynthesis would lead to the cessation of the pathway. One solution to this problem in archaea would be to reductively carboxylate a portion of the acetyl-CoA to pyruvate using reduced ferredoxin and the pyruvate oxidoreductase. Pyruvate could in turn replenish the citric acid cycle intermediate, oxaloacetate, via pyruvate carboxylase, or be converted to PEP via PEP synthase, the gene for which is found in all archaeal genomes sequenced. PEP could undergo gluconeogenesis or itself be used to replenish the citric acid cycle through PEP carboxylase (42).

A second solution would be the use of the glyoxylate cycle (Fig. 5). The key enzymes, isocitrate lyase and malate synthase, have been found in *H. volcanii*, *H. marismortui*, *P. aerophilum*, *S. solfataricus* and *S. acidocaldarius*. The synthesis of these enzymes is induced by growth on acetate in *H. volcanii* (143), and in *H. marismortui* acetyl-CoA synthetase is coordinately upregulated in an acetate-specific fashion (13). Acetate-induced induction of isocitrate lyase and malate synthase also occurs in *S. acidocaldarius* (162). Another halophile, *Halobacterium* NRC1, lacks the genes for isocitrate lyase and malate synthase and accordingly cannot grow on acetate as sole carbon source (107).

In archaea, the route to glucose for carbon introduced into the citric acid cycle by the glyoxylate cycle has not been defined but may be via PEP carboxykinase and gluconeogenesis as in many bacteria. PEP carboxykinase catalyzes the reaction:

$$\text{Oxaloacetate} + \text{GTP/ATP} \rightarrow \text{PEP} + CO_2 \\ + \text{GDP/ADP}$$

Within the *Archaea*, a GTP-dependent enzyme has been characterized in *T. kodakaraensis* (51), although this organism does not appear to have the glyoxylate cycle enzymes, isocitrate lyase and malate synthase. The transcription and activity levels of the enzyme in *T. kodakaraensis* were higher under gluconeogenic conditions than under glycolytic conditions, consistent with a role in the supply of PEP from oxaloacetate when gluconeogenesis is operational. A gene for PEP carboxykinase has been tentatively identified in the genomes of *Sulfolobus* and *Aeropyrum*, but no enzymological data are available. A second exit possibility is via malic enzyme, which reductively decarboxylates malate to pyruvate; this enzyme has also been detected in *T. kodakaraensis* and the recombinant protein characterized (52).

Catabolism of Amino Acids

Many archaea can take up and catabolize peptides and amino acids, and some, notably the halophilic archaea and *Pyrococcus*, can use these as their sole carbon and energy sources. In the sequenced archaeal genomes, aminotransferase genes have been identified. The archaeal aminotransferases presumably serve a similar role to aminotransferases of bacteria and eucarya, namely to convert amino acids to their corresponding 2-oxoacids by a transamination reaction with another 2-oxoacid. For example, alanine can be transaminated to pyruvate, aspartate to oxaloacetate, and glutamate to 2-oxoglutarate. The products of the transamination reactions can then directly enter the pathways of central metabolism. Other 2-oxoacids have to undergo a series of chemical transformations to convert them to central metabolic intermediates, although not all these pathways in the *Archaea* have been defined by enzymic studies.

The deamination of the branched-chain amino acids, valine, leucine, and isoleucine, yields the branched-chain 2-oxoacids 3-methyl-2-oxo-butanoate, 4-methyl-2-oxo-pentanoate, and 3-methyl-2-oxopentanoate, respectively. In archaea, these, in turn, are converted to their corresponding acyl-CoA derivatives by the branched-chain BCOR (140). Further metabolism converts them to acetyl-CoA and/or succinyl-CoA, both of which can enter the citric acid cycle.

As mentioned above (see "The metabolic fate of pyruvate"), BCOR is a member of a family of Fd-linked oxidoreductases that also includes the pyruvate (POR) and 2-oxoglutarate (KGOR) enzymes. The presence of this family of active oxidoreductase enzymes in the aerobic archaea might be considered unexpected as the reactions they carry out in aerobic bacteria and eucarya are catalyzed by 2-oxoacid dehydrogenase multienzyme complexes. However, 2-oxoacid dehydrogenase complex activity has not been detected in any members of the *Archaea*, supporting the view that their oxidoreductases are sufficient for these metabolic steps. While these observations appear definitive, genes that potentially encode the components of 2-oxoacid dehydrogenase complexes have been identified in aerobic archaea. To understand the significance of these findings, a brief discussion of the mechanism and structure of the bacterial and eucaryal complexes is necessary.

The 2-oxoacid dehydrogenase complexes catalyze the general reaction:

$$\text{2-Oxoacid} + \text{CoASH} + NAD^+ \rightarrow \text{acyl-SCoA} \\ + CO_2 + \text{NADH} + H^+$$

Similar to the oxidoreductases, members of this family include the pyruvate dehydrogenase complex (PDHC, catalyzes the conversion of pyruvate to acetyl-SCoA), the 2-oxoglutarate dehydrogenase complex (OGDHC, 2-oxoglutatarate to succinyl-SCoA), and the branched-chain 2-oxoacid dehydrogenase complex (BCODHC, oxidatively decarboxylates the 2-oxoacids produced by the transamination of amino acids valine, leucine, and isoleucine). The complexes are all three-component systems consisting of multiple copies of enzymes E1 (2-oxoacid decarboxylase), E2 (dihydrolipoyl acyltransferase), and E3 (dihydrolipoamide dehydrogenase) (111, 112). E2 forms the structural core of the complex, to which copies of E1 and E3 are noncovalently bound. The number of copies of each component can vary between the different complexes and between phylogenetic groups of any one system (68, 112). For example, in the PDHC from gram-negative bacteria there are 24 polypeptide chains of the E2 component in each core molecule, whereas in the complex from gram-positive bacteria and eucarya there are 60 E2 chains. Most OGDHCs and BCODHCs have 24 E2 chains in their core structures. E2 also forms the catalytic core of these multienzyme complexes, with each E2 polypeptide chain having at least one covalently bound acyl-carrying cofactor, lipoic acid, which serves to connect the three active sites and to channel substrate through the enzyme complex (Fig. 7). Substrate specificity is determined by the E1 and E2 components, while E3 serves a common role in reoxidizing the enzyme-bound dihydrolipoamide produced by acyl-SCoA formation; consequently, it is often the same E3 gene product that can serve in the different 2-oxoacid dehydrogenase complexes.

The first indications that aerobic archaea might contain a 2-oxoacid dehydrogenase complex came with the surprising finding that the third component of the 2-oxoacid complexes, dihydrolipoamide dehydrogenase (DHLipDH), was active in halophilic archaea (29) and in the thermophiles, *T. acidophilum* (152) and *A. pernix* (H.C. Aass, D. W. Hough, and M. J. Danson, unpublished data). The identification of lipoic acid in *H. halobium* by a combined gas-chromatographic and mass-spectrometric procedure added to the significance of the enzymological studies (114), in that the only known physiological function

of DHLipDH and lipoic acid is as part of the 2-oxoacid dehydrogenase complexes.

Genes with significant sequence identities and conserved motifs to the E1α, E1β, E2, and E3 genes of bacterial and eucaryal 2-oxoacid complexes were identified in *H. volcanii* (Fig. 8) (31, 73, 169). The four ORFs are tightly spaced or overlapping, and a single ribosome-binding site and TATA box upstream of the ORF 1 start codon, and a transcriptional stop signal (poly(dT) tract) downstream of ORF 4, have been putatively identified. A single cluster with the same gene content and arrangement is found in the genome sequences of all aerobic archaea, with the exception of *S. solfataricus* (Fig. 8). In this organism, the cluster comprises E1α, E1β, and E2 genes, with the latter two separated by a gene of unknown function; the gene encoding the putative E3 component is 560 bp upstream of this cluster. Furthermore, the *S. solfataricus* E2 gene is interrupted by a frameshift, although this may be a case of programmed −1 frameshift recoding events

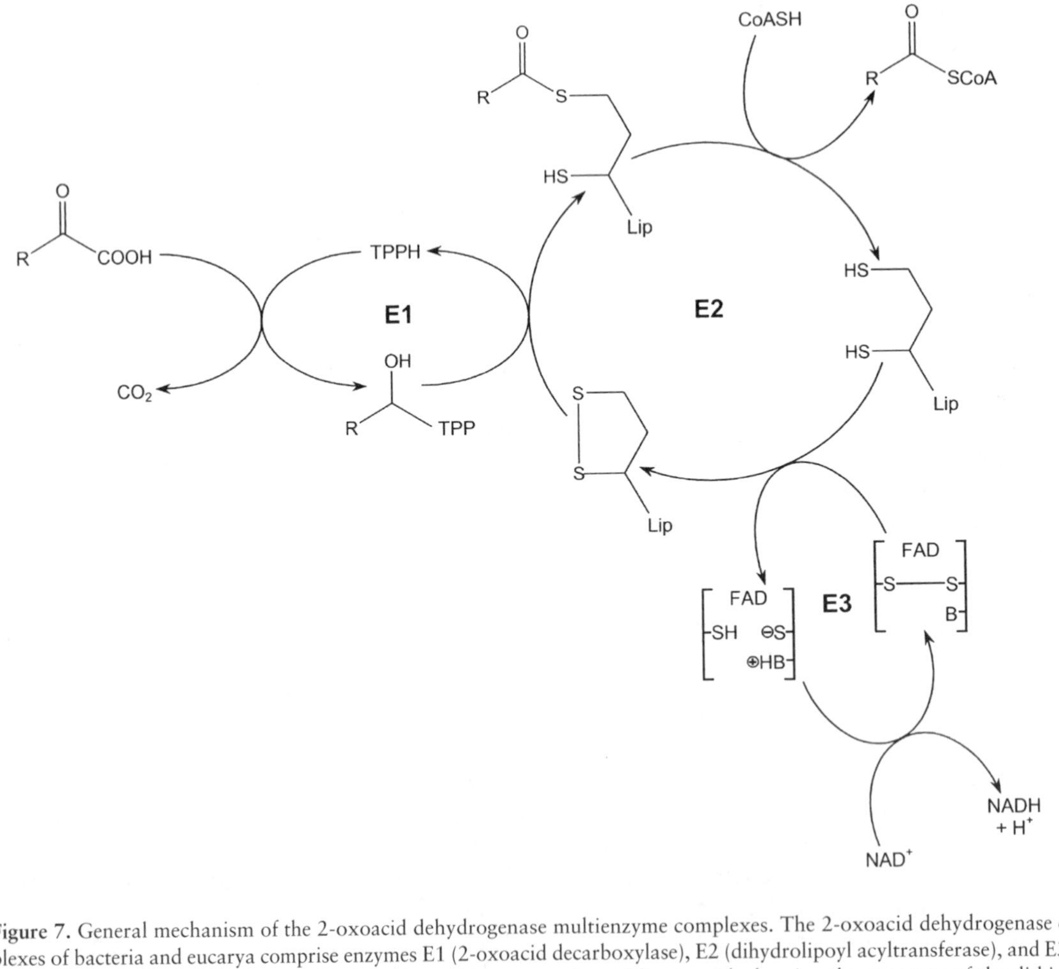

Figure 7. General mechanism of the 2-oxoacid dehydrogenase multienzyme complexes. The 2-oxoacid dehydrogenase complexes of bacteria and eucarya comprise enzymes E1 (2-oxoacid decarboxylase), E2 (dihydrolipoyl acyltransferase), and E3 (dihydrolipoamide dehydrogenase). B, histidine base; Lip, enzyme-bound lipoic acid, showing the structure of the dithiolane ring; S—S, protein disulfide bond; TPPH, thiamine pyrophosphate.

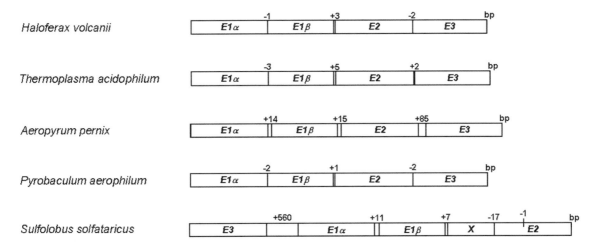

Figure 8. Gene clusters encoding the components of a putative archaeal 2-oxoacid dehydrogenase complex. The arrangement and intergene distances (bp) of the ORFs constituting the E1α, E1β, E2, and E3 genes of the proposed archaeal 2-oxoacid dehydrogenase are shown. The proposed direction of transcription (left to right, as drawn) is the same for all the genes. See text for details of the (−1) frameshift in the E2 gene of *S. solfataricus*.

that have been discovered in several genes of this hyperthermophilic archaeon (21).

The E1α and E1β genes from *T. acidophilum* have been coexpressed in *E. coli*, and the recombinant proteins form an $α_2β_2$-enzyme that decarboxylates the branched-chain 2-oxoacids 4-methyl-2-oxopentanoate, 3-methyl-2-oxopentanoate, and 3-methyl-2-oxobutanoate (63). A low catalytic activity is found with pyruvate, but the enzyme appears to be inactive toward 2-oxoglutarate. Similarly, the E2 gene has been recombinantly expressed as a 50% lipoylated product, and the E3 gene has been expressed as an active dihydrolipoamide dehydrogenase. Preliminary data indicate that the components assemble into an active complex with the same substrate specificity as the isolated E1 enzyme (C. Heath, H. C. Aass, D. W. Hough, and M. J. Danson, unpublished observations). The findings support a role for the complex in the metabolism of branched-chain amino acids.

In *H. volcanii*, the transcript levels for the genes encoding the putative 2-oxoacid dehydrogenase complex decreased when growth was changed from amino acid-based to glucose-based metabolism, supporting a role for the enzyme complex in the catabolism of amino acids but not glucose (178). However, insertional inactivation of the DHLipDH gene resulted in no detectable phenotypic difference from the wild-type organism when grown on a variety of central metabolic intermediates (73, 74). The catalytic activity of the halophilic complex is presently being examined using homologous expression of the components.

In addition to catabolism of amino acids via their 2-oxoacid derivatives, some halophilic archaea are able to ferment arginine to support anaerobic growth.

The consumption of arginine is coupled to the equimolar production of ornithine, indicating that this may occur via the arginine deiminase pathway, with ATP being generated by substrate-level phosphorylation (62). The three enzymes in this pathway are arginine deiminase, ornithine transcarbamylase, and carbamate kinase, which respectively catalyze the following reactions:

$$\text{L-arginine} + H_2O \rightarrow \text{L-citrulline} + NH_3$$

$$\text{L-citrulline} + P_i \rightarrow \text{carbamoyl-phosphate} + \text{L-ornithine}$$

$$\text{carbamoyl-phosphate} + ADP \rightarrow ATP + NH_3 + CO_2$$

The genes for the enzymes are present in a cluster in *Halobacterium salinarum* (128), and the identity of the encoded enzymes has been confirmed.

PERSPECTIVE: THE NEXT FIVE YEARS

Recent Developments

Although the study of microbial metabolism dates back to the foundations of the disciplines of microbiology and biochemistry, the advent of genome sequencing in the past decade has resulted in a new approach to the study of metabolic pathways. Annotated genome sequences are currently available for 20 species of archaea, with additional sequencing projects in progress, and this information has enabled comparative and functional genomics approaches to be added to the traditional techniques of enzyme purification and characterization and of metabolic labeling.

Publication of a new archaeal genome sequence often includes an integrated metabolic overview of nutrient uptake, energy production, biosynthesis, and other aspects of the biology of the organism (see references 89, 105, 146). Such overviews are usually incomplete since not all genes can be assigned an unambiguous functional annotation. In fact, all genome sequences published to date include a substantial number of hypothetical genes of unknown function that typically comprise up to 50% of the entire genome. Despite these limitations, a comparative approach, integrating genomic and functional information across a range of archaeal species, has been used to identify archaeal glycolytic pathways and the enzymes involved (166). However, it is evident that most of the proposed gene functions await experimental confirmation, and this is an ambitious goal given the range of organisms and enzyme activities involved. The completion of further genome sequences can only add to this problem.

The advent of genome sequencing has provided the background for global analysis of cell function by following changes in mRNA levels using DNA microarrays (transcriptomics) and protein expression, using two-dimensional gel electrophoresis or high-performance liquid chromatography and mass spectrometry (proteomics) (see Chapter 20). These approaches, applied initially to eucaryal (35) and bacterial (87) systems, have subsequently been applied to archaea. The first archaeal whole-genome DNA microarray was constructed for *P. furiosus*, using the 2065 ORFs annotated in the genome sequence (139). The expression levels of 8% of expressed ORFs were found to vary substantially between peptide and maltose-grown cells, and most of these ORFs were members of 27 putative operons. Among the 18 operons upregulated in maltose-grown cells were those responsible for maltose transport and the biosynthesis of several amino acids. An operon encoding three enzymes of the citric acid cycle, citrate synthase, aconitase, and isocitrate dehydrogenase, was also upregulated, as were several ORFs encoding hypothetical proteins of unknown function. Several of the nine operons upregulated in peptide-grown cells were involved in transamination of amino acids and metabolism of the resulting 2-oxoacids. It was also shown in a number of cases that there was good correlation between microarray data and changes in enzyme activity levels.

In the absence of a complete genome sequence, metabolic adaptation of *H. volcanii* following a switch from growth on amino acids to growth on glu-

cose has been studied using a shotgun DNA microarray (178). Again there were significant changes in expression levels for about 10% of all genes, some of which were expected on the basis of metabolic studies. The expression of genes encoding enzymes of the part-phosphorylative ED pathway was upregulated on glucose, as was a potential glucose transporter. Among the genes repressed on glucose were a number encoding proteins involved in translation and ATP synthesis. Also, there was a decrease in transcription of the genes constituting the putative 2-oxoacid dehydrogenase complex operon (73). Enzymic activity corresponding to this complex has not been detected in *Haloferax*, and its inactivation does not alter the metabolic phenotype. However, the microarray data suggest a role in amino acid catabolism, and this is supported by recent work on the corresponding complex in *T. acidophilum* (63). Heterologous expression of the E1 component of the *Thermoplasma* complex yielded a product with decarboxylase activity typical of the first component of a branched-chain 2-oxoacid dehydrogenase complex. This and other examples indicate the power of microarray-based methods, in combination with an enzymological approach, in the discovery of unexpected gene functions and metabolic processes.

Proteomics studies of archaea are still at a preliminary stage, in particular in the case of halophiles, since the instability of their proteins at low ionic strength has required the development of new methods for two-dimensional gel electrophoresis (77) (see Chapter 20). At present the majority of published archaeal proteomics studies relate to methanogens. For example, in the hyperthermophilic methanogen, *Methanocaldococcus jannaschii*, 963 proteins, representing ~54% of the genome, have been identified from whole-cell extracts using liquid chromatography and mass spectrometry methods (180).

It is perhaps unsurprising that there are as yet few instances of an integrated microarray and proteomics study of an archaeal species. Baliga et al. (9) have analyzed mRNA and protein profiles of wild-type and mutant strains of *Halobacterium* NRC-1 and found that in many cases changes in protein expression are not reflected in changes at the mRNA level, suggesting that these proteins are regulated by posttranscriptional control mechanisms. Furthermore, they concluded that two major energy production pathways in this organism, phototropy and arginine fermentation, are inversely regulated to control ATP production under anaerobic conditions

The Next 5 Years and Beyond

In the postgenomic era there is an increasing trend toward understanding the mechanisms underlying biological responses in the context of the whole organism. This is the goal of Systems Biology, an approach that aims to integrate genomics and proteomics with quantitative kinetic information on enzyme activities and metabolite levels. The aim is then to use mathematical and computational models to construct reaction networks that can simulate cellular functions. In the context of metabolism this extends the concept of a metabolic pathway, leading to the definition of a complex metabolic network that exhibits properties that cannot be described simply in terms of its individual components (109). However, this approach requires a vast amount of information relating to the organism of interest, which cannot be derived entirely from genome sequencing. At present there are severe limitations on both the extent and accuracy of sequence annotation and, even with high-throughput techniques, we are far from a situation where the function of every open reading frame can be confirmed experimentally and the kinetic and regulatory properties of the encoded enzyme defined.

Progress has been made by focussing on one aspect of the system. For example, central metabolism in *E. coli* has been modeled in terms of in vitro enzyme kinetic measurements and in vivo measurements of intracellular metabolite concentrations (20). Models such as this can be used to predict the effects of changes in nutrient supply and are an essential starting point for rational metabolic engineering. Clearly we are far from the point at which such metabolic models can be generated for an archaeal species. There is no model archaeal organism analogous to *E. coli* for which we have the necessary background data, and to accumulate this information will take a substantial, well-focused effort. Although genomic and proteomic studies will certainly be carried out on an ever-increasing range of archaea, the generation of enzyme kinetic and metabolomic data is likely to follow much more slowly. In some cases, in particular, with the hyperthermophiles, there will be additional factors related to enzyme and metabolite stability that will add a further dimension to the already complex models derived for mesophilic organisms.

CONCLUDING REMARKS

Central metabolism represents one of the most fundamental aspects of the biochemistry of the cell and is commonly perceived as invariant and sacrosanct. However, in reality there is a considerable diversity to the pathways, enzymes, and cofactors that comprise it, and this is no better illustrated than by considering the situation in the *Archaea*. The preceding text has served to catalog and evaluate the current knowledge of central metabolism in the *Archaea*, and

a striking variability to the nature of the processes and pathways in different genera is revealed. Nevertheless, trends can be discerned and unusual or distinctive features of archaeal metabolism can be identified.

Perhaps the most obvious trend in the nature of central metabolism in archaea is that the distribution of pathways and enzymes in a particular organism is closely linked to its specific growth environment. For example, halophiles use the part-phosphorylative ED pathway, thermoacidophiles use the nonphosphorylative ED pathway, and hyperthermophiles use variants of the EM pathway. This situation appears to apply regardless of the position of an organism in the rRNA-based universal phylogenetic tree. For example, the crenarchaeaon *S. solfataricus* and the euryarchaeaon *T. acidophilum* both possess the nonphosphorylative ED pathway. Both organisms grow in thermoacidophilic environments, and there is evidence of considerable lateral transfer of genes between them (129). The hyperthermophilic crenarchaeon *A. pernix* possesses the EM pathway, as do euryarchaeal hyperthermophiles such as *A. fulgidus* and *P. furiosus*. This observation provides some support for the proposal that metabolic genes are well suited for lateral transfer (173), a process that is undoubtedly favored between organisms that occupy a similar environmental niche. However, note that the hyperthermophilic bacterium *T. maritima* possesses EM and ED pathways similar to the classical bacterial and eucaryal pathways, despite evidence of large-scale lateral transfer between bacterial and archaeal hyperthermophiles (7).

As described throughout this chapter, central metabolic enzymes of archaea have a number of unusual and unique features. Many times in the past, researchers have sought to speculate that these features of archaeal metabolism, in particular those found in hyperthermophiles, are evolutionarily ancient and provide an insight into the nature of central metabolism in a primitive organism. Examples of this include the prevalence of variants of the ED pathway (122), the existence of ADP-dependent kinases (83), the existence of ferredoxin-dependent enzymes (25), the production of alanine (118), and the existence of promiscuous pathways (98). It is perhaps more likely that, instead of representing ancestral metabolic characteristics, these features provide some selective advantage for survival in a particular environment. As such, the diversity of central metabolic pathways documented in the *Archaea* may be just a remarkable illustration of the evolutionary adaptation of microorganisms to survival in a variety of different hostile growth environments. Many of the "unusual" features of archaeal central metabolism are also found in a limited number of bacterial and eucaryal species.

However, certain enzymes, such as ADP-dependent phosphofructokinase and the nonphosphorylating glyceraldehyde-3-phosphate oxidoreductase, have not yet been found in any organism other than archaea. These enzymes may have evolved specifically in the *Archaea*, presumably because they confer a selective advantage or, alternatively, they may have been present prior to the divergence of the three domains of life and were replaced in bacteria and eucarya during evolution.

The discovery of the fascinating array of diverse central metabolic characteristics in the *Archaea* alone justifies the efforts that have been expended by a myriad of researchers over the past two decades. It is hoped that this text will help future workers in the field to draw together the different aspects of archaeal central metabolism and identify areas to target for future research.

Acknowledgments. We are indebted to Professors M.W.W. Adams (University of Georgia, Athens, Ga.), R.K. Thauer (Max Plank Institute for Terrestrial Microbiology, Marburg, Germany), M. Dyall-Smith (University of Melbourne, Parkville, Australia), and P. Schönheit (University of Keil, Germany) for helpful discussions and advice, and to Professor J. Banfield (University of California Berkeley, Calif.) for prepublication release of information on the genes encoding citric acid cycle enzymes in *Ferroplasma*.

We thank the UK Biotechnological and Biological Research Council, The Royal Society, and NATO for generous financial support of the authors' own research reported in this chapter.

REFERENCES

1. Ahmed, H., T. J. G. Ettema, B. Tjaden, A. C. M. Geerling, J. van der Oost, and B. Siebers. 2005. The semi-phosphorylative Entner-Doudoroff pathway in hyperthermophilic archaea—a re-evaluation. *Biochem. J.* **390:**529–540.
2. Aitken, D. M., and A. D. Brown. 1969. Citrate and glyoxylate cycles in the halophil, *Halobacterium salinarium. Biochim. Biophys. Acta* **177:**351–354.
3. Altekar, W., and V. Rangaswamy. 1990. Indication of a modified EMP pathway for fructose breakdown in a halophilic archaebacterium. *FEMS Microbiol. Lett.* **69:**139–144.
4. Altekar, W., and V. Rangaswamy. 1992. Degradation of endogenous fructose during catabolism of sucrose and mannitol in halophilic archaebacteria. *Arch. Microbiol.* **158:**356–363.
5. Andreesen, J. R., and G. Gottschalk. 1969. Occurrence of a modified Entner-Doudoroff pathway in *Clostridium aceticum. Arch. Mikrobiol.* **69:**160–170.
6. Angelov, A., O. Futterer, O. Valerius, G. H. Braus, and W. Liebl. 2005. Properties of the recombinant glucose/galactose dehydrogenase from the extreme thermoacidophile, *Picrophilus torridus. FEBS J.* **272:**1054–1062.
7. Aravind, L., R. L. Tatusov, Y. I. Wolf, D. R. Walker, and E. V. Koonin. 1998. Evidence for massive gene exchange between archaeal and bacterial hyperthermophiles. *Trends Genet.* **14:** 442–444.
8. Baliga, N. S., R. Bonneau, M. T. Facciotti, M. Pan, G. Glusman, E. W. Deutsch, P. Shannon, Y. Chiu, R. S. Weng, R. R. Gan, P. Hung, S. V. Date, E. Marcotte, L. Hood, and W. V. Ng. 2004. Genome sequence of *Haloarcula marismortui:* a halophilic archaeon from the Dead sea. *Genome Res.* **14:**2221–2234.
9. Baliga, N. S., M. Pan, Y. A. Goo, E. C. Yi, D. R. Goodlett, K. Dimitrov, P. Shannon, R. Aebersold, W. V. Ng, and L. Hood. 2002. Coordinate regulation of energy transduction modules in *Halobacterium* sp analyzed by a global systems approach. *Proc. Natl. Acad. Sci. USA* **99:**14913–14918
10. Blaut, M. 1994. Metabolism in methanogens. *Antonie van Leeuwenhoek* **66:**187–208.
11. Bonete, M-J., C. Pire, F. I. Llorca, and M. L. Camacho. 1996. Glucose dehydrogenase from the halophilic archaeon *Haloferax mediterranei:* Enzyme purification, characterisation and N-terminal sequence. *FEBS Lett.* **383:**227–229.
12. Bräsen, C., and P. Schönheit. 2004. Unusual ADP-forming acetyl-coenzyme A synthetases from the mesophilic halophilic euryarchaeon *Haloarcula marismortui* and from the hyperthermophilic crenarchaeon *Pyrobaculum aerophilum. Arch. Microbiol.* **182:**277–287.
13. Bräsen, C., and P. Schönheit. 2004. Regulation of acetate and acetyl-CoA converting enzymes during growth on acetate and/or glucose in the halophilic archaeon *Haloarcula marismortui. FEMS Microbiol. Lett.* **241:**21–26.
14. Brunner, N. A., H. Brinkmann, B. Siebers, and R. Hensel. 1998. NAD$^+$-dependent glyceraldehyde-3-phosphate dehydrogenase from *Thermoproteus tenax. J. Biol. Chem.* **273:** 6149–6156.
15. Brunner, N. A., B. Siebers, and R. Hensel. 2001. Role of two different glyceraldehyde-3-phosphate dehydrogenases in controlling the reversible Embden-Meyerhof-Parnas pathway in *Thermoproteus tenax:* regulation on protein and transcript level. *Extremophiles* **5:**101–109.
16. Buchanan, C. L., H. Connaris, M. J. Danson, C. D. Reeve, and D. W. Hough. 1999. An extremely thermostable aldolase from *Sulfolobus solfataricus* with specificity for non-phosphorylated substrates. *Biochem. J.* **343:**563–570.
17. Budgen, N., and M. J. Danson. 1985. Metabolism of glucose via a modified Entner-Doudoroff pathway in the thermoacidophilic archaebacterium *Thermoplasma acidophilum. FEBS Lett.* **196:**207–210.
18. Bult, C. J., O. White, G. J. Olsen, L. X. Zhou, R. D. Fleischmann, G. G. Sutton, J. A. Blake, L. M. FitzGerald, R. A. Clayton, J. D. Gocayne, A. R. Kerlavage, B. A. Dougherty, J. F. Tomb, M. D. Adams, C. I. Reich, R. Overbeek, E. F. Kirkness, K. G. Weinstock, J. M. Merrick, A. Glodek, J. L. Scott, N. S. M. Geoghagen, J. F. Weidman, J. L. Fuhrmann, D. Nguyen, T. R. Utterback, J. M. Kelley, J. D. Peterson, P. W. Sadow, M. C. Hanna, M. D. Cotton, K. M. Roberts, M. A. Hurst, B. P. Kaine, M. Borodovsky, H-P. Klenk, C. M. Fraser, H. O. Smith, C. R. Woese, and J. C. Venter. (1996) Complete genome sequence of the methanogenic Archaeon, *Methanococcus jannaschii. Science* **273:**1058–1073.
19. Cardona, S., F. Remonsellez, N. Guiliani, and C. A. Jerez. 2001. The glycogen-bound polyphosphate kinase from *Sulfolobus acidocaldarius* is actually glycogen synthase. *Appl. Environ. Microbiol.* **67:**4773–4780.
20. Chassagnole, C., N. Noisommit-Rizzi, J. W. Schmid, K. Mauch, and M. Reuss. 2002. Dynamic modeling of the central carbon metabolism of *Escherichia coli. Biotechnol. Bioeng.* **79:**53–73.
21. Cobucci-Ponzano, B., M. Rossi, and M. Moracci. 2005. Recoding in Archaea. *Mol. Microbiol.* **55:**339–348.
22. Cohen, G. N., V. Barbe, D. Flament, M. Galperin, R. Heilig, O. Lecompte, O. Poch, D. Prieur, J. Querellou, R. Ripp, J-C. Thierry, J. van der Oost, J. Weissenbach, Y. Zivanovic, and P. Forterre. 2003. An integrated analysis of the genome of the hyperthermophilic archaeon *Pyrococcus abyssi. Mol. Microbiol.* **47:**1495–1512.
23. Conway, T. 1992. The Entner-Doudoroff pathway: history, physiology and molecular biology. *FEMS Microbiol. Rev.* **9:**1–27.

24. **Cooper, R. A.** 1978. Intermediary metabolism of monosaccharides by Bacteria, *Int. Rev. Biochem.* **16:**37–73.

25. **Daniel, R. M., and M. J. Danson.** 1995. Did primitive microorganisms use nonhem iron proteins in place of NAD/P? *J. Mol. Evol.* **40:**559–563.

26. **Danson, M. J.** 1989. Central metabolism of the archaebacteria: an overview. *Can. J. Microbiol.* **35:**58–64.

27. **Danson, M. J.** 1993. Central metabolism of the Archaea, p. 1–24. *In* M. Kates, D. Kushner, and A. T. Matheson (ed.), *New Comp. Biochemistry [The Biochemistry of Archaea]*, vol 26. Elsevier/North Holland Biomedical Press, Amsterdam.

28. **Danson, M. J., S. C. Black, D. L. Woodland, and P. A. Wood.** 1985. Citric acid cycle enzymes of the archaebacteria: citrate synthase and succinate thiokinase. *FEBS Lett.* **179:**120–124.

29. **Danson, M. J., R. Eisenthal, S. Hall, S. R. Kessell, and D. L. Williams.** 1984. Dihydrolipoamide dehydrogenase from halophilic archaebacteria. *Biochem. J.* **218:**811–818.

30. **Danson, M. J., and D. W. Hough.** 1992. The enzymology of archaebacterial pathways of central metabolism. *Biochem. Soc. Symp.* **58:**7–21.

31. **Danson, M. J., D. J. Morgan, A. C. Jeffries, D. W. Hough, and M. L. Dyall-Smith.** 2004. Multienzyme complexes in the Archaea: predictions from genome sequences, p. 177–191. *In* A. Ventosa (ed.), *Halophilic Microorganisms*. Springer-Verlag, Berlin, Germany.

32. **Darland, G., T. D. Brock, W. Samsonoff, and S. F. Conti.** 1970. A thermophilic, acidophilic mycoplasma isolated from a coal refuse pile. *Science* **170:**1416–1418.

33. **De Ley, J., and M. Doudoroff.** 1957. The metabolism of D-galactose in *Pseudomonas saccharophila*. *J. Biol. Chem.* **227:**745–757.

34. **De Montigny, C., and J. Sygusch.** 1996. Functional characterization of an extreme thermophilic class II fructose-1,6-bisphosphate aldolase. *Eur. J. Biochem.* **241:**243–248.

35. **DeRisi, J. L., V. R. Iyer, and P. O. Brown.** 1997. Exploring the metabolic and genetic control of gene expression on a genomic scale. *Science* **278:**680–686.

36. **De Rosa, M., A. Gambacorta, B. Nicolaus, P. Giardina, E. Poerio, and V. Buonocore.** 1984. Glucose metabolism in the extreme thermoacidophilic archaebacterium *Sulfolobus solfataricus*. *Biochem. J.* **224:**407–414.

37. **Dhar, N. M., and W. Altekar.** 1986. Distribution of class I and class II fructose biphosphate aldolases in halophilic archaebacteria. *FEMS Microbiol. Lett.* **35:**177–181.

38. **Dörr, C., M. Zaparty, B. Tjaden, H. Brinkmann, and B. Siebers.** 2003. The hexokinase of the hyperthermophile *Thermoproteus tenax*. *J. Biol. Chem.* **278:**18744–18753.

39. **Elshafei, A. M., and O. M. Abdel-Fatah.** 2001. Evidence for a non-phosphorylated route of galactose breakdown in cell-free extracts of *Aspergillus niger*. *Enzyme Microb. Technol.* **29:**76–83.

40. **Elzainy, T. A., M. M. Hassan, and A. M. Allam.** 1973. New pathway for non-phosphorylated degradation of gluconate by *Aspergillus niger*. *J. Bacteriol.* **114:**457–459.

41. **Entner, N., and M. Doudoroff.** 1952. Glucose and gluconic acid oxidation by *Pseudomonas saccharophila*. *J. Biol. Chem.* **196:**853–862.

42. **Ettema, T. J. G., K. S. Makarova, G. L. Jellema, H. J. Gierman, E. V. Koonin, M. A. Huynen, W. M. de Vos, and J. van der Oost.** 2004. Identification and functional verification of archaeal-type phosphoenolpyruvate carboxylase, a missing link in archaeal central carbohydrate metabolism. *J. Bacteriol.* **186:**7754–7762.

43. **Evans, J. N. S., D. P. Raleigh, C. J. Tolman, and M. F. Roberts.** 1986. ¹³C NMR spectroscopy of *Methanobacterium thermoautotrophicum*. *J. Biol. Chem.* **261:**16323–16331.

44. **Evans, J. N. S., C. J. Tolman, S. Kanodia, and M. F. Roberts.** 1985. 2,3-Cyclopyrophosphoglycerate in methanogens: evidence by ¹³C NMR spectroscopy for a role in carbohydrate metabolism. *Biochemistry* **24:**5693–5698.

45. **Fiala, G., and K. O. Stetter.** 1986. *Pyrococcus furiosus* sp. nov. represents a novel genus of marine heterotrophic archaebacteria growing optimally at 100°C. *Arch. Microbiol.* **145:**56–61.

46. **Finn, M. W., and F. R. Tabita.** 2003. Synthesis of catalytically active form III ribulose 1,5-bisphosphate carboxylase/oxygenase in archaea. *J. Bacteriol.* **185:**3049–3059.

47. **Finn, M. W., and F. R. Tabita.** 2004. Modified pathway to synthesize ribulose 1,5-bisphosphate in methanogenic archaea. *J. Bacteriol.* **186:**6360–6366.

48. **Fothergill-Gilmore, L. A., and P. A. Michels.** 1993. Evolution of glycolysis. *Prog. Biophys. Mol. Biol.* **59:**105–235.

49. **Fuhrer, T., E. Fischer, and U. Sauer.** 2005. Experimental identification and quantification of glucose metabolism in seven bacterial species. *J. Bacteriol.* **187:**1581–1590.

50. **Fukuda, E., and T. Wakagi.** 2002. Substrate recognition by 2-oxoacid:ferredoxin oxidoreductase from *Sulfolobus* sp. strain 7. *Biochim. Biophys. Acta* **1597:**74–80.

51. **Fukuda, W., T. Fukui, H. Atomi, and T. Imanaka.** 2004. First characterization of an archaeal GTP-dependent phosphoenolpyruvate carboxykinase from the hyperthermophilic archaeon *Thermococcus kodakaraensis* KOD1. *J. Bacteriol.* **186:**4620–4627.

52. **Fukuda, W., Y. S. Ismail, T. Fukui, H. Atomi, and T. Imanaka.** 2005. Characterisation of an archaeal malic enzyme from the hyperthermophilic archaeon *Thermococcus kodakaraensis* KOD1. *Archaea* **1:**293–301.

53. **Fukui, T., H. Atomi, T. Kanai, R. Matsumi, S. Fujiwara, and T. Imanaka.** 2005. Complete genome sequence of the hyperthermophilic archaeon *Thermococcus kodakaraensis* KOD1 and comparison with *Pyrococcus* genomes. *Genome Res.* **15:**352–363.

54. **Futterer, O., A. Angelov, H. Liesegang, G. Gottschalk, C. Schleper, B. Schepers, C. Dock, G. Antranikian, and W. Liebl.** 2004. Genome sequence of *Picrophilus torridus* and its implications for life around pH 0. *Proc. Natl. Acad. Sci. USA* **101:**9091–9096.

55. **Glasemacher, J., A.-K. Bock, R. Schmid, and P. Schönheit.** 1997. Purification and properties of acetyl-CoA synthetase (ADP-forming), an archaeal enzyme of acetate formation and ATP synthesis, from the hyperthermophile *Pyrococcus furiosus*. *Eur. J. Biochem.* **244:**561–567.

56. **Graham, D. E., H. Xu, and R. H. White.** 2002. A divergent archaeal member of the alkaline phosphatase binuclear metalloenzyme superfamily has phosphoglycerate mutase activity. *FEBS Lett.* **517:**190–194.

57. **Grogan, D. W.** 1989. Phentopyic characterization of the archaebacterial genus *Sulfolobus*: comparison of five wild-type strains. *J. Bacteriol.* **171:**6710–6719.

58. **Hansen, T., M. Oehlmann, and P. Schönheit.** 2001. Novel type of glucose-6-phosphate isomerase in the hyperthermophilic archaeon *Pyrococcus furiosus*. *J. Bacteriol.* **183:**3428–3435.

59. **Hansen, T., B. Reichstein, R. Schmid, and P. Schönheit.** 2002. The first archaeal ATP-dependent glucokinase, from the hyperthermophilic crenarchaeon *Aeropyrum pernix*, represents a monomeric, extremely thermophilic ROK glucokinase with broad hexose specificity. *J. Bacteriol.* **184:**5955–5965.

60. **Hansen, T., and P. Schönheit.** 2004. ADP-dependent 6-phosphofructokinase, an extremely thermophilic, non-allosteric enzyme from the hyperthermophilic, sulfate-reducing archaeon *Archaeoglobus fulgidus* strain 7324. *Extremophiles* **8:**29–35.

61. Hansen, T., D. Wendorff, and P. Schönheit. 2001. Bifunctional phosphoglucose/phosphomannose isomerases from the Archaea *Aeropyrum pernix* and *Thermoplasma acidophilum* constitute a novel enzyme family within the phosphoglucose isomerase superfamily. *J. Biol. Chem.* 279:2262–2272.

62. Hartmann, R., H-D. Sickinger, and D. Oesterhelt. 1980. Anaerobic growth of Halobacteria. *Proc. Natl. Acad. Sci. USA* 77:3821–3825.

63. Heath, C., A. C. Jeffries, D. W. Hough, and M. J. Danson. 2004. Discovery of the catalytic function of a putative 2-oxoacid dehydrogenase multienzyme complex in the thermophilic archaeon *Thermoplasma acidophilum*. *FEBS Lett.* 577:523–527.

64. Holden, H. M., I. Rayment, and J. B. Thoden. 2003. Structure and function of enzymes of the Leloir pathway for galactose metabolism. *J. Biol. Chem.* 278:43885–43888.

65. Holmes, M. L., and M. L. Dyall-Smith. 2000. Sequence and expression of a halobacterial β-galactosidase gene. *Mol. Microbiol.* 36:114–122.

66. Hügler, M., H. Huber, K. O. Stetter, and G. Fuchs. 2003. Autotrophic CO_2 fixation pathways in archaea (Crenarchaeota). *Arch. Microbiol.* 179:160–173.

67. Huynen, M. A., T. Dandekar, and P. Bork. 1999. Variation and evolution of the citric-acid cycle: a genomic perspective. *Trends Microbiol.* 7:281–291.

68. Izard, T., A. Ævarsson, M. D. Allen, A. H. Westphal, R. N. Perham, A. de Kok, and W. G. J. Hol. 1999. Principles of quasi-equivalence and Euclidean geometry govern the assembly of cubic and dodecahedral cores of pyruvate dehydrogenase complexes. *Proc. Natl. Acad. Sci. USA* 96:1240–1245.

69. Jansen, K., E. Stupperich, and G. Fuchs. 1982. Carbohydrate synthesis from acetyl CoA in the autotroph *Methanobacterium thermoautotrophicum*. *Arch. Microbiol.* 132:355–364.

70. Johnsen, U., T. Hansen, and P. Schönheit. 2003. Comparative analysis of pyruvate kinases from the hyperthermophilic archaea *Archaeoglobus fulgidus*, *Aeropyrum pernix*, and *Pyrobaculum aerophilum* and the hyperthermophilic bacterium *Thermotoga maritima*. *J. Biol. Chem.* 278:25417–25427.

71. Johnsen, U., and P. Schönheit. 2004. Novel xylose dehydrogenase in the halophilic archaeon *Haloarcula marimortui*. *J. Bacteriol.* 186:6198–6207.

72. Johnsen, U., M. Selig, K. B. Xavier, H. Santos, and P. Schönheit. 2001. Different glycolytic pathways for glucose and fructose in the halophilic archaeonn *Halococcus saccharolyticus*. *Arch. Microbiol.* 175:52–61.

73. Jolley, K. A., D. G. Maddocks, S. L. Gyles, Z. Mullan, S-L. Tang, M. L. Dyall-Smith, D. W. Hough, and M. J. Danson. 2000. 2-Oxoacid dehydrogenase multienzyme complexes in the halophilic Archaea? Gene sequences and protein structural predictions. *Microbiology* 146:1061–1069.

74. Jolley, K. A., E. Rapaport, D. W. Hough, M. J. Danson, W. G. Woods, and M. L. Dyall-Smith. 1996. Dihydrolipoamide dehydrogenase from the halophilic Archaeon, *Haloferax volcanii*: homologous over-expression of the cloned gene. *J. Bacteriol.* 178:3044–3048.

75. Jones, W. J., J. A. Leigh, F. Mayer, C. R. Woese, and R. S. Wolfe. 1983. *Methanococcus jannaschii* sp.nov., an extremely thermophilic methanogen from a submarine hydrothermal vent. *Arch. Microbiol.* 136:254–261.

76. Kandler, O. 1983. Carbohydrate metabolism in lactic acid bacteria. *Antonie van Leeuwenhock J. Microbiol.* 49:209–224.

77. Karadzic, I. M., and J. A. Maupin-Furlow. 2005. Improvement of two-dimensional gel electrophoresis proteome maps of the haloarchaeon *Haloferax volcanii*. *Proteomics* 5:354–359.

78. Kardinahl, S., C. L. Schmidt, T. Hansen, S. Anemüller, A. Petersen, and G. Schäfer. 1999. The strict molybdate-dependence of glucose-degradation by the thermoacidophile *Sulfolobus acidocaldarius* reveals the first crenarchaeotic molybdenum containing enzyme—an aldehyde oxidoreductase. *Eur. J. Biochem.* 260:540–548.

79. Kawarabayasi, Y., M. Sawada, H. Horikawa, Y. Haikawa, Y. Hino, S. Yamamoto, M. Sekine, S. Baba, H. Kosugi, A. Hosoyama, Y. Nagai, M. Sakai, K. Ogura, R. Otsuka, H. Nakazawa, M. Takamiya, Y. Ohfuku, T. Funahashi, T. Tanaka, Y. Kudoh, J. Yamazaki, N. Kushida, A. Oguchi, K. Aoki, T. Yoshizawa, Y. Nakamura, F.T. Robb, K. Horikoshi, Y. Masuchi, H. Shizuya, and H. Kikuchi. 1998. Complete sequence and gene organization of the genome of a hyperthermophilic archaebacterium, *Pyrococcus horikoshii* OT3. *DNA Res.* 5:55–76.

80. Kawarabayasi, Y., Y.Hino, H. Horikawa, S. Yamazaki, Y. Haikawa, K. Jin-No, M. Takahashi, M. Sekine, S. Baba, A. Ankai, H. Kosugi, A. Hosoyama, S. Fukui, Y. Nagai, K. Nishijima, H. Nakazawa, M. Takamiya, S. Masuda, T. Funahashi, T. Tanaka, Y. Kudoh, J. Yamazaki, N. Kushida, A. Oguchi, K. Aoki, K. Kubota, Y. Nakamura, N. Nomura, Y. Sako, and H. Kikuchi. 1999. Complete genome sequence of an aerobic hyperthermophilic crenarchaeon, *Aeropyrum pernix* K1. *DNA Res.* 6:83–101.

81. Kawashima, T., N. Amano, H. Koike, S. Makino, S. Higuchi, Y. Kawishima-Ohya, K. Watanabe, M. Yamazaki, K. Keiichi, T. Kawamoto, T. Nunoshiba, Y. Yamamoto, H. Aramaki, K. Makino, and M. Suzuki. 2000. Archaeal adaption to higher temperatures revealed by genomic sequence of *Thermoplasma volcanium*. *Proc. Natl. Acad. Sci. USA* 97:14257–14262.

82. Kay, J., and P. D. J. Weitzman (ed.). 1987. Krebs' citric acid cycle—half a century and still turning. *Biochem. Soc. Symp.* 54:1–195.

83. Kengen, S. W. M., F. A. M. de Bok, N-D. van Loo, C. Dijkema, A. J. M. Stams, and W. M. de Vos. 1994. Evidence for the operation of a novel Embden-Meyerhof pathway that involves ADP-dependent kinases during sugar fermentation by *Pyrococcus furiosus*. *J. Biol. Chem.* 269:17537–17541.

84. Kengen, S. W. M., and A. J. M. Stams. 1994. Formation of l-alanine as a reduced end product in carbohydrate fermentation by the hyperthermophilic archaeon *Pyrococcus furiosus*. *Arch. Microbiol.* 161:168–175.

85. Kengen, S. W. M., J. E. Tuininga, F. A. M. de Bok, A. J. M. Stams, and W. M. de Vos. 1995. Purification and characterization of a novel ADP-dependent glucokinase from the hyperthermophilic archaeon *Pyrococcus furiosus*. *J. Biol. Chem.* 270:30453–30457.

86. Kersters, K., and J. De Ley. 1968. The occurrence of the Entner-Doudoroff pathway in bacteria. *Antonie van Leeuwenhoek* 34:393–408.

87. Khodursky, A. B., B. J. Peter, N. R. Cozzarelli, D. Botstein, P. O. Brown, and C. Janofsky. 2000. DNA microarray analysis of gene expression in response to physiological and genetic changes that affect tryptophan metabolism in *Escherichia coli*. *Proc. Natl. Acad. Sci. USA* 97:12170–12175.

88. Kim, S., and S. B. Lee. 2005. Identification and characterisation of *Sulfolobus solfataricus* D-gluconate dehydratase: a key enzyme in the non-phosphorylated Entner-Doudoroff pathway. *Biochem. J.* 387:271–280.

89. Klenk, H. P., R. A. Clayton, J. F. Tomb, O. White, K. E. Nelson, K. A. Ketchum, R. J. Dodson, M. Gwinn, E. K. Hickey, J. D. Peterson, D. L. Richardson, A. R. Kerlavage, D. E. Graham, N. C. Kyrpides, R. D. Fleischmann, J. Quackenbush, N. H. Lee, G. G. Sutton, S. Gill, E. F. Kirkness, B. A. Dougherty, K. McKenney, M. D. Adams, B. Loftus, S. Peter-

son, C. I. Reich, L. K. McNeil, J. H. Badger, A. Glodek, L. X. Zhou, R. Overbeek, J. D. Gocayne, J. F. Weidman, L. McDonald, T. Utterback, M. D. Cotton, T. Spriggs, P. Artiach, B. P. Kaine, S. M. Sykes, P. W. Sadow, K. P. DAndrea, C. Bowman, C. Fujii, S. A. Garland, T. M. Mason, G. J. Olsen, C. M. Fraser, H. O. Smith, C. R. Woese, and J. C. Venter. 1997. The complete genome sequence of the hyperthermophilic, sulphate-reducing archaeon *Archaeoglobus fulgidus*. *Nature* 390:364–370.

90. Kletzin, A., and M. W. W. Adams. 1996. Molecular and phylogenetic characterization of pyruvate and 2- ketoisovalerate ferredoxin oxidoreductases from *Pyrococcus furiosus* and pyruvate ferredoxin oxidoreductase from *Thermotoga maritima*. *J. Bacteriol.* 178:248–257.

91. Kobayashi, T., S. Higuchi, K. Kimura, T. Kudo, and K. Horikoshi. 1995. Properties of glutamate-dehydrogenase and its involvement in alanine production in a hyperthermophilic archaeon, *Thermococcus profundus*. *J. Biochem.* 118:587–592.

92. Koga, S., I. Yoshioka, H. Sakuraba, M. Takahashi, S. Sakasegawa, S. Shimizu, and T. Ohshima. 2000. Biochemical characterisation, cloning, and sequencing of ADP-dependent (AMP-forming) glucokinase from two hyperthermophilic Archaea, *Pyrococcus furiosus* and *Thermococcus litoralis*. *J. Biochem.* 128:1079–1085.

93. König, H., E. Nusser, and K. O. Stetter. 1985. Glycogen in *Methanolobus* and *Methanococcus*. *FEMS Microbiol. Lett.* 28:265–269.

94. König, H., R. Skorto, W. Zillig, and W-D. Reiter. 1982. Glycogen in thermoacidophilic archaebacteria of the genera *Sulfolobus, Thermoproteus, Desulfurococcus* and *Thermococcus*. *Arch. Microbiol.* 132:297–303.

95. Kunow, J., D. Linder, and R. K. Thauer. 1995. Pyruvate:ferredoxin oxidoreductase from the sulfate-reducing *Archaeoglobus fulgidus*: molecular composition, catalytic properties, and sequence alignments. *Arch. Microbiol.* 163:21–28.

96. Labes, A., and P. Schönheit. 2001. Sugar utilisation in the hyperthermophilic, sulfate-reducing archaeon *Archaeoglobus fulgidus* strain 7324: starch degradation to acetate and CO_2 via a modified Embden-Meyerhof pathway and acetyl-CoA synthetase (ADP-forming). *Arch. Microbiol.* 176:329–338.

97. Labes, A., and P. Schönheit. 2003. ADP-dependent glucokinase from the hyperthermophilic sulfate-reducing archaeon *Archaeoglobus fulgidus* strain 7324. *Arch. Microbiol.* 180:69–75.

98. Lamble, H. J., N. I. Heyer, S. D. Bull, D. W. Hough, and M. J. Danson. 2003. Metabolic pathway promiscuity in the archaeon *Sulfolobus solfataricus* revealed by studies on glucose dehydrogenase and 2-keto-3-deoxygluconate aldolase. *J. Biol. Chem.* 278:34066–34072.

99. Lamble, H. J., C. C. Milburn, G. L. Taylor, D. W. Hough, and M. J. Danson. 2004. Gluconate dehydratase from the promiscuous Entner-Doudoroff pathway in *Sulfolobus solfataricus*. *FEBS Lett.* 576:133–136.

100. Maeder, D. L., R. B. Weiss, D. M. Dunn, J. L. Cherry, J. M. González, J. DiRuggiero, and F. T. Robb. 1999. Divergence of the hyperthermophilic archaea *Pyrococcus furiosus* and *P. horikoshii* inferred from complete genomic sequences. *Genetics* 152:1299–1305.

101. Mai, X., and M. W. W. Adams. 1996. Purification and characterization of two reversible and ADP-dependent acetyl coenzyme A synthetases from the hyperthermophilic archaeon *Pyrococcus furiosus*. *J. Bacteriol.* 178:5897–5903.

102. Makarova, K. S., and E. V. Koonin. 2003. Filling a gap in the central metabolism of archaea: prediction of a novel aconitase by comparative-genomic analysis. *FEMS Microbiol. Lett.* 227:17–23.

103. Musfeldt, M., and P. Schönheit. 2002. Novel type of ADP-forming acetyl coenzyme A synthetase in hyperthermophilic *archaea*: heterologous expression and characterization of isoenzymes from the sulfate reducer *Archaeoglobus fulgidus* and the methanogen *Methanococcus jannaschii*. *J. Bacteriol.* 184:636–644.

104. Nelson, K. E., R. A. Clayton, S. R. Gill, M. L. Gwinn, R. J. Dodson, D. H. Haft, E. K. Hickey, L. D. Peterson, W. C. Nelson, K. A. Ketchum, L. McDonald, T. R. Utterback, J. A. Malek, K. D. Linher, M. M. Garrett, A. M. Stewart, M. D. Cotton, M. S. Pratt, C. A. Phillips, D. Richardson, J. Heidelberg, G. G. Sutton, R. D. Fleischmann, J. A. Eisen, O. White, S. L. Salzberg, H. O. Smith, J. C. Venter, and C. M. Fraser. 1999. Evidence for lateral gene transfer between Archaea and Bacteria from genome sequence of *Thermotoga maritima*. *Nature* 399:323–329.

105. Ng, W. V., S. P. Kennedy, G. G. Mahairas, B. Berquist, M. Pan, H. D. Shukla, S. R. Lasky, N. S. Baliga, V. Thorsson, J. Sbrogna, S. Swartzell, D. Weir, J. Hall, T. A. Dahl, R. Welti, Y. A. Goo, B. Leithauser, K. Keller, R. Cruz, M. J. Danson, D. W. Hough, D. G. Maddocks, P. E. Jablonski, M. P. Krebs, C. M. Angevine, H. Dale, T. A. Isenbarger, R. F. Peck, M. Pohlschroder, J. L.Spudich, K. H. Jung, M. Alam, T. Freitas, S. B. Hou, C. J. Daniels, P. P. Dennis, A. D. Omer, H. Ebhardt, T. M. Lowe, R. Liang, M. Riley, L. Hood, and S. DasSarma. 2000. Genome sequence of *Halobacterium* species NRC-1. *Proc. Natl. Acad. Sci. USA* 97:12176–12181.

106. Nishizawa, Y., T. Yabuki, E. Fukuda, and T. Wakagi. 2005. Gene expression and characterization of two 2-oxoacid:ferredoxin oxidoreductases from *Aeropyrum pernix* K1. *FEBS Lett.* 579:2319–2322.

107. Oren, A., and P. Gurevich. 1995. Isocitrate lyase activity in halophilic archaea. *FEMS Microbiol. Lett.* 130:91–95.

108. Oren, A., and L. Mana. 2003. Sugar metabolism in the extremely halophilic bacterium *Salinibacter ruber*. *FEMS Microbiol. Lett.* 223:83–87.

109. Papin, J. A., N. D. Price, S. J. Wiback, D. A. Fell, and B. O. Palsson. 2003. Metabolic pathways in the post-genome era. *Trends Biochem. Sci.* 28:250–258.

110. Payton, M. A., and B. A. Haddock. 1985. Aerobic metabolism of glucose, p. 337–356. *In* A. T. Bull and H. Dalton (ed.), *Comprehensive Biotechnology*, vol. 1. Pergamon Press, Oxford.

111. Perham, R. N. 2000. Swinging arms and swinging domains in multifunctional enzymes: catalytic machines for multistep reactions. *Annu. Rev.Biochem.* 69:961–1004.

112. Perham, R. N., D. D. Jones, H. J. Chauhan, and M. J. Howard. 2002. Substrate channelling in 2-oxo acid dehydrogenase multienzyme complexes. *Biochem. Soc. Trans.* 30:47–51.

113. Plaga, W., F. Lottspeich, and D. Oesterhelt. 1992. Improved purification, crystallization and primary structure of pyruvate: ferredoxin oxidoreductase from *Halobacterium halobium*. *Eur. J. Biochem.* 205:391–397.

114. Pratt, K. J., C. Carles, T. J. Carne, M. J. Danson, and K. J. Stevenson. 1989. Detection of bacterial lipoic acid: a modified gas chromatographic-mass spectrometric procedure. *Biochem. J.* 258:749–754.

115. Racker, E. 1948. Enzymatic formation and breakdown of pentose phosphate. *Fed. Proc.* 7:180.

116. Racker, E. 1951. Enzymatic synthesis and breakdown of deoxyribose phosphate. *J. Biol. Chem.* 196:347–365.

117. Rashid, N., H. Imanaka, T. Kanai, T. Fukui, H. Atomi, and T. Imanaka. 2002. A novel candidate for the true fructose-1,6-bisphosphatase in archaea. *J. Biol. Chem.* 277:30649–30655.

118. Ravot, G., B. Ollivier, M. L. Fardeau, B. K. Patel, K. T. Andrews, M. Magot, and J. L. Garcia. 1996. L-alanine production from glucose fermentation by hyperthermophilic mem-

bers of the domains Bacteria and Archaea: a remnant of an ancestral metabolism? *Appl. Environ. Microbiol.* **62**:2657–2659.

119. Rawal, N., S. M. Kelkar, and W. Altekar. 1988. Alternative routes of carbohydrate metabolism in halophilic archaebacteria. *Ind. J. Biochem. Biophys.* **25**:674–686.

120. Ray, W. K., S. M. Keith, A. M. DeSantis, J. P. Hunt, T. J. Larson, R. F. Helm, and P. J. Kennelly. 2005. A phosphohexomutase from the archaeaon *Sulfolobus solfataricus* is covalently modified by phosphorylation on serine. *J. Bacteriol.* **187**:4270–4275.

121. Robb, F. T., D. L. Maeder, J. R. Brown, J. DiRuggiero, M. D. Stump, R. K. Yeh, R. B. Weiss, and D. M. Dunn. 2001. Genomic sequence of hyperthermophile *Pyrococcus furiosus*: implications for physiology and enzymology. *Methods Enzymol.* **330**:134–157.

122. Romano, A. H., and T. Conway. 1996. Evolution of carbohydrate metabolic pathways. *Res. Microbiol.* **147**:448–455.

123. Ronimus, R. S., Y. Kawarabayasi, H. Kikuchi, and H. W. Morgan. 2001. Cloning, expression and characterisation of a family B ATP-dependent phosphofructokinase activity from the hyperthermophilic crenarchaeon *Aeropyrum pernix*. *FEMS Microbiol. Lett.* **202**:85–90.

124. Ronimus, R. S., J. Koning, and H. W. Morgan. 1999. Purification and characterisation of an ADP-dependent phosphofructokinase from *Thermococcus zilligii*. *Extremophiles* **3**:121–129.

125. Ronimus, R. S., and H. W. Morgan. 2001. The biochemical properties and phylogenies of phosphofructokinases from extremophiles. *Extremophiles* **5**:357–373.

126. Ronimus, R. S., and H. W. Morgan. 2002. Distribution and phylogenies of enzymes of the Embden-Meyerhof-Parnas pathway from archaea and hyperthermophilic bacteria support a gluconeogenic origin of metabolism. *Archaea* **1**:199–221.

127. Rudolph, B., T. Hansen, and P. Schönheit. 2004. Glucose-6-phosphate isomerase from the hyperthermophilic archaeon *Methanococcus jannaschii*: characterization of the first archaeal member of the phosphoglucose isomerase superfamily. *Arch. Microbiol.* **181**:82–87.

128. Ruepp, A., and J. Soppa. 1996. Fermentative arginine degradation in *Halobacterium salinarium* (formerly *Halobacterium halobium*): genes, gene products, and transcripts of the arcRACB gene cluster. *J. Bacteriol.* **178**:4942–4947.

129. Ruepp, A., W. Graml, M-L. Santos-Martinez, K. K. Koretke, C. Volker, H. W. Mewes, D. Frishman, S. Stocker, A. N. Lupas, and W. Baumeister. 2000. The genome sequence of the thermoacidophilic scavenger *Thermoplasma acidophilum*. *Nature* **407**:508–513.

130. Sakuraba, H., I. Yoshioka, S. Koga, M. Takahashi, Y. Kitahama, T. Satomura, R. Kawakami, and T. Ohshima. 2002. ADP-Dependent glucokinase/phosphofructokinase, a novel bifunctional enzyme from the hyperthermophilic archaeon *Methanococcus jannaschi*. *J. Biol. Chem.* **277**:12495–12498.

131. Schäfer, G. 1996. Bioenergetics of the archaebacterium *Sulfolobus*. *Biochim. Biophys. Acta* **1277**:163–200.

132. Schäfer, S., C. Barkowski, and G. Fuchs. 1986. Carbon assimilation by the autotrophic thermophilic archaebacterium *Thermoproteus-neutrophilus*. *Arch. Microbiol.* **146**:301–308.

133. Schäfer, T., and P. Schönheit. 1993. Gluconeogenesis from pyruvate in the hyperthermophilic archaeon *Pyrococcus furiosus*: involvement of reactions of the Embden-Meyerhof pathway. *Arch. Microbiol.* **159**:354–363.

134. Schäfer, T., M. Selig, and P. Schönheit. 1993. Acetyl-CoA synthetase (ADP-forming) in archaea, a novel enzyme involved in acetate and ATP synthesis. *Arch. Microbiol.* **159**:72–83.

135. Schäfer, T., K. B. Xavier, H. Santos, and P. Schönheit. 1994. Glucose fermentation to acetate and alanine in resting cell suspensions of *Pyrococcus furiosus*: proposal of a novel glycolytic pathway based on ^{13}C labelling data and enzyme activities. *FEMS Microbiol. Lett.* **121**:107–114.

136. Schönheit, P., and T. Schäfer. 1995. Metabolism of hyperthermophiles. *World J. Microbiol. Biotechnol.* **11**:26–57.

137. Schramm, A., B. Siebers, B. Tjaden, H. Brinkmann, and R. Hensel. 2000. Pyruvate kinase of the hyperthermophilic crenarchaeote *Thermoproteus tenax*: physiological role and phylogenetic aspects. *J. Bacteriol.* **182**:2001–2009.

138. Schurig, H., N. Beaucamp, R. Ostendorp, R. Jaenicke, E. Adler, and J.R. Knowles. 1995. Phosphoglycerate kinase and triosephosphate isomerase from the hyperthermophilic bacterium *Thermotoga maritima* form a covalent bifunctional enzyme complex. *EMBO J.* **14**:442–451.

139. Schut, G. J., S. D. Brehm, S. Datta, and M. W. W. Adams. 2003. Whole-genome DNA microarray analysis of a hyperthermophile and an archaeon: *Pyrococcus furiosus* grown on carbohydrates or peptides. *J. Bacteriol.* **185**:3935–3947.

140. Schut, G. J., A .L. Menon, and M. W. W. Adams. 2001. 2-Keto acid oxidoreductases from *Pyrococcus furiosus* and *Thermococcus litoralis*. *Methods Enzymol.* **331**:144–158.

141. Segerer, A., T. A. Langworthy, and K. O. Stetter. 1988. *Thermoplasma acidophilum* and *Thermoplasma volcanium* sp. nov. from solfatara fields. *System. Appl. Microbiol.* **10**:161–171.

142. Selig, M., K. B. Xavier, H. Santos, and P. Schönheit. 1997. Comparative analysis of Embden-Meyerhof and Entner-Doudoroff glycolytic pathways in hyperthermophilic archaea and the bacterium *Thermotoga*. *Arch. Microbiol.* **167**:217–232.

143. Serrano, J. A., M. Camacho, and M. J. Bonete. 1998. Operation of glyoxylate cycle in halophilic archaea: presence of malate synthase and isocitrate lyase in *Haloferax volcanii*. *FEBS Lett.* **434**:13–16.

144. Severina, L. O., and N. V. Pimenov. 1988. Glucose metabolism in extreme halophic archaebacteria. *Microbiology* **57**:152–156.

145. Severina, L. O., and N. V. Pimenov. 1988. Glucose metabolism in extreme *Halococcus morrhuae*. *Microbiology* **57**:718–722.

146. She, Q., R. K. Singh, F. Confalonieri, Y. Zivanovic, G. Allard, M. J. Awayez, C. C. Y. Chan-Weiher, I. G. Clausen, B. A. Curtis, A. De Moors, G. Erauso, C. Fletcher, P. M. K. Gordon, I. Heikamp-de Jong, A. C. Jeffries, C. J. Kozera, N. Medina, X. Peng, H. P. Thi-Ngoc, P. Redder, M. E. Schenk, C. Theriault, N. Tolstrup, R. L. Charlebois, W. F. Doolittle, M. Duguet, T. Gaasterland, R. A. Garrett, M. A. Ragan, C. W. Sensen, and J. Van der Oost. 2001. The complete genome of the crenarchaeon *Sulfolobus solfataricus* P2. *Proc. Natl. Acad. Sci. USA* **98**:7835–7840.

147. Siebers, B., H. Brinkman, C. Dörr, B. Tjaden, H. Lilie, J. van der Oost, and C. H. Verhees. 2001. Archaeal fructose-1,6-bisphosphate aldolases constitute a new family of archaeal type class I aldolase. *J. Biol. Chem.* **276**:28710–28718.

148. Siebers, B., H-P. Klenk, and R. Hensel. 1998. PP$_i$-dependent phosphofructokinase from *Thermoproteus tenax*, an archaeal descendant of an ancient line in phosphofructokinase evolution. *J. Bacteriol.* **180**:2137–2143.

149. Siebers, B., B. Tjaden, K. Michalke, C. Dörr, H. Ahmed, M. Zaparty, P. Gordon, C. W. Sensen, A. Zibat, H-P. Klenk, S. C. Schuster, and R. Hensel. 2004. Reconstruction of the central carbohydrate metabolism of *Thermoproteus tenax* by use of genomic and biochemical data. *J. Bacteriol.* **186**:2179–2194.

150. Smith, D. R., L. A. Doucette-Stamm, C. Deloughery, H. Lee, J. Dubois, T. Aldredge, R. Bashirzadeh, D. Blakely, R. Cook,

K. Gilbert, D. Harrison, L. Hoang, P. Keagle, W. Lumm, B. Pothier, D. Qiu, R. Spadafora, R. Vicaire, Y. Wang, J. Wierzbowski, R. Gibson, N. Jiwani, A. Caruso, D. Bush, and J. N. Reeve. 1997. Complete genome sequence of *Methanobacterium thermoautotrophicum* deltaH: functional analysis and comparative genomics. *J. Bacteriol.* **179:**7135–7155.

151. Smith, L. D., N. Budgen, S. J. Bungard, M. J. Danson, and D. W. Hough. 1989. Purification and characterization of glucose dehydrogenase from the thermoacidophilic archaebacterium *Thermoplasma acidophilum*. *Biochem. J.* **261:**973–977.

152. Smith, L. D., S. J. Bungard, M. J. Danson, and D. W. Hough. 1987. Dihydrolipoamide dehydrogenase from the thermoacidophilic archaebacterium *Thermoplasma acidophilum*. *Biochem. Soc. Trans.* **15:**1097.

153. Soderberg, T. 2005. Biosynthesis of ribose-5-phosphate and erythrose-4-phosphate in archaea: a phylogenetic analysis of archaeal genomes. *Archaea* **1:**347–352.

154. Sonawat, H. M., S. Srivastava, S. Swaminathan, and G. Govil. 1990. Glycolysis and Entner-Doudoroff pathways in *Halobacterium halobium*: Some new observations based on ¹³C NMR spectroscopy. *Biochem. Biophys. Res. Commun.* **173:**358–362.

155. Stec, B., H. Yang, K. A. Johnson, L. Chen, and M. F. Roberts. 2000. MJ0109 is an enzyme that is both an inositol monophosphatase and the 'missing' archaeal fructose-1,6-bisphosphatase. *Nat. Struct. Biol.* **7:**1046–1050.

156. Steen, I. H., H. Hvoslef, T. Lien, and N-K. Birkeland. 2001. Isocitrate dehydrogenase, malate dehydrogenase, and glutamate dehydrogenase from *Archaeoglobus fulgidus*. *Methods Enzymol.* **331:**13–26.

157. Szymona, M., and M. Doudoroff. 1960. Carbohydrate metabolism in *Rhodopseudomonas spheroides*. *J. Gen. Microbiol.* **22:**167–183.

158. Tersteegen, A., D. Linder, R. K. Thauer, and R. Hedderich. 1997. Structures and functions of four anabolic 2-oxoacid oxidoreductases in *Methanobacterium thermoautotrophicum*. *Eur. J. Biochem.* **244:**862–868.

159. Tomlinson, G. A., and L. I. Hochstein. 1976. *Halobacterium saccharovorum* sp. nov., a carbohydrate-metabolizing, extremely halophilic bacterium. *Can. J. Microbiol.* **22:**587–591.

160. Tomlinson, G. A., T. K. Kock, and L. I. Hochstein. 1974. The metabolism of carbohydrates by extremely halophilic bacteria: glucose metabolism via a modified Entner-Doudoroff Pathway. *Can. J. Microbiol.* **20:**1085–1091.

161. Tuininga, J. E., C. H. Verhees, J. van der Oost, S. W. M. Kengen, A. J. M. Stams, and W. M. de Vos. 1999. Molecular and biochemical characterisation of the ADP-dependent phosphofructokinase from the hyperthermophilic archaeon *Pyrococcus furiosus*. *J. Biol. Chem.* **274:**21023–21028.

162. Uhrigshardt, H., M. Walden, H. John, A. Petersen, and S. Anemüller. 2002. Evidence for an operative glyoxylate cycle in the thermoacidophilic crenarchaeon *Sulfolobus acidocaldarius*. *FEBS Lett.* **513:**223–229.

163. van der Oost, J., M. A. Huynen, and C. H. Verhees. 2002. Molecular characterization of phosphoglycerate mutase in archaea. *FEMS Microbiol. Lett.* **212:**111–120.

164. van der Oost, J., G. Schut, S. W. M. Kengen, W. R. Hagen, M. Thomm, and W. M. de Vos. 1998. The ferredoxin-dependent conversion of glyceraldehyde-3-phosphate in the hyperthermophilic archaeon *Pyrococcus furiosus* represents a novel site of glycolytic regulation. *J. Biol. Chem.* **273:**28149–28154.

165. Verhees, C. H., M. A. Huynen, D. E. Ward, E. Schiltz, W. M. de Vos, and J. van der Oost. 2001. The phosphoglucose isomerase from the hyperthermophilic archaeon *Pyrococcus furiosus* is a unique glycolytic enzyme that belongs to the cupin superfamily. *J. Biol. Chem.* **276:**40926–40932.

166. Verhees, C. H., S. W. M. Kengen, J. E. Tuininga, G. J. Schut, M. W. W. Adams, W. M. de Vos, and J. van der Oost. 2003. The unique features of glycolytic pathways in Archaea. *Biochem. J.* **375:**231–246.

167. Verhees, C. H., D. G. M. Koot, T.J . G. Ettema, C. Dijkema, W. M. de Vos, and J. van der Oost. 2002. Biochemical adaptations of two sugar kinases from the hyperthermophilic archaeon *Pyrococcus furiosus*. *Biochem. J.* **366:**121–127.

168. Verhees, C. H., J. E. Tuininga, S. W. M. Kengen, A. J. M. Stams, J. van der Oost, and W. M. de Vos. 2001b. ADP-Dependent phosphofructokinase in mesophilic and thermophilic methanogenic Archaea. *J. Bacteriol.* **183:**7145–7153.

169. Vettakkorumakankav, N. N., and K. J. Stevenson. 1992. Dihydrolipoamide dehydrogenase from *Haloferax volcanii*: gene cloning, complete primary sequence and comparison to other dihydrolipoamide dehydrogenases. *Biochem. Cell Biol.* **70:**656–663.

170. Vreeland, R. H., S. Straight, J. Krammes, K. Dougherty, W. D. Rosenzweig, and M. Kamekura. 2002. *Halosimplex carlsbadense* gen. nov., sp. nov., a unique halophilic archaeon, with three 16S rRNA genes, that grows only in defined medium with glycerol and acetate or pyruvate. *Extremophiles* **6:**445–452.

171. Ward, D. E., S. W. M. Kengen, J. van der Oost, and W. M. de Vos. 2000. Purification and characterization of the alanine aminotransferase from the hyperthermophilic archaeon *Pyrococcus furiosus* and its role in alanine production. *J. Bacteriol.* **182:**2559–2566.

172. Weitzman, P. D. J. 1987. Patterns of diversity of citric acid cycle enzymes. *Biochem. Soc. Symp.* **54:**33–43.

173. Woese, C. R. 1998. The universal ancestor. *Proc. Natl. Acad. Sci. USA* **95:**6854–6859.

174. Wood, A. P., and D. P. Kelly. 1980. Carbohydrate degradation pathways in *Thiobacillus* A2 grown on various sugars. *J. Gen. Microbiol.* **120:**333–345.

175. Wood, A. P., D. P. Kelly, and P. R. Norris. 1987. Autotrophic growth of four *Sulfolobus* strains on tetrathionate and the effect of organic nutrients. *Arch. Microbiol.* **146:**382–389.

176. Xavier, K. B., M. S. Da Costa, and H. Santos. 2000. Demonstration of a novel glycolytic pathway in the hyperthermophilic archaeon *Thermococcus zilligii* by ¹³C-labeling experiments and nuclear magnetic resonance analysis. *J. Bacteriol.* **182:**4632–4636.

177. Yu, J-P., J. Ladapo, and W. B. Whitman. 1994. Pathway of glycogen metabolism in *Methanococcus maripaludis*. *J. Bacteriol.* **176:**325–332.

178. Zaigler, A., S. C. Schuster, and J. Soppa. 2003. Construction and usage of a onefold-coverage shotgun DNA microarray to characterize the metabolism of the archaeon *Haloferax volcanii*. *Mol. Microbiol.* **48:**1089–1105.

179. Zhang, Q., T. Iwasaki, T. Wakagi, and T. Oshima. 1996. 2-Oxoacid:ferredoxin oxidoreductase from the thermoacidophilic archaeon, *Sulfolobus* sp. strain 7. *J. Biochem.* **120:**587–599.

180. Zhu, W. H., C. I. Reich, G. J. Olsen, C. S. Giometti, and J. R. Yates. 2004. Shotgun proteomics of *Methanococcus jannaschii* and insights into methanogenesis. *J. Proteome Res.* **3:**538–548.

Archaea: Molecular and Cellular Biology
Edited by Ricardo Cavicchioli
© 2007 ASM Press, Washington, D.C.

Chapter 13

Methanogenesis

JAMES G. FERRY AND KYLE A. KASTEAD

INTRODUCTION

Methane-producing anaerobes (methanogens) were the first identified by Carl Woese to be phylogenetically distinct from all other cell types, and are the founding members of the *Archaea* (see Chapter 1). Methane-producing species were chosen based on their unusual morphological and biochemical characteristics (11, 38, 60, 114–116, 162, 165, 241, 271, 277) that guided comparisons of 16S ribosomal RNA sequences (64, 241) and discovery of the three-domain concept (270). Research on the methanogens, the largest group representing the *Archaea*, over the past three decades has had a major impact on our understanding of the ecology, physiology, biochemistry, and molecular biology of this domain. Indeed, the pathways for methanogenesis are one of the most extensively studied aspects of archaeal biology and one of the most prominent features that distinguish this domain of life from the *Bacteria* and *Eucarya*. Recent genomic sequencing, proteomic analyses, and development of genetic systems continue to expand our understanding of methanogenesis and the *Archaea*.

ECOLOGY

The Italian physicist Alessandro Volta is most often credited with the discovery of biological methane formation when he performed the "Volta experiment" at Lake Como, a freshwater lake in northern Italy. Disturbing the sediment with a pole, he released the trapped methane and collected the gas bubbles in an inverted funnel partly submerged in the water column. Lighting the gas produced a flame, prompting him to call the gas "combustible air." Figure 1 shows a modern-day Volta experiment in which gas collected from the sediment of a freshwater pond is ignited on release by tipping the funnel.

Including freshwater lakes and ponds, a significant fraction of the earth's biosphere contains vast and diverse oxygen-free environments where anaerobic microbes convert complex organic matter to methane and carbon dioxide, an essential link in the global carbon cycle (Fig. 2). The process occurs in habitats such as the rumen, the lower intestinal tract, sewage digesters, landfills, freshwater sediments of lakes and rivers, rice paddies, hydrothermal vents, coastal marine sediments, and the subsurface (264). A consortium of at least three interacting metabolic groups of anaerobes converts complex organic matter to the most oxidized (carbon dioxide) and reduced (methane) forms of carbon (Fig. 2). The process accounts for nearly one billion metric tons of biological methane produced annually (244). The first two groups (Fig. 2, step 3) are primarily from the *Bacteria*. The fermentative group decomposes complex organic matter to acetate, formate, higher volatile fatty acids, hydrogen, and carbon dioxide. The obligate hydrogen-producing group decomposes the higher volatile fatty acids to acetate, hydrogen, and carbon dioxide. The third group (Fig. 2, step 4), the methanogens, convert the metabolic products of the first two groups to methane by two major pathways. The conversion of the methyl group of acetate to methane (acetate fermentation pathway) produces about two-thirds of the annual production, whereas one-third derives from the reduction of carbon dioxide with electrons supplied from the oxidation of formate or hydrogen (carbon dioxide reduction pathway). Thus, the methanogens rely on the first two groups to supply substrates for growth and methanogenesis. Furthermore, the production of hydrogen by the fermentative and obligate hydrogen-producing groups is a thermodynamically unfavorable reaction, and growth of these groups depends on the hydrogen-utilizing methanogens to maintain low concentrations of hydrogen in the environment. Thus, the conversion of

James G. Ferry and Kyle A. Kastead • Department of Biochemistry and Molecular Biology, The Pennsylvania State University, University Park, PA 16802.

Figure 1. Reenactment of the Volta experiment illustrating methanogenesis in a freshwater pond.

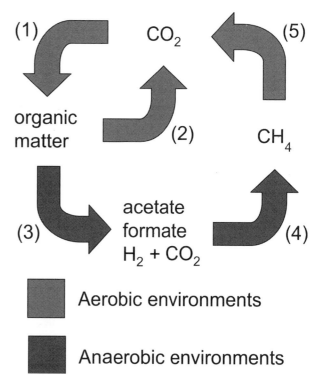

Aerobic environments

Anaerobic environments

Figure 2. The global carbon cycle. Steps: (1) Fixation of carbon dioxide into organic matter, (2) aerobic (oxygen-dependent) decomposition of organic matter to carbon dioxide, (3) deposition of organic matter into anaerobic (oxygen-free) environments and decomposition to metabolic end products by fermentative and obligate hydrogen-producing anaerobes, (4) conversion of the end products to methane by the methanoarchaea and escape of the methane to aerobic environments, (5) aerobic oxidation of methane to carbon dioxide by oxygen-requiring methylotrophs.

complex organic matter to methane requires a true syntrophic association of distinct metabolic groups of anaerobes.

PHYLOGENY

The domain *Archaea* is divided into four kingdoms, *Euryarchaeota*, *Crenarchaeota*, *Korarchaeota*, and the recently described *Nanoarchaeota* (see Chapter 2). Methanogens are the main constituency of the *Euryarchaeota* and are subdivided into five orders, each with distinctive characteristics.

The Order *Methanobacteriales*

The order *Methanobacteriales* comprises two families, *Methanobacteriaceae* and *Methanothermaceae*.

With thirty-two species in four genera, the *Methanobacteriaceae* is the largest and most diverse family of methanogens (8, 11, 17–19, 23, 26, 30, 33, 41, 44, 58, 100, 108, 121, 123–125, 138, 141, 142, 153, 169, 170, 171, 195, 209, 216, 223, 247, 253, 269, 273, 275, 278, 279, 281, 290). Cells range from coccoid, to filamentous rods, with most species being coccobacilliary or short rods. The cell wall structure is very similar to that of gram-positive bacteria, except that pseudomurein replaces muramic acid as the predominant peptidoglycan polymer. *Methanobacte-*

riaceae produce energy by reducing carbon dioxide with electrons generated by the oxidation of hydrogen, except for species of the genus *Methanosphaera* that reduce the methyl group of methanol with hydrogen (19, 68, 170). Many species also use formate, and two also use secondary alcohols as a source of reductant (44, 281). Utilizing CO_2 as their sole carbon and energy source, species of the genera *Methanobacterium* and *Methanothermobacter* are autotrophic. Most can also grow in the absence of any organic compounds, and some are even capable of fixing dinitrogen. Species of *Methanosphaera* and *Methanobrevibacter* are heterotrophic, as they require acetate as a source of cell carbon. Many species of *Methanobacteriaceae* can be considered enteric organisms, as they have been isolated from the digestive tracts and feces of animals, as well as from sewage sludge digesters. All species are mesophilic except for those of the genus *Methanothermobacter* that are thermophilic. Nearly all species are neutrophilic, having an optimum pH of 6.8 to 8.0. However, there are two alkalophilic and one acidophilic species of *Methanobacterium* as well as one acidophilic species of *Methanobrevibacter* (123, 195, 209, 273). Of particular interest is *Methanobacterium subterraneum*, which is not only alkalophilic but also the only species that is halotolerant, forms aggregates, and of all known methanogens, the only one to be isolated from granitic groundwater (123).

The family *Methanothermaceae* contains but a single genus, *Methanothermus*, composed of only two species (137, 236). Cells are rod shaped and motile by bipolar flagellar tufts. The cellular envelope is doubly layered, consisting of an inner pseudomurein layer similar to that of the family *Methanobacteriaceae* and an outer protein-containing S layer (see Chapter 14). *Methanothermus fervidus* cells are straight rods that occur in pairs or short chains (236). *Methanothermus sociabilis* cells are curved rods growing in large aggregates (137). Reduction of carbon dioxide to methane with hydrogen is the only means of energy production. Both organisms are autotrophic, although organic supplements such as yeast extract may stimulate growth. Both species were isolated from solfataric water sources, are neutralophilic, and as the name *Methanothermaceae* implies, are hyperthermophilic, having temperature optima of 83 and 88°C (137, 236).

The Order *Methanococcales*

Methanococcales is an order of coccoid marine species that contains two families, *Methanococcaceae* and *Methanocaldococcaceae*.

Methanococcaceae are a small family containing only two genera and five species (11, 43, 96, 107, 120, 159, 234, 239). Cells are irregular cocci, occurring singly and in pairs, motile by a polar tuft of flagella, and forming a protein cell wall or S layer. All *Methanococcaceae* reduce carbon dioxide to methane by using hydrogen and formate as electron donors. With the exception of *Methanococcus voltae*, which requires acetate for its cell carbon, all species are autotrophic (11, 234). Species of the genus *Methanococcus* can utilize sulfide and elemental sulfur as their sulfur source and ammonium as their nitrogen source; furthermore, *Methanococcus maripaludis* is also capable of fixing N_2 (107). And the impressive species *Methanothermococcus thermolithotrophicus* can grow on nearly any sulfur and nitrogen source, which includes sulfide, elemental sulfur, thiosulfate, sulfite, sulfate, ammonium, nitrate, and N_2 (96). All species of *Methanococcaceae* were isolated from marine or estuarine environments, with *Methanothermococcus* found in geothermally heated sea sediments or deep-sea hydrothermal vents. Despite these

habitats, most species are only halotolerant, preferring to grow at lower salinities.

The family *Methanocaldococcaceae* comprises only two genera and six species, most of which were transferred out of the family *Methanococcaceae*, differentiated primarily by their optimum growth temperatures (29, 35, 101, 102, 106, 132, 240, 287). Like the *Methanococcaceae*, cells are motile, irregular cocci occurring singly or in pairs. Another difference is that, while still using carbon dioxide as the sole carbon and energy source, *Methanocaldococcaceae* are only able to reduce it with hydrogen, with only one species capable of utilizing formate (240). Ammonium and sulfide serve as the nitrogen and sulfur sources and members of the genus *Methanocaldococcus* also require selenium. All species of *Methanocaldococcaceae* were isolated from deep-sea hydrothermal vents, and pursuant with this habitat, are hyperthermophiles, with all species exhibiting growth temperature optima between 80 and 88°C. These organisms are slightly acidophilic and prefer salt concentrations near seawater levels. The second genus, *Methanotorris*, contains only two species that are separated from all other *Methanocaldococcaceae* by the absence of both motility and the requirement for selenium, and by being slightly more halotolerant (NaCl range: 0.45 to 7.2%) than other species (35, 240).

The Order *Methanomicrobiales*

The order *Methanomicrobiales* contains the familes *Methanomicrobiaceae*, *Methanocorpusculaceae*, and *Methanospirillaceae*.

With twenty-three species in seven genera, *Methanomicrobiaceae* is one of the larger and more diverse families of methanogens (9, 11, 24, 40, 43, 50, 66, 85, 133–135, 155, 168, 174, 187–189, 198, 203–205, 233, 250, 266, 267, 274, 276, 282–285). Cell morphology ranges from cocci, to short rods, to the disc- or plate-shaped cells of the genus *Methanoplanus*. Cell walls are made of proteins and do not contain either peptidoglycan or pseudomurein. All species reduce carbon dioxide to form methane by using hydrogen and formate, and many species are also capable of using secondary alcohols. Nearly all species are heterotrophic, with acetate being the most common carbon source. *Methanomicrobiaceae* have been isolated from a wide range of anaerobic habitats, and their physiological diversity reflects this. While most species are mesophiles, *Methanoculleus thermophilicus* ("*Methanogenium frittonii*"), isolated from the effluent channel sediment of a nuclear power plant and from the sediment of Fritton Lake in England, is thermophilic (85, 204). There are two psychrophilic species, *Methanogen-*

ium marinum, isolated from Skan Bay, Alaska, and *Methanogenium frigidum,* isolated from sediments of Ace Lake in Antarctica (40, 66). *M. frigidum* is also one of two halophilic organisms, the other being *Methanocalculus halotolerans,* which was isolated from the well head of an oil field (66, 188). There are also three halotolerant organisms: *Methanoculleus submarinus* from deep-sea sediments of the Nankai Trough, *Methanofollis tationis* from a sulfataric field, and *Methanocalculus chunghsingensis* from a marine water aquaculture fish pond (135, 168, 276).

The *Methanocorpusculaceae* family, wherein *Methanocorpusculum* is the lone genus containing four species, is one of the smallest families of methanogens (29, 190, 275, 280, 286, 288). Cells are irregular cocci and occur singly or in aggregates. Cell walls are an S layer of hexagonally arranged units, and with the exception of *Methanocorpusculum labreanum,* all species of this family are motile by flagella (288). Growth occurs by the reduction of carbon dioxide to methane using hydrogen or formate, although *Methanocorpusculum parvum* and *Methanocorpusculum bavaricum* use secondary alcohols as reductants (280, 286). All species are heterotrophic, requiring acetate, peptones, or rumen fluid to serve as carbon and nitrogen sources. All species are mesophilic and neutralophilic. The type species *M. parvum* was isolated from an anaerobic sour whey digester inoculated with sewage sludge (280). *M. bavaricum* and *Methanocorpusculum sinense* were isolated from the industrial waste water from a sugar factory and a distillery, respectively, whereas *M. labreanum* was isolated from the surface sediments of Tar Pit Lake of the Labrea Tar Pits (286, 288).

Methanospirillaceae is one of two families of methanogens comprised of only a single genus and species (60). *Methanospirillum hungatei* cells are symmetrically curved rods that form wavy filaments. The cells are motile by polar tufts of flagella. The cell wall consists of an S layer of nonglycosylated polypeptides and is enveloped by a rigid paracrystalline outer sheath that may encase several cells. Formate or hydrogen serves as an electron donor for the reduction of carbon dioxide to methane. *M. hungatei* is best known for its association with syntrophic anaerobes that require the removal of hydrogen and formate, products of their metabolism, for continued growth (46). *M. hungatei* fixes N_2 and autotrophic growth is possible, although yeast extract and peptones stimulate growth. This organism is mesophilic with a temperature optimum of 37°C and neutralophilic with a very narrow pH range for growth (between 6.6 and 7.4). *M. hungatei* was isolated from sewage sludge, and a recently discovered strain, TM20-1, which may be designated as a new species, was isolated from a rice paddy field (246).

The Order *Methanosarcinales*

The order *Methanosarcinales* contains the families *Methanosarcinaceae* and *Methanosaetaceae.*

Currently, the family *Methanosarcinaceae* comprises eight genera and twenty-six species (3, 12, 21, 22, 28, 30, 32, 45, 56, 60, 67, 103, 109–111, 122, 148, 149, 151, 154, 156–160, 163, 178, 180, 181, 186, 191, 192, 196, 197, 207, 217, 220, 228, 229, 232, 235, 251, 268, 289, 291–293, 295, 296). Cells are coccoidal or pseudosarcinal, forming a protein cell wall devoid of peptidoglycan or pseudomurein and, in some cases, are surrounded by a heteropolysaccharide sheath to form large aggregates. All species can grow and produce methane by dismutating methyl compounds such as methanol, methyl amines, and methyl sulfides. Most also convert the methyl group of acetate to methane and reduce carbon dioxide with hydrogen as electron donor; thus, the family is the most metabolically versatile of all methanogens. Furthermore, CO can also be utilized by some species (207). Ammonium and sulfide serve as the major nitrogen and sulfur sources, and cofactor supplements such as biotin are required by some. Species of *Methanosarcinaceae* have been isolated from very diverse anaerobic habitats, and unsurprisingly, exhibit a wide range of physiological optima. Most species are mesophilic, but *Methanosarcina thermophila,* isolated from a 55°C anaerobic digestor, and *Methanomethylovorans thermophila,* isolated from a methanol-fed anaerobic sludge reactor, are thermophilic, while *Methanococcoides burtonii,* isolated from Ace Lake in Antarctica, *Methanosarcina baltica,* isolated from the Gotland Deep of the Baltic Sea, and *Methanococcoides alaskense,* isolated from Skan Bay, Alaska, are psychrophilic (67, 103, 220, 251, 296). Since many species have been isolated from marine environments, they prefer a salinity near that of seawater; however, *Methanohalobium evestigatum,* isolated from the sediments of a saline lagoon, is an extreme halophile, growing only at a NaCl concentration higher than 15.2% with an optimum of 25.1% (292). Furthermore, species of the genus *Methanohalophilus* are moderately halophilic ([NaCl] optimum between 5.0 and 12%), balancing their cytosolic osmolarity with the environment by synthesizing organic osmolytes such as glycine or betaine. With only three exceptions, all species of *Methanosarcinaceae* are neutralophilic, having pH optima between 6.5 and 7.5. The three exceptions are alkaliphilic (pH optimum greater than 8.0), and all three were isolated from saline environments; *Methanolobus oregonensis* is halotolerant (exhibiting highest growth rate at lower salinities, but continuing to grow in [NaCl] up to 9.0%), while *Methanolobus talorii* and

Methanosalsum zhilinae are moderately halophilic (148, 163, 191).

The family *Methanosaetaceae* is composed of only two species within a single genus (27, 97, 112, 113, 185, 193, 194, 248, 294). *Methanosaeta* species are nonmotile, straight rods that grow in chains and sometimes long filaments and are enclosed in a tubular sheath separated by structures called spacer plugs. Acetate serves as the only substrate for methanogenesis. Ammonium and sulfide serve as the nitrogen and sulfur sources, and both species also require cofactor supplements in the form of biotin or vitamins. *Methanosaeta concilii*, isolated from sewage sludge, is mesophilic, has an optimum pH of 7.0, and is inhibited by yeast extract (194). *Methanosaeta thermophila*, isolated from sludge of an anaerobic, thermophilic bioreactor, is thermophilic, with an optimum temperature of 55°C, is slightly acidophilic with a pH optimum of 6.0, and grows better in the presence of trace metals (185). This genus was previously identified as *Methanothrix*. However, the type species of this genus, *Methanothrix soehngenii*, was later discovered to be impure when initially characterized, and no cell culture collection contains a pure, viable culture. As the type species of the genus *Methanothrix* was invalidated, the rules of the International Code of Nomenclature of Bacteria state that the genus has no nomenclatural standing, and the name *Methanothrix* was rejected and *Methanosaeta* was adopted in its place (27, 194).

The Order *Methanopyrales*

The order *Methanopyrales* is composed of only one family, *Methanopyraceae*, and a single genus and species (130). *Methanopyrus kandleri* are motile rods that grow singly or in chains. The cell wall is composed of pseudomurein, and cells are also sheathed in a "fuzzy" coat of unknown composition. *Methanopyraceae* reduce carbon dioxide with hydrogen and are autotrophic, with ammonium and sulfide serving as nitrogen and sulfur sources. *M. kandleri*, isolated from hydrothermally heated deep-sea sediment, is a hyperthermophile with an optimum temperature of 98°C. It has a pH optimum of 6.5 and an optimum NaCl concentration of 2.0% (130). The taxonomic position of *M. kandleri* has been the subject of some controversy. Based on its 16S rDNA sequence, *M. kandleri* was placed in a deep branch of the *Euryarchaeota*, which is quite distant from all other methanogens. However, since the complete genome has been sequenced, there have been analyses of the translation machinery as well as the transcription machinery. These studies indicate that, while *M. kandleri* is still significantly different enough to be within its own family, it groups with the *Methanococcales* and *Methanobacteriales* (31).

PATHWAYS OF METHANOGENESIS

Coenzymes and Cofactors

Figure 3 shows the cofactors and coenzymes involved in methanogenic pathways, several of which are also found in organisms outside the *Archaea*. Methanofuran (MF) (143) and tetrahydromethanopterin (THMPT) (118) function as one-carbon carriers, the latter coenzyme also functioning in methylotrophic microbes from the *Bacteria* domain (39). Molybdopterin guanine dinucleotide, common in enzymes from the *Bacteria* and *Eucarya* domains, was first discovered in the *Archaea* as a cofactor of formate dehydrogenase (104, 164) from *Methanobacterium formicicum*, and later as a prosthetic group of formyl-MF dehydrogenase (117). Coenzyme F_{420} (F_{420}), a deazaflavin derivative (55), is an obligate two-electron carrier accepting or donating a hydride ion. Methanophenazine (MP) (1) is a 2-hydroxyphenazine derivative connected by ether linkage to a polyisoprenoid side chain that functions as a membrane electron carrier. Factor III is a cofactor of several methyltransferases and has a structure similar to vitamin B_{12}, a major exception being that factor III contains a 5-hydroxybenzimidazolyl base (211). Factor III functions in methyltransferases by accepting a methyl group as the upper axial ligand to the cobalt. Coenzyme M is the smallest cofactor known (241). The methylated form of (CH_3-S-CoM) is the substrate for the methylreductase, which catalyzes the reductive demethylation of CH_3-S-CoM to methane in all methanogenic pathways. Coenzyme B (CoB-SH) is the second substrate for methylreductase, providing electrons for the reaction (183). Cofactor F_{430} (F_{430}), the prosthetic group of methylreductase, was the first nickel-containing cofactor to be described (52, 265) and belongs to the family of macrocyclic tetrapyrroles (51), distinguished by a much higher degree of saturation than other members of the family (heme, siroheme, chlorophyll, bacteriochlorophyll, and corrinoids).

Considerable progress has been made on the biosynthesis of several of the above-mentioned cofactors, including the characterization of enzymes in the biosynthetic pathways (77, 79–82, 95, 202, 212, 225, 226, 255–262).

The Carbon Dioxide Reduction and Acetate Fermentation Pathways

The production of methane is the energy-yielding metabolism of methanogens. Two major pathways account for most of the methane produced biologically. Approximately two-thirds derive from the methyl group of acetate (reaction 1) by the "acetate fermentation" pathway (Fig. 4) and approximately

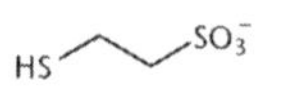

2-Mercaptoethanesulfonic acid
(coenzyme M, HS-CoM)

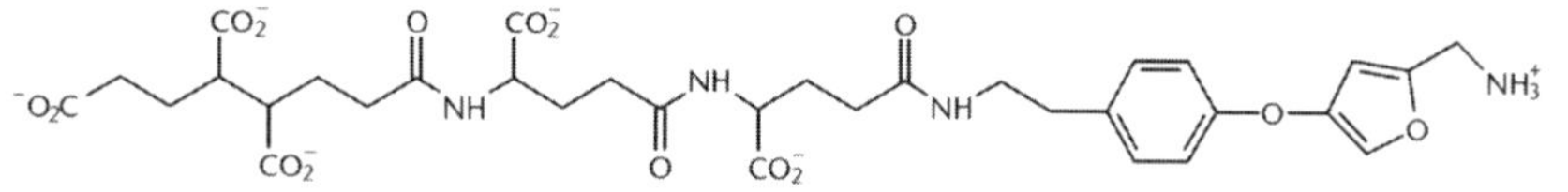

Methanofuran (MF)

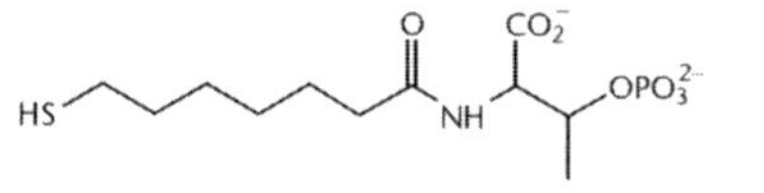

7-Mercaptoheptanoylthreonine phosphate
(coenzyme B, HS-CoB)

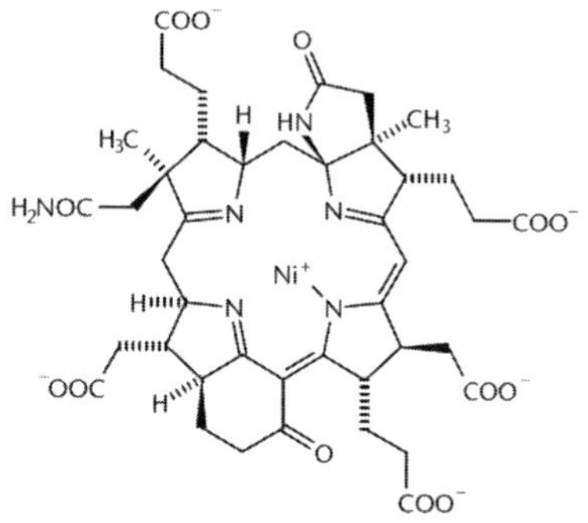

Factor F_{430} (F_{430})

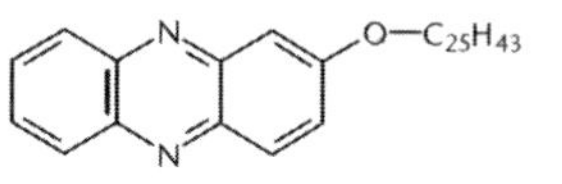

Methanophenazine

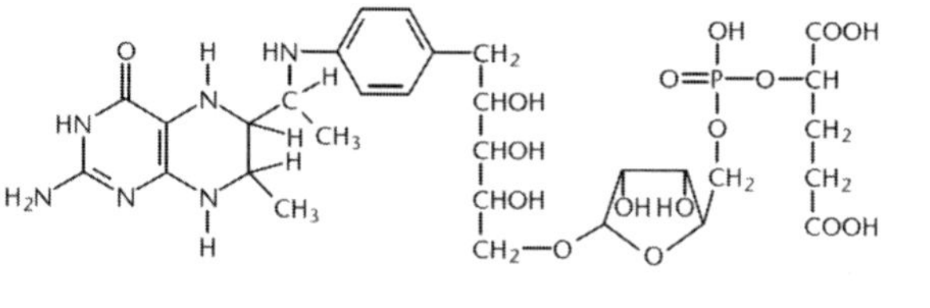

5,6,7,8-Tetrahydromethanopterin (H_4MPT)

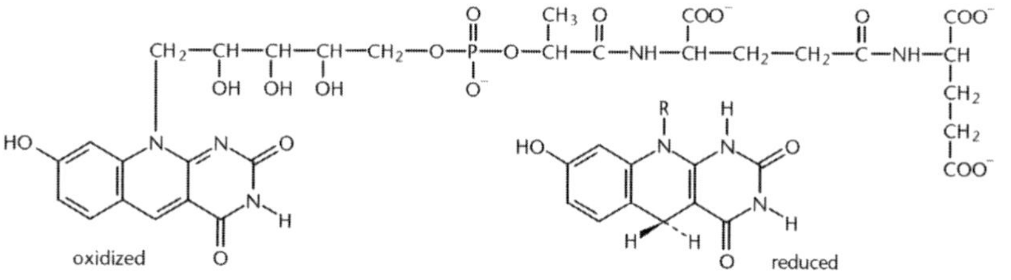

CoenzymeF_{420} (F_{420})

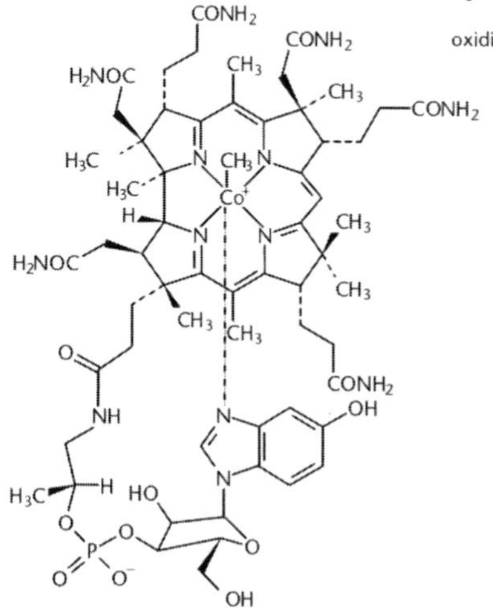

5-Hydroxybenzimidazolylcobamide
(factor III)

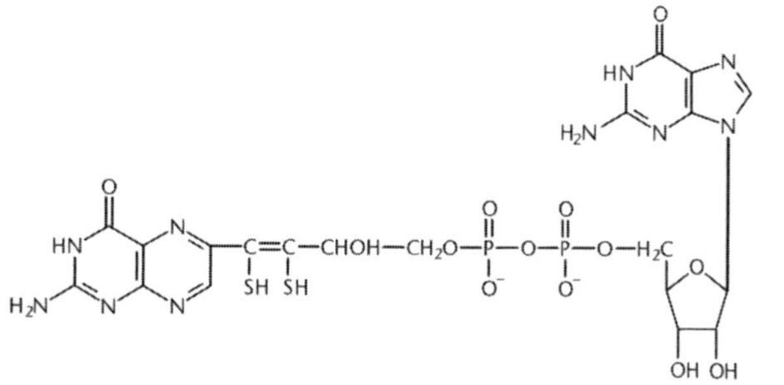

Molybdopterin guanine dinucleotide (MGD)

Figure 3. Cofactors and coenzymes utilized in methanogenic pathways.

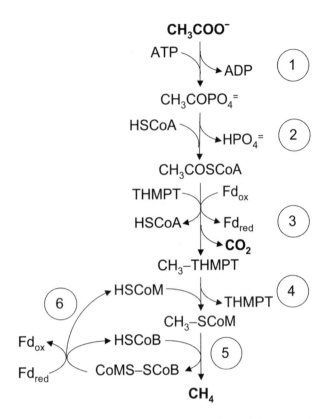

Figure 4. Steps in the acetate fermentation pathway. Substrates and products are shown in bold.

one-third by the reduction of carbon dioxide with electrons from either hydrogen or formate (reactions 2a or 2b) in the "carbon dioxide reduction" pathway (Fig. 5).

$$CH_3COO^- + H^+ \rightarrow CH_4 + CO_2 \quad (1)$$
$$CO_2 + 4H_2 \rightarrow CH_4 + 2H_2O \quad (2a)$$
$$4HCO_2H \rightarrow 3CO_2 + CH_4 + 2H_2O \quad (2b)$$

Several recent reviews describe both pathways in detail, including the description of other minor pathways (47, 48, 59, 78, 86, 215).

Reactions leading to methane that are common to both pathways

Reactions 3 to 5 are common to both major pathways (1 and 2a and b) that differ primarily in steps by which the methyl group is generated and passed to THMPT to form CH_3-THMPT (Figs. 4 and 5).

$$CH_3\text{-}THMPT + HS\text{-}CoM \rightarrow \\ CH_3\text{-}S\text{-}CoM + THMPT \quad (3)$$

$$CH_3\text{-}S\text{-}CoM + HS\text{-}CoB \rightarrow CoMS\text{-}SCoB + CH_4 \quad (4)$$

$$CoMS\text{-}SCoB + 2e^- + 2H^+ \rightarrow \\ HS\text{-}CoB + HS\text{-}CoM \quad (5)$$

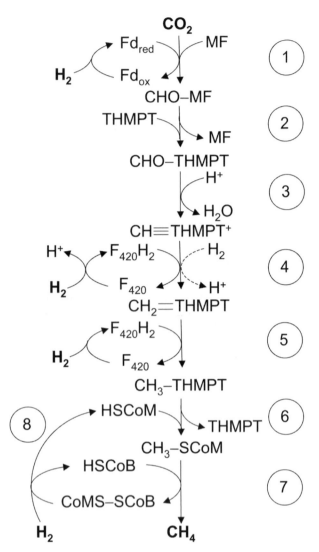

Figure 5. Steps in the carbon dioxide reduction pathway. Substrates and products are shown in bold.

Reaction 3 is catalyzed by ^5N-methyl-THMPT:CoM methyltransferase and is the subject of an excellent review (75). The methyltransferase is integral to the membrane and couples the exergonic methyl transfer reaction to generation of a sodium ion gradient across the membrane that could be used for various energy-requiring reactions. The enzyme contains eight non-identical subunits (MtrA to H) of which MtrA contains factor III and is thought to protrude into the cytoplasm. It is proposed that transfer of the methyl group from CH_3-THMPT to factor III is catalyzed by MtrH, whereas MtrE is postulated to demethylate factor III. It is hypothesized that conformational changes induced in MtrA by methylation and demethylation are transmitted to MtrE, which then drives the translocation of sodium.

The methyl-CoM reductase (Mcr) catalyzes reaction 4 utilizing HS-CoB as the electron donor, producing heterodisulfide CoMS-SCoB in addition to methane. Some species harbor genes encoding two isozymes designated MRI and MRII. The crystal structure of the MRI enzyme from the carbon dioxide-reducing species *Methanothermobacter marburgensis* (*Methanobacterium thermautotrophicum* strain Marburg) reveals an $\alpha_2\beta_2\gamma_2$-subunit structure that forms two active sites, each containing F_{430} (57). Based partly on arrangement of substrates in the structure, a mechanism has been proposed for which the first step is a nucleophilic attack of $[F_{430}]Ni(I)$ on CH_3-S-CoM, forming a $[F_{430}]Ni(III)$-CH_3 intermediate and HSCoM. In the next step, electrons are transferred from HS-CoM to Ni(III) producing the thiyl radical •S-CoM and $[F_{430}]Ni(II)$-CH_3. Methane is then produced by protonolysis of $[F_{430}]Ni(II)$-CH_3, and the thiyl radical is coupled to $^-$S-CoB to form CoB-S-S-CoM accompanied by a one-electron reduction of Ni(II). In the concerted mechanism, specific conformational changes ensure entry of CH_3-S-CoM adjacent to F_{430} and before entry of HS-CoB in the narrow active site channel (61, 76). Investigation of the dependence of temperature on activity indicates an alternating sites mechanism for release of CoMS-SCoB that is driven by conformational changes transmitted from the adjacent monomer (70). Under debate is an alternative to the first step (71), in which Ni(I) attacks the sulfur of CH_3-S-CoM, producing a free methyl radical reacting with HS-CoB to produce methane and the thiyl radical •S-CoB. A novel methyl-CoM reductase, containing a modified F_{430}, has been implicated in the first step of anaerobic methane oxidation by archaea present in anaerobic microbial communities that oxidize methane (214).

In both pathways, the disulfide bond of CoMS-SCoB is reduced by heterodisulfide reductase (Hdr), yielding the active sulfhydryl forms of the coenzymes (reaction 5). In both pathways, reduction of CoMS-SCoB is coupled to formation of an electrochemical proton gradient which drives ATP synthesis catalyzed by an A_1A_0-type ATP synthase (177). Two types of Hdr have been described, the two-subunit HdrDE and the three-subunit HdrABC. The HdrABC type functions in the carbon dioxide reduction pathway (87, 89), and the HdrDE type is found in acetate-grown cells of *M. thermophila* (218) and acetate- and methanol-grown *Methanosarcina barkeri* (90, 128). The HdrE contains cytochrome *b*, which is proposed to transfer electrons to HdrD. HdrD and HdrBC are highly conserved (89, 128), suggesting they are the catalytic subunits of their respective enzymes. Spectroscopic investigations indicate a 4Fe-4S center in HdrD and HdrB is the active site where reduction of

the disulfide occurs in two one-electron steps involving a thiyl radical intermediate (88).

Synthesis of CH$_3$-THMPT in the carbon dioxide reduction pathway

The six electrons required for carbon dioxide reduction to CH$_3$-THMPT derive from the oxidation of either hydrogen or formate (reactions 6 to 8).

$$H_2 + 2\ Fd^{ox} \rightarrow 2\ Fd^{red} + 2H^+ \qquad (6)$$
$$2H_2 + 2F_{420} \rightarrow 2F_{420}H_2 \qquad (7)$$
$$2HCOOH + 2F_{420} \rightarrow 2F_{420}H_2 + 2CO_2 \qquad (8)$$

Reaction 6 is coupled to the first step in the pathway where reduced ferredoxin (Fd^{red}) donates electrons for carbon dioxide reduction to the formyl level, an endergonic reaction. Thus, under low partial pressures of hydrogen encountered in the environment, the reduction of Fd^{ox} requires an input of energy. The enzymes catalyzing reaction 6 are different between *Methanosarcina* species (47) and "obligate carbon dioxide reducing" species such as *M. marburgensis* and *Methanothermobacter thermautotrophicus* (*Methanobacterium thermautotrophicum* strain ΔH) (237). In *Methanosarcina* species, reaction 6 is catalyzed by the Ech hydrogenase that utilizes Fd as the electron acceptor. The enzyme is a six-subunit complex with sequence identity to multisubunit hydrogenases in the *Bacteria* domain and subunits of the complex I energy-conserving NADH:quinone oxidoreductases. Thus, it is proposed that Ech hydrogenase depends on the proton gradient to drive the reduction of Fd by reverse electron transport (86). The 4Fe-4S clusters of the Ech from *M. barkeri* were assigned to subunits EchC and EchF (63), for which the electron paramagnetic resonance signals are pH dependent (129), suggesting a role in proton translocation. The enzyme catalyzing the energy-dependent reaction 6 in "obligate carbon dioxide reducing" species has not been established. The genomes do not encode an Ech hydrogenase, although these species encode two multisubunit membrane-bound hydrogenases (Eha and Ehb). These enzymes have sequence identity to Ech, and the genes are clustered with ferredoxins, leading to the proposal that Eha and Ehb function in analogy to Ech. The only formate dehydrogenases characterized from the methanogens are the F_{420}-dependent enzymes from *M. formicicum* (210) and *Methanococcus vannielii* (105), although the mechanism by which electrons are transferred from $F_{420}H_2$ to Fd has not been investigated. Reaction 7 is catalyzed by an F_{420}-dependent hydrogenase (4, 65) with properties distinct from the Ech hydrogenase.

The steps in involving conversion of carbon dioxide to CH_3-THMPT are shown in reactions 9 to 13.

$$CO_2 + MF + 2Fd^{red} + 2H^+ \rightarrow \\ 2Fd^{ox} + \text{formyl-MF} + H_2O \qquad (9)$$

$$\text{formyl-MF} + \text{THMPT} \rightarrow \\ N^5\text{-formyl-THMPT} + MF \qquad (10)$$

$$N^5\text{-formyl-THMPT} + H^+ \rightarrow \\ N^5, N^{10}\text{-methenyl-THMPT}^+ + H_2O \qquad (11)$$

$$N^5, N^{10}\text{-methenyl-THMPT}^+ + F_{420}H_2 \rightarrow \\ N^5, N^{10}\text{-methylene-THMPT} \qquad (12a) \\ + F_{420} + H^+$$

$$N^5, N^{10}\text{-methenyl-THMPT}^+ + H_2 \rightarrow \\ N^5, N^{10}\text{-methylene-THMPT} + H^+ \qquad (12b)$$

$$N^5, N^{10}\text{-methylene-THMPT} + F_{420}H_2 \rightarrow \\ N^5\text{-methyl-THMPT} + F_{420} \qquad (13)$$

In reaction 9, carbon dioxide attaches to MF and is then reduced to formyl-MF, with Fd^{red} catalyzed by formyl-MF dehydrogenase. Reaction 10 is catalyzed by formyl-MF:THMPT formyltransferase, and reaction 11 by methenyl-THMPT cyclohydrolase (Mch). Mutants of *M. barkeri* with *mch* disrupted hold promise for a genetic analysis to further probe the physiological function of this enzyme (83). Two mechanistically distinct N^5,N^{10}-methylene-THMPT dehydrogenases catalyze reaction 12, the first of which oxidizes reduced F_{420} ($F_{420}H_2$) (reaction 12a), and the second utilizes hydrogen as the electron donor (reaction 12b). The hydrogen-utilizing enzyme is essentially a hydrogenase that catalyzes the stereo-specific transfer of a hydride ion from hydrogen into the methylene carbon. The enzyme contains no iron-sulfur clusters or nickel in contrast to all known hydrogenases, although it contains a functional iron (213) and a low-molecular-mass cofactor of unknown composition (152). N^5,N^{10}-Methylene-THMPT reductase catalyses reaction 13. The crystal structure of the reductase (10) shows the isoalloxazine ring of F_{420} present in a pronounced butterfly conformation induced from the Re-face of F_{420} by a bulge containing a nonprolyl *cis*-peptide bond.

Synthesis of CH_3-THMPT in the acetate fermentation pathway

Two genera are known that ferment acetate to methane, *Methanosaeta* and *Methanosarcina*, of which the latter has been extensively investigated. In *Methanosarcina* species, CH_3-THMPT is synthesized by reactions 14 to 17. These reactions are widespread in diverse anaerobes; thus, an understanding of the enzymes has broad significance outside methanogenesis.

$$CH_3COO^- + ATP \rightarrow \\ CH_3CO_2PO_3^{2-} + ADP \qquad (14)$$

$$CH_3CO_2PO_3^{2-} + \text{HS-CoA} \rightarrow \\ CH_3COSCoA + P_i \qquad (15)$$

$$CH_3COSCoA + \text{THMPT} \\ + H_2O + Fd^{ox} > CH_3\text{-THMPT} \qquad (16) \\ + Fd^{red} + CO_2 + \text{HS-CoA}$$

$$CO_2 + H_2O \rightarrow HCO_3^- + H^+ \qquad (17)$$

Acetate kinase and phosphotransacetylase catalyze reactions 14 and 15, functioning in concert to convert acetate to acetyl-CoA. Features of the crystal structure (36) of acetate kinase from *M. thermophila* suggest the enzyme belongs to the *Acetate and Sugar Kinase/Hsc70/Actin* (ASKHA) superfamily of phosphotransferases and is possibly the founding member. Kinetic and biochemical analyses of site-specific amino acid variants of acetate kinase have identified residues essential for catalysis and established a direct in-line mechanism for transfer of the phosphate from ATP to acetate (74, 98). The crystal structure of phosphotransacetylase complexed with CoA-SH (99, 140) has revealed the active site architecture that, together with kinetic analyses of the wild-type (139) and site-specific amino acid variants, has identified essential residues leading to a proposed mechanism for catalysis. The concerted mechanism proceeds through base-catalyzed generation of $^-$S-CoA, followed by nucleophilic attack of the thiolate anion on the carbonyl carbon of acetyl phosphate. The five-subunit CO dehydrogenase/acetyl-CoA synthase (Cdh) complex cleaves the C—C and C—S bonds of acetyl-CoA, transfers the methyl group to THMPT, oxidizes the carbonyl group to carbon dioxide, and reduces Fd (reaction 16). The active site of the enzyme contains nickel in an unusual metal complex (201). A carbonic anhydrase, Cam, converts carbon dioxide to bicarbonate outside the cell (reaction 17), facilitating removal of the product from the cytoplasm. Cam is the prototype of an independently evolved class of carbonic anhydrases with a left-handed β-helical fold and contains iron in the active site (249).

Electron transport and energy conservation

Two recent reviews (47, 86) describe this topic in detail. In the carbon dioxide reduction pathway, the only reactions yielding enough energy for ATP synthesis are reactions 3 and 5, which are exergonic by -29 and -40 kJ/mol, respectively. The H_2:CoMS-SCoB oxidoreductase systems are different between *Methanosarcina* species (47) and "obligate carbon dioxide reducing" species such as *M. marburgensis* and *M. thermautotrophicus* (237). In *Methanosarcina mazei*, a

F_{420}-nonreducing hydrogenase (VhoAGC) oxidizes hydrogen and reduces MP (Fig. 6). The VhoAG subunits catalyze the oxidation of H_2, whereas VhoC contains a *b*-type heme that is proposed to donate electrons to MP. The reduced MP transfers electrons to the HdrDE-type heterodisulfide reductase, which reduces CoMS-SCoB to the sulfhydryl forms of the cofactors. The lipid-soluble quinonelike MP functions to translocate 2 protons to outside the cell membrane, generating a gradient that drives ATP synthesis catalyzed by a A_1A_0-type ATP synthase. An additional 2 protons are translocated by the Vho hydrogenase for a total of 4 protons translocated by the H_2:CoMS-SCoB oxidoreductase system. The "obligate carbon dioxide reducing" species do not contain cytochromes or MP, and the H_2:CoMS-SCoB oxidoreductase system (Fig. 7) is apparently only composed of a cytoplasmic F_{420}-nonreducing hydrogenase (MvhAGD) that is tightly bound to the soluble HdrABC type of heterodisulfide reductase (237). Evidence is reported that HdrABC is anchored to the membrane by the HdrB subunit (89), suggesting the H_2:CoMS-SCoB oxidoreductase system is loosely bound to the membrane. The MvhD subunit is proposed to transfer electrons to the HdrA subunit of the heterodisulfide reductase (237). Conclusive evidence that this H_2:

CoMS-SCoB oxidoreductase system is coupled to energy conservation has yet to be reported. However, the genome of *Methanosphaera stadtmanae*, which is only able to reduce CH_3-SCoM with hydrogen (see below), contains genes encoding HdrABC and MvhAGD of the H_2:CoMS-SCoB oxidoreductase system consistent with a role in the generation of a proton gradient that drives ATP synthesis (68).

In the acetate fermentation pathway of *Methanosarcina* species, a Fd^{red}:CoMS-SCoB oxidoreductase system pumps protons, providing the gradient which drives ATP synthesis. The Fd^{red} is generated by oxidation of the carbonyl group of acetyl-CoA, catalyzed by the Cdh complex (reaction 16). In the freshwater species *M. barkeri*, it is proposed that Fd^{red} donates electrons to the Ech hydrogenase (Fig. 8), which produces hydrogen and pumps protons (86). This contention is supported by gene knockout experiments (167). It is presumed that the H_2:CoMS-SCoB oxidoreductase then oxidizes H_2 and reduces the heterodisulfide, thus adding two additional proton-translocating segments. However, it is unlikely that the amount of energy available from the fermentation of acetate to methane ($\Delta G^{\circ\prime} = -36$ kJ/CH_4) can support three proton-translocating segments, especially when considering that one ATP is expended in con-

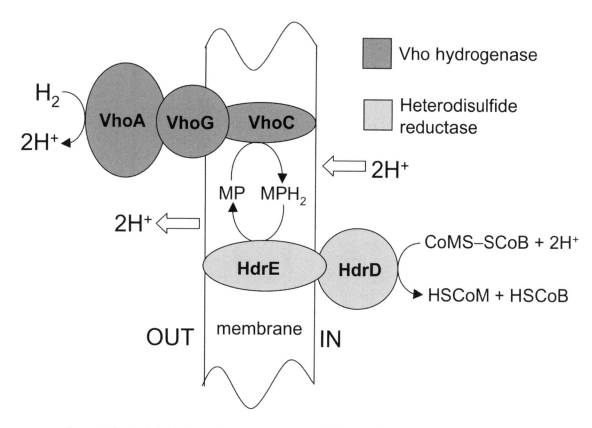

Figure 6. The H_2:CoMS-SCoB oxidoreductase system in *Methanosarcina* species. MP, methanophenazine.

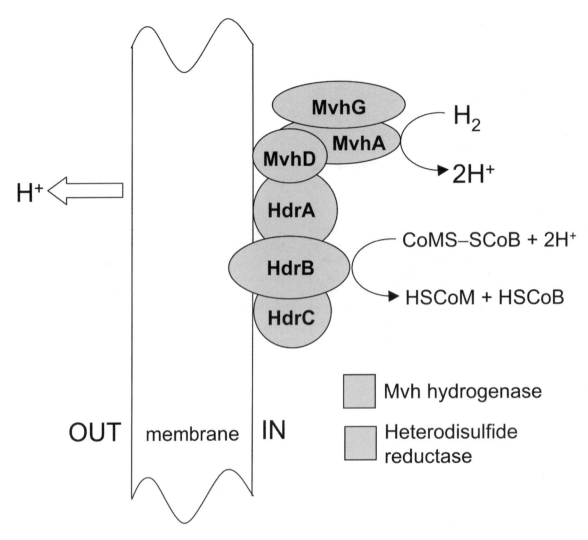

Figure 7. Predicted topology of the H$_2$:CoMS-SCoB oxidoreductase system in obligate carbon dioxide-reducing species.

verting acetate to acetyl-CoA in the first step of the pathway. Although the Ech hydrogenase has not been identified in the freshwater isolate *M. thermophila*, a HdrDE type of heterodisulfide reductase has been purified from membranes of acetate-grown cells with associated hydrogenase activity (218), suggesting a H$_2$:CoMS-SCoB oxidoreductase system. A role for the soluble electron-transferring protein Isf (iron-sulfur flavoprotein) has been proposed for *M. thermophila*, in which Isf mediates electron transfer from Fd^{red} to the membrane-bound electron transport chain (136), but the electron acceptor to Isf has not been identified. The crystal structure of Isf shows an unusual compact cysteine motif ligating the 4Fe4S cluster that is proposed to accept electrons from Fd^{red} and donate to the noncovalently bound FMN (5). In summary, the Fd^{red}:CoMS-SCoB oxidoreductase system of freshwater marine *Methanosarcina* species is not fully understood.

Recent evidence suggests the marine species *Methanosarcina acetivorans* employs a different strategy for oxidizing Fd^{red} and reducing CoMS-SCoB (145) that is more compatible with the thermodynamic constrains of the pathway (Fig. 9). The genome of *M. acetivorans* does not encode a functional Ech (69), and evidence (179) suggests hydrogen is not an obligatory intermediate in electron transport. Furthermore, gene knockout experiments suggest *M. acetivorans* utilizes an electron transport chain unique from *M. barkeri* (83). Proteomic analyses (145) indicate that acetate-grown *M. acetivorans* preferentially synthesizes subunits encoded by a six-gene cluster with sequence identity to the energy-converting Rnf complex described in species from the *Bacteria* domain (127). Gene knockouts confirm that the complex is essential for growth on acetate (W. Metcalf, personal communication). Transcriptional analysis shows that the six-gene cluster encoding the complex

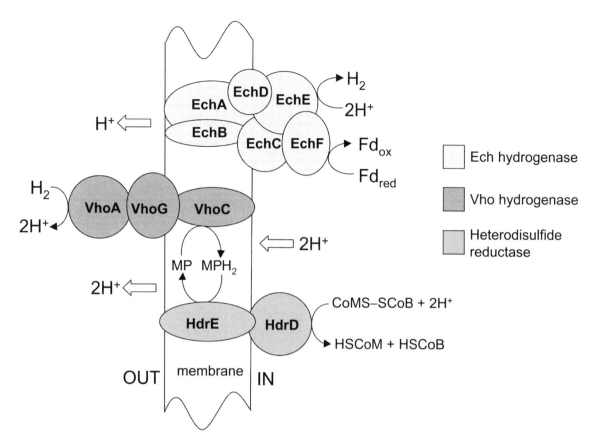

Figure 8. Predicted topology and function of the Fd:CoMS-SCoB oxidoreductase system in *Methanosarcina barkeri*. MP, methanophenazine.

is cotranscribed with two flanking genes, one of which encodes a cytochrome *c* (MA0658 in Fig. 9) shown to dominate in acetate-grown cells (145). It is proposed (but not experimentally verified) that Fd^{red} donates electrons to the Rnf complex containing cytochrome *c*, and that MP mediates electron transfer between the complex and HdrDE, which pumps two protons outside the cell (Fig. 9). It is also hypothesized that the Rnf complex pumps an unknown number of protons, contributing further to the proton gradient which drives ATP synthesis.

Pathways for Methanogenesis by Dismutation of the Methyl Groups of One-Carbon Compounds

In addition to the carbon dioxide reduction and acetate fermentation pathways, *Methanosarcina* and *Methanococcoides* species (230) convert the methyl groups of simple one-carbon compounds, such as methanol and methylamines, to methane and carbon dioxide for growth. Investigations of these pathways have revealed novel biochemistry and molecular biology.

Reactions leading to methane

Unlike the carbon dioxide and acetate fermentation pathways, the methyl group is transferred to HS-CoM by a system composed of two methyltransferases specific for methanol and monomethylamine (Fig. 10). The first (MT1) transfers the methyl group to the corrinoid cofactor of a protein that is the substrate for the second methyltransferase (MT2) that transfers the methyl group from the corrinoid protein to HS-CoM. The CH$_3$-S-CoM is reductively demethylated by methyl-CoM reductase producing methane and CoB-S-S-CoM as described above for the two major pathways. Electrons for the reduction are supplied by oxidation of the methyl group of CH$_3$-S-CoM by reversal of the carbon dioxide reduction pathway (Fig. 5). One methyl group is oxidized to carbon dioxide to yield six electrons for the conversion of three methyl groups to methane, as shown for methanol in reactions 18 and 19. Thus, the pathways are a dismutation of methyl groups to carbon dioxide and methane.

$$CH_3OH + H_2O \rightarrow CO_2 + 6\,e^- + 6\,H^+ \quad (18)$$
$$3CH_3OH + 6\,e^- + 6\,H^+ \rightarrow 3CH_4 + 3H_2O \quad (19)$$

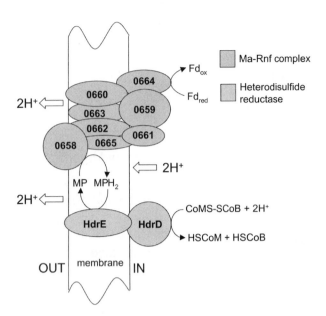

Figure 9. Postulated topology and function of the Fd:CoMS-SCoB oxidoreductase system in *Methanosarcina acetivorans*. Please see reference 145 for a detailed understanding. The numbers correspond to open reading frames of the gene cluster encoding the Ma-Rnf complex. MP, methanophenazine.

In the first step of the oxidation arm of the dismutation pathways, methyl-THMPT:CoM methyltransferase transfers the methyl group of methyl-CoM to THMPT that is an endergonic reaction driven by a sodium ion gradient. However, deletion of the operon encoding the methyltransferase in *M. barkeri* produces a mutant still able to produce methane from methanol, suggesting an alternative to the first step in the oxidation of the methyl group of methanol involving an unknown methyltransferase (254). As the mutant is unable to grow with methanol, the methyl group oxidation pathway involving the novel methyl-

transferase is unlikely to be physiologically relevant in the native environment. Furthermore, the mutant is able to grow and produce methane from methanol with electrons supplied by the oxidation of acetate to carbon dioxide and formate. However, the role of acetate in the metabolism of methanol in nature is in doubt since wild-type *Methanosarcina* species preferentially utilize methanol to the exclusion of acetate when both are present.

M. stadtmanae is a methanol-utilizing species with an unusual methanogenesis pathway (170). The genome sequence is missing 37 open reading frames that are present in the complete genome sequences of all other methanogens. Genes are absent for molybdopterin synthesis (Fig. 3), which is required for formyl-MF dehydrogenase. This absence explains why the organism is unable to oxidize the methyl group of methanol to supply electrons for reduction of the methyl group of methanol to methane and relies on hydrogen as an electron donor (68).

Electron transport and energy conservation

Studies on the mechanism of energy conservation in the pathway of methanogenesis from methyl-containing growth substrates have revealed novel electron transport components in the *Archaea* (47). As mentioned above (Reactions leading to methane), one-fourth of the methyl groups are oxidized to carbon dioxide by a reversal of the carbon dioxide reduction pathway (Fig. 5) to supply electrons for conversion of three-fourths of the methyl groups to methane (reactions 18 and 19). The electron acceptor in this oxidation is F_{420}, and the membrane-bound $F_{420}H_2$ dehydrogenase is the first component of the $F_{420}H_2$:CoMS-SCoB oxidoreductase system (Fig. 11), generating a proton gradient that drives ATP synthesis (13). As for all other systems in *Methanosarcina* species, MP mediates electron transfer to the HdrDE heterodisulfide reductase (14). Genomic analysis indicates the $F_{420}H_2$ dehydrogenase is a thirteen-subunit complex (FpoABCDHIJKLMNOF) with identity to the proton-translocating Complex I NADH:quinone oxidoreductases found in organisms from the *Bacteria* and the *Eucarya* (13, 49). Comparison of the deduced sequences with Complex I (13) suggest FpoA-JKMN constitutes the proton transporter module, and FpoBCDIHLOF couples electron transfer from $F_{420}H_2$ to MP (Fig. 11). FpoF and FpoO have no identity to subunits of Complex I, suggesting functions unique to the $F_{420}H_2$ dehydrogenase. FpoF, encoded separately from the *fpoABCDHIKLMNO* operon, is postulated to accept electrons directly from $F_{420}H_2$, and FpoO to interact with MP (47).

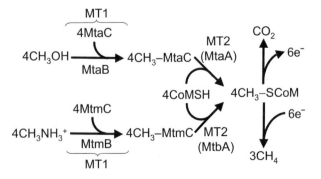

Figure 10. Steps in the pathway for dismutation of methanol and monomethylamine to carbon dioxide and methane.

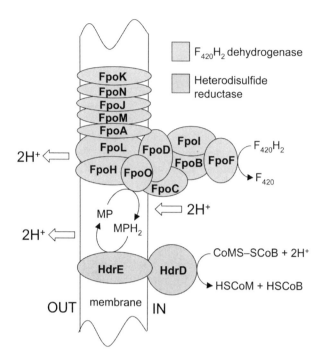

Figure 11. Postulated topology and function of the $F_{420}H_2$:CoMS-SCoB oxidoreductase system in *Methanosarcina mazei*. MP, methanophenazine.

MOLECULAR BIOLOGY OF METHANOGENESIS PATHWAYS

As for all archaea, gene expression in the methanogens is a hybrid of features from both the *Bacteria* and *Eucarya* (227). As in the *Bacteria*, genes encoding proteins in methanogenic pathways are present in operons transcribed as polycistronic mRNA, and a Shine-Dalgarno-like sequence directs translation initiation (see Chapter 8). However, the RNA polymerase is structurally homologous to that of the *Eucarya* and recognizes promoters that contain an AT-rich TATA box element similar to the *Eucarya* (15) (see Chapter 6). The TATA box is often preceded by a purine-rich B-recognition element (BRE) that facilitates the proper orientation of transcription (16). TBP (TATA box-binding protein) and TFB (transcription factor B), homologs of eucaryal general transcription factors, bind the TATA box and BRE elements and recruit the archaeal RNA polymerase to the promoter (15). Genes encoding three distinct TBPs are reported in the *M. acetivorans* genome (69), suggesting the possibility that pairings of different TBPs with the TFB is a novel mechanism for gene regulation in this metabolically diverse species. Although the transcription apparatus of the *Archaea* and *Eucarya* are similar, the sequencing of genomes from methanogens reveals several transcription factors homologous to regulators in the *Bacteria* (7, 34, 49, 68, 69, 91, 131,

221, 222). Thus, a major question is how the bacterial-like regulators interact with the eucaryal-like basal transcription apparatus. However, regulatory proteins other than bacterial homologs also function in the *Archaea*. In in vitro experiments, the eucaryal-like regulator, Tfx, was shown to bind downstream of the promoter for the *fmdECB* operon which encodes the molybdenum form of the formyl-MF dehydrogenase that functions in the carbon dioxide reduction pathway (92). Furthermore, a novel repressor (NrpR), regulating genes for nitrogen assimilation in *M. maripaludis*, represents a new family of regulators unique to the *Archaea* (146, 147). Thus, regulation of transcription in the methanogens complex and invites further investigation.

Regulation of Genes in the Carbon Dioxide Reduction Pathway

The expression of several genes in the carbon dioxide reduction pathway depends on the levels of hydrogen during growth. Northern blot analyses indicate that fed-batch cultures of *M. thermautotrophicus* supplied with excess hydrogen transcribe genes encoding the hydrogen-dependent methenyl-THMPT reductase (reaction 12b) and isozyme II of methyl-CoM reductase (MRII) (173). In contrast, under conditions where hydrogen levels were reduced and growth was limited, the genes encoding the $F_{420}H_2$-dependent methenyl-THMPT reductase (reaction 12a) and isozyme I of methyl-CoM reductase (MRI) were transcribed and the growth yields were greater than for cultures supplied with excess hydrogen. The results are consistent with a lower apparent K_m for hydrogen for the F_{420}-reducing hydrogenase compared with that of the hydrogen-dependent methylene-THMPT dehydrogenase (245). Experiments conducted with chemostat-cultured *M. marburgensis* revealed the same expression pattern for the gene encoding the $F_{420}H_2$-dependent methylene-THMPT dehydrogenase; however, expression of the gene encoding the hydrogen-dependent methylene-THMPT dehydrogenase was not dependent on the levels of hydrogen, suggesting the regulation of this gene is different in *M. thermautotrophicus* and *M. marburgensis* (2). In a more ecologically relevant experiment, *M. thermautotrophicus* was cocultured with obligate hydrogen-producing microbes in which the hydrogen partial pressure was maintained at a measurable low level (20 to 80 Pa) (150). Here again, only the gene encoding MRI was expressed under hydrogen-limiting growth conditions. The expression of genes encoding the formate dehydrogenase (reaction 8) of *M. thermautotrophicus* strain Z-245 increases when hydrogen becomes limiting in cultures growing on hydrogen and carbon dioxide (184). Consistent with these results is the report that *M. maripaludis* contains two sets of genes encoding formate dehydrogenase, and quantitative *lacZ* fusion experiments show both sets are downregulated when cells are grown with hydrogen plus carbon dioxide (272). Furthermore, when both formate and hydrogen plus carbon dioxide were present, expression of both sets were downregulated compared with growth on formate alone, which indicates that the regulation of transcription is controlled by hydrogen and not formate. At least for these two species, the results are consistent with a response that allows growth on formate when hydrogen is limiting in the environment. Hydrogen partial pressures also control the expression of genes encoding flagella in *Methanocaldococcus jannaschii* (*Methanococcus jannaschii*) (175). Flagella synthesis occurred when hydrogen was reduced from 178 kPa to 650 Pa, consistent with a role for this chemotactic response in adaptation to changing concentrations of hydrogen in the environment. In summary, although details of the transcriptional regulation are unknown, it appears that the energy-yielding metabolisms of carbon dioxide-reducing species adjust to different hydrogen concentrations in the environment to maintain an ecological advantage. Finally, the regulation of genes in response to the hydrogen partial pressure suggests a signal-sensing and transmission mechanism that has yet to be investigated.

Several lines of evidence indicate that the availability of metals regulate the expression of genes encoding metalloproteins essential for the carbon dioxide reduction pathway. The regulation of hydrogenase-encoding genes in *M. voltae* depends on the levels of selenium in the growth medium. *M. voltae* synthesizes two types of hydrogenases, F_{420}-reducing and non-F_{420}-reducing, for which there are selenium-containing and selenium-free isozymes of each type. The genes encoding both types of selenium-free hydrogenases are only transcribed under selenium limitation and linked by a common intergenic region containing activator and repressor binding sites (182). A 55-kDa putative regulatory protein binds to the positive element (176), and a LysR-type regulator (HrsM) binds to the negative element (238). Northern blot analysis indicates that *hrsM* is autoregulated.

A similar pattern of regulation of selenocysteine-containing and selenium-free hydrogenases by selenium is observed in *M. maripaludis* (206). However, when the *selB* gene essential for selenocysteine insertion is inactivated, selenium no longer exerts a regulatory effect, indicating that free selenium is not the effector molecule but rather a selenium-containing molecule. The genome sequence of *M. kandleri* is an-

notated with two types of tungsten-containing formyl-MF dehydrogenases (FwuB and FwcB) whose active sites contain either selenocysteine (FwuB) or cysteine (FwcB). Northern blot and primer-extension analyses indicate that, when grown in selenium-supplemented medium, only *fwuB* is transcribed; however, in media with low concentrations of selenium, both *fwuB* and *fwcB* are transcribed (252). The role of selenium or selenium-containing effector molecules in the regulation of selenoproteins of *M. voltae* has yet to be investigated. The presence of molybdenum in the growth medium is essential for transcription of genes encoding the molybdenum-containing formate dehydrogenase of *M. formicicum* (263) and the molybdenum-containing formyl-MF dehydrogenase (reaction 9) of *M. thermautotrophicus* (93), results consistent with a role for molybdenum as an effector molecule in regulation of transcription. Finally, when *M. marburgensis* is cultured in a chemostat with low concentrations of nickel that limit growth, there is an increase in gene expression and enzyme activity levels of the hydrogen-dependent methylene-THMPT dehydrogenase (2). Furthermore, levels of the nickel-containing $F_{420}H_2$-dependent hydrogenase decreased. The results are consistent with the inability of the nickel-containing hydrogenase to supply $F_{420}H_2$ for the $F_{420}H_2$-dependent methylene-THMPT dehydrogenase that is compensated by increased levels of the hydrogen-dependent methylene-THMPT dehydrogenase.

Regulation of Genes in the Acetate Fermentation Pathway

Development of a plasmid-mediated *lacZ* fusion reporter system in *M. acetivorans* has revealed that the operon encoding the Cdh complex (reaction 16) of *M. thermophila* is 54-fold downregulated in *M. acetivorans* grown on methanol compared with acetate, a result consistent with a unique role for this complex in the acetate fermentation pathway and preference for growth on methanol (6). The results are consistent with an earlier report on the regulation of *cdhA* in *M. thermophila* examined by Northern blotting (231). The genome sequences of *M. acetivorans* (69) and *M. mazei* (49) contain duplicate *cdh* operons with greater than 95% identity. The presence of two operons with high sequence identity raises the question as to whether both are transcribed during growth on acetate. A proteomic analysis of *M. acetivorans* indicates that one complex is expressed at least 16-fold over the other (145). The results are consistent with the predominance of a single Cdh complex in acetate-grown *M. mazei* strains Go1 and C-16 and expression of the cognate operon examined by Northern blotting. Although *M. mazei* strains Go1 and C-16 encode a duplicate *cdh* operon with high identity, evidence for expression was not reported (54). In contrast, global transcriptional profiling of *M. mazei* Go1 suggests both *cdh* operons are transcribed at approximately equal levels (94). Thus, expression of the duplicate operons in *M. mazei* strain Go1 is unresolved.

Ferredoxin accepts electrons from the Cdh complex of acetate-grown *Methanosarcina* species (62, 242, 243), and the gene encoding the ferredoxin from *M. thermophila* is downregulated in methanol- versus acetate-grown *M. thermophila* (42). Also essential and unique to the acetate fermentation pathway are acetate kinase and phosphotransacetylase (reactions 14 and 15), for which the encoding genes form an operon that is downregulated in methanol-grown compared with acetate-grown *M. thermophila* (219). Global proteomic and DNA microarray analyses of *M. mazei* and *M. acetivorans* (94, 144, 145) confirm the downregulation of genes essential and unique to the acetate fermentation pathway when cells are presented with methanol. This regulation in response to the growth substrate is consistent with the greater amount of energy available during growth with methanol ($\Delta G^{\circ\prime} = -106$ kJ/mol CH_4) versus acetate ($\Delta G^{\circ-} = -36$ kJ/mol) (244), providing cells with an ecological advantage.

Regulation of Genes in One-Carbon Dismutation Pathways

M. burtonii is a cold-adapted species that utilizes methanol and methylamines for growth and methanogenesis. A proteomic and transcriptional analysis of cold adaptation has revealed the thermal regulation of several genes essential for methanogenesis by dismutation of the methyl group of trimethylamine (72, 73). In particular, compared with growth at 23°C, cells grown at 4°C have elevated levels of $F_{420}H_2$ dehydrogenase (Fig. 11) which generates a proton gradient driving ATP synthesis. A sodium motive force is also generated by the methyl-THMPT:CoM methyltransferase (reaction 3) during growth on methylamines. However, at low temperatures, a proton motive force is easier to maintain, which is one explanation for elevated levels of $F_{420}H_2$ dehydrogenase when cells are grown at 4°C. Cells also upregulate synthesis of the F_{420}-dependent hydrogenase at the higher growth temperature, suggesting a role for hydrogen in electron transport. One possibility is that the hydrogen-dependent methylene-THMPT dehydrogenase (reaction 12b) functions at the higher temperature during oxidation of the methyl group, and the hydrogenase oxidizes the hydrogen with reduction of F_{420}, which supplies electrons to the proton-

translocating $F_{420}H_2$ dehydrogenase. The extent of regulation of genes essential for methanogenesis and other fundamental processes in response to temperature is consistent with a role in providing the cell with an ecological advantage in cold environments.

As mentioned above (see "Regulation of genes in the acetate fermentation pathway," above), growth on methanol is preferred over acetate. Thus, it is not surprising that proteomic and DNA microarray analyses show that genes essential for methanogenesis from methanol are upregulated when *Methanosarcina* species are cultured with this growth substrate (94, 144). Of particular interest is the regulation of multiple homologs of genes encoding methanol-specific methyltransferases MT1 and MT2 (Fig. 10). All *Methanosarcina* species contain an unusually large number of duplicated genes, which raises questions regarding expression and function of the gene products. The genome sequences of *M. acetivorans*, *M. mazei*, and *M. thermophila* harbor two homologs of *mtaA*, three homologs of *mtaB*, and three homologs of *mtaC* encoding enzymes specific for methanogenesis from methanol. Proteomic analyses of *M. thermophila* have established that none of the genes are silent and that methanol-grown cells express *mtaA-1*, *mtaB-1*, *mtaB-2*, *mtaB-3*, *mtaC-1*, *mtaC-2*, and *mtaC-3*, whereas acetate-grown cells express *mtaB-3* and *mtaC-3* (53). The authors speculate that synthesis of the methyltransferases in acetate-grown cells facilitates the switch from acetate to methanol as the growth substrate. A genetic analysis of multiple homologs of genes encoding the methyltransferases in *M. acetivorans* confirm results obtained with *M. thermophila* (200). Genetic analyses further showed that the Δ*mtaCB1* Δ*mtaCB2* deletion mutant of *M. acetivorans* cultured with methanol grew more slowly than wild-type, suggesting all three MT1 methyltransferases are required for wild-type growth on methanol.

Pyrrolysine, the 22nd Amino Acid

A most interesting outcome of the biochemical characterization of the methyltransferase systems is the discovery of the 22nd amino acid, pyrrolysine (Fig. 12), that was first found in the MtmB subunit of MT1 catalyzing transfer of the methyl group of monomethylamine to the corrinoid cofactor of the MtmC subunit of MT1 (84) (see Chapter 9). Since then, the cognate enzymes for dimethylamine and trimethylamine (MtbB and MttB) are reported to have UAG-encoded pyrrolysine residues (73, 224). Although a function for pyrrolysine has yet to be determined (126), multiple copies of *mtmB*, *mtbB*, and *mttB* containing in-frame UAG codons are reported

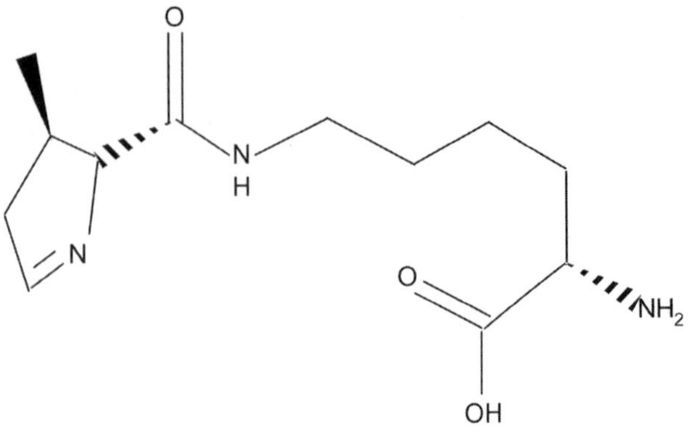

Figure 12. Pyrrolysine ((4R,5R)-4-methyl-pyrroline-5-carboxylate) present in MtmB, MtbB, and MttB.

in *M. barkeri*, *M. mazei*, *M. acetivorans*, and *M. thermophila*, supporting an essential function in the first step of the pathway for growth with methylated amines (49, 53, 69, 208). A mutant, unable to translate UAG, grows on methanol and acetate but not methylamines, indicating that the role for pyrrolysine is restricted to growth on methylamines (161). Furthermore, computational prediction of sequences-encoding pyrrolysine has only identified a small number of genes in selected genomes with the potential to encode pyrrolysine (37). Thus, it would appear that the 22nd amino acid is not widely used in nature and may be specific to the metabolism of methylamines. Although pyrrolysine is encoded by a potential stop codon, similar to selenocysteine (UGA), the mechanism of incorporation for these two unusual amino acids into proteins may be different (126). Whereas selenocysteine is synthesized attached to tRNA, pyrrolysine is charged directly onto a dedicated tRNA (199). The mechanism of translating UAG is unknown and under investigation. A role for UAG in stopping translation has not been reported for *Methanosarcina* species, consistent with UAG functioning as a sense codon, the same as for all other standard amino acids; however, a role for *cis-* and *trans*-acting factors cannot be ruled out.

PERSPECTIVE: THE NEXT FIVE YEARS

Although a wealth of information has accumulated over the past few decades on the enzymology of one-carbon reactions in methanogenesis, less is known about the details of membrane-bound electron transport and the structure/function of electron transfer proteins. The unusual structures revealed in the past for enzymes involved in one-carbon transformations portend novel structures and functions for electron transfer proteins and electron carriers. Advances in determining the crystal structures of membrane proteins are likely to be applied to the methanogens, facilitating discovery. Genomic sequencing of several methanogens has revealed that a large percentage of the proteome contains hypothetical proteins, many of which are likely to be involved in methanogenic pathways (34, 49, 68, 69, 91, 221, 222). Together with robust genetic systems paving the way for gene knockouts (20, 166), the discovery of novel proteins and their functions is at a threshold. Another understudied area ripe for discovery is gene regulation, in particular, the discovery and role of regulatory proteins in the context of a Eucarya-like basal transcription system. The finding that species related to *Methanosarcina* and *Methanococcoides* are present in anaerobic methane-oxidizing consortia (25), and that the process may be a reversal of methanogenesis pathways involving similar enzymes (172, 214), opens an entirely new field of research. Anaerobic methane oxidation, together with the exciting finding that plants produce methane aerobically (119), promises an unmatched flood of discovery revealing still more novel proteins and biochemical principals.

Acknowledgments. Research in the laboratory of J.G.F. has been supported by the National Institutes of Health, the Department of Energy, the National Science Roundation, and the National Aeronautics and Space Administration.

REFERENCES

1. **Abken, H.-J., M. Tietze, J. Brodersen, S. Baumer, U. Beifuss, and U. Deppenmeier.** 1998. Isolation and characterization of methanophenazine and the function of phenazines in membrane-bound electron transport of *Methanosarcina mazei* Go1. *J. Bacteriol.* **180:**2027–2032.
2. **Afting, C., E. Kremmer, C. Brucker, A. Hochheimer, and R. K. Thauer.** 2000. Regulation of the synthesis of H_2-forming methylenetetrahydromethanopterin dehydrogenase (Hmd) and of HmdII and HmdIII in *Methanothermobacter marburgensis*. *Arch. Microbiol.* **174:**225–232.
3. **Ahring, B. K., F. Alatristemondragon, P. Westermann, and R. A. Mah.** 1991. Effects of cations on *Methanosarcina thermophila* TM-1 growing on moderate concentrations of acetate: production of single cells. *Appl. Microbiol. Biotechnol.* **35:**686–689.
4. **Alex, L. A., J. N. Reeve, W. H. Orme-Johnson, and C. T. Walsh.** 1990. Cloning, sequence determination, and expression of the genes encoding the subunits of the nickel-containing 8-hydroxy-5- deazaflavin reducing hydrogenase from *Methanobacterium thermoautotrophicum* H. *Biochemistry* **29:**7237–7244.
5. **Andrade, S. L. A., F. Cruz, C. L. Drennan, V. Ramakrishnan, D. C. Rees, J. G. Ferry, and O. Einsle.** 2005. Structures of the iron-sulfur flavoproteins from *Methanosarcina thermophila* and *Archaeoglobus fulgidus*. *J. Bacteriol.* **187:**3848–3854.
6. **Apolinario, E. E., K. M. Jackson, and K. R. Sowers.** 2005. Development of a plasmid-mediated reporter system for *in vivo*

monitoring of gene expression in the archaeon *Methanosarcina acetivorans*. *Appl. Environ. Microbiol.* **71**:4914–4918.

7. **Aravind, L., and E. V. Koonin.** 1999. DNA-binding proteins and evolution of transcription regulation in the archaea. *Nucleic Acids Res.* **27**:4658–4670.

8. **Asakawa, S., H. Morii, M. Akagawamatsushita, Y. Koga, and K. Hayano.** 1993. Characterization of *Methanobrevibacter arboriphilicus* SA isolated from a paddy field soil and DNA-DNA hybridization among *M. arboriphilicus* strains. *Int. J. Syst. Bacteriol.* **43**:683–686.

9. **Asakawa, S., and K. Nagaoka.** 2003. *Methanoculleus bourgensis, Methanoculleus olentangyi* and *Methanoculleus oldenburgensis* are subjective synonyms. *Int. J. Syst. Evol. Microbiol.* **53**:1551–1552.

10. **Aufhammer, S. W., E. Warkentin, U. Ermler, C. H. Hagemeier, R. K. Thauer, and S. Shima.** 2005. Crystal structure of methylenetetrahydromethanopterin reductase (Mer) in complex with coenzyme F_{420}: Architecture of the F_{420}/FMN binding site of enzymes within the nonprolyl cis-peptide containing bacterial luciferase family. *Protein Sci.* **14**:1840–1849.

11. **Balch, W. E., G. E. Fox, L. J. Magrum, C. R. Woese, and R. S. Wolfe.** 1979. Methanogens: reevaluation of a unique biological group. *Microbiol. Rev.* **43**:260–296.

12. **Barker, H. A.** 1936. Studies upon the methane-producing bacteria. *Arch. Mikrobiol.* **7**:420–438.

13. **Baumer, S., T. Ide, C. Jacobi, A. Johann, G. Gottschalk, and U. Deppenmeier.** 2000. The $F_{420}H_2$ dehydrogenase from *Methanosarcina mazei* is a redox-driven proton pump closely related to NADH dehydrogenases. *J. Biol. Chem.* **275**:17968–17973.

14. **Baumer, S., E. Murakami, J. Brodersen, G. Gottschalk, S. W. Ragsdale, and U. Deppenmeier.** 1998. The $F_{420}H_2$:heterodisulfide oxidoreductase system from *Methanosarcina species*. 2-Hydroxyphenazine mediates electron transfer from $F_{420}H_2$ dehydrogenase to heterodisulfide reductase. *FEBS Lett.* **428**:295–298.

15. **Bell, S. D., and S. P. Jackson.** 2001. Mechanism and regulation of transcription in archaea. *Curr. Opin. Microbiol.* **4**:208–213.

16. **Bell, S. D., P. L. Kosa, P. B. Sigler, and S. P. Jackson.** 1999. Orientation of the transcription preinitiation complex in archaea. *Proc. Natl. Acad. Sci. USA* **96**:13662–13667.

17. **Belyaev, S. S., A. Y. Obraztsova, K. S. Laurinavichus, and L. V. Bezrukova.** 1986. Characteristics of rod-shaped methane-producing bacteria from an oil pool and description of *Methanobacterium ivanovii* sp. nov. *Microbiology* **55**:821–826.

18. **Belyaev, S. S., R. Wolkin, W. R. Kenealy, M. J. DeNiro, E. S., and J. G. Zeikus.** 1983. Methanogenic bacteria from the Bondyuzhskoe oil field: general characterization and analysis of stable-carbon isotopic fractionation. *Appl. Environ. Microbiol.* **45**:691–697.

19. **Biavati, B., M. Vasta, and J. G. Ferry.** 1988. Isolation and characterization of *Methanosphaera cuniculi* sp. nov. *Appl. Environ. Microbiol.* **54**:768–771.

20. **Blank, C. E., P. S. Kessler, and J. A. Leigh.** 1995. Genetics in methanogens: transposon insertion mutagenesis of a *Methanococcus maripaludis nifH* gene. *J. Bacteriol.* **177**:5773–5777.

21. **Blotevogel, K. H., and U. Fischer.** 1989. Transfer of *Methanococcus frisius* to the Genus *Methanosarcina* as *Methanosarcina frisia* comb. nov. *Int. J. Syst. Bacteriol.* **39**:91–92.

22. **Blotevogel, K. H., U. Fischer, and K. H. Lupkes.** 1986. *Methanococcus frisius* sp. nov., a new methylotrophic marine methanogen. *Can. J. Microbiol.* **32**:127–131.

23. **Blotevogel, K. H., U. Fischer, M. Mocha, and S. Jannsen.** 1985. *Methanobacterium thermoalcaliphilum* spec. nov., a

24. **Blotevogel, K. H., R. Gahljanssen, S. Jannsen, U. Fischer, F. Pilz, G. Auling, A. J. L. Macario, and B. J. Tindall.** 1991. Isolation and characterization of a novel mesophilic, freshwater methanogen from river sediment *Methanoculleus oldenburgensis* sp. nov. *Arch. Microbiol.* **157**:54–59.

25. **Boetius, A., K. Ravenschlag, C. J. Schubert, D. Rickert, F. Widdel, A. Gieseke, R. Amann, B. B. Jorgensen, U. Witte, and O. Pfannkuche.** 2000. A marine microbial consortium apparently mediating anaerobic oxidation of methane. *Nature* **407**:623–626.

26. **Boone, D. R.** 1987. Replacement of the type strain of *Methanobacterium formicicum* and reinstatement of *Methanobacterium bryantii* sp. nov. nom. rev. (ex Balch and Wolfe, 1981) with M.o.H. (DSM 863) as the Type Strain. *Int. J. Syst. Bacteriol.* **37**:172–173.

27. **Boone, D. R., and Y. Kamagata.** 1998. Rejection of the species *Methanothrix soehngenii*[VP] and the genus *Methanothrix*[VP] as *nomina confusa*, and transfer of *Methanothrix thermophila*[VP] to the genus *Methanosaeta*[VP] as *Methanosaeta thermophila* comb. nov. Request for an Opinion. *Int. J. Syst. Bacteriol.* **48**:1079–1080.

28. **Boone, D. R., I. M. Mathrani, Y. Liu, J. Menaia, R. A. Mah, and J. E. Boone.** 1993. Isolation and characterization of *Methanohalophilus portucalensis* sp. nov. and DNA reassociation study of the genus *Methanohalophilus*. *Int. J. Syst. Bacteriol.* **43**:430–437.

29. **Boone, D. R., W. B. Whitman, and P. E. Rouviere.** 1993. Diversity and taxonomy of methanogens, p. 35–80. *In* J. G. Ferry (ed.), *Methanogenesis: Ecology, Physiology, Biochemistry & Genetics*. Chapman & Hall, New York, N.Y.

30. **Boone, D. R., S. Worakit, I. M. Mathrani, and R. A. Mah.** 1986. Alkaliphilic methanogens from high-pH lake sediments. *Syst. Appl. Microbiol.* **7**:230–234.

31. **Brochier, C., P. Forterre, and S. Gribaldo.** 2004. Archaeal phylogeny based on proteins of the transcription and translation machineries: tackling the *Methanopyrus kandleri* paradox. *Genome Biol.* **5**:R17.

32. **Bryant, M. P., and D. R. Boone.** 1987. Emended description of strain MS[T] (DSM800[T]), the type strain of *Methanosarcina barkeri*. *Int. J. Syst. Bacteriol.* **37**:169–170.

33. **Bryant, M. P., and D. R. Boone.** 1987. Isolation and characterization of *Methanobacterium formicicum* MF. *Int. J. Syst. Bacteriol.* **37**:171–171.

34. **Bult, C. J., O. White, G. J. Olsen, L. X. Zhou, R. D. Fleischmann, G. G. Sutton, J. A. Blake, L. M. Fitzgerald, R. A. Clayton, J. D. Gocayne, A. R. Kerlavage, B. A. Dougherty, J. F. Tomb, M. D. Adams, C. I. Reich, R. Overbeek, E. F. Kirkness, K. G. Weinstock, J. M. Merrick, A. Glodek, J. L. Scott, N. S. M. Geoghagen, J. F. Weidman, J. L. Fuhrmann, D. Nguyen, T. R. Utterback, J. M. Kelley, J. D. Peterson, P. W. Sadow, M. C. Hanna, M. D. Cotton, K. M. Roberts, M. A. Hurst, B. P. Kaine, M. Borodovsky, H. P. Klenk, C. M. Fraser, H. O. Smith, C. R. Woese, and J. C. Venter.** 1996. Complete genome sequence of the methanogenic archaeon, *Methanococcus jannaschii*. *Science* **273**:1058–1073.

35. **Burggraf, S., H. Fricke, A. Neuner, J. Kristjansson, P. Rouvier, L. Mandelco, C. R. Woese, and K. O. Stetter.** 1990. *Methanococcus igneus* sp. nov., a novel hyperthermophilic methanogen from a shallow submarine hydrothermal system. *Syst. Appl. Microbiol.* **13**:263–269.

36. **Buss, K. A., D. R. Cooper, C. Ingram-Smith, J. G. Ferry, D. A. Sanders, and M. S. Hasson.** 2001. Urkinase: structure of acetate kinase, a member of the ASKHA superfamily of phosphotransferases. *J. Bacteriol.* **183**:680–686.

37. Chaudhuri, B. N., and T. O. Yeates. 2005. A computational method to predict genetically encoded rare amino acids in proteins. *Genome Biol.* **6:**R79.

38. Cheeseman, P. A., A. Toms-Wood, and R. S. Wolfe. 1972. Isolation and properties of a fluorescent compound, factor 420, from *Methanobacterium* strain M.o.H. *J. Bacteriol.* **112:**527–531.

39. Chistoserdova, L., J. A. Vorholt, R. K. Thauer, and M. E. Lidstrom. 1998. C-1 transfer enzymes and coenzymes linking methylotrophic bacteria and methanogenic Archaea. *Science* **281:**99–102.

40. Chong, S. C., Y. T. Liu, M. Cummins, D. L. Valentine, and D. R. Boone. 2002. *Methanogenium marinum* sp. nov., a H_2-using methanogen from Skan Bay, Alaska, and kinetics of H_2 utilization. *Antonie Van Leeuwenhoek Int. J. Gen. Mol. Microbiol.* **81:**263–270.

41. Ciulla, R., C. Clougherty, N. Belay, S. Krishnan, C. Zhou, D. Byrd, and M. F. Roberts. 1994. Halotolerance of *Methanobacterium thermoautotrophicum* ΔH and Marburg. *J. Bacteriol.* **176:**3177–3187.

42. Clements, A. P., and J. G. Ferry. 1992. Cloning, nucleotide sequence, and transcriptional analyses of the gene encoding a ferredoxin from *Methanosarcina thermophila*. *J. Bacteriol.* **174:**5244–5250.

43. Corder, R. E., L. A. Hook, J. M. Larkin, and J. I. Frea. 1983. Isolation and characterization of two new methane-producing cocci: *Methanogenium olentangyi*, sp. nov., and *Methanococcus deltae*, sp. nov. *Arch. Microbiol.* **134:**28–32.

44. Cuzin, N., A. S. Ouattara, M. Labat, and J. L. Garcia. 2001. *Methanobacterium congolense* sp. nov., from a methanogenic fermentation of cassava peel. *Int. J. Syst. Evol. Microbiol.* **51:**489–493.

45. Davidova, I. A., H. J. M. Harmsen, A. J. M. Stams, S. S. Belyaev, and A. J. B. Zehnder. 1997. Taxonomic description of *Methanococcoides euhalobius* and its transfer to the *Methanohalophilus* genus. *Antonie Van Leeuwenhoek* **71:**313–318.

46. de Bok, F. A. M., H. J. M. Harmsen, C. M. Plugge, M. C. de Vries, A. D. L. Akkermans, W. M. de Vos, and A. J. M. Stams. 2005. The first true obligately syntrophic propionate-oxidizing bacterium, *Pelotomaculum schinkii* sp. nov., co-cultured with *Methanospirillum hungatei*, and emended description of the genus *Pelotomaculum*. *Int. J. Syst. Evol. Microbiol.* **55:**1697–1703.

47. Deppenmeier, U. 2004. The membrane-bound electron transport system of *Methanosarcina* species. *J. Bioenerg. Biomembr.* **36:**55–64.

48. Deppenmeier, U. 2002. The unique biochemistry of methanogenesis. *Prog. Nucleic Acid Res. Mol. Biol.* **71:**223–283.

49. Deppenmeier, U., A. Johann, T. Hartsch, R. Merkl, R. A. Schmitz, R. Martinez-Arias, A. Henne, A. Wiezer, S. Baumer, C. Jacobi, H. Bruggemann, T. Lienard, A. Christmann, M. Bomeke, S. Steckel, A. Bhattacharyya, A. Lykidis, R. Overbeek, H. P. Klenk, R. P. Gunsalus, H. J. Fritz, and G. Gottschalk. 2002. The genome of *Methanosarcina mazei*: evidence for lateral gene transfer between Bacteria and Archaea. *J. Mol. Microbiol. Biotechnol.* **4:**453–461.

50. Dianou, D., T. Miyaki, S. Asakawa, H. Morii, K. Nagaoka, H. Oyaizu, and S. Matsumoto. 2001. *Methanoculleus chikugoensis* sp. nov., a novel methanogenic archaeon isolated from paddy field soil in Japan, and DNA-DNA hybridization among *Methanoculleus* species. *Int. J. Syst. Evol. Microbiol.* **51:**1663–1669.

51. Diekert, G., R. Jaenchen, and R. K. Thauer. 1980. Biosynthetic evidence for a nickel tetrapyrrole structure of factor F_{430} from *Methanobacterium thermoautotrophicum*. *FEBS Lett.* **119:**118–120.

52. Diekert, G., B. Klee, and R. K. Thauer. 1980. Nickel, a component of factor F_{430} from *Methanobacterium thermoautotrophicum*. *Arch. Microbiol.* **124:**103–106.

53. Ding, Y. H., S. P. Zhang, J. F. Tomb, and J. G. Ferry. 2002. Genomic and proteomic analyses reveal multiple homologs of genes encoding enzymes of the methanol:coenzyme M methyltransferase system that are differentially expressed in methanol- and acetate-grown *Methanosarcina thermophila*. *FEMS Microbiol. Lett.* **215:**127–132.

54. Eggen, R. I. L., R. Vankranenburg, A. J. M. Vriesema, A. C. M. Geerling, M. F. J. M. Verhagen, W. R. Hagen, and W. M. Devos. 1996. Carbon monoxide dehydrogenase from *Methanosarcina frisia* Go1. Characterization of the enzyme and the regulated expression of two operon-like *cdh* gene clusters. *J. Biol. Chem.* **271:**14256–14263.

55. Eirich, L. D., G. D. Vogels, and R. S. Wolfe. 1978. Proposed structure for coenzyme F_{420} from *Methanobacterium*. *Biochemistry* **17:**4583–4593.

56. Elberson, M. A., and K. R. Sowers. 1997. Isolation of an aceticlastic strain of *Methanosarcina siciliae* from Marine Canyon sediments and emendation of the species description for *Methanosarcina siciliae*. *Int. J. Syst. Bacteriol.* **47:**1258–1261.

57. Ermler, U., W. Grabarse, S. Shima, M. Goubeaud, and R. K. Thauer. 1997. Crystal structure of methyl-coenzyme M reductase: the key enzyme of biological methane formation. *Science* **278:**1457–1462.

58. Ferrari, A., T. Brusa, A. Rutili, E. Canzi, and B. Biavati. 1994. Isolation and characterization of *Methanobrevibacter oralis* sp. nov. *Curr. Microbiol.* **29:**7–12.

59. Ferry, J. G. 2003. One-carbon metabolism in methanogenic anaerobes, p. 143–156. *In* L. G. Ljungdahl, M. W. Adams, L. L. Barton, J. G. Ferry, and M. K. Johnson (ed.), *Biochemistry and Physiology of Anaerobic Bacteria*. Springer-Verlag, New York, N.Y.

60. Ferry, J. G., P. H. Smith, and R. S. Wolfe. 1974. *Methanospirillum*, a new genus of methanogenic bacteria, and characterization of *Methanospirillum hungatii* sp. nov. *Int. J. Syst. Bacteriol.* **24:**465–469.

61. Finazzo, C., J. Harmer, C. Bauer, B. Jaun, E. C. Duin, F. Mahlert, M. Goenrich, R. K. Thauer, S. Van Doorslaer, and A. Schweiger. 2003. Coenzyme B induced coordination of coenzyme M via its thiol group to Ni(I) of F_{430} in active methyl-coenzyme M reductase. *J. Am. Chem. Soc.* **125:**4988–4989.

62. Fischer, R., and R. K. Thauer. 1990. Ferredoxin-dependent methane formation from acetate in cell extracts of Methanosarcina barkeri (strain MS). *FEBS Lett.* **269:**368–372.

63. Forzi, L., J. Koch, A. M. Guss, C. G. Radosevich, W. W. Metcalf, and R. Hedderich. 2005. Assignment of the [4Fe-4S] clusters of Ech hydrogenase from *Methanosarcina barkeri* to individual subunits via the characterization of site-directed mutants. *FEBS J.* **272:**4741–4753.

64. Fox, G. E., L. J. Marrum, W. E. Balch, R. S. Wolfe, and C. R. Woese. 1977. Classification of methanogenic bacteria by 16S ribosomal RNA characterization. *Proc. Natl. Acad. Sci. USA* **74:**4537–4541.

65. Fox, J. A., D. J. Livingston, W.-H. Orme-Johnson, and C. T. Walsh. 1987. 8-hydroxy-5-deazaflavin-reducing hydrogenase from *Methanobacterium thermoautotrophicum*: 1. Purification and characterization. *Biochemistry* **26:**4219–4227.

66. Franzmann, P. D., Y. T. Liu, D. L. Balkwill, H. C. Aldrich, E. C. de Macario, and D. R. Boone. 1997. *Methanogenium frigidum* sp. nov., a psychrophilic, H_2-using methanogen from Ace Lake, Antarctica. *Int. J. Syst. Bacteriol.* **47:**1068–1072.

67. Franzmann, P. D., N. Springer, W. Ludwig, E. C. de Macario, and M. Rohde. 1992. A methanogenic archaeon from Ace

Lake, Antarctica: *Methanococcoides burtonii* sp. nov. *Syst. Appl. Microbiol.* **15:**573–581.

68. Fricke, W. F., H. Seedorf, A. Henne, M. Kruer, H. Liesegang, R. Hedderich, G. Gottschalk, and R. K. Thauer. 2006. The genome sequence of *Methanosphaera stadtmanae* reveals why this human intestinal archaeon Is restricted to methanol and H_2 for methane formation and ATP synthesis. *J. Bacteriol.* **188:**642–658.

69. Galagan, J. E., C. Nusbaum, A. Roy, M. G. Endrizzi, P. Macdonald, W. FitzHugh, S. Calvo, R. Engels, S. Smirnov, D. Atnoor, A. Brown, N. Allen, J. Naylor, N. Stange-Thomann, K. DeArellano, R. Johnson, L. Linton, P. McEwan, K. McKernan, J. Talamas, A. Tirrell, W. Ye, A. Zimmer, R. D. Barber, I. Cann, D. E. Graham, D. A. Grahame, A. M. Guss, R. Hedderich, C. Ingram-Smith, H. C. Kuettner, J. A. Krzycki, J. A. Leigh, W. Li, J. Liu, B. Mukhopadhyay, J. N. Reeve, K. Smith, T. A. Springer, L. A. Umayam, O. White, R. H. White, E. C. de Macario, J. G. Ferry, K. F. Jarrell, H. Jing, A. J. Macario, I. Paulsen, M. Pritchett, K. R. Sowers, R. V. Swanson, S. H. Zinder, E. Lander, W. W. Metcalf, and B. Birren. 2002. The genome of *M. acetivorans* reveals extensive metabolic and physiological diversity. *Genome Res.* **12:**532–542.

70. Goenrich, M., E. C. Duin, F. Mahlert, and R. K. Thauer. 2005. Temperature dependence of methyl-coenzyme M reductase activity and of the formation of the methyl-coenzyme M reductase red2 state induced by coenzyme B. *J. Biol. Inorg. Chem.* **10:**333–342.

71. Goenrich, M., F. Mahlert, E. C. Duin, C. Bauer, B. Jaun, and R. K. Thauer. 2004. Probing the reactivity of Ni in the active site of methyl-coenzyme M reductase with substrate analogues. *J. Biol. Inorg. Chem.* **9:**691–705.

72. Goodchild, A., M. Raftery, N. F. W. Saunders, M. Guilhaus, and R. Cavicchioli. 2005. Cold adaptation of the antarctic archaeon, *Methanococcoides burtonii*, assessed by proteomics using ICAT. *J. Proteome Res.* **4:**473–480.

73. Goodchild, A., N. F. W. Saunders, H. Ertan, M. Raftery, M. Guilhaus, P. M. G. Curmi, and R. Cavicchioli. 2004. A proteomic determination of cold adaptation in the antarctic archaeon, *Methanococcoides burtonii*. *Mol. Microbiol.* **53:**309–321.

74. Gorrell, A., S. H. Lawrence, and J. G. Ferry. 2005. Structural and kinetic analyses of arginine residues in the active-site of the acetate kinase from *Methanosarcina thermophila*. *J. Biol. Chem.* **280:**10731–10742.

75. Gottschalk, G., and R. K. Thauer. 2001. The Na^+ translocating methyltransferase complex from methanogenic archaea. *Biochim. Biophys. Acta* **1505:**28–36.

76. Grabarse, W., F. Mahlert, E. C. Duin, M. Goubeaud, S. Shima, R. K. Thauer, V. Lamzin, and U. Ermler. 2001. On the mechanism of biological methane formation: Structural evidence for conformational changes in methyl-coenzyme M reductase upon substrate binding. *J. Mol. Biol.* **309:**315–330.

77. Graham, D. E., M. Graupner, H. M. Xu, and R. H. White. 2001. Identification of coenzyme M biosynthetic 2-phosphosulfolactate phosphatase—a member of a new class of Mg^{2+}-dependent acid phosphatases. *Eur. J. Biochem.* **268:**5176–5188.

78. Grahame, D. A., and S. Gencic. 2000. Methane biochemistry, p. 188–198. *In Encyclopedia of Microbiology*, 2 ed, vol. 3. Academic Press, New York, N.Y.

79. Graupner, M., and R. H. White. 2001. Biosynthesis of the phosphodiester bond in coenzyme F-$_{420}$ in the methanoarchaea. *Biochemistry* **40:**10859–10872.

80. Graupner, M., H. M. Xu, and R. H. White. 2002. Characterization of the 2-phospho-L-lactate transferase enzyme involved in coenzyme F-420 biosynthesis in Methanococcus jannaschii. *Biochemistry* **41:**3754–3761.

81. Graupner, M., H. M. Xu, and R. H. White. 2000. Identification of the gene encoding sulfopyruvate decarboxylase, an enzyme involved in biosynthesis of coenzyme M. *J. Bacteriol.* **182:**4862–4867.

82. Graupner, M., H. M. Xu, and R. H. White. 2002. The pyrimidine nucleotide reductase step in riboflavin and F-420 biosynthesis in Archaea proceeds by the eukaryotic route to riboflavin. *J. Bacteriol.* **184:**1952–1957.

83. Guss, A. M., B. Mukhopadhyay, J. K. Zhang, and W. W. Metcalf. 2005. Genetic analysis of *mch* mutants in two Methanosarcina species demonstrates multiple roles for the methanopterin-dependent C-1 oxidation/reduction pathway and differences in H(2) metabolism between closely related species. *Mol. Microbiol.* **55:**1671–80.

84. Hao, B., W. Gong, T. K. Ferguson, C. M. James, J. A. Krzycki, and M. K. Chan. 2002. A new UAG-encoded residue in the structure of a methanogen methyltransferase. *Science* **296:**1462–1466.

85. Harris, J. E., P. A. Pinn, and R. P. Davis. 1984. Isolation and characterization of a novel thermophilic, freshwater methanogen. *Appl. Environ. Microbiol.* **48:**1123–1128.

86. Hedderich, R. 2004. Energy-converting [NiFe] hydrogenases from Archaea and extremophiles: ancestors of complex I. *J. Bioenerg. Biomembr.* **36:**65–75.

87. Hedderich, R., A. Berkessel, and R. K. Thauer. 1990. Purification and properties of heterodisulfide reductase from *Methanobacterium thermoautotrophicum* (strain Marburg). *Eur. J. Biochem.* **193:**255–261.

88. Hedderich, R., N. Hamann, and M. Bennati. 2005. Heterodisulfide reductase from methanogenic archaea: a new catalytic role for an iron-sulfur cluster. *Biol. Chem.* **386:**961–970.

89. Hedderich, R., J. Koch, D. Linder, and R. K. Thauer. 1994. The heterodisulfide reductase from *Methanobacterium thermoautotrophicum* contains sequence motifs characteristic of pyridine-nucleotide-dependent thioredoxin reductases. *Eur. J. Biochem.* **225:**253–261.

90. Heiden, S., R. Hedderich, E. Setzke, and R. K. Thauer. 1993. Purification of a cytochrome-b containing H_2-heterodisulfide oxidoreductase complex from membranes of *Methanosarcina barkeri*. *Eur. J. Biochem.* **213:**529–535.

91. Hendrickson, E. L., R. Kaul, Y. Zhou, D. Bovee, P. Chapman, J. Chung, E. Conway de Macario, J. A. Dodsworth, W. Gillett, D. E. Graham, M. Hackett, A. K. Haydock, A. Kang, M. L. Land, R. Levy, T. J. Lie, T. A. Major, B. C. Moore, I. Porat, A. Palmeiri, G. Rouse, C. Saenphimmachak, D. Soll, S. Van Dien, T. Wang, W. B. Whitman, Q. Xia, Y. Zhang, F. W. Larimer, M. V. Olson, and J. A. Leigh. 2004. Complete genome sequence of the genetically tractable hydrogenotrophic methanogen *Methanococcus maripaludis*. *J. Bacteriol.* **186:**6956–6969.

92. Hochheimer, A., R. Hedderich, and R. K. Thauer. 1999. The DNA binding protein Tfx from *Methanobacterium thermoautotrophicum*: structure, DNA binding properties and transcriptional regulation. *Mol. Microbiol.* **31:**641–650.

93. Hochheimer, A., D. Linder, R. K. Thauer, and R. Hedderich. 1996. The molybdenum formylmethanofuran dehydrogenase operon and the tungsten formylmethanofuran dehydrogenase operon from *Methanobacterium thermoautotrophicum*—structures and transcriptional regulation. *Eur. J. Biochem.* **242:**156–162.

94. Hovey, R., S. Lentes, A. Ehrenreich, K. Salmon, K. Saba, G. Gottschalk, R. P. Gunsalus, and U. Deppenmeier. 2005. DNA microarray analysis of *Methanosarcina mazei* Go1 reveals adaptation to different methanogenic substrates. *Mol. Genet. Genomics* **273:**225–239.

95. Howell, D. M., M. Graupner, H. M. Xu, and R. H. White. 2000. Identification of enzymes homologous to isocitrate dehydrogenase that are involved in coenzyme B and leucine biosynthesis in methanoarchaea. *J. Bacteriol.* **182**:5013–5016.

96. Huber, H., M. Thomm, H. Konig, G. Thies, and K. O. Stetter. 1982. *Methanococcus thermolithotrophicus*, a novel thermophilic lithotrophic methanogen. *Arch. Microbiol.* **132**:47–50.

97. Huser, B. A., K. Wuhrmann, and A. J. B. Zehnder. 1982. *Methanothrix soehngenii* gen. nov. sp. nov., a new acetotrophic non-hydrogen-oxidizing methane bacterium. *Arch. Microbiol.* **132**:1–9.

98. Ingram-Smith, C., A. Gorrell, S. H. Lawrence, P. Iyer, K. Smith, and J. G. Ferry. 2005. Identification of the acetate binding site in the *Methanosarcina thermophila* acetate kinase. *J. Bacteriol.* **187**:2386–2394.

99. Iyer, P. P., S. H. Lawrence, K. B. Luther, K. R. Rajashankar, H. P. Yennawar, J. G. Ferry, and H. Schindelin. 2004. Crystal structure of phosphotransacetylase from the methanogenic archaeon *Methanosarcina thermophila*. *Structure* **12**:559–567.

100. Jain, M. K., T. E. Thompson, E. C. de Macario, and J. G. Zeikus. 1987. Speciation of *Methanobacterium* Strain Ivanov as *Methanobacterium ivanovii*, sp. nov. *Syst. Appl. Microbiol.* **9**:77–82.

101. Jeanthon, C., S. L'Haridon, A. L. Reysenbach, E. Corre, M. Vernet, P. Messner, U. B. Sleytr, and D. Prieur. 1999. Methanococcus vulcanius sp. nov., a novel hyperthermophilic methanogen isolated from East Pacific Rise, and identification of Methanococcus sp. DSM 4213^T as Methanococcus fervens sp. nov. *Int. J. Syst. Bacteriol.* **49**:583–589.

102. Jeanthon, C., S. L'Haridon, A. L. Reysenbach, M. Vernet, P. Messner, U. B. Sleytr, and D. Prieur. 1998. *Methanococcus infernus* sp. nov., a novel hyperthermophilic lithotrophic methanogen isolated from a deep-sea hydrothermal vent. *Int. J. Syst. Bacteriol.* **48**:913–919.

103. Jiang, B., S. N. Parshina, W. van Doesburg, B. P. Lomans, and A. J. Stams. 2005. *Methanomethylovorans thermophila* sp. nov., a thermophilic, methylotrophic methanogen from an anaerobic reactor fed with methanol. *Int. J. Syst. Evol. Microbiol.* **55**:2465–2470.

104. Johnson, J. L., N. R. Bastian, N. L. Schauer, J. G. Ferry, and K. V. Rajagopalan. 1991. Identification of molybdopterin guanine dinucleotide in formate dehydrogenase from *Methanobacterium formicicum*. *FEMS Microbiol. Lett.* **77**:2–3.

105. Jones, J. B., and T. C. Stadtman. 1981. Selenium-dependent and selenium-independent formate dehydrogenases of *Methanococcus vannielii*. *J. Biol. Chem.* **256**:656–663.

106. Jones, W. J., J. A. Leigh, F. Mayer, C. R. Woese, and R. S. Wolfe. 1983. *Methanococcus jannaschii* sp. nov., an extremely thermophilic methanogen from a submarine hydrothermal vent. *Arch. Microbiol.* **136**:254–261.

107. Jones, W. J., M. J. B. Paynter, and R. Gupta. 1983. Characterization of *Methanococcus maripaludis* sp. nov., a new methanogen isolated form a salt marsh sediment. *Arch. Microbiol.* **135**:91–97.

108. Joulian, C., B. K. C. Patel, B. Ollivier, J. L. Garcia, and P. A. Roger. 2000. *Methanobacterium oryzae* sp. nov., a novel methanogenic rod isolated from a Philippines ricefield. *Int. J. Syst. Evol. Microbiol.* **50**:525–528.

109. Kadam, P. C., and D. R. Boone. 1996. Influence of pH on ammonia accumulation and toxicity in halophilic, methylotrophic methanogens. *Appl. Environ. Microbiol.* **62**:4486–4492.

110. Kadam, P. C., and D. R. Boone. 1995. Physiological characterization and emended description of *Methanolobus vulcani*. *Int. J. Syst. Bacteriol.* **45**:400–402.

111. Kadam, P. C., D. R. Ranade, L. Mandelco, and D. R. Boone. 1994. Isolation and characterization of *Methanolobus bombayensis* sp. nov., a methylotrophic methanogen that requires high concentrations of divalent cations. *Int. J. Syst. Bacteriol.* **44**:603–607.

112. Kamagata, Y., H. Kawasaki, H. Oyaizu, K. Nakamura, E. Mikami, G. Endo, Y. Koga, and K. Yamasato. 1992. Characterization of three thermophilic strains of *Methanothrix* ("*Methanosaeta*") *thermophila* sp. nov. and rejection of *Methanothrix* ("*Methanosaeta*") *thermoacetophila*. *Int. J. Syst. Bacteriol.* **42**:463–468.

113. Kamagata, Y., and E. Mikami. 1991. Isolation and characterization of a novel thermophilic *Methanosaeta* strain. *Int. J. Syst. Bacteriol.* **41**:191–196.

114. Kandler, O. 1981. Cell wall structures and their phylogenetic implications. *Zbl. Bakt. Hyg., Abt. Orig.*149–160.

115. Kandler, O., and H. Hippe. 1977. Lack of peptidoglycan in the cell walls of *Methanosarcina barkeri*. *Arch. Microbiol.* **113**:57–60.

116. Kandler, O., and H. Konig. 1978. Chemical composition of the peptidoglycan-free cell walls of methanogenic bacteria. *Arch. Microbiol.* **118**:141–152.

117. Karrasch, M., G. Borner, and R. K. Thauer. 1990. The molybdenum cofactor of formylmethanofuran dehydrogenase from *Methanosarcina barkeri* is a molybdopterin guanine dinucleotide. *FEBS Lett.* **274**:48–52.

118. Keltjens, J. T., M. J. Huberts, W. H. Laarhoven, and G. D. Vogels. 1983. Structural elements of methanopterin, a novel pterin present in *Methanobacterium thermoautotrophicum*. *Eur. J. Biochem.* **130**:537–544.

119. Keppler, F., J. T. Hamilton, M. Brass, and T. Rockmann. 2006. Methane emissions from terrestrial plants under aerobic conditions. *Nature* **439**:187–191.

120. Keswani, J., S. Orkand, U. Premachandran, L. Mandelco, M. J. Franklin, and W. B. Whitman. 1996. Phylogeny and taxonomy of mesophilic *Methanococcus* spp. and comparison of rRNA, DNA hybridization, and phenotypic methods. *Int. J. Syst. Bacteriol.* **46**:727–735.

121. Konig, H. 1984. Isolation and characterization of *Methanobacterium uliginosum* sp. nov. from a marshy soil. *Can. J. Microbiol.* **30**:1477–1481.

122. Konig, H., and K. O. Stetter. 1982. Isolation and characterization of *Methanolobus tindarius*, sp. nov., a coccoid methanogen growing only on methanol and methylamines. *Zbl. Bakt. Hyg., I. Abt. Orig.* **C3**:478–490.

123. Kotelnikova, S., A. J. L. Macario, and K. Pedersen. 1998. *Methanobacterium subterraneum* sp. nov., a new alkaliphilic, eurythermic and halotolerant methanogen isolated from deep granitic groundwater. *Int. J. Syst. Bacteriol.* **48**:357–367.

124. Kotelnikova, S. V., A. Y. Obraztsova, K. H. Blotevogel, and I. N. Popov. 1993. Taxonomic analysis of thermophilic strains of the genus *Methanobacterium*: reclassification of *Methanobacterium thermoalcaliphilum* as a synonym of *Methanobacterium thermoautotrophicum*. *Int. J. Syst. Bacteriol.* **43**:591–596.

125. Kotelnikova, S. V., A. Y. Obraztsova, G. M. Gongadze, and K. S. Laurinavichius. 1993. *Methanobacterium thermoflexum* sp. nov. and *Methanobacterium defluvii* sp. nov., thermophilic rod-shaped methanogens isolated from anaerobic digester sludge. *Syst. Appl. Microbiol.* **16**:427–435.

126. Krzycki, J. A. 2005. The direct genetic encoding of pyrrolysine. *Curr. Opin. Microbiol.* **8**:706–712.

127. Kumagai, H., T. Fujiwara, H. Matsubara, and K. Saeki. 1997. Membrane localization, topology, and mutual stabilization of the rnfABC gene products in *Rhodobacter capsulatus* and im-

plications for a new family of energy-coupling NADH oxidoreductases. *Biochemistry* **36:**5509–5521.

128. **Kunkel, A., M. Vaupel, S. Heim, R. K. Thauer, and R. Hedderich.** 1997. Heterodisulfide reductase from methanol-grown cells of *Methanosarcina barkeri* is not a flavoenzyme. *Eur. J. Biochem.* **244:**226–234.

129. **Kurkin, S., J. Meuer, J. Koch, R. Hedderich, and S. P. Albracht.** 2002. The membrane-bound [NiFe]-hydrogenase (Ech) from *Methanosarcina barkeri:* unusual properties of the iron-sulphur clusters. *Eur. J. Biochem.* **269:**6101–6111.

130. **Kurr, M., R. Huber, H. Konig, H. W. Jannasch, H. Fricke, A. Trincone, J. K. Kristjansson, and K. O. Stetter.** 1991. *Methanopyrus kandleri,* gen. and sp. nov. represents a novel group of hyperthermophilic methanogens, growing at 110°C. *Arch. Microbiol.* **156:**239–247.

131. **Kyrpides, N. C., and C. A. Ouzounis.** 1999. Transcription in archaea. *Proc. Natl. Acad. Sci. USA* **96:**8545–8550.

132. **L'Haridon, S., A. L. Reysenbach, A. Banta, P. Messner, P. Schumann, E. Stackebrandt, and C. Jeanthon.** 2003. *Methanocaldococcus indicus* sp. nov., a novel hyperthermophilic methanogen isolated from the Central Indian Ridge. *Int. J. Syst. Evol. Microbiol.* **53:**1931–1935.

133. **Lai, M. C., and S. C. Chen.** 2001. *Methanofollis aquaemaris* sp. nov., a methanogen isolated from an aquaculture fish pond. *Int. J. Syst. Evol. Microbiol.* **51:**1873–1880.

134. **Lai, M. C., S. C. Chen, C. M. Shu, M. S. Chiou, C. C. Wang, M. J. Chuang, T. Y. Hong, C. C. Liu, L. J. Lai, and J. J. Hua.** 2002. *Methanocalculus taiwanensis* sp. nov., isolated from an estuarine environment. *Int. J. Syst. Evol. Microbiol.* **52:**1799–1806.

135. **Lai, M. C., C. C. Lin, P. H. Yu, Y. F. Huang, and S. C. Chen.** 2004. *Methanocalculus chunghsingensis* sp. nov., isolated from an estuary and a marine fishpond in Taiwan. *Int. J. Syst. Evol. Microbiol.* **54:**183–189.

136. **Latimer, M. T., M. H. Painter, and J. G. Ferry.** 1996. Characterization of an iron-sulfur flavoprotein from *Methanosarcina thermophila. J. Biol. Chem.* **271:**24023–24028.

137. **Lauerer, G., J. K. Kristjansson, T. A. Langworthy, H. Konig, and K. O. Stetter.** 1986. *Methanothermus sociabilis* sp. nov., a second species within the *Methanothermaceae* growing at 97°C. *Syst. Appl. Microbiol.* **8:**100–105.

138. **Laurinavichyus, K. S., S. V. Kotelnikova, and A. Y. Obraztsova.** 1988. New species of thermophilic methane-producing bacteria *Methanobacterium thermophilum. Microbiology* **57:**832–838.

139. **Lawrence, S. H., and J. G. Ferry.** 2006. Steady-state kinetic analysis of phosphotransacetylase from *Methanosarcina thermophila. J. Bacteriol.* **188:**1155–1158.

140. **Lawrence, S. H., K. B. Luther, H. Schindelin, and J. G. Ferry.** 2006. Structural and functional studies suggest a catalytic mechanism for the phosphotransacetylase from *Methanosarcina thermophila. J. Bacteriol.* **188:**1143–1154.

141. **Leadbetter, J. R., and J. A. Breznak.** 1996. Physiological ecology of *Methanobrevibacter cuticularis* sp. nov. and *Methanobrevibacter curvatus* sp. nov., isolated from the hindgut of the termite *Reticulitermes flavipes. Appl. Environ. Microbiol.* **62:**3620–3631.

142. **Leadbetter, J. R., L. D. Crosby, and J. A. Breznak.** 1998. *Methanobrevibacter filiformis* sp. nov., a filamentous methanogen from termite hindguts. *Arch. Microbiol.* **169:**287–292.

143. **Leigh, J. A., K. L. Rinehart, Jr., and R. S. Wolfe.** 1984. Structure of methanofuran, the carbon dioxide reduction factor of *Methanobacterium thermoautotrophicum. J. Am. Chem. Soc.* **106:**3636–3640.

144. **Li, Q., L. Li, T. Rejtar, B. L. Karger, and J. G. Ferry.** 2005. The proteome of *Methanosarcina acetivorans.* Part II. Comparison of protein levels in acetate- and methanol-grown cells. *J. Proteome Res.* **4:**129–136.

145. **Li, Q., L. Li, T. Rejtar, D. J. Lessner, B. L. Karger, and J. G. Ferry.** 2006. Electron transport in the pathway of acetate conversion to methane in the marine archaeon *Methanosarcina acetivorans. J. Bacteriol.* **188:**702–710.

146. **Lie, T. J., and J. A. Leigh.** 2003. A novel repressor of *nif* and *glnA* expression in the methanogenic archaeon *Methanococcus maripaludis. Mol. Microbiol.* **47:**235–246.

147. **Lie, T. J., G. E. Wood, and J. A. Leigh.** 2005. Regulation of nif expression in *Methanococcus maripaludis*—roles of the euryarchaeal repressor NrpR, 2-oxoglutarate, and two operators. *J. Biol. Chem.* **280:**5236–5241.

148. **Liu, Y., D. R. Boone, and C. Choy.** 1990. *Methanohalophilus oregonense* sp. nov., a methylotrophic methanogen from an alkaline, saline aquifer. *Int. J. Syst. Bacteriol.* **40:**111–116.

149. **Lomans, B. P., R. Maas, R. Luderer, H. den Camp, A. Pol, C. van der Drift, and G. D. Vogels.** 1999. Isolation and Characterization of *Methanomethylovorans hollandica* gen. nov., sp nov., isolated from freshwater sediment, a methylotrophic methanogen able to grow on dimethyl sulfide and methanethiol. *Appl. Environ. Microbiol.* **65:**3641–3650.

150. **Luo, H. W., H. Zhang, T. Suzuki, S. Hattori, and Y. Kamagata.** 2002. Differential expression of methanogenesis genes of *Methanothermobacter thermoautotrophicus* (formerly *Methanobacterium thermoautotrophicum*) in pure culture and in cocultures with fatty acid-oxidizing syntrophs. *Appl. Environ. Microbiol.* **68:**1173–1179.

151. **Lyimo, T. J., A. Pol, H. J. M. den Camp, H. R. Harhangi, and G. D. Vogels.** 2000. *Methanosarcina semesiae* sp. nov., a dimethylsulfide-utilizing methanogen from mangrove sediment. *Int. J. Syst. Evol. Microbiol.* **50:**171–178.

152. **Lyon, E. J., S. Shima, G. Buurman, S. Chowdhuri, A. Batschauer, K. Steinbach, and R. K. Thauer.** 2004. UV-A/blue-light inactivation of the 'metal-free' hydrogenase (Hmd) from methanogenic archaea—the enzyme contains functional iron after all. *Eur. J. Biochem.* **271:**195–204.

153. **Ma, K., X. L. Liu, and X. Z. Dong.** 2005. *Methanobacterium beijingense* sp. nov., a novel methanogen isolated from anaerobic digesters. *Int. J. Syst. Evol. Microbiol.* **55:**325–329.

154. **Maestrojuan, G. M., and D. R. Boone.** 1991. Characterization of *Methanosarcina barkeri* MS^T and 227, *Methanosarcina mazei* S-6^T, and *Methanosarcina vacuolata* Z-761^T. *Int. J. Syst. Bacteriol.* **41:**267–274.

155. **Maestrojuan, G. M., D. R. Boone, L. Y. Xun, R. A. Mah, and L. F. Zhang.** 1990. Transfer of *Methanogenium bourgense, Methanogenium marisnigri, Methanogenium olentangyi,* and *Methanogenium thermophilicum* to the genus *Methanoculleus* gen. nov., emendation of *Methanoculleus marisnigri* and *Methanogenium,* and description of new strains of *Methanoculleus bourgense* and *Methanoculleus marisnigri. Int. J. Syst. Bacteriol.* **40:**117–122.

156. **Maestrojuan, G. M., J. E. Boone, R. A. Mah, J. Menaia, M. S. Sachs, and D. R. Boone.** 1992. Taxonomy and halotolerance of mesophilic *Methanosarcina* strains, assignment of strains to species, and synonymy of *Methanosarcina mazei* and *Methanosarcina frisia. Int. J. Syst. Bacteriol.* **42:**561–567.

157. **Mah, R. A.** 1980. Isolation and Characterization of *Methanococcus mazei. Curr. Microbiol.* **3:**321–326.

158. **Mah, R. A., and D. A. Kuhn.** 1984. Rejection of the type species *Methanosarcina methanica* (Approved Lists 1980), conservation of the genus *Methanosarcina* with *Methanosarcina barkeri* (Approved Lists 1980) as the type species, and emendation of the genus *Methanosarcina. Int. J. Syst. Bacteriol.* **34:**266–267.

159. Mah, R. A., and D. A. Kuhn. 1984. Transfer of the type species of the genus *Methanococcus* to the genus *Methanosarcina*, naming it *Methanosarcina mazei* (Barker 1936) comb. nov. et emend. and conservation of the genus *Methanococcus* (Approved Lists 1980) with *Methanococcus vannielii* (Approved Lists 1980) as the type specis. *Int. J. Syst. Bacteriol.* **34**:263–265.

160. Mah, R. A., M. R. Smith, and L. Baresi. 1978. Studies on an acetate-fermenting strain of *Methanosarcina*. *Appl. Environ. Microbiol.* **35**:1174–1184.

161. Mahapatra, A., A. Patel, J. A. Soares, R. C. Larue, J. K. Zhang, W. W. Metcalf, and J. A. Krzycki. 2006. Characterization of a *Methanosarcina acetivorans* mutant unable to translate UAG as pyrrolysine. *Mol. Microbiol.* **59**:56–66.

162. Makula, R. A., and M. E. Singer. 1978. Ether-containing lipids of methanogenic bacteria. *Biochem. Biophys. Res. Commun.* **82**:716–722.

163. Mathrani, I. M., D. R. Boone, R. A. Mah, G. E. Fox, and P. P. Lau. 1988. *Methanohalophilus zhilinae* sp. nov., an alkaliphilic, halophilic, methylotrophic methanogen. *Int. J. Syst. Bacteriol.* **38**:139–142.

164. May, H. D., N. L. Schauer, and J. G. Ferry. 1986. Molybdopterin cofactor from *Methanobacterium formicicum* formate dehydrogenase. *J. Bacteriol.* **166**:500–504.

165. McBride, B. C., and R. S. Wolfe. 1971. A new coenzyme of methyl transfer, coenzyme M. *Biochemistry* **10**:2317–2324.

166. Metcalf, W. W., J. K. Zhang, E. Apolinario, K. R. Sowers, and R. S. Wolfe. 1997. A genetic system for Archaea of the genus *Methanosarcina*. Liposome-mediated transformation and construction of shuttle vectors. *Proc. Natl. Acad. Sci. USA* **94**:2626–2631.

167. Meuer, J., H. C. Kuettner, J. K. Zhang, R. Hedderich, and W. W. Metcalf. 2002. Genetic analysis of the archaeon *Methanosarcina barkeri* Fusaro reveals a central role for Ech hydrogenase and ferredoxin in methanogenesis and carbon fixation. *Proc. Natl. Acad. Sci. USA* **99**:5632–5637.

168. Mikucki, J. A., Y. T. Liu, M. Delwiche, F. S. Colwell, and D. R. Boone. 2003. Isolation of a methanogen from deep marine sediments that contain methane hydrates, and description of *Methanoculleus submarinus* sp. nov. *Appl. Environ. Microbiol.* **69**:3311–3316.

169. Miller, T. L., and C. Z. Lin. 2002. Description of *Methanobrevibacter gottschalkii* sp. nov., *Methanobrevibacter thaueri* sp. nov., *Methanobrevibacter woesei* sp. nov. and *Methanobrevibacter wolinii* sp. nov. *Int. J. Syst. Evol. Microbiol.* **52**:819–822.

170. Miller, T. L., and M. J. Wolin. 1985. *Methanosphaera stadtmaniae* gen. nov. sp. nov.: a species that forms methane by reducing methanol with hydrogen. *Arch. Microbiol.* **141**:116–122.

171. Miller, T. L., M. J. Wolin, and E. A. Kusel. 1986. Isolation and characterization of methanogens from animal feces. *Syst. Appl. Microbiol.* **8**:234–238.

172. Moran, J. J., C. H. House, K. H. Freeman, and J. G. Ferry. 2005. Trace methane oxidation studied in several Euryarchaeota under diverse conditions. *Archaea* **1**:303–309.

173. Morgan, R. M., T. D. Pihl, J. Nolling, and J. N. Reeve. 1997. Hydrogen regulation of growth, growth yields, and methane gene transcription in *Methanobacterium thermoautotrophicum* ΔH. *J. Bacteriol.* **179**:889–898.

174. Mori, K., H. Yamamoto, Y. Kamagata, M. Hatsu, and K. Takamizawa. 2000. *Methanocalculus pumilus* sp. nov., a heavy-metal-tolerant methanogen isolated from a waste-disposal site. *Int. J. Syst. Evol. Microbiol.* **50**:1723–1729.

175. Mukhopadhyay, B., E. F. Johnson, and R. S. Wolfe. 2000. A novel p(H₂) control on the expression of flagella in the hyperthermophilic strictly hydrogenotrophic methanarchaeon *Methanococcus jannaschii*. *Proc. Natl. Acad. Sci. USA* **97**:11522–11527.

176. Muller, S., and A. Klein. 2001. Coordinate positive regulation of genes encoding [NiFe] hydrogenases in *Methanococcus voltae*. *Mol. Genet. Genomics* **265**:1069–1075.

177. Muller, V., C. Ruppert, and T. Lemker. 1999. Structure and function of the A₁A₀-ATPases from methanogenic Archaea. *J. Bioenerg. Biomembr.* **31**:15–27.

178. Murray, P. A., and S. H. Zinder. 1985. Nutritional requirements of *Methanosarcina* sp. strain TM-1. *Appl. Environ. Microbiol.* **50**:49–55.

179. Nelson, M. J. K., and J. G. Ferry. 1984. Carbon monoxide-dependent methyl coenzyme M methylreductase in acetotrophic *Methanosarcina* spp. *J. Bacteriol.* **160**:526–532.

180. Ni, S. S., and D. R. Boone. 1991. Isolation and characterization of a dimethyl sulfide-degrading methanogen, *Methanolobus siciliae* HI350, from an oil-well, characterization of *M. siciliae* T4/Mᵀ, and emendation of *M. siciliae*. *Int. J. Syst. Bacteriol.* **41**:410–416.

181. Ni, S. S., C. R. Woese, H. C. Aldrich, and D. R. Boone. 1994. Transfer of *Methanolobus siciliae* to the genus *Methanosarcina*, naming it *Methanosarcina siciliae*, and emendation of the genus *Methanosarcina*. *Int. J. Syst. Bacteriol.* **44**:357–359.

182. Noll, I., S. Muller, and A. Klein. 1999. Transcriptional regulation of genes encoding the selenium-free [NiFe]- hydrogenases in the archaeon *Methanococcus voltae* involves positive and negative control elements. *Genetics* **152**:1335–1341.

183. Noll, K. M., K. L. Rinehart, Jr, R. S. Tanner, and R. S. Wolfe. 1986. Structure of component B (7-mercaptoheptanoylthreonine phosphate) of the methylcoenzyme M methylreductase system of *Methanobacterium thermoautotrophicum*. *Proc. Natl. Acad. Sci. USA* **83**:4238–4242.

184. Nolling, J., and J. N. Reeve. 1997. Growth- and substrate-dependent transcription of the formate dehydrogenase (*fdhCAB*) operon in *Methanobacterium thermoformicicum* Z-245. *J. Bacteriol.* **179**:899–908.

185. Nozhevnikova, A. N., and V. I. Chudina. 1985. Morphology of the thermophilic acetate methane bacterium *Methanothrix thermoacetophila* sp. nov. *Microbiology* **53**:618–624.

186. Obraztsova, A. Y., O. V. Shipin, L. V. Bezrukova, and S. S. Belyaev. 1987. Properties of the coccoid methylotrophic methanogen, *Methanococcoides euhalobius* sp. nov. *Microbiology* **56**:523–527.

187. Ollivier, B., J. L. Cayol, B. K. C. Patel, M. Magot, M. L. Fardeau, and J. L. Garcia. 1997. *Methanoplanus petrolearius* sp. nov., a novel methanogenic bacterium from an oil-producing well. *FEMS Microbiol. Lett.* **147**:51–56.

188. Ollivier, B., M. L. Fardeau, J. L. Cayol, M. Magot, B. K. C. Patel, G. Prensier, and J. L. Garcia. 1998. *Methanocalculus halotolerans* gen. nov., sp. nov., isolated from an oil-producing well. *Int. J. Syst. Bacteriol.* **48**:821–828.

189. Ollivier, B. M., R. A. Mah, J. L. Garcia, and D. R. Boone. 1986. Isolation and characterization of *Methanogenium bourgense* sp. nov. *Int. J. Syst. Bacteriol.* **36**:297–301.

190. Ollivier, B. M., R. A. Mah, J. L. Garcia, and R. R. 1985. Isolation and characterization of *Methanogenium aggregans* sp. nov. *Int. J. Syst. Bacteriol.* **35**:127–130.

191. Oremland, R. S., and D. R. Boone. 1994. *Methanolobus taylorii* sp. nov., a new methylotrophic, estuarine methanogen. *Int. J. Syst. Bacteriol.* **44**:573–575.

192. Oremland, R. S., R. P. Kiene, I. Mathrani, M. J. Whiticar, and D. R. Boone. 1989. Description of an estuarine methylotrophic methanogen which grows on dimethyl sulfide. *Appl. Environ. Microbiol.* **55**:994–1002.

193. Patel, G. B. 1984. Characterization and nutritional properties of *Methanothrix concilii* sp. nov., a mesophilic, aceticlastic methanogen. *Can. J. Microbiol.* **30:**1383–1396.

194. Patel, G. B., and G. D. Sprott. 1990. *Methanosaeta concilii* gen. nov., sp. nov. ("*Methanothrix concilii*") and *Methanosaeta thermoacetophila* nom. rev., comb. nov. Int. *J. Syst. Bacteriol.* **40:**79–82.

195. Patel, G. B., G. D. Sprott, and J. E. Fein. 1990. Isolation and characterization of *Methanobacterium espanolae* sp. nov., a mesophilic, moderately acidiphilic methanogen. *Int. J. Syst. Bacteriol.* **40:**12–18.

196. Paterek, J. R., and P. H. Smith. 1985. Isolation and characterization of a halophilic methanogen from Great Salt Lake. *Appl. Environ. Microbiol.* **50:**877–881.

197. Paterek, J. R., and P. H. Smith. 1988. *Methanohalophilus mahii* gen. nov., sp. nov., a methylotrophic halophilic methanogen. *Int. J. Syst. Bacteriol.* **38:**122–123.

198. Paynter, M. J. B., and R. E. Hungate. 1968. Characterization of *Methanobacterium mobilis*, sp. n., isolated from the bovine rumen. *J. Bacteriol.* **95:**1943–1951.

199. Polycarpo, C., A. Ambrogelly, A. Berube, S. M. Winbush, J. A. McCloskey, P. F. Crain, J. L. Wood, and D. Soll. 2004. An aminoacyl-tRNA synthetase that specifically activates pyrrolysine. *Proc. Natl. Acad. Sci. USA* **101:**12450–12454.

200. Pritchett, M. A., and W. W. Metcalf. 2005. Genetic, physiological and biochemical characterization of multiple methanol methyltransferase isozymes in *Methanosarcina acetivorans* C2A. *Mol. Microbiol.* **56:**1183–1194.

201. Ragsdale, S. W. 2004. Life with carbon monoxide. *Crit. Rev. Biochem. Mol. Biol.* **39:**165–95.

202. Rasche, M. E., and R. H. White. 1998. Mechanism for the enzymatic formation of 4-(beta-D-ribofuranosyl)aminobenzene 5′-phosphate during the biosynthesis of methanopterin. *Biochemistry* **37:**11343–11351.

203. Rivard, C. J., J. M. Henson, M. V. Thomas, and P. H. Smith. 1983. Isolation and characterization of *Methanomicrobium paynteri* sp. nov., a mesophilic methanogen isolated from marine sediments. *Appl. Environ. Microbiol.* **46:**484–490.

204. Rivard, C. J., and P. H. Smith. 1982. Isolation and characterization of a thermophilic marine methanogenic bacterium, *Methanogenium thermophilicum* sp. nov. *Int. J. Syst. Bacteriol.* **32:**430–436.

205. Romesser, J. A., R. S. Wolfe, F. Mayer, E. Spiess, and A. Walther-Mauruschat. 1979. *Methanogenium*, a new genus of marine methanogenic bacteria, and characterization of *Methanogenium cariaci* sp. nov. and *Methanogenium marisnigri* sp. nov. *Arch. Microbiol.* **121:**147–153.

206. Rother, M., I. Mathes, F. Lottspeich, and A. Bock. 2003. Inactivation of the selB gene in *Methanococcus maripaludis*: effect on synthesis of selenoproteins and their sulfur-containing homologs. *J. Bacteriol.* **185:**107–114.

207. Rother, M., and W. W. Metcalf. 2004. Anaerobic growth of *Methanosarcina acetivorans* C2A on carbon monoxide: an unusual way of life for a methanogenic archaeon. *Proc. Natl. Acad. Sci. USA* **101:**16929–16934.

208. Saunders, N. F., T. Thomas, P. M. Curmi, J. S. Mattick, E. Kuczek, R. Slade, J. Davis, P. D. Franzmann, D. Boone, K. Rusterholtz, R. Feldman, C. Gates, S. Bench, K. Sowers, K. Kadner, A. Aerts, P. Dehal, C. Detter, T. Glavina, S. Lucas, P. Richardson, F. Larimer, L. Hauser, M. Land, and R. Cavicchioli. 2003. Mechanisms of thermal adaptation revealed from the genomes of the Antarctic Archaea *Methanogenium frigidum* and *Methanococcoides burtonii*. *Genome Res.* **13:**1580–1588.

209. Savant, D. V., Y. S. Shouche, S. Prakash, and D. R. Ranade. 2002. *Methanobrevibacter acididurans* sp. nov., a novel methanogen from a sour anaerobic digester. *Int. J. Syst. Evol. Microbiol.* **52:**1081–1087.

210. Schauer, N. L., and J. G. Ferry. 1986. Composition of the coenzyme F_{420}-dependent formate dehydrogenase from *Methanobacterium formicicum*. *J. Bacteriol.* **165:**405–411.

211. Scherer, P., V. Hollriegel, C. Krug, M. Bokel, and P. Renz. 1984. On the biosynthesis of 5-hydroxybenzimidazolyl-cobamide (vitamin B_{12}-factor III) in *Methanosarcina barkeri*. *Arch. Microbiol.* **138:**354–359.

212. Scott, J. W., and M. E. Rasche. 2002. Purification, overproduction, and partial characterization of beta-RFAP synthase, a key enzyme in the methanopterin biosynthesis pathway. *J. Bacteriol.* **184:**4442–4448.

213. Shima, S., E. J. Lyon, R. K. Thauer, B. Mienert, and E. Bill. 2005. Mossbauer studies of the iron-sulfur cluster-free hydrogenase: the electronic state of the mononuclear Fe active site. *J. Am. Chem. Soc.* **127:**10430–10435.

214. Shima, S., and R. K. Thauer. 2005. Methyl-coenzyme M reductase and the anaerobic oxidation of methane in methanotrophic Archaea. *Curr. Opin. Microbiol.* **8:**643–648.

215. Shima, S., E. Warkentin, R. K. Thauer, and U. Ermler. 2002. Structure and function of enzymes involved in the methanogenic pathway utilizing carbon dioxide and molecular hydrogen. *J. Biosci. Bioeng.* **93:**519–530.

216. Shlimon, A. G., M. W. Friedrich, H. Niemann, N. B. Ramsing, and K. Finster. 2004. *Methanobacterium aarhusense* sp. nov., a novel methanogen isolated from a marine sediment (Aarhus Bay, Denmark). *Int. J. Syst. Evol. Microbiol.* **54:**759–763.

217. Simankova, M. V., S. N. Parshina, T. P. Tourova, T. V. Kolganova, A. J. B. Zehnder, and A. N. Nozhevnikova. 2001. *Methanosarcina lacustris* sp. nov., a new psychrotolerant methanogenic archaeon from anoxic lake sediments. *Syst. Appl. Microbiol.* **24:**362–367.

218. Simianu, M., E. Murakami, J. M. Brewer, and S. W. Ragsdale. 1998. Purification and properties of the heme- and iron-sulfur-containing heterodisulfide reductase from *Methanosarcina thermophila*. *Biochemistry* **37:**10027–10039.

219. Singh-Wissmann, K., and J. G. Ferry. 1995. Transcriptional regulation of the phosphotransacetylase-encoding and acetate kinase-encoding genes (*pta* and *ack*) from *Methanosarcina thermophila*. *J. Bacteriol.* **177:**1699–1702.

220. Singh, N., M. M. Kendall, Y. Liu, and D. R. Boone. 2005. Isolation and characterization of methylotrophic methanogens from anoxic marine sediments in Skan Bay, Alaska: description of *Methanococcoides alaskense* sp. nov., and emended description of *Methanosarcina baltica*. *Int. J. Syst. Evol. Microbiol.* **55:**2531–2538.

221. Slesarev, A. I., K. V. Mezhevaya, K. S. Makarova, N. N. Polushin, O. V. Shcherbinina, V. V. Shakhova, G. I. Belova, L. Aravind, D. A. Natale, I. B. Rogozin, R. L. Tatusov, Y. I. Wolf, K. O. Stetter, A. G. Malykh, E. V. Koonin, and S. A. Kozyavkin. 2002. The complete genome of hyperthermophile *Methanopyrus kandleri* AV19 and monophyly of archaeal methanogens. *Proc. Natl. Acad. Sci. USA* **99:**4644–4649.

222. Smith, D, R, L. A. Doucette-Stamm, C. Deloughery, H. Lee, J. Dubois, T. Aldredge, R. Bashirzadeh, D. Blakely, R. Cook, K. Gilbert, D. Harrison, L. Hoang, P. Keagle, W. Lumm, B. Pothier, D. Qiu, R. Spadafora, R. Vicaire, Y. Wang, J. Wierzbowski, R. Gibson, N. Jiwani, A. Caruso, D. Bush, and J. N. Reeve. 1997. Complete genome sequence of *Methanobacterium thermoautotrophicum* ΔH: functional analysis and comparative genomics. *J. Bacteriol.* **179:**7135–7155.

223. Smith, P. H., and R. E. Hungate. 1958. Isolation and characterization of *Methanobacterium ruminatium* n. sp. *J. Bacteriol.* **75:**713–718.

224. Soares, J. A., L. Zhang, R. L. Pitsch, N. M. Kleinholz, R. B. Jones, J. J. Wolff, J. Amster, K. B. Green-Church, and J. A. Krzycki. 2005. The residue mass of L-pyrrolysine in three distinct methylamine methyltransferases. *J. Biol. Chem.* **280:** 36962–36969.

225. Solow, B., and R. H. White. 1997. Biosynthesis of the peptide bond in the coenzyme n-(7-mercaptoheptanoyl)-l-threonine phosphate. *Arch. Biochem. Biophys.* **345:**299–304.

226. Solow, B. T., and R. H. White. 1997. Absolute stereochemistry of 2-hydroxyglutaric acid present in methanopterin. *Chirality* **9:**678–680.

227. Soppa, J. 1999. Transcription initiation in Archaea: facts, factors and future aspects. *Mol. Microbiol.* **31:**1295–1305.

228. Sowers, K. R., S. F. Baron, and J. G. Ferry. 1984. *Methanosarcina acetivorans* sp. nov., an acetotrophic methane-producing bacterium isolated from marine sediments. *Appl. Environ. Microbiol.* **47:**971–978.

229. Sowers, K. R., and J. G. Ferry. 1983. Isolation and characterization of a methylotrophic marine methanogen, *Methanococcoides methylutens* gen. nov., sp. nov. *Appl. Environ. Microbiol.* **45:**684–690.

230. Sowers, K. R., and J. G. Ferry. 2002. Methanogenesis in the marine environment, p. 1913–1923. *In* G. Bitton (ed.), *The Encyclopedia of Environmental Microbiology.* John Wiley & Sons, Inc., New York, N.Y.

231. Sowers, K. R., T. T. Thai, and R. P. Gunsalus. 1993. Transcriptional regulation of the carbon monoxide dehydrogenase gene (*cdhA*) in *Methanosarcina thermophila.* *J. Biol. Chem.* **268:**23172–23178.

232. Sprenger, W. W., M. C. van Belzen, J. Rosenberg, J. H. P. Hackstein, and J. T. Keltjens. 2000. *Methanomicrococcus blatticola* gen. nov., sp. nov., a methanol- and methylamine-reducing methanogen from the hindgut of the cockroach *Periplaneta americana.* *Int. J. Syst. Evol. Microbiol.* **50:**1989–1999.

233. Spring, S., P. Schumann, and C. Sproer. 2005. *Methanogenium frittonii* (Harris *et al.* 1996) is a later synonym of *Methanoculleus thermophilus* (Rivard and Smith 1982) Maestrojuan *et al.* 1990. *Int. J. Syst. Evol. Microbiol.* **55:** 1097–1099.

234. Stadtman, T. C., and H. A. Barker. 1951. Studies on the Methane Fermentation X. A New Formate-Decomposing Bacterium, *Methanococcus vannielii. J. Bacteriol.* **62:**269–280.

235. Stetter, K. O. 1989. Genus II. Methanolobus, p. 2205–2207. *In* J. T. Staley, M. P. Bryant, N. Pfennig, and J. G. Holt (ed.), *Bergey's Manual of Systematic Bacteriology*, 1st ed., vol. 3. Williams & Wilkins Co., Baltimore, Md..

236. Stetter, K. O., M. Thomm, J. Winter, G. Wildgruber, H. Huber, W. Zillig, D. Jane-Covic, H. Konig, P. Palm, and S. Wunderl. 1981. *Methanothermus fervidus*, sp. nov., a novel extremely thermophilic methanogen isolated from an Icelandic hot spring. *Zbl. Bakt. Hyg. I. Abt. Orig. C* **2:**166–178.

237. Stojanowic, A., G. J. Mander, E. C. Duin, and R. Hedderich. 2003. Physiological role of the F_{420}-non-reducing hydrogenase (Mvh) from *Methanothermobacter marburgensis. Arch. Microbiol.* **180:**194–203.

238. Sun, J., and A. Klein. 2004. A LysR-type regulator is involved in the negative regulation of genes encoding selenium-free hydrogenases in the archaeon *Methanococcus voltae. Mol. Microbiol.* **52:**563–571.

239. Takai, K., A. Inoue, and K. Horikoshi. 2002. *Methanothermococcus okinawensis* sp. nov., a thermophilic, methane-producing archaeon isolated from a Western Pacific deep-sea hydrothermal vent system. *Int. J. Syst. Evol. Microbiol.* **52:** 1089–1095.

240. Takai, K., K. H. Nealson, and K. Horikoshi. 2004. *Methanotorris formicicus* sp. nov., a novel extremely thermophilic, methane-producing archaeon isolated from a black smoker chimney in the Central Indian Ridge. *Int. J. Syst. Evol. Microbiol.* **54:**1095–1100.

241. Taylor, C. D., and R. S. Wolfe. 1974. Structure and methylation of coenzyme M ($HSCH_2CH_2SO_3$). *J. Biol. Chem.* **249:** 4879–4885.

242. Terlesky, K. C., and J. G. Ferry. 1988. Ferredoxin requirement for electron transport from the carbon monoxide dehydrogenase complex to a membrane-bound hydrogenase in acetate-grown *Methanosarcina thermophila. J. Biol. Chem.* **263:** 4075–4079.

243. Terlesky, K. C., and J. G. Ferry. 1988. Purification and characterization of a ferredoxin from acetate-grown *Methanosarcina thermophila. J. Biol. Chem.* **263:**4080–4082.

244. Thauer, R. K. 1998. Biochemistry of methanogenesis: a tribute to Marjory Stephenson. *Microbiology* **144:**2377–2406.

245. Thauer, R. K., A. R. Klein, and G. C. Hartmann. 1996. Reactions with molecular hydrogen in microorganisms: evidence for a purely organic hydrogenation catalyst. *Chem. Rev.* **96:**3031–3042.

246. Tonouchi, A. 2002. Isolation and characterization of a motile hydrogenotrophic methanogen from rice paddy field soil in Japan. *FEMS Microbiol. Lett.* **208:**239–243.

247. Touzel, J. P., E. C. de Macario, J. Nolling, W. M. Devos, T. Zhilina, and A. M. Lysenko. 1992. DNA relatedness among some thermophilic members of the genus *Methanobacterium*: emendation of the species *Methanobacterium thermoautotrophicum* and rejection of *Methanobacterium thermoformicicum* as a synonym of *Methanobacterium thermoautotrophicum. Int. J. Syst. Bacteriol.* **42:**408–411.

248. Touzel, J. P., G. Prensier, J. L. Roustan, I. Thomas, H. C. Dubourguier, and G. Albagnac. 1988. Description of a new strain of *Methanothrix soehngenii* and rejection of *Methanothrix concilii* as a synonym of *Methanothrix soehngenii. Int. J. Syst. Bacteriol.* **38:**30–36.

249. Tripp, B. C., C. B. Bell, F. Cruz, C. Krebs, and J. G. Ferry. 2004. A role for iron in an ancient carbonic anhydrase. *J. Biol. Chem.* **279:**6683–6687.

250. van Bruggen, J. J. A., K. B. Zwart, J. G. F. Hermans, E. M. Vanhove, C. K. Stumm, and G. D. Vogels. 1986. Isolation and characterization of *Methanoplanus endosymbiosus* sp. nov., an endosymbiont of the marine sapropelic ciliate *Metopus contortus* Quennerstedt. *Arch. Microbiol.* **144:**367–374.

251. von Klein, D., H. Arab, H. Volker, and M. Thomm. 2002. *Methanosarcina baltica*, sp. nov., a novel methanogen isolated from the Gotland Deep of the Baltic Sea. *Extremophiles* **6:**103–110.

252. Vorholt, J. A., M. Vaupel, and R. K. Thauer. 1997. A selenium-dependent and a selenium-independent formylmethanofuran dehydrogenase and their transcriptional regulation in the hyperthermophilic *Methanopyrus kandleri. Mol. Microbiol.* **23:**1033–1042.

253. Wasserfallen, A., J. Nolling, P. Pfister, J. Reeve, and E. C. de Macario. 2000. Phylogenetic analysis of 18 thermophilic *Methanobacterium* isolates supports the proposals to create a new genus, *Methanothermobacter* gen. nov., and to reclassify several isolates in three species, *Methanothermobacter thermautotrophicus* comb. nov., *Methanothermobacter wolfeii* comb. nov., and *Methanothermobacter marburgensis* sp. nov. *Int. J. Syst. Evol. Microbiol.* **50:**43–53.

254. Welander, P. V., and W. W. Metcalf. 2005. Loss of the mtr operon in *Methanosarcina* blocks growth on methanol, but not methanogenesis, and reveals an unknown methanogenic pathway. *Proc. Natl. Acad. Sci. USA* **102:**10664–10669.

255. White, R. 1988. Characterization of the enzymatic conversion of sulfoacetaldehyde and L-cysteine into coenzyme M (3-Mercaptoethanesulfonic acid). *Biochemistry* **27:**7458–7462.

256. **White, R., H.** 1988. Biosynthesis of the 2-(Aminomethyl)-4-(hydroxymethyl) furan subunit of methanofuran. *Biochemistry* 27:4415–4420.

257. **White, R. H.** 1989. Biosynthesis of the 7-mercaptoheptanoic acid subunit of component B [(7-mercaptoheptanoyl)threonine phosphate] of methanogenic bacteria. *Biochemistry* 28:860–865.

258. **White, R. H.** 1994. Biosynthesis of (7-mercaptoheptanoyl)threonine phosphate. Biochemistry 33:7077–7081.

259. **White, R. H.** 1990. Biosynthesis of methanopterin. *Biochemistry* 29:5397–5404.

260. **White, R. H.** 1996. Biosynthesis of methanopterin. *Biochemistry* 35:3447–3456.

261. **White, R. H.** 1998. Methanopterin biosynthesis: methylation of the biosynthetic intermediates. *Bba Gen Subjects* 1380:257–267.

262. **White, R. H.** 1989. Steps in the conversion of alpha-ketosuberate to 7- mercaptoheptanoic acid in methanogenic bacteria. *Biochemistry* 28:9417–9423.

263. **White, W. B., and J. G. Ferry.** 1992. Identification of formate dehydrogenase-specific mRNA species and nucleotide sequence of the fdhC gene of *Methanobacterium formicicum*. *J. Bacteriol.* 174:4997–5004.

264. **Whitman, W. B., D. C. Coleman, and W. J. Wiebe.** 1998. Prokaryotes: the unseen majority. *Proc. Natl. Acad. Sci. USA* 95:6578–6583.

265. **Whitman, W. B., and R. S. Wolfe.** 1980. Presence of nickel in factor F_{430} from *Methanobacterium bryantii*. *Biochem. Biophys. Res. Commun.* 92:1196–1201.

266. **Widdel, F., P. E. Rouviere, and R. S. Wolfe.** 1988. Classification of secondary alcohol-utilizing methanogens including a new thermophilic isolate. *Arch. Microbiol.* 150:477–481.

267. **Wildgruber, G., M. Thomm, H. Konig, K. Ober, T. Ricchiuto, and K. O. Stetter.** 1982. *Methanoplanus limicola*, a plate-shaped methanogen representing a novel family, the Methanoplanaceae. *Arch. Microbiol.* 132:31–36.

268. **Wilharm, T., T. N. Zhilina, and P. Hummel.** 1991. DNA-DNA hybridization of methylotrophic halophilic methanogenic bacteria and transfer of *Methanococcus halophilus*[VP] to the genus *Methanohalophilus* as *Methanohalophilus halophilus* comb. nov. *Int. J. Syst. Bacteriol.* 41:558–562.

269. **Winter, J., C. Lerp, H.-P. Zabel, F. X. Wildenauer, H. Konig, and F. Schindler.** 1984. *Methanobacterium wolfei*, sp. nov., a new tungsten-requiring, thermophilic, autotrophic methanogen. *Syst. Appl. Microbiol.* 5:457–466.

270. **Woese, C. R., O. Kandler, and M. L. Wheelis.** 1990. Towards a natural system of organisms. Proposal for the domains archaea, bacteria, and eucarya. Proc. Natl. Acad. Sci. USA 87:4576–4579.

271. **Wolfe, R. S.** 1979. Methanogens: a surprising microbial group. *Antonie van Leeuwenhoek J. Microbiol. Serol.* 45:353–364.

272. **Wood, G. E., A. K. Haydock, and J. A. Leigh.** 2003. Function and regulation of the formate dehydrogenase genes of the methanogenic archaeon *Methanococcus maripaludis*. *J. Bacteriol.* 185:2548–2554.

273. **Worakit, S., D. R. Boone, R. A. Mah, M. E. Abdelsamie, and M. M. Elhalwagi.** 1986. *Methanobacterium alcaliphilum* sp. nov., an H_2-utilizing methanogen that grows at high pH values. *Int. J. Syst. Bacteriol.* 36:380–382.

274. **Wu, S. Y., S. C. Chen, and M. C. Lai.** 2005. *Methanofollis formosanus* sp. nov., isolated from a fish pond. *Int. J. Syst. Evol. Microbiol.* 55:837–842.

275. **Xun, L. Y., D. R. Boone, and R. A. Mah.** 1989. Deoxyribonucleic acid hybridization study of *Methanogenium* and *Methanocorpusculum* species, emendation of the genus *Methanocorpusculum*, and transfer of *Methanogenium aggregans* to the genus *Methanocorpusculum* as *Methanocorpusculum aggregans* comb. nov. *Int. J. Syst. Bacteriol.* 39:109–111.

276. **Zabel, H. P., H. Konig, and J. Winter.** 1984. Isolation and characterization of a new coccoid methanogen, *Methanogenium tatii* spec. nov. from a solfataric field on Mount Tatio. *Arch. Microbiol.* 137:308–315.

277. **Zeikus, J. G., and V. G. Bowen.** 1975. Comparative ultrastructure of methanogenic bacteria. *Can. J. Microbiol.* 21:121–129.

278. **Zeikus, J. G., and D. L. Henning.** 1975. *Methanobacterium arbophilicum* sp.nov. An obligate anaerobe isolated from wetwood of living trees. *Antonie van Leeuwenhoek* 41:543–552.

279. **Zeikus, J. G., and R. S. Wolfe.** 1972. *Methanobacterium thermoautotrophicum* sp. n., an anaerobic, autotrophic, extreme thermophile. *J. Bacteriol.* 109:707–713.

280. **Zellner, G., C. Alten, E. Stackebrandt, E. C. de Macario, and J. Winter.** 1987. Isolation and characterization of *Methanocorpusculum parvum*, gen. nov, spec. nov., a new tungsten requiring, coccoid methanogen. *Arch. Microbiol.* 147:13–20.

281. **Zellner, G., K. Bleicher, E. Braun, H. Kneifel, B. J. Tindall, E. C. de Macario, and J. Winter.** 1989. Characterization of a new mesophilic, secondary alcohol-utilizing methanogen, *Methanobacterium palustre* spec. nov. from a peat bog. *Arch. Microbiol.* 151:1–9.

282. **Zellner, G., D. R. Boone, J. Keswani, W. B. Whitman, C. R. Woese, A. Hagelstein, B. J. Tindall, and E. Stackebrandt.** 1999. Reclassification of *Methanogenium tationis* and *Methanogenium liminatans* as *Methanofollis tationis* gen. nov., comb. nov. and *Methanofollis liminatans* comb. nov. and description of a new strain of *Methanofollis liminatans*. *Int. J. Syst. Bacteriol.* 49:247–255.

283. **Zellner, G., P. Messner, H. Kneifel, B. J. Tindall, J. Winter, and E. Stackebrandt.** 1989. *Methanolacinia* gen. nov., incorporating *Methanomicrobium paynteri* as *Methanolacinia paynteri* comb. nov. *J. Gen. Appl. Microbiol.* 35:185–202.

284. **Zellner, G., P. Messner, J. Winter, and E. Stackebrandt.** 1998. *Methanoculleus palmolei* sp. nov., an irregularly coccoid methanogen from an anaerobic digester treating wastewater of a palm oil plant in North-Sumatra, Indonesia. *Int. J. Syst. Bacteriol.* 48:1111–1117.

285. **Zellner, G., U. B. Sleytr, P. Messner, H. Kneifel, and J. Winter.** 1990. *Methanogenium liminatans* spec. nov., a new coccoid, mesophilic methanogen able to oxidize secondary alcohols. *Arch. Microbiol.* 153:287–293.

286. **Zellner, G., E. Stackebrandt, P. Messner, B. J. Tindall, E. C. de Macario, H. Kneifel, U. B. Sleytr, and J. Winter.** 1989. *Methanocorpusculaceae* fam. nov., represented by *Methanocorpusculum parvum*, *Methanocorpusculum sinense* spec. nov. and *Methanocorpusculum bavaricum* spec. nov. *Arch. Microbiol.* 151:381–390.

287. **Zhao, H. X., A. G. Wood, F. Widdel, and M. P. Bryant.** 1988. An extremely thermophilic *Methanococcus* from a deep sea hydrothermal vent and Its plasmid. *Arch. Microbiol.* 150:178–183.

288. **Zhao, Y. Z., D. R. Boone, R. A. Mah, J. E. Boone, and L. Y. Xun.** 1989. Isolation and characterization of *Methanocorpusculum labreanum* sp. nov. from the Labrea Tar Pits. *Int. J. Syst. Bacteriol.* 39:10–13.

289. **Zhilina, T. N.** 1983. A new obligate halophilic methane-producing bacterium. *Mikrobiologiya* 52:375–382.

290. **Zhilina, T. N., and S. A. Ilarinov.** 1985. Characteristics of formate-assimilating methane bacteria and description of *Methanobacterium thermoformicicum* sp. nov. *Microbiology* 53:647–651.

291. **Zhilina, T.N., and G. A. Zavarzin.** 1979. Comparative cytology of methanosarcinae and description of *Methanosarcina vacuolata* sp. nova. *Microbiology* 48:223–228.

292. **Zhilina, T. N., and G. A. Zavarzin.** 1987. *Methanohalobium evestigatus* N. gen., N. sp.—extremely halophilic methane-forming archebacterium. *Doklady Akademii Nauk SSSR* **293:**464–468.

293. **Zhilina, T. N., and G. A. Zavarzin.** 1987. *Methanosarcina vacuolata* sp. nov., a vacuolated methanosarcina. *Int. J. Syst. Bacteriol.* **37:**281–283.

294. **Zinder, S. H., T. Anguish, and A. L. Lobo.** 1987. Isolation and characterization of a thermophilic acetotrophic strain of *Methanothrix. Arch. Microbiol.* **146:**315–322.

295. **Zinder, S. H., and R. A. Mah.** 1979. Isolation and characterization of a thermophilic strain of *Methanosarcina* unable to use H_2-CO_2 for methanogenesis. *Appl. Environ. Microbiol.* **38:**996–1008.

296. **Zinder, S. H., K. R. Sowers, and J. G. Ferry.** 1985. *Methanosarcina thermophila* sp. nov., a thermophilic, acetotrophic, methane-producing bacterium. *Int. J. Syst. Bacteriol.* **35:**522–523.

Archaea: Molecular and Cellular Biology
Edited by Ricardo Cavicchioli
© 2007 ASM Press, Washington, D.C.

Chapter 14

Proteinaceous Surface Layers of *Archaea*: Ultrastructure and Biochemistry

HELMUT KÖNIG, REINHARD RACHEL, AND HARALD CLAUS

INTRODUCTION

At the end of the 1970s a group of "prokaryotes" was recognized as a third domain of life distinct from the *Bacteria* and *Eucarya* (159, 160) (see Chapter 1). This domain was named *Archaea*, reflecting the fact that many species were found to live in habitats resembling the conditions of the early Earth (i.e., archaic). They comprise the methanogens, the extreme halophiles, the thermoacidophiles, sulfate- and/or sulfur-metabolizing thermophiles and hyperthermophiles (see Chapter 2). *Archaea* thrive in anaerobic niches, salt lakes, marine hydrothermal systems, and continental solfataras. However, by using 16S rDNA oligonucleotides as specific probes, they have also been found in many so-called moderate ecosystems, such as soil (13), freshwater (124), seawater, and marine sediments (34, 35). It has been estimated that *Archaea* make up about 20% of the marine picoplankton (67). According to our current knowledge, the archaeal domain consists of four main phylogenetic lineages: the *Crenarchaeota*, the *Euryarchaeota*, the *Nanoarchaeota*, and the *Korarchaeota*.

In early studies (see below) it became obvious that the phylogenetic diversity of the *Archaea* was also reflected in a remarkable diversity of cell envelope types (Fig. 1) (9, 39, 63, 64, 89, 100). All *Archaea* lack muramic acid and a lipopolysaccharide-containing outer membrane (see the notable exception of *Ignicoccus* in "The outer membrane of *Ignicoccus* and the surface layer of *Nanoarchaeum*," below). They also lack a universal cell wall polymer. The cell walls of the *Archaea* are composed of different polymers such as glutaminylglycan, heterosaccharide, methanochondroitin, pseudomurein, protein, glycoprotein, or glycocalyx. The most common archaeal cell envelope is composed of a single protein or glycoprotein surface layer (S layer), which is directly associated with the cytoplasmic membrane. The thermoplasmas lack a cell wall, but they possess a glycocalyx instead. Some cell envelope types are restricted to the *Archaea*. However, similar building blocks may be found in natural polymeric compounds (e.g., connective tissue) in members of the other two domains of life.

As early as the 1950s, significant differences between typical bacterial cell walls and those of *Archaea* were established during cell envelope investigations of *Halobacterium* (54). Subsequent work with cells of *Sulfolobus* (150, 151), *Halococcus* (18), and *Methanosarcina* (61) also showed unusual structures. The S-layer glycoprotein of *Halobacterium salinarum* was the first glycoprotein discovered in bacteria and archaea (97, 98). Initially, these novel cell wall structures were viewed as curiosities, and their taxonomic significance was not realized until the concept of the *Archaea* was published (159). At this time, the results of cell wall studies supported the new view of the phylogeny of the *Bacteria* and *Archaea*.

Many archaea possess proteinaceous surface layers (S layers), which form two-dimensional regular arrays (6, 11, 63, 64, 101, 139). The chemical structure of archaeal S-layer glycoproteins has been determined in detail for a few archaeal species, e.g., *Methanothermus fervidus* (64, 66), *H. salinarum* and *Haloferax volcanii* (88, 138, 139), and *Staphylothermus marinus* (116).

Helmut König • Institut für Mikrobiologie und Weinforschung, Johannes Gutenberg-Universität, Becherweg 15, D-55099 Mainz, Germany. **Reinhard Rachel** • Lehrstuhl für Mikrobiologie, Universität Regensburg, Universitätsstraße. 31, D-93053 Regensburg, Germany. **Harald Claus** • Institut für Mikrobiologie und Weinforschung, Johannes Gutenberg-Universität, Becherweg 15, D-55099 Mainz, Germany.

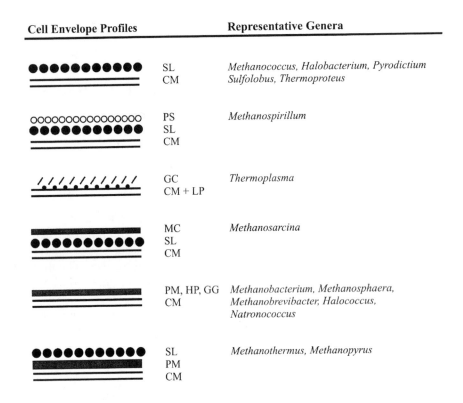

Figure 1. Cell wall profiles of *Archaea*. CM, cytoplasmic membrane; GC, glycocalyx; GG, glutaminylglycan; HP, heteropolysaccharide; LP, lipoglycan; MC, methanochondroitin; PM, pseudomurein; PS, protein sheath; SL, S layer.

GLYCOSYLATED S LAYERS OF (HYPER-) THERMOPHILIC ARCHAEA

Most of the known (hyper-)thermophilic and also many mesophilic species of the *Archaea* have been shown to possess S layers (64). The morphological building blocks are composed of six, four, three, or two subunits of one type of a (glyco-)protein. Accordingly, the symmetry axis with the highest symmetry is p6, p4, p3, or p2 (Fig. 2; Table 1).

In most species, the S layer is the sole cell envelope component outside the cytoplasmic membrane, to which it is anchored by an elongated, filamentous protrusion, spanning the quasi-periplasmic space between the S layer and the membrane. The center-to-center spacing varies among the known genera and species between about 11 nm in several species of the order *Methanococcales* (*M. vannielii* and *M. thermolithotrophicus* [85, 109]) and 30 nm, as found in several species of the order *Thermoproteales*, such as *Thermoproteus tenax* (99) or *Pyrobaculum islandicum* (117). The molecular masses of the surface proteins range from 40 to 325 kDa (101).

The two- and three-dimensional structures of isolated S-layer sheets have been studied in numerous species by electron crystallography: *Sulfolobus acidocaldarius* (31, 92, 141), *Sulfolobus solfataricus* (119), *Sulfolobus shibatae* (91), *Acidianus brierleyi* (8), *T. tenax* (99, 157), *P. islandicum* (117), *Pyrobaculum organotrophum* (118), *Desulfurococcus mobilis* (158), *Pyrodictium occultum* (51), *Pyrodictium brockii* (37), *Hyperthermus butylicus* (7), and *Archaeoglobus fulgidus* (73). Complementary structural information was obtained by studying freeze-fractured or freeze-etched cells, ultrathin sections, and freeze-dried and heavy-metal shadowed isolated S-layer sheets. Structural information at lower resolution is available for some newer isolates such as *Pyrobaculum aerophilum* (146), *Pyrodictium abyssi* (121), *Picrophilus oshimae* (125), *Ferroglobus placidus* (48), *Pyrolobus fumarii* (14), and *Archaeoglobus veneficus* (57).

Figure 2. Scheme showing the arrangement of S-layer subunits. From the left: p2, p3, p4, and p6.

Table 1. Characteristic structural features of archaeal S layers

Genus/species	Symmetry	Center-to-center (nm)	Width of periplasm (nm)	Source
Crenarchaeota				
Sulfolobales				
Sulfolobus sp.	p3	20–21	20–25	6, 91, 92
Metallosphaera sp.	p3	21	25	45
Acidianus brierleyi	p3	21	25	8
Thermoproteales				
Pyrobaculum organotrophum	p6	~30	25	118
Pyrobaculum islandicum	p6	~30	25	117
Thermoproteus tenax	p6	~30	25	99, 157
Thermofilum sp.	p6	27	25	Rachel, unpublished results
Desulfurococcales				
Desulfurococcus mobilis	p4	18	30	157
Staphylothermus marinus	p4	36	60–70	116
Aeropyrum pernix	p4	19	30	Rachel, unpublished results
Pyrolobus fumarii	p4	18.5	45	14
Pyrodictium sp.	p6	21	35	37, 51, 121
Hyperthermus butylicus	p6	25	~20	7
Ignicoccus sp.	–	–	20–400	120
Nanoarchaeota				
Nanoarchaeum equitans	p6	16	20	56
Euryarchaeota				
Thermococcales				
Thermococcus sp., *Pyrococcus* sp.	p6	~15–20	~10	7, 36, 58, 71, 84
Methanobacteriales				
Methanothermus fervidus	p6	20	15–20	This work[a]
Methanococcales				
Methanothermococcus thermautotrophicus	p6	11	≦10	109
Methanocaldococcus jannaschii	p6	11.5	10	109
Methanococcus vannielii, M. voltae	p6	10–11	~10	85, 109
Archaeoglobales				
Archaeoglobus sp.	p6	17.5	10	57, 73
Ferroglobus placidus	p4	23	~10	48
Thermoplasmatales				
Picrophilus sp.	p4	20	40	125
Halobacteriales				
Halobacterium salinarum	p6	18	6.5	77, 144
Haloferax volcanii	p6	18	6.5	144
Methanomicrobiales				
Methanoplanus limicola	p6	14.7	5–10	24

[a]See text and Fig. 12.

In some studies, the arrangement of the subunits is characteristic of the genus (e.g., *Pyrodictium* versus *Pyrolobus* versus *Hyperthermus*; *Archaeoglobus* versus *Ferroglobus*; *Thermoplasma* versus *Picrophilus*). In other cases, it is similar or identical in all species of the genera belonging to a family (*Thermoproteus* and *Pyrobaculum* within the *Thermoproteaceae*; *Sulfolobus*, *Acidianus*, and *Metallosphaera* within the *Sulfolobaceae*).

These results indicate that, at least to a limited extent, the S-layer structure correlates with the organism's phylogeny (6), as determined by 16S rDNA sequencing (see Chapter 2). In the following sections, the S-layer structures of (hyper-) thermophilic archaea are described in a phylogenetic context.

Crenarchaeota

Sulfolobales

The S layer of *S. acidocaldarius* (150, 151) was first isolated by lysing the cells with sodium dodecyl sulfate (SDS; 0.15%), digestion with DNase, and repeated treatment of the cell-shaped S-layer sheets with SDS. The isolated S layers were disintegrated with phosphate buffer pH 9 at 60°C, and the solubilized protein was purified by chromatography on Sepharose (102). Chemical analysis revealed that the S layer was composed of a single glycoprotein occurring in two modifications of apparent molecular masses of about 140 and 170 kDa, respectively. The glycoproteins contained high levels of serine and threonine and low levels of basic amino acids and dicaroxylic amino acids (102). Evidence was provided that a small protein subunit may anchor the S layer to the cell membrane of *S. acidocaldarius* (47).

The first electron micrographs (31, 141) of purified S layers of *S. acidocaldarius* were interpreted as showing that the subunits were arranged on a hexagonal lattice, with a two-sided plane group p6, and a 22-nm unit cell dimension. The three-dimensional structure of the S layer was elucidated by electron crystallography to about 2-nm resolution (31, 141). The three-dimensional (3D) model showed dimeric building blocks, arranged to form a series of hexagonal and triangular holes. This first 3D structure of an archaeal S layer already identified a feature that has subsequently been found to occur in many archaeal S layers; the external surface was fairly smooth, while the surface facing to the interior of the cell appeared sculptured, with large dome-shaped cavities and protruding "pedestals," which are now known to anchor the S layer into the lipid bilayer of the cytoplasmic membrane (6). The protein substructure consists of three types of globular domains, diad (D), triad (T), and ring region (R), connected by narrow bridges. These may act as "hinges," allowing the S layer to form a curved surface (lobes) and to follow the movements

of the cell surface during growth. Note that the larger pores of the S layer have the same diameter as pili (150). Pili have been observed that attach the cells to sulfur crystals (150). Similar results were subsequently obtained for the S layer of *S. solfataricus* (119).

The interpretation that the S layers of *Sulfolobus* species had sixfold symmetry was later shown to be incorrect (6; 91, 92). By improving the technique of sample preparation, image recording using cryoelectron microscopy, and refining the image-processing methods of the electron micrographs (including image classification), it became clear that the S layer of *S. acidocaldarius* had "only" threefold symmetry (92). The S layer of *S. shibatae* was also found to have p3 symmetry and the same fine structure (distribution of protein mass) as determined for *S. acidocaldarius*, with a mosaic arrangement of the protein complexes and with the frequent occurrence of twin boundaries (where neighboring unit cells are observed to be rotated by 120 degrees) and distortions (where unit cells on one lattice line are slightly displaced and rotated to each other) (91).

It has now been established that the S layers of the phylogenetically related organisms (Fig. 3), *S. acidocaldarius*, *S. solfataricus*, *S. shibatae*, *Acidianus infernus* (Fig. 4a; R. Rachel and H. Huber, unpublished

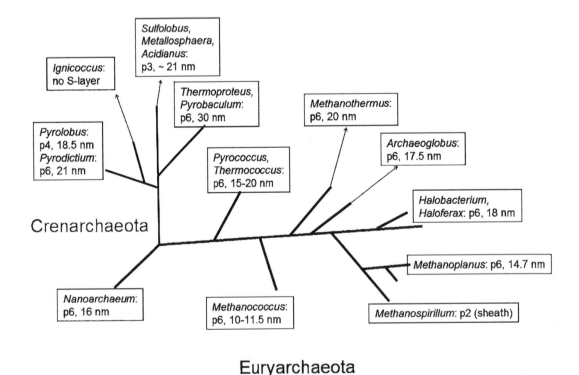

Figure 3. Phylogenetic tree of *Archaea*. The relative phylogenetic positions of the 16S rDNA sequences of archaea (*Crenarchaeota*, *Euryarchaeota*, and *Nanoarcheum*) discussed in this chapter are depicted in the tree. The arrangement and unit cell dimensions of S-layer subunits are shown.

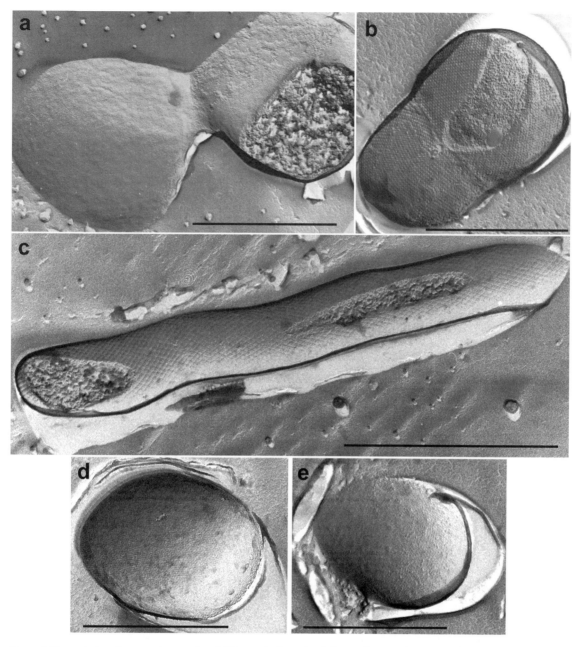

Figure 4. Transmission electron micrographs of different *Archaea*. (a–c) *Crenarchaeota*; (d, e) *Euryarchaeota*: *Thermococcales*. All cells prepared by freeze etching. (a) *Acidianus infernus*: p3 symmetry; (b) *Aeropyrum pernix*: p4 symmetry; (c) *Thermoproteus tenax*: p6 symmetry; (d) *Thermococcus chitonophagus*; (e) *Thermococcus acidaminovorans*. All bars = 1 μm.

results), *Acidianus brierlyi* (8), *Metallosphaera sedula* and *Metallosphaera prunae* (45) all have a high degree of similarity to each other. Their fine structure and mass distribution, the unique p3 symmetry, the center-to-center distance of about 21 nm, and the width of their periplasm (about 25 nm) are almost identical to each other (6). These structural characteristics appear to be common to many, if not all, species belonging to the crenarchaeal order *Sulfolobales*.

The putative S-layer gene of *S. acidocaldarius* DSM 639 was identified in the recently published whole genome sequence (25). The protein Slpsa1 consists of 1,424 amino acids, corresponding to a molecular mass of 151,040 daltons (Table 2). A putative leader peptide of 29 amino acids would be cleaved after membrane translocation to obtain the mature protein. Cysteine is present as found for other hyperthermophilic S-layer (glyco)proteins.

Table 2. Putative S-layer glycoproteins and membrane glycoproteins in three *Sulfolobus* species

Strain	N terminus[a]	Annotation	Accession no. (NCBI)	Size (aa)	MW (Da)[b]	MW (kDa) SDS-PAGE	Leader peptide[c] (aa)	pI[d]	PAS staining (N-glycosylation sites)[e]
S. acidocaldarius DSM 639	LTPEVSAGGIQAYLL	S-layer glycoprotein (experimental) Slp 1	CAJ30479	1,424	151,040	150[b]	29	5.17	+ (29)
S. solfataricus P2	TVPVILYAPFIF[f] AAIYTIPSVT	S-layer glycoprotein (experimental) Slp 1[g]	CAJ31324	1,231	131,868	130	30	6.56	+ (37)
S. tokodaii strain 7	n.d.	S-layer glycoprotein (putative) Slp 1[h]	BAB67302	1,440	156,240	n.d.	28	4.58	n.d (40)
S. acidocaldarius DSM 639	AVTINGITFYSPV	Membrane protein	YP_256928	475	49,546	60	24	9.47	+ (13)
S. solfataricus P2	ISKTLVAVIIVVVI	Maltose ABC transporter protein	NP_342629	523	56,856	60	–	5.98	+ (10)

[a]Determined by Edman degradation (B. Schlott, IMB, Jena).
[b]Matrix-assisted laser desorption/ionization (MALDI) determination: 176,974 Da.
[c]Determined with Signal P.
[d]Determined with Protparam.
[e]Determined with Prosite.
[f]Internal fragment obtained with proteinase K.
[g]Sequence alignments (Blast 2) with the S-layer glycoprotein Slp 1 show 25% identical residues.
[h]Sequence alignments (Blast 2) with the S-layer glycoprotein Slp 1 show 21% identical residues.

Desulfurococcales

For *D. mobilis*, the S layer was shown to exhibit p4 symmetry and a rather open meshwork of protein, with a lattice of 18 nm (158). Two other species in this order also have an S layer with p4 symmetry, although with a different fine structure. *P. fumarii*, an archaeon, which grows up to 113°C, has an S layer that encloses a 40-nm-wide "quasi-periplasmic space" and has a lattice of 18.5 nm (14). *S. marinus* has a unique glycoprotein S layer, named "tetrabrachion," which was intensively studied by biochemical methods, gene sequencing, and electron microscopy (115, 116). The morphological subunits of the S layer appear as a branched filiform meshwork and form a canopy with a distance of about 70 nm from the cell membrane, which encloses a "quasi-periplasmic space" (94). A morphological subunit is a tetramer of polypeptides (M_r, 92 kDa) that form a parallel, four-stranded α-helical rod, 70 nm long. It separates at one end into four strands, or straight arms (hence the name "tetra-brachion," "four arms"). Each polypeptide arm (M_r, 85 kDa; 24 nm in length) is composed of β-sheets and provides lateral connectivity to neighboring morphological subunits by end-to-end contacts (115). Attached to the middle of each stalk are two copies of a protease that provide an exodigestive function related to the heterotrophic energy metabolism of the organism (94). The tetrabrachion structure (that includes the protease) exhibits an unusual thermal stability (115, 116).

Recently, freeze etching was used to investigate the fine structure of cells of *Aeropyrum pernix*, a closely related *Aeropyrum* isolate, and three other isolates (CB9, HVE1, CH11), which were obtained from hot springs (R. Rachel, I. Wyschkony, H. Schmidt, and H. Huber; unpublished results). The physiological characterization of the three new isolates is incomplete. According to their 16S rRNA gene sequence, they all belong to the *Desulfurococcales*. Electron micrographs of freeze-etched cells (Fig. 4b) revealed that they have S layers with p4 symmetry and a lattice of about 18 to 19 nm, with obvious differences in their surface relief and in the mass distribution of the protein complexes. An S layer with p4 symmetry, an open network of protein, and a comparably large quasi-periplasmic space (30 to 70 nm) is a common feature of these organisms and of three others, *Desulfurococcus mobilis*, *Staphylothermus marinus*, and *Pyrolobus fumarii*.

In contrast to the *Aeropyrum*-related strains, cells of the genera *Hyperthermus* and *Pyrodictium* have S layers with p6 symmetry. For all species of the genus *Pyrodictium* that have been isolated, *P. occultum*, *P. abyssi*, *P. abyssi* strain TAG11, and *P. brockii*,

the center-to-center distance of the S layer is about 21 nm, and their mass distribution, surface reliefs, and 3D structures are almost identical (37, 51, 121). However, distribution of protein mass and surface relief of the *H. butylicus* S layer is clearly different from that of the *Pyrodictium* species (7); also, its center-to-center distance of 25 nm is significantly larger than in *Pyrodictium*. This shows that the phylogeny in this family, as determined by the sequence of the 16S rDNA, correlates with S-layer ultrastructure.

Thermoproteales

T. tenax (Fig. 4c) and *Thermofilum pendens* possess extraordinarily rigid S-layer sacculi of hexagonally arranged subunits that are resistant to detergent and protease treatment (166, 167). The S layer could be isolated by disrupting the cells by using sonication followed by incubation with DNase and RNase, SDS treatment (2% SDS, 100°C, 30 min), and differential centrifugation (81, 157). In *T. tenax*, the 25-nm-wide interspace is due to long protrusions that extend from the relatively thin (3 to 4 nm) filamentous network of the outer surface of the S layer toward the cytoplasmic membrane. The distal ends of these pillarlike protrusions appear to penetrate the membrane, thus serving as membrane anchors (99, 157).

The S layers of at least three species belonging to the genus *Pyrobaculum*, *P. islandicum* (117), *P. organotrophum*, and *P. aerophilum* (146), all have the same characteristic S-layer structure, which is almost identical to the S layer of *Thermoproteus*. This consists of a delicate protein meshwork, with long protrusions serving as membrane anchors, that encloses a 25-nm-wide periplasm, a center-to-center distance of about 30 nm, and a common chemical and thermal stability. Preliminary investigations with cells of two *Thermofilum* species showed a similar S-layer fine structure, with p6 symmetry and a center-to-center distance of 27 nm (R. Rachel, unpublished results).

Euryarchaeota

Methanococcales

The S layers of various genera of this order do not show a high level of detail when examined by freeze etching, indicating that they do not have elaborate surface relief (Fig. 5). The only common feature presently identified is the center-to-center distance of only 11 nm (85, 109) (K. Schuster and R. Rachel, unpublished results). The periplasmic space is narrow, with a width about 5 to 10 nm (20).

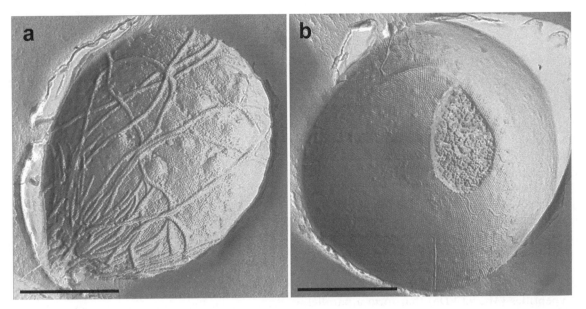

Figure 5. Transmission electron micrographs of species of the *Methanococcales*. (a) *M. thermolithotrophicus*; (b) *M. jannaschii*; freeze-etching. Bar = 0.5 μm.

Thermococcales

Similar to the *Methanococcales*, only limited information is available on the 3D structure of these S layers (Fig. 4d, e). The S layers of two genera of this order, *Thermococcus* and *Pyrococcus*, could not be isolated successfully by detergent extraction. In addition, they did not show much detail when examined by freeze etching because their surface relief is shallow (36, 58, 71). The only common feature is a lattice constant of ~15 to 20 nm (9, 84). The width of the periplasmic space in thin sectioned cells prepared by quick freezing and freeze substitution is about 10 nm (58).

Archaeoglobales

Cells of two *Archaeoglobus* species have been investigated. *A. fulgidus* (73) and *A. veneficus* (57) have an S layer with a center-to-center distance of 18 nm. When the 3D structure of the *A. fulgidus* S layer was successfully determined by electron crystallography, it showed a remarkable structural instability. It could not be isolated by detergent extraction of whole cells, as the S layers of many *Crenarchaeota*, but it could only be imaged after absorbing the cells to the carbon support and on-grid detergent extraction for a few seconds only (73). Based on its 16S rRNA gene sequence and physiology, *F. placidus* is related to the genus *Archaeoglobus*. In contrast to *Archaeoglobus*, *F. placidus* exhibits a different type of S layer, with a square lattice and a spacing of 23 nm (48).

Thermoplasmatales

Species of the genera *Thermoplasma* and *Ferroplasma* do not show any sign of an S layer. The only known genus of this order that exhibits an S layer is *Picrophilus*, which is known for its ability to grow at pH 0. The S layer has p4 symmetry and a center-to-center distance of 20 nm. The width of the periplasmic space is about 40 nm (125).

General remarks

In summary, there are many ultrastructural features of S layers that characterize the (hyper-) thermophiles. Within the *Crenarchaeota* that have been investigated, most S layers are stable and robust. They can be isolated by detergent extraction of whole cells or cell envelopes, indicating that the interaction between the protein complexes is strong. The distance between the cytoplasmic membrane and S layer is wide and varies between 20 and 25 nm (*Sulfolobales*, *Thermoproteales*), 40 nm (e.g., *Pyrolobus*), and 70 nm (*Staphylothermus*). In the S-layer-deficient *Ignicoccus*, the periplasm has been observed to vary between 20 and 500 nm in a single cell (120) (see "The outer membrane of *Ignicoccus* and the surface layer of *Nanoarchaeum*," below). These features are in contrast to many species that belong to the *Euryarchaeota*. Intact S-layer sheets of the *Euryarchaeota* tend to be difficult to obtain because the S layers are labile and can not be isolated by detergent extraction. The width of the periplasm in most cells is 15 nm or less and tends to be in the range 5 to 10 nm. The close spatial association between membrane and S layer correlates with the fact that their physicochemical interaction appears to be tight, while the interaction between the protein complexes is weak; detergent extraction of the membrane lipids results in dissolution of the S layer. There are notable exceptions within the *Euryarchaeota*: (i) *Picrophilus* exhibits an S layer that is strong enough to withstand detergent extraction, and the periplasm is 40 nm wide; (ii) in addition to an S layer, all species of the genera *Methanothermus* and *Methanopyrus* have a pseudomurein sacculus as a second cell wall polymer that is highly stable and serves to maintain cell shape maintenance and may protect the cells.

The characteristics of the structures of S layers are consistent with the primary structures of S-layer genes (2, 26). These studies revealed significant similarities of S-layer genes between closely related species of *Euryarchaeota*, including genes from members of the orders *Methanococcales*, *Thermococcales*, and *Halobacteriales*.

PROTEINACEOUS SHEATHS COVER THE CELLS OF *METHANOSPIRILLUM* AND *METHANOSAETA*

The filamentous chains of *Methanospirillum hungatei* and *Methanosaeta concilii* (formerly *Methanothrix soehngenii*) (111) are held together by a proteinaceous fibrillary sheath (163). Each single cell of *Methanospirillum* is surrounded by a separate electron-dense layer. Standard techniques for cell wall isolation provide pure preparations of sheath material (62). Freeze-etched specimens showed each fibril to be composed of two subfibrils. The isolated sheath material is resistant to detergent, chaotropic agents, and common proteases. Isolated sheaths (62, 132) are composed of amino acids and neutral sugars. Computer image processing of tilted-view electron micrographs of isolated, negatively stained sheaths revealed a two-dimensional S-layer-like paracrystalline structure. The structure consists of subunits with P1 symmetry with cellular dimensions of $\alpha = 12.0$ nm, $\beta = 2.9$ nm, and $\gamma = 93.7°$ (126), or subunits are arranged on a lattice with P2 symmetry and cells with $\alpha = 5.66$ nm, $\beta = 2.81$ nm, and $\gamma = 85.6°$ (135). Treatment of isolated sheaths at 90°C with a combination of β-mercaptoethanol and sodium dodecyl sulfate under alkaline conditions results in the solubilization of "glue peptides" and the liberation of single hoops, the essential structural component of the cylindrical sheaths (133). Imaging the inner and outer surfaces of isolated sheaths with a bimorph scanning tunneling electron microscope confirmed that the sheaths form a paracrystalline structure and that they consist of a series of stacked hoops of ca. 2.5 nm in width (12). This study also showed that the sheath possesses minute pores and therefore is impervious to solutes with a hydrated radius of >0.3 nm.

In cross sections of *M. concilii*, the envelope appears as a double track about 25 to 30 nm in width, with a very dark inner and a more electron-transparent outer layer. Only the inner layer participates in septum formation during cell division. Hence, it may be assumed that only the outer, electron-transparent layer represents a striated sheath that embraces many cells, whereas the inner layer represents a rigid cell wall sacculus surrounding individual cells (59, 162). They also show striation not only at the cylindric part of the sacculi, but also at the septa. This indicates that both layers seen in cross sections of whole cells belong to the same morphological entity, which therefore may not fit the definition of a sheath in the strict sense of the word. Chemical analysis of isolated envelopes revealed a complex amino acid pattern and the presence of neutral sugars (mainly glucose, mannose), resembling the composition of the sheath of *M. hungatei* (62, 132). After hydrozinolysis of sheath preparations from *M. concilii* strain FE, an asparagine-rich glycoprotein fraction was obtained (114). This is indicative of the presence of asparaginylrhamnose linkages on Asn-X-Ser glycosylation sites in the sheath glycoprotein.

PROTEIN SUBUNITS FORM THE CELL WALL OF METHANOCOCCI

All species belonging to the order *Methanococcales* possess hexagonal S layers as exclusive cell wall components outside the cytoplasma membrane (Fig. 5). As the *Methanococcales* include mesophilic, thermophilic, and extreme thermophilic species, they represent an ideal model system for studying thermal adaptation of S layers. The special features of these S layers are described below with reference to other bacterial and archaeal S-layer proteins (2, 5, 26, 77, 79, 109).

Gene Sequences

The first *Methanococcales* gene sequence for an S-layer protein was for the mesophile, *Methanococcus voltae* (82). Additional putative genes for S-layer proteins have been identified in the complete genome sequences of the mesophile *Methanococcus maripaludis* (165) and the hyperthermophile, *Methanocaldococcus jannaschii* (19). Based on the conserved N-terminal region, primers were constructed for PCR amplification and sequencing of previously unidentified S-layer genes from the mesophile *Methanococcus vannielii*, the thermophile *Methanothermococcus thermolithotrophicus*, and the hyperthermophile *Methanotorris igneus* (Table 3). The genes were confirmed to be S-layer genes by purification of the proteins they encoded and determination of their N termini by Edman degradation (1, 2).

Signal sequences for secretion

Most S-layer proteins contain N-terminal signal sequences that allow their secretion across the cytoplasmic membrane by the general secretory pathway (15, 42, 123) (See Chapter 17). The putative 28-amino-acid leader peptides of proteins from the *Methanococcales* (Fig. 6) are highly conserved (1, 2). They display typical characteristics of a signal sequence with a nonhydrophobic (n)-region, a hydrophobic (h)-core, and a charged (c)-region with an alanine residue at the peptide cleavage site (10).

The 22- and 30-amino-acid-long signal peptides of *M. fervidus* (17) and *M. mazei* (161), respectively, are not homologous with signal peptides of the *Methanococcales*. In addition to S-layer proteins, archaeal flagellins possess leader peptides (3, 5, 29). However, these short positively charged leader peptides consist of only 4 to 14 (in many cases 12) amino acids, with an invariant glycine at the cleavage site, and have little similarity to the signal peptides of S-layer proteins.

Several bacteria produce excess amounts of S-layer proteins to ensure complete coverage of the cells during all phases of growth. The thermophilic bacterium, *Geobacillus stearothermophilus*, appears to produce an S-layer protein pool in the peptidoglycan layer (16). Several bacteria have been reported to shed S-layer material into the culture medium (15). For the bacterium, *Acinetobacter*, this occurs as a result of an overproduction of new S-layer protein (143). Considerable quantities of S-layer proteins are released into the culture medium from *M. vannielii*, *M. thermolithotrophicus*, and *M. jannaschii* (between about 14% and 50% of the total S-layer protein). For *M. thermolithotrophicus* this commences early and continues throughout the growth phase (Fig. 7). A similar pattern was observed with the other two methanogens (data not shown).

Regulatory sequences controlling high-level gene expression

Analysis of the codon usage, ribosome-binding sites, and transcriptional promoters of S-layer genes indicates they are highly expressed (15). This is consistent with fast regeneration after S layers have been extracted by nonlethal methods. *M. voltae* protoplasts completely regenerated S layers within 60 min, although the mean generation time of growing cells is 16 h (43, 110). Large quantities of sheetlike S-layer patches were observed in the regeneration medium. Because of their efficient expression S-layer genes might be suitable for constructing vectors for heterologous overproduction of proteins.

Regulatory sequences involved in transcription (see Chapter 6) and translation (see Chapter 8) were examined in S-layer genes from members of the four genera, *Methanococcus*, *Methanothermococcus*, *Methanocaldococcus*, and *Methanotorris* (Table 4). For *M. jannaschii* genes, a TATA box and the "factor B recognition element" (BRE) are located 19 and 29 nucleotides upstream from the transcription start, respectively. The translation start and the Shine-Dalgarno sequence are located downstream from the transcription start site beginning at nucleotides 45 and 33, respectively. The TATA box perfectly matches the consensus sequence for methanogenic *Archaea* (142).

In contrast to *M. jannaschii*, several promotors have been described for the S-layer gene of *M. voltae* (65). The ribosome-binding site, usually localized 3 to 9 nucleotides upstream of the translation start site (30), was found to be complementary to a region at the 3' terminus of the 16 S rRNA of *M. jannaschii*. Translation is terminated with a series of stop codons, which is a common feature of methanogenic *Archaea* (30). They are followed by poly(A)/poly(T) sequence,

Table 3. S-layer genes from mesophilic and (hyper)thermophilic *Methanococcales*

Species	Opt. growth temp. (°C)	Gene	Nucleotide accession no.	Protein accession no.
Methanotorris igneus	88	*slmi 1*	AJ564995[a]	Q6KEQ4[b]
Methanocaldococcus jannaschii	85	*slmj 1*	AJ311636[a]	Q58232[b]
Methanothermococcus thermolithotrophicus	65	*slmt 1*	AJ308554[a]	Q8X235[b]
Methanococcus vannielii	37	*slmv 1*	AJ308553[a]	Q8X234[b]
Methanococcus voltae	37	*sla*	M59200[a]	Q50833[b]
Methanococcus maripaludis	37	*slp*	NC005791[c]	NP987503[c]

[a]EMBL Database.
[b]UniProt/Swiss Prot.
[c]GenBank/NCBI.

```
1    MAMSLXKIGAI/AVGGAMVASA/LASGVMA!ATTIG
2    MAMSLKKIGAI/AVGGAMVATA/LASGVAA!EVTTS
3    MAMSLKKIGAI/AVGGAMVASA/LASGVMA!ATTSG
4    MAMSMKKIGAI/AVGGAMVASA/LATGALA!AEKVG
5    MAMSLKKIGAI/AAGSAMVASA/LATGVFA!VEKIG
6    MAMSMKKIGAI/AVGGAMVASA/LATGAFA!AEKVG
     ****  ***** * ****** * ** * *
     n-region   h-region   c-region
```

Figure 6. Comparison of the leader peptides of the S-layer proteins of *Methanococcales*. (1) *M. igneus*; (2) *M. jannaschii*; (3) *M. thermolithotrophicus*; (4) *M. vannielii*; (5) *M. voltae*; (6) *M. maripaludis*.

which probably leads to formation of a hairpin structure and terminates transcription. Similar promoter regions were suggested for the other *Methanococcales*; however, it seems they are located at different positions within the gene sequence (Table 4).

Amino Acid Composition and Primary Structure

Molecular characteristics deduced from the gene sequences of S-layer proteins from *Methanococcales* are compiled in Table 5. They are all slightly acidic proteins with molecular masses ranging from 56 to 61 kDa.

Glycosylation

Thermal stabilization of S-layer proteins may be attributed to posttranslational modifications (e.g., glycosylation, phosphorylation), covalent cross-linking, or salt bridging (15, 40) (see Chapter 11). Glycosylation of S-layer proteins is in general well characterized (101), e.g., for the hyperthermophilic methanogenic species *M. fervidus* and *Methanothermus sociabilis* (17, 66, 108).

Table 4. Putative regulatory sequences for S-layer genes (25)

Region	Sequence	Position[a]
Promoter		
a. BRE box	-CGAAA-[b]	
M. jannaschii[c]	-CGTAA-	−33 to −29
M. igneus	-CGTAA-	−35 to −31
M. thermolithotrophicus	-CGTAA-	−35 to −31
M. vannielii	-CGTAA-	−40 to −36
M. voltae I	-CGTAA-	−34 to −30
M. voltae II	-CGTAA-	−33 to −29
b. TATA box	-AA/TTTATATA-[b]	
M. jannaschii	-TTTATATA-	−26 to −19
M. igneus	-TTTATATA-	−28 to −29
M. thermolithotrophicus	-TATATATA-	−28 to −21
M. vannielii	-TATAATAA-	−32 to −25
M. voltae I	-TATATATA-	−27 to −20
M. voltae II	-AATAAAA-	−26 to −19
c. Transcription start	-A/TTGC-[b]	
M. jannaschii	-ATAC-	1
M. igneus	-ATCG-	1
M. thermolithotrophicus	-ATCC-	1
M. vannielii	-ATAC-	1
M. voltae I	-ATTT-	1
M. voltae II	-ATAC-	1
Shine-Dalgarno sequence		
M. jannaschii	-AGGTGAT-	33–39
M. igneus	-AGGTGAT-	37–43
M. thermolithotrophicus	-AGGGTGA-	64–70
M. vannielii	-AGGTGAA-	60–66
M. voltae I	-AGGTGAT-	444–450
M. voltae II	-AGGTGAT-	204–210
Translation start		
M. jannaschii	-ATG-	45
M. igneus	-ATG-	49
M. thermolithotrophicus	-ATG-	77
M. vannielii	-ATG-	72
M. voltae I	-ATG-	456
M. voltae II	-ATG-	216

[a]Position relative to the transcription start.
[b]Consensus sequence.
[c]Full name of organisms: *Methanocaldococcus jannaschii*, *Methanotorris igneus*, *Methanothermococcus thermolithotrophicus*, *Methanococcus vannielii*, and *Methanococcus voltae*.

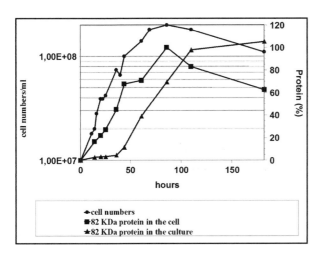

Figure 7. Distribution of S-layer protein in cells and in the culture medium during growth of *M. thermolithotrophicus*.

A larger number of *N*-glycan sites are predicted in the primary amino acid sequences of S-layer proteins from the hyperthermophilic *M. jannaschii* compared with their mesophilic relatives (2, 26, 27, 77). The same was found for the S-layer protein of the hyperthermophilic *M. igneus* (1, 25), suggesting a role for glycosylation in the thermostabilization of these proteins. Although conventional staining methods for the detection of glycoproteins (periodic acid-Schiff [PAS]) were negative, positive signals were obtained with more sensitive immuno blots (Fig. 8). In addition the S-layer proteins of *Methanococcales* revealed apparent higher molecular masses in SDS-polyacrylamide gel electrophoresis (PAGE) than expected from their gene sequences. Additional indirect evidence for posttrans-

Table 5. S-layer proteins from selected mesophilic and (hyper)thermophilic *Methanococcales*

Species[a]	Size (aa)	Molecular mass (daltons)	Isoelectric point	N-Glycosylation sites	Cys (mol %)	Ala (mol %)	Asp+Glu (mol %)	Arg+Lys (mol %)
M. igneus	519	55,669	4.68	8	0.4	9.4	15.4	11.2
M. jannaschii	558	60,547	4.27	8	0.4	9.9	20.4	10.8
M. thermolithotrophicus	559	59,225	4.30	5	–	12.2	18.2	10.2
M. vannielii	566	59,064	4.29	–	–	16.3	14.3	8.1
M. voltae	565	59,707	4.15	2	–	14.0	18.6	8.7
M. maripaludis	575	58,948	3.90	–	–	17.6	16.9	5.7

[a]Full name of organisms: *Methanocaldococcus jannaschii*, *Methanotorris igneus*, *Methanothermococcus thermolithotrophicus*, *Methanococcus vannielii*, *Methanococcus voltae*, and *Methanococcus maripaludis*.

lational modification is indicated by the smaller size of the S-layer protein from *M. jannaschii* when heterologously expressed in *Escherichia coli* (Fig. 9).

Recently, it has been demonstrated that flagellin proteins, and probably the S-layer protein of *M. voltae*, are glycosylated. The glycan structure of the flagellin elucidated by NMR analysis was shown to be a novel trisaccharide composed of β-Man*p*NAcA6Thr-(1-4)-β-Glc*p*NAc3NacA-(1-3)-β-Glc*p*Nac linked to Asn (145). This low degree of glycosylation might not be detected by less sensitive glycoprotein-staining methods.

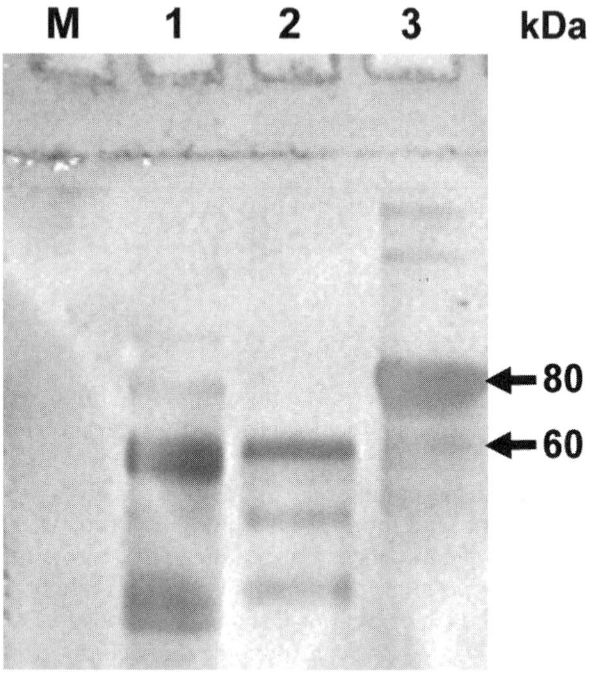

Thermostabilization

In addition to posttranslational modifications, intrinsic features of the polypeptide chain contribute to thermostabilization of proteins. Sequences of 115 proteins (S-layer proteins were not included) from *M. jannaschii* were compared with homologs from mesophilic *Methanococcales* (49). Characteristics of the proteins from thermophiles included higher residue volume and hydrophobicity, a higher percentage of charged amino acids (especially Glu, Lys, and Arg), and a lower percentage of uncharged polar residues (Ser, Thr, Asn, and Gln). In a similar study a large number of proteins from mesophilic and thermophilic to extreme thermophilic *Bacillus* and *Methanococcales*

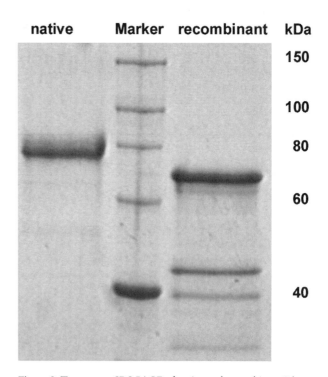

Figure 8. Immunoblot for glycoprotein detection. M, Marker proteins (negative control); S-layer proteins of (1) *M. thermolithotrophicus* (80 kDa); (2) *M. vannielii* (60 kDa); (3) *M. jannaschii* (80 kDa protein); S-layer proteins were electroeluted from SDS polyacrylamide gels and 5 µg applied to the gel for subsequent immunoblotting.

Figure 9. Ten percent SDS-PAGE of native and recombinant S-layer protein of *M. jannaschii*.

species were compared (95). An increase of Ile, Glu, Lys, and Arg and a decrease in Met, Asn, Gln, Ser, and Thr was observed in the thermophilic methanococcal proteins. In recent studies the complete genome sequences of mesophiles and thermophiles were analyzed (21). A large difference between the proportions of charged versus polar (noncharged) amino acids was found to be a common signature of all hyperthermophilic organisms. Ionic interactions may provide a mechanism for thermostabilization (49), and the proportional increase of oppositely charged residues in hyperthermophiles may provide a thermodynamic advantage due to the increased stability of coulombic interactions with temperature (21). Electrostatic interactions are important for S-layer stability in extreme halophiles (137) as well as in several bacteria (122, 123, 128). Sequences of putative soluble proteins from complete genomes of 8 thermophiles and 12 mesophiles were analyzed to gain insight into determinants of protein thermostability. The stabilizing factors include reduced protein size, increases in number of residues involved in hydrogen bonding, β-strand content, and helix stabilization through ion pairs. There are also significant increases in the relative amounts of charged and hydrophobic β-branched amino acids and decreases in uncharged polar amino acids in proteins from thermophiles relative to mesophilic organisms. Factors such as the relative proportion of residues in loops, proline and glycine content, and helix capping do not appear to be important (22).

Similarly, the histones from mesophilic (*Methanobacterium formicicum*), thermophilic (*Methanothermobacter thermautotrophicus*), and hyperthermophilic (*M. fervidus*) archaea, which have similar amino acid sequences but very different thermodynamic stabilities (93, 140), are believed to be stabilized by buried intramolecular arginine-aspartate interactions and intramolecular salt bridges on the surface of histone dimers. These structural features

are also present in S-layer proteins of *Methanococcales*. The S-layer proteins of the thermophilic and hyperthermophilic methanococci exhibited an increase in basic amino acids and a reduction of some amino acids with nonpolar side chains, e.g., alanine compared with their mesophilic counterparts (Table 5). The overall hydrophobicity is higher for the S-layer proteins from the mesophilic strains, indicating that it does not play a major role for adaptation to high temperatures in *Methanococcales*. As shown in Fig. 10, hydrophilic residues dominate in the S-layer protein of *M. jannaschii*, except in the N-terminal region, where hydrophobic residues may be involved in transmembrane transport or binding.

An increase of solvent-accessible surfaces in proteins of hyperthermophiles has been described (21). Thus, the increase in charged amino acids, especially lysine, in the S-layer proteins of *M. thermolithotrophicus*, *M. jannaschii*, and *M. igneus* could contribute to their increased thermal stability (2, 26). An increase in charged residues is present in the S-layer proteins of *M. mazei* (mesophilic) < *M. thermautotrophicus* (thermophilic) < *M. fervidus* (hyperthermophilic) and *A. fulgidus* (hyperthermophilic). The S-layer glycoprotein of *M. fervidus* contains high amounts of Asn and a basic isoelectric point (17). The relevance for the high amount of Asn is unknown.

Another significant feature of the S-layer proteins from the hyperthermophiles *M. jannaschii* and *M. igneus* is the occurrence of cysteine, which has only been detected in a few S-layer proteins (123, 128, 129). Intramolecular disulfide bridges may be another factor involved in the thermal stability of this surface protein.

Secondary Structures

A higher content of ordered structures (e.g., helical conformations), and fewer loops are predicted in

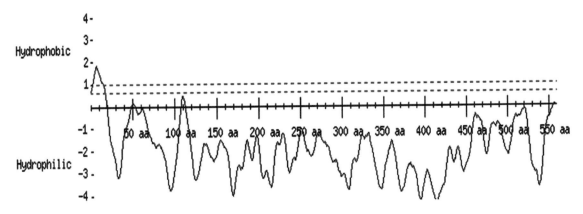

Figure 10. Hydropathy profile of the S-layer protein of *M. jannaschii*.

S-layer proteins from the mesophilic *M. voltae* and *M. vanniellii* compared with *M. thermolithotrophicus* and *M. jannaschii* (Table 6). In contrast to most common conceptions of factors determining protein thermostability, the relatively low extent of ordered secondary structures in the S-layer proteins of the hyperthermophilic members of the *Methanococcales* suggests that they are flexible molecules. This might explain their unusual behavior in SDS-PAGE, where they often appear in several conformational states (25). The migration of the purified S-layer protein from *M. jannaschii* varied in response to cations, pH, and temperature (Fig. 11). A similar temperature-dependent electrophoretic migration has been observed for the S-layer protein of the hyperthermophile *Nanoarchaeum equitans* (Schuster and Rachel, unpublished). Whether conformational adaptations of S-layer proteins may occur in vivo on the cell surface of *Archaea* living under extreme thermophilic conditions has to be investigated.

Sequence Comparison

In general, minor structural differences were observed in the primary and secondary structures of the S-layer proteins of mesophilic and hyperthermophilic *Methanococcales*. Most striking differences were found with respect to the occurrence of cysteine, the amount of basic amino acids residues, and the degree of hydrophobicity.

Apart from the leader peptides, sequence alignments revealed a notable degree of homology between the S-layer proteins of the mesophilic up to the (hyper)thermophilic *Methanococcales*, especially at the N and C termini (Table 7). The S-layer genes of the *Methanococcales* shared a significant homology with the presumptive S-layer genes of the hyperthermophilic heterotrophs *Pyrococcus abyssi* and *Pyrococcus horikoshii*. No significant similarity was found to any other archaeal S-layer protein.

Another group is formed by the S-layer proteins of *Methanosarcina mazei* and the gram-positive methanogens *M. thermautotrophicus*, *M. fervidus*, and *M. sociabilis*, which possess a significant degree of similarity to each other and to the sulfate-reducing hyperthermophilic *A. fulgidus*. The S-layer proteins of halobacteria and of the hyperthermophilic crenarchaeon *S. marinus* shared no homologies with the methanogens (26, 27).

TWO-LAYERED CELL WALLS CHARACTERIZE THE HYPERTHERMOPHILES *METHANOTHERMUS* AND *METHANOPYRUS*

The hyperthermophilic methanogens *M. fervidus* (134) and *Methanopyrus kandleri* (87) have double-layered cell envelopes that are recognizable in ultrathin sections (134). The pseudomurein has a thickness of ~15 to 20 nm and is covered by an external protein layer. This layer exhibits a hexagonal pattern with a center-to-center spacing of about 20 nm in freeze-etched cells of *M. fervidus* (Fig. 12a, b). In contrast, a crystalline arrangement of the outermost layer has not been observed for *Methanopyrus* cells (Fig. 12c, d). The S-layer glycoprotein of *M. fervidus* can be extracted with trichloroacetic acid and purified by reverse-phase chromatography using aqueous formic acid as eluant (17, 66, 108).

The mature protein consists of 593 amino acids. Compared with mesophilic S-layer proteins, it has a significantly higher content of isoleucine, asparagines, and cysteine, and a 14% higher content of β-sheet structure. The glycoprotein also possesses a leader peptide and 20 sequon structures (e.g., asparagine-X-serine/threonine) as potential N-glycosylation sites (17). One type of oligosaccharide is present and consists of D-3-O-MetMan, D-Man, and d-GalNAc (66). It is linked via N-acetylgalactosamine to asparagine residues of the peptide moiety (Fig. 13). In the biosynthesis of the glycan chains, nucleotide-activated oligosaccharides seem to be involved (50).

SULFATED PROTEOGLYCAN-LIKE S LAYERS IN NEUTROPHILIC HALOPHILES

The ultrastructure of the S layers of two genera of the order *Halobacteriales*, *Halobacterium* and *Haloferax*, has been studied in great detail (72, 74, 144). Members of both genera require similarly extreme osmotic conditions for growth. The surface glycoprotein of *H. salinarum* (former name: *H. halobium* [139]) was the first glycosylated protein detected in bacteria and archaea (97, 98). The S layers of both

Table 6. Secondary structures

Species[b]	Predicted structural features[a]		
	Helices	Sheets	Loops
M. voltae	36	27	36
M. vanniellii	45	19	36
M. thermolithotrophicus	27	28	46
M. jannaschii	22	25	51

[a]Predicted by the PHD program, represented at percentage of total structural features.
[b]Full name of organisms: *Methanocaldococcus jannaschii*, *Methanothermococcus thermolithotrophicus*, *Methanococcus vannielii*, and *Methanococcus voltae*.

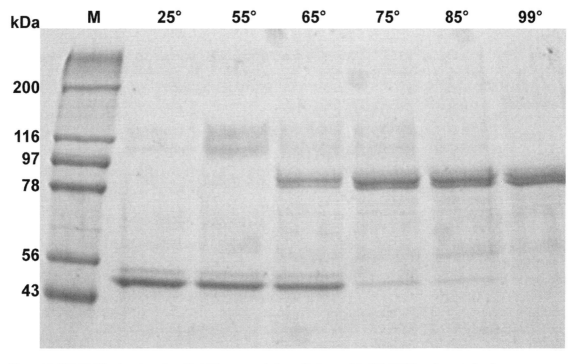

Figure 11. SDS-PAGE migration of purified S-layer protein from *M. jannaschii.* SDS-PAGE performed after heat treatment at different temperatures for 5 min in denaturing buffer.

microorganisms completely cover the cells. At present levels of resolution (2 nm in projection, ~2.5 to 3 nm in three-dimensional reconstructions), they show the same symmetry (p6) and center-to-center spacing of 15 nm and a high degree of similarity in the subunit organization of the oligomeric complexes (144). The width of the periplasmic space in both cases is about 10 nm.

The glycoprotein of *H. salinarum* possesses a stretch of 12 hydrophobic amino acids at the C terminus, which functions as a membrane anchor. Three different saccharide chains are linked to the peptide (Fig. 14). A sulfated pentasaccharide-repeating unit forms a glycosaminoglycan chain, which is linked to

the asparagine residue at the second N-terminal position of the peptide chain via an *N*-acetylgalactosamine. Ten sulfated glucose, glucuronic acid, and iduronic acid containing oligosaccharides are linked via glucose to an asparagine residue. In addition to the two types of *N*-linked glycan chains, about 15 *O*-glycosidically linked glucosylgalactose disaccharides occur. The disaccharides form a cluster close to the transmembrane domain. The glycoprotein of *H. salinarum* is acidic because of the occurrence of more than 20% aspartic and glutamic acid residues and up to 50 mol of uronic acids and 50 mol of sulfate residues per mol of peptide.

The mature polypeptide of the S-layer glycoprotein of *H. volcanii* is composed of 794 amino acids. Similar to the surface glycoprotein of *H. salinarum*, a hydrophobic stretch of about 20 amino acids at the C terminus probably functions as a transmembrane domain. Glucosyl-$(1\rightarrow2)$-galactose disaccharides are assumed to be linked to threonine residues clustering close to the membrane anchor. Sulfated and amino sugar-containing oligosaccharides are absent in the surface glycoprotein of *H. volcanii*. Newly synthesized S-layer glycoproteins of *H. volcanii* undergo a maturation step following translocation of the protein across the plasma membrane. The processing step is unaffected by inhibition of protein synthesis and is apparently unrelated to glycosylation of the protein (38). The S-layer proteins of moderate and

Table 7. Sequence homology of selected S-layer proteins[a]

Species	M. vol.	M. van.	M. lit.	M. jan.	M. ign.	P. aby.	P. hor.
1. *M. voltae*		44	48	38	31	23	28
2. *M. vannielii*	47		49	44	31	24	31
3. *M. thermolithotrophicum*	50	49		53	40	26	29
4. *M. jannaschii*	40	44	53		35	25	33
5. *M. igneus*	34	35	43	39		n.d.	n.d.
6. *P. abyssi*	24	25	26	26	n.d.		79
7. *P. horikoshii*	27	32	29	29	n.d.	79	

[a] Alignments of amino acid sequences (BLAST 2.0), represented as percent amino acid identity. 1, 2, *Methanococcus*; 3, *Methanothermococcus*; 4, *Methanocaldococcus*; 5, *Methanotorris*; 6, 7, *Pyrococcus*.

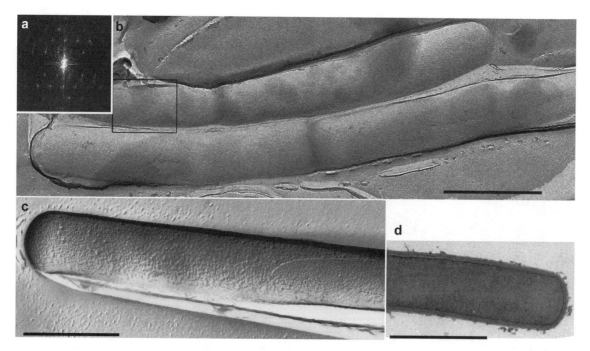

Figure 12. Transmission electron micrographs of *M. fervidus* and *M. kandleri* prepared by freeze etching. (a) Computed diffractogram, or power spectrum, of the image area marked in b. (b) *M. fervidus*, original micrograph of a freeze-etched cell. (c, d) *M. kandleri*. (c) Freeze-etched; (d) ultrathin section. All bars = 0.5 μm.

extreme halophilic archaea differ in the composition of the glycan chains (96).

To maintain a rod shape, halobacteria require a high concentration of extracellular NaCl (ca. >8% or >12%). Mg^{2+} ions may form salt bridges between sulfate and uronic acid residues of the oligosaccharides of S-layer glycoproteins of *H. salinarum* (156), because salt bridges are required for maintaining the integrity of the S-layer subunits. In the absence of Mg^{2+} ions, the S layer will disintegrate.

The biosynthesis of the glycan of the glycoprotein from *H. salinarum* has been studied in greater detail (90, 112, 136, 139, 152–156), while in the case of the neutrophilic methanogen *M. fervidus*, only activated oligosaccharides have been isolated from cell extracts (50, 78). In both species dolichol derivatives serve as lipid carriers, while in the case of the pseudomurein and methanochondroitin, undecaprenyl pyrophosphate functions as lipid carrier. According to the current knowledge, dolichyl-P-(P) is the universal lipid carrier in glycoprotein synthesis in *Archaea*, *Bacteria*, and *Eucarya* (78).

Depending on the glycan chain, C_{60}-dolichyl-monophosphate and dolichylpyrophosphate serve as the lipid carriers for the two sulfated glycan chains, the glycosaminoglycan and the second sulfated oligosaccharide, respectively. The complete glycosaminoglycan, including sulfation, is synthesized inside the cytoplasmic membrane at the lipid carrier dolychyl-P and then transferred to the nascent protein. The linkage to the protein takes place at the cell surface. The acceptor peptide is Asn-Ala-Ser. Replacement of the serine residue of the consensus sequence by valine, leucine, and asparagine did not prevent N-glycosylation. N-Glycosylation did not occur at Asn-479, when Ser-481 was removed (164), which indicates the presence of two different N-glycosyltransferases. In the case of the second sulfated oligosaccharide (c.f. above), completely sulfated lipid-linked precursors are formed before transfer to the protein. Prior to the transfer of this saccharide chain to the cell surface, some glucose residues are transiently methylated at carbon 3 (139, 156). Dolichol-linked oligosaccharides are also involved in the biosynthesis of the S-layer

α-D-3-O-MetManp-(1->6)-α-D-3-O-MetManp-[(1->2)-α-D-Manp]₃-(1->4)-D-GalNAc

Figure 13. Proposed structure of the oligosaccharide of the S-layer glycoprotein of *M. fervidus*. Modified from *The Journal of Biological Chemistry* (66) with permission of the publisher.

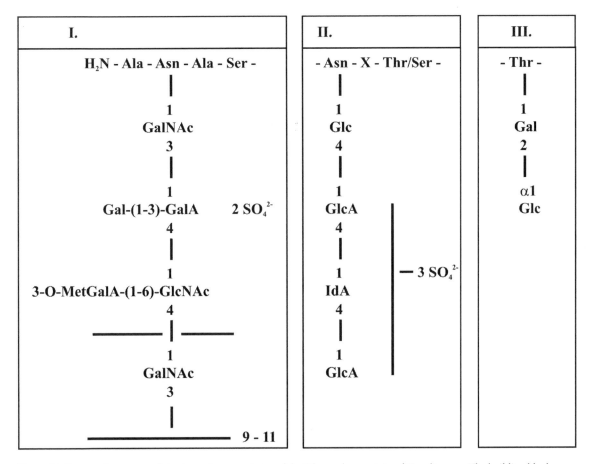

Figure 14. Proposed structure of the three glycan moieties of the S-layer glycoprotein of *H. salinarum*. The building block composed of five different sugars (glycan I) is linearly repeated 10 to 12 times. Modified from *Halophilic Bacteria*, vol. 2, CRC Press, Inc. (156), with permission of the publisher.

glycoprotein of *H. volcanii* (86). Along with protein glycosylation, additional posttranslational modification of the S-layer glycoprotein occurs at the external face of the plasma membrane of *H. volcanii* (38) (see Chapter 11).

Tunicamycin and bacitracin interfer with glycosylation in *H. salinarum*, while the glycan synthesis of the S-layer glycoprotein of *H. volcanii* is not affected (38). In contrast to *H. salinarum*, where C_{60}-dolichyl-monophosphate and dolichylpyrophosphate serve as the lipid carriers, only dolichyl monophosphate oligosaccharides were found in *H. volcanii* (86, 89). For comparison, *Sulfolobus acidocaldarius* is also sensitive to tunicamycin (38). Experiments with exogenously added peptides with an *N*-glycosylation site, which can not penetrate the cytoplasmic membrane, indicated an oligosaccharide transfer outside of the cytoplasmic membrane (89), because oligosaccharides were linked to these peptides by the cells. The proteins can also undergo further posttranslational modifications, which may include isoprenylation (83) or linkage of diphytanylglyceryl phosphate

residues (76). In addition to *H. salinarum* and *H. volcanii*, the S-layer protein from *Haloarcula japonica* has also been purified and heterologously expressed (103, 147, 148).

A glycoprotein S-layer with a three-dimensional structure and some properties similar to that of *Halobacteria* was observed and described for *Methanoplanus limicola* (24). Its dome-shaped complexes are arranged on a lattice with p6 symmetry and a center-to-center distance of 14.7 nm. In the cell envelope, they are in proximity to the cell membrane; the periplasm is 5 to 10 nm wide. The protein mass was determined to be at 135 kDa (native, glycosylated) and 115 kDa (after deglycosylation).

GLUTAMINYLGLYCAN IN NATRONOCOCCI

The majority of bacterial and archaeal exopolymers are polysaccharides, but exopolymers composed of L- or D-glutamate are also formed. In *Bacteria*, in general, α-linked polymers composed of glutamyl or glutaminyl residues possess the L-configuration, while in γ-linked exopeptides the glutamic acid residues have the D-configuration. The poly-γ-D-glutamyl polymers occur in the phylogenetically related genera *Bacillus*, *Sporosarcina*, and *Planococcus*. Similar polymers have been found in the archaeal genus *Natronococcus* (107), which are extremely halophilic cocci that grow optimally in alkaline and saline biotopes. The cell polymer of *N. occultus* is formed by a polyglutamine, which, in contrast to the bacterial polymer, is glycosylated. The cell wall polymer is composed of one amidated amino acid (L-glutamine), two amino sugars (glucosamine and galactosamine), two uronic acids (galacturonic acid and glucuronic acid), and a hexose sugar (glucose). In the intact polymer, the glutamine residues are linked via their γ-carboxylic group, forming a chain of about 60 monomers. Two types of oligosaccharides are linked to the polyglutamine backbone (Fig. 15). One oligosaccharide consists of an *N*-acetylglucosamine pentasaccharide at the reducing end and a galacturonic acid oligosaccharide at the nonreducing end. The second oligosaccharide has an *N*-acetylgalactosamine disaccharide at the reducing end and a maltose unit at the nonreducing end.

N-linked oligosaccharides of bacterial and archaeal cell wall glycoconjugates are usually linked via galactosamine, glucose, or rhamnose to the β-amide group of asparagine, while in *N. occultus* the oligosaccharides are linked to the α-carboxylic groups of the polyglutamine backbone via an *N*-amide linkage. Therefore, the exopolymer of *N. occultus* represents a novel type of naturally occurring glycoconjugates.

In *Bacillus anthracis* the polyglutamate polymer plays an important role in pathogenesis. It has not been determined whether the structurally related glutaminylglucan of the natronococci is a pathogenic factor.

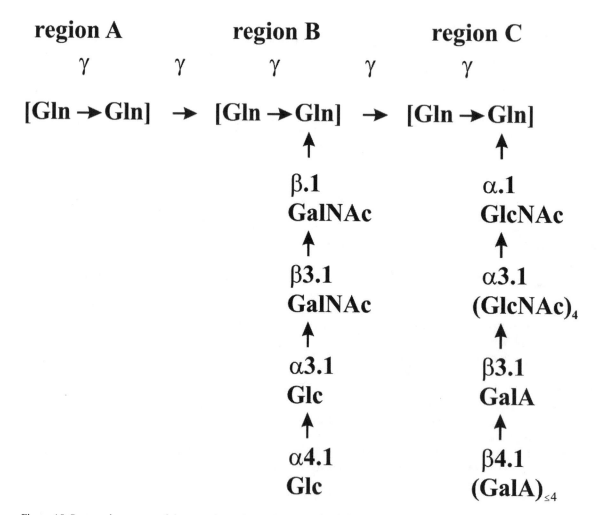

Figure 15. Proposed structure of the repeating units (regions A to C) of the cell wall polymer of *N. occultus*. Modified from the *European Journal of Biochemistry* (107) with permission of the publisher.

EXTRACELLULAR CANNULAE ARE A SPECIFIC FEATURE OF THE GENUS *PYRODICTIUM*

All known *Pyrodictium* species are able to grow at temperatures between 75 and 110°C under strictly anoxic conditions. The cells are covered by an S layer, i.e., a two-dimensional protein array, with hexagonal symmetry and a lattice constant of ~21 nm (Fig. 16a). The dome-shaped complexes are anchored to the cytoplasmic membrane by filiform stalks and span the periplasmic space with a constant width of approximately 35 nm (6, 37, 51). During growth on elemental sulfur, *Pyrodictium* cells form an extracellular matrix, in which the cells are entrapped. This matrix is an extended network of hollow cylinders consisting of helically arranged subunits (80, 121). Each of the cannulae has an outer diameter of 25 nm (Fig. 16b, c) and is made up of (at least) three homologous glycoproteins. In vivo observations of growing *Pyrodictium* cells at 90°C under anoxic conditions demonstrated that this network is dynamic; cell division and synthesis of the cannulae are directly linked (53). After cell division, the daughter cells remain interconnected by cannulae loops. Multiple generations result in the formation of a colony of cells entrapped in a dense cannulae network, in which each cell has connections with its neighbors. Analysis of dual-axis tilt series in cryoelectron tomography helped to reveal that the cannulae interconnect individual cells with each other, but only on the level of the periplasmic spaces of the cells; the cannulae do not enter the cytoplasms (Fig. 16c) (106).

THE OUTER MEMBRANE OF *IGNICOCCUS* AND THE SURFACE LAYER OF *NANOARCHAEUM*

From a sample taken at a hydrothermal vent of the Kolbeinsey Ridge, north of Iceland, a coculture of two coccoid, hyperthermophilic archaea was obtained (55). Based on 16S rRNA gene sequences, one belongs to the genus *Ignicoccus*, while the other was so different that it was attributed to a new phylum, *Nanoarchaeota*. Cells of *Ignicoccus* can be cultivated independently under strictly anaerobic conditions and thrive by sulfur-hydrogen autotrophy. However, *Nanoarchaeum equitans* can only be grown in coculture with, and in close contact to, cells of *Ignicoccus* sp. strain KIN4I. *Ignicoccus* cells are unique among the *Archaea* in two important ways: (i) they have a huge periplasm with variable width (20 to 500 nm), which contains vesicles; (ii) cells do not possess an S layer or any other cell wall polymer but have a unique outer membrane. Freeze-etch experiments revealed that the

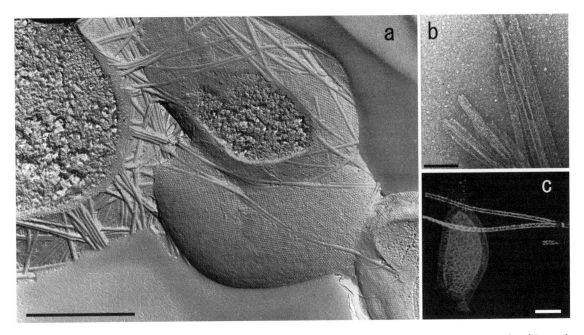

Figure 16. (*See the separate color insert for the color version of this illustration.*) Transmission electron micrographs of *P. occultum*. (a) *Pyrodictium* cells after freeze etching, exhibiting an S-layer with p6 symmetry. Bar = 0.5 μm. (b) Extracellular cannulae, a specific feature of the genus *Pyrodictium*; negative staining with uranyl acetate; bar = 0.1 μm. (c) 3D tomogram of a frozen-hydrated *Pyrodictium* cell with two cannulae; bar = 0.2 μm. Modified from the *Journal of Structural Biology* (106) with permission of the publisher.

Ignicoccus outer membrane fractures into two halves. This type of behavior is similar to the way in which many biological lipid membranes respond during freeze etching. The outer membrane is a highly dynamic structure: periplasmic vesicles can be observed in various stages of a fusion process, and vesicles are also released into the culture medium. The outer membrane is tightly packed with protein complexes, although there is no indication for crystallinity and the proteins are presently being investigated (104).

Ignicoccus sp. strain KIN4I and *N. equitans* both contain qualitatively identical amounts of glycerol ether lipids, archaeol and, to a minor degree, caldarchaeol (60). Cells of *N. equitans* are the smallest archaeal cocci presently known, and their ability to reproduce relies on the direct interaction with *Ignicoccus* cells. The reliance on a host is reflected in its genome sequence, which lacks genes for many important metabolic pathways (149). The ultrastructure of *N. equitans* is similar to many archaea. The cytoplasmic membrane is surrounded by a quasi-periplasmic space of constant width (20 nm) and covered by an S layer with sixfold symmetry and a lattice constant of about 15 nm (56). Future investigations of the unusual symbiosis of these two hyperthermophilic archaea aim at elucidating which proteins of both cell envelopes are directly involved in the physical interaction and in the exchange of metabolites from one cell to the other.

PERSPECTIVE: THE NEXT FIVE YEARS

The cell envelopes of the *Archaea* are often directly exposed to extreme environmental conditions, and they cannot be stabilized by cellular factors. Adaptation to the same type of extreme environment has not led to the evolution of similar cell surface structures. For example, *Halobacterium* sp., *Halococcus* sp., and *Natronococcus* sp. thrive in saturated salt brines, but they possess quite different cell wall polymers (e.g., glycoprotein, heterosaccharide, or glutaminylglycan). Differences in cell surface structures also exist for hyperthermophilic archaea. In this way extremophilic archaea provide a range of models for elucidating survival strategies of natural compounds and may give clues about molecular mechanisms of resistance against high temperature, low and high pH, and high-salt concentrations (41). Through studies of their cell surfaces, these investigations may lead to applications of new biomaterials. S layers represent the most common cell surface layer of *Archaea*. They are the simplest biological membranes found in nature. A wide spectrum of applications for S layers has already emerged (see Chapter 22). Isolated S-layer subunits assemble into monomolecular crystalline arrays in suspension, on surfaces and interfaces. These lattices have functional groups on the surface in an identical position and orientation in a nanometer range. These characteristics have led to their application as ultrafiltration membranes, immobilization matrices for functional molecules, affinity microcarriers and biosensors, conjugate vaccines, carriers for Langmuir-Blodgett films, and reconstituted biological membranes and pattering elements in molecular nanotechnology (130). In the past, applied studies have almost exclusively been performed with bacterial S layers. Since archaeal S-layer (glyco-)proteins are often resistant under extreme conditions, a new spectrum of future developments with archaeal S-layer glycoproteins should be found.

The genes of some archaeal S-layer proteins have been characterized (2, 17, 25–27, 88, 138, 161), and complete genome sequences have been published for many of the species described in this chapter (4, 19, 23, 28, 44, 46, 52, 68–70, 105, 127, 131, 149). To date, cell wall biosynthesis genes have not been unambiguously defined. Knowledge of the complete genome may be helpful for identifying special enzymes involved in the biosynthesis and degradation of cell wall polymers.

With respect to the evolution of methanococci, it seems that the common ancestor of the *Methanococcales* was probably a thermophile [75] and that mesophily is a modern adaptation. An exchange of an amino acid in mesophilic proteins may be the result of a relaxation of selection against this amino acid, which may be of importance in the extreme thermophilic counterparts.

The resolution of the 3D structure will be necessary to get a better knowledge of the molecular stabilization mechanisms of hyperthermophilic S-layer proteins. The first successful crystallizations indicate that this goal may be achievable (27, 32, 41, 113).

Acknowledgments. We thank the Deutsche Forschungsgemeinschaft, the "Ministerium für Bildung, Wissenschaft, Forschung und Technologie" (Bonn), and "Deutsches Zentrum für Luft- und Raumfahrt" (Bonn), Germany, for supporting this work and the European Space Agency for providing the flight opportunity for performing experiments under microgravity.

REFERENCES

1. Akca, E. 2004. Charakterisierung von S-Layer-Proteinen bei Prokaryoten. Ph.D. thesis. Johannes Gutenberg-Universität Mainz, Germany.

2. Akca, E., H. Claus, N. Schultz, G. Karbach, B. Schlott, T. Debaerdemaeker, J. P. Declercq, and H König. 2002. Genes and derived amino acid sequences of S-layer proteins from mesophilic, thermophilic and extremely methanococci. *Extremophiles* 6:351–358.

3. Albers, S.-V., and A. J. M. Driessen. 2002. Signal peptides of secreted proteins of the archaeon Sulfolobus solfataricus: a genomic survey. *Arch. Microbiol.* 177:209–216.

4. Baliga, N. S., R. Bonneau, M. T. Facciotti, M. Pan, G. Glusman, E. W. Deutsch, P. Shannon, Y. Chiu, R. S. Weng, R. R. Gan, P. Hung, S. V. Date, E. Marcotte, L. Hood, and W. V. Ng. 2004. Genome sequence of *Haloarcula marismortui*: a halophilic archaeon from the Dead Sea. *Genome Res.* **14:** 2221–2234.

5. Bardy, S. L., S. Y. M. Ng, and K. F. Jarrell. 2003. Prokaryotic motility structures. *Microbiology* **149:**295–304.

6. Baumeister, W., and G. Lembcke. 1992. Structural features of archaebacterial cell envelopes. *J. Bioenerg. Biomembr.* **24:** 567–575.

7. Baumeister, W., U. Santarius, S. Volker, R. Dürr, G. Lemcke, and H. Engelhardt. 1990. The surface protein of *Hyperthermus butylicus*: three dimensional structure and comparison with other archaebacterial surface proteins. *Syst. Appl. Microbiol.* **13:**105–111.

8. Baumeister, W., S. Volker, and U. Santarius. 1991. The three-dimensional structure of the surface layer protein of *Acidianus brierleyi* determined by electron crystallography. *Syst. Appl. Microbiol.* **14:**103–110.

9. Baumeister, W., I. Wildhaber, and B. M. Phipps. 1989. Principles and organization in eubacterial and archaebacterial surface proteins. *Can. J. Microbiol.* **35:**215–227.

10. Bendtsen, J. D., H. Nielsen, G. von Heijne, and S. Brunak. 2004. Improved prediction of signal peptides: SignalP 3.0. *J. Mol. Biol.* **16:**783–795.

11. Beveridge, T. J., and L. L. Graham. 1991. Surface layers of bacteria. *Microbiol. Rev.* **55:**684–705.

12. Beveridge, T. J., G. Southam, M. H. Jericho, and B. L. Blackford. 1990. High resolution topography of the S-layer sheath of the archaebacterium *Methanospirillum hungatei* provided by scanning tunneling microscopy *J. Bacteriol.* **172:**6589–6595.

13. Bintrim, S. B., T. J. Donohue, J. Handelsman, G. P. Roberts, and R. M. Goodman. 1997. Molecular phylogeny of *Archaea* from soil. *Proc. Natl. Acad. Sci. USA* **94:**277–282.

14. Blöchl, E., R. Rachel, S. Burggraf, D. Hafenbradl, H. W. Jannasch, and K. O. Stetter. 1997. *Pyrolobus fumarii*, gen. and sp.nov. represents a novel group of *Archaea*, extending the upper temperature limit for life to 113 °C. *Extremophiles* **1:**14–21

15. Boot, H. J., and P. H. Pouwels. 1996. Expression, secretion and antigenic variation of bacterial S-layer proteins. *Mol. Microbiol.* **21:**1117–1123.

16. Breitwieser, A., K. Gruber, and U. B. Sleytr. 1992. Evidence of an S-layer protein pool in the peptidoglycan of *Bacillus stearothermophilus*. *J. Bacteriol.* **174:**8008–8015.

17. Bröckl, G., M. Behr, S. Fabry, R. Hensel, H. Kaudewitz, E. Biendl, and H. König. 1991. Analysis and nucleotide sequence of the genes encoding the surface-layer glycoproteins of the hyperthermophilic methanogens *Methanothermus fervidus* and *Methanothermus sociabilis*. *Eur. J. Biochem.* **199:**147–152.

18. Brown, A. D., and K. J. Cho. 1970. The walls of extremely halophilic cocci. gram-positive bacteria lacking muramic acid. *J. Gen. Microbiol.* **62:**267–270.

19. Bult, C. J., O. White, G. J. Olsen, L. Zhou, R. D. Fleischmann, G. G. Sutton, J. A. Blake, L. M. FitzGerald, R. A. Clayton, J. D. Gocayne, A. R. Kerlavage, B. A. Dougherty, J.-F. Tomb, M. D. Adams, C. I. Reich, R. Overbeek, E. F. Kirkness, K. G. Weinstock, J. M. Merrick, A. Glodek, J. D. Scott, N. S. Geoghagen, J. F. Weidman, J. L. Fuhrmann, D. T. Nguyen, T. Utterback, J. M. Kelley, J. D. Peterson, P. W. Sadow, M. C. Hanna, M. D. Cotton, M. A. Hurst, K. M. Roberts, B. B. Kaine, M. Borodovsky, H. P. Klenk, C. M. Fraser, H. O. Smith, C. R. Woese, and J. C. Venter. 1996. Complete genome sequence of the methanogenic archaeon, *Methanococcus jannaschii*. *Science* **273:**1058–1073.

20. Burggraf, S., H. Fricke, A. Neuner, J. Kristjansson, P. Rouvier, L. Mandelco, C. R. Woese, and K. O. Stetter. 1990. *Methanococcus igneus* sp. nov., a novel hyperthermophilic methanogen from a shallow submarine hydrothermal system. *Syst. Appl. Microbiol.* **13:**263–269.

21. Cambillau, C., and J. M. Claverie. 2000. Structural and genomic correlates of hyperthermostabilty. *J. Biol. Chem.* **275:** 32383–32386.

22. Chakravarty, S., and R. Varadarajan. 2000. Elucidation of determinants of protein stability through genome sequence analysis. *FEBS Lett.* **470:**65–69.

23. Chen, L., K. Brugger, M. Skovgaard, P. Redder, Q. She, E. Torarinsson, B. Greve, M. Awayez, A. Zibat, H.-P. Klenk, and R. A. Garrett. 2005. The genome of *Sulfolobus acidocaldarius*, a model organism of the Crenarchaeota. *J. Bacteriol.* **187:**4992–4999.

24. Cheong, G.-W., Z. Cejka, J. Peters, K. O. Stetter, and W. Baumeister. 1991. The surface protein layer of *Methanoplanus limicola*: Three-dimensional structure and chemical characterization. *Syst. Appl. Microbiol.* **14:**209–217.

25. Claus, H., E. Akca, T. Debaerdemaeker, C. Evrard, J. P. Declercq, J. R. Harris, B. Schlott, and H. König. 2005. Molecular organization of selected prokaryotic S-layer proteins. *Can. J. Microbiol.* **51:**731–743.

26. Claus, H., E. Akca, T. Debaerdemaeker, C. Evrard, J. P. Declercq, and H. König. 2002. Primary structure of selected archaeal mesophilic and extremely thermophilic outer surface layer proteins. *Syst. Appl. Microbiol.* **25:** 3–12.

27. Claus, H., E. Akca, N. Schultz, G. Karbach, B. Schlott, T. Debaerdemaeker, J. P. Declercq, and H. König. 2001. Surface (glyco-)proteins: primary structure and crystallization under microgravity conditions, p. 806–809. *In* Proceedings of First European Workshop on Exo-/Astro-Biology, 21–23 May 2001, ESA SP-496. Frascati, Italy.

28. Cohen, G. N., V. Barbe, D. Flament, M. Galperin, R. Heilig, O. Lecompte, O. Poch, D. Prieur, J. Querellou, R. Ripp, J. C. Thierry, J. Van der Oost, J. Weissenbach, Y. Zivanovic, and P. Forterre. 2003. An integrated analysis of the genome of the hyperthermophilic archaeon *Pyrococcus abyssi*. *Mol. Microbiol.* **47:**1495–1512.

29. Correia, J., and K. F. Jarrell. 2000. Posttranslational processing of *Methanococcus voltae* preflagellin by preflagellin peptidases of M. *voltae* and other methanogens. *J. Bacteriol.* **182:**855–858.

30. Dalgaard, J. Z., and R. A. Garrett. 1993. Archaeal hyperthermophile genes, p. 535–563. *In* M. Kates, D. J. Kushner, and A. T. Matheson (ed.), *The Biochemistry of Archaea (Archaebacteria)*. Elsevier, Amsterdam.

31. Deatherage, J. F., K. A. Taylor, and L. A. Amos. 1983. Three-dimensional arrangement of the cell wall protein of *Sulfolobus acidocaldarius*. *J. Mol. Biol.* **167:**823–852.

32. Debaerdemaeker, T., C. Evrard, J. P. Declercq, H. Claus, E. Akca, and H. König. 2002. The first crystallization of the outer surface (S-layer) glycoprotein of the mesophilic bacterium *Bacillus sphaericus* and the hyperthermophilic archaeon *Methanothermus fervidus*, p. 411–412. *In* Proceedings of the Second European Workshop on Exo/Astro-Biology, 16–19 September 2002, ESA SP-518. Graz, Austria.

34. DeLong, E. F. 1992. *Archaea* in coastal marine environments. *Proc. Natl. Acad. Sci. USA* **89:**5685–5689.

35. DeLong, E. F., K. Y. Wu, B. B. Prezelin, and R. V. Jovine. 1994. High abundance of *Archaea* in Antarctic marine picoplancton. *Nature* **371:**695–697.

36. Dirmeier, R., M. Keller, D. Hafenbradl, F.-J. Braun, R. Rachel, S. Burggraf, and K. O. Stetter. 1998. *Thermococcus acidaminovorans* sp. nov., a new hyperthermophilic

alcalophilic archaeon growing on amino acids. *Extremophiles* 2:109–114.

37. **Dürr, R., R. Hegerl, S. Volker, U. Santarius, and W. Baumeister.** 1991. Three-dimensional reconstruction of the surface protein of *Pyrodictium brockii*: comparing two image processing strategies. *J. Struct. Biol.* 106:181–190.

38. **Eichler, J.** 2001. Post-translational modification of the S-layer glycoprotein occurs following translocation across the plasma membrane of the haloarchaeon *Haloferax volcanii*. *Eur. J. Biochem.* 268:4366–4373.

39. **Eichler, J.** 2003. Facing extremes: archaeal surface-layer (glyco)proteins. *Microbiology* 149:3347–3351.

40. **Engelhardt, H., and J. Peters.** 1998. Structural research on surface layers: a focus on stability, surface layer homology domains, and surface layer-cell wall interactions. *J. Struct. Biol.* 124:276–302.

41. **Evrard, C., J. P. Declercq, T. Debaerdemaeker, and H. König.** 1999. The first successful crystallization of a prokaryotic extremely thermophilic outer surface layer glycoprotein. *Z. Kristallogr.* 214:427–429.

42. **Fernandez, L. A., and J. Berenguer.** 2000. Secretion and assembly of regular surface structures in gram-negative bacteria. *FEMS Microbiol. Rev.* 24:21–44.

43. **Firtel, M., G. B. Patel, and T. Beveridge.** 1995. S-layer regeneration in *Methanococcus voltae* protoplasts. *J. Microbiol.* 14:817–824.

44. **Fitz-Gibbon, S. T., H. Ladner, U.-J. Kim, K. O. Stetter, M. I. Simon, and J. H. Miller.** 2002. Genome sequence of the hyperthermophilic crenarchaeon *Pyrobaculum aerophilum*. *Proc. Natl. Acad. Sci. USA* 99:984–989.

45. **Fuchs, T., H. Huber, K. Teiner, S. Burggraf, and K. O. Stetter.** 1995. *Metallosphaera prunae*, sp.nov., a novel metal-mobilizing, thermoacidophilic archaeum, isolated from a Uranium mine in Germany. *Syst. Appl. Microbiol.* 18:560–566.

46. **Fütterer, O., A. Angelov, H. Liesegang, G. Gottschalk, C. Schleper, B. Schepers, C. Dock, G. Antranikian, and W. Liebl.** 2004. Genome sequence of *Picrophilus torridus* and its implications for life around pH 0. *Proc. Natl. Acad. Sci. USA* 101:9091–9096.

47. **Grogan, D. W.** 1996. Organization and interactions of cell envelope proteins of the extreme thermoacidophile *Sulfolobus acidocaldarius*. *Can. J. Microbiol.* 42:1163–1171.

48. **Hafenbradl, D., M. Keller, R. Dirmeier, R. Rachel, P. Roßnagel, S. Burggraf, H. Huber, and K. O. Stetter.** 1996. *Ferroglobus placidus* gen. nov., sp. nov., a novel hyperthermophilic archaeum that oxidizes Fe^{2+} at neutral pH under anoxic conditions. *Arch. Microbiol.* 166:308–314.

49. **Haney, P. J., J. H. Badger, G. L. Buldak, C. I. Reich, C. R. Woese, and G. J. Olsen.** 1999. Thermal adaption analyzed by comparison of protein sequences from mesophilic and extremely thermophilic *Methanococcus* species. *Proc. Natl. Acad. Sci. USA* 36:3578–3583.

50. **Hartmann. E., and H. König.** 1989. Uridine and dolichyl diphosphate activated oligosaccharide intermediates are involved in the biosynthesis of the surface layer glycoprotein of *Methanothermus fervidus*. *Arch. Microbiol.* 151:274–281.

51. **Hegerl, R., and W. Baumeister.** 1988. Correlation averaging of a badly distorted lattice: the surface protein of *Pyrodictium occultum*. *Electron. Microsc. Tech.* 9:413–419.

52. **Hendrickson, E. L., R. Kaul, Y. Zhou, D. Bovee, P. Chapman, J. Chung, E. Conway de Macario, J. A. Dodsworth, W. Gillett, D. E. Graham, M. Hackett, A. K. Haydock, A. Kang, M. L. Land, R. Levy, T. J. Lie, T. A. Major, B. C. Moore, I. Porat, A. Palmeiri, G. Rouse, C. S. Saenphimmachak, S. Van Dien, T. Wang, W. B. Whitman, Q. Xia, Y. Zhang, F. W. Larimer, M. V. Olson, and J. A. Leigh.** 2004. Complete genome sequence of the genetically tractable hydrogenotrophic methanogen *Methanococcus maripaludis*. *J. Bacteriol.* 186:6956–6969.

53. **Horn, C., B. Paulmann, B. Kerlen, N. Junker, and H. Huber.** 1999. *In vivo* observation of cell division of anaerobic hyperthermophiles by using a high-intensity dark-field microscope. *J. Bacteriol.* 181:5114–5118.

54. **Houwink, A. L.** 1956. Flagella, gas vacuoles and cell-wall structure in *Halobacterium halobium*; an electron microscope study. *J. Gen. Microbiol.* 15:146–150.

55. **Huber, H., M. J. Hohn, R. Rachel, T. Fuchs, V. C. Wimmer, and K. O. Stetter.** 2002. A new phylum of *Archaea* represented by a nanosized hyperthermophilic symbiont. *Nature* 417:63–67.

56. **Huber, H., M. J. Hohn, K. O. Stetter, and R. Rachel.** 2003. The phylum *Nanoarchaeota*: present knowledge and future perspectives of a unique form of life. *Res. Microbiol.* 154:165–171.

57. **Huber, H., H. Jannasch, R. Rachel, T. Fuchs, and K. O. Stetter.** 1997. *Archaeoglobus veneficus* sp. nov., a novel facultative chemolithoautotrophic hyperthermophilic sulfite reducer, isolated from abyssal black smokers. *Syst. Appl. Microbiol.* 20:374–380.

58. **Huber, R., J. Stöhr, S. Hohenhaus, R. Rachel, S. Burggraf, H. W. Jannasch, and K. O. Stetter.** 1995. *Thermococcus chitonophagus* sp. nov., a novel, chitin-degrading hyperthermophilic archaeum from a deep-sea hydrothermal vent environment. *Arch. Microbiol.* 164:255–264.

59. **Huser, B. A., K. Wuhrmann, and A. J. B. Zehnder.** 1982. *Methanotrix soehngenii* gen. nov. sp. nov., a novel acetotrophic non-hydrogen-oxidizing methane bacterium. *Arch. Microbiol.* 132:1–9.

60. **Jahn, U., R. Summons, H. Sturt, E. Grossjean, and H. Huber.** 2004. Composition and source of the lipids of *Nanoarchaeum equitans* and their origin in the cytoplasmic membrane of its host *Ignicoccus* sp. KIN4I. *Arch. Microbiol.* 182:404–413.

61. **Kandler, O., and H. Hippe.** 1977. Lack of peptidoglycan in the cell walls of *Methanosarcina barkeri*. *Arch. Microbiol.* 113:57–60.

62. **Kandler, O., and H. König.** 1978. Chemical composition of the peptidoglycan-free cell walls of methanogenic bacteria. *Arch. Microbiol.* 118:141–152.

63. **Kandler, O., and H. König.** 1985. Cell envelopes of archaebacteria, p. 413–457. *In*: C. R. Woese and R. S. Wolfe. (ed.), *The Bacteria. A Treatise on Structure and Function. Archaebacteria*, vol. VIII. Academic Press, New York.

64. **Kandler, O., and H. König.** 1993. Cell envelopes of *Archaea*: structure and chemistry, p. 223–259. *In* M. Kates et al. (ed.), *The Biochemistry of Archaea (Archaebacteria)*. Elsevier Science, New York, N.Y.

65. **Kansy, W. J., M. E. Carinato, L. M. Monteggia, and J. Konisky.** 1994. In vivo transcripts of the S-layer encoding structural gene of the archaeon *Methanococcus voltae*. *Gene* 148:131–135.

66. **Kärcher, U., H. Schröder, E. Haslinger, G. Allmeier, R. Schreiner, F. Wieland, A Haselbeck, and H. König.** 1993. Primary structure of the heterosaccharide of the surface glycoprotein of *Methanothermus fervidus*. *J. Biol. Chem.* 268:26821–26826.

67. **Karner, M. B., E. F. DeLong, and D. M. Karl.** 2001. Archaeal dominance in the mesopelagic zone of the Pacific Ocean. *Nature* 409:507–510.

68. **Kawarabayasi, Y., Y. Hino, H. Horikawa, K. Jin-no, M. Takahashi, M. Sekine, S. Baba, A. Ankai, H. Kosugi, A. Hosoyama, S. Fukui, Y. Nagai, K. Nishijima, R. Otsuka,**

H. Nakazawa, M. Takamiya, Y. Kato, T. Yoshizawa, T. Tanaka, Y. Kudoh, J. Yamazaki, N. Kushida, A. Oguchi, K. Aoki, S. Masuda, M. Yanagii, M. Nishimura, A. Yamagishi, T. Oshima, and H. Kikuchi. 2001. Complete genome sequence of an aerobic thermoacidophilic crenarchaeon, *Sulfolobus tokodaii* strain7. *DNA Res.* 8:123–140.

69. Kawarabayasi, Y., Y. Hino, H. Horikawa, S. Yamazaki, Y. Haikawa, K. Jin-no, M. Takahashi, M. Sekine, S. Baba, A. Ankai, H. Kosugi, A. Hosoyama, S. Fukui, Y. Nagai, K. Nishijima, H. Nakazawa, M. Takamiya, S. Masuda, T. Funahashi, T. Tanaka, Y. Kudoh, J. Yamazaki, N. Kushida, A. Oguchi, K. Aoki, K. Kubota, Y. Nakamura, N. Nomura, Y. Sako, and H. Kikuchi. 1999. Complete genome sequence of an aerobic hyperthermophilic crenarchaeon, *Aeropyrum pernix* K1. *DNA Res.* 6: 83–101.

70. Kawarabayasi, Y., M. Sawada, H. Horikawa, Y. Haikawa, Y. Hino, S. Yamamoto, M. Sekine, S. Baba, H. Kosugi, A. Hosoyama, Y. Nagai, M. Sakai, K. Ogura, R. Otuka, H. Nakazawa, M. Takamiya, Y. Ohfuku, T. Funahashi, T. Tanaka, Y. Kudoh, J. Yamazaki, N. Kushida, A. Oguchi, K. Aoki, Y. Nakamura, T. F. Robb, K. Horikoshi, Y. Masuchi, H. Shizuya, and H. Kikuchi. 1998. Complete sequence and gene organization of the genome of a hyper-thermophilic archaebacterium, *Pyrococcus horikoshii* OT3. *DNA Res.* 5:55–76.

71. Keller, M., F.-J. Braun, R. Dirmeier, D. Hafenbradl, S. Burggraf, R. Rachel, and K.O. Stetter. 1995. *Thermococcus alcaliphilus*, sp.nov., a new hyperthermophilic archaeum growing on polysulfide at alkaline pH. *Arch. Microbiol.* 164:390–395.

72. Kessel, M., E. L. Buhle, Jr., S. Cohen, and U. Aebi. 1988. The cell wall structure of a magnesium-dependent halobacterium, *Halobacterium volcanii* CD-2, from the Dead Sea. *J. Ultrastruct. Mol. Struct. Res.* 100:94–106.

73. Kessel, M., S. Volker, U. Santarius, R. Huber and W. Baumeister. 1990. Three-dimensional reconstruction of the surface protein of the extremely thermophilic archaebacterium *Archaeoglobus fulgidus*. *Syst. Appl. Microbiol.* 13: 207–213.

74. Kessel, M., I. Wildhaber, S. Cohen, and W. Baumeister. 1988. Three-dimensional structure of the regular surface glycoprotein layer of *Halobacterium volcanii* from the Dead Sea. *EMBO J.* 7:1549–1554.

75. Keswani, J., S. Orkand, U. Premachandran, L. Mandelco, M. J. Franklin, and W. B. Whitman. 1996. Phylogeny and taxonomy of mesophilic *Methanococcus spp.* and comparison of rRNA, DNA hybridization, and phenotypic methods. *Int. J. Syst. Bacteriol.* 46:727–735.

76. Kikuchi, A., H. Sagami, and K. Ogura. 1999. Evidence for the covalent attachment of diphytanylglycerol phosphate to the cell-surface glycoprotein of *Halobacterium halobium*. *J. Biol. Chem.* 274:18011–18016.

77. König, H., H. Claus, and E. Akca. 2004. Cell wall structures of mesophilic, thermophilic and hyperthermophilic *Archaea*, p. 281–298. *In* J. Seckbach (ed.), *Origins*, Kluwer Academic, Dordrecht, The Netherlands.

78. König, H., E. Hartmann, and U. Kärcher. 1994. Pathways and principles of the biosynthesis of methanobacterial cell wall polymers. *Syst. Appl. Microbiol.* 16:510–517.

79. König, H., and P. Messner (ed.). 1997. International Workshop on Structure, Biochemistry, Molecular Biology and Applications of Microbial S-Layers (Special Issue). *FEMS Microbiol. Rev.* 20:1–178.

80. König, H., P. Messner, and K. O. Stetter. 1988. The fine structure of the fibers of *Pyrodictium occultum*. *FEMS Microbiol. Lett.* 49:207–212.

81. König, H., and K. O. Stetter. 1986. Studies on archaebacterial S-layers. *Syst. Appl. Microbiol.* 7:300–309.

82. Konisky, J., D. Lynn, M. Hoppert, F. Mayer, and P. Haney. 1994. Identification of the *Methanococcus voltae* S-layer structure gene. *J. Bacteriol.* 176:1790–1792.

83. Konrad, Z., and J. Eichler. 2002. Lipid modification of proteins in *Archaea*: attachment of a mevalonic acid-based lipid moiety to the S-layer glycoprotein of *Haloferax volcanii* follows protein translocation. *Biochem. J.* 366:959–964.

84. Kostyukova, A., G. Gongadze, Y. Polosina, E. Bonch-Osmolovskaya, M. Miroshnichenko, N. Chernyh, M. Obraztsova, V. Svetlichny, P. Messner, U. Sleytr, S. L'Haridon, C. Jeanthon, and D. Prieur. 1999. Investigation of structure and antigenic capacities of Thermococcales cell envelopes and reclassification of "*Caldococcus litoralis*" Z-1301 as *Thermococcus litoralis* Z-1301. *Extremophiles* 3:239

85. Koval, S., and K. Jarrell. 1987. Ultrastructure and biochemistry of the cell wall of *Methanococcus voltae*. *J. Bacteriol.* 169:1298–1306

86. Kuntz, C., J. Sonnenbichler, I. Sonnenbichler, M. Sumper, and R. Zeitler. 1997. Isolation and characterization of dolichol-linked oligosaccharides from *Haloferax volcanii*. *Glycobiology* 7:897–904.

87. Kurr, M., R. Huber, H. König, H. W. Jannasch, H. Fricke, A. Trincone, J. K. Kristjannson, and K. O. Stetter. 1991. *Methanopyrus kandleri*, gen. and sp. nov. represents a novel group of hyperthermophilic methanogens, growing at 110° C. *Arch. Microbiol.* 156:239–247.

88. Lechner, J., and Sumper. 1987. The primary structure of a procaryotic glycoprotein. Cloning and sequencing of the cell surface glycoprotein gene of halobacteria. *J. Biol. Chem.* 262: 9724–9729.

89. Lechner, J., and F. Wieland. 1989. Structure and biosynthesis of prokaryotic glycoproteins. *Annu. Rev. Biochem.* 58: 173–194.

90. Lechner, J., F. Wieland, and M. Sumper.1985. Biosynthesis of sulfated saccharides N-glycosidically linked to the protein via glucose. Purification and identification of sulfated dolichyl monophosphoryl tetrasaccharides from halobacteria. *J. Biol. Chem.* 260:860–866.

91. Lembcke, G., W. Baumeister, E. Beckmann, and F. Zemlin. 1993. Cryo-electron microscopy of the surface protein of *Sulfolobus shibatae*. *Ultramicroscopy* 49:397–406.

92. Lembcke, G., R. Dürr, R. Hegerl, and W. Baumeister. 1990. Image analysis and processing of an imperfect two-dimensional crystal. The surface layer of *Sulfolobus acidocaldarius* reinvestigated. *J. Microsc.* 161:263–278.

93. Li, W.T., J. W. Shriver, and J. N. Reeve. 2000. Mutational analysis of differences in thermostability between histones from mesophilic and hyperthermophilic *Archaea*. *J. Bacteriol.* 182:812–817.

94. Mayr, J., A. Lupas, J. Kellermann, C. Eckerskorn, W. Baumeister, and J. Peters. 1996. A hyperthermostable protease of the subtilisin family bound to the surface layer of the archaeon *Staphylothermus marinus*. *Curr. Biol.* 6:739–749.

95. McDonald, J. H., A. M. Grasso, and L. K. Rejto. 1999. Patterns of temperature adaptation in proteins from *Methanococcus* and *Bacillus*. *Mol. Biol. Evol.* 16:1785–1790.

96. Mengele, R., and M. Sumper. 1992. Drastic differences in glycosylation of related S-layer glycoproteins from moderate and extreme halophiles. *J. Biol. Chem.* 267:8182–8185.

97. Mescher, M. F., and J. L. Strominger. 1976. Structural (shape-maintaining) role of the cell surface glycoprotein of *Halobacterium salinarium*. *Proc. Natl. Acad. Sci. USA* 73:2687–2691.

98. Mescher, M. F., and J. L. Strominger. 1976. Purification and characterisation of a prokaryotic glycoprotein from the cell

envelope of *Halobacterium salinarium. J. Biol. Chem.* **251:** 2005–2014.

99. **Messner, P., D. Pum, M. Sara, K. O. Stetter, and U. B. Sleytr.** 1986. Ultrastructure of the cell envelope of the Archaebacteria *Thermoproteus tenax* and *Thermoproteus neutrophilus. J. Bacteriol.* **166:**1046–1054.

100. **Messner, P., and C. Schäffer.** 2003. Prokaryotic glycoproteins, p. 51–124. *In* H. Herz, H. Falk, and G. W. Kirby (ed.), *Progress in the Chemistry of Organic Natural Products.* Springer Verlag, Heidelberg, Germany.

101. **Messner, P., and U. B. Sleytr.** 1992. Crystalline bacterial cell-surface layers. *Adv. Microbial. Physiol.* **33:**213–274.

102. **Michel, H., D.-C. Neugebauer, and D. Oesterhelt.** 1980. The 2-d crystalline cell wall of *Sulfolobus acidocaldarius:* Structure, solubilization, and reassembly, p. 27–35. *In* W. Baumeister and W. Vogell (ed.), *Electron Microscopy at Molecular Dimensions,* Springer-Verlag, Berlin, Germany.

103. **Nakamura, S., S. Mizutani, H. Wakai, H. Kawasaki, R. Aono, and K. Horikoshi.** 1995. Purification and partial characterization of the cell surface glycoprotein from extremely halophilic *Archaeon Haloarcula japonica* strain TR-1. *Biotechnol. Lett.* **17:**705–706.

104. **Näther, D.J., and R. Rachel.** 2004. The outer membrane of the hyperthermophilic archaeon *Ignicoccus:* dynamics, ultrastructure and composition. *Biochem. Soc. Trans.* **32:**199–203.

105. **Ng, W. V., S. P. Kennedy, G. G. Mahairas, B. Berquist, M. Pan, H. D. Shukla, S. R. Lasky, N. Baliga, V. Thorsson, J. Sbrogna, S. Swartzell, D. Weir, J. Hall, T. A. Dahl, R. Welti, Y. A. Goo, B. Leithauser, K. Keller, R. Cruz, M. J. Danson, D. W. Hough, D. G. Maddocks, P. E. Jablonski, M. P. Krebs, C. M. Angevine, H. Dale, T. A. Isenbarger, R. F. Peck, M. Pohlschrod, J. L. Spudich, K.-H. Jung, M. Alam, T. Freitas, S. Hou, C. J. Daniels, P. P. Dennis, A. D. Omer, H. Ebhardt, T. M. Lowe, P. Liang, M. Riley, L. Hood, and S. DasSarma.** 2000. Genome sequence of *Halobacterium* species NRC-1. *Proc. Natl. Acad. Sci. USA* **97:**12176–12181.

106. **Nickell, S., R. Hegerl, W. Baumeister, and R. Rachel.** 2003. *Pyrodictium* cannulae enter the periplasmic space but do not enter the cytoplasm, as revealed by cryo-electron tomography. *J. Struct. Biol.* **141:**34–42.

107. **Niemetz, R., U. Kärcher, O. Kandler, B. Tindall, and H. König.** 1997. The cell wall polymer of the extremely halophilic archaeon *Natronococcus occultus. Eur. J. Biochem.* **249:**905–911.

108. **Nußer, E., E. Hartmann, H. Allmeier, H. König, G. Paul, and K. O. Stetter.** 1988. A glycoprotein surface layer covers the pseudomurein sacculus of the extreme thermophile *Methanothermus fervidus,* p. 21–25. *In* U. B. Sleytr P. Messner, D. Pum, and M. Sàra (ed.), *Crystalline Bacterial Cell Surface Layers.* Springer Verlag, Berlin, Germany.

109. **Nußer, E., and H. Köng.** 1987. S-layer studies on three species of *Methanococcus* living at different temperatures. *Can. J. Microbiol.* **33:**256–261.

110. **Patel, G. B., C. G. Choquet, J. H. E. Nash, and G. D. Sprott.** 1993. Formation and regeneration of *Methanococcus voltae* protoplasts. *Appl. Environ. Microbiol.* **59:**27–33.

111. **Patel, G. B., and G. O. Sprott.** 1990. *Methanosaeta concilii* gen. nov., sp. nov. ("*Methanothrix concilii*") and *Methanosaeta thermoacetophila* nom. rev., comb. nov. *Int. J. Bacteriol.* **40:**79–82.

112. **Paul, G., F. Lottspeich, and F. Wieland.** 1986. Asparaginyl-N-acetylgalactosamine. Linkage unit of halobacterial glycosaminoglycan. *J. Biol. Chem.* **261:**1020–1024.

113. **Pavkov, T., M. Oberer, E. M. Egelseer, M. Sara, U. B. Sleytr, and W. Keller.** 2003. Crystallization and preliminary structure determination of the C-terminal truncated domain of the S-layer protein SbsC. *Acta Crystallogr D Biol Crystallogr.* **59:**1466–1468.

114. **Pellerin, P., B. Fournet, and P. Debeire.** 1990. Evidence for the glycoprotein nature of the cell sheath of *Methanosaeta*-like cells in the culture of *Methanothrix soehngenii* strain FE. *Can. J. Microbiol.* **36:**631–636.

115. **Peters, J., W. Baumeister, and A. Lupas.** 1996. Hyperthermostable surface layer protein tetrabrachion from the archaebacterium *Staphylothermus marinus:* evidence for the presence of a right-handed coiled coil derived from the primary structure. *J. Mol. Biol.* **257:**1031–1041.

116. **Peters, J., M. Nitsch, B. Kühlmorgen, R. Golbik, A.Lupas, J. Kellermann, H. Engelhardt, J.-P. Pfander, S. Müller, K. Goldie, A. Engel, K. O. Stetter, and W. Baumeister.** 1995. Tetrabrachion: a filamentous archaebacterial surface protein assembly of unusual structure and extreme stability. *J. Mol. Biol.* **245:**385–401.

117. **Phipps, B. M., H. Engelhardt, R. Huber, and W. Baumeister.** 1990. Three-dimensional structure of the crystalline protein envelope layer of the hyperthermophilic archaebacterium *Pyrobaculum islandicum. J. Struct. Biol.* **103:**152–163.

118. **Phipps, B. M., R. Huber, and W. Baumeister.** 1991. The cell envelope of the hyperthermophilic archaebacterium *Pyrobaculum organotrophum* consists of two regularly arrayed protein layers: three-dimensional structure of the outer layer. *Mol. Microbiol.* **5:**253–265.

119. **Prüschenk, R., and W. Baumeister.** 1987. Three-dimensional structure of the surface protein of *Sulfolobus solfataricus. Eur. J. Cell Biol.* **45:**185–191.

120. **Rachel, R., I. Wyschkony, S. Riehl, and H. Huber.** 2002. The ultrastructure of *Ignicoccus:* Evidence for a novel outer membrane and for intracellular vesicle budding in an archaeon. *Archaea* **1:**9–18.

121. **Rieger, G, R. Rachel, R. Hermann, and K. O. Stetter.** 1995. Ultrastructure of the hyperthermophilic archaeon *Pyrodictium abyssi. J. Struct. Biol.* **115:**78–87.

122. **Sára, M., and U. B. Sleytr.** 1987. Charge distribution on the S-layer of *Bacillus stearothermophilus* NRS 1536/3c and importance of charged groups for morphogenesis and function. *J. Bacteriol.* **169:**2804–2809.

123. **Sára, M., and U. B. Sleytr.** 2000. S-layer proteins. *J. Bacteriol.* **182:**859–868.

124. **Schleper, C., W. Holben, and H. P. Klenk.** 1997. Recovery of crenarchaeotal ribosomal DNA sequences from freshwater-lake sediments. *Appl. Environm. Microbiol.* **63:**321–323.

125. **Schleper, C., G. Puehler, I. Holz, A. Gambacorta, D. Janekovic, U. Santarius, H.-P. Klenk, and W. Zillig.** 1995. *Picrophilus* gen. nov., fam. nov.: a novel aerobic, heterotrophic, thermoacidophilic genus and family comprising *Archaea* capable of growth around pH 0. *J. Bacteriol.* **177:**7050–7059.

126. **Shaw, P.J., G. J. Hills, J. A. Henwood, J. E. Harris, and O. B. Archer.** 1985. Three-dimensional architecture of the cell sheath and septa of *Methanospirillum hungatei. J. Bacteriol.* **161:**750–757.

127. **She, Q., R. K. Singh, F. Confalonieri, Y. Zivanovic, G. Allard, M. J. Awayez, C. C. Chan-Weiher, I. G. Clausen, B. A. Curtis, A. De Moors, G. Erauso, C. Fletcher, P. M. Gordon, I. Heikamp-de Jong, A. C. Jeffries, C. J. Kozera, N. Medina, X. Peng, H. P. Thi-Ngoc, P. Redder, M. E. Schenk, C. Theriault, N. Tolstrup, R. L. Charlebois, W. F. Doolittle, M. Duguet, T. Gaasterland, R. A. Garrett, M. A. Ragan, C. W. Sensen, and J. Van der Oost.** 2001. The complete genome of the crenarchaeon *Sulfolobus solfataricus* P2. *Proc. Natl. Acad. Sci. USA* **98:**7835–7840.

128. **Sleytr, U. B.** 1997. Basic and applied S-layer research: an overview. *FEMS Microbiol. Rev.* **20:**5–12.

129. Sleytr, U. B., and T. J. Beveridge. 1999. Bacterial S-layers. *Trends Microbiol.* 7:253–260.

130. Sleytr, U. B., M. Sára, P. Messner, and D. Pum. 1994. Two-dimensional protein crystals (S-layers): fundamentals and applications. *J. Cell. Biochem.* 56:171–176.

131. Smith D. R., L. A. Doucette-Stamm, C. Deloughery, H. Lee, J. Dubois, T. Aldredge, R. Bashirzadeh, D. Blakely, R. Cook, K. Gilbert, D. Harrison, L. Hoang, P. Keagle, W. Lumm, B. Pothier, D. Qiu, R. Spadafora, R. Vicaire, Y. Wang, J. Wierzbowski, R. Gibson, N. Jiwani, A. Caruso, D. Bush, H. Safer, D. Patwell, S. Prabhakar, S. McDougall, G. Shimer, A. Goyal, S. Pietrokovski, G. M. Church, C. J. Daniels, J.-I. Mao, P. Rice, J. Nölling and J. N. Reeve, 1997. Complete genome sequence of *Methanobacterium thermoautotrophicum* ΔH: functional and comparative genomics. *J. Bacteriol.* 179:7135–7155.

132. Sprott, G. D., and R. C. McKellar. 1980. Composition and properties of the cell wall of *Methanospirillum hungatei. Can. J. Microbiol.* 26:115–120.

133. Sprott, G. O., T. J. Beveridge, B. G. Patel, and G. Ferrante. 1986. Sheath disassembly in *Methanospirillum hungatei* strain GP1. *Can. J. Microbiol.* 32:847–854.

134. Stetter, K. O., M. Thomm, J. Winter, G. Wildgruber, H. Huber, W. Zillig, D. Janecovic, H. König, P. Palm, and S. Wunderl. 1981. *Methanothermus fervidus*, sp. nov., a novel extremely thermophilic methanogen isolated from an Icelandic hot spring. *Zbl. Bakt. Hyg., I. Abt. Orig.* C 2:166–178.

135. Stewart, M., T. J. Beveridge, and G. O. Sprott. 1985. Crystalline order to high resolution in the sheath of *Methanospirillum hungatei*: a cross beta structure. *J. Mol. Biol.* 183:509–515.

136. Sumper, M. 1987. Halobacterial glycoprotein biosynthesis. *Biochim. Biophys. Acta* 906:69–79.

137. Sumper, M. 1993. S-layer glycoproteins from moderately and extremely halophilic archeobacteria, p. 109–117. *In* T. J. Breveridge and S. F. Koval (ed.), *Advances in Bacterial Paracrystalline Surface Layers*. NATO ASI Series A: Life Sciences, vol. 252. Plenum Press, New York, N.Y.

138. Sumper, M., E. Berg, R. Mengele, and I. Strobel. 1990. Primary structure and glycosylation of the S-layer protein of *Haloferax volcanii. J. Bacteriol.* 172:7111–7118.

139. Sumper, M., and F. T. Wieland. 1995. Bacterial glycoproteins, p. 455–473. *In* J. Montreuil, J. F. G. Vliegenthart, and H. Schachter (ed.), *Glycoproteins*. Elsevier, Amsterdam, The Netherlands.

140. Tabassum, R., K. M. M. Sandman, and J. N. Reeve. 1990. HMt, a histone-related protein from *Methanobacterium thermoautotrophicum. J. Bacteriol.* 172:7111–7118.

141. Taylor, K. A., J. F. Deatherage, and L. A. Amos. 1982. Structure of the S-layer of *Sulfolobus acidocaldarius. Nature* 299:840–842.

142. Thomm, M. 1996. Archaeal transcription factors and their role in transcription inititation. *FEMS Microbiol. Rev.* 18:159–171.

143. Thorne, K. J. I., R. C. Oliver, and A. M. Glauert. 1976. Synthesis and turnover of the regularly arranged surface protein of *Acinetobacter sp.* relative to the other components of the cell envelope. *J. Bacteriol.* 127:440–450.

144. Trachtenberg, S., B. Pinnick, and M. Kessel. 2000. The cell surface glycoprotein layer of the extreme halophile *Halobacterium salinarum* and its relation to *Haloferax volcanii*: cryo-electron tomography of freeze-substituted cells and projection studies of negatively stained envelopes. *J. Struct. Biol.* 130:10–26.

145. Voisin, S., R. S. Houliston, J. Kelly, J. B. Brisson, D. Watson, S. L. Bardy, K. F. Jarrell, and S. M. Logan. 2005. Identification and characterization of the unique N-linked glycan common to the flagellins and S-layer glycoprotein of *Methanococcus voltae. J. Biol. Chem.* 280:16586–16593.

146. Völkl, P., R. Huber, E. Drobner, R. Rachel, S. Burggraf, A. Trincone, and K. O. Stetter. 1993. *Pyrobaculum aerophilum* sp. nov., a novel nitrate-reducing hyperthermophilic Archaeum. *Appl. Env. Microbiol.* 59:2918–2926.

147. Wakai, H., S. Nakamura, H. Kawasaki, K. Takada, S. Mizutani, R. Aono, and K. Horikoshi. 1997. Cloning and sequencing of the gene encoding the cell surface glycolprotein of *Haloarcula japonica* strain TR-1. *Extremophiles* 1:29–35.

148. Wakai, H., K. Takada, S. Nakamura, and K. Horikoshi. 1995. Structure and heterologous expression of the gene encoding the cell surface glycoprotein from *Haloarcula japonica* strain TR-1. *Nucleic Acids Symp. Ser.* 95:101–102.

149. Waters, E., M. J. Hohn, I. Ahel, D. E. Graham, M. B. Adams, M. Barnstead, K. Y. Beeson, L. Bibbs, R. Bolanos, M. Keller, K. Kretz, X. Lin, E. Mathur, J. Ni, M. Podar, T. H. Richardson, G. S. Sutton, M. Simon, D. Soll, K. O. Stetter, J. M. Short, and M. Noorderwier. 2003. The genome of *Nanoarchaeum equitans*: insights into early archaeal evolution and derived parasitism. *Proc. Natl. Acad. Sci. USA* 100:12984–12988.

150. Weiss, L. R. 1973. Attachment of bacteria to sulphur in extreme environments. *J. Gen. Microbiol.* 77:501–507.

151. Weiss, L. R. 1974. Subunit cell wall of *Sulfolobus acidocaldarius. J. Bacteriol.* 118:275–284.

152. Wieland, F., J. Lechner, G. Bernhardt, and M. Sumper. 1980. Halobacterial glycolprotein saccharides contain covalently-linked sulphate. *FEBS Lett.* 120:110–114.

153. Wieland, F., W. Dompert, G. Bernhardt, and M. Sumper. 1981. Sulphatation of a repetitive saccharide in halobacterial cell wall glycoprotein. Occurrence of a sulphated lipid-linked precursor. *FEBS Lett.* 132:319–323.

154. Wieland, F., R. Heitzer, and W. Schaefer. 1983. Asparaginyl-glucose: novel type of carbohydrate linkage. *Proc. Natl. Acad. Sci. USA* 80:5470–5474.

155. Wieland, F., J. Lechner, and M. Sumper. 1982. The cell wall glycoprotein of halobacteria: structural, functional and biosynthetic aspects. *Zbl. Bakt. Hyg. I. Abt. Orig.* C 3:161–170.

156. Wieland, F. T. 1988. The cell surfaces of halobacteria, p. 55–65. *In* F. Rodriguez-Valera (ed.), *Halophilic Bacteria*, vol. 2, CRC Press Inc., Boca Raton, Fla.

157. Wildhaber, I., and W. Baumeister. 1987. The cell envelope of *Thermoproteus tenax*: three-dimensional structure of the surface layer and its role in shape maintenance *EMBO J.* 6:1475–1480.

158. Wildhaber, I., U. Santarius, and W. Baumeister. 1987. Three-dimensional structure of the surface protein of *Desulfurococcus mobilis. J. Bacteriol.* 169:5563–5568.

159. Woese, C. R. 1987. Bacterial evolution. *Microbiol. Rev.* 51:221–271.

160. Woese, C. R., O. Kandler, and M. L. Wheelis. 1990. Towards a natural system of organisms: Proposal for the domains Archaea, Bacteria, Eucarya. *Proc. Natl. Acad. Sci. USA* 87:4576–4579.

161. Yao, R., A. J. Macario, and E. Conway de Macario. 1994. An archaeal S-layer gene homolog with repetitive units. *Biochim. Biophys. Acta* 1219:697–700.

162. Zehnder, A. J., B. A. Huser, T. D. Brock, and K. Wuhrman. 1980. Characterization of and acetate-decarboxylating, non-hydrogen-oxidizing methane bacterium. *Arch. Microbiol.* 124:1–11.

163. Zeikus, J. G., and V. G. Bowen. 1975. Fine structure of *Methanospirillum hungatei. J. Bacteriol.* 121:373–380.

164. Zeitler, R., E. Hochmuth, R. Deutzmann, and M. Sumper. 1998. Exchange of Ser-4 for Val, Leu and Asn in the sequon Asn-Ala-Ser does not prevent N-glycosylation of the cell sur-

face glycoprotein from *Halobacterium halobium*. *Glycobiology* 8:1157–1164.

165. Zhang, Y., F. W. Larimer, M. V. Olson, and J. A. Leigh. 2004. Complete genome sequence of the genetically tractable hydrogenotrophic methanogen *Methanococcus maripaludis*. *J. Bacteriol.* 186:6956–6969.

166. Zillig, W., K. O. Stetter, D. Prangishvilli, W. Schäfer, S. Wunderl, D. Janékovic, I. Holz, and P. Palm. 1982. Desulfurococcaceae the second family of the extremely thermophilic, anaerobic, sulfur respiring Thermoproteales. *Zentbl. Bakteriol., Mikrobiol. Hyg. Abt. I. Orig.* C 3:304–317.

167. Zillig, W., K. O. Stetter, W. Schäfer, D. Janékovic, S. Wunderl, I. Holz, and P. Palm. 1981. Thermoproteales: a novel type of extremely thermoacidophilic anaerobic archaebacteria isolated from Icelandic solfataras. *Zentralbl. Bakteriol., Mikrobiol. Hyg. Abt. 1. Orig.* C 2:205–227.

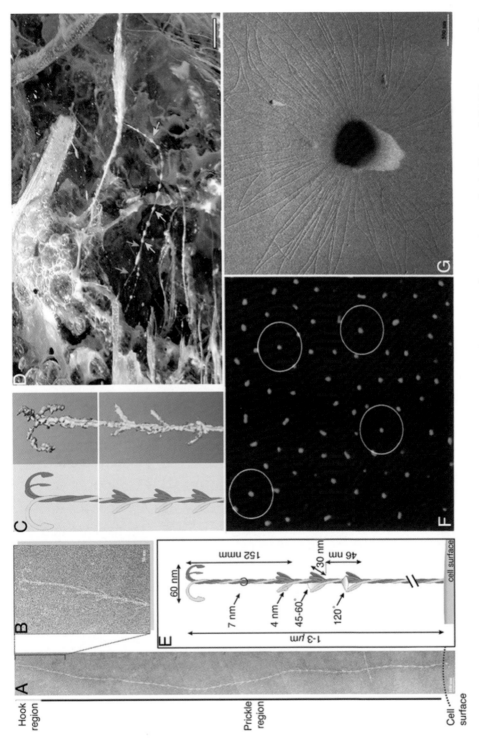

Figure 7 (Chapter 2). The euryarchaeon SM1 and its extracellular appendages ("hami"). (A) Electron micrograph of a "hamus." (B) Enlargement of the hook region. (C) Simplified model of a hamus with the three filaments shown in different colors and 3D reconstruction from cryoelectron microscopy. (D) "String of pearls," archaeal/bacterial community in cold, sulfurous spring water. (E) Hamus model with dimensions. (F) Natural biofilm hybridized with an SM1-specific fluorescent probe; circle diameter, 4 μm. (G) Pt-shadowed electron micrograph of a single SM1 cell with appendages. Figure compiled, modified, and reproduced from *Biospektrum* (264) and *Molecular Microbiology* (265) with permission of the publishers.

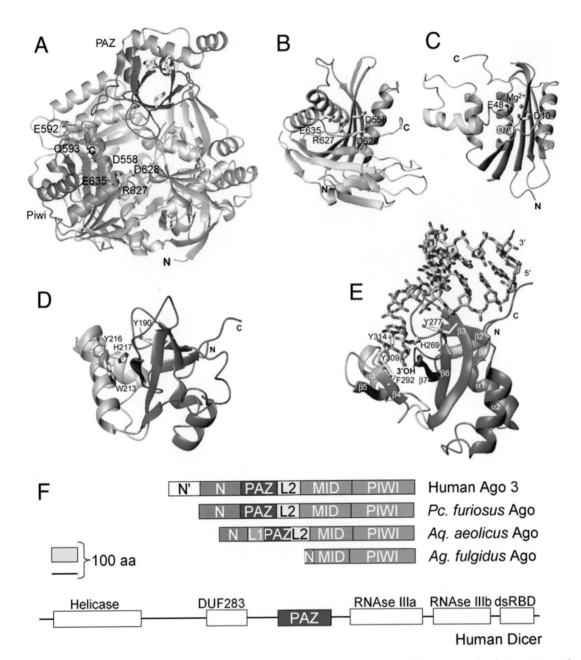

Figure 8 (Chapter 2). Three-dimensional structures of archaeal Argonaute proteins. (A) 3D structure of *P. furiosus* Ago with the PAZ domain (blue) and the PIWI domain (green/yellow) (PDB code 1U04). (B, C) Similarity of the *P. furiosus* PIWI domain (B) with the catalytic core of the *E. coli* RNase H1 (C) (PDB code 1RDD) with the catalytic DDE triad and bound Mg²⁺ ion highlighted. The *P. furiosus* PIWI domain has a putative, similar catalytic DDE triad and a conserved Arg (position 627). (D, E) 3D structure of the *P. furiosus* PAZ domain (D) and comparison with the homologous domain of human Ago1 bound to an siRNA mimic (E) (PDB code 1SI3). (F) Domain structures of Ago proteins, including N-terminal, linker (L1 and L2), PAZ, Mid, and PIWI domains and of human dicer comprising a DEXH helicase, a PAZ, two RNase III, and dsRBD domains and a conserved domain of unknown function (DUF). Panels A to E reproduced with modifications from *Current Opinion in Structural Biology* (241) with permission of the publisher.

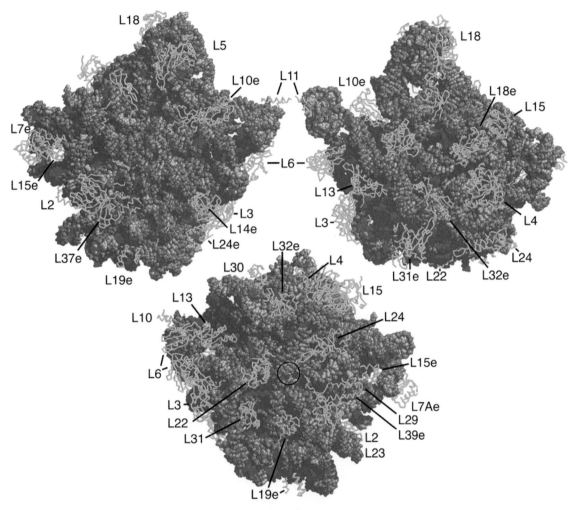

Figure 10 (Chapter 2). Three-dimensional structure of the 50S subunit of the *Haloarcula marismortui* ribosome. The ribosome arm around ribosomal protein L1 was omitted (for a more complete picture see reference 200). Figure drawn from the coordinates from PDB entry 1QVF (200); ribosomal RNAs are displayed in red (backbone) and gray (bases), proteins are displayed as yellow backbone ribbons. Top left, crown view; top right, back view; bottom, bottom view; the circle indicates the position of the polypeptide exit tunnel.

Figure 11 (Chapter 2). Haloarchaea in liquid cultures and within salt crystals. (A) Cultures of *Haloferax* and *Halorubrum*: first flask (front), *H. volcanii* WFD11 wild type; second flask, *H. volcanii* WFD11 gas vesicle ΔD mutant (see Fig. 14); third flask, *H. volcanii* WFD11 gas vesicle ΔD mutant complemented with the *gvpD* gene; fourth flask, *Halorubrum vacuolatum* wild type. (B) Himalayan rock salt ("Eubiona"; Claus, GmbH, Baden-Baden, Germany). (C, D) Crystals formed from dried *Halobacterium* cultures (cells trapped within). Bars, 1 cm. Crystals courtesy of F. Pfeifer, Darmstadt, Germany. Photographs by F. Pfeifer, Darmstadt, Germany (panel A), and A. Kletzin (panels B to D).

Figure 15 (Chapter 2). Solfatara and Pisciarelli fumaroles. (Left) Fumaroles in the Solfatara caldera (Pozzuoli near Naples, Italy) with deposition of sulfur, mercury, and arsenic salts. (Right) Fumarole-heated hole with boiling water, typical of habitats for *Sulfolobales* (Pisciarelli, near Naples, Italy). Photos taken by A. Kletzin.

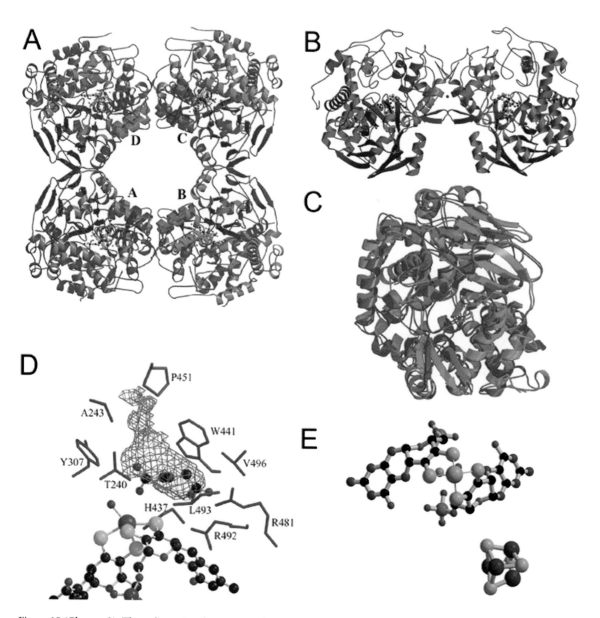

Figure 18 (Chapter 2). Three-dimensional structures of tungsten-containing aldehyde:ferredoxin oxidoreductases from *Pyrococcus furiosus*. (A) Cartoon of the formaldehyde:ferredoxin oxidoreductase (FOR), homotetrameric holoenzyme (150). (B) Cartoon of the aldehyde:ferredoxin oxidoreductase (AOR) homodimeric holoenzyme (53). (C) Peptide chains of AOR (cyan) superimposed on FOR (magenta) showing close structural similarity (150). (D) Active-site cavity of the FOR with surrounding residues and glutarate shown (150). (E) [4Fe-4S] cluster and the W-(bis-tungstopterin) cofactor of the AOR (53). FOR images reproduced from the *Journal of Molecular Biology* with permission of the publisher (150); AOR images reproduced from *Science* (53) with permission of the publisher.

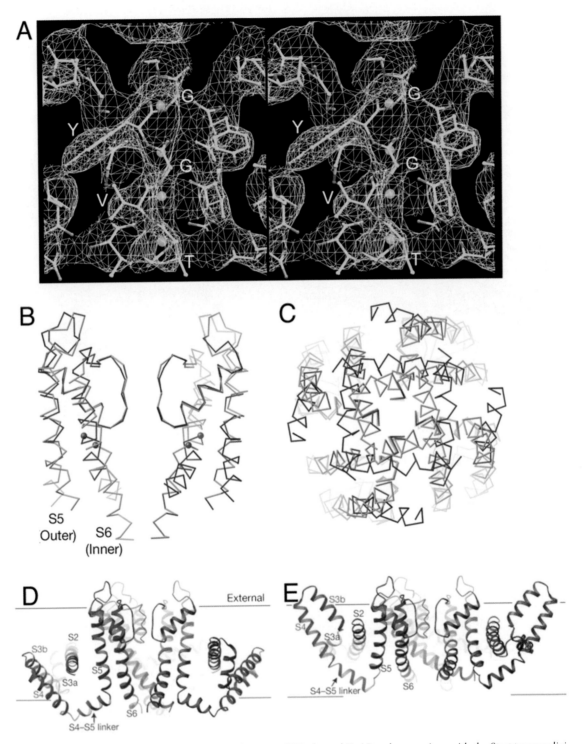

Figure 19 (Chapter 2). Model of the *Aeropyrum* voltage-gated K⁺-channel KvAP and comparison with the *Streptomyces lividans* KcsA K⁺ channel. (A) Stereo view of the KvAP pore with electron density map contoured at 1.0 Δ_ carbon (yellow), nitrogen (blue), oxygen (red), potassium (green). (B, C) α-Carbon traces of the KvAP pore (blue) and the *Streptomyces lividans* KcsA K⁺ channel (green) shown as a side view (B) and end-on from the intracellular side (C); S5, S6, outer and inner helices; glycine-gating hinges (red spheres). (D, E) Models of the closed (D) and open (E) KvAP structures based on the positions of the paddles (red), the pore and the S5 and S6 helices of KcsA. Reproduced with modifications from *Nature* (168, 169) with permission of the publisher.

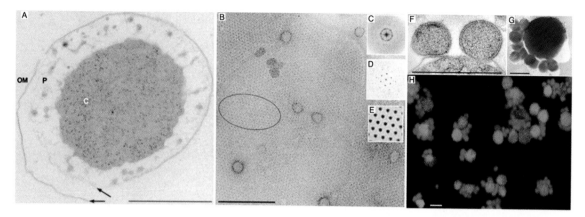

Figure 20 (Chapter 2). Electron micrograph and fluorescence images of *Ignicoccus* and *Nanoarchaeum*. (A) Transmission electron micrograph of thin-sectioned *Ignicoccus* cell with broad periplasmic space (P) and budded vesicles; OM, outer membrane, C, cytoplasm, bar, 1 μm. (B) Negative stained *Ignicoccus* outer membrane, highlighting power spectra of image field (C to E) (275). Panels A to E reproduced from *Biochemical Society Symposia* (275) with permission of the publisher. (F) Ultrathin section of *Nanoarchaeum* cells attached to the outer membrane of *Ignicoccus* sp. KIN/4. (G) Platinum shadowing of *Ignicoccus* cell with several *Nanoarchaeum* cells attached (left side of photograph). (H) Confocal laser-scanning micrograph using *Nanoarchaeum* (red) and *Ignicoccus*-specific probes (green). Panels F to H reproduced from *Nature* (152) with permission of the publisher.

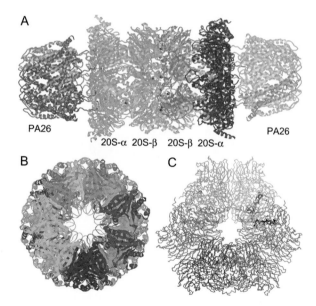

Figure 23 (Chapter 2). Three-dimensional structures of the *T. acidophilum* proteasome and tricorn protease. (A) Side view of the 26S proteasome/activator particle with the two sets of seven terminal PA26 subunits and the two $\alpha_7\beta_7\beta_7\alpha_7$ rings (PDB code 1YA7) (93). (B) Top view of the 20S proteasome core particle showing the sevenfold symmetry (PDB code 1PMA) (245). (C) Top view of the homohexameric tricorn protease complexed with a tridecameric peptide derivative (PDB code 1N6E) (195).

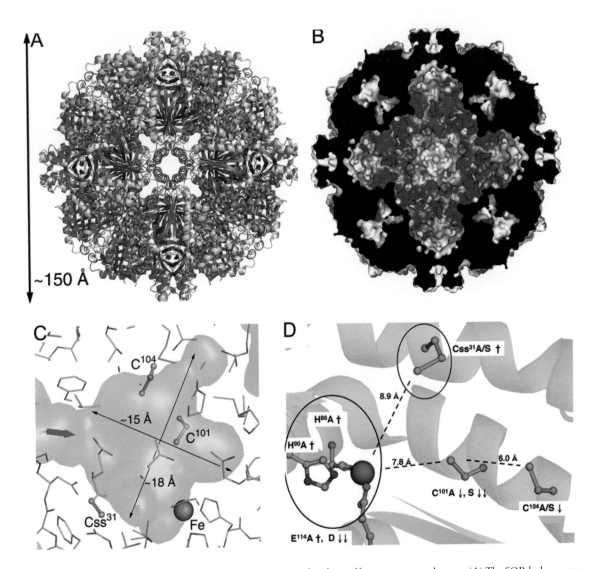

Figure 27 (Chapter 2). Three-dimensional structure of the *A. ambivalens* sulfur oxygenase reductase. (A) The SOR holoenzyme. Cartoon representation viewed along the crystallographic fourfold axis; cyan, α-helices; purple, β-sheets; red spheres, Fe ions. (B) Molecular accessible surface representation in the same orientation of inner surface of the sphere, color-coded according to the calculated electrostatic potentials: red, $\leq -10 \pm 1$ κT/e; white, neutral; blue, $\geq +10 \pm 1$ κT/e. (C) Cavity surface representation of the catalytic pocket, with conserved cysteines and iron highlighted; gray arrow, cavity entrance. (D) Effect of mutants on SOR activity; †, zero activity; ↓ reduced activity; ⇓ strongly reduced activity. The core active site composed of the Fe site and the persulfide-modified Css31 is highlighted within ellipsoids. Reproduced with minor modifications from *Science* (411) with permission from the publisher.

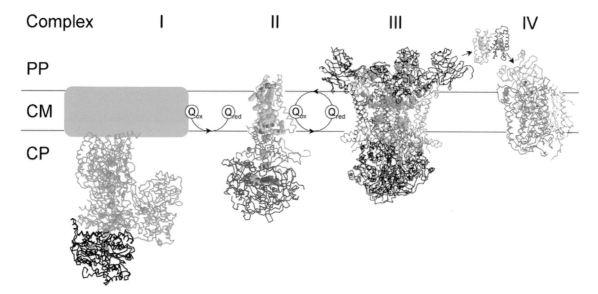

Figure 28 (Chapter 2). Canonical respiratory chain in bacteria and mitochondria. Scheme based on 3D structures with the exception of the membrane domain of complex I, for which a structure is not available. Domains that have not been identified in *Archaea* are shown in black. PP, periplasm; CM, cytoplasmic membrane; CP, cytoplasm; Q, quinols/quinones. The figure was prepared from the coordinates of PDB entries 1FUG (complex I, *Thermus thermophilus*), 1NEK (complex II, *E. coli*), 1KYO (complex III, *Saccharomyces cerevisiae*), 1EHK (complex IV, *Thermus thermophilus*), and 2CCY (cytochrome *c*, *Rhodospirillum molischianum*).

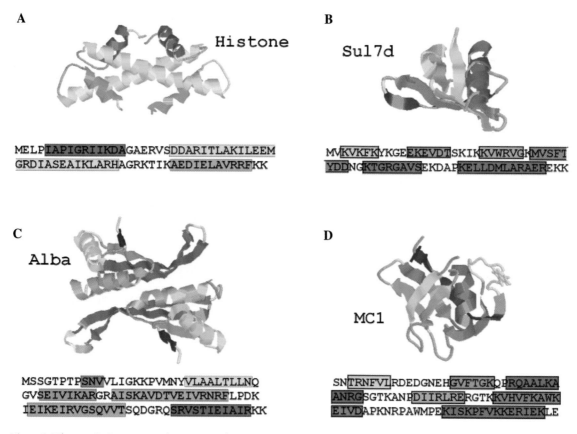

Figure 2 (Chapter 4). Sequences and structures of representative archaeal chromatin proteins. Primary sequences of HMfB from *M. fervidus* (A), Sul7d (Sac7d) from *S. acidocaldarius* (B), Alba (Sso10b1) from *S. solfataricus* (C), and MC1 from *Methanosarcina* sp. CHTI55 (D) are shown below the corresponding protein structure. The figure was constructed using structures available from the Protein Data Bank (11). Regions with α-helical and β-strand structures are colored identically in the sequence and in the corresponding structure.

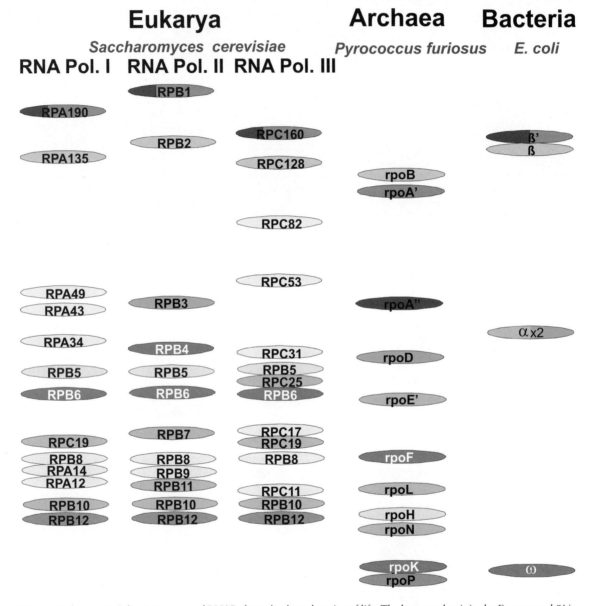

Figure 3 (Chapter 6). Subunit structure of RNAPs from the three domains of life. The largest subunit in the *Eucarya* and β′ in the *Bacteria* is split into two subunits, A1 and A2, in the *Archaea*. In methanogens, subunit B is also split into two polypeptides, B′ and B″. Different parts of bacterial subunit α are encoded by the genes for the archaeal subunits D and L. Subunits E1, F, H, N, and P are only shared between the *Archaea* and *Eucarya*. The pattern shown is based on separation of subunits by polyacrylamide gel electrophoresis under denaturing conditions. The numbers in the subunits of the eucaryal RNAP A (I), B (II), and C (III) indicate the molecular mass.

A

B

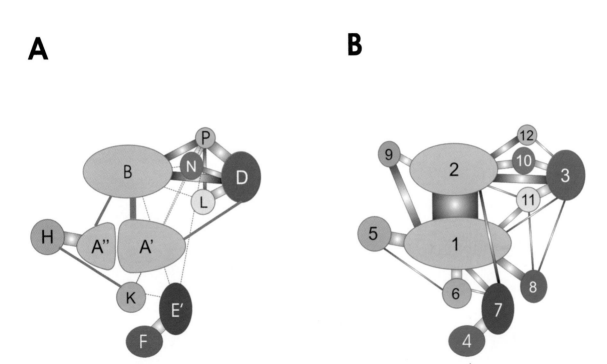

Figure 4 (Chapter 6). Structural similarity of *Pyrococcus* RNAP (A) and yeast RNAPII (B). Comparison of interactions of an archaeal RNAP inferred from Far-Western analysis with interactions of yeast RNAPII observed in the crystal structure of the enzyme. The width of the lines connecting subunits is a measure of the intensity of the interaction. Modified from *Science* (27) with additional data from *Proceedings of the National Academy of Sciences USA* (2).

Organism	(S10)	L3	L4	L23	L2	S19	L22	S3	L29	SUI	HY	S17	L14	L24	S4	L5	S14	S8	L6	L32	L19	L18	S5	L30	L15
SSO		L3	L4	L23	L2	S19	L22	S3	L29			S17	L14	L24	S4	L5	S14	S8	L6	L32	L19	L18	S5	L30	L15
STO		L3	L4	L23	L2	S19	L22	S3	L29			S17	L14	L24	S4	L5	S14	S8	L6	L32	L19	L18	S5	L30	L15
AFU		L3	L4	L23	L2	S19	L22	S3	L29		HY	S17	L14	L24	S4	L5	S14	S8	L6	L32	L19	L18	S5	L30	L15
APE		L3	L4	L23	L2	S19	L22	S3	L29		HY	S17	L14	L24	S4	L5	S14	S8	L6	L32	L19	L18	S5	L30	L15
PFU		L3	L4	L23	L2	S19	L22	S3	L29	SUI	HY	S17	L14	L24	S4	L5	S14	S8	L6	L32	L19	L18	S5	L30	L15
PHO		L3	L4	L23	L2	S19	L22	S3	L29	SUI	HY	S17	L14	L24	S4	L5	S14	S8	L6	L32	L19	L18	S5	L30	L15
PAB		L3	L4	L23	L2	S19	L22	S3	L29	SUI	HY	S17	L14	L24	S4	L5	S14	S8	L6	L32	L19	L18	S5	L30	L15
TKO		L3	L4	L23	L2	S19	L22	S3	L29	SUI	HY	S17	L14	L24	S4	L5	S14	S8	L6	L32	L19	L18	S5	L30	L15
PAE		L3	L4	L23	L2			S3	L29									S8		L32	L19	L18			
MKA		L3	L4	L23	L2		L22	S3	L29	SUI		S17	L14	L24	S4	L5	S14	S8	L6	L32	L19	L18	S5	L30	L15
MMA		L3	L4	L23	L2	S19	L22	S3	L29			S17	L14	L24	S4	L5	S14	S8	L6	L32	L19	L18	S5	L30	L15
MAC		L3	L4	L23	L2	S19	L22	S3	L29			S17	L14	L24	S4	L5	S14	S8	L6	L32	L19	L18	S5	L30	L15
MTH		L3	L4	L23	L2	S19	L22	S3	L29	SUI	HY	S17	L14	L24	S4	L5	S14	S8	L6	L32	L19	L18	S5	L30	L15
MJA		L3	L4	L23	L2			S3	L29			S17	L14	L24	S4	L5	S14	S8	L6	L32	L19	L18	S5	L30	L15
MMP		L3	L4	L23	L2	S19	L22	S3	L29			S17	L14	L24	S4	L5	S14	S8	L6	L32	L19	L18	S5	L30	L15
HMA		L3	L4	L23	L2	S19	L22	S3	L29			S17	L14	L24	S4	L5	S14	S8	L6	L32	L19	L18	S5	L30	L15
H-sp		L3	L4	L23	L2	S19	L22	S3	L29		HY	S17	L14	L24	S4	L5	S14	S8	L6	L32	L19	L18	S5	L30	L15
TAC		L3	L4	L23	L2	S19	L22	S3	L29	SUI	HY	S17	L14	L24	S4	L5	S14	S8	L6	L32	L19	L18	S5	L30	L15
TVO		L3	L4	L23	L2	S19	L22	S3	L29	SUI	HY	S17	L14	L24	S4	L5	S14	S8	L6	L32	L19	L18	S5	L30	L15
ECO	S10	L3	L4	L23	L2	S19	L22	S3		L16	L29	S17	L14	L24		L5	S14	S8	L6			L18	S5	L30	L15

Figure 3 (Chapter 8). Organization of the main ribosomal protein gene clusters in archaeal genomes. SSO, *Sulfolobus solfataricus*; STO, *Sulfolobus tokodaii*; AFU, *Archaeoglobus fulgidus*; APE, *Aeropyrum pernix*; PFU, *Pyrococcus furiosus*; PHO, *Pyrococcus horikoshii*; PAB, *Pyrococcus abyssi*; TKO, *Thermococcus kodakaraensis*; PAE, *Pyrobaculum aerophilum*; MKA, *Methanopyrus kandleri*; MMA, *Methanosarcina mazei*; MAC, *Methanosarcina acetivorans*; MTH, *Methanothermobacter thermautotrophicus*; MJA, *Methanococcus jannaschii*; MMP, *Methanococcus maripaludis*; HMA, *Haloarcula marismortui*; H-sp, *Halobacterium* sp. NRC1; TAC, *Thermoplasma acidophilum*; TVO, *Thermoplasma volcanii*. The last line (ECO) shows for comparison the organization of the same genes in *E. coli* that is also present in most bacteria. Genes that are within 50 bp of each other, and may therefore be cotranscribed, are indicated in the same color. Domain-specific genes are underlined.

M. thermautotrophicus GatDE:tRNA^{Gln} Complex

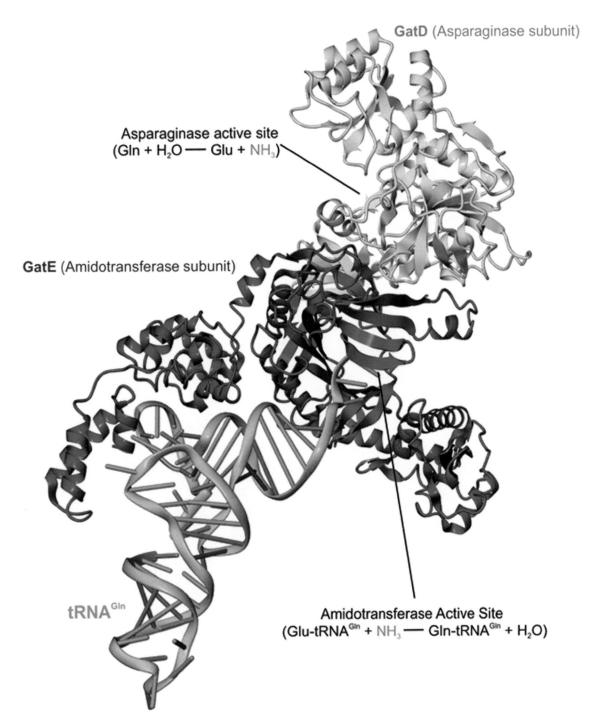

GatD (Asparaginase subunit)

Asparaginase active site
(Gln + H_2O —— Glu + NH_3)

GatE (Amidotransferase subunit)

tRNA^{Gln}

Amidotransferase Active Site
(Glu-tRNA^{Gln} + NH_3 —— Gln-tRNA^{Gln} + H_2O)

Figure 4 (Chapter 9). Crystal structure of the *M. thermautotrophicus* GatDE complexed with tRNA^{Gln}. The dimer of the heterodimeric GatDE (thus forming a heterotetramer) binds two tRNA molecules. The asparaginase active site of GatD and the kinase/amidotransferase active site of GatE are distantly separated, connected with a "molecular tunnel."

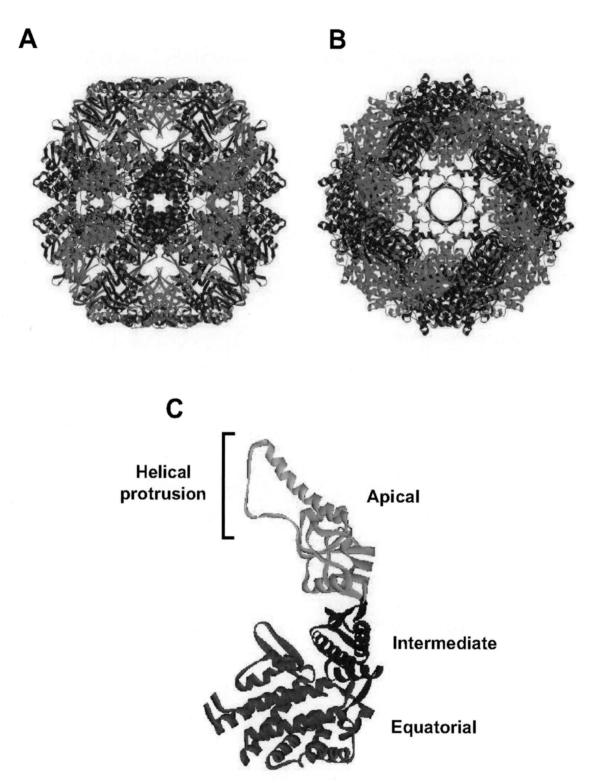

Figure 3 (Chapter 10). Structure of archaeal group II chaperonin from *Thermoplasma acidophilum*. (A and B) The side view and top view of the crystal structure of *T. acidophilum* chaperonin, respectively. The α subunits are shown in dark green, and the β subunits are shown in dark blue. The hexadecameric structure was drawn using MOLSCRIPT (48). (C) The subunit structure of *T. acidophilum* chaperonin. Apical, intermediate, and equatorial domains are represented by green, blue, and red, respectively. The helical protrusion is highlighted by yellow. The figure was drawn with the Viewer Light 5.0 software (Accelrys).

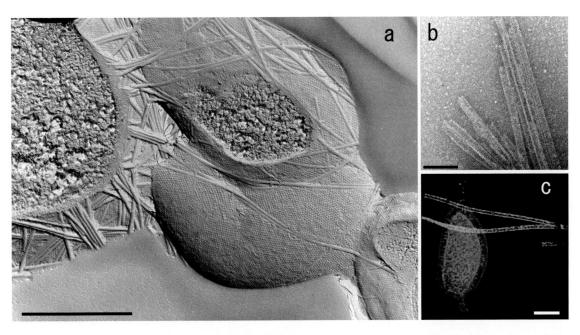

Figure 16 (Chapter 14). Transmission electron micrographs of *P. occultum*. (a) *Pyrodictium* cells after freeze etching, exhibiting an S-layer with p6 symmetry. Bar = 0.5 μm. (b) Extracellular cannulae, a specific feature of the genus *Pyrodictium*; negative staining with uranyl acetate; bar = 0.1 μm. (c) 3D tomogram of a frozen-hydrated *Pyrodictium* cell with two cannulae; bar = 0.2 μm. Modified from the *Journal of Structural Biology* (106) with permission of the publisher.

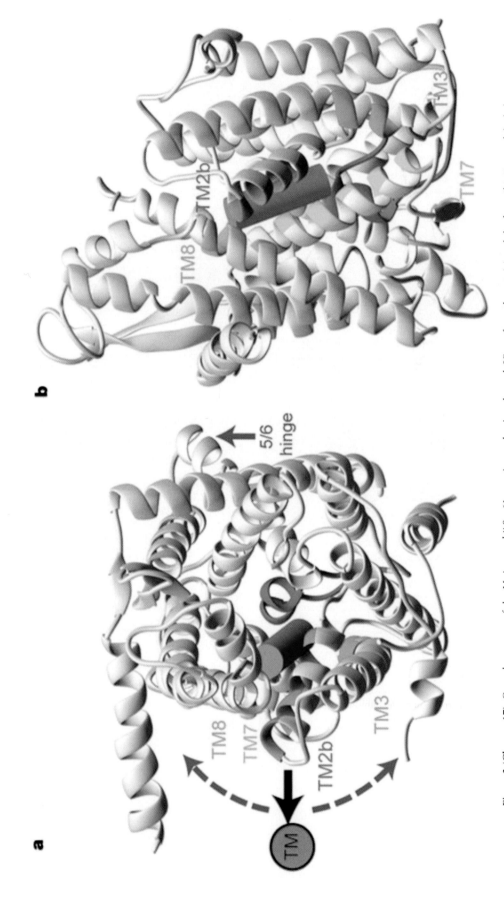

Figure 3 (Chapter 17). Crystal structure of the *M. jannaschii* Sec61 protein-conducting channel. Views from the top (a) and the front (b). Faces of the helices that form the signal-sequence-binding site and the lateral gate through which TMs of nascent membrane proteins exit the channel into lipid are colored. The plug, which gates the pore, is green. The hydrophobic core of the signal sequence probably forms a helix, modeled as a magenta cylinder, which intercalates between TM2b and TM7 above the plug. Intercalation requires opening the front surface as indicated by the broken arrows, with the hinge for the motion being the loop between TM5 and TM6 at the back of the molecule (5/6 hinge). A solid arrow pointing to the magenta circle in the top view indicates schematically how a TM of a nascent membrane protein would exit the channel into lipid. Structure and legend reprinted from *Nature* (107) with permission from the publisher.

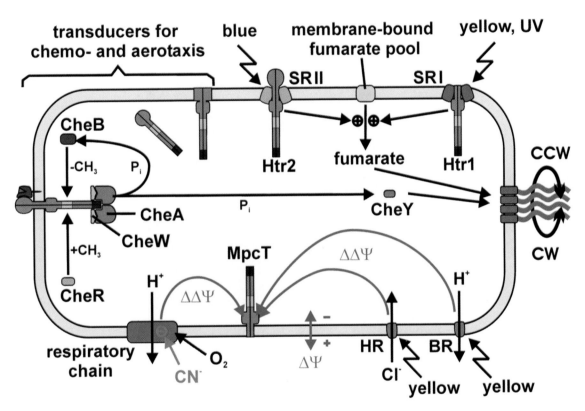

Figure 8 (Chapter 18). Overview of halobacterial signal transduction. Transducer proteins (Htr proteins) are depicted as dimers (brown) and shown in their expected topology. The Htr regions involved in adaptation (yellow) and in signal relay (dark gray) to the flagellar motor via various Che proteins are indicated. The actions of the Che-protein machinery are illustrated for only one of the Htr proteins, shown on the left, for which an interaction with a substrate-loaded, membrane-anchored binding protein is indicated. CheD and CheJ (CheC) proteins are omitted for clarity. Htr1 and Htr2 transduce light signals via direct interaction with their corresponding receptors SRI and SRII. Repellent light signals mediated by SRI and SRII elicit the release of switch factor fumarate from a membrane-bound fumarate pool. MpcT senses changes in membrane potential ($\Delta\Psi$) generated via light-dependent changes in ion transport activity of BR and HR. The relative sizes of receptors, binding proteins, transducers, and Che proteins approximately reflect their corresponding molecular masses. Reproduced from *Molecular Microbiology* (54) with permission of the publisher and D. Oesterhelt.

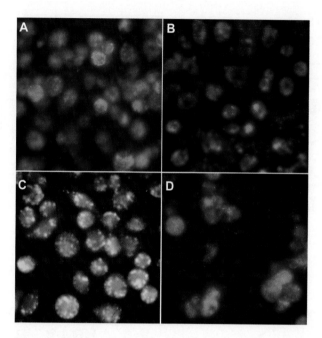

Figure 5 (Chapter 23). Uptake of fluorescent archaeosomes by phagocytic cells. Archaeosomes composed of total polar lipids were prepared either by incorporating a small amount of the fluorescent lipid rhodamine-phosphatidylethanolamine (62) or by entrapping 1.5 mM carboxyfluorescein (66). Uptake was performed in 1 ml of RPMI medium added to 0.5 million adhered cells (62). Panels show 30-min uptakes: (A) *M. smithii* rhodamine-archaeosomes (100 μg) by thioglycollate-activated mouse peritoneal macrophages; (B) uptake of *M. smithii* rhodamine-archaeosomes (25 μg) by bone marrow-derived DCs; (C) uptake of *M. mazei* carboxyfluorescein-archaeosomes (40 μg) by macrophages culture J774A.1; and (D) uptake of *H. salinarum* rhodamine-archaeosomes (100 μg) by thioglycollate-activated mouse peritoneal macrophages.

Archaea: Molecular and Cellular Biology
Edited by Ricardo Cavicchioli
© 2007 ASM Press, Washington, D.C.

Chapter 15

Lipids: Biosynthesis, Function, and Evolution

YAN BOUCHER

INTRODUCTION

One of the fundamental ways in which the *Archaea* differ from the *Bacteria* and *Eucarya* is the composition of their cell membrane. Core membrane lipids, in all domains of life, consist of two hydrocarbon side chains linked to a glycerol moiety. In bacteria and eucarya, fatty acid side chains are ester linked to an *sn*-glycerol-3-phosphate (G3P) moiety. In archaea, on the other hand, isoprenoid side chains are ether linked to an *sn*-glycerol-1-phosphate (G1P) backbone. The side chains and the type of bond that links them to the glycerol moiety differ only quantitatively between the *Archaea* and the other two domains of life. Indeed, a minor proportion of ether-linked lipids can be found in some eucarya (27) and bacteria (16), and fatty acid side chains are part of some archaeal lipids (11). The *sn*-1 stereochemistry of the glycerol backbone, on the other hand, is truly unique to *Archaea* and remains the most distinctive feature of their lipids.

A great diversity of archaeal lipids are derived from the *sn*-2,3-diacylglycerol diether basic structure. The most common is *sn*-2,3-diphytanylglycerol diether (archaeol) (Fig. 1), which is a core lipid in most archaea (Table 1) (8). Variations of this structure include the length of either or both side chains (25-carbon sesterterpanyl chain instead of a 20-carbon phytanyl chain), introduction of a hydroxy group at the C-3 of the phytanyl chain on the *sn*-3 position (hydroxyarchaeol), and linking of the terminal portion of the phytanyl chains to create a macrocyclic diether (Fig. 1). One of the most frequent and functionally important variations in structure is the head-to-head condensation of two diether lipid molecules to form a glycerol-dialkylglycerol tetraether lipid (GDGT) (Fig. 2). The most widespread of these is caldarchaeol (GDGT-0), which does not contain any modification to its alkyl core. Several variations of this basic tetraether lipid also exist, including the introduction of a number of penta- or hexacyclic rings, the replacement of one of the glycerol moieties by nonitol, a break in one of the C40 alkyl chains, and the introduction of a bridge between the alkyl chains (Fig. 2). Diether and tetraether lipids usually have a polar head group or sugar attached to the *sn*-1 position of their glycerol moieties, and can display various levels of unsaturation in their side chains (41).

The biosynthetic steps leading to this wide variety of core lipids are not all well understood, but many enzymes involved in assembling backbone structure have been identified and characterized. The synthesis of the isoprenoid building blocks, isopentenyl diphosphate (IPP) and dimethylallyl diphosphate (DMAPP), their assembly in side chains, and attachment to the *sn*-glycerol-1-phosphate moiety are relatively well understood. However, the order and nature of subsequent steps, such as attachment of head groups or sugars, saturation of the side chains and synthesis of tetraether lipids, are not as clearly defined.

The variety of lipid structures observed in archaea is underlined by a variation in the subset of biosynthetic genes present in different organisms as well as the functional flexibility of those genes. The mevalonate pathway describes the synthesis of the IPP and DMAPP building blocks that form the isoprenoid side chains of archaeal lipids. This pathway evolved through a combination of processes that included orthologous and nonorthologous gene displacement, integration of components from eucarya and bacteria, and lateral gene transfer between members of the *Archaea*. Isoprenyl phosphate synthases are responsible for elongating side chains using the end products of the mevalonate pathway. In some archaeal phyla they have gained the capacity to make chains longer than the basic 20 carbons through amino acid substitutions at a few specific positions. However, the most intriguing evolutionary development is the archaeal invention of three stereospecific enzymes, from broader protein families, that led to the unique chirality of their lipids:

Yan Boucher • Department of Chemistry and Biomolecular Sciences, Macquarie University, Sydney, NSW 2109, Australia.

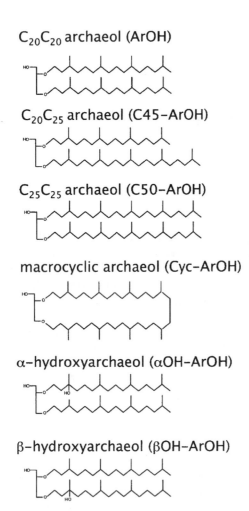

C₂₀C₂₀ archaeol (ArOH)

C₂₀C₂₅ archaeol (C45–ArOH)

C₂₅C₂₅ archaeol (C50–ArOH)

macrocyclic archaeol (Cyc–ArOH)

α–hydroxyarchaeol (αOH–ArOH)

β–hydroxyarchaeol (βOH–ArOH)

Figure 1. Basic structure of glycerol diether isoprenoid lipids. The sugars or polar head groups that are frequently attached to the *sn*-1 position of the glycerol moiety in archaeal diether lipids are not shown.

sn-glycerol-1-phosphate dehydrogenase (synthesis of the *sn*-1-glycerol phosphate backbone), geranylgeranylglyceryl phosphate synthase, and digeranylgeranylglyceryl phosphate synthase (addition of the *sn*-2 and *sn*-3 side chains to the glycerol moiety, respectively).

The function of the structurally diverse core diether lipids is just beginning to be addressed. Some types of modifications, such as the introduction of pentacyclic rings or side-chain unsaturation, have been hypothesized to play a role in adaptation to different growth temperatures (12, 33). Likewise, some lipid-profile studies have suggested that the proportion of different lipids in the cell membrane can vary with environmental conditions such as pressure, or physiological conditions such as growth phase (19, 47).

This chapter summarizes the different biosynthetic steps of isoprenoid ether lipid biosynthesis in archaea, describing the underlying enzymatic reactions that have been characterized. The evolution of this lipid

biosynthesis apparatus in a variety of archaea is discussed. The effect of the environment on the nature of the lipids present in archaeal cell membranes is also scrutinized in an attempt to link structure and function.

BIOSYNTHETIC PATHWAYS

Synthesis of the Isoprenoid Building Blocks: IPP and DMAPP

Similar to other isoprenoids, archaeal lipid side chains are assembled from two universal precursors: isopentenyl diphosphate (IPP) and its isomer dimethylallyl diphosphate (DMAPP). Most eucarya and some bacteria synthesize these precursors through the mevalonate pathway (1, 26). This pathway has been described in detail in these organisms and is always composed of five steps, as illustrated in Fig. 3: (i) conversion of acetyl-CoA and acetoacetyl-CoA to 3-hydroxy-3-methylglutaryl-CoA (HMG-CoA); (ii) reduction of HMG-CoA to mevalonate; (iii) phosphorylation of mevalonate; (iv) phosphorylation of phosphomevalonate; and (v) conversion of diphosphomevalonate to IPP. Archaea also derive their isoprenoids from mevalonate, as shown by tracer studies more than two decades ago (21). Orthologs of the enzymes catalyzing the first three steps of the pathway in bacteria and eucarya can also be found in archaea (45). However, the last two enzymes of the standard bacterial/eucaryal mevalonate pathway (phosphomevalonate kinase [PMK] and diphosphomevalonate decarboxylase [PPMD]) are missing in most archaea (Table 1). A few exceptions are known, including the PPMD gene present in the *Halobacteriales* and *Thermoplasmatales*, and PMK and PPMD in *Sulfolobales*. The conversion of the PPMD product, IPP, to the other essential isoprenoid precursor, DMAPP, is performed by isopentenyl diphosphate isomerase (IDI), of which two analogous types exist. All archaea harbor IDI2, but the *Halobacteriales* also possess an IDI1, the latter being the enzyme used in most eucarya and bacteria for the biosynthesis of DMAPP.

Elongation of the Isoprenoid Side Chains

The products of the mevalonate pathway (IPP and DMAPP) are used by archaea as building blocks for lipid side chains. These side chains are composed of 20 or 25 carbons (C20 or C25). The full length is reached through the sequential condensation of IPP to a growing allylic polyisoprenoid diphosphate (Fig. 3). The first molecule in the chain is the IPP isomer, DMAPP, to which IPP molecules are successively added to obtain GPP (geranyl diphosphate, 10 carbons), FPP

Table 1. Distribution of lipid biosynthesis enzymes and core lipids in the *Archaea*[a]

Archaeal groups	Enzymes[b]												Core lipids[c]									
	HMGS	HMGR	MK	PMK	PPMD	IDI1	IDI2	GGPPS	FGPPS	GGGPS	DGGGPS	G1PDH	ArOH	C25 chains	Cyc-ArOH	OH-ArOH	CAOH	GDGT-(1-8)	GDNT	GTGT	FU	CRe-OH
Euryarchaeota																						
Halobacteriales	+	+	+	−	B	B	+	+	+	D	+	+	+	+	−	−	−	−	−	−	−	−
Methanomicrobiales	+	+	+	−	−	−	+	+	−	+	+	+	+	−	−	−	+	−	−	−	−	−
Methanosarcinales	+	+	+	−	−	−	+	+	−	+	+	+	+	−	+	+	+	−	−	−	−	−
Methanococcales	+	+	+	−	−	−	+	+	−	+	+	+	+	−	+	+	+	−	−	−	−	−
Methanobacteriales	+	+	+	−	−	−	+	+	−	+	+	+	+	−	−	−	+	−	−	−	+	−
Methanopyrales	+	+	+	−	−	−	+	+	−	+	+	+	+	−	−	+	+	−	−	−	−	−
Thermococcales	+	+	+	−	−	−	+	+	−	D	+	+	+	−	−	−	+	−	−	−	−	−
Archaeoglobales	+	B	+	−	−	−	+	+	−	D	+	+	+	−	−	−	+	−	−	−	−	−
Thermoplasmatales	+	B	+	−	B	−	+	+	−	+	+	+	+	−	−	+	+	+	−	−	−	−
Crenarchaeota																						
Sulfolobales	+	+	+	E	E	−	+	+	+	+	+	+	+	−	−	+	+	+	+	+	−	−
Desulfurococcales	+	+	+	−	−	−	+	+	+	+	+	+	+	−	−	−	+	+	−	−	−	−
Thermoproteales	+	+	?	?	?	?	+	+	−	+	+	+	+	+	−	−	+	+	−	−	−	−
Marine group I	?	?	?	?	?	?	?	?	?	?	?	?	+	−	−	−	+	+	−	−	−	+
Nanoarchaeota	?	−	?	−	−	−	−	−	−	−	−	−	+	−	−	−	+	−	−	−	−	−

[a] +, present; −, absent or too divergent to be detected; B, bacterial type; E, eucaryal type; D, divergent from main archaeal type but origin uncertain; ?, no information available on presence or absence (uncultured archaeal group).
[b] HMGS, 3-hydroxy-3-methylglutaryl coenzyme A synthase; HMGR, 3-hydroxy-3-methylglutaryl coenzyme A reductase; MK, mevalonate kinase; PMK, phosphomevalonate kinase; PPMD, diphosphomevalonate decarboxylase; IDI1/IDI2, isopentenyl diphosphate isomerase type 1 and 2; GGPPS, geranylgeranyl diphosphate synthase; FGPPS, farnesylgeranyl diphosphate synthase; GGGPS, geranylgeranylglyceryl phosphate synthase; DGGGPS, digeranylgeranylglyceryl phosphate synthase; G1PDH, sn-glycerol-1-phosphate dehydrogenase.
[c] ArOH, archaeol; C25 chains, presence of one or two sesterpanyl side-chains; Cyc-ArOH, macrocyclic archaeol; OH-ArOH, hydroxyarchaeol; CAOH, caldarchaeol (GDGT-0); GDGT (1–8), glycerol-dialkyl-glycerol tetraether (1–8 pentacyclic rings); GDNT, glycerol-dialkyl-nonitol tetraether; GTGT, glycerol-trialkyl-glycerol tetraether; FU, H-shaped caldarchaeol derivative; CreOH, crenarchaeol.

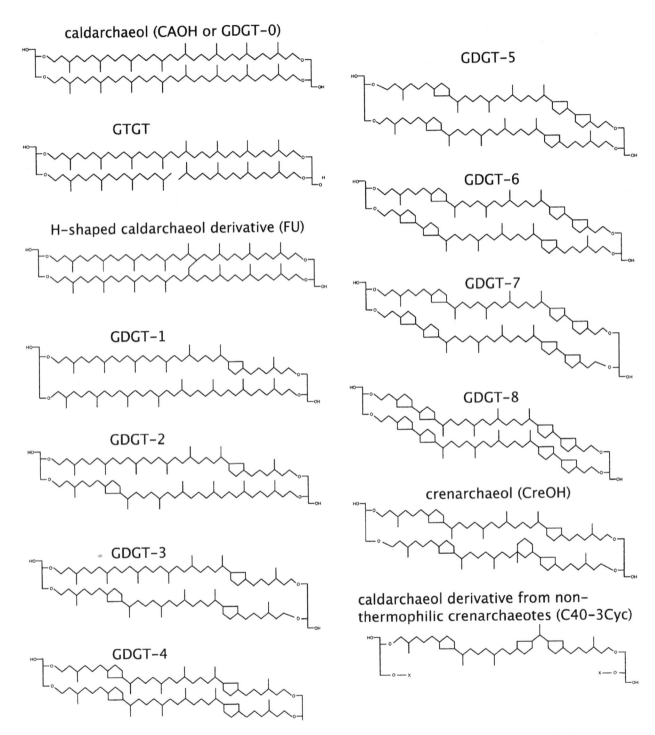

Figure 2. Basic structure of glycerol tetraether isoprenoid lipids. GDGT (glycerol-diakyl-glycerol tetraether) can display a number of cyclic rings (0 to 8) in its alkyl core. In GTGT (glycerol-triakyl-glycerol tetraether), only two of the four phytanyl side chains from the precursor diether lipids are linked by a C—C bond. Although only the antiparallel configuration is shown for the two glycerols forming the backbone of the tetraether lipids, both isomers are likely to be found in archaeal cells (43). GDNT (glycerol-dialkyl-nonitol tetraether) versions of most GDGTs, where one of the glycerol moieties is replaced by nonitol, are also found in a variety of archaea. The sugars or polar head groups usually attached to the hydroxy of the glycerol moieties in archaeal lipids are not shown.

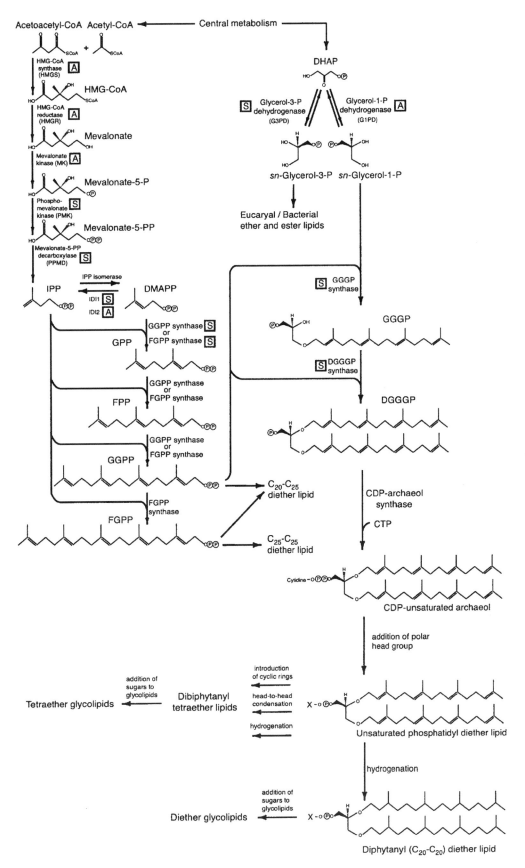

Figure 3. Pathways for the biosynthesis of archaeal glycerol ether isoprenoid lipids. Boxed "A" beside an enzyme name indicates that the enzyme is found in all archaea, and boxed "S" indicates that the enzyme is found in some archaea.

(farnesyl diphosphate, 15 carbons), GGPP (geranylgeranyl diphosphate, 20 carbons), and FGPP (farnesylgeranyl diphosphate, 25 carbons). These chain-elongation reactions are catalyzed by isoprenyl diphosphate synthases, such as GGPP and FGPP synthases, which differ from one another by the allylic substrate they accept to start the elongation process, and the chain length of their product(s) (22). In archaea, GGPP synthase can elongate DMAPP to obtain both FPP and GGPP, the latter being the isoprenyl forming the side chains of C20-C20 diether lipids. Archaea harboring C20-C25 or C25-C25 diether lipids require an FGGP synthase, which can either elongate directly from DMAPP or from longer allylic substrates like GGPP. Among the *Crenarchaeota*, only *Aeropyrum pernix* is known to display C25 side chains, C25-C25 archaeol being the sole core lipid found in this species. The only *Euryarchaeota* known to have one or two C25 chains in their core lipids are haloalkaliphilic archaea from the order *Halobacteriales* (17).

Synthesis of the Glycerol Phosphate Backbone

Although the unique stereoconfiguration of archaeal lipids has been known for decades, there have been questions about the precursor used to synthesize the glycerol phosphate backbone. The *sn*-glycerol-1-phosphate (G1P) dehydrogenase enzyme from *Methanothermobacter thermautotrophicus* was identified a decade ago to be responsible for the synthesis of the *sn*-glycerol-1-phosphate phospholipid backbone from dihydroxyacetone phosphate (DHAP) (Fig. 3) (34). It was subsequently demonstrated that G1P dehydrogenase activity was exhibited by other archaea (36) and that a G1P dehydrogenase gene is present in all sequenced archaeal genomes (3).

Linking Backbone and Side Chains

The enzyme responsible for linking the first side chain to the glycerol phosphate backbone of C20-C20 diether lipids was first characterized from *M. thermautotrophicus* (46). This enzyme, termed geranylgeranylglyceryl phosphate (GGGP) synthase, is encoded in all sequenced archaeal genomes, with the exception of the obligate parasite/symbiont *Nanoarchaeum equitans* (15). GGGP synthase strongly favors *sn*-glycerol-1-phosphate as a substrate, and therefore plays a role in defining the stereoconfiguration of archaeal lipids. The enzymatic activity responsible for adding the second C20 isoprenoid side chain in *M. thermautotrophicus* diether lipids, digeranylgeranylglyceryl phosphate (DGGGP) synthase, was identified in a cell extract that separately contained activity for GGGP synthase (5). The enzymatic activity of a recombinant

DGGGP synthase from *Sulfolobus acidocaldarius* has also been characterized in vitro (14).

Addition of Polar Head Groups

The product of DGGGP synthase, *sn*-2,3-digeranylgeranylglyceryl phosphate, is the substrate of CDP-archaeol synthase, which adds a cytidine group at the *sn*-1 position of the glycerol moiety to produce unsaturated CDP-archaeol (Fig. 3). As GGGP synthase, DGGGP synthase, and CDP-archaeol synthase all require fully unsaturated isoprenyl side chains as substrate (23), the saturation of double bonds must occur after the formation of unsaturated CDP-archaeol. By analogy to the fatty acid biosynthesis pathway in bacteria, unsaturated CDP-archaeol is thought to be a precursor for the addition of polar head groups or sugars at the *sn*-1 position (30). The nature of the head groups or sugars found in archaeal lipids varies greatly in different taxa (20, 24). The head groups of phospholipids can be a variety of polar compounds (glycerol, serine, inosine, ethanolamine, myo-inositol, aminopentanetretols), and sugar moieties of glycolipids can take many forms (glucose, mannose, galactose, gulose, *N*-acetylglucosamine, or combinations thereof) (13, 17, 20, 25, 31, 44). None of the enzymes responsible for the addition of sugar moieties are known, and only one enzyme catalyzing the addition of a polar head group has been characterized (archaetidylserine synthase, responsible for the addition of L-serine) (29).

Modification of the Side Chains: Hydrogenation and Cyclization

None of the enzymes involved in the saturation of isoprenoid side chains, introduction of cyclic rings in the alkyl core of tetraether lipids, or cyclization of diether lipids (formation of macrocyclic diether lipids or tetraether lipids) are currently known. However, some experiments have revealed information about the order in which these biosynthetic reactions might take place. Nemoto et al. (32) demonstrated that the diether lipid precursors of *Thermoplasma acidophila* tetraether lipids contained a modified polar head group (glycerophosphate). Addition of polar head groups was suggested to precede the formation of a tetraether molecule in *S. acidocaldarius*, where two molecules of archaetidylinositol would be precursors to the tetraether lipids (48). In *T. acidophila*, the sugar moieties of glycolipids only seem to be attached after the synthesis of the tetraether lipid core (32), and this may also be the case in *S. acidocaldarius*.

It is unclear whether saturation of side chains occurs during or after the head-to-head condensation reaction leading to tetraether lipids. An unusual double-bond migration can occur along the unsaturated side chains of lipid precursors in *M. thermautotrophicum* and *Methanococcus jannaschii*, leading to an isomerized intermediate with a terminal double bond. According to Eguchi et al. (9), this isomer could be the precursor for the creation of C—C bonds leading to macrocyclic and tetraether lipids. In this proposed mechanism, the formation of a C—C bond is triggered by the addition of a proton, which suggests that it would occur simultaneously with the saturation of the side chains. On the other hand, Nemoto et al. (32) have shown from mass spectrometry experiments of the lipids of *T. acidophila* that the diether precursors of the tetraether lipids are fully saturated. They suggest that saturation/unsaturation reactions could be reversible and that a terminal double bond could be reintroduced before the condensation reaction leading to the tetraether lipid.

The presence of ether lipids with partially saturated side chains in the methanogens, *Methanococcoides burtonii* and *Methanopyrus kandleri*, and the halophiles, *Halobacterium salinarum* and *Natronobacterium magadii*, adds a level of complexity by demonstrating that saturation reactions can also be partial (33, 35, 41). Moreover, the introduction of cyclic rings in the side chains, including head-to-head condensation of diether lipids, is thought to require the presence of some double bonds. However, similar to the studies on partially saturated side chains (34, 36, 42), there is little experimental evidence on when these structures are formed during the process of archaeal lipid biosynthesis.

EVOLUTION OF LIPID BIOSYNTHESIS IN *ARCHAEA*

Origins of the Archaeal Mevalonate Pathway

All organisms harboring a functional mevalonate pathway possess homologous enzymes catalyzing the first three steps: HMG-CoA synthase (HMGS), HMG-CoA reductase (HMGR), and mevalonate kinase (MVK) (Table 1). As a rule, Life's three domains (*Bacteria*, *Archaea*, and *Eucarya*) exhibit monophyly for these enzymes (enzymes within a domain are more similar to each other than between domains). However, there are some striking and well-supported exceptions. Some of the archaeal genes in the mevalonate pathway appear to have been acquired by lateral gene transfer (LGT). For example, all representatives of the *Euryarchaeota* from the *Archaeoglobales* and *Thermoplasmatales* harbor an HMGR gene of seemingly bacterial origin (2).

In the mevalonate pathway of bacteria and fungi, the phosphorylation of mevalonate is catalyzed by

mevalonate kinase (MVK), and the further phosphorylation of phosphomevalonate is catalyzed by a homologous protein, phosphomevalonate kinase (PMK). In archaea, PMK is only found in the genus *Sulfolobus* and is most closely related to the PMKs from fungi (3). A protein with high similarity to PPMD from eucarya is encoded adjacent to the PMK in the genome sequences of *S. tokadaii* and *S. solfataricus*. In phylogenetic analyses, this protein clusters with eucaryal homologs (3). This suggests that the PMK and PPMD found in *Sulfolobus* were acquired from eucarya by one of their ancestors. The presence of these two enzymes only in *Sulfolobus* signifies that this is the sole genus of the *Archaea* known to possess all five enzymes of the standard mevalonate pathway. However, *Sulfolobus* are not the only archaea to harbor a PPMD. This enzyme is also found in the *Halobacteriales* and the *Thermoplasmatales* (Table 1). Phylogenetic analysis indicates that the PPMD in these archaea are likely to have a different origin than the *Sulfolobus* enzymes. Indeed, these PPMD genes cluster with bacterial homologs and have distinct sequences from the eucaryal enzymes (3).

The isoprenoid biosynthesis enzyme isopentenyl diphosphate isomerase (IDI) type 1 (IDI1) is found in multiple genera of *Halobacteriales* and is possibly ubiquitous in this group (3). However, it is not found in any other archaea (Table 1). The fact that this enzyme is common in *Eucarya* and *Bacteria* but appears to be restricted to *Halobacteriales* in *Archaea* indicates that it was acquired by an ancestral *Halobacteriales* from either of these domains. All archaea, including the *Halobacteriales*, harbor the analogous type 2 IDI (IDI2) described by Kaneda et al. (18). Type 2 IDI is therefore likely to have been present in the ancestors of archaea, with a more recent acquisition of the type 1 IDI in the *Halobacteriales*.

Variation of Isoprenoid Side-Chain Length

The extent to which isoprenyl diphosphate synthases can synthesize isoprenoid chains of varying length was demonstrated by mutational studies. A single amino acid substitution changed *S. acidocaldarius* GGPP synthase into an enzyme that synthesized FGPP as its main product, with small amounts of hexaprenyl diphosphate as a secondary product (30 carbons isoprenyl diphosphate) (37). The same enzyme with two or three amino acid substitutions was capable of converting DMAPP, FPP, or GPP to a main product of 35 or 40 carbons in length (heptaprenyl and octaprenyl diphosphates) and secondary products of 65 to 120 carbons in length (38).

Such a change in the length of synthesized products has occurred for the isoprenyl diphosphate synthase of *A. pernix*. This archaeon only synthesizes disestert-erpanyl (C25-C25) diether lipids. This is rare among archaea as most only produce diphytanyl C20-C20 diether lipids (8). *A. pernix* does not possess a GGPP synthase, but instead encodes a single FGPP synthase (49). The FGPP synthase was probably derived from an ancestral archaeal GGPP synthase; phylogenetic analysis places the *A. pernix* FGPP synthase in a group with archaeal GGPP synthases (3). A similar change in isoprenyl diphosphate synthase polyisoprenoid chain elongation properties must have occurred for haloalkiliphilic species of *Halobacteriales*, as they are the only other archaea to harbor sesterterpanyl side chains in their lipids. However, these archaea maintained GGPP synthase activity and synthesize C20-C20, C20-C25, and/or C25-C25 core lipids (17). It is not known whether these *Halobacteriales* have separate GGPP and FGPP synthases or an enzyme that displays plasticity in the chain length of its products.

Emergence of the Specific Stereochemistry of Archaeal Lipids

Three enzymes involved in the biosynthesis of archaeal lipids show stereospecificity. *sn*- glycerol-1-phosphate (G1P) dehydrogenase introduces stereospecificity into archaeal lipids by specifically synthesizing glycerol phosphate with the *sn*-1 stereoconfiguration from dihydroxyacetone phosphate (DHAP) (34). Geranylgeranylglyceryl phosphate (GGGP) synthase strongly favors this glycerol phosphate stereoisomer when attaching the first isoprenoid side chain, yielding *sn*-3-geranylgeranylglyceryl phosphate. The attachment of the second side chain is also stereospecific, as DGGGP synthase will only recognize the *sn*-1 phosphate monoether as opposed to an *sn*-3 phosphate monoether substrate, yielding only unsaturated archaeol (*sn*-2,3-digeranylgeranylglyceryl phosphate or DGGGP) as a product (5). CDP-archaeol synthase then catalyzes the condensation of a CTP molecule with the unsaturated archaeol released by DGGGP synthase to obtain unsaturated CDP-archaeol (30). This enzyme does not recognize the stereochemical structure of the glycerol phosphate backbone or the linkage between glycerol and the isoprenoid side chains (ester or ether linkage). It therefore seems that the specific stereoconfiguration of archaeal lipids is established by G1P dehydrogenase, as well as the GGGP and DGGGP synthases.

G1P dehydrogenase and glycerol dehydrogenase catalyze similar reactions; the substrate (dihydroxyacetone phosphate or dihydroxyacetone) and product (*sn*-glycerol-1-phosphate or glycerol) being phosphorylated for the former, but not the latter, enzyme. These enzymes are also homologous (sharing about 20 to 25% amino acid identity) members of the NAD-dependent dehydrogenase superfamily (Fig. 3). Although G3P and G1P dehydrogenases are functionally equivalent (with the exception of their stereospecificity), they share little sequence similarity. Phylogenetic analysis presents G1P dehydrogenases as a monophyletic cluster among the larger NAD-dependent dehydrogenase superfamily (Fig. 4). This suggests that G1P dehydrogenase is an archaeal invention derived from an enzyme of the NAD-dependent dehydrogenase superfamily, possibly glycerol dehydrogenase, which shares similar sequence, substrate, and product.

The enzyme adding the first isoprenoid side chain to the glycerol backbone, GGGP synthase, is selective for *sn*-glycerol-1-phosphate as an acceptor (5). A surprising finding from phylogenetic analysis is the existence of two distinct but homologous types of GGGP synthases. Several *Halobacteriales* as well as *Archaeoglobus fulgidus* possess divergent enzymes that cluster with homologs from various species of the bacterial order *Bacillales* (Fig. 4). Two paralogs of this divergent enzyme are present in a few species of *Halobacteriales*. The role of these two paralogs is unclear. It is possible that one of them encodes a farnesylgeranylglyceryl phosphate synthase, as some *Halobacteriales* (including *Haloterrigena, Halococcus, Natronobacterium, Natrinema, Natrialba*, and *Natronomonas*) produce C20-C25 and/or C25-C25 diether lipids (17). However, some *Halobacteriales* that only have C20-C20 lipids (*Haloferax* and *Halobacterium*) also exhibit two paralogs.

Due to its presence in several *Halobacteriales* genera, the divergent type of GGGP synthase was most likely present in the ancestors of this order (and possibly of the order *Archaeoglobales* as well) (Fig. 4). How this divergent type of GGGP synthase arose is unclear. As GGGP synthase homologs are ubiquitous in archaea and only found in the order *Bacillales*, a *Chlorobium* species and a *Cytophaga* species in the Bacteria (Fig. 4), it is likely that this enzyme is an archaeal invention that was later acquired by bacteria through LGT. The *Halobacteriales, Archaeoglobus*, and *Bacillales* homologs are closely related to each other and distinct from other archaeal homologs and the *Cytophaga/Chlorobium* genes. The *Bacillales* GGGP synthase homologs would therefore have originated from the *Archaeoglobales* or the *Halobacteriales*, while the *Cytophaga/Chlorobium* homolog would descend from the enzymes of some other archaeal group. Since no ether lipids of the *sn*-2,3-stereoconfiguration are found in bacteria, the enzymes present in the *Bacillales* and *Cytophaga/Chlorobium* were probably co-opted for a different function, possibly DNA replication (3, 15). Lateral transfer of GGGP synthase homologs is also likely to have happened

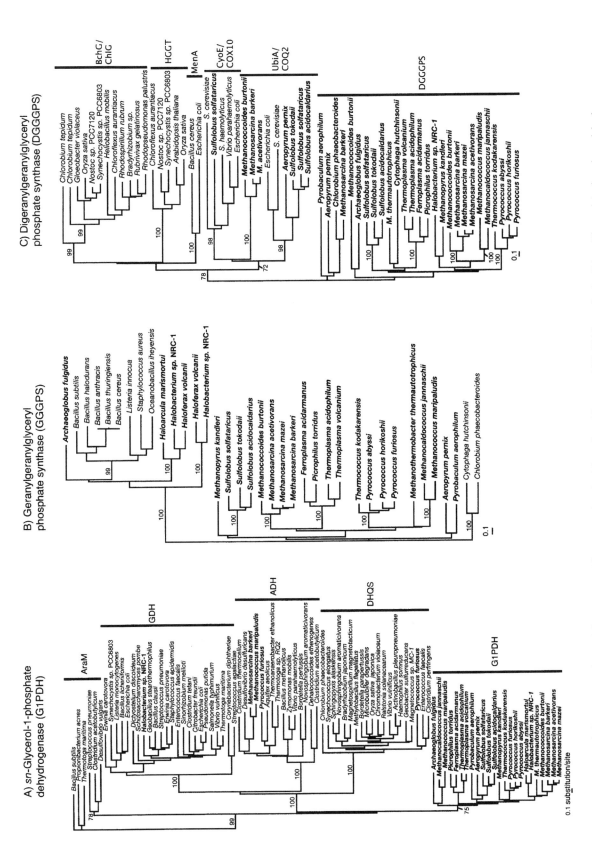

Figure 4. Phylogenetic analysis of the stereospecific enzymes involved in archaeal isoprenoid lipid biosynthesis. (A) *sn*-Glycerol-1-phosphate dehydrogenase (G1PDH) and related protein families; dehydroquinate synthase (DHQS), L-arabinose isomerase (AraM), glycerol dehydrogenase (GDH), alcohol dehydrogenase (ADH). (B) Geranylgeranylglyceryl phosphate synthase (GGGPS). (C) Digeranylgeranylglyceryl phosphate synthase (DGGGPS) and related protein families; Bacteriochlorophyll/chlorophyll synthase (BchG/ChlG), homogentisic acid geranylgeranyl transferase (HGGT), 1,4-dihydroxy 2-naphtoate octaprenyl-transferase (MenA), heme biosynthesis farnesyltransferase (CyoE/COX10), ubiquinone biosynthetic polyprenyl transferase (UbiA/COQ2). The trees presented are based on maximum likelihood amino acid distances under the minimum evolution model and were obtained using PROTDIST. Bootstrap values represent the consensus of 100 trees obtained from pseudo-replicates of the original dataset. Taxon names of archaea are highlighted in bold.

among archaea, as the specific relationship found between *Archaeoglobus* and *Halobacteriales* GGGP synthases (Fig. 4) is not consistent with the archaeal phylogeny derived from proteins involved in translation or small-subunit ribosomal proteins (28). The incongruence between well-known phylogenetic markers and GGGP synthase suggests that the latter enzyme was laterally transferred between the ancestors of the *Halobacteriales* and *Archaeoglobus*.

Similar to G1P dehydrogenase, DGGGP synthase does not display a history of LGT (Fig. 4). It is related to multiple families of prenyltransferases (e.g., chlorophyll biosynthesis, heme biosynthesis) and has probably originated from a large prenyltransferase superfamily. Therefore, with the exception of GGGP synthase, the stereospecific enzymes of archaeal lipid biosynthesis appear to have been derived from a homologous enzyme with a broadly similar function in the same protein superfamily. These changes of function in archaeal proteins might have been a critical step in the origin of this domain of life.

FUNCTION OF ARCHAEAL LIPIDS

Most of the microorganisms adapted to life in "extreme" environments (high salinity, high or low pH, high or low temperature, high pressure) are members of the *Archaea*. The survival ability of many members of the *Archaea* may be attributed to their unique cell membranes. Indeed, membranes composed of ether lipids have a higher stability than those made of ester lipids (10, 39) (see Chapters 14 and 23). In addition to the ether linkage between backbone and side chains, other features of archaeal lipids have been proposed to facilitate the growth of archaea in "extreme" environments. For example, cyclic lipids (such as macrocyclic diether lipids or tetraether lipids) and pentacyclic rings in side chains are thought to facilitate life at higher temperatures. The composition of the *M. jannaschii* cell membrane is greatly altered depending on its growth temperature; the proportion of macrocyclic and tetraether lipids increases compared with acyclic diether lipids when the organism is grown at higher temperatures (47). Cyclization of the side chains decreases their freedom of motion and leads to an appropriate level of membrane fluidity at higher temperatures. Pentacyclic rings are similarly believed to reduce membrane fluidity. In response to increased growth temperature, *Sulfolobus solfataricus* increases the proportion of membrane lipids that contain six to eight pentacyclic rings, at the expense of those containing none to two rings (12).

Members of the *Crenarchaeota* which live at low temperature also display cyclic rings in the side-chains of their lipids (7). The presence of such cyclic rings in their membrane lipids is likely to be a consequence of their shared ancestry with thermophilic *Crenarchaeota*. Given the evolutionary success of this cold-adapted group of archaea (i.e., they are present in great numbers in oceans), cyclic rings have limited effects on growth at lower temperature and/or their impact on membrane fluidity is compensated for by other structural characteristics. Some of these nonthermophilic marine *Crenarchaeota*, such as the sponge symbiont *Cenarchaeum symbiosum*, contain crenarchaeol in their cell membrane; a lipid that harbors a hexacyclic ring in one of its side-chains (Fig. 2). This hexacyclic ring was thought to prevent dense packing of lipid membranes, thereby increasing their fluidity (6). However, crenarchaeol has recently been found in hot springs, where increased membrane fluidity would have a negative impact on fitness for bacterial and archaeal organisms (41). Following this discovery, it was suggested that the presence of hexacyclic rings might instead increase membrane porosity and provide some biological advantage associated with permeability (40).

Archaeal lipids may contain varying degrees of saturation. The saturation level of ether lipid side chains is apparently not a purely genotypically determined trait, as it can vary with growth conditions. In the methanogen *Methanococcoides burtonii*, the proportion of unsaturated lipids from cells grown at 4°C is significantly higher than for cells grown at 23°C (33). It has been shown for the fatty acids of bacterial cell membranes that membrane fluidity is directly related to the degree of unsaturation (42). Higher degrees of unsaturation can maintain a more fluid membrane at lower temperatures. The presence of a high proportion of unsaturated lipids in *M. burtonii*'s cell membrane seems to be an adaptation to growth at low temperature, and its ability to regulate the proportion of such lipids in its cellular envelope would allow it to survive fluctuations in the temperature of its environment.

Factors other than growth temperature might influence the type of lipids found in cell membranes. For example, it has been suggested that the amount of solutes found in the environment could regulate the type of lipids composing the cellular membrane, which in turn would modify membrane permeability (40). The fact that C25 sesterterpanyl side chains are only found in alkaliphiles among the Halobacteriales could indicate that the thicker cell membrane formed by these core lipids helps them tolerate a high pH environment (17). The composition of cell membranes can also vary with the growth phase. For example, in *M. jannaschii*, the ratio of tetraether lipids to diether lipids increases when cells progress from logarithmic to stationary-phase growth (47). Atmospheric pressure was also shown to vary the proportion of different lipids present in the cell membrane of this archaeon (19). We are just beginning to unravel environmental factors that influence the composition of cell membranes in archaea and to link particular lipid structures with their function.

PERSPECTIVE: THE NEXT FIVE YEARS

The enzymes catalyzing most of the steps of lipid biosynthesis up to the addition of head groups have been characterized. However, for two steps of the mevalonate pathway (which synthesizes the isporenoid precursors IPP and DMAPP; Fig. 3), the archaeal analogs have not been identified (PMK and PPMD; Table 1). It should be possible to identify the genes for these enzymes through complementation experiments, using genetically engineered *Escherichia coli* that harbors specific genes of the mevalonate pathway (4).

The detection of CDP-archaeol synthase and archaetidylserine synthase activities in *M. thermautotrophicus* suggests that the biochemical steps for the addition of polar head groups on archaeal lipid precursors, might proceed in a manner analogous to fatty acid biosynthesis in bacteria. In bacteria, the addition of a cytidyl group at the *sn*-1 position of the glycerol backbone is a precursor step to the addition of a polar head group (which replaces the cytidyl group). The situation might be similar in archaea, where CDP-archaeol synthase could perform this cytidilation step, and enzymes such as archaetidylserine synthase may subsequently add polar head groups. This potential analogy (and possible homology) between bacterial and archaeal pathways could guide the discovery of enzymes involved in the addition of other important polar head groups (in addition to serine). Once these enzymes have been characterized, the door to the production of molecules such as archaetidylglycerol or archaetidylinositol is opened. These have been proposed as precursors for subsequent biosynthesis steps, such as the head-to-head condensation that produces tetraether lipids, or the addition of sugars leading to glycolipids (32, 48).

Recent introduction of different combinations of techniques has increased our capacity to purify and do a structural analysis of the polar membrane lipids of archaea. High-performance liquid chromotography (HPLC) combined with evaporative light-scattering detection (ELSD) allowed the determination of the complete polar lipid composition of *T. acidophila* (using gas chromatography, mass spectrometry, and NMR to get structural data on isolated peaks) (44). HPLC, in combination with electrospray mass spectrometry

(ES-MS) was used to characterize the membrane phospholipids and glycolipids of halophilic archaea and the cold-adapted methanogen *M. burtonii* (using tandem mass spectrometry for structural data) (34, 42). An interesting finding of the latter studies was that lipids with partially saturated side chains were present in cell membranes. Determining how this occurs may provide insight into the process of side-chain hydrogenation. Such lipid analyses can be used to look at the variation of membrane lipid composition under different growth conditions, as was illustrated for *M. burtonii*, in this case, where the proportion of unsaturated lipids was found to vary with growth temperature (33). This type of analysis has so far only been performed on a few different archaea and under a limited number of environmental variables (pressure and temperature). Examination of a wider range of taxa and environmental factors using these methods could help to correlate lipid structure with cell membrane function.

REFERENCES

1. **Boucher, Y., and W. F. Doolittle.** 2000. The role of lateral gene transfer in the evolution of isoprenoid biosynthesis pathways. *Mol. Microbiol.* **37:**703–716.
2. **Boucher, Y., H. Huber, S. L'Haridon, K. O. Stetter, and W. F. Doolittle.** 2001. Bacterial origin for the isoprenoid biosynthesis enzyme HMG-CoA reductase of the archaeal orders Thermoplasmatales and Archaeoglobales. *Mol. Biol. Evol.* **18:**1378–1388.
3. **Boucher, Y., M. Kamekura, and W. F. Doolittle.** 2004. Origins and evolution of isoprenoid lipid biosynthesis in archaea. *Mol. Microbiol.* **52:**515–527.
4. **Campos, N., M. Rodriguez-Concepcion, S. Sauret-Gueto, F. Gallego, L. M. Lois, and A. Boronat.** 2001. *Escherichia coli* engineered to synthesize isopentenyl diphosphate and dimethylallyl diphosphate from mevalonate: a novel system for the genetic analysis of the 2-C-methyl-d-erythritol 4-phosphate pathway for isoprenoid biosynthesis. *Biochem. J.* **353:**59–67.
5. **Chen, A., D. Zhang, and C. D. Poulter.** 1993. (S)-geranylgeranylglyceryl phosphate synthase. Purification and characterization of the first pathway-specific enzyme in archaebacterial membrane lipid biosynthesis. *J. Biol. Chem.* **268:**21701–21705.
6. **Damste, J. S., S. Schouten, E. C. Hopmans, A. C. van Duin, and J. A. Geenevasen.** 2002. Crenarchaeol: the characteristic core glycerol dibiphytanyl glycerol tetraether membrane lipid of cosmopolitan pelagic crenarchaeota. *J. Lipid Res.* **43:**1641–1651.
7. **DeLong, E. F., L. L. King, R. Massana, H. Cittone, A. Murray, C. Schleper, and S. G. Wakeham.** 1998. Dibiphytanyl ether lipids in nonthermophilic crenarchaeotes. *Appl. Environ. Microbiol.* **64:**1133–1138.
8. **De Rosa, M., and A. Gambacorta.** 1988. The lipids of Archaebacteria. *Prog. Lipid Res.* **27:**153–175.
9. **Eguchi, T., H. Takyo, M. Morita, K. Kakinuma, and Y. Koga.** 2000. Unusual double-bond migration as a plausible key reaction in the biosynthesis of the isoprenoidal membrane lipids of methanogenic archaea. *Chem. Commun.* 1545–1546.
10. **Elferink, M. G., J. G. de Wit, A. J. Driessen, and W. N. Konings.** 1994. Stability and proton-permeability of liposomes

composed of archaeal tetraether lipids. *Biochim. Biophys. Acta* **1193**:247–254.

11. Gattinger, A., M. Schloter, and J. C. Munch. 2002. Phospholipid etherlipid and phospholipid fatty acid fingerprints in selected euryarchaeotal monocultures for taxonomic profiling. *FEMS Microbiol. Lett.* **213**:133–139.

12. Gliozzi, A., G. Paoli, M. De Rosa, and A. Gambacorta. 1983. Effect of isoprenoid cyclization on the transition temperature of lipids in thermophilic archaebacteria. *Biochim. Biophys Acta* **735**:234–242.

13. Hafenbradl, D., M. Keller, and K. O. Stetter. 1996. Lipid analysis of *Methanopyrus kandleri*. *FEMS Microbiol. Lett.* **136**:199–202.

14. Hemmi, H., K. Shibuya, Y. Takahashi, T. Nakayama, and T. Nishino. 2004. (S)-2,3-Di-O-geranylgeranylglyceryl phosphate synthase from the thermoacidophilic archaeon *Sulfolobus solfataricus*. Molecular cloning and characterization of a membrane-intrinsic prenyltransferase involved in the biosynthesis of archaeal ether-linked membrane lipids. *J. Biol. Chem.* **279**:50197–50203.

15. Jahn, U., R. Summons, H. Sturt, E. Grosjean, and H. Huber. 2004. Composition of the lipids of *Nanoarchaeum equitans* and their origin from its host *Ignicoccus sp.* strain KIN4/I. *Arch. Microbiol.* **182**:404–413.

16. Jahnke, L. L., W. Eder, R. Huber, J. M. Hope, K. U. Hinrichs, J. M. Hayes, D. J. Des Marais, S. L. Cady, and R. E. Summons. 2001. Signature lipids and stable carbon isotope analyses of Octopus Spring hyperthermophilic communities compared with those of Aquificales representatives. *Appl. Environ. Microbiol.* **67**:5179–5189.

17. Kamekura, M., and M. Kates. 1999. Structural diversity of membrane lipids in members of the Halobacteriaceae. *Biosci. Biotechnol. Biochem.* **63**:969–972.

18. Kaneda, K., T. Kuzuyama, M. Takagi, Y. Hayakawa, and H. Seto. 2001. An unusual isopentenyl diphosphate isomerase found in the mevalonate pathway gene cluster from *Streptomyces sp.* strain CL190. *Proc. Natl. Acad. Sci. USA* **98**:932–937.

19. Kaneshiro, S. M., and D. S. Clark. 1995. Pressure effects on the composition and thermal behaviour of lipids from the deep-sea thermophile *Methanococcus jannaschii*. *J. Bacteriol.* **177**:3668–3672.

20. Kates, M. 1992. Archaebacterial lipids: structure, biosynthesis and function. *In* M. J. Danson, D. W. Hough, and G. G. Lunt (ed.), *The Archaebacteria: Biochemistry and Biotechnology.* Portland Press, London, United Kingdom.

21. Kates, M., and N. Kushwaha. 1978. Biochemistry of the lipids of extremely halophilic bacteria, p. 461–480. *In* S. R. Caplan and M. Ginzburg (ed.), *Energetics and Structure of Halophilic Microorganisms.* Elsevier, Amsterdam, The Netherlands.

22. Kellog, B., and C. D. Poulter. 1997. Chain elongation in the isoprenoid biosynthetic pathway. *Curr. Opin. Chem. Biol.* **1**:570–578.

23. Koga, Y., and H. Morii. 2006. Special methods for the analysis of ether lipid structure and metabolism in archaea. *Anal. Biochem.* **348**:1–14.

24. Koga, Y., H. Morii, A. M. Masayo, and M. Ohga. 1998. Correlation of polar lipid composition with 16S rRNA phylogeny in Methanogens. Further analysis of lipid component parts. *Biosci. Biotechnol. Biochem.* **62**:230–236.

25. Koga, Y., M. Nishihara, H. Morii, and M. Akagawa-Matsuhita. 1993. Ether polar lipids of methanogenic bacteria: structures, comparative aspects, and biosyntheses. *Microbiol. Rev.* **57**:164–182.

26. Lange, B. M., T. Rujan, W. Martin, and R. Croteau. 2000. Isoprenoid biosynthesis: the evolution of two ancient and distinct pathways across genomes. *Proc. Natl. Acad. Sci. USA* **97**:13172–1317.

27. Mangold, H. K., and F. Paltauf. 1983. *Ether Lipids: Biochemical and Biomedical Aspects.* Academic Press, New York, N.Y.

28. Matte-Tailliez, O., C. Brochier, P. Forterre, and H. Philippe. 2002. Archaeal phylogeny based on ribosomal proteins. *Mol. Biol. Evol.* **19**:631–639.

29. Morii, H., and Y. Koga. 2003. CDP-2,3-Di-O-geranylgeranyl-sn-glycerol:L-serine O-archaetidyltransferase (archaetidylserine synthase) in the methanogenic archaeon *Methanothermobacter thermautotrophicus*. *J. Bacteriol.* **185**:1181–1189.

30. Morii, H., M. Nishihara, and Y. Koga. 2000. CTP:2,3-di-O-geranylgeranyl-sn-glycero-1-phosphate cytidyltransferase in the methanogenic archaeon *Methanothermobacter thermoautotrophicus*. *J. Biol. Chem.* **275**:36568–36574.

31. Morii, H., H. Yagi, H. Akutsu, N. Nomura, Y. Sako, and Y. Koga. 1999. A novel phosphoglycolipid archaetidyl(glucosyl)inositol with two sesterterpanyl chains from the aerobic hyperthermophilic archaeon *Aeropyrum pernix* K1. *Biochim. Biophys. Acta* **1436**:426–436.

32. Nemoto, N., Y. Shida, H. Shimada, T. Oshima, and A. Yamagishi. 2003. Characterization of the precursor of tetraether lipid biosynthesis in the thermoacidophilic archaeon *Thermoplasma acidophilum*. *Extremophiles* **7**:235–243.

33. Nichols, D. S., M. R. Miller, N. W. Davies, A. Goodchild, M. Raftery, and R. Cavicchioli. 2004. Cold adaptation in the Antarctic Archaeon *Methanococcoides burtonii* involves membrane lipid unsaturation. *J. Bacteriol.* **186**:8508–8515.

34. Nishihara, M., and Y. Koga. 1995. sn-glycerol-1-phosphate dehydrogenase in *Methanobacterium thermoautotrophicum*: key enzyme in biosynthesis of the enantiomeric glycerophosphate backbone of ether phospholipids of archaebacteria. *J. Biochem. (Tokyo)* **117**:933–935.

35. Nishihara, M., H. Morii, K. Matsuno, M. Ohga, K. O. Stetter, and Y. Koga. 2002. Structural analysis by reductive cleavage with LiAlH4 of an allyl ether choline-phospholipid, archaetidylcholine, from the hyperthermophilic methanoarchaeon *Methanopyrus kandleri*. *Archaea* **1**:123–131.

36. Nishihara, M., T. Yamazaki, T. Oshima, and Y. Koga. 1999. sn-glycerol-1-phosphate-forming activities in Archaea: separation of archaeal phospholipid biosynthesis and glycerol catabolism by glycerophosphate enantiomers. *J. Bacteriol.* **181**:1330–1333.

37. Ohnuma, S., K. Hirooka, C. Ohto, and T. Nishino. 1997. Conversion from archaeal geranylgeranyl diphosphate synthase to farnesyl diphosphate synthase. Two amino acids before the first aspartate-rich motif solely determine eukaryotic farnesyl diphosphate synthase activity. *J. Biol. Chem.* **272**:5192–5198.

38. Ohnuma, S., K. Hirooka, N. Tsuruoka, M. Yano, C. Ohto, H. Nakane, and T. Nishino. 1998. A pathway where polyprenyl diphosphate elongates in prenyltransferase. Insight into a common mechanism of chain length determination of prenyltransferases. *J. Biol. Chem.* **273**:26705–26713.

39. Patel, G. B., and G. D. Sprott. 1999. Archaeabacterial ether lipid liposomes (archaeosomes) as novel vaccine and drug delivery systems. *Crit. Rev. Biotechnol.* **19**:317–357.

40. Pearson, A., Z. Huang, A. E. Ingalls, C. S. Romanek, J. Wiegel, K. H. Freeman, R. H. Smittenberg, and C. L. Zhang. 2004. Nonmarine crenarchaeol in Nevada hot springs. *Appl. Environ. Microbiol.* **70**:5229–5237.

41. Qiu, D., M. P. Games, X. Xiao, D. E. Games, and T. J. Walton. 2000. Characterisation of membrane phospholipids and glycolipids from a halophilic archaebacterium by high-performance liquid chromatography/electrospray mass spectrometry. *Rapid Commun. Mass Spectrom.* **14**:1586–1591.

42. Russell, N. J., and D. S. Nichols. 1999. Polyunsaturated fatty acids in marine Bacteria-a dogma rewritten. *Microbiology* 145:767–779.

43. Schouten, S., E. C. Hopmans, R. D. Pancost, and J. S. Damste. 2000. Widespread occurrence of structurally diverse tetraether membrane lipids: evidence for the ubiquitous presence of low-temperature relatives of hyperthermophiles. *Proc. Natl. Acad. Sci. USA* 97:14421–14426.

44. Shimada, H., N. Nemoto, Y. Shida, T. Oshima, and A. Yamagishi. 2002. Complete polar lipid composition of *Thermoplasma acidophilum* HO-62 determined by high-performance liquid chromatography with evaporative light-scattering detection. *J. Bacteriol.* 184:556–563.

45. Smit, A., and A. Mushegian. 2000. Biosynthesis of isoprenoids via mevalonate in Archaea: the lost pathway. *Genome Res.* 10:1468–1484.

46. Soderberg, T., A. Chen, and C. D. Poulter. 2001. Geranylgeranylglyceryl phosphate synthase. Characterization of the recombinant enzyme from *Methanobacterium thermoautotrophicum*. *Biochemistry* 40:14847–14854.

47. Sprott, G. D., M. Meloche, and J. C. Richards. 1991. Proportions of diether, macrocyclic diether, amd tetraether lipids in *Methanococcus jannaschii* grown at different temperatures. *J. Bacteriol.* 173:3907–3910.

48. Sugai, A., I. Uda, K. Kon, S. Ando, Y. Itoh, and T. Itoh. 1996. Structural identification of minor phosphoinositol lipids in *Sulfolobus acidocaldarius* N-8. *J. Jpn. Oil Chem. Soc.* 45:327–333.

49. Tachibana, A., Y. Yano, S. Otani, N. Nomura, Y. Sako, and M. Taniguchi. 2000. Novel prenyltransferase gene encoding farnesylgeranyl diphosphate synthase from a hyperthermophilic archaeon, *Aeropyrum pernix*. Molecularevolution with alteration in product specificity. *Eur. J. Biochem.* 267:321–328.

Archaea: Molecular and Cellular Biology
Edited by Ricardo Cavicchioli
© 2007 ASM Press, Washington, D.C.

Chapter 16

Solute Transport

Sonja V. Albers, Wil N. Konings, and Arnold J. M. Driessen

SOLUTE TRANSPORT IN *ARCHAEA*

Living cells are surrounded by a lipid membrane that forms a barrier between the cytoplasm and the extracellular environment. These membranes are impermeable for most ions and solutes, thereby enabling cells to control the ionic and nutrient composition of their cytoplasm. The membrane permeability barrier is essential for keeping optimal conditions for metabolic and energy-transducing reactions. Nevertheless, specific ions and solutes are necessary for cellular processes, and these molecules need to be transported into the cell. On the other hand, waste products need to be removed from the cell. Both the uptake and excretion processes involve specialized integral membrane proteins, i.e., transport proteins. Membrane proteins are embedded in the lipid layer of the cytoplasmic membrane. Typically, the lipid molecules form a bilayer, but some hyperthermophilic archaea, such as *Sulfolobus*, contain bipolar membrane-spanning lipids that form a lipid monolayer with the same thickness as a lipid bilayer (see Chapter 15). The fluidity and permeability properties of the membranes are mainly determined by their lipid composition. An optimal fluidity of the membrane is required for the barrier function of the membrane, and also for the activity of the membrane proteins. Likewise, the proton permeability of the membrane needs to be controlled to permit efficient energy transduction (94). Organisms can respond to changes in the environment, such as temperature, salinity, or pH, and adjust the composition of the cytoplasmic membrane by alteration of the lipid composition (94). Within limits, adjustment of the lipid composition allows homeostasis of membrane fluidity, proton permeability, and the phasic properties of the lipids.

THE CYTOPLASMIC MEMBRANE AND BIOENERGETICS

Cells can generate metabolic energy by distinct mechanisms. One is substrate-level phosphorylation, whereby chemical energy released in catabolic processes is stored in ADP or ATP. The second mechanism utilizes energy-transducing systems located in the cytoplasmic membrane that generate ATP or an electrochemical ion gradient, which in turn is used to produce ATP. Electron transfer systems or membrane-bound ATPases are examples of such systems. They translocate protons or sodium ions across the cytoplasmic membrane into the medium and thereby create an electrochemical ion gradient across the membrane. When protons are extruded, these mechanisms yield an electrochemical gradient of protons that is composed of a transmembrane pH gradient (ΔpH) that is "inside alkaline versus acidic outside," and a transmembrane electrical potential ($\Delta\Psi$) that is "inside negative versus positive outside." Both the ΔpH and $\Delta\Psi$ exert an inward-directed force on the protons, the proton motive force (PMF). The PMF can be expressed in millivolts according to the following formula:

$$\text{PMF} = \Delta\Psi - 2.303 \, (RT/F) \, \Delta\text{pH}$$

where R is the gas constant, T is the absolute temperature (K), and F is the Faraday constant. The effect of 1 pH unit difference between cytoplasm and external medium is 59 mV at 25°C, and 70 mV at 80°C. In most organisms, the PMF has a negative value and the driving force of the protons is directed into the cell. However, in many organisms (e.g., halophiles), sodium ions are used as coupling ions instead of protons. By analogy to the PMF, extrusion of sodium

Sonja V. Albers, Wil N. Konings, and Arnold J. M. Driessen • Department of Microbiology, Groningen Biomolecular Sciences and Biotechnology Institute and Materials Science Center Plus, University of Groningen, Kerklaan 30, 9751 NN Haren, The Netherlands.

ions from the cytoplasm into the medium results in the formation of an electrochemical sodium gradient that exerts a sodium motive force (SMF). The latter is composed of a $\Delta\Psi$ and the transmembrane chemical gradient of the sodium ions:

$$SMF = \Delta\Psi + 2.303 \, RT/F \, \log[Na_{in}^+]/[Na_{out}^+]$$

The PMF and SMF can be used for energy-requiring processes such as ATP synthesis, substrate transport across the membrane, flagellar rotation, and maintenance of the intracellular pH. Evidently, these processes can take place only if the electrochemical gradients of protons and sodium ions remain intact. This is only possible if the cytoplasmic membrane has a limited permeability for these ions.

Extreme acidophiles such as *Picrophilus* and *Sulfolobus* maintain an intracellular pH that is close to neutrality. As a consequence, these organisms experience a very high ΔpH across their cell membrane. This ΔpH can be up to 4 pH units, which it is for *Picrophilus* (82, 93). Such a high ΔpH can only be maintained with a membrane that has a very low proton permeability. On the other hand, the very high ΔpH in these archaea is partially compensated by an inverted $\Delta\Psi$ (negative outside versus positive inside) to keep the PMF within viable values (63). The inversion of the $\Delta\Psi$ is mainly realized by the uptake of potassium ions. Alkalophiles also maintain their intracellular pH close to neutrality. Thus, in these organisms, the ΔpH is reversed, i.e., alkaline outside versus acidic inside. To keep the PMF at viable values, these organisms maintain a large $\Delta\Psi$ (positive outside versus negative inside) across their membrane.

LIPIDS IN ARCHAEAL MEMBRANES AND ADAPTATIONS TO ENVIRONMENTAL CHANGES

In contrast to eucaryal and bacterial ester lipids, in archaea membrane lipids consist of two phytanyl chains that are linked via an ether bond to a glycerol or other alcohols like nonitol (see Chapter 15). The structure of the archaeal lipids has been reviewed extensively (18, 41, 100). The archaeal lipid chain contains isoprenoid units where every fourth carbon atom is linked to a methyl group. Most of the archaeal phytanyl chains are fully saturated isoprenoids (18, 40, 48, 100). Halobacteria and most mesophilic archaea contain lipids with a C_{20} diether lipid core, which form a bilayer membrane similar to ester lipids (40, 42, 91). In extreme thermophiles and acidophiles

monolayer-forming tetraether lipids are found which consist of C_{40} isoprenoid acyl chains (18).

Archaea respond to changes of the environmental temperature by the adaptation of the lipid composition of the cytoplasmic membrane, and as discussed above, these changes are necessary to keep the membrane in a liquid crystalline state and to limit its proton permeability. In *Thermoplasma* and *Sulfolobus solfataricus* lipids the degree of cyclization of the C_{40} isoprenoid in the tetraether lipid increases with the growth temperature (17, 23, 55). In the euryarchaeote *Methanococcus jannaschii*, an increase in temperature induces a change from diether lipids to tetraether lipids, which are more thermostable (86). Cyclization of the C_{40} isoprenoid chains and the use of tetraether lipids results in a tighter packing of the lipid acyl chains and a restricted mobility of the lipids in the membrane. This prevents an intolerable increase of membrane fluidity and enables these archaea to tolerate elevated growth temperatures. The arctic archaeon *Methanococcoides burtonii* employs unsaturation of ether lipids to adapt to low-growth temperatures. The fraction in the membrane of unsaturated lipids was significantly higher at 4°C than at 23°C (68). The unsaturation appears to be caused by an incomplete reduction of the lipid precursor instead of by a desaturase mechanism (68). In marine crenarchaeota a novel tetraether lipid (crenarchaeol) was found. The characteristic glycerol dibiphytanyl glycerol tetraether (GDGT) membrane lipid contains one cyclohexane and four cyclopentane rings formed by internal cyclization of the biphytanyl chains. Its structure is similar to that of GDGTs biosynthesized by (hyper)thermophilic crenarchaeota apart from the cyclohexane ring (15). The degree of cyclization of the core of the lipid increased within a temperature range from 0 to 30°C (84).

GENERAL PRINCIPLES OF TRANSPORT: MECHANISMS AND CLASSES

Solute transport systems can be classified in different groups according to their molecular architecture and their energy requirement (Fig. 1). Five classes of transport system have been identified: (i) channels, including the well-known examples of mechanosensitive channels, which can exhibit either large or small conductance; (ii) secondary transporters, which make use of the electrochemical gradient of either protons, sodium ions, or substrates to drive transport of substrates across the membrane. Secondary transporters are subdivided into three classes: (1) uniporters, which transport solutes without a coupling ion; (2) symporters, which translocate solutes together with a coupling ion, such as protons or sodium ions; and (3) anti-

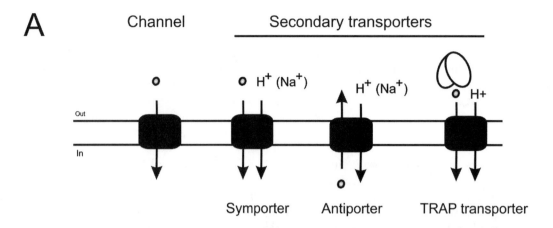

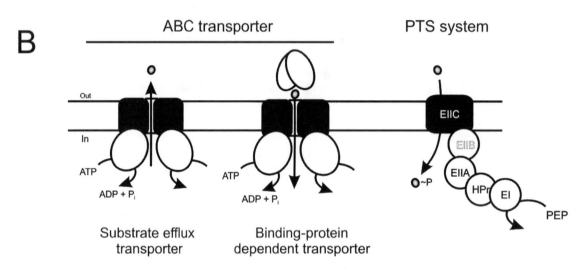

Figure 1. Classes of transporters. (A) Channels and different modes of secondary transporters. (B) ABC transporters and the PTS system. Integral membrane subunits of transporters (filled); cytoplasmic or extracellular subunits (open oval); transported substrate (open circle). The names of the subunits of the PTS system refer to the mannitol transporter of *E. coli*.

porters, which transport two substrates/ions in opposite directions. Secondary transporters usually consist of one polypeptide chain containing 4 to 14 transmembrane segments; (iii) binding-protein-dependent secondary transporters (TRAP transporters), which consist of a periplasmic binding protein and a membrane protein. These systems use the PMF or SMF to drive uptake of solutes; (iv) group translocation systems, i.e., the phosphoenolpyruvate (PEP)-dependent phosphotransferase systems (PTS), which couple transport of sugars to phosphorylation; and finally (v) primary transporters, which use light or chemical energy such as ATP or other compounds to drive substrate translocation. Well-studied examples are ion-translocating respiratory chains, the light-driven proton pump (bacteriorhodopsin), various types of ion-translocating ATPases, and the ABC (ATP-binding cassettes) trans-

porters. ABC transporters have a typical modular domain structure which usually comprises two integral membrane proteins that form the permease domain, and two cytoplasm-located ATPases which drive the transport of the substrate by the hydrolysis of ATP. In contrast to bacteria and archaea, ABC transporters in eucarya usually consist of a single polypeptide that encompasses all four domains. However, in bacteria and archaea, the membrane and ATPase domains may either exist as homo- or heterodimers, or even as separately fused membrane or ATPase domains. The ABC export systems are often homo- or heterodimers of a membrane domain fused with an ATPase domain. Bacterial and archaeal ABC uptake systems comprise a fifth subunit which is an extracellular substrate-binding-protein that binds the substrate at the outside of the cell and delivers it to the permease.

DISTRIBUTION OF TRANSPORT SYSTEMS IN ARCHAEA

Archaeal membrane proteins have only rarely been studied. The exception is bacteriorhodopsin, which has been studied in great detail by various biochemical, structural, and biophysical methods. This light-induced proton pump was first isolated in the early 1970s from membranes of *Halobacterium halobium* and *H. cutirubrum* (53, 70). The protein exists in a semicrystalline state in the purple membrane; membranes are purple due to the color of bacteriorhodopsin. The ease of its purification from purple membranes has made it a very suitable membrane system to study. Bacteriorhodopsin contains a nonprotein cofactor; the retinal that absorbs light is mechanistically involved in the proton-pumping mechanism. The structural and molecular mechanism of light-induced proton pumping has been elucidated and discussed in many excellent reviews (21, 32, 66).

During recent years, the genomic sequences of a large number of *Archaea* have been determined. The transport proteins were identified in 18 archaeal genome sequences (72) and described in a relational database (78) (see TransportDB at http://www.mem branetransport.org/). All classes of transporters can be found in archaea except for PTS systems which are completely absent. Strikingly, PTS systems are also absent in the hyperthermophilic bacteria *Thermotoga maritima* and *Aquifex aeolicus* that deeply branch in the universal phylogenetic tree, indicating that PTS-systems may have arisen relatively late in evolution.

The number of transporters present in the different archaeal genomes can be expressed as the number of transporters per Mb of genome. The value for this parameter varies from 21.3 in *Methanopyrus kandleri* to 73 in *Picrophilus torridus* and *Thermoplasma volcanicum*, and in most archaea it is about 40. Because there are no PTS systems present in archaea, only four classes of transporters exist, i.e., ion channels, ATP-dependent transporters and secondary transporters, and TRAP transporters, which so far have been identified in only three archaeal species. The ion channels constitute the smallest class and amount to only 5 to 8% of all transporters. Only *M. kandleri* and *Nanoarchaeum equitans* are exceptions, with 14% and 19% ion channels, respectively. Whereas *N. equitans* contains two small conductance mechanosensitive ion channels, *M. kandleri* encodes three candidates of voltage-gated ion channels. On average, about 50% of the transporters belong to the class of secondary transporters, while the remainder (~37%) are ATP-dependent transport systems. The ATP-dependent transport systems are the predominant class, only in

Aeropyrum pernix, *N. equitans*, and *Pyrobaculum aerophilum*. In *A. pernix* and *P. aerophilum* this is mainly caused by a high number of ATP-binding cassette (ABC) transporters, whereas in *N. equitans* (which has a small genome), 5 of the 8 genes belong to an F-type ATPase. In this organism, only three ABC transport genes are found, of which only one encodes a protein with a predicted membrane domain. This suggests that the other two, which have ATP-binding domains but not membrane domains, have other, maybe cytoplasmic, functions.

In the group of acidophiles (*P. torridus, S. solfataricus, Sulfolobus tokodaii, Thermoplasma acidophilum,* and *T. volcanicum*), the fraction of secondary transporters (~68%) is very large. The genomes of these organisms encode a multitude of members of the amino acid-polyamine-organocation (APC) family of secondary transporters. These systems mostly function as solute:cation symporters or solute:solute antiporters (77). Moreover, the analysis of all archaeal genome sequences indicates that secondary transporters of the NiCoT (Ni^{2+}/Co^{2+}) and Nramp (metal-ion transporter) family are only found in acidophiles. These systems are often involved in the uptake of heavy metals or iron. Due to the selective pressures that metal-rich acidic environments pose, acidophiles need to be resistant to heavy metals and some of these transporters might be involved in the resistance mechanisms.

Channels

Mechanosensitive (MS) channels have been extensively studied in eucarya and bacteria (30). In eucarya, MS channels play an important role in organ functions, such as hearing, touching, and in cell swelling. In bacteria, MS channels mainly play a role in the survival of osmotic stresses experienced under hypotonic conditions. The physiological function of MS channels in archaea has not been studied, but it is likely to be similar to bacteria. Interestingly, the growth of an *Escherichia coli* strain that lacks the endogenous MS channels can be partially restored by expression of the MS channel of *M. jannaschii*. However, this can only occur in a medium with high osmolarity (45). MS channels were first described in *Haloferax volcanii* (56), and systems from *T. acidophilum*, *T. volcanicum*, and *Methanocaldococcus jannaschii* were cloned and characterized (44–46). These channels share many mechanistic features, such as gating by osmotic stress, a large conductance and low ion selectivity, and weak voltage dependence. The two MS channels from *M. jannaschii*, MscMJ and MscMJLR, differ significantly in conductance (0.3 nS and 2.0 nS, respectively) (46), but they appear highly selective for cations. In this respect, these channels functionally resemble the Ca^{2+}-permeable channels of skeletal and heart muscle (60).

Two classes of ion channels can be discriminated that are opened either by ligand binding (e.g., neurotransmitters) or by the transmembrane potential (voltage-gated channels). Structural information for both classes of channels has been gathered from archaeal counterparts, i.e., the calcium-gated K^+-channel, MthK, of *Methanothermobacter thermautotrophicus* (36) and the voltage-dependent K^+-channel, KvAP, of *A. pernix* (37).

MthK is a tetramer, and each subunit contains two integral membrane domains and one C-terminal RCK (regulators of K^+ conductance) domain, which is located in the cytoplasm. The RCK domains form a gating ring. When Ca^+ binds, the pore of the channel opens to allow K^+ to be transported (Fig. 2). The voltage-dependent K^+-channel from *A. pernix*, KvAP, was first shown to be a functional and structural homolog of eucaryal voltage-gated K^+-channels (79). Subsequent structural studies revealed the presence of "voltage-sensor paddles" (37, 38), which may carry the charge across the membrane along the inface of the transporter and the lipid membrane to open the channel. However, this model (37, 38) is being debated as it opposes the conventional model in which the charge is transported along a helix in the core of the protein. Recent data suggest that the movement of the sensor paddles might only be very small (11, 78) and insufficient to carry ions across the membrane.

Secondary Transporters

Although secondary transporters are the most abundant class in archaeal genomes, few studies have been performed on members of this class. The earliest report on secondary transporters in archaea is from 1977 and describes the characterization of the uptake systems for amino acids in membrane vesicles of *H. halobium* in response to a light-induced $\Delta\psi$ and an artificially imposed sodium-ion gradient (59). This study reports the kinetics of all amino acid transporters, except for cysteine and aspartate. The aspartate transporter was purified from membranes of *H. halobium* and reconstituted in proteoliposomes (26). This system utilizes the sodium-ion gradient to drive transport and is highly specific for L-aspartate (26). A sodium-dependent glucose transporter has been described in *H. volcanii* (89). Expression of this system was shown to be induced by growth of the cells in the presence of glucose. Transport of glucose was blocked by inhibitors of mammalian glucose transporters (89), suggesting common structural and functional features of both systems.

A lactose transporter was identified in *S. solfataricus* by functionally complementing a mutant strain that was unable to grow on lactose (9). Complementation required both the *lacS* gene (β-galactosidase) and a gene located upstream of *lacS* that encodes a secondary transporter (73). The lactose transporter belongs to the major facilitator family (MFS) and is predicted to contain 12 transmembrane segments.

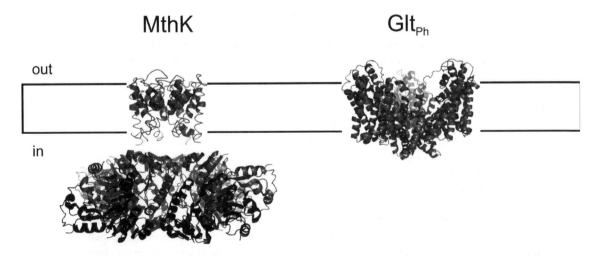

Figure 2. Channel structures. The calcium-gated K^+-channel, MthK of *M. thermautotrophicus*, and the secondary transport structure of the glutamate transporter, Glt_{Ph} of *P. horikoshii*, are shown. The orientation of both transporters in the membrane is depicted. The membrane-inserted part of MthK is relatively small, in comparison with the large complex of cytoplasmically located RCK domains. The structure of Glt_{Ph} shows the large cavity directed toward the extracellular side where the substrate is bound.

Methanosarcina thermophila TM-1 *and Methanohalophilus portucalensis* strain FDF1 were both shown to actively accumulate compatible solutes such as glycine-betaine. The glycine-betaine transporter of *M. thermophila* is highly specific for glycine-betaine and transports this molecule with a high affinity (K_t of 10 µM) (74). Studies with inhibitors suggest that transport depends on a PMF or SMF. However, since the gene(s) have not yet been identified, the exact mechanism remains to be determined. In *M. portucalensis*, growth on trimethylamine in the presence of 2.1 M NaCl was dramatically stimulated by the presence of glycine-betaine (54). This transporter is also specific for glycine-betaine and transports the molecule with a K_t of 23 µM (54).

Recently, the structure of a putative glutamate transporter from *Pyrococcus horikoshii* was elucidated (101). This is the first structural report of a secondary transporter from a member of the *Archaea*. This protein is homologous to high-affinity neurotransmitter transporters of the synaptic cleft in mammals. Although active transport of glutamate could not be demonstrated, the structure suggests a transport mechanism for this trimeric system that involves a large hydrophilic pore exposed to the outside that easily accommodates the substrate (Fig. 2) (101).

ABC Transporters

ABC transporters are a major class of transporters in archaea, and several systems have been studied in a great detail. Similar to bacterial counterparts, the subunits of these transporters exhibit the typical consensus motifs that classify them as ABC transporters. The ATPase subunits contain the typical Walker sequences, the Q-loop and the H-region. Archaeal ABC transporters can be divided into at least two classes, the binding-protein-dependent uptake systems and the multidrug/antibiotic transporter, which probably function as exporters. The latter class is typically composed of two subunits, each with one integral membrane part and a cytoplasmic nucleotide-binding domain. In some cases, these two domains of multidrug transporters are fused into a single polypeptide chain. In addition to the two permease and cytoplasmic ATPase domains, which form either homo- or heterodimers, the binding-protein-dependent ABC transporters constitute a substrate-binding protein. The membrane domains contain the EAAAx3Gx9IxLP motif, which has been shown to be the site of interaction between the membrane domain and the cytoplasmic ATPase subunit (16). Table 1 lists the ABC transporters in archaea that have been studied in some detail. ABC sugar transporters were identified and characterized in *Thermococcus litoralis*, *Pyrococcus furiosus*, and *S. solfataricus*. Transport of osmoprotectants has been studied primarily in methanogens (Table 1).

Bacterial sugar ABC uptake systems are divided into two main classes: the carbohydrate uptake transporters (the CUT class) and the di/oligopeptide transport class (83). The archaeal systems involved in the uptake of mono- or disaccharides belong to the CUT class. However, ABC transporters that transport di- and oligosaccharides, such as the cellobiose/β-glucoside transport system of *P. furiosus* (50) and the maltose/maltodextrin transporter of *S. solfataricus* (22), are homologous to the di/oligopeptide class. The

Table 1. Characterized ABC transporters in the *Archaea*

ABC transporter	Substrate	K_m for uptake (nM)	K_d for solute binding[a] (nM)	Reference
T. litoralis	Maltose/trehalose	22/17	160	34
S. solfataricus	Glucose	2,000	480	2
	Cellobiose + cellooligomers	–[b]	–	22
	Trehalose	–	–	22
	Maltose/maltotriose	–	–	22
	Arabinose	–	130	22
P. furiosus	Cellobiose + cellooligomers	175	45	50
	Maltose/trehalose			51
	Maltotrios, maltodextrin	–	270	24, 51
H. volcanii	Glucose (anaerobic)	–	–	98
	Molybdate	–	–	98
	Inorganic anions	–	–	98
A. fulgidus	Glycine betaine, proline betaine	–	–	33
M. mazei Göl	Glycine betaine	–	–	18
M. portucalensis	Glycine betaine	23,000	–	54
M. thermophila TM-1	Glycine betaine	10,000	–	70

[a]Solute binding to binding protein.
[b]–, not determined.

similarity of the two systems includes the substrate-binding proteins which have shown homology on the primary sequence level to bacterial di/oligopeptide-binding proteins, through to the subunit composition of the transporter, which consists of two permeases and two cytoplasmic ATPases. By analogy to bacterial di/oligopeptide-binding proteins, the homologous archaeal sugar-binding proteins recognize a large range of oligosaccharides (50). Operons encoding subunits of putative di/oligopeptide transporters are highly abundant in hyperthermophilic archaea and bacteria. Strikingly, these operons are positioned in the vicinity of genes that encode sugar-degrading enzymes (65). In *Thermotoga maritima*, some of these transporters were shown to be upregulated when the cells are grown on specific sugar substrates (14). It is advantageous for archaea and bacteria to accumulate shorter sugar oligomers in a single transport step as the uptake of a sugar monomer and oligomer presumably takes similar amounts of ATP. Uptake of oligomers thus minimizes the overall energy requirements for the uptake processes.

In *H. volcanii*, the function of several ABC transporters has been studied by using genetics. Complementation studies of mutants unable to grow under nitrogen limitation yielded three ABC transporters that are involved in the transport of glucose, molybdate, or inorganic ions (98). In a recent study, an ABC transporter for a corrinoid (vitamin B_{12} precursor), was genetically characterized in *Halobacterium* sp. strain NRC-1 (99).

The best-studied example of an archaeal ABC uptake system is the trehalose/maltose transporter of *T. litoralis*. The trehalose/maltose-binding protein (TMBP) binds its substrate with a very high affinity (160 nM at 80°C) (34); transport is also a high-affinity process. The ATPase subunit of this transporter, MalK, was expressed in *E. coli* and characterized (27), the structure was elucidated for both MalK and TMBP (19, 20), and the entire transporter was heterologously expressed in *E. coli* and purified (28). Although the solubilized and purified transporter displayed an intrinsic ATPase activity, the activity was not stimulated by the addition of binding protein, in contrast to studies with bacterial systems (76).

The substrate-binding protein

The binding protein captures the substrate at the outside of the cell and delivers it to the membrane-embedded permease domains, whereupon it is translocated across the membrane. Substrate-binding proteins contain two large domains (lobes) that are linked by a flexible hinge region. Upon binding of the substrate, the two lobes close like a "venus-fly trap" mechanism (75). The structures have been solved for TMBP of *T. litoralis* (20), the maltodextrin-binding protein (*Pfu*MBP) of *P. furiosus* (25), and ProX, the glycine-betaine- and proline-betaine-binding protein of *Archaeoglobus fulgidus* (81). Both TMBP and *Pfu*MBP have high structural similarity to the maltose-binding protein of *E. coli* (*Ec*MBP), even though the primary amino acid sequence identity is low. Structurally equivalent, but not identical, amino acids are involved in sugar binding in TMBP and the *Ec*MBP (20). In all of these binding proteins, two elongated patches of hydrophobic amino acids outline the binding pocket.

No significant heat release or absorption occurs when maltose binds to *Pfu*MBP, in contrast to the endothermic binding of maltose to *Ec*MBP. Both enzymes have the same binding constant. On the other hand, binding of maltotriose to *Pfu*MBP was strongly exothermic and occurs with an affinity ($K_d \sim 2.7 \times 10^7$ M^{-1}) that is 20-fold higher than observed for maltotriose binding to the *Ec*MBP ($K_d \sim 1.3 \times 10^6$ M^{-1}) (25). In comparison with *Ec*MBP, the bound sugar seems less deeply buried in the binding cleft of *Pfu*MBP. The latter structure is much less flexible, indicating that the sugar-bound form occurs more in an "open" than in an "occluded" state. It has been suggested that this binding mode resembles more a "lock-and-key" mode than the "venus-fly trap" mode (25).

ProX from *A. fulgidus* binds glycine-betaine and proline-betaine with a K_d of 60 and 50 nM at room temperature, respectively (33). These compounds act as thermoprotectants in this organism. The structure of ProX was determined without an apo form and in complex with glycine-betaine, proline-betaine, and trimethylammonium (81). Upon binding of these compounds, the two lobes move closer to each other. The binding of the substrates was mediated by cation-pi interactions and nonclassical hydrogen bonds between four tyrosine residues, a main-chain carbonyl oxygen, and the substrates. The mechanism of binding of the ligands occurs in a manner that is similar to the *E. coli* homolog. However, the *A. fulgidus* and *E. coli* proteins have low sequence identity, and the residues involved in binding are structurally equivalent but not conserved (81).

Structural studies on archaeal substrate-binding proteins have been performed with heterologously expressed proteins from *E. coli*. A common feature of the substrate-binding proteins and other extracellular, especially membrane-bound, proteins from archaea is that they are extensively glycosylated in their native host (24, 28, 31, 88) (see Chapter 11). Binding proteins isolated from *P. furiosus* contain glucose moieties (51), whereas mannose, glucose, galactose, and *N*-acetylglucosamine have been identified in the binding protein of *S. solfataricus* (22). Due to the extensive glycosylation, the archaeal binding proteins are readily isolated from solubilized membrane fractions by lectin affinity chromatography (2, 22, 28, 50). However, glycosylation does not appear to be essential for functionality as the heterologous, nonglycosylated proteins normally bind sugar (20, 25, 34, 50, 51, 81). Glycosylation may affect the stability of the proteins, protect them from proteolytic degradation, or influence their interaction with the cell envelope.

The distinction between the CUT and di/oligopeptide classes of ABC transporters is also evident for the protein-domain organization of the binding proteins (Fig. 3A and B). Binding proteins belonging to the CUT class contain an N-terminal signal peptide followed by a stretch of hydroxylated amino acids that is up to 60 residues long (the ST linker). The ST linker might act as a flexible region between the membrane anchor and the binding domain. ST linkers were also identified in a pullulanase (24) and a haloarchaeal S-layer protein (88). In the latter case, this region was the site of O-linked glycosylation.

As archaeal cells are surrounded by an S layer with pores of 4 to 5 nm size (47) (see Chapter 14), small molecules and proteins can easily diffuse into the medium. Therefore, extracellular proteins need to be attached to the archaeal cell envelope. In gram-positive bacteria, sugar-binding proteins are attached to the membrane by a lipid anchor that is covalently attached to the protein after translocation across the membrane. In bacteria, a cysteine residue at the +1 position is first lipidated prior to signal sequence removal by a specific lipoprotein signal peptidase (see Chapter 17). However, archaeal genomes do not appear to contain a homolog of the bacterial enzyme, and it is currently unclear how lipidation takes place

in archaea. Nevertheless, the N termini of many euryarchaeal binding proteins contain the same "lipobox" motif as bacterial proteins, a cysteine residue at the +1 position that is preceded by small hydrophobic residues (4). Analysis of the halocyanin of the archaeon, *Natronobacter pharaonis*, by mass spectroscopy, revealed that the N-terminal cysteine residue is covalently linked to a C_{20} diphytanyl diether lipid (61) (see Chapter 15).

Members of the CUT class of binding proteins from *S. solfataricus* contain type IV pilinlike signal peptides and do not contain secretory signal peptides (2, 5, 22). The signal peptides are usually found in the subunits of bacterial pili and archaeal flagellins (90), and typically consist of a positively charged N terminus followed by a hydrophobic domain (see Chapter 17). The hydrophobic domain acts as a scaffold for the assembly of the translocated subunits into multimeric structure. However, this process requires the positively charged N terminus to be removed by a type IV pilinlike signal peptidase. Because of the presence of type IV pilinlike signal peptides, it has been hypothesized that the binding proteins of *S. solfataricus* multimerize into a structure, referred to as the "bindosome" (1). In vitro assays demonstrated that the signal peptides of these binding proteins are processed by a specialized type IV prepilin signal peptidase, PibD, and not by the typical signal peptidase I (6). This enzyme also processes the flagellin subunits before assembly into the flagellum (7).

The binding proteins of the di/oligopeptide class contain typical bacterial signal peptides that are processed by an archaeal homolog of signal peptidase I (8, 67). In contrast to the CUT class binding proteins, these proteins harbor a hydrophobic region at the

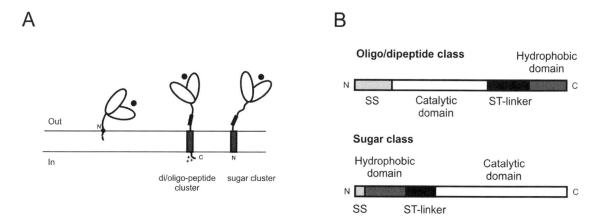

Figure 3. Mode of anchoring of archaeal substrate-binding proteins and their domain structure. (A) Archaeal substrate-binding proteins are either bound to the membrane by a fatty acid modification of their N terminus or a hydrophobic domain at the C or N terminus. (B) Domain organization of substrate-binding proteins. N, N terminus; C, C terminus; SS, signal sequence; ST-linker, serine/threonine-rich amino acid stretch; filled circle, substrate of binding protein.

C terminus that is preceded by a ST linker. Therefore, it appears that the CUT class binding proteins have a domain organization that mirrors the di/oligopeptide class sugar-binding proteins. Because of the presence of the C-terminal hydrophobic domain, which is likely to function as a transmembrane domain to anchor these proteins to the cytoplasmic membrane, the di/oligopeptide class binding proteins do not need to be lipid modified at their N termini. However, some pyrococcal binding proteins contain a GGICG sequence motif immediately downstream of the hydrophobic domain at their C terminus. Similar to some halophilic S-layer proteins that contain this motif, C-terminal lipid modification (52) may occur in a similar manner.

Many archaea contain binding proteins of the CUT class (4). *S. solfataricus* is an exception that has more binding proteins from the di/oligopeptide class (rather than the CUT class), and methanogens do not contain any member of the di/oligopeptide class. This might imply that *S. solfataricus* grows on carbon sources derived from higher oligomeric substrates without the need of extracellular enzymatic "pre-digestion," whereas methanogens do need to enzymatically cleave substrates before they can be transported into the cell.

Halobacterium contains a class of binding proteins that function as receptors in chemotaxis rather than in substrate uptake (49) (see Chapter 18). Although some bacterial substrate-binding proteins are also involved in extracellular substrate concentration sensing, their main function is transport (87). In general, substrate-binding proteins cluster in operon-like structures together with the other structural genes of the ABC transporter. However, in *Halobacterium*, the substrate binding-proteins, BasB (branched amino acids) and CosB (compatible solutes), are found in transcriptional units with their cognate transducer protein (49).

The ATP-binding protein

ATP-binding proteins of ABC transporters provide the driving force for substrate translocation. These proteins typically exist as homo- or heterodimers. Binding of ATP stabilizes the dimeric form of the isolated subunits, whereas hydrolysis of ATP causes the destabilization of the dimer. In short, upon ATP binding, the conformational change in the ATPase dimer affects the permease domain and results in a conformational change of the membrane domains and the passage of the substrate. ABC ATPases are highly conserved proteins in all three domains of life and share several motifs, such as the Walker A/B motifs (98) and the ABC signature motif, LSGGQ (10). The residues of the Walker motifs have been shown to be important

for binding of ATP and hydrolysis, whereas the ABC signature motif is essential for the dimerization process following ATP binding (64).

Many archaeal ATP-binding proteins have been crystallized, such as LolD (MJ0769) and LivG (MJ1267) from *M. jannaschii* (102), MalK from the trehalose/maltose transporter of *T. litoralis* (19), and GlcV, the ATPase subunit of the glucose transporter of *S. solfataricus* (95). Studies on LivG showed that the ABC signature motif is essential for dimer formation. Dimerization results in the completion of the catalytic ATP-binding site, as both monomers contribute residues for the active site (64). Although the ABC ATPase dimer contains two nucleotide-binding sites, hydrolysis of ATP at only one of the sites appears sufficient for transport (96). The overall fold of the archaeal ATP-binding domains is structurally the same as that of bacterial and eucaryal proteins.

MalK and GlcV contain an additional domain at their C-terminus (Fig. 4) that predominantly consists of β-sheets that are organized in an OB (oligonucleotide/oligosaccharide) fold. This C-terminal extension is also present in MalK; the ATPase of the *E. coli* maltose ABC transporter (12). This domain interacts with a positive regulator of the *mal* operon, MalT (71). When MalT is bound by MalK, activation of the *mal* operon cannot occur. However, when maltose is present in the medium, ATP is hydrolyzed by MalK, and MalT is released into the cytosol, where it can activate the transcription of the *mal* operon. This is the only example of an ATPase subunit that is involved in the regulation of expression of its own operon. MalT homologs have not been identified in

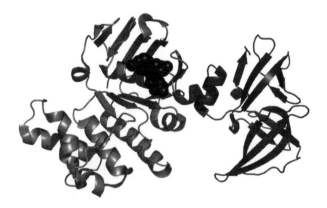

ATP-binding domain   C-terminal domain

Figure 4. Structure of the ATP-binding subunit, GlcV, of the glucose ABC transporter *of S. solfataricus*. The ATP-binding domain contains all the necessary residues for ATP hydrolysis. Bound ATP is shown in the structure (ball model). The function of the C-terminal domain is unknown.

archaea, and at this stage it is unclear if the C-terminal extension of the archaeal ABC ATPases are involved in regulation. Expression of the trehalose/maltose transporter of *T. litoralis* is repressed by the negative regulator, TrmB (57) (see Chapter 6). In *P. furiosus*, TrmB regulates both the trehalose/maltose transporter and the maltodextrin transporter (51, 58). This argues against the presence of a MalT-like regulation system in archaea. DNA microarray studies in *P. furiosus* revealed that the 16-kb region in the genome that is nearly identical with a portion of the *T. litoralis* genome showed that all the transporter genes are upregulated during growth on maltose, whereas *malK* and *trmB* retain high levels of expression, even when cells are grown on peptides. This implies that there might be a regulatory role for MalK beside the direct regulation of the operon by TrmB, but how this regulation is conferred remains to be investigated.

Comparison of ABC transporters present in *Sulfolobus* species

The genome sequences of three *Sulfolobus* species have been completed: *S. tokodaii*, *S. solfataricus* (43, 86), and *S. acidocaldarius* (13). *S. solfataricus* contains 28 ABC transporters, while *S. tokodaii* and *S. acidocaldarius* contain only 18 and 13 systems, respectively (Table 2). The larger number of ABC transporters in *S. solfataricus* can be mainly attributed to a subset of binding-protein-dependent ABC transporters. Of the three *Sulfolobus* species, *S. solfataricus* can utilize the largest number of carbon sources for growth. It can grow heterotrophically in a basal salt medium on a wide range of sugars, including glucose, cellobiose, maltose, arabinose, lactose, and others (22, 29). This diversity of carbon source utilization is reflected by the diversity of ABC transporters. Substrates of five ABC transporters in *S. solfataricus* have been identified (2, 22). In *S. tokodaii*, only a close homolog of the cellobiose transporter is present, whereas no homologs of the arabinose, glucose, maltose, or trehalose transporters have been identified either in *S. tokodaii* or *S. acidocaldarius*.

As described above (see "The substrate-binding protein"), a group of substrate-binding proteins of *S. solfataricus* contain type IV pilinlike signal peptides (5, 22). Most of the ABC transporters with a binding protein with type IV pilin signal peptides are not present in the other two strains. However, the predicted sugar transporter in *S. solfataricus* SSO1171-SSO1168 has homologs in the other strains, and these binding proteins all possess these unusual signal peptides. Moreover, the homologs of the predicted di/oligopeptide transporter SSO2619-SSO2616, Saci1038-1034

and Sto2539-2543, both contain binding proteins with type IV pilin signal peptides, whereas the equivalent transporter from *S. solfataricus* contains a typical type I signal peptide. All three strains contain a homolog of the type IV pili signal peptidase, PibD, which has been described and characterized from *S. solfataricus*. However, in *S. acidocaldarius* the peptidase has not been correctly annotated. The substrate-binding proteins with type IV pilin signal peptides appear to be unique to *Sulfolobales*. By analogy to the type IV signal sequence containing proteins, flagella and pili, their role in *Sulfolobus* species is likely to enable the binding proteins to assemble into multimeric structures on the membrane or S-layer surface.

MULTIDRUG TRANSPORTERS IN *ARCHAEA*

A general feature of bacteria and archaea is that they are equipped with defense systems against toxic compounds from the environment. This protection mechanism involves multiple drug transport systems that extrude toxic compounds from the cell. In general, the drug transport systems belong to the class of secondary transporters or ABC transporters. A P-glycoprotein-like drug efflux pump was described in *H. volcanii* (62). An anthracycline-resistant mutant of *H. volcanii* showed an improved ability to extrude a range of drugs (e.g., Rhodamine123), compared with wild-type cells (39). Similar to the human multidrug transporter, P-glycoprotein, the drug transport in the *H. volcanii* mutant could be reversed by the addition of an uncoupler, FCCP, or the $Ca(2+)$-channel antagonist, nifedipine (NDP) (62). This indicated that the improved resistance was mediated by an ABC transporter (62).

Another multidrug transporter has been characterized in *H. salinarum* (69). Hsmr is a small transporter that belongs to the group of small multidrug resistance proteins that are typically only 110 amino acids in length. These systems function as drug/proton antiporters. Hsmr was expressed in *E. coli* and was shown to actively extrude ethidium from cells. Detergent-solubilized Hsmr binds the substrate tetraphenylphosphonium (TPP^+) with high affinity, in a salt-dependent manner, a K_D of 200 and 40 nM at low and high salt, respectively. Hsmr has a highly unusual sequence, with over 40% of the amino acids being valine and alanine residues. These are distributed throughout the protein and are often concentrated in regions where there is little or no sequence conservation with other transporters. The authors suggested that this phenomenon is the outcome of a natural process of alanine and valine preference that is caused by the high GC content of the gene (69).

Table 2. Comparison of the predicted ABC-transport clusters in the three *Sulfolobus* species

Family[a]	*S. solfataricus* (SSO)			*S. acidocaldarius* (Saci)			*S. tokodaii*		
	Binding protein	Permease(s)	ATPase(s)	Binding protein	Permease(s)	ATPase(s)	Binding protein	Permease(s)	ATPase(s)
Amino acid	2835[b,c]	2839, 2841	2836, 3838	–	–	–	–	–	–
Antibiotic	–	1080	1078	–	–	–	–	–	–
Antibiotic	–	1936	1934	2304	2305	–	–	2348	2437
Di/oligopeptide	1273	1274, 1275	1276, 1277	1760	1762, 1763	1764, 1765	0706	0702, 0704	0700, 0701
Di/oligopeptide	1288	1283, 1284	1281, 1282	–	–	–	–	–	–
Di/oligopeptide[d]	3053	3058, 3059	3055	–	–	–	–	–	–
Di/oligopeptide[e]	2669	2668, 2671	2670, 2672	–	–	–	2534	2535, 2532	2536, 2537
Di/oligopeptide	3043	3047, 3048	3045, 3046	–	–	–	–	1648, 1649	1647
Di/oligopeptide	2619	2617, 2618	2615, 2616	1038[f]	1037, 1036	1035, 1034	2539[f]	2540, 2541	2542, 2543
Inorganic ion	2029	2031	2030	1811	1809	1810	1137	1135	1136
Inorganic ion	1894	1992	1893	1220	1221	1222	–	–	–
Iron	0485	0486	0487	–	–	–	0840	0841	0840
Multidrug	–	0051	0053	–	–	–	–	0609	0608
Multidrug	–	1318	1319	–	–	–	–	1771	1772
Multidrug	–	2135	2137	–	0945	0946	–	0536	0535
Multidrug	–	2404	2402	1007	1006	–	–	–	–
Multidrug	–	2601	2600	–	1800	1799	–	0755	0756
Multidrug	–	2646	2647	–	2123	−[f]	–	1099	–
Multidrug	–	3012	–	–	2303	2302	–	2445	2446
Multidrug	–	3168	3169	–	–	−[f]	–	2510	2509
Nitrate	–	2468	2469	–	0258	0257	–	0594	0595
Phosphate	0489[f]	0490, 0491	0488	–	–	–	–	–	–
Sugar	1171[f]	1169, 1170	1168	1165[f]	1163, 1164	1166	1103[f]	1104, 1105	1106
Sugar	2712[f]	2714	2713	–	–	–	1915	1918	1917
Sugar (glucose)	2847[f]	2848, 2849	2850	–	–	–	–	–	–
Sugar (arabinose)	3066[f]	3067, 3068	3069	–	–	–	–	–	–
Sugar (trehalose)	0999[f]	1000, 1001	1003	–	–	–	–	–	–
Sulfate	–	1032	1034	–	0114	–	–	2033	2035

[a] As assigned by the COG database.
[b] Numbers indicate the ORF numbers as annotated in the different genome sequences (13, 43, 85).
[c] The entries along the same row indicate homologous sets of genes in the respective organisms.
[d] Maltose transporters.
[e] Cellobiose transporters.
[f] Binding proteins with type IV pilin signal peptides.

PERSPECTIVE: THE NEXT FIVE YEARS

Membrane proteins from (hyper)thermophiles are very amendable to structural studies, because they tend to crystallize much more easily than their mesophilic counterparts (see Chapter 20). The crystallization of membrane proteins has caused problems for many years as it has been difficult to obtain sufficiently large amounts of purified membrane proteins. Moreover, the need for lipids or detergents to prevent precipitation of these hydrophobic proteins in the crystallization solutions causes additional problems. Expression of archaeal membrane proteins remains a bottleneck. Whereas most archaeal cytoplasmic proteins are easily produced in bacterial hosts, archaeal membrane proteins are difficult to express heterologously. The recent development of expression systems for *Sulfolobus* (3) and *Thermococcus* (80) may provide new opportunities for producing large quantities of functional archaeal membrane proteins. To date, only five structures of archaeal membrane proteins have been solved. Bacteriorhodopsin was the first that was crystallized; a process that was greatly facilitated by the presence of these proteins in a semicrystalline state in the halobacterial membrane. Other examples of archaeal membrane protein structures are for the voltage-gated channel of *M. thermautotrophicus*, MthK (36), the voltage-dependent K^+-channel from *A. pernix* (35), the glutamate transporter of *P. horikoshii* (101), and the protein-secretion (SecY) complex from *M. jannaschii* which is involved in the translocation of proteins across the cytoplasmic membrane (92) (see Chapter 17).

REFERENCES

1. **Albers, S. V., and A. J. Driessen.** 2005. Analysis of ATPases of putative secretion operons in the thermoacidophilic archaeon *Sulfolobus solfataricus*. *Microbiology* **151:**763–773.
2. **Albers, S. V., M. G. Elferink, R. L. Charlebois, C. W. Sensen, A. J. Driessen, and W. N. Konings.** 1999. Glucose transport in the extremely thermoacidophilic *Sulfolobus solfataricus* involves a high-affinity membrane-integrated binding protein. *J. Bacteriol.* **181:**4285–4291.
3. **Albers, S. V., M. Jonuscheit, S. Dinkelaker, T. Urich, A. Kletzin, R. Tampe, A. J. M. Driessen, and C. Schleper.** 2006. Production of recombinant and tagged proteins in the hyperthermophilic archaeon *Sulfolobus solfataricus*. *Appl. Environ. Microbiol.* **72:**102–111.
4. **Albers, S. V., S. M. Koning, W. N. Konings, and A. J. Driessen.** 2004. Insights into ABC transport in archaea. *J. Bioenerg. Biomembr.* **36:**5–15.
5. **Albers, S. V., W. N. Konings, and A. J. M. Driessen.** 1999. A unique short signal sequence in membrane anchored proteins of Archaea. *Mol. Microbiol.* **31:**1595–1596.
6. **Albers, S. V., Z. Szabó, and A. J. Driessen.** 2003. Archaeal homolog of bacterial type IV prepilin signal peptidases with broad substrate specificity. *J. Bacteriol.* **185:**3918–3925.
7. **Bardy, S. L., and K. F. Jarrell.** 2002. FlaK of the archaeon *Methanococcus maripaludis* possesses preflagellin peptidase activity. *FEMS Microbiol. Lett.* **208:**53–59.
8. **Bardy, S. L., S. Y. Ng, D. S. Carnegie, and K. F. Jarrell.** 2005. Site-directed mutagenesis analysis of amino acids critical for activity of the type I signal peptidase of the archaeon *Methanococcus voltae*. *J. Bacteriol.* **187:**1188–1191.
9. **Bartolucci, S., M. Rossi, and R. Cannio.** 2003. Characterization and functional complementation of a nonlethal deletion in the chromosome of a beta-glycosidase mutant of *Sulfolobus solfataricus*. *J. Bacteriol.* **185:**3948–3957.
10. **Boos, W., and H. Shuman.** 1998. Maltose/maltodextrin system of *Escherichia coli*: transport, metabolism, and regulation. *Microbiol. Mol. Biol. Rev.* **62:**204–229.
11. **Chanda, B., O. Kwame Asamoah, R. Blunck, B. Roux, and F. Bezanilla.** 2005. Gating charge displacement in voltage-gated ion channels involves limited transmembrane movement. *Nature* **436:**852–856.
12. **Chen, J., G. Lu, J. Lin, A. L. Davidson, and F. A. Quiocho.** 2003. A tweezers-like motion of the ATP-binding cassette dimer in an ABC transport cycle. *Mol. Cell* **12:**651–661.
13. **Chen, L., K. Brugger, M. Skovgaard, P. Redder, Q. She, E. Torarinsson, B. Greve, M. Awayez, A. Zibat, H. P. Klenk, and R. A. Garrett.** 2005. The genome of *Sulfolobus acidocaldarius*, a model organism of the Crenarchaeota. *J. Bacteriol.* **187:**4992–4999.
14. **Chhabra, S. R., K. R. Shockley, S. B. Conners, K. L. Scott, R. D. Wolfinger, and R. M. Kelly.** 2003. Carbohydrate-induced differential gene expression patterns in the hyperthermophilic bacterium *Thermotoga maritima*. *J. Biol. Chem.* **278:**7540–7552.
15. **Damsté, J. S., S. Schouten, E. C. Hopmans, A. C. van Duin, and J. A. Geenevasen.** 2002. Crenarchaeol: the characteristic core glycerol dibiphytanyl glycerol tetraether membrane lipid of cosmopolitan pelagic crenarchaeota. *J. Lipid Res.* **43:**1641–1651.
16. **Dassa, E., and M. Hofnung.** 1985. Sequence of gene *malG* in *E. coli* K12: homologies between integral membrane components from binding protein-dependent transport systems. *EMBO J.* **4:**2287–2293.
17. **De Rosa, M., and A. Gambacorta.** 1988. The lipids of archaebacteria. *Prog. Lipid Res.* **27:**153–175.
18. **De Rosa, M., A. Trincone, B. Nicolaus, and A. Gambacorta.** 1991. Archaebacteria: lipids, membrane structures, and adaptations to environmental stresses, p. 61–87. *In* G. di Prisco (ed.), *Life under Extreme Conditions*. Springer-Verlag, Berlin, Germany.
19. **Diederichs, K., J. Diez, G. Greller, C. Muller, J. Breed, C. Schnell, C. Vonrhein, W. Boos, and W. Welte.** 2000. Crystal structure of MalK, the ATPase subunit of the trehalose/maltose ABC transporter of the archaeon *Thermococcus litoralis*. *EMBO J.* **19:**5951–5961.
20. **Diez, J., K. Diederichs, G. Greller, R. Horlacher, W. Boos, and W. Welte.** 2001. The crystal structure of a liganded trehalose/maltose-binding protein from the hyperthermophilic archaeon *Thermococcus litoralis* at 1.85 Å. *J. Mol. Biol.* **305:**905–915.
21. **Edmonds, B. W., and H. Luecke.** 2004. Atomic resolution structures and the mechanism of ion pumping in bacteriorhodopsin. *Front. Biosci.* **9:**1556–1566.
22. **Elferink, M. G. L., S.-V. Albers, W. N. Konings, and A. J. M. Driessen.** 2001. Sugar transport in *Sulfolobus solfataricus* is mediated by two families of binding protein-dependent ABC transporters. *Mol. Microbiol.* **39:**1494–1503.
23. **Elferink, M. G. L., J. G. De Wit, R. Demel, A. J. M. Driessen, and W. N. Konings.** 1992. Functional reconstitution of membrane proteins in monolayer liposomes from bipolar lipids of *Sulfolobus acidocaldarius*. *J. Biol. Chem.* **267:**1375–1381.

24. Erra-Pujada, M., P. Debeire, F. Duchiron, and M. J. O'Dono-hue. 1999. The type II pullulanase of *Thermococcus hydrothermalis*: molecular characterization of the gene and expression of the catalytic domain. *J. Bacteriol.* **181:**3284–3287.

25. Evdokimov, A. G., D. E. Anderson, K. M. Routzahn, and D. S. Waugh. 2001. Structural basis for oligosaccharide recognition by *Pyrococcus furiosus* maltodextrin-binding protein. *J. Mol. Biol.* **305:**891–904.

26. Greene, R. V., and R. E. MacDonald. 1984. Partial purification and reconstitution of the aspartate transport system from *Halobacterium halobium*. *Arch. Biochem. Biophys.* **229:**576–584.

27. Greller, G., R. Horlacher, J. DiRuggiero, and W. Boos. 1999. Molecular and biochemical analysis of MalK, the ATP-hydrolyzing subunit of the trehalose/maltose transport system of the hyperthermophilic archaeon *Thermococcus litoralis*. *J. Biol. Chem.* **274:**20259–20264.

28. Greller, G., R. Riek, and W. Boos. 2001. Purification and characterization of the heterologously expressed trehalose/maltose ABC transporter complex of the hyperthermophilic archaeon *Thermococcus litoralis*. *Eur. J. Biochem.* **268:**4011–4018.

29. Grogan, D. W. 1989. Phenotypic characterization of the Archaebacterial genus *Sulfolobus*: comparison of five wild-type strains. *J. Bacteriol.* **171:**6710–6719.

30. Hamill, O. P., and B. Martinac. 2001. Molecular basis of mechanotransduction in living cells. *Physiol. Rev.* **81:**685–740.

31. Hettmann, T., C. L. Schmidt, S. Anemuller, U. Zahringer, H. Moll, A. Petersen, and G. Schafer. 1998. Cytochrome b558/566 from the archaeon *Sulfolobus acidocaldarius*. A novel highly glycosylated, membrane-bound b-type hemoprotein. *J. Biol. Chem.* **273:**12032–12040.

32. Hirai, T., and S. Subramaniam. 2003. Structural insights into the mechanism of proton pumping by bacteriorhodopsin. *FEBS Lett.* **545:**2–8.

33. Holtmann, G. 2004. Ph.D. Thesis. University of Marburg, Germany.

34. Horlacher, R., K. B. Xavier, H. Santos, J. DiRuggiero, M. Kossmann, and W. Boos. 1998. Archaeal binding protein-dependent ABC transporter: molecular and biochemical analysis of the trehalose/maltose transport system of the hyperthermophilic archaeon *Thermococcus litoralis*. *J. Bacteriol.* **180:**680–689.

35. Jiang, Q. X., D. N. Wang, and R. MacKinnon. 2004. Electron microscopic analysis of KvAP voltage-dependent K+ channels in an open conformation. *Nature* **430:**806–810.

36. Jiang, Y., A. Lee, J. Chen, M. Cadene, B. T. Chait, and R. MacKinnon. 2002. Crystal structure and mechanism of a calcium-gated potassium channel. *Nature* **417:**515–522.

37. Jiang, Y., A. Lee, J. Chen, V. Ruta, M. Cadene, B. T. Chait, and R. MacKinnon. 2003. X-ray structure of a voltage-dependent K+ channel. *Nature* **423:**33–41.

38. Jiang, Y., V. Ruta, J. Chen, A. Lee, and R. MacKinnon. 2003. The principle of gating charge movement in a voltage-dependent K+ channel. *Nature* **423:**42–48.

39. Kaidoh, K., S. Miyauchi, A. Abe, S. Tanabu, T. Nara, and N. Kamo. 1996. Rhodamine 123 efflux transporter in *Haloferax volcanii* is induced when cultured under 'metabolic stress' by amino acids: the efflux system resembles that in a doxorubicin-resistant mutant. *Biochem. J.* **314**(Pt 1):355–359.

40. Kates, M. 1996. Structural analysis of phospholipids and glycolipids in extremely halophilic archaebacteria. *J. Microbiol. Methods* **25:**113–128.

41. Kates, M. 1993. Membrane lipids of Archaea, p. 261–295. *In* M. Kates, D. J. Kushner, and A. T. Matheson (ed.), *The Biochemistry of Archaea (Archaebacteria)*. Elsevier, London, United Kingdom.

42. Kates, M., N. Moldoveanu, and L. C. Stewart. 1993. On the revised structure of the major phospholipid of *Halobacterium salinarium*. *Biochim. Biophys. Acta* **1169:**46–53.

43. Kawarabayasi, Y., Y. Hino, H. Horikawa, K. Jin-no, M. Takahashi, M. Sekine, S. Baba, A. Ankai, H. Kosugi, A. Hosoyama, S. Fukui, Y. Nagai, K. Nishijima, R. Otsuka, H. Nakazawa, M. Takamiya, Y. Kato, T. Yoshizawa, T. Tanaka, Y. Kudoh, J. Yamazaki, N. Kushida, A. Oguchi, K. Aoki, S. Masuda, M. Yanagii, M. Nishimura, A. Yamagishi, T. Oshima, and H. Kikuchi. 2001. Complete genome sequence of an aerobic thermoacidophilic crenarchaeon, *Sulfolobus tokodaii* strain7. *DNA Res.* **8:**123–140.

44. Kloda, A., and B. Martinac. 2001. Mechanosensitive channel of *Thermoplasma*, the cell wall-less archaea: cloning and molecular characterization. *Cell Biochem. Biophys.* **34:**321–347.

45. Kloda, A., and B. Martinac. 2001. Molecular identification of a mechanosensitive channel in archaea. *Biophys. J.* **80:**229–240.

46. Kloda, A., and B. Martinac. 2001. Structural and functional differences between two homologous mechanosensitive channels of *Methanococcus jannaschii*. *EMBO J.* **20:**1888–1896.

47. Koenig, H. 1988. Archaeobacterial cell envelopes. *Can. J. Microbiol.* **34:**395–406.

48. Koga, Y., M. Nishihara, H. Morii, and M. Kagawa-Matsushita. 1993. Ether polar lipids of methanogenic bacteria: structures, comparative aspects, and biosyntheses. *Microbiol. Rev.* **57:**164–182.

49. Kokoeva, M. V., K. F. Storch, C. Klein, and D. Oesterhelt. 2002. A novel mode of sensory transduction in archaea: binding protein-mediated chemotaxis towards osmoprotectants and amino acids. *EMBO J.* **21:**2312–2322.

50. Koning, S. M., M. G. Elferink, W. N. Konings, and A. J. Driessen. 2001. Cellobiose uptake in the hyperthermophilic archaeon *Pyrococcus furiosus* is mediated by an inducible, high-affinity ABC transporter. *J. Bacteriol.* **183:**4979–4984.

51. Koning, S. M., W. N. Konings, and A. J. M. Driessen. 2002. Biochemical evidence for the presence of two alpha-glucoside ABC-transport systems in the hyperthermophilic archaeon *Pyrococcus furiosus*. *Archaea* **1:**19–25.

52. Konrad, Z., and J. Eichler. 2002. Lipid modification of proteins in Archaea: attachment of a mevalonic acid-based lipid moiety to the surface-layer glycoprotein of *Haloferax volcanii* follows protein translocation. *Biochem. J.* **366:**959–964.

53. Kushwaha, S. C., and M. Kates. 1973. Isolation and identification of "bacteriorhodopsin" and minor C40-carotenoids in *Halobacterium cutirubrum*. *Biochim. Biophys. Acta* **316:**235–243.

54. Lai, M. C., T. Y. Hong, and R. P. Gunsalus. 2000. Glycine betaine transport in the obligate halophilic archaeon *Methanohalophilus portucalensis*. *J. Bacteriol.* **182:**5020–5024.

55. Langworthy, T. A. 1982. Lipids of *Thermoplasma*. *Methods Enzymol.* **88:**369–406.

56. Le Dain, A. C., N. Saint, A. Kloda, A. Ghazi, and B. Martinac. 1998. Mechanosensitive ion channels of the archaeon *Haloferax volcanii*. *J. Biol. Chem.* **273:**12116–12119.

57. Lee, S. J., A. Engelmann, R. Horlacher, Q. Qu, G. Vierke, C. Hebbeln, M. Thomm, and W. Boos. 2003. TrmB, a sugar-specific transcriptional regulator of the trehalose/maltose ABC transporter from the hyperthermophilic archaeon *Thermococcus litoralis*. *J. Biol. Chem.* **278:**983–990.

58. Lee, S. J., C. Moulakakis, S. M. Koning, W. Hausner, M. Thomm, and W. Boos. 2005. TrmB, a sugar sensing regulator of ABC transporter genes in *Pyrococcus furiosus* exhibits dual promoter specificity and is controlled by different inducers. *Mol. Microbiol.* **57:**1797–1807.

59. MacDonald, R. E., R. V. Greene, and J. K. Lanyi. 1977. Light-activated amino acid transport systems in *Halobac-

terium halobium envelope vesicles: role of chemical and electrical gradients. *Biochemistry* **16:**3227–3235.

60. **Martinac, B.** 2004. Mechanosensitive ion channels: molecules of mechanotransduction. *J. Cell Sci.* **117:**2449–2460.

61. **Mattar, S., B. Scharf, S. B. Kent, K. Rodewald, D. Oesterhelt, and M. Engelhard.** 1994. The primary structure of halocyanin, an archaeal blue copper protein, predicts a lipid anchor for membrane fixation. *J. Biol. Chem.* **269:**14939–14945.

62. **Miyauchi, S., M. Komatsubara, and N. Kamo.** 1992. In archaebacteria, there is a doxorubicin efflux pump similar to mammalian P-glycoprotein. *Biochim. Biophys. Acta* **1110:**144–150.

63. **Moll, R., and G. Schäfer.** 1988. Chemiosmotic H+ cycling across the plasma membrane of the thermoacidophilic archaebacterium *Sulfolobus acidocaldarius. FEBS Lett.* **232:**359–363.

64. **Moody, J. E., L. Millen, D. Binns, J. F. Hunt, and P. J. Thomas.** 2002. Cooperative, ATP-dependent association of the nucleotide binding cassettes during the catalytic cycle of ATP-binding cassette transporters. *J. Biol. Chem.* **277:**21111–21114.

65. **Nelson, K. E., R. A. Clayton, S. R. Gill, M. L. Gwinn, R. J. Dodson, D. H. Haft, E. K. Hickey, J. D. Peterson, W. C. Nelson, K. A. Ketchum, L. McDonald, T. R. Utterback, J. A. Malek, K. D. Linher, M. M. Garrett, A. M. Stewart, M. D. Cotton, M. S. Pratt, C. A. Phillips, D. Richardson, J. Heidelberg, G. G. Sutton, R. D. Fleischmann, J. A. Eisen, and C. M. Fraser.** 1999. Evidence for lateral gene transfer between Archaea and bacteria from genome sequence of *Thermotoga maritima. Nature* **399:**323–329.

66. **Neutze, R., E. Pebay-Peyroula, K. Edman, A. Royant, J. Navarro, and E. M. Landau.** 2002. Bacteriorhodopsin: a high-resolution structural view of vectorial proton transport. *Biochim. Biophys. Acta* **1565:**144–167.

67. **Ng, S. Y., and K. F. Jarrell.** 2003. Cloning and characterization of archaeal type I signal peptidase from *Methanococcus voltae. J. Bacteriol.* **185:**5936–5942.

68. **Nichols, D. S., M. R. Miller, N. W. Davies, A. Goodchild, M. Raftery, and R. Cavicchioli.** 2004. Cold adaptation in the Antarctic Archaeon *Methanococcoides burtonii* involves membrane lipid unsaturation. *J. Bacteriol.* **186:**8508–8515.

69. **Ninio, S., and S. Schuldiner.** 2003. Characterization of an archaeal multidrug transporter with a unique amino acid composition. *J. Biol. Chem.* **278:**12000–12005.

70. **Oesterhelt, D., M. Meentzen, and L. Schuhmann.** 1973. Reversible dissociation of the purple complex in bacteriorhodopsin and identification of 13-cis and all-trans-retinal as its chromophores. *Eur. J. Biochem.* **40:**453–463.

71. **Panagiotidis, C. H., W. Boos, and H. A. Shuman.** 1998. The ATP-binding cassette subunit of the maltose transporter MalK antagonizes MalT, the activator of the *Escherichia coli* mal regulon. *Mol. Microbiol.* **30:**535–546.

72. **Paulsen, I. T., L. Nguyen, M. K. Sliwinski, R. Rabus, and M. H. J. Saier.** 2000. Microbial genome analyses: comparative transport capabilities in eighteen prokaryotes. *J. Mol. Biol.* **301:**75–100.

73. **Prisco, A., M. Moracci, M. Rossi, and M. Ciaramella.** 1995. A gene encoding a putative membrane protein homologous to the major facilitator superfamily of transporters maps upstream of the β-glycosidase gene in the Archaeon *Sulfolobus solfataricus. J. Bacteriol.* **177:**1614–1619.

74. **Proctor, L. M., R. Lai, and R. P. Gunsalus.** 1997. The methanogenic archaeon Methanosarcina thermophila TM-1 possesses a high-affinity glycine betaine transporter involved in osmotic adaptation. *Appl. Environ. Microbiol.* **63:**2252–2257.

75. **Quiocho, F. A., and P. S. Ledvina.** 1996. Atomic structure and specificity of bacterial periplasmic receptors for active transport and chemotaxis: variation of common themes. *Mol. Microbiol.* **20:**17–25.

76. **Reich-Slotky, R., C. Panagiotidis, M. Reyes, and H. A. Shuman.** 2000. The detergent-soluble maltose transporter is activated by maltose binding protein and verapamil. *J. Bacteriol.* **182:**993–1000.

77. **Ren, Q., K. H. Kang, and I. T. Paulsen.** 2004. TransportDB: a relational database of cellular membrane transport systems. *Nucleic Acids Res.* **32:**D284–D288.

78. **Revell Phillips, L., M. Milescu, Y. Li-Smerin, J. A. Mindell, J. I. Kim, and K. J. Swartz.** 2005. Voltage-sensor activation with a tarantula toxin as cargo. *Nature* **436:**857–860.

79. **Ruta, V., Y. Jiang, A. Lee, J. Chen, and R. MacKinnon.** 2003. Functional analysis of an archaebacterial voltage-dependent K+ channel. *Nature* **422:**180–185.

80. **Sato, T., T. Fukui, H. Atomi, and T. Imanaka.** 2003. Targeted gene disruption by homologous recombination in the hyperthermophilic archaeon *Thermococcus kodakaraensis* KOD1. *J. Bacteriol.* **185:**210–220.

81. **Schiefner, A., G. Holtmann, K. Diederichs, W. Welte, and E. Bremer.** 2004. Structural basis for the binding of compatible solutes by ProX from the hyperthermophilic archaeon *Archaeoglobus fulgidus. J. Biol. Chem.* **279:**48270–48281.

82. **Schleper, C., G. Pühler, I. Holz, A. Gambacorta, D. Janekovic, U. Santarius, H.-P. Klenk, and W. Zillig.** 1995. *Picrophilus* gen.nov., fam. nov.: a novel aerobic, heterotrophic, thermoacidophilic genus and family comprising archaea capable of growth around pH 0. *J. Bacteriol.* **177:**7050–7059.

83. **Schneider, E.** 2001. ABC transporters catalyzing carbohydrate uptake. *Res. Microbiol.* **152:**303–310.

84. **Schouten, S., E. C. Hopmans, J. S. Schefuss, and J. S. Damste.** 2002. Distributional variations in marine crenarchaeotal membrane lipids: a new tool for reconstructing ancient sea water temperatures? *Earth Planet. Sci. Lett.* **204:**265–274.

85. **She, Q., R. K. Singh, F. Confalonieri, Y. Zivanovic, G. Allard, M. J. Awayez, C. C. Chan-Weiher, I. G. Clausen, B. A. Curtis, A. De Moors, G. Erauso, C. Fletcher, P. M. Gordon, I. Heikamp-De Jong, A. C. Jeffries, C. J. Kozera, N. Medina, X. Peng, H. P. Thi-Ngoc, P. Redder, M. E. Schenk, C. Theriault, N. Tolstrup, R. L. Charlebois, W. F. Doolittle, M. Duguet, T. Gaasterland, R. A. Garrett, M. A. Ragan, C. W. Sensen, and O. J. van der.** 2001. The complete genome of the crenarchaeon *Sulfolobus solfataricus* P2. *Proc. Natl. Acad. Sci. USA* **98:**7835–7840.

86. **Sprott, G. D., M. Meloche, and J. C. Richards.** 1991. Proportions of diether, macrocyclic diether, and tetraether lipids in *Methanococcus jannaschii* grown at different temperatures. *J. Bacteriol.* **173:**3907–3910.

87. **Stock, J., M. Surette, and P. Park.** 1994. Chemosensing and signal transduction in bacteria. *Curr. Opin. Neurobiol.* **4:**474–480.

88. **Sumper, M., E. Berg, R. Mengele, and I. Strobel.** 1990. Primary structure and glycosylation of the S-layer protein of *Haloferax volcanii. J. Bacteriol.* **172:**7111–7118.

89. **Tawara, E., and N. Kamo.** 1991. Glucose transport of *Haloferax volcanii* requires the Na(+)-electrochemical potential gradient and inhibitors for the mammalian glucose transporter inhibit the transport. *Biochim. Biophys. Acta* **1070:**293–299.

90. **Thomas, N. A., S. L. Bardy, and K. F. Jarrell.** 2001. The archaeal flagellum: a different kind of prokaryotic motility structure. *FEMS Microbiol. Rev.* **25:**147–174.

91. **Upasani, V. N., S. G. Desai, N. Moldoveanu, and M. Kates.** 1994. Lipids of extremely halophilic archaeobacteria from saline environments in India: a novel glycolipid in *Natronobacterium* strains. *Microbiology* **140**(Pt 8):1959–1966.

92. Van den Berg. B., W. M. Clemons, Jr., I. Collinson, Y. Modis, E. Hartmann, S. C. Harrison, and T. A. Rapoport. 2004. X-ray structure of a protein-conducting channel. *Nature* **427**:36–44.

93. van de Vossenberg, A. J. Driessen, W. Zillig, and W. N. Konings. 1998. Bioenergetics and cytoplasmic membrane stability of the extremely acidophilic, thermophilic archaeon *Picrophilus oshimae*. *Extremophiles* **2**:67–74.

94. Van de Vossenberg, J. L. C. M., T. Ubbink-Kok, M. G. L. Elferink, A. J. M. Driessen, and W. N. Konings. 1995. Ion permeability of the cytoplasmic membrane limits the maximum growth temperature of bacteria and archaea. *Mol. Microbiol.* **18**:925–932.

95. Verdon, G., S. V. Albers, B. W. Dijkstra, A. J. Driessen, and A. M. Thunnissen. 2003. Crystal structures of the ATPase subunit of the glucose ABC transporter from *Sulfolobus solfataricus*: nucleotide-free and nucleotide-bound conformations. *J. Mol. Biol.* **330**:343–358.

96. Verdon, G., S. V. Albers, N. van Oosterwijk, B. W. Dijkstra, A. J. Driessen, and A. M. Thunnissen. 2003. Formation of the productive ATP-Mg2+-bound dimer of GlcV, an ABC-ATPase from *Sulfolobus solfataricus*. *J. Mol. Biol.* **334**:255–267.

97. Walker, J. E., M. Saraste, M. J. Runswick, and N. J. Gay. 1982. Distantly related sequences in the alpha- and beta-subunits of ATP synthase, myosin, kinases and other ATP-requiring enzymes and a common nucleotide binding fold. *EMBO J.* **1**:945–951.

98. Wanner, C., and J. Soppa. 1999. Genetic identification of three ABC transporters as essential elements for nitrate respiration in *Haloferax volcanii*. *Genetics* **152**:1417–1428.

99. Woodson, J. D., A. A. Reynolds, and J. C. Escalante-Semerena. 2005. ABC Transporter for corrinoids in *Halobacterium* sp. strain NRC-1. *J. Bacteriol.* **187**:5901–5909.

100. Yamauchi, K., and M. Kinoshita. 1995. Highly stable lipid membranes from archaebacterial extremophiles. *Prog. Polym. Sci.* **18**:763–804.

101. Yernool, D., O. Boudker, Y. Jin, and E. Gouaux. 2004. Structure of a glutamate transporter homologue from *Pyrococcus horikoshii*. *Nature* **431**:811–818.

102. Yuan, Y. R., S. Blecker, O. Martsinkevich, L. Millen, P. J. Thomas, and J. F. Hunt. 2001. The crystal structure of the MJ0796 ATP-binding cassette: Implications for the structural consequences of ATP hydrolysis in the active site of an ABC-transporter. *J. Biol. Chem.* **276**:32313–32321.

Archaea: Molecular and Cellular Biology
Edited by Ricardo Cavicchioli
© 2007 ASM Press, Washington, D.C.

Chapter 17

Protein Translocation into and across Archaeal Cytoplasmic Membranes

MECHTHILD POHLSCHRÖDER AND KIERAN C. DILKS

INTRODUCTION

The cytoplasmic membrane is an amphiphatic structure that separates the interior of the cell from the external environment. However, interactions between the cytoplasm and the extracytoplasmic environment are essential for cellular life. For example, cells must communicate signals across the membrane, secrete toxins, and take up nutrients. The latter process not only requires integration of transporters, but also the secretion of substrate-binding proteins across the membrane (see Chapter 16). Additional proteins that need to be secreted include, among others, polymer-degrading enzymes and subunits of extracytoplasmic cellular structures, such as the cell wall, flagella, or pili (see Chapters 14 and 18). Thus, the ability to translocate proteins into and through the hydrophobic membranes that provide the semipermeable barrier to the cytoplasm is required by all forms of life.

A few proteins have been reported to spontaneously insert into the membrane without the aid of protein "machineries." However, most membrane proteins and substrates translocated across the membrane require such machineries to catalyze the translocation process. Several pathways have been identified that are involved in the translocation of unfolded proteins into or across the endoplasmic reticulum (ER) and cytoplasmic membranes, in eucarya, bacteria, and archaea, respectively (49, 79, 98). The secretion of the majority of these proteins is thought to depend on the universally conserved Sec pathway (79). Substrates translocated via the Sec pore can be translocated cotranslationally or posttranslationally, after much of the protein has been synthesized and extruded from the ribosome. The mechanisms of co- and posttranslational translocation through this pore

vary among organisms and require distinct components for substrate recognition and protein translocation (53, 86). Many bacteria and archaea also encode a Sec-independent translocation machinery, the Tat pathway, dedicated to the secretion of cytoplasmically folded proteins (11).

While the most extensive studies of these translocation pathways have been conducted in bacteria and eucarya, recent analyses of the archaeal Sec and Tat pathways have revealed novel and crucial information about archaeal protein translocation, as well as protein translocation in general.

THE Sec TRANSLOCATION PATHWAY

The Sec pathway is the only known universally conserved protein translocation pathway. To translocate proteins into and across cytoplasmic and ER membranes in bacteria, archaea, and eucarya, respectively, the Sec pathway must: (i) distinguish Sec substrates from other proteins; (ii) distinguish proteins that are integrated into the membrane as opposed to those translocated across the lipid bilayer; (iii) transport these proteins across or into the bilayer; and finally, (iv) allow for the translocation of proteins without disrupting the integrity of the membrane.

Sec Substrates, Their N-Terminal Signal Peptides, and Signal Peptidases

Integral membrane proteins

Essential cellular processes, such as the communication between the cytoplasm and the extracellular environment, uptake of nutrients, or the generation of ATP, require proteins that are embedded in the lipid

Mechthild Pohlschröder and Kieran C. Dilks • Biology Department, University of Pennsylvania, Philadelphia, PA 19104.

bilayer. In archaea, these include well-characterized membrane proteins, such as bacteriorhodopsin or subunits of voltage-gated channels. It is believed that most of these proteins are inserted into the lipid bilayer via the Sec pathway. While little is known about archaeal membrane protein insertion, it is thought that, similar to bacteria and eucarya, targeting of membrane proteins to the Sec pore involves the recognition of the first hydrophobic transmembrane segment (TM) by the cytoplasmic signal recognition particle (SRP) (see "Protein targeting" below) as it is being translated (26, 35).

An exception seems to be bacterioopsin (BO), a seven-TM polypeptide in the *Halobacterium salinarum* cytoplasmic membrane (104). BO, the apoprotein of bacteriorhodopsin (BR), is a light-driven proton pump that has been attractive for membrane insertion studies, as high-resolution structural analysis has revealed its topology in the native membrane. BO is synthesized with an N-terminal 13-amino-acid presequence that is distinct from a TM or any of the

known Sec signal sequences (see below and Fig. 1). Several studies are consistent with a model in which BO is inserted into the membrane by the Sec pathway (40). In this model, the SRP is localized to the membrane upon induction of BO synthesis (40). However, reports of posttranslational translocation of BO conflict with this model and highlight the need for additional experiments to determine its mechanism of translocation.

Secreted substrates and their N-terminal signals

Many Sec substrates, such as substrate-binding proteins, polymer-degrading enzymes, or cell wall subunits, are secreted and are either anchored to the membrane or released into the external environment. These substrates possess N-terminal signal peptides that resemble a TM, but also contain charged amino acids at the N terminus that interact with negatively charged phospholipids at the cytoplasmic face of the membrane, thus orienting the N terminus of the sig-

Sec signal sequences

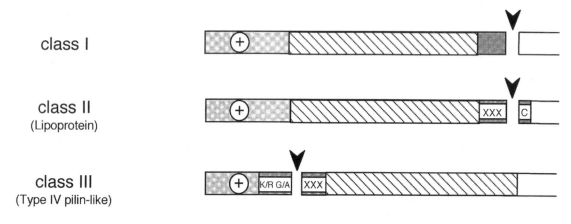

Tat signal sequences

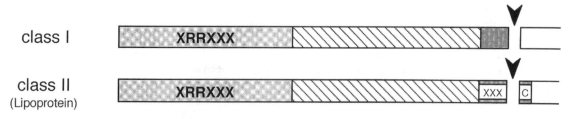

Figure 1. Schematic representation of different classes of Sec and Tat signal sequences. Grey and hatched boxes represent N-terminally charged and hydrophobic (H region) domains, respectively. Arrows indicate the signal peptide cleavage sites. Cleavage of predicted class 2 signal peptides of Tat substrates by SPII has not yet been confirmed experimentally. Modified from *FEMS Reviews* (85) with permission of the publisher.

nal sequence into the cytoplasm as the substrate is being translocated (108) (Fig. 1). Moreover, unlike sig-
nal-anchor sequences in membrane proteins, the signal sequences are often removed upon translocation
by a signal peptidase (SPase) (80). Depending on the
peptidase cleavage site, signal sequences are divided
into three classes (Fig. 1).

Class I signal peptides are cleaved from the preprotein by the universally conserved, membrane-associated, type I SPase (SPI). Consequently, the mature proteins are either released from the membrane
or anchored to the lipid bilayer via a C-terminal
membrane anchor. In bacteria, the SPI consists of a
catalytic core (domain I) and, in some cases, a second domain (domain II) whose role is not defined
(80). The eucaryal homolog of this SPI, SPC18, is part
of the ER membrane-bound heterooligomeric signal
peptidase complex (SPC) (29). In contrast to the bacterial SPI, SPC18 does not contain a domain II, and
its catalytic site involves a Ser-His-Arg triad rather
than the bacterial Ser-Lys dyad (29, 80, 100). It is intriguing that, while all known archaeal SPIs possess
a putative eucaryal Ser-His-Arg triad, some of them
also possess a domain that has similarity to the bacterial domain II (3, 27). Site-directed mutagenesis of
the *Methanococcus voltae* SPI revealed the requirement of the conserved Ser^{52} and His^{122} for activity
but, surprisingly, the presumed active-site Asp^{142}
(corresponding to Asp^{273} in *Escherichia coli*) was not
essential (8). Conversely, mutation of a second aspartate residue (Asp^{148}, equivalent to *E. coli* Asp^{280}) had
a severe negative effect on enzyme activity (8).

Computational analyses of signal sequences
from the euryarchaeon *Methanococcus jannaschii*
(74), and the crenarchaeon, *Sulfolobus solfataricus*
(3), suggest that these signal sequences possess a
charge distribution similar to those of bacteria, but
their SPase recognition site is similar to that of eucarya. However, despite these observed differences
among the signal peptides and SPIs, class I signal peptides from both eucarya and archaea are properly targeted and processed when expressed in bacteria (85,
105, 108).

While some of these substrates are released into
the extracellular environment, others are anchored
to the membrane via C-terminal hydrophobic domains. For example, several archaeal S-layer glycoproteins contain typical class I signal peptides and are
associated with the membrane via a C-terminal anchor (see Chapter 14).

Certain archaeal Sec signal sequences resemble
bacterial class II signal peptides, which contain a
lipobox motif ([I/L/G/A]-[A/G/S]-C) and are recognized by the type II SPase (SPII) (Fig. 1). In bacteria,
the terminal cysteine of the lipobox is modified upon
translocation by the addition of a diacylglycerol moiety via a thioester linkage (48). Only acylated substrates are recognized and cleaved by the SPII, which
temporally ensures that the substrate is able to be anchored in the membrane via lipid prior to cleavage of
the membrane-anchoring signal sequence (see Chapters 11, 14, and 15). The new amino terminus is further acylated, resulting in the mature lipoprotein that
will be anchored to the cytoplasmic- or outer membrane (95). While many secreted archaeal proteins do
possess signal peptides with a motif that resembles a
lipobox (13, 59, 60), and recent data suggest that
lipid modification occurs at the conserved cysteine
(65), homology searches have not resulted in the identification of an archaeal SPII homolog. This may in
part be due to the addition of slightly distinct lipid
moieties ($C_{20}C_{20}$ diphytanyl diether) to the lipobox
cysteine in archaea, which may require unique features of an SPII for recognition.

Extracellular protein structures

Archaea have evolved distinct mechanisms to assemble subunits of extracytoplasmic structures after
translocation via the Sec pathway. For example, archaeal flagellins are synthesized as preproteins with
signal sequences that are cleaved before the incorporation of the protein into the flagellar filament (9,
106). However, distinct from class I and class II signal
peptides, flagellin signal sequences (class III), initially
identified in bacterial type IV prepilin subunits, contain a SPase (prepilin-peptidase)-processing site that
precedes the hydrophobic stretch (Fig. 1 and Table 1).
This hydrophobic stretch constitutes part of the mature protein and is essential for initial membrane-
anchoring and subunit-subunit interactions that are
critical for the biosynthesis of type IV pili and piluslike structures. Note that archaeal flagellins do not resemble bacterial flagellins, and bacteria have evolved
a Sec-independent pathway for flagellin translocation
and flagella biosynthesis (62) (see Chapter 18).

Archaea also appear to utilize class III signal peptides for the translocation and biosynthesis of other
extracytoplasmic structures. In addition to the archaeal
flagellins, several sugar-binding proteins in *S. solfataricus* have been shown to contain typical class III signal peptides (4). While the reason these binding proteins possess class III signal peptides is currently
unknown, it is tempting to postulate that they also assemble into surface structures ("bindosomes") that
may provide a selective advantage under certain growth
conditions (2).

Recent in silico analysis of all available completely sequenced archaeal genomes using a *perl* program that predicts archaeal class III signal peptides

Table 1. Signal peptide classes in *Archaea*

Organism/protein	Signal sequence[a]	Processing peptidase
Sec signal sequences		
Class 1		SPI
Halobacterium salinarum Csg [59]	MTDTTGKLRAVLLTALMVGSVIGAGVAFTGG**AAA** AN	
Methanothermus fervidus S1gA [15]	MRKFTLLMLLLIVISMSGI**AGA** AQ	
Methanococcus maripaludis S1p[b]	MAMSMKKIGAIAVGGAMVASALATG**AFA** AE	
Class 2		SPII[c]
Archaeoglobus fulgidus G1nH [4]	MKKVVPILVLLAALLLLG CT	
Methanocaldococcus jannaschii BraC [4]	MPYYWGAILIGGVFLAG CT	
Pyrococcus abyssi Ma1E [4]	MKRGIYAVLLVGVLIFSVV**ASG** CI	
Class 3		FlaK/PibD
Methanococcus voltae FlaB1 [15]	MNIKEFLSNK**KG** AS	
Methanococcus vannielii FlaB1 [51]	MSVKNFMNNK**KG** DS	
Sulfolobus solfataricus TreS [3]	MSRSDKFSNKEKMR**RG** LS	
Tat signal sequences		
Class 1		SPI
Natronococcus sp. ←Amy [88][d]	MRRNHSHTSDSAGI<u>DRRTVLR</u>SSAAAGALALTGVTIGSTS**AAA** RS	
Sulfolobus solfataricus Oxyr [88][e]	MLKL<u>SRRDFLK</u>ISGATAVATAFILGGNS**VA** KR	
Archaeoglobus fulgidus Se1A [88][e]	MRKVMNSPDDGN<u>GRRRFLQ</u>FSMAALASAAAPSSVW**AFS** KI	
Class 2		SPII[1]
Halobacterium salinarum BasB [57][e]	MHST<u>TRREWLG</u>AIGATAATGLG CA	
Halobacterium salinarum CosB [57][e]	MMDTPEHASTS<u>SRRQLLG</u>MLAAGGTTA**VAG** CT	
Natronomonas pharaonis Hcy [64][e]	MKDIS<u>RRRFVL</u>GTGATVAAA**TLA** CN	

[a]Cleavage site is indicated by arrowhead. Conserved amino acids at the cleavage site are in bold. The Tat motif is underlined.
[b]D. VanDyke and K. F. Jarrell, unpublished data.
[c]SPII homolog not found in archaea; SPII analog predicted.
[d]Signal peptidase cleavage site identified but no direct evidence for SPI cleavage.
[e]Predicted signal peptidase cleavage site.

revealed the presence of a diverse set of proteins with class III signal peptides in archaea other than *Sulfolobus*. These include, among others, binding proteins as well as putative archaeal type IV pilins. Genes encoding proteins identified by the perl program, FLAFIND (http://signalfind.org), were often located in an operon with other FLAFIND positives, prepilin peptidases, and/or bacterial ATPases that may provide energy for assembly of the extracellular structures (104a). Further investigation will provide a better understanding of the diversity of bacterial and archaeal substrates containing such signal peptides, presumably for the biosynthesis of multimeric cell surface-associated structures.

Protein Targeting

The initial TM or H region of many Sec substrates is recognized by the SRP as it emerges from translating ribosomes (26) (Fig. 2). After the SRP-ribosome nascent chain (RNC) complex is targeted to the Sec pore, the substrate is translocated cotranslationally (26, 79, 99). SRP-independent substrates are translocated posttranslationally, requiring SRP-independent

targeting factors, chaperones to keep the precursor in a translocation-competent conformation, and an ATPase to drive the translocation process (85). While translocation across membranes can either be co- or posttranslational, membrane protein insertion is thought to mainly occur cotranslationally.

SRP-dependent targeting to the Sec pore

The SRP is composed of an RNA molecule and one or more protein subunits (26). In eucarya, where this pathway has been studied extensively, the SRP binds to the ribosome next to the peptide exit site. At this site on the ribosome, the S domain of the SRP RNA and the universally conserved SRP54 protein component sample nascent peptide and recognize signal sequences as they are translated (110) (Fig. 2). Once bound to an appropriate signal sequence, the SRP then undergoes a conformational change that allows for the binding of the second domain within the RNA (termed the Alu domain) to adhere to the rRNA that overlaps with the elongation factor-binding site of the ribosome (A site) (57, 63, 111). As the Alu domain blocks both elongation factors and tRNAs from

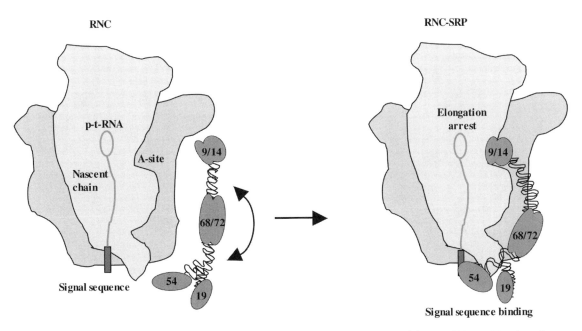

Figure 2. Mammalian SRP interaction with the ribosome. Upon cytoplasmic exposure of the initial TM or H region of many Sec substrates, the SRP interacts with the ribosome nascent chain (RNC) complex via several points of interaction. The SRP54 protein recognizes and binds the nascent polypeptide, while SRP9/14 bind and block the A site (see text). The bending of the SRP RNA molecule required for both of these interactions to occur simultaneously may be facilitated by SRP68/72. SRP proteins are represented by their corresponding numbers. Modified from *Current Opinion in Structural Biology* (25) with permission of the publisher.

entering the ribosome, translational elongation is likely to be arrested (42). The RNC-SRP complex is then targeted to the membrane, mediated by the affinity of SRP for the membrane-associated SRP receptor (SR) (73). The SR is anchored to the ER membrane by its SRβ-subunit, while the soluble SRα-subunit is responsible for the interaction with SRP54 (66) (Fig. 2). Once the SR has escorted the RNC-SRP complex to the translocon, SR and SRP disengage in a GTP-dependent manner. Concurrent with the resumption of translation, the Sec substrate is transported through the Sec pore (see "The Sec pore," below) (114).

Most known bacterial SRP RNA homologs possess a 7S RNA with both the S and the Alu domains (77). Despite this prevalence, only one bacterial SRP protein that interacts with the Alu domain, the *Bacillus subtilis* protein Hbsu, has been identified (71, 81). Moreover, some phylogenetically diverse bacteria (e.g., *E. coli* and *Streptomyces lividans*) possess a minimalistic, yet essential, SRP. Their SRP consists only of an RNA molecule (that lacks the Alu domain) and the universally conserved SRP54 homolog.

The archaeal SRP is essential and most closely resembles the eucaryal SRP. The 7S RNA and SRP54 homologs are more similar to their eucaryal than to their bacterial counterparts, and they contain a homolog of the eucaryal SRP19 subunit (68, 69, 91, 92). While no additional archaeal homologs of either eucaryal or bacterial SRP proteins have been identified, a recent analysis of archaeal small RNAs suggests an interaction of the ribosomal protein L7Ae with the *S. solfataricus* 7S RNA (118) (see Chapter 8). It has been postulated that such an interaction may facilitate the bending of the RNC-bound SRP to allow the Alu domain of the RNA to interact with the ribosomal exit tunnel and its A site to arrest translation (Fig. 2). Since homology searches have not identified homologs of the eucaryal proteins binding to this domain (14, 85), it is intriguing to speculate that the archaeal ribonuclear Alu domain by itself may be sufficient for pausing translation. Alternatively, biochemical characterization of bacterial and archaeal SRPs may reveal the presence of additional protein components involved in translational arrest.

Similar to the apparently more simple bacterial and archaeal SRP, the essential bacterial and archaeal SRP receptor homologs (FtsY) have a less complex composition, as they lack the eucaryal SRβ-subunit (Fig. 3). Consistent with the absence of the SRβ-subunit in bacteria and archaea (which is required for tight association of the SRP receptor to the membrane in eucarya), cell-fractionation experiments showed that at least 50% of the FtsY is localized in the cytoplasm (41, 61, 67). Recent copurification studies of the membrane-associated *E. coli* FtsY suggest that the bacterial and archaeal FtsY interacts directly with the Sec pore (7).

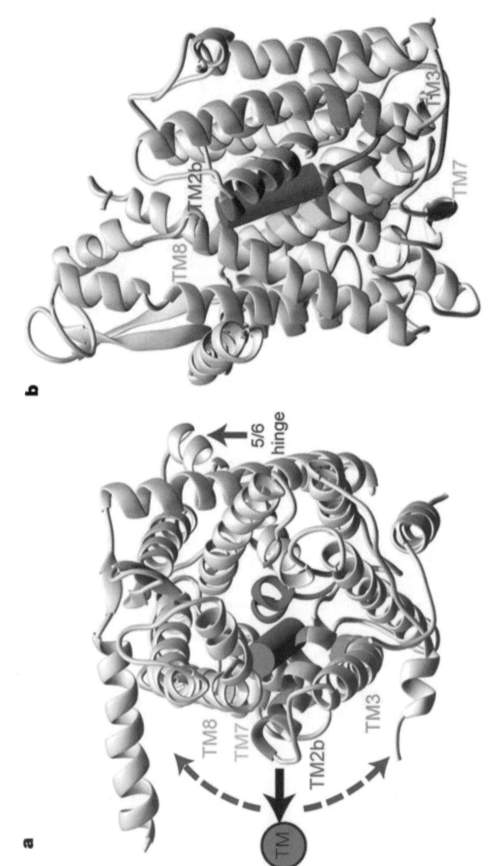

Figure 3. (*See the separate color insert for the color version of this illustration.*) Crystal structure of the *M. jannaschii* Sec61 protein-conducting channel. Views from the top (a) and the front (b). Faces of the helices that form the signal-sequence-binding site and the lateral gate through which TMs of nascent membrane proteins exit the channel into lipid are colored. The plug, which gates the pore, is green. The hydrophobic core of the signal sequence probably forms a helix, modeled as a magenta cylinder, which intercalates between TM2b and TM7 above the plug. Intercalation requires opening the front surface as indicated by the broken arrows, with the hinge for the motion being the loop between TM5 and TM6 at the back of the molecule (5/6 hinge). A solid arrow pointing to the magenta circle in the top view indicates schematically how a TM of a nascent membrane protein would exit the channel into lipid. Structure and legend reprinted from *Nature* (107) with permission from the publisher.

SRP-independent targeting to the Sec pore

The translation of Sec substrates emerging from the ribosome that are SRP independent are translocated after much of the protein has been translated. Thus, these substrates not only require SRP-independent targeting factors, but also require chaperones to keep them in an unfolded, translocation-competent conformation. One such chaperone in *E. coli*, the SecB chaperone, binds to slowly folding Sec substrates (89). The bound substrate is then targeted to the Sec pore by way of the affinity of SecB for SecA, the bacterial translocation ATPase that is crucial for providing energy for bacterial posttranslational protein translocation (109; and see "The Sec pore," below). SecA is also able to interact with Sec substrates in a SecB-independent manner. In fact, recent data suggest a direct interaction of SecA with the ribosome, where it may also be involved in the selection of emerging posttranslationally translocated nascent chains (55). While no SecB homolog has been identified in eucarya, universally conserved chaperones, such as Hsp70, are required for the targeting of unfolded proteins to the eucaryal Sec pore. Similarly, neither a SecB nor a SecA homolog has been identified in archaea, and distinct cytoplasmic chaperones may be involved in this process (14, 86). However, it is unclear whether posttranslational protein translocation occurs in archaea (see "Energetics," below).

The Sec Pore

The Sec pore consists of two universally conserved components, Sec61α and Sec61γ (SecY and SecE, respectively, in bacteria). A third component, Sec61β in archaea and eucarya or SecG in bacteria is also involved (44, 46, 58, 107) (Fig. 3). In both bacteria and eucarya, SecY/Sec61α and SecE/Sec61γ are essential, while the third pore subunit (SecG or Sec61β, respectively) is dispensable (32, 33). The archaeal Sec61αβγ homologs are most similar to the eucaryal counterparts. A high-resolution X-ray crystal structure of the Sec pore from the archaeon, *M. jannaschii* (the first and only such Sec pore structure for any organism), revealed that it is a 1:1:1 heterotrimer, with SecY forming a channel-like structure through which the nascent secretion substrate is likely to be translocated (Fig. 4) (107). The hourglass-shaped channel possesses a 5- to 8-Å diameter constriction between cytoplasmic and external hydrophilic funnels in the center of the membrane, which is lined with hydrophobic residues. It is blocked by a small helical segment on the extracytoplasmic side that has been called the "plug," and in that configuration, marks the closed channel. While the structural data of this archaeal Sec pore were generally consistent with previous topology predictions based on genetic analyses of the *E. coli* pore, it also led to a revision of previous interpretations of some data. Combining mutant signal sequence recognition data with the structure allowed for the proposal of a "clamshell" model for the transition from a closed inactive form of the pore to an open active form. These analyses suggest that the plug is being moved from the 5- to 8-Å constriction to the external side of the cell. The structure also revealed the site of the complex that is likely to open laterally toward the lipid phase to allow for TM integration into the membrane.

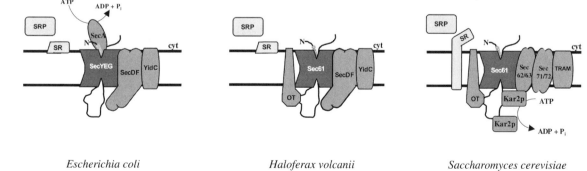

Escherichia coli   *Haloferax volcanii*   *Saccharomyces cerevisiae*

Figure 4. Sec machinery in the three domains of life. Components of the Sec machinery in representatives of bacteria (*E. coli*), archaea (*H. volcanii*), and eucarya (*S. cerevisiae*). Sec substrates are translocated into or across hydrophobic membranes via the universally conserved heterotrimeric Sec61 (SecYEG in bacteria) pore. Translocation through this protein-conducting channel requires distinct sets of additional Sec components in bacteria, archaea, and eucarya. YidC and TRAM are only involved in the insertion of proteins into the bacterial cytoplasmic and the ER membrane, respectively. While ATP hydrolysis by SecA and Kar2p are involved in energizing Sec translocation in bacteria and eucarya, respectively, no archaeal translocation ATPases have been identified. Cyt, cytoplasm. Reprinted from *Current Opinions in Microbiology* (84) with permission of the publisher.

Sec Pore-Associated Components

In vitro reconstitution studies suggest that the mammalian Sec-pore components are the only membrane components required for cotranslational translocation of proteins across lipid bilayers (39) (Fig. 3). However, it is thought that in bacteria and eucarya additional pore-associated components are necessary for efficient co- and posttranslational protein translocation. None of these accessory components are universally conserved. For example, eucarya contain Kar2p, an ER luminal ATPase, which binds to numerous Sec substrates as they emerge from the Sec pore and prevents their retrograde movement into the pore. Kar2p may also be involved in the gating of the translocon (5, 43). This ATPase is associated with the pore via its interaction with the membrane protein, Sec62, which together with Sec63 forms a subcomplex of the Sec pore (83, 93, 94). While Kar2p is not thought to be essential for mammalian cotranslational protein translocation, both co- and posttranslational protein translocation in yeast requires this ATP-dependent chaperone. Another eucaryal Sec-pore-associated component, the oligosyltransferase (OT), may be involved in giving the substrate directionality by glycosylating proteins emerging from the pore, thus preventing their retraction. Several additional proteins interact with the Sec pore, including TRAM, TRAP, and RAMP, although their function in translocation is not clear (31, 34, 37–39, 54, 101, 103, 113).

The E. coli SecYEG pore copurifies with the four distinct protein components, YidC, and the heterotrimeric complex SecDFYajC (Fig. 3) (23, 24, 76, 98). In addition, it transiently associates with the cytoplasmic translocation ATPase, SecA (109). This ATPase binds substrate, inserts into the Sec pore with the substrate, releases the substrate, and exits back out of the pore. This ATP-dependent cycle effectively results in "pushing" the substrate through the pore. While initially it was thought that a "push" through the pore was only required for the translocation of proteins across the membrane, recent studies suggest that the insertion of certain membrane proteins also depends on SecA (18, 88). A second component that is critical for the biogenesis of a subset of Sec-dependent (as well as Sec-independent) membrane proteins is YidC (117). Eucaryal homologs of YidC have only been identified in the mitochondrial inner membrane and the thylakoid membranes of chloroplasts (Oxa1 and Alb3, respectively) (116). While its exact function is not clear, it has been suggested to be involved in lateral integration, folding, and assembly of membrane proteins in bacteria and eucarya (17). The interaction of YidC with the SecYEG pore in bacteria may be mediated by SecDFYajC (76).

In addition to promoting the interaction of YidC with the bacterial translocon, the SecDFYajC complex may promote the membrane cycling of SecA and may stabilize SecG (24, 56). Moreover, both SecD and SecF contain large periplasmic domains that are crucial for the function of the protein. This indicates these membrane proteins may play an extracytoplasmic role late in translocation by facilitating the release of Sec substrates from the pore (64).

All complete archaeal genome sequences contain homologs of the eucaryal OT but apparently lack homologs of the other eucaryal Sec accessory components (14, 86). Furthermore, all genome sequences for Euryarchaeota encode a distant homolog of the bacterial YidC (116). However, it is not clear whether this protein is involved in archaeal membrane protein insertion (116). Many Euryarchaeota also contain homologs of SecD and SecF (14, 86) (Fig. 3). The amino acid similarity among bacterial and archaeal SecD and SecF homologs is more significant than among bacterial and archaeal YidC homologs, and the predicted membrane topologies of these archaeal membrane proteins are conserved with the bacterial counterparts (28).

Archaeal SecD and SecF have been studied in vivo in the model archaeon, Haloferax volcanii (45). Similar to E. coli, the H. volcanii SecD and SecF proteins form a membrane complex and a deletion of the secFD operon leads to a strong cold-sensitive phenotype. Consistent with their involvement in Sec translocation, the secFD knockout strain exhibits a specific Sec-protein export defect (45). The structural and functional conservation among the bacterial and archaeal homologs suggests a common function. However, the proposed functions of the bacterial SecDFYajC complex include SecA cycling and SecG stability. Since the archaeal Sec pore does not contain a SecG homolog, and no archaeal SecA homolog has been identified, the mechanism has to be independent of SecA and SecG. Additional studies are necessary to reveal the role of SecDF in facilitating bacterial and archaeal protein translocation.

Energetics

Co- and posttranslational Sec translocation in S. cerevisiae and E. coli requires a luminal and cytoplasmic ATPase, respectively (79). As archaea have a similar morphology to bacteria, it is unlikely that they possess an extracytoplasmic translocation ATPase analogous to eucaryal Kar2p, since ATP is not readily available in the extracellular environment. However, it is intriguing that archaea also lack a homolog of the bacterial SecA. Since no functionally equivalent ATPases have been identified in archaea, mechanisms by which energy is generated for archaeal protein translocation remain elusive.

Considering what is known about energetics of translocation in bacteria and eucarya, at least four distinct mechanisms for archaeal protein translocation may be proposed. It is also possible that any number of these energy-coupling mechanisms may act synergistically to drive translocation through the archaeal Sec pore (Fig. 5).

As noted in "The Sec pore," above, in vitro reconstitution studies of the mammalian translocation system suggest that proteins can be translocated across lipid bilayers in an ATP-independent manner. This indicates that translation elongation may be the sole driving force of unidirectional movement of proteins through the Sec pore (39) (Fig 5a). This is an attractive model for archaeal protein translocation, in view of the absence of a SecA or Kar2p homolog. However, while this may be one route of archaeal protein translocation, recent secretion studies across the *H. volcanii* cytoplasmic membrane suggest that two substrates are translocated across the archaeal cytoplasmic membrane, entirely by a posttranslational mechanism (50). Moreover, the translocation of a *Halobacterium* sp. NRC-1 bacterioopsin fusion construct was demonstrated to occur posttranslationally (78). While these studies indicate that posttranslational protein translocation is possible in archaea, it should be noted that the substrates used in both studies were heterolo-

gous fusion constructs. In addition, studies of BO in its native host are contradictory (see "Integral membrane proteins," above), suggesting that its translocation occurs cotranslationally (76).

Data supporting posttranslational protein translocation in archaea may indicate that an archaeal cytoplasmic SecA-like protein exists and that it lacks significant amino acid sequence conservation with the bacterial SecA. Alternatively, a cytoplasmic nucleotide-hydrolyzing enzyme unrelated to SecA in structure and/or mechanism may be driving translocation through the Sec pore (Fig. 5b).

Protein translocation may also be driven by one or several extracytoplasmic activities that provide directionality by preventing movement of the polypeptide chain back into the cytoplasm (Fig. 5c). In support of this, all complete genome sequences of archaea contain an OT, which may prevent the substrate from reentering the pore by glycosylating the translocating polypeptide chains (see "Sec pore-associated components," above). Similarly, the folding of the proteins aided by non-ATP-dependent extracytoplasmic chaperones may result in the unidirectional movement of the polypeptide. SecD and SecF may play such a role in the mechanism of translocation (Fig. 5c).

Finally, in vitro studies suggest that the proton motive force (PMF), in concert with the action of SecA, facilitates bacterial secretion via the Sec pore (70, 75). Furthermore, the PMF is apparently sufficient to drive translocation of proteins via the Tat pore (see "Tat machinery," below). Thus, it is possible that an ion gradient across the archaeal membrane is the sole source of energy for protein translocation (Fig. 5d).

THE Tat PATHWAY

Many bacteria and archaea possess an additional general secretion pathway, described as the twin-arginine translocation (Tat) pathway. Unlike the Sec pathway, secretion across this pore is, apparently, solely driven by the proton gradient across the membrane, and substrates of the Tat pathway can be secreted after significant, if not complete, folding of the protein has occurred in the cytoplasm (85). Since the Tat pathway does not require extracellular protein folding, organisms that have this pathway possess a powerful alternative mechanism to achieve extracellular protein activity. Similar to the Sec pathway, recent analyses of the Tat pathway in archaea have provided insights into how the Tat pathway functions in all bacteria and archaea and revealed features of this secretory route that are ostensibly specific to archaeal species.

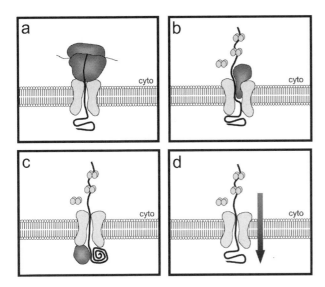

Figure 5. Models of putative archaeal protein translocation energetics. See text for details. (a) Cotranslational translocation. (b) Posttranslational translocation with a cytoplasmic energy-coupling protein. (c) Posttranslational translocation with extracytoplasmic activity. (d) Posttranslational translocation harnessing a gradient (e.g., ΔpH) across the cytoplasmic membrane. Figure reprinted from *FEMS Reviews* (85) with permission of the publisher.

Tat Signals and Substrates

Signals

Although the Tat and Sec translocation machineries are distinct, the features of the signals directing substrates to each pathway are remarkably similar. Similar to class I/II Sec signal peptides, Tat signals are located at the N terminus of the precursor and contain the same structure: a charged N terminus, followed by a stretch of hydrophobic amino acids (Fig. 1). In addition, Tat signal peptides often have a Sec-like, C-terminal signal cleavage site that is apparently processed by SPI (115). Despite the similarity of Sec and Tat signal peptides, general biochemical differences between the two must exist to prevent futile targeting of all Sec and Tat substrates to both pathways.

The N-terminal charged region of a Tat signal sequence is represented by the conserved S/T-R-R-X-F-L-K motif (11). The core twin-arginine residues are essentially universally conserved, and replacement of these arginines with other amino acids often results in a significant inhibition of substrate secretion, presumably due to the loss of substrate recognition by the appropriate Tat machinery (1, 87, 102). The surrounding residues are much less conserved. However, substitution of the surrounding residues with unfavorable residues can lead to a decrease in translocation efficiency (102). In addition, the hydrophobic stretch of Tat signal peptides generally has lower hydrophobicity than that found in Sec signal peptides (16).

The presence of the twin-arginine motif in the Tat signal sequence provided a means of identifying novel Tat substrates by computational pattern-matching techniques. Initial searches for secreted proteins with the twin-arginine residues proved fruitful (51). However, a large number of falsely predicted Tat substrates were made by using this approach. Analysis of the genome of one halophilic archaeon, *Halobacterium* NRC-1, revealed that nearly all of the putative secretory proteins possessed typical Tat signal peptides (90). This large pool of putative Tat signal peptides was used to develop a rule-based perlscript (TATFIND) that predicts these peptides based on numerous characteristics (90). TATFIND positively identifies all confirmed Tat substrates in organisms such as *E. coli* and *B. subtilis* (22, 52, 90). Moreover, due to the large number of criteria required for a successful match by TATFIND, the number of falsely predicted Tat substrates decreased dramatically compared with early pattern-matching efforts. Since the development of TATFIND, additional programs have been developed for Tat substrate identification, such as TatP, which utilizes both pattern matching and a neural

network (10). Future experimental verification of bacterial and archaeal Tat signal peptides that were predicted by computational methods (as well as those the programs failed to identify) will provide useful information to further optimize the sensitivity and specificity of these programs.

Substrates

Many of the earliest Tat substrates that were described were redox proteins from *E. coli* that incorporate their cofactor in the cytoplasm (11). After synthesis, many of these proteins are bound by chaperones that prevent interaction with the secretion machinery until the substrate accepts its cofactor (82). Maturation to the holoenzyme is likely to render the substrates incompatible with secretion through the 5- to 8-Å Sec pore constriction (107). Hence, the Tat pathway was originally viewed as a specialized transport system required mainly for a small class of apoenzymes whose prosthetic group is incorporated prior to translocation.

The availability of the TATFIND program made it possible to search entire genomes very rapidly for the presence of putative Tat substrates. By using this program, it became apparent that the Tat pathway was being used to vastly different extents, with the number of predicted substrates ranging from 0 to 145 in different bacteria and archaea (22). Similar to *E. coli*, many bacteria and archaea appear to employ the Tat pathway for the export of cofactor-containing redox proteins. However, a large number (in some cases, the majority) of putative Tat substrates were not cofactor-containing redox proteins, and included proteases, virulence factors, substrate-binding proteins, and polymer-degrading enzymes. Thus, an important finding of the signal peptide predictions from the largely archaeal-based program, TATFIND, was that the Tat system appeared to be a general secretion pathway, rather than a specialized translocation system (22).

Many Tat substrates are released into the periplasm or extracellular space. However, numerous substrates are attached to the membrane via one of three distinct mechanisms: (i) interaction with membrane proteins, (ii) a C-terminal membrane anchor, or (iii) lipid modification. Five different Tat substrates have been experimentally shown in *E. coli* to possess a C-terminal transmembrane segment that integrates into the membrane in a YidC-independent fashion (47). Analysis of TATFIND results indicated that this class of substrate also exists in archaea. Comprehensive bioinformatic examination of putative Tat substrates from halophilic archaea indicates that the majority of their Tat signal peptides contained typical lipoboxes (13). It has since been demonstrated experimentally that mny of these proteins are secreted via the Tat pathway and that their membrane localization depends on the critical cysteine residue in the lipobox (K.C. Dilks and M. Pohlschröder, unpublished results).

Sec substrates in *E. coli* that are anchored to the membrane in either of these ways require lateral integration of the C-terminal transmembrane segment into the membrane or signal anchoring of the lipoprotein in the membrane until it is processed by the proper peptidase and transferase/transacylase (72, 107). Tat substrates may also achieve proper localization by means of lateral integration and signal anchoring. However, it is not clear how these mechanisms would be compatible with translocation through the Tat apparatus.

Fourteen of the 23 complete archaeal genome sequences available from NCBI possess the necessary genes (*tatA* and *tatC*) for a functional Tat pathway (Table 2). Examination of archaeal genome sequences with TATFIND predicted that most of these 14 organisms encode very few Tat substrates, mostly redox proteins (Table 2). For example, TATFIND identified 9 of the 14 putative Tat substrates in the genome of *Picrophilus torridus* as likely to be cofactor-containing redox proteins. In sharp contrast, the three members of the *Halobacteriaceae* (*Haloarcula marismortui*, *Halobacterium* NRC-1, and *Natronomonas pharaonis*) encode a large number of putative Tat substrates, the majority of which are not redox proteins (Table 2). Computational analyses of the secretory proteins from these haloarchaea have revealed that this group of organisms uses the Tat pathway as the major translocation route, rather than the Sec pathway, which is used by most other organisms (13, 20, 30, 90). The preferential routing of proteins to the Tat pathway in haloarchaea may be in response to the high cytoplasmic and extracellular salt concentrations

(13, 90). Such salt conditions could increase the rate of spontaneous protein folding in the cytoplasm, or alternatively, necessitate a high level of chaperone-mediated protein folding in the cytoplasm. The Tat pathway, as opposed to the Sec pathway, would be able to accommodate the secretion of these folded proteins.

Tat Machinery

Three functionally distinct Tat machinery proteins have been identified: TatA, TatB, and TatC (Fig. 6A) (96, 112). TatA and TatB possess a single N-terminal transmembrane segment and a cytoplasmic C-terminal region composed of an amphipathic helix domain and a region that has not been structurally characterized. TatC typically contains six TM segments with N and C termini localized in the cytoplasm.

In *E. coli*, TatA, TatB, and TatC are integral membrane proteins that differentially interact to form two distinct oligomeric structures. A complex of multiple heterodimers of TatB and TatC act as the initial site of membrane interaction for cytoplasmic Tat substrates (Fig. 6B) (1, 15). Tat substrates that have had the twin-arginine residues in the signal peptide altered lose the ability to interact with the TatBC structure (1). Once the TatBC complex is occupied by a substrate, it engages the second oligomeric structure (a ring-shaped TatA multimer) in a PMF-dependent manner (1, 15) (Fig. 6B). The cytoplasmic region of TatA presumably plugs one end of the aqueous membrane pore created by the TatA multimer (36). Due to its structure and late-stage interaction with Tat substrates, the TatA multimer is believed to act as the protein-conducting channel of the Tat pathway. Biochemical and microscopic data indicate that the size of the TatA structure varies from approximately 10 to 40 copies (36, 77). The different sizes of TatA structures would help rationalize how this pathway is able

Table 2. The Tat pathway in *Archaea* deduced from complete genome sequences

Organism	Kingdom[a]	No. of ORFs	Putative Tat substrates	TatA and TatC present
Archaeoglobus fulgidus	Eury	2,420	10	Yes
Halobacterium sp. NRC-1	Eury	2,622	68	Yes
Methanocaldococcus jannaschii	Eury	1,786	0	No
Methanosarcina acetivorans	Eury	4,540	5	Yes
Methanothermobacter thermautotrophicus	Eury	1,873	1	No
Natronomonas pharaonis	Eury	2,622	106	Yes
Pyrobaculum aerophilum	Cren	2,605	14	Yes
Pyrococcus furiosus	Eury	2,125	3	No
Sulfolobus acidocaldarius	Cren	2,223	5	Yes
Sulfolobus tokodaii	Cren	2,825	4	Yes
Thermoplasma acidophilum	Eury	1,482	3	Yes

[a]Cren, *Crenarchaeota*; Eury, *Euryarchaeota*.

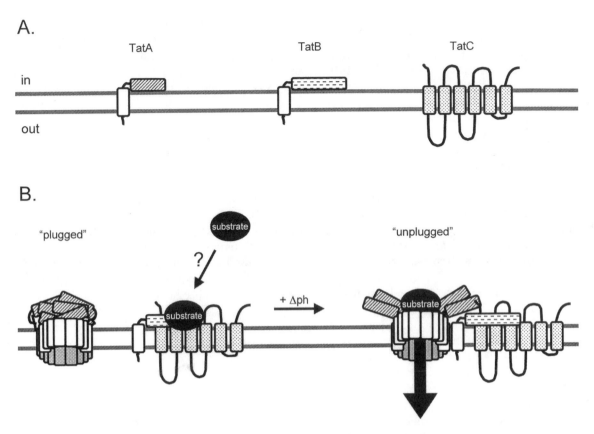

Figure 6. Tat components and model of Tat secretory mechanism. (A) Typcial structure of Tat machinery components in bacteria and archaea. The postamphipathic helical C terminus for TatA and TatB has been excluded for visual simplicity. (B) Model of Tat substrate translocation in *E. coli*. Tat substrates (oval) obtain tertiary structure in the cytoplasm and are targeted to the membrane TatBC complex in an unknown manner. Once bound to substrate, the TatBC complex interacts with a multimeric TatA ring in a ΔpH-dependent manner. The plugged inactive TatA ring likely alters to an active unplugged confirmation upon engaging substrate. There are insufficient data describing points of protein interactions, and the depicted points of interaction between proteins are not meant to be completely accurate.

to accommodate the passage of substrates of vastly different sizes, while maintaining a large degree of membrane impermeability.

While many archaea encode multiple *tatA* and *tatC* genes, all archaea (and several bacteria) lack a copy of *tatB* (22). The absence of TatB from many organisms was surprising in view of the requirement of TatB for Tat translocation in *E. coli* (97). However, recent studies in *E. coli* demonstrated that mutant alleles of *tatA* could suppress the translocation defect of a *tatB* deletion (12). Thus, TatA homologs may provide the function of both TatA and TatB in bacteria and archaea that do not have TatB.

Analysis of the *B. subtilis* Tat pathway revealed the presence of TatA homologs in the cytoplasm, in addition to its presence in the membrane (87). Cytoplasmic localization of TatA homologs was also observed in *Streptomyces lividans* and the haloarchaeon *H. volcanii* (19, 21). The *B. subtilis* TatA homolog (TatAd) coimmunoprecipitates with its cytoplasmic

Tat substrate, PhoD (87). Note that, despite extensively investigating the mechanism of the Tat pathway in *E. coli*, it remains unclear how substrates are directed to the membrane TatBC complex (Fig. 6B). Although substrate interaction with cytoplasmic TatA may represent a membrane-targeting mechanism, it is inconsistent with the absence of this protein from the cytoplasm and the late translocation role of this protein in *E. coli*. Future investigations of the nature and role of cytoplasmic and membrane TatA in bacteria and archaea will be useful in reconciling the localization and function of this protein.

The most well-defined archaeal Tat pathway is that of the model archaeon *H. volcanii*. As described above ("Tat signals and substrates") computational predictions and recent experimental analyses of putative Tat substrates indicate that the haloarchaea use the Tat pathway for the export of the majority of secreted proteins. Genetic studies in *H. volcanii*, demonstrating the essential nature of this pathway, are consistent with its predicted extensive use. The two TatC homologs in *H. volcanii* and other haloarchaea have atypical features. Specifically, one of the TatC homologs has an extended cytoplasmic N terminus, while its paralog contains 14 predicted TM segments, in which the sequence resembles two independent TatC proteins fused by 2 TMs (13, 21). These modifications of the TatC components have only been identified in the haloarchaea and may represent a functional adaptation to either high salt or increased usage of the Tat pathway in this group of archaea.

PERSPECTIVE: THE NEXT FIVE YEARS

Recent in vivo, in vitro, and in silico studies have led to a better understanding of archaeal protein translocation. Moreover, the elucidation of an archaeal Sec-pore X-ray crystal structure strikingly demonstrates how analysis of this pathway in archaea can significantly advance the field of protein translocation as a whole. In addition to standard molecular and biochemical approaches, it is now crucial to develop in vitro Sec and Tat protein translocation systems that will more clearly define the mechanisms of these pathways and reveal the energetics of these cellular processes in archaea. Moreover, with the growing number of auxotrophic strains, selectable markers, recombinatory techniques, and self-replicating vectors, the future use of genetic selections should help to identify potential archaea-specific aspects of protein translocation (6). Such genetic selections may also reveal translocation components conserved in bacteria and/or eucarya that have not yet been identified.

Acknowledgments. Support was provided to M.P. by a grant from the National Science Foundation (reference no. MCB-0239215) and the Department of Energy (reference no. DE-FG02-01ER15169), and to K.D. by a predoctoral fellowship from the American Heart Association (reference no. 0415376U).

REFERENCES

1. **Alami, M., I. Luke, S. Deitermann, G. Eisner, H.G. Koch, J. Brunner, and M. Muller.** 2003. Differential interactions between a twin-arginine signal peptide and its translocase in Escherichia coli. *Mol. Cell* 12:937–946.
2. **Albers, S. V., and A. J. Driessen.** 2005. Analysis of ATPases of putative secretion operons in the thermoacidophilic archaeon Sulfolobus solfataricus. *Microbiology* 151:763–773.
3. **Albers, S. V., and A. M. Driessen.** 2002. Signal peptides of secreted proteins of the archaeon Sulfolobus solfataricus: a genomic survey. *Arch. Microbiol.* 177:209–216.
4. **Albers, S. V., Z. Szabo, and A. J. Driessen.** 2003. Archaeal homolog of bacterial type IV prepilin signal peptidases with broad substrate specificity. *J. Bacteriol.* 185:3918–3925.
5. **Alder, N. N., Y. Shen, J. L. Brodsky, L. M. Hendershot, and A. E. Johnson.** 2005. The molecular mechanisms underlying BiP-mediated gating of the Sec61 translocon of the endoplasmic reticulum. *J. Cell Biol.* 168: 389–99.
6. **Allers, T., and M. Mevarech.** 2005. Archaeal genetics—the third way. *Nat. Rev. Genet.* 6:58–73.
7. **Angelini, S., S. Deitermann, and H. G. Koch.** 2005. FtsY, the bacterial signal-recognition particle receptor, interacts functionally and physically with the SecYEG translocon. *EMBO Rep.* 6:476–481.
8. **Bardy, S. L., S. Y. Ng, D. S. Carnegie, and K. F. Jarrell.** 2005. Site-directed mutagenesis analysis of amino acids critical for activity of the type I signal peptidase of the archaeon Methanococcus voltae. *J. Bacteriol.* 187:1188–1191.
9. **Bardy, S, L., S. Y. Ng, and K. F. Jarrell.** 2004. Recent advances in the structure and assembly of the archaeal flagellum. *J. Mol. Microbiol. Biotechnol.* 7:41–51.
10. **Bendtsen, J. D., H. Nielsen, D. Widdick, T. Palmer, and S. Brunak.** 2005. Prediction of twin-arginine signal peptides. *BMC Bioinformatics* 6:167.
11. **Berks, B. C.** 1996. A common export pathway for proteins binding complex redox cofactors? *Mol. Microbiol.* 22:393–404.
12. **Blaudeck, N., P. Kreutzenbeck, M. Muller, G. A. Sprenger, and R. Freudl.** 2005. Isolation and characterization of bifunctional Escherichia coli TatA mutant proteins that allow efficient tat-dependent protein translocation in the absence of TatB. *J. Biol. Chem.* 280:3426–3432.
13. **Bolhuis, A.** 2002. Protein transport in the halophilic archaeon Halobacterium sp. NRC-1: a major role for the twin-arginine translocation pathway? *Microbiology* 148:3335–3346.
14. **Cao, T. B., and M. H. Saier, Jr.** 2003. The general protein secretory pathway: phylogenetic analyses leading to evolutionary conclusions. *Biochim. Biophys. Acta* 1609:115–125.
15. **Cline, K., and H. Mori.** 2001. Thylakoid DeltapH-dependent precursor proteins bind to a cpTatC-Hcf106 complex before Tha4-dependent transport. *J. Cell Biol.* 154:719–729.
16. **Cristobal, S., J. W. de Gier, H. Nielsen, and G. von Heijne.** 1999. Competition between Sec- and TAT-dependent protein translocation in Escherichia coli. *EMBO J.* 18:2982–2990.
17. **Dalbey, R. E., and M. Chen.** 2004. Sec-translocase mediated membrane protein biogenesis. *Biochim. Biophys. Acta* 1694: 37–53.
18. **Deitermann. S., G. S. Sprie, and H. G. Koch.** 2005. A dual function for SecA in the assembly of single spanning membrane proteins in Escherichia coli. *J. Biol. Chem.* 280:39077–39085.
19. **De Keersmaeker, S., L. Van Mellaert, K. Schaerlaekens, W. Van Dessel, Vrancken K., et al.** 2005. Structural organization of the twin-arginine translocation system in Streptomyces lividans. *FEBS Lett.* 579:797–802.
20. **Dilks, K., M. I. Gimenez, and M. Pohlschröder.** 2005. Genetic and biochemical analysis of the twin-arginine translocation pathway in halophilic archaea. *J. Bacteriol.* 187:8104–8113.
21. **Dilks, K., M. I. Gimenez, and M. Pohlschröder.** 2005. Genetic and biochemical analysis of the twin-arginine translocation pathway in halophilic archaea. *J. Bacteriol.* 187:8104–8113.
22. **Dilks, K., R. W. Rose, E. Hartmann, and M. Pohlschröder.** 2003. Prokaryotic utilization of the twin-arginine translocation pathway: a genomic survey. *J. Bacteriol.* 185:1478–1483.
23. **Duong, F., and W. Wickner.** 1997. Distinct catalytic roles of the SecYE, SecG and SecDFyajC subunits of preprotein translocase holoenzyme. *EMBO J.* 16:2756–2768.
24. **Duong, F., and W. Wickner.** 1997. The SecDFyajC domain of preprotein translocase controls preprotein movement by regulating SecA membrane cycling. *EMBO J.* 16:4871–4879.
25. **Egea, P. F., S. O. Shan, J. Napetschnig, D. F. Savage, P. Walter, and R. M. Stroud.** 2004. Substrate twinning activates the signal recognition particle and its receptor. *Nature* 427:215–221.

26. Egea, P. F., R. M. Stroud, and P. Walter. 2005. Targeting proteins to membranes: structure of the signal recognition particle. *Curr. Opin. Struct. Biol.* **15**:213–220.

27. Eichler, J. 2002. Archaeal signal peptidases from the genus Thermoplasma: structural and mechanistic hybrids of the bacterial and eukaryal enzymes. *J. Mol. Evol.* **54**:411–415.

28. Eichler, J. 2003. Evolution of the prokaryotic protein translocation complex: a comparison of archaeal and bacterial versions of SecDF. *Mol. Phylogenet. Evol.* **27**:504–509.

29. Evans, E. A., R. Gilmore, and G. Blobel. 1986. Purification of microsomal signal peptidase as a complex. *Proc. Natl. Acad. Sci. USA* **83**:581–585.

30. Falb, M., F. Pfeiffer, P. Palm, K. Rodewald, V. Hickmann, J. Tittor, and D. Oesterhelt. 2005. Living with two extremes: conclusions from the genome sequence of *Natronomonas pharaonis*. *Genome Res.* **15**:1336–1343.

31. Fang, H., and N. Green. 1994. Nonlethal *sec71-1* and *sec72-1* mutations eliminate proteins associated with the Sec63p-BiP complex from *S. cerevisiae*. *Mol. Biol. Cell* **5**:933–942.

32. Finke, K., K. Plath, S. Panzner, S. Prehn, T. A. Rapoport, E. Hartmann, and T. Sommer. 1996. A second trimeric complex containing homologs of the Sec61p complex functions in protein transport across the ER membrane of *S. cerevisiae*. *EMBO J.* **15**:1482–1494.

33. Flower, A. M., L. L. Hines, and P. L. Pfennig. 2000. SecG is an auxiliary component of the protein export apparatus of *Escherichia coli*. *Mol. Gen. Genet.* **263**:131–136.

34. Fons, R. D., B. A. Bogert, and R. S. Hegde. 2003. Substrate-specific function of the translocon-associated protein complex during translocation across the ER membrane. *J. Cell Biol.* **160**:529–539.

35. Gier, J. W. 2005. Biogenesis of inner membrane proteins in Escherichia coli. *Annu. Rev. Microbiol.* **59**:329–355.

36. Gohlke, U., L. Pullan, C. A. McDevitt, I. Porcelli, E. de Leeuw, et al. 2005. The TatA component of the twin-arginine protein transport system forms channel complexes of variable diameter. *Proc. Natl. Acad. Sci. USA* **102**:10482–10486.

37. Görlich, D., E. Hartmann, S. Prehn, and T. A. Rapoport. 1992. A protein of the endoplasmic reticulum involved early in polypeptide translocation. *Nature* **357**:47–52.

38. Görlich, D., S. Prehn, E. Hartmann, K. U. Kalies, and T. A. Rapoport. 1992. A mammalian homolog of SEC61p and SECYp is associated with ribosomes and nascent polypeptides during translocation. *Cell* **71**:489–503.

39. Görlich, D., and T. A. Rapoport. 1993. Protein translocation into proteoliposomes reconstituted from purified components of the endoplasmic reticulum membrane. *Cell* **75**:615–630.

40. Gropp, R., F. Gropp, and M. C. Betlach. 1992. Association of the halobacterial 7S RNA to the polysome correlates with expression of the membrane protein bacterioopsin. *Proc. Natl. Acad. Sci. USA* **89**:1204–1208.

41. Haddad, A., R. W. Rose, and M. Pohlschröder. 2005. The Haloferax volcanii FtsY homolog is critical for haloarchaeal growth but does not require the A domain. *J. Bacteriol.* **187**:4015–4022.

42. Halic, M., T. Becker, M. R. Pool, C. M. Spahn, R. A. Grassucci, et al. 2004. Structure of the signal recognition particle interacting with the elongation-arrested ribosome. *Nature* **427**:808–814.

43. Hamman, B. D., L. M. Hendershot, and A. E. Johnson. 1998. BiP maintains the permeability barrier of the ER membrane by sealing the lumenal end of the translocon pore before and early in translocation. *Cell* **92**:747–758.

44. Hanada, M., K. I. Nishiyama, S. Mizushima, and H. Tokuda. 1994. Reconstitution of an efficient protein translocation machinery comprising SecA and the three membrane proteins, SecY, SecE, and SecG (p12). *J. Biol. Chem.* **269**:23625–23631.

45. Hand, N. J., A. Laskewitz, R. Klein, and M. Pohlschröder. 2006. Archaeal and Bacterial SecD and SecF homologs exhibit striking structural and functional conservation. *J. Bacteriol.* **188**:1251–1259.

46. Hartmann, E., T. Sommer, S. Prehn, D. Gorlich, S. Jentsch, and T. A. Rapoport. 1994. Evolutionary conservation of components of the protein translocation complex. *Nature* **367**:654–657.

47. Hatzixanthis, K., T. Palmer, and F. Sargent. 2003. A subset of bacterial inner membrane proteins integrated by the twin-arginine translocase. *Mol. Microbiol.* **49**:1377–1390.

48. Hayashi, S., and H. C. Wu. 1990. Lipoproteins in bacteria. *J. Bioenerg. Biomembr.* **22**:451–471.

49. Houben, E. N., P. A. Scotti, Q. A. Valent, J. Brunner, J. L. de Gier, B. Oudega, and J. Luirink. 2000. Nascent Lep inserts into the *Escherichia coli* inner membrane in the vicinity of YidC, SecY and SecA. *FEBS Lett.* **476**:229–233.

50. Irihimovitch, V., and J. Eichler. 2003. Post-translational secretion of fusion proteins in the halophilic archaea *Haloferax volcanii*. *J. Biol. Chem.* **278**:12881–12887.

51. Jongbloed, J. D., H. Antelmann, M. Hecker, R. Nijland, S. Bron, et al. 2002. Selective contribution of the twin-arginine translocation pathway to protein secretion in Bacillus subtilis. *J. Biol. Chem.* **277**:44068–44078.

52. Jongbloed, J. D., U. Grieger, H. Antelmann, M. Hecker, R. Nijland, S. Bron, and J. M. van Dijl. 2004. Two minimal Tat translocases in Bacillus. *Mol. Microbiol.* **54**:1319–1325.

53. Jungnickel. B., T. A. Rapoport, and E. Hartmann. 1994. Protein translocation: common themes from bacteria to man. *FEBS Lett.* **346**:73–347.

54. Kabani, M., J. M. Beckerich, and C. Gaillardin. 2000. Sls1p stimulates Sec63p-mediated activation of Kar2p in a conformation-dependent manner in the yeast endoplasmic reticulum. *Mol. Cell. Biol.* **20**:6923–6934.

55. Karamyshev, A. L., and A. E. Johnson. 2005. Selective SecA association with signal sequences in ribosome-bound nascent chains: A potential role for SecA in ribosome targeting to the bacterial membrane. *J. Biol. Chem.* **280**:40489–40493.

56. Kato, Y., K. Nishiyama, and H. Tokuda. 2003. Depletion of SecDF-YajC causes a decrease in the level of SecG: implication for their functional interaction. *FEBS Lett.* **550**:114–118.

57. Keenan, R. J., D. M. Freymann, R. M. Stroud, and P. Walter. 2001. The signal recognition particle. *Annu. Rev. Biochem.* **70**:755–775.

58. Kinch, L. N., M. H. Saier, Jr., and N. V. Grishin. 2002. Sec61beta—a component of the archaeal protein secretory system. *Trends Biochem. Sci.* **27**:170–171.

59. Kokoeva, M. V., K. F. Storch, C. Klein, and D. Oesterhelt. 2002. A novel mode of sensory transduction in archaea: binding protein-mediated chemotaxis towards osmoprotectants and amino acids. *EMBO J.* **21**:2312–2322.

60. Koning, S. M., S. V. Albers, W. N. Konings, and A. J. Driessen. 2002. Sugar transport in (hyper)thermophilic archaea. *Res. Microbiol.* **153**:61–67.

61. Lichi, T., G. Ring, and J. Eichler. 2004. Membrane binding of SRP pathway components in the halophilic archaea *Haloferax volcanii*. *Eur. J. Biochem.* **271**:1382–1390.

62. Macnab, R. M. 2004. Type III flagellar protein export and flagellar assembly. *Biochim. Biophys. Acta* **1694**:207–217.

63. Mason, N., L. F. Ciufo, and J. D. Brown. 2000. Elongation arrest is a physiologically important function of signal recognition particle. *EMBO J.* **19**:4164–4174.

64. Matsuyama, S., Y. Fujita, and S. Mizushima. 1993. SecD is involved in the release of translocated secretory proteins from the cytoplasmic membrane of *Escherichia coli*. *EMBO J.* 12:265–270.

65. Mattar, S., B. Scharf, S. B. Kent, K. Rodewald, D. Oesterhelt, and M. Engelhard. 1994. The primary structure of halocyanin, an archaeal blue copper protein, predicts a lipid anchor for membrane fixation. *J. Biol. Chem.* 269:14939–14945.

66. Miller, J. D., S. Tajima, L. Lauffer, and P. Walter. 1995. The beta subunit of the signal recognition particle receptor is a transmembrane GTPase that anchors the alpha subunit, a peripheral membrane GTPase, to the endoplasmic reticulum membrane. *J. Cell Biol.* 128:273–282.

67. Moll, R., S. Schmidtke, and G. Schaefer. 1996. A putative signal recognition particle receptor alpha subunit (SR alpha) homologue is expressed in the hyperthermophilic crenarchaeon Sulfolobus acidocaldarius. *FEMS Microbiol. Lett.* 137:51–56.

68. Moll, R. G. 2003. Protein-protein, protein-RNA and protein-lipid interactions of signal-recognition particle components in the hyperthermoacidophilic archaeon *Acidianus ambivalens*. *Biochem. J.* 374:247–254.

69. Moll, R. G. 2004. The archaeal signal recognition particle: steps toward membrane binding. *J. Bioenerg. Biomembr.* 36:47–53.

70. Mori, H., and K. Ito. 2003. Biochemical characterization of a mutationally altered protein translocase: proton motive force stimulation of the initiation phase of translocation. *J. Bacteriol.* 185:405–412.

71. Nakamura, K., S. Yahagi, T. Yamazaki, and K. Yamane. 1999. *Bacillus subtilis* histone-like protein, HBsu, is an integral component of a SRP-like particle that can bind the Alu domain of small cytoplasmic RNA. *J. Biol. Chem.* 274:13569–13576.

72. Narita, S., S. Matsuyama, and H. Tokuda. 2004. Lipoprotein trafficking in Escherichia coli. *Arch. Microbiol.* 182:1–6.

73. Neuhof, A., Rolls, M. M., B. Jungnickel, K. U. Kalies, and T. A. Rapoport. 1998. Binding of signal recognition particle gives ribosome/nascent chain complexes a competitive advantage in endoplasmic reticulum membrane interaction. *Mol. Biol. Cell* 9:103–115.

74. Nielsen, H., S. Brunak, and G. von Heijne. 1999. Machine learning approaches for the prediction of signal peptides and other protein sorting signals. *Protein Eng.* 12:3–9.

75. Nishiyama, K., A. Fukuda, K. Morita, and H. Tokuda. 1999. Membrane deinsertion of SecA underlying proton motive force-dependent stimulation of protein translocation. *EMBO J.* 18:1049–1058.

76. Nouwen, N., and A. J. Driessen. 2002. SecDFyajC forms a heterotetrameric complex with YidC. *Mol. Microbiol.* 44:1397–1405.

77. Oates, J., C. M. Barrett, J. P. Barnett, K. G. Byrne, A. Bolhuis, and C. Robinson. 2005. The Escherichia coli twin-arginine translocation apparatus incorporates a distinct form of TatABC complex, spectrum of modular TatA complexes and minor TatAB complex. *J. Mol. Biol.* 346:295–305.

78. Ortenberg, R., and M. Mevarech. 2000. Evidence for post-translational membrane insertion of the integral membrane protein bacterioopsin expressed in the heterologous halophilic archaeon *Haloferax volcanii*. *J. Biol. Chem.* 275:22839–22846.

79. Osborne, A. R., T. A. Rapoport, and B. van den Berg. 2005. Protein translocation by the Sec61/SecY channel. *Annu. Rev. Cell Dev. Biol.* 21:529–550.

80. Paetzel, M., A. Karla, N. C. Strynadka, and R. E. Dalbey. 2002. Signal peptidases. *Chem. Rev.* 102:4549–4580.

81. Palacín, A., R. de la Fuente, L. A. Valle I, Rivas, and R. P. Mellado. 2003. *Streptomyces lividans* contains a minimal functional signal recognition particle that is involved in protein secretion. *Microbiology* 149:2435–2442.

82. Palmer, T., F. Sargent, and B. C. Berks. 2005. Export of complex cofactor-containing proteins by the bacterial Tat pathway. *Trends Microbiol.* 13:175–180.

83. Panzner, S., L. Dreier, E. Hartmann, S. Kostka, and T. A. Rapoport. 1995. Posttranslational protein transport in yeast reconstituted with a purified complex of Sec proteins and Kar2p. *Cell* 81:561–570.

84. Pohlschröder, M., M. I. Gimenes, and K. F. Jarrell. 2005. Protein transport in Archaea: Sec and Twin arginine translocation pathways. Invited review. *Curr. Opin. Microbiol.* 8:713–719.

85. Pohlschröder, M., K. Dilks, N. J. Hand, and R. Wesley Rose. 2004. Translocation of proteins across archaeal cytoplasmic membranes. *FEMS Microbiol. Rev.* 28:3–24.

86. Pohlschröder, M., W. A. Prinz, E. Hartmann, and J. Beckwith. 1997. Protein translocation in the three domains of life: variations on a theme. *Cell* 91:563–566.

87. Pop, O. I., M. Westermann, R. Volkmer-Engert, D. Schulz, C. Lemke, et al. 2003. Sequence-specific binding of prePhoD to soluble TatAd indicates protein-mediated targeting of the Tat export in Bacillus subtilis. *J. Biol. Chem.* 278:38428–38436.

88. Qi, H. Y., and H. D. Bernstein. 1999. SecA is required for the insertion of inner membrane proteins targeted by the Escherichia coli signal recognition particle. *J. Biol. Chem.* 274:8993–8997.

89. Randall, L. L., and S. J. Hardy. 2002. SecB, one small chaperone in the complex milieu of the cell. *Cell. Mol. Life Sci.* 59:1617–1623.

90. Rose, R. W., T. Bruser, J. C. Kissinger, and M. Pohlschröder. 2002. Adaptation of protein secretion to extremely high-salt conditions by extensive use of the twin-arginine translocation pathway. *Mol. Microbiol.* 45:943–950.

91. Rose, R. W., and M. Pohlschröder. 2002. In vivo analysis of an essential archaeal signal recognition particle in its native host. *J. Bacteriol.* 184:3260–3267.

92. Rosendal, K. R., K. Wild, G. Montoya, and I. Sinning. 2003. Crystal structure of the complete core of archaeal signal recognition particle and implications for interdomain communication. *Proc. Natl. Acad. Sci. USA* 100:14701–14706.

93. Rothblatt, J. A., R. J. Deshaies, S. L. Sanders, G. Daum, and R. Schekman. 1989. Multiple genes are required for proper insertion of secretory proteins into the endoplasmic reticulum in yeast. *J. Cell Biol.* 109:2641–2652.

94. Sadler, I., A. Chiang, T. Kurihara, J. Rothblatt, J. Way, and P. Silver. 1989. A yeast gene important for protein assembly into the endoplasmic reticulum and the nucleus has homology to DnaJ, an *Escherichia coli* heat shock protein. *J. Cell Biol.* 109:2665–2675.

95. Sankaran, K., S. D. Gupta, and H. C. Wu. 1995. Modification of bacterial lipoproteins. *Methods Enzymol.* 250:683–697.

96. Sargent, F., E. G. Bogsch, N. R. Stanley, M. Wexler, C. Robinson, et al. 1998. Overlapping functions of components of a bacterial Sec-independent protein export pathway. *EMBO J.* 17:3640–3650.

97. Sargent, F., N. R. Stanley, B. C. Berks, and T. Palmer. 1999. Sec-independent protein translocation in Escherichia coli. A distinct and pivotal role for the TatB protein. *J. Biol. Chem.* 274:36073–36082.

98. Scotti, P. A., M. L. Urbanus, J. Brunner, J. W. de Gier, G. von Heijne, et al. 2000. YidC, the *Escherichia coli* homologue of mitochondrial Oxa1p, is a component of the Sec translocase. *EMBO J.* 19:542–549.

99. Shan, S. O., and P. Walter. 2005. Co-translational protein targeting by the signal recognition particle. *FEBS Lett.* **579:** 921–926.

100. Shelness, G. S., Y. S. Kanwar, and G. Blobel. 1988. cDNA-derived primary structure of the glycoprotein component of canine microsomal signal peptidase complex. *J. Biol. Chem.* **263:**17063–17070.

101. Snapp, E. L., G. A. Reinhart, B. A. Bogert, J. Lippincott-Schwartz, and R. S. Hegde. 2004. The organization of engaged and quiescent translocons in the endoplasmic reticulum of mammalian cells. *J. Cell Biol.* **164:**997–1007.

102. Stanley, N. R., T. Palmer, and B. C. Berks. 2000. The twin arginine consensus motif of Tat signal peptides is involved in Sec-independent protein targeting in Escherichia coli. *J. Biol. Chem.* **275:**11591–11596.

103. Steel, G. J., D. M. Fullerton, J. R. Tyson, and C. J. Stirling. 2004. Coordinated activation of Hsp70 chaperones. *Science* **303:**98–101.

104. Stoeckenius, W., R. H. Lozier, and R. A. Bogomolni. 1979. Bacteriorhodopsin and the purple membrane of halobacteria. *Biochim. Biophys. Acta* **505:**215–278.

104a. Szabó, Z., A. Oliveira Stahl, S.-V Albers, J. C. Kissinger, A. J. M. Driessen, and M. Pohlschröder. 2007. Identification of diverse archaeal proteins with class III signal peptides cleaved by distinct archaeal prepilin peptidases. *J. Bacteriol.* **189:**772–778.

105. Talmadge, K., S. Stahl, and W. Gilbert. 1980. Eukaryotic signal sequence transports insulin antigen in Escherichia coli. *Proc. Natl. Acad. Sci. USA* **77:**3369–3373.

106. Thomas, N. A., S. L. Bardy, and K. F. Jarrell. 2001. The archaeal flagellum: a different kind of prokaryotic motility structure. *FEMS Microbiol. Rev.* **25:**147–174.

107. Van den Berg, B., W. M. Clemons, Jr., I. Collinson, Y. Modis, E. Hartmann, et al. 2004. X-ray structure of a protein-conducting channel. *Nature* **427:**36–44.

108. von Heijne, G. 1990. The signal peptide. *J. Membr. Biol.* **115:** 195–201.

109. Vrontou, E., and A. Economou. 2004. Structure and function of SecA, the preprotein translocase nanomotor. *Biochim. Biophys. Acta* **1694:**67–80.

110. Walter, P., and G. Blobel. 1981. Translocation of proteins across the endoplasmic reticulum III. Signal recognition protein (SRP) causes signal sequence-dependent and site-specific arrest of chain elongation that is released by microsomal membranes. *J. Cell Biol.* **91:**557–561.

111. Weichenrieder, O., K. Wild, K. Strub, and S. Cusack. 2000. Structure and assembly of the Alu domain of the mammalian signal recognition particle. *Nature* **408:**167–173.

112. Weiner, J. H., P. T. Bilous, G. M. Shaw, S. P. Lubitz, L. Frost, et al. 1998. A novel and ubiquitous system for membrane targeting and secretion of cofactor-containing proteins. *Cell* **93:** 93–101.

113. Wiedmann, M., T. V. Kurzchalia, E. Hartmann, and T. A. Rapoport. 1987. A signal sequence receptor in the endoplasmic reticulum membrane. *Nature* **328:**830–833.

114. Wild, K., M. Halic, I. Sinning, and R. Beckmann. 2004. SRP meets the ribosome. *Nat. Struct. Mol. Biol.* **11:**1049–1053.

115. Yahr, T. L., and W. T. Wickner. 2001. Functional reconstitution of bacterial Tat translocation in vitro. *EMBO J.* **20:** 2472–2479.

116. Yen, M. R., K. T. Harley, Y. H. Tseng, and M. H. Saier, Jr. 2001. Phylogenetic and structural analyses of the oxa1 family of protein translocases. *FEMS Microbiol. Lett.* **204:**223–231.

117. Yi, L., and R. E. Dalbey. 2005. Oxa1/Alb3/YidC system for insertion of membrane proteins in mitochondria, chloroplasts and bacteria (review). *Mol. Membr. Biol.* **22:**101–111.

118. Zago, M. A., P. P. Dennis, and A. D. Omer. 2005. The expanding world of small RNAs in the hyperthermophilic archaeon Sulfolobus solfataricus. *Mol. Microbiol.* **55:**1812–1828.

Archaea: Molecular and Cellular Biology
Edited by Ricardo Cavicchioli
© 2007 ASM Press, Washington, D.C.

Chapter 18

Flagellation and Chemotaxis

Ken F. Jarrell, Sandy Y. M. Ng, and Bonnie Chaban

INTRODUCTION

Flagellation is a characteristic that is found widely throughout the *Archaea* (Fig. 1) (45, 111). While the gross observations of a rotating structure with a hook responsible for swimming motility indicate an affinity of the archaeal flagellum with the bacterial flagellum, biochemical, genetic, and ultrastructural evidence has accumulated over the years indicating that the archaeal flagellum is a unique motility structure. It is distinct from the bacterial flagellum but with several similarities to another bacterial motility structure, namely the type IV pilus, an organelle responsible for a type of surface translocation termed twitching (11, 12, 23, 84, 111). The presence of flagella on archaea inhabiting extremes of temperature, salinity, and pH (Fig. 2) indicate a remarkable stability of the structure. Novel posttranslational processing of the flagellins (unusual type IV pilin-like signal peptides, novel N-linked glycosylation) (5, 8, 9, 13, 24, 116, 118) is known, and insights into the assembly of the flagellins are slowly emerging (see Chapter 11), aided by the continuous improvement in the genetic tools for various archaea (see Chapter 21). Studies of the preflagellin processing have also aided our understanding of protein export in these organisms (7, 86, 89) (see Chapter 17).

One of the better-studied aspects of archaea physiology is the understanding of various types of taxis (phototaxis, chemotaxis), especially in halobacteria (64, 90, 107). While the components and molecular mechanisms are similar in bacteria and archaea, novel findings in archaea have been reported (56, 64). Both the bacterial and archaeal sensory systems have a two-component signaling mechanism at their heart. However, while the response regulator, CheY-P, is known to bind to the switch component FliM in bacteria (107), a functionally equivalent protein to FliM

has yet to be identified in archaea (29, 79). Indeed, the method of linkage of the unusual archaeal motility organelle with the bacterial-like chemotaxis system remains a compelling mystery. Continued study of this fascinating motility structure and associated sensing system will no doubt add to our knowledge of archaeal physiology and genetics well beyond the limited area of bacterial and archaeal movement.

ULTRASTRUCTURE

Phenomenologically, the archaeal flagellum resembles the bacterial flagellum in being a rotating structure responsible for swimming motility. However, the absence of detectable homologs in completely sequenced genomes from any archaeon of genes involved in bacterial flagella structure, function, and assembly indicate that the archaeal organelle is likely composed of archaeal-specific gene products (29). Electron microscopic examination of flagella from numerous diverse archaea indicates several common features. Archaeal flagella are typically much thinner than their bacterial counterparts, usually 10 to 13 nm (17, 27, 33, 46, 99), compared with bacterial flagellum diameters that are typically about 20 nm (48). In the case of *Halobacterium salinarum* filaments, there are ~3.3 subunits per turn of a ~19-Å pitch left-handed helix compared with ~5.5 subunits per turn of an approximately 26-Å pitch right-handed helix for plain bacterial filaments (23). *H. salinarum* cells alternate between forward and reverse swimming due to clockwise (CW) and counterclockwise (CCW) rotation of the right-handed flagella bundle, respectively (3). CW rotation of the halobacterial flagella exerts a pushing force, propelling the cell forward, while CCW rotation results in the flagella pulling the cell, so that the cell swims with the flagella

Ken F. Jarrell, Sandy Y. M. Ng, and Bonnie Chaban • Department of Microbiology and Immunology, Queen's University, Kingston, Ontario, Canada K7L 3N6.

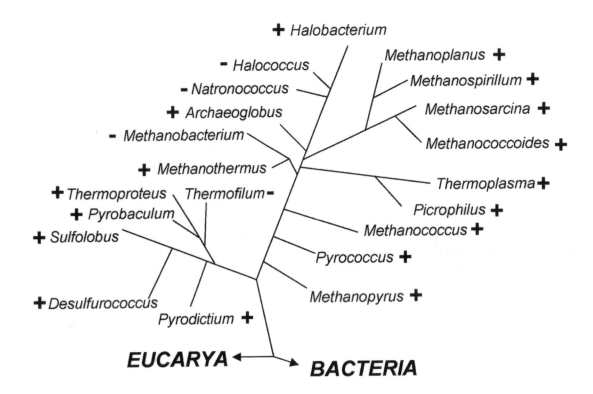

Figure 1. Distribution of flagellation throughout the domain *Archaea*. +, at least some species of the genus are flagellated; −, no members of the genus are reported to be flagellated. Reproduced with modifications from *FEMS Microbiology Reviews* (111).

at the front (54). This is unlike most bacteria, such as *Escherichia coli*, which alternate between swimming (CCW rotation of the left-handed flagella bundle) and tumbling (CW rotation). Also unlike most bacterial flagella, the archaeal flagella filaments are composed of multiple different flagellins (with the apparent exception of *Sulfolobus* species) (5), which are usually, perhaps universally, posttranslationally modified. N-Linked glycosylation structures attached to flagellins have been determined for *H. salinarum* (60, 118) and *Methanococcus voltae* (116). In addition, there is indirect evidence from specific staining (28, 32, 51, 87, 96, 99) and proteomic analyses (38, 74) of further, as yet uncharacterized, modifications on the flagellins from other archaeal species. The core structure of the flagella from *H. salinarum* is similar to that of type IV pili (23, 115), and in sharp contrast to bacterial flagella, the archaeal filament has no detectable central cavity (23). This observation implies that the assembly of archaeal flagella is distinct from that of bacterial flagella and is consistent with a long-standing hypothesis (36, 45). Flagella dissociation experiments in *Natrialba magadii* indicate that flagella filaments are composed of thinner filaments termed protofilaments and that the flagella can be untwined into protofilaments upon a decrease in NaCl concen-

tration (33). Immunolabeling experiments indicated that the *N. magadii* flagella are composed of several longitudinal rows of protofilaments, each consisting of one flagellin type (88). Cross-sectional analysis of the density map of *H. salinarum* flagella indicates the presence of 10 protofilaments (Fig. 3) (23).

Anchoring Structure

A cell-proximal hook region has been detected in stained samples of various archaeal flagella preparations. Staining with phosphotungstic acid routinely results in better resolution of the hook region than staining with uranyl acetate (10, 27, 52). Compared with the defined hook length in bacteria, archaeal hooks vary considerably in length and are observed to be longer, in general (10, 27). In addition, the junction between the filament and hook in archaeal flagella is indistinguishable, in contrast to the case in bacterial flagella (10). In *M. voltae*, there is a large enrichment in flagellin FlaB3 in flagella stub samples. These preparations are enriched in cell-proximal filament fragments (10), suggesting that in archaea the hook region may be composed primarily of one of the multiple flagellins.

The anchoring structure or basal body equivalent of archaeal flagella has proven to be difficult to visu-

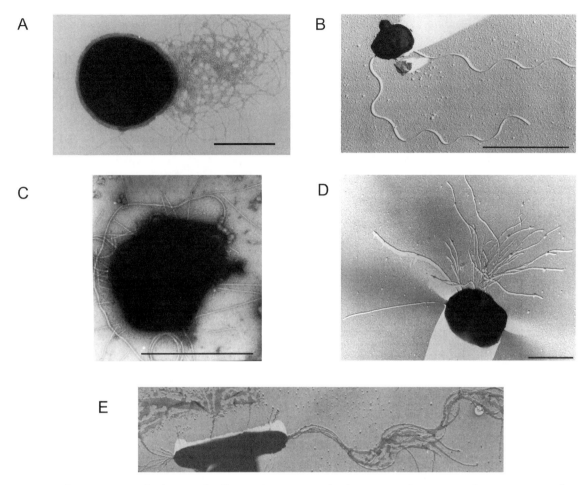

Figure 2. Electron micrographs depicting flagella on a diverse range of archaea. (A) *Methanococcus voltae*, negative stain, bar = 1 μm. Reproduced from the *Journal of Bacteriology* (10) with permission of the publisher. (B) *Archaeoglobus veneficus*, platinum shadowed, bar = 1 μm. Picture courtesy of Reinhard Rachel. (C) *Sulfolobus tengchongensis*, negative stain, bar = 1 μm. Picture courtesy of Li Huang. (D) *Pyrococcus furiosus*, platinum shadowed, bar = 1 μm. Picture courtesy of Reinhard Rachel. (E) *Halobacterium salinarum*, carbon/platinum shadow. Reproduced from the *Journal of Molecular Biology* (3) with permission of the publisher and D. Oesterhelt.

alize. This may be complicated by the novel cell envelopes of archaea, which lack murein. Indeed, in the best-studied cases, flagella have been isolated from organisms (*Methanococcus, Halobacterium*) that have an extremely simple envelope structure with a protein or glycoprotein S layer overlying a cytoplasmic membrane—an envelope structure that is simple in comparison with bacteria (57). This simple wall structure may preclude the presence of the typical basal body rings that are associated with the peptidoglycan layer and the outer membrane in bacteria flagella. In archaea, such rings have, on occasions, been reported (*Methanococcus thermolithotrophicus, Methanospirillum hungatei*) (27), although more typically archaeal flagella have an ill-defined knob at their cell proximal end (10, 47, 52, 59). It is unclear whether these observations indicate that the anchoring lacks rings typical of bacterial basal bodies, or whether the archaeal

equivalents are simply more delicate structures that are lost upon flagella preparation.

It is apparent from several lines of evidence that further anchoring of the flagella of archaea occurs by a structure that may be located beneath the cytoplasmic membrane. This was predicted more than 20 years ago (3), based on electron microscopy of halobacterial cells. Kupper et al. (59) were able to isolate halobacterial flagella inserted into a polar cap structure, where the filaments of an entire flagella bundle were attached. These authors could not determine whether the polar cap was part of the wall, membrane, or subcytoplasmic membrane. However, thin sections of many archaea, such as *M. voltae* (46, 58) and *H. salinarum* (69, 100), have revealed the presence of polar membrane-like structures. In addition, a second distinct, disk-like lamellar structure has been detected below the cytoplasmic membrane in *H. salinarum* (69,

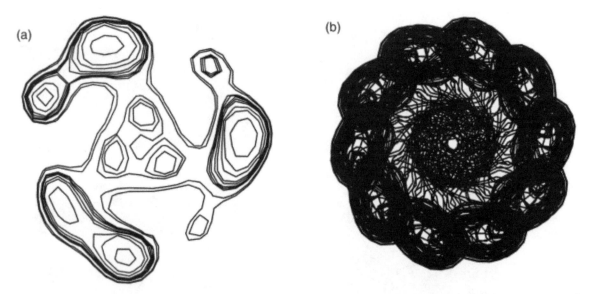

Figure 3. Density map of the *H. salinarum* flagellar filament. (a) Cross-section through the three-dimensional density map of a *H. salinarum* flagellum. The triskelion-like shape and lack of a central channel is shown. (b) An axial projection of a stack of ten sections as shown in (a) spaced 5.8 Å apart and rotated by 108° relative to each other. Reproduced from the *Journal of Molecular Biology* (23) with permission of the publisher and S. Trachtenberg.

100). This complex structure is always located at the cell pole exactly where the flagellar bundle is located and lies about 20 nm beneath the cytoplasmic membrane (69).

ARCHAEAL FLAGELLA GENES

One genetic locus, containing up to 13 flagella-related genes, is present in the genome sequences of the majority of flagellated archaea (Table 1). This locus normally contains multiple flagellin genes arranged in tandem, followed by a number of conserved flagella-associated genes (*flaC-flaJ* or a subset thereof; Fig. 4). A gene for the preflagellin peptidase is often present outside of this cluster (8, 9). In *M. voltae*, genes involved in the transfer of a novel *N*-linked glycan to the flagellins have recently been identified, separate from the main *fla* cluster, although their role is not limited to flagellin modification (21a).

Flagellins

The best-studied genes in the archaeal flagella loci are the flagellin genes. Among the 25 completely sequenced genomes available at the NCBI website (http://www.ncbi.nlm.nih.gov/), 19 have readily identified flagellin genes. These include three species of *Methanosarcina*, which have never been reported to

be flagellated or motile, although another *Methanosarcina* species, *M. baltica*, can be flagellated. On the other hand, two species reported to have flagella and/or be motile (*Pyrobaculum aerophilum* and *Methanopyrus kandleri*) do not have any genes currently annotated as flagellins. Examination of the *M. kandleri* genome does reveal the presence of genes that encode several flagellin-like proteins, located very close to apparent *flaHIJ* homologs (K. F. Jarrell, S. Y. M. Ng, and B. Chaban, unpublished results). The predicted proteins encoded by these genes have short signal peptides ending in Lys-Gly, followed by a hydrophobic stretch, typical of archaeal flagellins. Between these small genes lie two genes predicted to be prepilin peptidase homologs. However, these flagellin-like proteins are very small, less than 80 amino acids. Whether these are the proteins responsible for the appendages observed on *M. kandleri* cells remains to be determined.

Flagellin genes are present as multigene families in all archaea, with the exception of *Sulfolobus* species, where only a single flagellin gene is present. Two to six flagellin genes are usually present in tandem. However, in some archaea, including *H. salinarum*, one group is present at a separate locus on the chromosome from another tandem set (36, 77). In *Haloarcula marismortui*, one flagellin gene is located on a plasmid (TIGR gene annotation NT01HM0026; http://www.tigr.org/). The flagellin genes are known to be cotranscribed with the downstream *fla* genes in the several

Table 1. Distribution of *fla* gene clusters in archaea[a]

Organism	Flagellin	fla genes								
		C	D	E	F	G	H	I	J	K
Crenarchaeota										
Aeropyrum pernix	14601712 14601713	–	–	–	–	14601709 14601708	14601707	14601706 14601914 14600901	14601705	–
Pyrobaculum aerophilum	–	–	–	–	–	–	–	18313579 18312170 18313110	18313580	–
Sulfolobus acidocaldarius	70606948	–	–	–	70606945	70606946	70606944	70606943 70608025 70607238	70606942	70605985
Sulfolobus solfataricus	15899081	–	–	–	15899078	15897078	15899077	15899076	15899075	15897089
Sulfolobus tokodaii	15922853	–	–	–	15922856	15922855	15922857	15922858 15920739 1521686	15922859	15922589
Euryarchaeota										
Archaeoglobus fulgidus	11498659 11498660	–		11498658	11498656	11498657	11498655	11498654	11498653	11498541
Ferroplasma acidarmanus	55376132 55380089 55378891	–	–	–	–	–	–	–	–	–
Haloarcula marismortui			55378885		55378884	55378883	55378882	55378881 55378048	55378880 55378047	55379184
Halobacterium salinarum	15790079 15790080 15790081 15790120 15790121 15790252	–	15790076 15790075	–	15790074	16554480	16554479	15790072 15789975 15789521	15790071	15791095
Methanocaldococcus jannaschii	15669081 15669082 15669083	15669084	15669085	15669086	15669087	15669088	15669089	15669090 1591924 1591480	15669091 1591923	15669092
Methanococcoides burtonii	68185311 68185312 68185254		68186332 68185854			68186333	68186334 68185307	68186336 68185306	68186337 68185305	68184794
Methanococcus maripaludis	45359229 45359230 45359231	45359232	45359233	45359234	45359235	45359236	45359237	45359238	45359239	45335818
Methanopyrus kandleri	–	–	–	–	–	-	–	20094144 20089132 20092769	19887149	20094140 20089408
Methanosarcina acetivorans	20091879 20091880 20091895	–	20091878 20091896		20091876 20091898	20091877 20091897	20091875 20091899	20091874 20091900 20091914	20091873 20091901	

Continued on following page

Table 1. *Continued*

| Organism | Flagellin | | fla genes | | | | | | | | |
		C	D	E	F	G	H	I	J	K
Methanosarcina barkeri	68080180	–	68080179	–	68080177	68080178	68080176	68080175 68081047	68080174	68133328
Methanosarcina mazei	21226424 21226425 21226520	–	21226423 21226519	–	21226421 21226517	21226422 21226518	21226420 21227041 21226516	21226419 21226515 21227618 21227040	21226418 21227039 21226514	21227780
Methanothermobacter thermautotrophicus	–	–	–	–	–	–	–	2622833	2622834	2621485
Nanoarchaeum equitans	–	–	–	–	–	–	–	41614965 41615211	–	–
Picrophilus torridus	–	–	–	–	–					
Pyrococcus abyssi	14521692 14521693 14521694	14521691	14521690	–	14521689	14521688	14521687	14521686 14521589	14521685 14521587	14521786
Pyrococcus furiosus	18976709 18976710	18976708	18976707	–	18976706	18976705	18976704	18976703 18977366	18976702 18977364	18976843
Pyrococcus horikoshii	14590447 14590448 14590449 14590450 14590451	14590452	14590453	–	33359301	14590455	14590456	14590457 14590539	14590458 14590540 14590541	14590360
Thermococcus kodakaraensis	57639973 57639974 57639975 57639976 57639977	57639978	57639979	–	57639980	57639981	57639982	57639983 57639935 57641788	57639984	57639988
Thermoplasma acidophilum	16082620 16081658	16081659	16081660	–	16081661	16082646	16081662	16081663 16081953 16081950	16081664 16081951	–
Thermoplasma volcanium	13542257 13541438	13541439	13541440	–	13541441	13541442	13541443	13541444 13541846 13541843	13541445 13541844	–

aGI numbers are for proteins from genome sequences at NCBI Genomes (www.ncbi.nlm.nih.gov/genomes/static/a.html).

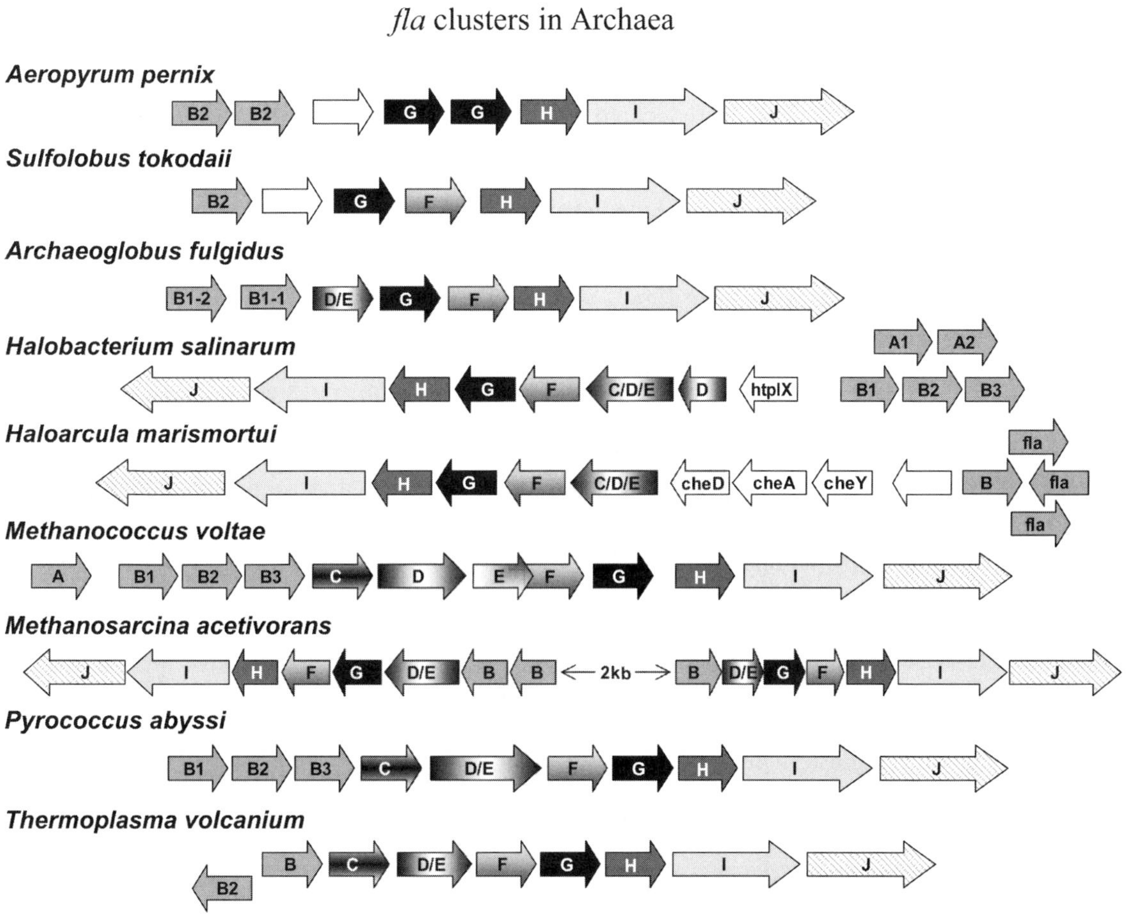

fla clusters in Archaea

Figure 4. Flagella gene families in selected archaeal species. Similar shadings indicate homologs shared among families. Genes are transcribed in the direction of the arrows. In *H. salinarum*, the B flagellin genes are adjacent to the accessory genes while the A flagellin genes are located elsewhere on the chromosome. One of the flagellin genes of *H. marismortui* is located on a plasmid.

instances where this has been studied (various methanococci [50, 113], *Thermococcus* (formerly *Pyrococcus*) *kodakaraensis* [75]). In *H. salinarum*, the flagella-associated *fla* genes are present next to tandem flagellin genes. However, the orientation of the two gene clusters is inverted, thereby precluding cotranscription, though not necessarily coregulation.

Proteins from all flagellin genes are predicted to be made with short signal peptides, usually ending with a basic amino acid and a glycine, similar to type IV pilins (111). Typical signal peptides of about 12 amino acids are found in many flagellins, especially from halophiles and methanogens, while even shorter ones of 4 amino acids are predicted for flagellins from *Pyrococcus* sp. Analysis of the deduced amino acid sequence also indicates the presence of *N*-linked glycosylation sequons (Asn-Xaa-Ser/Thr) in the vast majority of archaeal flagellins (76). The number of such sequons ranges up to 16 for one flagellin of *M. thermolithotrophicus*. The presence of glycosylation is believed to be widespread among archaeal

flagellins and may account for most of the discrepancies observed between the predicted molecular weights and that observed in sodium dodecyl sulfate-polyacrylamide gel electrophoresis (SDS-PAGE) analyses (15, 45, 104, 111). The N termini of archaeal flagellins is very hydrophobic (50), and there is considerable similarity in sequence to the conserved N termini of type IV pilins in bacteria (30).

Evidence from mutagenesis studies in several archaea has indicated that the flagellins are not interchangeable. In *M. voltae*, insertional inactivation of the minor flagellin, *flaA*, resulted in cells that looked to be normally flagellated. However, they were not as motile as wild-type cells in semi-swarm plate analysis (44). In *H. salinarum*, the two flagellin gene clusters that encode FlgA and FlgB flagellin proteins are transcribed separately. Each gene cluster has been insertionally inactivated, and the ensuing effect on flagellation has been studied (110). The *flgA* mutants were flagellated but less motile than the wild-type cells. The flagella themselves were curved and spiral, similar to the wild type. However, the flagella of the mutant cells were shortened and located at both the poles and sides of the cells. In addition, a mutant that had an insertion in *flgA2* only had straight flagella that were located mainly on the cell's poles. Mutants that had disrupted *flgB* genes were also less motile than the wild-type cells and had both spiral-shaped flagella at the poles of the cells and outgrowths from the cell surface. These outgrowths are believed to be membrane sacs filled with basal body-like structures. Some of these outgrowths were observed to have a flagellum at their ends (110). From these data it was interpreted that the A flagellins form the main component of the filament and are incorporated at the initial stages of assembly. The less abundant B flagellins are most likely located near the base. The phenotype of the *flgB* mutants with outgrowths to which flagella may be attached suggest a role for the B flagellins in proper anchoring of the flagella structure. In the absence of the A flagellins, it appears that improper incorporation of the B flagellins can occur, leading to truncated filaments not located at the poles.

Preflagellin Peptidase Genes

The enzyme responsible for cleavage of the signal peptide from the archaeal flagellins was identified as FlaK, the preflagellin peptidase in methanococci (8, 9) and as PibD (peptidase involved in biogenesis of prepilin-like proteins) in *Sulfolobus* (5). PibD processes several nonflagella-related proteins involved in sugar binding and therefore appears to have broader substrate specificity than the methanococcal FlaK enzymes (5). FlaK is usually located distant from the flagella gene locus, although in *M. jannaschii* it follows *flaJ* and is likely to be cotranscribed. FlaK/PibD is a member of the same COG (1989: www.ncbi.nlm.nih .gov/COG/) that includes the bacterial prepilin peptidases, such as PilD of *Pseudomonas aeruginosa*, even though the amino acid sequence similarity of the bacterial and archaeal enzymes is very low. An in vitro assay was developed that allowed for the identification of key amino acids in the signal peptide required for proper processing (5, 112). In vitro analysis revealed that the −1 glycine and a basic amino acid at the −2 position are extremely important for proper processing of the preflagellin. Replacement of this conserved glycine with alanine by site-directed mutagenesis still allowed processing of the archaeal preflagellin; this replacement also allows processing of the type IV prepilin in the bacterial system. The in vitro assay was also used to show that two conserved aspartic acid residues were critical for the activity of the peptidase. This clearly identifies the archaeal peptidase as a member of the aspartic acid peptidases, a family that

includes the prepilin peptidases (9). Disruption of the *flaK* gene in *M. voltae* leads to larger-molecular-weight flagellins being produced, consistent with retention of the signal peptides. Furthermore, the mutant cells are not flagellated, indicating that the flagellins need to be N-terminally processed to be assembled into a flagellar filament (9).

Other *fla* Genes

Among the *fla*-associated genes, the most conserved are the *flaHIJ* clusters. The genes are present in the genome sequences of all archaea that possess flagellin genes and are believed to be involved in the export of the flagellins. Similar to *flaK*, *flaI* and *flaJ* have homologs in the bacterial type IV pili system. FlaI contains a Walker box involved in NTP binding (14), and the similarity of FlaI to ATPases involved in type IV pili extension and retraction has been recognized for many years (PilT/PilB in *P. aeruginosa* [14]; TadA in *Actinobacillus actinomycetemcomitans* [49]). Recently, Albers and Driessen (4) demonstrated that FlaI from *Sulfolobus solfataricus* possesses divalent cation-dependent ATPase activity. Whether the archaeal FlaI forms homohexamers similar to the bacterial PilT (34), and likely for BfpF (the PilT equivalent in the bundle forming pilus system [26]), is still to be determined. FlaJ is an integral membrane protein with numerous predicted transmembrane domains, and its similarity to the conserved membrane PilC/TadB component of type IV pili has been reported (49, 84). Since PilC may interact with the corresponding ATPase in the type IV pilus system (6), it may be that in the archaeal flagella system, FlaI and FlaJ interact. FlaH also contains a Walker Box A (113), but it does not have strong similarity to any bacterial proteins.

In a comprehensive study of secretion ATPases in *Sulfolobus*, it was suggested that FlaI-FlaJ might form the minimal core of the secretion systems (4). However, the universal association of FlaH, also a potential ATPase (4, 113) with FlaI and FlaJ in the archaeal flagella operons, suggests that it too is likely to be a required component for the export/assembly system of this organelle. Some archaea contain two or more copies of some, or all, of the *flaHIJ* genes. The set located in the immediate vicinity of the flagellins genes may be dedicated solely to the export of the flagellins, while the other set may be involved in export of other substrates. It has been shown in both *H. salinarum* (83) and in *M. voltae* (114) that the *flaI* and *flaJ* genes in the vicinity of the flagellins are required for flagellation, since either inframe deletions of *flaI* in *H. salinarum* (83) or insertional inactivation into *flaJ* in *M. voltae* (114) results in nonflagellated cells. In these two cases it is clear that the presence of other *flaI* or *flaJ* homologs elsewhere on the chromosome did not compensate for disruptions of the *flaI* and *flaJ* genes located at the flagella locus.

The involvement of the *flaC-flaG* genes in archaeal motility is much more of a mystery. Because of their location between the required flagellin genes and *flaHIJ* genes and their cotranscription with these genes in at least some archaea (50, 113), they have been assumed to be involved in archaeal flagellation. In *H. salinarum*, the orientation of the flagella-associated genes is opposite to that of neighboring flagellin genes, illustrating that they are not cotranscribed with flagellin genes in this species, although coregulation is still possible. While all of these genes have homologs in most of the flagellated *Euryarchaeota*, including members of methanococci, pyrococci, halophiles, and *Thermoplasma* species, in *Crenarchaeota*, no similarity, or only weak similarity exists to these genes. Indeed, in most members of the *Crenarchaeota*, the number of putative open reading frames (ORFs) located between the flagellin genes and the *flaHIJ* set simply does not leave room for all of *flaC-F* (76), clearly indicating that different subsets of the flagella-associated genes are present in different organisms. Homologs to *flaC-F* have not been identified in bacteria, and they therefore encode archaea-specific proteins involved in flagella structure, function, or assembly. None of the genes have been reported in archaea that do not also have flagellin genes (76).

Some interesting observations have been made about the *flaC-F* genes. The *flaD* gene has been shown to encode two proteins in *M. voltae*; one full-length gene product of about 52 kDa and a smaller 15-kDa protein that initiates at an inframe methionine downstream of a strong ribosome binding site (rbs) toward the 3′ end of the gene. Both versions of FlaD are observed when the gene is expressed in *Escherichia coli*, and both versions are detected in *M. voltae* cells by using anti-FlaD antisera (113). The truncated FlaD is very similar to the full length of FlaE (42.7% identity over a 127-amino-acid overlap). A potential start codon downstream of a strong rbs at a position similar to that of the *M. voltae flaD* was identified in other methanococci, suggesting that these archaea also make two proteins from the single gene (113). A 16-kDa C-terminal portion of FlaD was detected in a proteomic study of *M. jannaschii* (74), providing proof that the translation of *flaD* in this hyperthermophile was similar to that observed in *M. voltae*. Furthermore, western blots using anti-*M. voltae* FlaD revealed two cross-reacting bands in *Methanococcus maripaludis* membrane fractions, at similar molecular masses to those observed in *M. voltae* (S. Y. M. Ng and K. F. Jarrell, unpublished

data). Studies are currently underway to address the role, if any, that this truncated FlaD has in flagellation.

Another observation suggests that the FlaC, FlaD, and FlaE proteins might interact. In several archaea, notably *H. salinarum* and *Methanococcoides burtonii*, there is one large gene with similarity to *flaC*, *flaD*, and *flaE* from *Methanococcus* spp. In *H. salinarum*, this protein is predicted to be 504 amino acids long, and it aligns with the *M. voltae* FlaC (188 amino acids) over amino acids 30 to 233, and with the *M. voltae* FlaD (362 amino acids) over the remainder of its length. Because of the strong sequence similarity between FlaD and FlaE, the halophile protein also aligns with FlaE from amino acid position 380 to the end of the protein. If this one large protein is fulfilling the roles of FlaC, FlaD, and FlaE, it could indicate that in those archaea that have the three separate proteins, that these proteins may interact with each other. Using a recently developed method to create markerless inframe deletions in *Methanococcus maripaludis* (72), the requirement for, and possible role of, each of these *fla*-associated genes in archaeal flagellation is currently being studied (B. Chaban, S. Y. M. Ng, and K. F. Jarrell, unpublished data).

Glycosylation Genes

A regular finding in many, if not all, archaeal flagellins is an aberrantly high apparent molecular mass determined by SDS-PAGE compared with the predicted mass from the gene sequence (111). In fact, this finding is so common that posttranslational modification of some sort may be the rule for the flagellins from these organisms. In many cases, this modification has been shown either directly or indirectly through specific staining to be glycosylation. In general, glycans can be attached to proteins by either an N linkage or an O linkage (see Chapter 11). N linkages involve glycan attachment to an Asn residue within the consensus sequence Asn-Xaa-Ser/Thr, where Xaa can be any amino acid except Pro. This sequon is regarded as necessary but not sufficient for N-glycosylation in eucarya (35) as well as in bacteria (78). On the other hand, the attachment sites for O linkages are less clearly defined, with the glycan covalently linked to a Ser or Thr residue. Thus far, glycans found on archaeal flagellins have all been of the N-linkage type (104, 116). This contrasts sharply with glycans that have been detected on both bacterial flagella and type IV pili, which to date have all been reported to be O linked (109).

From archaeal flagellins, N-linked glycosylation structures have been determined in *H. salinarum* (104, 118) and *M. voltae* (116); preliminary evidence for modification of flagellins has been presented for *M. jannaschii* (38). The *H. salinarum* glycan is a sulfated oligosaccharide linked to Asn residues via glucose followed by 2 or 3 glucuronic acids, with approximately one-third of these replaced by iduronic acid. Furthermore, a sulfate residue is attached to each hexuronic acid at position 3 and the terminal position has been shown to contain some variability, being occupied by either a hexuronic acid or a glucose residue (104, 105). In comparison, the *M. voltae* glycan appears to be less variable. The same trisaccharide structure was detected a total of 14 times among the four *M. voltae* flagellins. Linked to Asn via *N*-acetylglucosamine, the structure is followed by a di-*N*-acetylglucuronic acid and finally an *N*-acetylated mannuronic acid with a threonine attached at position 6 (116). In both cases, the same glycan detected on the flagellins was also found on S layer proteins within the cells, suggesting that a common glycosylation pathway exists for the posttranslational modification of both types of proteins.

For the first time, genes involved in the glycosylation of archaeal flagellins have been identified in *M. voltae* (21a). Using SDS-PAGE, it was found that mutations that altered the *M. voltae* glycan caused corresponding alterations in the apparent molecular mass of both the flagellins and S layer protein. Two genes critical for glycosylation were identified. One is a homolog of the oligosaccharyltransferase STT3 subunit. Homologs of this gene are found among the *pgl* genes (*pglB*) in bacteria such as *Campylobacter*, and as part of a complex of proteins collectively known as the oligosaccharyltransferase (OTase) in eukarya, where they are critical in *N*-linked glycosylation systems (21, 108, 109). The STT3 homologs are responsible for the transfer of the completed glycan from a lipid carrier to the protein target in a reaction that occurs on the periplasmic side of the cytoplasmic membrane in bacteria or the lumenal side of the endoplasmic reticulum in eukarya (109). It is hypothesized that the STT3 protein in archaea behaves similarly to its bacterial counterpart, being membrane bound with its activity on the periplasmic side of the cytoplasmic membrane. Searches of archaeal genomes reveal the presence of STT3 homologs in most archaea, including nonflagellated ones. The second gene demonstrated to be necessary for glycosylation in *M. voltae* is one that has strong similarity to glycosyltransferases in many bacteria, such as *Thermus* and *Pseudomonas*. In *M. voltae*, this gene product appears to be involved in the transfer of the third, terminal sugar to the glycan structure of the flagellins and S layer protein (21a). The *M. voltae* glycosyltransferase is also predicted to be membrane bound (www.ch.embnet.org/software/TMPRED_form.html), with its active site on the

cytoplasmic side of the cytoplasmic membrane. Genes with significant similarity to this gene are also found in a variety of archaea, again including nonflagellated species. As with the STT3, this glycosyltransferase is likely to be involved in general glycan synthesis not specific to flagellins and appears to play a role in posttranslational modification of the S layer protein and perhaps other cell surface proteins.

Given the similarities between the genes involved in archaeal and bacterial *N*-glycosylation, the current model for glycan assembly and attachment in archaea is based on the bacterial model. For bacterial *N*-glycosylation, the glycan is proposed to be assembled on the interior side of the cytoplasmic membrane through the sequential addition of nucleotide-activated sugars onto a lipid carrier, followed by a "flipping" of the glycan to the periplasmic side of the membrane and finally a transferring en block to the target protein (108, 109). At the present time, the same sequence of events is proposed for archaeal *N*-linked glycosylation and research is continuing to identify further genes involved in this pathway in methanococci.

REGULATION OF FLAGELLATION

Many archaea have been observed to possess flagella under certain growth conditions and be nonflagellated if those growth conditions vary. Growth at suboptimal temperature often results in nonflagellated cells: this has been observed with *Methanospirillum hungatei* (31) and *M. jannaschii* (Jarrell, Ng, and Chaban, unpublished). However, *M. maripaludis* strain LL is more heavily flagellated at 30°C than at 37°C in Balch III medium (K. F. Jarrell, S. Y. M. Ng, and B. Chaban, unpublished observation).The regulation in *M. hungatei* appeared to be at the level of subunit assembly, as similar levels of flagellins were detected by western blotting in both flagellated and nonflagellated cells.

More recently, the regulation of flagellation in *M. jannaschii* has been studied at the proteomic level. Significant changes in the flagellins and other proteins believed to be critical for flagellation were observed in response to decreased H_2 availability, limited ammonium availability, and stage of growth (38, 74). In addition, the flagellins appear to be modified with an undetermined modification(s), and the protein variants (in p*I* and molecular weight) vary with growth conditions (38). Growth under limited ammonium resulted in only minimal amounts of flagellins FlaB1 and FlaB2. An earlier study revealed that flagellins FlaB2 and FlaB3, as well as flagella-associated proteins, FlaD and FlaE, were present in low amounts when the cells were grown in hydrogen excess conditions and the cells lacked flagella. Flagella synthesis occurred when hydrogen became limited. This represented the first case of flagella regulation by hydrogen in any domain of life. This may be relevant to *M. jannaschii* in its natural habitat of deep-sea hydrothermal vents. It has been postulated that a 10-base direct repeat GTGTGGGGGAattGTGTGGGGGA located 183 nucleotides downstream of the putative promoter and 49 nucleotides upstream of the presumed translation start site may be a possible regulatory element (113). Recent transcriptome analysis of the mesophilic *M. maripaludis* grown under hydrogen-limited conditions in a chemostat indicated genes for flagella synthesis were upregulated under such conditions (E. L. Hendrickson, A. K. Haydock, I. Porat, W. B. Whitman, and J. A. Leigh, 105th ASM General Meeting, abstract K-086, 2005), consistent with the observations made on the hyperthermophilic *M. jannaschii*.

ARCHAEAL FLAGELLA, BACTERIAL FLAGELLA, AND TYPE IV PILI

Bacteria and archaea have developed many very different mechanisms that enable them to move in their environments (12). Bacterial and archaeal flagella are responsible for swimming, while type IV pili are involved in surface translocation or twitching. A comparison of archaeal flagella to bacterial flagella and type IV pili is presented in Table 2. In gross appearance and performance, the archaeal flagellum resembles the bacterial flagellum but, at the genetic level, there appears to be no conserved genes between these two organelles. On the other hand, several conserved proteins are present in the archaeal flagellum and type IV pilus systems (11, 14, 84, 111). The notion that the archaeal flagellum shared similarities to the type IV pilus began with the observation that archaeal flagellins and type IV pilins possess significant

Table 2. Comparison of archaeal flagella to bacterial flagella and type IV pili

Bacterial flagella	Archaeal flagella	Type IV pili
• Usually single flagellin with a few exceptions	• Usually multiple flagellins with rare exception	• Single major pilin
		• Several minor pilins
• No sequence similarities between bacterial and archaeal flagellins		• N-terminal sequence similarities with archaeal flagellins
• Flagellins rarely posttranslationally modified	• Flagellins posttranslationally modified, usually through glycosylation	• Pilins may be glycosylated or phosphorylated depending on the species
• If glycosylated, *O*-linked	• Glycosylation *N*-linked	• Glycosylation *O*-linked
• Flagellins do not have N-terminal leader peptides	• Flagellins have N-terminal signal peptides	• Prepilins have N-terminal signal peptides
	• Cleaved by FlaK/PibB	• Cleaved by PilD
• No archaeal homologues of genes involved in bacterial flagellation (flagellins, rod, hook, hook-associated, ring, switch, mot)	• No archaeal homologues of genes involved in bacterial flagellation (flagellins, rod, hook, hook-associated, ring, switch, mot)	• FlaI is homologous to PilB, PilT, and TadA, ATPases involved in the assembly/disassembly process of type IV pili
		• FlaJ has sequence similarity with TadB and PilC, integral membrane proteins involved in type IV pili biogenesis
• Rod and ring containing basal body	• Knoblike anchoring structure	• No anchoring structure observed
• Hook region	• Curved hooklike region	• No hook
• Well-defined hook region length of about 60 nm	• Great variability in hook length (72–265 nm)	
• Filament is usually a left-hand helix	• Filament is a right-hand helix in *H. salinarum*	
• ~20 nm diameter	• ~10–15 nm diameter	• ~5–7 nm diameter
• 2 nm channel allowing for passage of flagellin subunits	• Channel not detected	• Channel too narrow for proteins to pass through
• Swimming motility	• Swimming motility	• Surface translocation or twitching motility
• Rotating filament	• Rotating filament	• Movement by pilus retraction/elongation at base
• Switching of rotation	• Switching of rotation	
• Swim/tumble motion	• Push/pull motion	
• Growth at distal end	• Presumed growth at proximal end (base)	• Growth at proximal end (base)
• Associated chemotaxis system	• Associated chemotaxis system	• Often associated chemotaxis system

amino acid sequence similarity at their N termini (30). This very hydrophobic region has been demonstrated to be the oligomerization domain for type IV pilins. Several observations have strengthened this initial hypothesis, including the lack of archaeal genes that have detectable homology to genes in bacteria involved in flagella structure or assembly (29, 79). Perhaps the most significant observation is that both archaeal flagellins and type IV pilins are made as preproteins with unusual, atypically short signal peptides processed by specific signal peptidase homologs (5, 8, 9, 24). As the flagella gene loci of more archaea were released, it became obvious that besides the flagellin/pilin similarity and the homologous signal peptidases, two other proteins were shared in the two systems: an ATPase and a poorly described, conserved integral membrane protein.

In addition to the abovementioned similarities, the archaeal flagellar filament has been shown to share structural similarities to type IV pili, rather than to bacterial flagella filaments (23, 115). In bacterial flagella assembly, newly synthesized flagellins must first travel from the cytoplasm through the hollow, growing flagella structure via a specialized type III secretion system located at the base of the MS ring, before incorporating under the filament cap located at the distal tip of the structure (62, 63). Distal growth would seem to be precluded in archaeal flagella if there is no central channel for subunits to pass through. By default, assembly must occur at the base, similar to type IV pili (68, 73), a mechanism that was suggested for archaeal flagella several years ago (45).

PROPOSED MODEL OF ASSEMBLY OF ARCHAEAL FLAGELLA

Even in very early studies of archaeal flagella, the unusual nature of these structures led researchers to speculate that the likely mechanism of assembly would be distinct from the bacterial flagella assembly mechanism (36). Since then, different laboratories have reported similarities of archaeal flagella genes to type IV pili genes while homologs to bacterial flagella genes remain undetected (14, 23, 84, 115). These observations resulted in a proposed mechanism for assembly of archaeal flagella with its striking prediction of incorporation of new subunits at the base (45), as in type IV pili (73). The work of Trachtenberg (23, 115), which indicates, through various electron microscopy techniques, that archaeal flagella lack a hollow channel large enough to accommodate the passage of flagellins, appears to support a model of assembly at the base. Flagellin mutant studies in

H. salinarum have been interpreted with this mechanism of incorporation in mind (110). However, it must be emphasized that direct proof of polarity of growth in the archaeal filaments is still lacking.

Our current speculative model for assembly of archaeal flagella is shown in Fig. 5. Since the total number of genes identified in the archaeal flagella system seems very low compared with either bacterial flagella or type IV pili systems, it is believed that many more genes involved in flagellation remain to be identified: for instance, no anchoring or motor components or genes involved in gene regulation are yet identified. In the proposed model, preflagellins interact with chaperones in the cytoplasm, preventing aggregation through the hydrophobic N termini (87). The preproteins are delivered to the cytoplasmic membrane where at least three events must occur. The signal peptide is removed from the N terminus by the preflagellin peptidase FlaK. In most cases, the flagellin is modified by attachment of an *N*-linked glycan. This glycan is assembled on a lipid carrier at the cytoplasmic membrane through the activity of various glycosyltransferases. The completed glycan is flipped through the membrane and then transferred to an asparagine residue present as part of an *N*-linked sequon. The oligosaccharyltransferase, STT3 homolog, is responsible for this final transfer of the completed glycan. This *N*-linked system is not flagellin-specific and in certain archaea is shared with at least the modification of S layer proteins. The order of these two posttranslational events is unknown, but the two processes can occur independently of each other. In support of this, a mutant in which *flaK* was disrupted still produced flagellins that migrated as completely glycosylated (9), and in mutants disrupted in the glycosylation pathway, flagellins still had their signal peptides removed (21a). Following signal peptide removal and *N*-linked glycan addition, the newly synthesized flagellins are incorporated into the filament. Because they are made as preproteins, unlike bacterial flagellins, we believe it likely that the new subunits are added from the cytoplasmic membrane to the base of the structure. This could be achieved through the combined activity of the FlaHIJ ATPase-conserved membrane protein complex that has type IV pilus homologs. Strongly supporting this unorthodox assembly mechanism is the observation that, unlike bacterial flagella, the archaeal flagella (at least in the case of *H. salinarum*) do not have a hollow channel for passage of subunits to the distal tip (23). However, this assembly mechanism also raises numerous fundamental questions. What is the nature of the anchoring structure and its assembly in relationship to the filament, and what is the role for the subcytoplasmic membrane complex? Is there an instigator or

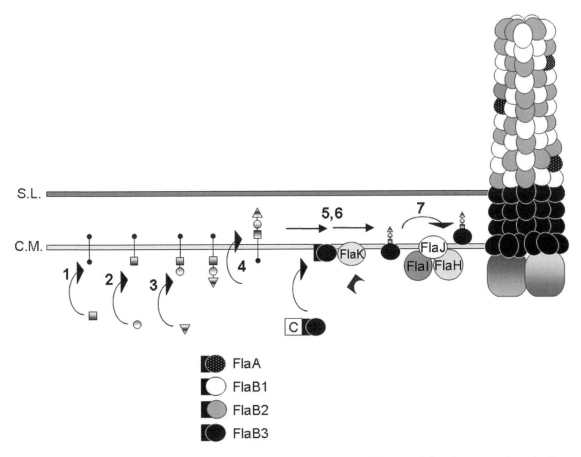

Figure 5. Model of assembly of archaeal flagella. The assembly of the glycan modification is believed to occur independently on a lipid carrier in the cytoplasmic membrane (steps 1 to 3 for the three-sugar glycan of *M. voltae*) before transfer via an SST3 homolog to an Asn residue within the N-linked sequon present on the flagellins (step 4). Glycosylation, as well as removal of the signal peptide by FlaK (steps 5 to 6, although the order of the two steps is unknown), occurs in the cytoplasmic membrane prior to incorporation of the flagellins into the filament (step 7). The incorporation step is presumed to involve a FlaHIJ complex of ATPases and a conserved membrane protein that has homologs in the type IV pili system. Incorporation of new flagellin subunits is hypothesized to occur at the base of the structure since no hollow center for passage of subunits has been observed.

terminator of flagella growth, and how do archaeal flagella regenerate when they are sheared off if growth occurs at the base? Since many flagellated archaea lack a murein equivalent, to what is the archaeal flagellar stator bound? Is this part of the function of the subcytoplasmic membrane structure? These and many other intriguing questions can only be answered by further study.

CHEMOTAXIS

Analysis of the many sequenced genomes from flagellated archaea indicates the presence of easily detected homologs of bacterial genes involved in chemotaxis. Indeed, the presence of a highly conserved, bacterial-like chemotaxis system interacting with the unusual archaeal flagellum is a novel feature

of the archaeal sensory system. Nonetheless, studies of potential archaeal taxis (chemotaxis, aerotaxis, phototaxis, thermotaxis, and osmotaxis) are rare and often limited to a preliminary demonstration of taxis with no follow-up genetic work; no doubt partly due to a lack of genetic tools available at that time.

Capillary assays, from Adler's classic studies of bacterial chemotaxis (1), were adapted for use in anaerobic chambers to demonstrate chemotaxis of *M. hungatei* to acetate (70) and *M. voltae* to acetate, isoleucine, and leucine (all requirements for growth) but not histidine (97). *Natronomonas pharaonis* is both phototactic and chemotactic (94). *H. salinarum* is chemotactic to all 20 common amino acids (120) and compatible osmolytes (56). Storch et al. (103) performed a systematic screening of >80 chemicals to test for their ability to act as potential attractants or repellents in *H. salinarum*. The strongest attractants included several essential amino acids (leucine, iso-leucine, valine, methionine, and arginine) as well as the nonessential amino acid cysteine and several peptides. No evidence for taxis towards sugars was found, consistent with the inability of this organism to utilize carbohydrates. *H. salinarum* is also aerotactic, attracted to orange light while repelled by UV and blue light (95, 122) and, most recently, has been shown to sense a change in membrane potential (54).

The study of phototaxis and chemotaxis in *H. salinarum* is a rare instance where significant biochemical and genetic studies on taxis in an archaeon have been reported (64, 93). The purpose of various taxis systems (chemotaxis, phototaxis, aerotaxis, etc.) in bacteria and archaea is to allow the organisms to swim away from harmful conditions and/or toward ones favorable for growth and survival. For *H. salinarum*, maximum growth rate occurs under aerobic conditions chemoheterotrophically (41). *H. salinarum* has the ability to use light as an energy source, in addition to using aerobic respiration and arginine fermentation (54). Bacteriorhodopsin (BR) and halorhodopsin (HR) pump out protons or import chloride ions, respectively, setting up an electrochemical gradient across the cytoplasmic membrane. Two sensory rhodopsins (SRI and SRII) affect swimming migration of the halophiles by influencing the flagella motor rotation in response to various wavelengths of light (82). SRI absorbs at ~600 nm and triggers attraction to green/orange light but avoidance to harmful UV light, while SRII (absorbing light at ~500 nm) results in avoidance of blue light (40, 71). The dual role of SRI means that cells migrate toward green/orange light where the ion pumps will be fully activated but not into full spectrum solar radiation containing damaging short wavelength photons (41, 101).

When oxygen and nutrients available for respiration are plentiful, SRI, BR, and HR synthesis in *H. salinarum* are inhibited, but induction of SRII occurs. The resulting migration of the cells away from high luminescence toward the dark mediated by SRII presumably occurs to prevent light-induced damage to cells. Under conditions of low oxygen and limited nutrients, induction of BR, HR, and SRI all occur, with the synthesis of SRI resulting in migration of the cells toward orange light where BR and HR act to establish a proton motive force to drive synthesis of ATP (82).

The presence of a dedicated chemotaxis system to detect trimethylammonium compounds (used as osmolytes) in *H. salinarum* may provide an important selective advantage in its high-salt environment. However, when cells were grown in the presence of such compounds, growth and survival were not greatly affected (56).

CHEMOTAXIS GENES

In recent years, it has become obvious that the *Archaea* possess a chemotaxis system that is similar to the *Bacteria*. In most flagellated archaea, analysis of completely sequenced genomes reveals the presence of a complete set of chemotaxis genes homologous to those of bacteria (Table 3) (107). These include proteins involved in all processes of bacterial chemotaxis, namely signal recognition and transduction (various methyl-accepting chemotaxis proteins [MCPs] and CheD), excitation (CheA, CheW, and CheY), adaptation (CheR and CheB), and signal removal (i.e., CheY-P dephosphorylation, CheC). Some *che* genes found only in a subset of bacteria (*cheV*, *cheX*, and *cheZ*) have not been reported in archaeal genome annotations. To date, all *che* gene clusters have been found within organisms belonging to the *Euryarcheaota*, with no *che* genes or MCPs detected within members of the *Crenarchaeota*, despite the availability of complete genome sequences for several flagellated members of the *Crenarchaeota*. Similar to their subcellular localization in many bacteria, MCPs are physically located at the poles in *H. salinarum*, suggesting that this feature has been conserved in evolution (37). Since the location of MCPs in bacteria may play a role in regulation of chemotaxis, especially in regard to the role of MCP clustering in regulating ligand binding and signaling, the similar localization of MCPs in at least one archaeon could indicate a similar regulation of chemotaxis in all bacteria and archaea.

Unlike the *fla* operon, *che* clusters in archaea are much more variable in their composition and gene order. Genes within *che* clusters are often inverted, precluding cotranscription in many cases. With the exception of *M. maripaludis*, *cheY*, *cheB*, and *cheA* tend to be clustered together in that order (Fig. 6). Surrounding these genes are *cheC*, *cheD*, *cheR*, *cheW*, and at least one MCP. Tandem copies of *cheC* are observed, and in many cases small genes of unknown function are found within the cluster. In addition, clusters from *Halobacterium* and *Methanosarcina* include several flagellin genes. *Methanosarcina* species, *M. acetivorans* and *M. mazei* have two distinct, complete *che* gene clusters (designated Cluster 1 and 2). Many other species contain additional copies of *cheA*, *cheY*, and/or *cheC* and rarely *cheR* or *cheW*. These genes are located outside the *che* cluster elsewhere on the chromosome. It has not been determined whether the additional *che* genes serve functions related to chemotaxis or are involved in other processes. Strangely, *P. furiosus* lacks detectable *che* genes while the genomes of the two other *Pyrococcus* species contain a complete set. It is also unusual that *M. jannaschii* lacks *che* genes, since a complete set is found in the mesophilic relative, *M. maripaludis*.

Table 3. Distribution of *che* gene clusters in archaea[a]

Organism	Flagellin	*che* genes							MCP
		CheA	CheB	CheC	CheD	CheR	CheW	CheY	
Crenarchaeota									
Aeropyrum pernix	2	–	–	–	–	–	–	–	–
Pyrobaculum aerophilum	0	–	–	–	–	–	–	–	–
Sulfolobus acidocaldarius	1	–	–	–	–	–	–	–	–
Sulfolobus solfataricus	1	–	–	–	–	–	–	–	–
Sulfolobus tokodaii	1	–	–	–	–	–	–	–	–
Euryarchaeota									
Archaeoglobus fulgidus	2	11498645	11498646	11498644	11498643	11498642	11498649	11498647	11498650, 11498639
Ferroplasma acidarmanus	0	–	–	–	–	–	–	–	–
Haloarcula marismortui	3	55378887 55378897 55379264	55378896	55378056 55377404	55378886	55378898	55378265 55378895	55378888 55377368	55379092, 55377553, 55377801, 55377770, 55379379, 55378800, 55376954, 55378507, 55376654, 55377425, 55379831, 55379155, 55379349, 55377103, 55379274, 55377296, 55379260, 55377270, 55377812
Halobacterium salinarum	6	15790089	15790090	15790087 15790088 15790567	15790086	15790085	15790067 15790092	15790091	15790682, 15790681, 15790508, 15790077, 15790413, 15789966, 15789953, 15789618, 15790664, 15790685, 15790497, 15789962, 15790756, 15789818, 15790609, 15790124, 15790447
Methanocaldococcus jannaschii	3	–	–	–	–	–	–	–	–
Methanococcoides burtonii	3	68211075	68211076	68211074	68185296	68079876	68185302	68185300	68185303, 68184724, 68184796
Methanococcus maripaludis	3	45358490	45358489	45358495 45358494	45358491	45358493	45358488	45358496 45358867	45358492, 45357976, 45358050, 45358351
Methanopyrus kandleri	0	–	–	–	–	–	–	–	–

Continued on following page

Table 3. *Continued*

Organism	Flagellin	che genes							MCP
		CheA	CheB	CheC	CheD	CheR	CheW	CheY	
Methanosarcina acetivorans	3	20088913 20091884	20088914 20091885	200888911 20091883	20088910 20091882	20088912 20091881 20090837 20092349	20088919 20091888	20091886 20088915 20091276	20088918, 20091889
Methanosarcina barkeri	1	68134062	68079874	68134061	68079877	68079876	68079923	68079873 68079486	68079922, 68079922, 68081123
Methanosarcina mazei	3	21226430 21227427	21226431 21227428	21226429 21227425	21226428 21227424	21226427 21227426	21226434 21227432	21226432 21227429	21226435, 21227431, 21227760
Methanothermobacter thermautotrophicus	0	–	–	–	–	–	–	–	
Nanoarchaeum equitans	0	–	–	–	–	–	–	–	–
Picrophilus torridus	0	–	–	–	–	–	–	–	–
Pyrococcus abyssi	3	14521750 14521752	14521753	14521750 14521751 33356802	14521749	14521755	14521757	14521754	14520639, 14521748, 14521756, 14521788, 14520639
Pyrococcus furiosus	2	–		–	–	–	–	–	18976805
Pyroccoccus horikoshii	5	14590396	14590395	14590398 14590397	14590399	14590393	14590390	14590394	14590358, 14590391, 14591598, 14590400, 14591707
Thermococcus kodakaraensis	5	57640570 57640569	57640568	57640571 57640572 57158895 57158896	57640574	57640566	57640564	57640567	57640091, 57640573, 57642082, 57640565, 15790447
Thermoplasma acidophilum	2	–	–	–	–	–	–	–	–
Thermoplasma volcanium	2	–	–	–	–	–	–	–	–

[a]GI numbers are proteins from genome sequences at NCBI Genomes (www.ncbi.nlm.nih.gov/genomes/static/a.html).

To date, the only direct genetic study of archaeal *che* genes was undertaken by Rudolph and Oesterhelt (90) in *H. salinarum*, an archaeon that possesses a *che* operon with greatest similarity to *B. subtilis* (54). This investigation involved the deletion of the *H. salinarum cheA, cheB, cheY*, and *cheJ* (a gene with significant similarity to *cheC*) genes. All four genes were inactivated and then complemented in different combinations of the three genes. The loss of *cheA, cheB*, or *cheY* led to the complete loss of chemotaxis and phototaxis, whereas the absence of *cheJ* resulted in a reduction of chemotactic and phototactic ability. The genes *cheA* and *cheY* were found to be required for reverse swimming and counterclockwise rotation of the flagella. *cheB* mutants displayed wild-type distributions of forward and reverse swimming (50:50), while *cheJ* mutants were skewed to an 88:12 distribution of forward:reverse swimming. Although the exact role of *cheJ* remains unknown, it was suggested to play a role in halobacterial signal transduction.

ARCHAEAL TRANSDUCERS

The detection of various external signals is accomplished through a variety of receptor proteins. The majority of research in this area has been performed in extreme halophiles. In *H. salinarum*, 18 transducer genes have been identified in the sequenced genome (54, 77). Archaeal transducers belong to three distinct families (Fig. 7). Family A consists of bacterial-type chemotaxis transducers with periplasmic and cytoplasmic domains interconnected by two transmembrane segments. Family B consists of transducers with two or more transmembrane segments and no periplasmic domain, such as the SRI transducer HtrI. In family C, transducers are soluble cytoplasmic proteins. Transducer proteins are responsible for the detection of the external signal that is transmitted to the internal components of the chemotaxis system. Light receptors such as SRI and SRII interact with cognate transducer proteins, HtrI and HtrII, respectively. In

che clusters in Archaea

Archaeoglobus fulgidus

Halobacterium salinarum

Methanococcus maripaludis

Methanosarcina acetivorans and Methanosarcina mazei (2 operons each)

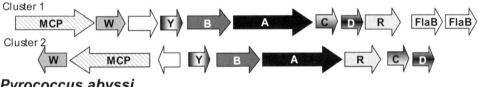

Pyrococcus abyssi

Pyrococcus horikoshii

Figure 6. Chemotaxis gene families in selected archaeal species. Similar shadings indicate homologs shared among families. Genes are transcribed in the direction of the arrows.

H. salinarum, HtrII also acts as a chemotransducer for serine via its large periplasmic domain (42) while the equivalent protein in *N. pharaonis*, which lacks the periplasmic ligand-binding domain, has no known role in chemotaxis (71). In other cases, chemicals can be detected through binding proteins such as BasB and CosB, which interact with their cognate transducer proteins BasT and CosT (55, 56). Other receptors, such as HemAT (for aerotaxis) (43), and probably Car (arginine taxis) (103) and MpcT (for detection of changes in membrane potential) (54), also act as transducers. Many of these transducers have unusual features. For example, Car and HemAT are cytoplasmic transducers while BasT and CosT are both associated with membrane-attached binding proteins (BasB and CosB) that have a role in chemotaxis but not in transport. The binding proteins have features characteristic of lipoproteins, including the LIPO box at the N terminus (56). MpcT is the first transducer identified that responds to a change in the membrane

potential, and HemAT is the first myoglobin-like heme-containing protein reported in the *Archaea*.

Co-localization of receptor and transducer genes on the chromosome may be a general principle in archaea (56). This is observed with the *bas*B/T and *cos*B/T systems where the substrate-binding and transducer genes are organized in operon-like structures in the genome. Furthermore, the genes for SRI and SRII are apparently transcribed with their respective transducer partners, HtrI and HtrII.

Phototaxis Transducers

In *H. salinarum*, the photoreceptors responsible for phototactic behavior are SRI and SRII. SRII is well characterized for its role in *H. salinarum*'s repellant response to blue light (66). SRI, on the other hand, mediates both an attractant response to orange light and a repellant response to UV light.

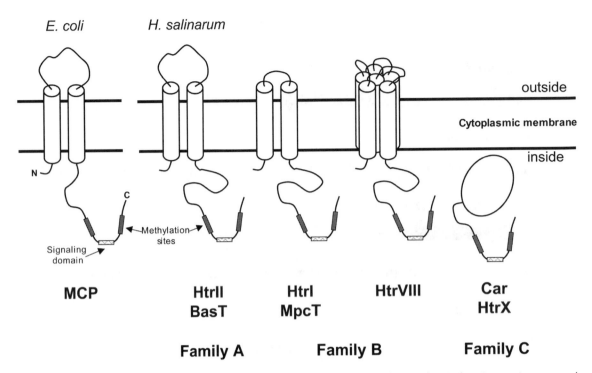

Figure 7. Archaeal transducer families involved in taxis. Similar to bacterial MCPs, the cytoplasmic domain contains conserved modules involved in methylation/demethylation for signal adaptation as well as for signal transmission to the histidine kinase, CheA. Family A transducers are similar to bacterial chemotaxis transducers while Family B transducers lack the periplasmic domain. Family C transducers are soluble. Aerotaxis transducer Htr VIII is unusual in having six transmembrane helices rather than two. Depiction of HtrII is from *H. salinarum*. *N. pharaonis* HtrII lacks the periplasmic domain. Figure based on data presented in Hoff et al. (41) and Marwan and Oesterhelt (64).

The halobacterial transducers for phototaxis are homologous in sequence, structure, and function to the bacterial methyl-accepting chemotaxis proteins. However, they lack an extracellular ligand-binding domain and are physically attached to the rhodopsins. Similar to bacterial transducers (methyl-accepting chemotaxis proteins), HtrI contains two transmembrane helices, a strongly conserved cytoplasmic region involved in binding of a histidine kinase, and flanking regions containing carboxylmethylation sites. Crystallization of the receptor-transducer complex has been achieved using a shortened transducer (residues 1 to 114, *N. pharaonis* HtrII) comprising the two transmembrane helices (TM1 and TM2) and an additional small cytoplasmic fragment (39). One intriguing aspect of sensory rhodopsin is that, in its transducer-free state, it acts as a light-driven proton pump. Interaction with transducer interrupts the transport cycle through closure of the cytoplasmic half-channel. Through the generation of a series of mutants truncated in the membrane-proximal region of an SRI-HtrI fusion protein, it was demonstrated that residues 62 to 66 were nearly completely responsible for channel closure by converting transducer-free rhodopsin from a proton pump into a signal-relaying device (22). These findings were consistent with fluorescent probe-labeling and cysteine cross-linking studies, which revealed that the five residues were within energy-transferring distances (119). Site-directed mutagenesis of these five residues (K62A, E63A, I64A, A65S, and A66S) demonstrated that channel closure requires the presence of amino acids at these five positions, but surprisingly, closure does not depend on a specific amino acid type (22).

Chemotaxis Transducers

Several chemotactic transducers were identified in *H. salinarum* by Western-blot analysis with antisera to bacterial transducers (2). Using a probe specific to a conserved region in transducers, 13 putative transducers were subsequently identified in *H. salinarum* (122). With the completion of the *H. salinarum* genome sequence, as many as 18 putative transducers were identified (54, 77). They share highly conserved C-terminal regions that are likely to be involved in signaling to the histidine kinase and methylation-dependent adaptation. Although different transducers have been characterized to varying degrees (see below), many others remain uncharacterized.

The phototransducer HtrII of *H. salinarum* is a member of transducer family A. One fundamental difference between the phototransducers HtrI and HtrII is that HtrII is more like its bacterial counterpart and contains a large periplasmic domain at the N terminus. Subsequent functional analysis using HtrII deletion and overexpression strains confirmed the hypothesis that, in addition to being a phototransducer, HtrII also functions as a chemotransducer (42). Deletion mutants are nonchemotactic toward serine and are devoid of methyltransferase activities following serine chemostimulation, while responses to aspartic acid and glutamic acid remain comparable to wild type.

Several potential halobacterial transducer genes were inactivated in *H. salinarum*, and the mutants were tested against a wide range of compounds. This analysis led to the identification of Car, a specific transducer for arginine (103). Car is rare in that it is a soluble, cytoplasmic transducer (rather than a cytoplasmic membrane-located transducer) that is able to be methylated. Its function is to sense the arginine level in response to arginine:ornithine antiporter activity, through either direct binding or interaction with an unknown, cytoplasmic receptor component. Another halobacterial cytoplasmic transducer is HtrXI from halobacterial strain Flx15 (20). HtrXI differs from Car at only 9 amino acid residues. An *htrXI* deletional mutant is defective in chemotaxis toward glutamic acid and aspartic acid and devoid of methyltransferase activity. It is not known whether HtrXI interacts with a membrane protein or directly binds to its intracellular chemoeffector.

Analysis of the *H. salinarum* genome led to the identification of binding protein orthologs immediately upstream of potential transducers (BasB/BasT and CosB/CosT pairs) (56). These binding proteins are membrane bound and are presumably attached to the archaeal membrane through lipid anchors. No promoter-like elements were identifiable upstream of *basT* and *cosT*; Northern blot analysis confirmed that *basB/basT* belong to a single transcriptional unit. Functional analysis of deletion mutants revealed that BasB/BasT mediates chemotaxis behavior toward branched-chain amino acids (leucine, valine, isoleucine, methionine, and cysteine), while CosB/CosT mediates chemotaxis toward compatible solutes. Cells carrying a truncated BasT transducer were found to be nonchemotactic toward BasT-specific solutes, thereby demonstrating a function for the extracellular signaling domain. The homologous binding proteins in bacteria serve the dual function of mediating chemotactic responses and initiating solute uptake through interaction with the ABC transport system. However, since the halobacterial genome does not encode genes for an equivalent ABC transporter system

for branched-chain amino acids, in *H. salinarum* both BatB and CosB have roles exclusively limited to chemotaxis and not to actual transport of substrates (56). It has been speculated that the chromosomal co-localization of the substrate-binding and transducer genes may be due to a need for precise regulation of the two modules (56).

Aerotaxis Transducers

The *H. salinarum* aerotactic response was initially observed as motile cells accumulating around an air bubble (16, 102). The first aerotaxis transducer was characterized by Brooun and colleagues (19). The *htrVIII* gene was originally cloned as one of 13 putative transducers with homology to methyl-accepting proteins (122). HtrVIII was speculated to have a role in oxygen sensing, based on its six transmembrane segments being homologous to the heme-binding sites of the eucaryal cytochrome *c* oxidase (19). The HtrVIII C-terminal domain shares homology with the bacterial methyl-accepting chemotaxis protein. Aerotaxis was demonstrated through microslide capillary assays; an *htrVIII* deletion strain lost aerotactic response while an overexpression strain exhibited enhanced aerotaxis. The aerotactic behavior was found to be methylation dependent.

HemAT-Hs (previously HtrX) is another newly identified myoglobin-like, heme-containing aerotactic transducer in *H. salinarum*. It possesses N-terminal homology to myoglobin, as well as similarity to the cytoplasmic signaling domain of Tsr (*E. coli* MCP). Analysis of deletion and overexpression strains confirmed that HemAT-Hs is responsible for the aerophobic response in *H. salinarum* (43). The signaling is a methylation-dependent process, elicited by the two methylation sites at the K1 region of the C terminus. HemAT-Hs is a soluble transducer that appears to function by binding diatomic oxygen at its heme when the heme is in the ferrous state, with the oxygen binding triggering a conformational change in the N-terminal sensor domain that alters the C-terminal signaling domain. The similarity of the C-terminal signaling domain to the MCP family of bacterial chemoreceptors implies that the subsequent signal transduction events leading to aerotaxis in *H. salinarum* are likely to be similar to those involved in chemotaxis in bacteria.

Transducer Responding to Membrane Potential

Recently, MpcT (formerly Htr14) was identified in *H. salinarum* as the transducer responsible for communicating changes in the membrane potential to the taxis system (54). In *H. salinarum* mutants containing only one of the light-driven ion pumps (bacteriorhodpsin or halorhodopsin) as their sole retinal protein, a decrease in irradiance causes a phototactic response in the absence of respiration. The decrease in the proton motive force (PMF) is detected, and the cell responds with a reversal of flagellum rotation. By systematically knocking out each of the 18 Htr-encoding genes, *htr14* was identified as the putative PMF sensor and was the only Htr gene essential for the photoresponse in these cells. Responsiveness was restored by complementation with *htr14*. Based on calculations of the cytoplasmic buffering capacity, it was determined that MpcT was responding to changes in the membrane potential component of the PMF and not the change in internal pH.

GENERAL MECHANISM
OF TAXIS IN *ARCHAEA*

Extreme halophiles remain the only archaea in which significant chemotaxis/phototaxis studies have been reported. Similar to bacteria, in the absence of a gradient of stimuli, halobacteria exhibit a random walk as cells intermittently change the direction of their swimming (81). The presence of a gradient of attractant results in suppression of spontaneous motor switching, and as a result extends swimming, while increasing the level of repellents leads to activated switching and a resultant change in direction. Work by Oesterhelt's group on *H. salinarum* established that archaea possess an chemotaxis system analogous to bacteria (90–92). Moreover, the identification and function of various transducers have been reported (19, 43, 54, 56, 71, 103). It appears that once the stimulus has been detected, a bacterial-like response is triggered (Fig. 8). As in bacteria, the heart of the chemotaxis mechanism in archaea lies in a two-component regulatory system in which the ligand occupancy state of the receptors controls autophosphorylation of the histidine kinase, CheA. CheA is inhibited by attractants in most bacteria, such as *E. coli*. *B. subtilis* is an exception where attractants stimulate CheA activity. Similar to most bacteria, chemical and light attractants decrease CheA activity in archaea (107).

Because many chemoreceptors remain unidentified, the first step in transmembrane signaling is not very well characterized. Limited information has come from studying the *N. pharaonis* SRII transducer complex. The four retinal proteins, BR, Hr, SRI, and SRII, share a common topology with membrane helices arranged in two arcs (80). The inner arc is composed of helices B, C, and D, while the outer arc comprises helices A, E, F, and G. A transmembrane pore is formed mainly between helices B, C, F, and G, and the retinal is attached to a lysine residue in helix G. Using the crystal structure of the *N. pharaonis* receptor-transducer complex as a model, a working model has been proposed in which light excitation triggers an outward movement of the cytoplasmic part of helix F, which in turn induces a rotation of one of the transmembrane helices of the transducer, thereby triggering it into active state (39). It has also been suggested that as SRII-HtrII of *H. salinarum* displays dual functionality (in both phototaxis and chemotaxis; see "Phototaxis transducers and Chemotaxis transducers," above), the transmembrane-signaling mechanism involved in both types of taxis is also likely shared.

Interaction of the MCP-like transducers initially occurs with CheW, and the signal is passed to CheA. Following autophosphorylation, CheA transfers the phosphate to CheY. Interaction of the phosphorylated CheY (CheY-P) with a switch component of the archaeal flagellum is assumed to then occur, resulting in a change in the rotation of the flagellum. In the bacterial system, CheY binds to the switch component, FliM (18), but a homolog or functionally equivalent protein has not been identified in any archaea (29, 79). Hence, the necessary connection of the sensing system to the motility system remains an enigma. For phototaxis in *H. salinarum* at least and presumably for other sensory systems in addition to the two-component signaling system, there is also signaling through fumarate. Both CheY and fumarate are required for switching of flagella rotation in phototaxis; a similar role for fumarate as a switch factor was later shown in *E. coli*. Activated transducer causes the release of fumarate from a fumarate-binding protein located in the cytoplasmic membrane (64, 65, 67). As many as 60,000 molecules of fumarate may be released per second in a cell responding to light stimulus. Evidence suggests that fumarate and CheY act cooperatively at the level of the flagellar switch (64). Models for the mechanism of various taxes must be able to integrate not only data on detection of external stimuli but also more recent data that clearly show cells engaging in fumarate-mediated signaling as well as sensing changes in membrane potential and intracellular arginine levels. These observations clearly suggest that, in addition to changing external stimuli, the metabolic state of a bacterial or archaeal cell is important in regulating tactic behavior (64).

For the cells to detect changes in the signal intensity over a wide range of signal strength (light intensity or chemical concentration), flagellated archaea are equipped with an adaptation mechanism. Similar to bacterial MCPs, archaeal transducers contain glutamate residues in conserved regions of their cytoplasmic domains that can be methylated. These residues are substrates for the methylesterase CheB and

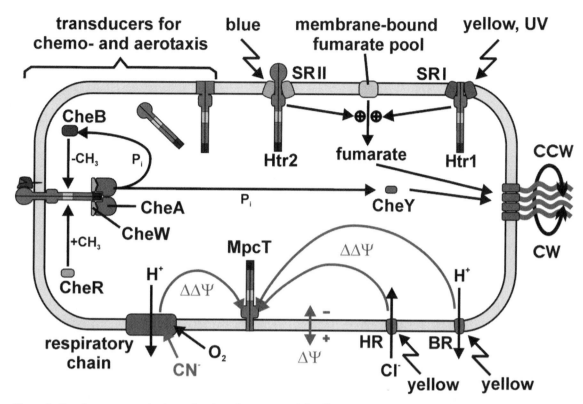

Figure 8. (*See the separate color insert for the color version of this illustration.*) Overview of halobacterial signal transduction. Transducer proteins (Htr proteins) are depicted as dimers (brown) and shown in their expected topology. The Htr regions involved in adaptation (yellow) and in signal relay (dark gray) to the flagellar motor via various Che proteins are indicated. The actions of the Che-protein machinery are illustrated for only one of the Htr proteins, shown on the left, for which an interaction with a substrate-loaded, membrane-anchored binding protein is indicated. CheD and CheJ (CheC) proteins are omitted for clarity. Htr1 and Htr2 transduce light signals via direct interaction with their corresponding receptors SRI and SRII. Repellent light signals mediated by SRI and SRII elicit the release of switch factor fumarate from a membrane-bound fumarate pool. MpcT senses changes in membrane potential ($\Delta\Psi$) generated via light-dependent changes in ion transport activity of BR and HR. The relative sizes of receptors, binding proteins, transducers, and Che proteins approximately reflect their corresponding molecular masses. Reproduced from *Molecular Microbiology* (54) with permission of the publisher and D. Oesterhelt.

the methyltransferase CheR. In the HtrI and HtrII sequences in *H. salinarum*, Perazzona and Spudich (85) identified three glutamate pairs containing potential methylation sites. Mutagenic substitution of these glutamate pairs by alanine subsequently identified Glu265-Glu266 (HtrI) and Glu513-Glu514 (HtrII) as being responsible for signal transduction. Furthermore, it was revealed that *H. salinarum* adaptation to chemostimuli is different from the *E. coli* chemotaxis paradigm but is instead similar to the *B. subtilis* case, in which turnover rates increase regardless of the nature of the stimulus. The methylation state of the Htr proteins results in adaptation to signals, or a resetting of the system so that cells can continue to respond to changes in the signal rather than absolute signal strength. Thus it appears that the same mode of adaptation to stimulus occurs in archaea as in bacteria.

CheY-P determines the direction of flagella rotation: in *E. coli* it causes CW rotation and in *B. subtilis*

CCW rotation. When *cheY* or *cheA* was deleted in *H. salinarum*, no reversals in swimming were seen and cells showed a preferential forward swimming due to CW flagella rotation (90). Thus CheY-P is responsible for the reversal of rotation, i.e., CCW. The default direction of flagella rotation in *H. salinarum* is CW, and CheY-P is required to cause rotation in the CCW direction (107). Since quick responses to changing signals in the environment are a necessary part of the chemotaxis system, the half-life of the response regulator, CheY-P, is usually very short (<1 min for various bacterial CheY-P); dephosphorylation is often aided by specific phosphatases. While the best-studied phosphatase is CheZ in *E. coli*, most chemotactic bacteria and archaea lack the gene for this protein (117). In *B. subtilis*, CheY-P dephosphorylation is carried out by a combination of FliY, CheC, and possibly CheX (106). FliY is particularly interesting as it is a component of the flagellar switch in *Bacillus*. In archaea, the candidate for CheY-P dephosphorylation appears to be CheC, which is present in all chemotactic archaea (107). Since many archaea appear to contain multiple copies of *cheY*-like genes, it may be that some archaea use alternate versions of CheY to act as phosphate sinks in signal removal from CheY-P, as is the case in *Sinorhizobium meliloti* (98).

PERSPECTIVE: THE NEXT FIVE YEARS

The study of flagella, type IV pili, and chemotaxis in bacteria has generated a wealth of knowledge on a variety of fundamental concepts, including gene regulation, protein export, and protein-protein interactions, that far supersede the limited direct application to bacterial and archaeal motility. Continued study of archaeal flagellation and chemotaxis is expected to yield the same far-ranging information about the much less well-studied archaea. In the next five years, it is predicted that significant advances will occur to answer fundamental questions about how the archaeal flagella is assembled, its component proteins, and its interaction with the chemotaxis system.

New genes are slowly being identified that are involved with flagellin posttranslational modification, i.e., the processing peptidase and glycosylation genes. The latter also play a role in modification of the S layer. Genes involved in posttranslational modification may be unique to each organism, and many of these are yet to be identified. Structural studies of the glycans are a necessary prerequisite for seeking out the required genes and identifying the precise roles of the gene products in glycan composition, assembly, and attachment. However, additional flagella structural genes still need to be identified. These must include genes for anchoring proteins (basal body equivalents), *mot* genes (needed for flagella motor function, i.e., rotation of a completed flagellum) and switch genes (needed for torque generation and reversal of flagella rotation). None of these have yet been identified by homology searches based on their bacterial counterparts and so other methods must be used. Methods may include random mutagenesis by insertion mutagenesis using cloned genomic fragments, or possibly transposon mutagenesis. These studies will be facilitated by the methodological development of systems for genetic manipulation of archaea, particularly for methanogens (53, 72, 61, 121; see Chapter 13). Significant progress in understanding motility, flagellation, and chemotaxis will occur as the genetic tools continue to improve in the various model organisms. Study of the flagella-associated genes which are often cotranscribed with flagellins will, it is hoped, yield important information about their, so far completely unknown, role in archaeal flagellation.

Another issue that, it is hoped, will be resolved in the next 5 years is proof of the polarity of growth of archaeal flagella. While similarities of archaea flagella to type IV pili have been described, these are limited to the flagellins, associated peptidase, ATPase, and one conserved membrane protein. These similarities, especially the signal peptide-containing flagellins and peptidase, led to the proposal many years ago that the incorporation of the new subunits would be at the base rather than the distal tip, as is the case with bacterial flagellins (45). The lack of a detectable central channel in archaeal flagella that was reported recently strongly supports the hypothesized assembly mechanism (23). However, evidence that clearly demonstrates polarity is still desirable and necessary. How a rotary apparatus would be assembled from the base and the nature of the anchoring structure for this organelle are important protein secretion/assembly questions that need answering.

The connection between the motility apparatus and the environmental sensing system needs to be identified. Identification of the switch protein that binds CheY-P would be a big step forward since the gene for such a protein may be located in an operon of other flagella structural protein genes. In the same vein, the site of interaction of fumarate at the flagella motor should be identified. Continued deletion analysis of the 18 homologs of the MCPs in *H. salinarum* (54, 103) will undoubtedly lead to the identification of the role each of them has in sensing the environment in this extreme halophile. This will allow for a complete picture and integration of various types of taxis in one archaeon.

Finally, the structure of the flagellum itself has only been tackled recently by Tractenberg's group (23, 115). The properties of the filament proteins and how they interact, especially in the most extreme habitats of hyperthermophily and saturated salt, to create such stable structures is so far largely unexplored and ripe for analysis.

It is expected that the continued study of archaeal flagellation and chemotaxis will lead to novel discoveries about these structures and processes in archaea, and these may in turn lead to insights into the understanding of bacterial chemotaxis and type IV pili assembly, structure, and function (25).

Acknowledgments. Research in our laboratory is supported by a grant from the Natural Sciences and Engineering Research Council of Canada (NSERC) to K.F.J. Both S.Y.M. Ng and B. Chaban are supported by postgraduate awards from NSERC.

We are grateful to all colleagues who kindly allowed use of the figures used in this review.

REFERENCES

1. Adler, J. 1973. A method for measuring chemotaxis and use of the method to determine optimum conditions for chemotaxis by *Escherichia coli*. *J. Gen. Microbiol.* **74**:77–91.

2. Alam, M., and G. L. Hazelbauer. 1991. Structural features of methyl-accepting taxis proteins conserved between archaebacteria and eubacteria revealed by antigenic cross-reaction. *J. Bacteriol.* **173**:5837–5842.

3. Alam, M., and D. Oesterhelt. 1984. Morphology, function and isolation of halobacterial flagella. *J. Mol. Biol.* **176**:459–475.

4. Albers, S.-V., and A. J. M. Driessen. 2005. Analysis of ATPases of putative secretion operons in the thermoacidophilic archaeon *Sulfolobus solfataricus*. *Microbiology* **151**:763–773.

5. Albers, S.-V., Z. Szabo, and A. J. M. Driessen. 2003. Archaeal homolog of bacterial type IV prepilin signal peptidases with broad substrate specificity. *J. Bacteriol.* **185**:3918–3925.

6. Alm, R. A., and J. S. Mattick. 1997. Genes involved in the biogenesis and function of type-4 fimbriae in *Pseudomonas aeruginosa*. *Gene* **192**:89–97.

7. Bardy, S. L., J. Eichler, and K. F. Jarrell. 2003. Archaeal signal peptides—a comparative survey at the genome level. *Protein Sci.* **12**:1833–1843.

8. Bardy, S. L., and K. F. Jarrell. 2002. FlaK of the archaeon *Methanococcus maripaludis* possesses preflagellin peptidase activity. *FEMS Microbiol. Lett.* **208**:53–59.

9. Bardy, S. L., and K. F. Jarrell. 2003. Cleavage of preflagellins by an aspartic acid signal peptidase is essential for flagellation in the archaeon *Methanococcus voltae*. *Mol. Microbiol.* **50**:1339–1347.

10. Bardy, S. L., T. Mori, K. Komoriya, S.-I. Aizawa, and K. F. Jarrell. 2002. Identification and localization of flagellins FlaA and FlaB3 within flagella of *Methanococcus voltae*. *J. Bacteriol.* **184**:5223–5233.

11. Bardy, S. L., S. Y. M. Ng, and K. F. Jarrell. 2004. Recent advances in the structure and assembly of the archaeal flagellum. *J. Mol. Microbiol. Biotechnol.* **7**:41–51.

12. Bardy, S. L., S. Y. M. Ng, and K. F. Jarrell. 2003. Prokaryotic motility structures. *Microbiology* **149**:295–304.

13. Bayley, D. P., and K. F. Jarrell. 1999. Overexpression of *Methanococcus voltae* flagellins subunits in *Escherichia coli* and *Pseudomonas aeruginosa*: a source of archaeal preflagellin. *J. Bacteriol.* **181**:4146–4153.

14. Bayley, D. P., and K. F. Jarrell. 1998. Further evidence to suggest that archaeal flagella are related to bacterial type IV pili. *J. Mol. Evol.* **46**:370–373.

15. Bayley, D. P., M. L. Kalmokoff, and K. F. Jarrell. 1993. Effect of bacitracin on flagellar assembly and presumed glycosylation of the flagellins of *Methanococcus deltae*. *Arch. Microbiol.* **160**:179–185.

16. Bibikov, S. I., and V. P. Skulachev. 1989. Mechanisms of phototaxis and aerotaxis in *Halobacterium halobium*. *FEBS Lett.* **243**:303–306.

17. Black, F. T., E. A. Freund, O. Vinther, and C. Christiansen. 1979. Flagellation and swimming motility of *Thermplasma acidophilum*. *J. Bacteriol.* **137**:456–460.

18. Bren, A., and M. Eisenbach. 1998. The N terminus of the flagellar switch protein, FliM, is the binding domain for the chemotactic response regulator, CheY. *J. Mol. Biol.* **278**:507–514.

19. Brooun, A., J. Bell, T. Freitas, R.W. Larsen, and M. Alam. 1998. An archaeal aerotaxis transducer combines subunit I core structures of eukaryotic cytochrome c oxidase and eubacterial methyl-accepting chemotaxis proteins. *J. Bacteriol.* **180**:1642–1646.

20. Brooun, A., W. Zhang, and M. Alam. 1997. Primary structure and functional analysis of the soluble transducer protein HtrXI in the archaeon *Halobacterium salinarium*. *J. Bacteriol.* **179**:2963–2968.

21. Burda, P., and M. Aebi. 1999. The dolichol pathway of N-linked glycosylation. *Biochim. Biophys. Acta* **1426**:239–257.

21a. Chaban, B., S. Voisin, J. Kelly, S. M. Logan, and K. F. Jarrell. 2006. Identification of genes involved in the biosynthesis and attachment of *Methanococcus voltae* N-linked glycans: insight into N-linked glycosylation pathways in *Archaea*. *Mol. Microbiol.* **61**:259–268.

22. Chen, X., and J. Spudich. 2004. Five residues in the HtrII transducer membrane-proximal domain close the cytoplasmic proton-conducting channel of sensory rhodopsin I. *J. Biol. Chem.* **279**:42964–42969.

23. Cohen-Krausz, S., and S. Trachtenberg. 2002. The structure of the archaebacterial flagellar filament of the extreme halophiles *Halobacterium salinarum* R1M1 and its relation to eubacterial flagellar filaments and type IV pili. *J. Mol. Biol.* **321**:383–395.

24. Correia, J. D., and K. F. Jarrell. 2000. Posttranslational processing of *Methanococcus voltae* preflagellin by preflagellin peptidases of *M. voltae* and other methanogens. *J. Bacteriol.* **182**:855–858.

25. Craig, L., M. E. Pique, and J. A. Trainer. 2004. Type IV pilus structure and bacterial pathogenicity. *Nat. Rev. Microbiol.* **2**:363–378.

26. Crowther, L. J., R. P. Anantha, and M. S. Donnenberg. 2004. The inner membrane subassembly of the enteropathogenic *Escherichia coli* bundle-forming pilus machine. *Mol. Microbiol.* **52**:67–79.

27. Cruden, D., R. Sparling, and A. J. Markovetz. 1989. Isolation and ultrastructure of the flagella of *Methanococcus thermolithotrophicus* and *Methanospirillum hungatei*. *Appl. Environ. Microbiol.* **55**:1414–1419.

28. Faguy, D. M., D. S. Bayley, A. S. Kostyukova, N. A. Thomas, and K. F. Jarrell. 1996. Isolation and characterization of flagella and flagellin proteins from the thermoacidophilic archaea *Thermoplasma volcanium* and *Sulfolobus shibatae*. *J. Bacteriol.* **178**:902–905.

29. Faguy, D. M., and K. F. Jarrell. 1999. A twisted tale: the origin and evolution of motility and chemotaxis in prokaryotes. *Microbiology* **145**:279–281.

30. Faguy, D. M., K. F. Jarrell, J. Kuzio, and M. L. Kalmokoff. 1994. Molecular analysis of archaeal flagellins: similarity to the type IV pilin-transport superfamily widespread in bacteria. *Can. J. Microbiol.* **40**:67–71.

31. Faguy, D. M., S. F. Koval, and K. F. Jarrell. 1993. Effects of changes in mineral concentration and growth temperature on filament length and flagellation in the archaeon *Methanospirillum hungatei*. *Arch. Microbiol.* **159**:512–520.

32. Faguy, D. M., S. F. Koval, and K. F. Jarrell. 1992. Correlation between glycosylation of flagellin proteins and sensitivity of flagellar filaments to Triton X-100 in methanogens. *FEMS Microbiol. Lett.* **90**:129–134.

33. Fedorov, O. V., M. G. Pyatibratov, A. S. Kostyukova, N. K. Osina, and V. Y. Tarasov. 1994. Protofilament as a structural element of flagella of haloalkalophilic archaebacteria. *Can. J. Microbiol.* **40**:45–53.

34. Forest, K. T., K. A. Satyshur, G. A. Worzalla, J. K. Hansen, and T. J. Herderdorf. 2004. The pilus retraction protein PilT:ultrastructure of the biological assembly. *Acta Crystallogr. Sect. D Biol Crystallogr.* **60**:978–982.

35. Gavel, Y., and G. von Heijne. 1990. Sequence differences between glycosylated and non-glycosylated Asn-X-Thr/Ser ac-

ceptor sites: implications for protein engineering. *Protein Eng.* 3:433–442.

36. Gerl, L., and M. Sumper. 1988. Halobacterial flagellins are encoded by a multigene family. Characterization of five flagellins genes. *J. Biol. Chem.* 263:13246–13251.

37. Gestwicki, J. E., A. C. Lamanna, R. M. Harshey, L. L. McCarter, L. L. Kiessling, and J. Adler. 2000. Evolutionary conservation of methyl-accepting chemotaxis protein location in *Bacteria* and *Archaea. J. Bacteriol.* 182:6499–6502.

38. Giometti, C. S., C. I. Reich, S. L. Tollaksen, G. Babnigg, H. Lim, J. R. Yates III, and G. J. Olsen. 2001. Structural modifications of *Methanococcus jannaschii* flagellin proteins revealed by proteome analysis. *Proteomics* 1:1033–1042.

39. Gordeliy, V. I., J. Ilabahn, R. Moukhametzlanov, R. Efremov, J. Granzin, R. Schlesinger, G. Buldt, T. Savopol, A. J. Scheidig, J. P. Klare, and M. Engelhard. 2002. Molecular basis of transmembrane signaling by sensory rhodopsin II-transducer complex. *Nature* 419:484–487.

40. Hellingwerf, K. J. 2002. The molecular basis of sensing and responding to light in microorganisms. *Antonie van Leeuwenhoek* 81:51–59.

41. Hoff, W. D., K.-H. Jung, and J. L. Spudich. 1997. Molecular mechanism of photosignalling by archaeal sensory rhodopsins. *Annu. Rev. Biophys. Biomol. Struct.* 26:223–258.

42. Hou, S., A. Brooun, H. S. Yu, T. Freitas, and M. Alam. 1998. Sensory rhodopsin II transducer HtrII is also responsible for serine chemotaxis in the archaeon *Halobacterium salinarium. J. Bacteriol.* 180:1600–1602.

43. Hou, S., R. W. Larsen, D. Boudko, C. W. Riley, E. Karatan, M. Zimmer, G. W. Ordal, and M. Alam. 2000. Myoglobin-like aerotaxis transducers in Archaea and Bacteria. *Nature* 403:540–544.

44. Jarrell, K. F., D. P. Bayley, V. Florian, and A. Klein. 1996. Isolation and characterization of insertional mutations in flagellins genes in the archaeon *Methanococcus voltae. Mol. Microbiol.* 20:657–666.

45. Jarrell, K. F., D. P. Bayley, and A. S. Kostyukova. 1996. The archaeal flagellum: a unique motility structure. *J. Bacteriol.* 178:5057–5064.

46. Jarrell, K. F. and S. F. Koval. 1989. Ultrastructure and biochemistry of *Methanococcus voltae. CRC Crit. Rev. Microbiol.* 17:53–87.

47. Jaschke, M., H.-J. Butt, and E. K. Wolff. 1994. Imaging flagella of halobacteria by atomic force microscopy. *Analyst* 119:1943–1946.

48. Jones, C. J. and S.-I. Aizawa. 1991. The bacterial flagellum and flagellar motor: structure, function and assembly. *Adv. Microb. Physiol.* 32:109–172.

49. Kachlany, S. C., P. J. Planet, M. K. Bhattacharjee, E. Kollia, R. deSalle, D. H. Fine, and D. H. Figurski. 2000. Nonspecific adherence by *Actinobacillus actinomycetemcomitans* requires genes widespread in *Bacteria* and *Archaea. J. Bacteriol.* 182:6169–6176.

50. Kalmokoff, M. L., and K. F. Jarrell. 1991. Cloning and sequencing of a multigene family encoding the flagellins of *Methanococcus voltae. J. Bacteriol.* 173:7113–7125.

51. Kalmokoff, M. L., S. F. Koval, and K. F. Jarrell. 1992. Relatedness of the flagellins from methanogens. *Arch. Microbiol.* 157:481–487.

52. Kalmokoff, M. L., S. F. Koval, and K. F. Jarrell. 1988. Isolation of flagella of the archaebacterium *Methanococcus voltae* by phase separation with Triton X-114. *J. Bacteriol.* 170:1752–1758.

53. Kim, W., and W. B. Whitman. 1999. Isolation of acetate auxotrophs of the methane-producing archaeon *Methanococcus maripaludis* by random insertional mutagenesis. *Genetics* 152:1429–1437.

54. Koch, M. K., and D. Oesterhelt. 2005. MpcT is the transducer for membrane potential changes in *Halobacterium salinarum. Mol. Microbiol.* 55:1681–1694.

55. Kokoeva, M.V., and D. Oesterhelt. 2000. BasT, a membrane-bound transducer for protein for amino acid detection in *Halobacterium salinarum. Mol. Microbiol.* 35:647–656.

56. Kokoeva, M. V., K.-F. Storch, C. Klein, and D. Oesterhelt. 2002. A novel mode of sensory transduction in archaea: binding protein-mediated chemotaxis towards osmoprotectants and amino acids. *EMBO J.* 21:2312–2322.

57. Konig, H. 1988. Archaeobacterial cell envelopes. *Can. J. Microbiol.* 34:395–406.

58. Koval, S. F., and K. F. Jarrell. 1987. Ultrastructure and biochemistry of the cell wall of *Methanococcus voltae. J. Bacteriol.* 169:1298–1306.

59. Kupper, J., W. Marwan, D. Typke, H. Grunberg, U. Uwer, M. Gluch, and D. Oesterhelt. 1994. The flagella bundle of *Halobacterium salinarium* is inserted into a distinct polar cap structure. *J. Bacteriol.* 176:5184–5187.

60. Lechner, J., and F. Wieland. 1989. Structure and biosynthesis of prokaryotic glycoproteins. *Annu. Rev. Biochem.* 58:173–194.

61. Luo, Y., and A. Wasserfallen. 2001. Gene transfer systems and their applications in Archaea. *Syst. Appl. Microbiol.* 24:15–25.

62. Macnab, R. M. 2004. Type III flagellar protein export and flagellar assembly. *Biochim. Biophys. Acta* 1694:207–217.

63. Macnab, R. M. 1999. The bacterial flagellum: reversible rotary propellor and type III export apparatus. *J. Bacteriol.* 181:7149–7153.

64. Marwan, W., and D. Oesterhelt. 2000. Archaeal vision and bacterial smelling. *ASM News* 66:83–89.

65. Marwan. W., and D. Oesterhelt. 1991. Light-induced release of the switch factor during photophobic responses of *Halobacterium halobium. Naturwissenschaften* 78:127–129.

66. Marwan, W., and D. Oesterhelt. 1987. Signal formation in the halobacterial photophobic response mediated by a fourth retinal protein (P480). *J. Mol. Biol.* 195:333–342.

67. Marwan, W., W. Schafer, and D. Oesterhelt. 1990. Signal transduction in *Halobacterium* depends on fumarate. *EMBO J.* 9:355–362.

68. Mattick, J. S. 2002. Type IV pili and twitching motility. *Annu. Rev. Microbiol.* 56:289–314.

69. Metlina, A. L. 2004. Bacterial and archaeal flagella as prokaryotic motility organelles. *Biochemistry (Moscow)* 69:1203–1212.

70. Migas, J., K. L. Anderson, D. L. Cruden, and A. J. Markovetz. 1989. Chemotaxis in *Methanospirillum hungatei. Appl. Environ. Microbiol.* 55:264–265.

71. Mironova, O .S., R. G. Efremov, B. Person, J. Heberle, I. L. Budyak, G. Buldt, and R. Schlesinger. 2005. Functional characterization of sensory rhodopsin II from *Halobacterium salinarum* expressed in *Escherichia coli. FEBS Lett.* 579:3147–3151.

72. Moore, B. C., and J. A. Leigh. 2005. Markerless mutagenesis in *Methanococcus maripaludis* demonstrates roles for alanine dehydrogenase, alanine racemase, and alanine permease. *J. Bacteriol.* 187:972–979.

73. Morand, P. C., E. Bille, S. Morelle, E. Eugene, J.-L. Beretti, M. Wolfgang, T. F. Meyer, M. Kooney, and X. Nassif. 2004. Type IV pilus retraction in pathogenic *Neisseria* is regulated by the PilC protein. *EMBO J.* 23:2009–2017.

74. Mukhopadhyay, B., E. F. Johnson, and R. S. Wolfe. 2000. A novel p_{H2} control on the expression of flagella in the hyperthermophilic strictly hydrogenotrophic methanarchaeon *Methanococcus jannaschii. Proc. Natl. Acad. Sci. USA* 97:11522–11527.

75. Nagahisa, K., S. Ezaki, S. Fujiwara, T. Imanaka, and M. Takagi. 1999. Sequence and transcriptional studies of five clus-

tered flagellin genes from hyperthermophilic archaeon *Pyrococcus kodakaraensis* KOD1. *FEMS Microbiol. Lett.* **178:** 183–190.

76. Ng, S. Y. M., B. Chaban, and K. F. Jarrell. 2006. Archaeal flagella, bacterial flagella and type IV pili: a comparison of genes and posttranslational modification. *J. Mol. Microbiol. Biotechnol.*, **11:**167–191.

77. Ng, W. V., S. P. Kennedy, G. G. Mahairas, B. Berquist, M. Pan, H. D. Shukla, S. R. Lasky, N. S. Baliga, V. Thorsson, J. Sbrogna, S. Swartzell, D.Weir, J. Hall, T. A. Dahl, R. Welti, Y. A. Goo, B. Leithauser, K. Keller, R. Cruz, M. J. Danson, D. W. Hough, D. G. Maddocks, P. E. Jablonski, M. P. Krebs, C. M. Angevine, H. Dale, T. A. Isenbarger, R. F. Peck, M. Pohlschroder, J. L. Spudich, K. W. Jung, M. Alam, T. Freitas, S. Hou, C. J. Daniels, P. P. Dennis, A. D. Omar, H. Ebhardt, T. M. Lowe, P. Liang, M. Riley, L. Hood, and S. DasSarma. 2000. Genomic sequence of *Halobacterium* species NRC-1. *Proc. Natl. Acad. Sci. USA* **97:**12176–12181.

78. Nita-Lazar, M., M. Wacker, B. Schegg, S. Amber, and M. Aebi. 2004. The N-X-S/T consensus sequence is required but not sufficient for bacterial N-linked protein glycosylation. *Glycobiology* **15:**361–367.

79. Nutsch, T., W. Marwan, D. Oesterhelt, and E. D. Gilles. 2003. Signal processing and flagellar motor switching during phototaxis of *Halobacterium salinarum*. *Genome Res.* **13:**2406–2412.

80. Oesterhelt, D. 1998. The structure and mechanism of the family of retinal proteins from halophilic archaea. *Curr. Opin. Struct. Biol.* **8:**489–500.

81. Oesterhelt, D., and W. Marwan. 1993. Signal transduction in halobacteria, p. 173–187. *In* M. Kates et al. (ed.), *The Biochemistry of Archaea (Archaebacteria)*. Elsevier Science Publications, New York, N.Y.

82. Oprian, D. D. 2003. Phototaxis, chemotaxis and the missing link. *Trends Biochem. Sci.* **28:**167–169.

83. Patenge, N., A. Berendes, H. Engelhardt, S. C. Schuster, and D. Oesterhelt. 2001. The *fla* gene cluster is involved in the biogenesis of flagella in *Halobacterium salinarum*. *Mol. Microbiol.* **41:**653–663.

84. Peabody, C. R., Y. J. Chung, M.-R. Yen, D. Vidal-Ingigliardi, A. P. Pugsley, and M. H. Saier, Jr. 2003. Type II protein secretion and its relationship to bacterial type IV pili and archaeal flagella. *Microbiology* **149:**3051–3072.

85. Perazzona, B., and J. L. Spudich. 1999. Identification of methylation sites and effects of phototaxis stimuli on transducer methylation in *Halobacterium salinarium*. *J. Bacteriol.* **181:**5676–5683.

86. Pohlschroder, M., K. Dilks, N. J. Hand, and R. W. Rose. 2004. Translocation of proteins across archaeal cytoplasmic membranes. *FEMS Microbiol. Rev.* **28:**3–24.

87. Polosina, Y. Y., K. F. Jarrell, O. V. Fedorov, and A. S. Kostyukova. 1998. Nucleotide diphosphate kinase from haloalkaliphilic archaeon *Natronobacterium magadii*: purification and characterization. *Extremophiles* **2:**333–338.

88. Pyatibratov, M. G., K. Leonard, V. Y. Tarasov, and O. V. Fedorov. 2002. Two immunologically distinct types of protofilaments can be identified in *Naltrialba magadii* flagella. *FEMS Microbiol. Lett.* **212:**23–27.

89. Ring, G., and J. Eichler. 2004. Extreme secretion: protein translocation across the archaeal plasma membrane. *J. Bioenerg. Biomembr.* **36:**35–45.

90. Rudolph, J., and D. Oesterhelt. 1996. Deletion analysis of the *che* operon in *Halobacterium salinarium*. *J. Mol. Biol.* **258:** 548–554.

91. Rudolph, J., and D. Oesterhelt. 1995. Chemotaxis and phototaxis require a CheA histidine kinase in the archaeon *Halobacterium salinarium*. *EMBO J.* **14:**667–673.

92. Rudolph, J., N. Tolliday, C. Schmitt, S. C. Schuster, and D. Oesterhelt. 1995. Phosphorylation in halobacterial signal transduction. *EMBO J.* **14:**4249–4257.

93. Schafer, G., M. Engelhard, and V. Muller. 1999. Bioenergetics of the archaea. *Microbiol. Mol. Biol. Rev.* **63:**570–620.

94. Scharf, B., and E. K. Wolff. 1994. Phototactic behaviour of the archaebacterial *Natronobacterium pharaonis*. *FEBS Lett.* **340:**114–116.

95. Seidel, R., B. Scharf, M. Gautel, K. Kleine, D. Oesterhelt, and M. Engelhardt. 1995. The primary structure of sensory rhodopsin II: a member of an additional retinal protein subgroup is coexpressed with its transducer, the halobacterial transducer of rhodopsin II. *Proc. Natl. Acad. Sci. USA* **92:** 3036–3040.

96. Serganova, I. S., Y. Y. Polosina, A. S. Kostyukova, A. L. Metlina, M. G. Pyatibratov, and O. V. Fedorov. 1995. Flagella of halophilic archaea: biochemical and genetic analysis. *Biochemistry (Moscow)* **60:**953–957.

97. Sment, K. A., and J. Konisky. 1989. Chemotaxis in the archaebacterium *Methanococcus voltae*. *J. Bacteriol.* **171:** 2870–2872.

98. Sourjik, V., and R. Schmitt. 1998. Phosphotransfer between CheA, CheY1, and CheY2 in the chemotaxis signal transduction chain of *Rhizobium meliloti*. *Biochemistry* **37:**2327–2335.

99. Southam, G., M. L. Kalmokoff, K. F. Jarrell, S. F. Koval, and T. J. Beveridge. 1990. Isolation, characterization and cellular insertion of the flagella from two strains of the archaebacterium *Methanospirillum hungatei*. *J. Bacteriol.* **172:**3221–3228.

100. Speranskii, V. V., A. L. Metlina, T. M. Novikova, and L. Y. Bakeyeva. 1996. Disk-like lamellar structure as part of the archaeal flagellar apparatus. *Biophysics* **41:**167–173.

101. Spudich, J. L., and H. Luecke.2002. Sensory rhodopsin II: functional insights from structure. *Curr. Opin. Struct. Biol.* **12:**540–546.

102. Stoeckenius, W., E. K. Wolff, and B. Hess. 1988. A rapid population method for action spectra applied to *Halobacterium halobium*. *J. Bacteriol.* **170:**2790–2795.

103. Storch, K. F., J. Rudolph, and D. Oesterhelt. 1999. Car: a cytoplasmic sensor responsible for arginine chemotaxis in *Halobacterium salinarum*. *EMBO J.* **18:**1146–1158.

104. Sumper, M. 1987. Halobacterial glycoprotein synthesis. *Biochim. Biophys. Acta* **906:**69–79.

105. Sumper, M., and F. T. Wieland. 1995. Bacterial glycoproteins. *In* J. Montreuil, H. Schachter, and J. F. G. Vliegenthart (ed.), *Glycoproteins*. Elsevier Science, New York, N.Y.

106. Szurmant, H., T. J. Muff, and G. W. Ordal. 2004. *Bacillus subtilis* CheC and FliY are members of a novel class of CheY-P-hydrolyzing proteins in the chemotactic signal transduction cascade. *J. Biol. Chem.* **279:**21787–21792.

107. Szurmant, H., and G. Ordal. 2004. Diversity in chemotaxis mechanisms among the Bacteria and Archaea. *Microbiol. Mol. Biol. Rev.* **68:**301–319.

108. Szymanski, C. M., S. M. Logan, D. Linton, and B. W. Wren. 2003. *Campylobacter*—a tale of two protein glycosylation systems. *Trends Microbiol.* **11:**233–238.

109. Szymanski, C. M., and B. W. Wren. 2005. Protein glycosylation in bacterial mucosal pathogens. *Nat. Rev. Microbiol.* **3:**225–237.

110. Tarasov, V. Y., M. G. Pyatibratov, S. Tang, M. Dyall-Smith, and O. V. Fedorov. 2000. Role of flagellins from A and B loci in flagella formation of *Halobacterium salinarum*. *Mol. Microbiol.* **35:**69–78.

111. Thomas, N. A., S. L. Bardy, and K. F. Jarrell. 2001. The archaeal flagellum: a different kind of prokaryotic motility structure. *FEMS Microbiol. Rev.* **25:**147–174

112. Thomas, N. A., E. D. Chao, and K. F. Jarrell. 2001. Identification of amino acids in the leader peptide of *Methanococ-*

cus voltae preflagellin that are important in posttranslational processing. *Arch. Microbiol.* **175**:263–269.

113. **Thomas, N. A., and K. F. Jarrell.** 2001. Characterization of flagellum gene families of methanogenic archaea and localization of novel accessory proteins. *J. Bacteriol.* **183**:7154–7164.

114. **Thomas, N. A., S. Mueller, A. Klein, and K. F. Jarrell.** 2002. Mutants in *flaI* and *flaJ* of the archaeon *Methanococcus voltae* are deficient in flagellum assembly. *Mol. Microbiol.* **46**:879–887.

115. **Trachtenberg, S., V. E. Galkin, and E. H. Egelman.** 2005. Refining the structure of the *Halobacterium salinarum* flagellar filament using the iterative helical real space reconstruction method: insights into polymorphism. *J. Mol. Biol.* **346**:665–676.

116. **Voisin, S., R. S. Houliston, J. Kelly, J.-R. Brisson, D. Watson, S. L. Bardy, K. F. Jarrell, and S. M. Logan.** 2005. Identification and characterization of the unique N-linked glycan common to the flagellins and S-layer glycoprotein of *Methanococcus voltae*. *J. Biol. Chem.* **280**:16586–16593.

117. **Wadhams, G. H., and J. P. Armitage.** 2004. Making sense of it all: bacterial chemotaxis. *Nat. Rev. Mol. Cell Biol.* **5**:1024–1037.

118. **Wieland, F., G. Paul, and M. Sumper.** 1985. Halobacterial flagellins are sulfated glycoproteins. *J. Biol. Chem.* **260**:15180–15185.

119. **Yang, C.-S., O. Sineshchekov, E. N. Spudich, and J. L. Spudich.** 2004. The cytoplasmic membrane-proximal domain of the HtrII transducer interacts with the E-F loop of photoactivated *Natronomonas pharoanis* sensory rhodopsin II. *J. Biol. Chem.* **279**:42970–42976.

120. **Yu, H. S., and M. Alam.** 1997. An agarose-in-plug bridge method to study chemotaxis in the Archaeon *Halobacterium salinarum*. *FEMS Microbiol. Lett.* **156**:265–269.

121. **Zhang, J. K., M. A. Pritchett, D. J. Lampe, H. M. Robertson, and W. W. Metcalf.** 2000. In vivo transposon mutagenesis of the methanogenic archaeon *Methanosarcina acetivorans* C2A using a modified version of the insect mariner-family transposable element Himar1. *Proc. Natl. Acad. Sci. USA* **97**:9665–9670.

122. **Zhang, W., A. Brooun, J. McCandless, P. Banda, and M. Alam.** 1996. Signal transduction in the archaeon *Halobacterium salinarium* is processed through three subfamilies of 13 soluble and membrane-bound transducer proteins. *Proc. Natl. Acad. Sci. USA* **93**:4649–4654.

Archaea: Molecular and Cellular Biology
Edited by Ricardo Cavicchioli
© 2007 ASM Press, Washington, D.C.

Chapter 19

Structure and Evolution of Genomes

Patrick Forterre, Yvan Zivanovic, and Simonetta Gribaldo

INTRODUCTION

At the archaeal meeting held in Munich in 1993 for the official retirement of Professor Wolfram Zillig (who continued working and publishing until 2004) a small informal gathering was organized by Roger Garrett to discuss the possibility of sequencing an archaeal genome (two years before TIGR sequenced the first bacterial genome). The main topic discussed was: which archaeal genome should we sequence? Everyone in the room started to defend his (her) favorite organism (*Sulfolobus* or *Halobacterium*, why not a methanogen?). After some discussion, Carl Woese asked to speak, and everybody wondered what would be his preferred archaeon. "We need to sequence at least five archaeal genomes!" were his words. This was nonsense for most participants, for whom it would have been already good to have at least one, besides the human genome! Ten years later, twenty archaeal genomes were indeed available in public databases. So Carl was right after all, as it was for the realization that the *Archaea* are not simply "strange bacteria," but a domain of life on their own (see Chapter 1).

The sequencing of archaeal genomes has been critical for archaeal research, especially because of the limited availability of easy-to-handle genetic systems (see Chapter 21). Moreover, it has permitted the existence of *Archaea* to be advertized outside the immediate fan club. After genome data became available, people who had never considered working with such exotic microbes ("God, we have to grow them!") could jump onto the bandwagon by just ordering DNA and cloning it into *Escherichia coli*. The complete sequence of the genome of *Methanococcus jannaschii* was made available by TIGR (17) only one year after the historical *Haemophilus influenzae* publication. Since then, a large number of proteins from

M. jannaschii have been cloned and sequenced by people who would have never considered dealing with a methanogen before. With the yeast genome becoming accessible the same year (1996), the sequencing of the *M. jannaschii* genome immediately opened up Pandora's box for a completely new research field: large-scale comparative genomics.

The first eight sequenced archaeal genomes were all from thermophiles and hyperthermophiles. This bias toward high temperatures was the product of both an academic interest (biologists have been fascinated by hyperthermophiles since their discovery in the early eighties) and of funding necessity (thanks to PCR and *Taq* polymerase it was easier in the 1990s to ask for money for the genome sequencing of a hyperthermophile). More recently, other types of extremophiles (halophiles, thermoacidophiles, psychrophiles) have entered the pipeline. Eventually, the genomes of several mesophilic archaea have appeared, and more exciting ones from previously uncultivated archaea are to come in the near future.

The availability of archaeal genomes from organisms living in such diverse environments has provided a unique opportunity to look at how DNA sequences are shaped by environmental factors. However, and primarily, the sequencing of archaeal genomes has allowed the concept of the three domains to be tested. In fact, although the existence of three distinct domains of life was already firmly established through rRNA sequence comparison, and supported by the discovery of unique features in archaeal biochemistry and molecular biology (see Chapter 2), it remained important to determine how the uniqueness of *Archaea* translated at the whole-genome level. Not so surprisingly, some people have found the resemblance of the structure and organization of archaeal and bacterial genomes to be a temptation to get back to the old

Patrick Forterre • Unité Biologie Moléculaire du Gène chez les Extremophiles, Institut de Génétique et Microbiologie, UMR CNRS 8621, Université Paris-Sud, 91405 Orsay, and Institut Pasteur, 25 rue du Dr. Roux, 75724 Paris Cedex 15, France. **Yvan Zivanovic** • Institut de Génétique et Microbiologie, UMR CNRS 8621, Université Paris-Sud, 91405 Orsay, France. **Simonetta Gribaldo** • Unité Biologie Moléculaire du Gène chez les Extremophiles, Institut Pasteur, 25 rue du Dr. Roux, 75724 Paris Cedex 15, France.

"prokaryote/eukaryote" dichotomy. However, the three-domains concept is not based on phenotypic traits, but relies on the existence of three different sequence spaces for the macromolecules that are central to gene expression and reproduction in the living world. This trichotomy was nicely reemphasized by comparative genomics, especially when several archaeal genomes became available. The vast majority of genes in archaeal genomes have indeed their most closely related homologs in other archaeal genomes, rather than bacterial or eucaryal genomes. A general principle that has emerged is that all archaeal housekeeping proteins (especially those for information processing) cluster together in phylogenetic trees, distinct from bacterial and eucaryal homologs (when they exist). Hence, they correspond to archaeal versions (sensu Woese), in the same way that informational proteins from bacteria cluster together and eucaryal proteins cluster together.

Archaeal genomics has also confirmed and extended the evolutionary linkage between *Archaea* and *Eucarya* that was previously uncovered by the pioneering work of Wolfram Zillig and coworkers on archaeal RNA polymerases. This was especially striking for DNA replication, since the majority of bacterial DNA replication genes have no homologs in *Archaea*, whereas most essential eucaryal DNA replication proteins have closely related archaeal homologs. The existence of many informational proteins that are common to *Archaea* and *Eucarya* and absent from *Bacteria* is not only true for DNA replication, but also for translation (especially the initiation step), transcription, and RNA and protein processing (i.e., informational proteins). This distribution is well illustrated by the phylogenetic affinities of the core proteins common to all archaeal genomes. This includes 118 proteins with homologs in both *Eucarya* and *Bacteria*; of these, 63 have only eucaryal homologs while just 13 have only bacterial homologs (74). If the core of archaeal genomes has a profound eucaryal flavor, its periphery (the proteins only present in one or a few archaeal genomes) has a rather bacterial affinity. In fact, archaeal genomes as a whole always encode more bacterial-like than eucaryal-like proteins. However, most of these are operational proteins (such as proteins involved in metabolism or transport) that show a long history of horizontal gene transfer (HGT) between archaea and bacteria.

Many reviews have summarized insight from newly sequenced archaeal genomes and have usually focused on aspects of comparative genomics (35, 73, 74, 91). This chapter focuses on the description of archaeal genomes and what can be learned from archaeal genomics about the mechanisms of genome evolution, and the history of the *Archaea* domain itself.

GENERAL FEATURES OF ARCHAEAL GENOMES

Compared with *Bacteria* (231 completely sequenced genomes and 413 in progress at the end of August 2005), the number of completely sequenced archaeal genomes is much smaller. Currently, only 24 archaeal genome sequences are available (of which only five are for *Crenarchaeota*) (Table 1) and 25 are in progress (of which eleven are for *Crenarchaeota*) (Table 2).

All archaeal genomes sequenced to date are circular and relatively compact, with sizes ranging from 490 kb for *Nanoarchaeum equitans* (the smallest cellular genome ever sequenced) (113), up to the 5.75 Mb for *Methanosarcina acetivorans* (42). The larger genomes are from mesophilic archaea and contain a high proportion of genes recruited from bacteria by HGT (up to a third in *Methanosarcina mazei* [31]). As a rule, genomes from hyperthermophiles appear to be smaller than genomes from mesophiles. However, this might be biased, since most sequenced genomes of mesophilic archaea belong to species that are quite distantly related from hyperthermophiles. Indeed, the genomes from the closely related *Methanococcus maripaludis* (mesophilic) and *M. jannaschii* (hyperthermophilic) have similar sizes (1.6 Mb and 1.74 Mb, respectively) (17, 48).

The average gene density is about one gene per kilobase, with usually short intergenic regions. Similar to bacteria, gene density slightly decreases and the length of intergenic regions increases with genome size (42). Generally speaking, little work has been done to efficiently mine intergenic regions of archaeal genomes, and more systematic and exhaustive approaches should be very fruitful for retrieving new, important information (e.g., RNA genes, regulatory signals, and so forth). In addition to the main chromosome, several archaeal genome sequences include one or several plasmids of varying sizes (from 3444 bp of *Pyrococcus abyssi* pGT5 up to 410,554 bp of *Haloarcula marismortui*).

Many archaeal genes are grouped in clusters of various sizes, some of them representing bona fide operons (i.e., two or more genes translated from a single transcript and under the regulation of a repressor and an inducer). The clustering of archaeal genes is particularly useful to make functional predictions (55, 73, 81), since genes encoding proteins that interact with each other (especially subunits of multiprotein complexes) or participate in a similar molecular mechanism (functional operons) are often clustered in archaeal genomes. For example, the conserved genomic localization allowed the detection and biochemical characterization of specific archaeal helicases and nucleases

Table 1. Complete and in-progress archaeal genomes in public databases[a]

Organism name	Group	Size	#chr[b]	#pl[c]	GC%	Inteins[d]	Released
Aeropyrum pernix K1	*Crenarchaeota*	1.67	1	0	67	1	06/26/1999
Archaeoglobus fulgidus DSM 4304	*Euryarchaeota*	2.18	1	0	46	0	12/06/1997
Haloarcula marismortui ATCC 43049	*Euryarchaeota*	4.27	1	7	61.1	4	11/03/2004
Halobacterium sp. NRC-1	*Euryarchaeota*	2.57	1	2	65.9	1	10/01/2000
Methanocaldococcus jannaschii DSM 2661	*Euryarchaeota*	1.74	1	3	31.3	19	08/23/1996
Methanococcus maripaludis S2	*Euryarchaeota*	1.66	1	0	33.1	0	10/07/2004
Methanopyrus kandleri AV19	*Euryarchaeota*	1.69	1	0	60	5	04/04/2002
Methanosarcina acetivorans C2A	*Euryarchaeota*	5.75	1	0	42.7	0	04/05/2002
Methanosarcina barkeri strain fusaro	*Euryarchaeota*	4.87	1	1	39.2	0	06/26/2002
Methanosarcina mazei Go1	*Euryarchaeota*	4.1	1	0	41.5	0	07/20/2002
Methanothermobacter thermautotrophicus strain Delta H	*Euryarchaeota*	1.75	1	0	49.5	0	11/26/1997
Nanoarchaeum equitans Kin4-M	*Nanoarchaeota*	0.49	1	0	31.6	1	10/21/2003
Natronomonas pharaonis DSM 2160	*Euryarchaeota*	2.75	1	2	63.1	0	09/28/2005
Picrophilus torridus DSM 9790	*Euryarchaeota*	1.55	1	0	36	1	06/09/2004
Pyrobaculum aerophilum strain IM2	*Crenarchaeota*	2.22	1	0	52	0	01/17/2002
Pyrococcus abyssi GE5	*Euryarchaeota*	1.77	1	1	42	14	06/01/2001
Pyrococcus furiosus DSM 3638	*Euryarchaeota*	1.91	1	0	42	10	08/03/1999
Pyrococcus horikoshii OT3	*Euryarchaeota*	1.74	1	0	42	14	07/29/1998
Sulfolobus acidocaldarius DSM 639	*Crenarchaeota*	2.23	1	0	36.7	0	07/05/2005
Sulfolobus solfataricus P2	*Crenarchaeota*	2.99	1	0	35.8	0	06/28/2001
Sulfolobus tokodaii strain 7	*Crenarchaeota*	2.69	1	0	32.8	0	09/27/2001
Thermococcus kodakaraensis KOD1	*Euryarchaeota*	2.09	1	0	52.0	8	01/13/2005
Thermoplasma acidophilum DSM 1728	*Euryarchaeota*	1.56	1	0	50	1	10/12/2000
Thermoplasma volcanium GSS1	*Euryarchaeota*	1.58	1	0	50	1	12/20/2000

[a]Data adapted from http://www.ncbi.nlm.nih.gov/genomes/lproks.cgi as of 11 October 2005.
[b]Number of chromosomes.
[c]Number of plasmids.
[d]From (http://bioinfo.weizmann.ac.il/~pietro/inteins/Inteins_table.html June 2004) and SWISSPROT release 48.2 of 11 October 2005 (http://www.expasy.org/cgi-bin/lists?intein.txt).

that are coexpressed with the universal DNA recombination proteins, Rad50 and Mre11 (27). A few gene clusters are conserved in most archaeal genomes, such as the superoperon of ribosomal proteins, the RNA polymerase operon, or the Mre11 cluster. However, most clusters tend to be conserved only in a subset of genomes, indicating that their association is constantly challenged by genome disruption (see "Mechanisms of genome evolution," below), and subsequently selected for in different gene arrangements for functional reasons. As a rule, gene clusters, including operons, tend to be less abundant in small genomes; they are very rare in the genome of *N. equitans* (74, 113). This suggests that the genomes have been streamlined by rearrangements leading to gene losses and cluster disruption.

Similar to bacteria, some archaeal rRNA and tRNA genes contain introns, and some protein-coding genes are interrupted by inteins. These mobile elements, especially inteins, appear to be more abundant in archaea than in bacteria (see Chapters 5 and 22). Inteins are especially abundant in *Thermococcales* and in *M. jannaschii*, but they are rare or even absent in other archaea (Table 1) (95). There are no introns in archaeal protein-coding genes, consistent with the absence of genes encoding homologs of eucaryal spliceo-

somes. In addition to rRNA and tRNA genes that are present in all cellular organisms, archaeal genomes contain a plethora of eucaryal-like noncoding RNAs (snoRNAs) that are involved in rRNA processing, and probably many other types of RNA genes that are yet to be analyzed (92) (see Chapter 7).

The analysis of complete archaeal genome sequences was especially important for identifying some major features of the mechanism of translation initiation in archaea (62) (see Chapter 8). Preliminary analyses of a few archaeal genes led to the conclusion that archaea, like bacteria, mainly use Shine-Dalgarno (SD) sequences for the recognition of mRNAs by the ribosome. However, this has not been confirmed by analyses at the whole-genome level (62). For example, mRNA with Shine-Dalgarno sequences are less abundant than "leaderless" mRNAs in *Sulfolobus acidocaldarius* (23), and apparently, completely absent in *Pyrobaculum aerophilum* (33). Archaeal genomes encode homologs of all the eucaryal genes that are involved in translation initiation (a few of them being in all three domains of life), except for the factor that recognizes the cap structure present at the 5′ end of most eucaryal mRNA (a structure that is absent in archaea). Analysis of archaeal genomes also confirmed

Table 2. Archaeal genomes in progress in public databases[a]

Organism name	Group	Size[b]	GC%
Acidianus brierleyi	*Crenarchaeota*	1.8	–
Caldivirga maquilingensis	*Crenarchaeota*	–	43
Cenarchaeum symbiosum	*Crenarchaeota*	–	–
Ferroplasma acidarmanus Fer1	*Euryarchaeota*	1.87	36.8
Halobacterium salinarum	*Euryarchaeota*	4	–
Halobaculum gomorrense	*Euryarchaeota*	–	70
Haloferax volcanii DS2	*Euryarchaeota*	4.03	–
Halorubrum lacusprofundi	*Euryarchaeota*	–	–
Hyperthermus butylicus	*Crenarchaeota*	2	56
Methanococcoides burtonii DSM 6242	*Euryarchaeota*	2.56	40.8
Methanococcus voltae	*Euryarchaeota*	–	31
Methanocorpusculum labreanum	*Euryarchaeota*	–	50
Methanoculleus marisnigri	*Euryarchaeota*	–	61
Methanogenium frigidum Ace-2	*Euryarchaeota*	2.5	–
Methanosarcina thermophila	*Euryarchaeota*	–	42
Methanospirillum hungatei	*Euryarchaeota*	–	–
Methanothermococcus thermolithotrophicus	*Euryarchaeota*	–	34
Natrialba asiatica	*Euryarchaeota*	–	62.3
Pyrobaculum arsenaticum DSM 13514	*Crenarchaeota*	–	58.3
Pyrobaculum calidifontis	*Crenarchaeota*	–	51
Pyrobaculum islandicum	*Crenarchaeota*	–	–
Pyrolobus fumarii	*Crenarchaeota*	–	53
Staphylothermus marinus	*Crenarchaeota*	–	35
Thermofilum pendens	*Crenarchaeota*	–	57.4
Thermoproteus neutrophilus	*Crenarchaeota*	–	56.2

[a]Data adapted from http://www.ncbi.nlm.nih.gov/genomes/lproks.cgi as of 11 October 2005.
[b]Size estimates in Mb.

the eucaryal nature of the archaeal transcription machinery and revealed a complex network of regulatory factors with both bacterial and eucaryal features (3). Whereas promoters are now well defined in archaea, the identification of transcription terminators is still elusive. Analyses of a few transcripts in the 1980s identified the presence of sequences resembling "bacterium-like" terminators at the 3′ end of a few archaeal genes. However, these observations were not confirmed by analyses at the whole-genome level.

CHROMOSOME ORGANIZATION

Replication Origins

The impact of genomics on archaeal research was especially critical for the field of DNA replication. Prior to genome sequencing, nothing was known about chromosomal replication origins in archaea and only a handful of proteins that were possibly involved in DNA replication had been identified. The availability of complete genome sequences allowed the localization of DNA replication origins in several archaeal genomes to be predicted by in silico analyses (see Chapter 3). In bacterial genomes, the leading strand is enriched in guanine (G) residues in compar-

ison with the lagging strand (Fig. 1a) (61). It has been proposed that GC skews may compensate for cytosine deamination on single-stranded DNA at replication forks (38, 103). Short, specific sequences of a few nucleotides (words) are also unevenly distributed between the two strands (9). In *E. coli*, octamers that exhibit a strong bias in their distribution between the two strands are recognition signals for the primase DnaG and for the RecBCD recombination system (Chi sequences) (9). Since the leading strand becomes the lagging strand both at the origin and terminus of DNA replication (*oriC* and *terC*, respectively) (Fig. 1b), several studies showed that it was possible to localize the origin and terminus of DNA replication in bacteria by scanning (from an arbitrary position) any complete genome sequence for either GC or specific word skews (Fig. 1c) (60).

The first direct application of GC skew analysis to the genome of an archaeon, *M. jannaschii*, suggested many origins (86). However, by using a more refined method (cumulative GC or word skews), a single origin was predicted in the genomes of *Methanothermobacter thermautotrophicus* and *Pyrococcus horikoshii* (63). In both cases, graphs of cumulative GC or word skews exhibited two inverted peaks on each side of the base line (with the putative *oriC* located at the bottom of the negative peak in GC

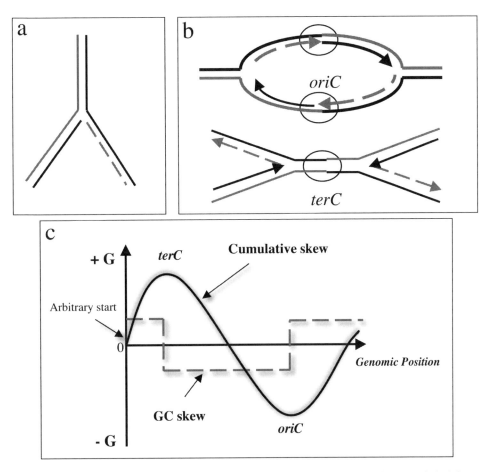

Figure 1. (a) A schematic drawing of a DNA replication fork in *Bacteria* and *Archaea*. The leading strand (thick line) is enriched in guanine residues or specific nucleotide words with respect to the lagging strand (thin line). (b) At the origin and terminus of DNA replication (*oriC* and *terC*, respectively) the leading strand becomes the lagging strand, and vice versa. (c) It is possible to localize the origin and terminus of DNA replication by scanning from an arbitrary position any complete genome sequence for either GC or specific word skews. Cumulative skews exhibit two inverted peaks on each side of the baseline, which correspond to the origin and terminus of replication.

skew analysis) (Fig. 2), indicating that these archaea replicate their genome from a single origin with two replication forks moving in opposite directions, as in bacteria. Bidirectional replication from a single origin was experimentally validated in another *Pyrococcus* species, *P. abyssi* (80, 87). Collectively, these results indicate that, although archaea possess a eucarya-like replication machinery, the selection pressure responsible for GC skews has affected them in a manner similar to bacteria.

Computational studies using cumulative GC skews, or a more exhaustive method (Z curves), allowed identifying additional *oriC* sites in archaea. Z curves (their name is derived from their zigzag shape) are three-dimensional curves that take into account all possible skews in base composition (purine versus pyrimidine [RY], amino versus keto [MK], or weak versus strong hydrogen bond bases [WS]). For

a particular genome, the Z curve produces several different graphs (disparity curves) with peaks that correspond to putative origins or termini of replication (cumulative GC skews can be seen as a special case of Z curves). As a consequence, several origins that are not detected by GC skew analyses can be discovered by other skews revealed by Z curves. This method was especially successful for identifying a previously undetected, putative *oriC* in the genome of *M. jannaschii* (119). Software to draw and manipulate Z curves is available free online at http://tubic.tju.edu.cn/zcurve/.

Putative *oriC* sites have been detected in 15 of 19 completely sequenced archaeal genomes tested (117) (Table 3). Computational methods have failed to predict any *oriC* sites in the genomes of *Methanopyrus kandleri*, *N. equitans*, *Archaeoglobus fulgidus*, and *Sulfolobus tokodaii* (117). This is probably due to recent rearrangements that have perturbed the

Pyrococcus horikoshii

Methanobacterium
thermoautotrophicum

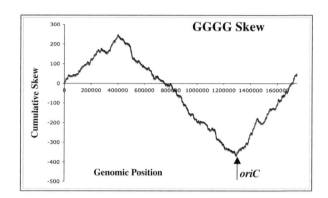

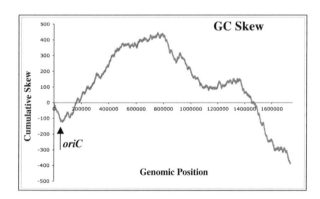

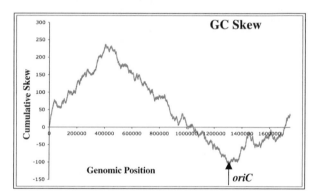

Figure 2. Graphs of cumulative skew and GC skew for *Methanobacterium thermautotrophicum* and *Pyrococcus horikoshii*. Putative *oriC* in both species are indicated by arrows.

skew in these particular genomes. A putative *oriC* was predicted in *A. fulgidus* by marker frequency analysis (70). This method is based on the fact that the copy number of a particular gene in an exponentially growing cell population is directly related to its position relative to the replication origin (Fig. 3). Surprisingly, two origins of replication were predicted in the genome of *Halobacterium* NRC1 (54), and three origins were recently detected in the genome of *Sulfolobus solfataricus* and *S. acidocaldarius* by marker frequency analysis on DNA chips (67). The three origins of *S. solfataricus* have been validated by two-dimensional gel electrophoresis (99).

The identification and experimental validation of archaeal *oriC* sites allowed analysis of their sequences and genomic contexts. Archaeal *oriC* sites are characterized by the presence of 13mer repeats, AT-rich

regions, and long inverted repeats (63, 87, 99, 119). Most archaeal *oriC* sites are located in intergenic regions just upstream of the genes encoding the homologs of eucaryal initiation factors Cdc6 and Orc1 (see Chapter 3). These proteins (hereafter referred to as Orc1/Cdc6) bind to archaeal *oriC*, both in vivo and in vitro (80, 99). Several archaeal genomes (and haloarchaeal plasmids) encode more than one Orc1/Cdc6 protein (up to 9 *in Halobacterium* NRC1), and the presence of an *orc1/cdc6* gene is not necessarily correlated with the presence of an origin of replication (and vice versa). For example, although *Sulfolobus* genomes contain three *orc1/cdc6* genes, only two of the three *oriC* are located close to an *orc1/cdc6* gene (66). Surprisingly, in *Halobacterium* NRC1, two *orc1/cdc6* genes are located at the peaks that normally correspond to *terC* positions in cumulative GC skew

Table 3. Complete archael genome replication origin identification

Organism name	Group	Status of replication origin identification
Aeropyrum pernix K1	*Crenarchaeota*	Unknown
Archaeoglobus fulgidus DSM 4304	*Euryarchaeota*	Approximate location known based on marker frequency (70)
Halobacterium sp. NRC-1	*Euryarchaeota*	Two replication origins predicted (54), one identified in vivo (8)
Methanocaldococcus jannaschii DSM 2661	*Euryarchaeota*	One replication origin identified with Z-curve method (119)
Methanococcus maripaludis S2	*Euryarchaeota*	Unknown
Methanopyrus kandleri AV19	*Euryarchaeota*	Uncertain
Methanosarcina acetivorans C2A	*Euryarchaeota*	One replication origin identified with GC skew analysis (D. Cortez, personal communication)
Methanosarcina mazei Go1	*Euryarchaeota*	One replication origin identified with Z-curve method (119)
Methanothermobacter thermautotrophicus strain Delta H	*Euryarchaeota*	One replication origin identified with word skew analysis (63)
Nanoarchaeum equitans Kin4-M	*Nanoarchaeota*	Unknown
Picrophilus torridus DSM 9790	*Euryarchaeota*	Unknown
Pyrobaculum aerophilum strain IM2	*Crenarchaeota*	Unknown
Pyrococcus abyssi GE5	*Euryarchaeota*	One replication origin identified with word skew analysis (63), and confirmed in vivo by 2D gel analysis (87)
Pyrococcus furiosus DSM 3638	*Euryarchaeota*	One replication origin identified with word skew analysis (63) and comparison with close homolog *P. abyssi*
Pyrococcus horikoshii OT3	*Euryarchaeota*	One replication origin identified with word skew analysis (63) and comparison with close homolog *P. abyssi*
Sulfolobus solfataricus P2	*Crenarchaeota*	Two replication origins identified in vivo (99), third replication origin identified by markers frequency analysis (66)
Sulfolobus tokodaii strain 7	*Crenarchaeota*	Unknown
Thermoplasma acidophilum DSM 1728	*Euryarchaeota*	One replication origin identified with GC skew analysis (D. Cortez, personal communication)
Thermoplasma volcanium GSS1	*Euryarchaeota*	Unknown

analyses (54). One of these peaks is a bona fide origin that has now been experimentally validated in vivo (8). The GC skew is therefore inverted in *Halobacterium salinarium* compared with other organisms (i.e., the leading strand being enriched in C instead of G). The same GC skew was observed in the bacterium *Streptomyces coelicolor* (6). It is unclear if these inversions are related to the very high GC content of both genomes (71% in *S. coelicolor* and 66% in *Halobacterium* NRC1) or to both genomes having experienced frequent rearrangements triggered by deletions and insertion of large plasmids. In a few archaeal genomes, the genes around *oriC* include genes encoding putative DNA replication proteins other than *orc1/cdc6*. The most striking example is a replication island in *Pyrococcus* that contains, in addition to *orc1/cdc6* genes, genes encoding the two subunits of the DNA polymerase D family, and the two subunits of the clamp-loading factor, RFC (87). Other putative ori-

gins have been localized close to genes encoding DNA primase or DNA polymerases subunits (63, 118).

Replication Termini and Replichore Organization

In archaea with a single replication origin, the chromosome is divided in two replichores (left and right) of equal sizes by *oriC* and *terC*, indicating that the two replication forks move at the same rate. The four replichores of the *H. salinarium* NRC1 genome also have the same size (54), but this is not the case for the six replichores of *Sulfolobus* species (66). The three *oriC* of *Sulfolobus* are initiated simultaneously in a population of synchronized cells and move at the same rate. Therefore, termination is not simultaneous, and short replichores are completely replicated before larger ones. This feature partially resembles the eucaryal mode of replication, where multiple replicons are replicated at different times of the cell cycle.

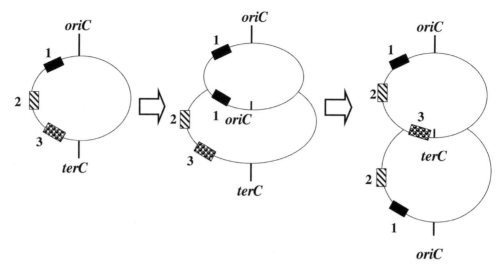

Figure 3. Marker frequency analysis. In a population of nonsynchronized cells, the copy number of a particular gene is directly proportional to its closeness to the origin of replication (*oriC*). Boxes represent genes. Gene 1 is the closest to *oriC* and the most represented.

The regions corresponding to archaeal *terC* have not yet been characterized at the sequence level. They seem to be less well defined than *oriC,* since the peaks of cumulated GC skew or Z-curve graphs corresponding to *terC* are usually broader than those corresponding to *oriC* (Fig. 2). Proteins known to be involved in the termination of DNA replication in *E. coli* and *Bacillus subtilis* (Tus and RTP, respectively) are not homologous to each other and have homologs only in closely related species. The mechanism of DNA replication termination is completely unknown in eucarya, and it remains a mystery in archaea (see Chapter 2).

All archaeal genomes encode a single protein that is homologous to bacterial paralogs XerC and XerD. In bacteria, these proteins are involved in the resolution of dimeric chromosomes that are produced by recombination during bacterial DNA replication (60). XerCD proteins recognize specific sites, referred to as *dif* (deletion induces filamentation) sites, which are located near *terC*. Sequences with similarity to *dif* sites, or other putative Xer-binding sites, have not been identified in archaeal genomes. It would be especially interesting to purify archaeal Xer proteins in vitro to search for their specific binding sequences.

The chromosome terminus appears to be a hot spot of recombination in archaea, as it is in bacteria. This was clearly shown from a genome comparison of the two closely related species, *P. abyssi* and *P. horikoshii.* The terminus region has in fact experienced several rearrangements after the divergence of these two species, including the translocation of two regions of synteny (conservation of gene order) (Fig. 4A, fragments E and F) (120).

Organization of Transcription Units at the Whole-Genome Level

Similar to bacteria, highly expressed genes (in particular, those encoding rRNA and ribosomal proteins) in archaea are frequently located close to *oriC* to increase their copy number in actively replicating cells (see references 100 and 101 and references therein). There is also a general tendency to have more genes orientated in the direction of DNA replication than in the opposite direction (see references 100, 101 and references therein). This phenomenon induces a skew of transcription (more genes are transcribed on the leading strand). Cumulative transcription skew analysis in bacteria and archaea often produces graphs that are quite similar to those obtained with GC skews, since this skew also changes orientation at *oriC* and *terC* (63, 120). The strength of the colinearity between transcription and replication is very variable in different bacteria (82), and this is probably also the case for archaea. This trend is even reversed in *P. furiosus,* where slightly more genes are transcribed in the opposite direction to DNA replication (120). This could be due to recent, extensive rearrangements in this archaeon (see "Mechanisms of genome evolution," below). However, a clear colinearity between replication and transcription was observed in the three *Pyrococcus* species when the analysis was restricted to highly expressed genes (120).

The correlation between colinearity in transcription and replication and a high level of gene expression has also been observed in many bacterial genomes (102). Since a replication fork has more chance to en-

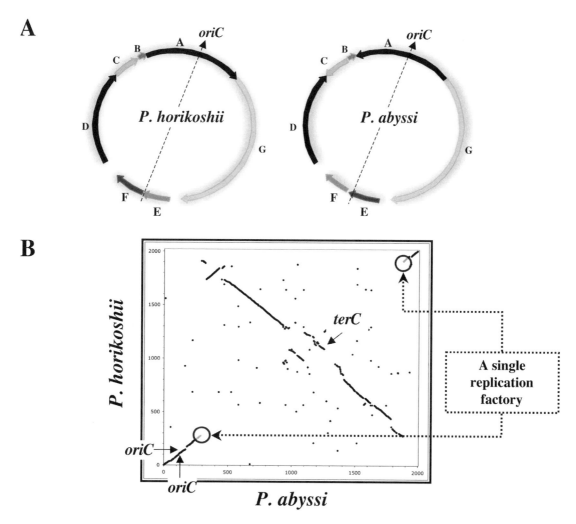

Figure 4. (A) A schematic view of the genome organization of the two closely related species *Pyrococcus abyssi* and *Pyrococcus horikoshii*. Fragments E and F in the terminus region show a translocation between the two species, while fragment A, containing *oriC*, and fragment C show an inversion. (B) Comparative BLAST-hit plot of the two genomes. (Adapted from reference 120 with permission.)

counter an RNA polymerase transcribing a highly expressed gene in its trajectory along the chromosome, it was thought that the main purpose of this organization was to avoid head-to-head collisions between replication forks and transcription forks to prevent the loss of transcripts, the reduction of replication rate, and/or the collapse of replication forks (10). However, from the systematic gene disruption analysis of protein function in *B. subtilis*, it was recently shown that the strongest trend was for transcription of essential genes in the direction of DNA replication (102). Therefore, the purpose of colinearity in bacteria (as represented by *B. subtilis*) is probably to prevent the abortive transcription of essential genes to avoid the accumulation of truncated mRNA or proteins that could be deleterious to the cell. As essential genes are often highly transcribed (but not always), the correlation between the level of transcription and transcription/replication colinearity could be indirect. It will be very interesting to determine if the same rule exists in archaea. This will require the systematic disruption of all genes in at least one archaeon.

Repeated Sequences

Archaeal genomes contain various types of repeated sequences. Most archaeal genomes contain insertion sequences (IS) that may represent active mobile elements or relics of past invasion by transposable elements (see Chapter 5). Active IS are especially abundant in *S. solfataricus* and *Halobacterium* NRC1 (201 and 48 copies, respectively) (15), where they are responsible for the well-known genetic instability of these strains. In the case of *S. solfataricus*, they constitute more than 10% of the genome sequence (23). These IS elements belong to various families (some of them including bacterial homologs), indicating that they may play a major role in promoting HGT between the two domains via illegitimate recombination between the archaeal chromosome and incoming bacterial DNA fragments (or vice versa).

A second class of mobile elements that are only present in a few archaeal genomes (mainly *Sulfolobus*), are MITEs (*Miniature Inverted Repeat Transposable Elements*) (15) (see Chapter 5). MITEs lack open reading frames (ORFs), indicating that they depend on host-encoded transposases for their mobility. These small elements (~50 to 400 bp) are common in eucaryal and bacterial genomes and could be derived from ISs as a result of deletion within an IS element or acquisition of terminal inverted repeats. The high abundance of MITEs in *S. solfataricus* and *S. toko-dai* genomes (143 and 61, respectively) correlates with the large number of IS elements in these species (201 and 34, respectively) (15).

A very different group of nonautonomous repeated sequences found in most archaeal genomes (both *Euryarchaeota* and *Crenarchaeota*) are clusters of short repeated sequences (20 to 40 bp) separated by spacers of similar size and variable noncoding sequences (Fig. 5). These clusters, that are also present in several bacterial lineages, have been described as SRSRs (*Short Regularly Spaced Repeats*) (84), LCTRs (*Large Cluster of Tandem Repeats*) (120), or CRISPRs (*Clustered Regularly Interspaced Palindromic Short Repeats*) (52). The term CRISPR has been adopted for this chapter. These repeats are conserved between closely related species but are quite variable between different archaeal and bacterial lineages. The number of CRISPRs and their location in the genome can vary. For instance, 4, 6, and 7 CRISPRs are present at different locations in the genomes of *P. abyssi*, *P. hori-koshii*, and *P. furiosus*, respectively (Fig. 4) (120).

CRISPRs are transcribed, suggesting a regulatory role via antisense RNA, and a CRISPR-binding protein has been purified from *S. solfataricus* (94). Clues to the origin of CRISPRs came from the finding that some spacer sequences are homologous to sequences from plasmids, viruses, or integrated prophages (97). Furthermore, it has been observed that archaeal viruses or conjugative plasmids harboring CRISPRs, apparently cannot infect cells containing CRISPRs with homologous spacer sequences (83). This suggests that CRISPRs in cellular genomes are of viral (plasmid) origin and are now part of an immunity system that prevents invasion by further foreign elements. This hypothesis is supported by experimental evidence in *Haloarchaea* that exhibit incompatibility between replicons that contain similar CRISPRs (83). Genes that are present in the vicinity of CRISPRs (referred to as *cas* genes, for CRISPR *a*ssociated genes) (52) belong to a cluster of genes predicted to encode a new and "mysterious" DNA repair system (72, 97). These proteins, which include a putative helicase and a RecB homolog, are probably involved in the mobility of their associated CRISPRs. It has also been suggested that CRISPRs are important for chromosome partitioning (85). This hypothesis is supported by the presence of CRISPRs located symmetrically with respect to *oriC* in *P. abyssi* and *P. horikoshii*, or to *terC* in the three *Pyrococcus* species (Fig. 4) and *S. acidocaldarius* (23, 120). The evolutionary origin of CRISPR is mysterious since they are present in completely unrelated viruses, such as bacteriophages and viruses of hyperthermophilic archaea (97).

MECHANISMS OF GENOME EVOLUTION

Replication-Driven Genome Rearrangements

The sequencing of several genomes from closely related species is a very powerful approach to analyze the mechanism of genome evolution. For example, it was very informative to compare the three *Pyrococcus* genomes that were sequenced at the end of the 1990s. The genomes of *P. abyssi* and *P. horikoshii* are sufficiently closely related to allow the identification of long regions of synteny, thereby enabling a precise analysis of the genomic rearrangements that occurred after the separation of these two species. The major rearrangement is the inversion of a large fragment containing *oriC* (Fig. 4A, fragment A). As *oriC* is located roughly at the middle of the inverted fragment, this inversion produces an X-shaped figure in a comparative BLAST-hit plot of the two genomes (Fig. 4B). X-shaped figures often arise from the comparative analyses of closely related bacterial genomes (111). This indicates that inversions around *oriC* are more frequent than others that may occur and that they frequently occur between closely related bacteria. These types of inversions are probably favored since they maintain the same direction of transcription and

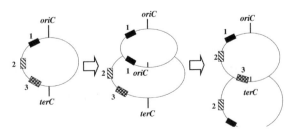

Figure 5. Schematic structure of a typical CRISPR.

replication (i.e., colinear) for all the genes present on these fragments. This suggests that the inversions take place between the two replication forks, moving at the same speed in opposite directions from the origin (111). If this model is correct, the observation of X-shaped figures in archaea suggests that a single replication "factory" exists that couples two replication forks (similar to bacteria) (120). It has been proposed that type II DNA topoisomerases that are associated with replication forks may be involved in this process (76). Alternatively, recombination repair of a broken replication fork may use the daughter strand of the second fork that is present in the replication "factory" as template, instead of the daughter strand of the same fork. Irrespective of the mechanism, the X-shaped figures derived from closely related genomes, and the presence of extensive rearrangement at *terC* (see "Chromosome organization," above), indicate that replication is a major force driving genome rearrangement (111) that is common to both archaea and bacteria.

Remarkably, comparative analysis of the cumulative skews of transcription between *P. horikoshii* and *P. abysii* allowed attributing specific rearrangements to a particular species (120). It was in fact inferred that the inversion of a small fragment inside the left replichore (Fig. 4A, fragment C), and a large translocation at the terminus (Fig. 4A, fragments E and D), both oc-

curred in the *P. abyssi* lineage (120). This conclusion is based on the observation that the cumulative transcription skew lines are disturbed at these two locations in the *P. abyssi* graph, whereas they are smooth in the graph for *P. horikoshii* (Fig. 6). This means that, in *P. abyssi*, the genes present on these two fragments are transcribed in a direction opposite to that of transcription, while in *P. horikoshii* they kept the same direction that they had in the genome of the ancestor of these two *Thermococcales*. Note that the nucleotide skews for these two fragments in *P. abyssi* are smooth, in contrast to the marked alteration of the transcriptional skew line (Fig. 6). This indicates that compositional skews were fully restored in *P. abyssi* (at the time of genome sequencing). This rapid restoration is consistent with the general observation that strand bias in nucleotide composition is probably neutral and evolves fast (103). The rapid restoration of base composition skews implies a huge modification in the genome sequence at the nucleotide level (with possible accumulation of neutral mutations at the protein level). Genome shuffling thus appears to be a critical force in modifying DNA sequences, via selection pressure acting for the restoration of nucleotide skews.

The main sequence of events that may occur in archaeal and bacterial genome evolution would thus be as follows: (i) replication-driven rearrangements inducing skew inversions, (ii) restoration of normal

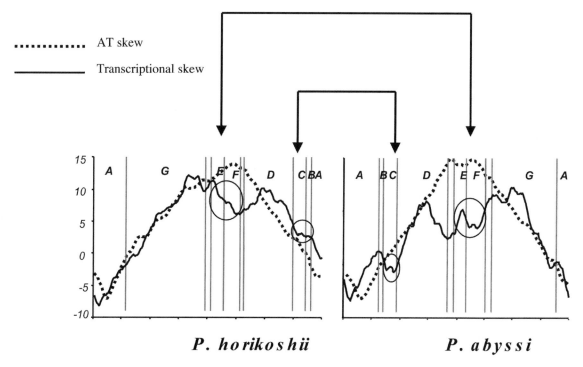

Figure 6. Cumulative transcription skew lines are smooth at fragments E, F, and C in *P. horikoshii*, while they are disturbed in *P. abyssi*. (Adapted from reference 120 with permission.)

skews inducing sequence evolution, (iii) selection of sequence changes compatible with optimal function for an organism at a given time. It is important to test the validity of these assumptions by comparing other closely related archaeal genomes and to check the prediction that genome rearrangements may increase the rate of protein evolution.

Recombination Induced by Transposition

Besides rearrangements driven by DNA replication, a major force shaping genome evolution in archaeal and bacterial genomes is recombination induced by mobile elements (see Chapter 5). The presence of repeated sequences can also promote homologous recombination inside the same chromosome, or between different chromosomes in species containing multiple chromosomal copies. Comparing the three *Pyrococcus* species, and extending this to the recently published sequence of *Thermococcus kodakaraensis*, has been very informative for unraveling the mechanisms of genome evolution. The extensive studies of *Sulfolobales* genomes and their extrachromosomal elements have also been extremely enlightening for this purpose (16, 23, 107). Many relics of tRNA genes and insertion sequences have been detected at the border of recombined fragments between different *Pyrococcus* and *Sulfolobus* species, indicating that the majority of small rearrangements were mediated by mobile elements that integrated in tRNA genes (59, 69, 120). In the case of *Pyrococcus*, many rearrangements are also associated with genes encoding enzymes involved in restriction/modification (RM) systems, suggesting that RM genes themselves behave as mobile elements and cause genome rearrangements (24). Genome shuffling by transposon-driven chromosomal rearrangements can induce a strong perturbation in the usual patterns of genome organization. The genomes of *S. solfataricus* and *S. tokodai* contain many ISs and are both extensively shuffled. Similarly, chromosomal rearrangements have been more important in the evolution of the genome of *P. furiosus* than in the two other *Pyrococcus* species (120), and this could be related to the higher number of transposases in *P. furiosus*, than in *P. abyssi* and *P. horikoshii*. Recent invasion by transposases could explain the inverted transcriptional skew in *P. furiosus* (120). The major rearrangements that occurred between *P. furiosus* and the two other pyrococci (four inversions and four transpositions), occurred within one replichore (120). This could reflect the existence of a physical or topological barrier that could prevent recombination by transposition between DNA fragments present on different replichores (120). However, this could also simply reflect the fact that these transposi-

tions preferentially occur at relatively short distances on the same chromosome, for structural or mechanistic reasons.

Gene Loss and Acquisition

Comparative genomics has shown that gene loss, gene duplication, and integration of foreign DNA are major forces shaping genome evolution. Although gene loss is difficult to quantify, it may be a frequent and ongoing process in archaeal genomes. Indeed, it was recently shown that as much as 15% of archaeal ribosomal proteins have been lost in different archaeal lineages (59). Gene loss and nonorthologous replacement seem to be frequent in the components of archaeal metabolic pathways (71). Several protein family expansions identified in complete archaeal genomes have been linked to different physiological requirements (71). For example, nearly half of all annotated genes in the metabolically versatile methanogen, *M. acetivorans*, belong to one of 539 multigene families (42).

Integration of foreign elements (plasmids, viruses) containing ISs, or activation of resident transposons, can directly induce gene loss, acquisition, or inversion. The continuous evolution of archaeal genomes by gene loss and acquisition is obvious from the comparative analyses of the proteomes of closely related species (23, 25, 39). There are always a number of proteins that are absent in close relatives. This clearly indicates either recent acquisition by HGT or recent loss in these relatives. The genes that appear to have recently integrated in archaeal genomes sometimes encode for proteins with homologs in more distantly related archaeal or bacterial species. In the genome of *P. furiosus*, a 16-kb fragment is flanked by two IS elements (32) (see Chapters 16 and 20). This region is absent from the two other *Pyrococcus* species, but a very similar fragment is present in *Thermococcus littoralis*, indicating a recent transfer between *Thermococcales* (32). The identification of bacterial-like genes in *P. abyssi*, which are absent in the two other *Pyrococcus* species, suggests a very recent transfer from bacteria (26).

Despite these examples, most species-specific proteins have no homologs in sequence databases, and many are likely to be of plasmid or viral origin. This is consistent with the extraordinary abundance of these extrachromosomal elements in nature, and with the general observation that most proteins encoded by viral or plasmid DNA (both in archaea and bacteria) have no homologs in databases. This trend is extreme for some archaeal viruses. For instance, the 40-kb genome of PSV (*Pyrobaculum* Spherical Virus) contains 49 putative orphan ORFs (none of them with a predicted function) (47). There is likely to be a huge reservoir of viral and plasmid genes in the biosphere that are continuously providing new genes to archaeal and bacterial genomes. The time of residence of these orphan genes in cellular genomes may be very brief, and they may be rapidly lost. They would rarely provide a substantial selective advantage to the host and then become fixed in the cellular genome. The same fate is expected for genes of cellular origin that are carried by extrachromosomal elements and transferred to a new genome by transduction, conjugation, or transformation.

HGT has occurred between different archaeal lineages, between *Archaea* and *Bacteria*, and also possibly from *Archaea* to *Eucarya*; the latter is suggested by the presence of a homolog of archaeal DNA topoisomerase VI that is otherwise exclusively found in plants (41), as well as homologs of archaeal alanyl- and prolyl-tRNA synthetases in two protist lineages (2). The percentage of genes acquired by HGT greatly varies from one archaeon to the other. *Methanosarcinales*, *Haloarchaea*, and marine *Crenarchaeota* appear to contain a large proportion of genes of bacterial origin that may have been used to expand their metabolic repertoire and for adaptation to mesophily (31, 54, 64). The important role that HGT has played in the adaptation of archaea to extreme environments is discussed below (see "Genomics and adaptation to extremophily").

Fast Genome Evolution

In some cases, specific mechanisms (and/or selection pressure) can greatly accelerate the evolution of a particular genome. The most spectacular case of rapid genome evolution in the *Archaea* is that of *N. equitans*. The genome of *N. equitans* is highly reduced, probably following adaptation to its symbiotic/parasitic lifestyle. It has lost all enzymes involved in metabolic pathways, including those required for lipid biosynthesis. *N. equitans* thus relies completely on the metabolism of its host *Ignicoccus* (113). Most spectacularly, 100 of the 301 genes otherwise present in all archaeal genomes are missing in *N. equitans* (74). This reduction was coupled to a complete loss of operon structure and the split of several genes including those coding for tRNAs. The presence of several tRNA genes split in two at the position of an intron insertion in the anticodon loop, and of an intein in one of the split protein genes, suggests that these types of insertion elements may have made important contributions to the evolution of the *N. equitans* genome. However, the *N. equitans* genome appears today stable since no pseudogenes are found (113).

Another case of rapid genome evolution in the *Archaea* is for *M. kandleri*. Although *M. kandleri* is a free-living organism (unlike *N. equitans*), its genome is highly unusual. In addition to the absence of a normal genomic skew pattern (Table 3 and "Chromosome organization," above), it contains many orphan genes, as well as many split and fused genes (109). Phylogenetic analyses have shown that *M. kandleri* RNA polymerase subunits have evolved more rapidly than in other archaea. This was inferred from the long branches displayed by *M. kandleri* in RNA polymerase phylogenies, and from the presence of a higher number of indels (insertions/deletions) in their sequences than any other archaeal species that was analyzed (13). It was suggested that the rapid evolution of the *M. kandleri* genome may be related to the loss in this archaeon of transcription elongation factor, TFS (otherwise present in all archaea) (13). This factor is evolutionarily related to the eucaryal factor, TFIIS, which is involved in the fidelity of transcription and in the bypass of DNA lesions. TFIIS and TFS stimulate an intrinsic RNase activity of RNA polymerase that cleaves the 3′ end of nascent messenger RNAs that are extruded at stalled forks, thus resuming transcription. The absence of TFS may influence the rate of genome evolution by an as-yet-unknown mechanism. For example, it may increase homologous recombination to promote transcription by stalled transcription machinery, and/or somehow affect DNA repair. Such an unsuspected link between transcription and evolutionary rate, inferred from purely in silico analyses, highlights the need for experimental investigation.

GENOMICS AND ADAPTATION TO EXTREMOPHILY

Insight from Protein Content and Comparative Genomics

Most archaea that have been studied experimentally live in environments that are considered to be extreme (such as high-salt concentration, low pH, high or low temperature) (see Chapter 2). For many years, the molecular mechanisms of adaptation to these conditions could only be studied at the biochemical level. The recent availability of archaeal genome sequences has greatly expanded the capacity to look for hallmarks of extremophily.

Members of the *Archaea* are the only organisms that can thrive in hydrothermal habitats where temperatures range from 95 to 113°C. In general, although hyperthermophilic bacteria exist, archaea are predominant in all biotopes with temperatures above 80°C. The connection between archaea and hyperthermophily is thus pervasive, and it possibly dates back to a hyperthermophilic archaeal ancestor. Much attention has therefore been given to possible clues of adaptation to hyperthermophily in archaeal genomes. Surprisingly, comparative genomics analysis has shown that all hyperthermophiles (archaea and bacteria) share a single protein that is absent in all non-hyperthermophiles: reverse gyrase (34). This unique hyperthermophile-specific protein is formed by the fusion of a classical type I DNA topoisomerase and a large helicase-like domain. Reverse gyrase is the only DNA topoisomerase that can produce positive supercoiling into circular DNA in vitro. Its specific phylogenetic pattern suggests a crucial role in thermoadaptation. Consistent with this hypothesis, a mutant of *Thermococcus kodakaraensis* lacking reverse gyrase grows more slowly than the wild type at 80°C and cannot grow above 90°C (4).

It has been purported that reverse gyrase controls intracellular DNA topology in hyperthermophiles, and that positive supercoiling is essential for life at high temperature. This hypothesis was supported by experiments showing that viral or plasmid DNA isolated from hyperthermophiles were either relaxed or positively supercoiled, compared with the negatively supercoiled plasmids isolated from mesophilic bacteria or archaea (22). This unusual topology was thought to reflect the topology of the chromosome itself. Consistent with this idea, the genome sequences of archaeal hyperthermophiles exhibit a sequence periodicity of ~10 bp (much lower that the 11 bp typically observed in bacterial genome sequences), suggesting a shorter helical path induced by positive supercoiling (49). However, further experimental work showed that *A. fulgidus*, which displays a sequence periodicity of 10, contains DNA gyrase, in addition to reverse gyrase, and harbors a negatively supercoiled plasmid (65). It was also found that *M. kandleri* (an archaeon with only reverse gyrase) has a sequence periodicity of 10.9 (105). The sequence periodicity of a given genome is therefore unlikely to correlate with the level of DNA supercoiling. On the other hand, the presence of negatively supercoiled plasmids in hyperthermophiles harboring DNA gyrase, such as *A. fulgidus* and *Thermotoga maritima* (45), indicates that hyperthermophiles can live with this type of DNA topology and that gyrase dominates over reverse gyrase in determining intracellular DNA topology. In summary, it seems that reverse gyrase does not play a major role in determining intracellular DNA topology.

The assumption that reverse gyrase helps to stabilize DNA at high temperature is also probably wrong, since intracellular DNA is highly resistant to denaturation (77). Recent data suggest that reverse gyrase may be involved in some form of DNA protection against thermodegradation, possibly coupled to DNA repair (53, 88). The relationships between such activity, and the positive supercoiling activity observed in vitro, is presently unclear.

By relaxing the protein identification criteria, it is possible to identify proteins that are not unique to hyperthermophiles but do exhibit a phylogenomic profile with preference for hyperthermophiles (72, 75). In particular, a cluster of genes that is overrepresented in hyperthermophilic archaea and bacteria and is predicted to form a novel DNA repair system important for growth at very high temperature (72). However, as described above (Repeated sequences), this cluster includes the Cas proteins that appear to be linked to the propagation of CRISPR sequences. Since CRISPRs are also overrepresented in hyperthermophilic genomes, it is not clear whether Cas proteins are really involved in high-temperature adaptation.

Comparative genomics has shown that, for adaptation to extreme environments, microorganisms have benefited to a great extent from HGT. This is exemplified by the high number of archaea-like proteins that are present in hyperthermophilic bacteria, including reverse gyrase (37, 89). If the common ancestor of all archaea was a hyperthermophile (as suggested by phylogenetic analyses of reverse gyrase, ribosomal proteins, and RNA polymerase subunits [12, 37]), low-temperature environments can be considered extreme for archaea! Indeed, archaea have probably recruited many proteins from mesophilic bacteria during their adaptation to mesophily. This is strongly suggested by the existence of a nearly unidirectional flux of HGT from mesophilic bacteria to mesophilic archaea (115). Proteins of bacterial origin are especially abundant in mesophilic methanogens and in members of the *Halobacteriales*.

HGT appears to have been especially prevalent between thermoacidophiles. For example, the euryarchaeon *Picrophilus oshimae* (*Thermoplasmatales*) shares 35% of its genes with pyrococci (another euryarchaeon), but 58% with the crenarchaeon, *S. solfataricus*. Even more strikingly, 13% of the proteins shared by *P. oshimae* and *S. solfataricus* are absent in *Thermoplasma acidophilum*, indicating that genes have recently been transferred from *Sulfolobus* to *Picrophilus* (reviewed in reference 25).

A core of 690 genes common to all thermoacidophiles (*Thermoplasmatales* and *Sulfolobales*) have been identified from a comparative genomic analysis (40). Many proteins with specific affinity for thermoacidophiles are transporters that could exploit the large transmembrane pH gradient present in these microorganisms (see Chapter 16). Transporters may also function at low pH as uncouplers of the respiratory chain and may therefore be involved in the removal of organic acids that are harmful to acidophiles (25, 40). In particular, *P. oshimae* has an intracellular pH close to 4.6 and contains many enzymes for the degradation and transformation of these compounds. Proteins apparently linked to the thermoacidophile phenotype tend to be more similar to enzymes from *Crenarchaeota* or bacteria than to proteins of other *Euryarchaeota*, again highlighting an important role for HGT in adaptation to new environments.

Haloarchaea exhibit the extraordinary property of accumulating very high concentrations of intracellular K$^+$ (up to 3 to 4 M). Consistent with this, the genome of *Halobacterium* NRC1 has a high number of active K$^+$ transporters and potential pumps involved in Na$^+$ efflux (90) (see Chapter 16). Three genomes of halophilic archaea are now available, and a comparative analysis of genomes from halophilic microorganisms should be available soon. It will be particularly interesting to obtain the genome sequence of an extremely halophilic bacterium such as *Salinibacter ruber* to compare the mechanisms of high-salt adaptation between archaea and bacteria.

Insights from Amino Acid Composition

Analysis of the whole proteome of *Halobacterium* NRC1 confirmed that proteins of halophilic archaea are extremely acidic (basic proteins are essentially absent), with acidic amino acids concentrated on the protein surface (54). In contrast, analysis of the predicted proteome of the thermoacidophile, *Picrophilus torridus*, failed to detect any significant bias in amino acid composition, even though its proteins are exposed to an intracellular pH of about 4.6 (40).

With respect to thermal adaptation, the proteomes of hyperthermophiles appear to be especially enriched in glutamate, lysine, valine, isoleucine, and tyrosine, and depleted in asparagine, glutamine, histidine, alanine, and threonine (46, 108). The decrease in asparagine, glutamine, and histidine is probably correlated to the sensitivity of these amino acids (especially Asn) to thermodegradation (112). A comparison between the complete, predicted proteomes of 53 mesophiles, 9 thermophiles, and 9 hyperthermophiles revealed a clear trend in the polar index of amino acids, increasing charged (aspartate, glutamate, lysine, arginine) versus polar (noncharged) amino acids, (asparagine, glutamine, serine, threonine) (19, 110) (Fig. 7). The simultaneous increase of positively and negatively charged amino acids fits a priori well with previous biochemical data obtained from the comparative structural analysis of hyperthermophilic proteins with their mesophilic homologs. These biochemical studies have indicated that a major determinant in the stabilization of proteins at very high temperature is the formation of networks of ion pairs at their surface (98). However, this interpretation based on the polar

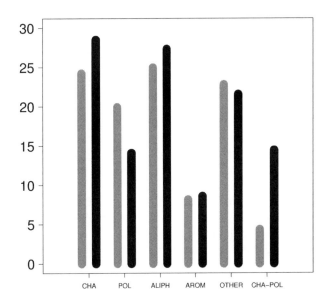

Figure 7. The percentage of various classes of amino acids in gene coding sequences from hyperthermophiles and mesophiles. Percentages were calculated for 9 fully sequenced hyperthermophilic and 53 mesophilic bacteria and archaea. Grey and black vertical bars indicate the mesophiles and the hyperthermophiles, respectively. The *y* axis indicates percentages. CHA, charged; POL, polar; ALIPH, aliphatic; AROM, aromatic.

index of amino acids was partially challenged by a recent study using unfolding simulation experiments that showed that hyperthermophilic proteins are structurally adapted to high temperature through compactness and higher residue hydrophobicity (7). In this study, the number of charged residues in hyperthermophiles was reported to be much greater than it would need to be for the stabilization by ion pairs (7). The high polar index observed in the genome of hyperthermophilic archaea could thus have an alternative purpose (see "Nucleotide composition and codon index," below). The trends between temperature and amino acid polarity were confirmed in a comparative analysis of genomes from psychrophilic archaea, which showed that proteins from cold-adapted archaea are characterized by a high content of noncharged polar amino acids (104).

Nucleotide Composition and Codon Index

Genomics data have confirmed preliminary observations that G+C content of tRNA and rRNA increases, as expected, with higher optimal growth temperature (OGT), whereas the genomic G+C content is not correlated to OGT (43). In a recent review, Hickey and Singer stated that "the obvious question that comes to mind is why we observed the expected correlation between nucleotide content and growth temperature in the paired region of RNA molecules (rRNA, tRNA) but not in double-stranded DNA?" (50). The answer is that, differently from tRNA and rRNA, which are topologically open structures (77–79), intracellular DNA can be assumed to be a topologically closed DNA, because the two DNA strands cannot freely rotate around each other in the cell (for instance, because of the inertia of large transcription complexes bound by ribosomes, especially those translating membrane proteins anchored to the cytoplasmic membrane). As a consequence, intracellular DNA is intrinsically resistant to thermodenaturation and does not need to be stabilized by high GC. In particular, it was shown that a circular double-stranded plasmid (topologically closed DNA) remains double stranded up to at least 107°C, whereas the same plasmid, once linearized, denatures at 78°C (77).

A high genomic GC content could also stabilize codon-anticodon interaction. Two important papers have clarified the role of GC content in codon usage by looking at the possible preference for synonymous codons in relation to OGT. It was recently shown that hyperthermophiles exhibit a preference for C and G ending codons, but only when the first two positions are AT base pairs (5). In particular, a strong preference was noticed for AAG among arginine codons. In contrast, C or G is avoided at third positions when the second base pair is G or C. This can be explained by a general rule preventing avoidance of adjacent GC base pairs in anticodon because they are probably too sticky for correct codon-anticodon interactions (5). It was independently reported that synonymous codon usage is subject to selection in hyperthermophiles, and a strong preference exists for AGR, CGN (arginine) and ATH (isoleucine) codons (68). The difference in codon usage between hyperthermophiles and mesophiles was more pronounced in highly expressed genes (68). A conclusion of the study was that codon preferences in hyperthermophiles may be related to mRNA stability, rather than specific codon-anticodon base pairing (68).

A global analysis has shown that archaea and bacteria tend to acquire A and lose C at high temperature, while keeping T and G relatively constant (57). As a consequence, the genome sequences of hyperthermophiles are enriched in purine (purine loading) with a marked preference for adenine and a decrease in cytosine (57, 58, 108). It has been speculated that the increase in purine could minimize unnecessary RNA-RNA interactions and prevent the formation of "self"-double-stranded RNA. Since RNA-RNA interactions have a strong entropy-driven component, the need to minimize them should increase as temperature increases. Adenine could also help stabilizing mRNA against thermodegradation at high temperature, as it would stabilize the structure of mRNA (20). Avoid-

ance of RNA thermodegradation may indeed represent a major survival strategy in hyperthermophiles (36). A recent, exhaustive study on purine loading reported that mRNAs corresponding to highly expressed genes were especially enriched in adenine in hyperthermophiles (93). Furthermore, the study reported a strong bias in favor of polypurine tracts, with runs of more than five As in the mRNAs of hyperthermophiles. This correlates with the observed glutamate + lysine/(glutamine + histidine) ratio (110) and could explain why the number of charged residues in hyperthermophiles is much greater than would be necessary for the stabilization by ion pairs (7). In conclusion, life at high temperature appears to involve a complex web of different mechanisms acting at the genome level.

INSIGHTS INTO THE HISTORY OF THE *ARCHAEA* FROM COMPARATIVE GENOMICS

From the time of their "discovery," archaea have been at the center of heated debates about the evolution of early life. Despite wide expectations, comparative archaeal genomics has not been very helpful in settling debates concerning the topology of the universal tree, and the root of the tree. Proponents of various contradictory hypotheses often used to select from genomic data what apparently helped to support their favorite scenario. The eucaryal character of archaeal informational proteins was used to confirm the sisterhood of archaea and eucarya, previously inferred from the rooting of the tree of life by universal couples of anciently duplicated paralogs (116). However, this interpretation can be misleading, since it cannot be determined from comparative genomics alone whether the features common to archaea and eucarya are primitive or derived traits. Comparative genomics can only provide evidence of characteristics shared between domains, but it does not help polarizing these characters. (For example, a trait uniquely shared between archaea and eucarya [AE trait] may have originated in the branch leading to their common ancestor, and thus be considered to be derived. However, it is also equally likely that the AE trait is ancestral and was lost in the branch leading to bacteria.) The bacterial rooting of the universal tree remains controversial, and some genomewide analyses even support the rooting of the tree of life in the eucaryal branch (18), as it was suggested from a careful analysis of the universal couple of paralogs composed of the signal recognition particle and its receptor (11).

Despite the eucaryal affinity of their core genes, the general features (such as operon structure) and mechanisms of evolution of archaeal genomes (such as replication-driver rearrangements) are very similar to those of bacteria. If these features have a common ancestry, this indicates that they were already present in a common ancestor of archaea and bacteria. However, some of these features could also be due to convergent evolution because of similar lifestyles, or else, they could reflect a common mechanism at the origin of the archaeal and bacterial domains. For instance, it has been suggested that both archaeal and bacterial genomes may have been derived from larger genomes by streamlining, with traits being retained in extant eucarya (96). Streamlining may have been triggered by adaptation to thermophilic environments (36).

The sequencing of archaeal genomes has been extremely fruitful for unraveling the evolutionary history of the *Archaea*. Crenarchaeal genomes do not contain more "eucarya-like genes" than euryarchaeal genes. This argues against the "eocyte theory," which postulated a specific phylogenetic affinity between eucarya and *Crenarchaeota* and implied that archaea are not monophyletic (56). Archaeal comparative genomics has confirmed the importance of the *Euryarchaeota-Crenarchaeota* divide. Members of these two archaeal kingdoms exhibit striking differences in their DNA replication and cell division machineries. Crucial cell division proteins, such as MinD and FtsZ, appear to be exclusive to *Euryarchaeota* and *Nanoarchaeota* and systematically absent from all current crenarchaeal genomes. The same is true for DNA polymerase of the D family and for histones, whose functions are likely performed by nonhomologous proteins in *Crenarchaeota* (44, 114). On the other hand, some of these differences between *Euryarchaeota* and *Crenarchaeota* may have to be reevaluated after the sequencing of marine *Crenarchaeota*; archaeal histone genes were recently discovered in several contigs of uncultured marine *Crenarchaeota* and in *Cenarchaeum symbiosum* (28), although it cannot be excluded at present that they may have been acquired by HGT from *Euryarchaeota*. The present sequencing of a member of the tentative phylum, *Korarchaeota*, should provide especially important information on the early diversification in the *Archaea*.

The genomic era has led to an explosion of "whole genome trees," where different characteristics, such as the number of shared genes, shared protein folds, or even metabolic pathways, replace nucleotide or amino acid substitutions to quantify evolutionary distances. A variant is supertree methods, which combine phylogenetic trees obtained from different genes (for a review see reference 30). All genome trees cluster all archaea together as a coherent group, distinct from bacteria and eucarya, thereby confirming the three-domain concept. Except for a universal supertree (29), all of them failed to recover the *Euryarchaeota-Crenarchaeota* divide. Moreover, depending on the method of reconstruction, the whole-genome trees usually place *Haloarchaea* and/or *Thermoplasma* (two *Euryarchaeota*) either at the base of the *Archaea* or as sister groups to *Crenarchaeota*. This indicates that archaeal whole-genome trees are strongly affected by the extensive HGT that occurred within archaea, as well as between archaea and bacteria. In fact, the different archaeal genomes have not been affected to the same extent by HGT, and this may explain why only some archaea are clearly misplaced. In particular, the genomes of *Haloarchaea* and *Thermoplasmatales* contain a high number of genes of bacterial origin, explaining why they are attracted toward the *Bacteria* in whole-genome trees and, as a consequence, emerge at the base of the *Archaea*. For *Thermoplasmatales*, biased placement is increased by the presence of many genes in their genomes that appear to have been recruited from thermoacidophilic *Crenarchaeota* (*Sulfolobales*).

Whole-genome trees based on shared characteristics are therefore not reliable for obtaining a good phylogeny of the *Archaea*. Presently, the best methods are those that focus on selected sets of proteins that are less prone to HGT over large evolutionary distances, such as ribosomal proteins and RNA polymerase subunits. Several precautions have to be considered to generate a "good" archaeal tree. In particular, outgroup sequences (eucaryal and/or bacterial) that are normally used to root archaeal trees should be excluded from analysis to avoid possible long-branch attraction artifacts. Furthermore, individual trees should be analyzed carefully before concatenation to detect evident cases of HGT and remove the corresponding genes from the dataset. Following this strategy, the construction of datasets of concatenated ribosomal proteins and RNA polymerase subunits that are common to all archaea and have not been involved in HGT, and the accurate construction of the corresponding "translation" and "transcription" trees, has allowed a few questions to be solved that had remained confusing in 16 rRNA trees (12).

The translation and transcription trees based on the analysis of 23 complete or draft genomes (Fig. 8, A and B, respectively) are largely congruent, with one exception (see below), confirming the existence of a core of proteins that can be used to reconstruct the history of the *Archaea* domain (12). The results obtained have validated most of the evolutionary relationships previously obtained from 16S rRNA sequence comparison. In particular, they have confirmed that *Halobacteriales* are sister groups of *Methanomicrobiales*, and that *Thermoplasmatales* belong to a large clade comprising *Halobacteriales*, *Methanomicrobiales*, and *Archaeoglobales*. The only difference

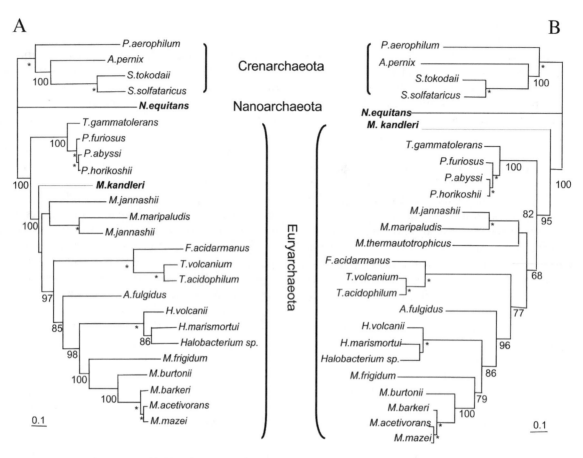

Figure 8. Unrooted maximum likelihood (ML) trees based on concatenation of 53 ribosomal proteins (A) and 12 RNA polymerase subunits and two transcription factors (B). Numbers at nodes are bootstrap values (BVs). The scale bars represent the number of changes per position for a unit branch length. Trees were produced by exhaustive searches performed by PROTML (1). Branch lengths and likelihood values were calculated by TREE-PUZZLE (JJT model including a (-correction [8 categories of sites]) (106). Numbers at nodes are bootstrap values computed with PUZZLEBOOT (51) from 1,000 replications. Asterisks indicate constrained nodes (supported by BV = 100% in preliminary NJ and heuristic ML analyses). The names of groups showing incongruent positions between the two trees are in bold. (Adapted from reference 12 with permission.)

between the translation and transcription trees is the position of *M. kandleri. M. kandleri* branches with *Methanobacteriales* in the translation tree (Fig. 8A), whereas it is located at the base of *Euryarchaeota* in the transcription tree (Fig. 8B), as it is in the 16S rRNA tree. However, the position of *M. kandleri* in the transcription tree is most likely an artifact of long-branch attraction induced by the rapid evolution of its RNA polymerase (see "Mechanisms of genome evolution" above). This interpretation is apparent from the very long branch displayed by *M. kandleri* in the transcription tree, compared with the translation tree (Fig. 8) (12).

For some organisms, the congruence of transcription and translation trees is not sufficient to reconstruct the history of a particular lineage. This is illustrated by *N. equitans*, which branches independently from *Euryarchaeota* and *Crenarchaeota* in

both trees, although it may be a distant relative of *Thermococcales*. Indeed, a specific phylogenetic proximity of *N. equitans* and *Thermococcales* is indicated by several single-gene phylogenies (elongation factors, reverse gyrase, DNA topoisomerase VI, tyrosyl-tRNA synthetase) and by the concatenation of only the proteins of the ribosome small subunit (14). The abnormal branching of *N. equitans* in the rRNA, transcription and translation trees is probably due to a mixture of long-branch attraction artifact (due to the rapid rate of evolution of its proteins, under pressure for reductive evolution) and HGT from the *Ignicoccus* host (especially for the proteins of the large subunit). In conclusion, the analysis of the components of two independent molecular systems, and the critical evaluation of the position of *M. kandleri* and *N. equitans*, allow a reasonable view of archaeal history to be proposed (depicted in Fig. 9).

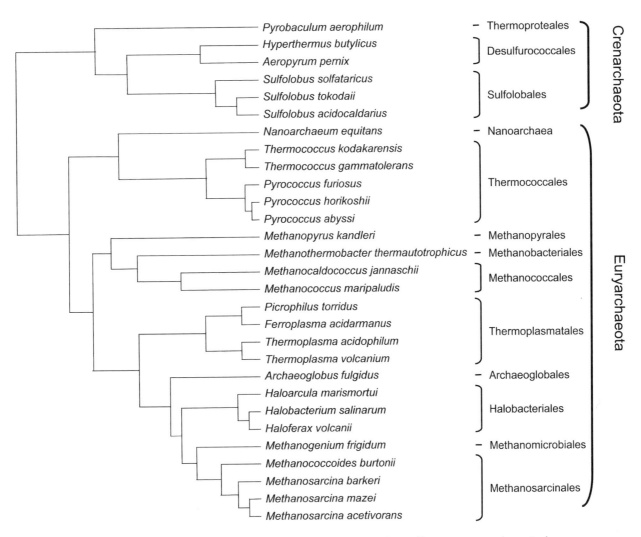

Figure 9. An ideal tree of the *Archaea* based on recent phylogenetic analyses of large concatenated protein data sets.

PERSPECTIVE: THE NEXT FIVE YEARS

Nearly ten years after the first archaeal genome was sequenced, a good knowledge of the structure and evolution of archaeal chromosomes has been obtained. Nevertheless, the gap between the number of complete genome sequences for archaea versus bacteria remains important, and this bias will probably continue because archaea are not considered a threat for human health. However, this may change in the near future, with the development of metagenomics projects aimed at identifying the complete flora of human commensals. Moreover, archaea may reveal themselves to be important players in human illness (21).

Ongoing metagenomics projects will continue to broaden the understanding of archaeal diversity and evolution. The cultivation of species that have only been detected in sequences of environmental samples will be a key step for understanding the features of the organisms that occupy critical positions in archaeal phylogeny; these include the deep-branching *Korarchaeota*, and the *Crenarchaeota* thriving in cold environments. This will help to gain a better understanding of the nature of the archaeal ancestor itself, and the origin of the major differences between the two major archaeal phyla, the *Euryarchaeota* and the *Crenarchaeota*.

Several features remain to be determined about archaeal chromosome structure and evolution. For instance, not much is known about termination of DNA replication and their associated sequence sites. Increasing the number of complete archaeal genomes will permit a thorough comparative analysis that will surely help clarify this issue in the near future. Comparative genomics will also allow a more complete investigation of the nature and extent of HGT in the *Archaea* and will possibly provide clues on the mechanisms responsible for the passage of genetic information between archaeal and bacterial species. Since archaeal viruses may be key players in this process, ongoing studies will shortly shed light onto this issue by providing a better understanding of their diversity, biology, and survival strategies.

Acknowledgments. We thank the editor for accurate revision of the manuscript and precious comments, Karsten Shure and Jean-Michel Claverie for kindly providing Figure 7, Celine Brochier for kindly providing Figure 9, and Diego Cortez for kindly providing Figure 2 and for artwork on Figures 1 and 4.

REFERENCES

1. Adachi, J., and M. Hasegawa. 1996. MOLPHY version 2.3: programs for molecular phylogenetics based on maximum likelihood. *Comput. Sci. Monogr.* **28**:1–150.

2. Andersson, J. O., S. W. Sarchfield, and A. J. Roger. 2005. Gene transfers from nanoarchaeota to an ancestor of diplomonads and parabasalids. *Mol. Biol. Evol.* **22**:85–90.

3. Aravind, L., and E. V. Koonin. 1999. DNA-binding proteins and evolution of transcription regulation in the archaea. *Nucleic Acids Res.* **27**:4658–4670.

4. Atomi, H., R. Matsumi, and T. Imanaka. 2004. Reverse gyrase is not a prerequisite for hyperthermophilic life. *J. Bacteriol.* **186**:4829–4833.

5. Basak, S., and T. C. Ghosh. 2005. On the origin of genomic adaptation at high temperature for prokaryotic organisms. *Biochem. Biophys. Res. Commun.* **330**:629–632.

6. Bentley, S. D., K. F. Chater, A. M. Cerdeno-Tarraga, G. L. Challis, N. R. Thomson, K. D. James, D. E. Harris, M. A. Quail, H. Kieser, D. Harper, A. Bateman, S. Brown, G. Chandra, C. W. Chen, M. Collins, A. Cronin, A. Fraser, A. Goble, J. Hidalgo, T. Hornsby, S. Howarth, C. H. Huang, T. Kieser, L. Larke, L. Murphy, K. Oliver, S. O'Neil, E. Rabbinowitsch, M. A. Rajandream, K. Rutherford, S. Rutter, K. Seeger, D. Saunders, S. Sharp, R. Squares, S. Squares, K. Taylor, T. Warren, A. Wietzorrek, J. Woodward, B. G. Barrell, J. Parkhill, and D. A. Hopwood. 2002. Complete genome sequence of the model actinomycete Streptomyces coelicolor A3(2). *Nature* **417**:141–147.

7. Berezovsky, I. N., and E. I. Shakhnovich. 2005. Physics and evolution of thermophilic adaptation. *Proc. Natl. Acad. Sci. USA* **102**:12742–12747.

8. Berquist, B. R., and S. DasSarma. 2003. An archaeal chromosomal autonomously replicating sequence element from an extreme halophile, Halobacterium sp. strain NRC-1. *J. Bacteriol.* **185**:5959–5966.

9. Blattner, F. R., G. Plunkett III, C. A. Bloch, N. T. Perna, V. Burland, M. Riley, J. Collado-Vides, J. D. Glasner, C. K. Rode, G. F. Mayhew, J. Gregor, N. W. Davis, H. A. Kirkpatrick, M. A. Goeden, D. J. Rose, B. Mau, and Y. Shao. 1997. The complete genome sequence of Escherichia coli K-12. *Science* **277**:1453–1474.

10. Brewer, B. J. 1988. When polymerases collide: replication and the transcriptional organization of the E. coli chromosome. *Cell* **53**:679–686.

11. Brinkmann, H., and H. Philippe. 1999. Archaea sister group of Bacteria? Indications from tree reconstruction artifacts in ancient phylogenies. *Mol. Biol. Evol.* **16**:817–825.

12. Brochier, C., P. Forterre, and S. Gribaldo. 2005. An emerging phylogenetic core of Archaea: phylogenies of transcription and translation machineries converge following addition of new genome sequences. *BMC Evol. Biol.* **5**:36.

13. Brochier, C., P. Forterre, and S. Gribaldo. 2004. Archaeal phylogeny based on proteins of the transcription and translation machineries: tackling the Methanopyrus kandleri paradox. *Genome Biol.* **5**:R17.

14. Brochier, C., S. Gribaldo, Y. Zivanovic, F. Confalonieri, and P. Forterre. 2005. Nanoarchaea: representatives of a novel archaeal phylum or a fast-evolving euryarchaeal lineage related to Thermococcales? *Genome Biol.* **6**:R42.

15. Brügger, K., P. Redder, Q. She, F. Confalonieri, Y. Zivanovic, and R. A. Garrett. 2002. Mobile elements in archaeal genomes. *FEMS Microbiol. Lett.* **206**:131–141.

16. Brugger, K., E. Torarinsson, P. Redder, L. Chen, and R. A. Garrett. 2004. Shuffling of Sulfolobus genomes by autonomous and non-autonomous mobile elements. *Biochem. Soc. Trans.* **32**:179–183.

17. Bult, C. J., O. White, G. J. Olsen, L. Zhou, R. D. Fleischmann, G. G. Sutton, J. A. Blake, L. M. FitzGerald, R. A. Clayton, J. D. Gocayne, A. R. Kerlavage, B. A. Dougherty, J. F. Tomb, M. D. Adams, C. I. Reich, R. Overbeek, E. F. Kirkness, K. G. Weinstock, J. M. Merrick, A. Glodek, J. L. Scott, N. S. M. Geoghagen, and J. C. Venter. 1996. Complete genome sequence of the methanogenic archaeon, Methanococcus jannaschii. *Science* **273**:1058–1073.

18. Caetano-Anolles, G., and D. Caetano-Anolles. 2005. Universal sharing patterns in proteomes and evolution of protein fold architecture and life. *J. Mol. Evol.* **60**:484–498.

19. Cambillau, C., and J. M. Claverie. 2000. Structural and genomic correlates of hyperthermostability. *J. Biol. Chem.* **275**:32383–32386.

20. Cate, J. H., A. R. Gooding, E. Podell, K. Zhou, B. L. Golden, A. A. Szewczak, C. E. Kundrot, T. R. Cech, and J. A. Doudna. 1996. RNA tertiary structure mediation by adenosine platforms. *Science* **273**:1696–1699.

21. Cavicchioli, R., P. M. Curmi, N. Saunders, and T. Thomas. 2003. Pathogenic archaea: do they exist? *Bioessays* **25**:1119–1128.

22. Charbonnier, F., and P. Forterre. 1994. Comparison of plasmid DNA topology among mesophilic and thermophilic eubacteria and archaebacteria. *J. Bacteriol.* **176**:1251–1259.

23. Chen, L., K. Brugger, M. Skovgaard, P. Redder, Q. She, E. Torarinsson, B. Greve, M. Awayez, A. Zibat, H. P. Klenk, and R. A. Garrett. 2005. The genome of Sulfolobus acidocaldarius, a model organism of the Crenarchaeota. *J. Bacteriol.* **187**:4992–4999.

24. Chinen, A., I. Uchiyama, and I. Kobayashi. 2000. Comparison between Pyrococcus horikoshii and Pyrococcus abyssi genome sequences reveals linkage of restriction-modification genes with large genome polymorphisms. *Gene* **259**:109–121.

25. Ciaramella, M., A. Napoli, and M. Rossi. 2005. Another extreme genome: how to live at pH 0. *Trends Microbiol.* **13**:49–51.

26. Cohen, G. N., V. Barbe, D. Flament, M. Galperin, R. Heilig, O. Lecompte, O. Poch, D. Prieur, J. Querellou, J. Ripp, J. C. Thierry, J. Van der Oost, J. Weissenbach, Y. Zivanovic, and P. Forterre. 2003. An integrated analysis of the genome of the hyperthermophilic archaeon Pyrococcus abyssi. *Mol. Microbiol.* **47**:1495–1512.

27. Constantinesco, F., P. Forterre, E. V. Koonin, L. Aravind, and C. Elie. 2004. A bipolar DNA helicase gene, herA, clusters with rad50, mre11 and nurA genes in thermophilic archaea. *Nucleic Acids Res.* **32**:1439–1447.

28. Cubonova, L., K. Sandman, S. J. Hallam, E. F. Delong, and J. N. Reeve. 2005. Histones in crenarchaea. *J. Bacteriol.* **187**:5482–5485.

29. Daubin, V., M. Gouy, and G. Perriere. 2002. A phylogenomic approach to bacterial phylogeny: evidence of a core of genes sharing a common history. *Genome Res.* **12**:1080–1090.

30. Delsuc, F., H. Brinkmann, and H. Philippe. 2005. Phylogenomics and the reconstruction of the tree of life. *Nat. Rev. Genet.* 6:361–375.

31. Deppenmeier, U., A. Johann, T. Hartsch, R. Merkl, R. A. Schmitz, R. Martinez-Arias, A. Henne, A. Wiezer, S. Baumer, C. Jacobi, H. Bruggemann, T. Lienard, A. Christmann, M. Bomeke, S. Steckel, A. Bhattacharyya, A. Lykidis, R. Overbeek, H. P. Klenk, R. P. Gunsalus, H. J. Fritz, and G. Gottschalk. 2002. The genome of Methanosarcina mazei: evidence for lateral gene transfer between bacteria and archaea. *J. Mol. Microbiol. Biotechnol.* 4:453–461.

32. Diruggiero, J., D. Dunn, D. L. Maeder, R. Holley-Shanks, J. Chatard, R. Horlacher, F. T. Robb, W. Boos, and R. B. Weiss. 2000. Evidence of recent lateral gene transfer among hyperthermophilic archaea. *Mol. Microbiol.* 38:684–693.

33. Fitz-Gibbon, S. T., H. Ladner, U. J. Kim, K. O. Stetter, M. I. Simon, and J. H. Miller. 2002. Genome sequence of the hyperthermophilic crenarchaeon Pyrobaculum aerophilum. *Proc. Natl. Acad. Sci. USA* 99:984–989.

34. Forterre, P. 2002. A hot story from comparative genomics: reverse gyrase is the only hyperthermophile-specific protein. *Trends Genet.* 18:236–237.

35. Forterre, P. 1997. Archaea: what can we learn from their sequences? *Curr. Opin. Genet. Dev.* 7:764–770.

36. Forterre, P. 1995. Thermoreduction, a hypothesis for the origin of prokaryotes. *C. R. Acad. Sci. Ser. III* 318:415–422.

37. Forterre, P., C. Bouthier De La Tour, H. Philippe, and M. Duguet. 2000. Reverse gyrase from hyperthermophiles: probable transfer of a thermoadaptation trait from archaea to bacteria. *Trends Genet.* 16:152–154.

38. Frank, A. C., and J. R. Lobry. 1999. Asymmetric substitution patterns: a review of possible underlying mutational or selective mechanisms. *Gene* 238:65–77.

39. Fukui, T., H. Atomi, T. Kanai, R. Matsumi, S. Fujiwara, and T. Imanaka. 2005. Complete genome sequence of the hyperthermophilic archaeon Thermococcus kodakaraensis KOD1 and comparison with Pyrococcus genomes. *Genome Res.* 15:352–363.

40. Futterer, O., A. Angelov, H. Liesegang, G. Gottschalk, C. Schleper, B. Schepers, C. Dock, G. Antranikian, and W. Liebl. 2004. Genome sequence of Picrophilus torridus and its implications for life around pH 0. *Proc. Natl. Acad. Sci. USA* 101:9091–9096.

41. Gadelle, D., J. Filee, C. Buhler, and P. Forterre. 2003. Phylogenomics of type II DNA topoisomerases. *Bioessays* 25:232–242.

42. Galagan, J. E., C. Nusbaum, A. Roy, M. G. Endrizzi, P. Macdonald, W. FitzHugh, S. Calvo, R. Engels, S. Smirnov, D. Atnoor, A. Brown, N. Allen, J. Naylor, N. Stange-Thomann, K. DeArellano, R. Johnson, L. Linton, P. McEwan, K. McKernan, J. Talamas, A. Tirrell, W. Ye, A. Zimmer, R. D. Barber, I. Cann, D. E. Graham, D. A. Grahame, A. M. Guss, R. Hedderich, C. Ingram-Smith, H. C. Kuettner, J. A. Krzycki, J. A. Leigh, W. Li, J. Liu, B. Mukhopadhyay, J. N. Reeve, K. Smith, T. A. Springer, L. A. Umayam, O. White, R. H. White, E. Conway de Macario, J. G. Ferry, K. F. Jarrell, H. Jing, A. J. Macario, I. Paulsen, M. Pritchett, K. R. Sowers, R. V. Swanson, S. H. Zinder, E. Lander, W. W. Metcalf, and B. Birren. 2002. The genome of M. acetivorans reveals extensive metabolic and physiological diversity. *Genome Res.* 12:532–542.

43. Galtier, N., and J. R. Lobry. 1997. Relationships between genomic G+C content, RNA secondary structures, and optimal growth temperature in prokaryotes. *J. Mol. Evol.* 44:632–636.

44. Grabowski, B., and Z. Kelman. 2003. Archeal DNA replication: eukaryal proteins in a bacterial context. *Annu. Rev. Microbiol.* 57:487–516.

45. Guipaud, O., E. Marguet, K. M. Noll, C. B. de la Tour, and P. Forterre. 1997. Both DNA gyrase and reverse gyrase are present in the hyperthermophilic bacterium Thermotoga maritima. *Proc. Natl. Acad. Sci. USA* 94:10606–10611.

46. Haney, P. J., J. H. Badger, G. L. Buldak, C. I. Reich, C. R. Woese, and G. J. Olsen. 1999. Thermal adaptation analyzed by comparison of protein sequences from mesophilic and extremely thermophilic Methanococcus species. *Proc. Natl. Acad. Sci. USA* 96:3578–3583.

47. Haring, M., X. Peng, K. Brugger, R. Rachel, K. O. Stetter, R. A. Garrett, and D. Prangishvili. 2004. Morphology and genome organization of the virus PSV of the hyperthermophilic archaeal genera Pyrobaculum and Thermoproteus: a novel virus family, the Globuloviridae. *Virology* 323:233–242.

48. Hendrickson, E. L., R. Kaul, Y. Zhou, D. Bovee, P. Chapman, J. Chung, E. Conway de Macario, J. A. Dodsworth, W. Gillett, D. E. Graham, M. Hackett, A. K. Haydock, A. Kang, M. L. Land, R. Levy, T. J. Lie, T. A. Major, B. C. Moore, I. Porat, A. Palmeiri, G. Rouse, C. Saenphimmachak, D. Soll, S. Van Dien, T. Wang, W. B. Whitman, Q. Xia, Y. Zhang, F. W. Larimer, M. V. Olson, and J. A. Leigh. 2004. Complete genome sequence of the genetically tractable hydrogenotrophic methanogen Methanococcus maripaludis. *J. Bacteriol.* 186:6956–6969.

49. Herzel, H., O. Weiss, and E. N. Trifonov. 1998. Sequence periodicity in complete genomes of archaea suggests positive supercoiling. *J. Biomol. Struct. Dyn.* 16:341–345.

50. Hickey, D. A., and G. A. Singer. 2004. Genomic and proteomic adaptations to growth at high temperature. *Genome Biol.* 5:117.

51. Holder, M. E., and A. J. Roger. 2002. A shell-script program called "puzzleboot" that allows the analysis of multiple data sets with PUZZLE even though PUZZLE lacks the "M" option of many PHYLIP programs. http://hades.biochem.dal.ca/Rogerlab/Software/software.html#puzzleboot.

52. Jansen, R., J. D. van Embden, W. Gaastra, and L. M. Schouls. 2002. Identification of a novel family of sequence repeats among prokaryotes. *Omics* 6:23–33.

53. Kampmann, M., and D. Stock. 2004. Reverse gyrase has heat-protective DNA chaperone activity independent of supercoiling. *Nucleic Acids Res.* 32:3537–3545.

54. Kennedy, S. P., W. V. Ng, S. L. Salzberg, L. Hood, and S. DasSarma. 2001. Understanding the adaptation of Halobacterium species NRC-1 to its extreme environment through computational analysis of its genome sequence. *Genome Res.* 11:1641–1650.

55. Koonin, E. V., Y. I. Wolf, and L. Aravind. 2001. Prediction of the archaeal exosome and its connections with the proteasome and the translation and transcription machineries by a comparative-genomic approach. *Genome Res.* 11:240–252.

56. Lake, J. A., E. Henderson, M. Oakes, and M. W. Clark. 1984. Eocytes: a new ribosome structure indicates a kingdom with a close relationship to eukaryotes. *Proc. Natl. Acad. Sci. USA* 81:3786–3790.

57. Lambros, R. J., J. R. Mortimer, and D. R. Forsdyke. 2003. Optimum growth temperature and the base composition of open reading frames in prokaryotes. *Extremophiles* 7:443–450.

58. Lao, P. J., and D. R. Forsdyke. 2000. Thermophilic bacteria strictly obey Szybalski's transcription direction rule and politely purine-load RNAs with both adenine and guanine. *Genome Res.* 10:228–236.

59. Lecompte, O., R. Ripp, J. C. Thierry, D. Moras, and O. Poch. 2002. Comparative analysis of ribosomal proteins in complete genomes: an example of reductive evolution at the domain scale. *Nucleic Acids Res.* 30:5382–5390.

60. Lewis, P. J. 2001. Bacterial chromosome segregation. *Microbiology* **147**:519–526.

61. Lobry, J. R. 1996. Asymmetric substitution patterns in the two DNA strands of bacteria. *Mol. Biol. Evol.* **13**:660–605.

62. Londei, P. 2005. Evolution of translational initiation: new insights from the archaea. *FEMS Microbiol. Rev.* **29**:185–200.

63. Lopez, P., H. Philippe, H. Myllykallio, and P. Forterre. 1999. Identification of putative chromosomal origins of replication in Archaea. *Mol. Microbiol.* **32**:883–886.

64. Lopez-Garcia, P., C. Brochier, D. Moreira, and F. Rodriguez-Valera. 2004. Comparative analysis of a genome fragment of an uncultivated mesopelagic crenarchaeote reveals multiple horizontal gene transfers. *Environ. Microbiol.* **6**:19–34.

65. Lopez-Garcia, P., P. Forterre, J. van der Oost, and G. Erauso. 2000. Plasmid pGS5 from the hyperthermophilic archaeon Archaeoglobus profundus is negatively supercoiled. *J. Bacteriol.* **182**:4998–5000.

66. Lundgren, M., A. Andersson, L. Chen, P. Nilsson, and R. Bernander. 2004. Three replication origins in Sulfolobus species: synchronous initiation of chromosome replication and asynchronous termination. *Proc. Natl. Acad. Sci. USA* **101**:7046–7051.

67. Luscombe, N. M., D. Greenbaum, and M. Gerstein. 2001. What is bioinformatics? A proposed definition and overview of the field. *Methods Inf. Med.* **40**:346–358.

68. Lynn, D. J., G. A. Singer, and D. A. Hickey. 2002. Synonymous codon usage is subject to selection in thermophilic bacteria. *Nucleic Acids Res.* **30**:4272–4277.

69. Maeder, D. L., R. B. Weiss, D. M. Dunn, J. L. Cherry, J. M. Gonzalez, J. DiRuggiero, and F. T. Robb. 1999. Divergence of the hyperthermophilic archaea *Pyrococcus furiosus* and *P. horikoshii* inferred from complete genomic sequences. *Genetics* **152**:1299–1305.

70. Maisnier-Patin, S., L. Malandrin, N. K. Birkeland, and R. Bernander. 2002. Chromosome replication patterns in the hyperthermophilic euryarchaea Archaeoglobus fulgidus and Methanocaldococcus (Methanococcus) jannaschii. *Mol. Microbiol.* **45**:1443–1450.

71. Makarova, K. S., L. Aravind, M. Y. Galperin, N. V. Grishin, R. L. Tatusov, Y. I. Wolf, and E. V. Koonin. 1999. Comparative genomics of the Archaea (Euryarchaeota): evolution of conserved protein families, the stable core, and the variable shell. *Genome Res.* **9**:608–628.

72. Makarova, K. S., L. Aravind, N. V. Grishin, I. B. Rogozin, and E. V. Koonin. 2002. A DNA repair system specific for thermophilic Archaea and bacteria predicted by genomic context analysis. *Nucleic Acids Res.* **30**:482–496.

73. Makarova, K. S., and E. V. Koonin. 2003. Comparative genomics of Archaea: how much have we learned in six years, and what's next? *Genome Biol.* **4**:115.

74. Makarova, K. S., and E. V. Koonin. 2005. Evolutionary and functional genomics of the Archaea. *Curr. Opin. Microbiol.* **117**:52–67.

75. Makarova, K. S., Y. I. Wolf, and E. V. Koonin. 2003. Potential genomic determinants of hyperthermophily. *Trends Genet.* **19**:172–176.

76. Makino, S., and M. Suzuki. 2001. Bacterial genomic reorganization upon DNA replication. *Science* **292**:803.

77. Marguet, E., and P. Forterre. 1994. DNA stability at temperatures typical for hyperthermophiles. *Nucleic Acids Res.* **22**:1681–1686.

78. Marguet, E., and P. Forterre. 1998. Protection of DNA by salts against thermodegradation at temperatures typical for hyperthermophiles. *Extremophiles* **2**:115–122.

79. Marguet, E., and P. Forterre. 2001. Stability and manipulation of DNA at extreme temperatures. *Methods Enzymol.* **334**:205–115.

80. Matsunaga, F., P. Forterre, Y. Ishino, and H. Myllykallio. 2001. In vivo interactions of archaeal Cdc6/Orc1 and minichromosome maintenance proteins with the replication origin. *Proc. Natl. Acad. Sci. USA* **98**:11152–11157.

81. Matte-Tailliez, O., Y. Zivanovic, and P. Forterre. 2000. Mining archaeal proteomes for eukaryotic proteins with novel functions: the PACE case. *Trends Genet.* **16**:533–536.

82. McLean, M. J., K. H. Wolfe, and K. M. Devine. 1998. Base composition skews, replication orientation, and gene orientation in 12 prokaryote genomes. *J. Mol. Evol.* **47**:691–696.

83. Mojica, F. J., C. Diez-Villasenor, J. Garcia-Martinez, and E. Soria. 2005. Intervening sequences of regularly spaced prokaryotic repeats derive from foreign genetic elements. *J. Mol. Evol.* **60**:174–182.

84. Mojica, F. J., C. Diez-Villasenor, E. Soria, and G. Juez. 2000. Biological significance of a family of regularly spaced repeats in the genomes of Archaea, Bacteria and mitochondria. *Mol. Microbiol.* **36**:244–6.

85. Mojica, F. J., C. Ferrer, G. Juez, and F. Rodriguez-Valera. 1995. Long stretches of short tandem repeats are present in the largest replicons of the Archaea Haloferax mediterranei and Haloferax volcanii and could be involved in replicon partitioning. *Mol. Microbiol.* **17**:85–93.

86. Mrazek, J., and S. Karlin. 1998. Strand compositional asymmetry in bacterial and large viral genomes. *Proc. Natl. Acad. Sci. USA* **95**:3720–3725.

87. Myllykallio, H., P. Lopez, P. Lopez-Garcia, R. Heilig, W. Saurin, Y. Zivanovic, H. Philippe, and P. Forterre. 2000. Bacterial mode of replication with eukaryotic-like machinery in a hyperthermophilic archaeon. *Science* **288**:2212–2215.

88. Napoli, A., A. Valenti, V. Salerno, M. Nadal, F. Garnier, M. Rossi, and M. Ciaramella. 2004. Reverse gyrase recruitment to DNA after UV light irradiation in Sulfolobus solfataricus. *J. Biol. Chem.* **279**:33192–33198.

89. Nelson, K. E., R. A. Clayton, S. R. Gill, M. L. Gwinn, R. J. Dodson, D. H. Haft, E. K. Hickey, J. D. Peterson, W. C. Nelson, K. A. Ketchum, L. McDonald, T. R. Utterback, J. A. Malek, K. D. Linher, M. M. Garrett, A. M. Stewart, M. D. Cotton, M. S. Pratt, C. A. Phillips, D. Richardson, J. Heidelberg, G. G. Sutton, R. D. Fleischmann, J. A. Eisen, and C. M. Fraser. 1999. Evidence for lateral gene transfer between Archaea and bacteria from genome sequence of Thermotoga maritima. *Nature* **399**:323–329.

90. Ng, W. V., S. P. Kennedy, G. G. Mahairas, B. Berquist, M. Pan, H. D. Shukla, S. R. Lasky, N. S. Baliga, V. Thorsson, J. Sbrogna, S. Swartzell, D. Weir, J. Hall, T. A. Dahl, R. Welti, Y. A. Goo, B. Leithauser, K. Keller, R. Cruz, M. J. Danson, D. W. Hough, D. G. Maddocks, P. E. Jablonski, M. P. Krebs, C. M. Angevine, H. Dale, T. A. Isenbarger, R. F. Peck, M. Pohlschroder, J. L. Spudich, K. H. Jung, M. Alam, T. Freitas, S. Hou, C. J. Daniels, P. P. Dennis, A. D. Omer, H. Ebhardt, T. M. Lowe, P. Liang, M. Riley, L. Hood, and S. DasSarma. 2000. Genome sequence of Halobacterium species NRC-1. *Proc. Natl. Acad. Sci. USA.* **97**:12176–12181.

91. Olsen, G. J., and C. R. Woese. 1997. Archaeal genomics: an overview. *Cell* **89**:991–994.

92. Opalka, N., M. Chlenov, P. Chacon, W. J. Rice, W. Wriggers, and S. A. Darst. 2003. Structure and function of the transcription elongation factor GreB bound to bacterial RNA polymerase. *Cell* **114**:335–345.

93. Paz, A., D. Mester, I. Baca, E. Nevo, and A. Korol. 2004. Adaptive role of increased frequency of polypurine tracts in mRNA sequences of thermophilic prokaryotes. *Proc. Natl. Acad. Sci. USA* **101**:2951–2956.

94. Peng, X., K. Brugger, B. Shen, L. Chen, Q. She, and R. A. Garrett. 2003. Genus-specific protein binding to the large

clusters of DNA repeats (short regularly spaced repeats) present in Sulfolobus genomes. *J. Bacteriol.* **185:**2410–2417.

95. Pietrokovski, S. 2001. Intein spread and extinction in evolution. *Trends Genet.* **17:**465–472.

96. Poole, A., D. Jeffares, and D. Penny. 1999. Early evolution: prokaryotes, the new kids on the block. *Bioessays* **21:**880–889.

97. Pourcel, C., G. Salvignol, and G. Vergnaud. 2005. CRISPR elements in Yersinia pestis acquire new repeats by preferential uptake of bacteriophage DNA, and provide additional tools for evolutionary studies. *Microbiology* **151:**653–663.

98. Querol, E., J. A. Perez-Pons, and A. Mozo-Villarias. 1996. Analysis of protein conformational characteristics related to thermostability. *Protein Eng.* **9:**265–271.

99. Robinson, N. P., I. Dionne, M. Lundgren, V. L. Marsh, R. Bernander, and S. D. Bell. 2004. Identification of two origins of replication in the single chromosome of the archaeon Sulfolobus solfataricus. *Cell* **116:**25–38.

100. Rocha, E. P. 2004. Order and disorder in bacterial genomes. *Curr. Opin. Microbiol.* **7:**519–527.

101. Rocha, E. P. 2004. The replication-related organization of bacterial genomes. *Microbiology* **150:**1609–1627.

102. Rocha, E. P., and A. Danchin. 2003. Essentiality, not expressiveness, drives gene-strand bias in bacteria. *Nat. Genet.* **34:**377–378.

103. Rocha, E. P., and A. Danchin. 2001. Ongoing evolution of strand composition in bacterial genomes. *Mol. Biol. Evol.* **18:**1789–1799.

104. Saunders, N. F., T. Thomas, P. M. Curmi, J. S. Mattick, E. Kuczek, R. Slade, J. Davis, P. D. Franzmann, D. Boone, K. Rusterholtz, R. Feldman, C. Gates, S. Bench, K. Sowers, K. Kadner, A. Aerts, P. Dehal, C. Detter, T. Glavina, S. Lucas, P. Richardson, F. Larimer, L. Hauser, M. Land, and R. Cavicchioli. 2003. Mechanisms of thermal adaptation revealed from the genomes of the Antarctic Archaea Methanogenium frigidum and Methanococcoides burtonii. *Genome Res.* **13:**1580–1588.

105. Schieg, P., and H. Herzel. 2004. Periodicities of 10-11bp as indicators of the supercoiled state of genomic DNA. *J. Mol. Biol.* **343:**891–901.

106. Schmidt, H. A., K. Strimmer, M. Vingron, and A. von Haeseler. 2002. TREE-PUZZLE: maximum likelihood phylogenetic analysis using quartets and parallel computing. *Bioinformatics* **18:**502–504.

107. She, Q., R. K. Singh, F. Confalonieri, Y. Zivanovic, G. Allard, M. J. Awayez, C. C. Chan-Weiher, I. G. Clausen, B. A. Curtis, A. De Moors, G. Erauso, C. Fletcher, P. M. Gordon, I. Heikamp-de Jong, A. C. Jeffries, C. J. Kozera, N. Medina, X. Peng, H. P. Thi-Ngoc, P. Redder, M. E. Schenk, C. Theriault, N. Tolstrup, R. L. Charlebois, W. F. Doolittle, M. Duguet, T. Gaasterland, R. A. Garrett, M. A. Ragan, C. W.

Sensen, and J. Van der Oost. 2001. The complete genome of the crenarchaeon *Sulfolobus solfataricus* P2. *Proc. Natl. Acad. Sci. USA* **98:**7835–7840.

108. Singer, G. A., and D. A. Hickey. 2003. Thermophilic prokaryotes have characteristic patterns of codon usage, amino acid composition and nucleotide content. *Gene* **317:**39–47.

109. Slesarev, A. I., K. V. Mezhevaya, K. S. Makarova, N. N. Polushin, O. V. Shcherbinina, V. V. Shakhova, G. I. Belova, L. Aravind, D. A. Natale, I. B. Rogozin, R. L. Tatusov, Y. I. Wolf, K. O. Stetter, A. G. Malykh, E. V. Koonin, and S. A. Kozyavkin. 2002. The complete genome of hyperthermophile Methanopyrus kandleri AV19 and monophyly of archaeal methanogens. *Proc. Natl. Acad. Sci. USA* **99:**4644–4649.

110. Shure, K., and J. M. Claverie. 2003. Genomic correlates of hyperthermostability, an update. *J. Biol. Chem.* **278:**17198–17202.

111. Tillier, E. R., and R. A. Collins. 2000. Genome rearrangement by replication-directed translocation. *Nat. Genet.* **26:**195–197.

112. Vieille, C., and G. J. Zeikus. 2001. Hyperthermophilic enzymes: sources, uses, and molecular mechanisms for thermostability. *Microbiol. Mol. Biol. Rev.* **65:**1–43.

113. Waters, E., M. J. Hohn, I. Ahel, D. E. Graham, M. D. Adams, M. Barnstead, K. Y. Beeson, L. Bibbs, R. Bolanos, M. Keller, K. Kretz, X. Lin, E. Mathur, J. Ni, M. Podar, T. Richardson, G. G. Sutton, M. Simon, D. Soll, K. O. Stetter, J. M. Short, and M. Noordewier. 2003. The genome of Nanoarchaeum equitans: insights into early archaeal evolution and derived parasitism. *Proc. Natl. Acad. Sci. USA* **100:**12984–12988.

114. White, M. F., and S. D. Bell. 2002. Holding it together: chromatin in the Archaea. *Trends Genet.* **18:**621–626.

115. Wiezer, A., and R. Merkl. 2005. A comparative categorization of gene flux in diverse microbial species. *Genomics* **86:**462–475.

116. Woese, C. R., O. Kandler, and M. L. Wheelis. 1990. Towards a natural system of organisms: proposal for the domains Archaea, Bacteria, and Eucarya. *Proc. Natl. Acad. Sci. USA* **87:**4576–4579.

117. Zhang, J. 1999. Performance of likelihood ratio tests of evolutionary hypotheses under inadequate substitution models. *Mol. Biol. Evol.* **16:**868–875.

118. Zhang, R., and C. T. Zhang. 2005. Identification of replication origins in archaeal genomes based on the Z-curve method. *Archaea* **1:**335–346.

119. Zhang, R., and C. T. Zhang. 2004. Identification of replication origins in the genome of the methanogenic archaeon, Methanocaldococcus jannaschii. *Extremophiles* **8:**253–258.

120. Zivanovic, Y., P. Lopez, H. Philippe, and P. Forterre. 2002. Pyrococcus genome comparison evidences chromosome shuffling-driven evolution. *Nucleic Acids Res.* **30:**1902–1910.

Archaea: Molecular and Cellular Biology
Edited by Ricardo Cavicchioli
© 2007 ASM Press, Washington, D.C.

Chapter 20

Functional Genomics

Francis E. Jenney, Jr., Sabrina Tachdjian, Chung-Jung Chou,
Robert M. Kelly, and Michael W. W. Adams

INTRODUCTION

The substantial amount of genome sequence information available for members of the *Archaea* facilitates the use of functional genomics tools to investigate issues related to genetics, metabolism, physiology, and ecology of these microorganisms. Indeed, over 20 genome sequences have been completed for archaea (see Chapter 19). In some cases, genome sequences for more than one member of a genus are available, e.g., *Pyrococcus* (39, 94, 146) and *Sulfolobus* (34, 95, 162). Given the phylogenetic placement of the domain *Archaea* relative to the *Bacteria* and *Eucarya*, use of functional genomics to probe the relationship between genotype and phenotype for these fascinating microorganisms can provide a unique perspective on both evolutionary processes and mechanisms in extant biological systems. As genetic systems for archaea become further developed and implemented (see Chapter 21), functional genomics tools can be expanded to enable a full systems biology approach to studying archaea.

The current status of functional genomics efforts to investigate archaea are reviewed in this chapter. In this discussion, functional genomics refers to transcriptional response (transcriptomics), protein inventory and differential abundance (proteomics), and protein structural attributes (structural genomics) examined in the context of entire genomes. Ultimately, the goal is to integrate information gained from these three perspectives to discern biological mechanisms, although most efforts to date have focused on developing and implementing effective protocols for each methodology for specific archaea. However, significant progress has been made through the use of functional genomics tools in understanding the biology of ar-

chaea, examples of which are presented. In addition, specific challenges and opportunities that relate to the use of functional genomics to study archaeal biology are considered.

Choice of Model Organisms

The choice of model archaea for functional genomics studies arises to a large extent from the prevailing interests within the scientific community. As has been the case in other areas of biology, model systems for functional genomics studies will emerge from long-term efforts focused on the biology of specific archaea. Such efforts have generated information related to specific biomolecules, metabolic pathways, regulatory processes, and physiological patterns that can be investigated in a more complete way through functional genomics. Furthermore, model archaea are obviously those for which complete genome sequence information is available, although this may soon be a moot point when bacterial and archaeal genomes are sequenced within hours (116). Without genome sequence data, the use of functional genomics tools, though not impossible, is problematic. Model archaea should be readily cultivated in laboratory settings for a range of environmental and nutritional conditions; this is paramount to gaining functional genomics information on regulation, physiology, and ecology. The choice of model archaea for functional genomics studies may also depend on the availability of useful genetic systems. Despite the challenges in developing such systems for anaerobic and extremophilic archaea, much progress has been made along these lines (see Chapter 21). On the other hand, functional genomics tools can be used to circumvent some of the limitations. The combination of both modern and clas-

Francis E. Jenney, Jr., and Michael W. W. Adams • Department of Biochemistry & Molecular Biology, Davison Life Sciences Complex, Green Street, University of Georgia, Athens, GA 30602-7229. Sabrina Tachdjian, Chung-Jung Chou, and Robert M. Kelly • Department of Chemical and Biomolecular Engineering, North Carolina State University, EB-1, Box 7905, 911 Partners Way, Raleigh, NC 27695-7905.

sical tools for studying biological processes is the best situation for systems biology approaches.

Currently, certain archaea are emerging as model systems for functional genomics studies. Table 1 lists several possible model archaea chosen for the reasons discussed above. Within the hyperthermophilic archaea, *Pyrococcus furiosus* is among the most extensively studied; indeed, over 600 PubMed literature citations exist for this hyperthermophile at the time of writing (September 2005). *P. furiosus* was isolated from geothermal waters near Vulcano Island, Italy (58). It grows anaerobically near 100°C on a number of glucans in the presence or absence of elemental sulfur (52) but will grow on peptides only if sulfur is present in the medium. Primary metabolic products are acetate, H_2, and H_2S (when sulfur is present). Not only has the *P. furiosus* genome been sequenced (146), but genomes for two other members of the genus are also available (39, 94). Although no genetic system is currently available for *P. furiosus*, efforts have been made in this respect for members of the genus *Pyrococcus* (134). *P. furiosus* can be cultured in large-scale batch (181) and continuous culture (141) to relatively high biomass concentrations for a wide variety of growth conditions, thus facilitating functional genomics efforts. *P. furiosus* has been used in laboratory studies for almost twenty years, is relatively easy to cultivate, and has a versatile growth physiology. As a result, a large number of proteins from *P. furiosus* have been characterized with respect to microbial biochemistry and biotechnology (2–4). In light of the these factors, several functional genomics efforts have focused on *P. furiosus*, including transcriptomics, proteomics, and structural genomics, as summarized in Table 2.

Sulfolobus solfataricus and *P. furiosus* represent the two most-studied hyperthermophilic archaea; over 400 PubMed citations exist for *S. solfataricus*. Isolated from acidic thermal hot spring near Naples, Italy, in 1980 (199), *S. solfataricus* has been extensively studied, for example, to understand the basis for protein thermostability (158), adaptation to low pH environments, biotechnology (190), and features of eucaryal genetics from the perspective of the archaea (50). *S. solfataricus* grows fastest aerobically at pH 3.5 and 80°C on a range of glucans and peptides as carbon sources (199). Its utilization of elemental sulfur (S^0) during growth has been questioned despite initial reports of this capability at the time of isolation. The *S. solfataricus* genome has been sequenced (162), as have those of *S. tokodaii* (95) and *S. acidocaldarius* (34). Given its aerobic lifestyle and the large number of mobile genetic elements in its genome, a genetic system for *S. solfataricus* has been demonstrated (41). As the results from functional genomics are obtained, the availability of a genetic system will be invaluable for further studying the functions of specific genes in this archaeon.

Thermococcus kodakaraensis was isolated in 1994 from a shallow marine vent off the coast of Japan (13). Similar in many ways to *P. furiosus* in terms of its growth physiology, albeit growing optimally at 85°C rather than near 100°C, the biochemistry of *T. kodakaraensis* has been extensively studied using recombinant versions of its proteins produced with information derived from its recently reported genome sequence (61). Although functional genomics efforts have yet to be reported for this archaeon, the recent development of a versatile genetic system is very ex-

Table 1. Possible model archaeal systems for functional genomics studies

Parameter	*Pyrococcus furiosus*	*Sulfolobus solfataricus* P2	*Halobacterium* NRC-1	*Thermococcus kodakaraensis*	*Methanocaldococcus jannaschii*
Literature citations	>600	~400	~35	~85	~160
Isolation	Vulcano Island, Italy (1986)	Pisciarelli Solfatara, Italy (1980)	Salt mine, Bad Ischl, Austria (1994)	Kodakara, Japan (1994)	White Smoker Chimney, East Pacific Rise (1985)
T_{opt} for growth	100°C	80°C	37°C	85°C	85°C
Physiology	Fermentative heterotroph	Heterotrophic thermoacidophile	Extreme halophile	Fermentative heterotroph	Autotroph
Oxygen requirement	Anaerobe	Aerobe	Facultative aerobe	Anaerobe	Anaerobe
Easily plated	No	Yes	Yes	No	No
Year genome was sequenced	2001	2001	2000	2005	1996
Genome size (Mb)	1.91	2.99	2.57	2.09	1.74
Protein-coding ORFs	2,125	2,977	2,075	2,306	1,786
Chromosomal elements	1	1	1	1	3
Plasmids	0	0	2	0	0
Genetic system	No	Yes	Yes	Yes	Some tools available
Closely related species	*P. horikoshii* *P. abyssi* *T. kodakaraensis*	*S. tokodaii* *S. acidocaldarius*	*H. salinarum* *H. volcanii*	*P. horikoshii* *P. abyssi* *P. furiosus*	Related methanogens are listed in Table 2.

Table 2. Archaeal "omics": proteomic, transcriptomic, and structural genomic studies

Organism	Proteome[a]				Transcriptome		Structural genome[b]
	Single (S) or multiple (M) growth conditions	No. of proteins identified	Separation, identification technique	Reference	Growth conditions	Reference	No. of recombinant protein structures[c]
Methanocaldococcus jannaschii	M (H$_2$)	10	2D, MT	123	65/85/95°C	22	119, 19
	M (H$_2$, NH$_4^+$)	2	2D, MT	66			
	S	170	2D, MT and LC-MS/MS	65			
	S	954	LC, LC-MS/MS	198			
	S	74	1D/LC, ESI/Q-FTMS	60			
Methanococcus maripaludis		–		–		–	18, 2
Methanococcoides burtonii	M (4/23°C)	54	2D, LC-MS/MS	69		–	–
	S	528	LC, LC-MS/MS	71, 151		–	
Methanosarcina acetivorans	M (4/23°C)	163	ICAT/LC, LC-MS/MS	70			–
	M (methanol/acetate)	412	2D, MT	107, 108		–	–
Methanosarcina thermophila	M (methanol/acetate)	7	2D, comigration	87		–	
	M (methanol/acetate)	6	2D, N-term	49		–	
Methanosarcina mazei		–	–	–	Methanol/acetate	82	37, 8
Methanothermobacter thermautotrophicus		–	–	–		–	>212, 36[d]
Halobacterium salinarum	S (cytoplasm)	>800	2D, MT	173		–	
	S (membrane)	114	1D, LC-MS/MS	102			
	M (salinity)	>50	2D, ESI-Q TOF-MS/MS	132			3, 3
	M (salinity)	29	2D, ESI-Q TOF-MS/MS	37			
	M (hydrocarbons)	5	2D, ESI-Q TOF-MS/MS	92			
Haloferax volcanii					Peptides/glucose	193	None
Halobacterium NRC1	M (membrane mutants)	272	ICAT, µLC-ESI-MS/MS	14	M (membrane mutants)	14	None
	S (membrane, cytoplasm)	426	µLC-ESI-MS/MS	68	UV radiation	14	8, 0
					DMSO/TMAO	124	
Pyrococcus furiosus	S (membrane, cytoplasm)	66	2D, MT and µLC-ESI-MS/MS	80	Sulfur (partial array) 90/105°C	157	425, 39[e]
	S	62	2D, MT and µLC-ESI-MS/MS	109		165	
					Peptides/maltose	156	
					95/72°C	188	
Pyrococcus horikoshii		–		–		–	116, 87
Thermoplasma acidophilum	S	1	2D, N-term	189		–	74, 17
Archaeoglobus fulgidus		–		–	78/89°C	148	122, 31
Sulfolobus solfataricus		–		–		–	52, 9
Aeropyrum pernix		–		–		–	39, 15

[a]The proteome list does not include studies where patterns were presented but individual proteins were not identified (e.g., 46, 121). The abbreviations are: 2D, 2D-gel electrophoresis; MT, MALDI-TOF.

[b]The recombinant proteins and structures are those made available from structural genomics projects that target the individual organisms.

[c]Data are derived from the Protein Data Bank (http://targetdb.pdb.org/) representing data voluntarily deposited by structural genomics centers worldwide. These data do not include individual research projects outside of the structural genomics groups. Recombinant proteins refers to purified proteins (not simply targets for expression). Structures refer to both X-ray and NMR-based structures, and not all structures counted have been released.

[d]For more information, see reference 154.

[e]For more information, see reference 185.

citing and will pave the way for comprehensive systems biology studies (12).

Several methanogens have been studied to various extents using functional genomics tools and, thus, represent prospective model systems. Among these, members of the *Methanococcales* (179), such as *Methanocaldococcus jannaschii*, whose sequence was released in 1996 (73), have already been the object of several proteomic (60, 65, 198) and transcriptional response studies (22). In addition, the genome of *Methanococcus maripaludis* was recently sequenced (78) and the availability of mutagenesis tools for this methanogen will facilitate functional genomics studies (122). The *Methanosarcinales* have been the focus of genomics studies; no less than three complete genome sequences (47, 63) and one shotgun sequence (42, 152) have become available over the last few years. A genetic system has been developed for the genus *Methanosarcina* (119), and additional species-specific tools have been reported for *Methanosarcina acetivorans* (9, 135, 195, 196), *Methanosarcina mazei* (54), and *Methanosarcina thermophila* (49, 104, 120). Furthermore, transcriptional data generated by DNA microarray experiments were recently reported for *M. mazei* (82). The proteomes of *M. acetivorans* (60, 107, 108), *Methanococcoides burtonii* (69–71, 151) and *M. thermophila* (49) have also been examined. In addition, members of the *Methanobacteriales*, whose flagship organism *Methanobacterium thermautotrophicum* (now *Methanothermobacter thermautotrophicus*) was sequenced in 1997 (166), are currently the object of structural proteomics initiatives (38, 192).

Last, but not least, are the halophilic archaea, some members of which were the first archaeal model systems for "extreme" environments. *Halobacterium* NRC1 is one of three halophiles that have been studied to a significant extent with respect to physiology, metabolism, and genetics (68, 96, 124, 126, 186), *H. halobium* (*salinarum*) (19, 36, 102, 169, 173) and *Haloferax volcanii* being the others (7, 8, 18, 93, 169). Any or all of these can be viewed as a model system for functional genomic analyses, with pioneering studies in *Halobacterium* NRC1 having already been reported (68, 92).

Merits and Challenges of Functional Genomics Approaches

The new field of systems biology involves the rationalization of the behavior of an organism in terms of the interplay of all of its genes and the proteins that they encode. This approach has been a driving force for and has greatly facilitated genomewide and multigenome studies, thereby generating unprecedented amounts of data that need to be sifted through for specific insights. Initial efforts with transcriptomics and proteomics indicated that, while the potential to know how the influence of specific stimuli on every gene and protein associated with an organism's genome could be monitored, such information may or may not be readily useful. This realization was sobering given the need for highly sophisticated, analytical, and statistical skills, as well as the significant expense, that are typically part and parcel of functional genomics approaches. Thus, the substantial promise offered by functional genomics can be tempered at present by the investment in terms of time and money required and the actual outcome in terms of useful biological information.

Table 2 summarizes current activities with respect to the use of "omics" to study archaea. Not unlike other areas of biological research, "omics" tools have been used to various extents and with varying effectiveness. In some cases, studies have shown that certain tools have been especially useful in exploring biological phenomena. In other cases, beyond the demonstration that a specific "omics" tool can be applied to a particular archaeon, not much new and useful information has resulted. It remains to be seen how these new tools impact the study of archaea and how to be of best use to a systems biology perspective.

As functional genomics tools become more available and are implemented for the study of archaea, it will be important to make strategic use of the technologies. Initial efforts have often been "look and see," with the expectation that genomewide responses to environmental perturbations will lead to interesting insights. While this has been true in some instances, hypothesis-driven efforts will be increasingly important so that the end result of functional genomics studies provides definitive answers to biological questions.

PROTEOMICS

The term "proteomics" is used herein to describe the experimental determination of protein species within a given cell on a genomewide basis. Such approaches obviously include the identification of proteins and can also include quantitation relative to other proteins in the cell or quantitation relative to that observed when the cell is grown under different conditions. Purely computational analyses of the proteins encoded by a genome are not considered, unless coupled with experimental investigation. The proteomic analyses of archaeal species have followed the methodologies that have been developed for bacteria and eucarya in general. In fact, some of the earlier pioneering studies using two-dimensional gel electrophoresis (2-DGE) were carried out with the halophilic archaea (46) and the methanogenic archaeon *M. jannaschii* (123), where changes in protein patterns were evident in response to high temperature and pressure, respectively. Spot identification subsequently became routine, initially by N-terminal Edman sequencing, and then by mass spectrometry (MS), typically of tryptic peptides. The effiency of identifying proteins in 2-DGE spots has benefited from advances in developments in analytical liquid chromatography (LC) as well as in MS, with archaea providing some samples for method's development (109). The 2-DGE approach has several severe limitations, due to the nature of the separation medium and the staining methods. It is slowly being replaced in the more well-studied bacterial systems by direct high-throughput LC of complex mixtures coupled to more sophisticated MS applications (see, for example, references 84, 111, 197). Such advances are also being utilized by the research community in general, including those working with archaea, and some examples are included herein. The next frontier is environmental proteomics; archaea have been involved, although they have not yet played a dominant role (for example, reference 143). As is the case with many facets of archaeal biology, the field of proteomics can be divided into accomplishments achieved with three different groups of archaea: methanogens, halophiles, and organisms that fit into neither category. However, this will no doubt change significantly in the near future, as the proteomes of more and more novel organisms are examined by increasingly rapid, high-throughput (HTP) techniques. Nevertheless, these categories are currently convenient and will be used here. The reader is also referred for additional information to an excellent summary of the proteomics techniques that have been applied to archaea (32).

Proteomics of Methanogens

M. jannaschii

M. jannaschii was isolated from a deep-sea hydrothermal vent and is regarded as a hyperthermophile, with an optimum growth temperature near 90°C. Its genome sequence was published in 1996, the fourth genome to be sequenced (27) and the first from an archaeon. It is, perhaps, not surprising that *M. jannaschii* has been the subject of many genome-related analyses (73). These include a seminal proteomics study (123) using 2-DGE to monitor changes in the cellular protein content when cells of this strict hydrogenotroph were grown under high (178 kPa) and ultralow (650 Pa) partial pressures of hydrogen. While numerous changes were seen in the 2-DGE pat-

terns of the two growth conditions, attention was focused on representatives of each: five from low pH_2- and one from high pH_2-grown cells. Low pH_2 caused increased levels of F420-dependent methylenetetrahydromethanopterin dehydrogenase (MTD) and four flagellar proteins, while high pH_2 favored higher levels of H_2-dependent methylenetetrahydromethanopterin dehydrogenase (HMDX). The response of the two enzymes was anticipated, but regulation of expression of flagella by hydrogen had not been seen in any domain of life before this study. This was also the first example of the regulation of flagella synthesis in the *Archaea* and demonstrated that *M. jannaschii* does exhibit a chemotactic response.

A subsequent proteomic analysis by 2-DGE of *M. jannaschii* also focused on two flagellar proteins B1 and B2 (encoded by MJ0891 and MJ0892, respectively). Their cellular concentrations changed in response to H_2 concentration, availability of ammonia, and the stage of cell growth (66). Multiple protein spots were observed after separation by 2-DGE that were not observed in the earlier study (123). These yielded peptides that corresponded to these two proteins, even though the spots were at different isoelectric points and mass values. Moreover, the abundance of the spots varied with the growth conditions. Given the extreme stability of proteins from organisms like *M. jannaschii* that thrive at temperatures close to 90°C, it is also possible that multiple protein spots after 2-DGE could arise from the incomplete dissociation and denaturation of protein-protein complexes. Alternatively, or in addition, the flagellar proteins might be proteolytically cleaved or have varying degrees of glycosylation (88), phosphorylation (see below), or deamidation (66). It was concluded that the flagellar proteins are structurally modified in some fashion in response to the cellular environment and growth stage, although this complex regulatory response is, as yet, not understood.

The same authors subsequently reported a more complete proteome analysis of *M. jannaschii* using 2-DGE and matrix-assisted laser desorption ionization-time of flight mass spectrometry (MALDI-TOF-MS) and liquid chromatography coupled to tandem mass spectrometry (LC/MS/MS) to analyze tryptic digests (65). In an effort to separate and identify as many proteins as possible, changes were made to the typical separation approach, including the use of two different protein stains and the use of two different parameters to resolve proteins in the first dimension (nonequilibrium pH electrophoresis as well as isoelectric focusing). A total of 170 proteins of the 1,738 proteins annotated in the genome were identified after the analyses of extracts of cells grown under "optimal" conditions. Almost a third of them were annotated as (conserved) hypothetical proteins, demonstrating the power of the approach in showing that previously uncharacterized proteins were produced within the cell. Several proteins behaved anomalously during 2-DGE, indicating that not all protein complexes are dissociated by the sample preparation procedure. For example, the three subunits ($\alpha\beta\gamma$) of methyl coenzyme M reductase isoenzyme I and one subunit of isoenzyme II were all identified in the same spot on the 2-DGE. In addition, some proteins were detected in more than one spot, suggesting differences in degrees of phosphorylation in the case of an elongation factor (MJ0822), or in glycosylation in the case of an S-layer protein (MJ0324). It was also apparent that glycosylation might interfere with protein staining. All in all, this study nicely demonstrated that, even for relatively small genomes such as that of *M. jannaschii*, which at 1.7 Mb (27) is close to a third that of *Escherichia coli* (4.6 Mb [20]), there was far more to rationalizing proteomic analyses than merely identifying proteins.

The most recent report on the proteome of *M. jannaschii* employed a non-gel method of protein separation (198). This so-called shotgun technique, which had been successfully applied to identify proteins in lower eucarya, involves digestion of proteins in a complex mixture without prior separation, and then their separation and identification using multidimensional chromatographic capillary columns and LC/LC/tandem MS. Hundreds of proteins can be identified in the same complex mixture. For *M. jannaschii*, more than 13,000 peptides were identified and assigned to 954 proteins, more than half the proteins encoded by the complete genome. They ranged in size from less than 10 to more then 100 kDa, and almost all identifications were based on more than two peptides. More than 40% were identified over 50% of their sequence, and these are assumed to represent the more abundant proteins in the cell. About 60% of the genes in the genome encode (conserved) hypothetical proteins, and about 27% of them were identified. A total of 32 of the 36 proteins involved in the conversion of CO_2 to methane were identified. Several steps are catalyzed by multiple enzymes, and their relative abundance in the cells (grown under high H_2 partial pressure) indicated unique regulatory control of methane production. The proteome analysis also identified self-splicing intein peptides and new proteins created by splicing, once more demonstrating the complexity of even a "simple" genome.

A more targeted, non-2-DGE approach with *M. jannaschii* was recently reported (60) that used a continuous elution denaturing one-dimensional gel electrophoresis approach, coupled to subsequent reverse-phase liquid chromatography and electro-

spray ionization quadrupole mass spectrometry (ESI/Q-FTMS) to fragment and identify intact proteins. For a ribosomal-enriched fraction, 24 of the 68 proteins predicted to be in the ribosome were identified. In addition, 50 proteins were identified from fractionated samples of cell-extracts, and four were found to have significant mass shifts due to methylation, acetylation, and/or incorrect start sites. Analysis of a histone-enriched fraction gave no evidence for posttranslational modification of this DNA-binding protein, in agreement with the absence of the relevant enzymes from the *M. jannaschii* genome. However, the same rather surprising conclusion was reached by a similar analysis of histone proteins in *Methanosarcina acetivorans*, whose genome does contain acetyl transferase enzymes predicted to modify histones (60).

Methanococcoides burtonii

Methanococcoides burtonii was isolated from permanently cold (2°C), methane-rich lake water in Antarctica, although in the laboratory it grows fastest (T_{opt}) at 23°C. The organism uses C-1 compounds such as methylamines and methanol, but not formate, H_2 and CO_2, or acetate, as growth substrates. The draft genome of the organism indicates that the genome is ~2.9 Mb in size and encodes 2,676 proteins (152). A proteomics approach to investigate protein levels in cells grown at low (4°C) and optimal temperatures was recently reported by Goodchild et al. (71). This study represented the first global analysis of proteins involved in cold adaptation (rather than examining the cold shock response). The results were particularly of interest since this archaeon does not contain homologs of the well-characterized cold shock protein (Csp) family found in bacteria. Comparison of 2-DGE from the two cell types revealed 54 proteins that appeared to be differentially regulated, and these were identified by LC/ESI-MS/MS. Reverse-transcriptase PCR analyses were conducted for the expression of the genes encoding 33 of the 54 proteins, and for about half of them there was good agreement between the observed changes in protein and mRNA levels. Of course, an exact match is unlikely, given that the stability of RNA and protein products will vary from gene to gene. There were also examples of posttranslationally modified proteins. One example was methyl coenzyme M reductase, whose three subunits were identified in eight separate spots. This made quantitation difficult, but an examination of mRNA levels showed that all three genes were up-regulated at the lower temperature by approximately 3-fold.

The results showed that adaptation to temperatures significantly below the optimum involves transcription and RNA polymerase subunit E and protein folding with peptidyl prolyl *cis/trans* isomerase. There was also evidence the optimum growth temperature is a stressful condition, and this required production of the heat shock protein DnaK. In a companion study the same authors identified approximately 20% of the proteins encoded in the genome (528 of 2676) by direct separation and analysis of trypsin-treated cell-free extracts by LC-MS/MS (2-DGE was not utilized). Of those 528 identified, about 25% were previously annotated as hypothetical proteins. The remainder were assigned a variety of anticipated physiological roles, including the complete set of proteins of the methanogenic pathway (see also reference 151). One unanticipated result was the identification of two transposases, indicating the potential for active genetic exchange at these cold temperatures.

The proteome of *M. burtonii* has also been examined (70) by a second generation MS-based technique using isotope labeling, the so-called isotope-coded affinity tag (ICAT) approach (74). This involves labeling cysteine residues already present in proteins in complex mixtures with either a ^{12}C- or ^{13}C-labeled reagent. The two protein mixtures are then mixed and are digested with trypsin. The peptides are purified by affinity chromatography and analyzed by MS/MS. The essence of this approach is that the relative abundance of the ^{12}C and ^{13}C isotopes allows accurate quantitation as well as the identification of the same protein in the two samples. In the case of *M. burtonii*, cells were grown at 4 and at 23°C, the same as were used for the 2-DGE study, and a total of 163 proteins were identified using the double-labeling approach (70). Labeling consistency of each sample was assessed by two independent labelings, and the relative abundance of the peptides that were identified was comparable. However, note that, although statistical analyses showed a high level of confidence in protein identification, only 48 of the 163 were identified from two or more peptides. A comparison of the results from this study with those from the 2-DGE showed that the two approaches are very complementary. Thus, the number of proteins identified by the 2-DGE approach that were differentially expressed by at least 2-fold (16%) was 4-fold higher than in the ICAT study. This is assumed to be a result of the higher stringency used for the ICAT analysis, where the relative abundances must be consistent in all labeling experiments. Similarly, the 2-DGE and ICAT studies each identified unique proteins (21 and 11, respectively), further emphasizing the need to carry out both techniques to optimize coverage. In general, the biological implications of the ICAT study confirmed and extended those made from the 2-DGE analysis, with a key regulatory role for proteins involved in methanogenesis and transcription, as well as hypothetical proteins of unknown function.

Methanosarcina acetivorans

One of the surprises of the genome sequencing projects was the large size of the genome of *M. acetivorans*. At 5.75 Mb, it is the largest yet reported for any member of the *Archaea* (63). Clearly, this organism has a considerable metabolic capacity that is well beyond that of, for example, *M. jannaschii*, whose genome is less than one third the size. One striking factor is the large number of duplicated genes in the *M. acetivorans* genome; the role of these paralogs, and how their expression is regulated is largely unknown. A recent proteomic study sought to address this issue (107, 108). Using 2-DGE coupled with MS analysis (MALDI-TOF-TOF) after in-gel tryptic digestion, a total of 412 proteins were identified, 30% of which were annotated as encoding hypothetical or conserved hypothetical proteins. Many of the proteins were detected only in acetate- (122 proteins) or methanol-grown cells (102), rather than both (188). Many of them included paralogs involved in the methanogenesis pathway. Some, such as the three operons encoding the methanol-utilizing methyltransferases, appeared to be differentially regulated, while others, such as the two operons encoding the heterodisulfide redutase, were not. Other genes apparently differentially regulated included those involved in ATP synthesis, protein synthesis, protein-folding, and stress responses. One of the most intriguing findings concerned the presence in acetate-grown cells of proteins that are part of a NADH-dependent, sodium-transporting membrane oxidoreductase. The results suggest that different methanogens may have different mechanisms for energy conservation during methanogenesis. This proteomics study has given the first glimpse into the extremely complex metabolism of a fascinating but poorly understoood methanogen.

Methanosarcina thermophila

Methanosarcina thermophila is a moderately thermophilic methanogen and was one of the first archaea to be studied using a proteomic approach. The protein content of acetate- and methanol-grown cells was examined using 2-DGE, and the study resolved more than 400 protein spots from each growth condition, with more than 100 being exclusive to one or the other (87). Spots corresponding to three enzymes (acetate kinase, phosphotransacetylase, and carbon monoxide dehydrogenase) were identified in acetate-grown cells using purified proteins as markers, and these were either absent or were at a lower level in methanol-grown cells. These studies were extended by Ding et al. (49) who identified the most abundant proteins in methanol- and acetate-grown cells by N-terminal Edman sequencing and estimated the relative abundances of subunits of the methanol: coenzyme M methyltransferase (MTase) system. MTase I contains two subunits (MtaB and MtaC), while MTase II contains only one (MtaA), and there are two homologs of MtaA and three for both MtaB and MtaC. Some of these were shown to be at higher levels, or only detected in methanol-grown cells, and some only in acetate-grown cells. Global proteomic analyses have not been reported for this organism, and its genome sequence is not yet publicly available, in contrast to those of the related species *M. mazei*, *M. acetivorans,* and *M. barkeri* (47, 63; see www .genomesonline.org).

Proteomics of Halophiles

Extensive proteomic analyses have been carried out with the halophilic archaea (92). In a seminal paper, 2-DGE was used to analyze cell extracts of several *Halobacterium* species and demonstrated that they exhibit a response to heat shock (60°C for 3 min) by synthesizing a limited number of proteins of defined molecular weight range (46). Cells rapidly assumed their "normal" pattern of protein expression when returned to the 'normal' temperature (37°C). Some proteins that appeared in response to heat were also observed in gel patterns when cells were exposed to a change in salinity, suggesting that, like the other two domains of life, archaea, or at least the halophilic species, exhibited a general stress response. Specific changes in protein patterns on 2-DGE were subsequently observed with *Haloferax volcanii* when it was grown at high and low salinities, although the identities of specific proteins were not determined (121). The problems caused by the need for high-salt concentrations to prevent aggregation and precipitation of proteins from halophiles have recently been addressed by using filtration methods in the case of *Halobacterium salinarum* (36) and various washing methods with *H. volcanii* (93), although in these studies, proteins were not identified and differential expression was not assessed.

The first study reporting the large-scale identification of expressed proteins of a haloarchaeon from an identification perspective was for *Halobacterium* sp. NRC-1 (14). The genome sequence of this organism encodes 2682 putative protein-coding genes (126). The ambitious proteome analysis was part of a global systems approach that also included transcriptional profiling and utilized the wild-type strain and three mutants that either overproduced or did not produce the light-harvesting, membrane-associated bacteriorhodopsin. The proteomic approach involved direct

analysis of differentially labeled cell-extracts using the ICAT technique as described above for *M. burtonii*. A total of 272 different proteins were identified from 1,120 tryptic peptides, showing that 50 of the proteins were differentially produced in the wild-type and mutant strains. Note that the ICAT approach can detect quite minor changes (<50%) in relative protein abundance (76). Only seven of the 50 regulated proteins were of unknown function, compared with 41% predicted from the genome sequence. The rest of the proteins were associated with phototrophy, membrane and carotenoid biosynthesis, and primary N metabolism (arginine, glutamate metabolism, and pyrimidine). Perhaps surprisingly, corresponding changes in the transcriptional levels were observed for only 17 of these 50 differentially regulated proteins. Nevertheless, it could be concluded that at the systems level, there was reciprocal regulation of phototrophy and arginine fermentation, which are the two major pathways of energy conservation in *Halobacterium* sp.

The protein content of the insoluble membrane and soluble cytoplasmic fractions of cell extracts of *Halobacterium* NRC-1 have also been examined directly using a simple shotgun approach. The tryptic digests of the two fractions were separated by microcapillary HPLC, and peptides were identified by ESI-MS/MS. A total of 426 proteins were identified, with 401 encoded in the chromosome and 25 in the minichromosomes of this organism. Of these, 232 were soluble, 165 insoluble, and 29 in both fractions. In this case, only one growth condition was used and regulatory effects were not assessed.

A global proteomic analysis of *H. salinarum* (strain R1, DSM681) was recently reported (173). As most halophilic proteins are acidic (82% have pI values between 3.5 and 5.5), one of the goals was to resolve proteins that migrate over a narrow pH range. The use of overlapping 2D gels and subsequent MALDI-TOF MS peptide mass fingerprinting (PMF) led to the identification of over 800 proteins in the more than 1,800 spots resolved in the cytoplasmic fraction. A semiautomated protocol was utilized which included automated spot excision, 96-well plate-based in-gel digestions, and automated spectral acquisition and peak annotation generation. The genome sequence of the organism has not been released (www.genomes online.org) but is reported (173) to contain at least 2784 open reading frames (ORFs) that encode proteins. Approximately 40% of the proteins encoded by these ORFs were therefore identified by 2-DGE analysis. It was estimated that the gels contained 1,600 unique proteins, indicating that a high percentage of the number of predicted cytoplasmic proteins (a total of 1,959) are present in cells grown under "standard" conditions. Almost 100 proteins were found in more than one spot on the 2D gel, indicating that a significant fraction may be posttranslationally modified, although some might be artifacts (deamidation, carbamylation, etc.) of the separation procedure (173). The correlation between predicted and observed protein migration on the 2D gels enabled reassignment of start codons for some proteins based on their apparently anomalous positions. The results showed two general biases in the types of proteins identified. First, proteins with a mass above 25 kDa were more readily identified than those smaller (50% versus 20% of those predicted), and this is presumably because of the smaller number of tryptic peptides from smaller proteins. Second, chromosomally encoded proteins were much more readily detected (48% of predicted) than those encoded by the four megaplasmids (41 to 284 kb; 14% of predicted) in this organism, suggesting that in general the latter are not significantly expressed under the growth conditions. Tebbe et al. (173) speculated that a higher percentage of *H. salinarum* proteins were not identified by the 2-DGE analyses because of a lack of gene expression, loss during sample preparation, poor staining, and lack of identification by MALDI-TOF.

The membrane proteomes of *H. halobium* and *H. salinarum* have also been analyzed. The problem with membrane proteins is to solubilize the proteins under conditions that allow separation while maintaining solubility. This was achieved with the membrane proteins of *H. halobium* by a rapid and simple technique involving solubilization with, and tryptic digestion in, aqueous methanol and separation and peptide identification using a micro-LC-MS/MS approach (21). A total of 41 proteins were identified, 80% of which were predicted to contain transmembrane domains. These included all tryptic peptides of bacteriorhodopsin, even those posttranslationally modified. In an extension of the proteomics analysis of *H. salinarum* (173), an LC, rather than 2D approach, was used to analyze membrane proteins (102). Initial studies using various approaches showed that 2D analysis was not very productive because the membrane proteins precipitated at their pI and could not be solublized, and peptides that are obtained from resolved spots are not suited to identification by MALDI-TOF analysis. Nevertheless, one-dimensional-sodium dodecyl sulfate (SDS) gel separation coupled with LC-MS/MS resulted in the identification of 114 proteins that were predicted to contain transmembrane α-helical regions, which is about 20% of the number estimated from the genome sequence. In combination with the study of the cytoplasmic proteins, a total of almost 950 *H. salinarum* proteins were identified, representing 34% of the predicted gene products. This archaeon

is therefore one of the best characterized of all organisms by proteomic approaches.

The use of proteomics to identify proteins of biotechnological relevance has recently been utilized in halophilic archaea. A pioneering study (132) examined cell extracts of *H. salinarum* grown with 3.5, 4.3, or 6.0 M NaCl by 2-DGE, and observed 773 spots, 56 of which appeared to be differentially regulated. As the authors pointed out, protein identification using (MALDI-TOF) analysis of tryptic peptides is not efficient using halophilic organisms as their proteins contain a low proportion of the basic residues targeted by trypsin; thus, the more sophisticated electrospray ionization quadrupole (ESI-Q-) TOF-MS/MS approach was used to identify more than 50 proteins from spots. These included inosine monophosphate dehydrogenase, which was more abundant in cells grown at high-salt concentration. Because of the biotechnological application of such a salt-tolerant enzyme, the recombinant form was obtained and further characterized. In a subsequent extension of this work, the same approach was used to identify 29 proteins from spots after 2-DGE that appeared to be differentially regulated after growing cells at different salinities (37). Of these, six were of unknown function and, based on detailed bioinformatic analyses, recombinant expression and immunological pull-down assays, one of them was proposed to catalyze the acetylation of a ribosomal protein (L13). This approach to identify halophilic enzymes of biotechnological significance was recently applied to *H. salinarum* grown in the presence of petroleum hydrocarbons (4% diesel). Five spots on 2-DGE were shown to vary dramatically after analysis of cell extracts from diesel-grown and control cells, although the proteins involved were not reported (92).

Proteomics of Nonmethanogenic, Nonhalophilic Archaea

Sulfolobus species

A complete proteomic analysis of a *Sulfolobus* species has not been reported. However, some seminal proteomic work has been performed. In the first proteomics study (178), 2-DGE analysis was performed on *Sulfolobus* strain B12 and demonstrated that thermotolerance was accompanied by the synthesis of three heat shock proteins. This extremely thermophilic acidophile grows optimally at 70°C and pH 3.5 and acquired thermotolerance at 92°C after exposure at 88°C for one hour. A heat shock response was also observed in *S. acidocaldarius* by 2-DGE (130). The same approach was also used with this organism to show phosphorylation of a specific protein in response to phosphate starvation, and this led to the first proposal for a two-component regulatory system in the *Archaea* (130). The use of 2-DGE to identify phosphoproteins in *Sulfolobus* sp. was subsequently utilized by others (e.g., 112). Very recently, a double-labeling procedure involving *S. solfataricus* growth on ^{15}N- and ^{13}C-enriched media was reported. The approach improves the efficiency of protein identification as well as enabling accurate quantitation of protein abundance, although the biology of the organism was not specifically addressed (167).

Metallosphaera sedula

The approach used to investigate the response to heat shock in *Sulfolobus* species has been applied to examine the proteome of another extreme thermoacidophile, *M. sedula* (75). However, in this case, continuous culture was used to minimize growth phase and growth rate effects. *M. sedula* was grown in a 10-liter chemostat at 74°C, pH 2.0, at a dilution rate of 0.04 h^{-1}, and was subjected to both abrupt and gradual temperature shifts. Growth in continuous culture for 100 h at the supraoptimal temperature of 81°C (the stress condition) led to an approximately sevenfold lower steady-state cell density than that observed for growth at or below 79°C. SDS-polyacrylamide gel electrophoresis (PAGE) (both one and two dimensional) revealed significantly higher levels (sixfold increase) of a 66-kDa stress-response protein (MseHSP60), immunologically related to Thermophilic Factor 55 from *Sulfolobus shibatae* (178), which was later determined to be the archaeal thermosome. The thermosome was clearly the most abundant intracellular protein apparent by 2D SDS-PAGE. If the culture that had been acclimated to 81°C was returned to a lower temperature (74°C), the amount of thermosome reverted to a level observed prior to thermal acclimation. Furthermore, when the previously acclimated culture (at 81°C) was shifted back from 74 to 81°C, without going through gradual acclimation steps, the result was the immediate washout of cells from the chemostat, indicative of a loss of thermotolerance. This study showed that gradual thermal acclimation of *M. sedula* could only extend the upper growth temperature limit of stable growth for this organism by 2°C (and this occurred at reduced cell densities).

Pyrococcus furiosus

In one of the few proteomics studies directed toward membrane-bound proteins (80), cell extracts of *Pyrococcus furiosus* were separated into soluble and insoluble fractions and analyzed by 2-DGE, and

MALDI-TOF and micro-LC-ESI-MS/MS; a total of 32 membrane proteins and 34 cytoplasmic proteins were identified. Based on bioinformatic analyses for signal peptides (SignalP, TargetP, and SOSUISignal) and transmembrane-spanning α-helices (TSEG, SOSUI, and PRED-TMR2), it was concluded that 23 of the 32 proteins (72%) from the membrane fraction should be membrane associated, and that all of the proteins from the cytoplasmic fraction should be in the cytoplasm. In a separate study, 62 cytoplasmic proteins were separated by 2-DGE, and these were utilized for a systematic comparision to demonstrate the much greater efficiency of the micro-LC-MS/MS approach in identifying proteins, and particularly for smaller ones (<25 kDa), compared with the routinely used MALDI-TOF methods (109). However, the two approaches are complementary, and both should be utilized if possible.

Thermoplasma acidophilum

Proteomic-type, 2-DGE analyses have been applied to *T. acidophilum* and demonstrated some unexpected eucarya-like properties of the organism. Cell-free extracts were fractionated, and low-molecular-weight proteins were systematically analyzed by N-terminal Edman sequencing. This led to the discovery of a ubiquitin-labeled peptide, suggesting that ATP-ubiquitin-dependent proteolysis occurs in this organism (189). In a related study from the same group, 2-DGE was used to analyze purified proteasomes from *T. acidophilum* (200).

TRANSCRIPTOMICS

The relatively small sizes (<2.5 Mb) of several archaeal genomes, and the increasing availability of full genome sequences for these organisms, make DNA microarrays (155) a strategic tool for their study. While present knowledge of the biology of the third domain of life is still lagging in comparison with the *Bacteria*, the rapid maturation of microarray technology, in conjunction with improvements in bioinformatics and data analysis tools, should close this gap in the near future. Efforts to determine the transcriptome of archaea using DNA microarrays are reviewed in this section.

Considerations for Genomewide Transcriptional Response Experiments

One of the first realizations that arise from the use of DNA microarrays to investigate genomewide transcriptional responses is the sensitivity of gene expression to even minor (and sometimes unintended) perturbations in growth conditions. Thus, cultures must be grown, whether in batch or continuous mode, under carefully controlled conditions to reliably assess responses to intended environmental and nutritional perturbations. This may prove to be difficult for archaea that grow under extremes of temperature, pH, salinity, or strict anaerobic conditions. There are many issues to consider when choosing between batch and continuous culture for differential expression experiments with archaea. For example, a key advantage of chemostat culture is the ability to regulate cell density and maintain nutrient-limited, steady-state growth for extended periods (164). However, chemostats are more difficult to operate than batch reactors. Fortuitously, the extreme growth conditions characteristic of many archaea can eliminate issues with culture contamination that can plague continuous growth of conventional mesophilic microorganisms. Long-term chemostat operation is known to be affected by significant wall growth; the seeding of the culture by sessile biofilm-associated cells can be problematic in interpreting transcriptional response data (139). While biofilm formation has not been extensively examined in the *Archaea*, *P. furiosus* (140) and *Thermococcus litoralis* (144, 145) have been observed to form biofilms in chemotat culture. While batch cultures are typically easier to operate than chemostats, sampling times are a critical issue. Preferably, cells are perturbed and sampled during exponential phase in batch culture. If these steps are performed too early or late, it may be almost impossible to separate growth phase and growth rate effects from the intended transcriptional response. Cell density should be carefully controlled since microbial behavior, such as cell signaling or aggregation, can be cell density dependent, as has been noted for cocultures of the hyperthermophilic bacterium *Thermotoga maritima* and *M. jannaschii* (90).

Although not often considered in functional genomics experiments, gene expression can be growth phase dependent; the extent to which this is important for archaea compared with bacteria is not yet known. To illustrate the effect of growth phase on the transcriptome, a culture of *S. solfataricus* grown on peptide-based media at 80°C/pH 4.0, was examined at mid-exponential, late-exponential, and early-stationary phases. Overall, almost 600 ORFs (approximately 20% of the genome) were differentially expressed 2-fold or more between growth phases (Fig. 1) (S. Tachdjian and R. M. Kelly, unpublished results).

To determine the time course response of *S. solfataricus* to heat shock, a 2-liter batch culture was grown to mid-exponential phase at 80°C, and the temperature was then shifted to 90°C within 10 min.

The culture was sampled 10 minutes before temperature shift was initiated, and then at 5, 30, and 60 min after the temperature had reached 90°C. As anticipated, the results demonstrated that many genes respond dynamically to heat shock. For example, more than one third of the genome was found to be differentially expressed within 5 min of the culture reaching 90°C, but very little differential expression occurred between 30 and 60 min (see Fig. 1). Similar results were obtained after heat shock of the hyperthermophilic bacterium *T. maritima*, where differential gene expression patterns could be categorized into immediate, short-term, and long-term responses (142). As discussed below, the same was true for the cold shock response (from 95 to 72°C) of *P. furiosus*, although the low temperature slowed down the acclimation time, which occurred over a 5-h period (188). Failure to track time-dependent gene expression patterns can lead to inaccurate interpretation of biological phenomena. For example, in a *S. solfataricus* acid shock experiment (pH 4.0 to 2.0) similar to the heat shock experiment described in this section, less than 2-fold changes were observed for the hypothetical proteins SSO0337 and SSO2887 after 60 min, in comparison with pre-acid shock levels (Fig. 2). However, SSO0337 was more than 8-fold upregulated at 5 min and 2-fold upregulated at 30 min compared with pre-acid shock values. On the other hand, SSO2887 was unaffected after 5 min, downregulated 3.6-fold and 2.2-fold after 30 and 60 min, respectively. Clearly, transcriptional response is significant for both ORFs although this was not clear from "before-and-after" samples comparing the 60-min and pre-acid shock expression levels.

Transcriptional Response Analysis of Archaea

Although DNA microarrays have only begun to be used for the study of archaea, interesting insights into specific aspects of the biology of these organisms have already been obtained from representatives of most of the major types, including the methanogens, halophiles, and hyperthermophiles. Studies are described, commencing with organisms that have been the most studied to date, the heterotrophic anaerobe, *P. furiosus*.

Pyrococcus furiosus

Influence of sulfur on *P. furiosus* metabolism. The first application of DNA microarrays to study the metabolism of either an archaeon or a hyperthermophile was reported in 2001 using *P. furiosus* (157). An array was constructed using full-length PCR products to 271 of the 2,065 genes annotated in its genome.

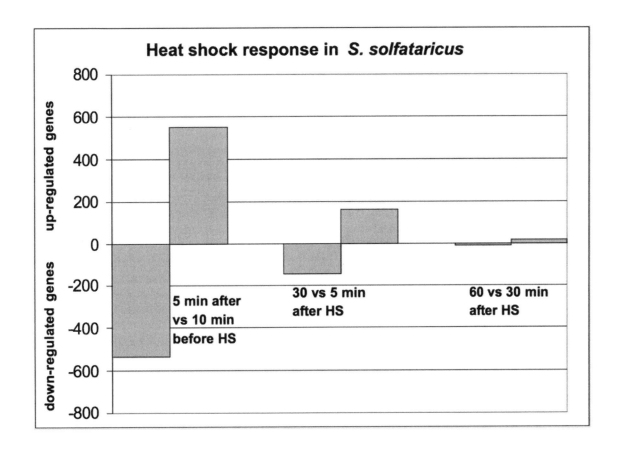

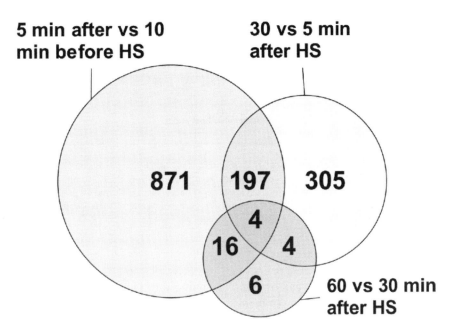

Figure 1. Time-dependent response of *Sulfolobus solfataricus* to a temperature shift from 80°C to 90°C (Tachdjian and Kelly, unpublished). Histogram (top) and Venn diagram (bottom) represent the number of genes differentially expressed more than 2-fold at each stage of the heat shock. HS, heat shock.

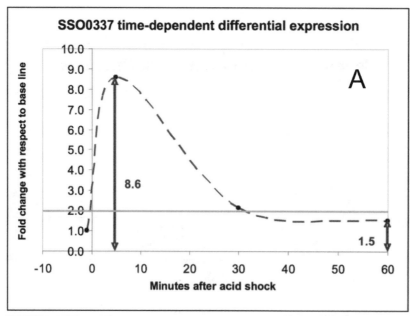

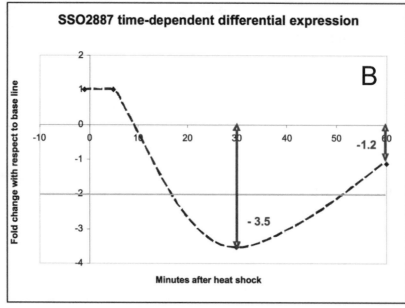

Figure 2. Time-dependent differential expression of ORFs SSO0337 and SSO2887 from *Sulfolobus solfataricus* subjected to a pH shift. ORFs SSO0337 (A) and SSO2887 (B) were subjected to a pH shift from pH 4.0 to pH 2.0 at 80°C (Tachdjian and Kelly, unpublished). Samples were taken 1 min before pH shift was initiated, then 5, 30, and 60 min after reaching pH 2.0 (lines drawn through points). Significant levels of differential expression occurred at intermediate sampling times, although no substantial changes were observed for the pre-acid shock versus 60-min contrast.

The genes included those that are predicted to encode proteins mainly involved in the pathways of sugar and peptide catabolism and in the utilization of metals (such as Fe, Ni, W, and Mo) and the biosynthesis of cofactors, amino acids, lipids, polyamines, ribosomes, and nucleotides, together with putative chaperonins, ATPases, and transcriptional regulators. *P. furiosus* obtains energy for growth by converting either sugars or peptides to organic acids, CO_2, and hydrogen. Cells grow well in the absence of S^0 but if it is present, it is reduced to H_2S. For the array analysis, a batch comparison was carried out where RNA was derived from cells grown in the presence and absence of S^0. The expression of 21 of the 271 genes decreased dramatically (>5-fold) in the presence of S^0, and of these, 18 encoded subunits associated with the three different hydrogenase systems that have been characterized from *P. furiosus*. The other three encode homologs of ornithine carbamoyltransferase and HypF, both of which appear to be involved in hydrogenase biosynthesis, as well as a conserved hypothetical protein. The presence of S^0 resulted in the upregulation by more than 2-fold of the expression of two previously uncharacterized genes. Their products were termed SipA and SipB (for sulfur-induced proteins), and it was proposed that they are part of a novel S^0-reducing, membrane-associated, iron-sulfur cluster-containing complex. For example, SipB was predicted to contain two [4Fe-4S]] clusters that might be involved in providing electrons for S^0 reduction. Other genes whose expression was upregulated by S^0 encoded a putative flavoprotein and a second putative iron-sulfur protein, both of which may also be involved in S^0 reduction. However, note that the mechanism by which *P. furiosus* generates H_2S is still largely unknown. Although the DNA microarray approach generates strictly comparative results, the fluorescence intensities associated with a given spot on the array gives a qualitative indication of the degree to which the relevant gene is expressed. Consequently, it was shown that, of the twenty (S^0-independent) genes that appeared to be the most highly expressed (at more than 20 times the detection limit), twelve of them encoded enzymes that had been previously purified from *P. furiosus* biomass, indicating that their protein products are among the most abundant within the cell. Conversely, none of the products of the thirty-four (S^0-independent ORFs) that were not expressed above the detection limit had been characterized biochemically. As yet, the effects of S^0 on the expression on all genes encoded by the *P. furiosus* genome has not been reported.

***P. furiosus* transcriptional response to thermal stress.** The effect of supraoptimal temperatures on gene expression in *P. furiosus* was examined with a targeted DNA microarray containing 201 genes (165). These were predicted to encode proteins involved in proteolysis, the stress response, proteolytic fermentation, and glycoside hydrolysis. When a culture growing with peptides as the carbon source was shifted from 90 to 105°C for one hour, differential expression (4-fold or higher) of several genes was noted, including the thermosome (PF1974), small heat shock protein (PF1883), and two other putative molecular chaperones (PF0963 and PF1882, CDC-48 homologs, VAT-related). Of the 42 protease-related genes monitored, only two were differentially expressed more than 5-fold; pyrolysin (PF0287) was downregulated, while a subtilisin-like protease (PF0688) was upregulated. The two ATP-dependent proteases in *P. furiosus* responded to heat shock in different ways. The Lon protease (PF0467) was downregulated (3-fold), and while the expression of the genes encoding the

two proteasome β-subunits (PF0159 and PF1404) were slightly stimulated (2-fold), that encoding the α-subunit (PF1571) was downregulated (4-fold), and the proteasome ATP-dependent regulatory subunit (PAN PF0115) was not affected by heat shock. This result raises questions concerning 20S proteasome assembly and function during thermal stress, and cellular abundance of this complex. Genes related to proteolytic fermentation were among the most highly expressed under normal growth conditions and were either not affected or were downregulated to varying extents upon heat shock. Compatible solute formation under thermal stress was indicated by the induction of genes encoding a putative trehalose synthase (PF1742) and L-*myo*-inositol 1-phosphate synthase (PF1616), the latter result being obtained through Northern analysis. Mannosyl glycerate formation (e.g., PF0591) was not induced. Several glycosidase genes were significantly induced upon heat shock (e.g., PF0073, PF0076, PF0442), as were genes encoding maltose/trehalose-binding proteins (e.g., PF1739 and PF1938). Saccharide recruitment involving glycosidases during thermal stress may serve to meet the increased bioenergetic needs of the cell or be recruited for compatible solute synthesis.

Influence of carbon source on *P. furiosus* metabolism. The first complete genome DNA microarray to be constructed for a hyperthermophile or for a non-halophilic archaeon was also for *P. furiosus* (156). The array contained PCR products for the 2,065 genes that have been annotated in the genome; all but 105 of which were full-length (the exceptions were approximately 1 kb). A batch comparison was carried out where cells were grown at 95°C using either peptides (hydrolyzed casein) or the disaccharide maltose as the primary carbon source in the presence of elemental sulfur. In addition to providing information on changes in gene expression, the DNA microarray approach can be used to assess which genes are not significantly expressed, and this was the case for approximately 20% (398 of 2065) of genes in both growth conditions. For the remaining 80%, the focus was on genes whose expression appeared to be strongly regulated (signal intensity changed by at least 5-fold) by the presence of peptides or maltose. As illustrated in Fig. 3, expression of these genes is essentially shut down under one or the other condition. This applied to a total of 125 genes, most of which (82 of 125) were part of 27 clusters of two or more adjacent genes, indicating the extensive and coordinated regulation of the expression of putative operons. A total of 18 operons were upregulated by growth on maltose and 9 by growth on peptides, although 5 of the 27 operons encoded only (conserved) hypothetical proteins and so

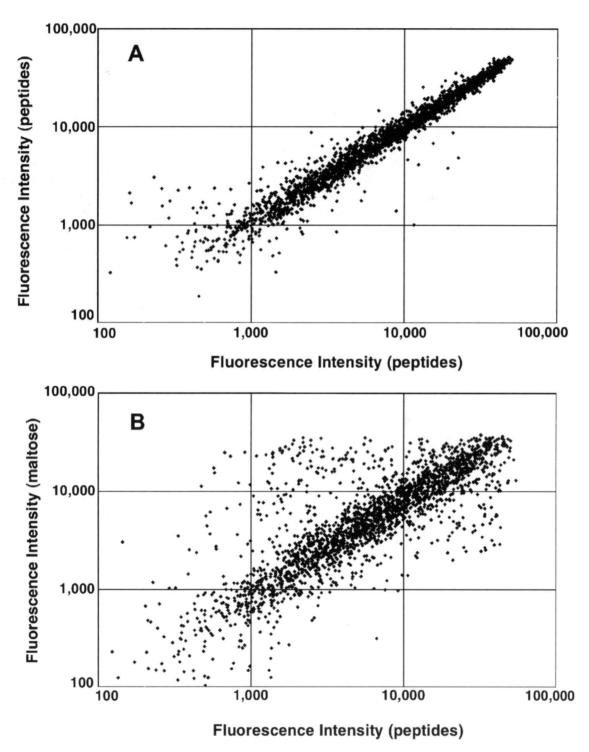

Figure 3. Fluorescence intensities of *Pyrococcus furiosus* DNA microarrays. (A) cDNA versus cDNA derived from two independent cultures of cells grown with peptides as the carbon source. (B) cDNA versus cDNA derived from two independent cultures of cells grown with peptides or maltose as the carbon source. In A, the upper and lower diagonal line pairs indicate twofold and fivefold changes in the signal intensities, respectively, while only the lines indicating fivefold changes are given in B. See text for details. Reproduced from the *Journal of Bacteriology* (156) with permission of the publisher.

their functions could not be assessed. Those operons regulated by maltose and its metabolites include those that are responsible for the biosynthesis of 12 amino acids (Glu, Arg, Leu, Val, Ile, Ser, Thr, Met, His, Phe, Trp, and Try), of ornithine, of citric acid cycle intermediates, and for maltose transport. On the other hand, operons that are upregulated in peptide-grown cells included those encoding enzymes involved in the production of organic acids and 2-ketoacids from the amino acids serving as carbon and energy sources. These findings were consistent with the pathways that had been proposed for amino acid degradation involving transamination and production of energy-yielding CoA derivatives (6). This was confirmed experimentally by finding that acetate was the major acid in spent media for maltose-grown cells, in contrast to isovalerate, butyrate, isobutyrate, and phenylacetate in spent media of peptide-grown cells (157). Several sugar metabolism genes that did not appear to be part of operons were highly regulated. These included three in peptide-grown cells unique to the gluconeogenic pathway, and three in maltose-grown cells that are unique to the novel glycolytic pathway found in this organism (182). The microarray results also provided an unexpected insight into the regulation of the hydrogenases of *P. furiosus* (156). In maltose-grown cells, the expression and the activities of the two cytoplasmic (I and II) and one membrane-bound enzyme decrease if S^0 is present in the growth medium (6, 156). However, the genes encoding the four subunits of one of the two cytoplasmic hydrogenases, hydrogenase I, were expressed in peptide-grown cultures, even though S^0 was present, and the hydrogenase activity in the cell extracts remained very low. The role of what appears to be inactive hydrogenase I in peptide-grown cells is still not understood.

Response of *P. furiosus* to cold shock. While the response of hyperthermophilic organisms to supraoptimal growth temperatures has been the subject of several investigations, as discussed herein, the effects of suboptimal temperatures on archaeal physiology and metabolism has received little attention. This is of some interest as genomes of archaea generally lack the canonical cold shock protein A (CspA) and ribosomal-binding factor A (RbfA) found in bacteria, proteins that are thought to enhance ribosomal function at low temperature (56). The first such investigation of a member of the *Archaea* using DNA microarrays monitored the effects of growing *P. furiosus* at the lower end of its temperature range for significant growth, 72°C (doubling time ~5 h) and at 95°C, which is near its optimal growth temperature (doubling time ~1 h) (188). The study included both a batch comparison of cells grown at the two temperatures and a kinetic or

shock experiment where cultures were shocked by rapidly dropping the temperature from 95 to 72°C. This shock resulted in a 5-h lag or acclimation phase, during which time little growth occurred. Transcriptional analyses showed that cells undergo three very different responses at the suboptimal temperature: an early shock (over ~2 h), a late shock (at 5 h), and an adapted response (seen in the batch comparison, occurring after many generations growing at 72°C).

In general, at the suboptimal growth temperature, much less mRNA was present for certain cellular processes (188). For example, after 1, 2, and 5 h, there were 171, 69, and 152 genes, respectively, downregulated by >2.5-fold. On the other hand, there were several genes whose expression was significantly upregulated at 72°C, and their products are assumed to be involved in the processes by which the cells become acclimated to the lower temperature. After 1, 2, and 5 h at 72°C, and in cells adapted to 72°C, there were 49, 35, 30, and 59 ORFs, respectively, that were significantly upregulated (>2.5-fold). These encoded proteins involved in translation, solute transport, amino acid biosynthesis, and tungsten and intermediary carbon metabolism, as well as numerous conserved/hypothetical and/or membrane-associated proteins. The overall response to suboptimal temperatures is summarized in Fig. 4. A priority for the cell appears to be the production of amino acids, by transport, intracellular proteolysis and biosynthesis, for the production of cold-specific proteins. Some are involved in translation and influencing DNA topology, but a major problem in understanding the overall response is that many of the genes are of unknown function. For example, of the 22 putative cold-responsive operons that were identified, 9 contain exclusively or predominantly conserved hypothetical proteins.

Another characteristic of the cold response is the upregulation of a significant number of genes predicted to encode membrane-bound proteins (188). For example, this includes 21 of the 59 genes upregulated in cells adapted to 72°C (and 34 of the 59 fall into the conserved hypothetical category). Analysis of such cells by one-dimensional SDS-gel electrophoresis revealed two major membrane proteins, which staining indicated were glycoproteins. They were termed CipA (PF0190) and CipB (PF1408), and their cold-induced expression was evident from the DNA microarray data and was confirmed by real-time (RT-) PCR. Both appear to be solute-binding transport proteins that are related phylogenetically to some cold-responsive genes identified in certain bacteria (23). It was postulated that the Cip proteins may represent a general prokaryotic-type cold shock response mechanism that is present in bacteria and archaea, and

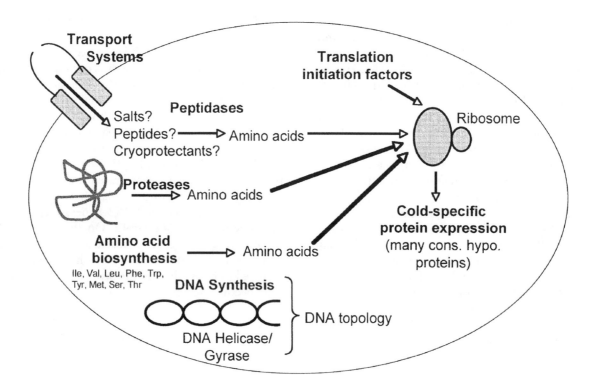

Figure 4. Cellular processes involved in the cold shock response of *Pyrococccus furiosus* when grown at 72°C rather than at the optimal temperature near 100°C. Modified from the Ph.D. thesis (187) with permission of the author.

even in hyperthermophilic archaea, although further evidence is needed to substantiate this (188).

Methanogens

It is perhaps surprising that, despite the considerable amount of proteomic analyses that have been carried out with methanogenic species (see "Proteomics of methanogens" above), there have been only two transcriptional profiling studies, both reported in 2005. One (22) examined a temperature response of *M. jannaschii,* while the other (82) evaluated the response of *M. mazei* to different carbon sources. While these data should allow comparisons to be made with proteomic analyses, unfortunately, the corresponding experiments have not as yet been reported.

M. jannaschii. *M. jannaschii* grows optimally at 85°C, and its response to nonlethal cold shock (65°C) and lethal heat shock (95°C) was studied using microarrays made of PCR products representing 99% of the 1,738 genes annotated on the genome (22). A total of 95 genes were differentially regulated by the heat shock, including those encoding both ATP-dependent and -independent chaperones. In contrast, a much more general response was seen to a decreased temperature, with 345 genes showing differential reg-

ulation, more than 170 of which are conserved/hypothetical. Eight of the temperature-responsive genes were examined by RT-PCR, which showed on average a threefold greater change in the response than in the array data. The known genes regulated by the cold shock included many involved in the processes noted above in the general cold response of *P. furiosus* (188), such as maintenance of DNA topology, transcription, and translation. The proteins included topoisomerase, DNA helicase, RNA helicase, ribosomal subunits, prolyl isomerase, and proteases. However, in contrast to *P. furiosus,* an increase in amino acid biosynthesis was not observed, and specific changes to membrane-bound proteins were not identified. Direct comparison between these two archaea is not straightforward, however, as the *M. jannaschii* data represent one time point after the cold shock, and acclimation and adaptation phases were not distinguished. In *M. jannaschii,* the genes involved in methanogenesis and an ATP synthase, which represent the primary pathway for energy conservation, were downregulated about 2.5-fold at the lower temperature.

M. mazei. The genome of *M. mazei* is 4.1 Mb in size and encodes 3,371 genes (47). PCR products (200 to 100 bp) were used to construct a microarray representing 3,269 (97%) of the genes. Of these, 2,480

(76%) showed signficant expression in cells grown on methanol or acetate. This is similar (80%) to the number of genes that were reported to be expressed in *P. furiosus* when assessed by the microarray approach (156). Comparison of cells grown with methanol or acetate as the carbon source identified a total of 317 genes that displayed either an increased level (155 genes) or decreased level (162 genes) of expression (>2.5 fold). More than 70% of them were organized into potential operons. Many reflected known or suspected differences in the pathways of methanogensis and energy conservation using the two carbon sources, and the up- and downregulation of several key enzymes was confirmed by RT-PCR. Some unexpected findings included the higher requirement for aromatic amino acids and the upregulation of peptide transporters, flavodoxins, and ferredoxins in acetate-grown cells, as well as the regulation of more than 100 genes of unknown or undefined function. This is the first comprehensive analysis of how gene expression in a methanogenic archaeon varies with carbon source as determined by mRNA levels, and a corresponding proteomic analysis would be extremely informative.

Halophiles

Halobacterium **sp. NRC-1.** The first global analysis of the proteome of a haloarchaeon from an identification perspective was for *Halobacterium* sp. NRC-1 (14). This seminal study also included the first whole-genome transcriptional analysis for an archaeon using arrays containing PCR products representing 2,413 genes. Expression profiles were compared for the wild type and for three mutant strains that either lacked or overproduced bacteriorhodopsin. Each sample pair was analyzed four times independently, and it was stated that statistically significant changes in expression level of 50% could be detected, although this seems a somewhat optimistic assertion. The mutations that led to a greater amount of purple membrane resulted in the differential regulation of 151 of the 2,682 genes examined (66 upregulated and 85 downregulated) relative to the wild type. A surprising number of regulated genes (98 of 151, or 65%) were of unknown function (compared with 41% of unknown function in the complete genome). The genes with annotated functions were involved in phototrophy and membrane synthesis, which were upregulated. In contrast, genes involved in arginine fermentation (the alternative pathway by which this organism conserves energy), were downregulated. Very few (7 of 50) of the genes that appeared to be regulated according to the proteome analysis showed corresponding changes at the transcriptional level.

Very recently, whole-genome DNA microarrays were also used to investigate the discovery that *Halobacterium* NRC-1 can respire anaerobically using N- and S-oxides as terminal electron acceptors (124). Each of the 2,473 genes was represented on the array by up to three 60-mer oligonucleotides. A total of 104 genes were upregulated, and 137 were downregulated at least twofold during anaerobic respiration. These included 24 of the 109 genes in this organism that are predicted to be involved in energy metabolism and included a 3-fold change in expression of the *dms* operon, which encodes the proteins involved in DMSO reduction. The expression of almost all of the genes involved in aerobic respiration were found to be unaffected by anaerobic growth, indicating that the organism is prepared to metabolize oxygen, even under anaerobic conditions.

H. volcanii. The metabolism of *H. volcanii* was investigated by a novel variation of the DNA microarray approach that did not utilize the genome sequence (which at the time of writing is still not available; www.genomesonline.org). A PCR-product library was generated from a genomic library, and this was used to construct a 2,880-spot, one-fold-coverage array (where each spot represented less than an entire gene) (193). About 10% of the spots (273 of 2,880) indicated a more than 2.5-fold change in expression of the representative gene when the carbon source for growth was changed from peptides to glucose. The clones represented by the 273 spots were sequenced, and the genes were identified by comparison with available databases, which included the genome sequence of *H. salinarum*. The results for many genes were rationalized in terms of their predicted roles in transport, the central metabolic pathways, and growth phase. However, many regulatory patterns were unanticipated, including several genes encoding enzymes involved in metabolizing acetate and 2-ketoacids. It is estimated that, by this onefold coverage approach, there is a one-in-three chance that a gene will not be represented. Nevertheless, this study does demonstrate that a shotgun genomic library can be utilized to give important insights into global regulatory pathways and can do so in the absence of a complete genome sequence for the target organism.

A. fulgidus

Regulation of heat shock response in *A. fulgidus*. The heat shock response of *A. fulgidus* (which grows optimally at 83°C) to a temperature shift from 78 to 89°C was studied with a whole-genome cDNA microarray (based on 500 to 2,000 bp PCR products)

(148). Significant changes in expression between 5 and 60 min after the temperature change were observed for 350 genes, 189 of which were up-regulated and 161 down-regulated (although the fold-changes and the statistical significance were not stated). One of the most highly expressed ORFs, AF1298, appeared to be upregulated ~10-fold; this showed similarity to a transcriptional regulator identified previously in *P. furiosus* (PF1790) (183). Electrophoresis mobility shift and footprinting assays demonstrated that (the recombinant form of) AF1298 was a DNA-binding protein and bound to regions upstream of two *A. fulgidus* genes, AF1298 and AF1971. It was proposed that AF1298 is part of an operon that encodes a small heat shock protein, Hsp20 (AF1296) and a cell division control protein (cdc48, AF1297). By sequence analysis, AF1298 is only distantly related to the *P. furiosus* protein (PF1790), and it is thought that they may be members of a diverse protein family that mediates a heat shock response in hyperthermophilic archaea. *A. fulgidus* is known to respond to heat shock by increased production of organic solutes, such as di-myo-inositol-1,1' (3,3')-phosphate (DIP), and to change the composition of membrane isoprenoid lipids. However, with one exception (3-hydroxy-3-methylglutaryl-CoA reductase, AF1736), the genes encoding enzymes relevant to these pathways did not show significant changes in expression (148).

STRUCTURAL GENOMICS

What Is Structural Genomics?

In the 1990s, DNA-sequencing efforts leapt forward, beginning symbolically in 1995 with the determination of the first complete genome from a free-living organism, the bacterium *Haemophilus influenzae* (59). Ten years later, there are currently ~300 complete bacterial or archaeal genome sequences available, with more than twenty from the *Archaea* (85) (see Chapter 19). In the late 1990s, discussions began on one of the next logical extensions of genomic research, understanding the proteome at the structural level. This evolved into the concept of "structural genomics" (SG), large-scale, rapid determination of the three-dimensional structures of proteins

using nuclear magnetic resonance (NMR) and X-ray crystallography techniques, with the lofty goal of determining the structure of a representative fold from all possible families of protein folds in the living world (26, 62, 98, 149, 161, 175). It was estimated that approximately 5,000 to 15,000 different structures may be necessary to cover the estimated 1,000 to 3,000 unique folds that may be represented in living organisms (62), although this estimate is being constantly revised (110). Such an undertaking would require extensive development of novel high-throughput (HTP) technologies at the levels of bioinformatics, gene cloning, heterologous protein expression (both screening for expression and milligram level production), crystallization, structure determination (NMR and X-ray), and data processing. In the United States, beginning in 2000, the National Institute of General Medical Sciences (NIGMS/NIH) created a five-year Protein Structure Initiative (PSI) (128, 138), devoted to development of new tools and approaches for automating and increasing the rate of all these methods required for rapid production of protein structures. Nine centers in the United States (17, 25, 28, 128, 138, 174), along with other international group efforts (67, 77, 86, 170), set out to develop these new technologies in projects that focus in some cases on particular target organisms, and in some, on particular target protein families using homologs from multiple organisms.

It was well known that recombinant protein production presents several significant obstacles. Initially, the bacterium *E. coli* was the host organism of choice due to the many well-developed tools available for heterologous protein expression (15, 16). High-throughput gene cloning turned out to be remarkably simple, with virtually 100% success rates (17) whether using standard ligation-based cloning (89, 150) or commercially available recombination-based cloning systems (117, 184). However, there is a remarkable attrition rate following cloning (Table 3) independent of research group, protocols used, and target list (33, 38, 105, 160). About 20% of targeted proteins are successfully expressed and purified, 37% of purified proteins crystallize, half of these give diffraction quality crystals, and 25% of these diffracting crystals fail to yield a structure. However, this is not to say the SG efforts have not yielded tremendous amounts of use-

Table 3. Success rates for the various steps in the SG protocol from gene to protein structure (137) in the Protein Data Bank (17)

Status	Cloned	Expressed	Soluble	Purified	Crystallized	Diffraction	NMR assigned	Crystal structure	NMR structure	In PDB
No. of targets	54,170	31,388	13,241	11,164	4,126	2,123	610	1,551	508	1,985
% Success	100.0	57.9	24.4	20.6	7.6	3.9	1.1	2.9	0.9	3.7

ful information. As of August 2005 (after almost five years of effort), more than 2,100 structures at atomic resolution have been deposited in the Protein Data Bank (PDB) by SG groups throughout the world (24, 137, 177).

Why do so many proteins fail along this pathway? One class of proteins is well known to give significant problems for overexpression and crystallization: the membrane proteins (91). Initially most centers removed membrane proteins from their target list (these typically represent about 25% of the ORFs in a genome). However, more recently experiments have been initiated for HTP expression and purification of these difficult proteins (51, 57, 103, 114). In addition to problems with membrane proteins, a large number of nonmembrane proteins still fail to be heterologously expressed in *E. coli* (5) due to a lack of various posttranslational modifications. Important posttranslational modifications include formation of heteromeric protein complexes, proteolytic processing, modifications such as phosphorylation or acetylation, and addition of organic (such as flavin) or inorganic (for example, the metals Fe and Zn) cofactors (5, 89, 171). As many as a third of the ORFs in a given bacterial or archaeal genome could be part of an operon, and thus potentially the protein products may form a heteromeric protein complex, or at least coexpression may be required for assembly. This is compounded by the difficulty in predicting operons (55). Considering metal cofactors alone, approximately one-third of all structurally characterized proteins contain a metal cofactor, and perhaps as many as half of all proteins could contain metal (81, 89). Given these caveats, the success rate in producing folded, stable proteins from a random selection of ORFs from a genome was originally predicted to be ~20% (5), which is consistent with the observed success rate (Table 3). In the United States, NIGMS has instituted a second five-year Protein Structure Initiative, with centers focused on HTP protein and structure production, and centers specifically tasked with addressing these difficult expression problems (often termed the "high-hanging fruit") (138, 159).

Archaeal Targets

Many SG groups (not just those examining archaea) have targeted specific organisms (5, 72), although it is clear that when expression of a gene and/or purification or crystallization of the protein fails, an ortholog from another organism may succeed (131, 154). Thus, specific archaea have been targeted (5, 38, 115), in addition to numerous proteins being chosen from archaea solely because they possess an ortholog of the protein of interest. Genes from hyper-

thermophilic archaea (and bacteria) are particularly useful because, even if the proteins they encode do not crystallize more easily than those from mesophiles (an anecdotal belief), they are certainly more stable than mesophilic proteins and thus easier to work with. Some of the pioneering work in SG was performed using archaeal target organisms, in particular, the hyperthermophiles *Pyrobaculum aerophilum* (115) and *Methanothermobacter thermautotrophicus* (38). Some of the earliest work on target selection was performed using the *P. aerophilum* genome by attempting to assign protein folds to each ORF and predict which may contain novel folds (115). *P. aerophilum* is no longer a primary target as the center working on this organism moved on to the disease-causing organism *Mycobacterium tuberculosis*, and more recently, to a more technology-based approach to focus on producing properly folded proteins regardless of source (Integrated Center for Structure and Function Innovation) (138). Nonetheless this has resulted in a number of structures (e.g., 11, 44, 125, 136).

One of the first uses of SG was to assign a biochemical function to a hypothetical protein based on a structure determination. This was performed using a *M. jannaschii* protein (194). *M. jannaschii* was originally used as a test case by the Berkeley center (194), which has since shifted focus to structures of proteins from *Mycoplasma* species (99). The *M. jannaschii* project is still active, and 144 proteins are listed as SG targets in the PDB, mainly targeted by the Northeast Consortium (191). The PDB lists 33 *M. jannaschii* structures completed by SG groups, for example (83, 97, 118). There are also several structures from the related species *M. maripaludis* deposited in the PDB but as yet unpublished (1YVC, 1YWX).

The first (and most complete) archaeal structural genomics project was derived from pioneering work by a group based at the Ontario Cancer Institute at the University of Toronto. Using *M. thermautotrophicus* as the model organism, the goal was to test and begin development of the tools needed for HTP protein production and structure determination (38). A group of 424 (nonmembrane) proteins was targeted, and ~20% were produced in sufficient quantity and quality for structural analyses by either NMR or crystallography (38). The PDB TargetDB lists 213 current *M. thermautotrophicus* targets and 49 structures derived from SG groups (e.g., 30, 45).

One of the most significant archaeal SG efforts involves the hyperthermophilic archaeon *P. furiosus* (5). As one of three target organisms (along with *Caenorhabditis elegans* and *Homo sapiens*) of the Southeast Collaboratory for Structural Genomics (SECSG) (168), the goal of this laboratory was to target all the ORFs in this organism and to express them to yield proteins in a fully folded, functional form. As a free-living nonnucleated cell with a small genome (1.9 Mb representing ~2,000 ORFs), *P. furiosus* is an excellent candidate for attempting a genomewide SG project (5). To ensure that proteins that contain cofactors and/or are part of multiprotein complexes are expressed under conditions that enable them to fold properly, cells are grown, for example, in the presence of excess Fe or Zn for cofactors, or by coexpression of multiple ORFs for complexes. One of the first key issues addressed by the SECSG was precisely how many ORFs to target for cloning and expression. The original annotation of the *P. furiosus* genome deposited in Genbank defined 2,065 putative ORFs. However, depending on the annotation, the number of ORFs predicted was as high as 2,261 (133). As a result, the initial effort targeted 2,192 ORFs for cloning. Considering the difficulties in obtaining recombinant forms of membrane proteins, cofactor-containing proteins, and proteins that may only be stable when coexpressed with their partners from a multiprotein complex, it was anticipated that only about 20% of the target genes would yield stable, properly folded proteins (i.e., the "low-hanging fruit"). To date, 1,909 ORFs (168) have been cloned into an expression vector containing an amino-terminal affinity tag (MAH-HHHHGS-), which allows purification by immobilized metal affinity chromatography (IMAC), as well as detection using an enzyme-linked immunosorbent assay (ELISA) with a commercial antibody against the affinity tag (171).

Screening thousands of clones at a preparative scale (1 liter culture) is prohibitive in terms of time and cost. As a result, screening was performed in a small scale (1 ml) expression system (SSE) using robotics to automate screening for protein expression (5, 171). Screens were performed using several different growth conditions known to affect protein expression, such as culture medium, temperature, *E. coli* strain, etc. Using the SSE, 1-ml cultures are grown and heterologous protein expression is induced in deep-well 96-well plates overnight. The cultures are harvested by centrifugation, and the cells are lysed and then fractionated using vacuum filtration into a soluble fraction and an insoluble inclusion body fraction (solubilized with 6 M guanidine), with a whole-cell extract fraction retained for screening for membrane proteins. Using an antibody against the affinity tag, expression

of the target protein (under a specific growth condition) can be detected and the fraction in which it is expressed can be determined (water-soluble form, or insoluble, presumably as an inclusion body) (89). A number of recent reports demonstrate the usefulness of similar types of HTP screens for protein expression (e.g., 1, 29, 40, 79, 113, 127).

These results for the expression screens are used to scale up production to (typically) a 1-liter scale for purification of milligram amounts of protein, which can then be used for both X-ray crystallography and NMR structural analyses. For clones that fail to express (or express an insoluble, presumably unfolded protein), increasingly complex protocols are followed to attempt to obtain suitable expression and protein quality; for example, alternative *E. coli* expression strains, recloning with different expression vectors or affinity tags, different host organisms, etc. For *P. furiosus*, 2,373 cultures representing 1,006 unique ORFs (and 1,367 additional attempts to express these unique ORFs under different growth conditions) have so far been scaled up to 1 liter, of which 57% (575) successfully expressed some protein (as determined by SDS-PAGE after one affinity column). Three hundred eighty-five unique proteins have been purified, and 259 of these have been analyzed by mass spectrometry analysis and match the predicted mass of the target protein (i.e., not degraded, or posttranslationally modified by *E. coli*). A total of 240 have been submitted for X-ray crystallography screening and 137 have been sent for NMR screening (Table 4) (168). A summary of output from all SG groups can be found at the Protein Data Bank (137). At the time of writing (August 2005), 89,721 targets have resulted in 54,530 clones, 11,632 purified proteins, and 1,926 X-ray and 782 NMR structures.

In addition to *P. furiosus*, 31 targets from *P. abyssi* have been registered in the PDB and 11 structures have been completed but not yet released (136). *P. horikoshii* has 124 targets registered by a number of different centers and 45 structures available (e.g., 10, 53, 172), and a large number of presently unpublished and unreleased structures (136). Other major archaeal SG projects (again, typically using archaea as sources of orthologous genes) are *Sulfolobus solfataricus* (328 ORFs targeted) and other *Sulfolobus* spp. (25 ORFs targeted), which are primarily being performed at the Northeast Consortium (191) and the Joint Center for

Table 4. August 2005 production statistics for *Pyrococcus furiosus* proteins from gene to protein structure (137, 168)

Genes cloned	Expression attempt	Expression success	Pure protein	Crystal	Diffracts	X-ray structure	NMR folded	NMR structure
1,909	1,106	575	385	108/259	59	29	112/137	2

Structural Genomics (JCSG) (105); only a few structures are available (35, 129), although many are completed but unreleased (136). *Archaeoglobus fulgidus* SG (197 targets) is occurring primarily at the JCSG (105), the Midwest Center for Structural Genomics (MCSG) (100), and the New York Consortium (NYS-GXRC) (176). In addition, 50 proteins from other *Archaeoglobus* sp. have been targeted with 41 structures available (e.g., 106, 153) and a large number unreleased (136).

More than 300 ORFs from *Thermoplasma acidophilum* and *T. volcanium* have been targeted by at least five different groups, with 20 structures produced so far (e.g., 101, 163). *Aeropyrum pernix* has been targeted by a large number of groups (344 targets), and six structures are complete. A few ORFs have been targeted from *H. salinarum*, *Halobacterium* sp. NRC-1, *M. mazei*, and *Ferroplasma acidarmanus* (137). In addition to archaea, some extremophilic bacteria are major SG targets; for example, *T. maritima* (one of the pioneering projects) (48, 105) and *Aquifex aeolicus*. The above-mentioned SG projects represent a snapshot of a rapidly changing, ongoing process that will continue to target archaea until the primary SG goal of "filling structure space" has been completed.

It is therefore clear that, within a remarkably short period of five years, tremendous strides have been made in developing HTP protocols for every step in the structure determination pipeline. For the low-hanging fruit, this has resulted in a dramatic reduction in time and cost for determination of high-resolution structures. In addition to the structures determined, these initiatives have also resulted in creation of large banks of knowledge and materials that are available to the scientific community (138). This includes novel software tools, cloning and expression tools and vectors, and many purified proteins that can be used for a range of purposes, for example, producing antibodies and protein arrays (64), and most importantly, for functional analyses. However, despite these rapid advances, all the techniques currently in use are still empirical in nature. No clear predictive rules have emerged that can guarantee protein production leading to structure determination, although relationships between contact order and thermostability (147) and biophysical properties versus crystallization have emerged from structural studies (31). Essentially, each of these steps depends on the unique chemistry of each individual protein. The field is therefore at a stage of having a vastly enhanced, but nonetheless still brute force, approach to structural genomics. As the available data expand, rules for predicting how well an individual ORF will behave at each stage of the process should ultimately become apparent. In the new U.S. Protein Structure Initiative, HTP production centers will combine and optimize the tools that have been developed for rapid screening and production of targets, and orthologous genes will be used for recalcitrant targets; this will be facilitated by technology that uses a large degree of automation for very rapid cloning and screening. The smaller research-based centers will tackle the much more difficult to express, high-hanging fruit, in particular, represented by membrane and posttranslationally modified proteins (138, 159).

PERSPECTIVE: THE NEXT FIVE YEARS

As "omics" tools become more accessible to the research community for the study of archaeal biology, significant progress in the confident annotation (unequivocal evidence of functional/structural characteristics ORFs/genes/proteins) of many of the archaeal genomes will likely result in the next five years. Coupled with useful genetic systems (see Chapter 21), strategic use of functional genomics approaches will form the basis for complete and accurate genome annotation. Archaeal genome sequence data will swell due to the rapidly falling sequencing time and costs. One outcome of this is the prospect of sequencing and functional genomics analyses of environmental samples, studies focusing on community interactions in situ, and comprehensive insights into archaeal ecology of extreme environments (180). This could provide clues to the roles that seemingly uncultured and "unculturable" archaea play in various ecosystems. One of the current frontiers that may come into clearer focus is the relationship between the membrane transcriptome and proteome in archaea; how the unique archaeal membrane composition and structure impacts this remains to be seen. While the archaeal biology research community will no doubt benefit from generic advances in functional genomics tools, it will be important to develop genetic systems that complement the use of these tools. Information gained from ongoing and emerging functional genomics efforts focusing on archaea will provide clues and insights that will accelerate the genetic system's development. Archaea have also played an important role in the recently defined field of structural genomics. In the past five years, several species have been the targets of such endeavors. However, that field is now more directed toward representatives of protein families rather than specific organisms. Consequently, in the next five years, the inclusion of genes from archaea and the subsequent determination of protein structures will be more by accident than by design. In contrast, it is clear that the use of proteomic and transcriptomic tools is at an early stage in the study of archaeal biology. As analytical techniques become more robust and exquisite, a comprehensive cataloging of the functional components of the cell at a molecular level, coupled with the functionality elucidated by "omics" protocols, will give amazing new insights into archaeal biology.

Acknowledgments. Work reported here from the authors' laboratories was supported in part by grants from the National Science Foundation, the Department of Energy, the National Institutes of Health, the National Aeronautics and Space Administration, the University of Georgia and Georgia Research Alliance.

REFERENCES

1. Acton, T. B., K. C. Gunsalus, R. Xiao, L. C. Ma, J. Aramini, M. C. Baran, Y. W. Chiang, T. Climent, B. Cooper, N. G. Denissova, S. M. Douglas, J. K. Everett, C. K. Ho, D. Macapagal, P. K. Rajan, R. Shastry, L. Y. Shih, G. V. Swapna, M. Wilson, M. Wu, M. Gerstein, M. Inouye, J. F. Hunt, and G. T. Montelione. 2005. Robotic cloning and Protein Production Platform of the Northeast Structural Genomics Consortium. *Methods Enzymol.* **394:**210–243.

2. Adams, M. W., and R. M. Kelly (ed.). 2001. *Hyperthermophilic Enzymes Part A*, vol. 330. Academic Press, San Diego, Calif.

3. Adams, M. W., and R. M. Kelly (ed.). 2001. *Hyperthermophilic Enzymes Part B*, vol. 331. Academic Press, San Diego, Calif.

4. Adams, M. W., and R.M. Kelly (ed.). 2001. *Hyperthermophilic Enzymes Part C*, vol. 334. Academic Press, San Diego, Calif.

5. Adams, M. W., H. A. Dailey, L. J. DeLucas, M. Luo, J. H. Prestegard, J. P. Rose, and B. C. Wang. 2003. The Southeast Collaboratory for Structural Genomics: a high-throughput gene to structure factory. *Acc. Chem. Res.* **36:**191–198.

6. Adams, M. W., J. F. Holden, A. L. Menon, G. J. Schut, A. M. Grunden, C. Hou, A. M. Hutchins, F. E. Jenney, Jr., C. Kim, K. Ma, G. Pan, R. Roy, R. Sapra, S. V. Story, and M. F. Verhagen. 2001. Key role for sulfur in peptide metabolism and in regulation of three hydrogenases in the hyperthermophilic archaeon *Pyrococcus furiosus*. *J. Bacteriol.* **183:**716–724.

7. Allers, T., and H. P. Ngo. 2003. Genetic analysis of homologous recombination in Archaea: *Haloferax volcanii* as a model organism. *Biochem. Soc. Trans.* **31:**706–710.

8. Allers, T., H. P. Ngo, M. Mevarech, and R. G. Lloyd. 2004. Development of additional selectable markers for the halophilic archaeon *Haloferax volcanii* based on the *leuB* and *trpA* genes. *Appl. Environ. Microbiol.* **70:**943–953.

9. Apolinario, E. E., K. M. Jackson, and K. R. Sowers. 2005. Development of a plasmid-mediated reporter system for in vivo monitoring of gene expression in the archaeon *Methanosarcina acetivorans*. *Appl. Environ. Microbiol.* **71:**4914–4918.

10. Aramini, J. M., Y. J. Huang, J. R. Cort, S. Goldsmith-Fischman, R. Xiao, L. Y. Shih, C. K. Ho, J. Liu, B. Rost, B. Honig, M. A. Kennedy, T. B. Acton, and G. T. Montelione. 2003. Solution NMR structure of the 30S ribosomal protein S28E from *Pyrococcus horikoshii*. *Protein Sci.* **12:**2823–2830.

11. Arcus, V. L., K. Backbro, A. Roos, E. L. Daniel, and E. N. Baker. 2004. Distant structural homology leads to the functional characterization of an archaeal PIN domain as an exonuclease. *J. Biol. Chem.* **279:**16471–16478.

12. Atomi, H. 2005. Recent progress towards the application of hyperthermophiles and their enzymes. *Curr. Opin. Chem. Biol.* **9:**166–173.

13. Atomi, H., T. Fukui, T. Kanai, M. Morikawa, and T. Imanaka. 2004. Description of *Thermococcus kodakaraensis* sp. nov., a well studied hyperthermophilic archaeon previously reported as *Pyrococcus* sp. KOD1. *Archaea* **1:**263–267.

14. Baliga, N. S., M. Pan, Y. A. Goo, E. C. Yi, D. R. Goodlett, K. Dimitrov, P. Shannon, R. Aebersold, W. V. Ng, and L. Hood. 2002. Coordinate regulation of energy transduction modules in *Halobacterium* sp analyzed by a global systems approach. *Proc. Natl. Acad. Sci. USA* **99:**14913–14918.

15. Baneyx, F. 1999. Recombinant protein expression in *Escherichia coli*. *Curr. Opin. Biotechnol.* **10:**411–421.

16. Baneyx, F., and M. Mujacic. 2004. Recombinant protein folding and misfolding in *Escherichia coli*. *Nat. Biotechnol.* **22:**1399–1408.

17. Berman, H. M., T. N. Bhat, P. E. Bourne, Z. Feng, G. Gilliland, H. Weissig, and J. Westbrook. 2000. The Protein Data Bank and the challenge of structural genomics. *Nat. Struct. Biol.* **7**(Suppl.):957–959.

18. Bidle, K. A. 2003. Differential expression of genes influenced by changing salinity using RNA arbitrarily primed PCR in the archaeal halophile *Haloferax volcanii*. *Extremophiles* **7:**1–7.

19. Blaseio, U., and F. Pfeifer. 1990. Transformation of *Halobacterium halobium*: development of vectors and investigation of gas vesicle synthesis. *Proc. Natl. Acad. Sci. USA* **87:**6772–6776.

20. Blattner, F. R., G. Plunkett, C. A. Bloch, N. T. Perna, V. Burland, M. Riley, J. ColladoVides, J. D. Glasner, C. K. Rode, G. F. Mayhew, J. Gregor, N. W. Davis, H. A. Kirkpatrick, M. A. Goeden, D. J. Rose, B. Mau, and Y. Shao. 1997. The complete genome sequence of *Escherichia coli* K-12. *Science* **277:**1453–1474.

21. Blonder, J., T. P. Conrads, L. R. Yu, A. Terunuma, G. M. Janini, H. J. Issaq, J. C. Vogel, and T. D. Veenstra. 2004. A detergent- and cyanogen bromide-free method for integral membrane proteomics: application to *Halobacterium* purple membranes and the human epidermal membrane proteome. *Proteomics* **4:**31–45.

22. Boonyaratanakornkit, B. B., A. J. Simpson, T. A. Whitehead, C. M. Fraser, N. M. El-Sayed, and D. S. Clark. 2005. Transcriptional profiling of the hyperthermophilic methanarchaeon *Methanococcus jannaschii* in response to lethal heat and non-lethal cold shock. *Environ. Microbiol.* **7:**789–797.

23. Borezee, E., E. Pellegrini, and P. Berche. 2000. OppA of *Listeria monocytogenes*, an oligopeptide-binding protein required for bacterial growth at low temperature and involved in intracellular survival. *Infect. Immun.* **68:**7069–7077.

24. Bourne, P. E., C. K. Allerston, W. Krebs, W. Li, I. N. Shindyalov, A. Godzik, I. Friedberg, T. Liu, D. Wild, S. Hwang, Z. Ghahramani, L. Chen, and J. Westbrook. 2004. The status of structural genomics defined through the analysis of current targets and structures. *Pac. Symp. Biocomput.* **2004:**375–386.

25. Brenner, S. E. 2000. Target selection for structural genomics. *Nat. Struct. Biol.* **7**(Suppl.):967–969.

26. Brenner, S. E., and M. Levitt. 2000. Expectations from structural genomics. *Protein Sci.* **9:**197–200.

27. Bult, C. J., O. White, G. J. Olsen, L. Zhou, R. D. Fleischmann, G. G. Sutton, J. A. Blake, L. M. FitzGerald, R. A. Clayton, J. D. Gocayne, A. R. Kerlavage, B. A. Dougherty, J. F. Tomb, M. D. Adams, C. I. Reich, R. Overbeek, E. F. Kirkness, K. G. Weinstock, J. M. Merrick, A. Glodek, J. L. Scott, N. S. Geoghagen, and J. C. Venter. 1996. Complete genome sequence of the methanogenic archaeon, *Methanococcus jannaschii*. *Science* **273:**1058–1073.

28. Burley, S. K. 2000. An overview of structural genomics. *Nat. Struct. Biol.* **7**(Suppl.):932–934.

29. Busso, D., R. Kim, and S. H. Kim. 2004. Using an *Escherichia coli* cell-free extract to screen for soluble expression of recombinant proteins. *J. Struct. Funct. Genomics* **5:**69–74.

30. Canaves, J. M. 2004. Predicted role for the archease protein family based on structural and sequence analysis of TM1083 and MTH1598, two proteins structurally characterized through structural genomics efforts. *Proteins* 56:19–27.

31. Canaves, J. M., R. Page, I. A. Wilson, and R. C. Stevens. 2004. Protein biophysical properties that correlate with crystallization success in *Thermotoga maritima*: maximum clustering strategy for structural genomics. *J. Mol. Biol.* 344:977–991.

32. Cavicchioli, R., A. Goodchild, and M. Raftery. 2006. Proteomics of archaea, pp. 57–72. *In* I. Humphery-Smith and M. Hecker (ed.), *Microbial Proteomics: Functional Biology of Whole Organisms*. John Wiley & Sons, Hoboken, N.J.

33. Chayen, N. E. 2003. Protein crystallization for genomics: throughput versus output. *J. Struct. Funct. Genomics* 4:115–120.

34. Chen, L., K. Brugger, M. Skovgaard, P. Redder, Q. She, E. Torarinsson, B. Greve, M. Awayez, A. Zibat, H. P. Klenk, and R. A. Garrett. 2005. The genome of *Sulfolobus acidocaldarius*, a model organism of the Crenarchaeota. *J. Bacteriol.* 187:4992–4999.

35. Chen, L., L. R. Chen, X. E. Zhou, Y. Wang, M. A. Kahsai, A. T. Clark, S. P. Edmondson, Z. J. Liu, J. P. Rose, B. C. Wang, E. J. Meehan, and J. W. Shriver. 2004. The hyperthermophile protein Sso10a is a dimer of winged helix DNA-binding domains linked by an antiparallel coiled coil rod. *J. Mol. Biol.* 341:73–91.

36. Cho, C. W., S. H. Lee, J. Choi, S. J. Park, D. J. Ha, H. J. Kim, and C. W. Kim. 2003. Improvement of the two-dimensional gel electrophoresis analysis for the proteome study of *Halobacterium salinarum*. *Proteomics* 3:2325–2329.

37. Choi, J., W. A. Joo, S. J. Park, S. H. Lee, and C. W. Kim. 2005. An efficient proteomics based strategy for the functional characterization of a novel halophilic enzyme from *Halobacterium salinarum*. *Proteomics* 5:907–917.

38. Christendat, D., A. Yee, A. Dharamsi, Y. Kluger, A. Savchenko, J. R. Cort, V. Booth, C. D. Mackereth, V. Saridakis, I. Ekiel, G. Kozlov, K. L. Maxwell, N. Wu, L. P. McIntosh, K. Gehring, M. A. Kennedy, A. R. Davidson, E. F. Pai, M. Gerstein, A. M. Edwards, and C. H. Arrowsmith. 2000. Structural proteomics of an archaeon. *Nat. Struct. Biol.* 7:903–909.

39. Cohen, G. N., V. Barbe, D. Flament, M. Galperin, R. Heilig, O. Lecompte, O. Poch, D. Prieur, J. Querellou, R. Ripp, J. C. Thierry, J. Van der Oost, J. Weissenbach, Y. Zivanovic, and P. Forterre. 2003. An integrated analysis of the genome of the hyperthermophilic archaeon *Pyrococcus abyssi*. *Mol. Microbiol.* 47:1495–1512.

40. Coleman, M. A., V. H. Lao, B. W. Segelke, and P. T. Beernink. 2004. High-throughput, fluorescence-based screening for soluble protein expression. *J. Proteome Res.* 3:1024–1032.

41. Contursi, P., R. Cannio, S. Prato, G. Fiorentino, M. Rossi, and S. Bartolucci. 2003. Development of a genetic system for hyperthermophilic Archaea: expression of a moderate thermophilic bacterial alcohol dehydrogenase gene in *Sulfolobus solfataricus*. *FEMS Microbiol. Lett.* 218:115–120.

42. Copeland, A., S. Lucas, A. Lapidus, K. Barry, C. Detter, T. Glavina, N. Hammon, S. Israni, S. Pitluck, and P. Richardson. 2005. Sequencing of the draft genome and assembly of *Methanococcoides burtonii* DSM 6242. Released. U.S. Department of Energy Joint Genome Institute, Walnut Creek, Calif.

43. Reference deleted.

44. Cort, J. R., S. V. Mariappan, C. Y. Kim, M. S. Park, T. S. Peat, G. S. Waldo, T. C. Terwilliger, and M. A. Kennedy. 2001. Solution structure of *Pyrobaculum aerophilum* DsrC, an archaeal homologue of the gamma subunit of dissimilatory sulfite reductase. *Eur. J. Biochem.* 268:5842–5850.

45. Cort, J. R., A. Yee, A. M. Edwards, C. H. Arrowsmith, and M. A. Kennedy. 2000. Structure-based functional classification of hypothetical protein MTH538 from *Methanobacterium thermoautotrophicum*. *J. Mol. Biol.* 302:189–203.

46. Daniels, C. J., A. H. Z. McKee, and W. F. Doolittle. 1984. Archaebacterial heat-shock proteins. *EMBO J.* 3:745–749.

47. Deppenmeier, U., A. Johann, T. Hartsch, R. Merkl, R. A. Schmitz, R. Martinez-Arias, A. Henne, A. Wiezer, S. Baumer, C. Jacobi, H. Bruggemann, T. Lienard, A. Christmann, M. Bomeke, S. Steckel, A. Bhattacharyya, A. Lykidis, R. Overbeek, H. P. Klenk, R. P. Gunsalus, H. J. Fritz, and G. Gottschalk. 2002. The genome of *Methanosarcina mazei*: evidence for lateral gene transfer between bacteria and archaea. *J. Mol. Microbiol. Biotechnol.* 4:453–461.

48. DiDonato, M., A. M. Deacon, H. E. Klock, D. McMullan, and S. A. Lesley. 2004. A scaleable and integrated crystallization pipeline applied to mining the *Thermotoga maritima* proteome. *J. Struct. Funct. Genomics* 5:133–146.

49. Ding, Y. H., S. P. Zhang, J. F. Tomb, and J. G. Ferry. 2002. Genomic and proteomic analyses reveal multiple homologs of genes encoding enzymes of the methanol:coenzyme M methyltransferase system that are differentially expressed in methanol- and acetate-grown *Methanosarcina thermophila*. *FEMS Microbiol. Lett.* 215:127–132.

50. Dionne, I., N. P. Robinson, A. T. McGeoch, V. L. Marsh, A. Reddish, and S. D. Bell. 2003. DNA replication in the hyperthermophilic archaeon *Sulfolobus solfataricus*. *Biochem. Soc. Trans.* 31:674–676.

51. Dobrovetsky, E., M. L. Lu, R. Andorn-Broza, G. Khutoreskaya, J. E. Bray, A. Savchenko, C. H. Arrowsmith, A. M. Edwards, and C. M. Koth. 2005. High-throughput production of prokaryotic membrane proteins. *J. Struct. Funct. Genomics* 6:33–50.

52. Driskill, L. E., K. Kusy, M. W. Bauer, and R. M. Kelly. 1999. Relationship between glycosyl hydrolase inventory and growth physiology of the hyperthermophile *Pyrococcus furiosus* on carbohydrate-based media. *Appl. Environ. Microbiol.* 65:893–897.

53. Du, X., I. G. Choi, R. Kim, W. Wang, J. Jancarik, H. Yokota, and S. H. Kim. 2000. Crystal structure of an intracellular protease from *Pyrococcus horikoshii* at 2-A resolution. *Proc. Natl. Acad. Sci. USA* 97:14079–14084.

54. Ehlers, C., K. Weidenbach, K. Veit, U. Deppenmeier, W. W. Metcalf, and R. A. Schmitz. 2005. Development of genetic methods and construction of a chromosomal glnK1 mutant in *Methanosarcina mazei* strain Go1. *Mol. Genet. Genomics* 273:290–298.

55. Ermolaeva, M. D., O. White, and S. L. Salzberg. 2001. Prediction of operons in microbial genomes. *Nucleic Acids Res.* 29:1216–1221.

56. Ermolenko, D. N., and G. I. Makhatadze. 2002. Bacterial cold-shock proteins. *Cell. Mol. Life Sci.* 59:1902–1913.

57. Eshaghi, S., M. Hedren, M. I. Nasser, T. Hammarberg, A. Thornell, and P. Nordlund. 2005. An efficient strategy for high-throughput expression screening of recombinant integral membrane proteins. *Protein Sci.* 14:676–683.

58. Fiala, G., and K. O. Stetter. 1986. *Pyrococcus furiosus* sp-nov represents a novel genus of marine heterotrophic Archaebacteria growing optimally at 100-degrees C. *Arch. Microbiol.* 145:56–61.

59. Fleischmann, R. D., M. D. Adams, O. White, R. A. Clayton, E. F. Kirkness, A. R. Kerlavage, C. J. Bult, J. F. Tomb, B. A. Dougherty, J. M. Merrick, et al. 1995. Whole-genome random sequencing and assembly of *Haemophilus influenzae* Rd. *Science* 269:496–512.

60. Forbes, A. J., S. M. Patrie, G. K. Taylor, Y. B. Kim, L. Jiang, and N. L. Kelleher. 2004. Targeted analysis and discovery of posttranslational modifications in proteins from methanogenic

archaea by top-down MS. *Proc. Natl. Acad. Sci. USA* **101:** 2678–2683.

61. Fukui, T., H. Atomi, T. Kanai, R. Matsumi, S. Fujiwara, and T. Imanaka. 2005. Complete genome sequence of the hyperthermophilic archaeon *Thermococcus kodakaraensis* KOD1 and comparison with *Pyrococcus* genomes. *Genome Res.* **15:**352–363.

62. Gaasterland, T. 1998. Structural genomics: bioinformatics in the driver's seat. *Nat. Biotechnol.* **16:**625–627.

63. Galagan, J. E., C. Nusbaum, A. Roy, M. G. Endrizzi, P. Macdonald, W. FitzHugh, S. Calvo, R. Engels, S. Smirnov, D. Atnoor, A. Brown, N. Allen, J. Naylor, N. Stange-Thomann, K. DeArellano, R. Johnson, L. Linton, P. McEwan, K. McKernan, J. Talamas, A. Tirrell, W. Ye, A. Zimmer, R. D. Barber, I. Cann, D. E. Graham, D. A. Grahame, A. M. Guss, R. Hedderich, C. Ingram-Smith, H. C. Kuettner, J. A. Krzycki, J. A. Leigh, W. Li, J. Liu, B. Mukhopadhyay, J. N. Reeve, K. Smith, T. A. Springer, L. A. Umayam, O. White, R. H. White, E. Conway de Macario, J. G. Ferry, K. F. Jarrell, H. Jing, A. J. Macario, I. Paulsen, M. Pritchett, K. R. Sowers, R. V. Swanson, S. H. Zinder, E. Lander, W. W. Metcalf, and B. Birren. 2002. The genome of *M. acetivorans* reveals extensive metabolic and physiological diversity. *Genome Res.* **12:**532–542.

64. Gilbert, M., T. C. Edwards, and J. S. Albala. 2004. Protein expression arrays for proteomics. *Methods Mol. Biol.* **264:** 15–23.

65. Giometti, C. S., C. Reich, S. Tollaksen, G. Babnigg, H. Lim, W. Zhu, J. Yates, and G. Olsen. 2002. Global analysis of a "simple" proteome: *Methanococcus jannaschii*. *J. Chromatogr. B Anal. Technol. Biomed. Life Sci.* **782:**227–243.

66. Giometti, C. S., C. I. Reich, S. L. Tollaksen, G. Babnigg, H. Lim, J. R. Yates, and G. J. Olsen. 2001. Structural modifications of *Methanococcus jannaschii* flagellin proteins revealed by proteome analysis. *Proteomics* **1:**1033–1042.

67. Gong, W. M., H. Y. Liu, L. W. Niu, Y. Y. Shi, Y. J. Tang, M. K. Teng, J. H. Wu, D. C. Liang, D. C. Wang, J. F. Wang, J. P. Ding, H. Y. Hu, Q. H. Huang, Q. H. Zhang, S. Y. Lu, J. L. An, Y. H. Liang, X. F. Zheng, X. C. Gu, and X. D. Su. 2003. Structural genomics efforts at the Chinese Academy of Sciences and Peking University. *J. Struct. Funct. Genomics* **4:** 137–139.

68. Goo, Y. A., E. C. Yi, N. S. Baliga, W. A. Tao, M. Pan, R. Aebersold, D. R. Goodlett, L. Hood, and W. V. Ng. 2003. Proteomic analysis of an extreme halophilic archaeon, *Halobacterium* sp. NRC-1. *Mol. Cell. Proteomics* **2:**506–524.

69. Goodchild, A., M. Raftery, N. F. Saunders, M. Guilhaus, and R. Cavicchioli. 2004. Biology of the cold adapted archaeon, *Methanococcoides burtonii* determined by proteomics using liquid chromatography-tandem mass spectrometry. *J. Proteome Res.* **3:**1164–1176.

70. Goodchild, A., M. Raftery, N. F. Saunders, M. Guilhaus, and R. Cavicchioli. 2005. Cold adaptation of the Antarctic archaeon, *Methanococcoides burtonii* assessed by proteomics using ICAT. *J. Proteome Res.* **4:**473–480.

71. Goodchild, A., N. F. Saunders, H. Ertan, M. Raftery, M. Guilhaus, P. M. Curmi, and R. Cavicchioli. 2004. A proteomic determination of cold adaptation in the Antarctic archaeon, *Methanococcoides burtonii*. *Mol. Microbiol.* **53:**309–321.

72. Goulding, C. W., L. J. Perry, D. Anderson, M. R. Sawaya, D. Cascio, M. I. Apostol, S. Chan, A. Parseghian, S. S. Wang, Y. Wu, V. Cassano, H. S. Gill, and D. Eisenberg. 2003. Structural genomics of *Mycobacterium tuberculosis*: a preliminary report of progress at UCLA. *Biophys. Chem.* **105:**361–370.

73. Graham, D. E., N. Kyrpides, I. J. Anderson, R. Overbeek, and W. B. Whitman. 2001. Genome of *Methanocaldococcus* (*Methanococcus*) *jannaschii*. *Methods Enzymol.* **330:**40–123.

74. Gygi, S. P., B. Rist, S. A. Gerber, F. Turecek, M. H. Gelb, and R. Aebersold. 1999. Quantitative analysis of complex protein mixtures using isotope-coded affinity tags. *Nat. Biotechnol.* **17:**994–999.

75. Han, C. J., S. H. Park, and R. M. Kelly. 1997. Acquired thermotolerance and stressed-phase growth of the extremely thermoacidophilic archaeon *Metallosphaera sedula* in continuous culture. *Appl. Environ. Microbiol.* **63:**2391–2396.

76. Han, D. K., J. Eng, H. L. Zhou, and R. Aebersold. 2001. Quantitative profiling of differentiation-induced microsomal proteins using isotope-coded affinity tags and mass spectrometry. *Nat. Biotechnol.* **19:**946–951.

77. Heinemann, U. 2000. Structural genomics in Europe: slow start, strong finish? *Nat. Struct. Biol.* **7**(Suppl.):940–942.

78. Hendrickson, E. L., R. Kaul, Y. Zhou, D. Bovee, P. Chapman, J. Chung, E. Conway de Macario, J. A. Dodsworth, W. Gillett, D. E. Graham, M. Hackett, A. K. Haydock, A. Kang, M. L. Land, R. Levy, T. J. Lie, T. A. Major, B. C. Moore, I. Porat, A. Palmeiri, G. Rouse, C. Saenphimmachak, D. Soll, S. Van Dien, T. Wang, W. B. Whitman, Q. Xia, Y. Zhang, F. W. Larimer, M. V. Olson, and J. A. Leigh. 2004. Complete genome sequence of the genetically tractable hydrogenotrophic methanogen *Methanococcus maripaludis*. *J. Bacteriol.* **186:**6956–6969.

79. Hewitt, L., and J. M. McDonnell. 2004. Screening and optimizing protein production in *E. coli*. *Methods Mol. Biol.* **278:**1–16.

80. Holden, J. F., F. L. Poole, S. L. Tollaksen, C. S. Giometti, H. Lim, J. R. Yates, and M. W. W. Adams. 2001. Identification of membrane proteins in the hyperthermophilic archaeon *Pyrococcus furiosus* using proteomics and prediction programs. *Comp. Funct. Genomics* **2:**275–288.

81. Holm, L., and C. Sander. 1996. Mapping the protein universe. *Science* **273:**595–603.

82. Hovey, R., S. Lentes, A. Ehrenreich, K. Salmon, K. Saba, G. Gottschalk, R. P. Gunsalus, and U. Deppenmeier. 2005. DNA microarray analysis of *Methanosarcina mazei* Go1 reveals adaptation to different methanogenic substrates. *Mol. Genet. Genomics* **273:**225–239.

83. Hwang, K. Y., J. H. Chung, S. H. Kim, Y. S. Han, and Y. Cho. 1999. Structure-based identification of a novel NTPase from *Methanococcus jannaschii*. *Nat. Struct. Biol.* **6:**691–696.

84. Ihling, C., and A. Sinz. 2005. Proteome analysis of *Escherichia coli* using high-performance liquid chromatography and Fourier transform ion cyclotron resonance mass spectrometry. *Proteomics* **5:**2029–2042.

85. Institute for Genomic Research. 2005. http://www.tigr.org/. The Institute for Genomic Research. [Online.]

86. International Structural Genomics Organization. 2005. International Structural Genomics Organization. http://www.isgo.org/. [Online.]

87. Jablonski, P. E., A. A. Dimarco, T. A. Bobik, M. C. Cabell, and J. G. Ferry. 1990. Protein content and enzyme activities in methanol grown and acetate grown *Methanosarcina thermophila*. *J. Bacteriol.* **172:**1271–1275.

88. Jarrell, K. F., D. P. Bayley, and A. S. Kostyukova. 1996. The archaeal flagellum: A unique motility structure. *J. Bacteriol.* **178:**5057–5064.

89. Jenney, F. E., Jr., P. S. Brereton, M. Izumi, F. L. Poole II, C. Shah, F. J. Sugar, H. S. Lee, and M. W. Adams. 2005. High-throughput production of *Pyrococcus furiosus* proteins: considerations for metalloproteins. *J. Synchrotron Radiat.* **12:**8–12.

90. Johnson, M. R., C. I. Montero, S. B. Conners, K. R. Shockley, S. L. Bridger, and R. M. Kelly. 2005. Population density-dependent regulation of exopolysaccharide formation in the hyperthermophilic bacterium *Thermotoga maritima*. *Mol. Microbiol.* **55**:664–674.

91. Jones, D. T., and W. R. Taylor. 1998. Towards structural genomics for transmembrane proteins. *Biochem .Soc. Trans.* **26**:429–438.

92. Joo, W. A., and C. W. Kim. 2005. Proteomics of halophilic archaea. *J. Chromatogr. B Anal.Technol. Biomed. Life Sci.* **815**:237–250.

93. Karadzic, I. M., and J. A. Maupin-Furlow. 2005. Improvement of two-dimensional gel electrophoresis proteome maps of the haloarchaeon *Haloferax volcanii*. *Proteomics* **5**:354–359.

94. Kawarabayasi, Y. 2001. Genome of *Pyrococcus horikoshii* OT3. *Methods Enzymol.* **330**:124–134.

95. Kawarabayasi, Y., Y. Hino, H. Horikawa, K. Jin-no, M. Takahashi, M. Sekine, S. Baba, A. Ankai, H. Kosugi, A. Hosoyama, S. Fukui, Y. Nagai, K. Nishijima, R. Otsuka, H. Nakazawa, M. Takamiya, Y. Kato, T. Yoshizawa, T. Tanaka, Y. Kudoh, J. Yamazaki, N. Kushida, A. Oguchi, K. Aoki, S. Masuda, M. Yanagii, M. Nishimura, A. Yamagishi, T. Oshima, and H. Kikuchi. 2001. Complete genome sequence of an aerobic thermoacidophilic crenarchaeon, *Sulfolobus tokodaii* strain7. *DNA Res.* **8**:123–140.

96. Kennedy, S. P., W. V. Ng, S. L. Salzberg, L. Hood, and S. DasSarma. 2001. Understanding the adaptation of *Halobacterium* species NRC-1 to its extreme environment through computational analysis of its genome sequence. *Genome Res.* **11**:1641–1650.

97. Kim, J. S., A. DeGiovanni, J. Jancarik, P. D. Adams, H. Yokota, R. Kim, and S. H. Kim. 2005. Crystal structure of DNA sequence specificity subunit of a type I restriction-modification enzyme and its functional implications. *Proc. Natl. Acad. Sci. USA* **102**:3248–3253.

98. Kim, S. H. 1998. Shining a light on structural genomics. *Nat. Struct. Biol.* **5**(Suppl.):643–645.

99. Kim, S. H., D. H. Shin, I. G. Choi, U. Schulze-Gahmen, S. Chen, and R. Kim. 2003. Structure-based functional inference in structural genomics. *J. Struct. Funct. Genomics* **4**:129–135.

100. Kim, Y., I. Dementieva, M. Zhou, R. Wu, L. Lezondra, P. Quartey, G. Joachimiak, O. Korolev, H. Li, and A. Joachimiak. 2004. Automation of protein purification for structural genomics. *J. Struct. Funct. Genomics* **5**:111–118.

101. Kim, Y., A. F. Yakunin, E. Kuznetsova, X. Xu, M. Pennycooke, J. Gu, F. Cheung, M. Proudfoot, C. H. Arrowsmith, A. Joachimiak, A. M. Edwards, and D. Christendat. 2004. Structure- and function-based characterization of a new phosphoglycolate phosphatase from *Thermoplasma acidophilum*. *J. Biol. Chem.* **279**:517–526.

102. Klein, C., C. Garcia-Rizo, B. Bisle, B. Scheffer, H. Zischka, F. Pfeiffer, F. Siedler, and D. Oesterhelt. 2005. The membrane proteome of *Halobacterium salinarum*. *Proteomics* **5**:180–197.

103. Laible, P. D., H. N. Scott, L. Henry, and D. K. Hanson. 2004. Towards higher-throughput membrane protein production for structural genomics initiatives. *J. Struct. Funct. Genomics* **5**:167–172.

104. Leartsakulpanich, U., M. L. Antonkine, and J. G. Ferry. 2000. Site-specific mutational analysis of a novel cysteine motif proposed to ligate the 4Fe-4S cluster in the iron-sulfur flavoprotein of the thermophilic methanoarchaeon *Methanosarcina thermophila*. *J. Bacteriol.* **182**:5309–5316.

105. Lesley, S. A., P. Kuhn, A. Godzik, A. M. Deacon, I. Mathews, A. Kreusch, G. Spraggon, H. E. Klock, D. McMullan, T. Shin, J. Vincent, A. Robb, L. S. Brinen, M. D. Miller, T. M. McPhillips, M. A. Miller, D. Scheibe, J. M. Canaves, C. Guda, L. Jaroszewski, T. L. Selby, M. A. Elsliger, J. Wooley, S. S. Taylor, K. O. Hodgson, I. A. Wilson, P. G. Schultz, and R. C. Stevens. 2002. Structural genomics of the *Thermotoga maritima* proteome implemented in a high-throughput structure determination pipeline. *Proc. Natl. Acad. Sci. USA* **99**:11664–11669.

106. Levin, I., R. Schwarzenbacher, R. Page, P. Abdubek, E. Ambing, T. Biorac, L. S. Brinen, J. Campbell, J. M. Canaves, H. J. Chiu, X. Dai, A. M. Deacon, M. DiDonato, M. A. Elsliger, R. Floyd, A. Godzik, C. Grittini, S. K. Grzechnik, E. Hampton, L. Jaroszewski, C. Karlak, H. E. Klock, E. Koesema, J. S. Kovarik, A. Kreusch, P. Kuhn, S. A. Lesley, D. McMullan, T. M. McPhillips, M. D. Miller, A. Morse, K. Moy, J. Ouyang, K. Quijano, R. Reyes, F. Rezezadeh, A. Robb, E. Sims, G. Spraggon, R. C. Stevens, H. van den Bedem, J. Velasquez, J. Vincent, F. von Delft, X. Wang, B. West, G. Wolf, Q. Xu, K. O. Hodgson, J. Wooley, and I. A. Wilson. 2004. Crystal structure of a PIN (PilT N-terminus) domain (AF0591) from *Archaeoglobus fulgidus* at 1.90 A resolution. *Proteins* **56**:404–408.

107. Li, Q., L. Li, T. Rejtar, B. L. Karger, and J. G. Ferry. 2005. Proteome of *Methanosarcina acetivorans* Part I: an expanded view of the biology of the cell. *J. Proteome Res.* **4**:112–128.

108. Li, Q., L. Li, T. Rejtar, B. L. Karger, and J. G. Ferry. 2005. Proteome of *Methanosarcina acetivorans* Part II: comparison of protein levels in acetate- and methanol-grown cells. *J. Proteome Res.* **4**:129–135.

109. Lim, H., J. Eng, J. R. Yates, S. L. Tollaksen, C. S. Giometti, J. F. Holden, M. W. W. Adams, C. I. Reich, G. J. Olsen, and L. G. Hays. 2003. Identification of 2D-gel proteins: A comparison of MALDI/TOF peptide mass mapping to mu LC-ESI tandem mass spectrometry. *J. Am. Soc. Mass Spectrom.* **14**:957–970.

110. Liu, X., K. Fan, and W. Wang. 2004. The number of protein folds and their distribution over families in nature. *Proteins* **54**:491–499.

111. Lopez-Campistrous, A., P. Semchuk, L. Burke, T. Palmer-Stone, S. J. Brokx, G. Broderick, D. Bottorff, S. Bolch, J. H. Weiner, and M. J. Ellison. 2005. Localization, annotation, and comparison of the *Escherichia coli* K-12 proteome under two states of growth. *Mol. Cell. Proteom.* **4**:1205–1209.

112. Lower, B. H., M. Ben Potters, and P. J. Kennelly. 2004. A phosphoprotein from the archaeon *Sulfolobus solfataricus* with protein-serine/threonine kinase activity. *J. Bacteriol.* **186**:463–472.

113. Luan, C. H., S. Qiu, J. B. Finley, M. Carson, R. J. Gray, W. Huang, D. Johnson, J. Tsao, J. Reboul, P. Vaglio, D. E. Hill, M. Vidal, L. J. Delucas, and M. Luo. 2004. High-throughput expression of *C. elegans* proteins. *Genome Res.* **14**:2102–2110.

114. Lundstrom, K. 2004. Structural genomics on membrane proteins: mini review. *Combinator. Chem. High Throughput Screen* **7**:431–439.

114a. Maeder, D. L., I. Anderson, T. S. Brettin, D. C. Bruce, P. Gilna, C. S. Han, A. Lapidus, W. W. Metcalf, E. Saunders, R. Tapia, and K. R. Sowers. 2006. The *Methanosarcina barkeri* genome: comparative analysis with *Methanosarcina acetivorans* and *Methanosarcina mazei* reveals extensive rearrangement with methanosarcinal genomes. *J. Bacteriol.* **188**:7922–7931.

115. Mallick, P., K. E. Goodwill, S. Fitz-Gibbon, J. H. Miller, and D. Eisenberg. 2000. Selecting protein targets for structural genomics of *Pyrobaculum aerophilum*: validating automated fold assignment methods by using binary hypothesis testing. *Proc. Natl. Acad. Sci. USA* **97**:2450–2455.

116. Margulies, M., M. Egholm, W. E. Altman, S. Attiya, J. S. Bader, L. A. Bemben, J. Berka, M. S. Braverman, Y. J. Chen, Z. Chen, S. B. Dewell, L. Du, J. M. Fierro, X. V. Gomes, B. C. Godwin, W. He, S. Helgesen, C. H. Ho, G. P. Irzyk, S. C. Jando, M. L. Alenquer, T. P. Jarvie, K. B. Jirage, J. B. Kim, J. R. Knight, J. R. Lanza, J. H. Leamon, S. M. Lefkowitz, M. Lei, J. Li, K. L. Lohman, H. Lu, V. B. Makhijani, K. E. McDade, M. P. McKenna, E. W. Myers, E. Nickerson, J. R. Nobile, R. Plant, B. P. Puc, M. T. Ronan, G. T. Roth, G. J. Sarkis, J. F. Simons, J. W. Simpson, M. Srinivasan, K. R. Tartaro, A. Tomasz, K. A. Vogt, G. A. Volkmer, S. H. Wang, Y. Wang, M. P. Weiner, P. Yu, R. F. Begley, and J. M. Rothberg. 2005. Genome sequencing in microfabricated high-density picolitre reactors. *Nature* **437:**376–380.

117. Marsischky, G., and J. LaBaer. 2004. Many paths to many clones: a comparative look at high-throughput cloning methods. *Genome Res.* **14:**2020–2028.

118. Martinez-Cruz, L. A., M. K. Dreyer, D. C. Boisvert, H. Yokota, M. L. Martinez-Chantar, R. Kim, and S. H. Kim. 2002. Crystal structure of MJ1247 protein from *M. jannaschii* at 2.0 A resolution infers a molecular function of 3-hexulose-6-phosphate isomerase. *Structure (Camb.)* **10:**195–204.

119. Metcalf, W. W., J. K. Zhang, E. Apolinario, K. R. Sowers, and R. S. Wolfe. 1997. A genetic system for Archaea of the genus *Methanosarcina*: liposome-mediated transformation and construction of shuttle vectors. *Proc. Natl. Acad. Sci. USA* **94:**2626–2631.

120. Miles, R. D., P. P. Iyer, and J. G. Ferry. 2001. Site-directed mutational analysis of active site residues in the acetate kinase from *Methanosarcina thermophila. J. Biol. Chem.* **276:**45059–45064.

121. Mojica, F. J. M., E. Cisneros, C. Ferrer, F. RodriguezValera, and G. Juez. 1997. Osmotically induced response in representatives of halophilic prokaryotes: the bacterium *Halomonas elongata* and the Archaeon *Haloferax volcanii. J. Bacteriol.* **179:**5471–5481.

122. Moore, B. C., and J. A. Leigh. 2005. Markerless mutagenesis in *Methanococcus maripaludis* demonstrates roles for alanine dehydrogenase, alanine racemase, and alanine permease. *J. Bacteriol.* **187:**972–979.

123. Mukhopadhyay, B., E. F. Johnson, and R. S. Wolfe. 2000. A novel p(H2) control on the expression of flagella in the hyperthermophilic strictly hydrogenotrophic methanarchaeaon *Methanococcus jannaschii. Proc. Natl. Acad. Sci. USA* **97:**11522–11527.

124. Muller, J. A., and S. DasSarma. 2005. Genomic analysis of anaerobic respiration in the archaeon *Halobacterium* sp. strain NRC-1: dimethyl sulfoxide and trimethylamine N-oxide as terminal electron acceptors. *J. Bacteriol.* **187:**1659–1667.

125. Mura, C., J. E. Katz, S. G. Clarke, and D. Eisenberg. 2003. Structure and function of an archaeal homolog of survival protein E (SurEalpha): an acid phosphatase with purine nucleotide specificity. *J. Mol. Biol.* **326:**1559–1575.

126. Ng, W. V., S. P. Kennedy, G. G. Mahairas, B. Berquist, M. Pan, H. D. Shukla, S. R. Lasky, N. S. Baliga, V. Thorsson, J. Sbrogna, S. Swartzell, D. Weir, J. Hall, T. A. Dahl, R. Welti, Y. A. Goo, B. Leithauser, K. Keller, R. Cruz, M. J. Danson, D. W. Hough, D. G. Maddocks, P. E. Jablonski, M. P. Krebs, C. M. Angevine, H. Dale, T. A. Isenbarger, R. F. Peck, M. Pohlschroder, J. L. Spudich, K. W. Jung, M. Alam, T. Freitas, S. Hou, C. J. Daniels, P. P. Dennis, A. D. Omer, H. Ebhardt, T. M. Lowe, P. Liang, M. Riley, L. Hood, and S. DasSarma. 2000. Genome sequence of *Halobacterium* species NRC-1. *Proc. Natl. Acad. Sci. USA* **97:**12176–12181.

127. Nguyen, H., B. Martinez, N. Oganesyan, and R. Kim. 2004. An automated small-scale protein expression and purification screening provides beneficial information for protein production. *J. Struct. Funct. Genomics* **5:**23–27.

128. Norvell, J. C., and A. Z. Machalek. 2000. Structural genomics programs at the US National Institute of General Medical Sciences. *Nat. Struct. Biol.* **7**(Suppl.)**:**931.

129. Oku, Y., A. Ohtaki, S. Kamitori, N. Nakamura, M. Yohda, H. Ohno, and Y. Kawarabayasi. 2004. Structure and direct electrochemistry of cytochrome P450 from the thermoacidophilic crenarchaeon, *Sulfolobus tokodaii* strain 7. *J. Inorg. Biochem.* **98:**1194–1199.

130. Osorio, G., and C. A. Jerez. 1996. Adaptive response of the archaeon *Sulfolobus acidocaldarius* BC65 to phosphate starvation. *Microbiology* **142:**1531–1536.

131. O'Toole, N., M. Grabowski, Z. Otwinowski, W. Minor, and M. Cygler. 2004. The structural genomics experimental pipeline: insights from global target lists. *Proteins* **56:**201–210.

132. Park, S. J., W. A. Joo, J. Choi, S. H. Lee, and C. W. Kim. 2004. Identification and characterization of inosine monophosphate dehydrogenase from *Halobacterium salinarum. Proteomics* **4:**3632–3641.

133. Poole, F. L., II, B. A. Gerwe, R. C. Hopkins, G. J. Schut, M. V. Weinberg, F. E. Jenney, Jr., and M. W. Adams. 2005. Defining genes in the genome of the hyperthermophilic archaeon *Pyrococcus furiosus*: implications for all microbial genomes. *J. Bacteriol.* **187:**7325–7332.

134. Prieur, D., G. Erauso, C. Geslin, S. Lucas, M. Gaillard, A. Bidault, A. C. Mattenet, K. Rouault, D. Flament, P. Forterre, and M. Le Romancer. 2004. Genetic elements of Thermococcales. *Biochem. Soc. Trans.* **32:**184–187.

135. Pritchett, M. A., J. K. Zhang, and W. W. Metcalf. 2004. Development of a markerless genetic exchange method for *Methanosarcina acetivorans* C2A and its use in construction of new genetic tools for methanogenic archaea. *Appl. Environ. Microbiol.* **70:**1425–1433.

136. Protein Data Bank, P. 2004. http://www.rcsb.org/pdb/. [Online.]

137. Protein Data Bank, T. 2005. http://targetdb.pdb.org/statistics/TargetStatistics.html. [Online.}

138. Protein Structure Initiative, P. 2005. NIGMS/NIH Protein Structure Initiative. http://www.nigms.nih.gov/psi/. [Online.]

139. Pysz, M. A., S. B. Conners, C. I. Montero, K. R. Shockley, M. R. Johnson, D. E. Ward, and R. M. Kelly. 2004. Transcriptional analysis of biofilm formation processes in the anaerobic, hyperthermophilic bacterium *Thermotoga maritima. Appl. Environ. Microbiol.* **70:**6098–6112.

140. Pysz, M. A., C. I. Montero, S. R. Chabra, and R. M. Kelly. 2004. Significance of polysaccharides in the microbial physiology and ecology of hot deep-sea subsurface biotopes, p. 213–226. *In* W. Wilcock, C. Cary, E. DeLong, D. Kelley, and J. Baross (ed.), *The Subseafloor Biosphere at Mid-Ocean Ridges.* American Geophysical Union Monograph Series 144. American Geophysical Union, Washington, D.C.

141. Pysz, M. A., K. D. Rinker, K. R. Shockley, and R. M. Kelly. 2001. Continuous cultivation of hyperthermophiles. *Methods Enzymol.* **330:**31–40.

142. Pysz, M. A., D. E. Ward, K. R. Shockley, C. I. Montero, S. B. Conners, M. R. Johnson, and R. M. Kelly. 2004. Transcriptional analysis of dynamic heat-shock response by the hyperthermophilic bacterium *Thermotoga maritima. Extremophiles* **8:**209–217.

143. Ram, R. J., N. C. Verberkmoes, M. P. Thelen, G. W. Tyson, B. J. Baker, R. C. Blake II, M. Shah, R. L. Hettich, and J. F.

Banfield. 2005. Community proteomics of a natural microbial biofilm. *Science* 308:1915–1920.

144. Rinker, K. D., and R.M. Kelly. 1996. Growth physiology of the hyperthermophilic archaeon *Thermococcus litoralis*: Development of a sulfur-free defined medium, characterization of an exopolysaccharide, and evidence of biofilm formation. *Appl. Environ. Microbiol.* 62:4478–4485.

145. Rinker, K. D., and R. M. Kelly. 2000. Effect of carbon and nitrogen sources on growth dynamics and exopolysaccharide production for the hyperthermophilic archaeon *Thermococcus litoralis* and bacterium *Thermotoga maritima*. *Biotechnol. Bioeng.* 69:537–547.

146. Robb, F. T., D. L. Maeder, J. R. Brown, J. DiRuggiero, M. D. Stump, R. K. Yeh, R. B. Weiss, and D. M. Dunn. 2001. Genomic sequence of hyperthermophile, *Pyrococcus furiosus*: implications for physiology and enzymology. *Methods Enzymol.* 330:134–157.

147. Robinson-Rechavi, M., and A. Godzik. 2005. Structural genomics of *Thermotoga maritima* proteins shows that contact order is a major determinant of protein thermostability. *Structure (Camb.)* 13:857–860.

148. Rohlin, L., J. D. Trent, K. Salmon, U. Kim, R. P. Gunsalus, and J. C. Liao. 2005. Heat shock response of *Archaeoglobus fulgidus*. *J. Bacteriol.* 187:6046–6057.

149. Rost, B. 1998. Marrying structure and genomics. *Structure* 6:259–263.

150. Sambrook, J., and D. W. Russell. 2001. *Molecular Cloning: A Laboratory Manual*, 3rd ed. Cold Spring Harbor Laboratory Press, Cold Spring Harbor, N.Y.

151. Saunders, N. F., A. Goodchild, M. Raftery, M. Guilhaus, P. M. Curmi, and R. Cavicchioli. 2005. Predicted roles for hypothetical proteins in the low-temperature expressed proteome of the Antarctic archaeon *Methanococcoides burtonii*. *J. Proteome Res.* 4:464–472.

152. Saunders, N. F., T. Thomas, P. M. Curmi, J. S. Mattick, E. Kuczek, R. Slade, J. Davis, P. D. Franzmann, D. Boone, K. Rusterholtz, R. Feldman, C. Gates, S. Bench, K. Sowers, K. Kadner, A. Aerts, P. Dehal, C. Detter, T. Glavina, S. Lucas, P. Richardson, F. Larimer, L. Hauser, M. Land, and R. Cavicchioli. 2003. Mechanisms of thermal adaptation revealed from the genomes of the Antarctic Archaea *Methanogenium frigidum* and *Methanococcoides burtonii*. *Genome Res.* 13:1580–1588.

153. Savchenko, A., N. Krogan, J. R. Cort, E. Evdokimova, J. M. Lew, A. A. Yee, L. Sanchez-Pulido, M. A. Andrade, A. Bochkarev, J. D. Watson, M. A. Kennedy, J. Greenblatt, T. Hughes, C. H. Arrowsmith, J. M. Rommens, and A. M. Edwards. 2005. The Shwachman-Bodian-Diamond syndrome protein family is involved in RNA metabolism. *J. Biol. Chem.* 280:19213–19220.

154. Savchenko, A., A. Yee, A. Khachatryan, T. Skarina, E. Evdokimova, M. Pavlova, A. Semesi, J. Northey, S. Beasley, N. Lan, R. Das, M. Gerstein, C. H. Arrowmith, and A. M. Edwards. 2003. Strategies for structural proteomics of prokaryotes: Quantifying the advantages of studying orthologous proteins and of using both NMR and X-ray crystallography approaches. *Proteins* 50:392–399.

155. Schena, M., D. Shalon, R. W. Davis, and P. O. Brown. 1995. Quantitative monitoring of gene expression patterns with a complementary DNA microarray. *Science* 270:467–470.

156. Schut, G. J., S. D. Brehm, S. Datta, and M. W. Adams. 2003. Whole-genome DNA microarray analysis of a hyperthermophile and an archaeon: *Pyrococcus furiosus* grown on carbohydrates or peptides. *J. Bacteriol.* 185:3935–3947.

157. Schut, G. J., J. Zhou, and M. W. Adams. 2001. DNA microarray analysis of the hyperthermophilic archaeon *Pyro-coccus furiosus*: evidence for a new type of sulfur-reducing enzyme complex. *J. Bacteriol.* 183:7027–7036.

158. Sehgal, A. C., R. Tompson, J. Cavanagh, and R. M. Kelly. 2002. Structural and catalytic response to temperature and cosolvents of carboxylesterase EST1 from the extremely thermoacidophilic archaeon *Sulfolobus solfataricus* P1. *Biotechnol. Bioeng.* 80:784–793.

159. Service, R. 2005. Structural biology. Structural genomics, round 2. *Science* 307:1554–1558.

160. Service, R. F. 2002. Structural genomics. Tapping DNA for structures produces a trickle. *Science* 298:948–950.

161. Shapiro, L., and C. D. Lima. 1998. The Argonne Structural Genomics Workshop: Lamaze class for the birth of a new science. *Structure* 6:265–267.

162. She, Q., R. K. Singh, F. Confalonieri, Y. Zivanovic, G. Allard, M. J. Awayez, C. C. Chan-Weiher, I. G. Clausen, B. A. Curtis, A. De Moors, G. Erauso, C. Fletcher, P. M. Gordon, I. Heikamp-de Jong, A. C. Jeffries, C. J. Kozera, N. Medina, X. Peng, H. P. Thi-Ngoc, P. Redder, M. E. Schenk, C. Theriault, N. Tolstrup, R. L. Charlebois, W. F. Doolittle, M. Duguet, T. Gaasterland, R. A. Garrett, M. A. Ragan, C. W. Sensen, and J. Van der Oost. 2001. The complete genome of the crenarchaeon *Sulfolobus solfataricus* P2. *Proc. Natl. Acad. Sci. USA* 98:7835–7840.

163. Shin, D. H., N. Oganesyan, J. Jancarik, H. Yokota, R. Kim, and S. H. Kim. 2005. Crystal structure of a nicotinate phosphoribosyltransferase from *Thermoplasma acidophilum*. *J. Biol. Chem.* 280:18326–18335.

164. Shockley, K. R., K. L. Scott, M. A. Pysz, S. B. Conners, M. R. Johnson, C. I. Montero, R. D. Wolfinger, and R. M. Kelly. 2005. Genome-Wide Transcriptional Variation within and between Steady States for Continuous Growth of the Hyperthermophile *Thermotoga Maritima*. *Appl. Environ. Microbiol.* 71:5572–5576.

165. Shockley, K. R., D. E. Ward, S. R. Chhabra, S. B. Conners, C. I. Montero, and R. M. Kelly. 2003. Heat shock response by the hyperthermophilic archaeon *Pyrococcus furiosus*. *Appl. Environ. Microbiol.* 69:2365–2371.

166. Smith, D. R., L. A. Doucette-Stamm, C. Deloughery, H. Lee, J. Dubois, T. Aldredge, R. Bashirzadeh, D. Blakely, R. Cook, K. Gilbert, D. Harrison, L. Hoang, P. Keagle, W. Lumm, B. Pothier, D. Qiu, R. Spadafora, R. Vicaire, Y. Wang, J. Wierzbowski, R. Gibson, N. Jiwani, A. Caruso, D. Bush, et al. 1997. Complete genome sequence of *Methanobacterium thermoautotrophicum* deltaH: functional analysis and comparative genomics. *J. Bacteriol.* 179:7135–7155.

167. Snijders, A. P. L., M. G. J. de Vos, and P. C. Wright. 2005. Novel approach for peptide quantitation and sequencing based on N-15 and C-13 metabolic labeling. *J. Proteome Res.* 4:578–585.

168. Southeast Collaboratory for Structural Genomics, S. 2005. http://www.secsg.org. [Online.]

169. St Jean, A., and R. L. Charlebois. 1996. Comparative genomic analysis of the *Haloferax volcanii* DS2 and *Halobacterium salinarium* GRB contig maps reveals extensive rearrangement. *J. Bacteriol.* 178:3860–3868.

170. Stevens, R. C., S. Yokoyama, and I. A. Wilson. 2001. Global efforts in structural genomics. *Science* 294:89–92.

171. Sugar, F. J., F. E. Jenney, Jr., F. L. Poole, 2nd, P. S. Brereton, M. Izumi, C. Shah, and M. W. Adams. 2005. Comparison of small- and large-scale expression of selected *Pyrococcus furiosus* genes as an aid to high-throughput protein production. *J. Struct. Funct. Genomics* 6:149–158.

172. Tajika, Y., N. Sakai, T. Tamura, M. Yao, N. Watanabe, and I. Tanaka. 2005. Crystal structure of PH0010 from *Pyrococ-*

cus horikoshii, which is highly homologous to human AM-MECR 1C-terminal region. *Proteins* **58**:501–503.

173. Tebbe, A., C. Klein, B. Bisle, F. Siedler, B. Scheffer, C. Garcia-Rizo, J. Wolfertz, V. Hickmann, F. Pfeiffer, and D. Oesterhelt. 2005. Analysis of the cytosolic proteome of *Halobacterium salinarum* and its implication for genome annotation. *Proteomics* **5**:168–179.

174. Terwilliger, T. C. 2000. Structural genomics in North America. *Nat. Struct. Biol.* **7**(Suppl.):935–939.

175. Terwilliger, T. C., G. Waldo, T. S. Peat, J. M. Newman, K. Chu, and J. Berendzen. 1998. Class-directed structure determination: foundation for a protein structure initiative. *Protein Sci.* **7**:1851–1856.

176. Thirumuruhan, R., S. Almo, A. Bresnick, R. Huang, F. Nagajyothi, T. Dodatko, A. Sharp, Y. Patskovsky, M. R. Chance, and S. Burley. 2003. Progress report from the New York structural Genomics research consortium. *Biophys. J.* **84**:175a.

177. Todd, A. E., R. L. Marsden, J. M. Thornton, and C. A. Orengo. 2005. Progress of structural genomics initiatives: an analysis of solved target structures. *J. Mol. Biol.* **348**:1235–1260.

178. Trent, J. D., J. Osipiuk, and T. Pinkau. 1990. Acquired thermotolerance and heat shock in the extremely thermophilic archaebacterium *Sulfolobus* sp. strain B12. *J. Bacteriol.* **172**:1478–1484.

179. Tumbula, D. L., and W. B. Whitman. 1999. Genetics of *Methanococcus*: possibilities for functional genomics in Archaea. *Mol. Microbiol.* **33**:1–7.

180. Venter, J. C., K. Remington, J. F. Heidelberg, A. L. Halpern, D. Rusch, J. A. Eisen, D. Wu, I. Paulsen, K. E. Nelson, W. Nelson, D. E. Fouts, S. Levy, A. H. Knap, M. W. Lomas, K. Nealson, O. White, J. Peterson, J. Hoffman, R. Parsons, H. Baden-Tillson, C. Pfannkoch, Y. H. Rogers, and H. O. Smith. 2004. Environmental genome shotgun sequencing of the Sargasso Sea. *Science* **304**:66–74.

181. Verhagen, M. F., A. L. Menon, G. J. Schut, and M. W. Adams. 2001. *Pyrococcus furiosus*: large-scale cultivation and enzyme purification. *Methods Enzymol.* **330**:25–30.

182. Verhees, C. H., S. W. Kengen, J. E. Tuininga, G. J. Schut, M. W. Adams, W. M. De Vos, and J. Van Der Oost. 2003. The unique features of glycolytic pathways in Archaea. *Biochem. J.* **375**:231–246.

183. Vierke, G., A. Engelmann, C. Hebbeln, and M. Thomm. 2003. A novel archaeal transcriptional regulator of heat shock response. *J. Biol. Chem.* **278**:18–26.

184. Walhout, A. J., G. F. Temple, M. A. Brasch, J. L. Hartley, M. A. Lorson, S. van den Heuvel, and M. Vidal. 2000. GATEWAY recombinational cloning: application to the cloning of large numbers of open reading frames or ORFeomes. *Methods Enzymol.* **328**:575–592.

185. Wang, B. C., M. W. Adams, H. Dailey, L. Delucas, M. Luo, J. Rose, R. Bunzel, T. Dailey, J. Habel, P. Horanyi, F. E. Jenney, Jr., I. Kataeva, H. S. Lee, S. Li, T. Li, D. Lin, Z. J. Liu, C. H. Luan, M. Mayer, L. Nagy, M. G. Newton, J. Ng, F. L. Poole II, A. Shah, C. Shah, F. J. Sugar, and H. Xu. 2005. Protein Production and Crystallization at SECSG—An Overview. *J. Struct. Funct. Genomics* **6**:233–243.

186. Wang, G., S. P. Kennedy, S. Fasiludeen, C. Rensing, and S. DasSarma. 2004. Arsenic resistance in *Halobacterium* sp. strain NRC-1 examined by using an improved gene knockout system. *J. Bacteriol.* **186**:3187–3194.

187. Weinberg, M. V. 2004. Oxidative and cold stress responses in the hyperthermophilic archaeon *Pyrococcus furiosus*. Ph.D. dissertation. University of Georgia, Athens, Ga.

188. Weinberg, M. V., G. J. Schut, S. Brehm, S. Datta, and M. W. Adams. 2005. Cold shock of a hyperthermophilic archaeon: *Pyrococcus furiosus* exhibits multiple responses to a suboptimal growth temperature with a key role for membrane-bound glycoproteins. *J. Bacteriol.* **187**:336–348.

189. Wolf, S., F. Lottspeich, and W. Baumeister. 1993. Ubiquitin found in the archaebacterium *Thermoplasma acidophilum*. *FEBS Lett.* **326**:42–44.

190. Worthington, P., V. Hoang, F. Perez-Pomares, and P. Blum. 2003. Targeted disruption of the a-amylase gene in the hyperthermophilic archaeon *Sulfolobus solfataricus*. *J. Bacteriol.* **185**:482–488.

191. Wunderlich, Z., T. B. Acton, J. Liu, G. Kornhaber, J. Everett, P. Carter, N. Lan, N. Echols, M. Gerstein, B. Rost, and G. T. Montelione. 2004. The protein target list of the Northeast Structural Genomics Consortium. *Proteins* **56**:181–187.

192. Yee, A., K. Pardee, D. Christendat, A. Savchenko, A. M. Edwards, and C. H. Arrowsmith. 2003. Structural proteomics: toward high-throughput structural biology as a tool in functional genomics. *Acc. Chem. Res.* **36**:183–189.

193. Zaigler, A., S. C. Schuster, and J. Soppa. 2003. Construction and usage of a onefold-coverage shotgun DNA microarray to characterize the metabolism of the archaeon *Haloferax volcanii*. *Mol. Microbiol.* **48**:1089–1105.

194. Zarembinski, T. I., L. W. Hung, H. J. Mueller-Dieckmann, K. K. Kim, H. Yokota, R. Kim, and S. H. Kim. 1998. Structure-based assignment of the biochemical function of a hypothetical protein: a test case of structural genomics. *Proc. Natl. Acad. Sci. USA* **95**:15189–15193.

195. Zhang, J. K., M. A. Pritchett, D. J. Lampe, H. M. Robertson, and W. W. Metcalf. 2000. In vivo transposon mutagenesis of the methanogenic archaeon *Methanosarcina acetivorans* C2A using a modified version of the insect mariner-family transposable element Himar1. *Proc. Natl. Acad. Sci. USA* **97**:9665–9670.

196. Zhang, J. K., A. K. White, H. C. Kuettner, P. Boccazzi, and W. W. Metcalf. 2002. Directed mutagenesis and plasmid-based complementation in the methanogenic archaeon *Methanosarcina acetivorans* C2A demonstrated by genetic analysis of proline biosynthesis. *J. Bacteriol.* **184**:1449–1454.

197. Zhong, H. Y., S. L. Marcus, and L. Li. 2004. Two-dimensional mass spectra generated from the analysis of N-15-labeled and unlabeled peptides for efficient protein identification and de novo peptide sequencing. *J. Proteome Res.* **3**:1155–1163.

198. Zhu, W., C. I. Reich, G. J. Olsen, C. S. Giometti, and J. R. Yates III. 2004. Shotgun proteomics of *Methanococcus jannaschii* and insights into methanogenesis. *J. Proteome Res.* **3**:538–548.

199. Zillig, W., K. O. Stetter, A. Wunderl, W. Schultz, H. Priess, and I. Scholtz. 1980. The *Sulfolobus*-"*Caldariella*" group: taxonomy on the basis of the structure of DNA-dependent RNA polymerases. *Arch. Mikrobiol.* **125**:259–269.

200. Zwickl, P., A. Grziwa, G. Puhler, B. Dahlmann, F. Lottspeich, and W. Baumeister. 1992. Primary structure of the *Thermoplasma* proteasome and its implications for the structure, function, and evolution of the multicatalytic proteinase. *Biochemistry* **31**:964–972.

Archaea: Molecular and Cellular Biology
Edited by Ricardo Cavicchioli
© 2007 ASM Press, Washington, D.C.

Chapter 21

Molecular Genetics of *Archaea*

KEVIN SOWERS AND KIMBERLY ANDERSON

INTRODUCTION

The *Archaea* are a phylogenetic lineage of microorganisms genotypically distinct from the *Bacteria* and *Eucarya* (see Chapters 1 and 19). These microorganisms are morphologically similar to monocellular, nonnucleated bacteria and possess many biosynthesis pathways typically found in bacteria (see Chapter 2). In contrast, nucleic acid processing, including both transcription and translation, in general occurs by mechanisms similar to those in eucarya (see Chapters 6 and 8). Like the *Bacteria* this domain of microorganisms exhibits a wide range of morphological and physiological diversity, which includes heterotrophs and autotrophs, acidophiles and alkaliphiles, psychrophiles and thermophiles, high-solute-intolerant nonhalophiles and halophiles. However, many archaea require extreme conditions for growth. These include halophiles that require saline concentrations up to saturation, methanogens that require highly reduced anoxic environments poised below −350 mv, and hyperthermophiles that grow at temperatures up to 113°C. The unique characteristics of these microorganisms have largely precluded attempts to apply bacterial or eucaryal genetic protocols to members of this domain. Traditional colony growth of many of the fastidious anaerobes in traditional anaerobic roll tubes is not as practical as plating for screening large numbers of clones; unique cell wall structures prevent the use of commonly used antibiotic genetic markers that target cell wall synthesis; bacterial gene promoters associated with many of the commonly used genetic markers are not recognized by the archaeal transcriptional apparatus; bacterial plasmids and phages will not replicate in archaeal species. In recent years several laboratories have overcome these difficulties by developing effective plating techniques, identifying genetic markers that do not target cell wall synthesis, fusing archaeal promoters with recombinant genes, and isolating native archaeal vectors and identifying promiscuous nonarchaeal vectors. Gene transfer systems now exist for species within all three physiological groups of archaea, halophiles, methanogens, and nonmethanogenic hyperthermophiles, and development of these systems has been reviewed (3, 72, 94, 103). Despite varying degrees of difficulty applying these protocols they are routinely used by laboratories conducting research on archaeal genetics and can be mastered by anyone with a fundamental knowledge of microbial genetic techniques.

MEDIA AND GROWTH

One of the initial challenges faced by investigators in the early development of archaeal genetic systems was the difficulty associated with growing many members of this domain on solidified medium for selection of recombinant clones. Fortunately, techniques have been developed to obtain efficient surface colonization for species of both the *Crenarchaeota* and *Euryarchaeota* (Table 1).

Halophilic euryarchaeota are grown on standard agar-solidified medium supplemented with 12 to 25% NaCl and Mg^{2+} salts, which serve as osmolytes for growth (91). As these microorganisms are aerobic mesophiles, they do not require specialized equipment used for growing hyperthermophiles and anaerobic methanogens. Because of the ease of growing these extremophiles, methods for genetic manipulation of haloarchaea have been developed and standardized over the past twenty years (see reference 30).

Methods have been developed for efficient growth of fastidiously anaerobic methanogens on conventional methanogen medium solidified with agar in Petri dishes, including *Methanobacterium* spp.

Kevin Sowers and Kimberly Anderson • Center of Marine Biotechnology, University of Maryland Biotechnology Institute, 701 E. Pratt Street, Baltimore, MD 21202.

Table 1. Physiological characteristics and growth efficiency of archaea on solidified medium

Organism	Salinity (%)	Growth temp (°C)	Aerobic/anaerobic	Mean efficiency[a] (%)	Reference
Halophiles					
Halobacterium spp.	20–25	40–50	Facultative anaerobe	100	91
Haloferax volcanii	12.5–20	40–50	Aerobe	100	91
Haloarcula spp.	20	40–50	Aerobe	100	91
Methanogens					
Methanococcus spp.	2–3	37	Strict anaerobe	90	54, 58, 74
Methanosarcina spp.	2–3	35	Strict anaerobe	55	6, 74
Methanothermobacter spp.	0.6	37	Strict anaerobe	80–100	43, 58
Methanobrevibacter spp.	0.6	30–37	Strict anaerobe	90–100	43, 58
Thermophiles					
Sulfolobus spp.	0.01	65–78	Aerobe	100	65
Pyrobaculum spp.	1.5	92	Facultative anaerobe	100	110
Pyrococcus spp.	2–4	80–95	Strict anaerobe	ND	34
Thermococcus celer	2–4	80–95	Strict anaerobe	ND	34

[a]ND, not determined.

and *Methanobrevibacter* spp. (43, 58), *Methanococcus* spp. (54, 74) and *Methanosarcina* spp. (6, 31, 52, 54, 74, 108) (Table 1). However, owing to the requirement by methanogens for both anoxic and highly reduced growth conditions, solidified medium must be prepared and inoculated in an anaerobic glove box. Anaerobic medium containing agar is melted in sealed serum vials outside the glove box, then transferred into the glove box and poured into Petri dishes. Although methanogens will tolerate the oxygen concentrations in an anaerobic glove box (<1 ppm) for purposes of inoculation, they will not grow without a reduced atmosphere. Inoculated media must be transferred to a gas-tight vessel or chamber containing a reduced atmosphere generated by hydrogen sulfide for incubation and growth (6, 108). The long incubation periods of 5 to 30 days require that the jars selected are not wholly composed of plastics or other polymers, which are slightly permeable to oxygen (104). Metal culture jars are available from commercial sources (Medica Instrument Mfg. Co., Don Whitley Scientific Ltd.), and they can be constructed from a modified pressure cooker (11), paint pressure canister (108), canning jar (6), or glove box airlock (74) or custom-made by a machine shop (10). Several commercially manufactured glass and/or metal anaerobe jars suitable for growing methanogens from TORBAL, Oxoid, and BBL are no longer available but often can be found in storage or through used-equipment distributors.

In addition to a reduced atmosphere the moisture content of the medium is often a critical factor. If the plates are too moist, the closed environment of the anaerobe jar results in confluence of the colonies due to surface condensate; too low a moisture content is inhibitory to many osmotically sensitive species with S-layer cell walls. The moisture content of the plates is controlled by preincubating the plates in an anaerobic glove box at a known relative humidity for a set period of time, then storing the plates in a sealed container until they are inoculated. Both the hydrogen sulfide concentration and moisture content required for optimal plating efficiencies must be determined empirically for each species. Another factor that can critically affect plating efficiency is exposure of the cells to trace amounts of oxygen during plating in the anaerobic glove box. When cells are spread onto the surface of the solidified medium they are extremely sensitive to oxygen as they are no longer protected by the reduced medium. Glassware and spreading implements should be equilibrated in the glove box overnight to remove adsorbed oxygen. Care must be taken to ensure that the chamber atmosphere is at its minimum oxygen tension by using freshly charged palladium catalyst and by not opening the gas-lock for a period of time before inoculation to allow the chamber atmosphere to become anoxic. Inoculating cells in a molten top agar (0.5% wt/vol in medium) will often improve colonization efficiencies by protecting the cells from trace oxygen exposure (58). The growth efficiency of *Methanosarcina mazei* Gö1, which grows poorly on agar-solidified medium, was improved by selection for a spontaneous mutant that yielded greater plating efficiencies (31).

Colonization of methanogenic and nonmethanogenic hyperthermophiles on solidified medium is equally problematic, as agar is rapidly dehydrated at high temperatures, especially at the concentrations required for it to remain solidified. Therefore, gellan gum (Gelrite) is used as the solidifying agent for growth of thermophiles and hyperthermophiles, which are incubated in a plastic bag or anaerobe jar to minimize dehydration (34, 65). Conventional polystyrene disposable Petri plates will melt above 60°C;

therefore, glass or higher-temperature-resistant polyethylene plates (Greiner Bio One International AG) are used for thermophiles. Growth medium is inoculated at room temperature, and then the medium is incubated at the desired temperature. For anaerobic hyperthermophiles media are prepared and inoculated by the methods described for methanogens.

STRAIN CHARACTERISTICS

Archaea generate cell walls that range from pliable paracrystalline protein and glycoprotein S-layers to rigid forms that include a pseudomurein monolayer analogous to bacterial gram-positive cell walls and heteropolysaccharide with chemical properties of chondroitin (56) (see Chapter 14). Most of the species for which tractable gene transfer systems have been developed have cell walls composed of a paracrystalline S-layer rather than a rigid cell wall (Table 1). For DNA-uptake procedures, spheroplasts are generated by suspending S-layer cells in a Mg^{2+}-free sucrose buffer, in some instances with the addition of EDTA, which disrupts the integrity of the S-layer protein subunits (35). After transformation, the S-layer is regenerated by resuspension of the cells in medium that contains divalent cations.

Genetic systems have been developed for archaeal species with rigid cell walls. *Methanosarcina* spp. grown in low-saline medium synthesize a rigid chondroitin-like (methanochondroitin) outer layer in addition to a glycoprotein S-layer, which causes cells to grow as multicellular aggregates. For genetic studies these species are adapted to, and grown at, marine saline concentrations, which prevents synthesis of methanochondroitin, enabling the cells to be grown as individual S-layer cells (101). Spheroplasts of these cells can be generated in Mg^{2+}-free sucrose buffer for transformation, and the S-layer can be regenerated as described above (73). *Methanothermobacter* spp. have a rigid pseudomurein cell wall, and spheroplasts can be generated with a pseudomurein endopeptidase

from *Methanothermobacter wolfei*. Although the spheroplasts remain actively methanogenic in a sucrose buffer, methods for regenerating a cell wall have not yet been developed, which currently limits the tractability of these species for genetics (57, 77).

Besides cell structure, cell physiology has also influenced the development of genetic systems for the *Archaea*. Most of the currently available genetic systems were developed for mesophilic and moderately thermophilic archaea (Table 2). In the extreme temperature ranges there has been limited success with hyperthermophilic archaea (2, 67), and there are no published reports for psychrophilic archaea.

SELECTABLE GENE MARKERS

Vectors have been developed for halophiles, methanogens, and nonmethanogenic hyperthermophiles (Table 3). Plasmids isolated thus far from archaea do not possess phenotypes that can be readily used for selection; the unique physiology of archaea precludes the use of many common genetic markers. As examples, antibiotics that act by disrupting cell wall synthesis are ineffective because archaea lack bacterial cell walls; high temperatures used to grow hyperthemophiles degrade many antibiotics; high temperatures or high-salt concentrations degrade many of the gene products that detoxify the selection marker; in the case of methanogens, chloramphenicol effectively inhibits protein synthesis, but the product of chloramphenicol acetyltransferase inhibits methanogenesis (14). However, there are several archaeal and orthologous genes from bacterial and eucaryal systems that have been used as effective genetic markers in the *Archaea*, and these are described below.

The high intracellular salt concentration found in the moderate and extreme halophiles will denature proteins generated by many nonhalotolerant resistance loci currently in use. This problem has been circumvented by the isolation of a genetic determinant

Table 2. Gene transfer methods for archaea

Type	Method	Species	Reference
Transformation	Polyethylene glycol (PEG)-mediated	*Haloferax volcanii, Haloarcula* spp., *Halobacterium* spp., *Methanococcus maripaludis, Pyrococcus abyssi*	24–26, 67, 109
	Electroporation	*Methanococcus voltae, Sulfolobus* spp.	22, 83, 98
	Liposome-mediated	*Methanosarcina* spp.	73
	$CaCl_2$ with heat shock	*Sulfolobus solfataricus, Pyrococcus furiosus*	2
Transduction	ΨM1-mediated	*Methanothermobacter thermautotrophicus*	71
Conjugation	Cell mating	*Haloferax volcanii, Sulfolobus acidocaldarius*	39, 92
	Plasmid mediated	*Sulfolobus solfataricus*	97

Table 3. Shuttle vectors for use in archaea and *E. coli*

Type of organism	Vector	Host organism(s)	Genetic marker(s)[a]	Size (kb)	Characteristics	Reference
Halophiles	pHRZ	*Halobacterium* spp.	Ani/Thi	12.6	Unknown origin of replication	69
	pMDS20	*Haloferax volcanii*	Mev/Amp	10	Derived from pMDS10	47
	pMLH3	*Haloferax volcanii*	Nov/Mev/Amp	11.3	Contains 2 selective markers (Nov/Mev)	47
	pMPK54, 62	*Halobacterium salinarium*	Mev/Amp	8.6, 10.2	Used for expression of bacteriorhodopsin gene, *bop*	60
	pNB102	*Haloarcula* spp., *Halobacterium* spp.	Mev/Amp	9.1	Derived from pNB101	120
	pNG168	*Halobacterium* spp.	Mev/Amp	8.9	Multiple cloning sites; blue/white screening in *E. coli*	30
	pTA230–233	*Haloferax volcanii*	Pyr, Trp, Leu, Hdr/Amp	7.4, 7.7, 7.8, 7.4	Blue/white screening in *E. coli*; requires Δ*pyrE2*, Δ*trpA*, Δ*leuB*, or Δ*hdrB* host	4
	pUBP2	*Halobacterium salinarium*, *Haloferax volcanii*	Mev/Amp	12.3	Derived from pHH1 from *Halobacterium*	18, 25
	pWL102, pWL104	*Haloferax volcanii*, *Haloarcula* spp.	Mev/Amp	10.5, 8.7	Derived from pHV2 from *Haloferax*	18, 25, 62
	pWL-Nov	*Haloferax volcanii*	Nov/Amp	9.4	Derived from pWL102	82
	pWL204	*Haloferax volcanii*	Mev/Amp	10.4	Expression vector based on pWL102	81
Methanogens	pDTL44	*Methanococcus maripaludis*	Pur/Amp	12.7	Used for gene expression and complementation	107
	pEA103	*Methanosarcina acetivorans*	Pur/Amp	12.4	Based on pWM313; used for gene expression; requires *E. coli pir* host for replication	5
	pJK89	*Methanosarcina acetivorans*	Pro/Amp	9.1	Contains *proC* from *E. coli*; used for complementation of Δ*proC* host	89
	pPB31–35	*Methanosarcina acetivorans*	PA/Amp	11.3	Blue-white screening in *E. coli*; requires *E. coli pir* host for replication	19, 119
	pWLG30	*Methanococcus maripaludis*	Pur/Amp	12.7	Used for gene expression	36
	pWM309–321	*Methanosarcina* spp.	Pur/Amp	8.2–8.9	Based on pC2A; multiple cloning sites; blue/white screening in *E. coli* (except pWM321)	73
Thermophiles	pAG21	*Pyrococcus furiosus*, *Sulfolobus acidocaldarius*	Adh/Amp	6.5	More stable in *E. coli* due to low copy number	8
	pCSV1	*Pyrococcus furiosus*, *Sulfolobus acidocaldarius*	Amp	6.1	Based on pGT5 from *P. abyssi*	2
	pEXADH	*Sulfolobus solfataricus*	Hyg, Adh/Amp	7.7	Contains 2 selective markers (Hyg, Adh) for use in *S. solfataricus*	29
	pEX*lacOP*	*Sulfolobus solfataricus*	Hyg/Amp	11.7	Contains *lacS* and *lacTr* from *S. solfataricus*	13
	pEXSs	*Sulfolobus solfataricus*	Hyg/Amp	6.4	Uses SSV1 ORI for replication in *S. solfataricus*	22
	pYS2	*Pyrococcus abyssi*	Pyr/amp	6.4	Based on pGT5 from *P. abyssi*; used with Δ*pyrE* strains	67

[a]Adh, alcohol resistance; Amp, ampicillin; Ani, anisomycin; Hdr, trimethoprim; Hyg, hygromycin B; Leu, leucine; Mev, mevinolin; Nov, novobiocin; PA, pseudomonic acid; Pro, proline; Pur, puromycin; Pyr, 5-fluoroorotic acid; Thi, thiostrepton; Trp, trytophan.

encoding resistance to the 3-hydroxy-3-methylglu-taryl coenzyme A reductase inhibitor mevinolin, which is a cholesterol-reducing drug for humans that inhibits synthesis of isoprenoid lipid side chains in archaea (62). A promoter point mutation leads to over-production of the reductase, which is the target of mevinolin. The *Haloferax volcanii* Mev^R determinant has been shown to confer mevinolin resistance in *Haloferax* spp., *Haloarcula* spp., and *Halobacterium salinarium* (formerly *Halobacterium halobium*) (18, 42, 59, 80, 115). Another haloarchaeal genetic locus isolated from spontaneous gyrase mutants of *H. volcanii* (48) provides resistance to the gyrase inhibitor novobiocin. Several hybrid vectors have been constructed that use both markers for transformant selection and insertional inactivation (Table 3). Resistance to the broad-spectrum protein inhibitors anisomycin, sparsomycin, and thiostrepton has been reported in *H. salinarium* using a plasmid-borne, altered 23S rRNA gene (69, 105), and deletion of dihydrofolate reductase (*hdr*A) in *H. volcanii* confers resistance to trimethoprim (82). However, these latter resistance markers have not been exploited for haloarchaeal selection systems. Counter-selectable gene knockout systems have been developed with spontaneous 5-fluoroorotic acid-resistant mutants of *Halobacterium* spp. and *H. volcanii* using *ura3* and *pyrE*, respectively (4, 17, 85). Auxotrophic mutants have also been developed that can be complemented with genes encoding for histidine, leucine, thymidine, and tryptophan biosynthesis as selectable markers (4, 28, 82).

Mesophilic methanogenic archaea are sensitive to several antibiotics that inhibit protein synthesis, including puromycin, neomycin, and pseudomonic acid (20, 88). The most commonly used antibiotic for genetic selection is puromycin. In this system the structural gene from the bacterium *Streptomyces alboniger* that encodes puromycin transacetylase (*pac*) has been used as a selective marker for *Methanococcus voltae*, *Methanococcus maripaludis*, and *Methanosarcina* spp. As with all archaea the *S. alboniger* bacterial promoter is not recognized and transcribed; therefore, the *pac* gene is flanked with the constitutively expressed methyl coenzyme M reductase (MCR) promoter and terminator from *M. voltae* (plasmid pMEB.2) or *Methanosarcina barkeri* (plasmid pJK3) (37, 73). The *pac* transcription cassette is flanked by either EcoRI sites or multiple restriction sites in plasmids pMEB.2 and pJK3, respectively, for excision and insertion into other vectors. Another antibiotic resistance marker for the methanogens utilizes aminoglycoside phosphotransferase genes APH3′I and APH3′II from bacterial transposons Tn*903* and Tn*5*, respectively, engineered with flanking archaeal *mcr* promoter and terminator sequences (9). When introduced into

Methanococcus maripaludis on plasmids, these genes confer resistance to the protein synthesis inhibitor neomycin at frequencies equivalent to those achieved for puromycin resistance with *pac*. However, these markers are ineffective for *Methanosarcina* spp. because these species are not inhibited by neomycin (19). Spontaneous mutations in the gene encoding isoleucyl t-RNA synthetase, *ileS*, confer resistance to pseudomonic acid by *Methanothermobacter thermautotrophicus* (53). Pseudomonic acid was developed also as a selectable marker in *M. barkeri* by directed mutagenesis of the methanosarcinal *ileS* (19). Although this is a tractable genetic system, pseudomonic acid, also known as mupirocin, is the active ingredient in a topical antimicrobial ointment and is currently available in pure form only by request from the manufacturer (GlaxoSmithKline). Bacitracin, an inhibitor of bacterial murein synthesis, has been reported to inhibit *M. thermautotrophicus*, presumably by inhibiting synthesis of pseudomurein, but a tractable gene transfer system has not been developed for this genus (44). In addition to antibiotics, auxotrophic mutants have been developed with defects in the biosynthesis of histidine, leucine, and proline (45, 86, 89). A shuttle vector encoding *Escherichia coli* *proC* is available as a selectable marker for a *Methanosarcina acetivorans* ΔproC host (89). Disruption in the purine salvage pathway gene *hpt*, which encodes hypoxanthine phosphoribosyltransferase, has been used for counter selection with either 8-aza-2,6-diaminopurine or 8-aza-hypoxanthine in *M. acetivorans* and *M. maripaludis* (76, 89).

For both methanogenic and nonmethanogenic hyperthermophiles, thermostability of both the marker gene product and its target has largely precluded the use of mesophilic bacterial and archaeal markers. However, several high-temperature-resistant markers have been described for use in these microorganisms. *Sulfolobus solfataricus* is sensitive to the thermostable antibiotic hygromycin with a low spontaneous reversion frequency. A plasmid that contains a gene for a thermostable mutant form of hygromycin phosphotransferase (*hph*) isolated from *E. coli* was constructed with flanking *S. solfataricus* aspartate aminotransferase promoter and terminator sequences for expression in thermophilic archaea (22, 23). Another selectable genetic marker is based on butanol sensitivity of hyperthermophilic species that lack alcohol dehydrogenase (ADH) (2). The ADH structural gene from alcohol-tolerant *S. solfataricus*, together with its promoter and terminator sequences, confers resistance when transformed into *Sulfolobus acidocaldarius* and *Pyrococcus furiosus* (8, 29). Two other selectable markers have been developed based on the 23S rRNA gene of *S. acidocaldarius* and *P. furiosus* where a point mutation in one position confers resistance to caromycin, chloramphenicol, and celesticetin, and a point mutation at a second site confers resistance to thiostrepton (2). However, the rate of spontaneous reversion ($>10^{-7}$) at either location is too high for use as a selection marker and efforts are under way to use a 23S rRNA gene possessing both point mutations to reduce the probability of spontaneous reversion of *P. furiosus* grown in the presence of both antibiotics. Spontaneous novobiocin-resistant mutants of *S. acidocaldarius* have also been reported but must be maintained with novobiocin because of high reversion rates (40). Counter-selectable gene knockouts have been reported for 5-fluoroorotic acid-resistant mutants of *Sulfolobus* spp., *Pyrococcus abyssi*, and *Thermococcus kodakaraensis* using *pyrE* and *pyrF* (40, 55, 67, 70, 96, 112). *S. solfataricus* auxotrophic mutants for the β-linked disaccharide lactose (*lacS*) (117) and the α-linked oligosaccharides starch and glycogen (*malA*) also have the potential to be used for targeted mutagenesis. An auxotrophic mutant for tryptophan *trpE* in *T. kodakaraensis* has been reported, but it has not been tested with a selectable marker (96).

GENE TRANSFER SYSTEMS

Transformation

Haloarchaea are transformed via polyethylene glycol (PEG) mediated transformation, which was first described in *H. volcanii* (24). This method requires the formation of spheroplasts generated by removing the surface glycoprotein layer (S-layer). The primary complication encountered with transformation in halophilic species is that most halophiles have restriction systems that lower the efficiency of the transformation (25). However, transformation efficiencies have been improved significantly by using restriction-minus mutants and passing the DNA through a Dam$^+$ strain of *E. coli* (50).

Within the methanogenic archaea, there are only two genera for which transformation protocols have been developed: *Methanococcus* and *Methanosarcina*. *M. voltae* spheroplasts can be transformed using electroporation and can also take up circular and linear DNA without any further manipulation (83). However, the transformation efficiency for these methods is low, at around 10^1 to 10^3 transformants μg^{-1} DNA (83). PEG-mediated transformation is a very efficient method for use with *M. maripaludis* and is capable of producing 10^7 transformants μg^{-1} DNA (109). In contrast PEG-mediated transformation is ineffective for the *Methanosarcina* species, but liposome-mediated transformation has proven an effi-

cient method for transformation (31, 73). Transformation via this method is similar to PEG-mediated transformation of *M. maripaludis*, producing $>10^7$ transformants μg^{-1} DNA for *Methanosarcina* species.

There are several effective methods for transformation of the thermophilic archaea. Similar to *M. voltae*, *T. kodakaraensis* is capable of taking up DNA without any modification or mediation (96). PEG-mediated transformation with spheroplasts has been shown to be effective for *P. abyssi* (67). *S. solfataricus* and *P. furiosus* have both been successfully transformed with circular DNA using CaCl$_2$ treatment followed by heat shock (2). *S. acidocaldarius* and *S. solfataricus* can be transformed with linear and circular DNA using electroporation (1, 8, 22, 32, 98).

Transduction

Phages and viruses have been discovered within all three groups of archaea, but transduction currently is not a tractable method for gene transfer. *Methanothermobacter marburgensis* (formerly *Methanobacterium thermoautotrophicum* Marburg) has been directly transduced with ΨM1, a phage isolated from this organism (71). However, this phage has a limited burst size, which makes it impractical for use in gene transfer experiments.

Conjugation

Gene transfer by conjugation has been documented within the halophiles and the hyperthermophiles. In *H. volcanii*, direct cell-cell contact and fusion with the transfer of genetic material has been shown (92). The transfer efficiency is stimulated by treatment with DNase (75). The DNA transfer is bidirectional and does not appear to require any plasmid or transposon-encoded genes. Cell mating has also been reported for two members of the hyperthermophilic archaea, *S. acidocaldarius* and *S. solfataricus*. With *S. acidocaldarius*, genetic marker exchange has been reported between two auxotrophic mutants (39). *S. solfataricus* uses plasmid-mediated conjugation requiring direct cell-cell contact in which the conjugative plasmid pNOB8 is transferred unidirectionally (97, 99).

GENE VECTORS

Hybrid vectors contain both bacterial and archaeal components that enable them to be replicated and selected in both systems (Fig. 1). A large number of shuttle vectors are available for the halophiles in

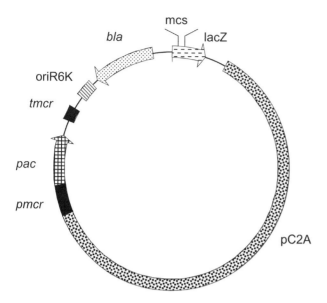

Figure 1. Recombinant plasmid showing construction typical for an *E. coli*/archaeal shuttle vector. The construct includes the *pir*-dependent R6K *ori* for replication of the plasmid in *pir*⁺ *E. coli*, the *bla* gene for selection of *E. coli* transformants with ampicillin, pC2A *ori* and *repA* for replication in *Methanosarcina* spp., and *pac* under transcriptional control of the archaeal *mcrB* gene for selection of methanosarcinal recombinants on puromycin. pC2A, autonomously replicating plasmid pC2A from *M. acetivorans*; *pac*, puromycin N-acetyltransferase flanked by the methylCoM reductase archaeal promoter (pmcr) and terminator (tmcr); oriR6K, *pir*-dependent R6K origin of replication; *bla*, β-lactamase; mcs, multiple cloning site in the *lacZ* gene encoding β-galactosidase for blue-white screening of DNA insertion.

comparison to the methanogens or the thermophiles (Table 3). *H. volcanii* contains a naturally occurring miniplasmid, pHV2. Using the origin from this vector, shuttle vectors for halophilic archaea were developed containing the gene encoding mevinolin resistance for selection within the halophile and a separate resistance gene for selection within *E. coli* (62). The shuttle vectors derived from pHV2 can be used in *H. volcanii* as well as *Halobacterium* spp. Other shuttle vectors for use in the halophiles have also been developed from naturally occurring plasmids, such as pNRC100, pHH1, and pGRB1 (18, 59, 78, 79).

Within the methanogens, shuttle vectors for *Methanosarcina* spp. have been constructed from the naturally occurring plasmid pC2A from *M. acetivorans* C2A (73, 102). These vectors contain the origin of replication from pC2A, a ColE1 origin for replication in *E. coli*, the *pac* gene for puromycin resistance within the methanogen, and a gene encoding ampicillin resistance for selection in *E. coli* (Fig. 1). The pWM family of plasmids, which are all derivatives of pC2A, are capable of transformation into all *Methanosarcina* species (31, 73). Shuttle vectors for *M. maripaludis* have been constructed using the cryp-

tic plasmid pURB500 although, unlike derivatives of methanosarcinal pC2A, these vectors can only be used with *M. maripaludis* (107).

There is a relative dearth of shuttle vectors available for use within the hyperthermophilic archaea. One type of shuttle vector currently used is based on pGT5, an autonomously replicating plasmid from *P. abyssi* GE9 (33). These plasmids can be used in both *P. abyssi* and *S. acidocaldarius* (2, 8, 67). Another shuttle vector uses a replication origin from *Sulfolobus* phage SSV1 (22). This vector, pEXSs, contains the *hph* gene for selection of transformants of *S. solfataricus* with hygromycin.

GENE DELETION SYSTEMS

Random Mutagenesis

Gene disruption is required to identify and confirm the function of genes within the archaea. Random mutagenesis using chemical and UV radiation has been successfully used for *H. volcanii*, *M. voltae*, *M. maripaludis*, and *P. abyssi* (16, 61, 75, 113) (also see Chapter 5). However, a significant problem with these approaches is determining the mutation loci. A transposition system has been developed for *M. acetivorans* using a modified *mariner*-family *Himar1* transposon that utilizes only the cognate transposase expressed from methanosarcinal promoters (118). The transposon elements flank both archaeal and bacterial selection markers and an *E. coli* origin of replication. After transformation of the vector into the archaeon, transposed genomic DNA is restriction digested, ligated, and propagated as a plasmid by transformation into *E. coli*. The mutated gene is readily identified by sequencing the genomic DNA flanking the transposon elements. By use of this approach, *mariner*-induced *M. acetivorans* mutants were identified for genes that encode proteins for heat shock, nitrogen fixation, and cell wall structures (118).

Targeted Gene Knockout Systems

Gene knockout systems that disrupt specific loci have been developed for all three groups of archaea. These systems typically include disruption with retention of a genetic marker and "markerless systems" by which the genetic marker is removed after disruption by a second homologous recombination event, enabling reuse of the markers for subsequent deletions (Fig. 2). Within the halophilic archaea, a commonly used method is based on restoration of uracil prototrophy in a *ura* mutant or *pyrE2* strain as a genetic marker (4, 17, 85, 111). Circular nonreplicating plas-

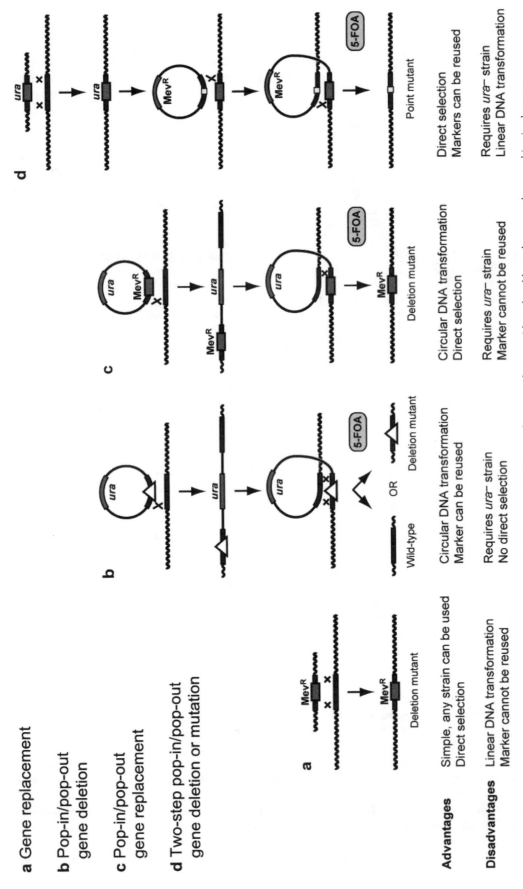

a Gene replacement

b Pop-in/pop-out
gene deletion

c Pop-in/pop-out
gene replacement

d Two-step pop-in/pop-out
gene deletion or mutation

Advantages	Simple, any strain can be used Direct selection	Circular DNA transformation Marker can be reused	Circular DNA transformation Direct selection	Direct selection Markers can be reused
Disadvantages	Linear DNA transformation Marker cannot be reused	Requires *ura⁻* strain No direct selection	Requires *ura⁻* strain Marker cannot be reused	Requires *ura⁻* strain Linear DNA transformation

Figure 2. Gene disruption methods that are used in archaeal genetics. (a) Direct replacement of a gene with a selectable marker occurs by recombination between linear DNA flanked by regions of the target gene and the wild-type chromosomal gene. (b) The "pop-in pop-out" method uses circular DNA and selection for transformation to uracil prototrophy using a *ura⁻* strain (17, 85, 96). Recombinants that have lost the plasmid are counter-selected using 5-fluoroorotic acid (5-FOA), which inhibits growth of *ura⁺* cells. Deletion mutants must be screened by Southern hybridization. (c) A variant of the "pop-in pop-out" method for gene deletion utilizes a genetic marker for gene disruption that allows direct selection (4). (d) Another variant used for generating point mutations employs gene replacement with a *ura* marker and subsequent replacement of the *ura* disrupted target gene with a gene containing the desired point mutation (85). Mutants with the desired point mutation are counter-selected with 5-FOA. Reprinted from *Nature Reviews Genetics* (3) with permission of the publisher.

mids that contain the *ura3* or *pyrE2* genes and the disrupted target gene are integrated into the target gene loci by homologous recombination and selected in uracil-free medium. Counter-selection with 5-fluoroorotic acid selects for clones that have lost the *ura3* or *pyrE2* gene and either lost the wild-type or mutated the target gene by a second homologous recombination event. Recombinant strains must then be screened by Southern analysis or PCR to identify clones with single copies of the disrupted gene. A modification of this method utilizes the gene encoding mevinolin resistance (*mev^R*) to disrupt the target gene, which allows direct selection of the disrupted clone by adding mevinolin to the medium (4). Another variation employs insertion of *ura3* by homologous insertion into the genomic target gene followed by substitution of the genomic target gene with a copy of the gene with a point mutation on a plasmid (85). The point mutants are counter-selected with 5-fluoroorotic acid. *H. volcanii* Δ*leuB* and Δ*trpA* mutants are also available for positive selection of gene deletions (4).

Within the methanogenic archaea, the gene knockout systems that were developed for *Methanococcus* spp. and *Methanosarcina* spp. utilize selective markers for puromycin, pseudomonic acid, and neomycin resistance (9, 19, 37). The simplest means of gene disruption is by homologous recombination with the genetic marker distally flanked by sequences from the target gene on a nonreplicating shuttle vector or linear DNA fragment. Mutants are then selected by growing the transformant with the corresponding antibiotic. This approach has been used in *Methanococcus* spp. by disruption with neomycin (APH3′I/II) and puromycin (*pac*) resistance markers (12, 46, 63, 64, 66, 87, 106, 116) and in *Methanosarcina* spp. with *pac* and pseudomonic acid (*ileS*) resistance markers using homologous recombination (41, 68, 114) and transposon insertion (119). Directed gene disruption is also possible by complementing histidine auxotrophy in a *hisA* mutant strain of *M. voltae* and proline auxotrophy in a *proC* mutant strain of *M. acetivorans* (15, 119).

A disadvantage of gene disruption with an antibiotic or by complementation of an auxotrophic mutant is that the genetic marker cannot be reused for multiple disruptions. As an alternative, a markerless gene disruption system for the methanogens has been developed using a mutant defective for hypoxanthine phosphoriboxyltransferase (*hpt*), an enzyme involved in the purine salvage pathway, which is resistant to inhibition by the purine base analog 8-*azo*-2,6-diamino-purine (8-ADP) (21, 89, 76). In the first step of disruption, wild-type *hpt* and *pac* flanked by target DNA is inserted into the wild-type gene by homologous recombination to create an unstable merodiploid that is selected for puromycin resistance and screened for resistance to 8-ADP. In the second step, selected transformants are grown under nonselective conditions without puromycin to promote plasmid excision, which creates a mixed population of clones with either the wild-type gene or a disrupted gene. However, clones with the mutant allele must be selected by screening for puromycin sensitivity either by Southern hybridization or PCR.

A modified markerless disruption method has recently been devised that eliminates the requirement for screening (Fig. 3) (94). This method first involves disruption of the target gene by a double-recombination event with a linear DNA fragment containing the *pac-hpt* cassette, flanked by *flp* recombinase recognition sites and sequence from the target gene. Insertion of this construct yields a gene disruption that is selected by puromycin resistance and 8-ADP sensitivity. The mutant is then transformed with a nonreplicating plasmid that encodes the gene for *flp* recombinase under archaeal promoter control, which, when expressed, will delete the region between the *flp* recognition sites, generating a deletion mutant that is puromycin sensitive and 8-ADP resistant. As an alternative to gene deletions, conditional gene inactivation has also been used to test for essential genes (93). In this approach the target gene is fused to a highly regulated promoter, and the effects of this fusion are assayed under expressing and nonexpressing growth conditions. This approach has also been proposed for identification of *trans*-acting regulatory factors that control multiple genes in a regulatory system (94).

Gene disruption protocols exist for two species of thermophilic archaea. A *lacS* mutant of *S. solfataricus* created through transposon-mediated mutagenesis is disrupted by *lacS* gene distally flanked by sequences of the target gene via homologous recombination. *T. kodakaraensis* genes can also be disrupted via recombination, but this method has been modified to allow for the reuse of the selective marker (95, 96). Using either *pyrF* or *trpE* as selective markers with endogeneous and exogenous sequences that flank the selective marker gene and act as tandem repeats, the genetic marker is excised via recombination under nonselective growth conditions. Subsequent deletions can be made in the mutant strain using the same marker.

GENE REPORTERS

Several phenotypic markers are available for the archaea that have been exploited largely for in vivo gene expression studies rather than genetic markers. A salt-tolerant β-galactosidase from *Haloferax alicantei* has been expressed in *Halobacterium* spp. and

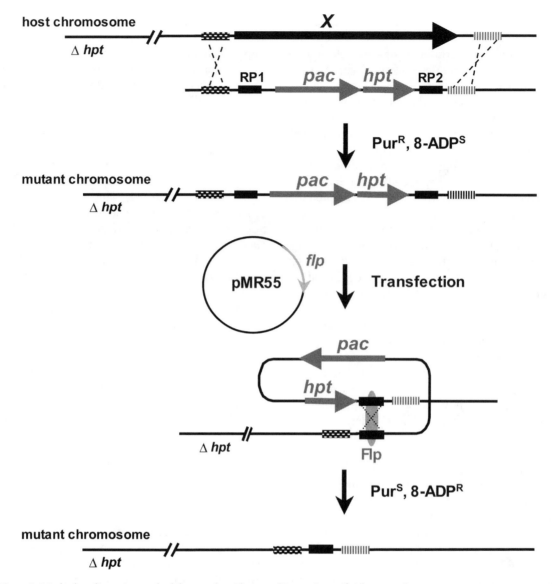

Figure 3. Markerless disruption method that employs Flp recombinase. An artificial operon that expresses puromycin N-acetyl-transferase (*pac*) and hypoxanthine phosphoribosyltransferase (*htp*) is flanked by Flp recombinase recognition sites (RP1 and RP2) and regions homologous to the target gene. The linear DNA is transformed into an *M. acetivorans* Δhpt strain that is resistant to 8-aza-2,6-diamino-purine (8-ADP). The target gene is replaced by homologous recombination, and recombinants are selected by resistance to puromycin. The deletion mutant is subsequently transformed with the nonreplicating plasmid pMR55 encoding Flp recombinase, which removes the *pac–hpt* operon by site-specific recombination between RP1 and RP2. Reprinted from *Current Opinion in Microbiology* (94) with permission of the publisher.

H. volcanii, which lack detectable β-galactosidase background activity (38, 49, 51, 84). A modified derivative of green fluorescent protein has also been expressed in *H. volcanii* as a gene reporter (90). A phenotype with potential for use in genetic studies is the purple color associated with bacteriorhodopsin in *H. salinarium*, which is encoded by the *bop* gene. When present on a plasmid, bacteriorhodopsin production is capable of complementing a *bop* insertion mutant, which is detected by purple-colored colony formation (59). Constructs containing functional *bop*

genes are potentially useful not only for studies employing *bop* mutant strains but also when propagated in naturally occurring *bop*-less halophiles (e.g., *H. volcanii*), provided that the organism is capable of synthesizing the retinal chromophore required for bacteriorhodopsin function or it is supplemented to the medium.

The *E. coli uidA* (encodes β-glucuronidase) and *lacZ* (encodes β-galactosidase) genes and *Bacillus subtilis treA* (encodes trehalase) have been expressed as phenotypic markers in the methanogenic archaea,

Methanococcus spp. and *Methanosarcina* spp. (7, 15, 27, 89, 100). When flanked by archaeal promoter and terminator sequences, these three markers can be used to quantify archaeal gene expression. Since the chromophores are colorless under anaerobic conditions, the assays are conducted aerobically, or filter blots of the transformed colonies must be exposed to air to develop the color.

One phenotypic marker, a thermostable gene encoding *lacS* from *S. solfataricus*, has been shown to complement mutants in transformed cells (32). However, plasmid pNOB8 containing the *lacS* gene was eventually lost in transformants with restored β-galactosidase activity because of inadvertent integration of *lacS* into the genome. A plasmid mediated *lacS* reporter has also been developed for directed integration into the *S. solfataricus* genome based on counterselection with *pyrEF* (55). Two other thermostable β-galactosidase genes isolated and cloned from *P. furiosus* that exhibit maximum activity at 95°C could potentially function as phenotypic markers for hyperthermophilic species (2).

PERSPECTIVE: THE NEXT FIVE YEARS

Progress in the development of methodologies for archaeal genetics has rapidly accelerated in the past decade. Additional methods are currently under development for all three archaeal phyla, including additional systems for markerless exchange, gene expression, topological mapping, protein tagging and expression, as well as others. The intense interest in developing these methods has in large part been promoted by the availability of archaeal genome sequences. As the earliest archaeal genomes became available in the mid-1990s it was recognized that the lack of tractable gene transfer systems for many of these species limited the ability to confirm the function of annotated genes and determine the function of genes encoding unidentified proteins. The choice of archaeal genomes for sequencing is now largely driven by the availability of genetic systems, which at present include complete genomes of the halophiles *Haloarcula marismortui* and *Halobacterium* sp. NRC-1, the methanogens *M. maripaludis*, *M. acetivorans*, *M. barkeri*, and *M. mazei*, and the hyperthermophiles *S. solfataricus* P2 and *T. kodakaraensis* KOD1. The development of genetic tools for the *Archaea* is ongoing, and more sophisticated techniques are currently under development that will enable researchers to use genetic analysis on a routine basis to understand the unique abilities of these microorganisms to proliferate in extreme environments.

Acknowledgments. This work was supported in part by grants to K.S. from the Department of Energy, Energy Biosciences Program (DE-FG02-93ER20106), and the National Science Foundation, Division of Molecular & Cellular Biosciences (MCB0110762).

REFERENCES

1. Aagaard, C., J. Z. Dalgaard, and R. A. Garrett. 1995. Intercellular mobility and homing of an archaeal rDNA intron confers a selective advantage over intron⁻ cells of *Sulfolobus acidocalderius. Proc. Natl. Acad. Sci. USA* **92**:12285–12289.

2. Aagaard, C., I. Leviev, R. N. Aravalli, P. Forterre, D. Prieur, and R. A. Garrett. 1996. General vectors for archaeal hyperthermophiles: strategies based on a mobile intron and a plasmid. *FEMS Microbiol. Rev.* **18**:93–104.

3. Allers, T., and M. Mevarech. 2005. Archaeal genetics—the third way. *Nat. Rev. Genet.* **6**:58–73.

4. Allers, T., H. P. Ngo, M. Mevarech, and R. G. Lloyd. 2004. Development of additional selectable markers for the halophilic archaeon *Haloferax volcanii* based on the *leuB* and *trpA* genes. *Appl. Environ. Microbiol.* **70**:943–953.

5. Reference deleted.

6. Apolinario, E. A., and K. R. Sowers. 1996. Plate colonization of *Methanococcus maripaludis* and *Methanosarcina thermophila* in a modified canning jar. *FEMS Microbiol. Lett.* **145**:131–137.

7. Apolinario-Smith, E., K. M. Jackson, and K. R. Sowers. 2005. Development of a plasmid-mediated reporter system for in vivo monitoring of gene expression in the archaeon *Methanosarcina acetivorans. Appl. Envion. Microbiol.* **71**: 4914–4918.

8. Aravalli, R. N., and R. A. Garrett. 1997. Shuttle vectors for hyperthermophilic archaea. *Extremophiles* **1**:183–191.

9. Argyle, J. L., D. L. Tumbula, and J. A. Leigh. 1996. Neomycin resistance as a selectable marker in *Methanococcus maripaludis. Appl. Environ. Microbiol.* **62**:4233–4237.

10. Balch, W. E., G. E. Fox, L. J. Magrum, C. R. Woese, and R. S. Wolfe. 1979. Methanogens: reevaluation of a unique biological group. *Microbiol. Rev.* **43**:260–296.

11. Balch, W. E., and R. S. Wolfe. 1976. New approach to the cultivation of methanogenic bacteria: 2-mercaptoethanesulfonic acid (HS-CoM)-dependent growth of *Methanobacterium ruminantium* in a pressurized atmosphere. *Appl. Environ. Microbiol.* **32**:781–791.

12. Bardy, S. L., and K. F. Jarrell. 2003. Cleavage of preflagellins by an aspartic acid signal peptidase is essential for flagellation in the archaeon *Methanococcus voltae. Mol. Microbiol.* **50**:1339–1347.

13. Bartolucci, S., M. Rossi, and R. Cannio. 2003. Characterization and functional complementation of a nonlethal deletion in the chromosome of a beta-glycosidase mutant of *Sulfolobus solfataricus. J. Bacteriol.* **185**:3948–3957.

14. Beckler, G. S., L. A. Hook, and J. N. Reeve. 1984. Chloramphenicol acetyltransferase should not provide methanogens with resistance to chloramphenicol. *Appl. Environ. Microbiol.* **47**:868–869.

15. Beneke, S., H. Bestgen, and A. Klein. 1995. Use of the *Escherichia coli uidA* gene as a reporter in *Methanococcus voltae* for the analysis of the regulatory function of the intergenic region between the operons encoding selenium-free hydrogenases. *Mol. Gen. Genet.* **248**:225–228.

16. **Bertani, G., and L. Baresi.** 1987. Genetic transformation in the methanogen *Methanococcus voltae* PS. *J. Bacteriol.* l69:2730–2738.

17. **Bitan-Banin, R. O. G, and M. Mevarech.** 2003. Development of a gene knockout system for the halophilic archaeon *Haloferax volcanii* by use of the *pyrE* gene. *J. Bacteriol.* 185:772–778.

18. **Blaseio, U., and F. Pfeifer.** 1990. Transformation of *Halobacterium halobium:* development of vectors and investigation of gas vesicle synthesis. *Proc. Natl. Acad. Sci. USA* 87:6772–6776.

19. **Boccazzi, P., K. J. Zhang, and W. W. Metcalf.** 2000. Generation of dominant selectable markers for resistance to pseudomonic acid by cloning and mutagenesis of the *ileS* gene from the archaeon *Methanosarcina barkeri* Fusaro. *J. Bacteriol.* 182:2611–2618.

20. **Bock, A., and O. Kandler.** 1985. Antibiotic sensitivity of archaebacteria. *In* C. R. Woese and R. S. Wolfe (ed.), *The Bacteria. A Treatise on Structure and Function*, vol. 8. Academic Press, Inc., New York, N.Y.

21. **Bowen, T. L., and W. B. Whitman.** 1987. Incorporation of exogenous purines and pyrimidines by *Methanococcus voltae* and isolation of analog-resistant mutants. *Appl. Environ. Microbiol.* 53:1822–1826.

22. **Cannio, R., P. Contursi, M. Rossi, and S. Bartolucci.** 1998. An autonomously replicating transforming vector for *Sulfolobus solfataricus*. *J. Bacteriol.* 180:3237–3240.

23. **Cannio, R., P. Contursi, M. Rossi, and S. Bartolucci.** 2001. Thermoadaptation of a mesophilic hygromycin B phosphotransferase by directed evolution in hyperthermophilic Archaea: selection of a stable genetic marker for DNA transfer into *Sulfolobus solfataricus*. *Extremophiles* 5:153–159.

24. **Charlebois, R. L., W. L. Lam, S. W. Cline, and W. F. Doolittle.** 1987. Characterization of pHV2 from *Halobacterium volcanii* and its use in demonstrating transformation of an archaebacterium. *Proc. Natl. Acad. Sci. USA* 84:8530–8534.

25. **Cline, S., and W. F. Doolittle.** 1992. Transformation of members of the genus *Haloarcula* with shuttle vectors based on *Halobacterium halobium* and *Haloferax volcanii* plasmid replicons. *J. Bacteriol.* 174:1076–1080.

26. **Cline, S. W., and W. F. Doolittle.** 1987. Efficient transfection of the archaebacterium *Halobacterium halobium*. *J. Bacteriol.* 169:1341–1344.

27. **Cohen-Kupiec, R., C. Blank, and J. A. Leigh.** 1997. Transcriptional regulation in Archaea: *In vivo* demonstration of a repressor binding site in a methanogen. *Proc. Natl. Acad. Sci. USA* 94:1316–1320.

28. **Conover, R. K., and W. F. Doolittle.** 1990. Characterization of a gene involved in histidine biosynthesis in *Halobacterium (Haloferax) volcanii*: isolation and rapid mapping by transformation of an auxotroph with cosmid DNA. *J. Bacteriol.* 172:3244–3249.

29. **Contursi, P., R. Cannio, S. Prato, G. Fiorentino, M. Rossi, and S. Bartolucci.** 2003. Development of a genetic system for hyperthermophilic Archaea: expression of a moderate thermophilic bacterial alcohol dehydrogenase gene in *Sulfolobus solfataricus*. *FEMS Microbiol. Lett.* 218:115–120.

30. **DasSarma, S., and E. M. Fleischmann (ed.).** 1995. *Halophiles.* Cold Spring Harbor Laboratory Press, Plainview, N.Y.

31. **Ehlers, C., K. Weidenbach, K. Veit, U. Deppenmeier, W. W. Metcalf, and R. A. Schmitz.** 2005. Development of genetic methods and construction of a chromosomal *glnK1* mutant in *Methanosarcina mazei* strain Go1. *Mol. Genet. Genomics* 273:290–298.

32. **Elferink, M. G., C. Schleper, and W. Zillig.** 1996. Transformation of the extremely thermoacidophilic archaeon *Sulfolobus solfataricus* via a self-spreading vector. *FEMS Microbiol. Lett.* 137:31–35.

33. **Erauso, G., S. Marsin, N. Benbouzid-Rollet, M. F. Baucher, T. Barbeyron, Y. Zivanovic, D. Prieur, and P. Forterre.** 1996. Sequence of plasmid pGT5 from the archaeon *Pyrococcus abyssi*: evidence for rolling-circle replication in a hyperthermophile. *J. Bacteriol.* 178:3232–3237.

34. **Erauso, G., D. Prieur, A. Godfroy, and G. Raguenes.** 1995. Plate cultivation techniques for strictly anaerobic, thermophilic, sulfur-metabolizing Archaea, p. 25–29. *In* F. T. Robb, A. R. Place, K. R. Sowers, H. J. Schreier, S. DasSarma, and E. M. Fleischmann (ed.), *Archaea—A Laboratory Manual.* Cold Spring Harbor Laboratory Press, Plainview, N.Y.

35. **Firtel, M., G. B. Patel, and T. J. Beveridge.** 1995. S layer regeneration in *Methanococcus voltae* protoplasts. *Microbiology* 141:817–824.

36. **Gardner, W. L., and W. B. Whitman.** 1999. Expression vectors for *Methanococcus maripaludis*: overexpression of acetohydroxyacid synthase and beta-galactosidase. *Genetics* 152:1439–1447.

37. **Gernhardt, P., O. Possot, M. Foglino, L. Sibold, and A. Klein.** 1990. Construction of an integration vector for use in the archaebacterium *Methanococcus voltae* and expression of a eubacterial resistance gene. *Mol. Gen. Genet.* 22:273–279.

38. **Gregor, D., and F. Pfeifer.** 2001. Use of a halobacterial *bgaH* reporter gene to analyse the regulation of gene expression in halophilic archaea. *Microbiology* 147:1745–1754.

39. **Grogan, D. W.** 1996. Exchange of genetic markers at extremely high temperatures in the archaeon *Sulfolobus acidocaldarius*. *J. Bacteriol.* 178:3207–3211.

40. **Grogan, D. W.** 1991. Selectable mutant phenotypes of the extremely thermophilic archaebacterium *Sulfolobus acidocaldarius*. *J. Bacteriol.* 173:7725–7727.

41. **Guss, A. M., B. Mukhopadhyay, J. K. Zhang, and W. W. Metcalf.** 2005. Genetic analysis of *mch* mutants in two *Methanosarcina* species demonstrates multiple roles for the methanopterin-dependent C-1 oxidation/reduction pathway and differences in H-2 metabolism between closely related species. *Mol. Microbiol.* 55:1671–1680.

42. **Hackett, N. R., and S. DasSarma.** 1989. Characterization of the small endogenous plasmid of *Halobacterium* strain SB3 and its use in transformation. *Can. J. Microbiol.* 35:86–91.

43. **Harris, J. E.** 1985. Gelrite as an agar substitute for the cultivation of mesophilic *Methanobacterium* and *Methanobrevibacter* species. *Appl. Environ. Microbiol.* 50:1107–1109.

44. **Harris, J. E., and P. A. Pinn.** 1985. Bacitracin-resistant mutants of a mesophilic *Methanobacterium* species. *Arch. Microbiol.* 143:151–153.

45. **Haydock, A. K., I. Porat, W. B. Whitman, and J. A. Leigh.** 2004. Continuous culture of *Methanococcus maripaludis* under defined nutrient conditions. *FEMS Microbiol. Lett.* 238:85–91.

46. **Heinicke, I., J. Muller, M. Pittelkow, and A. Klein.** 2004. Mutational analysis of genes encoding chromatin proteins in the archaeon *Methanococcus voltae* indicates their involvement in the regulation of gene expression. *Mol. Genet. Genomics* 272:76–87.

47. **Holmes, M., F. Pfeifer, and M. Dyall-Smith.** 1994. Improved shuttle vectors for *Haloferax volcanii* including a dual-resistance plasmid. *Gene* 146:117–121.

48. **Holmes, M. L., and M. L. Dyall-Smith.** 1990. A plasmid vector with a selectable marker for halophilic archaebacteria. *J. Bacteriol.* 172:756–761.

49. **Holmes, M. L., and M. L. Dyall-Smith.** 2000. Sequence and expression of a halobacterial β-galactosidase gene. *Mol. Microbiol.* 36:114–122.

50. Holmes, M. L., S. D. Nuttall, and M. L. Dyall-Smith. 1991. Construction and use of halobacterial shuttle vectors and further studies on *Haloferax* DNA gyrase. *J. Bacteriol.* **173:**3807–3813.

51. Holmes, M. L., R. K. Scopes, R. L. Moritz, R. J. Simpson, C. Englert, F. Pfeifer, and M. L. Dyall-Smith. 1997. Purification and analysis of an extremely halophilic β-galactosidase from *Haloferax alicantei*. *BBA Prot. Struct. Mol. Enzymol.* **1337:**276–286.

52. Hook, L. A., R. E. Corder, P. T. Hamilton, J. I. Frea, and J. N. Reeve (ed.). 1984. *Development of a Plating System for Genetic Exchange Studies in Methanogens Using a Modified Ultra-Low Oxygen Chamber.* The Ohio State University Press, Columbus, Ohio.

53. Jenal, U., T. Rechsteiner, P. Y. Tan, E. Buhlmann, L. Meile, and T. Leisinger. 1991. Isoleucyl-tRNA synthetase of *Methanobacterium thermoautotrophicum* Marburg. Cloning of the gene, nucleotide sequence, and localization of a base change conferring resistance to pseudomonic acid. *J. Biol. Chem.* **266:**10570–10577.

54. Jones, W. J., W. B. Whitman, F. D. Fields, and R. S. Wolfe. 1983. Growth and plating efficiency of methanococci on agar media. *Appl. Environ. Microbiol.* **46:**220–226.

55. Jonuscheit, M., E. Martusewitsch, K. M. Stedman, and C. Schleper. 2003. A reporter gene system for the hyperthermophilic archaeon *Sulfolobus solfataricus* based on a selectable and integrative shuttle vector. *Mol. Microbiol.* **48:**1241–1252.

56. Kandler, O., and H. Konig. 1998. Cell wall polymers in Archaea (Archaebacteria). *Cell Mol. Life Sci.* **54:**305–308.

57. Kiener, A., H. Konig, J. Winter, and T. Leisinger. 1987. Purification and use of *Methanobacterium wolfei* pseudomurein endopeptidase for lysis of *Methanobacterium thermoautotrophicum*. *J. Bacteriol.* **169:**1010–1016.

58. Kiener, A., and T. Leisinger. 1983. Oxygen sensitivity of methanogenic bacteria. *Syst. Appl. Microbiol.* **4:**305–312.

59. Krebs, M. P., T. Hauss, M. P. Heyn, and U. L. RajBhandary. 1991. Expression of the bacterioopsin gene in *Halobacterium halobium* using a multicopy plasmid. *Proc. Natl. Acad. Sci. USA* **88:**859–863.

60. Krebs, M. P., R. Mollaaghababa, and H. G. Khorana. 1993. Gene replacement in *Halobacterium halobium* and expression of bacteriorhodopsin mutants. *Proc. Natl. Acad. Sci. USA* **90:**1987–1991.

61. Ladapo, J., and W. B. Whitman. 1990. Method for isolation of auxotrophs in the methanogenic archaebacteria: role of the acetyl-CoA pathway of autotrophic CO_2 fixation in *Methanococcus maripaludis*. *Proc. Natl. Acad. Sci. USA* **87:**5598–5602.

62. Lam, W. L., and W. F. Doolittle. 1989. Shuttle vectors for the archaebacterium *Halobacterium volcanii*. *Proc. Natl. Acad. Sci. USA* **86:**5478–5482.

63. Lie, T., and J. Leigh. 2003. A novel repressor of *nif* and *glnA* expression in the methanogenic archaeon *Methanococcus maripaludis*. *Mol. Microbiol.* **47:**235–246.

64. Lie, T. J., G. E. Wood, and J. A. Leigh. 2005. Regulation of *nif* expression in *Methanococcus maripaludis*: roles of the euryarchaeal repressor NrpR, 2-oxoglutarate, and two operators. *J. Biol. Chem.* **280:**5236–5241.

65. Lindstrom, E. B., and H. M. Sehlin. 1989. High efficiency of plating of the thermophilic sulfur-dependent archaebacterium *Sulfolobus acidocaldarius*. *Appl. Environ. Microbiol.* **55:**3020–3021.

66. Long, S. W., and D. M. Faguy. 2004. Anucleate and titan cell phenotypes caused by insertional inactivation of the structural maintenance of chromosomes (*smc*) gene in the archaeon *Methanococcus voltae*. *Mol. Microbiol.* **52:**1567–1577.

67. Lucas, S., L. Toffin, Y. Zivanovic, D. Charlier, H. Moussard, P. Forterre, D. Prieur, and G. Erauso. 2002. Construction of a shuttle vector for, and spheroplast transformation of, the hyperthermophilic archaeon *Pyrococcus abyssi*. *Appl. Environ. Microbiol.* **68:**5528–5536.

68. Mahapatra, A., A. Patel, J. A. Soares, R. C. Larue, J. K. Zhang, W. W. Metcalf, and J. A. Krzycki. 2006. Characterization of a *Methanosarcina acetivorans* mutant unable to translate UAG as pyrrolysine. *Mol. Microbiol.* **59:**56–66.

69. Mankin, A. S., I. M. Zyrianova, V. K. Kagramanova, and R. A. Garrett. 1992. Introducing mutations into the single-copy chromosomal 23S rRNA gene of the archaeon *Halobacterium halobium* by using an rRNA operon-based transformation system. *Proc. Natl. Acad. Sci. USA* **89:**6535–6539.

70. Martusewitsch, E., C. W. Sensen, and C. Schleper. 2000. High spontaneous mutation rate in the hyperthermophilic archaeon *Sulfolobus solfataricus* is mediated by transposable elements. *J. Bacteriol.* **182:**2574–2581.

71. Meile, L., P. Abendschein, and T. Leisinger. 1990. Transduction in the archaebacterium *Methanobacterium thermautotrophicum* Marburg. *J. Bacteriol.* **172:**3507–3508.

72. Metcalf, W. W. 1999. Genetic analysis in the domain Archaea, p. 277–326. *In* M. C. Smith and R. E. Sockett (ed.), *Genetic Methods for Diverse Prokaryotes*, vol. 29. Academic Press, New York, N.Y.

73. Metcalf, W. W., J. K. Zhang, E. Apolinario, K. R. Sowers, and R. S. Wolfe. 1997. A genetic system for Archaea of the genus *Methanosarcina*: Liposome-mediated transformation and construction of shuttle vectors. *Proc. Natl. Acad. Sci. USA* **94:**2626–2631.

74. Metcalf, W. W., J. K. Zhang, and R. S. Wolfe. 1998. An anaerobic, intrachamber incubator for growth of *Methanosarcina* spp. on methanol-containing solid media. *Appl. Environ. Microbiol.* **64:**768–770.

75. Mevarech, M., and R. Werczberger. 1985. Genetic transfer in *Halobacterium volcanii*. *J. Bacteriol.* **162:**461–462.

76. Moore, B. C., and J. A. Leigh. 2005. Markerless mutagenesis in *Methanococcus maripaludis* demonstrates roles for alanine dehydrogenase, alanine racemase, and alanine permease. *J. Bacteriol.* **187:**972–979.

77. Morii, H., and Y. Koga. 1992. An improved assay method for a pseudomurien-degrading enzyme of *Methanobacterium wolfei* and the protoplast formation of *Methanobacterium thermoautotrophicum* by the enzyme. *J. Ferment. Bioeng.* **73:**6–10.

78. Ng, W.-L., P. Arora, and S. DasSarma. 1994. Large deletions in class III gas-vesicles deficient mutants of *Halobacterium halobium*. *Syst. Appl. Microbiol.* **16:**560–568.

79. Ng, W.-L., and S. DasSarma. 1993. Minimal replication origin of the 200-kilobase *Halobacterium* plasmid pNRC100. *J. Bacteriol.* **175:**4584–4596.

80. Ng, W. L., and S. DasSarma. 1993. Minimal replication origin of the 200-kilobase *Halobacterium* plasmid pNRC100. *J. Bacteriol.* **175:**4584–4596.

81. Nieuwlandt, D. T., and C. J. Daniels. 1990. An expression vector for the archaebacterium *Haloferax volcanii*. *J. Bacteriol.* **172:**7104–7110.

82. Ortenberg, R., O. Rozenblatt-Rosen, and M. Mevarech. 2000. The extremely halophilic archaeon *Haloferax volcanii* has two very different dihydrofolate reductases. *Mol. Microbiol.* **35:**1493–1505.

83. Patel, G. B., J. H. E. Nash, B. J. Agnew, and G. D. Sprott. 1994. Natural and electroporation-mediated transformation of *Methanococcus voltae* protoplasts. *Appl. Environ. Microbiol.* **60:**903–907.

84. **Patenge, N., A. Haase, H. Bolhuis, and D. Oesterhelt.** 2000. The gene for a halophilic β-galactosidase (*bgaH*) of *Haloferax alicantei* as a reporter gene for promoter analysis in *Halobacterium salinarium. Mol. Microbiol.* **36:**105–113.

85. **Peck, R. F., S. DasSarma, and M. P. Krebs.** 2000. Homologous gene knockout in the archaeon *Halobacterium* with ura3 as a counterselectable marker. *Mol. Microbiol.* **35:**667–676.

86. **Pfeiffer, M., H. Bestgen, A. Burger, and A. Klein.** 1998. The *vhuU* gene encoding a small subunit of a selenium-containing [NiFe]-hydrogenase in *Methanococcus voltae* appears to be essential for the cell. *Arch. Microbiol.* **170:**418–426.

87. **Porat, I., B. W. Waters, Q. Teng, and W. B. Whitman.** 2004. Two biosynthetic pathways for aromatic amino acids in the archaeon *Methanococcus maripaludis. J. Bacteriol.* **186:**4940–4950.

88. **Possot, O., P. Gernhardt, A. Klein, and L. Sibold.** 1988. Analysis of drug resistance in the archaebacterium *Methanococcus voltae* with respect to potential use in genetic engineering. *Appl. Environ. Microbiol.* **54:**734–740.

89. **Pritchett, M. A., J. K. Zhang, and W. W. Metcalf.** 2004. Development of a markerless genetic exchange method for *Methanosarcina acetivorans* C2A and its use in construction of new genetic tools for methanogenic archaea. *Appl. Environ. Microbiol.* **70:**1425–1433.

90. **Reuter, C. J., and J. A. Maupin-Furlow.** 2004. Analysis of proteasome-dependent proteolysis in *Haloferax volcanii* cells, using short-lived green fluorescent proteins. *Appl. Environ. Microbiol.* **70:**7530–7538.

91. **Rodriguez-Valera, F.** 1995. Cultivation of halophilic *Archaea*, p. 13–16. *In* F. T. Robb, A. R. Place, K. R. Sowers, H. J. Schreier, S. DasSarma, and E. M. Fleischmann (ed.), *Archaea—A Laboratory Manual.* Cold Spring Harbor Laboratory Press, Plainview, N.Y.

92. **Rosenshine, I., R. Tchelet, and M. Mevarech.** 1989. The mechanism of DNA transfer in the mating system of an archaebacterium. *Science* **245:**1387–1389.

93. **Rother, M., P. Boccazzi, A. Bose, M. A. Pritchett, and W. W. Metcalf.** 2005. Methanol-dependent gene expression demonstrates that methyl-coenzyme M reductase is essential in *Methanosarcina acetivorans* C2A and allows isolation of mutants with defects in regulation of the methanol utilization pathway. *J. Bacteriol.* **187:**5552–5559.

94. **Rother, M., and W. W. Metcalf.** 2005. Genetic technologies for Archaea. *Curr. Opin. Microbiol.* **8:**745–751.

95. **Sato, T., T. Fukui, H. Atomi, and T. Imanaka.** 2005. Improved and versatile transformation system allowing multiple genetic manipulations of the hyperthermophilic archaeon *Thermococcus kodakaraensis. Appl. Environ. Microbiol.* **71:**3889–3899.

96. **Sato, T., T. Fukui, H. Atomi, and T. Imanaka.** 2003. Targeted gene disruption by homologous recombination in the hyperthermophilic archaeon *Thermococcus kodakaraensis* KOD1. *J. Bacteriol.* **185:**210–220.

97. **Schleper, C., I. Holz, D. Janekovic, J. Murphy, and W. Zillig.** 1995. A multicopy plasmid of the extremely thermophilic archaeon *Sulfolobus* effects its transfer to recipients by mating. *J. Bacteriol.* **177:**4417–4426.

98. **Schleper, C., K. Kubo, and W. Zillig.** 1992. The particle SSV1 from the extremely thermophilic archaeon *Sulfolobus* is a virus: demonstration of infectivity and of transfection with viral DNA. *Proc. Natl. Acad. Sci. USA* **89:**7645–7649.

99. **She, Q., H. Phan, R. A. Garrett, S.-V. Albers, K. M. Stedman, and W. Zillig.** 1998. Genetic profile of pNOB8 from *Sulfolobus*: the first conjugative plasmid form an archaeon. *Extremophiles* **2:**417–425.

100. **Sniezko, I., C. Dobson-Stone, and A. Klein.** 1998. The *treA* gene of *Bacillus subtilis* is a suitable reporter gene for the ar-

chaeon *Methanococcus voltae. FEMS Microbiol. Lett.* **164:**237–242.

101. **Sowers, K. R., J. E. Boone, and R. P. Gunsalus.** 1993. Disaggregation of *Methanosarcina* spp. and growth as single cells at elevated osmolarity. *Appl. Environ. Microbiol.* **59:**3832–3839.

102. **Sowers, K. R., and R. P. Gunsalus.** 1988. Plasmid DNA from the Acetotrophic Methanogen *Methanosarcina acetivorans. J. Bacteriol.* **170:**4979–4982.

103. **Sowers, K. R., and H. J. Schreier.** 1999. Gene transfer systems for the archaea. *Trends Microbiol.* **7:**212–219.

104. **Sowers, K. R., and H. J. Schreier.** 1995. Techniques for anaerobic growth, p. 15–55. *In* F. T. Robb, K. R. Sowers, S. DasSharma, A. R. Place, H. J. Schreier, and E. M. Fleischmann (ed.), *Archaea—A Laboratory Manual.* Cold Spring Harbor Laboratory Press, Plainview, N.Y.

105. **Tan, G. T., A. DeBlasio, and A. S. Mankin.** 1996. Mutations in the peptidyl transferase center of 23 S rRNA reveal the site of action of sparsomycin, a universal inhibitor of translation. *J. Mol. Biol.* **261:**222–230.

106. **Thomas, N. A., S. Mueller, A. Klein, and K. F. Jarrell.** 2002. Mutants in flaI and flaJ of the archaeon *Methanococcus voltae* are deficient in flagellum assembly. *Mol. Microbiol.* **46:**879–887.

107. **Tumbula, D. L., T. L. Bowen, and W. B. Whitman.** 1997. Characterization of pURB500 from the archaeon *Methanococcus maripaludis* and construction of a shuttle vector. *J. Bacteriol.* **179:**2976–2986.

108. **Tumbula, D. L., T. L. Bowen, and W. B. Whitman.** 1995. Growth of methanogens on solidified medium, p. 49–55. *In* F. T. Robb, A. R. Place, K. R. Sowers, H. J. Schreier, S. DasSarma, and E. M. Fleischmann (ed.), *Archaea—A Laboratory Manual.* Cold Spring Harbor Laboratory Press, Plainview, N.Y.

109. **Tumbula, D. L., R. A. Makula, and W. B. Whitman.** 1994. Transformation of *Methanococcus maripaludis* and identification of a PstI-like restriction system. *FEMS Microbiol. Lett.* **121:**309–314.

110. **Völkl, P., R. Huber, E. Drobner, R. Rachel, S. Burggraf, A. Trincone, and K. Stetter.** 1993. *Pyrobaculum aerophilum* sp. nov., a novel nitrate-reducing hyperthermophilic archaeum. *Appl. Environ. Microbiol.* **59:**2918–2926.

111. **Wang, G., S. P. Kennedy, S. Fasiludeen, C. Rensing, and S. DasSarma.** 2004. Arsenic resistance in *Halobacterium* sp. strain NRC-1 examined by using an improved gene knockout system. *J. Bacteriol.* **186:**3187–3194.

112. **Watrin, L., S. Lucas, C. Purcarea, C. Legrain, and D. Prieur.** 1999. Isolation and characterization of pyrimidine auxotrophs, and molecular cloning of the *pyrE* gene from the hyperthermophilic archaeon *Pyrococcus abyssi. Mol. Gen. Genet.* **262:**378–381.

113. **Watrin, L., and D. Prieur.** 1996. UV and ethyl methanesulfonate effects in hyperthermophilic archaea and isolation of auxotrophic mutants of *Pyrococcus* strains. *Curr. Microbiol.* **33:**377–382.

114. **Welander, P. V., and W. W. Metcalf.** 2005. Loss of the *mtr* operon in *Methanosarcina* blocks growth on methanol, but not methanogenesis, and reveals an unknown methanogenic pathway. *Proc. Natl. Acad. Sci. USA* **102:**10664–10669.

115. **Wendoloski, D., C. Ferrer, and M. L. Dyall-Smith.** 2001. A new simvastin (mevinolin)-resistant marker from *Haloarchula hispanica* and a new *Haloferax volcanii* strain cured of plasmid pHV2. *Microbiology* **147:**959–964.

116. **Wood, G. E., A. K. Haydock, and J. A. Leigh.** 2003. Function and regulation of the formate dehydrogenase genes of the methanogenic archaeon *Methanococcus maripaludis. J. Bacteriol.* **185:**2548–2554.

117. **Worthington, P., V. Hoang, P. Perez-Pomares, and P. Blum.** 2003. Targeted disruption of the α-amylase gene in the hyperthermophilic archaeon *Sulfolobus solfataricus. J. Bacteriol.* **185:**482–488.

118. **Zhang, J. K., M. A. Pritchett, D. J. Lampe, H. M. Robertson, and W. W. Metcalf.** 2000. *In vivo* transposon mutagenesis of the methanogenic archaeon *Methanosarcina acetivorans* C2A using a modified version of the insect *mariner*-family transposable element *Himar1. Proc. Natl. Acad. Sci. USA* **97:**9665–9670.

119. **Zhang, J. K., A. K. White, H. C. Kuettner, P. Boccazzi, and W. W. Metcalf.** 2002. Directed mutagenesis and plasmid-based complementation in the methanogenic archaeon *Methanosarcina acetivorans* C2A demonstrated by genetic analysis of proline biosynthesis. *J. Bacteriol.* **184:**1449–1454.

120. **Zhou, M., H. Xiang, C. Sun, and H. Tan.** 2004. Construction of a novel shuttle vector based on an RCR-plasmid from a haloalkaliphilic archaeon and transformation into other haloarchaea. *Biotechnol. Lett.* **26:**1107–1113.

Archaea: Molecular and Cellular Biology
Edited by Ricardo Cavicchioli
© 2007 ASM Press, Washington, D.C.

Chapter 22

Biotechnology

MARCO MORACCI, BEATRICE COBUCCI-PONZANO, GIUSEPPE PERUGINO, AND MOSÈ ROSSI

INTRODUCTION

The exploitation of natural catalysts as whole cells or as purified enzymes is the essence of biotechnology and was developed in an effort to obtain environmentally friendly and economically convenient industrial processes. Nevertheless, several industrial processes are damaging for most "conventional" organisms and biocatalysts. This may explain why, since their discovery in the 1950s, organisms living under extreme environmental conditions and the molecules extracted from them appeared to provide very promising tools for biotechnology. These expectations were motivated by the unusual physical conditions required by extremophiles for growth.

Extremophiles are mainly bacterial and archaeal microorganisms, with the majority from the *Archaea*, although they exist within the three domains of life since a eucaryal thermophile has also been reported (18). This biodiversity is a key feature that enables the isolation of organisms with peculiar or unexpected metabolic pathways and enzymes with different characteristics in terms of substrate specificity, selectivity, and reaction mechanism. As a consequence, biotechnologists can screen extremophiles for the most suitable biocatalyst to apply in a particular industrial field. The exploitation of the natural diversity of extremophiles is one of the missions of the Diversa Corporation (USA), which is committed to the rapid discovery of novel microbes isolated from the most diverse biotopes (www.diversa.com/).

Despite these premises and the extensive research, the number of extant applications of extremophiles is still limited. The reason for this stalling is that their commercial use is based not only on their peculiar properties but also on several other different criteria, including technology integration, intellectual property, regulatory compliance, and assurance of supply, that

must be satisfied. In other words, microorganisms or biomolecules can be successfully exploited in novel industrial processes if they introduce an innovation that outcompetes existing products; this commercial reality also applies to the exploitation of extremophiles.

The biotechnology of the *Archaea* has been reviewed in recent excellent articles (35, 117). The topics covered in this chapter include enzymes and molecules from archaea and the use of archaeal whole cells. Other topics relevant to the biotechnology of archaea, such as functional genomics, the development of molecular biology tools, and archaeosomes are covered in Chapters 20, 21 and 23, respectively. Most of the applications described here are common to extremophilic bacteria and archaea and have been described in many excellent reviews (9, 20, 45, 50, 123, 144, 145). The chapter updates the current state of biotechnology of the *Archaea*, paying special attention to distinguish between extant and potential applications, to provide a realistic overview of the impact of these organisms on biotechnology.

ENZYMES FROM *ARCHAEA*

The estimated value of the worldwide use of industrial enzymes grew from $1 billion in 1995 to $2 billion in 2004 and, depending on different estimates, is expected to rise $2.4 to 5.1 billion in 2009 (Business Communication Company, www.bccresearch.com; Freedonia group, www.freedoniagroup.com/world .html). Among industrial enzymes, the mesophilic ones are often not well suited for the harsh conditions adopted in industry because they are rapidly denaturated at the operational conditions; therefore, extremophilic industrial enzymes (extremozymes) provide interesting alternatives for working at high temperatures or in the presence of high ionic strength and or-

Marco Moracci, Beatrice Cobucci-Ponzano, Giuseppe Perugino, and Mosè Rossi • Institute of Protein Biochemistry, National Research Council, Via P. Castellino 111, 80131 Naples, Italy.

ganic solvents (28, 71). Enzymes from psychrophiles, thermophiles, halophiles, and piezophiles usually show the same characteristics of the organisms from which they have been extracted. With the exception of enzymes secreted in the medium, this is not true for acidophiles and alkaliphiles, whose enzymes are optimally active at the neutral intracellular pH. Nevertheless, the extreme acidophile *Picrophilus torridus* maintains an intracellular pH of 4.6, and therefore the enzymes from this organism hold particular interest for biotechnology (43).

The interest in the application of extremozymes resides in their surprising properties. There has been extensive research on the structure/function relationship in extremozymes, with the aim to uncover the molecular mechanisms of stabilization of these biocatalysts, and to engineer stabilized mutants of conventional enzymes (reviewed in references 56, 145). Enzymes from cold-adapted microorganisms offer considerable potential for biotechnological applications (review in references 20, 35, 45). However, very few psychrophilic archaea have been isolated (20, 35), and they will therefore not be covered extensively in this chapter.

Among the extremozymes, those from halophiles and thermophiles (thermozymes) are, by far, the most studied, and their application in biotechnology has been exploited for a long time. Enzymes from halophilic bacteria optimally working at high-salt concentrations, at which the water activity is greatly reduced, make them interesting biocatalysts in aqueous/organic and nonaqueous media (1). The general advantage of thermozymes in industrial applications is their outstanding stability to high temperature, proteases, and high concentration of organics. These catalysts have allowed the development of new processes (e.g., the polymerase chain reaction) and have been adopted as alternatives to enzymes from mesophiles since their prolonged life span reduces the cost of the continuous addition of fresh biocatalyst. Additional advantages of conducting biocatalysis and biotransformation processes at high temperatures include the reduced risk of contamination by common mesophilic organisms, an increased solubility and diffusion coefficient of organic compounds, and a reduced viscosity of the medium. These features lead to increased reaction rates and yield by favoring the extraction of volatile organic compounds, the accessibility of recalcitrant substrates to hydrolysis, and the equilibrium displacement in endothermic reactions (74, 88). In addition, their peculiar characteristics allow easier purification procedures for the recombinant enzymes (i.e., simple heating steps eliminating the unstable proteins of the host), resulting in more convenient downstream processing.

The following sections overview the most promising applications of extremozymes, emphasizing the archaeal examples. A short survey of examples of enzymes from extremophiles is shown in Table 1. The table is not exhaustive as the number of extremozymes is large. The reader is directed to the papers cited in this chapter and a number of excellent reviews on these biocatalysts (9, 28, 50, 56, 81, 123, 143–145), some of which deal specifically with archaeal extremozymes (35, 117).

Hydrolases

Protease

Among hydrolases, proteases, esterases/lipases, and glycoside hydrolases are important in industry. In particular, proteases are the major industrial enzymes covering more than 25% of the world market (Business Communication Company, www.bccresearch.com) and are extensively used in food, pharmaceutical, leather, and textile industries. (Hyper)thermophilic proteases are important components of the detergent formulations as they improve the cleaning ability of the detergent on protein based stains (37). The stability of the thermophilic proteases to the laundry components (in particular, detergents and alkaline pH) is the main reason for this interest (71). Several archaea harbor an abundant number of intracellular and extracellular proteases, such as *Pyrococcus furiosus*, which shows more than a dozen of such enzymes (26), *Sulfolobus solfataricus* (49), several thermococcales (42), for which also a patent has been filed in the United States (128), and *Aeropyrum pernix* (19). A detailed account of archaeal proteases has been described (35), including several articles in *Methods in Enzymology*, vol. 330, 2001. More recently, proteases working at low temperature in household laundry detergents have been used to maintain the colors of the clothes and to save energy. Therefore, archaeal psychrophilic proteases may also be exploited in the near future in laundry detergents.

Halophilic proteases, such as those from *Haloferax mediterranei* (65, 127) and *Haloarcula marismortui* (40), are of considerable interest for their ability to have greater activity in solvents of increased polarity (reviewed in references 81, 123). Under these conditions the hydrolytic activity is reduced, allowing the exploitation of proteases from halophiles in peptide synthesis (69, 111). An extracellular protease from *Halobacterium halobium* synthesized glycine containing peptides with yields of about 76% (111). The activity of the enzyme was maximal in 32% (vol:vol) dimethyl formamide while, under the same conditions, subtilisin Carlsberg did not exhibit the same level of increased activity.

Table 1. Applications of extremozymes

Enzyme	Type	Representative applications	Organism	T_{opt} (°C)	pH_{opt}	Remarks	References
Glycosyl-hydrolases	α-Amylases	Starch hydrolysis, brewing, baking, detergents	*Pyrococcus furiosus*	100	6.5–7.5	Extracellular, recombinant	32, 62
			Pyrococcus furiosus	100	7.0	Extracellular, recombinant	76
	Pullulanases type II	Production of linear small sugars	*Desulfurococcus mucosus*	85	5.0	Recombinant	33
			Pyrococcus woesei	100	6.0	Recombinant	119
	Pullulanases type III		*Thermococcus aggregans*	100	6.5	Recombinant	92
	Glucoamylases	Production of glucose	*Picrophilus oshimae*	90	2.0	–	124
			Picrophilus torridus	90	2.0	–	124
			Thermoplasma acidophilum	90	2.0	–	124
	α-Glucosidases	Final step of starch degradation	*Sulfolobus solfataricus*	120	4.5	Recombinant	108
	Endo-β-glucanase	Cotton products biopolishing	*Pyrococcus horikoshii*	85	5.6	Recombinant	5
		degradation of cellulose and production of cellooligomers	*Sulfolobus solfataricus*	80	1.8	Recombinant	58
	Xylanases	Paper bleaching	*Sulfolobus solfataricus*	90	7.0	Membrane associated	17
	α-Xylosidases		*Sulfolobus solfataricus*	90	5.5	Recombinant	86
	Chitinases	Food, cosmetics, pharmaceuticals, agrochemicals	*Thermococcus chitonophagus*	80	6.0	Extracellular part of multi-component enzymatic apparatus	6
	β-Galactosidases		*Haloferax alicantei*	25	7.2	Maximum activity in 4 M NaCl	54
	β-Glycosidases		*Pyrococcus horikoshii*	90	6.5	Membrane bound; recombinant	3
			Sulfolobus solfataricus	85	5.4	Recombinant; nucleophile mutant able to synthesize oligosaccharides	99, 139
Proteases	Lysine amino peptidases (KAP) Ion proteases	Baking, brewing, detergents, leather industry	*Pyrococcus furiosus*	100	8.0	First KAP from an archaeon	129
			Thermococcus kodakaraensis	70	9.0	Recombinant; membrane bound; ATP-independent Ion protease on unfolded protein substrates	42
	Aminopeptidases		*Haloarcula marismortui*	40	8.0	Active in 2 M KCl	40
	Cysteine protease		*Sulfolobus solfataricus*	70	7.5	Intracellular	49
	Serine protease		*Aeropyrum pernix* K1	90	8.0–9.0	Extracellular; recombinant	19

Continued on following page

Table 1. *Continued*

Enzyme	Type	Representative applications	Organism	T_{opt} (°C)	pH_{opt}	Remarks	References
			Thermococcus stetteri	85	8.5–9.0	Extracellular; 1% SDS-resistant	72
			Staphylothermus marinus	90	9.0	Membrane associated; resistant after a 135°C treatment	83
Esterases/ lipases	Carboxylesterase	Organic synthesis in industrial processes	*Sulfolobus solfataricus*	95	7.0–9.0	Recombinant	
			Pyrobaculum calidifontis	90	7.0	Recombinant	55
			Archaeoglobus fulgidus	80	7.0	Recombinant	80
	L-Aminoacylase	Production of L-amino acids from racemic solutions of *N*-acetyl amino acids	*Thermococcus litoralis*	85	8.0	Recombinant; enantiospecific for L-amino acids	135, 137
Alcohol dehydro-genases		Stereochemistry	*Sulfolobus solfataricus*	85	8.8–9.6	Recombinant; estreme enantio stereoselective	101
			Aeropyrum pernix	95	8.0–10.0	Recombinant	
			Pyrococcus furiosus	90	7.0–10.0	Recombinant	141
DNA poly-merases	*Pfu* pol	In PCR for DNA fragment up to 4 kb	*Pyrococcus furiosus*	72		Low processivity; uracil stalling	
	Vents pol	–	*Thermococcus litoralis*	72		Low processivity; uracil stalling	
	Deep Vent pol	–	*Pyrococcus* sp. GB-D	–		Low processivity; uracil stalling	
	Tgo pol	In PCR reaction for DNA fragment up to 3.5 kb	*Thermococcus gorgoniarus*	72		Low processivity; uracil stalling	
	PfuUltra pol	*Pfu* mutant with improved fidelity; combined with dUTPase (Archae Maxx; U.S. patent 2005,003,401) for PCR product up to 17 kb genomic	*Pyrococcus furiosus*	72		Low processivity; uracil stalling	53
	Platinum Pfx pol	In PCR reaction for DNA fragment up to 12 kb genomic and 20 kb vector	*Thermococcus kodakaraensis*	68		High processivity; uracil stalling	
DNA ligases			*Thermococcus kodakaraensis*	70–80	8.0	First DNA ligase from an archaeon	89, 90
		Stratagene (U.S. patent 6,280,998)	*Pyrococcus furiosus*	80	7.5	Higher melting temperature for ligase chain reaction	
		Stratagene (U.S. patent 6,280,998)	*Pyrococcus furiosus*	80	7.5	Higher melting temperature for ligase chain reaction	

Esterase/lipase

Esterases and lipases are some of the most versatile industrial enzymes as they have been exploited in the food industry, in lubricants and cosmetic formulations, in the pulp and paper industry (136, 149), and in organic synthesis in which they are the most widely used biocatalysts (28). In organic synthesis, esterases/lipases are exploited in hydrolysis, transesterification, alcoholysis, acidolysis, and aminolysis reactions. The application of these enzymes was extended after the discovery of thermophilic versions; for example, thermostable lipases have enhanced the physical refining of seed oils by enabling the separation of the lysophosphatide from oil at pH 5.0 and 75°C (50). Recently, several esterases/lipases have been described from archaea; the enantioselectivity and the catalytic efficiency in organic solvents of the esterase from *S. solfataricus* has been described in detail (120–122), and, more recently, three open reading frames (ORFs) producing enzymes with esterase activity and an additional phosphotriesterase have been cloned from the same organism (70, 85). Other enzymes have been identified in *Pyrobaculum calidifontis* (55), *P. furiosus* (60), and *Archaeoglobus fulgidus* (80) for which the three-dimensional structure of the carboxyl-esterase is available (29). Patents of esterase enzymes from various archaeal genera (*Pyrodictium*, *Archaeoglobus*, *Thermococcus*, and *Sulfolobus*) have been filed (106), but no extant application of the enzymes is known.

Glycosidases

Industrial glycoside hydrolases are widely used for the hydrolysis of starch and β-polysaccharides. Starch is composed of α-glucose units linked by α-1,4- and α-1,6-glucosidic bonds. The linear polymer of amylose consists of α-1,4-linked glucopyranose residues, and amylopectin has additional α-1,6-linked branch points every 17 to 26 glucose units along the linear polymer. Approximately 1.4×10^9 tons of starch is produced every year (67), and it is the main source of sugar for the food industry. One of the largest markets of starch enzymatic hydrolysis is the production of glucose/fructose syrups for soft drinks. Starch conversion can also produce anticariogenic sugars, such as isomaltooligosaccharides (67), antistaling agents in baking (46), and cyclic glucans (cyclodextrins) that have applications as carriers of small molecules, artificial protein chaperones, and thermoreversible starch gels (67).

Enzymatic hydrolysis of starch occurs through liquefaction, which includes gelatinization (swelling of starch at 105 to 110°C, pH 5.8 to 6.0) and saccharification (production of maltodextrines by an α-amylase at 95°C for 2 to 3 h). After liquefaction, glucose and maltose syrups are produced by the combined action of pullulanase with glucoamylases or β-amylases. In these processes, temperature and pH control is critical: gelatinization below 105°C produces incomplete swelling, making the refining of starch difficult, while above 105°C the enzymes become inactivated reducing the yields of saccharification. Similarly, a pH that is higher or lower than 5.5 can lead to by-products and unwanted color formation (145). To avoid these problems, pH adjustments are required before and after the liquefaction step, thereby increasing the cost of chemicals. Termamyl and Fungamyl are commercially available enzymes from nonarchaeal, moderately thermophilic organisms that are used in the degradation of starch. Their activity depends on calcium, and the formation of calcium oxalate can damage the industrial plant (50). An engineered version of Termamyl, (Termamyl LC) greatly reduces the amount of calcium that is required (50). Extremophilic amylases active and stable at 100°C, pH 4.0 to 5.0, and not requiring calcium, would be well suited for making starch conversion more economical. In addition, thermostable pullulanases, β-amylases, and glucoamylases could be used in a "one-pot" strategy during starch biodegradation.

Hyper/thermophiles, including members of the *Archaea*, synthesize amylolytic enzymes to hydrolyze glycogen. The enzymes include α-amylases, glucoamylases, α-glucosidases, and pullulanases (reviewed in reference 12). Enzymes that have potential biotechnological application include an α-amylase from *Pyrococcus woesei* that does not require calcium for activity and stability (41), and a novel type III pullulanase from *Thermococcus aggregans* that differs from type I and type II enzymes on the basis of its substrate specificity. This type III enzyme simultaneously attacks α-1,4- and α-1,6-glycosidic linkages in pullulan, producing a mixture of maltotriose, panose, maltose, and glucose (92). Glucoamylases from members of the genera *Picrophilus* and *Thermoplasma* that are active at 70 to 100°C and pH 0.7 to 3.0 may have biotechnological potential (12). The use of these enzymes for starch degradation requires the development of large-scale recombinant expression.

Cellulases and hemicellulases are also important in biotechnology. Cellulose is a linear polymer of β-1,4-linked glucose units and is the most abundant natural polymer on Earth. Cellulolytic enzymes are exploited in a variety of industrial fields (reviewed in reference 10). The crystalline structure of cellulose makes it a recalcitrant substrate for hydrolysis. Thermostable cellulases operating at high temperature allow the loosening of the wood fibers and provide access to the hydrolytic enzymes. Therefore, extremophilic cellulases would be very useful alternatives to mesophilic counterparts in the production of fermentable sugars for fuel ethanol (150), in the biostoning of denim, and in detergent formulas to improve the color brightening and softening by the biopolishing of the cotton fabrics. In particular, while commercial enzymes are active only at 50 to 55°C, this latter application requires cellulases that are stable at temperatures close to 100°C (5).

The dominating component of hemicelluloses is xylan, a linear polymer of β-1,4-linked xylose residues. Xylan degradation is one of the steps of the pulp and paper refining, since its hydrolysis facilitates the lignin removal during bleaching. Thermophilic xylanases would improve this process by disrupting the wood structure at elevated temperature, reducing the chlorine consumption and the consequent environmental pollution. However, these enzymes must be devoid of cellulolytic activity, should be of low molecular size to enable easier access to the pulp fibers, and should be available at very low costs (50).

Interesting reviews on hyperthermophilic cellulases and hemicellulases are available in the literature (11, 130). These enzymes are uncommon in archaea, although cellulases have been found in *P. furiosus* (4, 16), *P. horikoshii* (5), the archaeon AEPII1a (77), and in the thermoacidophile *S. solfataricus* (58, 79). Xylanase activity has been demonstrated in *Thermococcus zilligii* (140), although its unequivocal assignment is still disputed (110). Xylanases have also been reported in *P. abyssi* (7), *Halorhabdus utahensis* (146), and *S. solfataricus* (17). The xylanases from *P. abyssi* and *S. solfataricus* are highly thermostable and active up to 110°C and 90°C, respectively, while the enzyme from *H. utahensis* is, quite surprisingly, active over a broad range of NaCl concentrations (0 to 30%). Presently, the main drawback for the exploitation of archaeal cellulolytic and xylanolytic enzymes is the difficulty of expressing them in heterologous hosts (57), making their application to an industrial scale problematic.

Another class of glycosidases that has biotechnological value is chitinases. Chitin is a waste polymer produced from the exoskeleton of crabs and shrimp, and it is used to produce chitosan that has a variety of applications in cosmetics, paper and photographic products, and medical fields (50). Extremophilic chitinases would be attractive alternatives to the current use of corrosive solutions of 40% sodium hydroxide, for the hydrolysis of chitin (50). Three hypothetical proteins have been found to constitute a new pathway for the degradation of chitin in *Thermococcus kodakaraensis* (132–134). This is a nice example illustrating how the biodiversity of archaea has enhanced the potential for identifying novel biotechnological applications.

Glycosidases are attractive enzymes also for the synthesis of carbohydrates, which are becoming increasingly important as therapeutics in the pharmaceutical industry (154). Glycosidases synthesize oligosaccharides in reactions of reverse hydrolysis (equilibrium-controlled synthesis) or transglycosylation (kinetically controlled process) in which an alcohol or another sugar acts as acceptor instead of water (Fig. 1). The β-glycosidase from the hyperthermophilic archaeon *S. solfataricus* promotes transglycosylation reactions to pyranosides with a yield of 10 to 40% (25). However, since the product of the reaction has the same anomeric configuration of the substrate, it is hydrolyzed by the enzyme with a consequent reduction in the final yield. To avoid these problems, a new class of engineered glycosidase has been produced to promote the synthesis of sugars with almost quantitative yields; these novel enzymatic activities have been termed *glycosynthases* (reviewed in reference 97). Three hyperthermophilic glycosynthases have been produced by engineering the β-glycosidases from *S. solfataricus*, *Thermosphaera aggregans*, and *P. furiosus* (87, 96). The approach used to obtain hyperthermophilic glycosynthases differs from that described for mesophilic enzymes (Fig. 2A). The active site carboxylate (glutamic acid), acting as the nucleophile of the reaction, was changed in a nonnu-

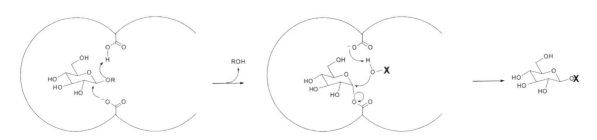

Figure 1. Reaction mechanism of β-glycosidases. Hydrolysis occurs if "X" is a hydrogen (H) atom, or transglycosylation occurs if "X" is a different "R" group.

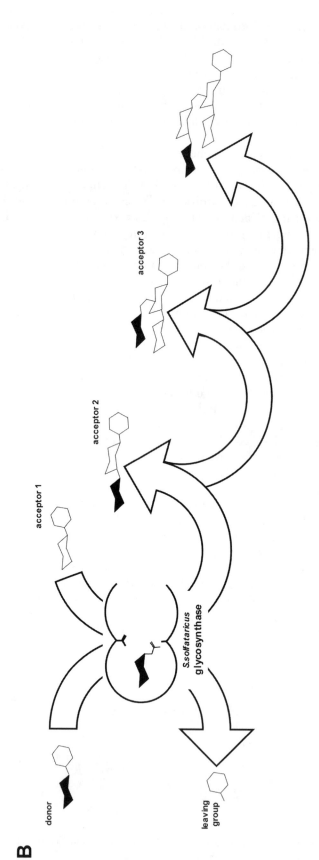

Figure 2. Reaction mechanism of hyperthermophilic glycosynthases. (A) Scheme of the glycosynthase mechanism: the transferred sugar is shown in bold. (B) Scheme of the processive glycosynthetic reaction: the transferred sugar in a chair conformation is shown with closed symbols; the aryl leaving group is shown as a hexagon.

cleophilic residue (glycine) by mutagenesis leading to a completely inactive enzyme. The mutant was re-activated at 65°C in the presence of sodium formate buffer, pH 4.0 (functioning as an external nucleophile), and aryl-glycoside (the activated substrate) (96). The mutant enzyme, assisted by the external nucleophile, formed the formyl-glucoside intermediate in situ. In the second step of the reaction, a substrate molecule, working as acceptor, resolves the intermediate leading to the product (Fig. 2A). The key to this approach is that the product accumulates in the reaction because it is a nonactivated disaccharide that cannot be hy-drolyzed by the mutant. When sufficient amounts of the disaccharide product are present, they compete with the initial substrate and act as new acceptors lead-ing to the formation of trisaccharides; the iteration of this process leads to the generation of oligosaccharides of up to four glycosidic residues containing branched chains with β-1,3 and β-1,6 bonds (138, 139) (Fig. 2B). This is a clear example of how the unique char-acteristics of stability to high temperatures and acidic pH of the glycosidases from hyperthermophilic ar-chaea allowed the development of a novel strategy for the exploitation of oligosaccharide synthesis.

Archaeal enzymes in molecular biology

The amplification of DNA fragments by the well-defined technique of the PCR has led to a burst in the knowledge of life sciences, comprising molecular and cellular biology, genetic engineering, and biotechnol-ogy. The use of the DNA polymerase from the ther-mophilic bacterium *Thermus aquaticus* (*Taq*) bypassed the problem of the addition of a mesophilic enzyme at each cycle after the denaturation step (38), leading to a rapid development of the PCR technique that is widely used (22, 64). *Taq* DNA polymerase is used for a large number of PCR applications. However, the enzyme exhibits poor fidelity due to a lack of 3′-5′ ex-onuclease activity (proofreading activity). For this rea-son it is not used when high-fidelity amplification is required, such as during the analysis of allelic poly-morphisms, allelic stages of single cells, or rare mu-tations in human cells.

Significant time and effort has been spent to im-prove the performance of *Taq* DNA polymerase (34). However, in the past decade, knowledge of archaeal DNA polymerase enzymes has led to their use in a range of PCR applications (Table 1). All archaeal DNA polymerases (B-type) (95) possess proofreading ac-tivity and lack 5′-3′ exonuclease activity. These fea-tures of the archaeal enzymes have facilitated the de-velopment of PCR techniques by eliminating the need for downstream error-correction steps and minimiz-ing the number of clones that must be sequenced to obtain error-free constructs. However, most archaeal DNA polymerases also have disadvantages compared with the bacterial *Taq* polymerase: (i) exhibit limited processivity in vitro (the polymerization rate is in the range 9 to 25 s^{-1}, compared with 47 to 61 s^{-1} for *Taq* polymerase) and (ii) uracil residues (dU), are formed by the temperature-dependent deamination of dCTP during the PCR reaction, which causes synthe-sis to stall, thereby decreasing their performance (48). These characteristics preclude amplifications of DNA fragments longer than 3-4 kb since prolonged ex-tension times (1-2 min kb^{-1} at 72°C) promote dUTP formation.

The problem with low processivity has been over-come with the isolation of an archaeal DNA poly-merase from *Thermococcus kodakaraensis*, which ex-hibits a high extension rate (106 to 138 s^{-1}) and low error rate (131), and is therefore able to perform longer PCR reactions similar to *Taq* DNA polymerase. The limitation of the dUTP formation was also overcome by the introduction of a new class of archaeal enzyme: dUTPases. These thermostable biocatalysts from *Pyro-coccus* spp. prevent the incorporation of dUTP in PCR products by transforming dUTP to dUMP, and conse-quently increasing the PCR final yield (53).

Most currently available commercial products for the amplification of DNA are composed of a mix-ture of archaeal and bacterial thermostable DNA polymerases to ensure that low error rates and high processivity are achieved. Examples include Hercu-lase and PicoMaxx (Stratagene, USA), in addition to the dUTPase (ArchaeMaxx Polymerase Enhancing Factor), which minimizes uracil poisoning, resulting in amplification products as long as 19 kb (14, 52).

Other enzymes

Transaminases and aminotransferases are widely used in the food industry to improve the viscosity of products and to produce L- or D-amino acids as food additives and precursors of antibiotics for the manu-facture of pharmaceuticals and agricultural products (71). Patents have been filed for this class of enzymes from thermophilic bacteria and the hyperthermo-philic crenarchaeon *P. aerophilum* (147); the trans-aminase provided improved stability at high tempera-ture and in organic solvents for the production of optically pure chiral compounds.

Alcohol dehydrogenases (ADHs) from hyperther-mophilic archaea have also found a role in biotech-nology industries (100). Hyperthermophilic ADHs can be useful for stereoselective transformation of ketones to alcohols, and vice versa. The ADH from

S. solfataricus is a zinc-containing, NAD(H)-dependent enzyme that is able to convert 3-methylbutan-2-one to (S)-3-methylbutan-2-ol with almost 100% stereoselectivity (39, 101). The ADH from *A. pernix* produces an enantiomeric excess of about 90% with 2-octanone, 2-nonanone, and 2-decanone substrates (51). In contrast, although the ADH from *P. furiosus* has broad substrate specificity, it exhibited only a slight preference for (S)-2-butanol (141).

Several other archaeal enzymes have attractive properties for chemical synthesis. A cysteine synthase from *A. pernix* has been used for the production of sulfur-containing organic compounds (61). A phosphatidylethanolamine N-methyltransferase from *P. furiosus* (82) catalyzes the synthesis of phosphatidylcholine, which is used in food as a digestible surfactant, and in the medical and pharmaceutical industries as a component of microcapsules for drugs. The stability of the enzyme in organic solvents and its ability to synthesize phosphatidylcholine of high optical purity make it a good alternative to phosphatidylethanolamine N-methyltransferases from bacteria and yeast.

An interesting example of how an archaeal extremozyme can be developed for use in industry is the L-aminoacylase from *Thermococcus litoralis* (135). The enzyme, which catalyzes the hydrolysis of a variety of N-acyl groups from several amino acids, was isolated and characterized in a research program involving Chirotech Technology (UK) and the University of Exeter (UK). The ability of the L-aminoacylase from *T. litoralis* to remove aromatic groups from N-acylamino acids, is a feature that distinguishes it from the commercially available L-aminoacylase from the Amano Enzyme Inc. (Japan). In addition, the archaeal enzyme was enantiospecific for L-amino acids, was active in a broad range of pH values (at least 70% of maximum activity in the pH range 6.5 to 9.5), showed maximum activity at 85°C, and a half-life of 25 h at 70°C. To produce a commercial product, the enzyme was expressed in *Escherichia coli* at a 500-liter scale and was partially purified to prevent contamination by its genetically modified host. The downstream process involved only a 2-fold purification with the final preparation being a cell-free extract containing about 80 million units from 30 to 40 kg of cells. Several commercial applications of the L-aminoacylase from *T. litoralis* are reported in the Dowpharma Chirotech Technology catalog (135). The archaeal enzyme is reported to improve the resolution of an N-acyl-protected amino acid: 240 units of enzyme g^{-1} of substrate allowed the production of more than 100 kg of product in 8 to 10 h at 70°C. The same process performed by heating the racemic N-acylamino acid to 50°C completely inactivated the mesophilic counterpart.

MOLECULES FROM *ARCHAEA*

Inteins

Since their discovery in 1990, inteins have been studied for their fascinating chemical mechanism of cleaving and rejoining proteins. Although their exploitation is still in their infancy, they provide enormous opportunities for developing novel tools for protein engineering and for understanding of the fundamental mechanism of protein splicing. This is an extraordinary posttranslational processing event that involves the precise removal of an internal polypeptide segment, termed intein, from a precursor protein, with the concomitant ligation of the flanking polypeptide sequences, termed exteins (Fig. 3) (94).

Due to rapid splicing in vivo, early in vitro studies of the chemical mechanism of protein splicing were hampered by the inability to purify precursor proteins or splicing intermediates. The discovery of inteins from hyperthermophilic archaea was fundamental to overcoming this obstacle. The *Pyrococcus* Psp intein-1 only undergoes splicing at high temperature. The unspliced precursor protein and the splicing process were able to be examined in vitro by inducing splicing using high temperature (152).

Inteins have been identified in 17 of 22 archaeal genome sequences (including *Nanoarchaeum equitans*) (http://bioinformatics.weizmann.ac.il/~pietro/inteins; 21, 93). They were not found in the genome sequences of *A. fulgidus*, *Methanosarcina acetivorans*, *P. aeropyrum*, *S. solfataricus*, and *S. acidocaldarius*. The crystal structure of the intein spliced out from a ribonucleotide reductase from *P. furiosus* (PI-*PfuI*) has been reported (59).

An intein-mediated affinity protein purification system has been developed by New England Biolabs (23, 24) and it is called IMPACT (Intein Mediated Purification with an Affinity Chitin binding Tag). This system allows the purification of a candidate protein by thiol-induced cleavage of a fusion protein that is bound to a chitin column. The intein-based purification technology avoids the use of exogenous proteases that can degrade the protein of interest, thereby offering a time-saving and economical approach.

Since the establishment of the first intein purification system, many advances have been made and several other intein fusion systems developed (153). For instance, after their discovery mini-inteins replaced the full-length inteins. Their smaller molecular weights can increase protein expression and final yield. One of the smallest mini-intein known has been found in the ribonucleoside diphosphate reductase gene of the archaeon *Methanothermobacter thermautotrophicus* (*Mth* RIR1 intein) (126). The *Mth* RIR1 intein, the in-

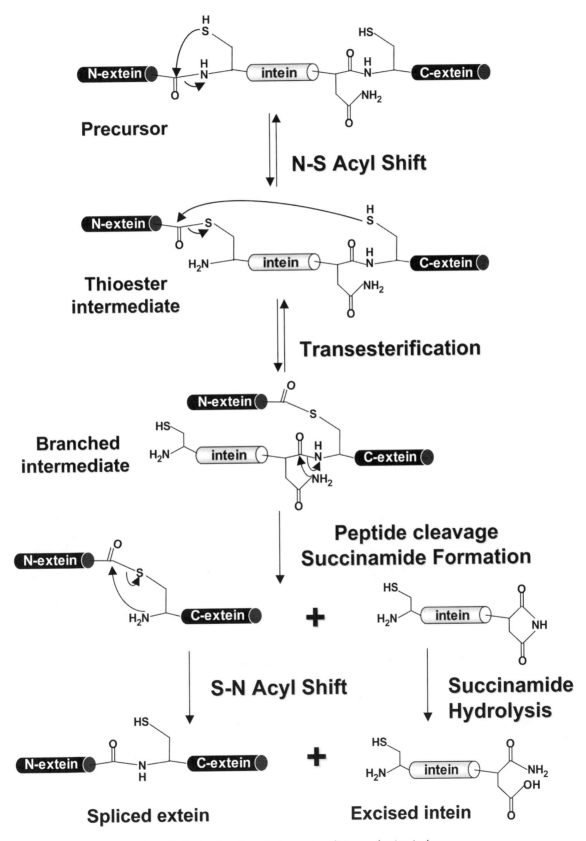

Figure 3. The intein self-catalytic protein-splicing mechanism is shown.

tein from the *Mycobacterium xenopi gyrA* gene (*Mxe* GyrA intein), and the mini-intein made from the *Synechocystis* spp. DnaB intein are commercially available from New England Biolabs as the IMPACT-TWIN (Intein Mediated Purifcation with an Affinity Chitin-binding Tag-Two Intein) *E. coli* expression vectors.

Polymers

Biopolymers have been identified in different groups of archaea, including bioplastics from halophiles and exopolysaccharides (EPS) from halophiles and thermophiles. *T. litoralis* secretes EPS when grown on maltose (104), and *Sulfolobus* sp. produces a sulfated exopolysaccharide containing glucose, mannose, glucosamine, and galactose when grown on glucose in stationary phase (91). Microbial exopolysaccharides are used as stabilizers, thickeners, and emulsifiers in several industries (e.g., pharmaceutical, textile, paper), and the major commercial EPS derives from the bacterium *Xanthomonas campestris* (143). The archaeon *Haloferax mediterranei* secretes up to 3 g liter^{-1} of a highly sulphated and acidic EPS that contains mannose as a major component (8).

In the past decade it was suggested that an acidic heteropolysaccharide from *H. mediterranei*, which showed excellent rheological properties, could have potential application as emulsifying agent in the oil industry (143). Residual oil in natural oil fields can be extracted by injection of pressurized water into a new well; in these circumstances, halobacterial membrane lipids and EPSs from this archaeon may be useful due to their biosurface activity and bioemulsifying properties (143).

Among biopolymers, bioplastics have, by far, the widest range of biotechnological applications; these compounds are produced from polyhydroxyalkanoates (PHAs), a heterogeneous family of polyesters of which poly-β-hydroxybutyrate (PHB) is the most common. Several marine archaea accumulate large amounts (25 to 80% of dry weight) of intracellular PHA that are subsequently metabolized for carbon and energy during periods of starvation (148). PHAs are used in industry for the production of thermoplastic polymers, the properties of which resemble polystyrene, polycarbonate, and polypropylene. However, these bioplastics are biodegradable. For instance, PHB has a structure similar to polypropylene and sinks to the sediment where it can be degraded in approximately one month, rather than in one year as is typically the case for polypropylene (148). The archaeon *H. mediterranei* produces large amounts of PHA when grown on starch or glucose (78). The carbon source (starch instead of glucose used by bacterial

halophiles) is relatively inexpensive, and yields are higher due to the large accumulation of polymer and ease of cell lysis (143).

Several efforts to commercialize PHA, notably by ICI in the 1980s and early 1990s under the trade name of Biopol, and by Monsanto in the mid-1990s, foundered due to problems associated with high cost and a very limited ability to process PHA. In 2001, Metabolix purchased Monsanto's Biopol assets to add to its current range of PHA. These products include molding resins, coatings for water-resistant paper bags, films, adhesives, and fibers for textiles and carpets (/www.metabolix.com/), thereby illustrating the real potential for application of archaeal PHA.

Compatible Solutes

Microorganisms cope with high osmolarity by uptaking of K^+ from the environment, or by synthesizing compatible solutes such as amino acids and their derivatives, sugars and their derivatives, polyols, betains, and ectoins (27, 44, 105, 107, 112) (see Chapter 16). Compatible solutes are small, highly soluble, organic molecules that do not interfere with central metabolism, even if they accumulate to high concentrations. Compatible solutes from archaea have biotechnological roles as cryoprotectants and preservatives.

Some thermophilic and hyperthermophilic organisms accumulate di-*myo*-inositol-phosphate, di-mannosyl-di-*myo*-inositol-phosphate, di-glycerol-phosphate, mannosylglycerate, and mannosylglyceramide. Compatible solutes from archaea, including *P. woesei*, *Pyrodictium* sp., *Thermococcus* sp., and *A. fulgidus*, tend to be negatively charged, although *P. aerophilum* and *Sulfolobus* sp. accumulate the neutral compatible solute, trehalose, which is widespread in nature. In addition to osmoadaptation, compatible solutes may protect cells from dehydration, freezing, desiccation, high temperature, and oxygen radicals (see 112 and references therein). Due to the wide range of protective effects offered by compatible solutes, they have several biotechnological applications, including the conservation of tissues and other biological products in medicine, food, and cosmetic industry, and to preserve cultures and cell lines in scientific research (116).

Trehalose and mannosylglycerate have found particular biotechnological applications. Trehalose (α-D-glucopyranosyl-α-D-glucopyranoside) is currently synthesized using two enzymes from *Arthrobacter* sp. The first enzyme, called Q36, is a trehalosyl-dextrin forming enzyme (EC 5.4.99.15) that converts dextrins to trehalosyl-dextrin, and a trehalose-forming enzyme

(EC 3.2.1.141) releases trehalose and low-molecular-weight dextrins from trehalosyl-dextrins (116). The two enzymes work sequentially at 45°C. The use of thermophilic enzymes may help to reduce viscosity and microbial contamination. The same two enzymes have been identified in *S. solfataricus* strains KM1 and P2 (73; www-archbac.u-psud.fr/Projects/Sulfolobus/Sulfolobus.html), in *S. shibatae* (30), and in *S. tokodaii* (www.bio.nite.go.jp/dogan/Top). When the *Sulfolobus* enzymes were expressed in *E. coli* in a high-density microfiltration bioreactor, the recombinant cells produced trehalose from dextrins at 75°C with a final conversion of 90% (31). This study illustrates the true potential for industrial application.

Mannosylglycerate has potential use as bioprotectant. This compound, together with di-*myo*-inositol-phosphate, accumulates in *Pyrococcus* sp. with increasing NaCl concentration of the medium (112). The use of mannosylglycerate as an enzyme stabilizer has been patented (113) and studied extensively (13). It is considerably better than trehalose as thermoprotectant for enzymes, and as good as trehalose for protecting enzymes against desiccation (113). For example, after incubation for 10 min at 50°C, the residual activity of rabbit muscle lactate dehydrogenase increases from 5% to 90% in the presence of mannosylglycerate and trehalose, respectively (113). The patent showed that mannosylglycerate could be purified from ethanolic extracts of several archaea, including *P. furiosus*, *P. woesei*, *Methanothermus* spp., and *Thermococcus* spp. grown in complex medium at supraoptimal temperatures and salinities. For *P. furiosus*, 2.5 g of mannosylglycerate was obtained from 100 g of cell paste (113).

Self-Assembling Components

Self-assembling S-layer glycoproteins (118, 125) and bacteriorhodopsins (15, 143, 151) have drawn interest for their potential in molecular nanotechnology. S layers are commonly observed as cell envelopes of bacteria and archaea (see Chapter 14). The S layers of these organisms are particularly appealing for biotechnology because they form a regular lattice that is very stable. In addition, because each unit of the S-layer lattice possesses identical physicochemical properties, they form unique structures. The mechanical and permeability properties of these structures has enabled S layers to be used as ultrafiltration membranes that exhibit precise molecular exclusion properties, and are therefore interesting alternatives to conventional membranes (125). Moreover, S layers from many archaea and some bacteria are glycosylated (36, 114), which might allow macromolecules to be immobilized not only on the protein moiety, but also on the carbohydrate residues.

Bacteriorhodopsin is a photoactive proton pump present in the purple membrane of halophilic archaea, including *Halobacterium salinarium*. At low oxygen availability, bacteriorhodopsin trimers are assembled in a two-dimensional hexagonal lattice in the purple membrane. The primary function of bacteriorhodopsin is to pump protons (see Chapter 16). The main biotechnology application of this protein resides in its ability to absorb light energy and convert it to chemical energy. Bacteriorhodopsin-based devices have been used since the 1980s for optical data processing (holography, information storage), as biochips, and as light sensors obtained by sandwiching the protein between an oxide electrode and an electrically conductive gel (143). In the late 1970s, the Soviet military explored the potential of bioelectronics in the computer technology in a program called "Project Rhodopsin." The project led to the production of *bio-chrom*, a real-time photochromic and holographic film containing bacteriorhodopsin (15). More recently, a new generation of bacteriorhodopsin-based devices were developed using genetic manipulation, which led to the generation of a mutant with a 700-fold improvement in volumetric data storage (reviewed in rerference 151).

ARCHAEAL WHOLE CELLS

The utilization of archaeal cells in biotechnological applications is limited by the low biomass yield of cells, and the availability of molecular biology tools (see Chapter 21). The biotechnology of archaeal whole cells is still in its infancy, and the applications are limited to the treatment of contaminated soil and wastewater (bioremediation) and hydrogen production.

Bioremediation

The treatment of contaminated soil and wastewater with microorganisms to remove toxic organic compounds, metals, etc., is an established practice that is undergoing continous expansion. Although bacteria have been exploited (e.g., treatment of the petroleum wastes; reviewed in reference 142), archaea have not been widely used. Nevertheless, studies on the microbial populations associated with treatment of contaminated soil and water revealed the presence of archaea. For example, methanogenic archaea were found as syntropic consortia in the leachate of a full-scale recirculating landfill (57), in anaerobic reactors for the treatment of industrial dye effluents (98) and domestic wastewater (84), in soil decontaminated by steam treatment (103), and in pharmaceutical and alcohol distillery wastewater (2). Most of the archaea identified belong to species of *Methanosaeta*, *Methanobacterium*, and *Methanospirillum*. In contrast to these examples, archaeal populations were generally reduced in oil-polluted beach ecosystems, suggesting that they play small role in oil degradation in comparison with bacteria (109).

The presence of methanogenic archaea in contaminated anaerobic environments is consistent with their anaerobic lifestyle. However, the importance of these microorganisms in aerobic wastewater has received little attention. More recently, anoxic microenvironments containing methanogenic archaea have been identified in industrial and domestic aerated activated sludge plants (47). These types of studies (47) are particularly useful as they provide an improved understanding of the ecological forces governing the wastewater treatment and may therefore help in optimizing the process. The study highlighted that the dynamics of microbial-driven processes are often determined by minority populations. One remarkable example of this type of regulation was recently reported for an activated sludge reactor (63). Periodic addition of solutions containing anaerobic methanogenic archaea to a conventional activated sludge system led to increased nitrogen removal and to sludge reduction. This method has been adopted by ArchaeaSolutions, Inc. (www.archaeasolutions.com, Tyrone, Ga.) in the treatment of wastewater from winery industries.

Hydrogen Production

Hydrogen gas is an attractive alternative to fossil fuel as it is a clean, nonpolluting source of energy. However, conventional production is based, at present, on the steam reforming of natural gas and petroleum, and microbial production of H_2 is gaining increasing interest. Many studies on the microbial production of H_2 have focused on the bacterial genera, *Clostridium* (68) and *Enterobacter* (75). In the early 1990s, *P. furiosus* was shown to efficiently produce H_2 when grown on rich medium (115), suggesting that archaea can also play a role in this field. More recently, *T. kodakaraensis* KOD1 was shown to produce H_2 at a rate ranging from 14 to 59.6 mmol per gram dry weight h^{-1} (depending on the dilution rate), when grown at 85°C on rich medium supplemented with pyruvate or starch (66). These values are comparable to those reported for *Enterobacter cloacae* (29.6 mmol per gram dry weight h^{-1}). The higher operational temperatures of the hyperthermophilic archaea provide the advantage of eliminating the risks of microbial contamination and the ability to perform starch liquefaction thereby increasing the solubility of the substrate. These preliminary reports illustrate a high potential for using carbohydrate-degrading archaea for H_2 production.

PERSPECTIVE: THE NEXT FIVE YEARS

Extremophilic archaea have been considered an interesting source of molecules for novel biotechnological applications. Their stability and activity to extreme conditions make them useful alternatives to labile mesophilic counterparts. Therefore, it is not surprising that a number of companies worldwide exploit extremophilic organisms and their biomolecules. Nevertheless, products from archaea represent a small part of commerical catalogs. One of the leading biotechnology companies, Genencor (USA), has no products, either in development or on the market, that are derived from archaea, despite the fact that they deal with enzymes from extremophiles. Presently, only two companies are fully committed to the exploitation of archaea and their biomolecules: Archaezyme Ltd., Jerusalem, Israel, and ArchaeaSolutions Inc., Tyrone, Georgia, USA. However, this is a rapidly evolving field. In the past year, numerous exciting findings that have potential relevance to biotechnology have been reported, two of which are the genome sequence of the extreme acidophile *P. torridus* (43) and a new mechanism of gene expression in *N. equitans* (102). Such discoveries hold great expectations for the future. There is much to be gained from the exploitation of archaea, and this will be fostered not only by scientific advances, but also by the willingness of industry and government to make suitable investment.

REFERENCES

1. **Adams, M. W. W., F. B. Perler, and R. M. Kelly.** 1995. Extremozymes—expanding the limits of biocatalysis. *BioTechnology* 13:662–668.
2. **Akarsubasi, A. T., O. Ince, B. Kirdar, N. A. Oz, D. Orhon, T. P. Curtis, I. M. Head, and B. K. Ince.** 2005. Effect of wastewater composition on archaeal population diversity. *Water Res.* 39:1576–1584.
3. **Akiba, T., M. Nishio, I. Matsui, and K. Harata.** 2004. X-ray structure of a membrane-bound beta-glycosidase from the hyperthermophilic archaeon *Pyrococcus horikoshii. Proteins.* 57:422–431.
4. **Andersen, L. N., M. E. Bjoernvad, and N. Schuelein.** 1998. A thermostable endo-β-1,4-glucanase. Patent WO9833895.
5. **Ando, S., H. Ishida, Y. Kosugi, and K. Ishikawa.** 2002. Hyperthermostable endoglucanase from *Pyrococcus horikoshii. Appl. Environ. Microbiol.* 68:430–433.
6. **Andonopoulou, E., and C. E. Vorgias.** 2004. Multiple components and induction mechanism of the chitinolytic system of the hyperthermophilic archaeon *Thermococcus chitonophagus. Appl. Microbiol. Biotechnol.* 65:694–702.

7. Andrade, C. M., W. B. Aguiar, and G. Antranikian. 2001. Physiological aspects involved in production of xylanolytic enzymes by deep-sea hyperthermophilic archaeon *Pyrodictium abyssi*. *Appl. Biochem. Biotechnol.* **91-93**:655–669.

8. Anton, J., I. Meseguer, and F. Rodriguezvalera. 1988. Production of an extracellular polysaccharide by *Haloferax mediterranei*. *Appl. Environ. Microbiol.* **54**:2381–2386.

9. Atomi, H. 2005. Recent progress towards the application of hyperthermophiles and their enzymes. *Curr. Opin. Chem. Biol.* **9**:166–173.

10. Bayer, E. A., E. Morag, and R. Lamed. 1994. The cellulosome-a treasure-trove for biotechnology. *Trends Biotechnol.* **12**:379–386.

11. Bergquist, P. L., M. D. Gibbs, D. D. Morris, V. S. Te'o, D. J. Saul, and H. W. Moran. 1999. Molecular diversity of thermophilic cellulolytic and hemicellulolytic bacteria. *FEMS Microbiol. Ecol.* **28**:99–110.

12. Bertoldo, C., and G. Antranikian. 2002. Starch-hydrolyzing enzymes from thermophilic archaea and bacteria. *Curr. Opin. Chem. Biol.* **6**:151–160.

13. Borges N, A. Ramos, N. D. Raven, R. J. Sharp, and H. Santos. 2002. Comparative study of the thermostabilizing properties of mannosylglycerate and other compatible solutes on model enzymes. *Extremophiles* **6**:209–216.

14. Borns, M., and H. Hogrefe. 2000. Unique enhanced DNA polymerase delivers high fidelity and great PCR performance. *Strategies* **13**:76–79.

15. Burykin, N. N., E. Y. Korchemskaya, M. S. Soskin, V. B. Taranenko, T. V. Dukova, and N. N. Vsevolodov. 1985. Photoinduced Anisotropy In Bio-Chrom Films. *Opt. Commun.* **54**:68–70.

16. Cady, S. G., M. W. Bauer, W. Callen, M. A. Snead, E. J. Mathur, J. M. Short, and R. M. Kelly. 2001. β-Endoglucanase from *Pyrococcus furiosus*. *Methods Enzymol.* **330**:346–354.

17. Cannio, R., N. Di Prizito, M. Rossi, and A. Morana. 2004. A xylan-degrading strain of *Sulfolobus solfataricus*: isolation and characterization of the xylanase activity. *Extremophiles* **8**:117–124.

18. Cary S. C., T. Shank, and J. Stein. 1998. Worms bask in extreme temperatures. *Nature* **391**:545–546.

19. Catara, G., G. Ruggiero, F. La Cara, F. A. Digilio, A. Capasso, and M. Rossi. 2003. A novel extracellular subtilisin-like protease from the hyperthermophile *Aeropyrum pernix* K1: biochemical properties, cloning, and expression. *Extremophiles* **7**:391–399.

20. Cavicchioli, R., K. S. Siddiqui, D. Andrews, and K. R. Sowers. 2002. Low-temperature extremophiles and their applications. *Curr. Opin. Biotechnol.* **13**:253–261.

21. Chen, L., K. Brugger, M. Skovgaard, P. Redder, Q. She, E. Torarinsson, B. Greve, M. Awayez, A. Zibat, H. P. Klenk, and R. A. Garrett. 2005. The genome of *Sulfolobus acidocaldarius*, a model organism of the Crenarchaeota. *J. Bacteriol.* **187**:4992–4999.

22. Chien, A., D. Edgar, and J. Trela. 1976. Deoxyribonucleic acid polymerase from the extreme thermophilile *Thermus aquaticus*. *J. Bacteriol.* **127**:1550–1557.

23. Chong, S., F. B. Mersha, D. G. Comb, M. E. Scott, D. Landry, L. M. Vence, F. B. Perler, J. Benner, R. B. Kucera, C. A. Hirvonen, J. J. Pelletier, H. Paulus, and M. Q. Xu. 1997. Single-column purification of free recombinant proteins using a self-cleavable affinity tag derived from a protein splicing element. *Gene* **192**:271–281.

24. Chong, S., G. E. Montello, A. Zhang, E. J. Cantor, W. Liao, M. Q Xu, and J. Benner. 1998. Utilizing the C-terminal cleavage activity of a protein splicing element to purify recombinant proteins in a single chromatographic step. *Nucleic Acids Res.* **26**:5109–5115.

25. Cobucci-Ponzano, B., G. Perugino, A. Trincone, M. Mazzone, B. Di Lauro, A. Giordano, M. Rossi, and M. Moracci. 2003. Applications in biocatalysis of glycosyl hydrolases from the hyperthermophilic archaeon *Sulfolobus solfataricus*. *Biocat. Biotrans.* **21**:215–221.

26. Connaris, H., D. A. Cowan, and R. J. Sharp. 1991. Heterogeneity of proteinases from the hyperthermophilic archaeobacterium *Pyrococcus furiosus*. *J. Gen. Microbiol.* **137**:1193–1199.

27. da Costa, M. S., H. Santos, and E. A. Galinski. 1998. An overview of the role and diversity of compatible solutes in Bacteria and Archaea. *Adv. Biochem. Eng. Biotechnol.* **61**:117–153.

28. Demirjian, D. C., F. Moris-Varas, and C. S. Cassidy. 2001. Enzymes from extremophiles. *Curr. Opin. Chem. Biol.* **5**:144–151.

29. De Simone, G., V. Menchise, G. Manco, L. Mandrich, N. Sorrentino, D. Lang, M. Rossi, and C. Pedone. 2001. The crystal structure of a hyper-thermophilic carboxylesterase from the archaeon *Archaeoglobus fulgidus*. *J. Mol. Biol.* **314**:507–518.

30. Di Lernia, I., A. Morana, A. Ottombrino, S. Fusco, M. Rossi, and M. De Rosa. 1998. Enzymes from *Sulfolobus shibatae* for the production of trehalose and glucose from starch. *Extremophiles* **2**:409–416.

31. Di Lernia, I., C. Schiraldi, M. Generoso, and M. De Rosa. 2002. Trehalose production at high temperature exploiting an immobilized cell bioreactor. *Extremophiles* **6**:341–347.

32. Dong, G., C. Veille, A. Savchenko, and J. G. Zeikus. 1997. Cloning, sequencing and expression of the gene encoding extracellular α-amylase from *Pyrococcus furiosus* and biochemical characterization of the recombinant enzyme. *Appl. Environ. Microbiol.* **63**:3569–3576.

33. Duffner, F., C. Bertoldo, J. T. Andersen, K. Wagner, and G. Antranikian. 2000. A new thermoactive pullulanase from *Desulfurococcus mucosus*: cloning, sequencing, purification, and characterization of the recombinant enzyme after expression in *Bacillus subtilis*. *J. Bacteriol.* **182**:6331–6338.

34. Eckert, K. A., and T. A. Kunkel. 1991. DNA polymerase fidelity and the polymerase chain reaction. *PCR Methods Appl.* **1**:17–24.

35. Eichler, J. 2001. Biotechnological uses of archaeal extremozymes. *Biotechnol. Adv.* **19**:261–278.

36. Eichler, J. 2003. Facing extremes: archaeal surface-layer (glyco)proteins. *Microbiology* **149**:3347–3351.

37. Eriksen, N. 1996. Detergents, p.187–200. *In* T. Godfrey and S. West (ed.), *Industrial Enzymes*, 2nd ed. The Macmillan Press Ltd., Basingstoke, United Kingdom.

38. Erlich, H., D. Gelfand, and R. Saiki. 1988. Specific DNA amplification product review. *Nature* **33**:–461–462.

39. Esposito, L., I. Bruno, F. Sica, C, A, Raia, A. Giordano, M. Rossi, L. Mazzarella, and A. Zagari. 2003. Crystal structure of a ternary complex of the alcohol dehydrogenase from *Sulfolobus solfataricus*. *Biochemistry* **42**:–14397–14407.

40. Franzetti, B., G. Schoehn, J. F. Hernandez, M. Jaquinod, R. W. Ruigrok, and G. Zaccai. 2002. Tetrahedral aminopeptidase: a novel large protease complex from archaea. *EMBO J.* **21**:2132–2138.

41. Frillingos, S., A. Linden, F. Niehaus, C. Vargas, J. J. Nieto, A. Ventosa, G. Antranikian, and C. Drainas. 2000. Cloning and expression of alpha-amylase from the hyperthermophilic archaeon *Pyrococcus weosei* in the moderately halophilic bacterium *Halomonas elongata*. *J. Appl. Microbial.* **88**:495–503.

42. Fukui, T., T. Eguchi, H. Atomi, and T. Imanaka. 2002. A membrane-bound archaeal Lon protease displays ATP-independent proteolytic activity towards unfolded proteins and ATP-dependent activity for folded proteins. *J. Bacteriol.* **184**:3689–3698.

43. Futterer, O., A. Angelov, H. Liesegang, G. Gottschalk, C. Schleper, B. Schepers, C. Dock, G. Antranikian, and W. Liebl. 2004. Genome sequence of *Picrophilus torridus* and its implications for life around pH 0. *Proc. Natl. Acad. Sci. USA* **101**:9091–9096.

44. Galinski, E. A. 1995. Osmoadaptation in bacteria. *Adv. Microb. Physiol.* **37**:272–328.

45. Gerday, C., M. Aittaleb, M. Bentahir, J. P. Chessa, P. Claverie, T. Collins, S. D'Amico, J. Dumont, G. Garsoux, D. Georlette, A. Hoyoux, T. Lonhienne, M. A. Meuwis, and G. Feller. 2000. Cold-adapted enzymes: from fundamentals to biotechnology. *Trends Biotechnol.* **18**:103–107.

46. Godfrey, T. 1996. Baking, p. 87–101. *In* T. Godfrey and S. West (ed.), *Industrial Enzymes*, 2nd ed. The Macmillan Press Ltd., Basingstoke United Kingdom.

47. Gray, N. D., I. P. Miskin, O. Kornilova, T. P Curtis, and I. M. Head. 2002. Occurrence and activity of Archaea in aerated activated sludge wastewater treatment plants. *Environ. Microbiol.* **4**:158–168.

48. Greagg, M. A., M. J. Fogg, G. Panayotou, S. J. Evans, B. A. Connolly, and L. H. Pearl. 1999. A read-head function in archaeal DNA polymerases detects promutagenic template-strand uracil. *Proc. Natl. Acad. Sci. USA* **96**:9045–9050.

49. Guagliardi, A., L. Cerchia, and M. Rossi. 2002. An intracellular protease of the crenarchaeon *Sulfolobus solfataricus*, which has sequence similarity to eukaryotic peptidases of the CD clan. *Biochem. J.* **368**:357–363.

50. Haki, G. D., and S. K. Rakshit. 2003. Developments in industrially important thermostable enzymes: a review. *Bioresour. Technol.* **89**:17–34.

51. Hirakawa, H., N. Kamiya, Y. Kawarabayashi, and T. Nagamune. 2004. Properties of an alcohol dehydrogenase from the hyperthermophilic archaeon *Aeropyrum pernix* K1. *J. Biosci. Bioeng.* **97**:202–206.

52. Hogrefe, H., B. Scott, K. Nielson, V. Hedden, C. Hansen, J. Cline, F. Bai, J. Amberg, R. Allen, and M. Madden. 1997. Nover PCR enhancing factor improves performance of *Pfu* DNA polymerase. *Strategies* **10**:93–96.

53. Hogrefe, H. H., C. J. Hansen, B. R. Scott, and K. B. Nielson. 2002. Archaeal dUTPase enhances PCR amplifications with archaeal DNA polymerase by preventing dUTP incorporation. *Proc. Natl. Acad. Sci. USA* **99**:596–601.

54. Holmes, M. L., R. K. Scopes, R. L. Moritz, R. J. Simpson, C. Englert, F. Pfeifer, and M. L. Dyall-Smith. 1997. Purification and analysis of an extremely halophilic beta-galactosidase from *Haloferax alicantei*. *Biochim. Biophys. Acta* **1337**:276–286.

55. Hotta, Y., S. Ezaki, H. Atomi, and T. Imanaka. 2002. Extremely stable and versatile carboxylesterase from a hyperthermophilic archaeon. *Appl. Environ. Microbiol.* **68**:3925–3931.

56. Hough, D. W., and M. J. Danson. 1999. Extremozymes. *Curr. Opin. Chem. Biol.* **3**:39–46.

57. Huang, L. N., H. Zhou, Y. Q. Chen, S. Luo, C. Y. Lan, and L. H. Qu. 2002. Diversity and structure of the archaeal community in the leachate of a full-scale recirculating landfill as examined by direct 16S rRNA gene sequence retrieval. *FEMS Microbiol. Lett.* **214**:235–240.

58. Huang, Y., G. Krauss, S. Cottaz, H. Driguez, and G. Lipps. 2005. A highly acid-stable and thermostable endo-beta-glucanase from the thermoacidophilic archaeon *Sulfolobus solfataricus*. *Biochem. J.* **385**(Pt 2):581–588.

59. Ichiyanagi, K., Y. Ishino, M. Ariyoshi, K. Komori, and K. Morikawa. 2000. Crystal structure of an archaeal intein-encoded homing endonuclease PI-PfuI. *J. Mol. Biol.* **300**:889–901.

60. Ikeda, M., and D. S. Clark. 1998. Molecular cloning of extremely thermostable esterase gene from hyperthermophilic archaeon *Pyrococcus furiosus* in *Escherichia coli*. *Biotechnol. Bioeng.* **57**:624–629.

61. Ishikawa, K., and K. Mino. 2004. Heat-resistant cysteine synthase. U.S. Patent 2,004,002,075.

62. Jørgensen, S., C. E. Vorgias, and G. Antranikian. 1997. Cloning, sequencing and expression of an extracellular α-amylase from the hyperthermophilic archeon *Pyrococcus furiosus* in *Escherichia coli* and *Bacillus subtilis*. *J. Biol. Chem.* **272**:16335–16342.

63. Jun, H. B., S. M. Park, N. B. Park, and S. H. Lee. 2004. Nitrogen removal and sludge reduction in a symbiotic activated sludge system between anaerobic archaea and bacteria. *Water Sci. Technol.* **50**:189–197.

64. Kaledin, A., A. Sliusarenko, and S. Gorodetski. 1980. Isolation and properties of DNA polymerase from extreme thermophilic bacteria *Thermus aquaticus* YT-1. *Biochimiya* **45**:644–651.

65. Kamekura, M., Y. Seno, and M. Dyall-Smith. 1996. Halolysin R4, a serine proteinase from the halophilic archaeon *Haloferax mediterranei*; gene cloning, expression and structural studies. *Biochim. Biophys. Acta* **1294**:159–167.

66. Kanai, T, H. Imanaka, A. Nakajima, K. Uwamori, Y. Omori, T. Fukui, H. Atomi, and T. Imanaka. 2005. Continuous hydrogen production by the hyperthermophilic archaeon, *Thermococcus kodakaraensis* KOD1. *J. Biotechnol.* **116**:271–282.

67. Kaper, T., M. J. van der Maarel, G. J. Euverink, and L. Dijkhuizen. 2004. Exploring and exploiting starch-modifying amylomaltases from thermophiles. *Biochem. Soc. Trans.* **32**:279–282.

68. Kataoka, N., A. Miya, and K. Kiriyama. 1997. Studies on hydrogen production by continuous culture system of hydrogen-producing anaerobic bacteria. *Water Sci. Technol.* **36**:41–47.

69. Kim, J., and J. S. Dordick. 1997. Unusual salt and solvent dependence of a protease from an extreme halophile. *Biotechnol. Bioeng.* **55**:471–479.

70. Kim, S., and S. B. Lee. 2004. Thermostable esterase from a thermoacidophilic archaeon: purification and characterization for enzymatic resolution of a chiral compound. *Biosci. Biotechnol. Biochem.* **68**:2289–2298.

71. Kirk, O., T. V. Borchert, and C. C. Fuglsang. 2002. Industrial enzyme applications. *Curr. Opin. Biotechnol.* **13**:345–351.

72. Klingeberg, M., B. Galunsky, C. Sjoholm, V. Kasche, and G.Antranikian. 1995. Purification and properties of a highly thermostable, sodium dodecyl sulfate-resistant and stereospecific proteinase from the extremely thermophilic archaeon *Thermococcus stetteri*. *Appl. Environ. Microbiol.* **61**:3098–3104.

73. Kobayashi, K., M. Kato, Y. Miura, M. Kettoku, T. Komeda, and A. Iwamatsu. 1996. Gene cloning and expression of new trehalose-producing enzymes from the hyperthermophilic archaeum *Sulfolobus solfataricus* KM1. *Biosci. Biotechnol. Biochem.* **60**:1882–1885.

74. Krahe, M., G. Antranikian, and H. Markl. 1996. Fermentation of extremophilic microorganisms. *FEMS Microbiol. Rev.* **18**:271–285.

75. Kumar, N., and D. Das. 2001. Continuous hydrogen production by immobilized *Enterobacter cloacae* IIT-BT 08 using lignocellulosic materials as solid matrices. *Enzyme Microbiol. Technol.* **29**:280–287.

76. Laderman, K. A., K. Asada, T. Uemori, H. Mukai, Y. Taguchi, I. Kato, and C. B. Anfinsen. 1993. α-Amylase from the hyperthermophilic archaebacterium *Pyrococcus furiosus*. Cloning and sequencing of the gene and expression in *Escherichia coli*. *J. Biol. Chem.* **268**:24402–24407.

77. **Lam, D. E., and E. J. Mathur.** 1998. Endoglucanases U.S. Patent 5,789,228.

78. **Lillo, J. G., and F. Rodriguezvalera.** 1990. Effects of culture conditions on poly(beta-hydroxybutyric acid) production by *Haloferax-Mediterranei. Appl. Environ. Microbiol.* **56:**2517–2521.

79. **Limauro, D., R. Cannio, G. Fiorentino, M. Rossi, and S. Bartolucci.** 2001. Identification and molecular characterization of an endoglucanase gene, *celS,* from the extremely thermophilic archaeon *Sulfolobus solfataricus. Extremophiles* **5:**213–219.

80. **Manco, G., E. Giosuè, S. D'Auria, P. Herman, G. Carrea, and M. Rossi.** 2000. Cloning, overexpression, and properties of a new thermophilic and thermostable esterase with sequence similarity to hormone-sensitive lipase subfamily from the archaeon *Archaeoglobus fulgidus. Arch. Biochem. Byophys.,* **373:**182–192.

81. **Marhuenda-Egea, F. C., and M. J. Bonete.** 2002. Extreme halophilic enzymes in organic solvents. *Curr. Opin. Biotechnol.* **13:**385–389.

82. **Matsui, I., K. Ishikawa, H. Ishida, Y. Kosugi, and Y. Tahara.** 2002. Thermostable enzyme having phosphatidylethanolamine N-methyltransferase activity. U.S. Patent 6,391,604.

83. **Mayr, J., A. Lupas, J. Kellermann, C. Eckerskom, W. Baumeister, and J. Peters.** 1996. A hyperthermostable protease of the subtilisin family bound to the surface layer of the archaeon *Staphylothermus marinus. Curr. Biol.* **6:**739–749.

84. **Mendonca, N. M., C. L. Niciura, E. P. Gianotti, and J. R. Campos.** 2004. Full scale fluidized bed anaerobic reactor for domestic wastewater treatment: performance, sludge production and biofilm. *Water Sci. Technol.* **49:**319–325.

85. **Merone, L., L. Mandrich, M. Rossi, and G. Manco.** 2005. A thermostable phosphotriesterase from the archaeon *Sulfolobus solfataricus*: cloning, overexpression and properties. *Extremophiles* **9:**297–305.

86. **Moracci, M., B. Cobucci-Ponzano, A. Trincone, S. Fusco, M. De Rosa, J. Van der Oost, C. W. Sensen, R. L. Charlebois, and M. Rossi.** 2000. Identification and molecular characterization of the first α-xylosidase from an archaeon. *J. Biol. Chem.* **275:**22082–22089.

87. **Moracci, M., A. Trincone, G. Perugino, M. Ciaramella, and M. Rossi.** 1998. Restoration of the activity of active-site mutants of the hyperthermophilic beta-glycosidase from *Sulfolobus solfataricus*: dependence of the mechanism on the action of external nucleophiles. *Biochemistry* **37:**17262–17270.

88. **Mozhaev, V. V.** 1993. Mechanism-based strategies for protein thermostabilization. *Trends Biotechnol.* **11:**88–95.

89. **Nakatani, M., S. Ezaki, H. Atomi, and T. Imanaka.** 2000. A DNA ligase from a hyperthermophilic archaeon with unique cofactor specificity. *J. Bacteriol.* **182:**6424–6433.

90. **Nakatani, M., S. Ezaki, H. Atomi, and T. Imanaka.** 2002. Substrate recognition and fidelity of strand joining by an archaeal DNA ligase. *Eur. J. Biochem.* **269:**650–656.

91. **Nicolaus, B., M. C. Manca, I. Romano, and L. Lama.** 1993. Production of an exopolysaccharide from 2 thermophilic archaea belonging to the genus *Sulfolobus. FEMS Microbiol. Lett.* **109:**203–206.

92. **Niehaus, F., A. Peters, T. Groudieva, and G. Antranikian.** 2000. Cloning, expression and biochemical characterisation of a unique thermostable pullulan-hydrolysing enzyme from the hyperthermophilic archaeon *Thermococcus aggregans. FEMS Microbiol. Lett.* **190:**223–229.

93. **Perler, F. B.** 2002. InBase, the Intein Database. *Nucleic Acids Res.* **30:**383–384.

94. **Perler, F. B., E. O. Davis, G. E. Dean, F. S. Gimble, W. E. Jack, N. Neff, C. J. Noren, J. Thorner, and M. Belfort.** 1994. Protein splicing elements: inteins and exteins—a definition of terms and recommended nomenclature. *Nucleic Acids Res.* **22:**1125–1127.

95. **Perler, F. B., S. Kumar, and H. Kong.** 1996. Thermostable DNA polymerases. *Adv. Protein Chem.* **48:**377–435.

96. **Perugino, G., A. Trincone, A. Giordano, J. van der Oost, T. Kaper, M. Rossi, and M. Moracci.** 2003. Activity of hyperthermophilic glycosynthases is significantly enhanced at acidic pH. *Biochemistry* **42:**8484–8493.

97. **Perugino, G., A. Trincone, M. Rossi, and M. Moracci.** 2004. Oligosaccharide synthesis by glycosynthases. *Trends Biotechnol.* **22:**31–37.

98. **Plumb, J. J., J. Bell, and D. C. Stuckey.** 2001. Microbial populations associated with treatment of an industrial dye effluent in an anaerobic baffled reactor. *Appl. Environ. Microbiol.* **67:**3226–3235.

99. **Pouwels, J., M. Moracci, B. Cobucci-Ponzano, G. Perugino, J. Van der Oost, T. Kaper, J. H. G. Lebbink, V. W. De Vos, M. Ciaramella, and M. Rossi.** 2000. Activity and stability of hyperthermophilic enzymes: a comparative study on two archaeal β-glycosidases. *Extremophiles* **4:**157–164.

100. **Radianingtyas, H, and P. C. Wright.** 2003. Alcohol dehydrogenases from thermophilic and hyperthermophilic archaea and bacteria. *FEMS Microbiol Rev.* **27:**593–616.

101. **Raia, C. A., A. Giordano, and M. Rossi.** 2001. Alcohol dehydrogenase from *Sulfolobus solfataricus. Methods Enzymol.* **331:**176–195.

102. **Randau, L., R. Munch, M. J. Hohn, D. Jahn, and D. Soll.** 2005. *Nanoarchaeum equitans* creates functional tRNAs from separate genes for their 5′- and 3′-halves. *Nature* **433:**537–541.

103. **Richardson, R. E., C.A. James, V. K. Bhupathiraju, and L. Alvarez-Cohen.** 2002. Microbial activity in soils following steam treatment. *Biodegradation* **13:**285–295.

104. **Rinker, K. D., and R. M. Kelly.** 2000. Effect of carbon and nitrogen sources on growth dynamics and exopolysaccharide production for the hyperthermophilic archaeon *Thermococcus litoralis* and bacterium *Thermotoga maritima. Biotechnol Bioeng.* **69:**537–547.

105. **Roberts, M. F.** 2004. Osmoadaptation and osmoregulation in archaea: update 2004. *Front. Biosci.* **9:**1999–2019.

106. **Robertson, D. E., D. Murphy, J. Reid, A. M. Maffia, S. Link, R. V. Swanson, P. V. Warren, and A. Kosmotka.** 1999. Esterases. U.S. Patent 5,942,430.

107. **Roeßler, M., and V. Müller.** 2001. Osmoadaptation in bacteria and archaea: common principles and differences. *Environ. Microbiol.* **3:**743–754.

108. **Rolfsmeier, M., C. Haseltine, E. Bini, A. Clark, and P. Blum.** 1998. Molecular characterization of the α-glucosidase gene (*malA*) from the hyperthermophilic archaeon *Sulfolobus solfataricus. J. Bacteriol.* **180:**1287–1295.

109. **Roling, W. F., I R. de Brito Couto, R. P. Swannell, and I. M. Head.** 2004. Response of Archaeal communities in beach sediments to spilled oil and bioremediation. *Appl. Environ. Microbiol.* **70:**2614–2620.

110. **Rolland, J. L., Y. Gueguen, D. Flament, Y. Pouliquen, P. F. Street, and J. Dietrich.** 2002. Comment on "The first description of an archaeal hemicellulase: the xylanase from *Thermococcus zilligii* strain AN1": evidence that the unique N-terminal sequence proposed comes from a maltodextrin phosphorylase. *Extremophiles* **6:**349–350.

111. **Ryu, K., J. Kim, and J. S. Dordick.** 1994. Catalytic properties and potential of an extracellular protease from an extreme halophile. *Enzyme Microb. Technol.* **16:**266–275.

112. **Santos, H., and M. S. da Costa.** 2002. Compatible solutes of organisms that live in hot saline environments. *Environ. Microbiol.* **4:**501–509.

113. Santos, M. E. D., A. A. Martins, and M. da Costa. 1998. Thermostabilization, osmoprotection, and protection against desiccation of enzymes, cell components and cells by mannosyl-glycerate. Eur. Patent 0816509.

114. Schaffer, C., and P. Messner. 2001. Glycobiology of surface layer proteins. *Biochimie* **83**:591–599.

115. Schicho, R. N., K. Ma, M. W. Adams, and R. M. Kelly. 1993. Bioenergetics of sulfur reduction in the hyperthermophilic archaeon *Pyrococcus furiosus. J. Bacteriol.* **175**:1823–1830.

116. Schiraldi, C., I. Di Lernia, and M. De Rosa. 2002. Trehalose production: exploiting novel approaches. *Trends Biotechnol.* **20**:420–425.

117. Schiraldi, C., M. Giuliano, and M. De Rosa. 2002. Perspectives on biotechnological applications of archaea. *Archaea* **1**:75–86.

118. Schuster, B., E. Gyorvary, D. Pum, and U. B. Sleytr. 2005. Nanotechnology with S-layer proteins. *Methods Mol. Biol.* **300**:101–123.

119. Schwerdtfeger, R. M., R. Chiaraluce, V. Consalvi, R. Scandurra, and G. Antranikian. 1999. Stability, refolding and Ca2+ binding of pullulanase from the hyperthermophilic archaeon *Pyrococcus woesei. Eur. J. Biochem.* **264**:479–487.

120. Sehgal, A. C., W. Callen, E. J. Mathur, J. M. Short, and R. M. Kelly. 2001. Carboxylesterase from *Sulfolobus solfataricus* P1. *Methods Enzymol.* **330**:461–471.

121. Sehgal, A. C., and R. M. Kelly. 2002. Enantiomeric resolution of 2-aryl propionic esters with hyperthermophilic and mesophilic esterases: contrasting thermodynamic mechanisms. *J. Am. Chem. Soc.* **124**:8190–8191.

122. Sehgal, A. C., R. Tompson, J. Cavanagh, and R. M. Kelly. 2002. Structural and catalytic response to temperature and cosolvents of carboxylesterase EST1 from the extremely thermoacidophilic archaeon *Sulfolobus solfataricus* P1. *Biotechnol. Bioeng.* **80**:784–793.

123. Sellek, G. A., and J. B. Chaudhuri. 1999. Biocatalysis in organic media using enzymes from extremophiles. *Enzyme Microbiol. Technol.* **25**:471–482.

124. Serour, E., and G. Antranikian. 2002. Novel thermoactive glucoamylases from the thermoacidophilic Archaea *Thermoplasma acidophilum, Picrophilus torridus* and *Picrophilus oshimae. Antonie Van Leeuwenhoek* **81**:73–83.

125. Sleytr, U. B., and M. Sara. 1997. Bacterial and archaeal S-layer proteins: structure-function relationships and their biotechnological applications. *Trends Biotechnol.* **15**:20–26.

126. Smith, D. R., L. A. Doucette-Stamm, C. Deloughery, H. Lee, J. Dubois, T. Aldredge, R. Bashirzadeh, D. Blakely, R. Cook, K. Gilbert, D. Harrison, L. Hoang, P. Keagle, W. Lumm, B. Pothier, D. Qiu, R. Spadafora, R. Vicaire, Y. Wang, J. Wierzbowski, R. Gibson, N. Jiwani, A. Caruso, D. Bush, B. Pothier, D. Qiu, H. Safer, D. Patwell, S. Prabhakar, S. McDougall, G. Shimer, A. Goyal, S. Pietrokovski, G. M. Church, C. J. Daniels, J. Mao, P. Rice, J. Nölling, and J. N. Reeve. 1997. Complete genome sequence of *Methanobacterium thermautotrophicum* deltaH: functional analysis and comparative genomics. *J. Bacteriol.* **179**:7135–7155.

127. Stepanov, V. M., G. N. Rudenskaya, L. P. Revina, Y. B. Gryaznova, E. N. Lysogorskaya, I. Y. U. Filippova, and I. I. Ivanova. 1992. A serine proteinase of an archaebacterium *Halobacterium mediterranei*. A homologue of eubacterial subtilisins. *Biochem. J.* **285**:281–286.

128. Stetter, K. O. 1998. *Thermococcus* AV4 and enzymes produced by the same. U.S. Patent 5,714,373.

129. Story, S. V., C. Shah, F. E. Jr Jenney, and M. W. Adams. 2005. Characterization of a novel zinc-containing, lysine-specific aminopeptidase from the hyperthermophilic archaeon *Pyrococcus furiosus. J. Bacteriol.* **187**:2077–2083.

130. Sunna, A., M. Moracci, M. Rossi, and G. Antranikian. 1997. Glycosyl hydrolases from hyperthermophiles. *Extremophiles* **1**:2–13.

131. Takagi, M., M. Nishioka, H. Kakihara, M. Kitabayashi., H. Inoue, B. Kawakami, M. Oka, and T. Imanaka. 1997. Characterization of DNA polymerase from *Pyrococcus* sp. Strain KOD1 and its application to PCR. *Appl. Environ. Microbiol.* **63**:4504–4510.

132. Tanaka, T., T. Fukui, S. Fujiwara, H. Atomi, and T. Imanaka. 2003. Characterization of an exo-beta-D-glucosaminidase involved in a novel chitinolytic pathway from the hyperthermophilic archaeon *Thermococcus kodakaraensis* KOD1. *J. Bacteriol.* **185**:5175–5181.

133. Tanaka, T., T. Fukui, S. Fujiwara, H. Atomi, and T. Imanaka. 2004. Concerted action of diacetylchitobiose deacetylase and exo-beta-D-glucosaminidase in a novel chitinolytic pathway in the hyperthermophilic archaeon *Thermococcus kodakaraensis* KOD1. *J. Biol. Chem.* **279**:30021–30027.

134. Tanaka, T., T. Fukui, and T. Imanaka. 2001. Different cleavage specificities of the dual catalytic domains in chitinase from the hyperthermophilic archaeon *Thermococcus kodakaraensis* KOD1. *J. Biol. Chem.* **276**:35629–35635.

135. Taylor, I. N., R. C. Brown, M. Bycroft, G. King, J. A. Littlechild, M. C. Lloyd, C. Praquin, H. S. Toogood, and S. J. Taylor. 2004. Application of thermophilic enzymes in commercial biotransformation processes. *Biochem. Soc. Trans.* **32**:290–292.

136. Tolan, J. S. 1996. Pulp and paper, p. 327–338. *In* T. Godfrey and S. West (ed.), *Industrial Enzymes*, 2nd ed. The Macmillan Press Ltd., Basingstoke, United Kingdom.

137. Toogood, H. S., E. J. Hollingsworth, R. C. Brown, I. N. Taylor, S. J. C. Taylor, R. McCague, and J. A. Littlechild. 2002. A thermostable L-aminoacylase from Thermococcus litoralis: cloning, overespression, characterization, and applications in biotransformations. *Extremophiles* **6**:111–122.

138. Trincone, A., A. Giordano, G. Perugino, M. Rossi, and M. Moracci. 2003. Glycosynthase-catalysed syntheses at pH below neutrality. *Bioorg. Med. Chem. Lett.* **13**:4039–4042.

139. Trincone, A., A. Giordano, G. Perugino, M. Rossi, and M. Moracci. 2005. Highly productive autocondensation and transglycosylation reactions with *Sulfolobus solfataricus* glycosynthase. *Chembiochem* **6**:1431–1437.

140. Uhl, A. M., and R. M. Daniel. 1999. The first description of an archaeal hemicellulase: the xylanase from *Thermococcus zilligii* strain AN1. *Extremophiles* **3**:263–267.

141. van der Oost, J., W. G. Voorhorst, S. W. Kengen, A. C. Geerling, V. Wittenhorst, Y. Gueguen, and W. M. de Vos. 2001. Genetic and biochemical characterization of a short-chain alcohol dehydrogenase from the hyperthermophilic archaeon *Pyrococcus furiosus. Eur. J. Biochem.* **268**:3062–3068.

142. Van Hamme, J. D., A. Singh, and O. P. Ward. 2003. Recent advances in petroleum microbiology. *Microbiol. Mol. Biol. Rev.* **67**:503–549.

143. Ventosa, A., and J. J. Nieto. 1995. Biotechnological applications and potentialities of halophilic microorganisms. *World J. Microbiol. Biotechnol.* **11**:85–94.

144. Ventosa, A., J. J. Nieto, and A. Oren. 1998. Biology of moderately halophilic aerobic bacteria. *Microbiol. Mol. Biol. Rev.* **62**:504–544.

145. Vieille, C., and G. J. Zeikus. 2001. Hyperthermophilic enzymes: Sources, uses, and molecular mechanisms for thermostability. *Microbiol. Mol. Biol. Rev.* **65**:1–43.

146. **Waino, M., and K. Ingvorsen.** 2003. Production of beta-xylanase and beta-xylosidase by the extremely halophilic archaeon *Halorhabdus utahensis. Extremophiles* 7:87–93.

147. **Warren, P. V., and R. V. Swanson.** 1998. Transaminases and aminotransferases. U.S. Patent 5,814,473.

148. **Weiner, R. M.** 1997. Biopolymers from marine prokaryotes. *Trends Biotechnol.* **15**:390–394.

149. **West, S.** 1996. Flavour production with enzymes, p. 209–224. *In* T. Godfrey and S. West (ed.), *Industrial Enzymes*, 2nd ed. The Macmillan Press Ltd., Basingstoke, United Kingdom.

150. **Wheals, A. E., L. C. Basso, D. M. Alves, and H. V. Amorim.** 1999. Fuel ethanol after 25 years. *Trends Biotechnol.* **17**:482–487.

151. **Wise, K. J., N. B. Gillespie, J. A. Stuart, M. P. Krebs, and R. R. Birge.** 2002. Optimization of bacteriorhodopsin for bioelectronic devices. *Trends Biotechnol.* **20**:387–394.

152. **Xu, M. Q., M.W. Southworth, F. B. Mersha, L. J. Hornstra, and F. B. Perler.** 1993. *In vitro* protein splicing of purified precursor and the identification of a branched intermediate. *Cell* **75**:1371–1377.

153. **Xu, M. Q., H. Paulus, and S. Chong.** 2000. Fusions to self-splicing inteins for protein purification. *Methods Enzymol.* **326**:376–418.

154. **Zopf, D., and S. Roth.** 1996. Oligosaccharide anti-infective agents. *Lancet* **347**:1017–1021.

Archaea: Molecular and Cellular Biology
Edited by Ricardo Cavicchioli
© 2007 ASM Press, Washington, D.C.

Chapter 23

Archaeosome Vaccines

G. Dennis Sprott and Lakshmi Krishnan

INTRODUCTION

The discovery that hydration of polar lipids, such as phosphatidylglycerol and phosphatidylcholine, causes the spontaneous formation of liposomes (2), led to concepts for use of liposomes as biochemical tools (19), drug delivery systems (23), and antigen carrier systems for vaccines (1). A description of these important data that laid the foundation for liposome methodologies can be found elsewhere (39).

The polar membrane lipids from archaea are structurally unique (see Chapter 15) and appear to have uses in biotechnology. Predating the discovery of *Archaea* as a domain of life, novel nonsaponifiable lipids were discovered in the archaeon *Halobacterium salinarum* (52). Subsequently, it was established that the polar lipids of this extremely halophilic archaeon are based on archaeol (2,3-di-*O-sn*-phytanylglycerol), a novel lipid with a glycerol backbone linked at carbons 2 and 3 via ether bonds to two saturated, carbon-20, isopranoid chains (26) (see Chapter 15). It is now known that the archaeol lipid structure is a distinguishing and ubiquitous feature in *Archaea*. In addition, the dimer form of archaeol, caldarchaeol, and other more subtle variations to these core lipids are characteristics peculiar to each strain of archaeon (27).

The term liposome defines a bilayer arrangement of lipids and does not accurately describe archaeal lipid "liposomes" that contain caldarchaeol bipolar, membrane-spanning lipids. To account for the presence of unilayer membrane regions described below (see "Stability," below), as well as for simplicity, "archaeosome" was proposed to replace "archaeal lipid liposome" as a new term to describe the closed lipid vesicles prepared from archaeal lipids (55).

As members of the *Archaea* can thrive in extreme environments, their polar lipids are expected to be stable and therefore to be of value for a range of bio-technological applications. Archaeosomes have been developed from various archaea for use as drug delivery systems (11, 44, 49) and vaccine applications that utilize their adjuvant properties (29, 63, 66).

This chapter describes the mechanism of archaeosome adjuvants as self-adjuvanting antigen carrier systems that are taken up by specific receptor-mediated endocytosis to promote both CD4$^+$ and CD8$^+$ T-cell responses. The effects of lipid structure on the immune response are also reviewed.

PREPARATION OF ARCHAEOSOMES

Archaeal lipids possess several features that make them ideal for the preparation of archaeosomes. The first is the inherent stability of the polar lipids, allowing long-term storage of these lipids in air without oxidation or other chemical changes. This stability results from the fully saturated nature of the isopranoid chains and the stable ether linkages between the chains and the glycerol backbone. Exceptions to this rule of archaeal lipid saturation may be found, as the polar lipids of some cold-adapted archaea exhibit unsaturation of their isoprenoid chains (20, 40), and these sources have been excluded from most applied studies. The archaeal *sn*-2,3 stereochemistry versus *sn*-1,2 found in glycerolipids of the *Bacteria* and *Eucarya* may influence their susceptibility to enzymatic attack (27) and contribute to stability properties. Second, archaeal lipids form archaeosomes over physiological temperature ranges, allowing preparation of vaccines at ambient temperatures. Formation of liposomes from nonarchaeal lipids must be performed above the phase transition temperature, where heat sensitivity of the active ingredient being entrapped could be an issue. Third, once formed, archaeosomes of 50- to 250-nm diameters remain sus-

G. Dennis Sprott and Lakshmi Krishnan • Institute for Biological Sciences, 100 Sussex Drive, National Research Council of Canada, Ottawa, Ontario, K1A 0R6, Canada.

pended indefinitely and resist fusion or aggregation over long storage periods. The stability of archaeosome membranes accounts for the long retention times for entrapped compounds (11, 45, 49).

A historical perspective on formation of archaeosomes has been described in a review (47). Methods to prepare archaeosomes are similar to the methods used to prepare liposomes from nonarchaeal lipids (60). In brief, the preparation of archaeosomes involves growing the archaeon of interest, extracting total lipids, and precipitating the total polar lipid fraction (the bulk of the lipid) with cold acetone. Drying an aliquot of the total polar lipids and hydrating in the presence of antigen forms multilamellar archaeosomes containing entrapped antigen. In general, hydration occurs readily with lipid extracts high in archaeols but is more difficult and occurs with lower archaeosome yield if the content of caldarchaeols is exceptionally high (e.g., total polar lipids from *Thermoplasma acidophilum* or *Sulfolobus acidocaldarius*). The size of archaeosomes is usually reduced by brief sonication in a sonic water bath or by extrusion through membranes of defined porosity (3) prior to removal of any nonentrapped antigen, usually by centrifugation. The final steps are filtration through a sterile 0.45-μm-pore-size filter and quantification of entrapped antigen. Archaeosome size is measured in a submicron particle sizer and is typically 50 to 250 nm. Each archaeon has its own unique, total polar lipid composition that, in turn, imparts unique properties to the resulting archaeosomes.

To avoid the necessity of purifying lipids for archaeosome preparations and to potentially achieve multiple receptor interactions, most studies that have used archaeosomes as adjuvants included the total polar lipids from a chosen archaeon. Using total polar lipids also has the advantage of producing a mixture of polar lipids that form stable archaeosomes (10).

STABILITY

An important consideration relating to the stability of archaeosomes is the proportion of archaeol to caldarchaeol lipids in the preparation. This in turn may have profound effects on vaccine adjuvant properties.

Bipolar caldarchaeol lipids span lipid films and intact cell membranes with a polar group facing each side (21, 33) (see Chapter 15). Archaea, such as *H. salinarum*, are incapable of caldarchaeol synthesis. Many other archaea (e.g., *T. acidophilum*) generate a mixture of archaeols and caldarchaeols. As a result, three different membrane structures occur in archaeosomes depending on the archaeal lipid source (Fig. 1). A bilayer may form from archaeal lipid ex-

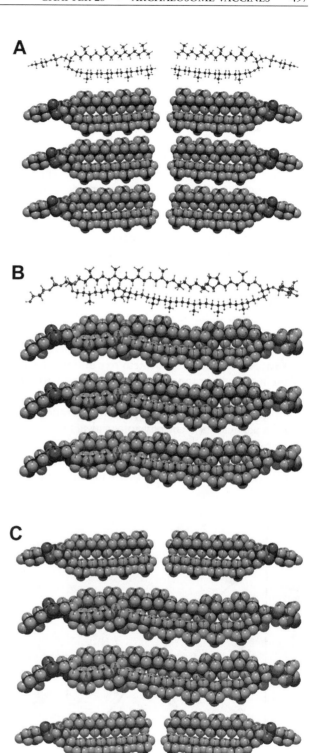

Figure 1. Archaeol and caldarchaeol polar lipid membrane models. (A) Model lipids are archaetidylglycerol arranged to depict a bilayer archaeosome membrane. Regular methyl branches of the isopranoid chains are shown. (B) The membrane is depicted as a unilayer composed of the main polar lipid of *T. acidophilum* (64). (C) This model of an archaeosome membrane consists of a mixture of A and B polar lipids.

tracts containing 100% archaeols (Fig. 1A). A membrane unilayer is typical for lipids from *T. acidophilum* with 90% caldarchaeol (Fig. 1B), while a combination of bilayer and unilayer arrangements may be expected for the total polar lipids from archaea such as *Methanobrevibacter smithii* (Fig. 1C).

Freeze-fracture results corroborate the membrane arrangements predicted for archaeosomes prepared from various total polar lipid sources (Fig. 2). The frequency of intramembrane fractures occurring along the hydrophobic plane of the membrane of archaeal cell plasma membranes (and archaeosomes) correlate with the amount of caldarchaeols present (Fig. 2) (4). Intramembrane fractures were essentially absent at a content of caldarchaeols greater than 50%. During freeze fracturing, intramembrane particles were seen when fractures occasionally occurred along the hydrophobic membrane domain. However, this only occurred in cases where archaeosomes consisted of mixtures of archaeols and caldarchaeols. As these archaeosomes were protein free and the density of the intramembrane particles correlated with caldarchaeol content, it was concluded that caldarchaeol complexes must form these (intramembrane) lipid raftlike particles (4). The physiological significance of these particles is unknown.

Archaeosomes consisting largely of caldarchaeol lipids are unusually stable to thermally induced leak-age of solutes, and exhibit low ion permeabilities (7, 16, 49). The thermal stability during autoclaving of archaeosomes with different proportions of caldarchaeol lipids was examined (Fig. 3). The method used was to first load archaeosomes with self-quenching amounts of carboxyfluorescein (CF). An increase in fluorescence occurred as CF leaked from the archaeosomes during autoclaving and diluted into the suspending fluid. It was found that CF leakage declined as the caldarchaeol content increased, reaching a plateau at about 50%. This plateau at about 50% caldarchaeol content is similar to the caldarchaeol content that prevented intramembrane cleavage during freeze fracturing (Fig. 2). Thermal stability was higher for *Methanococcus jannaschii* archaeosomes than for *M. smithii* archaeosomes, despite similar caldarchaeol content, but this is likely to reflect the additional stabilizing effect of macrocyclic archaeols from *M. jannaschii*. These data clearly indicate that caldarchaeol membrane-spanning lipids enhance thermostability.

Permeability has been examined for several total polar lipid compositions of archaeosomes, compared with liposomes prepared from *Escherichia coli* lipids and from commercial 1,2-di-*O*-phytanyl-*sn*-glycero-3-phosphocholine (ester linkages to phytanyl chains) (Fig. 4) (35). These and other data (35) support the conclusion that the ether bonds of archaeal lipids (versus ester bonds), and particularly caldarchaeol lipids, decrease the proton permeability of archaeosomes (Fig. 4A). Furthermore, limiting the mobility of

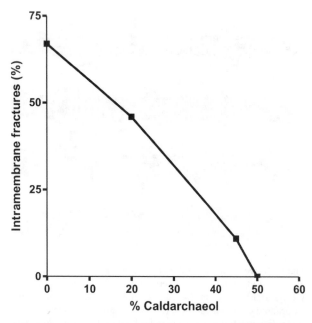

Figure 2. Correlation between the caldarchaeol content of archaeal membranes and the frequency of intramembrane fractures observed by freeze-fracture electron microscopy. Similar data were obtained for freeze-fractured archaeosomes prepared from the total polar lipids extracted from the same archaea. Modified from the *Journal of Bacteriology* (4) with permission of the publisher.

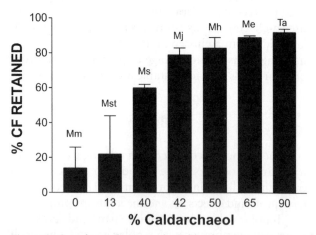

Figure 3. Correlation between the caldarchaeol content of archaeosomes and their thermal stabilities. Carboxyfluorescein (CF) was entrapped within archaeosomes made of total polar lipids from *M. mazei* (Mm), *Methanosphaera stadtmanae* (Mst), *M. smithii* (Ms), *M. jannaschii* (Mj), *Methanospirillum hungatei* (Mh), *Methanobacterium espanolae* (Me), and *T. acidophilum* (Ta). Retention of CF was measured following autoclaving for 15 min at 121°C. Modified from the *Canadian Journal of Microbiology* (12) with permission of the publisher.

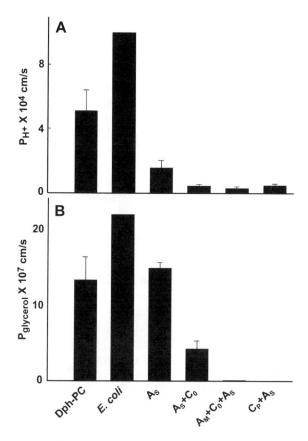

Figure 4. Proton and glycerol permeabilities of liposomes and archaeosomes. Permeability rates were measured in lipid vesicles prepared from the ester lipid diphytanylphosphatidylcholine (Dph-PC), total polar lipids of *E. coli* (*E. coli*), *H. salinarum* (As), *M. smithii* (As + Co), *M. jannaschii* (Am + Co + As), and *T. acidophilum* (Cp + As). Symbols indicate archaeols (As), caldarchaeols (Cs), macrocyclic archaeols (Am), and caldarchaeols with cyclopentane rings (Cp). Reproduced from the *Journal of Biological Chemistry* (35) with permission of the publisher.

hydrocarbon chains (caldarchaeols and macrocyclic archaeols) plays a role in decreasing permeability to glycerol (Fig. 4B) and other small molecules (35).

The extreme environmental condition in which an archaeon thrives might be expected to correlate with the stability of their archaeosomes. However, this is not always the case. For example, archaeosomes prepared from the polar lipids of alkaliphilic *Natronobacterium magadii* were sensitive to alkaline pH (11). As described above (Fig. 3), thermostability of archaeosomes correlates with caldarchaeol content. However, some hyperthermophiles, such as *Methanopyrus kandleri*, lack, or have a low content of, caldarchaeol lipids (53), indicating from a physiological perspective that the presence of a high amount of caldarchaeols is not necessarily a prerequisite to extreme thermophily. In addition, the total polar lipids from extremely thermophilic archaea do

not appear to be structurally very different from those from other archaea (53). In contrast to these examples, the lipids from extremely halophilic archaea form archaeosomes in salt concentrations up to 4 M, whereas liposomes from *E. coli* (67) or archaeosomes from total polar lipids of nonhalophilic archaea do not form in high-salt concentrations. The mechanism of salt stability for these archaeosomes is correlated with the content of the diacidic lipid, archaetidylglycerol phosphate-O-methyl (see Chapter 15), a lipid that is unique to and abundant in extremely halophilic archaea (65).

INTERACTION WITH ANTIGEN-PRESENTING CELLS

Macrophages, and especially dendritic cells (DCs), are the professional antigen-presenting cells (APCs) that initiate the immune response and are the critical cell types for adjuvant action. It has been known since the 1970s that liposomes are taken up by phagocytosis and adjuvant the immune response by antigen delivery (1). The concept of using archaeal lipids as adjuvants to deliver antigens with higher efficiency than conventional liposomes stemmed from the stability properties of archaeosomes. However, lower efficiency was also considered to be possible if the archaeosomes proved to be too stable to release their antigen cargo.

Much of the recent activity in adjuvant design has been centered on discovery of molecules associated with pathogens (pathogen-associated molecular patterns) that are agonists for Toll-like receptors found on APCs (51). Archaea are not known to be associated with humans except in anaerobic niches such as the intestinal tract, where *M. smithii* and *Methanosphaera stadtmanae* generate methane as part of the normal microbiota (36). Because archaea have not been linked to pathogenic activity and lack endotoxins (6, 22), interaction of archaeal molecules with mammalian receptors found on APCs, such as the Toll-like danger receptors designed to recognize and prompt the immune response to the presence of pathogens, would not be expected. Therefore it was a serendipitous discovery that archaeal lipids proved to be good adjuvants capable of activating the immune system of mammals.

Endocytosis

In macrophages cultures, the uptake of archaeosomes derived from several lipid compositions is 3 to 53 times higher than the uptake of conventional liposomes (66). Prior to this study, uptake of archaeosomes by mammalian cells had not been assessed. The results were unexpected and highlighted the potential utility of archaeosomes as antigen carriers. Non-phagocytic cell cultures took up comparatively little of the archaeosomes, indicating that uptake was through a phagocytic mechanism. Cytochalasins, inhibitors of phagocytosis, were used to confirm that uptake was by phagocytosis and to show that following internalization the structural integrity of the archaeosomes was lost over time (24, 66). This observation was important as it indicated that the archaeal delivery system was not too stable to release the antigen for processing and presentation. Incubation of fluorescent archaeosomes with macrophages and DCs resulted in the rapid accumulation of hot spots of archaeosomes (Fig. 5), indicating internalization of many archaeosomes within each phagosome.

Archaeosomes prepared from the total polar lipids of *M. smithii* have proven to be especially strong adjuvants. Recently, several reasons for this activity have been elucidated. First, the polar lipids are approximately 60% archaeols and 40% caldarchaeols,

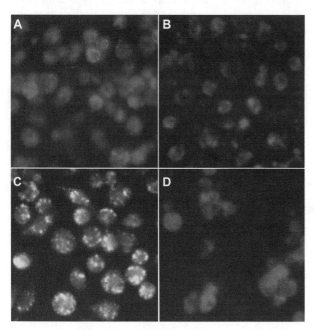

Figure 5. (*See the separate color insert for the color version of this illustration.*) Uptake of fluorescent archaeosomes by phagocytic cells. Archaeosomes composed of total polar lipids were prepared either by incorporating a small amount of the fluorescent lipid rhodamine-phosphatidylethanolamine (62) or by entrapping 1.5 mM carboxyfluorescein (66). Uptake was performed in 1 ml of RPMI medium added to 0.5 million adhered cells (62). Panels show 30-min uptakes: (A) *M. smithii* rhodamine-archaeosomes (100 μg) by thioglycollate-activated mouse peritoneal macrophages; (B) uptake of *M. smithii* rhodamine-archaeosomes (25 μg) by bone marrow-derived DCs; (C) uptake of *M. mazei* carboxyfluorescein-archaeosomes (40 μg) by macrophages culture J774A.1; and (D) uptake of *H. salinarum* rhodamine-archaeosomes (100 μg) by thioglycollate-activated mouse peritoneal macrophages.

thereby imparting enhanced stability. Second, the total polar lipids are exceptionally high in lipids with phosphoserine headgroups (Fig. 6). Evidence for endocytosis (receptor mediated phagocytosis) of *M. smithii* archaeosomes via a phosphatidylserine (PS) receptor present on APCs is described below (see "Presentation pathway," below). Third, the total polar lipids of *M. smithii* form archaeosomes readily, and good yields are recovered following filter sterilization. Studies examining the mechanism of action of archaeosome adjuvants have tended to use archaeosomes prepared from the total polar lipids of *M. smithii*.

Costimulation

Two signals are required to activate T cells. The first is antigen presentation in the context of MHC, and the second is costimulation of the specific T cell recognizing the presented antigen. Consequently, for an adjuvant to effectively augment an immune response, not only must the antigen be delivered appropriately to APCs, but it is also imperative that costimulation occurs. Dendritic cells mature and upregulate expression of key costimulatory molecules on their surface, thereby providing the important second signal for T-cell activation. Activated mature DCs also secrete specific cytokines that orchestrate the immune response and direct the CD4$^+$ T cells toward a Th1- or Th2-type response (37).

M. smithii archaeosomes provide a strong costimulation signal to macrophages (Fig. 7) and DCs (30). Conventional liposome composition (PC/PG/cholesterol) provides little costimulation (Fig. 7), consistent with a need to incorporate a coadjuvant to improve its vaccine potential (48). The striking upregulation of MHC class II, CD80, and CD86 observed with *M. smithii* archaeosomes, compares favorably with that obtained with potent lipopolysaccharide from *E. coli*. A similar striking upregulation of expression of costimulatory marker CD40 occurs in mouse bone marrow-derived DCs exposed to *M. smithii* archaeosomes (24). Furthermore, DCs were activated by *M. smithii* archaeosomes to secrete moderate amounts of cytokines interleukin 12 (IL-12) and tumor necrosis factor (TNF) (24, 30). Collectively, these studies highlight the value of archaeosomes when compared with nonarchaeal liposome preparations.

In contrast to archaeosomes from *M. smithii*, there is relatively little known about costimulatory effects of archaeosomes prepared from the total polar lipids of other archaea. However, it is known that costimulatory molecules are also upregulated in macrophages and DC cultures exposed to *H. salinarum* or *T. acidophilum* archaeosomes (L. Krishnan and G. D. Sprott, unpublished data). Furthermore,

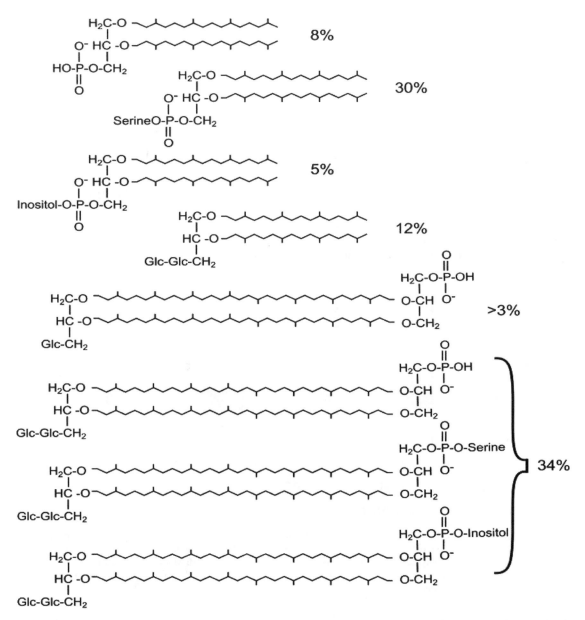

Figure 6. Structures and abundance of the polar lipids found in the total polar lipids of *M. smithii* (54). Phosphoserine head-groups are abundant in both archaeol (2,3-di-*O-sn*-phytanylglycerol) and its dimer, caldarchaeol, lipids. Other lipids present in minor amounts are not shown.

many archaeosome types evoke moderate amounts of IL-12 secretion by APCs (62). As the headgroups greatly vary in these different archaea, this suggests that the core lipid structure that is common to archaea may be important for the interaction of archaeosomes with APCs.

Presentation Pathway

Antigen presentation may be assessed by an in vitro assay designed to monitor the extent to which ovalbumin-derived MHC class I epitope (SIINFEKL)

is presented on the surface of an APC. For this to occur by the classical pathway, the ovalbumin entrapped within archaeosomes must undergo endocytosis by the DC or macrophages culture and be delivered to the cytosol for processing in the proteasome. In brief, the major MHC class I peptide SIINFEKL, so generated, may then be loaded on MHC class I molecules and the complex delivered to the cell surface for presentation to CD8$^+$ T cells. Thus the in vitro assay is based on incubating APCs with antigen entrapped in archaeosomes, allowing intracellular processing, and then incubating them with CD8$^+$

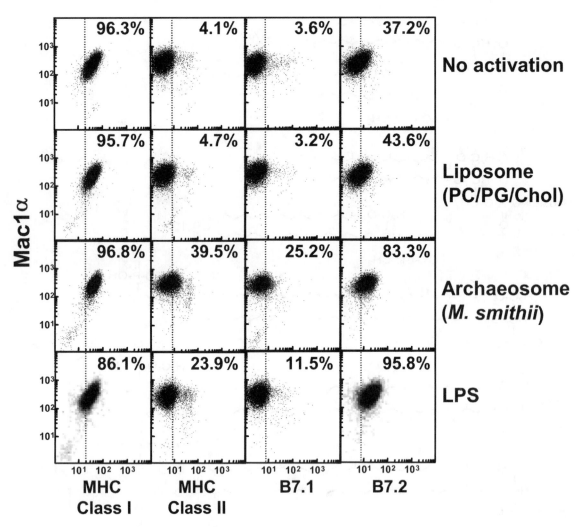

Figure 7. Upregulation of cell surface molecules on J774A.1 macrophages treated with archaeosomes. Macrophages were treated for 24 h, with no activator, 25 μg of antigen-free liposomes ml⁻¹ (PC/PG/cholesterol), 25 μg of antigen-free archaeosomes ml⁻¹, or 10 μg of lipopolysaccharide ml⁻¹. Cells were then double stained with Mac1α-PE (macrophages marker) and one of the cell surface markers shown. Data were acquired by flow cytometry. Reproduced from the *Journal of Immunology* (30) with permission of the publisher.

T cells that have the specific T-cell receptor to bind the presented MHC class I-SIINFEKL epitope. Binding activates the CD8⁺ T cells to secrete the cytokine interferon-γ (IFN-γ). Quantification of IFN-γ production by the T cells is a direct indication of the ability of archaeosomes to deliver ovalbumin to the cytosol for MHC class I loading (Fig. 8). Synthetic SIINFEKL added to a separate aliquot of the APC culture will bind to MHC class I molecules exogenously and is equated to 100% loading. In this assay, soluble ovalbumin fails to activate T cells, as it cannot gain access to the MHC class I pathway, and serves as a negative control.

The pathway of antigen presentation in DCs and macrophages has been assessed in some detail for ovalbumin entrapped in *M. smithii* archaeosomes and found to occur via a series of classical steps for both MHC class I and class II processing (Fig. 8). For *M. smithii* archaeosomes, the high content of archaetidylserine (phosphoserine-archaeol) and phosphoserine-caldarchaeol lipids indicates that the mechanism of endocytosis may be promoted by a PS receptor. PS receptors are commonly found in DCs and macrophages. Their role in clearing cells undergoing apoptosis has been studied in macrophages (25), but not their role in influencing antigen presentation. Several lines of evidence support a PS receptor-mediated endocytic mechanism. First, Annexin V labeling experiments indicate that PS headgroups in *M. smithii* archaeosomes are surface exposed and available for receptor interaction (62). Second, cytochalasin inhibitors of phagocytosis prevent uptake of these archaeosomes. Third, phosphatidylserine liposomes compete for uptake of *M. smithii* archaeo-

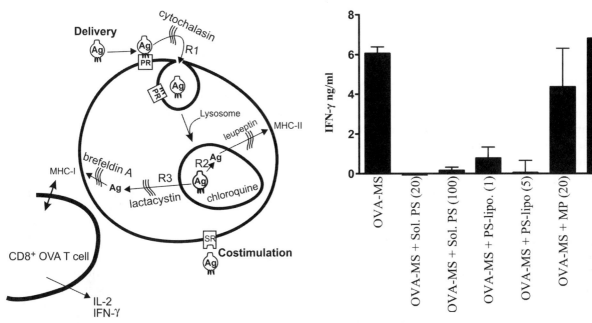

Figure 8. Scheme depicting processing of ovalbumin encapsulated in archaeosomes of *M. smithii* by an APC. Sites of inhibitor action are shown. Specific peptide fragments of the Ag (ovalbumin) are presented by MHC class I or MHC class II to CD8$^+$ T cells and CD4$^+$ T cells, respectively. A CD8$^+$ T cell specific to the MHC class I ovalbumin peptide complex is shown docking with the APC. Docking results in activation and excretion of IL-2 and IFN-γ. SR, signaling receptor; PR, phagocytosis receptor; R1, R2, and R3 are steps 1 to 3, the rate of which depend on the archaeosome composition and APC.

Figure 9. PS receptor-mediated endocytosis of *M. smithii* archaeosomes containing ovalbumin. Processing of ovalbumin and presentation of peptide by MHC class I is quantified by assay of IL-2 or IFN-γ by activated CD8 T cells, as described in the legend to Fig. 8. Competition for uptake of archaeosomes was assayed for soluble phospho-L-serine (Sol. PS), phosphatidylserine liposomes (PS lipo), or mannopentaose (MP), at the μg concentrations shown in parentheses. Reproduced from the *Journal of Immunology* (24) with permission of the publisher.

somes, whereas liposomes or archaeosomes lacking the phosphoserine headgroup are less effective (24, 62). This is illustrated by the results of the assay just described that was designed to measure antigen processing by DCs incubated with ovalbumin-archaeosomes (Fig. 9). Inhibition of antigen processing by either soluble phosphoserine or phosphatidylserine liposomes indicates competition for the endocytosis of *M. smithii* archaeosomes. Specificity for the PS receptor was indicated by lack of inhibition upon blocking either the mannose receptor by mannopentaose, or the Fc receptors by specific antibody (24). As the PS receptor is expressed primarily on APCs, engaging this receptor for antigen delivery may result in efficient antigen delivery for processing by the immune system.

Inhibition of MHC class I processing by chloroquine and monensin (inhibitors of acidification of the phagolysosome) indicates that acidification is a necessary event to promote the cytosolic translocation of the antigen from the phagolysosome (24). Inhibition of proteases in the phagolysosome by leupeptin and antipain did not interfere in MHC class I processing, whereas inhibition of cytosolic proteasome

processing by inhibitors such as lactacystin blocked MHC class I processing (Fig. 8). The implication of this finding is that the protein antigen must be processed for MHC class I presentation, subsequent to translocation to the cytosol. Furthermore, DCs from TAP (transporter associated with antigen processing) deficient mice did not process antigen carried in *M. smithii* archaeosomes, for MHC class I presentation (24). All of these findings support a classical MHC class I pathway for antigen entrapped in *M. smithii* archaeosomes. In addition, part of the antigen delivered was directed to MHC class II processing (24), indicating that MHC class I and II compete for antigen. Clearly, the properties of the specific archaeosome type used in a vaccine may influence this competition to favor either one or other pathway, and may be used to bias processing by either MHC class I (CD8$^+$ T-cell response) or class II (CD4$^+$ T-cell response).

IMMUNE RESPONSES

CD4$^+$ T-Cell Response

Antigen presentation in the context of MHC class II results in interaction with CD4$^+$ T cells and, depending on the cytokine environment, differentiate primarily toward either Th1 cells secreting type 1 cytokines (cell-mediated response) or Th2 cells secreting type 2 cytokines (antibody response). Initial studies revealed that a strong antiprotein antibody response developed in mice that had been immunized with various archaeosome types that contained a protein antigen (63). This response was higher than that obtained using conventional liposomes. In fact, the adjuvant activity decreased as the proportion of conventional lipids to archaeosomes was increased (28). This may have been caused by the lower stability and leakage of entrapped antigen in vivo, as adjuvant activity demands that antigen be retained within the archaeosome/liposome for delivery, or loss of the costimulation signal.

Further studies revealed that the archaeal lipid composition used to entrap the antigen has a marked effect on the antibody titers measured in the sera of immunized mice (Fig. 10). Highest mean titers (6 to 8 mice/group) occurred with archaeosomes prepared from the total polar lipids of *Halobacterium halobium* and were significantly higher than all other ar-

chaeosome types at day 10 ($P < 0.05$). Following a second injection, the antibody mean titer for the *H. halobium* archaeosome adjuvant was not significantly different from those obtained for *Halococcus morrhuae* or *Methanosarcina mazei* archaeosomes by day 49, but was significantly higher ($P < 0.05$) compared with either *M. smithii* or *M. jannaschii*. Significant differences were seen at 49 days between higher titers for *M. mazei* than for *M. smithii*, and for *M. smithii* than for *M. jannaschii*. It is tempting to link the somewhat lower titers obtained for *M. smithii*, and especially the low titers for *M. jannaschii*, with the higher archaeosome stability. Caldarchaeol lipids account for about 40 mol% of the polar lipids in *M. smithii* (54) and, in combination with macrocyclic archaeols, about 85 mol% in *M. jannaschii* (when cells are grown at 65°C) (58). A higher stability and lower antigen release into the phagolysosome (Step R2 in Fig. 8) may account for the reduced antigenicity of archaeosomes from these archaea (Fig. 10).

Structural differences in the total polar lipid headgroups present in the various archaeosome types may influence the magnitude of MHC class II re-

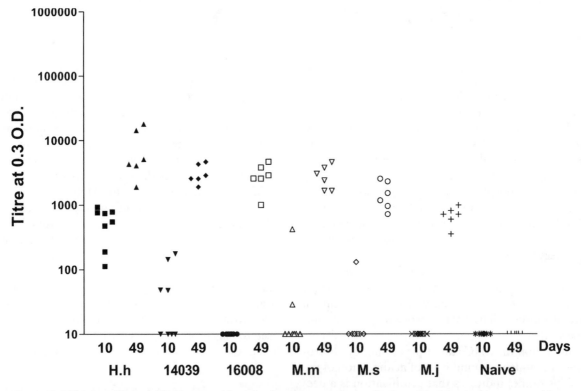

Figure 10. Influence of the polar lipid composition of archaeosomes on humoral adjuvant activity. Archaeosomes with entrapped ovalbumin were prepared from the total polar lipids extracted from *H. halobium* (H.h), *Halococcus morrhuae* strains 14039 and 16008, *Methanosarcina mazei* (M.m), *M. smithii* (M.s), and *Methanococcus jannaschii* (M.j). Injections, given subcutaneously at zero and 3 weeks, contained 15 μg ovalbumin. Titers of antiovalbumin antibody in sera are given for 10-day and 49-day bleeds. Titers in sera from mice immunized with 15 μg ovalbumin (no adjuvant) were below 10. Each data point is for a different mouse. Reproduced from *Archaea* (62) with permission of the publisher.

sponses. Headgroup structural features may be especially important in receptor-mediated endocytosis and signaling mechanisms. Several lines of evidence illustrate the importance of archaeal lipid headgroups to humoral adjuvant activity. Lipids from the extremely halophilic archaea (*H. halobium* and *H. morrhuae* strains 14039 and 16008) differ from the other archaeosome types by having a high content of archaetidylglycerol phosphate (PGP-CH_3) and sulfated-glycolipids (62). N-Acetylgalactosamine–4-SO_4 is recognized by the mannose receptor of phagocytic cells (17), and sulfation of archaeal glycolipids may promote interaction with this or other receptors. Antibody titers were high for *M. mazei* archaeosome adjuvant (Fig. 10), and these archaeosomes are composed almost exclusively of phospholipids. This suggests that in the presence of archaetidylserine or other bioactive phospholipids, archaeal glycolipids are not essential for an antibody response. *M. jannaschii* lipid extracts are high in glycolipids (the dominant sugar is N-acetylglucosamine) (62), and this archaeosome type produced a relatively weak antibody response, indicating that relatively low adjuvant activity is induced by this headgroup. Phagocytosis is promoted when conventional liposomes contain phosphatidylserine (34). Therefore, the presence of the phosphoserine archaeal lipids found in *M. smithii*, *M. mazei*, and *M. jannaschii* might account for some of the adjuvanticity observed for these archaeosomes.

Surprisingly, *M. smithii* archaeosomes induced both Th1 (IFN-γ) and Th2 (IL-4) cytokines in spleen cells from immunized mice, in contrast to conventional liposomes and Alum that induced only IL-4 (28). The finding that this archaeosome was capable of a mixed adjuvant activity may be highly beneficial for the development of new vaccines.

CD8$^+$ T-Cell Response

A CD8$^+$ T-cell response is an adaptive immune response largely evolved for protection against viral or other intracellular infections. As exogenous antigens generally do not gain access to the cytosol of APCs for processing, they do not promote a CD8$^+$ T-cell response. However, a cytotoxic T-cell (CTL) response is required for vaccines to be protective against many diseases, to clear host cells infected by intracellular pathogens or in a cancerous state. Several adjuvant systems are being developed to try and alleviate this problem and to deliver antigen for MHC class I processing (5, 38). Archaeosomes may serve well for this purpose as they are potent CTL adjuvants for entrapped protein and peptide antigens (29).

M. smithii archaeosomes deliver protein antigens from their location in phagolysosomes to the cytoso-

lic, classical processing pathway, and load MHC class I molecules with antigen-derived peptides. This process results in development and proliferation of cytotoxic T cells capable of killing target cells that express these same peptide epitopes on their surface. Various types of archaeosomes (containing ovalbumin as the test antigen) have been compared as CTL adjuvants in mice. Although all the archaeosome types demonstrate potential to adjuvant a CTL response, a single subcutaneous injection produced a primary (10 day) CTL response that was most strong for *H. halobium*, *H. morrhuae* 14039, and *M. smithii* archaeosomes (Fig. 11). The polar lipids from *H. morrhuae* 14039 are remarkably similar to strain 16008, except for the presence of an unidentified glycolipid in strain 14039 (62). Archaeosomes from strain 14039 were superior to 16008 in promoting a CTL response and in activating DCs to secrete IL-12, implicating the unknown glycolipid as an immune activator (62).

Immune Memory

To be effective, vaccine adjuvants must not only elicit strong primary immune responses to the protective antigens but also achieve long-lasting immunity with strong memory recall. Most published studies only focus on the short term. The importance of long-term studies was demonstrated recently using a liposome vaccine prepared from the total polar lipids of *Marinococcus* (*Planococcus* H8), a member of the *Bacteria* (56). This liposome type initially appeared to have promise as it activated DCs and produced a good primary CTL response, but unfortunately the response was only short lived. A liposome vaccine prepared from the total polar lipids of *Bacillus firmus* was a weak adjuvant for both antibody and CTL responses (56), further highlighting the value of archaeal lipids for vaccine applications.

Archaeosomes containing bovine serum albumin as test antigen produced long-term antibody titers in mice (28). Two total polar lipid archaeosome types were tested. *H. salinarum* archaeosomes resulted in a strong primary response, long-lasting titers over the life of the mice, and memory recall. Such a long-lasting response was not expected as these archaeosomes lack caldarchaeol lipids and was anticipated to contribute less antigen persistence than caldarchaeol-containing adjuvants. For *T. acidophilum* archaeosomes, the primary response was not as strong as for the *H. salinarum* archaeosomes, but titers were maintained, and a strong memory recall response occurred. Alum is the antibody-adjuvant approved for human use, and it produced a response comparable to *H. salinarum* archaeosomes, although memory recall was not as strong (28).

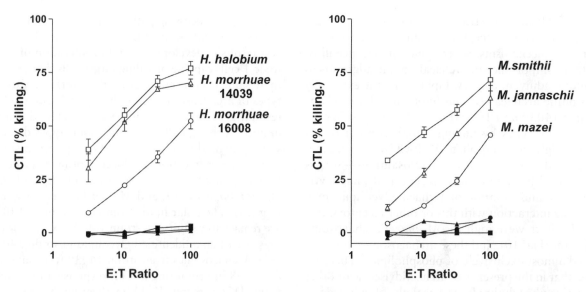

Figure 11. Influence of the polar lipid composition on cytotoxic T-cell responses (CTL) to ovalbumin entrapped in archaeosomes. C57Bl/6 mice were immunized subcutaneously with 15 μg ovalbumin entrapped in various total polar lipid archaeosomes. Ten days later, CTL activity was measured in splenic cell cultures. Killing of specific target cells (open symbols) by effector cells and lack of killing of nonspecific targets (closed symbols) are shown. The ratio of effector to target cells is shown as the E:T ratio. Reproduced from *Archaea* (62) with permission of the publisher.

When using lipid adjuvants, long-term CD8+ T-cell responses may be more difficult to achieve than B-cell (antibody) memory responses (56). The longevity of CTL responses in mice immunized with archaeosomes prepared from several archaea were examined over the long term (up to 50 to 65 weeks following vaccination) (32). In contrast to formulations of liposomes, a long-term memory response was observed for *M. smithii* and *T. acidophilum* archaeosomes. CTL memory was less evident for the other archaeosome preparations, which all lack membrane-spanning lipids, and it is tempting to speculate that the caldarchaeols promote these unusually long-lasting CTL responses. However, in an apparent contradiction, caldarchaeols and macrocyclic archaeols are present in *M. jannaschii* archaeosomes, and these did not contribute to long-term CTL responses. The relatively weak priming of immune responses by *M. jannaschii* archaeosomes may have contributed to the poorer long-term responses. Furthermore, the reason for an apparent absence of long-term CTL responses in mice immunized with some archaeosomes may relate to variable factors such as antigen loading, antigen dose, or vaccination regimen.

VACCINES

Data using test antigens indicate the strong potential for archaeosome vaccines to achieve immunity against intracellular bacteria. Because *Listeria monocytogenes* is an intracellular pathogen that is typically cleared by the host CD8+ T-cell immunity (42), this model has been used as a test for potency of archaeosome vaccines. The protective CTL epitope of the key immunogenic protein of *Listeria*, Listeriolysin, is known. This peptide was synthesized and coupled to palmitoyl chains to promote high-efficiency encapsulation (14). Mice immunized with the *M. smithii* vaccine developed a strong CTL response that resulted in specific immunity following infection of the mice with *L. monocytogenes*. Protection did not occur with mice immunized with antigen only, or with archaeosomes lacking antigen. Furthermore, prolonged protection was demonstrated (Fig. 12). Similar protection was observed with *T. acidophilum* archaeosomes, somewhat less effective with *H. salinarum* archaeosomes, and the least amount of protection with conventional liposomes (14).

Archaeosomes also have potency and utility as adjuvants for cancer vaccines (31, 32), where a CD8+ T-cell response is required for protection or regression of tumors. Solid tumor and metastasis mouse models have been used to evaluate both the therapeutic and prophylactic utility of archaeosomes. Using ovalbumin as the test antigen, effective protection of mice vaccinated with *M. smithii* archaeosomes was demonstrated in both solid tumor and metastasis models. Protection was dependent on the presence of CD8+ T cells and required IFN-γ (31). Protection also occurred in IL-12 knockout mice (31). These data indicated strong adjuvant activity of the *M. smithii* ar-

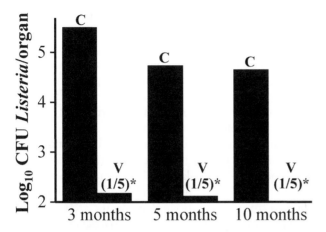

Figure 12. Protection of archaeosome-immunized mice against infection by the facultative intracellular bacterium, *Listeria monocytogenes*. BALB/c mice were vaccinated subcutaneously on days 0, 21, and 42 with 12.5 µg of a dipalmitoylated peptide entrapped in *M. smithii* archaeosomes. After 3, 5, and 10 months, mice were challenged with live *L. monocytogenes*, and the bacteria in spleens were enumerated 3 days later. C, mice not immunized; V, vaccinated mice. The number of spleens in each group of 5 mice that were still infected are shown as (*x*/5). Modified from *Vaccine* (14) with permission of the publisher, with the exception of the 10-month data (J. W. Conlan, unpublished data).

chaeosomes (31). Furthermore, in therapeutic tumor models, solid tumors regressed following the injection of *M. smithii* archaeosomes loaded with antigen. Some regression also occurred with injection of antigen-free *M. smithii* archaeosomes, but not with liposomes made from phosphatidylcholine, phosphatidylglycerol, and cholesterol (31). *T. acidophilum* antigen-free archaeosomes also had tumor-regressing properties, but *H. salinarum* archaeosomes did not (32). This intriguing effect of the *M. smithii* archaeosomes might relate to the high content of archaetidylserine. This is consistent with studies showing that liposomes containing phosphatidylserine damage the vascular system of tumors (9). However, this mechanism of tumor regression would not explain how the *T. acidophilum* archaeosomes exert their effect, as archaetidylserine is absent.

SAFETY

Archaea are nonpathogenic organisms (6, 22), not associated with endotoxin or other toxic metabolites, and it may therefore be anticipated that archaeosomes would be safe. The main polar lipid of *T. acidophilum* (64) has been administered to mice by intraperitoneal and oral routes and found to be free of toxicity (18). Furthermore, a series of total polar lipid archaeosome compositions were tested to as-

sess the possibility of toxicity related to their use in mammals (43, 46). In mice, significant levels of anti-archaeosome antibodies were not detected, and reactions or toxicity were absent in a standard series of toxicity tests (46). A similar conclusion was reached in studies using mice that were administered a high intravenous or oral dose of archaeosomes (43). In trials using hundreds of mice and various strains, ages, and sex, effects of toxicity have never been documented, suggesting that archaeosomes will prove to be a safe adjuvant suitable for trials in humans.

PERSPECTIVE: THE NEXT FIVE YEARS

The utility of archaeal polar lipids as adjuvants for mammalian vaccines is now well established. The stable physical properties of the polar lipids, the hydration properties of the lipids that result in archaeosome formation, the stability of archaeosomes, and the specific interactions that occur with APCs all contribute to their adjuvant potential. Patents are issued in various countries, including PCT applications covering the preparation of archaeosomes from total polar lipids of archaea, to those describing their adjuvant properties (59, 61). The intellectual property on adjuvant activity includes archaeosomes prepared from purified or synthetic archaeal lipids, and archaeal lipids mixed with conventional lipids. Although animal trials indicate that archaeosomes are safe with no toxicity issues, human clinical trials have yet to commence.

Archaeosomes act as mixed adjuvants augmenting Th1 and Th2 arms of immunity, as well as the CTL response. Delivery of antigen for presentation by MHC class I and MHC class II are competing pathways that are favored depending on the lipid composition of archaeosomes. For example, the total polar lipids from *H. salinarum* favor MHC class II presentation and antibody response, whereas the polar lipids from *M. smithii* and *T. acidophilum* favor MHC class I presentation and the CTL response. Further studies directed toward understanding archaeosome structure in relation to antigen presentation may allow design of an archaeosome type to meet the needs of a particular vaccine.

Archaeosomes are proving to be a valuable tool to study the interaction of lipid adjuvants with the mammalian immune system. Over the next several years we can anticipate that researchers will address many of the questions required to explain fully the mechanism of adjuvant action. To date, most of the published mechanistic work has used archaeosomes prepared from the total polar lipids of *M. smithii*, which are notable in their high content of phosphoserine lipids. It has not yet been determined whether other archaeosome compositions also interact with APCs to use the classical pathways of antigen presentation. While it is known that various archaeosome lipid compositions deliver protein antigen to the cytosol of APCs for MHC class I presentation, the mechanism of endosome to cytosol antigen translocation is unknown. This may occur by fusion of archaeosomes with the phagolysosome membrane triggered by the low pH and/or higher calcium content found in these compartments (13), or destabilization and leakage to the cytosol. With the exception of one study showing that nonphysiological amounts of calcium phosphate can promote slow fusion, little is known about membrane fusion events of archaeosomes, especially those high in caldarchaeol lipids (15).

M. smithii archaeosomes interact with the PS receptor on APCs and adjuvant the immune response (24) However, this observation is difficult to rationalize in view of the down-regulation of the immune response that occurs in DCs exposed to liposomes containing phosphatidylserine (8). M. smithii archaeosomes elicit a counterintuitive effect that includes capacity to stimulate allogeneic T-cell proliferation, to activate $CD4^+$ T cells, and to upregulate costimulatory molecules. The stereochemistry of archaeal lipids, or some other unique archaeal lipid feature, may explain this apparent contradiction.

Adjuvant development has centered on discovery of Toll-like receptor agonists to induce an inflammatory response by alerting the immune system to the presence of "danger signals" normally expressed by pathogens. Most archaeosomes are only mildly inflammatory (24, 29, 30, 57, 62), indicating that strong adjuvant activity need not necessitate the side reactions associated with strong inflammation. Indeed it has been suggested that overt inflammation may erode memory cells (56). Much remains unknown regarding the APC receptors and signaling pathways that may be activated by various archaeosome lipid compositions, or the relationship between adjuvant activity and inflammation.

M. smithii and T. acidophilum archaeosome lipid compositions have the unusual capacity to induce a long-lasting $CD8^+$ T-cell memory response in animals (29, 32). At present the mechanism is largely unknown but may relate to the induction of specific T-cell subsets during a strong primary response and/or to archaeosome-antigen persistence. The membrane-spanning caldarchaeol lipids are suspected to be required for this phenomenon. Furthermore, it is currently unknown whether the apparent inability of some archaeosome compositions to maintain a very long-lasting $CD8^+$ T-cell response in animals may be compensated for by varying parameters such as the antigen dose or injection schedule.

Antigen must be physically associated with the archaeosome for adjuvant activity to occur (28). However, studies have not compared antigen that are internalized versus being coupled to the surface of archaeosomes. Surface coupling may prove to be attractive for peptides in cancer vaccine applications (50), and more work is required in the area of antigen-archaeosome formulation.

To achieve the ultimate goal of an "optimized archaeosome" tailored to elicit the immune responses required for a particular vaccine, it may be necessary to use purified or synthetic archaeal lipids. This is indicated from observations that specific archaeal, and nonarchaeal, lipid structures have particular receptor recognition and APC-activating properties, whereas others do not (24, 56, 62). However, utility in animals of total polar lipid archaeosome vaccines has been demonstrated, and this approach is likely to have advantages of less cost and multiple receptor interactions and to contribute to the formation of stable vaccines. Success of archaeosomes as a human adjuvant now depends on successful clinical testing of an archaeosome vaccine.

Acknowledgments. We acknowledge the laboratory personnel who, over the past decade, have coauthored archaeosome publications and contributed to the development of archaeosome vaccines. Computer-generated models of archaeol and caldarchaeol lipids shown in Figure 1 were prepared by J.-R. Brisson, National Research Council of Canada. We are grateful to our photographer Tom Devecseri for his help with graphics.

This article is NRCC publication no. 42502.

REFERENCES

1. **Allison, A. G., and G. Gregoriadis.** 1974. Liposomes as immunological adjuvants. *Nature* 252:252.

2. **Bangham, A. D., M. M. Standish, and J. C. Watkins.** 1965. Diffusion of univalent ions across the lamellae of swollen phospholipids. *J. Mol. Biol.* 13:238–252.

3. **Berger, N., A. Sachse, J. Bender, R. Schubert, and M. Brandl.** 2001. Filter extrusion of liposomes using different devices: comparison of liposome size, encapsulation efficiency, and process characteristics. *Int. J. Pharm.* 223:55–68.

4. **Beveridge, T. J., C. G. Choquet, G. B. Patel, and G. D. Sprott.** 1993. Freeze-fracture planes of methanogen membranes correlate with the content of tetraether lipids. *J. Bacteriol.* 175:1191–1197.

5. **Brennan, F. R., and G. Dougan.** 2005. Non-clinical safety evaluation of novel vaccines and adjuvants: new products, new strategies. *Vaccine* 23:3210–3222.

6. **Cavicchioli, R., P. M. Curmi, N. Saunders, and T. Thomas.** 2003. Pathogenic archaea: do they exist? *Bioessays* 25:1119–1128.

7. **Chan, E. L.** 1994. Unusual thermal stability of liposomes made from bipolar tetraether lipids. *Biochem. Biophys. Res. Commun.* 202:673–679.

8. **Chen, X., K. Doffek, S. L. Sugg, and J. Shilyansky.** 2004. Phosphatidylserine regulates the maturation of human dendritic cells. *J. Immunol.* 173:2985–2994.

9. Chiu, G. N., M. B. Bally, and L. D. Mayer. 2003. Targeting of antibody conjugated, phosphatidylserine-containing liposomes to vascular cell adhesion molecule 1 for controlled thrombogenesis. *Biochim. Biophys. Acta* **1613**:115–121.

10. Choquet, C. G., G. B. Patel, T. J. Beveridge, and G. D. Sprott. 1992. Formation of unilamellar liposomes from total polar lipid extracts of methanogens. *Appl. Environ. Microbiol.* **58**:2894–2900.

11. Choquet, C. G., G. B. Patel, T. J. Beveridge, and G. D. Sprott. 1994. Stability of pressure-extruded liposomes made from archaeobacterial ether lipids. *Appl. Microbiol. Biotechnol.* **42**: 375–384.

12. Choquet, C. G., G. B. Patel, and G. D. Sprott. 1996. Heat sterilization of archaeal liposomes. *Can. J. Microbiol.* **42**:183–186.

13. Christensen, K. A., J. T. Myers, and J. A. Swanson. 2002. pH-dependent regulation of lysosomal calcium in macrophages. *J. Cell Sci.* **115**:599–607.

14. Conlan, J. W., L. Krishnan, G. E. Willick, G. B. Patel, and G. D. Sprott. 2001. Immunization of mice with lipopeptide antigens encapsulated in novel liposomes prepared from the polar lipids of various archaeobacteria elicits rapid and prolonged specific protective immunity against infection with the facultative intracellular pathogen, *Listeria monocytogenes*. *Vaccine* **19**:3509–3517.

15. Elferink, M. G., J. van Breemen, W. N. Konings, A. J. Driessen, and J. Wilschut. 1997. Slow fusion of liposomes composed of membrane-spanning lipids. *Chem. Phys. Lipids* **88**:37–43.

16. Elferink, M. G. L., J. G. deWit, A. J. M. Driessen, and W. N. Konings. 1994. Stability and proton-permeability of liposomes composed of archaeal tetraether lipids. *Biochim. Biophys. Acta* **1193**:247–254.

17. Fiete, D. J., M. C. Beranek, and J. U. Baenziger. 1998. A cysteine-rich domain of the "mannose" receptor mediates GalNAc-4-SO4 binding. *Proc. Natl. Acad. Sci. USA* **95**:2089–2093.

18. Freisleben, H. J., J. Bormann, D. C. Litzinger, F. Lehr, P. Rudolph, W. Schatton, and L. Huang. 1995. Toxicity and biodistribution of liposomes of the main phospholipid from the archaeobacterium *Thermoplasma acidophilum*. *J. Liposome Res.* **5**:215–223.

19. Freisleben, H. J., K. Zwicker, P. Jezek, G. John, A. Bettin-Bogutzki, K. Ring, and T. Nawroth. 1995. Reconstitution of bacteriorhodopsin and ATP synthase from *Micrococcus luteus* into liposomes of the purified main tetraether lipid from *Thermoplasma acidophilum*: proton conductance and light-driven ATP synthesis. *Chem. Phys. Lipids* **78**:137–147.

20. Gibson, J. A., M. R. Miller, N. W. Davies, G. P. Neill, D. S. Nichols, and J. K. Volkman. 2005. Unsaturated diether lipids in the psychrotrophic archaeon *Halorubrum lacusprofundi*. *Syst. Appl. Microbiol.* **28**:19–26.

21. Gliozzi, A., A. Relini, and P. L. G. Chong. 2002. Structure and permeability properties of biomimetic membranes of bolaform archaeal tetraether lipids. *J. Membr. Sci.* **206**:131–147.

22. Gophna, U., R. L. Charlebois, and W. F. Doolittle. 2004. Have archaeal genes contributed to bacterial virulence? *Trends Microbiol.* **12**:213–219.

23. Gregoriadis, G. and A. T. Florence. 1993. Recent advances in drug targeting. *Trends Biotechnol.* **11**:440–442.

24. Gurnani, K., J. Kennedy, S. Sad, G. D. Sprott, and L. Krishnan. 2004. Phosphatidylserine receptor-mediated recognition of archaeosome adjuvant promotes endocytosis and MHC class I cross-presentation of the entrapped antigen by phagosome-to-cytosol transport and classical processing. *J. Immunol.* **173**: 566–578.

25. Hoffmann, P. R., A. M. deCathelineau, C. A. Ogden, Y. Leverrier, D. L. Bratton, D. L. Daleke, A. J. Ridley, V. A. Fadok, and P. M. Henson. 2001. Phosphatidylserine (PS) induces PS receptor-mediated macropinocytosis and promotes clearance of apoptotic cells. *J. Cell Biol.* **155**:649–659.

26. Kates, M. 1978. The phytanyl ether-linked polar lipids and isoprenoid neutral lipids of extremely halophilic bacteria. *Prog. Chem. Fats. Other Lipids* **15**:301–342.

27. Kates, M. 1992. Archaebacterial lipids: structure, biosynthesis and function. Biochem.Soc.Symp. **58**:51–72.

28. Krishnan, L., C. J. Dicaire, G. B. Patel, and G. D. Sprott. 2000. Archaeosome vaccine adjuvants induce strong humoral, cell-mediated, and memory responses: comparison to conventional liposomes and alum. *Infect. Immun.* **68**:54–63.

29. Krishnan, L., S. Sad, G. B. Patel, and G. D. Sprott. 2000. Archaeosomes induce long-term CD8$^+$ cytotoxic T cell response to entrapped soluble protein by the exogenous cytosolic pathway, in the absence of CD4$^+$ T cell help. *J. Immunol.* **165**: 5177–5185.

30. Krishnan, L., S. Sad, G. B. Patel, and G. D. Sprott. 2001. The potent adjuvant activity of archaeosomes correlates to the recruitment and activation of macrophages and dendritic cells in vivo. *J. Immunol.* **166**:1885–1893.

31. Krishnan, L., S. Sad, G. B. Patel, and G. D. Sprott. 2003. Archaeosomes induce enhanced cytotoxic T lymphocyte responses to entrapped soluble protein in the absence of interleukin 12 and protect against tumor challenge. *Cancer Res.* **63**:2526–2534.

32. Krishnan, L., and G. D. Sprott. 2003. Archaeosomes as self-adjuvanting delivery systems for cancer vaccines. *J. Drug Target.* **11**:515–524.

33. Luzzati, V., A. Gambacorta, M. DeRosa, and A. Gulik. 1987. Polar lipids of thermophilic prokaryotic organisms: chemical and physical structure. *Annu. Rev. Biophys. Biophys. Chem.* **16**:25–47.

34. Makino, K., J. Tabata, T. Yoshioka, M. Fukuda, M. Ikekita, H. Ohshima, and H. Terada. 2003. Effects of liposomal phosphatidylserine on phagocytic uptake of liposomes by macrophage-like HL-60RG cells. *Colloids Surfaces B* **29**:277–284.

35. Mathai, J. C., G. D. Sprott, and M. L. Zeidel. 2001. Molecular mechanisms of water and solute transport across archaeobacterial lipid membranes. *J. Biol. Chem.* **276**:27266–27271.

36. Miller, T. L., and M. J. Wolin. 1982. Enumeration of *Methanobrevibacter smithii* in human feces. *Arch. Microbiol.* **131**:14–18.

37. Mosmann, T. R., and S. Sad. 1996. The expanding universe of T-cell subsets: Th1, Th2 and more. *Immunol. Today* **17**: 138–146.

38. Mossman, S. P., L. S. Evans, H. Fang, J. Staas, T. Tice, S. Raychaudhuri, K. H. Grabstein, M. A. Cheever, and M. E. Johnson. 2005. Development of a CTL vaccine for Her-2/neu using peptide-microspheres and adjuvants. *Vaccine* **23**:3545–3554.

39. New, R. C. C. (ed.) 1990. *Liposomes. A Practical Approach*. IRL Press, Oxford.

40. Nichols, D. S., M. R. Miller, N. W. Davies, A. Goodchild, M. Raftery, and R. Cavicchioli. 2004. Cold adaptation in the Antarctic Archaeon *Methanococcoides burtonii* involves membrane lipid unsaturation. *J. Bacteriol.* **186**:8508–8515.

41. Nishihara, M., H. Morii, and Y. Koga. 1987. Structure determination of a quartet of novel tetraether lipids from Methanobacterium thermoautotrophicum. *J. Biochem. (Tokyo)* **101**: 1007–1015.

42. North, R. J., P. L. Dunn, and J. W. Conlan. 1997. Murine listeriosis as a model of antimicrobial defense. *Immunol. Rev.* **158**:27–36.

43. Omri, A., B. J. Agnew, and G. B. Patel. 2003. Short-term repeated-dose toxicity profile of archaeosomes administered to mice via intravenous and oral routes. *Int. J. Toxicol.* **22**:9–23.

44. Omri, A., B. Makabi-Panzu, B. J. Agnew, G. D. Sprott, and G. B. Patel. 2000. Influence of coenzyme Q10 on tissue distribution of archaeosomes, and pegylated archaeosomes, administered to mice by oral and intravenous routes. *J. Drug Target.* 7:383–392.

45. Patel, G. B., B. J. Agnew, L. Deschatelets, L. P. Fleming, and G. D. Sprott. 2000. In vitro assessment of archaeosome stability for developing oral delivery systems. *Int. J. Pharm.* 194:39–49.

46. Patel, G. B., A. Omri, L. Deschatelets, and G. D. Sprott. 2002. Safety of archaeosome adjuvants evaluated in a mouse model. *J. Liposome Res.* 12:353–372.

47. Patel, G. B., and G. D. Sprott. 1999. Archaeobacterial ether lipid liposomes (archaeosomes) as novel vaccine and drug delivery systems. *Crit. Rev. Biotechnol.* 19:317–357.

48. Richards, R. L., M. Rao, N. M. Wassef, G. M. Glenn, S. W. Rothwell, and C. R. Alving. 1998. Liposomes containing lipid A serve as an adjuvant for induction of antibody and cytotoxic T-cell responses against RTS,S malaria antigen. *Infect. Immun.* 66:2859–2865.

49. Ring, K., B. Henkel, A. Valenteijn, and R. Gutermann. 1986. Studies on the permeability and stability of liposomes derived from a membrane spanning bipolar archaebacterial tetraetherlipid, p. 101–123. *In* K. H. Schmidt (ed.), *Liposomes as Drug Carriers.* Thieme, Stuttgart, Germany.

50. Roth, A., F. Rohrbach, R. Weth, B. Frisch, F. Schuber, and W. S. Wels. 2005. Induction of effective and antigen-specific antitumour immunity by a liposomal ErbB2/HER2 peptide-based vaccination construct. *Br. J. Cancer* 92:1421–1429.

51. Schwarz, K., T. Storni, V. Manolova, A. Didierlaurent, J. C. Sirard, P. Rothlisberger, and M. F. Bachmann. 2003. Role of Toll-like receptors in costimulating cytotoxic T cell responses. *Eur. J. Immunol.* 33:1465–1470.

52. Sehgal, S. N., M. Kates, and N. E. Gibbons. 1962. Lipids of *Halobacterium cutirubrum. Can. J. Biochem. Physiol.* 40:69–81.

53. Sprott, G. D., B. J. Agnew, and G. B. Patel. 1997. Structural features of ether lipids in the archaeobacterial thermophiles *Pyrococcus furiosis, Methanopyrus kandleri, Methanothermus fervidus,* and *Sulfolobus acidocaldarius. Can. J. Microbiol.* 43:467–476.

54. Sprott, G. D., J. Brisson, C. J. Dicaire, A. K. Pelletier, L. A. Deschatelets, L. Krishnan, and G. B. Patel. 1999. A structural comparison of the total polar lipids from the human archaea *Methanobrevibacter smithii* and *Methanosphaera stadtmanae* and its relevance to the adjuvant activities of their liposomes. *Biochim. Biophys. Acta* 1440:275–288.

55. Sprott, G. D., C. J. Dicaire, L. P. Fleming, and G. B. Patel. 1996. Stability of liposomes prepared from archaeobacterial lipids and phosphatidylcholine mixtures. *Cells Mater.* 6:143–155.

56. Sprott, G. D., C. J. Dicaire, K. Gurnani, L. A. Deschatelets, and L. Krishnan. 2004. Liposome adjuvants prepared from the total polar lipids of *Haloferax volcanii, Planococcus* spp. and *Bacillus firmus* differ in ability to elicit and sustain immune responses. *Vaccine* 22:2154–2162.

57. Sprott, G. D., S. Larocque, N. Cadotte, C. J. Dicaire, M. McGee, and J. R. Brisson. 2003. Novel polar lipids of halophilic eubacterium *Planococcus* H8 and archaeon *Haloferax volcanii. Biochim. Biophys. Acta* 1633:179–188.

58. Sprott, G. D., M. Meloche, and J. C. Richards. 1991. Proportions of diether, macrocyclic diether, and tetraether lipids in *Methanococcus jannaschii* grown at different temperatures. *J. Bacteriol.* 173:3907–3910.

59. Sprott, G. D., G. B. Patel, C. G. Choquet, and I. Ekiel. 1992. Formation of stable liposomes from lipid extracts of archaeobacteria (archaea). Patent PCT/CA/92/00464[WO 93/08202], 1–56.

60. Sprott, G. D., G. B. Patel, and L. Krishnan. 2003. Archaeobacterial ether lipid liposomes as vaccine adjuvants. *Methods Enzymol.* 373:155–172.

61. Sprott, G. D., G. B. Patel, B. Makabi-Panzu, and D. L. Tolson. 1997. Archaeosomes, archaeosomes containing coenzyme Q10, and other types of liposomes containing coenzyme Q10 as adjuvants and as delivery vehicles. Patent PCT/CA 96/00835 [WO 97/22333], 1–57.

62. Sprott, G. D., S. Sad, L. P. Fleming, C. J. Dicaire, G. B. Patel, and L. Krishnan. 2003. Archaeosomes varying in lipid composition differ in receptor-mediated endocytosis and differentially adjuvant immune responses to entrapped protein. *Archaea* 1:151–164.

63. Sprott, G. D., D. L. Tolson, and G. B. Patel. 1997. Archaeosomes as novel antigen delivery systems. *FEMS Microbiol. Lett.* 154:17–22.

64. Swain, M., J. R. Brisson, G. D. Sprott, F. P. Cooper, and G. B. Patel. 1997. Identification of beta-L-gulose as the sugar moiety of the main polar lipid *Thermoplasma acidophilum. Biochim. Biophys. Acta* 1345:56–64.

65. Tenchov, B., E. M. Vescio, G. D. Sprott, M. L. Zeidel, and J. C. Mathai. 2006. Salt tolerance of archaeal extremely halophilic lipid membranes. *J. Biol. Chem.* 281:10016–10023.

66. Tolson, D. L., R. K. Latta, G. B. Patel, and G. D. Sprott. 1996. Uptake of archaeobacterial liposomes and conventional liposomes by phagocytic cells. *J. Liposome Res.* 6:755–776.

67. van de Vossenberg, J. L., A. J. Driessen, W. D. Grant, and W. N. Konings. 1999. Lipid membranes from halophilic and alkali-halophilic *Archaea* have a low H+ and Na+ permeability at high salt concentrations. *Extremophiles* 3:253–257.

INDEX